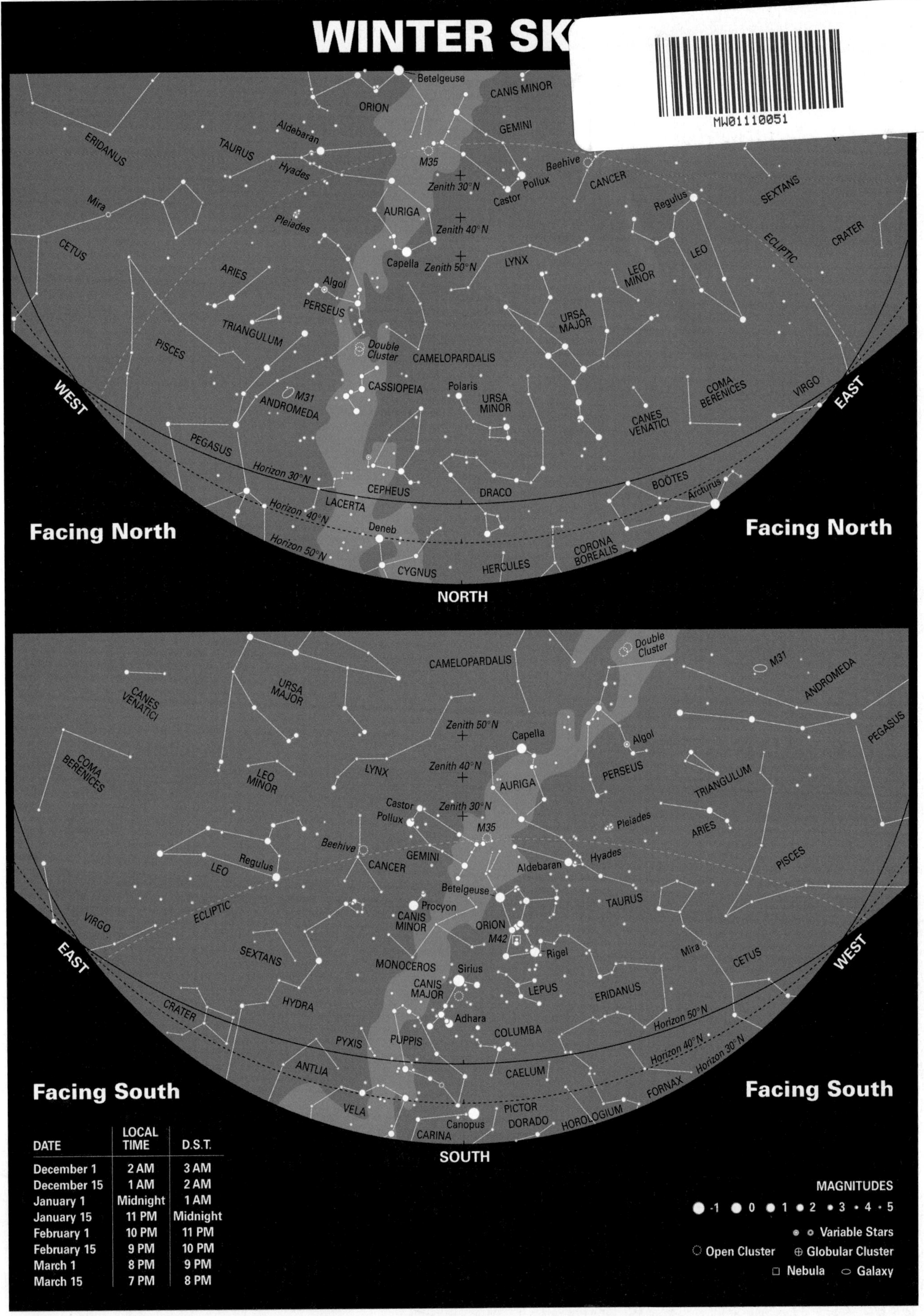

DATE	LOCAL TIME	D.S.T.
December 1	2 AM	3 AM
December 15	1 AM	2 AM
January 1	Midnight	1 AM
January 15	11 PM	Midnight
February 1	10 PM	11 PM
February 15	9 PM	10 PM
March 1	8 PM	9 PM
March 15	7 PM	8 PM

MAP BY WIL TIRION; FOR JAY M. PASACHOFF

A supernova remnant in the Large Magellanic Cloud, a satellite galaxy to our own.

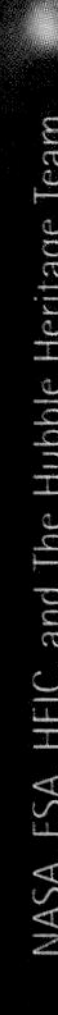

THIRD EDITION

THE COSMOS

Astronomy in the New Millennium

Jay M. Pasachoff
Williams College

Alex Filippenko
University of California, Berkeley

Australia · Brazil · Canada · Mexico · Singapore · Spain · United Kingdom · United States

The Cosmos: Astronomy in the New Millennium, Third Edition
Jay M. Pasachoff and Alex Filippenko

Astronomy Acquisitions Editor: Chris Hall
Publisher / Executive Editor: David Harris
Development Editor: Alyssa White
Editorial Assistants: Brandi Kirksey, Jessica Jacobs
Technology Project Manager: Samuel Subity
Senior Marketing Manager: Mark Santee
Marketing Assistant: Michele Colella
Marketing Communications Manager: Bryan Vann
Project Manager, Editorial Production: Teri Hyde
Art Director: Lee Friedman
Print Buyer: Doreen Suruki
Permissions Editor: Kiely Sisk
Production Service: Lachina Publishing Services
Text Designer: John Walker Design
Copy Editor: Lachina Publishing Services
Illustrator: Lachina Publishing Services
Cover Designer: John Walker Design
Cover Image: Courtesy of NASA/JPL/Space Science Institute
Cover Printer: Phoenix Color Corp—MD
Compositor: Lachina Publishing Services
Printer: Courier Corporation/Kendallville

Printed in the United States of America
1 2 3 4 5 6 7 10 09 08 07 06

Library of Congress Control Number: 2005931601

Student Edition: ISBN 0-495-01303-X

Thomson Higher Education
10 Davis Drive
Belmont, CA 94002-3098
USA

BRIEF CONTENTS

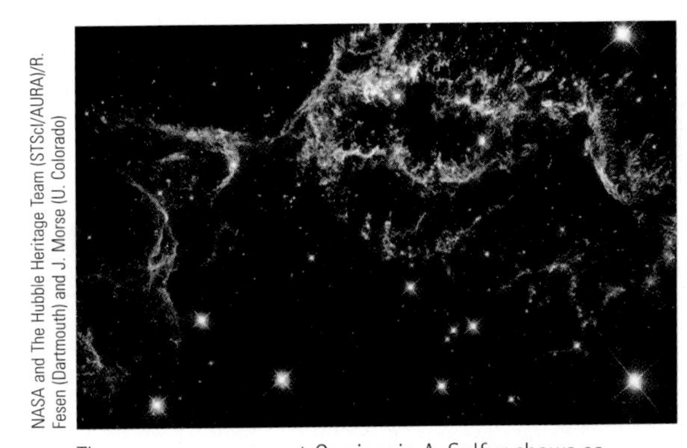

NASA and The Hubble Heritage Team (STScI/AURA)/R. Fesen (Dartmouth) and J. Morse (U. Colorado)

The supernova remnant Cassiopeia A. Sulfur shows as red and oxygen as blue in this Hubble image.

CONTENTS

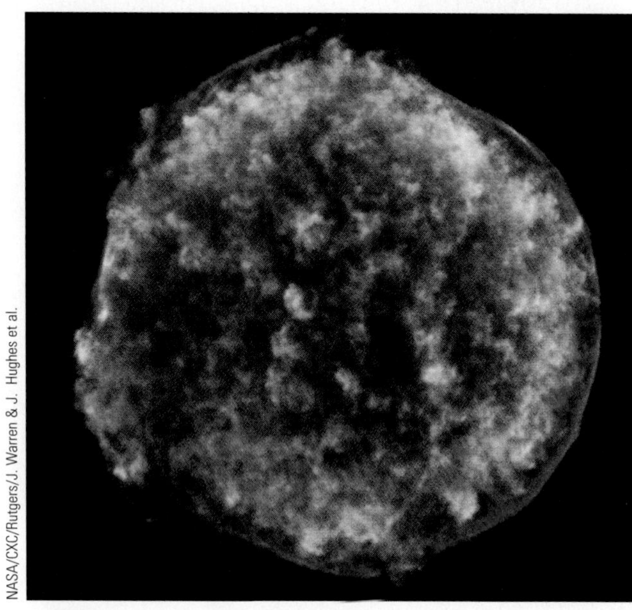
NASA/CXC/Rutgers/J. Warren & J. Hughes et al.

Tycho's supernova remnant, the remains of the explosion of a star in 1572.

Jay M. Pasachoff

The Swedish Solar Telescope in the Canary Islands, Spain.

Jay M. Pasachoff collection

The frontispiece of Galileo's *Dialogo*, published in 1632.

A 360° panoramic scan on Mars, from the Mars Exploration Rover named Spirit.

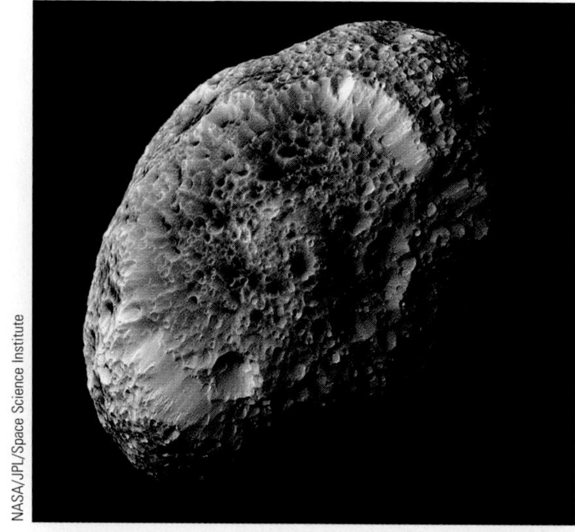

Saturn's moon Hyperion, viewed close up from NASA's Cassini mission.

NASA/JPL/Space Science Institute

Saturn's rings (edge-on at bottom) and its moon Dione, against a background of Saturn's disk with shadows of rings, imaged from NASA's Cassini mission.

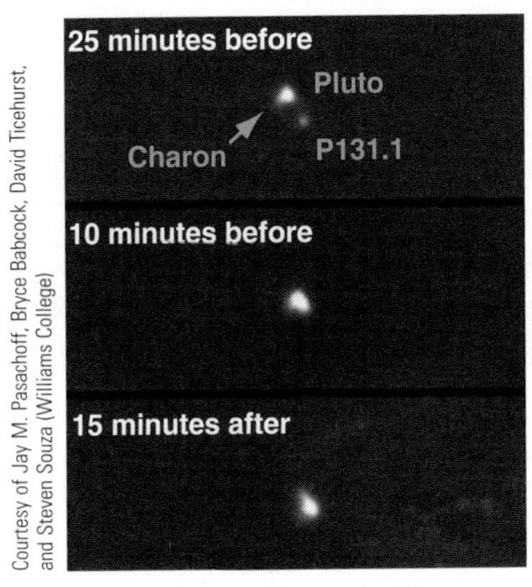

Courtesy of Jay M. Pasachoff, Bryce Babcock, David Ticehurst, and Steven Souza (Williams College)

The occultation of a star by Pluto, with Charon off to the side in exquisite seeing.

A composite image of the solar corona at the April 18, 2005, total solar eclipse, imaged from the mid-Pacific.

The Boomerang Nebula, gas reflecting starlight, imaged with the Hubble Space Telescope.

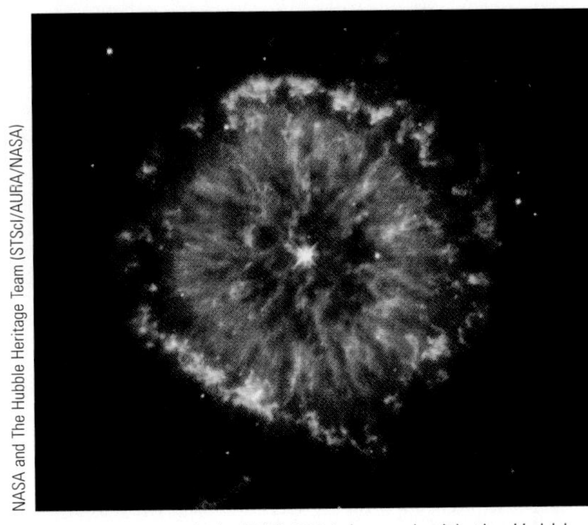

NASA and The Hubble Heritage Team (STScI/AURA/NASA)

The planetary nebula NGC 6751, imaged with the Hubble Space Telescope. We see gas thrown out by a dying star that once resembled our Sun.

NASE, ESA and The Hubble Heritage Team (STScI/AURA)/J.C. Green (U. Colorado) and the Cosmic Origins Spectograph (cos) GTO Team; NASA/CXO/SAO

A supernova remnant, N132D, in the Large Magellanic Cloud, a composite of images with the Hubble Space Telescope and the Chandra X-ray Observatory.

NASA, ESA, and The Hubble Heritage Team (STScI/AURA)/P. Knezek (WIYN)

The barred spiral galaxy NGC 1300, imaged with the Hubble Space Telescope.

NASA and the Early Release Observations Team: Andrew Fruchter, Sylvia Baggett, and Zoltan Levay (Space Telescope Science Institute), and Richard Hook (Space Telescope—European Coordinating Facility)

Gravitational lensing from the cluster of galaxies Abell 2218 bends light from even more-distant galaxies into circular arcs.

NASA/JPL/Caltech, Lockheed Martin, and Northrop Grumman Space Technologies

Artist's conception of SIM PlanetQuest, NASA's Space Interferometry Mission to measure star positions and to look for extrasolar planets.

PREFACE

Astronomy continues to flourish. The pair of Keck 10-meter-diameter telescopes in Hawaii and the quartet of 8-meter-diameter telescopes in Chile called the Very Large Telescope are among the instruments used to make important discoveries from mountaintop observatories. Still larger telescopes are on the drawing board, including the Thirty Meter Telescope (formerly called the California Extremely Large Telescope) and a European effort known, astonishingly, as the Overwhelmingly Large Telescope, though its 100-meter size may merit the name. The Hubble Space Telescope sends down exciting data all the time, though we worry about its future and about a gap before the James Webb Space Telescope is launched. The latest space observatories transmit images made with gamma rays, with x-rays, and with infrared radiation. The overall structure of the Universe is being mapped and analyzed, with catalogues of millions of objects being compiled. Cosmology has become a mathematical, and even a statistical, science. Further, new electronic instruments and computer capabilities, new space missions to Solar-System objects, and advances in computational astronomy and in theoretical work will continue to bring forth exciting results.

In *The Cosmos: Astronomy in the New Millennium,* we describe the current state of astronomy, both the fundamentals of astronomical knowledge that have been built up over decades and the incredible advances that are now taking place. We want simply to share with you the excitement and magnificence of the Universe.

We try to cover all the branches of astronomy without slighting any of them; each teacher and each student may well find special interests that are different from our own. One of our aims in writing this book is to educate citizens in the hope that they will understand the value and methods of scientific research in general and astronomical research in particular.

In writing this book, we share the goals of a commission of the Association of American Colleges, whose report on the college curriculum stated, "A person who understands what science is recognizes that scientific concepts are created by acts of human intelligence and imagination; comprehends the distinction between observation and inference and between the occasional role of accidental discovery in scientific investigation and the deliberate strategy of forming and testing hypotheses; understands how theories are formed, tested, validated, and accorded provisional acceptance; and discriminates between conclusions that rest on unverified assertion and those that are developed from the application of scientific reasoning." The scientific method permeates the book.

NASA/JPL-Caltech/R.A. Gutermuth (Harvard-Smithsonian CFA)

A reflection nebula, NGC 1333 in the constellation Perseus, imaged in the infrared with the Spitzer Space Telescope and displayed in false color. The blue-green shows shock waves in the gas, and the red shows glowing dust.

What is science? The following statement was originally drafted by the Panel on Public Affairs of the American Physical Society, in an attempt to meet the perceived need for a very short statement that would differentiate science from pseudoscience. This statement has been endorsed as a proposal to other scientific societies by the Council of the American Physical Society, and was endorsed by the Executive Board of the American Association of Physics Teachers:

Science is the systematic enterprise of gathering knowledge about the world and organizing and condensing that knowledge into testable laws and theories. The success and credibility of science is anchored in the willingness of scientists to:

1. *Expose their ideas and results to independent testing and replication by other scientists; this requires the complete and open exchange of data, procedures, and materials;*

2. *Abandon or modify accepted conclusions when confronted with more complete or reliable experimental evidence. Adherence to these principles provides a mechanism for self-correction that is the foundation of the credibility of science.*

Our book, through the methods it describes, should reveal this systematic enterprise of science to the readers.

Because one cannot adequately cover the whole Universe in a few months, we have therefore had to pick and choose topics from within the various branches of astronomy, while trying to describe a wide range, to convey the spirit of contemporary astronomy and of the scientists working in it. Our mix includes much basic astronomy and many of the exciting topics now at the forefront.

Organization

The Cosmos: Astronomy in the New Millennium is organized in an Earth-outward approach. Chapter 1 gives an overview of the Universe. Chapters 2 and 3 present, respectively, fundamental astronomical concepts about light, matter, and energy; and the various types of telescopes used to explore the electromagnetic spectrum. In Chapter 4, we discuss easily observed astronomical phenomena and the celestial sphere. Some professors prefer to take up that material at the very beginning of the course or at other points, and there is no problem with doing so.

An artist's conception of the James Webb Space Telescope, whose launch is now planned by NASA for 2013.

Chapter 5 examines the early history of the study of astronomy. Chapters 6 through 8 cover the Solar System and its occupants, although an in-depth discussion of the Sun is reserved for later, in Chapter 10. Chapter 6 compares Earth to the Moon and Earth's nearest planetary neighbors, Venus and Mars. Based largely on the Voyager, Galileo, and Cassini data, Chapter 7 compares and contrasts the Jovian gas/liquid giants—Jupiter, Saturn, Uranus, and Neptune. Chapter 8 looks at the outermost part of the Solar System, including Pluto and its moon, Charon; the chapter also spotlights comets and meteoroids, the Solar System's vagabonds. Chapter 9 discusses the formation of our own Solar System and describes the exciting discovery of over 160 planets around other stars.

Chapter 10 discusses the Sun, our nearest star. Moving outward, Chapters 11 through 14 examine all aspects of stars. Chapter 11 begins by presenting observational traits of stars—their colors and types—and goes on to show how we measure their distances, brightnesses, and motions. It also discusses binary stars, variable stars, and star clusters, in the process showing how we derive stellar masses and ages. Chapter 12 answers the question of how stars shine and reveals that all stars have life cycles. Chapter 13 tells what happens when stars die and describes some of the peculiar objects that violent stellar death can create, including neutron stars and pulsars. Black holes, the most bizarre objects to result from star death, are the focus of Chapter 14.

As we explore further, Chapter 15 describes the parts of the Milky Way Galaxy and our place in it. Chapter 16 pushes beyond the Milky Way to discuss galaxies in general, the fundamental units of the Universe, and evidence that they consist largely of dark matter. Ways in which we are studying the evolution of galaxies are also described. Chapter 17 looks at quasars, distant and powerful objects that are probably gigantic black holes swallowing gas in the central regions of galaxies.

Chapters 18 and 19 consider the ultimate questions of cosmological creation by analyzing recent findings and current theories. Evidence for an accelerating expansion of the Universe, possibilities for the overall geometry and fate of the Universe, ripples in the cosmic background radiation, the origin and phenomenally rapid early growth of the Universe, and the idea of multiple universes are among the fascinating (and sometimes very speculative) topics explored. Lastly, Chapter 20 discusses the always-intriguing search for extraterrestrial intelligence.

NASA/JPL/Space Science Institute

Saturn's rings, viewed from the Cassini spacecraft.

Features

The Cosmos: Astronomy in the New Millennium offers instructors a short text with concise coverage over a wide range of astronomical topics. An early discussion of the scientific method stresses its importance in the verification of observations. The text presents up-to-date coverage of many important findings and theories as well as the latest images, including observations of Jupiter and Saturn from the Cassini-Huygens mission (including the landing on Titan), close-up observations of Mars, images from the impact of Deep Impact on a comet in 2005, infrared images of stars in formation and of gas near them, and coverage of the 2005 total solar eclipse.

We provide, as well, a sampling of the many significant findings from the Hubble Space Telescope. Many images from the Chandra X-ray Observatory and the Spitzer Space Telescope also appear, as do results from the Swift spacecraft that is monitoring gamma-ray bursts, among the most powerful and violent phenomena in the Universe. Of particular interest is the recent advance in our understanding of the age and expansion of the Universe, both through better measurements of the current expansion rate and through discussion of the conclusion that the Universe's expansion is speeding up with time. We also present the recent exciting measurements that allow astronomers to determine the overall geometry of the Universe and the long-ago origin of its structure through detailed observations of the cosmic background radiation.

Origins

The study of our origins, whether it be ourselves as humans, our Earth as a planet, our Sun as a star, or our Galaxy as a whole, is as interesting to many of us as it is to look at our own baby pictures. Most of us have gazed at the stars and wondered how they came to be and what their relationship to us is. NASA has chosen *Origins* as one of its major themes for the organization of its missions and has several spacecraft planned in the Origins program discussed later in Chapter 1. We emphasize the study of origins in this text, first by singling out specifics on the first page of each chapter and then by dealing with a variety of relevant material in the text itself.

Pedagogy

Mixed in the book we have five kinds of features:

1. ***People in Astronomy.*** Each of these interviews presents a notable contemporary astronomer engaged in conversation about a variety of topics: current and future work in astronomy, what led that person to study and pursue astronomy as a career, and why learning about astronomy is an important scientific *and* human endeavor. We hope that you enjoy reading their comments as much as we enjoyed speaking with them and learning about their varied interests and backgrounds.
2. ***Star Parties.*** An occasional feature that shows students how to find things in the sky. These include observing exercises and links to the star maps that appear on the inside covers of the book.
3. ***Figure It Out.*** In some astronomy courses, it may be appropriate to elaborate on equations. Because we wrote *The Cosmos: Astronomy in the New Millennium* to be a descriptive presentation of modern astronomy for liberal arts students, we kept the use of mathematics to a minimum. However, we recognize that some instructors wish to introduce their students to more of the mathematics associated with astronomical phenomena. Consequently, we provide mathematical features, numbered so they can be assigned or not, at the instructor's option.

4. ***Lives in Science.*** These boxes provide biographies of important historical figures like Copernicus and Galileo.
5. ***A Closer Look.*** Using these boxes, students can further explore interesting topics, such as size scales in the Universe, observing with large telescopes, various celestial phenomena, mythology, and naming systems. We are pleased to supply some exciting close-ups of Mars, Titan, and a comet, based on 2005's spacecraft spectaculars.

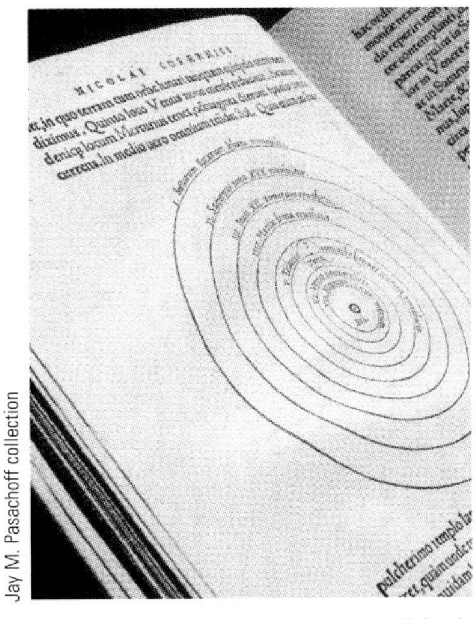

Jay M. Pasachoff collection

The first heliocentric diagram published, in Copernicus's 1543 *De Revolutionibus.*

We have provided aids to make the book easy to read and to study from. New vocabulary is boldfaced in the text and in the expanded summaries that are given in context at the end of each chapter, and defined in the glossary. The index provides further aid in finding explanations. An expanded set of end-of-chapter questions covers a range of material and includes some that are straightforward to answer from the text and others that require more thought. Appendices provide some information on planets, stars, constellations, and nonstellar objects. The exceptionally beautiful sky maps by Wil Tirion (inside the covers of the book) will help you find your way around the sky when you go outside to observe the stars. Be sure you do.

Recent educational research has shown that students often need to unlearn incorrect ideas in order to understand the correct ones. We have thus placed a list of such "Misconceptions," including common incorrect ideas and the correct alternatives, on the book's website. The book's website also has a list of many relevant URLs from the World Wide Web.

The Cosmos, 3rd edition, website (http://astronomy.brookscole.com/cosmos3) is updated frequently with links to information and photographs from a wide variety of sources. Students will find multiple-choice self-tests and other useful information. A section for professors, for which a password is required, provides various downloadable teaching aids, including most of this book's images. Both professors and students have access to news updates and new web links. Updates also appear at this book's other website at http://www.solarcorona.net. Each professor and each student should frequently look at these updates and links, which are organized in chapter order. Thus the updates should be checked for each lecture and each chapter read.

Ancillary Materials

Available to those who adopt this book are the following ancillaries:

Online Instructor's Manual with Test Bank

This resource manual, updated for *The Cosmos,* 3rd edition, contains the answers to all questions in the book. Additional instructor resources include suggested course syllabi, listings of main concepts, lists of relevant DVDs and other audiovisual resources, and a complete test bank with answers. It is available exclusively for download from the password-protected Instructor's Website.

NASA, ESA, A. Bolton (Harvard-Smithsonian CFA) and the SLACS team

An "Einstein ring," the blue ring showing light from an extremely distant galaxy focused toward us by an intermediate galaxy using concepts explained with Einstein's general theory of relativity.

Multimedia Manager Instructor's Resource CD

Almost all the photos and drawings from the text are available for professors to use in lecturing, supplied both as jpg images and as PowerPoint slides. The Multimedia Manager Instructor's Resource CD gives instructors a quick way to assemble art, photos, and multimedia with notes for their lectures. In addition to the images from *The Cosmos,* 3rd edition, the Multimedia Manager includes a collection of Active Figure animations and additional simulations. Preformatted PowerPoint™ presentations have been assembled for each text. It is easy to modify the PowerPoint™ presentations using the normal Microsoft® PowerPoint™ program. Also, the simple interface supplied on the CD provides an alternate way for professors to incorporate graphics, digital video, animations, and audio clips into lectures.

"Clicker Systems": JoinIn™ on TurningPoint®

"Clicker Systems," or "personal-response systems," are increasingly popular ways for students to provide feedback during lectures and for faculty to provide, in turn, feedback to students about their comprehension. Thomson Brooks/Cole now offers you book-specific JoinIn™ content for Response Systems specifically tailored to *The Cosmos,* 3rd edition. The system allows an instructor to assess students' progress with instant in-class quizzes and polls. Our publisher's exclusive agreement to offer TurningPoint® software lets instructors pose book-specific questions and display students' answers seamlessly within Microsoft® PowerPoint™ slides, in conjunction with the "clicker" hardware of an instructor's choice. For college and university adopters only. Instructors should contact their local Thomson representatives to learn more.

Transparency Acetates

One hundred twenty-five overhead transparency acetates containing art from *The Cosmos,* 3rd edition, including twenty-five new to this edition, are available upon request.

ExamView® Computerized Testing

Create, deliver, and customize tests and study guides (both print and online) in minutes with this easy-to-use assessment and tutorial system. ExamView offers both a Quick Test Wizard and an Online Test Wizard that guide instructors step-by-step through the process of creating tests, while "WYSIWYG" capability allows instructors to see the test being created on the screen exactly as it will print or display online. Instructors can build tests of up to 250 questions using up to 12 question types. Using ExamView's complete word-processing capabilities, instructors can enter an unlimited number of new questions or edit existing questions.

WebCT/Now Integration or Blackboard/Now Integration

Instructors can integrate the conceptual testing and multimedia tutorial features of AceAstronomy™ within a familiar WebCT/Blackboard environment by packaging the text with this special access code. Instructors can assign the AceAstronomy™ materials and have the results flow automatically to a WebCT/Blackboard grade book, creating a robust online course. Students access AceAstronomy™ via their WebCT/Blackboard course, without using a separate user name or password. For college and university adopters only. Instructors should contact their local Thomson representatives to learn more.

AceAstronomy™

This book is integrated with AceAstronomy™, an assessment-centered learning tool for astronomy. Seamlessly tied to *The Cosmos,* 3rd edition, through Active Figures (interactive animations of art from the text), this learning tool is web-based and is included with every new copy of most versions of the book. This interactive resource helps students gauge their individual study needs, then gives them a Personalized Learning Plan that focuses their study time on the concepts they most need to master.

By providing students with a better understanding of exactly what they need to focus on, AceAstronomy™ has the potential of helping students make the optimal use of their study time.

The Hubble Space Telescope, during its most recent repair mission.

Virtual Astronomy Labs

Automatically bundled at no additional cost with new copies of most versions of the text, the Virtual Astronomy Labs are an online, interactive way for students to learn. Focusing on twenty of the most important concepts in astronomy, the labs offer students hands-on exercises that complement the topics in the text. Instructors can set up classes online and view student results, or students can print their lab reports for submission, making the Virtual Astronomy Labs useful for homework assignments, lab exercises, or extra-credit work.

Planetarium CD-ROMs

Instructors can elect to have a CD-ROM of either Starry Night™ or TheSky Student Edition™ packaged with the textbook at no additional charge. The RedShift College Edition CD is available to bundle for an additional fee. Workbooks with exercises and examples are also available for TheSky™ and RedShift™. Contact your local sales representative for complete details.

Pasachoff's *Peterson Field Guide to the Stars and Planets* is a suggested accompaniment for those wanting monthly star maps, star charts, and other detailed observing aids. See http://www.solarcorona.com. Quarterly star maps appear on the inside covers in *The Cosmos,* 3rd edition.

The Cosmos Website

Adopters of *The Cosmos,* 3rd edition, have access to a rich array of teaching and learning resources at the book's web page (http://astronomy.brookscole.com /cosmos3). This site features chapter-by-chapter online tutorial quizzes, chapter outlines, chapter reviews, chapter-by-chapter web links, flashcards, a collection of common student misconceptions, and more. These items are available to qualified adopters. Please consult with your local sales representative for details. For more information about these ancillaries, visit the web page above or contact your local Brooks/Cole sales representative by e-mail at tl.support@thomson.com, by phone at 800-423-0563, or by fax at 859-647-5020. See the web page above for forms to order instructors' sample copies.

Acknowledgments

The publishers join us in placing a heavy premium on accuracy, and we have made certain that the manuscript has been read not only by students for clarity and style but also by other professional astronomers for scientific comments. As a result, you will find that the statements in this book, brief as they are, are authoritative.

This third edition of *The Cosmos* benefited from the advice of reviewers John Armstrong (Weber State University), Philip Hegenderfer (University of Akron), Lori Lubin (University of Calfornia, Davis), Brian Oetiker (Sam Houston State University), Ata Sarajedini (University of Florida), Anuj Sarma (DePaul University), Brian Scott (Bellevue Community College), Steinn Sigurdsson (Pennsylvania State University), and Jack Sulentic (University of Alabama). We are grateful for specific review comments and assistance from Michael Blanton (New York University), Neil Cornish (Montana State University), Anthony DelGenio (NASA), Michael Hanna, Andy Ingersoll (Caltech), Geoff Marcy (University of California, Berkeley), Carolyn Porco (Space

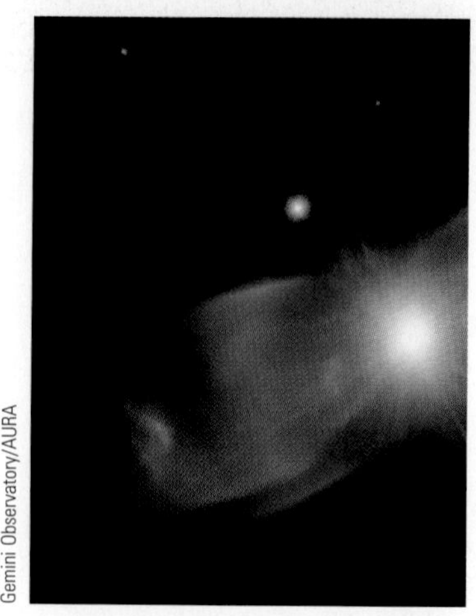

The planetary nebula M2-9, with the image sharpened using adaptive optics on the 8-m Gillett Gemini Telescope.

Science Institute), Cliff Rose, David Spergel (Princeton University), Keivan Stassun (Vanderbilt University), Richard Treffers (Diablo Valley College), Gerard van Belle (Jet Propulsion Laboratory/Caltech), Joshua Winn (MIT), and Rob Wittenmyer (University of Texas).

The second edition benefited from the advice of reviewers Gibor Basri (University of California, Berkeley), Orville Day (East Carolina University), John Eggert (Daytona Beach Community College), Thomas Hockey (University of Northern Iowa), Roger Kadala (Hawaii Pacific University), Robert Kren (University of Michigan, Flint), Claud Lacy (University of Arkansas), Philip Lockett (Centre College), John Mattox (Francis Marion University), and Charles Rogers (Southwestern Oklahoma State University). We thank Kelly Beatty *(Sky & Telescope)* and Heidi Hammel (Southwest Research Institute) for their comments on Solar-System material. We also found useful the extensive comments on a questionnaire filled out by A.F.'s students at UC Berkeley and past comments from J.M.P.'s students at Williams.

We thank the following reviewers for their help with the first edition of *The Cosmos:* Phillip Appleton (Iowa State University), Kelly Beatty (*Sky & Telescope* magazine), Robert Egler (North Carolina State University), Thomas Hockey (University of Northern Iowa), Scott Payson (Wayne State University), Steinn Sigurdsson (Pennsylvania State University), and Dan Wilkins (University of Nebraska, Omaha).

We are also grateful to Nancy Kutner for her excellent work on the index, Barbara Swanson and Madeline Kennedy for their assistance, and Samuel Subity and Milos Mladenovic for their computer work on this book's web pages.

At Brooks/Cole, we thank Chris Hall (Acquisitions Editor), Alyssa White (Development Editor), Teri Hyde (Senior Production Project Manager), Samuel Subity (Electronic Technology Manager), and Mark Santee and Julie Conover (Marketing Managers), as well as Brandi Kirksey for assistance with the Instructor's Manual and Test Bank and Kiely Sisk for assistance with permissions. At Lachina Publishing Services, we thank Ronn Jost (Project Manager).

J.M.P. thanks various members of his family, who have provided vital and valuable editorial services in addition to their general support. His wife, Naomi, has always been very helpful both personally and editorially. He is grateful for the family support of Deborah Pasachoff, Eloise Pasachoff, and Tom Glaisyer.

A.F. thanks his children, Zoe and Simon Filippenko, for the long hours they endured while he was busy working on this book rather than playing with them; they mean the Universe to him. A.F. also thanks his wife, Noelle Filippenko, for her incredible patience, advice, and support; he could not have succeeded without her, and he looks forward to sharing more of the wonders of the cosmos with her.

We are extremely grateful to all the individuals named above for their assistance. Of course, it is we who have put this all together, and we alone are responsible for the content. We would appreciate hearing from readers with suggestions for improved presentation of topics, with comments about specific points that need clarification, with typographical or other errors, or just to tell us how you like your astronomy course. We invite readers to write to us, respectively, at Williams College, Hopkins Observatory, 33 Lab Campus Drive, Williamstown, MA 01267–2630, and Astronomy Department, University of California, Berkeley, CA 94720–3411, by e-mail, or through the book's websites. We promise a personal response to each writer.

Jay M. Pasachoff
Williamstown, Massachusetts
jay.m.pasachoff@williams.edu

Alex Filippenko
Berkeley, California
alex@astro.berkeley.edu

Websites

http://astronomy.brookscole.com /cosmos3, http://www.solarcorona.net

ABOUT THE AUTHORS

Jay M. Pasachoff, during the 2005 annular solar eclipse.

Jay M. Pasachoff is Field Memorial Professor of Astronomy at Williams College, where he teaches the astronomy survey course and works with undergraduate students on a variety of astronomical research projects. He is also Director of the Hopkins Observatory and Chair of the Astronomy Department there. He received his undergraduate and graduate degrees from Harvard and was at the California Institute of Technology before going to Williams College.

Pasachoff pioneered the emphasis in textbooks on contemporary astronomy alongside the traditional bases. He has taken advantage of his broad experience with a wide variety of ground-based telescopes and spacecraft in writing his texts. He received the 2003 Education Prize of the American Astronomical Society.

Pasachoff's expedition with students to the 2006 total solar eclipse was his 42nd solar eclipse. His research is currently sponsored by the National Science Foundation, NASA, and the National Geographic Society. He is Chair of the Working Group on Eclipses of the International Astronomical Union. He is collaborating with colleagues to observe occultations of stars by Pluto, its largest moon (Charon), Triton, and other objects in the outer parts of the Solar System. He also works in radio astronomy of the interstellar medium, concentrating on deuterium and its cosmological consequences.

At this writing, Pasachoff is President of the Commission on Education and Development of the International Astronomical Union. He is co-editor of *Teaching and Learning Astronomy: Effective Strategies for Educators Worldwide* (2005).

Alex Filippenko is a Professor of Astronomy at the University of California, Berkeley. His teaching of an astronomy survey course is very popular on campus; he has won the most coveted teaching awards at Berkeley and has four times been voted "Best Professor" on campus. He received his undergraduate degree from the University of California, Santa Barbara, and his doctorate from the California Institute of Technology.

Filippenko has produced three video courses on college-level astronomy. The recipient of the 2004 Carl Sagan Prize for Science Popularization, he lectures widely, and he has appeared frequently on science newscasts and television documentaries.

Alex Filippenko, at the Keck II telescope.

Filippenko's primary areas of research are supernovae, gamma-ray bursts, active galaxies, black holes, and observational cosmology. He and his collaborators have obtained some of the best evidence for the existence of stellar-mass black holes in our Milky Way Galaxy. His robotic telescope, together with a large team that includes many undergraduate students, is conducting the world's most successful search for exploding stars in relatively nearby galaxies, having found over 500. He has made major contributions to the discovery that the expansion rate of the Universe is speeding up with time, propelled by a mysterious "dark energy." One of the world's most highly cited astronomers, his research has been recognized with several prestigious awards.

He has served as a Councilor of the American Astronomical Society and has been President of the Astronomical Society of the Pacific. He is an active member of the International Astronomical Union.

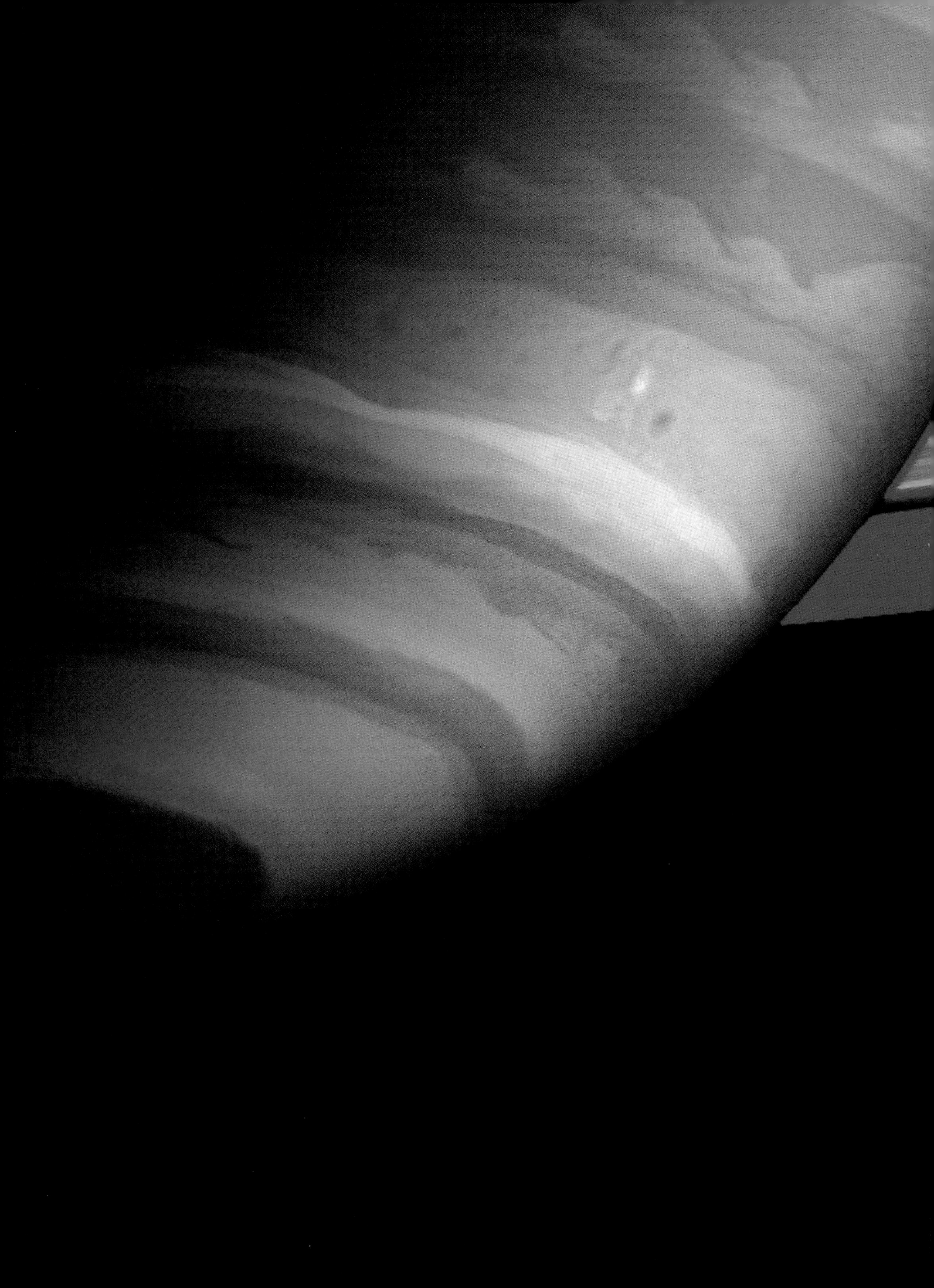

CHAPTER 1

A Grand Tour of the Heavens

Astronomy is in a golden age, filled with the excitement of new discoveries and a deeper understanding of the Universe, our home—and what an enthralling universe it is!

We have explored most of the planets in the Solar System, revealing an astonishingly wide variety of terrains and moons. We have discovered planets orbiting other stars, increasing our confidence that life exists elsewhere. We have solved many of the mysteries surrounding stellar birth and death, revealing among other things how the chemical elements inside our bodies, like calcium and oxygen, formed inside stars. With the Hubble Space Telescope, the Chandra X-ray Observatory, and the Spitzer Space Telescope we are examining galaxies shortly after their birth, deducing important clues to the origin and evolution of our own Milky Way Galaxy.

We have witnessed explosions of stars halfway across the visible Universe whose power is so tremendous that it rivals a galaxy containing ten billion normal stars. Indeed, we are lucky that none have recently occurred too close to Earth, for we wouldn't survive. We have detected black holes—strange objects whose gravitational pull is so strong that nothing, not even light, can escape. And, most recently, we have found strong evidence that what we thought was "empty space" actually contains a dark, gravitationally repulsive kind of energy that is causing the Universe to expand faster and faster with time. The origin of this "dark energy" is a complete mystery, but an understanding of it may revolutionize physics.

The study of astronomy enriches our view of the Universe and fills us with awe, increasing our appreciation of its sheer grandeur and beauty. We hope that your studies will inspire you to ask questions about the Universe all around us and show you how to use your detective skills to search for the answers. Get ready for a thrilling voyage unlike any that you've ever experienced!

ORIGINS

We comment on "Origins" and "Structure and Evolution" as organizing themes.

AIMS

1. Survey the Universe and the methods astronomers use to study it (Sections 1.1, 1.2, 1.4).
2. Learn the measurement units used by astronomers (Section 1.1).
3. See how the sky looks in different seasons (Section 1.3).
4. Understand the value of astronomy to humans (Section 1.5).
5. Assess the scientific method and show how pseudoscience fails scientific tests (Sections 1.6, 1.7).

AceAstronomy™ The AceAstronomy icon throughout this text indicates an opportunity for you to test yourself on key concepts and to explore animations and interactions of the AceAstronomy website at **http://astronomy.brookscole.com/cosmos3**

Saturn, imaged from the Cassini spacecraft, which went into orbit around that planet in 2004. This false-color view shows a storm, known as the "Dragon Storm," in Saturn's clouds and, at right, detail in Saturn's rings. Methane gas shows as red, so the clouds at right are under a lot of methane and are therefore low in Saturn's atmosphere.
NASA/JPL/Space Science Institute

1.1 Peering through the Universe: A Time Machine

Astronomers have deduced that the Universe began almost 14 billion years ago. Let us consider that the time between the origin of the Universe and the year 2006, or 14 billion years, is compressed into one day. If the Universe began at midnight, then it wasn't until slightly after 4 p.m. that the Earth formed; the first fossils date from 6 p.m. The first humans appeared only 2 seconds ago, and it is only 1/300 second since Columbus landed in America. Still, the Sun should shine for another 9 hours; an astronomical timescale is much greater than the timescale of our daily lives (■ Fig. 1–1).

One fundamental fact allows astronomers to observe what happened in the Universe long ago: Light travels at a finite speed, 300,000 km/sec (equal to 186,000 miles per second), or nearly 10 trillion km per year. As a result, if something happens far away, we can't know about it immediately. Light from the Moon takes about a second to reach us (1.3 seconds, more precisely), so we see the Moon as it was roughly one second ago. Light from the Sun takes about eight minutes to reach us. Light from the nearest of the other stars takes over four years to reach us; we say that it is over four "light-years" away. Once we look beyond the nearest stars, we are seeing much farther back in time (see *Figure It Out 1.1: Keeping Track of Space and Time*).

The Universe is so vast that when we receive light or radio waves from objects across our home, the Milky Way Galaxy (a collection of hundreds of billions of stars bound together by gravity), we are seeing back tens of thousands of years. Even for the most nearby other galaxies, light has taken hundreds of thousands, or even millions, of years to reach us. And for the farthest known galaxies, the light has been travelling to us for billions of years. New telescopes on high mountains and in orbit around the Earth enable us to study these distant objects much better than we could previously. When we observe these farthest objects, we see them as they were billions of years ago. How have they changed in the billions of years since? Are they still there? What have

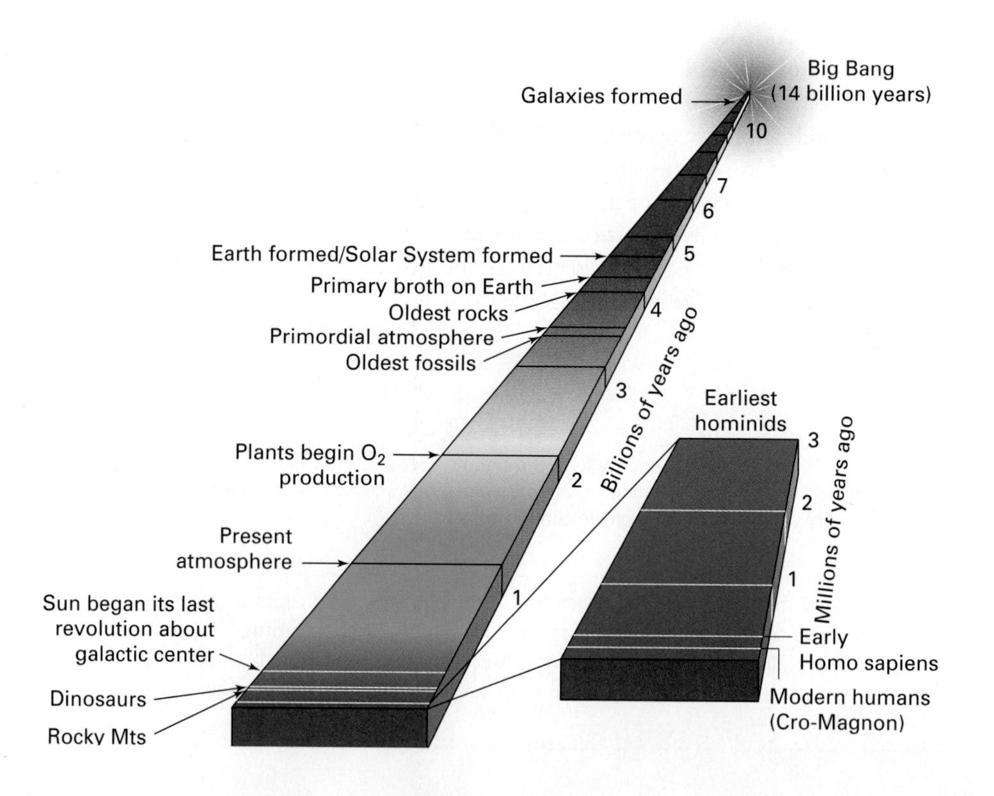

■ **FIGURE 1-1** A sense of time.

FIGURE IT OUT 1.1

Keeping Track of Space and Time

Throughout this book, we shall generally use the metric system, which is commonly used by scientists. The basic unit of length, for example, is the meter, which is equivalent to 39.37 inches, slightly more than a yard. Prefixes are used (Appendix 1) in conjunction with the word "meter," whose symbol is "m," to define new units. The most frequently used prefixes are "milli-," meaning 1/1000, "centi-," meaning 1/100, "kilo-," meaning 1000 times, and "mega-," meaning one million times. Thus 1 millimeter is 1/1000 of a meter, or about 0.04 inch, and a kilometer is 1000 meters, or about 5/8 of a mile.

As we describe in *Figure It Out 1.2: Scientific Notation,* we keep track of the powers of 10 by which we multiply 1 m by writing the number of tens we multiply together as an exponent; 1000 m (1000 meters), for example, is 10^3 m, since 1000 has 3 zeros following the 1 and thus represents three tens multiplying each other.

The standard symbol for "second" is "s," so km/s is kilometers per second. (We will generally use "sec" in this book for added clarity.) Astronomers measure mass in kilograms (kg), where each kilogram is 10^3 grams (1000 g).

We can keep track of distance not only in the metric system but also in units that are based on the length of time that it takes light to travel. The speed of light is, according to Einstein's special theory of relativity, the greatest speed that is physically attainable for objects travelling through space. Light travels at 300,000 km/sec (186,000 miles/sec), so fast that if we could bend it enough, it would circle the Earth over 7 times in a single second. (Such bending takes place only near a black hole, though; for all intents and purposes, light goes straight near the Earth.)

Even at that fantastic speed, we shall see that it would take years for us to reach the stars. Similarly, it has taken years for the light we see from stars to reach us, so we are really seeing the stars as they were years ago. Thus, we are looking backward in time as we view the Universe, with distant objects being viewed farther in the past than nearby objects.

The distance that light travels in a year is called a **light-year;** note that the light-year is a unit of *length* rather than a unit of time even though the term "year" appears in it. It is equal to 9.53×10^{12} km (nearly 10 trillion km)—an extremely large distance by human standards.

A "month" is an astronomical time unit, based on the Moon's orbit, and a "year" is an astronomical time unit, based on the Earth's orbit around the Sun. The measurement of time itself is now usually based on processes in atoms, which are used to define the second. So the second from atomic timekeeping is slightly different from the second that we think of as a sixtieth of a minute, which is a sixtieth of an hour, which is a twenty-fourth of a day, which is based on the rotation of the Earth on its axis. For some purposes, weird stars called pulsars keep the most accurate time in the Universe. In this book, when we are talking about objects billions of years old, it won't matter precisely how we define the second or the year.

they evolved into? The observations of distant objects that we make today show us how the Universe was a long, long time ago—peering into space, we watch a movie of the history of the Universe, allowing us to explore and eventually understand it.

Building on such observations, astronomers use a wide range of technology to gather information and construct theories to learn about the Universe, to discover what is in it, and to predict what its future will be. This book will show you how we look, what we have found, and how we interpret and evaluate the results.

1.2 How Do We Study Things We Can't Touch?

The Universe is a place of great variety—after all, it has everything in it! At times, astronomers study things of a size and scale that humans can easily comprehend: the planets, for instance. Most astronomical objects, however, are so large and so far away that we have trouble grasping their sizes and distances. Many of these distant objects

FIGURE IT OUT 1.2

Scientific Notation

In astronomy, we often find ourselves writing numbers that have strings of zeros attached, so we use what is called either scientific notation or exponential notation to simplify our writing chores. Scientific notation helps prevent making mistakes when copying long strings of numbers, and so aids astronomers (and students) in making calculations.

In scientific notation, which we use in *A Closer Look 1.1: A Sense of Scale,* included in this chapter, we merely count the number of zeros, and write the result as a superscript to the number 10. Thus the number 100,000,000, a 1 followed by 8 zeros, is written 10^8. The superscript is called the exponent. (In spreadsheets, like Microsoft Excel, the exponent is written after a caret, ^, as in 10^8.) We also say that "ten is raised to the eighth power."

When a number is not an integer power of 10, we divide it into two parts: a number between 1 and 10, and an integer power of 10. Thus the number 3645 is written as 3.645×10^3. The exponent shows how many places the decimal point was moved to the left.

We can represent numbers between zero and one by using negative exponents. A minus sign in the exponent of a number means that the number is actually one divided by what the quantity would be if the exponent were positive. Thus $10^{-2} = 1/10^2 = 1/100 = 0.01$. Instead of working with 0.00256, for a further example, we would move the decimal point three places to the right and write 2.56×10^{-3}.

Powers of 1000 beyond kilo- (a thousand) are mega- (a million), giga- (a billion), tera- (a trillion), peta-, exa-, zetta-, and yotta-. It has been suggested, not entirely seriously, to use groucho- and harpo-, after two of the Marx Brothers, for the next prefixes.

are fascinating and bizarre—ultra-dense pulsars that spin on their axes hundreds of times per second, exploding stars that light up the sky and incinerate any planets around them, giant black holes with a billion times the Sun's mass.

In addition to taking photographs of celestial objects, astronomers break down an object's light into its component colors to make a **spectrum** (see Chapter 2), much like a rainbow (■ Fig. 1–2). Today's astronomers, thanks to advances in telescopes and in devices to detect the incoming radiation, study not only the visible part of the spectrum, but also its gamma rays, x-rays, ultraviolet, infrared, and radio waves. We use telescopes on the ground and in space to observe astronomical objects in almost all parts of the spectrum. Combining views in the visible part of the spectrum with studies of invisible radiation gives us a more complete idea of the astronomical object we are studying than we could otherwise have (■ Fig. 1–3). Regardless of whether we are looking at nearby or very distant objects, the techniques of studying in various parts of the spectrum are largely the same.

■ **FIGURE 1–2** The visible spectrum, light from the Sun spread out in a band. The dark lines represent missing colors, which tell us about specific elements in space or in the Sun absorbing those colors.
National Optical Astronomy Observatories/Kitt Peak

■ **FIGURE 1–3** Objects often look very different in different parts of the spectrum. ⓐ The double cluster, two adjacent groups of stars, imaged from the ground in visible light. ⓑ One of the clusters imaged from space with the Chandra X-ray Observatory. Colors show images made in different parts of the x-ray spectrum.

© Robert Gendler 2004

Nancy Evans, Scott Wolk, Fred Seward, Tom Barnes, Scott Kenyon, and Jay Pasachoff, NASA/CXC/SAO

■ **FIGURE 1-4** Near the twin 10-m Keck telescopes (a), we find the 8.2-meter-diameter Gillett Gemini North telescope (b) on Mauna Kea volcano in Hawaii. Note the Milky Way above the Gemini North Telescope. A consortium of universities is planning to make the Giant Magellan Telescope, to be erected in Chile, out of several mirrors this size in order to have the equivalent of a 21-meter-diameter telescope.

■ **FIGURE 1-6** *(left)* The center of the Crab Nebula, the remnant of a star that exploded almost 1000 years ago, in visible light. This highly detailed image was taken with the Very Large Telescope in Chile. *(right)* An x-ray image of the heart of the Crab Nebula taken with the Chandra X-ray Observatory. The image reveals tilted 1-light-year rings of material lighted by electrons flowing from the central pulsar, plus jets of gas extending perpendicular to the rings.

■ **FIGURE 1-5** An image of part of a small, neighboring galaxy to our own, the Tarantula Nebula in the Large Magellanic Cloud. It was taken with the Hubble Space Telescope and a European Southern Observatory ground-based telescope. It was mosaicked and processed by a 23-year-old amateur astronomer using publicly available software. Shock waves from exploding stars have compressed the gas into the visible filaments and sheets.

The tools that astronomers use are bigger and better than ever. Giant telescopes on mountaintops collect visible light with mirrors as large as 10 meters across (■ Fig. 1–4). Up in space, above Earth's atmosphere, the Hubble Space Telescope sends back very clear images (■ Fig. 1–5). Many faraway objects are seen as clearly with Hubble as those closer to us appear with most ground-based telescopes. This accomplishment enables us to study a larger number of distant objects in detail. (The ground-based astronomers are developing methods of seeing very clearly, too, over limited areas of the sky.) The Chandra X-ray Observatory produces clear images of a wide variety of objects using the x-rays they emit (■ Fig. 1–6).

1.3 Finding Constellations in the Sky

When we look outward into space, we see stars that are at different distances from us. But our eyes don't reveal that some stars are much farther away than others of roughly the same brightness. People have long made up stories about groups of stars that appear in one part of the sky or another. The major star groups are called **constellations.** These constellations were given names, occasionally because they resembled

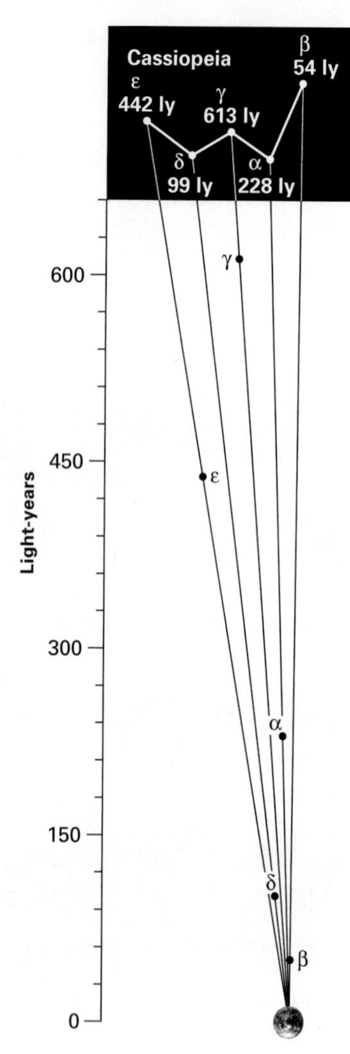

■ **FIGURE 1-7** The stars we see as a constellation are actually at different distances from us. In this case, we see the true relative distances of the stars in the "W" of Cassiopeia, as determined from the Hipparcos spacecraft. The stars' appearance projected on the sky is shown in the upper part.

something (for example, Scorpius, the Scorpion), but mostly to honor a hero or other subject of a story.

The International Astronomical Union put the scheme of constellations on a definite system in 1930. The sky was officially divided into 88 constellations (see Appendices) with definite boundaries, and every star is now associated with one and only one constellation. But the constellations give only the directions to the stars, and not the stars' distances. Individual stars in a given constellation generally have quite different distances from us (■ Fig. 1–7); these stars aren't physically associated with each other, and they were born at different times and locations.

Thus, the constellations show where things appear to be in the sky, but not what they are like or what they are made of. In most of this book, you will study the "how and why" of astronomy, not merely where things are located or what they are called. Still, it can be fun to look up at night and recognize the patterns in the sky.

Some groupings of stars are very familiar to many people but are not actually constellations. These configurations, made of parts of one or more constellations, are known as *asterisms.* The Big Dipper, for example, is an asterism but isn't a constellation, since it is but part of the constellation Ursa Major (the Big Bear). As we will see further in Chapter 4, some asterisms and constellations are sufficiently close to the celestial north pole in the sky that they are visible at all times of year, as seen from the United States. The Big Dipper is an example. But other asterisms and constellations, farther from celestial north, are visible at night for only part of the year. Let us now survey some of the prominent asterisms and constellations that you can see in each season; see also *Star Party 1.1: Using the Sky Maps.*

AceAstronomy™ Log into AceAstronomy and select this chapter to see the Active Figure called "Constellations from Different Latitudes."

1.3a The Autumn Sky

As it grows dark on an autumn evening, you will see the Pointers in the Big Dipper—the two end stars—point upward toward Polaris. Known as the "north star," Polaris is not one of the brightest or nearest stars in the sky, but is well known because it is close to the direction of the celestial north pole. As we will see in Chapter 4, that means it uniquely appears almost motionless in the sky over the night, and provides a bearing that can help you get safely out of the woods. Almost an equal distance on the other side of Polaris is a "W"-shaped constellation named Cassiopeia (■ Fig. 1–8). In Greek mythology, Cassiopeia was married to Cepheus, the king of Ethiopia (and the subject of the constellation that neighbors Cassiopeia to the west). Cassiopeia appears sitting on a chair.

As we continue across the sky away from the Pointers, we come to the constellation Andromeda, who in Greek mythology was Cassiopeia's daughter. In Andromeda, on a very dark night you might see a faint, hazy patch of light; this is actually the center of the nearest large galaxy to our own, and is known as the Great Galaxy in Andromeda, or the Andromeda Galaxy. Though it is one of the nearest galaxies, about 2.4 million light-years away, it is much farther away than any of the individual stars that we see in the sky, since they are all in our own Milky Way Galaxy.

Southwest in the sky from Andromeda, but still high overhead, are four stars that appear to make a square known as the Great Square of Pegasus. One of the corners of this asterism is actually in the constellation Andromeda.

If it is really dark outside (which probably means that you are far from a city and also that the Moon is not full or almost full), you will see the hazy band of light known as the "Milky Way" crossing the sky high overhead, passing right through Cassiopeia. This dim band with ragged edges, which marks the plane of our disk-shaped galaxy (see Chapter 16), has many dark patches that make rifts in its brightness.

■ **FIGURE 1-8** The constellation Cassiopeia is easily found in the sky from its distinctive "W" shape.

ASIDE 1.1: Observing the sky

Amateur astronomers often hold viewing sessions informally known as "star parties" during which they observe celestial objects. Likewise, the occasional "Star Party" boxes in this text highlight interesting observations that you can make.

Star Party 1.1 Using the Sky Maps

Because of Earth's motion around the Sun over the course of a year, the parts of the sky that are "up" after dark change slightly each day. A given star rises (and crosses the meridian, or highest point of its arc across the sky) about 4 minutes earlier each day. By the time a season has gone by, the sky has apparently slipped a quarter of the way around at sunset as the Earth has moved a quarter of the way around the Sun in its yearly orbit. Some constellations are lost in the afternoon and evening glare, while others have become visible just before dawn.

In December of each year, the constellation Orion crosses the meridian at midnight. Three months later, in March, when the Earth has moved through one quarter of its orbit around the Sun, the constellation Virgo crosses the meridian at midnight, when Orion is setting. Orion crosses the meridian at sunset (that is, 6 hours earlier than in December—consistent with 4 minutes/day × 90 days = 360 minutes = 6 hours). Another three months later, in June, Orion crosses the meridian an additional 6 hours earlier—that is, at noon. Hence, it isn't then visible at night. Instead, the constellation Ophiuchus crosses the meridian at midnight.

Because of this seasonal difference, inside the front and back covers of this book we have included four Sky Maps, one of which is best for the date and time at which you are observing. Suitable combinations of date and time are marked. Note also that if you make your observations later at night, it is equivalent to observing later in the year. Two hours later at night is the same as shifting over by one month.

Hold the map above your head while you are facing north or south, as marked on each map, and notice where your zenith is in the sky and on the map. The horizon for your latitude is also marked. Try to identify a pattern in the brightest stars that you can see. Finding the Big Dipper, and using it to locate the pole star, often helps you to orient yourself. Don't let any bright planets confuse your search for the bright stars—knowing that planets usually appear to shine steadily instead of twinkling like stars (see Chapter 4) may assist you in locating the planets.

Come back and look at Sections 1.3a–d at the appropriate time of year—even after you have finished with this course.

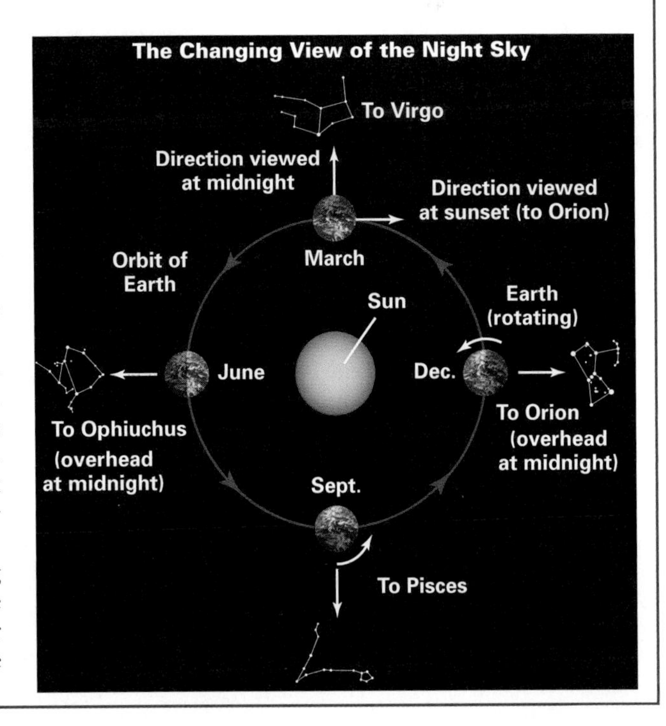

Moving southeast from Cassiopeia, along the Milky Way, we come to the constellation Perseus; he was the Greek hero who slew the Medusa. (He flew off on Pegasus, the winged horse, who is conveniently nearby in the sky, and saw Andromeda, whom he saved.) On the edge of Perseus nearest to Cassiopeia, with a small telescope or binoculars we can see two hazy patches of light that are really clusters of hundreds of stars called "open clusters," a type of grouping we will discuss in Chapter 11. This "double cluster in Perseus," also known as h and χ (the Greek letter "chi") Persei, provides two of the open clusters that are easiest to see with small telescopes. (They already appeared in Figure 1–3.) In 1603, Johann Bayer assigned Greek letters to the brightest stars and lower-case Latin letters to less-bright stars (■ Fig. 1–9), but in this case the system was applied to name the two clusters as well.

Along the Milky Way in the other direction from Cassiopeia (whose "W" is relatively easy to find), we come to a cross of bright stars directly overhead. This "Northern Cross" is an asterism marking part of the constellation Cygnus, the Swan (■ Fig. 1–10). In this direction, spacecraft detect x-rays whose brightness varies with time, and astronomers have deduced in part from that information that a black hole is located there. Also in Cygnus is a particularly dark region of the Milky Way, called the northern Coalsack. Dust in space in that direction prevents us from seeing as many stars as we see in other directions of the Milky Way.

Slightly to the west is another bright star, Vega, in the constellation Lyra (the Lyre). And farther westward, we come to the constellation Hercules, named for the mythological Greek hero who performed twelve great labors, of which the most famous was bringing back the golden apples. In Hercules is an older, larger type of star cluster called a "globular cluster," another type of grouping we will discuss in Chapter 11. It is known as M13, the great globular cluster in Hercules. It resembles a fuzzy mothball whether glimpsed with the naked eye or seen with small telescopes; larger telescopes have better clarity and so can reveal the individual stars.

1.3b The Winter Sky

As autumn proceeds and winter approaches, the constellations we have discussed appear closer and closer to the western horizon for the same hour of the night. By

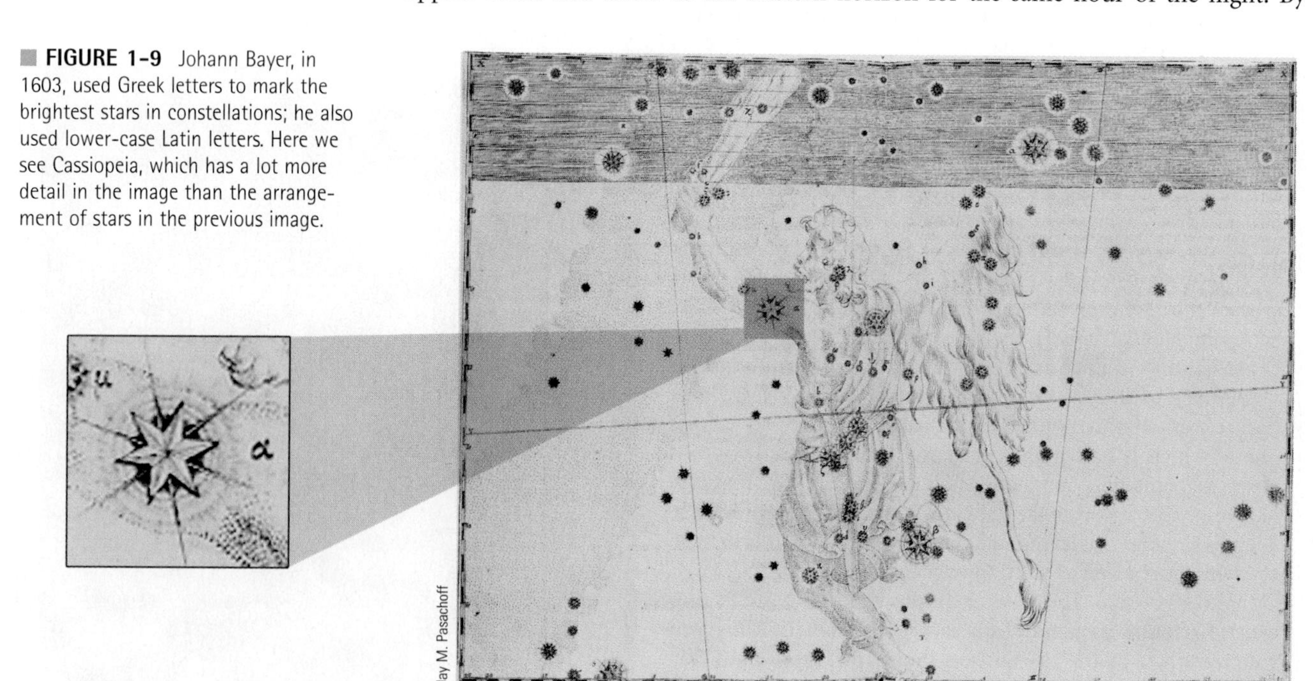

■ **FIGURE 1-9** Johann Bayer, in 1603, used Greek letters to mark the brightest stars in constellations; he also used lower-case Latin letters. Here we see Cassiopeia, which has a lot more detail in the image than the arrangement of stars in the previous image.

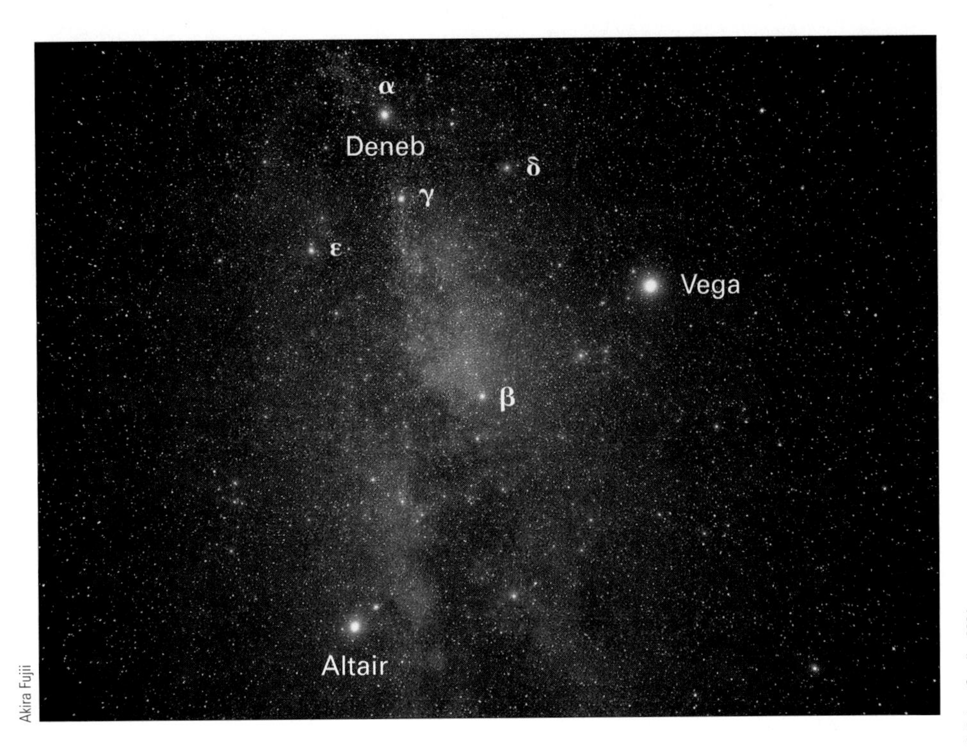

■ **FIGURE 1-10** The Northern Cross, composed of the brightest stars in the constellation Cygnus, the Swan. Deneb, also called alpha (α) Cygni; gamma (γ) Cygni; and beta (β) Cygni make the long bar; epsilon, gamma, and delta Cygni make the crossbar. The bright star Vega, in Lyra, is nearby. Also marked is the bright star Altair, in Aquila, the Eagle. These stars lie in the Milky Way, which shows clearly on the image.

■ **FIGURE 1-11** The Pleiades, the Seven Sisters, in the constellation Taurus, the Bull. It is a star cluster, and long exposures like this one show dust around the stars reflecting starlight, preferentially the bluish colors. When the Pleiades and the Hyades, another star cluster, rose just before dawn, ancient peoples in some parts of the world knew that the rainy season was about to begin.

early evening on January 1, Cygnus is setting in the western sky, while Cassiopeia and Perseus are overhead.

To the south of the Milky Way, near Perseus, we can now see a group of six stars close together in the sky (■ Fig. 1–11). The tight grouping tends to catch your attention as you scan the sky. It is the Pleiades (pronounced "plee′a-deez"), traditionally the Seven Sisters of Greek mythology, the daughters of Atlas. (We can usually see six stars with the unaided eye now, so either one of the stars has faded over the millennia or it was never visible and the association with the Pleiades myth was loose.) These stars are another example of an open cluster of stars. Binoculars or a small telescope will reveal dozens of stars there, whereas a large telescope will ordinarily show too small a region of sky for you to see the Pleiades well. So a bigger telescope isn't always better.

Farther toward the east, rising earlier every evening, is the constellation Orion, the Hunter (■ Fig. 1–12). Orion is perhaps the easiest constellation of all to pick out in the sky, for three bright stars close together in a line make up its belt. Orion is warding off Taurus, the Bull, whose head is marked by a large "V" of stars. A reddish star, Betelgeuse ("bee′tl-juice" would not be far wrong for pronunciation, though some say "beh′tl-jouz"), marks Orion's armpit, and symmetrically on the other side of his belt, the bright bluish star Rigel ("rye′jel") marks his heel. Betelgeuse is an example of a red supergiant star; it is hundreds of millions of kilometers across, far bigger itself than the Earth's *orbit* around the Sun!

Orion's sword extends down from his belt. A telescope, or a photograph, reveals a beautiful region known as the Great Nebula in Orion, or the Orion Nebula. Its general shape can be seen in even a smallish telescope; however, only photographs clearly reveal the vivid colors that long telescopic exposure show—though whether it is reddish or greenish in an image depends on what kind of film is used. It is a site where new stars are forming right now, as you read these words.

Rising after Orion is Sirius, the brightest star in the sky. Orion's belt points directly to it. Sirius appears blue-white, which indicates that its surface is very hot. Sirius is so much brighter than the other stars that it stands out to the naked eye. It is part of the constellation Canis Major, the Big Dog. (You can remember that it is near Orion by thinking of it as Orion's dog.)

■ **FIGURE 1-12** The constellation Orion, marked by reddish Betelgeuse, bluish Rigel, and a belt of three stars in the middle. The Orion Nebula is the reddish object below the belt. It is M42 on the Winter Sky Map at the end of the book. Sirius, the brightest star in the sky, is to the lower left of Orion in this view.

Back toward the top of the sky, between the Pleiades and Orion's belt, is a group of stars that forms the "V"-shaped head of Taurus. This open cluster is known as the Hyades ("hy′a-deez"). The stars of the Hyades mark the bull's face, while the stars of the Pleiades ride on the bull's shoulder. In a Greek myth, Jupiter turned himself into a bull to carry Europa over the sea to what is now called Europe.

1.3c The Spring Sky

We can tell that spring is approaching when the Hyades and Orion get closer and closer to the western horizon each evening, and finally are no longer visible shortly after sunset. Now Castor and Pollux, a pair of equally bright stars, are nicely placed for viewing in the western sky. Castor and Pollux were the twins in the Greek Pantheon of gods. The constellation is called Gemini, the twins.

On spring evenings, the Big Bear (Ursa Major) is overhead, and anything in the Big Dipper—which is part of the Big Bear—would spill out. Leo, the Lion, is just to the south of the overhead point, called the zenith (follow the Pointers backward). Leo looks like a backward question mark, with the bright star Regulus, the lion's heart, at its base. The rest of Leo, to the east of Regulus, is marked by a bright triangle of stars. Some people visualize a sickle-shaped head and a triangular tail.

If we follow the arc made by the stars in the handle of the Big Dipper, we come to a bright reddish star, Arcturus, another supergiant. It is in the kite-shaped constellation Boötes, the Herdsman.

Sirius sets right after sunset in the spring; however, a prominent but somewhat fainter star, Spica, is rising in the southeast in the constellation Virgo, the Virgin. It is farther along the arc of the Big Dipper through Arcturus. Vega, a star that is between Sirius and Spica in brightness, is rising in the northeast. And the constellation Hercules, with its notable globular cluster M13, is rising in the east in the evening at this time of year.

ASIDE 1.2: A capital idea

When we refer to our Milky Way Galaxy, we say "our Galaxy" or "the Galaxy" with an upper-case "G." When we refer to other galaxies, we use a lower-case "g."

1.3d The Summer Sky

Summer, of course, is a comfortable time to watch the stars because of the generally warm weather. Spica is over toward the southwest in the evening. A bright reddish star, Antares, is in the constellation Scorpius, the Scorpion, to the south. ("Antares" means "compared with Ares," another name for Mars, because Antares is also reddish.)

Hercules and Cygnus are high overhead, and the star Vega is prominent near the zenith. Cassiopeia is in the northeast. The center of our Galaxy is in the dense part of the Milky Way that we see in the constellation Sagittarius, the Archer, in the south (■ Fig. 1–13).

Akira Fujii

■ **FIGURE 1–13** The Milky Way—the band of stars, gas, and dust that marks the plane of our disk-shaped Galaxy—is easily recognized in a dark sky. It is shown here with Halley's Comet (*left, middle*, with a tail) passing by.

Around August 12 every summer is a wonderful time to observe the sky, because that is when the Perseid meteor shower occurs. (Meteors, or "shooting stars," are not stars at all, as we will discuss in Chapter 8.) In a clear, dark sky, with not much moonlight, one bright meteor a minute may be visible at the peak of the shower. Just lie back and watch the sky in general—don't look in any specific direction. The rate of meteors tends to be substantially higher after midnight than before midnight, since our part of Earth has then turned so that it is plowing through space, crossing through the paths of pebbles and ice chunks that streak through the sky as they heat up. Although the Perseids is the most observed meteor shower, partly because it occurs at a time of warm weather in the northern part of the country, many other meteor showers occur during various parts of the year.

The summer is a good time of year for observing a prime example that shows that stars are not necessarily constant in brightness. This "variable star," Delta Cephei, appears in the constellation Cepheus, which is midway between Cassiopeia and Cygnus. Delta Cephei varies in brightness with a 5.4-day period. As we will see in Chapters 11 and 16, studies of its variations have been important for allowing us to measure the distances to galaxies. This fact reminds us of the real importance of studying the sky—which is to learn what things are and how they work, and not just where they are. The study of the sky has led us to better understand the Universe, which makes astronomy beautiful and exciting to so many people.

AceAstronomy™ Log into AceAstronomy and select this chapter to see the Active Figure called "Constellations in Different Seasons."

1.4 How Do You Take a Tape Measure to the Stars?

It is easy to see the direction to an object in the sky but much harder to find its distance. Astronomers are always seeking new and better ways to measure distances of objects that are too far away to touch. Our direct ability to reach out to astronomical objects is limited to our Solar System. We can send people to the Moon and spacecraft to the other planets. We can even bounce radio waves off the Moon, most of the other planets, and the Sun, and measure how long the radio waves' round trip takes to find out how far they have travelled. The distance (d) travelled by light or by an object is equal to the constant rate of travel (its speed v) multiplied by the time (t) spent travelling ($d = vt$).

Once we go farther afield, we must be more ingenious. Repeatedly in this book, we will discuss ways of measuring distance. For the nearest hundreds of thousands of stars, we have recent results from a spacecraft that showed how much their apparent position shifts when we look at them from slightly different angles (see Chapter 11). For more distant stars, we find out how far away they are by comparing how bright they actually (intrinsically) are and how bright they appear to be. We often tell their intrinsic brightness from looking at their spectra (Chapter 11), though we will also see other methods.

Once we get to galaxies other than our own, we will see that our methods are even less precise. For the nearest galaxies, we search for stars whose specific properties we recognize. Some of these stars are thought to be identical in type to the same kinds of stars in our own Galaxy whose intrinsic brightnesses we know. Again, we can then compare intrinsic brightness with apparent brightness to give distance. For the farthest galaxies, as we shall discuss in Chapter 16, we find distances using the 1920s discovery that shifts in color of the spectrum of a galaxy reveal how far away it is. Of course, we continue to test these methods as best we can, and some of the most exciting investigations of modern astronomy are related to the determination of distances. For example, one of the Hubble Space Telescope's Key Projects has been devoted to finding distances of galaxies, and has made good progress toward resolving a long-term controversy over the size and age of the Universe.

ASIDE 1.3: Broken conversation

If we were carrying on a conversation by radio with someone at the distance of the Moon, there would be pauses of noticeable length after we finished speaking before we heard an answer. This is because radio waves, even at the speed of light, take over a second to travel each way. Astronauts on the Moon have to get used to these pauses when their messages travel by radio waves to people on Earth.

1.5 The Value of Astronomy

1.5a The Grandest Laboratory of All

Throughout history, observations of the heavens have led to discoveries that have had a major impact on people's ideas about themselves and the world around them. Even

A Closer Look 1.1 A Sense of Scale: Measuring Distances

Let us try to get a sense of scale of the Universe, starting with sizes that are part of our experience and then expanding toward the enormously large. Imagine a series of cubes, of which we see one face (a square).

© 1995 David Scharf

1 mm = 0.1 cm

We begin our journey through space with a view of something 1 millimeter across, an electron-microscope view of a fly. Every step we take will show a region 100 times larger in diameter than that in the previous picture.

Jay M. Pasachoff

10 cm = 100 mm

A square 100 times larger on each side is 10 centimeters × 10 centimeters. (Since the area of a square is the length of a side squared, the area of a 10 cm square is 10,000 times the area of a 1 mm square.) The area encloses a flower.

Neil Leifer for Sports Illustrated; © AOL Time Warner, Inc.

10 m = 1000 cm

Here we move far enough away to see an area 10 meters on a side, with Muhammad Ali triumphant.

Joseph Distefano, City of Boston, and Abrams Aerial Survey Corp

1 km = 10^3 m

A square 100 times larger on each side is now 1 kilometer square, about 250 acres. An aerial view of Boston shows how big this is.

NASA

100 km = 10^5 m

The next square, 100 km on a side, encloses the cities of Boston and Providence. Note that though we are still bound to the limited area of the Earth, the area we can see is increasing rapidly.

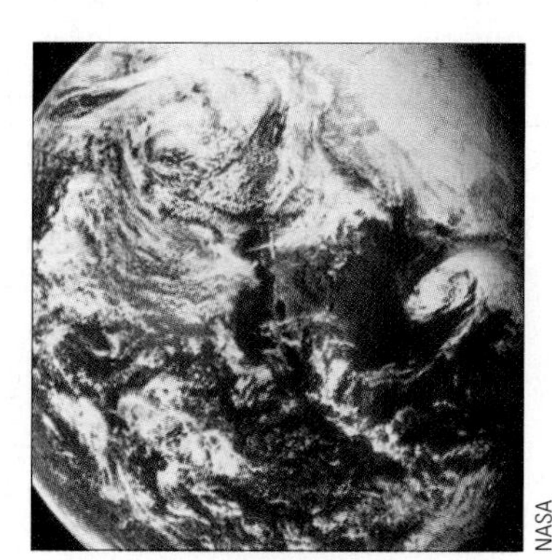

NASA

10,000 km = 10^7 m

A square 10,000 km on a side covers nearly the entire Earth. We see the southwestern United States, through the clouds, and northwestern Mexico, including the Baja California peninsula.

© NASA/JPL/Caltech

1,000,000 km = 10^9 m = 3.3 lt sec

When we have receded 100 times farther, we see a square 100 times larger in diameter: 1 million kilometers across. It encloses the orbit of the Moon around the Earth. We can measure with our wristwatches the amount of time that it takes light to travel this distance.

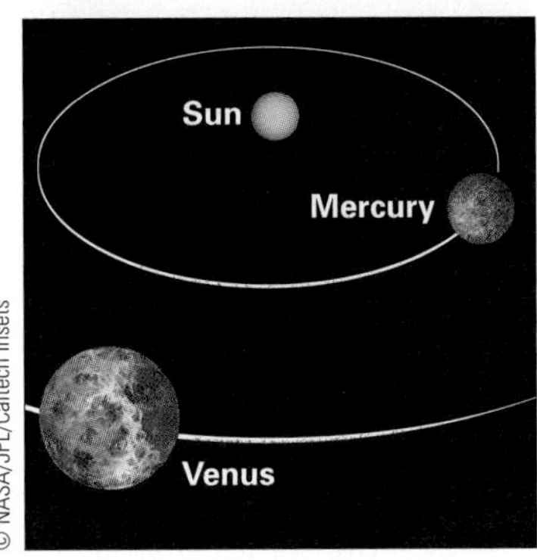

© NASA/JPL/Caltech Insets

10^{11} m = 5.5 lt min

When we look on from 100 times farther away still, we see an area 100 million kilometers across, 2/3 the distance from the Earth to the Sun. We can now see the Sun and the two innermost planets in our field of view.

Saturn image: NASA/JPL/Space Science Institute

10^{13} m = 9 lt hr

An area 10 billion kilometers across shows us the entire Solar System in good perspective. It takes light about 10 hours to travel across the Solar System. The outer planets have become visible and are receding into the distance as our journey outward continues.

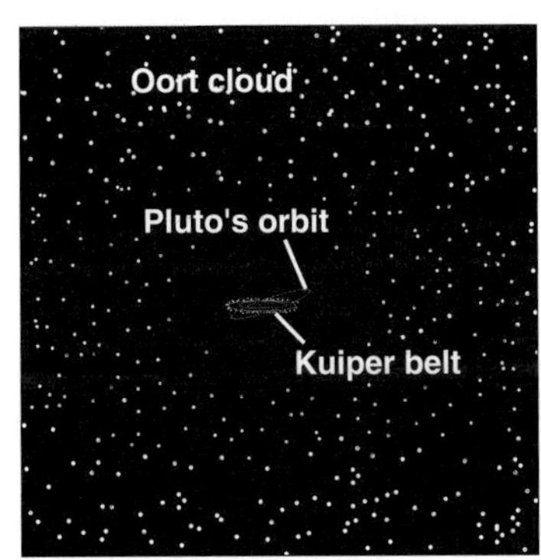

10^{15} m = 38 lt days

From 100 times farther away, 1 trillion km across, we see Pluto and other small, planet-like bodies in the Kuiper belt. At even larger distances from the Sun, the Oort cloud is a vast collection of comets. We have not yet reached the scale at which another star besides the Sun is in a cube of this size.

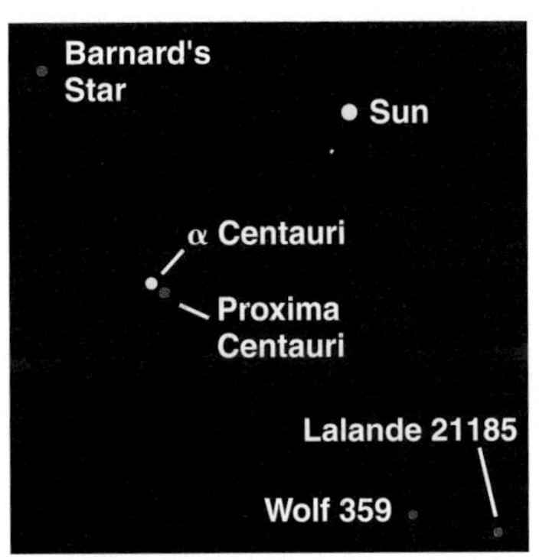

10^{17} m = 10 lt yr

As we continue to recede from the Solar System, the nearest stars finally come into view. We are seeing an area 10 light-years across, which contains only a few stars, most of whose names are unfamiliar (see the Appendices). The brightest stars tend to be intrinsically very powerful, but more distant than these dim, nearby stars.

Axel Mellinger, Universitat Potsdam

10^{19} m = 10^3 lt yr

By the time we are 100 times farther away, we can see a 1000 light-year fragment of our Galaxy, the Milky Way Galaxy. We see not only many individual stars but also many clusters of stars and many "nebulae"—regions of glowing, reflecting, or opaque gas or dust.

(box continued next page)

A Sense of Scale: Measuring Distances *(continued)*

Hubble Heritage Team (AURA/STScI/NASA)

10^{21} m = 10^5 lt yr

In a field of view 100 times larger in diameter, 100,000 light-years across, we would be able to view our entire Milky Way Galaxy, with its spiral arms, in one go. (The picture shown is a Hubble image of a galaxy similar to ours.)

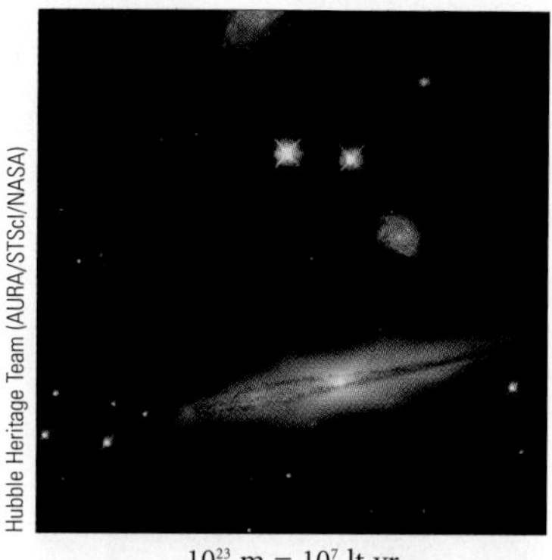

Hubble Heritage Team (AURA/STScI/NASA)

10^{23} m = 10^7 lt yr

Next we move sufficiently far away so that we can see an area 10 million light-years across. There are about as many centimeters across this image as there are grains of sand in all of Earth's beaches. The image shows part of a cluster of galaxies; our Galaxy is in such a grouping, the Local Group.

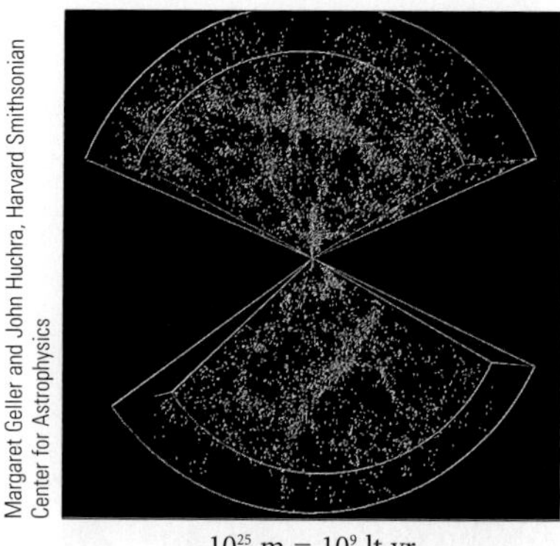

Margaret Geller and John Huchra, Harvard Smithsonian Center for Astrophysics

10^{25} m = 10^9 lt yr

If we could see a field of view 100 times larger, 1 billion light-years across, our Local Group of galaxies would appear as but part of a supercluster—a cluster of galaxy clusters. The image shows a mapping of a wedge of the Universe centered on Earth and containing hundreds of thousands of galaxies.

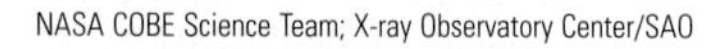

NASA COBE Science Team; X-ray Observatory Center/SAO

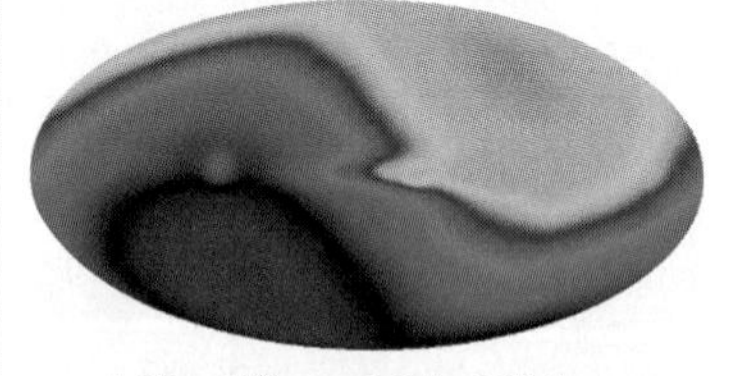

1.37 x 10^{26} m = 1.37 x 10^{10} lt yr

We have even detected radiation from the Universe's earliest years. This map of the whole sky shows a slight difference in the Universe's temperature in opposite directions (Chapter 19), resulting from our Sun's motion. In more detailed maps, we are seeing the seeds from which today's clusters of galaxies grew. A combination of radio, ultraviolet, x-ray, and optical studies, together with theoretical work and experiments with giant atom smashers on Earth, is allowing us to explore the past and predict the future of the Universe.

the dawn of mathematics may have stemmed from ancient observations of the sky, made in order to keep track of seasons and seasonal floods in the fertile areas of the Earth. Observations of the motions of the Moon and the planets, which are free of such complicating terrestrial forces as friction and which are massive enough so that gravity dominates their motions, led to an understanding of gravity and of the forces that govern all motion.

The regions of space studied by astronomers serve as a cosmic laboratory where we can investigate matter or radiation, often under conditions that we cannot duplicate on Earth (■ Fig. 1–14). These studies allow us to extend our understanding of the laws of physics, which govern the behavior and evolution of the Universe.

A new importance has been given to astronomy by the realization that large asteroids and comets have hit the Earth every few tens of millions of years with enough power to devastate our planet. We shall see in Chapter 8 how the dinosaurs and many other species were extinguished 65 million years ago by an asteroid or comet, according to a theory that has become widely favored. All those movies about asteroids and comets on collision courses with Earth aren't entirely science fiction!

Many of the discoveries of tomorrow—perhaps the control of nuclear fusion or the discovery of new sources of energy, or maybe even something so revolutionary

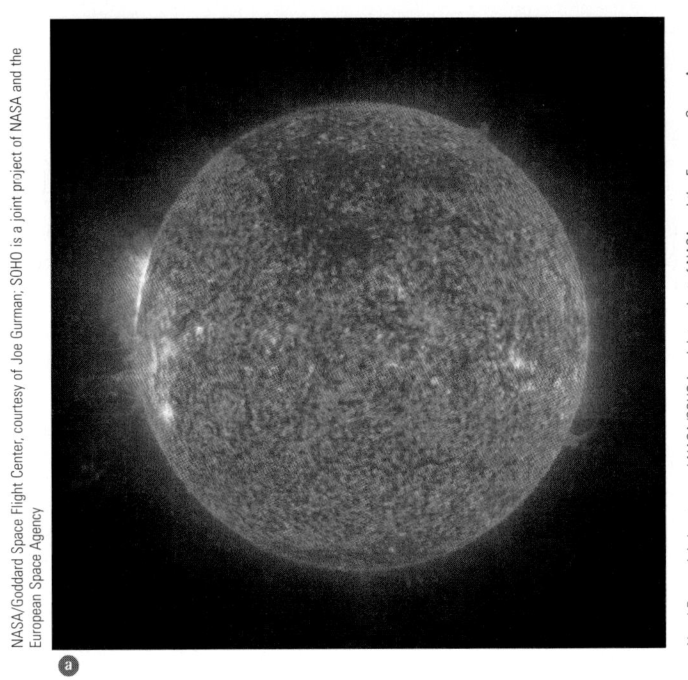

FIGURE 1-14 The Sun is made of plasma, a strange stage of matter that is gas so hot that it is ionized and subject to magnetic fields. Here we see an eruption on the Sun's edge that occurred in 2004. The images are from the Solar and Heliospheric Observatory (SOHO). (a) An image in the helium radiation that occurs in gas at temperatures around 60,000 K, showing the solar chromosphere. (b) An image with a "coronagraph" showing gas at temperatures of millions of degrees. The coronagraph blocks not only the ordinary solar disk, whose size is shown with a white circle, but also the inner corona, in order to limit scattering in the telescope that would fog the image.

that it cannot now be predicted—will undoubtedly be based on advances made through such basic research as the study of astronomical systems. Considered in this sense, astronomy is an investment in our future.

The impact of astronomy on our conception of the Universe has been strong through the years. Discoveries that the Earth is not at the center of the Universe, or that the Universe has been expanding for billions of years, affect our philosophical conceptions of ourselves and of our relations to space and time.

Yet most of us study astronomy not for its technological and philosophical benefits but rather for its beauty and inherent interest. We must stretch our minds to understand the strange objects and events that take place in the far reaches of space. The effort broadens us and continually enthralls us all. Ultimately, we study astronomy because of its fascination and mystery.

1.5b Origins

It is always fun to look at our own childhood pictures. Similarly, astronomers are increasingly looking at the baby pictures of the Universe and of various types of objects in it. NASA has made its "Origins" program a centerpoint of its strategy for investigation. Under this program, many scientists will use spacecraft, telescopes on Earth, computers, and just plain clear thought to study all kinds of origins: the origin of the Universe, of our Galaxy, of stars (Fig. 1–15), of the Sun, of our Earth and the elements in it, and of life itself—maybe someday not only life here on Earth but also on one of the planets around other stars that are being discovered at an increasing rate. After all, we now know of more planets outside our Solar System than we do inside, a big change in our view that occurred in the past decade.

FIGURE 1-15 A giant cloud of gas and dust, with thousands of stars forming. The view, with false color to bring out different wavelengths of infrared radiation, was taken with the Spitzer Space Telescope. It shows the Tarantula Nebula in the Large Magellanic Cloud. See Figure 1-5 for an optical view.

NASA's Origins program has three official major scientific goals: (1) to understand how galaxies formed early in the Universe and to determine the role of galaxies in the emergence of planetary systems and life; (2) to understand how stars and planetary systems form and to determine whether life-sustaining planets exist around other stars; and (3) to understand how life originated on Earth and to determine whether it began and may still exist elsewhere as well. Existing or planned missions relevant to the Origins program include the Hubble Space Telescope, the Spitzer Space Telescope, the Keck Interferometer on Mauna Kea (interferometers are ways of linking several telescopes together, as we shall see in Chapter 4, making them equivalent for some purposes to a single, huge telescope), various Mars missions, a Europa mission, and the airplane known as the Stratospheric Observatory for Infrared Astronomy (SOFIA).

Specific near-future space missions that are part of the Origins program are the Space Interferometry Mission and the Webb Space Telescope. In the longer term, perhaps two decades away, NASA is planning the Terrestrial Planet Finder, a telescope or set of telescopes in space that is to allow imaging of small planets like our own. The goal of the Terrestrial Planet Finder is to detect directly small, rocky planetary companions to other stars as well as to study spectra rich in features from molecules that might indicate the habitability of any planets that are discovered.

Another major theme of NASA's scientific investigations is Structure and Evolution. In this book, you will learn about the structure of the Solar System, of our Galaxy, and of space. You will learn the life stories of stars, and see how different kinds of stars evolve. And you will learn about the evolution of the Universe as a whole.

In this book, we will make these themes explicit as the central organizing structure. In particular, we will try to point out the links to Origins—to our Universe's childhood pictures—on a chapter-by-chapter basis.

ASIDE 1.4: NASA roadmap

The NASA "roadmap" for the structure and evolution of the Universe theme is "Beyond Einstein: From the Big Bang to Black Holes." NASA's former Physics and Astronomy Division is now the Universe Division.

1.6 What Is Science?

Science is not merely a body of facts; it is also a process of investigation. The standards that scientists use to assess their ideas and to decide which to accept—in some sense, those that are "true"—are the basis of much of our technological world. One of the guiding principles of science is that results should be *reproducible*; that is, other scientists should be able to get essentially the same result by repeating the same experiment or observation. Science is thus a self-checking way of carrying out investigations.

Though acceptable scientific investigations are actually carried out in many ways, there is a standard model for the **scientific method.** In this standard model, one first looks at a body of data and makes educated guesses as to what might explain them. An educated guess is a **hypothesis.** Then one thinks of consequences that would follow if the hypothesis were true and tries to carry out experiments or make observations that test the hypothesis. If, at any time, the results are contrary to the hypothesis, then the hypothesis is discarded or modified. (There may, though, be other assumptions that could be modified or discarded instead, because they are inappropriate. Also, one needs to check for experimental or observational errors.) If the hypothesis passes its tests and is established in some basic framework or set of equations, it can be called a **theory.**

ASIDE 1.5: Grammatical note

Note that "data" is a plural word; the singular form is "datum," a distinction not always followed these days.

If the hypothesis or theory survives test after test, it is accepted as being "true." Still, at any time a new experiment or observation could show it was false after all. This process of "falsification," being able to find out if a hypothesis or theory is false, is basic to the definition of science formulated by a leading set of philosophers of science.

Sometimes, you see something described as a "law" or a "principle" or a "fact." "Laws of nature"—like Kepler's laws of planetary motion, Newton's laws of motion, or Newton's law of gravitation (see Chapter 5)—are actually *descriptions* of how nature behaves, and they might in fact be incorrect. Newton used the word "law" (in Latin) over three hundred years ago. The words "law" and "principle" are historical usages associated with certain basic theories. Scientists consider something a "fact" when it is so well established that it would be extremely unlikely that it is incorrect.

What is an example of the scientific method? Albert Einstein advanced a theory in 1916 to explain how gravity works. (It was developed so completely that it was far beyond a hypothesis.) His "general theory of relativity" was based on mathematical equations he worked out from some theoretical ideas. His result turned out to explain some observations of the orbit of Mercury (the orbit's elliptical shape slowly rotates with time) that had been puzzling up to that time, though this hadn't been Einstein's motivation. Still, its basic tests lay ahead.

Einstein's theory predicted that starlight would appear to be bent by a certain amount if it could be observed to pass very near the Sun. Such bending would be visible only during a total solar eclipse, when the sky was dark but the Sun was still present in the sky. The theory seemed so important that expeditions were made to total solar eclipses to test it. When the first telegram came back that the strange prediction of Einstein's theory had been verified, the theory was quickly accepted. Successful predictions are usually given more weight than mere explanations of already known facts (■ Fig. 1–16). Einstein immediately gained a worldwide reputation as a great scientist. The theory's nearly complete acceptance by the scientific community was established as viable alternatives could not be found.

Most of the time, though, the scientific method does not work so straightforwardly. We can consider, for example, our understanding of how the stars shine. Until the 1930s, it was thought that the Sun got its energy from its contracting gas, but this idea seemed wrong, because it could not explain how the Sun could be as old as rocks on Earth whose ages we measured. (Presumably the Sun and Earth formed nearly simultaneously.) Then scientists suggested that nuclear fusion—the merging of 4 hydrogen nuclei to make a single helium nucleus, in particular—could provide the energy for the Sun to live 10 billion years (see Chapter 12), and even worked out the detailed ways in which the hydrogen could fuse. Thus a theory of nuclear energy as a source of power for the Sun and stars exists. But how could it be tested? The observed properties of a wide variety of stars around the Universe fit, in several ways, with the deductions of the theory. Still, a direct test had to wait.

Over the last thirty years, measurements have finally been made of individual subatomic particles called "neutrinos" that should be given off as the fusion process takes place. Neutrinos from fusion in the Sun were indeed measured, but only at one-third to one-half the rate expected, as we will describe in Chapter 12. Had our main theory of stellar energy failed? No, because explanations were discovered for why some of the neutrinos may not reach us in a detectable form. No physicists doubt that the Sun and stars get their energy from nuclear fusion, so our theory of the properties of neutrinos themselves is changing. Thus the astronomical study of neutrinos from the Sun has led to important changes in the most basic ideas of physics.

So the "scientific method" isn't cut-and-dried. But it does demand a rigor and honesty in scientific testing. The standards in science are high, especially those of evidence, and we hope that this book will provide enough examples to enable you to form an accurate impression of how the process actually works.

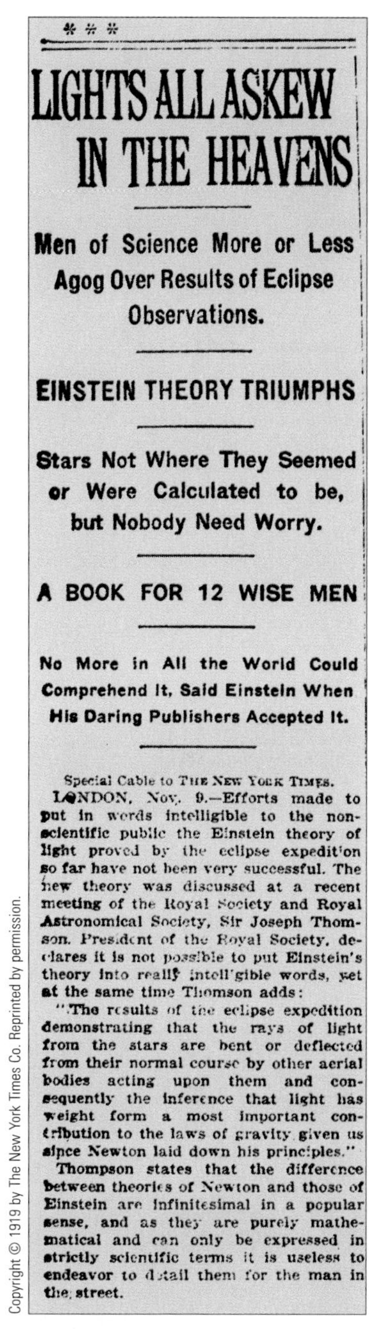

LIGHTS ALL ASKEW IN THE HEAVENS

Men of Science More or Less Agog Over Results of Eclipse Observations.

EINSTEIN THEORY TRIUMPHS

Stars Not Where They Seemed or Were Calculated to be, but Nobody Need Worry.

A BOOK FOR 12 WISE MEN

No More in All the World Could Comprehend It, Said Einstein When His Daring Publishers Accepted It.

Special Cable to THE NEW YORK TIMES.

LONDON, Nov. 9.—Efforts made to put in words intelligible to the non-scientific public the Einstein theory of light proved by the eclipse expedition so far have not been very successful. The new theory was discussed at a recent meeting of the Royal Society and Royal Astronomical Society, Sir Joseph Thomson, President of the Royal Society, declares it is not possible to put Einstein's theory into really intelligible words, yet at the same time Thomson adds:

"The results of the eclipse expedition demonstrating that the rays of light from the stars are bent or deflected from their normal course by other aerial bodies acting upon them and consequently the inference that light has weight form a most important contribution to the laws of gravity given us since Newton laid down his principles."

Thompson states that the difference between theories of Newton and those of Einstein are infinitesimal in a popular sense, and as they are purely mathematical and can only be expressed in strictly scientific terms it is useless to endeavor to detail them for the man in the street.

■ **FIGURE 1–16** This report was one of the early descriptions of general relativity's success that captured the public's fancy. From *The New York Times* of November 10, 1919.

1.7 Why Is Science Far Better Than Pseudoscience?

Though science is itself fascinating, too many people have beliefs that may seem related to science but either have no present verification or are false. Such beliefs, such as astrology or the idea that UFOs (unidentified flying objects) are now definitely bringing aliens from other planets, are pseudoscience rather than science. ("Pseudo" means that something is not authentic or sincere, in spite of it looking somewhat real.)

Astrology is not at all connected with astronomy, except in a historical context (they had similar origins, and hence the same root), so it does not really deserve a place in a text on astronomy. But since so many people incorrectly associate astrology with astronomy, and since astrologers claim to be using astronomical objects to make their predictions, let us use our knowledge of astronomy and of the scientific method

FIGURE 1-17 Twelve constellations through which the Sun, Moon, and planets pass make up the zodiac.

to assess astrology's validity. Millions of Americans—a number that shows no signs of decreasing—believe in astrology as more than just a source of amusement, so the topic is too widespread to ignore.

Astrology is an attempt to predict or explain our actions and personalities on the basis of the positions of the stars and planets now and at the instants of our births. Astrology has been around for a long time, but it has never definitively been shown to work. Believers may cite incidents that reinforce their faith in astrology, but statistical measurements do not compellingly show a real effect. If something happens to you that you had expected because of an astrological prediction, you would more certainly notice that this event occurred than you would notice the thousands of unpredicted things that also happened to you that day. Yet we do enough things, have sufficiently varied thoughts, and interact with enough people that if we make many predictions, some of them are likely to be at least partially fulfilled by chance. We simply forget the rest.

In fact, even the traditional astrological alignments are not accurately calculated, for the Earth's pole points in different directions in space as time passes over millennia. In truth, stars are overhead at different times of year compared with the case millennia ago, when the astrological tables that are often still in current use were computed. At a given time of year, the Sun is usually in a different sign of the zodiac from its traditional astrological one (Fig. 1–17). Further, we know that the constellations are illusions; they don't even exist as physical objects. They are merely projections of the positions of stars that are at very different distances from us.

ASIDE 1.6: Too little knowledge

In 2003, it was reported that hundreds of thousands of children in India could be paralyzed by polio because their parents are unknowledgeably rejecting the vaccine.

Studies have shown that superstition actively constricts the progress of science and technology in various countries around the world and is therefore not merely an innocent force. It is not just that some people harmlessly believe in astrology; their lack of understanding of scientific structure may actually impede the training or work of people needed to solve the problems of our age, such as AIDS, pollution, food shortages, and the energy crisis.

In addition to the lack of evidence corroborating astrological predictions, one minor reason why scientists in general and astronomers in particular doubt astrology is that they cannot conceive of a way in which it could possibly work. The human brain is so complex that it seems most improbable that any celestial alignment can affect people, including newborns, in an overall way. The celestial forces that are known are not sufficient to set personalities or influence daily events. Any unknown force must have truly remarkable properties to exert any noticeable effect on humans.

Let us mention a specific study showing that astrology doesn't work, one in which a psychologist tested certain values. Do Libras and Aquarians rank "Equality" highly? Do Sagittarians especially value "Honesty"? Do Virgos, Geminis, and Capricorns treasure the quality "Intellectual"? Several astrology books agreed that these and other similar examples are values typical of those signs. Although believers often criticize the objections of skeptics on the grounds that these "group horoscopes" are not as valuable or accurate as individualized charts, surely some general assumptions and rules hold in common, as claimed in newspaper horoscope columns. (Signs of the zodiac appear in Fig. 1–18.)

Jay M. Pasachoff

FIGURE 1-18 The symbols for the constellations of the zodiac shown on the historic 15th-century astronomical clock in Prague.

The subjects, 1600 psychology graduate students, did not know in advance what was being tested. They gave their birth dates, and the questioners determined their astrological signs. The results: no special correlation with the values they were supposed to hold was apparent for any of the signs. Also, when asked to what extent they shared the qualities of each given sign, as many subjects ranked themselves above average as below, regardless of their astrological signs.

From an astronomer's view, astrology is meaningless, unnecessary, and impossible to explain without contradicting the broad set of physical laws that we have developed and tested over the years and that very well explains what happens on the Earth and in the sky. Astrology snipes at the roots of all pure science. Let's learn from the stars, but let's learn the truth.

Science is more than just a set of facts, since a methodology of investigation and standards of proof are involved, but science is also more than just a methodology. In this course, you should not only learn certain facts about the Universe, but in addition appreciate the way that theories and facts come to be accepted.

We will be discussing such fascinating things as quasars, pulsars, and black holes. We will consider complex molecules that have spontaneously formed in interstellar space, and try to decide whether the Universe will expand forever or collapse. We have sent a rocket into interstellar space bearing a portrait of humans, and have beamed a radio message toward a cluster of stars 24,000 light-years away. These topics and actions are part of modern astronomy: what contemporary, often conservative, scientists are doing and thinking about the Universe. So, though it would be tremendously exciting if celestial alignments could affect individuals, the evidence is so strongly against this idea that it seems fruitless to spend time on astrology instead of the exciting aspects of the real astronomical Universe.

We will have more to say about pseudoscience in the final chapter of this book, where we discuss the notion that UFOs are currently bringing aliens to Earth.

ASIDE 1.7: Self-deception

A team of psychologists from the California State University at Long Beach arranged for a magician to perform three psychic-like stunts in front of psychology classes. Even when they emphasized to the students that the performer was a magician performing tricks, 50 per cent of the class still believed the magician to be psychic. Evidently, a lot of self-deception is involved with believing in pseudoscience.

CONCEPT REVIEW

This is an enormously exciting time to study astronomy. Telescopes and the instruments on them are better than ever, providing new clues to a wide variety of phenomena and deepening our understanding of the cosmos.

The Universe is about 14 billion years old. Compared with a 24-hour day scaled to match the age of the Universe, the Earth formed about 8 hours ago, and humans appeared only 2 seconds ago (Sec. 1.1).

Scientists often use a version of the metric system, in which prefixes like "kilo-" for one thousand and "mega-" for one million go with units like meter for distance and second for time, or are found in kilogram for mass. Astronomers also use special units, like the **light-year** for the distance light travels in one year (Sec. 1.1). By viewing objects that are at different distances from Earth, we look back to different times in the past, because light travels at a finite speed. Thus, the observation of very distant objects we see today provides a glimpse of the Universe as it was a long, long time ago.

Astronomers gain most of their information about the Universe by studying radiation from objects (Sec. 1.2). In addition to taking photographs, we can break down an object's light into its component colors to make a **spectrum,** like a rainbow. We study not only the visible part of the spectrum, but also its gamma rays, x-rays, ultraviolet, visible light, infrared, and radio waves.

The sky is divided into 88 regions known as **constellations** (Sec. 1.3). The individual stars in a given constellation are generally at quite different distances from Earth and are not physically associated with each other; they are simply apparent groupings of stars. As the Earth rotates about its axis and revolves around the Sun, different constellations appear in the night sky. It can be fun to look at the sky and know some of the constellations, as well as their related mythology (Sec. 1.3a–d).

Some of the most exciting studies in modern astronomy are related to the determination of distances (Sec. 1.4). We can measure how far away the Moon or a planet is by bouncing radio waves off it and measuring the total amount of time the waves were in flight. It is much more difficult to measure the distances of stars, galaxies, and other objects beyond our Solar System, but later in this book we explain the methods used by astronomers.

The study of astronomy has had a major impact on the development of science throughout human history (Sec. 1.5). For example, observations of the changing positions of the Moon and the planets led to an understanding of gravity and the laws governing the motions of bodies. Moreover, astronomy allows us to investigate human origins, including the creation of the chemical elements and the formation of the Sun and Earth. The Universe is a cosmic laboratory to test ideas of science in conditions often not available in laboratories on Earth. Observations of the skies may even help humans avoid or delay extinction, if we are able to discover (and subsequently deflect) large asteroids or comets that are on a collision course with Earth.

One of the main principles of science is that results should be reproducible; other scientists should be able to get essentially the same result by repeating the same experiment or observation (Sec. 1.6). Astronomers and other scientists follow the **scientific method,** which is difficult to define but provides a standard about which scientists agree. In the basic form of the scientific method, a **hypothesis** passes observational tests to become a **theory.**

Many people have beliefs that may seem related to science but either have no present verification or are false; they are based on pseudoscience rather than on authentic science (Sec. 1.7). Astrology (as opposed to astronomy), for example, passes no scientific tests and so is not a science. Furthermore, astrology has been shown not to work.

QUESTIONS

1. Why might we say that our senses have been expanded in recent years?

†2. The speed of light is 3×10^5 km/sec. Express this number in m/sec and in cm/sec.

3. Geologists study different strata to determine conditions on Earth long in the past. For example, dinosaur bones are found in strata dating from about 245 million to 65 million years ago. How is it that astronomers are able to see parts of the Universe appearing as they were in the past?

4. Distinguish between a hypothesis and a theory. Give a non-astronomical example of each.

5. Discuss an example of science and an example of pseudoscience, and distinguish between their respective values.

†6. **(a)** Write the following in scientific notation: 4642; 70,000; 34.7. **(b)** Write the following in scientific notation: 0.254; 0.0046; 0.10243. **(c)** Write out the following in an ordinary string of digits: 2.543×10^6; 2.0043×10^2; 7.673×10^{-4}.

†7. Suppose you wish to construct a scale model of the Solar System. The distance between the Earth and the Sun (one "Astronomical Unit"—1 A.U.) is 1.5×10^8 km, and you represent it by 15 km, about the size of a reasonably big city. **(a)** What is the scaled distance of the planet Pluto (39.5 A.U.)? (**Hint:** use *ratios* here and elsewhere when possible!) **(b)** The Earth is about 12,800 km in diameter. How large is Earth on your scale? Compare its size with that of a common object. **(c)** The Sun's diameter is about 1.4×10^6 km. How large is it relative to Earth? **(d)** The nearest star, Proxima Centauri, is 4.2 light-years away (i.e., 4.0×10^{13} km). How far is this in your scale model? **(e)** Compare your answer in part (d) with the true distance to the Moon, 3.84×10^5 km. Comment on whether this gives you some idea of the enormous distances of even the nearest stars.

†8. Suppose an asteroid is found that appears to be heading directly toward Earth (using Earth's location at the time of discovery). It is 4.2 A.U. away (i.e., 4.2 times the Earth-Sun distance), and travelling through the Solar System with a speed of 20 km/sec. **(a)** How long would it take the asteroid to reach Earth (at Earth's location at the time of discovery), assuming the asteroid's speed is constant? **(b)** Will the Earth be at approximately the same location in its orbit around the Sun after the time interval computed in part (a)? Discuss what could happen.

9. Look up the horoscopes for a given day in a wide variety of newspapers. (This can most easily be done over the Internet.) How well do you think they agree with each other? Do you think the predictions are sufficiently general or vague that quite a few people might think they came true, regardless of whether they had anything to do with celestial influences?

10. **True or false?** When we observe very distant stars and galaxies, we see them as they are at the present time (or nearly the present time), since light travels at the maximum possible speed.

11. **True or false?** Astronomers gain information about celestial objects by studying many forms of radiation, including x-rays and infrared, not just visible light.

12. **True or false?** Astrology is a pseudoscience, originally related to astronomy, that has been shown to be incorrect.

13. **True or false?** When added to a unit of measure, the prefixes "kilo," "mega," and "giga" mean one thousand, one million, and one billion (respectively).

14. **Fill in the blank:** The distance travelled by light or by an object moving with a constant speed is equal to speed multiplied by __________.

†15. **Fill in the blank:** If 14 billion years (the approximate current age of the Universe) were compressed to 1 year (i.e., 12 months), then on the same scale the Earth (whose true age is about 4.6 billion years) formed ___ months after the Universe was born.

16. **Fill in the blank:** Distinct groups of stars, called __________, were given names by ancient astronomers; stories were associated with them.

17. **Fill in the blank:** A light-year is a unit of _______.

†18. **Multiple choice:** Suppose the distance from the Sun to Pluto, 40 A.U. = 6×10^9 km, were compressed to the size of a pen (15 cm). On this scale, what would be the distance from the Sun to Aldebaran, a bright star (the Eye of Taurus, the Bull) whose true distance is roughly 60 light-years? (Note: 1 light-year is about 10^{13} km.) **(a)** 15 km. **(b)** 1.5×10^6 km. **(c)** 1.5×10^5 cm. **(d)** 6×10^{14} km. **(e)** 6×10^{14} cm.

†19. **Multiple choice:** An asteroid is found in space with a constant speed of 4.0×10^3 m/sec. How far does it travel in 2 minutes? **(a)** 33.3 m. **(b)** 4.8×10^2 m. **(c)** 2.0×10^3 m. **(d)** 8.0×10^3 m. **(e)** 4.8×10^5 m.

20. **Multiple choice:** Scientific results must be **(a)** hypothetical; **(b)** reproducible; **(c)** controversial; **(d)** believed by at least 50% of scientists; or **(e)** believed by 100% of scientists.

21. **Multiple choice:** The process of breaking light down into its component colors creates **(a)** a spectrum; **(b)** an image; **(c)** a pulse; **(d)** a hologram; or **(e)** a movie.

†This question requires a numerical solution.

TOPICS FOR DISCUSSION

1. What is the value of astronomy to you? How do you rank National Science Foundation (NSF) and National Aeronautics and Space Administration (NASA) funds for research with respect to other national needs? Reanswer this question when you have completed this course.
2. Weigh the risks of repairing and upgrading the Hubble Space Telescope with astronauts as opposed to building a robot capable of carrying out the repair.
3. Weigh the importance of maintaining the Hubble Space Telescope compared with other uses of money within NASA for other missions.
4. Discuss the role of pseudoscience, like astrology, in our modern society.

MEDIA

AceAstronomy™ Log into AceAstronomy at **http://astronomy.brookscole.com/cosmos3** to access quizzes and animations that will help you assess your understanding of this chapter's topics.

Log into the Student Companion Web Site at **http://astronomy.brookscole.com/cosmos3** for more resources for this chapter including a list of common misconceptions, news and updates, flashcards, and more.

CHAPTER 2

Light, Matter, and Energy: Powering the Universe

The light that reaches us from the stars and planets is only one type of radiation, a certain way in which energy moves through space. Radiation in this sense results from the continuous changes in electricity and magnetism at each point of space, so is more formally known as **electromagnetic radiation** or **electromagnetic waves.** Gamma rays, x-rays, ultraviolet, ordinary light, infrared, and radio waves are all merely electromagnetic radiation of different wavelengths.

In this chapter, we discuss the properties of radiation in what we call the spectrum and how analysis of this radiation enables scientists to study the Universe. After all, we cannot touch a star! Despite having brought bits of the Moon back to Earth for study, we cannot yet do the same for even the nearest planets, though we have found a few rocks from the Moon and from Mars in the form of meteorites (see Chapter 8). At various points in the book, we will discuss other types of contact that we on Earth have with the Universe beyond, including particles called cosmic rays and extremely elusive particles called neutrinos. Remember that this book's index can take you straight to any given topic, should you choose to learn about it before the text formally addresses it.

Studies of atoms are important for understanding radiation and how it is given off. The simplest atom is hydrogen, and we will see how it gives off or takes up radiation at certain colors, which are known as spectral lines. These spectral lines sometimes are visible as bright colors; they are known as emission lines. When seen from certain angles (to be discussed later), a gas's spectral lines can appear as gaps in a continuous band of color; they are then known as absorption lines. We will see how studying the spectral lines can even tell you whether an object is moving toward or away from us and how fast.

Finally, we will discuss the measurement scales most commonly used to describe temperature: Fahrenheit, Celsius, and kelvin. We also show how to convert from one temperature scale to another.

ORIGINS

We discuss the basic constituents of atoms and some of the techniques that allow us to explore the Universe nearly as far back as the beginning of time. We discover from studies of light that distant stars and galaxies are made of the same kinds of elements as those found on Earth.

AIMS

1. Learn about the spectrum, how spectral lines are formed by atoms, and what spectral lines tell us (Sections 2.1 to 2.4).
2. Explore the nature of the continuous radiation emitted by a body of a given temperature (Section 2.2).
3. See how the Doppler effect tells us about an object's motion (Section 2.5).

AceAstronomy™ The AceAstronomy icon throughout the text indicates an opportunity for you to test yourself on key concepts and to explore animations and interactions on the AceAstronomy website at **http://astronomy.brookscole.com/pf2e**

◀ The reddish color that shows in this visible-light image of gas known as the Trifid Nebula comes from emission by hydrogen, one of the topics covered in this chapter.
Canada-France-Hawaii Telescope/J.-C. Cuillandre/Coelum

■ **FIGURE 2–1** When white light passes through a prism, a full optical spectrum results, with different colors bent by different amounts.

2.1 Studying a Star Is Like Looking at a Rainbow

Over 300 years ago Isaac Newton showed that when ordinary sunlight is passed through a prism, a band of color like the rainbow comes out the other side. Thus "white light" is composed of all the colors of the rainbow (■ Fig. 2–1). A graph of color versus the brightness (we won't make technical distinctions in this book between "brightness" and "intensity") at each color is called a **spectrum** (plural: **spectra**), as is the actual display of color spread out. A very dense gas or a solid gives off a **continuous spectrum,** that is, a spectrum that changes smoothly in brightness from one color to the next.

Technically, each color corresponds to light of a specific **wavelength,** the distance between two consecutive wave crests or two consecutive wave troughs; see *Figure It Out 2.1: The Nature of Light.* This "wave theory" of light is not the only way we can consider light, but it does lead to very useful and straightforward explanations. Astronomers often measure the wavelength of light in angstroms (abbreviated Å, after A. J. Ångström, a Swedish physicist of the 19th century who mapped the solar spectrum). One angstrom is 10^{-10} m (that is, one ten-billionth of a meter, 1/10,000,000,000 meter).

Expanding the narrow rainbow of light in the center of Figure 2–2, the range of visible light from violet light through blue, green, yellow, orange, and red light extends only from about 4000 Å to about 6500 Å in wavelength. The entire visible region of the spectrum is in this narrow range of wavelength, which varies by less than a factor of 2. You can also see the wavelengths of visible light if you look ahead to

FIGURE IT OUT 2.1
The Nature of Light

Light can be described as an electromagnetic wave: self-propagating, oscillating (that is, getting larger and smaller) electric and magnetic fields.

The *wavelength,* denoted by the Greek lowercase λ (lambda), is the distance from one wave crest to the next; see the figure. This has units of length, such as cm.

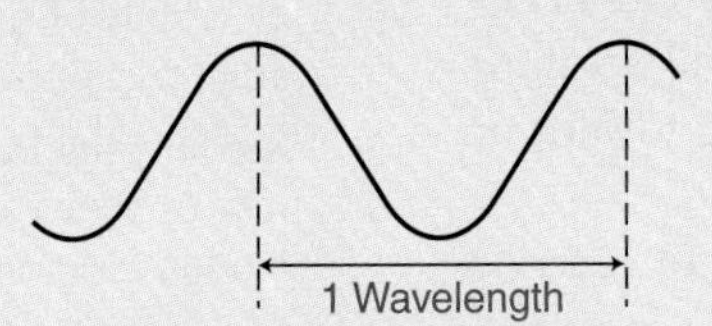

The *frequency,* denoted by the Greek lowercase ν (nu), is the number of times per second that a crest passes a fixed point. The unit is 1/sec, or hertz (Hz). Hence, the *period* of the wave, P (in seconds), is simply $1/\nu$.

In general, the length per wave (λ) multiplied by the number of waves per second (ν) gives the length per second traversed by the wave. This is its *speed* v, so we have

$$\lambda\nu = v.$$

In our case, $v = c$, the speed of light, so $\lambda\nu = c$.

Note that all electromagnetic waves in a vacuum travel with the same speed, c, regardless of their wavelength. The measured speed is independent of the relative speeds of the observer and the light source. This is admittedly counterintuitive, but it has been completely verified; indeed, it is one of the foundations of Einstein's theory of relativity. Electromagnetic waves do slow down in substances such as glass and water, and the speed generally depends on wavelength. This, in fact, is what leads to the spreading out of the colors when light passes through a prism.

Light can also behave as discrete particles known as *photons* (wave or energy "packets"). This is a fundamental aspect of quantum theory. With the right equipment, photons can be detected as discrete lumps of energy.

A photon has no mass, but its energy E is given by the product of Planck's constant h (named after the quantum physicist Max Planck) and its frequency ν: $E = h\nu$. Planck's constant is a very small number. (Its value is listed in Appendix 2 rather than here, to show that you shouldn't be memorizing it.)

Photons of higher energy, therefore, have higher frequency and shorter wavelength:

$$E = h\nu = hc/\lambda, \text{ since } \lambda\nu = c.$$

(All photons, regardless of their energy, travel through a vacuum with the same speed, c.)

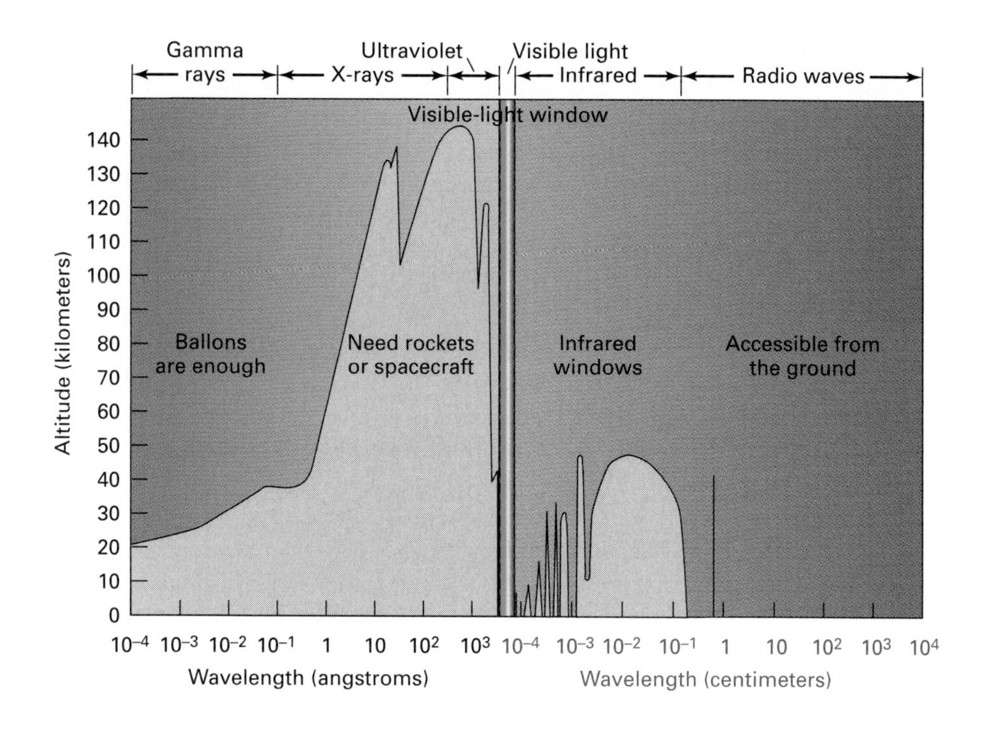

FIGURE 2-2 In a window of transparency in the terrestrial atmosphere, radiation can penetrate to the Earth's surface. The curve specifies the altitude at which the intensity of arriving radiation is reduced to half its original value. When this level of reduction is reached high in the atmosphere, little or no radiation of that wavelength reaches the ground.

Figure 2–4. These wavelengths are about half a millionth of a meter long, since 10,000 Å is 1 micrometer. We can remember the colors we perceive from the name of the friendly fellow **ROY G BIV: R**ed **O**range **Y**ellow **G**reen **B**lue **I**ndigo **V**iolet, going from longer to shorter wavelengths. (The name Indigo is no longer as commonly used.)

Astronomers still tend to use angstroms rather than the unit nanometers that is preferred in the metric system known as Système International (SI). Because 10 Å = 1 nm, divide all the numbers above by 10 to show their values in nanometers: violet light is about 400 nm, for example.

Only certain parts of the electromagnetic spectrum can penetrate the Earth's atmosphere. We say that our atmosphere has "windows" for the parts of the spectrum that can pass through it. The atmosphere is transparent (completely clear) at these windows and opaque (not passing light at all) at other parts of the spectrum. One window passes what we generally call "light," and what astronomers technically call visible light, or "the visible." (In this book, we will use the term "light" to refer to any form of electromagnetic radiation.) Another window falls in the radio part of the spectrum, and modern astronomers can thus base their "radio telescopes" on Earth and still detect that radiation (■ Fig. 2–2).

But we of the Earth are no longer bound to our planet's surface. Balloons, rockets, and satellites carry telescopes high up in or above our atmosphere. They can observe parts of the electromagnetic spectrum that do not reach the Earth's surface. In recent years, we have been able to make observations all across the spectrum. It may seem strange, in view of the longtime identification of astronomy with visible-light observations, to realize that optical studies no longer dominate astronomy.

ASIDE 2.1: A different syllabus

It is possible at any time to consider Chapter 4, which deals with phases of Solar-System objects, with motions and positions in the sky, and with notions of time and calendars. That material is independent of the fundamental concepts of understanding why the planets, stars, and galaxies are the way they are, which we discuss in this chapter and in most of the other chapters of this book.

2.2 "Black Bodies" and Their Radiation

Many things in astronomy seem simple to study. Stars, for example, are balls of gas, and they give off visible light and other electromagnetic radiation that on the whole follow an especially simple rule. If you have two stars of the same size and one is hotter than the other, it is also intrinsically brighter. Furthermore, the hotter star gives off more of its energy at shorter wavelengths.

ASIDE 2.2: How hot is it?

Also see *Figure It Out 2.4: Temperature Conversions,* on page 33.

Let us consider the simplest possible object that gives off radiation. It is called a **black body,** which indicates that it is, in principle, a simple thing—it isn't like a polka-dotted body, for example, that gives off radiation with different properties from different areas. A black body is an opaque object that absorbs all radiation that is incident upon it; absolutely no radiation is reflected or transmitted. Atomic motions within that object (because of its nonzero temperature) then cause it to emit (radiate) energy in a manner that depends only on its temperature, not on chemical composition or anything else. This result is called **black-body radiation** (or **thermal radiation**).

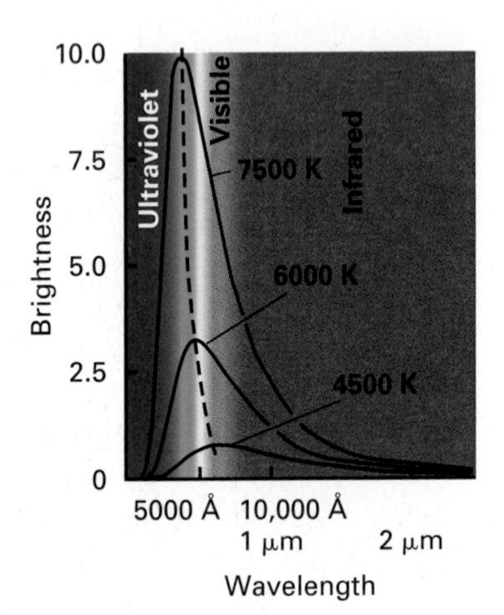

The brightness of radiation at different wavelengths for different temperatures. "Black bodies"—objects that give off (emit) radiation in an ideal manner—have spectra that follow these particular curves, which depend only on the temperature of the object. The spectra of stars are fairly similar to these curves; stars are nearly perfect black bodies.

Understanding black bodies and their radiation is a key to understanding stars. Each bit of gas in a star radiates as a black body. However, overall a star is not a perfect black body; its spectrum actually does depend in a minor way on the chemical composition of its outermost layers of gas. Moreover, these outer layers are partly transparent, so we may see gas having a small range of temperatures when we look at a star. Nevertheless, for many purposes we can approximate the spectrum of a star as being that of a black body. (There are, in fact, very few objects in nature that are perfect, idealized black bodies, yet the concept is useful in a wide range of circumstances.)

Spectra of hotter black bodies peak (that is, have their highest brightness) at shorter (bluer) wavelengths than spectra of colder black bodies (see *Figure It Out 2.2: Black-Body Radiation and Wien's Law*). Also, a hot black body emits much more energy per second than a cold black body of equal surface area (see *Figure It Out 2.3: Black-Body Radiation and the Stefan-Boltzmann Law*).

Planets are much cooler than stars, so they give off most of their radiation at long wavelengths, in the infrared. The Sun, on the other hand, gives off most of its radiation in visible light. When we look at the spectrum of a planet, we see that some of the radiation is strongest in the visible, which means it is reflected sunlight, while other radiation is strongest in the infrared and therefore emitted by the planet itself as thermal radiation, largely in the infrared. Thus a graph of the energy given off by a planet shows two peaks—one in the visible and one in the infrared (■ Fig. 2–3). Planets,

FIGURE IT OUT 2.2

Black-Body Radiation and Wien's Law

A black body is a perfect absorber: It absorbs all incident radiation, reflecting and transmitting none. It has a certain temperature, which is a measure of the average speed with which the particles within it jiggle around: the higher the temperature, the greater the average speed. The randomly moving particles, some of which are charged (that is, have electric charge), emit electromagnetic radiation. This emission is called thermal radiation, or black-body radiation.

A black body is also a perfect emitter: The shape of the object's emitted spectrum depends only on its temperature, not on its chemical composition or other properties. This spectral shape is called the *Planck curve,* in honor of the physicist Max Planck; its derivation was a fundamental problem in quantum physics.

As can be seen in the figure above, at all wavelengths the spectrum (Planck curve) of a hot black body is higher (that is, brighter) than that of a colder black body having the same surface area.

An important property is that the spectrum of a hot black body peaks at a shorter wavelength than that of a colder black body. The product of the temperature (T) and the peak wavelength (λ_{max}) is a constant:

$$\lambda_{max} T = 2.9 \times 10^7 \text{ Å K} = 0.29 \text{ cm K}.$$

This relation is known as *Wien's law.* It is mathematically derivable from the formula for the Planck curve. For example, the spectrum of the Sun, whose surface temperature is about 5800 K, peaks at a wavelength of

$$\lambda_{max} = (2.9 \times 10^7 \text{ Å K})/(5800 \text{ K}) = 5000 \text{ Å}.$$

A star that is twice as hot as the Sun has a spectrum that peaks at half this wavelength, or about 2500 Å; the product of temperature and peak wavelength must remain constant.

FIGURE IT OUT 2.3

Black-Body Radiation and the Stefan-Boltzmann Law

Another useful property of black-body radiation is that per unit of surface area, a hot black body emits much more energy per second than a cold black body. In fact, the energy emitted is proportional to the fourth power of the temperature: $T \times T \times T \times T$. This relation is known as the *Stefan-Boltzmann law,* which (like Wien's law) is derivable from the Planck curve. It can be expressed as

$$E = \sigma T^4,$$

where E is the energy emitted per unit area (for example, cm^2) per second, T is the surface temperature in kelvins (it is important that the temperature scale start at absolute zero), and σ (sigma) is a constant (known as the Stefan-Boltzmann constant).

For example, if two stars (labeled with subscripts 1 and 2) have the same surface area, but one is twice as hot as the other, the hotter star emits $2^4 = 16$ times as much energy (per second) as the colder star:

$$E_1/E_2 = (\sigma T_1^4)/(\sigma T_2^4) = T_1^4/T_2^4 = (T_1/T_2)^4 = 2^4 = 16.$$

However, in the above example, if the hotter star's radius (R) is $\frac{1}{4}$ that of the colder star, then its surface area is $(\frac{1}{4})^2 = \frac{1}{16}$ as large as that of the colder star, because the surface area of a sphere is given by $4\pi R^2$. This lesser area exactly balances the greater emission per unit area, making the two stars equally luminous.

In general, the luminosity (L) (that is, the energy emitted per second, also known as power or intrinsic brightness) of a black body is given by its surface area (S) multiplied by the energy emitted per unit area per second (E): $L = SE$. For a sphere of radius R and temperature T, we have

$$L = 4\pi R^2 \sigma T^4.$$

Thus, if we know the luminosity and surface temperature of a star, we can derive its radius R.

therefore, are not perfect black bodies, but by considering each of its spectral peaks and its black-body approximation, the concept is nonetheless often useful.

Similarly, humans reflect visible light from the Sun or from room lamps, and so are not perfect black bodies. But we do emit our own thermal (black-body) radiation, which is most intense at infrared wavelengths that are visible to certain infrared cameras but not to our eyes.

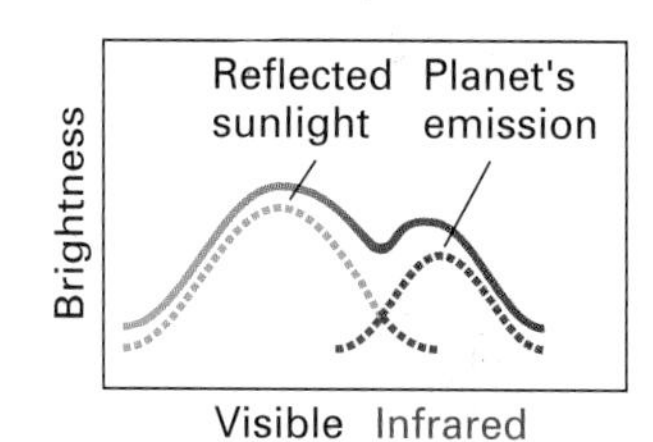

■ **FIGURE 2–3** The spectrum of a planet includes one peak in the visible reflected from ordinary sunlight and another peak in the infrared. The infrared peak is the thermal emission (nearly black-body radiation) from the planet itself, based on its surface temperature. The observed spectrum is the sum of the two curves; here, it is offset upward a little for clarity.

2.3 What Are Those Missing Colors and Where Are They?

In the early 1800s, when Joseph Fraunhofer looked in detail at the spectrum of the Sun, he noticed that the continuous range of colors in the Sun's light was crossed by dark gaps (■ Fig. 2–4). He saw a dozen or so of these "Fraunhofer lines"; astronomers have since mapped millions. The presence of such lines indicates that the Sun is not a perfect black body, but they allow astronomers to learn much more about the Sun (and other stars as well).

The dark Fraunhofer lines turn out to be from relatively cool gas absorbing radiation from behind it. We thus say that they are **absorption lines.** Atoms of each of the chemical elements in the gas absorb light at a certain set of wavelengths. By seeing what wavelengths are absorbed, we can tell what elements are in the gas, and the proportion of them present. We can also measure the temperature of the gas more accurately than from the peak of the nearly black-body spectrum.

Imagine the band of color shown in Figure 2–4 without the dark vertical lines crossing it. That continuous band of color would be the "continuous spectrum" we defined in Section 2.1. With the dark lines crossing the spectrum, however, we have an "absorption-line spectrum" or a "Fraunhofer spectrum."

How does the absorption take place? To understand it, we have to study processes inside the atoms themselves. **Atoms** are the smallest particles of a given chemical element. For example, all hydrogen atoms are alike, all iron atoms are alike, and all uranium

ASIDE 2.3: Spelling help

Note that "absorption" is spelled with a "p," not with a second "b."

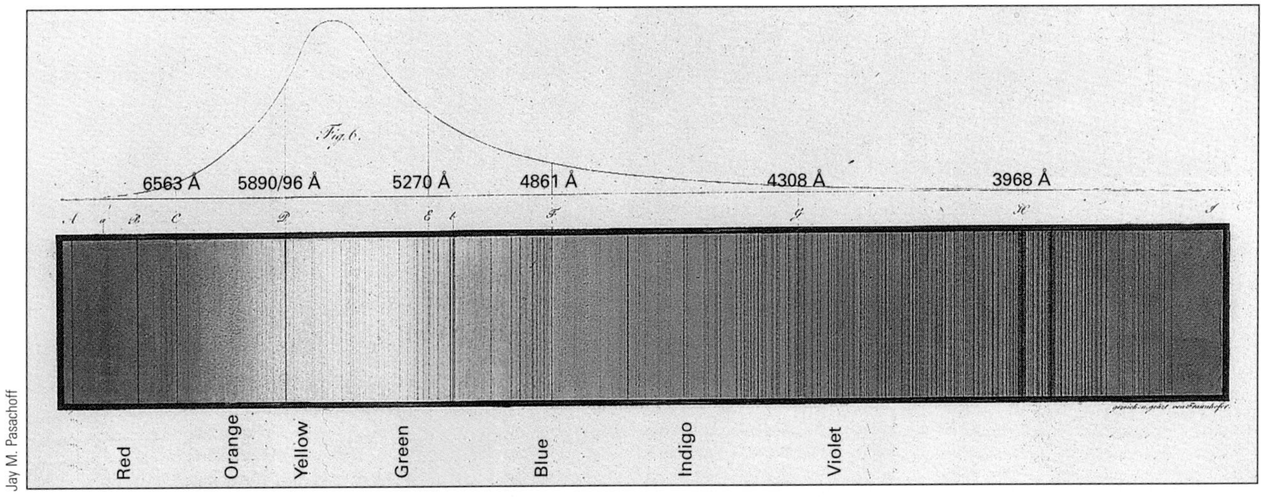

■ **FIGURE 2-4** The dark lines that extend from top to bottom, marking the absence of color at a specific wavelength, are the Fraunhofer lines. The underlying drawing is Fraunhofer's original. We still use the capital-letter notation today for the "D" line (sodium) and the "H" line (from ionized calcium, not hydrogen!). We now know that Fraunhofer's "C" and "F" lines are basic lines of hydrogen. The "E" line is from iron and the "G" line is from iron and calcium. Fraunhofer's lowercase "b" is magnesium. The features marked "A" and "B" are bands from oxygen in the Earth's atmosphere.

■ **FIGURE 2-5** A helium atom contains two protons and two neutrons in its nucleus and two "orbiting" electrons. The nucleus is drawn much larger than its real scale with respect to the overall atom; the actual radius of the electron cloud is over 10,000 times larger than that of the nucleus.

atoms are alike. As was discovered by Ernest Rutherford in 1911, atoms contain relatively massive central objects, which we call **nuclei** (■ Fig. 2–5). Distributed around the nuclei are relatively light particles, which we call **electrons.** The nucleus contains **protons,** which have a positive electric charge, and **neutrons,** which have no electric charge and so are neutral. Thus the nucleus—protons and neutrons together—has a positive electric charge. Electrons, on the other hand, have a negative electric charge. Each of the chemical elements has a different number of protons in its nucleus.

Why aren't the electrons pulled into the nucleus? In 1913, Niels Bohr (■ Fig. 2–6) made a suggestion that some rule (at that time the rule was arbitrary) kept the electrons at fixed distances, in what are often (inaccurately) thought of as orbits. His suggestion was an early step in the development of the **quantum theory,** which was elaborated in the following two decades. The theory incorporates the idea that light consists of individual packets—quanta of energy; the utility of considering such individual, non-continuous objects had been developed a decade earlier.

We can think of each quantum of energy as a massless particle of light, which is called a **photon;** see *Figure It Out 2.1: The Nature of Light.* Photons of blue light have higher energy than photons of red light, but they all travel at the same speed through a vacuum. For some purposes, it is best to consider light as waves, while for others it is best to consider light as particles (that is, as photons). Though we will not explore quantum theory in detail here, we should mention that it is one of the major intellectual advances of the 20th century.

Let us return to absorption lines. We mentioned that they are formed as light passes through a gas (■ Fig. 2–7). The atoms in the gas take up (absorb) some of the light (photons) at specific wavelengths. The energy of this light goes into giving the electrons in those atoms more energy, putting them in higher levels. Each change in energy corresponds to a fixed wavelength of absorbed light; the greater the energy absorbed, the shorter the absorbed photon's wavelength.

But energy can't pile up in the atoms. If we look at the atoms from the side, so that they are no longer seen in silhouette against a background source of light, we would see the atoms give off (emit) just as much energy as they take up (absorb), in random directions. Essentially, they emit this energy at the same set of wavelengths as the absorbed light, producing **emission lines** (abrupt spikes in the brightness of light) at these wavelengths; see ■ Figure 2–8.

■ **FIGURE 2-6** Niels Bohr and his wife, Margrethe. They have become widely known in the last few years through Michael Frayn's play *Copenhagen,* about different perceptions of a controversial meeting Bohr had with his former assistant Werner Heisenberg.

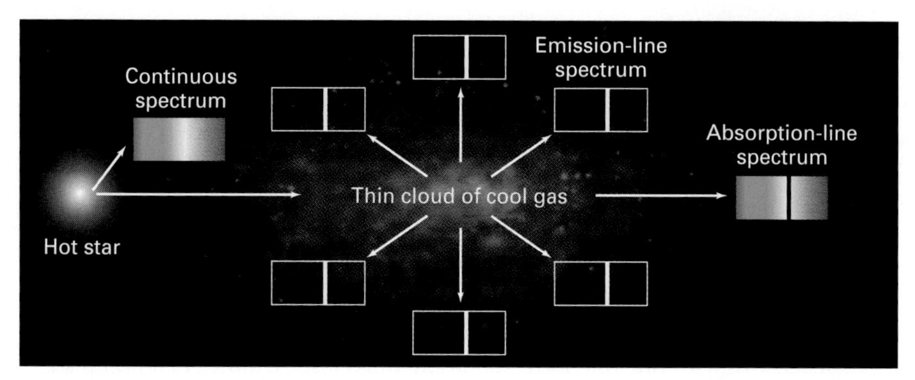

■ **FIGURE 2-7** When we view a hotter source that emits continuous radiation through a cooler gas, we may see absorption lines. These absorption lines appear at the same wavelengths at which the gas gives off emission lines when viewed with no background or against a cooler, darker background. Each of the spectra in the little boxes is the spectrum you would see looking back along the arrow. Note that the view from the right shows an absorption line. Only when you look through one source silhouetted against a hotter source do you see the absorption lines.

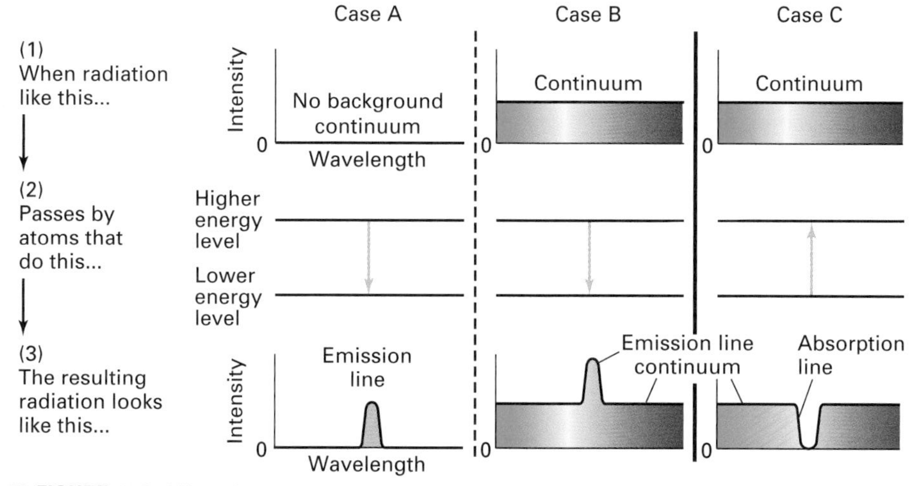

■ **FIGURE 2-8** When photons are emitted, we see an emission line. Continuous radiation may or may not be present (Cases A and B, respectively). An absorption line must absorb radiation from something; hence, an absorption line only appears when there is also continuous radiation (Case C). The top and bottom horizontal rows are graphs of the spectrum before and after the radiation passes by an atom. A schematic diagram of the atom's energy levels is shown in the center row (to simplify the situation, only two levels are given).

All stars have absorption lines. They are formed, basically, as light from layers just inside the star's visible surface passes through atoms right at the surface. The surface is cooler, so the atoms there absorb energy, making absorption lines. (Detailed models are much more complicated.) Emission lines occur only in special cases for stars. We see absorption lines on the Sun's surface, but we can see emission lines by looking just outside the Sun's edge, where only dark sky is the background. Extended regions of gas called "nebulae" (■ Fig. 2–9) give off emission lines, as we will explain later, because they are not silhouetted against background sources of light.

As we will discuss in more detail below, each of the chemical elements has its own distinctive, unique set of spectral lines, like a fingerprint, because of the different numbers of electrons and their allowed energy levels. So if you take the spectrum of a distant star, you may see several overlapping patterns of spectral lines. These patterns allow you to tell what chemical elements are in the star. We know what stars consist of from an analysis of their spectra!

AceAstronomy™ Log into AceAstronomy and select this chapter to see Astronomy Exercise "The Electromagnetic Spectrum."

■ **FIGURE 2-9** The Rosette Nebula, which has a reddish color because of its bright emission lines from hydrogen gas. It surrounds a cluster of hot, bluish stars.

2.4 The Story Behind the Bohr Atom

Bohr, in 1913, presented a model of the simplest atom—hydrogen—and showed how it produces emission and absorption lines. Stars often show the spectrum of hydrogen. Laboratories on Earth can also reproduce the spectrum of hydrogen; the same set of spectral lines appears both in emission and in absorption.

Hydrogen consists of a single central particle (a proton) in its nucleus with a single electron surrounding it (■ Fig. 2–10). The model for the hydrogen atom explained why hydrogen has only a few spectral lines (■ Fig. 2–11), rather than continuous bands of color. It postulated that a hydrogen atom could give off (emit) or take up (absorb) energy only in one of a fixed set of amounts, just as you can climb up stairs from step to step but cannot float in between. The position of the electron relative to the nucleus determines its specific, discrete **energy level,** the amount of energy in a hydrogen atom. So the electron can only jump from energy level to energy level, and not hover in between values of the fixed set of energies allowable.

When light hits an electron and makes it jump from a lower energy level to a higher energy level, a photon is absorbed, contributing to an absorption line. (When there are many absorbed photons, we see an absorption line in an otherwise continuous spectrum.) The photon that can be absorbed has an energy exactly equal to the difference between the higher and lower electronic energy levels. An atom cannot absorb a photon having too little energy, and it cannot take just part of the energy of a

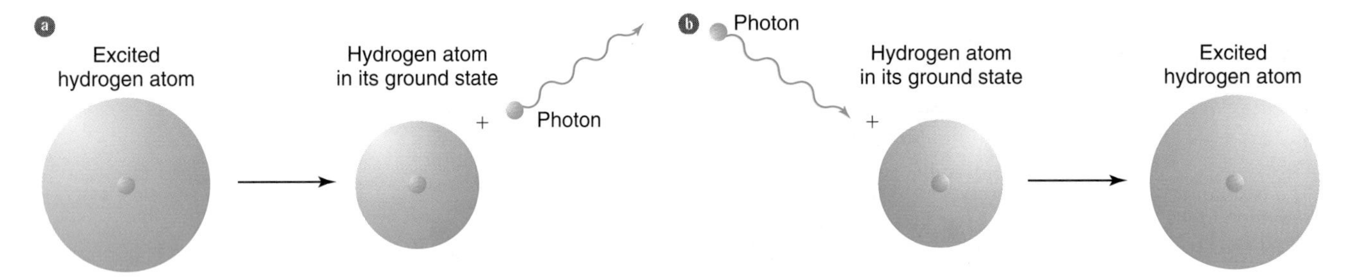

■ **FIGURE 2–10** Since the hydrogen atom has only a single electron, it is a particularly simple case to study. The lowest possible energy state of an atom is called its ground state; the electron is closest to the nucleus. All other energy states are called excited states; the electron is farther from the nucleus. (a) When an atom in an excited state gives off a photon, it drops back to a lower energy state, perhaps even to the ground state. We see the photons given off for a given transition as an emission line. (b) The formation of an absorption line, by taking up (absorbing) photons from incoming continuous radiation. Figure 2-12 shows more details for the specific transitions.

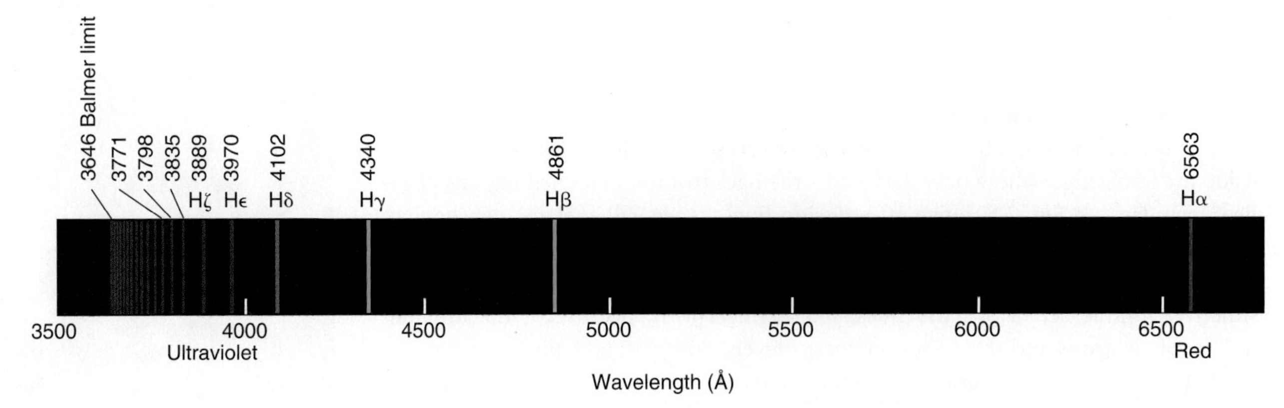

■ **FIGURE 2–11** The Balmer series, representing transitions down to (or up from) the second energy state of hydrogen. The strongest line in this series, Hα (H-alpha), is in the red. Note how the lines get closer and closer together as you go to shorter wavelengths. This distinct pattern makes the Balmer series, and thus hydrogen, easy to identify in a spectrum.

high-energy photon (releasing the excess energy as a lower-energy photon). Either all or none of the energy is absorbed.

When the electron jumps from a higher energy level to a lower one, a photon is given off (emitted). Such photons can go away from the atom in any direction, regardless of the direction of the original photon that was absorbed by the atom. Many photons together make an emission line.

The **Bohr atom** is Bohr's model that explains the hydrogen spectrum. In it, electrons can have orbits of different sizes. Each orbit corresponds to an energy level (■ Fig. 2–12). Only certain orbits are allowable. Bohr himself couldn't give a clear reason for this; he said that's just the way it is. (Note, however, that according to modern quantum theory, the electron orbits are not like planetary orbits; instead, the electron behaves as though it were distributed throughout its orbit. Also, we now understand the details of atoms in the context of quantum theory, but these are complicated.)

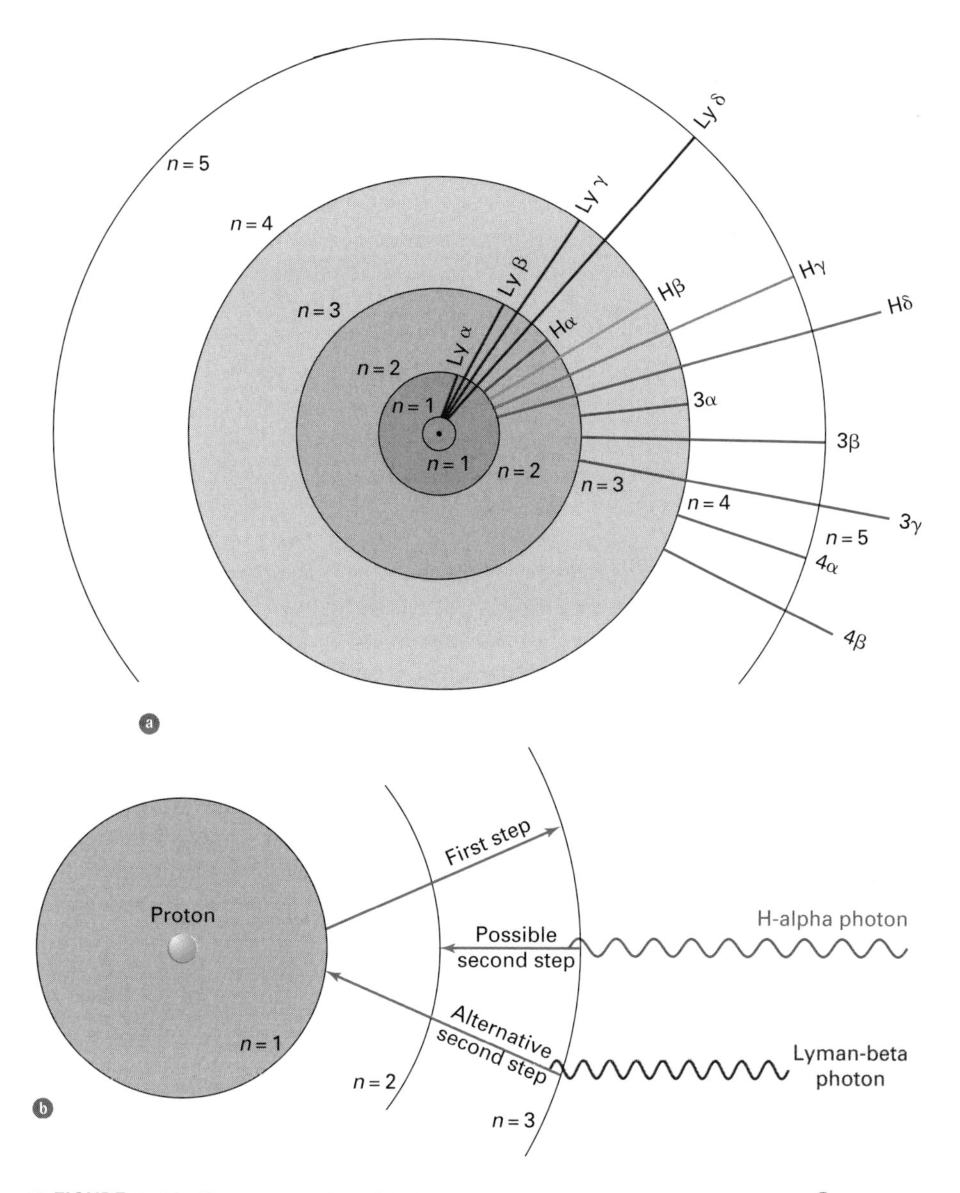

■ **FIGURE 2-12** The representation of hydrogen energy levels known as the Bohr atom. ⓐ Each of the circles shows the lone electron in a different energy level. ⓑ A schematic transition in which an electron is excited from level 1 to level 3 by an incoming photon. It can then jump down to level 2, emitting a red photon, or directly to level 1 (a greater difference in energy), emitting an ultraviolet photon, which has more energy than the red photon.

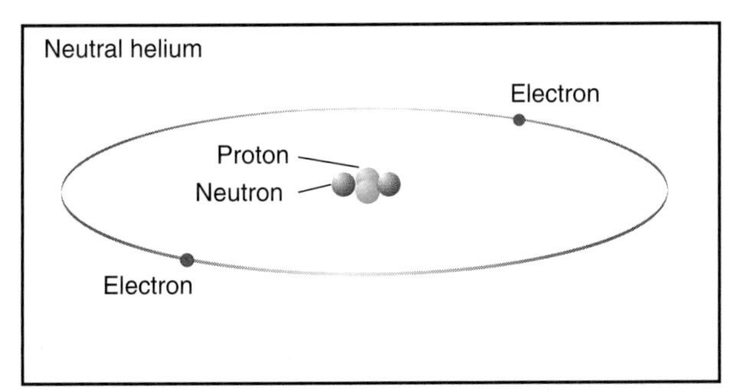

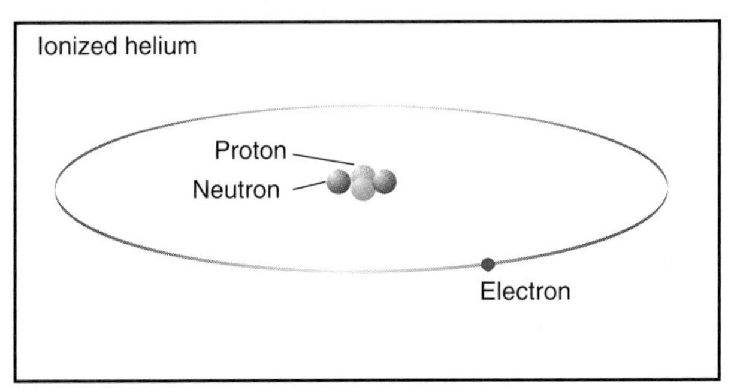

FIGURE 2-13 An atom missing one or more electrons is ionized. Neutral helium is shown at left; it has two electrons to balance the charge of the two protons. Ionized helium is shown at right. It has only one electron, so its net charge is +1.

We use the letter n to label the energy levels. We call the energy level for $n = 1$ the **ground level,** as it is the lowest possible energy state. **Excited levels** have $n = 2$ and higher. The hydrogen atom's series of transitions from or to the ground level is called the Lyman series, after the American physicist Theodore Lyman. The lines fall in the ultraviolet, at wavelengths far too short to pass through the Earth's atmosphere. Telescopes above the atmosphere now enable us to observe Lyman lines from the stars.

The series of transitions with $n = 2$ as the lowest level is the **Balmer series,** which is in the visible part of the spectrum. The bright red emission line in the visible part of the hydrogen spectrum, known as the Hα (H-alpha) line, corresponds to the transition from level 3 to level 2. The transition from $n = 4$ to $n = 2$ causes the Hβ (H-beta) line, and so on. Because the series falls in the visible where it is so well observed, we usually call the lines simply H-alpha, etc., instead of Balmer alpha, etc. So the Lyman series is transitions to or from lower level 1, and the Balmer series is transitions to or from lower level 2. Other series of hydrogen lines correspond to transitions with still different lower levels.

Since the higher energy levels have greater energy (they are higher above the ground state), a spectral line caused solely by transitions from a higher level to a lower level is an emission line. When, on the other hand, continuous radiation falls on cool hydrogen gas, some of the atoms in the gas can be raised to higher energy levels and absorption lines result.

Each of the chemical elements has its own set of discrete energy levels. Although the details are much more complex than in the simple Bohr model of the hydrogen atom, the general principles are similar. Thus, each of the chemical elements produces a unique pattern of spectral lines, like a fingerprint. The relative strength of the lines actually seen in the spectrum of a particular element depends on the distribution of electrons among the various energy levels, and this in turn depends on temperature and other factors.

Some atoms have lost one or more electrons, through either collisions with other particles or absorption of high-energy photons; they are said to be **ionized** (Fig. 2–13). The spectrum of an ionized element is different from the spectrum of the same element when it is not ionized, because the distribution of electron energy levels is different. Similarly, the spectrum of each molecule (two or more atoms bound to each other) is unique.

2.5 The Doppler Effect and Motion

Even though we can't reach out and touch a star, we can examine its light in great detail. Above, you saw how studying starlight reveals the temperature of a star (from its color) and what chemical elements it has near its surface (from its spectral lines).

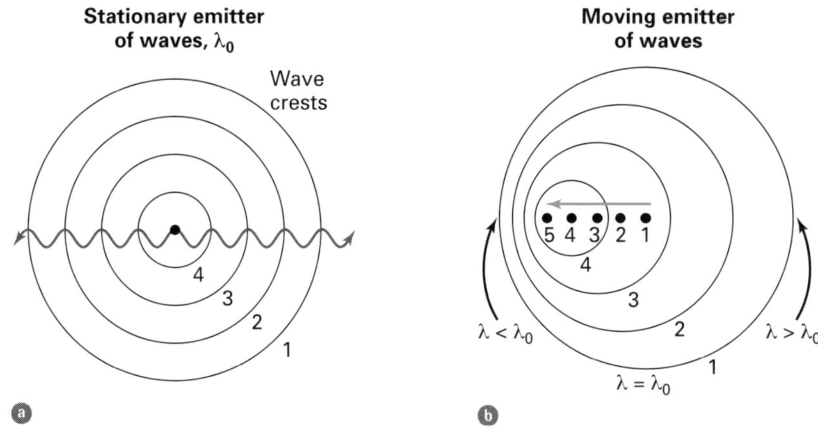

■ **FIGURE 2–14** ⓐ A stationary source is emitting waves of wavelength λ, in a circular pattern around it. Four wave crests, labeled 1 through 4, are shown. The observed wavelength is independent of the observer's position relative to the source. ⓑ The source is moving to the left (green arrow) with a constant speed. Along the direction of motion, the source partially keeps up with its most recently emitted wave crest before it emits another wave crest, so the crests get bunched closer together. From the left, the observed wavelength λ is shorter than λ_0; this is a "blueshift." Conversely, an observer on the right would see $\lambda > \lambda_0$; this is a "redshift." Perpendicular to the direction of motion, the spacing of the wave crests is unaffected, and an observer would therefore measure a wavelength $\lambda = \lambda_0$.

Studying starlight also tells us how fast the star is moving toward or away from us. The technique applies to all kinds of objects, including the planets.

The effect that motion has on waves is known as the **Doppler effect,** and was studied by Christian Doppler about 150 years ago (see ■ Fig. 2–14). It works for any waves—water waves, sound waves, or light waves. We will give more details about it when we study the motion of stars (Chapter 11).

For the moment, we need only realize that when an object is coming toward you, the waves you are getting from it are compressed, so the wavelength becomes shorter. When an object is going away from you, the waves you are getting from it are spread out, and the wavelength becomes longer.

Since the visible part of the spectrum has blue at short wavelengths and red at long wavelengths, the wavelengths are shifted toward the blue when objects are approaching and toward the red when objects are receding. Astronomers use the term **blueshift** to describe the shifts in wavelength of objects that are approaching and the term **redshift** to describe the shifts in wavelength of objects that are receding. Note that the light still travels at *c*, the speed of light in a vacuum; wavelengths and frequencies change, but the measured speed is independent of the motion of the source or of the observer.

One example of redshifts and blueshifts deals with planets in the Solar System. If you take spectra of several parts of a planet, and see that one side is blueshifted and the other side is redshifted, then you know that the first side is approaching you and the other side is receding (■ Fig. 2–15). You have found out that the planet is rotating! By measuring how much the light is shifted, you can even tell how fast the planet is rotating. The Doppler effect is a powerful tool for studying distant objects.

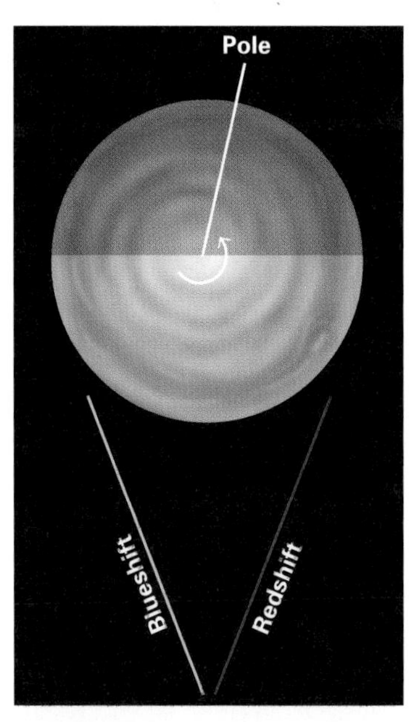

■ **FIGURE 2–15** If a planet is rotating, one side will be approaching us, blueshifting its spectral lines, while the other side will be receding from us, redshifting its spectral lines.

Perhaps the most exciting use of the Doppler effect has been the discovery, in 1999, of the first relatively normal planetary system other than our own Solar System. Since 1995, over 160 planets had been discovered around other stars. The discovery in a few cases of several planets around the same star confirms that at least most of these objects are indeed planets rather than some kind of failed companion star; we would not expect there to be so many failed stars in a system.

FIGURE IT OUT 2.4

Temperature Conversions

Though most Americans use the Fahrenheit temperature scale, in which water freezes at 32°F and boils at 212°F, most of the rest of the world uses the Celsius scale, in which water freezes at 0°C and boils at 100°C. Note that the difference between freezing and boiling points for each scale is 180°F and 100°C, respectively, so a change of 180°F equals a change of 100°C, or 9°F for every 5°C.

There is little, if any, water on the stars or on most planets, so astronomers use a more fundamental scale. Their scale, the kelvin scale (whose symbol is K), begins at absolute zero, the coldest temperature that can ever be approached. Since absolute zero is about −273.16°C, and a rise of one kelvin (1 K) is the same as a rise of 1°C, the freezing point of water (0°C) is about 273 K (see figure below).

To convert from kelvins to °C, simply subtract 273. To change from °C to °F, we must first multiply by 9/5 and then add 32. You may (or may not) find it easy to remember: times two, minus point two, plus thirty-two. That is, multiply by 2, subtract two-tenths of the original (which gives you 9/5), and then add 32, to get °F. For stellar temperatures, the 32° is too small to notice, and a sufficient level of approximation is often simply to multiply °C by 2 (see figure below).

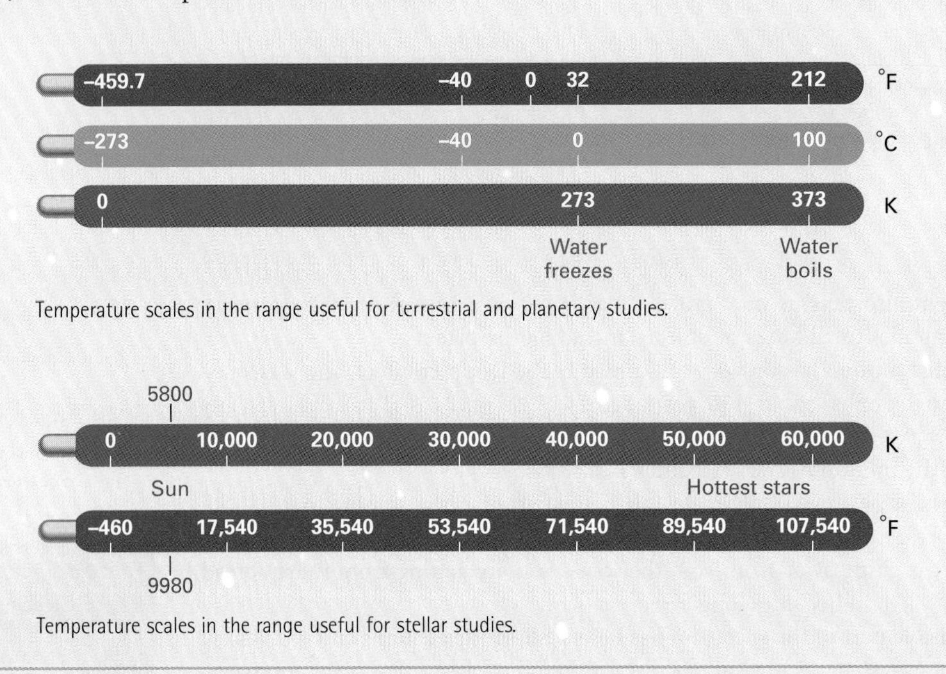

Temperature scales in the range useful for terrestrial and planetary studies.

Temperature scales in the range useful for stellar studies.

The technique used to find most of these planets relies upon the Doppler effect, since the planets are too faint and too close to their parent stars to be seen directly, even with our best telescopes on Earth or in space. The technique depends on the fact that the star isn't entirely steady in space, but rather moves to and fro as the planet around it moves fro and to.

Consider, for example, a pair of dancers waltzing: They usually rotate around a point between them. If one of the dancers were invisible, we could still see the other dancer moving around. Similarly, we can see the visible object, the star, moving slightly toward and away from us, and infer that there must be an invisible object, the planet, moving in step but in the opposite direction at every time. Since the star is so much more massive than the planet, it moves less than the planet (■ Fig. 2–16).

The breakthrough in 1995 that enabled astronomers to detect these planets came when they developed some extremely precise ways of measuring Doppler shifts, more precise than had been possible before. More planets are now known outside the Solar System than inside it. We will describe those "extrasolar planets," or "exoplanets," in Chapter 9.

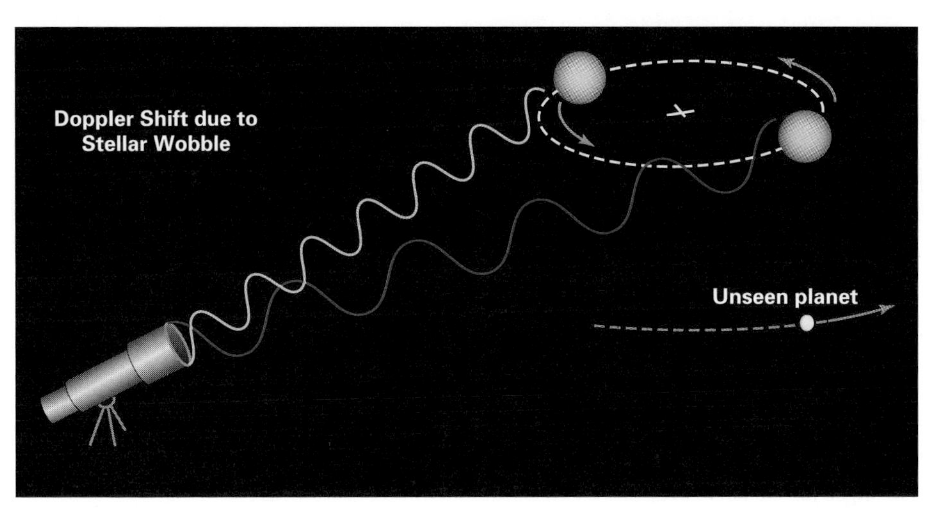

■ **FIGURE 2-16** The technique used by several teams of scientists to detect planets around Sun-like stars by searching for periodically changing Doppler shifts in the stars' spectra. The diagram shows a particular star at two positions in its orbit. The successive redshifts and blueshifts of that star's spectrum reveal the otherwise unseen planet. It must always be on the opposite side of the system's "center of mass" (like the fulcrum of a see-saw; Chapter 11 has more details) from the star.

AceAstronomy™ Log into AceAstronomy and select this chapter to see Astronomy Exercise "Doppler Shift."

CONCEPT REVIEW

Electromagnetic radiation is a combination of electricity and magnetism changing together in space; it is also known as **electromagnetic waves** (Introductory section). The **wavelength** is the distance between two consecutive crests (or troughs) of the wave, and different wavelengths correspond to different colors (Sec. 2.1). The rainbow produced by passing light through a prism, as well as the graph of color (wavelength) versus the brightness at each color, is called a **spectrum.** Such **spectra** are extremely valuable to astronomers; they can be used to determine the temperature, chemical composition, and other properties of the object being studied. From the Earth's surface, we view through atmospheric windows that pass only certain types of electromagnetic radiation, chiefly visible and radio.

A very dense gas or solid gives off a **continuous spectrum** that changes smoothly in brightness from one color to the next (Sec. 2.2). In the case of a **black body,** an object that absorbs all radiation that is incident upon it (none is reflected or transmitted), the shape of the emitted continuous spectrum depends only on the object's surface temperature, not on its chemical composition or other properties. This result is called **black-body radiation** or **thermal radiation.** Spectra of hotter black bodies are brightest at shorter (bluer) wavelengths than spectra of cooler black bodies. Also, a hot black body is intrinsically brighter than a cold black body of equal surface area.

Atoms are composed of positively charged **nuclei** "orbited" by negatively charged **electrons** (Sec. 2.3). The nuclei are made of positively charged **protons** (which distinguish each chemical element) and neutral **neutrons.**

Quantum theory explains how to describe atoms, and the rules that prevent electrons from spiraling into nuclei. When an atom loses one or more electrons, it is **ionized.** Two or more atoms bound together form a **molecule.**

Light sometimes acts as though it were made of particles, which are called **photons** (Sec. 2.3). **Absorption lines** are dark gaps in a spectrum caused by atoms (or molecules) absorbing some of the photons that are passing through them; the electrons jump from lower to higher energy levels. Conversely, **emission lines** are locations in a spectrum where extra photons are present. They are caused by electrons jumping from higher to lower energy levels, thereby emitting light.

The **Bohr atom,** with electrons at different radii centered on the nucleus, explains the spectrum of hydrogen in detail (Sec. 2.4). Transitions to the lowest **energy level,** called the **ground level,** do not cause lines in the visible part of the spectrum, but rather in the ultraviolet. Transitions between the second energy level and higher levels (called **excited levels**) lead to photons being added or subtracted in the visible part of the spectrum. The resulting spectral lines are known as the **Balmer series.**

The **Doppler effect** is a change in wavelength of light caused by an object's motion along a line toward or away from you (Sec. 2.5). If you and the object are receding from each other, the object's spectrum appears **redshifted,** while **blueshifts** are seen if you and the object are approaching each other. Larger relative speeds produce bigger Doppler shifts.

QUESTIONS

1. What is the difference between a gamma ray and a radio wave?
2. Why is electromagnetic radiation so important to astronomers?
3. Discuss the difference between the general concept of the spectrum and the particular spectrum we see as a rainbow.
4. Does white light contain red light? Green light? Discuss.
5. After Newton separated sunlight into its component colors with a prism, he reassembled the colors. What should he have found and why?
6. Identify Roy G Biv. Identify A. J. Ångström. What relation do they have to each other?
7. How can our atmosphere have a "window"? Can our air leak out through it? Explain.
8. Describe the relation of Fraunhofer lines to emission lines and to absorption lines.
9. Describe how the same atoms can sometimes cause emission lines and at other times cause absorption lines.
10. Sketch an atom, showing its nucleus and its electrons.
11. Sketch the energy levels of a hydrogen atom, showing how the Balmer series arises.
12. On a sketch of the energy levels of a hydrogen atom, show with an arrow the transition that matches Hα in absorption. Also show the transition that matches Hα in emission.
13. If you were to take an emission nebula and put it in front of a very hot, bright source of continuous radiation, what change would you see in the nebula's spectrum? Explain.
14. If the surface of a star were increasing in temperature as you went outward, would you see absorption lines or emission lines in its spectrum? Explain how this case differs from what we see in normal stars, which decrease in temperature as you go outward in the region from which they give off most of their light.
15. Does the Bohr atom explain iron atoms? Describe from this chapter what type of atom the Bohr atom explains.
16. If someone were to say that we cannot know the composition of distant stars, since there is no way to perform experiments on them in terrestrial laboratories, how would you respond?
17. A spectral line from the left side of Saturn's rings, as you see them, is at a wavelength slightly shorter than the same spectral line measured from the right side of Saturn's rings. Which way (direction) is Saturn rotating, and why?
18. †Give the temperature in degrees Celsius for a 60°F day.
19. †What Fahrenheit temperature corresponds to 30°C?
20. †The Earth's average temperature is about 27°C. What is its average temperature in kelvins?
21. †The Sun's surface is about 5800 K. What is its temperature in °F?
22. An iron rod is heated by a welder's torch. Initially it glows a dull red, then a brighter orange, and finally very bright white. Discuss this sequence in terms of black-body radiation.
23. Does the Doppler effect depend on the distance between the source of light and the observer? Explain.
24. †The Hβ line of hydrogen has a wavelength of 4861 Å. **(a)** What is its frequency? **(b)** What is the frequency of a line with twice the wavelength of Hα?
25. †Announcers at a certain radio station say that they are at "95 FM on your dial," meaning that they transmit at a frequency of 95 MHz (95 megahertz, or 95 million cycles per second). What is the wavelength of the radio waves from this station?
26. †**(a)** If one photon has 10 times the frequency of another photon, which photon is the more energetic, and by what factor? **(b)** Answer the same question for the case where the first photon has twice the wavelength of the second photon.
27. †Consider a black body whose temperature is 3 K. At what wavelength does its spectrum peak? (We will see the importance of this radiation in the discussion of cosmology in Chapter 19.)
28. †Suppose the peak of a particular star's spectrum occurs at about 6000 Å. **(a)** Use Wien's law to calculate the star's surface temperature. **(b)** If this star were a factor of four hotter, at what wavelength would its spectrum peak? In what part of the electromagnetic spectrum is this peak?
29. †**(a)** Compare the luminosity (amount of energy given off per second) of the Sun with that of a star the same size but three times hotter at its surface. **(b)** Answer the same question, but now assume the star also has twice the Sun's radius.
30. How many times hotter than the Sun's surface is the surface of a star the same size, but that gives off twice the Sun's energy per second (that is, is twice as luminous)?
31. **True or false?** When we see other people, our eyes are detecting the visible light that each of us radiates as an approximate "black body" (thermal emitter).
32. †**True or false?** If the surface temperature of Star Zeppo is 3 times that of Star Harpo, then the wavelength at which Star Zeppo's spectrum peaks is $\frac{1}{3}$ the wavelength at which Star Harpo's spectrum peaks.
33. **True or false?** The Doppler effect has been used to measure the rotation rates of planets in our Solar System, as well as to detect the presence of planets around some other stars.
34. **True or false?** The type of spectral feature usually observed from a hot gas with no star behind it along the line of sight is an absorption line.

35. **Multiple choice:** Which one of the following statements about atoms is *false*? (**a**) Electrons have discrete energy levels. (**b**) Each element produces a unique pattern of spectral lines, like a fingerprint. (**c**) Photon emission occurs randomly, in any direction. (**d**) An electron in an atom may absorb either part or all of the energy of a photon. (**e**) Absorption occurs when an electron in an atom jumps from a lower energy level to a higher energy level.
36. **Multiple choice:** In a vacuum, photons of higher energy (**a**) move faster than lower energy photons; (**b**) have higher frequencies and shorter wavelengths than lower energy photons; (**c**) have more mass than lower energy photons; (**d**) are not as likely to become redshifted as lower energy photons; or (**e**) travel less distance between their source and the observer than lower energy photons.

†37. **Multiple choice:** A local radio station broadcasts at 100.3 megahertz (megahertz = 10^6 cycles per second; symbol MHz). What is the approximate wavelength of the signal? (**a**) 3000 angstroms. (**b**) 30 meters. (**c**) 1.86 miles. (**d**) 30 centimeters. (**e**) 3 meters.

38. **Multiple choice:** Which one of the following statements about electromagnetic waves is *false*? (**a**) Human eyes are able to detect only a tiny fraction of all possible electromagnetic waves. (**b**) If electromagnetic wave Zoe has twice the wavelength of electromagnetic wave Simon, then Zoe also has twice the frequency of Simon. (**c**) "White light" such as sunlight actually consists of many electromagnetic waves, having different wavelengths, mixed together. (**d**) An electromagnetic wave consists of oscillating electric and magnetic fields that are perpendicular to each other and perpendicular to the direction of motion. (**e**) The measured speed of an electromagnetic wave is independent of the speed of its source relative to the observer.
39. **Fill in the blank:** If a neutral atom loses one or more electrons, the atom is said to be ______.
40. **Fill in the blank:** By observing the _____ of a star or planet, we can determine what kinds of atoms or molecules are present and their relative abundance.
41. **Fill in the blank:** Electromagnetic radiation behaves as though it has properties of both _______ and _______.

†42. **Fill in the blank:** According to the Stefan-Boltzmann law, the ratio of the energy emitted per second by two black bodies having the same surface area is proportional to the _____ power of the ratio of their temperatures.

†This question requires a numerical solution.

TOPICS FOR DISCUSSION

1. What are some examples in which you know that magnetic or electric fields play a prominent role? Is there evidence that one type of field produces or interacts with the other type?
2. Why is spectroscopy useful in astronomy? Describe and explain three possible applications.
3. Is any object truly a perfect "black body," absorbing absolutely all of the energy that hits it? If there are few such objects, why is the black-body concept useful?
4. What do you think of the dual nature of light, being both a wave and a particle? Is this contrary to your intuition?

MEDIA

Virtual Laboratories

- Units, Measure, and Unit Conversion
- Properties of Light and Its Interaction with Matter
- The Doppler Effect

AceAstronomy™ Log into AceAstronomy at **http://astronomy.brookscole.com/cosmos3** to access quizzes and animations that will help you assess your understanding of this chapter's topics.

Log into the Student Companion Web Site at **http://astronomy.brookscole.com/cosmos3** for more resources for this chapter including a list of common misconceptions, news and updates, flashcards, and more.

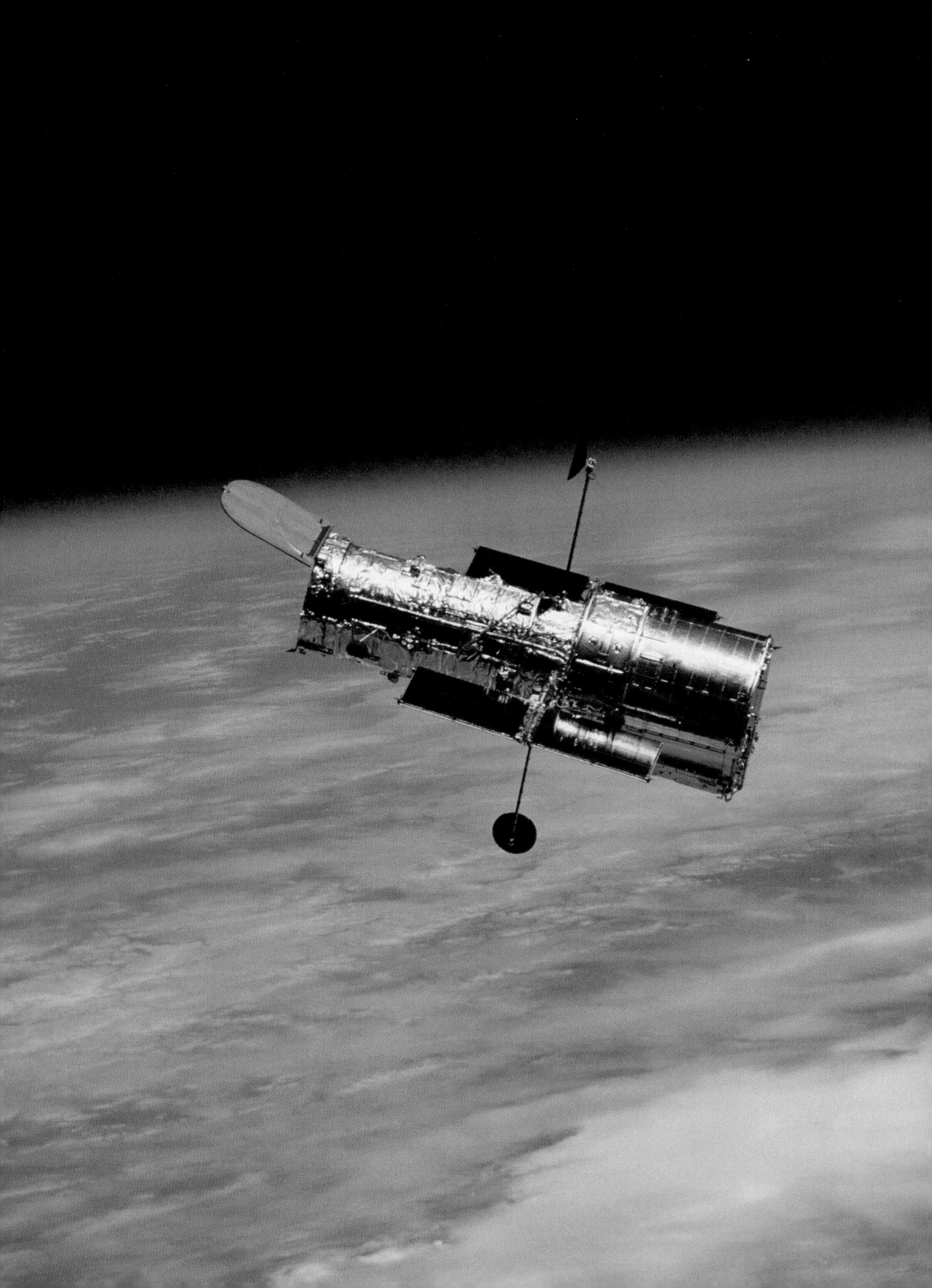

CHAPTER 3

Light and Telescopes: Extending Our Senses

Everybody knows that astronomers use telescopes, but not everybody realizes that the telescopes astronomers use are of very different types. Further, very few modern telescopes are used directly with the eye. In this chapter, we will first discuss the telescopes that astronomers use to collect visible light, as they have for hundreds of years. Then we will see how astronomers now also use telescopes to study gamma rays, x-rays, ultraviolet, infrared, and radio waves.

ORIGINS

We discuss the telescopes used by astronomers to see the farthest and faintest objects, including a new generation of huge telescopes on the ground and major visible, x-ray, and infrared space telescopes aloft. These instruments allow us to learn about the earliest epochs of the Universe, to study how stars form, and to search for other planetary systems, among other things.

AIMS

1. Discuss the historical importance of the development of telescopes (Section 3.1).
2. See how telescopes work, and understand the main uses of telescopes (Sections 3.2 and 3.3).
3. Learn about a variety of telescopes on the ground and in space for studying radiation inside and outside the visible spectrum (Sections 3.4 to 3.8).

3.1 The First Telescopes for Astronomy

Almost four hundred years ago, a Dutch optician put two eyeglass lenses together, and noticed that distant objects appeared closer (that is, they looked magnified). The next year, in 1609, the English scientist Thomas Harriot built one of these devices and looked at the Moon. But all he saw was a blotchy surface, and he didn't make anything of it.

Credit for first using a telescope to make astronomical studies goes to Galileo Galilei. In 1609, Galileo heard that a telescope had been made in Holland, so in Venice he made one of his own and used it to look at the Moon. Perhaps as a result of his training in interpreting light and shadow in drawings (he was surrounded by the Renaissance and its developments in visual perspective), Galileo realized that the light and dark patterns on the Moon meant that there were craters there (■ Fig. 3–1). With his tiny telescopes—using lenses only a few centimeters across and providing, with an eyepiece, a magnification of only 20 or 30, not much more powerful than a modern pair of binoculars and showing a smaller part of the sky—he went on to revolutionize our view of the cosmos, as will be further discussed in Chapter 5.

Whenever Galileo looked at Jupiter through his telescope, he saw that it was not just a point of light, but appeared as a small disk. He also spotted four points of light that moved from one side of Jupiter to another (■ Fig. 3–2). He eventually realized that the points of light were moons orbiting Jupiter, the first proof that not all bodies in the Solar System orbited the Earth.

AceAstronomy™ The AceAstronomy icon throughout the text indicates an opportunity for you to test yourself on key concepts and to explore animations and interactions on the AceAstronomy website at **http://astronomy.brookscole.com/cosmos3**

◀ The Hubble Space Telescope as seen from the space shuttle that had just brought astronauts to service it.
NASA's Johnson Space Center

■ **FIGURE 3-1** ⓐ An engraving of Galileo's observations of the Moon from his book *Sidereus Nuncius (The Starry Messenger)*, published in 1610. ⓑ A modern photo appears for comparison. Galileo was the first to report that the Moon has craters. It seems reasonable that Galileo had been sensitized to interpreting surfaces and shadows by the Italian Renaissance and by his related training in drawing. Aristotle and Ptolemy had held that the Earth was imperfect but that everything above it was perfect, so Galileo's observation contradicted them.

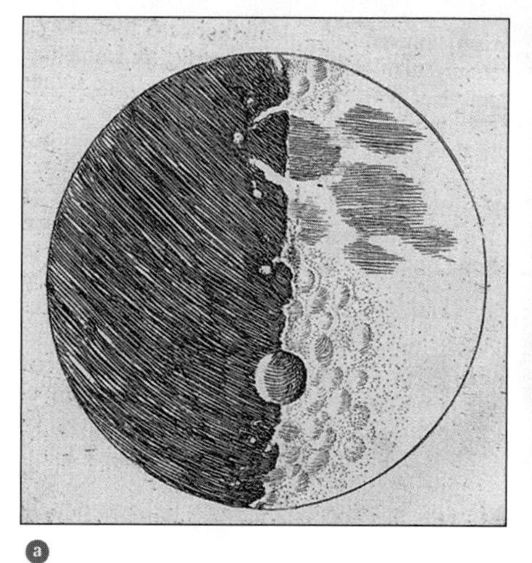

ⓐ

ⓑ

Jay M. Pasachoff, after Ewen Whitaker, Lunar and Planetary Laboratory, U. Arizona.

ⓐ

Mars Orbiter Camera on Mars Global Surveyor; NASA/JPL/Malin Space Science Systems; colorized with a Cassini image, NASA/JPL/Space Science Institute

■ **FIGURE 3-2** ⓐ Jupiter and three of its four Galilean satellites; from left to right: Callisto, Ganymede, Jupiter, Europa. Io was hidden behind Jupiter. The relative brightness of the satellites was enhanced. ⓑ Some of Galileo's notes about his first observations of Jupiter's moons. Simon Marius also observed the moons at about the same time; we now use the names Marius proposed for them: Io, Europa, Ganymede, and Callisto, though we call them the Galilean satellites. ⓒ An image of Jupiter (circled in purple) and four of its moons made in 2002 with a replica of one of Galileo's 2.5-cm (1-in) telescopes.

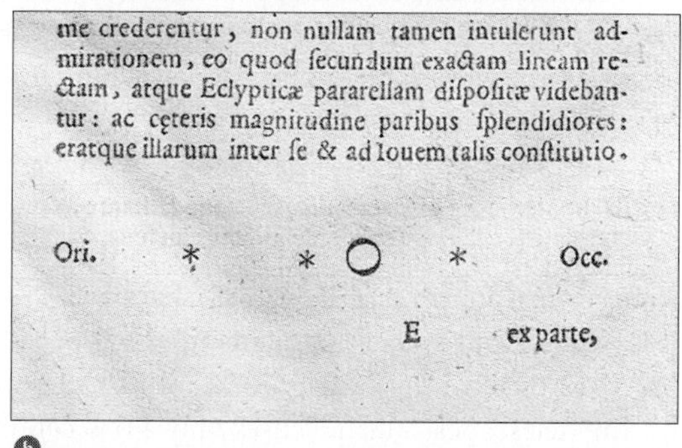

me crederentur, non nullam tamen intulerunt admirationem, eo quod ſecundum exactam lineam rectam, atque Eclypticæ pararellam diſpoſitæ videbantur: ac cęteris magnitudine paribus ſplendidiores: eratque illarum inter ſe & ad Iouem talis conſtitutio.

Ori. * * O * Occ.

E ex parte,

ⓑ

Jay M. Pasachoff

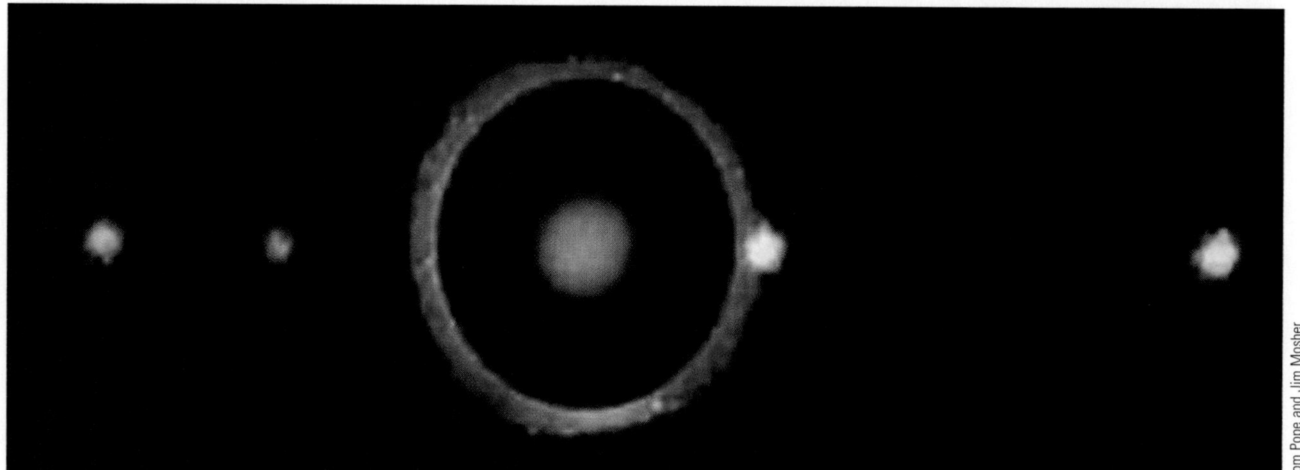

ⓒ

Tom Pope and Jim Mosher

■ **FIGURE 3-3** Venus goes through a full set of phases, from new to full (crescent to nearly full shown here).

The existence of Jupiter's moons contradicted the ancient Greek philosophers—chiefly Aristotle and Ptolemy—who had held that the Earth is at the center of all orbits (see Chapter 5). Further, the ancient ideas that the Earth could not be in motion because the Moon (and other objects) would be "left behind" was also wrong. Galileo's discovery of the moons thus backed the newer theory of Copernicus, who had said in 1543 that the Sun and not the Earth is at the center of the Universe. And Galileo's lunar discovery—that the Moon's surface had craters—had also endorsed Copernicus's ideas, since the Greek philosophers had held that celestial bodies were all "perfect." Galileo published these discoveries in 1610 in his book *Sidereus Nuncius (The Starry Messenger).*

■ **FIGURE 3-4** The position in the sky of Venus at different phases, in an artist's rendition of a computer plot. Venus is a crescent only when it is in a part of its orbit that is relatively close to the Earth, and so looks larger at those times.

Seldom has a book been as influential as Galileo's slim volume. He also reported in it that his telescope revealed that the Milky Way was made up of a myriad of individual stars. He drew many individual stars in the Pleiades, which we now know to be a star cluster. And he reported some stars in the middle of what is now known as the Orion Nebula. But one attempt made by Galileo in his *Sidereus Nuncius* didn't stick: he proposed to use the name "Medicean stars," after his financial backers, for the moons of Jupiter. Nowadays, recognizing Galileo's intellectual breakthroughs rather than the Medicis' financial contributions, we call them the "Galilean moons."

Galileo went on to discover that Venus went through a complete set of phases, from crescent to nearly full (■ Fig. 3–3), as it changed dramatically in size (■ Fig. 3–4). These variations were contrary to the prediction of the Earth-centered (geocentric) theory of Ptolemy and Aristotle that only a crescent phase would be seen (see Chapter 5). The Venus observations were thus the fatal blow to the geocentric hypothesis. He also found that the Sun had spots on it (which we now call "sunspots"), among many other exciting things.

3.2 How Do Telescopes Work?

The basic principles of telescopes are easy to understand. In astronomy we normally deal with light rays that are parallel to each other, which is the case for light from the stars and planets, since they are very far away (■ Fig. 3–5). Certain curved lenses and mirrors can bring starlight to a single point, called the **focus** (■ Fig. 3–6). The many different points of light coming from an extended object (like a planet) together form an image of the object in the focal plane. If an eyepiece lens is also included, then the image becomes magnified and can be viewed easily with the human eye. For example, each "monocular" in a pair of binoculars is a simple telescope of this kind, much like the ones made by Galileo.

But Galileo's telescopes had deficiencies, among them that white-light images were tinged with color, and somewhat out of focus. This effect, known as **chromatic**

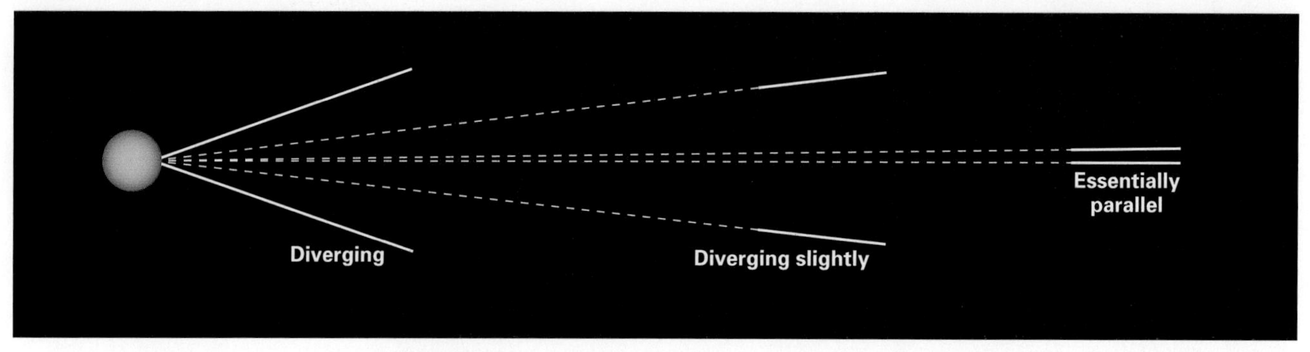

■ **FIGURE 3-5** The light we see from a star is almost parallel, diverging imperceptibly, because stars are so distant.

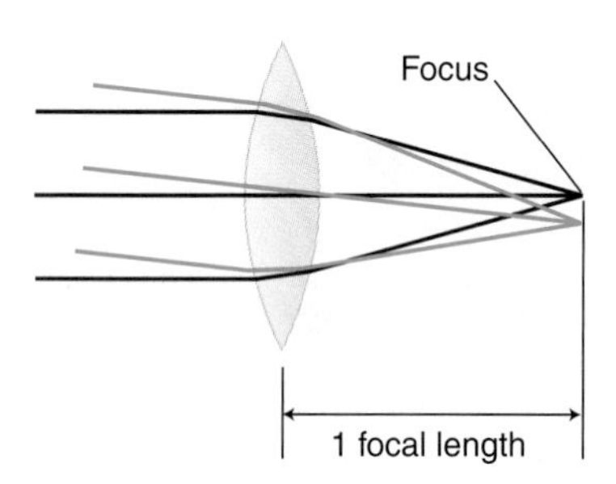

■ **FIGURE 3-6** The focal length is the distance behind a lens to the point at which a very distant object is focused. [The focal length of the human eye is about 2.5 cm (1 inch)]. An object in a slightly different part of the sky (indicated by parallel light rays tilted with respect to the horizontal ones) is imaged at a slightly different location in the focal plane, as shown.

aberration (■ Fig. 3–7), is caused by the fact that different colors of light are bent by different amounts as the light passes through a lens, similar to what happens to light in a prism (as discussed in Chapter 2). Each color ends up having a different focus.

Toward the end of the 17th century, Isaac Newton, in England, had the idea of using mirrors instead of lenses to make a telescope. Mirrors do not suffer from chromatic aberration. When your focusing mirror is only a few centimeters across, however, your head would block the incoming light if you tried to put your eye to this "prime" focus. Newton had the bright idea of putting a small, flat "secondary mirror" just in front of the focus to reflect the light out to the side, bringing the focus point outside the telescope tube. This **Newtonian telescope** (■ Fig. 3–8) is a design still in use by many amateur astronomers. But many telescopes instead use the **Cassegrain** design, in which a secondary mirror bounces the light back through a small hole in the middle of the primary mirror (■ Fig. 3–9).

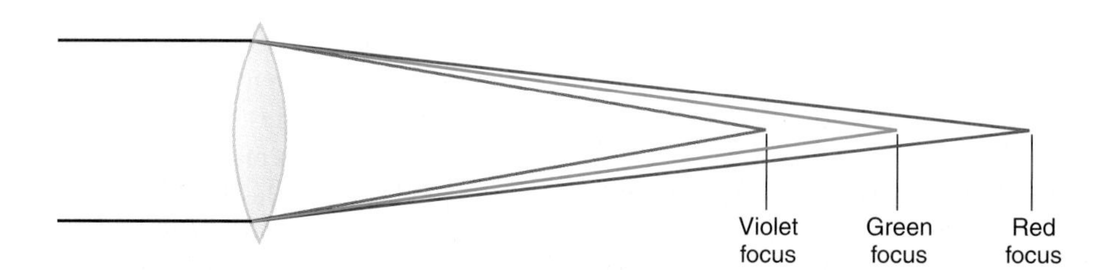

■ **FIGURE 3-7** Chromatic aberration occurs when white light passes through a simple lens. The various colors of light come to a focus at different distances from the lens; there is no single "best focus," and objects end up showing a tinge of color.

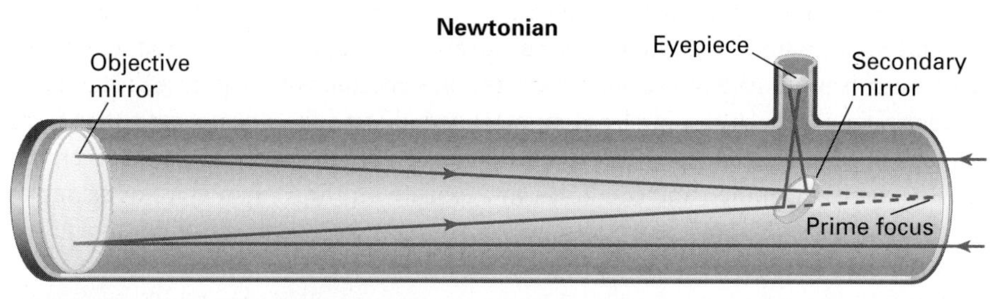

■ **FIGURE 3-8** The path of light in a Newtonian telescope; note the diagonal mirror that brings the focus out to the side. Many amateur telescopes are of this type.

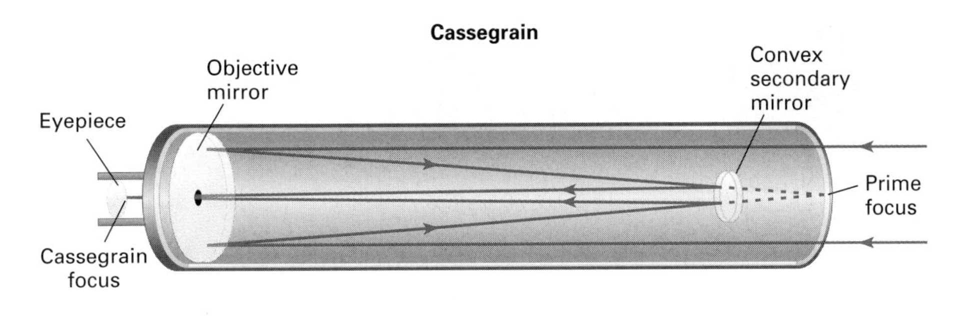

■ **FIGURE 3-9** A Cassegrain telescope, in which a curved secondary mirror bounces light through a hole in the center of the primary mirror.

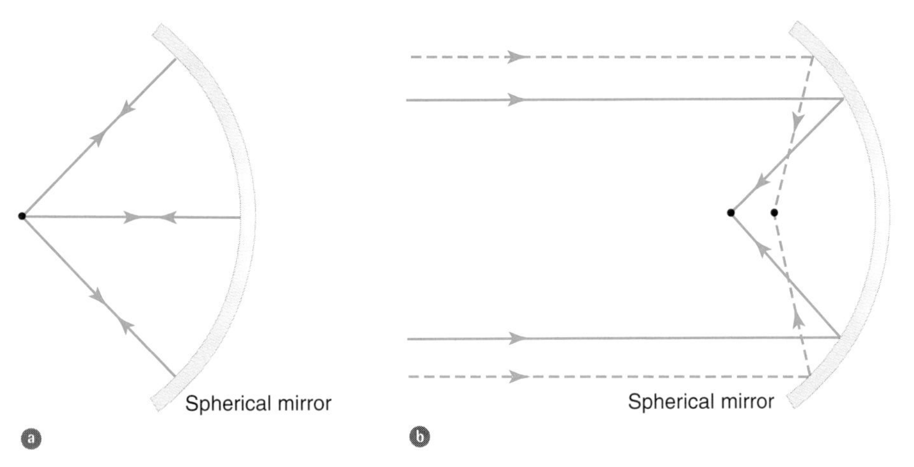

■ **FIGURE 3-10** ⓐ A spherical mirror (that is, part of the interior of a sphere) focuses light that originates at its center of curvature back on itself, as does the circular arc shown here. ⓑ A spherical mirror (or a circular arc, as shown here) suffers from spherical aberration in that it does not perfectly focus light from very large distances, for which the incoming rays are essentially parallel (recall Fig. 3-5). Lenses that don't have the proper shape can also show spherical aberration.

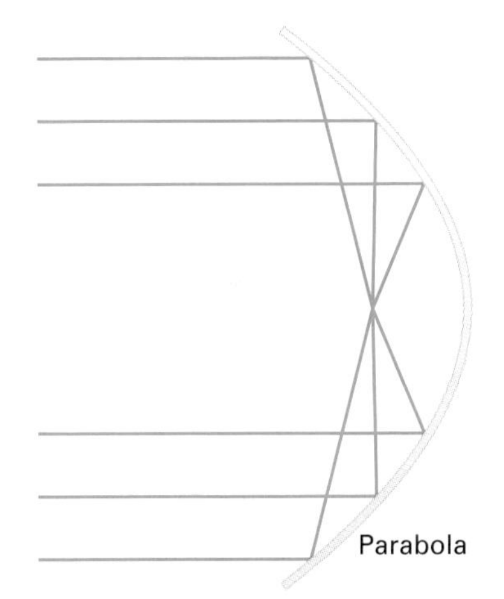

■ **FIGURE 3-11** A paraboloid is a three-dimensional curve (two-dimensional surface) created by spinning a parabola on its axis of symmetry. As shown here, a parabola focuses parallel light to a single point, as does a paraboloid.

Note that the hole, or an obstruction of part of the incoming light by the secondary mirror, only decreases the apparent brightness of the object; it does not alter its shape. Every part of the mirror forms a complete image of the object.

Spherical mirrors reflect light from their centers back onto the same point, but do not bring parallel light to a good focus (■ Fig. 3–10). This effect is called **spherical aberration.** We now often use mirrors that are in the shape of a **paraboloid** (a parabola rotated around its axis, forming a curved surface) since only paraboloids bring parallel light near the mirror's axis to a focus (■ Fig. 3–11). However, light that comes in from a direction substantially tilted relative to the mirror's axis is still out of focus; thus, simple reflecting telescopes generally have a narrow field of view.

Through the 19th century, telescopes using lenses (**refracting telescopes,** or refractors) and telescopes using mirrors (**reflecting telescopes,** or reflectors) were made larger and larger. The pinnacle of refracting telescopes was reached in the 1890s with the construction of a telescope with a lens 40 inches (1 m) across for the Yerkes Observatory in Wisconsin, now part of the University of Chicago (■ Fig. 3–12).

It was difficult to make a lens of clear glass thick enough to support its large diameter; moreover, such a thick lens may sag from its weight, absorbs light, and also suffers from chromatic aberration. And the telescope tube had to be tremendously long. Because of these difficulties, no larger telescope lens has ever been put into long-term service. (A 1.25-m refractor, mounted horizontally to point at a mirror that tracked the stars, was set up for a few months at an exposition in Paris in 1900.) In 2002, over 100 years later, a lens also 40 inches (1 m) across was put into use at the Swedish Solar Telescope on La Palma, Canary Islands.

The size of a telescope's primary lens or mirror is particularly important because the main job of most telescopes is to collect light—to act as a "light bucket." All the

Yerkes Observatory

■ **FIGURE 3-12** The Yerkes refractor, still one of the two largest in the world, has a 1-m-diameter lens.

FIGURE IT OUT 3.1
Light-Gathering Power of a Telescope

A telescope's main purpose is to collect light quickly and in large quantities. The mirror or lens of a telescope acts as a gigantic eye pupil that intercepts the light rays from a distant object. The larger its area, the more light it will collect in a given time, allowing fainter objects to be seen. (The length of the telescope tube is irrelevant to this "light-gathering power.")

The area of a circle is proportional to the square of its radius r (or diameter $D = 2r$):

$$A = \pi r^2 = \pi (D/2)^2 = \pi D^2/4.$$

Therefore, the ratio of the areas of two telescopes with circular mirrors of diameter D_1 and D_2 is given by

$$A_2/A_1 = D_2^2/D_1^2 = (D_2/D_1)^2.$$

Let's say that $D_2 = 8$ m (a typical size for a large optical telescope), and $D_1 = 8$ mm $= 0.008$ m (as for the pupil opening of a dilated eye). The ratio of areas is

$$(D_2/D_1)^2 = (8 \text{ m}/0.008 \text{ m})^2 = 1000^2 = 10^6.$$

Thus, looking through the eyepiece of such a telescope, one could see stars a million times fainter than with the unaided eye!

By attaching a detector to the telescope, the exposure time can be made very long, making even fainter stars visible. Some electronic detectors, such as charge-coupled devices (CCDs), are far more sensitive than eyes, and detect most of the photons that hit them. With large telescopes, long exposures, and high-quality CCDs, objects over 10^9 (a billion) times fainter than the limit of the unaided eye have been detected.

light is brought to a common focus, where it is viewed or recorded. (See *Figure It Out 3.1: Light-Gathering Power of a Telescope.*) The larger the telescope's lens or mirror, the fainter the objects that can be viewed or the more quickly observations can be made. A larger telescope would also provide better **resolution**—the ability to detect fine detail—if it weren't for the shimmering (turbulence) of the Earth's atmosphere, which limits all large telescopes to about the same resolution (technically, angular resolution). Only if you can improve the resolution is it worthwhile magnifying images. For the most part, then, the fact that telescopes magnify is secondary to their ability to gather light.

AceAstronomy™ Log into AceAstronomy and select this chapter to see Astronomy Exercise "Lenses-Focal Length."

AceAstronomy™ Log into AceAstronomy and select this chapter to see the Active Figure called "Resolution and Telescopes."

AceAstronomy™ Log into AceAstronomy and select this chapter to see Astronomy Exercise "Telescopes: Objective Lens and Eyepiece."

3.3 Modern Telescopes

From the mid-19th century onward, larger and larger reflecting telescopes were constructed. But the mirrors, then made of shiny metal, tended to tarnish. This problem was avoided by evaporating a thin coat of silver onto a mirror made of glass. More recently, a thin coating of aluminum turned out to be longer lasting, though silver with a thin transparent overcoat of tough material is now coming back into style. The 100-inch (2.5-m; see *Figure It Out 3.2: Changing Units*) "Hooker" reflector at the Mt. Wilson Observatory in California became the largest telescope in the world in 1917. Its use led to discoveries about distant galaxies that transformed our view of what the Universe is like and what will happen to it and us in the far future (Chapters 16 and 18).

In 1948, the 200-inch (5-m) "Hale" reflecting telescope opened at the Palomar Observatory, also in California, and was for many years the largest in the world. Current electronic imaging detectors, specifically **charge-coupled devices (CCDs)** similar to those in camcorders and digital cameras, have made this and other large telescopes many times more powerful than they were when they recorded images on film.

Some of the most interesting astronomical objects are in the southern sky, so astronomers need telescopes at sites more southerly than the continental United States. For example, the nearest galaxies to our own—known as the Magellanic Clouds—are not observable from the continental United States. The National Optical Astronomy Observatories, supported by the National Science Foundation, have a half-share in two telescopes, each with 8-m mirrors, the northern-hemisphere one in Hawaii and the southern-hemisphere one in Chile. The project is called Gemini, since "Gemini" are the twins in Greek mythology (and the name of a constellation) and these are twin telescopes. The other half-share in the project is divided among the United Kingdom, Canada, Chile, Australia, Argentina, and Brazil. By sharing, the United States has not only half the time on a telescope in the northern hemisphere, but also half the time on a telescope in the southern hemisphere, a better case than having a full telescope in the north and nothing in the south.

The observatory with the greatest number of large telescopes is now on top of the dormant volcano Mauna Kea in Hawaii, partly because its latitude is as far south as +20°, allowing much of the southern sky to be seen, and partly because the site is so high that it is above 40 per cent of the Earth's atmosphere. To detect the infrared part of the spectrum, telescopes must be above as much of the water vapor in the Earth's atmosphere as possible, and Mauna Kea is above 90 per cent of it. In addition, the peak is above the atmospheric inversion layer that keeps the clouds from rising, usually giving about 300 nights each year of clear skies with steady images. Consequently, several of the world's dozen largest telescopes are there (■ Fig. 3–13).

In particular, the California Institute of Technology (Caltech) and the University of California have built the two Keck 10-m telescopes (■ Fig. 3–14), whose mirrors are each twice the diameter and four times the surface area of Palomar's largest reflector. Hence, each one is able to gather light four times faster. When it was built, a single 10-m mirror would have been prohibitively expensive, so University of California scientists worked out a plan to use a mirror made of 36 smaller hexagons (■ Fig. 3–15).

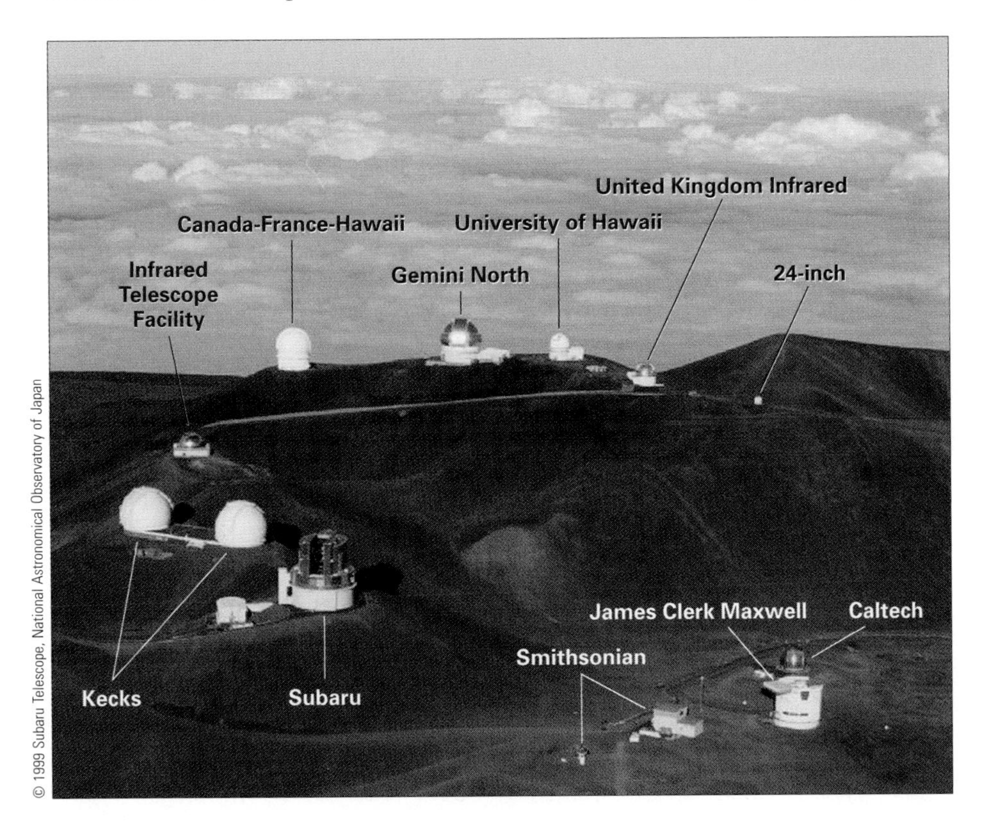

■ **FIGURE 3–13** Mauna Kea, with its many huge telescopes. The Keck twin domes with the Subaru dome near them are at the middle left; the NASA Infrared Telescope Facility is above them; and the Canada-France-Hawaii, Gemini North, University of Hawaii 88-inch, United Kingdom Infrared, and University of Hawaii 24-inch telescopes are on a ridge at top. In the "submillimeter valley" at lower right, we see the Caltech 10-m and James Clerk Maxwell 15-m submillimeter telescopes (Maxwell, one of the very greatest physicists of all time—one of the big three along with Newton and Einstein, in the view of many—unified electricity and magnetism theoretically), as well as the first to be installed of eight 6-m telescopes and the assembly building of the Smithsonian Astrophysical Observatory's submillimeter array.

■ **FIGURE 3-14** The dome of the Keck I 10-m telescope silhouetted against the Milky Way.

Richard Wainscoat, Institute for Astronomy, University of Hawaii

Jay M. Pasachoff

■ **FIGURE 3-15** Each 10-m mirror of the Keck telescopes is made of 36 contiguous hexagonal segments, which are continuously adjustable. Several outrigger telescopes are planned, though their construction is held up for local political reasons.

The first telescope worked so well that a twin (Keck II) was quickly built beside it. Not only the Gillett Gemini North 8-m telescope but also a Japanese 8-m telescope, named Subaru (for the star cluster known in English as the Pleiades), are on Mauna Kea.

The University of Texas and Pennsylvania State University have built a 9.2-m telescope in Texas, the largest optical telescope in the world after the Kecks. It is on an inexpensive mount and has limited mobility, but it is very useful for gathering a lot of light for spectroscopy. A clone has been built in South Africa—the South Africa Large Telescope (SALT)—by many international partners.

Another major project is the European Southern Observatory's Very Large Telescope, an array of four 8-m telescopes (■ Fig. 3–16) in Chile. Most of the time they

FIGURE IT OUT 3.2

Changing Units

When changing from one system of units to another, it is helpful to keep careful track of the conversion factors, to make sure you aren't getting them upside-down. For example, 1 inch = 2.54 cm. You can always divide both sides of an equation by the same number (except zero) without destroying the equality. So dividing both sides of the previous equation by 2.54 cm, we get (1 inch)/(2.54 cm) = 1.

We can also always multiply anything by 1 without changing its value. So if we are given, say, a telescope size that is 8 m = 800 cm, as for some in the current generation of large telescopes, we can convert that unit to inches by multiplying by the conversion factor just derived and making sure that units at the top and the bottom of the equation cancel properly. Thus

$$800 \text{ cm} \times \frac{(1 \text{ inch})}{(2.54 \text{ cm})} = \frac{(800) \text{ inches}}{(2.54)} = 300 \text{ inches}$$

(approximating 800/2.54 = 300).

European Southern Observatory

■ **FIGURE 3–16** Domes of the Very Large Telescope, in Chile. The entire project consists of four individual 8-m telescopes plus four smaller telescopes (a few of which are still under construction). One of these movable 1.8-m "auxiliary telescopes" is visible here. From left to right in this 2004 image, we see Antu, Kueyen, Melipal, an auxiliary telescope, the 2.6-m VLT Survey Telescope (a wide-field imaging telescope), and Yepun.

are used individually, but technology is advancing to allow them to be used in combination to give still more finely detailed images. (The Keck pair is also being used occasionally in that mode.) The images are superb (■ Fig. 3–17). A pair of 6.5-m telescopes on Las Campanas, another Chilean peak, is the Magellan project, a collaboration among the Carnegie Observatories, the University of Arizona, Harvard, the University of Michigan, and MIT. A compound Giant Magellan Telescope, with several mirrors making a surface equivalent to that of a 21.5-m telescope, is now in the planning stages by parts of the Magellan collaboration, as is a 30-m Keck-style telescope spearheaded by Caltech and the University of California.

The Large Binocular Telescope project is under way in Arizona with two 8.4-m mirrors on a common mount. Partners include the University of Arizona, Arizona State, Northern Arizona University, Ohio State, Notre Dame, Research Corporation, and European institutes. The Astrophysical Institute of the Canary Islands, with European membership, and with additional participation from Mexico and the University of Florida, is completing a 10.4-m telescope of the Keck design on La Palma, Canary Islands.

As we mentioned, the angular size of the finest details you can see (the resolution) is basically limited by turbulence in the Earth's atmosphere, but a technique called **adaptive optics** is improving the resolution of more and more telescopes. In adaptive optics, the light from the main mirror is reflected off a secondary mirror whose shape can be slightly distorted many times a second to compensate for the atmosphere's distortions. With this technology and other advances, the resolution from ground-based telescopes has been improving recently after a long hiatus. However, the images show fine detail only over very small areas of the sky, such as the disk of Jupiter.

European Southern Observatory

■ **FIGURE 3–17** The spiral galaxy NGC 1097, imaged with one of the telescopes of the Very Large Telescope in Chile. The president of Chile, on an official visit, helped operate the telescope for this image. This galaxy has a central black hole whose mass is tens of millions of times that of the Sun.

3.4 The Big Picture: Wide-field Telescopes

As mentioned above, ordinary optical telescopes see a fairly narrow field of view—that is, a small part of the sky is in focus. Even the most modern have images of less than about 1° × 1° (for comparison, the full moon appears roughly half a degree in diameter), which means it would take decades to make images of the entire sky (over 40,000 square degrees). The German optician Bernhard Schmidt, in the 1930s, invented a way of using a thin lens ground into a complicated shape together with a spherical mirror to image a wide field of sky (■ Fig. 3–18).

The largest **Schmidt telescopes,** except for one of interchangeable design, are at the Palomar Observatory in California and at the United Kingdom Schmidt site in Australia (■ Fig. 3–19); these telescopes have front lenses 1.25 m (49 inches) in diameter and mirrors half again as large. (The back mirror is larger than the front lens in order to allow study of objects off to the side.) They can observe a field of view some 7° × 7°—almost the size of your fist held at the end of your outstretched arm, compared with only the size of a grain of sand at that distance in the case of the Hubble Space Telescope!

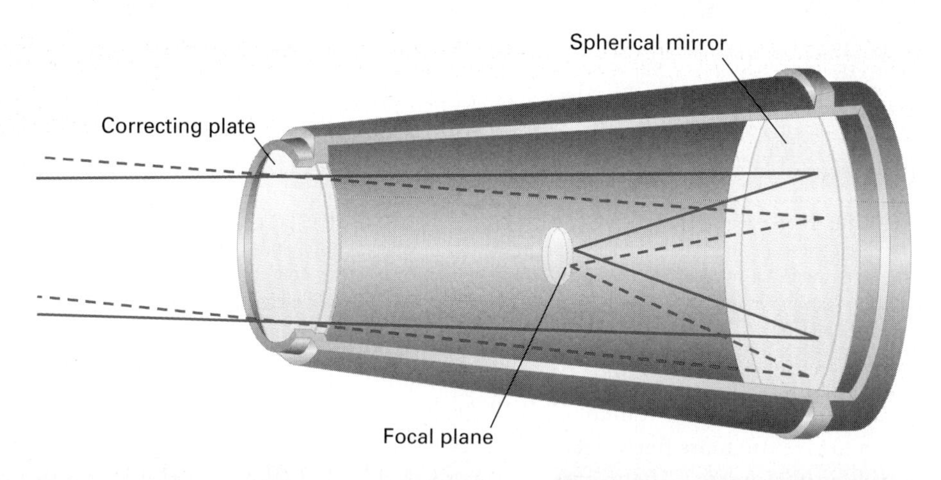

■ **FIGURE 3-18** By having a non-spherical thin lens called a correcting plate, a Schmidt camera is able to focus a wide angle of sky onto a curved piece of film. Since the image falls at a location where you cannot put your eye, the image is always recorded on film. Accordingly, this device is often called a Schmidt camera rather than a Schmidt telescope.

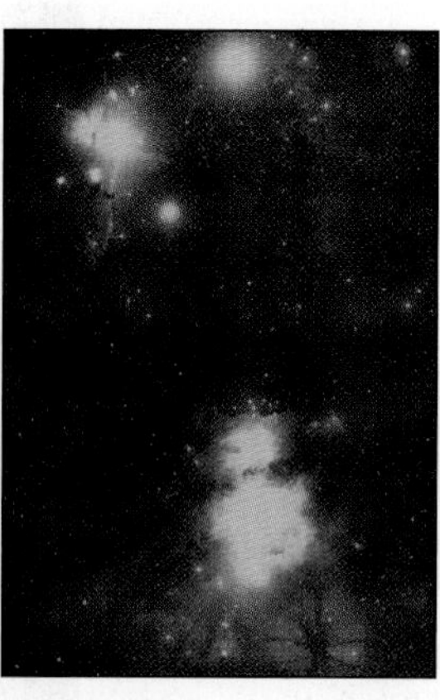

■ **FIGURE 3-19** (a) Three individual images, each taken with the U.K. Schmidt telescope (which is in Australia) through a filter in a different color, are printed together to make the full-color image. (b) This color view of the Horsehead Nebula in Orion is from the top-left quarter of the three monochromatic (single-color) images. The white circles show the size of the Moon.

The Palomar camera, now named the "Oschin" Schmidt telescope, was used in the 1950s to survey the whole sky visible from southern California with photographic film and filters that made pairs of images in red and blue light. This Palomar Observatory Sky Survey is a basic reference for astronomers. Hundreds of thousands of galaxies, quasars, nebulae, and other objects have been discovered on them. The Schmidt telescopes in Australia and Chile have extended this survey to the southern hemisphere. In the 1990s, Palomar carried out a newer survey with improved films and more overlap between adjacent regions. Among other things, it is being compared with the first survey to see which objects have changed or moved. Both surveys have been digitized, to improve their scientific utility. The Palomar Schmidt has now been converted to digital detection with CCDs, and is largely being used to search for Earth-approaching asteroids.

An interesting method of surveying wide regions of sky has been developed by the Sloan Digital Sky Survey team. They use a telescope (not of Schmidt design) at Apache Point, New Mexico, to record digital CCD images of the sky as the Earth turns, observing simultaneously in several different colors (■ Fig. 3–20). So many data were collected that developing new methods of data handling was an important part of the project. They have mapped hundreds of millions of galaxies and half a million "quasars" (extremely distant, powerful objects in the centers of galaxies; see Chapter 17). The telescope took spectra of 500,000 of the galaxies and 60,000 of the quasars that it mapped.

University of Chicago News Office

■ **FIGURE 3–20** The filters of the Sloan Digital Sky Survey; the charge-coupled devices (CCDs) are behind them. The color bands include ultraviolet, blue-green, red, far red, and near infrared.

3.5 Amateurs Are Catching Up

It is fortunate for astronomy as a science that so many people are interested in looking at the sky. Many are just casual observers, who may look through a telescope occasionally as part of a course or on an "open night," when people are invited to view through telescopes at a professional observatory, but others are quite devoted "amateur astronomers" for whom astronomy is a serious hobby. Some amateur astronomers make their own equipment, ranging up to quite large telescopes perhaps 60 cm in diameter. But most amateur astronomers use one of several commercial brands of telescopes.

Computer power and the techniques for CCD image processing have advanced so much that these days, amateur astronomers are producing pictures that professionals using the largest telescopes would have been proud of a decade ago. One interesting technique is to take thousands of photos very quickly, use a computer to throw out the blurriest, and combine the rest to form a sharp image. Some amateurs are even contributing significantly to professional research astronomers, obtaining high-quality complementary data.

Many of the amateur telescopes are Newtonian reflectors, with mirrors 15 cm in diameter being the most popular size (■ Fig. 3–21*a*). It is quite possible to shape your own mirror for such a telescope. The Dobsonian telescope is a variant of this type, made with very inexpensive mirrors and construction methods. Ordinary Dobsonians don't track the stars as the sky revolves overhead, but they are easy to turn by hand in the up-down direction (called "altitude") and in the left-right direction (called "azimuth") on cheap Teflon bearings. (Computerized tracking can be added.)

Compound telescopes that combine some features of reflectors with some of the Schmidt telescopes are very popular. This Schmidt–Cassegrain design (■ Fig. 3–21*b*) bounces the light, so that the telescope is relatively short, making it easier to transport and set up. Many of the current versions of these telescopes are now being provided with a computer-based "Go-To" function, where you press a button and the telescope "goes to" point at the object you have selected. Some have Global Positioning System (GPS) installed, so that the telescope knows where it is located. Then you point to at first one and then another bright star whose names you know, and the telescope's computer calculates the pointing for the rest of the sky.

Jay M. Pasachoff

a

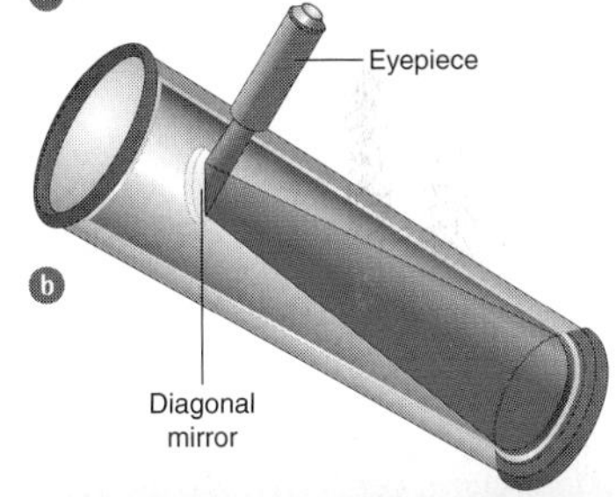

b

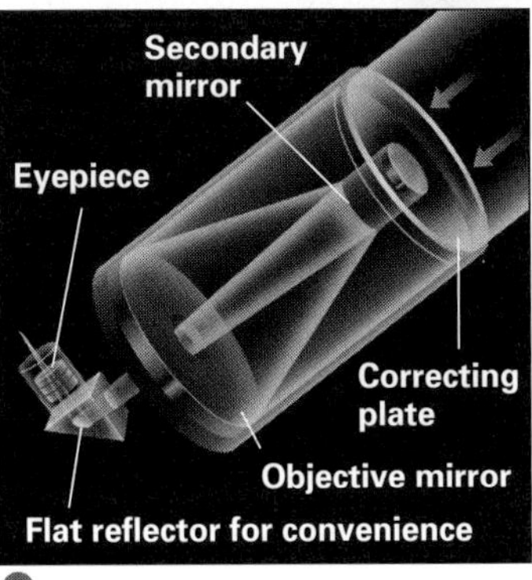

c

■ **FIGURE 3–21** (a) An amateur astronomer's Newtonian telescope. Note that the eyepiece is perpendicular to the tube near the tube's top. (b) A cutaway drawing of this Newtonian telescope. (c) A cutaway drawing of a compound Schmidt-Cassegrain design, now widely used by amateurs because of its light-gathering power and portability.

3.6 Glorious Hubble After Initial Trouble

The first moderately large telescope to be launched above the Earth's atmosphere is the **Hubble Space Telescope** (HST or Hubble, for short; ■ Fig. 3–22), built by NASA with major contributions from the European Space Agency. The set of instruments on board Hubble is sensitive not only to visible light but also to ultraviolet and infrared radiation that don't pass through the Earth's atmosphere. (These regions of the electromagnetic spectrum will be discussed in more detail below.)

The Hubble saga is a dramatic one. When launched in 1990, the 94-inch (2.4-m) mirror was supposed to provide images with about 10 times better resolution than ground-based images, because of Hubble's location outside of Earth's turbulent atmosphere. (But the latest advances in adaptive optics now eliminate, for at least some types of infrared observations involving narrow fields of view, the advantage Hubble formerly had over ground-based telescopes.) Unfortunately, the main mirror (■ Fig. 3–23) turned out to be made with slightly the wrong shape. Apparently, an optical system used to test it was made slightly the wrong size, and it indicated that the mirror was in the right shape when it actually wasn't. The result is some amount of spherical aberration, which blurred the images and caused great disappointment when it was discovered soon after launch.

Fortunately, the telescope was designed so that space-shuttle astronauts could visit it every few years to make repairs. A mission launched in 1993 carried a replacement for the main camera and correcting mirrors for other instruments, and brought the telescope to full operation. A second generation of equipment, installed in 1997, included a camera, the Near Infrared Camera and Multi-Object Spectrometer (NICMOS), that is sensitive to the infrared; however, it ran out of coolant sooner than expected. A mission in December 1999 replaced the gyroscopes that hold the telescope steady and made certain other repairs and improvements, such as installing better computers. A mission in 2002 installed the Advanced Camera for Surveys (ACS) and a cooler (refrigerator) that made the infrared camera work again. With increased sensitivity and a wider field of view than the best camera then on board, ACS is about 10 times more efficient in getting images, so it is taking Hubble to a new level.

Hubble's high resolution is able to concentrate the light of a star into an extremely small region of the sky—the star or galaxy isn't blurred out. This, plus the very dark background sky at high altitude, even at optical wavelengths but especially in the infrared (where Earth's warm atmosphere glows brightly), allows us to examine much fainter objects than we could formerly (■ Fig. 3–24). The combination of resolution

NASA's Johnson Space Center

■ **FIGURE 3-22** An astronaut (marked with an arrow) works on the Hubble Space Telescope during its 2002 servicing mission, when the Advanced Camera for Surveys was installed and the infrared camera was restored to working order. The telescope is named after Edwin Hubble, who discovered the true nature of galaxies and the redshift -distance relation for galaxies that led to our understanding of the expansion of the Universe (see Chapter 16).

Goodrich Corp., Optical and Space Systems Division

■ **FIGURE 3-23** The 2.4-m (94-inch) mirror of the Hubble Space Telescope, before launch but after it was ground and polished and covered with a reflective coating. This single telescope has already made numerous discoveries but is raising as many questions as it answers. So we need still more ground-based telescopes as partners in the enterprise, in addition to more telescopes in space.

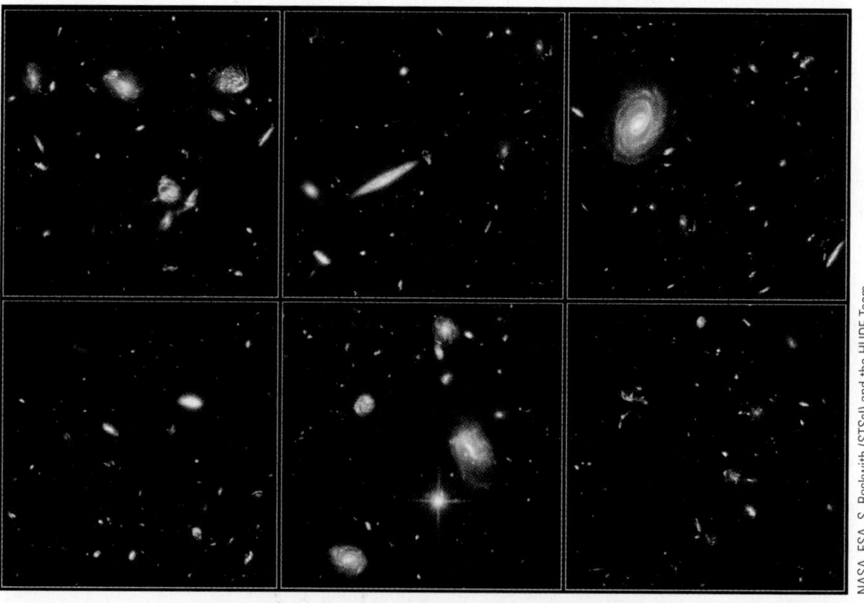

NASA, ESA, S. Beckwith (STScI) and the HUDF Team

■ **FIGURE 3-24** Sections of the Hubble Ultra Deep Field, a tiny region of the sky viewed over and over with the Hubble Space Telescope's Advanced Camera for Surveys. Many faint and distant galaxies are revealed, often interacting with each other.

and sensitivity has led to great advances toward solving several basic problems of astronomy. As we shall see later (Chapter 16), we were able to pin down our whole notion of the size and age of the Universe much more accurately than before. Thus, Hubble fills a unique niche, and was well worth its much larger cost compared with a ground-based telescope of similar size.

Still another upgrade of Hubble had been planned for many years, but it was put on hold after the space-shuttle *Columbia* disaster on February 1, 2003. At the time of this writing (fall 2005), it is unknown whether the upgrade can be accomplished (either with a robotic or a crewed mission) before the batteries fail or too many gyroscopes break. Thus, the future of Hubble is unclear, given limitations on our space-shuttle fleet and declines in Federal funding; consult our website or Hubble's for the latest information.

A Next Generation Space Telescope, named the James Webb Space Telescope after the Administrator of NASA in its moon-landing days, is under construction with a nominal launch date of 2013. Currently planned to have a 6.5-m mirror, the telescope will be optimized to work at infrared wavelengths, which will make it especially useful in studying the origins of planets and in looking for extremely distant objects in the Universe. But it will not have high-resolution visible-light capabilities, so it won't be a direct replacement for Hubble.

3.7 You Can't Look at the Sun at Night

Telescopes that work at night usually have to collect a lot of light. **Solar telescopes** work during the day, and often have far too much light to deal with. They have to get steady images of the Sun in spite of viewing through air turbulence caused by solar heating. So solar telescopes are usually designed differently from nighttime telescopes. The Swedish Solar Telescope on La Palma, with its 1-m mirror, uses adaptive optics to get fantastically detailed solar observations. Planning is under way for a 4-m (13-ft) Advanced Technology Solar Telescope, larger than any previous solar telescope, with the 10,000-foot (3050-m) altitude of Haleakala on Maui, Hawaii, selected as the site.

Solar observatories in space have taken special advantage of their position above the Earth's atmosphere to make x-ray and ultraviolet observations. The Japanese Yohkoh (Sunbeam) spacecraft, launched in 1991, carried x-ray and ultraviolet telescopes for study of solar activity. Yohkoh was killed by a solar eclipse in 2001; its trackers, used to find the edge of the round Sun, got confused, allowing the telescope to drift away and to start spinning. A successor with improved imaging detail, named Solar-B until it is launched, is to be its replacement.

The Solar and Heliospheric Observatory (SOHO), a joint European Space Agency/NASA mission, is located a million kilometers upward toward the Sun, for constant viewing of all parts of the Sun. The Transition Region and Coronal Explorer (TRACE) gives even higher-resolution images of loops of gas at the edge of the Sun (■ Fig. 3–25).

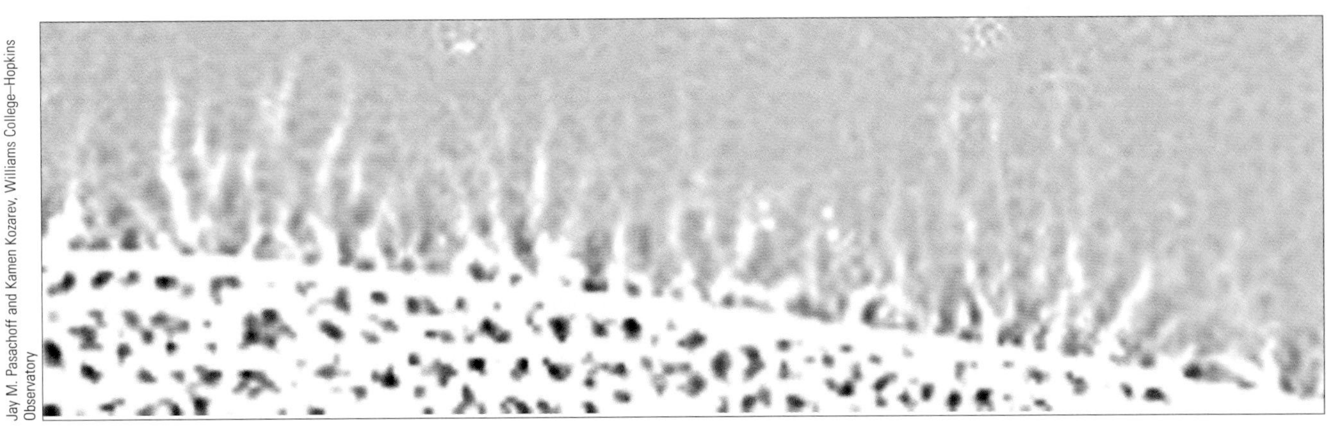

Jay M. Pasachoff and Kamen Kozarev, Williams College–Hopkins Observatory

■ **FIGURE 3-25** Spikes of hot chromospheric gas from the Transition Region and Coronal Explorer (TRACE) spacecraft imaged in the ultraviolet as part of one of the authors' (J.M.P.) research.

All these telescopes gave especially interesting data during the year 2000–2001 maximum of sunspot and other solar activity. A new set of high-resolution solar spacecraft should be in place for the sunspot maximum of 2010–2011.

3.8 How Can You See the Invisible?

As discussed more fully in Chapter 2, we can describe a light wave by its wavelength (■ Fig. 3–26). But a large range of wavelengths is possible, and visible light makes up only a small part of this broader spectrum (■ Fig. 3–27). **Gamma rays, x-rays,** and **ultraviolet** light have shorter wavelengths than visible light, and **infrared** and **radio waves** have longer wavelengths.

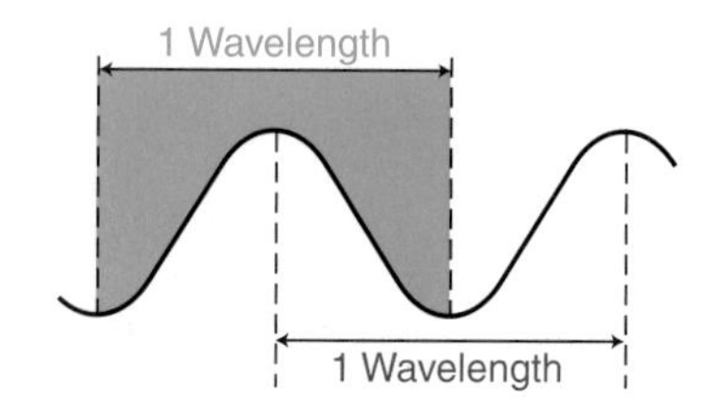

■ **FIGURE 3-26** Waves of electric and magnetic fields travelling across space are called "radiation." The wavelength is the length over which a wave repeats.

3.8a X-ray and Gamma-ray Telescopes

The shortest wavelengths would pass right through the glass or even the reflective coatings of ordinary telescopes, so special imaging devices have to be made to study them. And x-rays and gamma rays do not pass through the Earth's atmosphere, so they can be observed only from satellites in space. NASA's series of three High-Energy Astronomy Observatories (HEAOs) was tremendously successful in the late 1970s. One of them even made detailed x-ray observations of individual objects with resolution approaching that of ground-based telescopes working with ordinary light.

NASA's best x-ray telescope is called the **Chandra X-ray Observatory,** named after a scientist (S. Chandrasekhar; see Chapter 13) who made important studies of white dwarfs and black holes. It was launched in 1999. It makes its high-resolution images with a set of nested mirrors made on cylinders (■ Figs. 3–28 and ■ 3–29). Ordinary mirrors could not be used because x-rays would pass right through them. However, x-rays bounce off mirrors at low angles, just as stones can be skipped across a lake at low angles (■ Fig. 3–30). Chandra joins Hubble as one of NASA's "Great Observatories." The European Space Agency's XMM-Newton mission has more telescope area and so is more sensitive to faint sources than Chandra, but it doesn't have Chandra's high resolution.

ASIDE 3.1: X-ray black-hole satellite

The Japanese-American Suzaku satellite, launched in 2005, is capturing x-rays over a wide energy range to study violent phenomena such as those occuring near black holes, including those at the centers of many galaxies. Though one instrument failed, two others are working.

NASA's Swift spacecraft, named in part for the swiftness (within about a minute) with which it can turn its ultraviolet/visible-light and x-ray telescopes to point at gamma-ray burst positions, started its observations in 2005. NASA's major Constellation-X quartet of x-ray spacecraft is on the drawing board for 2019.

In the gamma-ray part of the spectrum, the Compton Gamma Ray Observatory was launched in 1991, also as part of NASA's Great Observatories program. NASA

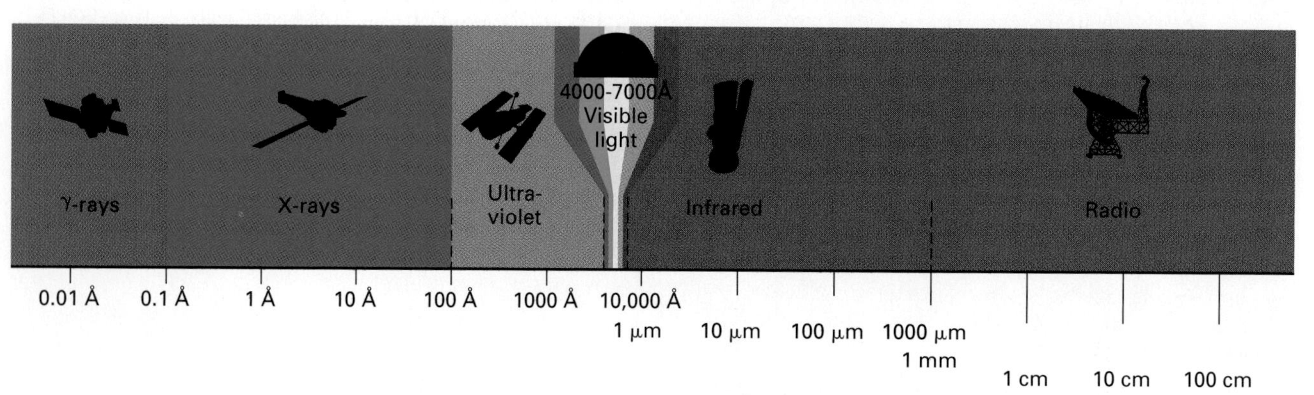

■ **FIGURE 3-27** The spectrum of electromagnetic waves. The silhouettes represent telescopes or spacecraft used or planned for observing that part of the spectrum: the Compton Gamma Ray Observatory, the Chandra X-ray Observatory, the Hubble Space Telescope, the dome of a ground-based telescope, the Spitzer Space Telescope, and a ground-based radio telescope.

■ **FIGURE 3–28** An outside view of the nested Chandra X-ray Observatory mirrors.

■ **FIGURE 3–29** One of the cylindrical mirrors from the Chandra X-ray Observatory. Its inside is polished to reflect x-rays by grazing incidence.

destroyed it in 2000, since the loss of one more of its gyros would have made it more difficult to control where its debris would land on Earth. It made important contributions to the study of exotic "gamma-ray bursts" (Chapter 14) and other high-energy objects. The European Space Agency launched its "Integral" (**Inte**rnational, **G**amma-**R**ay **A**strophysics **L**aboratory) telescope in 2002. Besides observing gamma rays, it can make simultaneous x-ray and visible-light observations.

Some huge "light buckets" are giant ground-based telescopes, bigger than the largest optical telescopes, but focusing only well enough to pick up (but not accurately locate) flashes of visible light in the sky (known as "Cerenkov radiation") that are caused by gamma rays hitting particles in our atmosphere. The MAGIC telescope (**M**ajor **A**tmospheric **G**amma-ray **I**maging **C**erenkov Telescope) on La Palma in the Canary Islands is such a device.

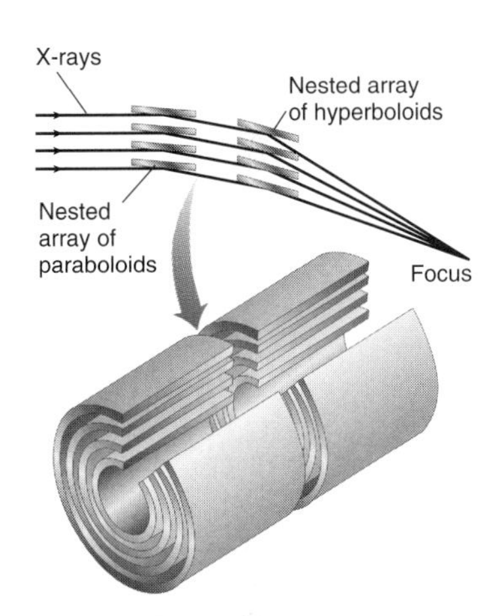

■ **FIGURE 3–30** Follow the arrows from the left on the light rays to see how each bounces off first a paraboloid and then a hyperboloid (the surface formed by spinning a hyperbola on its axis of symmetry) to come to a common focus. As the three-dimensional drawing in the lower part of the diagram shows, four similar paraboloid/hyperboloid mirror arrangements, used at low angles to incoming x-rays and all sharing the same focus, are nested within each other to increase the area of telescope surface that intercepts x-rays in NASA's Chandra X-ray Observatory. A photo of one segment was shown in the preceding figure (Fig. 3–29).

3.8b Telescopes for Ultraviolet Wavelengths

Ultraviolet wavelengths are longer than x-rays but still shorter than visible light. All but the longest wavelength ultraviolet light does not pass through the Earth's atmosphere, so must be observed from space. For about two decades, the 20-cm telescope on the International Ultraviolet Explorer spacecraft sent back valuable ultraviolet observations. Overlapping it in time, the 2.4-m Hubble Space Telescope was launched, as discussed above; it has a much larger mirror and so is much more sensitive to ultraviolet radiation.

Several ultraviolet telescopes have been carried aloft for brief periods aboard space shuttles, and brought back to Earth at the end of the shuttle mission. At present, NASA's Far Ultraviolet Spectrographic Explorer (FUSE) is taking high-resolution spectra largely in order to study the origin of the elements in the Universe. NASA's Galaxy Evolution Explorer (GALEX) is sending back ultraviolet views of distant and nearby galaxies to find out how galaxies form and change.

3.8c Infrared Telescopes

From high-altitude sites such as Mauna Kea, parts of the infrared can be observed from the Earth's surface. From high aircraft altitudes, even more can be observed, and NASA has refitted an airplane with a 2.5-m telescope to operate in the infrared. Cool objects such as planets and dust around stars in formation emit most of their radiation

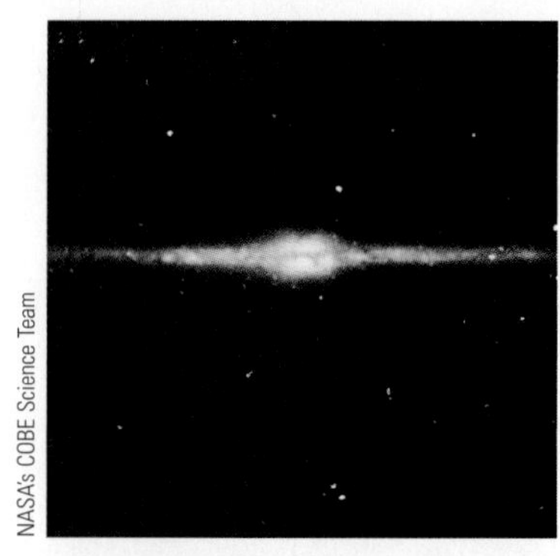
NASA's COBE Science Team

■ **FIGURE 3-31** The sky, mapped by the Cosmic Background Explorer spacecraft in 1990, reveals mainly the Milky Way and thus the shape of our Galaxy. The image here is false color, a translation into visible wavelengths of three infrared wavelengths that penetrate interstellar dust.

in the infrared, so studies of planets and of how stars form have especially benefited from infrared observations. This **S**tratospheric **O**bservatory **f**or **I**nfrared **A**stronomy (SOFIA) telescope should have its first scientific flights in 2006.

An international observatory, the **I**nfra**r**ed **A**stronomical **S**atellite (IRAS), mapped the sky in the 1980s, and then was followed by the European **I**nfrared **S**pace **O**bservatory (ISO) in the mid-1990s. Since the telescopes and detectors themselves, because of their warmth, emit enough infrared radiation to overwhelm the faint signals from space, the telescopes had to be cooled way below normal temperatures using liquid helium. These telescopes mapped the whole sky, and discovered a half-dozen comets, hundreds of asteroids, hundreds of thousands of galaxies, and many other objects. ISO took the spectra of many of these objects.

Since infrared penetrates the haze in space, its whole-sky view reveals our own Milky Way Galaxy. During 1990–1994, the **Co**smic **B**ackground **E**xplorer (COBE) spacecraft mapped the sky in a variety of infrared (■ Fig. 3–31) and radio wavelengths, primarily to make cosmological studies. NASA's **W**ilkinson **M**icrowave **A**nisotropy **P**robe (WMAP) was launched in 2001 to make higher-resolution observations. We shall discuss their cosmological discoveries in Chapter 19.

The whole sky was mapped in the infrared in the 1990s by 2MASS, a joint ground-based observation by the University of Massachusetts at Amherst and the Imaging Processing and Analysis Center at Caltech. (The "2" in the acronym is for its 2-micron wavelength and the initials of **2 M**icron **A**ll **S**ky **S**urvey are "Mass.")

A large, sensitive infrared mission, the **Spitzer Space Telescope** (■ Fig. 3–32*a*), was launched as the infrared Great Observatory in 2003. It has been producing phenomenal images (for example, ■ Fig. 3–32*b*). The European Space Agency plans its Herschel infrared telescope for launch in 2007. (Sir William Herschel discovered infrared radiation about two hundred years ago.)

3.8d Radio Telescopes

Since Karl Jansky's 1930s discovery (■ Fig. 3–33) that astronomical objects give off radio waves, radio astronomy has advanced greatly. Huge metal "dishes" are giant

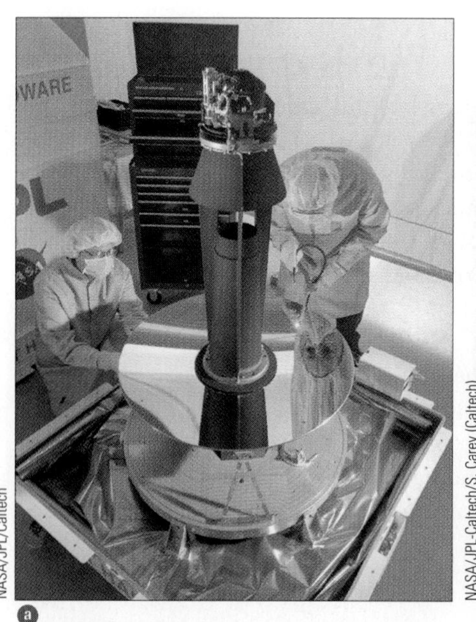

NASA/JPL/Caltech

a

NASA/JPL-Caltech/S. Carey (Caltech)

b

■ **FIGURE 3-32** ⓐ NASA's Spitzer Space Telescope before launch. We see its lightweight beryllium mirror. ⓑ A region of star formation, imaged with the Spitzer Space Telescope. This photograph is composed of images obtained at four wavelengths: 3.6 microns (shown as blue), 4.5 microns (shown as green), 5.8 microns (shown as orange), and 8 microns (shown as red). This object, DR6, is about as large across as the nearest star is to the Sun (4 light years).

■ **FIGURE 3-33** ⓐ Karl Jansky *(inset)* and the full-scale model of the rotating antenna with which he discovered radio waves from space. ⓑ A sculpture honoring the original Jansky telescope, erected at the original site in Holmdel, New Jersey, in 1998.

reflectors that concentrate radio waves onto antennae that enable us to detect faint signals from objects in outer space.

A still larger dish in Arecibo, Puerto Rico, is 1000 feet (330 m) across, but points only more or less overhead. Still, all the planets and many other interesting objects pass through its field of view. This telescope and one discussed below starred in the movie *Contact.*

Astronomers almost always convert the incoming radio signals to graphs or intensity values in computers and print them out, rather than converting the radio waves to sound with amplifiers and loudspeakers. If the signals are converted to sound, it is usually only so that the astronomers can monitor them to make sure no radio broadcasts are interfering with the celestial signals.

Radio telescopes were originally limited by their very poor resolution. The resolution of a telescope depends only on the telescope's diameter, but we have to measure the diameter relative to the wavelength of the radiation we are studying. For a radio telescope studying waves 10 cm long, even a 100-m telescope is only 1000 wavelengths across. A 10-cm optical telescope studying ordinary light is 200,000 wavelengths across, so it is effectively much larger and gives much finer images (■ Fig. 3–34); see *Figure It Out 3.3: Angular Resolution of a Telescope.*

ASIDE 3.2: Eyesight

The optical quality of the human eye, not the size of its lens, determines its final resolution.

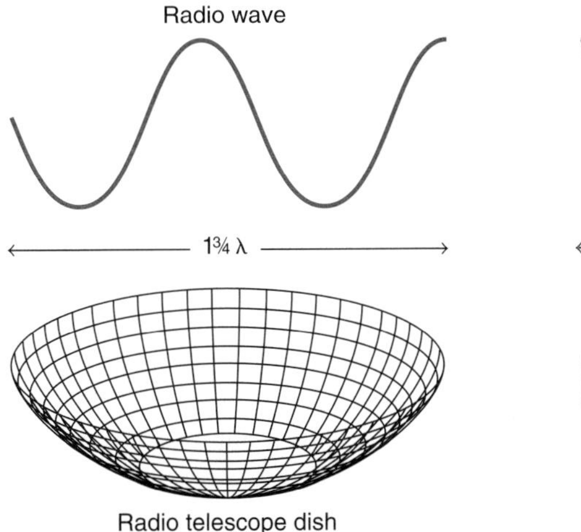

■ **FIGURE 3-34** For familiarity's sake, we often refer to telescopes by their physical size, though this is not the most significant measure. It is more meaningful to measure the diameter of telescope mirrors in terms of the wavelength of the radiation that is being observed than it is to measure them in terms of units like centimeters that have no particularly relevant significance. The radio dish at left is only $1\frac{3}{4}$ wavelengths across, whereas the mirror at right is 8 wavelengths across, making it effectively much bigger. The wavelengths are greatly exaggerated in this diagram relative to the size of any actual reflectors. For the Bonn radio telescope used to observe 10-cm waves, the 100-m diameter divided by 0.1 m per wave = 1000 wavelengths.

FIGURE IT OUT 3.3

Angular Resolution of a Telescope

Telescopes improve the clarity with which objects are seen: they have higher *angular resolution* (the ability to see fine detail) than the human eye.

Angular measure is important in astronomy. The full circle is divided into 360 degrees (360°). The Moon and the Sun each subtend (cover) about 1/2°. Each degree consists of 60 arc minutes (60'). Each minute of arc consists of 60 arc seconds (60"). A second of arc is very small—approximately the angle subtended by a dime viewed from a distance of 3.7 km. One can also use radians for angular measure. There are 2π radians in 360°, so 1" = 1/206265 radian.

If two point-like objects are closer together than about 1 arc minute, the unaided eye will perceive them as only one object because their individual "blur circles" merge together. With a telescope, the size of the individual blur circles decreases, and the objects become resolved (see the figure). The angular diameter of the blur circle is proportional to λ/D, where λ is the observation wavelength and D is the diameter of the lens or mirror. Hence, in principle, large telescopes are able to resolve finer details than small telescopes (at a given wavelength).

A rule of thumb found by the amateur astronomer Dawes in the 19th century is that the resolution (in arc seconds) equals $0.002\lambda/D$, if λ is given in angstroms (Å) and D is given in centimeters. This resolution is the angular separation at which a double star is detectable as having two separate components (if they have equal brightness).

As an example, the resolution of a telescope 1 m (i.e., 100 cm) in diameter for green light of approximately 5000 Å is given by 0.002(5000/100) = 0.1 arc second.

In practice, Earth's atmosphere blurs starlight: Layers of air with different densities move in a turbulent way relative to each other, and the rays of light bend in different directions. This is related to the twinkling of starlight. The angular resolution of telescopes larger than 20-30 cm is limited by the blurring effects of the atmosphere (typically 1 arc second, and rarely smaller than 1/3 arc second), not by the size of the mirror or lens. Thus, even bigger ground-based telescopes do not give clearer images, unless advanced techniques such as adaptive optics are used.

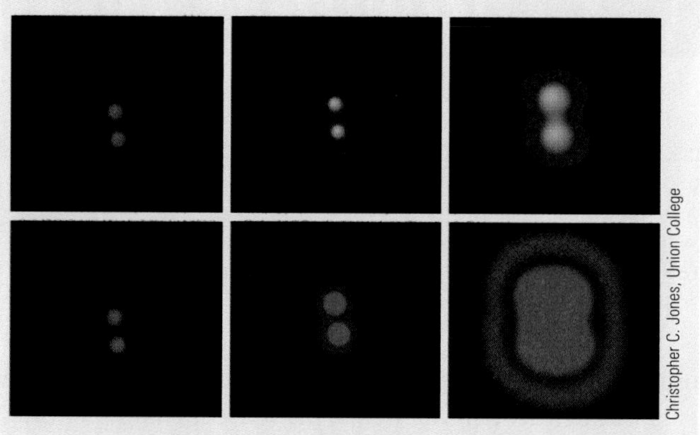

Christopher C. Jones, Union College

In addition to their primary use for gathering light from faint objects, telescopes are often used to increase the resolution, or clarity—our ability to distinguish details in an image. We see a pair of point sources (sources so small or so far away that they appear as points), in the right-hand column, that are not quite resolved in *(top row)* violet light (about 4000 Å) and *(bottom row)* red light (about 6500 Å). Note that the rightmost violet image is substantially better resolved than the rightmost red image. In this series, the images were obtained with lenses (or mirrors) of different diameter; from right to left, the diameter is larger by a factor of 2 in successive images. In astronomy, as long as we aren't limited by shimmering (turbulence) in the Earth's atmosphere, a smaller image disk is produced by a larger telescope or at a shorter wavelength.

National Radio Astronomy Observatory

FIGURE 3–35 The 100-m Robert C. Byrd Green Bank Telescope, which went into use in 2001.

A new radio telescope at the National Radio Astronomy's site at Green Bank, West Virginia, replacing one that collapsed, has a reflecting surface 100 m in diameter in an unusual design (Fig. 3–35). New technology has made it possible to observe radio waves well at relatively short wavelengths, those measured to be a few millimeters. Molecules in space are especially well studied at these wavelengths. The new Byrd Green Bank Telescope is useful for studying such molecules.

A breakthrough in providing higher resolution has been the development of arrays of radio telescopes that operate together and give the resolution of a single telescope spanning kilometers or even continents. The **Very Large Array (VLA)** is a set of 27 radio telescopes, each 26 m in diameter, and was also seen in the movie *Contact* (Fig. 3–36). All the telescopes operate together, and powerful computers analyze the joint output to make detailed pictures of objects in space. These telescopes are linked to allow the use of "interferometry" to mix the signals; analysis later on gives images of very high resolution. The VLA's telescopes are spread out over hundreds of square kilometers on a plain in New Mexico. An Expanded VLA (EVLA), with improved electronics and additional dishes, is under construction.

A Closer Look 3.1 A Night at Mauna Kea

An "observing run" with one of the world's largest telescopes is a highlight of the work of many astronomers. The construction of several huge telescopes at the top of Mauna Kea ("White Mountain") in Hawaii has made the mountain the site of one of the world's major observatories. It was chosen because of its clear and steady air. These conditions stem from the fact that the summit is so high—4200 m (13,800 ft) above sea level—and the mountain is surrounded by ocean. Though there are other tall mountains in the world, few have such favorable conditions for astronomy.

Let us imagine you have applied for observing time and have been chosen. During the months before your observing run, you make lists of objects to observe, prepare charts of the objects' positions in the sky based on existing star maps and photographs, and plan the details of your observing procedure. The air is very thin at the telescope; you may not think as clearly or rapidly as you do at lower altitudes. Therefore, it is wise and necessary to plan each detail in advance.

The time for your observing run comes, and you fly off to the island of Hawaii, the largest of the islands in the state of Hawaii. You drive up the mountain, as the scenery changes from tropical to relatively barren, crossing dark lava flows. First stop is the mid-level facility (Hale Pohaku) at an altitude of 2750 m (9000 ft). The Onizuka Center for International Astronomy at Hale Pohaku is named after Ellison Onizuka, the Hawaiian astronaut who died in the space-shuttle *Challenger* explosion in 1986.

The next night is yours. During the afternoon, you test the instrument with which the data will be obtained and determine the exact configuration in which it is to be used. For example, if you will be obtaining optical spectra of objects, the spectrograph needs to be focused properly and you must choose specific wavelength ranges to observe. An afternoon trip to the summit might be necessary.

After dinner you return to the summit to start your observing. You dress warmly, since the nighttime temperature approaches freezing at this altitude, even in the summertime. Although most telescope domes are equipped with warm rooms in which astronomers sit when collecting data, occasionally you might step outside to absorb the magnificence of the dark sky—or to monitor the motion of encroaching clouds.

During the night you are assisted by a Telescope Operator who knows the telescope and its systems very well, and who is responsible for the telescope, its operation, and its safety. The operator points the telescope to the object you have chosen to observe and makes sure the telescope is properly compensating for the Earth's rotation. A television camera mounted on the telescope shows objects within the telescope's field of view. You examine that region of the sky, carefully comparing it with your chart to identify the right object among many random, unwanted stars. That object is then centered at the proper position to take data.

Nearly everything else is computer-controlled. You measure the brightness of the object at the wavelength you are observing, or begin an exposure to get a spectrum, by starting a computer program on a console provided for the observer. The data are stored in digital form on computer disks and tapes, and you might even transfer them over the Internet to your university computer.

Suppose you are interested in what kinds of chemical elements are produced and ejected by supernovae (exploding stars) or how they reveal the overall structure of the Universe and its expansion. One of us (A.F.) is a regular user of the Keck Telescopes on supernova projects. Or, suppose you are studying Pluto's atmosphere by watching it pass in front of a star. One of us (J.M.P.) made such an observation at Mauna Kea in 2002. In either case, you labor through the night, occasionally seeing an obviously exciting result, but often just getting tantalizing hints that demand much more detailed and time-consuming analysis.

At some time during the night, you may go into the telescope dome to ponder the telescope at work. The telescope all but blocks your view of the sky. Hardly anyone ever actually looks through the telescope; indeed, there is usually no good way to do so. Still, you may sense a deep feeling for how the telescope is looking out into space.

You drive down the mountain to Hale Pohaku in the early morning sun. Over the next few days and nights, you repeat your new routine. When you leave Mauna Kea, it is a shock to leave the pristine air above the clouds. But you take home with you data about the objects you have observed. It may take you months or years to fully study the data that you gathered in a few brief days and nights at the top of the world. But you hope that your data will allow a more complete understanding of some aspect of the Universe. Sometimes major conceptual breakthroughs are made.

W.M. Keck Observatory

The view from one of the ridges of telescopes across Mauna Kea, with the submillimeter valley at left, the Subaru and twin Keck telescopes in the middle, and the NASA Infrared Telescope Facility at right.

Riccardo Giovanelli and Martha Haynes, Cornell U.

■ **FIGURE 3-36** The Very Large Array (VLA) is laid out in a giant "Y" with a diameter of 27 km, and sprawls across a plain in New Mexico. It contains 27 telescopes (plus a spare). It is so large that even the weather is sometimes different from one side to another. Its imaging capability has been important for understanding radio sources. An expansion to greater distances is under way.

To get even higher resolution, astronomers have built the **Very Long Baseline Array (VLBA),** spanning the whole United States. Its images are many times higher in resolution than even those of the VLA. Astronomers often use the technique of "very-long-baseline interferometry" to link telescopes at such distances, or even distances spanning continents, but the VLBA dedicates telescopes full-time to such high-resolution work.

AceAstronomy™ Log into AceAstronomy and select this chapter to see the Active Figure called "Different Telescopes."

CONCEPT REVIEW

About 400 years ago, Galileo was the first to use telescopes for extensive astronomical studies (Sec. 3.1). He discovered lunar craters, the disk and four main moons of Jupiter, countless stars in the Milky Way, the phases of Venus, sunspots, and other phenomena. These findings contributed to the Copernican revolution, to be further discussed in Chapter 5.

Refracting telescopes use lenses and **reflecting telescopes** use mirrors to collect light (Sec. 3.2). **Spherical aberration** occurs when parallel light is not reflected or refracted to a good **focus,** the place where the image is formed; it is solved for on-axis light with a mirror in the shape of a **paraboloid. Chromatic aberration,** a property only of lenses, occurs when light of different colors does not reach the same focus. **Newtonian telescopes** reflect the light from the main mirror diagonally out to the side; **Cassegrain** telescopes use a secondary mirror to reflect the light from the main mirror through a hole in the primary.

The larger the area of the telescope's main lens or mirror, the better its light-gathering power, the ability to detect faint objects (Sec. 3.2). Also, the larger the diameter, the higher the theoretical **resolution**—the ability to distinguish detail. In practice, the shimmering of the Earth's atmosphere limits the resolution of large telescopes, but a new technique called **adaptive optics** allows astronomers to achieve very clear images over small fields of view (Sec. 3.3). **Charge-coupled devices (CCDs)** are a common type of electronic detector that is extremely sensitive to light, allowing very faint objects to be studied.

Schmidt telescopes have a very wide field of view (Sec. 3.4). One of the most important astronomical surveys ever conducted was the Palomar Observatory Sky Survey; in the 1950s, it mapped the whole sky visible from southern California. Recently, the Sloan Digital Sky Survey carried out a CCD survey of a large portion of the sky.

Amateur astronomy has long been a popular hobby. Improvements in telescope design and computer technology have increased the portability and ease of operation of telescopes (Sec. 3.5). Equipped with CCDs and fancy image-processing computer programs, some amateur astronomers have even been contributing high-quality data to research projects conducted by professional astronomers.

The 2.4-m **Hubble Space Telescope** has been in the limelight since 1990 (Sec. 3.6). Although initially disappointing because its primary mirror suffered from spherical aberration, corrective optics were subsequently installed by space-shuttle astronauts and the telescope began delivering phenomenal results. For example, Hubble provides optical observations with resolution about ten times better than most ground-based data. It has been critical for ultraviolet observations, and many of its studies in the infrared have been superior as well. Ground-based **solar telescopes,** which collect enormous amounts of light and must deal with the effects of solar heating, have technical differences from nighttime telescopes (Sec. 3.7). Solar telescopes in space provide access to wavelengths not visible from Earth's surface and also obtain optical images with very high resolution. Telescopes to observe **x-rays, gamma rays,** and most **ultraviolet** light must be above the Earth's atmosphere. For nearly a decade, NASA's now-defunct Compton Gamma Ray Observatory provided data at the highest energies (Sec. 3.8a). The **Chandra X-ray Observatory,** one of NASA's "Great Observatories," has been an outstanding success. The Hubble Space Telescope has been NASA's premier ultraviolet telescope, but there have been other important ultraviolet telescopes as well (Sec. 3.8b). While some parts of the **infrared** can be observed with telescopes on the ground, telescopes in airplanes or in orbit, such as NASA's **Spitzer Space Telescope,** are also necessary (Sec. 3.8c).

Radio telescopes on the ground observe a wide range of wavelengths covered by **radio waves** (Sec. 3.8d). Arrays of radio telescopes, such as the **Very Large Array (VLA)** in New Mexico and the continent-spanning **Very Long Baseline Array (VLBA),** give high resolution through a process known as interferometry, linking the data from the different telescopes.

QUESTIONS

1. What are three discoveries that immediately followed the first use of the telescope for astronomy?
2. What advantage does a reflecting telescope have over a refracting telescope?
3. What limits a large ground-based telescope's ability to see detail?
4. List the important criteria in choosing a site for an optical observatory meant to study stars and galaxies.
5. Describe a method that allows us to make large optical telescopes more cheaply than simply scaling up designs of previous large telescopes.
6. Name some of the world's largest telescopes and their locations.
7. What are the similarities and differences between making radio observations and using a reflector for optical observations? Compare the path of the radiation, the detection of signals, and limiting factors.
8. Why is it sometimes better to use a small telescope in orbit around the Earth than it is to use a large telescope on a mountaintop?
9. Why is it better for some purposes to use a medium-size telescope on a mountain instead of a telescope in space?
10. Describe the problem that initially occurred with the Hubble Space Telescope, and how it was corrected.
11. Describe and compare the Hubble Space Telescope, the Spitzer Space Telescope, and the Chandra X-ray Observatory. Mention their uses, designs, and other relevant factors.
12. What are two reasons why the Hubble Space Telescope can observe fainter objects than we can now study from the ground?

†13. What is the light-gathering power of a telescope that is 3 meters in diameter, relative to a 1-m telescope? What is it relative to a dilated human eye (pupil 6 mm in diameter)?

†14. How many times more light is gathered by a single Keck telescope, whose mirror is equivalent in area to that of a circle 10 m in diameter, than by the Palomar Hale telescope, whose mirror is 5 m in diameter, in any given interval of time?

†15. Assume for simplicity that you have only one eye. Let's say that on a dark, clear night you can barely see Star Moe with your fully dilated (pupil 6 mm in diameter) naked eye. **(a)** How big an eye would you need to barely see Star Curly, which is 100 times fainter than Star Moe? Although human eyes aren't that large, one can buy small, inexpensive telescopes that have mirrors or lenses this size. **(b)** However, suppose you instead want to see Star Shemp, a million times fainter than Star Moe. How big would your telescope have to be? **(c)** Compare this with existing telescopes at major observatories. Do you think you could afford to buy one?

16. Why might some stars appear double in blue light, though they could not be resolved in red light with the same telescope?

†17. What mirror diameter gives 0.1 arc second resolution for infrared radiation of wavelength 2 micrometers?

†18. What mirror diameter gives 1 arc second resolution for radio radiation of wavelength 1 m? Compare this with the size of existing optical telescopes.

19. Why is light from stars and planets considered to be "parallel light"?
20. Why are optical and radio telescopes often built in groups, or arrays?
21. How is it possible to focus x-rays into images if x-rays pass right through most materials?

†22. **True or false?** All other things being equal, the faintest star one can see through a 10-m diameter telescope is 10 times fainter than the faintest star one can see through a 1-m diameter telescope.

23. **True or false?** In some cases, radio telescopes at many different locations have been used together, combining the light, to produce images with much higher resolution than that of the individual telescopes.
24. **True or false?** The primary purpose of an astronomical telescope is to magnify images of stars and other objects.
25. **True or false?** The clarity of images obtained with large ground-based optical telescopes is generally degraded by turbulence in Earth's atmosphere, unless special correction techniques are used.

26. **True or false?** Although Galileo was not the first to use two lenses in the form of a telescope, he was the first to conduct systematic astronomical studies with a telescope.
27. **Multiple choice:** Which one of the following is *not* an advantage of the Hubble Space Telescope (Hubble) over ground-based telescopes in gathering information about the Universe? (**a**) Hubble is better able to collect ultraviolet light. (**b**) Hubble detects fainter objects at optical wavelengths because it has a darker sky. (**c**) Hubble detects fainter objects at infrared wavelengths because it has a darker sky. (**d**) Hubble produces clearer images of celestial objects at optical wavelengths. (**e**) Hubble was cheaper to build than ground-based telescopes of comparable diameter.

†28. **Multiple choice:** Eloise's telescope has a mirror 10 m in diameter, while Deborah's telescope has a mirror 2 m in diameter. The tube of Eloise's telescope is twice the length of the tube of Deborah's telescope. When both telescopes are being used from the same location and in the same manner, the faintest stars Eloise can see are _______ times fainter than the faintest stars Deborah can see. (**a**) 5. (**b**) 10. (**c**) 12.5. (**d**) 25. (**e**) 50.

29. **Multiple choice:** A ground-based telescope to observe x-rays with x-ray detectors would (**a**) give astronomers a chance to study the hot interiors of stars and planets; (**b**) be worthless because x-rays cannot get through the Earth's atmosphere; (**c**) be worthless because astronomers have not yet devised detectors sensitive to x-rays; (**d**) be worthless because no astronomical objects emit x-rays; or (**e**) be useful only to Superman, who has x-ray vision.
30. **Multiple choice:** All other things being equal, and ignoring atmospheric effects, a reflecting telescope with a large primary mirror will have, compared with a telescope with a smaller primary mirror, (**a**) better light-gathering power only; (**b**) better resolution only; (**c**) better light-gathering power and better resolution; (**d**) better light-gathering power but poorer resolution; or (**e**) no clear advantage unless the secondary mirror is also larger.
31. **Multiple choice:** Chromatic aberration is a problem in ________ telescopes and refers to ________. (**a**) refracting : differing focal lengths for different wavelengths of light. (**b**) reflecting : differing focal points from different parts of the mirror. (**c**) refracting : the smearing of light due to atmospheric turbulence. (**d**) reflecting : the smearing of light due to atmospheric turbulence. (**e**) space : the deformation of the glass due to the absence of gravity.
32. **Fill in the blank:** About 300 years ago, ________ put a small diagonal mirror in front of the focal point of the primary mirror to produce the first workable reflecting telescope.
33. **Fill in the blank:** Electronic gadgets called _________ are now used much more frequently than photographic film to record images and spectra of astronomical objects.
34. **Fill in the blank:** The ability to distinguish between details or to distinguish two adjacent objects as separate is known as _________.

†35. **Fill in the blank:** A radio telescope 100 m in diameter, when used to measure 1-cm waves, is ________ wavelengths across.

36. **Fill in the blank:** Substantial improvements have recently been made to narrow-field images made with ground-based telescopes by using the technique of ________, which corrects for distortions in the waves of incoming light.

†This question requires a numerical solution.

TOPICS FOR DISCUSSION

1 Is it a coincidence that major improvements in astronomical detectors were made at nearly the same time as the electronics and computer revolution that began in the late 1970s?

2. Was the Hubble Space Telescope worth the roughly $2 billion that it cost? Reconsider this question after completing this textbook or your astronomy course.
3. Does it make sense that an obstruction inside of a telescope (such as the secondary mirror), or the hole in the center of the primary mirror of a Cassegrain telescope, doesn't produce a hole in the object that is being viewed?

MEDIA

AceAstronomy™ Log into AceAstronomy at **http://astronomy.brookscole.com/cosmos3** to access quizzes and animations that will help you assess your understanding of this chapter's topics.

Log into the Student Companion Web Site at **http://astronomy.brookscole.com/cosmos3** for more resources for this chapter including a list of common misconceptions, news and updates, flashcards, and more.

People in Astronomy

Jeff Hoffman

Jeffrey Hoffman is a NASA Scientist-Astronaut. He grew up near New York City, and went into the city every month or so to visit the American Museum of Natural History's Hayden Planetarium. After graduating from Amherst College (Amherst, Massachusetts) in 1966, he attended graduate school in astronomy at Harvard University, receiving his Ph.D. in 1971.

He then worked in England with the x-ray astronomy group at the University of Leicester for 3 years. While in England, he married, and his wife and he had their first child. Then he returned to the United States, working at the MIT Center for Space Research for 2 years. In 1978, he was selected by NASA as a scientist-astronaut, and moved to Houston. He flew on the space shuttle for the first time in 1985, making NASA's first unplanned space walk in an attempt to repair a malfunctioning satellite. In 1990 he was on the mission that carried aloft the "Astro" set of ultraviolet telescopes. During missions in 1992 and 1996, he conducted tests of a new type of space technology: the use of long tethers in space to generate electricity and to change the orbits of satellites. In 1993, he was one of the astronauts who repaired the Hubble Space Telescope. Hoffman is currently Professor of the Practice of Aeronautical Engineering at MIT. He is also director of the Massachusetts Space Grant Consortium.

How did you enjoy being in space?

Answering as an astronomer first, astronomers are used to working on mountaintops, and space is the ultimate mountaintop. Actually, for me, working on the Astro mission [a set of ultraviolet telescopes carried on a space shuttle] was unique, because my professional work had been x-ray astronomy using satellites, so I had never actually done anything with traditional telescopes. The first time I guided an actual optical-type telescope in my life was in space, but my three astronomer-astronaut colleagues let me do it anyway. Since I had spent most of my professional career building x-ray telescopes to fly in rockets and satellites, it was gratifying to fly with some telescopes on board. Of course, the fun of being in space goes beyond what your actual mission is. No matter what you are doing up there, it is an incredible view, an incredible feeling. But working as an astronomer on one of my space flights gave me a lot of professional satisfaction.

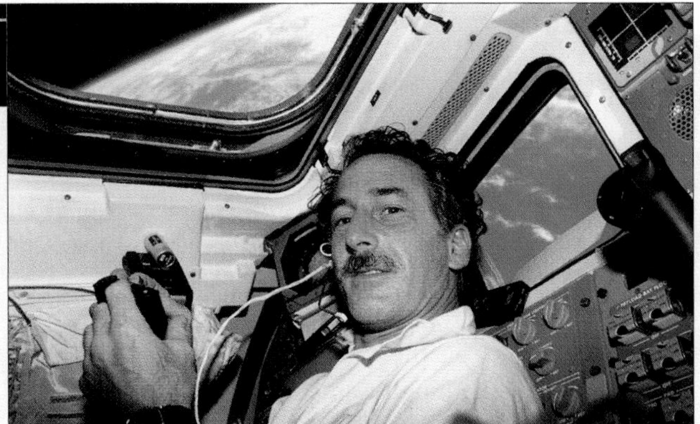

NASA's Johnson Space Center

What did it feel like to repair the Hubble Space Telescope?

Fixing the Hubble Space Telescope was the most important and challenging task of my entire astronaut career. It is easy in retrospect to forget the incredible shock produced by the discovery of Hubble's initial optical flaw, but in many ways the whole future of NASA's human spaceflight program rested on our ability to show that astronauts working in space suits could fix the problem. Had we not succeeded, it is quite possible that Congress would not have given NASA the go-ahead to build the new International Space Station. We worked extremely hard training for the mission, spending over 400 hours underwater and countless weeks in simulators. When we finally took off, I can honestly say that we had done everything we could think of to ensure the success of the mission. Of course, we knew that there were many unexpected things that could go wrong no matter how hard we trained. In fact, as things turned out, the most surprising thing about the mission was how few unpleasant surprises we had.

At the end of the fifth and final space walk, we were elated at having been able to accomplish every one of the tasks that we had set out to do. Of course, we would not know for several weeks whether the new optics we installed had actually corrected Hubble's vision. It was New Year's Eve when I finally got the news from some astronomer friends working at the Space Telescope Science Institute that Hubble was finally working flawlessly. What a great way to celebrate the new year!

How did you get to be an astronaut?

How did I originally decide I wanted to be an astronaut? That's been going on for a long time, ever since I was a little kid. I first got interested in astronomy back in the Hayden Planetarium in New York City. But I got interested not only in astronomy but also in anything having to do with space, including rockets. Of course, there was no such thing as a real space program back then, and I wanted to be a scientist. The first astronauts were all jet pilots, and that didn't appeal to me, though I was excited by the rocket part of it.

But when they announced that they needed scientists to be astronauts on the shuttle program, I always knew that this was something I wanted to do, so I applied.

"... MY INSPIRATION TO BECOME AN ASTRONOMER AND TO BECOME AN ASTRONAUT SPRING FROM THE SAME FASCINATION AT LOOKING BEYOND WHERE WE ARE NOW, LOOKING OUT FROM THE EARTH."

What do you do for NASA now?

After I left flight status as an astronaut, I spent four years representing NASA in Europe. This was a bit like being an ambassador. Spaceflight is increasingly an international activity, and it is useful to have someone "on the scene" to work with our partner space agencies. My office was in the American Embassy in Paris, since the headquarters of the European Space Agency is located in Paris. My technical background and experience with international projects during my various space missions gave me the qualifications for this job. In addition, I speak several European languages, which helps a lot.

Now, I have entered my fourth career, following being an astronomer, astronaut, and diplomat, as a faculty member of the Department of Aeronautics and Astronautics at the Massachusetts Institute of Technology. I am now working more as an engineer than an astrophysicist, a result of my extensive experience in space operations and design during my many years at NASA. We try to expose our students to all the phases involved in space projects: conceiving, designing, constructing, testing, and operating. I bring to the department special personal experience in operating space systems, which I try to share with students. In addition, I am involved in research projects using the International Space Station as a test bed for new satellite control technology and trying to develop more maneuverable space suits.

What would you most like to see NASA do next?

Exploration in all its aspects is NASA's primary mission. We need to develop space telescopes even more powerful than Hubble to continue our exploration of the astronomical universe. I think the search for extra-solar planetary systems and the search for life in the Universe is one of the most exciting scientific goals of the new millennium. Closer to home, we have a lot of exploring to do in our own Solar System. I am particularly excited about searching for signs of life on Mars and Europa. And of course I am interested in expanding human capability to travel and work in space. The International Space Station is the next step in this development. Making all this happen requires more reliable and cheaper space transportation, which is also one of NASA's main goals.

What message do you have for students?

First of all, I like to try to spread an ecological message that we have to take care of the Earth, our planetary home in space. We get a lot of responses to pictures that we take of the Earth from space, particularly where we can show the environmental changes taking place on the planet. Kids really seem to respond to that. We get disturbing sequences of pictures taken over the last 15 years showing the deforestation of the Amazon, the encroaching desert in sub-Saharan Africa, and land erosion in Madagascar, one environmental disaster after the other, which you can see better from space than from anywhere else.

I also like to talk to young people about the fact that you can study physics and astronomy and apply it in numerous different ways other than just becoming a professional astronomer. For instance, I can show my younger son, who is thinking about what he wants to do after university, two examples of friends of mine from graduate school. One of them started in physics and moved to biology and one was in applied mathematics and also moved to biology, and both do a lot of work in environmental science. Both developed the skills of mathematical analysis and facility with computers, which we use all the time to model complex systems in astronomy. I often find that my training as a physicist allows me to cut to the heart of problems in a way that some people who were trained as engineers sometimes don't do. The other thing that I often stress is the fact that my inspiration to become an astronomer and to become an astronaut spring from the same fascination at looking beyond where we are now, looking out from the Earth. I hope that we can keep the dream alive for the next generation so that they will be able to live out some of their dreams as well, whether they are studying through telescopes or travelling outside the Earth.

What comments do you have about the crewed space program?

I never felt that space flight was without risk. Space is a hazardous environment, unforgiving of human errors or mechanical failures. People have died exploring the oceans, the mountains, and the polar regions of the Earth, but it is part of the human spirit to push onward and outward. The exploration of space is one of the most exciting aspects of the past half-century. I wish we had a way to get into and back from space more safely, but I think it is important to continue. We owe it to the next generation.

NASA's Johnson Space Center

Jeff Hoffman fixing the Hubble Space Telescope.

C H A P T E R 4

Observing the Stars and Planets: Clockwork of the Universe

The Sun, the Moon, and the stars rise every day in the eastern half of the sky and set in the western half. If you leave your camera on a tripod with the lens open for a few minutes or hours in a dark place at night, you will photograph the "star trails"—the trails across the photograph left by the individual stars. In this chapter, we will discuss the phases of the Moon and planets, and how to find stars and planets in the sky. Stars twinkle; planets don't twinkle as much (■ Fig. 4–1). We will also discuss the motions of the Sun, Moon, and planets, as well as of the stars in the sky.

4.1 The Phases of the Moon and Planets

From the simple observation that the apparent shapes of the Moon and planets change, we can draw conclusions that are important for our understanding of the mechanics of the Solar System. In this section, we shall see how the positions of the Sun, Earth, and other Solar-System objects determine the appearance of these objects.

The **phases** of moons or planets are the shapes of the sunlighted areas as seen from a given vantage point. The fact that the Moon goes through a set of such phases approximately once every month is perhaps the most familiar astronomical observation, aside from the day/night cycle and the fact that the stars come out at night (■ Fig. 4–2*c*, 4–3). In fact, the name "month" comes from the word "moon." The actual period of the phases, the interval between a particular phase of the Moon and its next repetition, is approximately $29\frac{1}{2}$ Earth days.

The Moon is not the only object in the Solar System that goes through phases. Mercury and Venus both orbit inside the Earth's orbit, and so sometimes we see the

AceAstronomy™ The AceAstronomy icon throughout this text indicates an opportunity for you to test yourself on key concepts, and to explore animations and interactions of the AceAstronomy website at **http://astronomy.brookscole.com/cosmos3**

ORIGINS

Throughout their existence, humans have used the apparent positions of celestial objects to define the day, month, year, and seasons, as well as for navigation.

AIMS

1. Understand why objects in the Solar System often seem to go through phases (Section 4.1).
2. Learn about eclipses of the Sun and of the Moon (Section 4.2).
3. Learn why stars seem to twinkle (Section 4.3).
4. Become familiar with the scale used by amateur and professional astronomers to describe how bright stars and planets look (Section 4.4).
5. Explore the basic motions of celestial objects, and how they make us see and experience rising and setting (Section 4.5).
6. Discuss the celestial analogues of longitude and latitude (Section 4.6).
7. Understand how seasons occur (Section 4.7).
8. Learn about time and calendars (Sections 4.8 and 4.9).

This long exposure shows stars circling the celestial north pole; the Hawaiian volcano Mauna Kea, where many telescopes are located, is in the foreground. The shortest bright arc is Polaris, the "North Star," only about 1° from the pole. (Though only the 49th-brightest star in the sky, it is famous because of its location.)

Peter Michaud, Gemini Observatory

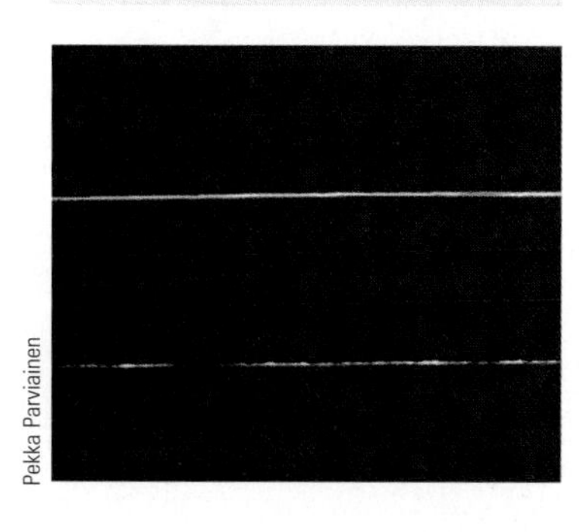

Pekka Parviainen

■ **FIGURE 4–1** A camera was moved steadily from left to right, with the bright star Sirius (*bottom trail*) in view. The star trail shows twinkling. Also, the camera was moved from left to right *(top trail)* when pointed at Jupiter, which left a more solid, less twinkling trail. Notice the contrast between the less-twinkling planet trail and the more-twinkling star trail. The fact that the star trail breaks up into bits of different colors was caused by part of the twinkling effect of the Earth's atmosphere, given that Sirius was low in the sky when the picture was taken.

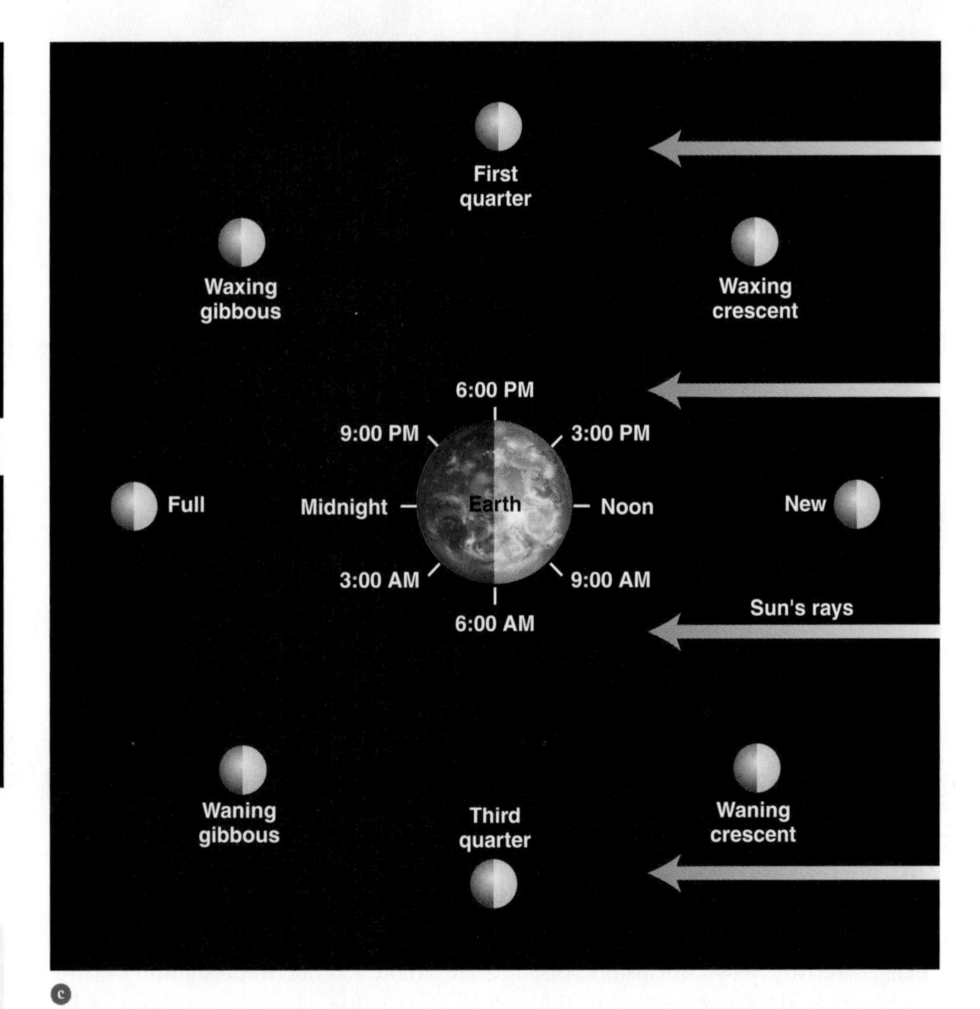

■ **FIGURE 4-2** (a) As the 2001 Mars Odyssey spacecraft left Earth, its visible-light camera recorded Earth as a crescent of reflected sunlight, since at that time the spacecraft was behind most of the hemisphere facing the Sun. (b) Though Saturn's disk always looks full when seen from Earth, the views backward toward the Sun from the Cassini spacecraft show Saturn and its moons as crescents. Here we see a 2005 view of Rhea. (c) The phases of the Moon depend on the Moon's position in orbit around the Earth. Here we visualize the situation as if we were looking down from high above the Earth's orbit. Each of the Moon images shows how the Moon is actually lighted. Note that the Sun always lights one hemisphere of the Moon—the one directly facing the Sun, which is generally not the one facing the Earth.

ASIDE 4.1: The rising Moon

The Moon rises an average of about 50 (actually 48.8) minutes later each day. Multiplying 48.8 minutes/day by $29\frac{1}{2}$ days (the lunar month) gives 1440 minutes, which is 24 hours. Thus, at a given phase, the Moon rises at the same time each month.

side that faces away from the Sun and sometimes we see the side that faces toward the Sun; see Chapter 5 for more details. Thus at times Mercury and Venus are seen as crescents, though it takes a telescope to observe their shapes. Spacecraft to the outer planets have looked back and seen the Earth as a crescent (■ Fig. 4–2*a*) and the other planets as crescents (■ Fig. 4–2*b*) as well.

■ **FIGURE 4-3** The phases of the Moon, as seen from Earth. Each phase is labeled to match the corresponding Sun-Moon-Earth relative position that Fig. 4-2c showed from a point of view high above the Solar System.

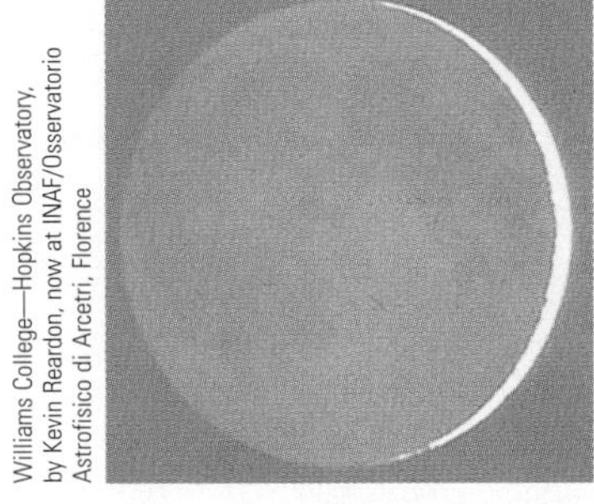

Waxing crescent

Waxing crescent

First quarter

Waxing gibbous

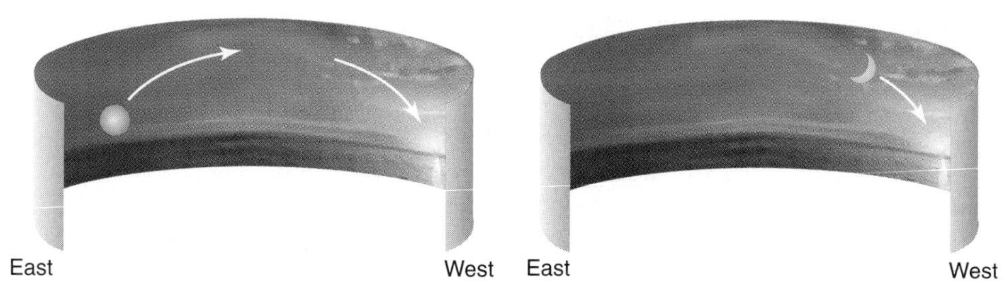

■ **FIGURE 4–4** Because the phase of the Moon depends on its position in the sky with respect to the Sun, a full moon always rises at sunset. A crescent moon is either setting shortly after sunset, as shown here, or rising shortly before sunrise. (These statements are true when viewing the Moon from latitudes relatively near Earth's equator. Close to either pole, the relationship between lunar phase and time of rising or setting is more complicated.)

The explanation of the phases is quite simple: the Moon is a sphere that shines by reflecting sunlight, and at all times the side that faces the Sun is lighted and the side that faces away from the Sun is dark. The phase of the Moon that we see from the Earth, as the Moon revolves (orbits) around us, depends on the relative orientation of the three bodies: Sun, Moon, and Earth. The situation is simplified by the fact that the plane of the Moon's revolution around the Earth is nearly, although not quite, the same as the plane of the Earth's revolution around the Sun; they are inclined relative to each other by 5°.

When the Moon is almost exactly between the Earth and the Sun, the dark side of the Moon faces us. We call this a "new moon." A few days earlier or later we see a sliver of the lighted side of the Moon, and call this a "crescent." As the month progresses, the crescent gets bigger (a "waxing crescent"), and about seven days after a new moon, half the face of the Moon that is visible to us is lighted. We are one quarter of the way through the cycle of phases, so we have the "first-quarter moon."

When over half the Moon's disk is visible, it is called a "gibbous moon" (pronounced "gibb'-us"). As the sunlighted portion visible to us grows, the gibbous moon is said to be "waxing." One week after the first-quarter moon, the Moon is on the opposite side of the Earth from the Sun, and the entire face visible to us is lighted. This is called a "full moon." Thereafter we have a "waning gibbous moon." One week after full moon, when again half the Moon's disk that we see appears lighted, we have a "third-quarter moon." This phase is followed by a "waning crescent moon," and finally by a "new moon" again. The cycle of phases then repeats.

Note that since the phase of the Moon is related to the position of the Moon with respect to the Sun, if you know the phase, you can tell approximately when the Moon will rise. For example, since the Moon is 180° across the sky from the Sun when it is full, a full moon rises just as the Sun sets (■ Fig. 4–4). Each day thereafter, the Moon rises an average of about 50 minutes later. The third-quarter moon, then, rises near midnight, and is high in the sky at sunrise. The new moon rises with the Sun in the east at dawn, and sets with the Sun in the west at dusk. The first-quarter moon rises near noon and is high in the sky at sunset.

It is natural to ask why the Earth's shadow doesn't generally hide the Moon during full moon, and why the Moon doesn't often block the Sun during new moon. These phenomena are rare because the Moon's orbit is tilted by about 5° relative to the

Full moon

Waning gibbous

Third quarter

Waning crescent

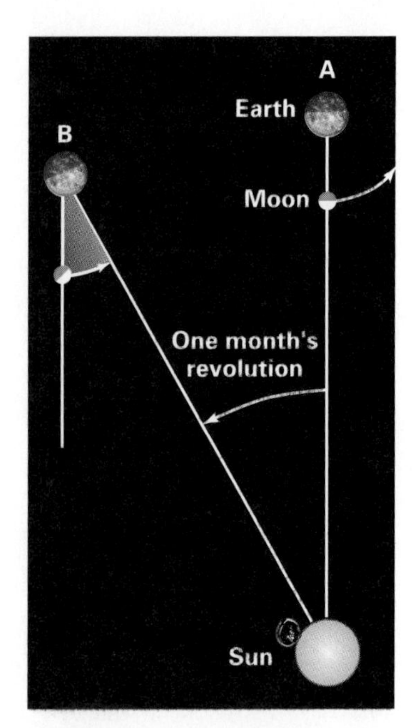

■ **FIGURE 4-5** After the Moon has completed its $27\frac{1}{3}$-day revolution around the Earth with respect to the stars, it has moved from point A to B. But the Earth has moved one month's worth around the Sun. It takes about an extra two days for the Moon to complete its revolution around the Earth as measured with respect to the Sun.

Earth–Sun plane, making it difficult for the Sun, Earth, and Moon to become exactly aligned. However, when they do reach the right configuration, eclipses occur, as we will soon discuss.

The Moon revolves around the Earth every $27\frac{1}{3}$ days with respect to the stars. But during that time, the Earth has moved partway around the Sun, so it takes a little more time for the Moon to complete a revolution with respect to the Sun (■ Fig. 4–5). Thus the cycle of phases that we see from Earth repeats with this $29\frac{1}{2}$-day period.

AceAstronomy™ Log into AceAstronomy and select this chapter to see the Active Figure called "Lunar Phases."

AceAstronomy™ Log into AceAstronomy and select this chapter to see Astronomy Exercise "Phases of the Moon."

AceAstronomy™ Log into AceAstronomy and select this chapter to see Astronomy Exercise "Moon Phase Calendar."

4.2 Celestial Spectacles: Eclipses

Because the Moon's orbit around the Earth and the Earth's orbit around the Sun are not precisely in the same plane (■ Fig. 4–6), the Moon usually passes slightly above or below the Earth's shadow at full moon, and the Earth usually passes slightly above or below the Moon's shadow at new moon. But up to seven times a year, full moons or new moons occur when the Moon is at the part of its orbit that crosses the Earth's orbital plane. At those times, we have a **lunar eclipse** or a **solar eclipse** (■ Fig. 4–7). Thus up to seven eclipses (mostly partial) can occur in a given year.

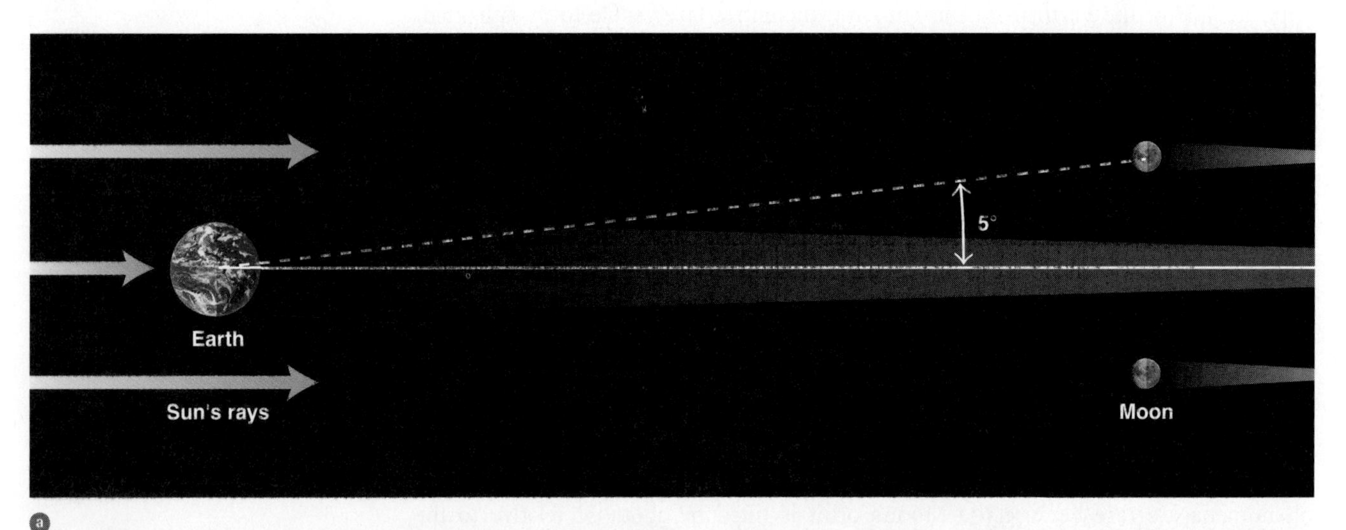

■ **FIGURE 4-6** (a) The plane of the Moon's orbit is tipped by 5° with respect to the plane of the Earth's orbit, so the Moon usually passes above or below the Earth's shadow. Therefore, we don't have lunar eclipses most months. (The diagram is not to scale.) (b) An actual image of the Earth and Moon, illustrating their true separation: about 30 Earth diameters. The image was obtained with the infrared camera on board the 2001 Mars Odyssey spacecraft. Taken at a wavelength of 9 μm, the image shows colder regions, such as the South Pole near the Earth's bottom, as dark, and Australia, which is warmer, as bright. Each pixel corresponds to 900 km across, about the size of Texas.

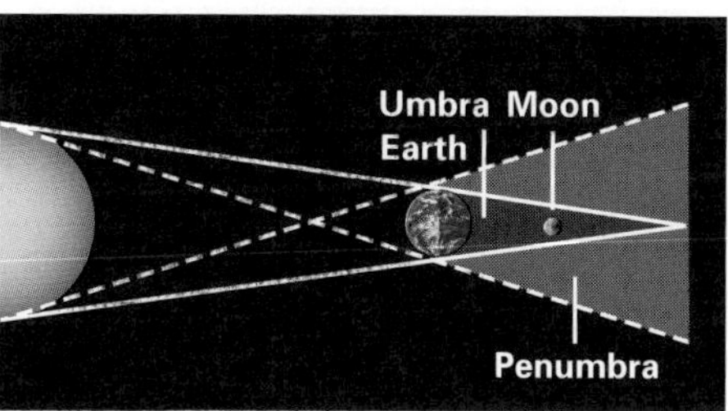

■ **FIGURE 4-7** When the Moon is between the Earth and the Sun, we observe an eclipse of the Sun. When the Moon is on the far side of the Earth from the Sun, we see a lunar eclipse. The part of the Earth's shadow that is only partially shielded from the Sun's view is called the "penumbra"; the part of the Earth's shadow that is entirely shielded from the Sun is called the "umbra." (The diagram is not to scale.)

Many more people see a total lunar eclipse than a total solar eclipse when one occurs. At a total lunar eclipse, the Moon lies entirely in the Earth's full shadow and sunlight is entirely cut off from it (neglecting the atmospheric effects we will discuss below). So for anyone on the entire hemisphere of Earth for which the Moon is up, the eclipse is visible. In a total solar eclipse, on the other hand, the alignment of the Moon between the Sun and the Earth must be precise, and only those people in a narrow band on the surface of the Earth see the eclipse. Therefore, it is much rarer for a typical person on Earth to see a total solar eclipse—when the Moon covers the whole surface of the Sun—than a total lunar eclipse.

We will discuss the science of the Sun in Chapter 10. Historically, many important things about the Sun, such as the hot "corona" (see below) and its spectrum, were discovered during eclipses. Now, satellites in space are able to observe the middle and outer corona on a daily basis, and to study the Sun in a variety of important ways. A few observatories on high mountains study some aspects of the inner corona, but what they can observe is limited. There are still gaps in what spacecraft and ground-based observatories can do to observe the corona, and eclipses remain scientifically useful for studies that fill in those gaps. For example, coronagraphs in space have to hide not only the Sun's surface but also a region around it, for technological reasons (■ Fig. 4–8). So images taken on the days of eclipses cover those physical gaps. In addition, eclipse images can be obtained with shorter intervals than images with current spacecraft.

ASIDE 4.2: Drama in the sky

The phenomenon of the darkening of the sky around you as totality approaches is dramatic. Watching on television loses most of the glory. For a few seconds, the chromosphere is visible as a pinkish band around the leading edge of the Moon. Then, as totality begins, the corona comes into view in all its magnificence (Fig. 4–12).

■ **FIGURE 4-8** A compound image of a solar eclipse. The disk of the Sun is from the Extreme-Ultraviolet Imaging Telescope on the Solar and Heliospheric Observatory (SOHO) spacecraft; the middle zone is from the Williams College Eclipse Expedition headed by one of the authors (J.M.P.), and the outer part is from the Large-Angle Spectrographic Coronagraph on SOHO. Note how features can be traced from lower levels in the corona to higher levels because of the way the images are used together.

Star Party 4.1 Observing Solar Eclipses

The next total solar eclipse visible from the United States isn't until 2017 and from Canada until 2024 (except for one far north in 2008). In general, one must travel far to witness totality. At a given location on Earth, the average interval between total solar eclipses is over 300 years.

Should you go out of your way to observe a total solar eclipse? The entire event is indescribably beautiful and moving. Photographs and words simply do not convey the drama, beauty, and thrill of a total solar eclipse: It must be witnessed in person. In our opinion, everyone should see at least one!

Only during the *total* part of a total eclipse can one look at the Sun directly without special filters. (During the Baily's beads that mark the last few seconds before or after totality, it is also safe to observe the eclipse directly with the naked eye, though not with binoculars or a telescope.) During other phases and types of eclipses, you can observe the Sun safely only through special filters or by projecting the image so that you are not looking directly at the Sun. Appropriate filters to use include welder's glass, shade 14, which is available at welding supply shops, or one of the special solar-viewing filters available from certain vendors (**http://www.eclipses.info**).

Alternatively, to make a pinhole camera (Fig. 4–11*b*) that projects the image of the Sun (so that you don't look directly at the Sun), punch a small hole (about 3 mm across) in a piece of thin cardboard. Face the cardboard toward the Sun and project the resulting image of the Sun onto a flat surface, such as another piece of cardboard, which you look at while you are facing away from the Sun. This method works best if the flat surface is in the shade, directly behind the cardboard with the hole in it. If you punch multiple holes, you will get an image of the Sun from each one; be creative by spelling your name, drawing a simple object, etc.

The total solar eclipse of April 8, 2005. (a) A wide-angle view, showing the eclipsed Sun. At the horizon, we see a 360° view of the sky glowing reddish from sunlight outside the Moon's shadow reflected into the shadow, where the camera was located. (b) The diamond-ring effect as totality began. (c) The solar corona surrounding the dark silhouette of the Moon in the middle of the 30-second period of totality. (d) Image of the partially eclipsed Sun projected by a monocular (half of a pair of binoculars), next to two round pieces of candy, showing the festive mood of successful eclipse viewers after totality.

4.2a Eerie Lunar Eclipses

A total lunar eclipse is a much more leisurely event to watch than a total solar eclipse. The partial phase, when the Earth's shadow gradually covers the Moon, usually lasts over an hour, similar to the duration of the partial phase of a total solar eclipse. But then the total phase of a lunar eclipse, when the Moon is entirely within the Earth's shadow, can also last for over an hour, in dramatic contrast with the few minutes of a total solar eclipse (see Section 4.2b).

During a total lunar eclipse, the sunlight is not entirely shut off from the Moon: A small amount is refracted (bent) around the edge of the Earth by our atmosphere. Much of the short-wavelength light (violet, blue, green) is scattered out by air molecules, or absorbed by dust and other particles, during the sunlight's passage through our atmosphere. The remaining light is reddish-orange, and this is the light that falls on the Moon during a total lunar eclipse. (For the same reason, the light from the Sun that we see near sunset or sunrise is orange or reddish; the blue light has been scattered away or absorbed by particles; see *A Closer Look 4.1: Colors in the Sky.*) Thus, the eclipsed Moon, though relatively dark, looks orange or reddish (■ Fig. 4–9); its overall appearance, in fact, is rather eerie and three-dimensional.

Weather permitting, people in the continental United States will be able to see a total lunar eclipse at or near moonrise on March 3, 2007. Everybody in the United States (including Alaska and Hawaii) and Canada will be able to see the total lunar eclipse of August 28, 2007, near moonset. The total lunar eclipse of February 21, 2008, will also be visible throughout the Americas. Try to see a lunar eclipse if you can!

4.2b Glorious Solar Eclipses

The hot, tenuous outer layer of the Sun, known as the **corona,** is fainter than the blue sky. To study it, we need a way to remove the blue sky while the Sun is up. A total solar eclipse does just that for us.

Akira Fujii

(a)

Dennis di Cicco

(b)

■ **FIGURE 4–9** (a) A total lunar eclipse, showing both partial and total phases. The camera was guided on the position of the Earth's umbral shadow. Note that the photograph shows that the Earth's shadow is round, and thus that the Earth is round. The totally eclipsed Moon appears faintly in the center of the shadow and looks reddish. (b) Another view of a total lunar eclipse, this one a longer exposure than in (*a*).

A Closer Look 4.1 COLORS IN THE SKY

Why is the sky blue? Air molecules in the Earth's atmosphere scatter (reflect) short-wavelength photons of light (violet, blue, and green) more effectively than they scatter longer-wavelength photons (orange or red), preferentially sending the short-wavelength light around in all directions, including toward our eyes. So, away from the Sun, the sky looks some shade of blue (see the figure). The response of the eye's color cones plays a role in making the sky blue instead of violet.

Why are sunsets and sunrises red? When the Sun is high in the sky, it appears white (not yellow, a common misconception). As the Sun moves progressively lower in the sky, we have to look more and more diagonally through the Earth's layer of air. The path of light through the atmosphere increases, so more of the short-wavelength photons get scattered away by air molecules before they reach us; the Sun then looks yellow, sometimes followed by orange (see the figure).

Also, various particles in the atmosphere, such as dust and smoke, tend to absorb blue light more than red light, further contributing to the Sun's changing color. As it approaches the horizon, it looks progressively more orange and red. Just before sunset (or just after sunrise), the Sun is sometimes very red if there is much dust aloft. Reflection of these orange and red rays of sunlight by air, dust, and clouds gives the sky its vivid sunset/sunrise colors.

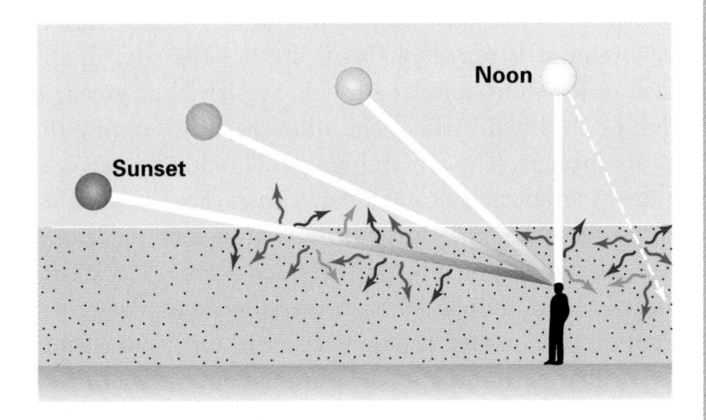

Air molecules preferentially scatter violet, blue, and green light, causing the sky away from the Sun to appear some shade of blue. The longer the path through the air, the more short-wavelength light is lost, so the setting Sun appears orange or red.

Solar eclipses arise because of a happy circumstance: Though the Moon is about 400 times smaller in diameter than the solar **photosphere** (the disk of the Sun we see every day, also known as the Sun's "surface"), it is also about 400 times closer to the Earth. Because of this coincidence, the Sun and the Moon cover almost exactly the same angle in the sky—about $^1/_2$° (■ Fig. 4–10).

Occasionally the Moon passes close enough to the Earth–Sun line that the Moon's shadow falls upon the surface of the Earth. The central part of the lunar shadow barely reaches the Earth's surface. This lunar shadow sweeps across the Earth's surface in a band up to 300 km wide. Only observers stationed within this narrow band can see the total eclipse.

From anywhere outside the band of totality, one sees only a partial eclipse. Sometimes the Moon, Sun, and Earth are not precisely aligned and the darkest part of the shadow—called the **umbra**—never hits the Earth. We are then in the intermediate

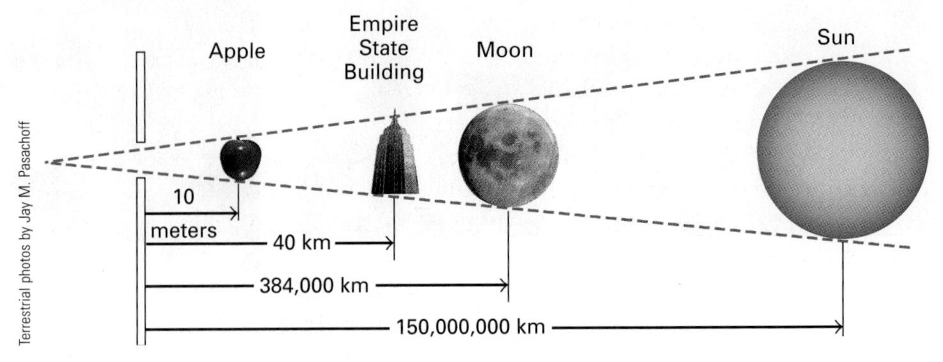

■ **FIGURE 4–10** An apple, the Empire State Building, the Moon, and the Sun are very different from each other in physical size, but here they cover the same angle because they are at different distances from us.

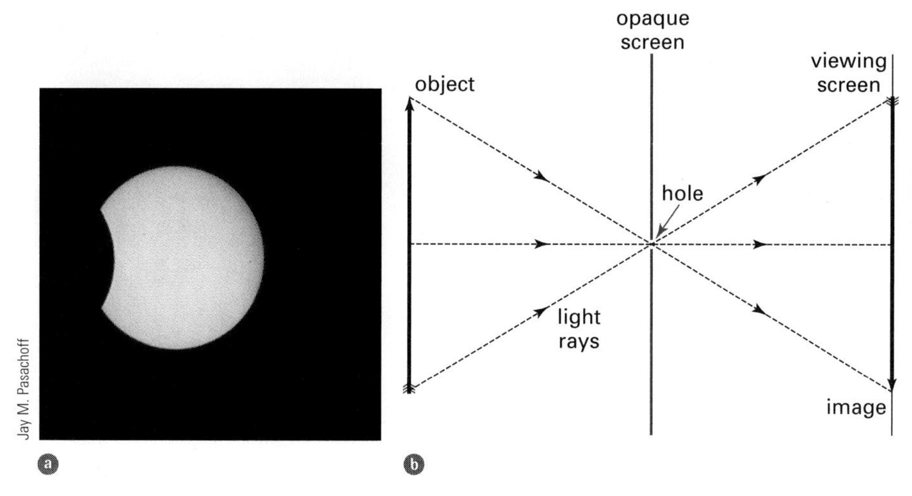

Jay M. Pasachoff

Alex Filippenko

■ **FIGURE 4–11** (a) The partially eclipsed Sun, as seen through a safe filter that eliminates all but one part in about 100,000 of the incident sunlight. (b) Diagram of a pinhole camera, essentially just a hole in an opaque screen such as cardboard. Two of the rays of light, in this case from the extreme ends of an arrow, show the formation of an inverted image of the arrow. (c) The multiple images of the partially eclipsed Sun of July 11, 1991, produced by the pinhole-like gaps between leaves in a tree.

part of the shadow, which is called the **penumbra.** Only a partial eclipse is visible on Earth under these circumstances (■ Fig. 4–11*a*).

As long as the slightest bit of photosphere is visible, even as little as 1 per cent, one cannot see the important eclipse phenomena—the faint outer layers of the Sun—that are both beautiful and the subject of scientific study. Thus partial eclipses are of little value for most scientific purposes. After all, the photosphere is 1,000,000 times brighter than the outermost layer, the corona; if 1 per cent of the photosphere is showing, then we still have 10,000 times more light from the photosphere than from the corona, which is enough to ruin our opportunity to see the corona.

To see a partial eclipse or the partial phase of a total eclipse, you should not look at the Sun except through a special filter, just as you shouldn't look at the Sun without such a special filter on a non-eclipse day (see *Star Party 4.1: Observing Solar Eclipses*). You need the filter to protect your eyes because the photosphere is visible throughout the partial phases before and after totality. Its direct image on your retina for an extended time could cause burning, heating, and blindness. Alternatively, you can project the image of the Sun with a telescope or binoculars onto a surface, being careful not to look up through the eyepiece. A home-made pinhole camera (■ Fig. 4–11*b*) is the simplest and safest device to use, forming an image of the Sun. Multiple images of the Sun appear with cameras having many pinholes, and are sometimes produced by natural phenomena such as holes between leaves in trees (■ Fig. 4–11*c*).

You still need the special filter or pinhole camera to watch the final minute of the partial phases. As the partial phase ends, the bright light of the solar photosphere passing through valleys on the edge of the Moon glistens like a series of bright beads, which are called **Baily's beads.** At that time, the eclipse becomes safe to watch unfiltered, but only with the naked eye, not with binoculars or telescopes. The last bit of the uneclipsed photosphere seems so bright that it looks like the diamond on a ring—the **diamond-ring effect** (■ Fig. 4–12*a*).

During totality (■ Fig. 4–12*b*), you see the full glory of the corona, often having streamers of gas near the Sun's equator and finer plumes near its poles. You can also see prominences (■ Fig. 4–12*c*), glowing pockets of hydrogen gas that appear red. The total phase may last just a few seconds, or it may last as long as about 7 minutes. (The next eclipse to be almost that long will cross China, including Shanghai, on July 22,

FIGURE 4–12 The total solar eclipse of April 8, 2005, viewed from a ship in the middle of the Pacific Ocean, since totality was not visible from land. (a) The diamond-ring effect, with just a little bit of the bright underlying solar disk not covered by the edge of the Moon, marks the beginning and the end (in this particular case) of the total phase of a solar eclipse. Here, sunlight shining through valleys or between mountains on the edge of the Moon led to the effect looking like a double diamond ring. (b) The *totally* eclipsed Sun is safe to view with the naked eye, binoculars, or a telescope; here we see the corona. (c) Some prominences (red), a thin layer known as the chromosphere (also red), and the inner corona during totality.

2009, though the maximum duration will be available only over the Pacific ocean, and people on land will have to settle for "only" about 5 minutes of totality.) Astronomers use their spectrographs and make images through special filters. Tons of equipment are used to study the corona during this brief time of totality.

At the end of the eclipse, the diamond ring appears on the other side of the Sun, followed by Baily's beads and then the more mundane partial phases.

Somewhere in the world, a total solar eclipse occurs about every 18 months, on average (■ Fig. 4–13). The most recent total solar eclipse, on March 29, 2006, crossed Africa from Ghana through Libya and northwestern Egypt, passed a tiny Greek island and the middle of Turkey, and continued on through Russia, Georgia, and Kazakhstan. The total solar eclipse of August 1, 2008, will cross Siberia, western Mongolia, and northern China. The total solar eclipse of July 22, 2009, will cross India and China, reaching its peak length over the Pacific Ocean southeast of Japan. Not until 2017 will a total eclipse cross the United States.

Sometimes the Moon covers a slightly smaller angle in the sky than the Sun, because the Moon is in the part of its elliptical orbit that is relatively far from the Earth. When a well-aligned eclipse occurs in such a circumstance, the Moon doesn't quite cover the Sun. An annulus—a ring—of the photosphere remains visible, so we call this special type of partial eclipse an **annular eclipse.** In the continental United States, we won't get an annular eclipse until May 20, 2012 (■ Fig. 4–14).

Rarely, a solar eclipse is annular over some parts of the eclipse path on Earth, and total over other parts; two such "hybrid" cases will occur in 2013 (Fig. 4–14). In Figure 4–12 and in *Star Party 4.1: Observing Solar Eclipse,* we show photographs taken during the total part of the hybrid eclipse of April 8, 2005. Totality was visible only in the middle of the Pacific Ocean, and was observed by just 1000 people who were on the three ships that sailed to observe it. Fortunately, both of the authors were included in that group. Contrast this with the over 10 million people who will have the opportunity to observe the 2009 eclipse when it passes over Shanghai!

We will discuss the scientific value of eclipses in the chapter on the Sun, Chapter 10.

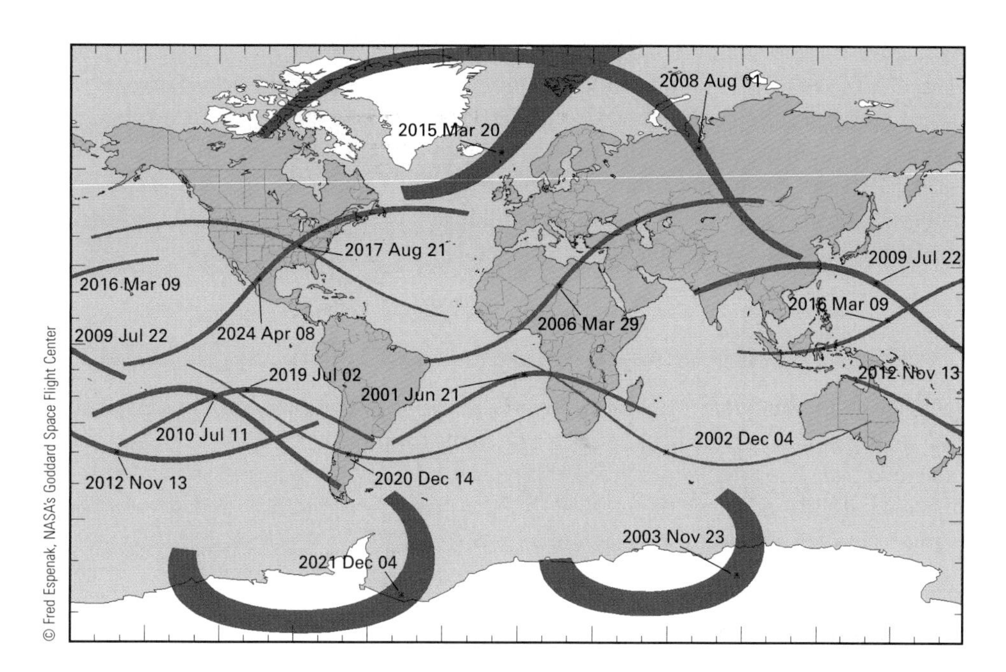

■ **FIGURE 4-13** Total eclipses of the Sun from 2001 to 2025. On April 8, 2005, totality was visible only in the middle of a band whose ends showed just an annular eclipse, so that path is marked (in blue) in the next figure (Fig. 4-14) rather than here. The two hybrid (annular/total) eclipses of 2013 are also shown only in Figure 4-14.

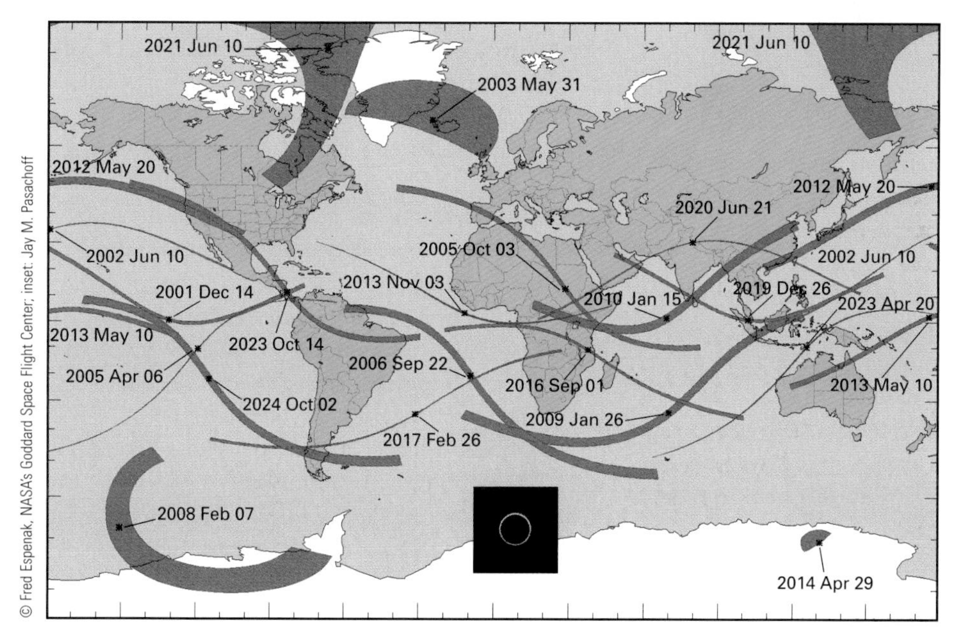

■ **FIGURE 4-14** Annular eclipses of the Sun from 2001 to 2025. The inset shows an image of the annular eclipse of May 10, 1994, photographed from New Hampshire. The parts of the very narrow paths of three annular/total "hybrid" eclipses (one in 2005, two in 2013) from which totality is seen are shown in blue.

4.3 Twinkle, Twinkle, Little Star . . .

If you look up at night in a place far from city lights, you may see hundreds of stars with the naked eye. They will seem to change in brightness from moment to moment—that is, to **twinkle.** This twinkling comes from the moving regions of air in the Earth's atmosphere. The air bends starlight, just as a glass lens bends light. As the air moves, the starlight is bent by different amounts and the strength of the radiation hitting your eye varies, making the distant, point-like stars seem to twinkle (as we saw in Figure 4–1).

ASIDE 4.3: Venus is no UFO

When a planet is low enough on the horizon, as Venus often is when we see it, it too can twinkle because there is so much turbulent air along the line of sight. Since the bending of light by air is different for different colors, we sometimes even see Venus or a bright star turning alternately reddish and greenish. On those occasions, professional astronomers sometimes get calls that UFOs have been sighted—especially when the crescent moon is nearby, drawing attention to the unusual configuration.

Unlike stars, planets are close enough to us that they appear as tiny disks when viewed with telescopes, though we can't quite see these disks with the naked eye. As the air moves around, even though the planets' images move slightly, there are enough points on the image to make the average amount of light we receive keep relatively steady. (At any given time, some points are brighter than average and some are fainter.) So planets, on the whole, don't twinkle as much as stars. Generally, a bright object in the sky that isn't twinkling is a planet.

4.4 The Concept of Apparent Magnitude

To describe the apparent brightness of stars in the sky, astronomers—professionals and amateurs alike—usually use a scale that stems from the ancient Greeks. Over two millennia ago, Hipparchus described the typical brightest stars in the sky as "of the first magnitude," the next brightest as "of the second magnitude," and so on. The faintest stars were "of the sixth magnitude."

The **magnitude scale**—in particular, **apparent magnitude,** since it is how bright the stars appear—is fixed by comparison with the historical scale (■ Fig. 4–15). The

FIGURE IT OUT 4.1
Using the Magnitude Scale

Thousands of years since the terminology was first used by Hipparchus, we still use a similar scale, the **magnitude scale,** though now it is on a mathematical basis. Its basic structure—characterizing the brightness of the naked-eye stars with easy-to-visualize numbers between about 0 and 6—is simple to use.

For example, if you are lucky enough to be outdoors at a telescope, you can tell in a moment that a 1st-magnitude star is brighter than a 3rd-magnitude star. Whether you use more complicated aspects of the magnitude scale (below) is optional, and later in this book we give brightnesses not only using magnitudes but also using the actual values for brightness or relative brightness that some people prefer.

Technically, each *difference* of 5 magnitudes is a *factor* of 100 times in brightness. A 1st-magnitude star is exactly 100 times brighter than a 6th-magnitude star—that is, we receive exactly 100 times more energy in the form of light. Sixth magnitude is still the faintest that we can see with the naked eye, though because of urban sprawl the dark skies necessary to see such faint stars are harder to find these days than they were long ago. Objects too faint to see with the naked eye have magnitudes greater than 6th.

A few stars and planets are brighter than 1st magnitude, so the scale has also been extended in the opposite direction into negative numbers. For example, Sirius (the brightest star in the sky) has a magnitude of −1.5, and Venus can reach magnitude −4.4.

Since each difference of 5 magnitudes is a factor of 100 times, each one magnitude is a number that, when five of them are multiplied together, equals 100. No integer has this property. After all, $1 \times 1 \times 1 \times 1 \times 1 = 1$, $2 \times 2 \times 2 \times 2 \times 2 = 32$, which is less than 100, and $3 \times 3 \times 3 \times 3 \times 3 = 243$, which is greater than 100, so the number we want is somewhere between 2 and 3. A number with this property is simply called "the fifth root of 100." So, each difference of 1 magnitude is a factor of the fifth root of 100 (which is approximately equal to 2.512, or even just 2.5) in brightness. This choice is necessary to make a *difference* of 5 magnitudes ($1 + 1 + 1 + 1 + 1$, an additive process) equal to a *factor* of 100 ($2.5 \times 2.5 \times 2.5 \times 2.5 \times 2.5$, a multiplicative process).

For those who would like the formula linking magnitudes m (which we add and subtract) with brightnesses b (which we multiply or divide), it is

$$b_A = 2.512^{(m_B - m_A)} \cdot b_B$$

Examples: If Star B is 1 magnitude fainter than Star A (say 3rd magnitude instead of 2nd magnitude), $m_B - m_A = 3 - 2 = 1$, $b_A/b_B = 2.512^1 = 2.512$, and $b_A = 2.512 b_B$, making Star A approximately 2.512 times brighter than Star B.

If Star B is 2 magnitudes fainter than Star A, $b_A/b_B = 2.512^2$, which is approximately 6, so Star A is about 6 times brighter than Star B.

A magnitude difference of 5 corresponds to $(2.512\ldots)^5 = 100$ times (exactly) in brightness.

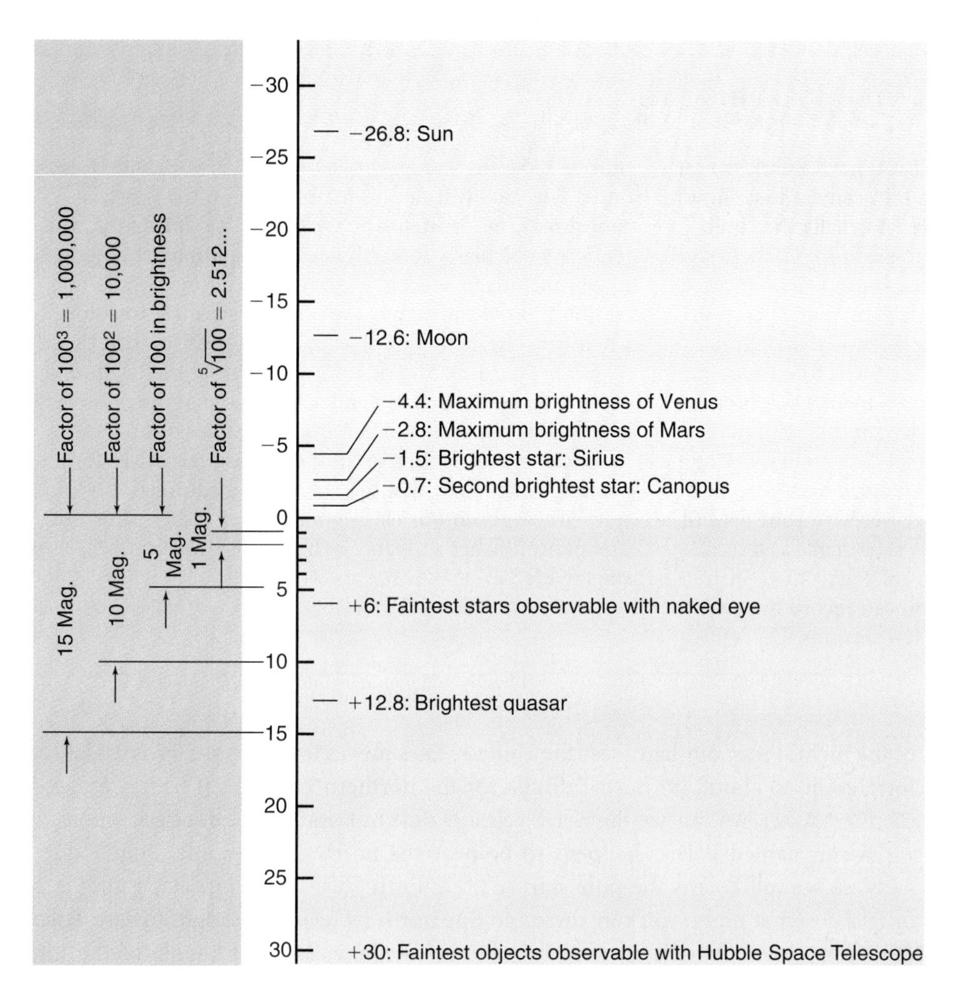

■ **FIGURE 4-15** The apparent magnitude scale is shown on the vertical axis. At the left, sample intervals of 1, 5, 10, and 15 magnitudes are marked and translated into multiplicative factors of 2.512 . . . , 100, 10,000, and 1,000,000, respectively. For example, a difference of 1 magnitude is a factor of about 2.5 times in brightness; a difference of 5 magnitudes is a factor of exactly 100 times in brightness.

higher the number, the fainter the star; see *Figure It Out 4.1: Using the Magnitude Scale.* If you hear about a star of 2nd or 4th magnitude, you should know that it is relatively bright and that it would be visible to your naked eye. If you read about a 13th-magnitude quasar, on the other hand, it is much too faint to see with the naked eye. If you read about a telescope on the ground or in space observing a 30th-magnitude galaxy, it is among the faintest objects we can currently study.

AceAstronomy™ Log into AceAstronomy and select this chapter to see the Active Figure called "Daily Motion in the Sky."

4.5 Rising and Setting Stars

Stars and planets are so distant that we have no depth perception: they all seem to be glued to an enormously large sphere surrounding the Earth, the **celestial sphere.** This sphere is imaginary; in reality, the stars and planets are at different distances from Earth.

Though stars seem to rise in the east, move across the sky, and set in the west each night, the Earth is actually turning on its axis and the celestial sphere is holding steady. If you extend the Earth's axis beyond the north pole and the south pole, these extensions point to the celestial poles. Stars appear to traverse circles or arcs around the celestial poles. (See *A Closer Look 4.2: Photographing the Stars.*)

Since the orientation of the Earth's axis doesn't change relative to the distant stars (at least on timescales of years), the celestial poles don't appear to move during the course

A Closer Look 4.2 PHOTOGRAPHING THE STARS

It is easy to photograph the stars if you have a dark sky far from city lights. You should use a film camera that allows the option of taking a long exposure (normally the "bulb" setting); most "point-and-shoot" cameras or digital cameras won't take such long exposures. Place your camera on a tripod, disable your flash unit, make the lens wide open (perhaps *f*/1.4 or *f*/2, where *f*/ is the "focal ratio" marked on many lenses and cameras), and set the focus to infinity.

Use a cable release to open the shutter for 4 minutes or more; don't take exposures of over 15 minutes without first testing shorter times to make sure that background skylight doesn't fog the film. You will have a picture of star trails. If the North Star (Polaris) is roughly centered in your field of view, you will see the circles that the stars make as a result of the Earth spinning on its axis.

With the new fast color films, you can record the stars in a constellation with exposures of a few seconds. Use a 50-mm or 135-mm lens and try a series: 1, 2, 4, 8, and 16 seconds. The stars will not noticeably trail on the shorter exposures. The constellation Orion, visible during the winter, is a particularly interesting constellation to photograph, since your image will show the reddish Orion Nebula in addition to the stars. Your eyes are not sensitive to color when viewing such faint objects, but film will record them.

Hint: Take a picture of a normal scene at the beginning of the roll, and one at the end as well, so that the photofinisher will know where to cut apart the slides or how to make the prints. Be prepared to send back negatives for printing (or request that they be printed automatically), in spite of the photofinisher's note that they didn't come out. The photofinisher probably didn't notice the tiny specks the stars made.

of the night. From our latitudes (the United States ranges from about 20° north latitude for Hawaii, to about 49° north latitude for the northern continental United States, to 65° for Alaska), we can see the north celestial pole but not the south celestial pole.

A star named Polaris happens to be near the north celestial pole, only about 1° away, so we call Polaris the **pole star** or the "North Star." If you are navigating at sea or in a desert at night, you can always go due north by heading straight toward Polaris (■ Fig. 4–16). Polaris is conveniently located at the end of the handle of the Little Dipper, and can easily be found by following the "Pointers" at the end of the bowl of the Big Dipper. Polaris isn't especially bright, but you can find it if city lights have not brightened the sky too much.

The north celestial pole is the one fixed point in our sky, since it never moves. To understand the motion of the stars, let us first consider two simple cases. If we were at the Earth's equator, then the two celestial poles would be on the horizon (due north and due south), and stars would rise in the eastern half of the sky, go straight up and across the sky, and set in the western half (■ Fig. 4–17). Only a star that rose due east of us would pass directly overhead and set due west.

If, on the other hand, we were at the Earth's north (or south) pole, then the north (or south) celestial pole would always be directly overhead. No stars would rise and set, but they would all move in circles around the sky, parallel to the horizon (■ Fig. 4–18).

We live in an intermediate case, where the stars rise at an angle relative to the horizon (■ Fig. 4–19). Close to the celestial pole, we can see that the stars are really circling

Peanuts reprinted by permission of United Feature Syndicate

(a)

■ **FIGURE 4-16** Misconceptions held by Linus and Lucy in *Peanuts.* (a) But there is no "East Star" or "West Star," since the Earth rotates on its axis. Only the positions of the north and south celestial poles are steady, and there is no bright "South Star." *(Cartoon by Charles Schulz. ©1970 United Feature Syndicate, Inc.)* (b) Lucy is still making things up. You can never see the north and south celestial poles in the sky at the same time, since they are 180° from each other in the sky (ignoring atmospheric refraction). Further, there is no "South Star"—that is, there is no obvious star near the south celestial pole. *(Cartoon by Charles Schulz. © 1970 United Feature Syndicate, Inc.)*

FIGURE IT OUT 4.2

Sidereal Time

Since a whole circle is divided into 360°, and the sky appears to turn completely around the Earth once every 24 hours, the sky appears to turn 15 degrees per hour (15°/hr). Astronomers therefore divide right ascension (r.a.) into hours, minutes, and seconds (of time) instead of degrees. At the equator, one hour of r.a. = 15°. Dividing by 60, 1 minute of r.a. = $\frac{1}{4}$° = 15 minutes of arc (since there are 60 minutes of arc in a degree). Dividing by 60 again, 1 second of r.a. = $\frac{1}{4}$ minute of arc = 15 arc seconds (since there are 60 seconds of arc in a minute of arc).

We can keep "star time" by noticing the right ascension of a star that is "crossing the meridian," where the meridian is the line that extends from the north celestial pole through the point right over our heads (the zenith) to the horizon due south of us. This star time is called "sidereal time"; astronomers keep special clocks that run on sidereal time to show them which celestial objects are crossing the meridian—that is, crossing the north-zenith-south line. A sidereal day (a day by the stars) is about 4 minutes shorter than a solar day (a day by the Sun), which is the one we usually keep track of on our watches.

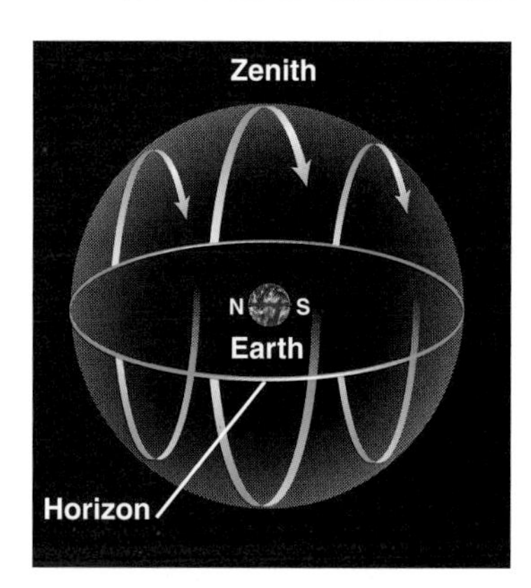

■ **FIGURE 4–17** Viewed from the equator, the stars rise straight up, pass across the sky, and set straight down.

Peter Michaud, Gemini Observatory

■ **FIGURE 4–19** Near the celestial equator, the star circles are so large that they appear almost straight. Here we are looking past Mauna Kea, and we see the rightward extension of the photograph that opened this chapter. Note how relatively straight the arcs are at the extreme right.

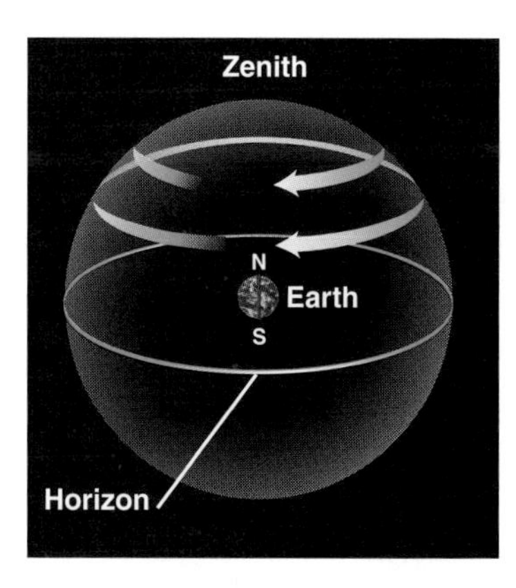

■ **FIGURE 4–18** Viewed from the pole, the stars move around the sky in circles parallel to the horizon, never rising or setting.

ⓑ

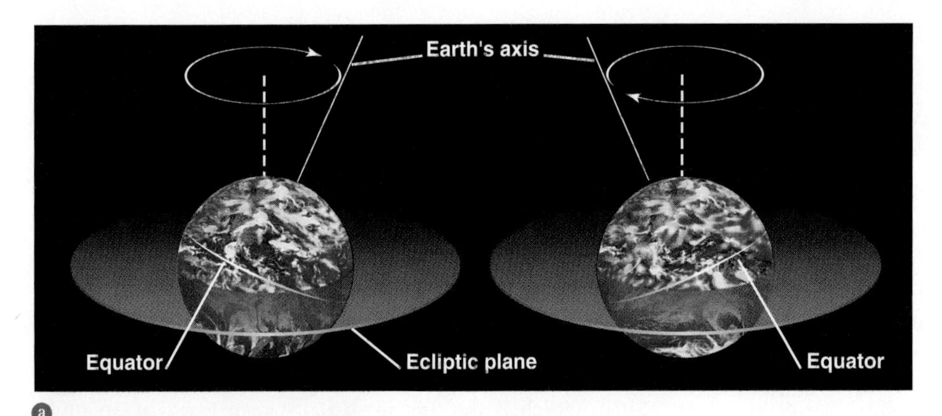

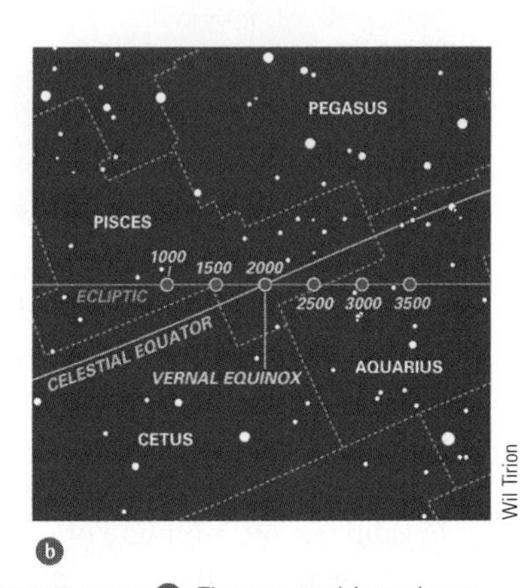

■ **FIGURE 4-21** The Earth's axis precesses with a period of 26,000 years. (a) The two positions shown are separated by 13,000 years. As the Earth's pole precesses, the equator moves with it (since the Earth is a rigid body). The celestial equator and the ecliptic—the Sun's apparent path through the stars—will always maintain the $23^1/_2$° angle between them, but the points of intersection (the equinoxes) will change. Thus, over the 26,000-year cycle, the vernal equinox will move through all signs of the zodiac. (b) The vernal equinox is now in the constellation Pisces and approaching Aquarius (and thus the celebration, in the musical *Hair,* of the "Age of Aquarius").

■ **FIGURE 4-20** When we look toward the north celestial pole, stars appear to move in giant circles about the pole. Here, we are looking past one of the telescopes of the Kitt Peak National Observatory in Arizona.

the pole (■ Fig. 4–20). Only the pole star (Polaris) itself remains relatively fixed in place, although it too traces out a small circle of radius about 1° around the north celestial pole.

As the Earth spins, it wobbles slightly, like a giant top, because of the gravitational pulls of the Sun and the Moon. As a result of this **precession,** the axis actually traces out a large circle in the sky with a period of 26,000 years (■ Fig. 4–21). So Polaris is the pole star at the present time, and generally there isn't a prominent star near either celestial pole when the orientation differs.

4.6 Celestial Coordinates to Label the Sky

Geographers divide the surface of the Earth into a grid, so that we can describe locations. The equator is the line halfway between the poles (■ Fig. 4–22). Lines of constant longitude run from pole to pole, crossing the equator perpendicularly. Lines of constant latitude circle the Earth, parallel to the equator.

Astronomers have a similar coordinate system in the sky. Imagine that we are at the center of the celestial sphere, looking out at the stars. The **zenith** is the point directly over our heads. The **celestial equator** circles the sky on the celestial sphere, halfway between the celestial poles. It lies right above the Earth's equator. Lines of constant **right ascension** run between the celestial poles, crossing the celestial equator perpendicularly. (See *Figure It Out 4.2: Sidereal Time.*) They are similar to terrestrial longitude. Lines of constant **declination** circle the celestial sphere, parallel to the celestial equator, similarly to the way that terrestrial latitude circles our globe (■ Fig. 4–23). The right ascension and declination of a star are essentially unchanging, just as each city on Earth has a fixed longitude and latitude. (Precession actually causes the celestial coordinates to change very slowly over time.)

From any point on Earth that has a clear horizon, one can see only half of the celestial sphere at any given time. This can be visualized by extending a plane that skims the Earth's surface so that it intersects the very distant celestial sphere (■ Fig. 4–24).

AceAstronomy™ Log into AceAstronomy and select this chapter to see the Active Figure called "Celestial Sphere."

Jay M. Pasachoff

■ **FIGURE 4–22** In Ecuador (which means "Equator" in Spanish), a monument marks a 19th-century French expedition to measure the length of a section of the equator. A monument on the equator itself is often visited. Though charlatans there claim to show that water swirls down a drain oppositely in the two hemispheres, this is not the case. (The "coriolis force" that causes hurricanes and tornados to rotate in opposite directions is much too small to affect water going down a drain.)

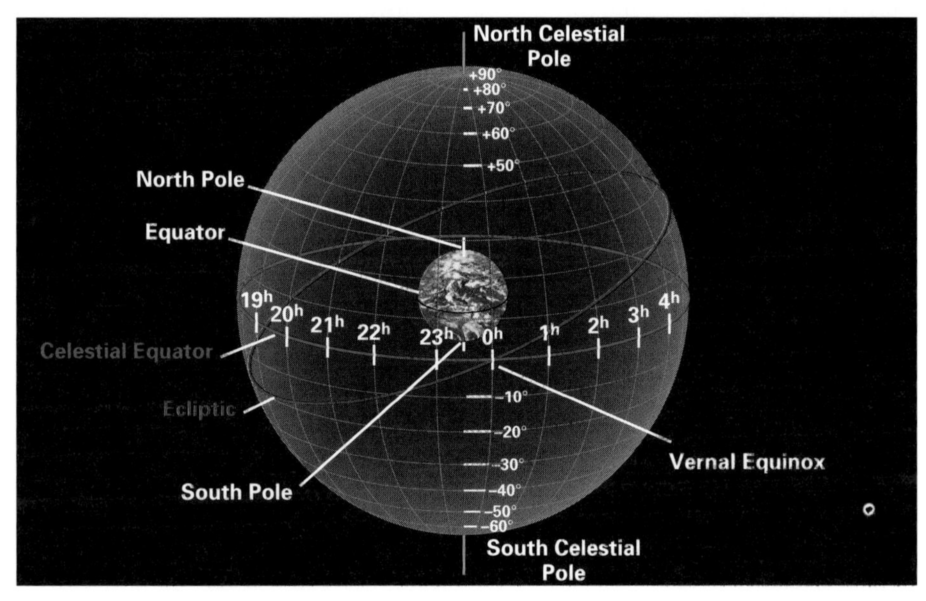

■ **FIGURE 4–23** The celestial equator is the projection of the Earth's equator onto the sky, and the ecliptic is the Sun's apparent path among the stars in the course of a year. The vernal equinox is one of the intersections of the ecliptic and the celestial equator, and is the zero-point of right ascension. From a given location at the latitude of the United States, the stars nearest the north celestial pole never set and the stars nearest the south celestial pole never rise above the horizon. Right ascension is measured along the celestial equator. Each hour of right ascension equals 15°. Declination is measured perpendicularly (+10°, +20°, etc.).

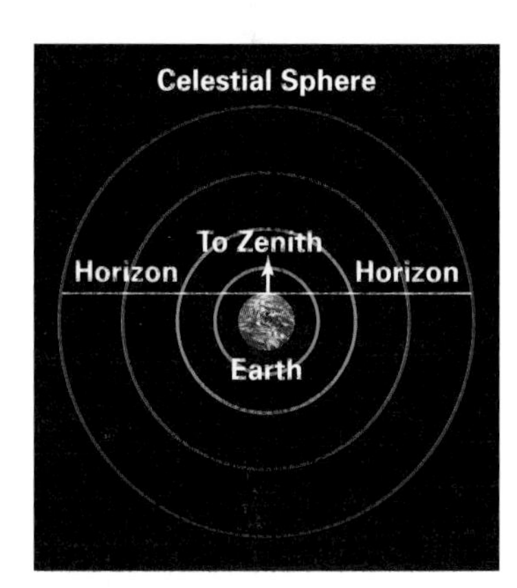

■ **FIGURE 4–24** The zenith is the point directly over your head; the nadir is the point straight down below your feet. The horizon marks as far as you can see. It is clear that if the celestial sphere were nearby (*innermost circle*), much less than half of it would be above the horizon. We can see progressively more of the celestial sphere at any given time if it is farther away, as shown (*middle and outer circles*). In fact, it is so far away that we see essentially half of it at a time.

4.7 The Reason for the Seasons

Though the stars appear to turn above the Earth at a steady rate, the Sun, the Moon, and the planets appear to slowly drift among the stars in the sky, as discussed in more detail in Chapter 5. The planets were long ago noticed to be "wanderers" among the stars, slightly deviating from the daily apparent motion of the entire sky overhead.

The path that the Sun follows among the stars in the sky is known as the **ecliptic.** We can't notice this path readily because the Sun is so bright that we don't see the stars

Star Party 4.2 The Paths of the Moon and Planets

Watch the Moon from night to night and note its position among the stars on each occasion. The Moon goes through its full range of right ascension and of declination each month as it circles the Earth.

The motion of the planets is much more difficult to categorize. For whatever planets are visible, plot their paths among the stars once a night every 3 days or so. Mercury and Venus circle the Sun more quickly than does Earth, and have smaller orbits than the Earth. They wiggle in the sky around the Sun over a period of weeks or months, and are never seen in the middle of the night. They are the "evening stars" or the "morning stars," depending on which side of the Sun they are on.

Mars, Jupiter, and Saturn (as well as the other planets too faint to see readily) circle the Sun more slowly than does Earth. So they drift across the sky with respect to the stars, and don't change their right ascension or declination rapidly. (On a given night, they seem to rise and set at the same rate as the stars; it takes observations over several nights to notice their motion.)

The U.S. and U.K. governments jointly put out a volume of tables each year, *The Astronomical Almanac.* It is an "ephemeris," a list of changing things; the word comes from the same root as "ephemeral." The book includes tables of the positions of the Sun, Moon, and planets. One of us (J.M.P.) has written the *Field Guide to the Stars and Planets,* which includes graphs and less detailed tables.

when it is up, and because the Earth's rotation causes another more rapid daily motion (the rising and setting of the Sun). The ecliptic is marked with a dotted line on the Sky Maps, on the inside covers of this book.

The Earth and the other planets revolve around the Sun in more or less a flat plane. So from Earth, the apparent paths of the other planets are all close to the ecliptic. (See *Star Party 4.2: The Paths of the Moon and Planets.*) But the Earth's axis is not perpendicular to the ecliptic. It is, rather, tipped from perpendicular by 23½°.

The ecliptic is therefore tipped with respect to the celestial equator. The two points of intersection are known as the **vernal equinox** and the **autumnal equinox.** The Sun is at those points at the beginning of our northern hemisphere spring and autumn, respectively. On the equinoxes, the Sun's declination is zero.

Three months after the vernal equinox, the Sun is on the part of the ecliptic that is farthest north of the celestial equator. The Sun's declination is then +23½°, and we say it is at the summer **solstice,** the first day of summer in the northern hemisphere. (Conversely, when the Sun's declination is −23½°, we say it is at the winter solstice, the first day of winter in the northern hemisphere.) The summer is hot because the Sun is above our horizon for a longer time and because it reaches a higher angle above the horizon when it is at high declinations (■ Fig. 4–25). A consequence of the latter effect is that a given beam of light intercepts a smaller area than it does when the light strikes at a glancing angle, so the heating per unit area is greater (■ Fig. 4–26).

The seasons (■ Fig. 4–27), thus, are caused by the variation in the declination of the Sun. This variation, in turn, is caused by the fact that the Earth's axis of spin is tipped by 23½°. Many, if not most, people misunderstand the cause of the seasons. Note that the seasons are not a consequence of the changing distance between the Earth and the Sun. If they were, then seasons would not be opposite in the northern and southern hemispheres, and there would be no seasonal changes in the number of daytime hours. In fact, the Earth is closest to the Sun each year around January 4, which falls in the northern hemisphere winter.

The word "equinox" means "equal night," implying that in theory the length of day and night is equal on those two occasions each year. The equinoxes mark the dates at which the center of the Sun crosses the celestial equator. But the Sun is not just a theoretical point, which is what is used to calculate the equinoxes. Since the top of the Sun obviously rises before its middle, the daytime is actually a little longer than the nighttime on the day of the equinox. Also, bending (refraction) of sunlight by the

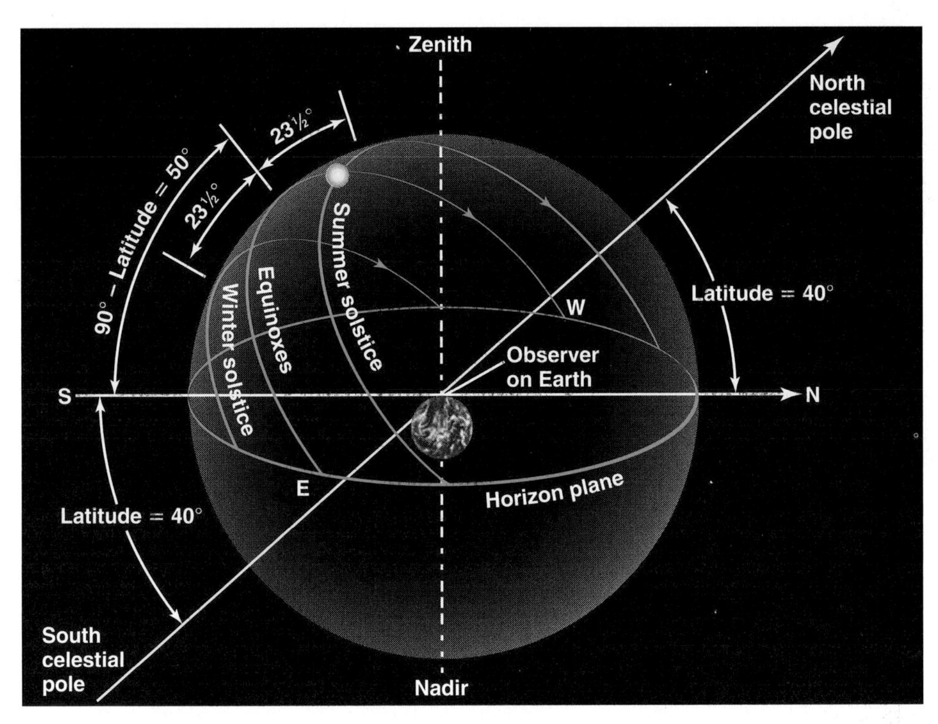

■ **FIGURE 4–25** The blue arcs from the equator up and over the top of the sphere shown are the path of the Sun at different times of the year. Around the summer solstice, June 21, the Sun (shown in yellow) is at its highest declination, rises highest in the sky, stays up longer (because, as shown, more of its path is above the horizon), and rises and sets farthest to the north. The opposite is true near the winter solstice, December 21. The diagram is drawn for latitude 40°.

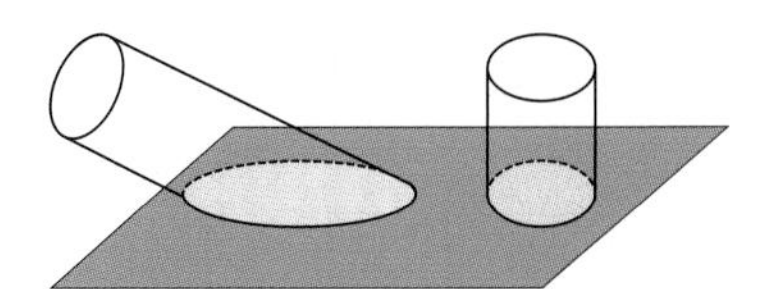

■ **FIGURE 4–26** There is more heating per unit area of the ground when the Sun is high than when it is low, because when the Sun is high a given beam of light intercepts a smaller area.

Earth's atmosphere allows us to see the Sun when it is really a little below our horizon, thereby lengthening the daytime. So the days of equal daytime and nighttime are displaced by a few days from the equinoxes.

Because of its apparent motion with respect to the stars, the Sun goes through the complete range of right ascension and between $-23\frac{1}{2}°$ and $+23\frac{1}{2}°$ in declination each year. As a result, its height above the horizon varies from day to day. If we were to take a photograph of the Sun at the same hour each day, over the year the Sun would sometimes be relatively low and sometimes relatively high.

If we were at or close to the north pole, we would be able to see the Sun whenever it was at a declination sufficiently above the celestial equator. This phenomenon is

■ **FIGURE 4–27** A schematic view from space of the Sun and the Earth (not to scale). The seasons occur because the Earth's axis is tipped with respect to the plane of the orbit in which it revolves around the Sun. The dotted line is drawn perpendicular to the plane of the Earth's orbit. When the northern hemisphere is tilted toward the Sun, it has its summer; at the same time, the southern hemisphere is having its winter. At both locations of the Earth shown, the Earth rotates through many 24-hour day-night cycles before its motion around the Sun moves it appreciably.

Emil Schulthess, Black Star

■ **FIGURE 4–28** In this series taken in June from northern Norway, above the Arctic Circle, one photograph was obtained each hour for an entire day. The Sun never set, a phenomenon known as the "midnight sun." Since the site was not quite at the north pole, the Sun and stars move somewhat higher and lower in the sky during the course of a day.

known as the **midnight sun** (■ Fig. 4–28). Indeed, from the north or south poles, the Sun is continuously visible for about six months of the year, followed by about six months of darkness.

AceAstronomy™ Log into AceAstronomy and select this chapter to see the Active Figures called "Seasons" and "Shadow and Seasons."

AceAstronomy™ Log into AceAstronomy and select this chapter to see Astronomy Exercise "The Seasons."

AceAstronomy™ Log into AceAstronomy and select this chapter to see Astronomy Exercise "Sunrise Through the Seasons."

4.8 Time and the International Date Line

Every city and town on Earth used to have its own time system, based on the Sun, until widespread railroad travel made this inconvenient. In 1884, an international conference agreed on a series of longitudinal time zones. Now all localities in the same zone have a standard time (■ Fig. 4–29).

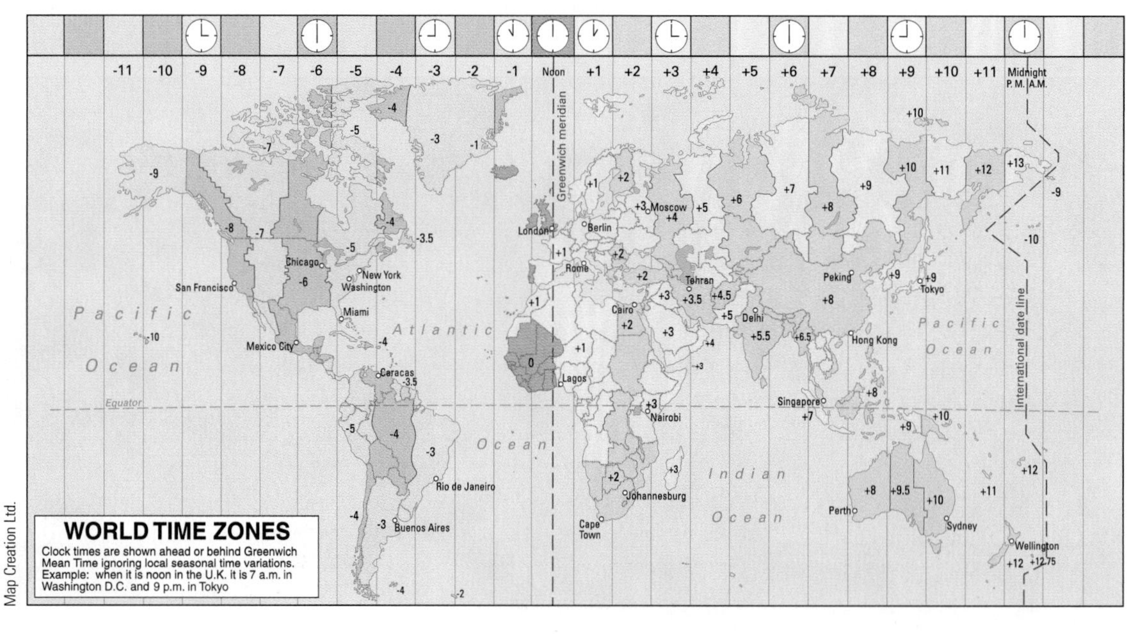

Map Creation Ltd.

■ **FIGURE 4–29** International time zones for standard time.

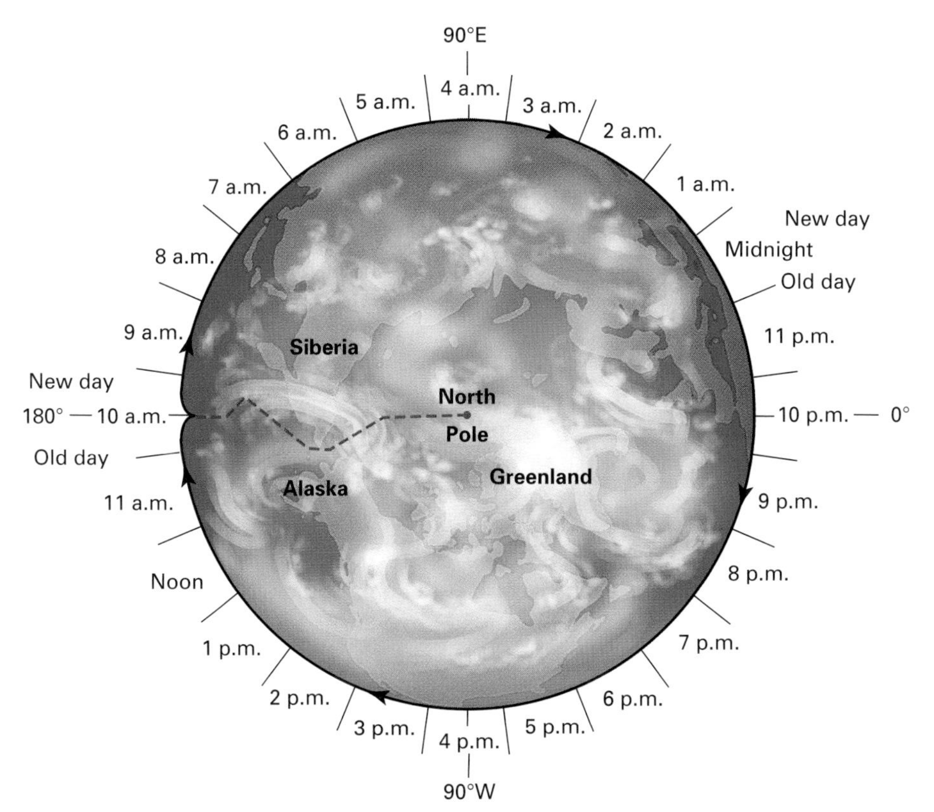

■ **FIGURE 4–30** This top view of the Earth, looking down from above the north pole, shows how days are born and die at the international date line. When you cross the international date line, your calendar changes by one day. To be among the first locations to see the new millennium, countries even changed their time zones, which shows how timekeeping can be arbitrary.

Since there are twenty-four hours in a day, the 360° of longitude around the Earth are divided into 24 standard time zones, each modified to some degree from a basic north-south swath that is 15° wide. Each of the standard time zones is centered, in principle, on a meridian of longitude exactly divisible by 15, though the actual zones are modified for geopolitical reasons. Because the time is the same throughout each zone, the Sun is not directly overhead at noon at each point in a given time zone, but in principle is less than about a half-hour off (though this varies considerably throughout the world). Standard time is based on the average length of a solar day. As the Sun seems to move in the sky from east to west, the time in any one place gets later.

We can visualize noon, and each hour, moving around the world from east to west, minute by minute. We get a particular time back 24 hours later, but if the hours circled the world continuously the date would not be able to change at midnight. So we specify a north-south line and have the date change there. We call it the **international date line** (■ Fig. 4–30). With this line present on the globe, as we go eastward

Jay M. Pasachoff

■ **FIGURE 4–31** The international zero circle of longitude at the Royal Observatory Greenwich is marked by a telescope that looked, historically, up and down along the north-south line. This site, in Greenwich, England, is a museum on the former site of the Royal Observatory there.

the hours get later and we go into the next day. At some time in our eastward trip, which you can visualize on Figure 4–30, we cross the international date line, and we go back one day. Thus we have only 24 hours on Earth at any one time.

A hundred years ago, England won the distinction of having the basic line of longitude, 0° (■ Fig. 4–31), for Greenwich, then the site of the Royal Observatory. Realizing that the international date line would disrupt the calendars of those who crossed it, that line was put as far away from the populated areas of Europe as possible—near or along the 180° longitude line. The international date line passes from north to south through the Pacific Ocean and actually bends to avoid cutting through continents or groups of islands, thus providing them with the same date as their nearest neighbor.

In the summer, in order to make the daylight last into later hours, many countries have adopted "daylight savings time." Clocks are set ahead 1 hour on a certain date in the spring. Thus if darkness falls at 7 p.m. Eastern Standard Time (E.S.T.), that time is called 8 p.m. Eastern Daylight Time (E.D.T.), and most people have an extra hour of daylight after work (but, naturally, one less hour of daylight in the morning). In most places, clocks are set back one hour in the fall, though some places have adopted daylight savings time all year. The phrase to remember to help you set your clocks is "fall back, spring forward." Of course, daylight savings time is just a bookkeeping change in how we name the hours, and doesn't result from any astronomical changes.

It is interesting to note that daylight savings time is of political interest. In 2005, the energy bill passed by Congress and signed by the president extended daylight savings time for an extra month, to save energy by providing more light in the evening. (Traditionally, this desire has been counterbalanced by the farmers' lobby, who didn't want more morning darkness, but their influence is waning. Children waiting for schoolbuses in the morning darkness, though, remain a counterargument for extending the period of daylight savings time.)

4.9 Calendars

The period of time that the Earth takes to revolve once around the Sun is called, of course, a year. This period is about 365¼ average solar days.

Roman calendars had, at different times, different numbers of days in a year, so the dates rapidly drifted out of synchronization with the seasons (which follow solar years). Julius Caesar decreed that 46 B.C. would be a 445-day year in order to catch up, and defined a calendar, the **Julian calendar,** that would be more accurate. This calendar had years that were normally 365 days in length, with an extra day inserted every fourth year in order to bring the average year to 365¼ days in length. The fourth years were, and still are, called leap years.

The name of the fifth month, formerly Quintillis, was changed to honor Julius Caesar; in English we call it July. The year then began in March; the last four months of our year still bear names from this system of numbering. (For example, October was the eighth month, and "oct" is the Latin root for "eighth.") Augustus Caesar, who carried out subsequent calendar reforms, renamed August after himself. He also transferred a day from February in order to make August last as long as July.

The Julian calendar was much more accurate than its predecessors, but the actual **solar year** is a few minutes shorter than 365¼ days. By 1582, the calendar was about 10 days out of phase with the date at which Easter had occurred at the time of a religious council 1250 years earlier, and Pope Gregory XIII issued a proclamation to correct the situation: he simply dropped 10 days from 1582. Many citizens of that time objected to the supposed loss of the time from their lives and to the commercial complications. Does one pay a full month's rent for the month in which the days were omitted, for example? "Give us back our fortnight," they cried.

In the **Gregorian calendar,** years that are evenly divisible by four are leap years, except that three out of every four century years—the ones not evenly divisible by 400—have only 365 days. Thus 1600 was a leap year; 1700, 1800, and 1900 were not; and 2000 was again a leap year. Although many countries immediately adopted the Gregorian calendar, Great Britain (and its American colonies) did not adopt it until 1752, when 11 days were skipped. (Current U.S. states that were then Spanish were already using the Gregorian calendar.) As a result, we celebrate George Washington's birthday on February 22, even though he was born on February 11 (■ Fig. 4–32). Actually, since the year had begun in March instead of January, the year 1752 was cut short; it began in March and ended the next January. Washington was born on February 11, 1731, then often written February 11, 1731/2, but since 1752, people have referred to his date of birth as February 22, 1732. The Gregorian calendar is the one in current use. It will be over 3000 years before this calendar is as much as one day out of step.

When did the new millennium start? It is tempting to say that it was on "January 1, 2000." That marks the beginning of the thousand years whose dates start with a "2." But, if you are trying to count millennia, since there was no "year zero," two thousand years after the beginning of the year 1 would be January 1, 2001. (Nobody called it "year one" then, however; the dating came several hundred years later.) If the first century began in the year 1, then the 21st century began in the year 2001! We, the authors, had two new millennium parties—a preliminary one on New Year's Eve 2000 and then the real one on New Year's Eve 2001.

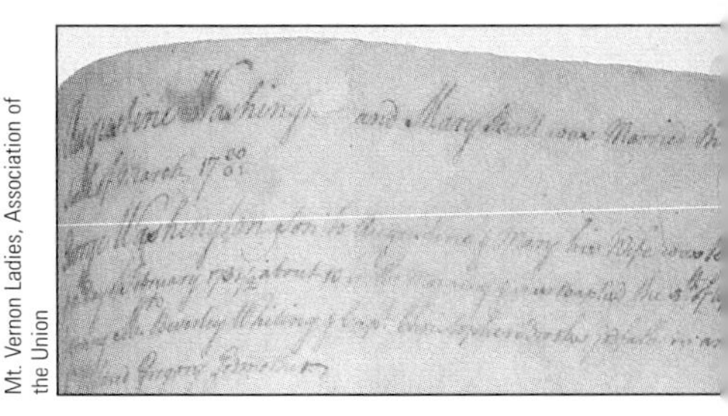

Mt. Vernon Ladies, Association of the Union

■ **FIGURE 4-32** In George Washington's family Bible, his date of birth is given as the "11th day of February 1731/2." Some contemporaries would have said 1731; we now say February 22, 1732.

ASIDE 4.4: Spelling Hint

Note that the word "millennium," in the title of this book and everywhere else, has two *l*'s and two *n*'s; many people miss that second *n*.

CONCEPT REVIEW

Planets and moons have detectable disks that show **phases,** depending on how much of the sunlighted side we see (Sec. 4.1). In particular, the phase of the Moon changes over the course of a month because of the changing angle formed by the Moon-Earth-Sun system.

During a **lunar eclipse,** the Moon is in the shadow cast by the Earth (Sec. 4.2a). Even when a lunar eclipse is total, however, the Moon can be faintly seen because Earth's atmosphere bends some sunlight toward it. The Moon then has a reddish-orange color for the same reason that sunsets and sunrises are reddish-orange: air molecules preferentially scatter blue light in random directions (causing the daytime sky to appear blue), and particles such as dust in the atmosphere absorb blue light more than red light.

At a total **solar eclipse,** the **umbra,** or darkest part, of the Moon's shadow hits the Earth (Sec. 4.2b). People in the **penumbra** see only a partial eclipse, which must be viewed through special filters or by using a projection technique like a pinhole camera. Just before an eclipse is total, the last bits of the Sun's surface, or photosphere, shine through lunar valleys and appear as **Baily's beads.** The last bit of the uneclipsed photosphere seems so bright that it looks like the diamond on a ring—the **diamond-ring effect.** During totality, you can see the Sun's hot, tenuous atmosphere, the **corona;** a total eclipse is perfectly safe to view with the naked eye, binoculars, or a telescope. When the Moon's disk appears too small to fully cover the Sun's photosphere, we have an **annular eclipse.** Because the Moon's orbit around the Earth and the Earth's orbit around the Sun are not precisely in the same plane, lunar and solar eclipses each occur only a few times a year, not every month.

The point-like images of stars **twinkle** because we see them through the Earth's turbulent atmosphere (Sec. 4.3). Planets don't twinkle as much as stars, because the disk-like images of planets consist of many independently twinkling points that tend to average out.

Astronomers use a simple scale of brightness, where the brightest visible stars are of the first magnitude, the next brightnest of the second magnitude, and so on (Sec. 4.4). Long ago this was set on a mathematical scale, the **magnitude scale,** in which each factor of 100 in brightness corresponds to a difference of 5 magnitudes. A star's **apparent magnitude** is how bright it looks to us.

Stars appear to rise and set each day because the Earth turns (Sec. 4.5). We see the stars as though they were on a giant **celestial sphere,** with **celestial poles** over the Earth's north and south poles. Polaris, now the **pole star,** marks the north celestial pole, but **precession,** the very slowly changing orientation of the Earth's axis, will change that in the future (Sec. 4.5).

Astronomical coordinate systems include one based on the Earth (Sec. 4.6). The **celestial equator** is the line on the celestial sphere above the Earth's equator. **Right ascension** corresponds to terrestrial longitude and **declination** corresponds to terrestrial latitude. The meridian is the north-south line extending through the **zenith,** the point overhead. Sidereal time, time by the stars, differs from solar time by about 4 minutes per day.

The planets "wander" among the stars as their coordinates change (Sec. 4.7). They travel close to the path of the Sun across the sky, the **ecliptic.** The ecliptic meets the celestial equator at the **vernal equinox** and at the **autumnal equinox.** The Sun is at those points at the beginning of our northern hemisphere spring and autumn, respectively. The Sun's highest and lowest declinations correspond to the summer **solstice** (the first day of summer in the northern hemisphere) and the winter solstice (the first day of winter), respectively. Thus, the seasons are caused by the tilt of the Earth's axis. When you are far enough north or south on the Earth so that the Sun never sets at the current season, you see the **midnight sun.**

Standard time is based on the average length of a solar day (Sec. 4.8). There are 24 time zones, each nominally 15° wide in longitude (though there are large variations throughout the Earth for geopolitical reasons). Days change at the **international date line.**

The **Julian calendar** included leap years, but the seasons drifted because the actual **solar year** is not exactly $365\frac{1}{4}$ days (Sec. 4.9). We now use the **Gregorian calendar,** in which leap years are sometimes omitted to adjust for the fact that the year is not composed of exactly $365\frac{1}{4}$ days.

QUESTIONS

†1. On the picture opening the chapter, estimate the angle covered by the star trails and deduce how long the exposure lasted.

2. Describe the phases of the Earth you would see over a month if you were on the Moon.

†3. If the Moon were placed at twice its current distance from the Earth, how large would it have to be so that total solar eclipses could still occur?

†4. Compare the average frequencies of total and annular eclipses over a 25-year interval, say, 2001–2025.

5. In what phases of a total eclipse must one use eye protection and in what phase do you look directly? Is it ever safe to look directly at an annular eclipse? Explain.

6. If you look toward the horizon, are the stars you see likely to be twinkling more or less than the stars overhead? Explain.

7. Is the planet Saturn, which is farther away and physically smaller than Jupiter, likely to twinkle more or less than Jupiter?

8. Explain how it is that some stars never rise in our sky, and others never set.

9. Using the Appendices and the Sky Maps, comment on whether Polaris is one of the twenty brightest stars in the sky.

†10. Compare a 6th-magnitude star and an 11th-magnitude star in brightness. Which is brighter, and by how many times?

†11. **(a)** Compare a 16th-magnitude star with an 11th-magnitude star in brightness, as in Question 10. **(b)** Now compare the 16th-magnitude with the 6th-magnitude star, specifying which is brighter and by how many times.

†12. Since the sky revolves once a day, how many degrees does it appear to revolve in 1 hour?

13. Use the suitable Sky Map to list the bright stars that are near the zenith during the summer.

14. Explain why Christmas comes in the summer in Australia.

15. Are the lengths of day and night exactly equal at the vernal equinox? Explain.

16. Does the Sun pass due south of you at exactly noon on your watch each day? Explain why or why not.

17. Explain why for an observer at the north pole the Sun changes its altitude in the sky over the year, but the stars do not.

†18. When it is 5 p.m. on November 1 in New York City, what time of day and what date is it in Tokyo?

†19. Describe how to use Figure 4–29 to find the date and time in England when it is 9 a.m. on February 21 in California. Carry out the calculation both by going eastward and by going westward, and describe why you get the same answer.

†20. List the leap years between 1800 and 2200. How many will there be?

21. **True or false?** Earth's atmosphere scatters (reflects) blue light more easily than red light, causing the daytime sky to appear blue and contributing to the reddish-orange appearance of sunsets.

22. **True or false?** If Earth's axis of rotation were perpendicular to Earth's orbital plane (i.e., if Earth's axis weren't tilted), there would be no major seasonal variations at a given location on Earth throughout the year.

23. **True or false?** Venus and Mercury are sometimes visible at a position in the celestial sphere diametrically opposite the Sun.

24. **True or false?** More "circumpolar stars" (i.e., stars that are always above the horizon) are visible from Alaska than from Hawaii.

25. **True or false?** As seen from Earth's surface, stars appear to twinkle because our eyes don't see photons outside visible wavelengths, which are emitted when the star appears to be dimming slightly.

26. **Multiple choice:** Which one of the following statements about lunar eclipses is *false*? (**a**) Lunar eclipses don't occur monthly due to the inclination of the Moon's orbit around the Earth relative to the Earth's orbit around the Sun. (**b**) Total lunar eclipses last longer than total solar eclipses. (**c**) Lunar eclipses are predictable. (**d**) At a given time, a total lunar eclipse is visible only from a small part of the Earth's surface. (**e**) The Moon is still visible during a total lunar eclipse because of light going through the Earth's atmosphere.
27. **Multiple choice:** If the Moon as seen from the Earth is in the waning crescent phase, what phase of the Earth would be seen from the near side of the Moon? (**a**) Full. (**b**) Waxing gibbous. (**c**) Waning crescent. (**d**) The Earth does not go through phases as seen from the Moon. (**e**) The Earth's phases as seen from the Moon are unrelated to the Moon's phases as seen from the Earth.
28. **Multiple choice:** Which one of the following is the main reason why Earth has seasons? (**a**) Earth's orbit is elliptical. We have summer when we are closer to the Sun and winter when we are farther from the Sun. (**b**) The tilt of Earth's axis of rotation causes one hemisphere of the planet to be closer to the Sun during the day than the other hemisphere. Because it is closer to the Sun, it receives much more solar energy per hour. (**c**) Over the course of the year, the tilt of Earth's axis of rotation varies from 23.5° to 0° in such a way as to bring more heating per hour in the summer than in the winter. (**d**) The tilt of Earth's axis of rotation causes the Sun to pass higher in the sky during the day in one hemisphere than in the other, thereby giving more daylight hours and more heating per hour. (**e**) It has seasons so that sunbathers will know when to go to the beach, and skiers will know when to go skiing.
29. **Multiple choice:** You are in New York City at 3 p.m. and you want to look at the full Moon. The best idea would be to (**a**) look toward the southern horizon; (**b**) look toward the western horizon; (**c**) look toward the eastern horizon; (**d**) look overhead, or nearly overhead; or (**e**) give up and try again later.
30. **Multiple choice:** If the Moon were half its current distance from the Earth, then (**a**) we would no longer see the Sun's photosphere fully blocked by the Moon during solar eclipses; (**b**) we would no longer see the entire corona at a given time during total solar eclipses; (**c**) solar prominences would be easier to see during solar eclipses; (**d**) partial solar eclipses would not be visible from far-northern and far-southern latitudes on Earth; or (**e**) there would never be partial lunar eclipses.

†31. **Fill in the blank:** A given star near the celestial equator rises about _____ minutes earlier each night because of Earth's revolution around the Sun.

32. **Fill in the blank:** When the Sun reaches the point in the sky known as the _______, it is the first day of spring in the northern hemisphere.

33. **Fill in the blank:** The celestial analogues of longitude and latitude are _______ and _______, respectively.

†34. **Fill in the blank:** A magnitude 3 star is ______ times brighter than a magnitude 10 star.

35. **Fill in the blank:** As viewed from the Earth's equator, the North Star (Polaris) appears to be located _________.

†This question requires a numerical solution.

TOPICS FOR DISCUSSION

1. Physically, how is it possible that a given solar eclipse can sometimes be both total and annular along different parts of the eclipse path?
2. During a total lunar eclipse, what would someone on the Moon see when looking toward the Sun?
3. When the Sun is on either of the equinoxes, observers at both the North Pole and the South Pole of the Earth can see it completely above the horizon. How is this possible?

MEDIA

AceAstronomy™ Log into AceAstronomy at **http://astronomy.brookscole.com/cosmos3** to access quizzes and animations that will help you assess your understanding of this chapter's topics.

Log into the Student Companion Web Site at **http://astronomy.brookscole.com/cosmos3** for more resources for this chapter including a list of common misconceptions, news and updates, flashcards, and more.

DIALOGO
di
GALILEO GALILEI LINCEO
AL SER.mo FERD. II. GRAN. DVCA DI
TOSCANA

CHAPTER 5

Gravitation and Motion: The Early History of Astronomy

ORIGINS

We explore the overall structure of the Solar System and how ideas about it were developed over many centuries, providing basic information for theories of its formation.

AIMS

1. Summarize theories of the Universe developed two millennia ago (Sections 5.1 and 5.2).
2. Follow how modern astronomy progressed through the ideas of Copernicus, Kepler, Galileo, and Newton (Sections 5.3 to 5.7).
3. Introduce the orbital motions of planets and how they provide clues to the formation of the Solar System (Section 5.8).

Ancient peoples knew of five planets—Mercury, Venus, Mars, Jupiter, and Saturn. When we observe these planets in the sky, we join the people of long ago in noticing that the positions of the planets vary from night to night with respect to each other and with respect to the stars. In this chapter, we will discuss the motions of the planets and you will learn how major figures in the history of astronomy explained these motions. These explanations have led to today's conceptions of the Universe and of our place in it.

5.1 A Brief Survey of the Solar System

Our ideas of the Solar System took shape before even the telescope was invented, but nowadays we have visited almost all the Solar System with spacecraft (■ Fig. 5–1). Let us first give an overview of the Solar System before discussing the historical figures who started us on our current path. Discussion on the individual Solar-System objects occurs later chapters.

Since 1957, when the first spacecraft were launched into orbit around the Earth, we have made giant strides. Satellites routinely study the Earth and even the Moon, mapping them in detail and watching how they change (■ Fig. 5–2). The series of crewed landings on the Moon during 1969–1972 revolutionized our understanding of that relatively nearby object.

The closest planet to Earth is Venus, only 70 per cent our distance from the Sun. Venus is covered with clouds, but radar has mapped its surface. Beyond Venus, Mercury orbits so close to the Sun that it has essentially no atmosphere left, given its small mass; the gravitational pull of Mercury is too weak to retain hot gases. No spacecraft has visited Mercury for decades, though NASA's MESSENGER will arrive there in 2008.

Farther outward than the Earth's orbit is Mars, a fascinating planet now the subject of extensive study. Its seasonal changes have long made it an object of interest. The signs (now photographed from orbiting spacecraft) that water once flowed on its surface make it all the more interesting, especially since liquid water is thought to be necessary for life to arise.

AceAstronomy™ The AceAstronomy icon throughout this text indicates an opportunity for you to test yourself on key concepts, and to explore animations and interactions of the AceAstronomy website at **http://astronomy.brookscole.com/cosmos3**

NASA/JPL/Caltech

■ **FIGURE 5–1** A composite of spacecraft images of the 8 innermost planets: Mercury, Venus, Earth, Mars, Jupiter, Saturn, Uranus, and Neptune.

The frontispiece of Galileo's *Dialogue on the Two Great World Systems,* published in 1632. According to the labels, Copernicus is to the right, with Aristotle and Ptolemy to the left; Copernicus was drawn with Galileo's face, however.
Jay M. Pasachoff collection

Star Party 5.1 Prograde and Retrograde Motions

If Mars or Jupiter is visible during times when you are awake, you can monitor the apparent position of either one in the sky to see prograde or even retrograde motion. Both planets, but especially Mars, move fairly quickly relative to the stars. Note that the planets move almost imperceptibly on a given night with respect to the stars; the stars and planets rise and set together at almost the same rate. (To locate the current positions of planets, consult magazines such as *Sky and Telescope* or *Astronomy,* consult the *Field Guide to the Stars and Planets,* or look them up with planetarium software.)

With your naked eye (or through binoculars), note the position of Mars or Jupiter relative to stars near them in the sky. Carefully sketch or photograph what you observe. (See the hints in Chapter 4 on photographing the sky at night.) Repeat this exercise every 6 to 9 days, or more often if you wish. The best results are obtained with at least 10 observations over the course of 2 to 3 months.

Examining your sequence of observations made over several days, can you see the planet in its prograde motion? (Prograde is from west to east among the stars.) Were you viewing at a time when it went through retrograde motion?

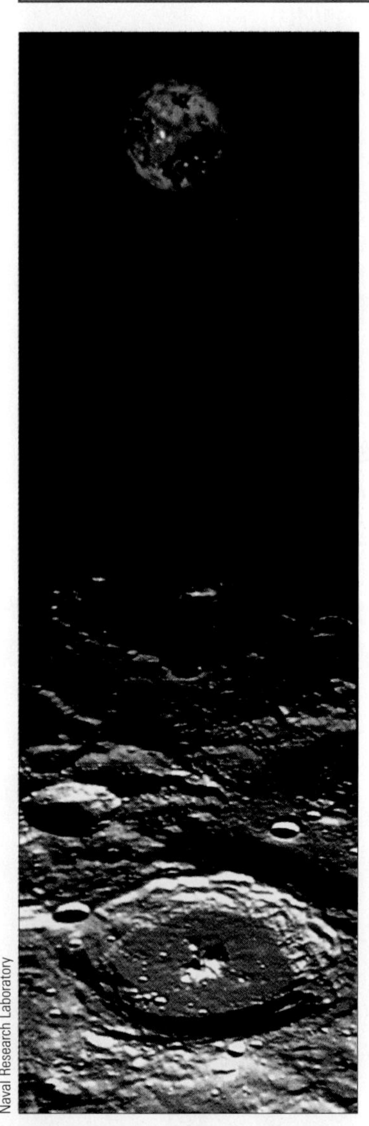
Naval Research Laboratory

■ **FIGURE 5-2** The Earth, with a lunar crater in the foreground, imaged by the Clementine spacecraft in orbit around the Moon. The angular distance between the images of the Earth and Moon was electronically reduced in this image.

The four planets beyond Mars are much larger than the Earth or Mars. Giant Jupiter is second only to the Sun in ruling the Solar System by its gravity. The Galileo spacecraft in orbit around it has sent back detailed images of Jupiter and most of its many moons. Saturn, with its extensive, beautiful rings, is always a pretty object to see in a telescope. Its rings are made of material broken up by the same kinds of factors that cause the tides on Earth, as we shall learn in the next chapter.

Uranus and Neptune were visited by a Voyager spacecraft after it passed Jupiter and Saturn, providing our only close-up views of these large planets. The Hubble Space Telescope is useful for monitoring changes in their atmospheres, as are large ground-based telescopes equipped with adaptive optics (recall the discussion in Chapter 3). These objects can also be studied in other ways, such as when they pass in front of stars.

Beyond the orbit of Neptune, tiny Pluto is strange in its properties. We will come to the debate as to whether it even deserves to be classified as a planet.

Along with the planets, we find comets and asteroids, as well as interplanetary dust. The planets and asteroids, but not most of the comets, orbit the Sun in nearly the same plane rather than having their orbits oriented every which way. Before we describe all these objects in more detail, let us tell an interesting chronological story.

5.2 The Earth-Centered Astronomy of Ancient Greece

Most of the time, planets appear to drift slowly in one direction (west to east) with respect to the background stars. This forward motion, moving slightly slower in the sky than the stars as they rise and set, is called **prograde motion.** But sometimes a planet drifts in the opposite direction (that is, east to west) with respect to the stars. We call this backward drift **retrograde motion** (■ Fig. 5–3). (See *Star Party 5.1: Prograde and Retrograde Motions.*)

When we make a mental picture of how something works, we call it a **model,** just as sometimes we make mechanical models. The ancient Greeks made theoretical models of the Solar System in order to explain the motion of the planets. For example, by comparing the length of the periods of retrograde motion of the different planets, they were able to discover the order of distance of the planets.

One of the earliest and greatest philosophers, Aristotle (■ Fig. 5–4), lived in Greece in about 350 B.C. Aristotle thought, and actually believed that he definitely knew, that the Earth was at the center of the Universe. Further, he thought that he knew that the planets, the Sun, and the stars revolved around it (■ Fig. 5–5).

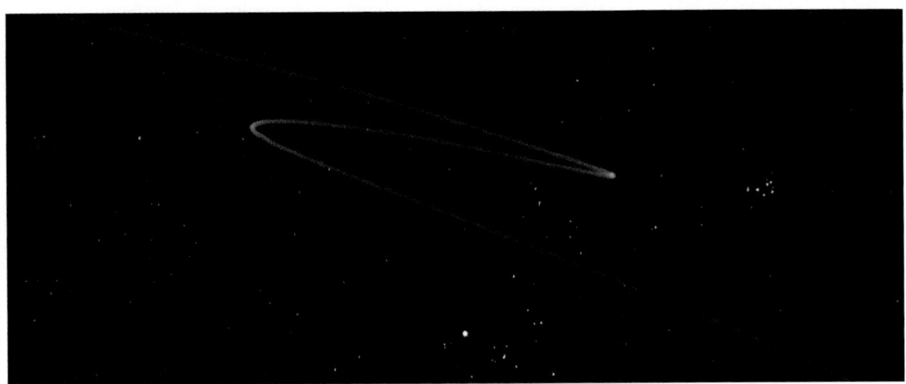

Simulation with the Zeiss ZKP3/B planetarium projector of Williams College's Hopkins Observatory by Jay M. Pasachoff, Yariv Pierce, and Megan Bruck.

■ **FIGURE 5-3** A planetarium simulation of the path of Mars over a nine-month period. In the middle we see a period of retrograde motion. Mars begins at the right (west) and proceeds toward the left (east) on this image. After it passes above the V-shaped Hyades star cluster (whose brightest star is Aldebaran), it goes into retrograde motion. Then it goes into prograde motion again. Because Mars and the Earth orbit in different planes, our perspective changes over time and the backward motion doesn't retrace the forward motion exactly. Rather, it makes a loop in the sky.

In Aristotle's model, the Universe was made up of a set of 55 celestial spheres that fit around each other. Each sphere's natural motion was rotation. The planets were carried around by some of the spheres, and the motion of the spheres affected the other spheres. Some retrograde motion could be accounted for in this way. The outermost sphere was that of the *fixed stars*. Outside this sphere was the "prime mover" that caused the rotation of the stars.

ASIDE 5.1: Wandering planets

The very slow drifting of planets ("the wanderers" in Greek) among the stars should not be confused with the much more rapid, daily rising and setting of stars and planets caused by Earth's rotation.

Photo Vatican Museums

■ **FIGURE 5-4** Aristotle *(pointing down)* and Plato *(pointing up)* in *The School of Athens* by Raphael (1483–1520), a fresco in the Vatican.

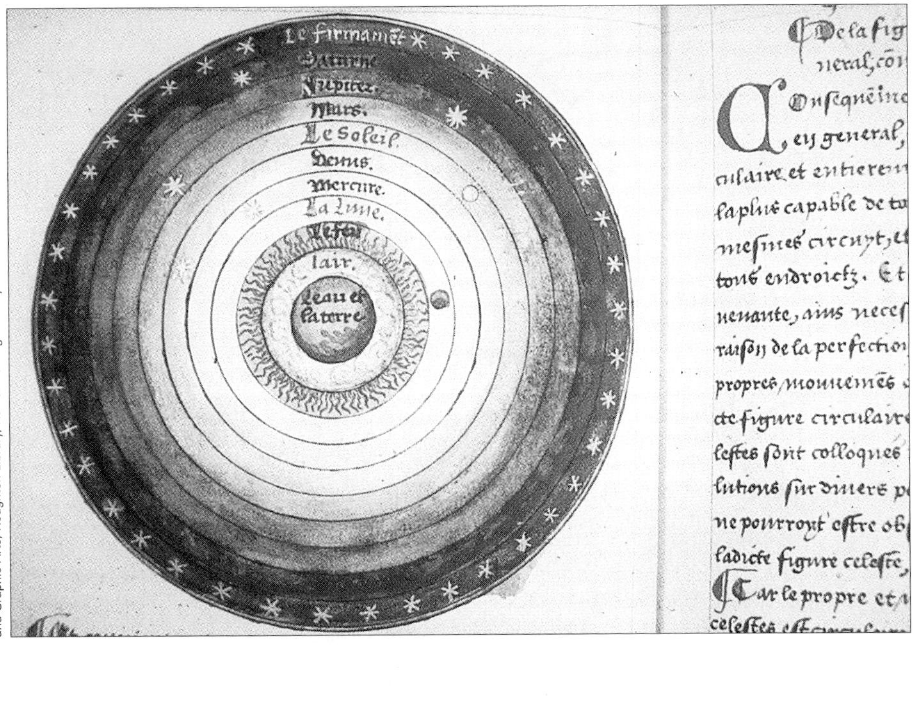

Folio 8 verso, from Oronce Fine, *Le Sphere du monde*, 1549. MS Typ 57. Department of Printing and Graphic Arts, Houghton Library, Harvard College Library

■ **FIGURE 5-5** Aristotle's cosmological system, with water and Earth at the center, surrounded by air, fire, the Moon, Mercury, Venus, the Sun, Mars, Jupiter, Saturn, and the firmament of fixed stars.

A Closer Look 5.1 PTOLEMAIC TERMS

deferent–A large circle centered approximately on the Earth, actually centered midway between the Earth and the equant (see below).

epicycle–A small circle whose center moves along a deferent. The planets move on the epicycles.

equant–The point around which the epicycle moves at a uniform angular rate.

■ **FIGURE 5-6** Ptolemy, in a 15th-century drawing.

Aristotle's theory dominated scientific thinking for about 1800 years, until the Renaissance. Unfortunately, his theories were accepted so completely that they may have impeded scientific work that might have led to new theories.

In about A.D. 140, almost 500 years after Aristotle, the Greek scientist Claudius Ptolemy (■ Fig. 5–6) flourished in Alexandria. He presented a detailed theory of the Universe that explained the retrograde motion. Ptolemy's model was **geocentric** (Earth-centered), as was Aristotle's. To account for the planets' retrograde motion, Ptolemy's model has the planets travelling along small circles that move on the larger circles of the planets' overall orbits. The small circles are called **epicycles,** and the larger circles are called **deferents** (■ Fig. 5–7). (See *A Closer Look 5.1: Ptolemaic Terms.*) The center of an epicycle moved with a constant angular speed relative to a point called the **equant** (Fig. 5–7). Since circles were thought to be "perfect" shapes, it seemed natural that planets should follow circles in their motion.

Ptolemy's views were very influential in the study of astronomy for a long time. His tables of planetary motions, which were reasonably accurate considering how long ago he developed them, were accepted for nearly 15 centuries. His major work became known as the *Almagest (Greatest).* It contained not only his ideas but also a summary of the ideas of his predecessors. Most of our knowledge of Greek astronomy comes from Ptolemy's *Almagest.*

AceAstronomy™ Log into AceAstronomy and select this chapter to see the Active Figure called "Epicycles."

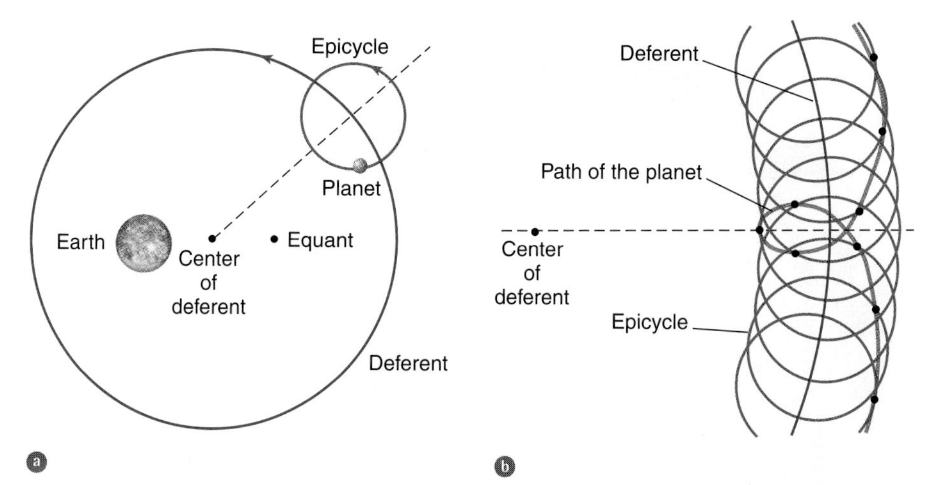

■ **FIGURE 5-7** (a) In the Ptolemaic system, a planet would move along an epicycle. The epicycle, in turn, moved on a deferent. The observed changing speed of planets in the sky would mean that the center of the epicycle did not move at a constant speed around the center of the deferent. Rather, it moved at a constant angular speed (that is, an angle changing with time at a constant rate) as seen from a point called the equant, which was not quite at the center of the deferent. (b) In the Ptolemaic system, the projected path in the sky of a planet in retrograde motion is shown.

5.3 A Heretical Idea: The Sun-Centered Universe

We credit Nicolaus Copernicus, a 16th-century Polish astronomer, with our modern view of the Solar System and of the Universe beyond. (See *Lives in Science: Copernicus.*) Copernicus suggested a **heliocentric** (Sun-centered) theory (■ Figs. 5–8, 5–9, and 5–10). He explained the retrograde motion with the Sun rather than the Earth at the center of the Solar System.

Aristarchus of Samos, a Greek scientist, had suggested a heliocentric theory 18 centuries earlier. We do not know, though, how detailed a picture he presented. His ancient heliocentric suggestion required the counterintuitive notion that the Earth itself moved. This idea seems contrary to our senses, and it contradicts the theories of Aristotle. If the Earth is rotating, for example, why aren't birds and clouds "left behind"? Only the 17th-century discovery by Isaac Newton of laws of motion that differed substantially from Aristotle's solved this dilemma, as we shall see later in this chapter.

Copernicus's heliocentric theory still assumed that the planets moved in circles, though the circles were not quite centered on the Sun. The notion that celestial bodies had to follow such "perfect" shapes (circles) shows that Copernicus had not broken entirely away from the old ideas; the observations that would later force scientists to do so were still over 100 years in the future. Copernicus's manuscript had the mechanisms that drove the planets housed in zones. The exact notion of "orbits" did not yet exist.

Copernicus still invoked the presence of some epicycles in order to have his predictions agree better with observations. Copernicus was proud, though, that by using circular orbits not quite centered on the Sun, he had eliminated the equant. In any case, the detailed predictions that Copernicus himself computed were not in much better agreement with the existing observations than tables based on Ptolemy's model. The heliocentric theory appealed to Copernicus mostly on philosophical rather than on observational grounds. Scientists did not then have the standards we now have for how to compare observations with theory.

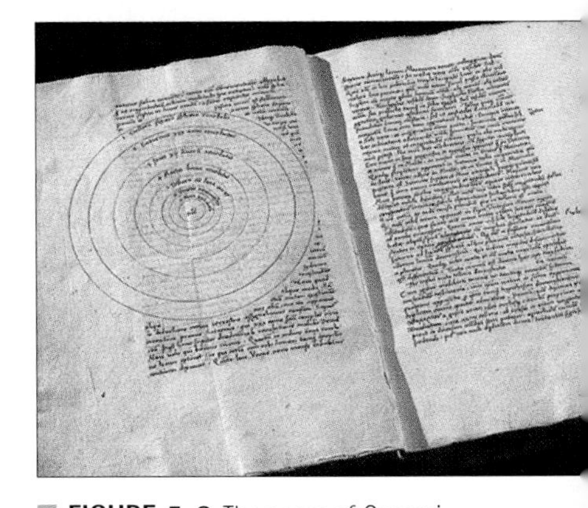

■ **FIGURE 5-8** The pages of Copernicus's original manuscript in which he drew his heliocentric system.

Photograph by Charles Eames, reproduced courtesy of Eames Office and Owen Gingerich, © 1990, 1998 Lucia Eames dba Eames Office

■ **FIGURE 5-9** Copernicus's heliocentric diagram as printed in his book *De Revolutionibus* (1543). The printed version was not an exact copy of the hand-drawn one in the manuscript.

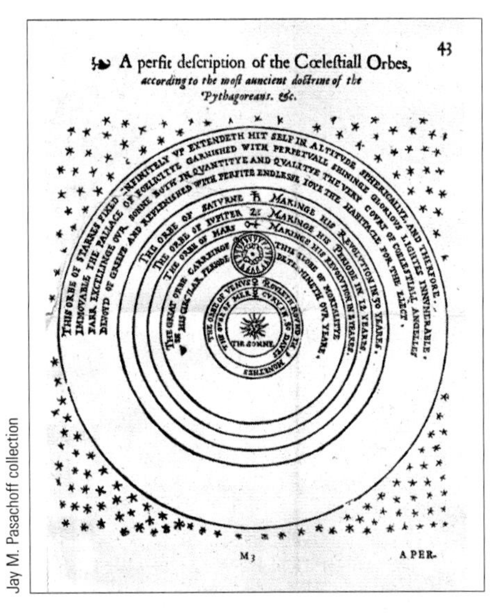

■ **FIGURE 5-10** The first English diagram of the Copernican system, which appeared in 1576.

LIVES IN SCIENCE

Copernicus

Nicolaus Copernicus was born in 1473. Around 1505–1510, he formulated his heliocentric ideas. He wrote a lengthy manuscript, and continued to develop his model even while carrying out official duties in Frauenberg (now Frombork, Poland).

He published his book *De Revolutionibus (Concerning the Revolutions)* in 1543. Though it is not certain why he did not publish his book earlier, his reasons for withholding the manuscript from publication probably included a general desire to perfect the manuscript and his realization that to publish the book would involve him in controversy.

Torun Museum, courtesy of Marek Demianski

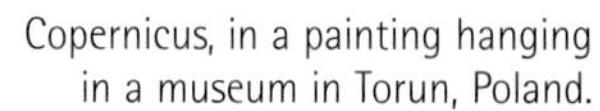
Copernicus, in a painting hanging in a museum in Torun, Poland.

When the book appeared, an unsigned preface by a theologian who had been drafted by the printer to proofread the book incorrectly implied that Copernicus disavowed any true belief in the physical reality of his hypothesis. The preface implied that Copernicus thought it was only a means for computation. Copernicus did not receive a full copy of his book until the day of his death.

Although some people, including Martin Luther, noticed a contradiction between the Copernican theory and the Bible, little controversy occurred at that time. In particular, the Pope—to whom the book had been dedicated—did not object.

The Harvard-Smithsonian astronomer and historian of science Owen Gingerich has, by valiant searching, located about 300 extant copies of the first edition of Copernicus's masterpiece and about 300 copies of the second edition. By studying which were censored and what notes were handwritten into the book by readers of the time, he has been able to study the spread of belief in Copernicanism.

Copernicus published his theory in 1543 in the book he called *De Revolutionibus (Concerning the Revolutions)*. The model explained the retrograde motion of the planets in terms of a projection effect, as follows (■ Fig. 5–11):

Let us consider, first, a planet that is farther from the Sun than is the Earth. For Mars and other such planets, notice what happens when the Earth approaches the part of its orbit that is closest to Mars. (The farther away from the Sun the planet, the more slowly it revolves around the Sun.) As the Earth passes Mars, the projection of the Earth-Mars line outward to the distant, apparently motionless stars moves backward

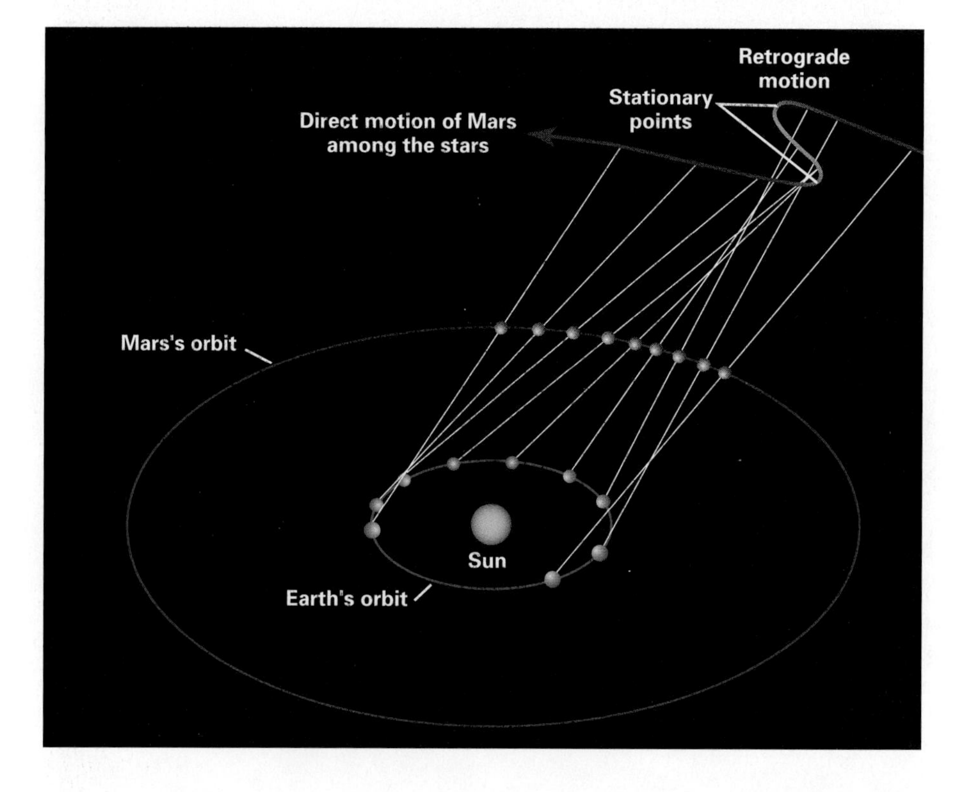

■ **FIGURE 5-11** The Copernican theory explains retrograde motion as an effect of projection. For each of the nine positions of Mars shown from right to left on the red line, follow the white line from Earth's position through Mars's position to see the projection of Mars against the sky. Mars's forward motion appears to slow down as the Earth overtakes it. Between the two "stationary points," Mars appears in retrograde motion; that is, it appears to move backward with respect to the stars. A similar argument works for planets closer to the Sun than the Earth.

compared with the way it had been moving. Then, as the Earth and Mars continue around their orbits, the projection of the Earth-Mars line appears to go forward again. A similar analysis holds for Mercury and Venus, which orbit the Sun at smaller distances than does the Earth.

The idea that the Sun was at the approximate center of the Solar System led Copernicus to two additional important results. First, he was able to work out the relative distances to the planets. Second, he was able to derive the length of time the planets take to orbit the Sun from observations of how long they take between appearances in the same place in our sky. His ability to derive these results played a large part in persuading Copernicus that his heliocentric system was better than the geocentric, Ptolemaic model.

AceAstronomy™ Log into AceAstronomy and select this chapter to see Astronomy Exercise "Aristarchus's Measurement."

5.4 The Keen Eyes of Tycho Brahe

In the last part of the 16th century, not long after Copernicus's death, Tycho Brahe (see *Lives in Science: Tycho Brahe*) began to observe Mars and the other planets to improve quantitative predictions of the positions of planets. Tycho, a Danish nobleman, set up an observatory on an island off the mainland of Denmark (■ Fig. 5–12). The first building there was called Uraniborg (after Urania, the muse of astronomy). You can still see its foundations, but the island is now Swedish.

Though the telescope had not yet been invented, Tycho used giant instruments to make observations that were unprecedented in their accuracy. In 1597, Tycho lost his financial support in Denmark. He arrived in Prague two years later. There, Johannes

Photograph courtesy of the Royal Ontario Museum

■ **FIGURE 5–12** Tycho's observatory at Uraniborg, Denmark. In this image from one of Tycho's books, a portrait of Tycho was drawn into the mural quadrant (a marked quarter-circle on a wall) that he used to measure the altitudes at which stars and planets crossed the meridian (their highest point in the sky, for all but the most northerly stars).

LIVES IN SCIENCE

Tycho Brahe

Tycho Brahe was born in 1546 to a Danish noble family. As a child he was taken away and raised by a wealthy uncle. In 1560, a total solar eclipse was visible in Portugal, and the young Tycho witnessed the partial phases in Denmark. Though the event itself was not visually spectacular—partial eclipses rarely are—Tycho, at the age of 14, was so struck by the ability of astronomers to predict the event that he devoted his life from then on to making an accurate body of observations.

When Tycho was 20, he dueled with swords with a fellow student over which of them was the better mathematician. During the duel, part of his nose was cut off. For the rest of his life he wore a gold and silver replacement and was frequently rubbing the remainder with ointment. Portraits made during his life and the relief on his tomb (see figure) show a line across his nose, though just how much was actually cut off is not now definitely known.

Jay M. Pasachoff

Tombstone of Tycho Brahe.

In 1572, Tycho was astounded to discover an apparently new star in the sky, so bright that it outshone Venus. It was what we now call a "supernova"; indeed, we call it "Tycho's supernova." It was the explosion of a star, and remained visible in the sky for 18 months. As Europe was emerging from the Dark Ages, Tycho's observations provided important evidence that the heavens were not immutable. He published a book about the supernova, and his fame spread. In 1576, the king of Denmark offered to set up Tycho on the island of Hveen with funds to build a major observatory, as well as various other grants.

Unfortunately for Tycho, a new king came into power in Denmark in 1588, and Tycho's influence waned. Tycho had always been an argumentative and egotistical fellow, and he fell out of favor in the countryside and in the court. Finally, in 1597, his financial support cut, he left Denmark. Two years later he settled in Prague, at the invitation of the Holy Roman Emperor, Rudolph II.

In 1601, Tycho attended a dinner where etiquette prevented anyone from leaving the table before the baron (or at least so Tycho thought). The guests drank freely and Tycho wound up with a urinary tract infection. Within two weeks he was dead of it.

Jay M. Pasachoff

■ **FIGURE 5–13** The plaque over Johannes Kepler's house in Prague.

Kepler came to work with him as a young assistant. (See *Lives in Science: Johannes Kepler.*) At Tycho's death in 1601, after some battles for access, Kepler was left to analyze all the observations that Tycho and his assistants had made (■ Fig. 5–13).

5.5 Johannes Kepler and His Laws of Orbits

Johannes Kepler studied with one of the first professors to believe in the Copernican view of the Universe. Kepler came to agree, and made some mathematical calculations involving geometrical shapes. His ideas were generally wrong, but the quality of his mathematical skill had impressed Tycho before Kepler arrived in Prague.

Tycho's new, more thorough and precise observational data showed that the tables of the positions of the planets then in use were not very accurate. When Kepler joined Tycho, he carried out detailed calculations to explain the planetary positions. In the years after Tycho's death, Kepler succeeded in explaining, first, the orbit of Mars. But he could do so only by dropping the idea that the planets orbited in circles, the perfect shape. (See *A Closer Look 5.2: Kepler's Laws.*)

5.5a Kepler's First Law

Kepler's first law, published in 1609, says that the planets orbit the Sun in ellipses, with the Sun at one focus. An ellipse is defined in the following way: Choose any two points on a plane; these points are called the **foci** (each is a **focus**). From any point on the ellipse, we can draw two lines, one to each focus. The sum of the lengths of these two lines is the same for each point on the ellipse.

The **major axis** of an ellipse is the line within the ellipse that passes through the two foci, or the length of that line (■ Fig. 5–14*a*). The **minor axis** is the perpendicular to the center of the major axis, or its length. The **semimajor axis** is half the length of the major axis, and the **semiminor axis** is half the length of the minor axis. A circle is the special case of an ellipse where the two foci are in the same place (the center of the circle).

It is easy to draw an ellipse (■ Fig. 5–14*b*). A given spacing of the foci and a given length of string define each ellipse. The shape of the ellipse will change if you change the length of the string or the distance between the foci.

ASIDE 5.2: Parts of an orbit

The farthest point from the Sun in a planet's orbit is called the "aphelion," in contrast to the perihelion (the closest point to the Sun).

A Closer Look 5.2 KEPLER'S LAWS

To this day, we consider Kepler's laws to be the basic description of the motions of Solar-System objects:

1. The planets orbit the Sun in ellipses, with the Sun at one focus.
2. The line joining the Sun and a planet sweeps through equal areas in equal times.
3. The square of the orbital period of a planet is proportional to the cube of the semimajor axis of its orbit—half the longest dimension of the ellipse. (Loosely: the squares of the periods are proportional to the cubes of the planets' distances from the Sun.)

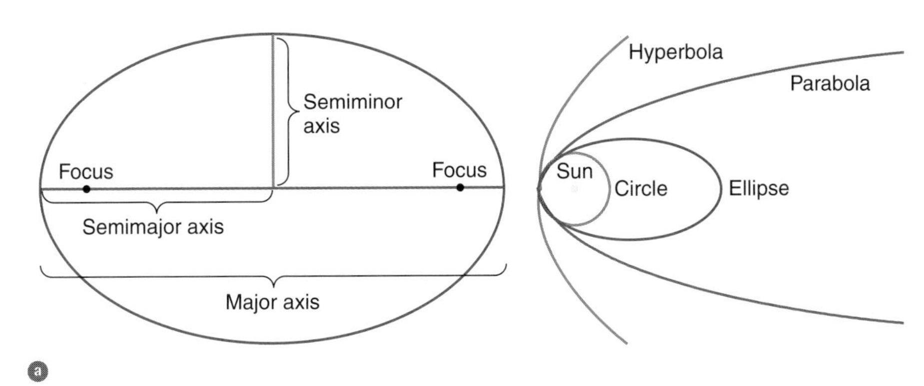

Jay M. Pasachoff

■ **FIGURE 5-14** ⓐ The parts of an ellipse, a closed curve. *(right)* The ellipse shown has the same perihelion distance (closest approach to the Sun) as does the circle. Its **eccentricity,** the distance between the foci divided by the major axis, is 0.8. If the perihelion distance remains constant but the eccentricity reaches 1, then the curve is a parabola, whose ends do not join. Curves for eccentricities greater than 1 are hyperbolas (which, like parabolas, are open curves). ⓑ It is easy to draw an ellipse. Put two nails or thumbtacks in a piece of paper, and link them with a piece of string that has some slack in it. If you pull a pen around while the pen keeps the string taut, the pen will necessarily trace out an ellipse. It does so since the string doesn't change in length. The sum of the lengths of the lines from the point to the two foci remains constant, and this is the defining property of an ellipse.

According to Kepler's first law, the Sun is at one focus of the elliptical orbit of each planet. What is at the other focus? Nothing special; we say that it is "empty."

5.5b Kepler's Second Law

Kepler's second law describes the speed with which the planets travel in their orbits. It states that *the line joining the Sun and a planet sweeps through equal areas in equal times.* It is thus also known as the **law of equal areas.**

When a planet is at its greatest distance from the Sun as it follows an elliptical orbit, the line joining it with the Sun sweeps out a long, skinny sector. This sector has two straight lines starting at a focus and extending to the ellipse. The curved part of the ellipse is relatively short for a planet at its great distance from the Sun. On the

LIVES IN SCIENCE

Johannes Kepler

Shortly after Tycho Brahe arrived in Prague, he was joined by a mathematically inclined assistant, Johannes Kepler. Kepler was born in 1571 in Weil der Stadt, near what is now Stuttgart, Germany. He was of a poor family, and had gone to school on scholarships.

In 1604, Kepler observed a supernova and soon wrote a book about it. It was the second exploding star to appear in 32 years. It was also seen by Galileo, but Kepler's study has led us to call it "Kepler's supernova." We have not seen another bright supernova in our Milky Way Galaxy since then.

Prague statue of Tycho Brahe and Johannes Kepler.

Jay M. Pasachoff

By the time of the supernova, through analysis of Tycho's data, Kepler had discovered what we now call his second law. He was soon to find his first law. It took four additional years, though, to arrange and finance publication. Much later on, he developed his third law.

The abdication of Kepler's patron, the Emperor, led Kepler to leave Prague and move to Linz. Much of his time had to be devoted to financing the printing of the planetary tables and to his other expenses—and to defending his mother against the charge that she was a witch. He died in 1630, while travelling to collect an old debt.

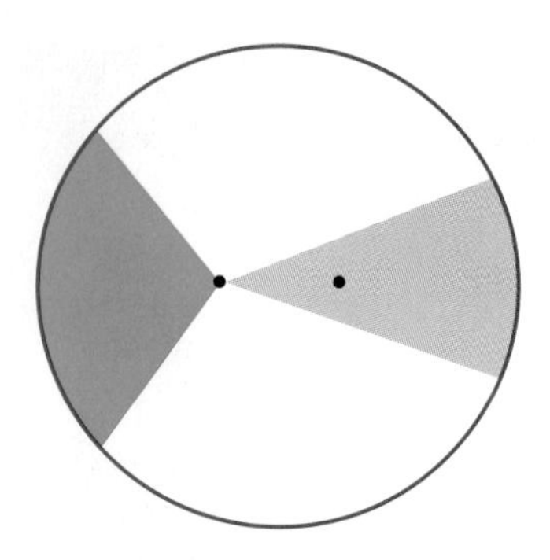

■ **FIGURE 5-15** Kepler's second law states that the two shaded sectors are equal in area. They represent the areas covered by a line drawn from the Sun, which is at a focus of the ellipse, to an orbiting planet in a given length of time.

other hand, according to Kepler's second law, the area of this long, skinny sector must be the same as that of any short, fat sector formed for the planet nearer to **perihelion,** the closest point to the Sun in a planet's orbit (■ Fig. 5–15). Thus, the planet moves most quickly when it is near the Sun.

Since the time of the Greeks, the idea that the planets travelled at a constant rate had been thought to be important. Kepler's second law replaced this old notion with the idea that the total area swept out changes at a constant rate.

Kepler's second law is especially noticeable for comets, which have very eccentric (that is, flattened) elliptical orbits. For example, Kepler's second law demonstrates why Halley's Comet sweeps so quickly through the inner part of the Solar System. It moves much more slowly when it is farther from the Sun, since the sector swept out by the line joining it to the Sun is skinny but long.

5.5c Kepler's Third Law

Kepler's third law deals with the length of time a planet takes to orbit the Sun, which is its **period** of revolution. Kepler's third law relates the period to some measure of the planet's distance from the Sun. Specifically, it states that *the square of the period of revolution is proportional to the cube of the semimajor axis of the ellipse.* (Mathematically, $P^2 = kR^3$, where k is a constant; or, $P^2 \propto R^3$, where "$\propto$" means "proportional to.") That is, if the cube of the semimajor axis of the ellipse goes up, the square of the period goes up by the same factor (see *Figure It Out 5.1: Kepler's Third Law).*

A terrestrial application of Kepler's third law is in "geostationary satellites," which are so high that they orbit Earth once a day as Earth rotates at the same rate underneath (■ Fig. 5–16). They thus appear to hover over the equator (■ Fig. 5–17), and are used for relaying television and telephone signals.

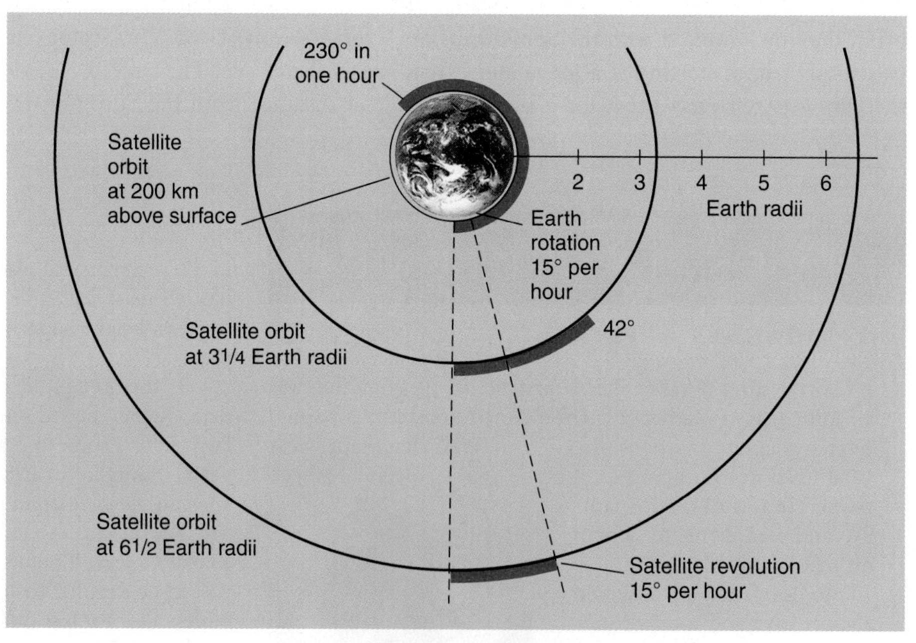

■ **FIGURE 5-16** Most satellites, including the space shuttles, orbit only 200 km or so above the Earth's surface. They orbit Earth in about 90 minutes. From Kepler's third law, we can see that the speed in orbit of a satellite decreases as the satellite gets higher; see *Figure It Out 5.3: Orbital Speed of Planets.* But distances are measured from the center of the Earth, so even a doubling of the orbital height above the Earth's surface is only a slight increase in the orbit's radius. When a satellite is as high as 6½ Earth radii from Earth's center, the satellite orbits in 24 hours. This 24-hour period is the same as the period with which a point on the Earth's surface below rotates, so the satellite appears from the ground to hover in one place. It is then called a "geostationary satellite."

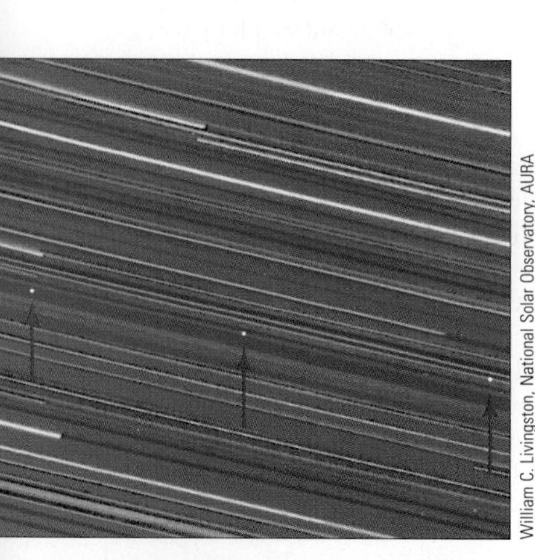

William C. Livingston, National Solar Observatory, AURA

■ **FIGURE 5-17** In this time exposure, the stars trail but the series of geostationary satellites hovering over the Earth's equator, many delivering TV or telephone signals, appear as points.

FIGURE IT OUT 5.1

Kepler's Third Law

It is often easiest to express Kepler's third law, $P^2 = kR^3$ (where P is the period of revolution, R is the semimajor axis, and k is a constant), by relating values for a planet to values for the Earth:

$$\frac{P^2_{\text{Planet}}}{P^2_{\text{Earth}}} = \frac{kR^3_{\text{Planet}}}{kR^3_{\text{Earth}}} = \frac{R^3_{\text{Planet}}}{R^3_{\text{Earth}}}.$$

We can choose to work in units that are convenient for us here on Earth. We use the term **1 Astronomical Unit (1 A.U.)** to mean the semimajor axis of the Earth's orbit. This semimajor axis can be shown to be the average distance from the Sun to the Earth. We use a unit of time based on the length of time the Earth takes to orbit the Sun: **1 year.** When we use these values, Kepler's law appears in a simple form, since the numbers on the bottom of the equation are just 1, and we have

$$P^2_{\text{Planet}} = R^3_{\text{Planet}} \text{ (distance in A.U.; period in Earth years).}$$

Example: We know from observation that Jupiter takes 11.86 years to revolve around the Sun. What is Jupiter's average distance from the Sun?

Answer: $(11.86)^2 = R^3_{\text{Jupiter}}$, so $R_{\text{Jupiter}} = 5.2$ A.U.

Kepler had to calculate by hand, but we can calculate R easily with a pocket calculator. Astronomers are often content with approximate values that can be calculated in their heads. For example, since 11.86 is about 12, $(11.86)^2$ is about $12^2 = 144$. Do you need a calculator to find the rough cube root of 144? No. Just try a few numbers. $1^3 = 1$, $2^3 = 8$, $3^3 = 9 \times 3 = 27$, and $4^3 = 16 \times 4 = 64$, which are all too small. But $5^3 = 125$ is closer. $6^3 = 216$ is too large, so the answer must be a little over 5 A.U.

The period with which objects revolve around bodies other than the Sun (for example, other planetary systems) follows Kepler's third law as well. The laws apply also to artificial satellites in orbit around the Earth (Figs. 5–16 and 5–17) and to the moons of other planets.

5.6 The Demise of the Ptolemaic Model: Galileo Galilei

Galileo Galilei flourished in what is now Italy at the same time that Kepler was working farther north. Galileo began to believe in the Copernican heliocentric system in the 1590s, and identified some of our basic laws of physics in the following years. In late 1609 or early 1610, Galileo was the first to use a telescope for systematic astronomical studies (■ Fig. 5–18).

Jay M. Pasachoff

■ **FIGURE 5-18** Galileo's telescopes.

In 1610, he reported in a book that with his telescope he could see many more stars than he could see with his unaided eye. Indeed, the Milky Way and certain other hazy-appearing regions of the sky actually contained numerous individual stars. He described views of the Moon, including the discovery of mountains, craters, and the relatively dark lunar "seas"; see *Star Party 5.2: Galileo's Observations.*

Galileo's discovery that small bodies revolved around Jupiter (■ Fig. 5–19) was very important because it proved that not all bodies revolved around the Earth. Moreover,

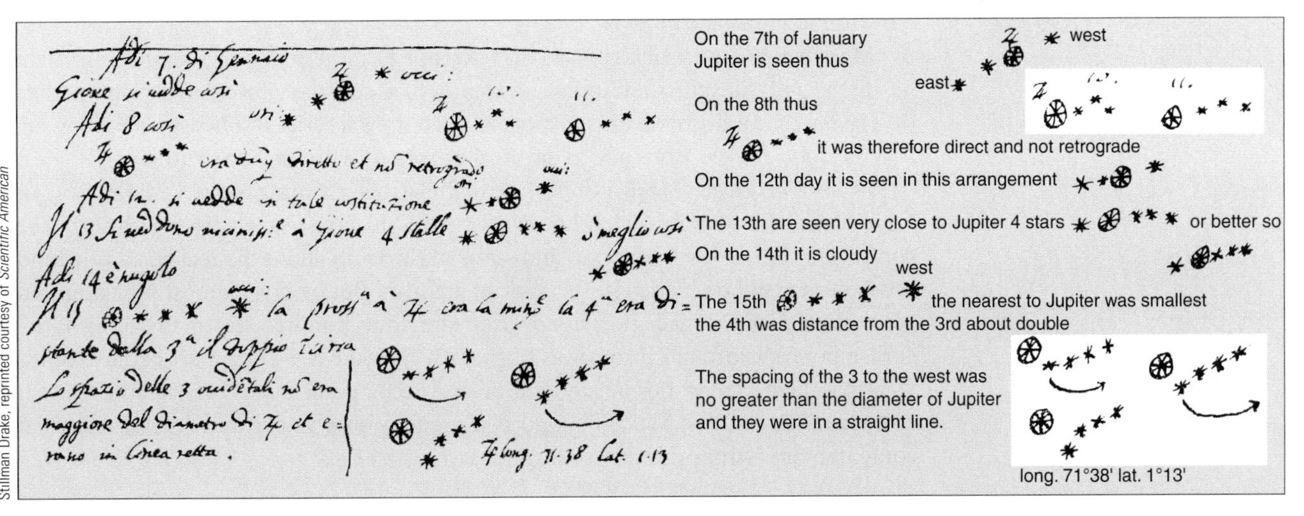

U. Michigan Library, Dept. of Rare Books and Special Collections; translation by Stillman Drake, reprinted courtesy of *Scientific American*

■ **FIGURE 5-19** A translation *(right)* of Galileo's original notes *(left)* summarizing his first observations of Jupiter's moons in 1610. The highlighted areas were probably added later. It had not yet occurred to Galileo that the objects were moons in revolution around Jupiter.

LIVES IN SCIENCE

Galileo Galilei

Galileo Galilei was born in Pisa, in northern Italy, in 1564. In the 1590s, he became a professor at Padua, in the Venetian Republic. At some point in this decade, he adopted Copernicus's heliocentric theory. After hearing that a device to magnify far-off objects had been made, he figured out the principles himself and made his own version. In late 1609 or early 1610, Galileo was among the first to turn a telescope toward the sky.

In 1613, Galileo published his acceptance of the Copernican theory. The book in which he did so was received and acknowledged with thanks by the Cardinal who, significantly, was Pope at the time of Galileo's trial. Just how Galileo got out of favor with the Church has been the subject of much study. His bitter quarrel with a Jesuit astronomer, Christopher Scheiner, over who deserved the priority for the discovery of sunspots did not put him in good stead with the Jesuits. Still, the disagreement on Copernicanism was more fundamental than personal, and the Church's evolving views against the Copernican system were undoubtedly more important than personal feelings. Also, the Church apparently objected to his belief that matter consists of atoms.

Galileo.

In 1615, Galileo was denounced to the Inquisition, and was warned against teaching the Copernican theory. There has been much discussion of how serious the warning was and how seriously Galileo took it.

Over the following years, Galileo was certainly not outspoken in his Copernican beliefs, though he also did not cease to hold them. In 1629, he completed his major manuscript, *Dialogue on the Two Great World Systems,* these systems being the Ptolemaic and Copernican. With some difficulty, clearance from the censors was obtained, and the book was published in 1632. It was written in contemporary Italian, instead of the traditional Latin, and could thus reach a wide audience. The book was condemned and Galileo was again called before the Inquisition.

Galileo's trial provided high drama, and has indeed been transformed into drama on the stage. He was convicted, and sentenced to house arrest. He lived nine more years, until 1642, his eyesight and his health failing. He devoted his time to the experimental study of the motions of falling bodies. Indeed, most of the principles in Newton's laws of motion are based on experimental facts determined by Galileo and documented in his important book on mechanics, written during this period. In 1992, Pope John Paul II acknowledged that Galileo had been correct to favor the Copernican theory, an implicit though not an explicit pardon.

the moons were obviously not "left behind" while Jupiter moved across the sky, suggesting that the Earth might move as well without leaving objects behind.

By displaying facts that Aristotle and Ptolemy obviously had not known, he showed that Aristotle and Ptolemy had not been omniscient, and opened the idea that more remained to be discovered. But not all of Galileo's contemporaries accepted that what was seen through the newfangled telescope was real, or were even willing to look through this contraption at all.

Critical to the support of the heliocentric model was Galileo's discovery that Venus went through an entire series of phases (■ Fig. 5–20). The non-crescent phases could not be explained with the Ptolemaic system (■ Fig. 5–21). After all, if Venus travelled on an epicycle located between the Earth and the Sun, Venus should always appear as a crescent (or "new"). Thus, the complete set of phases of Venus provided the fatal blow to the Ptolemaic model.

Moreover, in 1612, Galileo used his telescope to project an image of the Sun. Here he discovered sunspots, additional evidence that celestial objects were not perfect (■ Fig. 5–22). In his book on sunspots, Galileo drew a series of photographs showing how sunspots looked from day to day as they rotated with the Sun's surface.

In *Lives in Science: Galileo Galilei,* we discuss Galileo's controversy with the Roman Catholic Church, during which he was sentenced to house arrest (■ Fig. 5–23). The controversy echos down to today, and some of the truth about the real reasons behind the fight may still be hidden in the Vatican archives. But by now, almost four hundred years since Galileo made his discoveries and four hundred years since his near-contemporary Giordano Bruno was burned at the stake at least in part because of his view that there were worlds beyond our Solar System, peace reigns between the Church and scientists, and the Vatican supports a modern observatory and several respected contemporary astronomers.

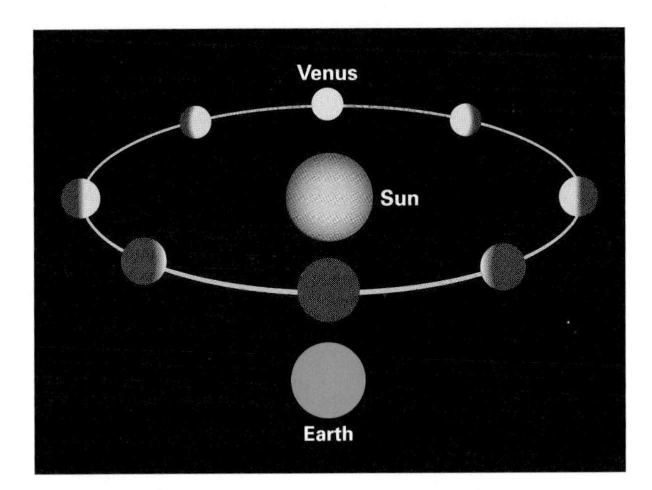

■ **FIGURE 5-20** When Venus is on the far side of the Sun as seen from the Earth, it is more than half illuminated. We thus see it go through a whole cycle of phases, from thin crescent (or even "new") through full. Venus thus matches the predictions of the Copernican system.

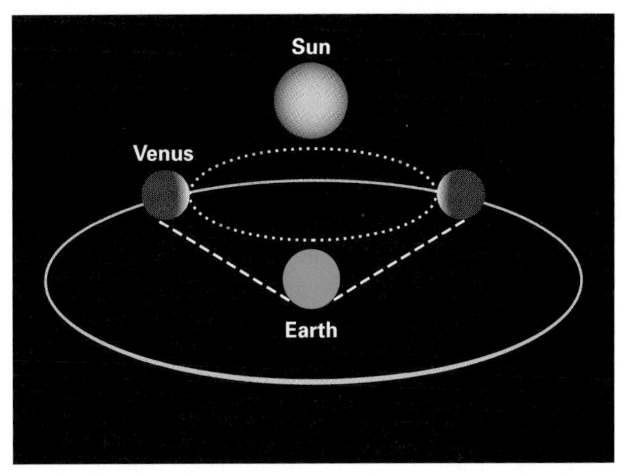

■ **FIGURE 5-21** In the Ptolemaic theory, Venus and the Sun both orbit the Earth. However, because Venus is never seen far from the Sun in the sky, Venus's epicycle (which is shown dotted) must be restricted to always fall on the line joining the center of the deferent (which is near the Earth) and the Sun. In this diagram, Venus could never get farther from the Sun than the dashed lines. It would therefore always appear as a crescent.

5.7 On the Shoulders of Giants: Isaac Newton

Kepler discovered his laws of planetary orbits by trial and error; they were purely empirical, and he had no physical understanding of them. Only with the work of Isaac Newton 60 years later did we find out why the laws existed.

Newton was born in England in 1642, the year of Galileo's death. He became the greatest scientist of his time and perhaps of all time. His work on optics, his invention of the reflecting telescope, and his discovery that visible light can be broken down into a spectrum of colors (see Chapter 2) would merit him a place in astronomy texts. But still more important was his work on motion and on gravity.

Newton set modern physics on its feet by deriving laws showing how objects move on Earth and in space, and by finding the law that describes the force of gravity. In order to work out the law of gravity, Newton had to invent calculus, a new branch of mathematics! Newton long withheld publishing his results, until Edmond Halley (whose name we now associate with the famous comet) convinced him to publish his work. Newton's *Principia* (pronounced "Prin-kip′ee-a"), short for the Latin form of *Mathematical Principles of Natural Philosophy,* appeared in 1687.

The Principia contains Newton's three laws of motion. The first is that bodies in motion tend to remain in motion, in a straight line with constant speed, unless acted upon by an external (that is, an outside) force. It is a law of inertia, which was really discovered by Galileo.

Newton's second law relates a force with its effect on accelerating (speeding up) a mass. A larger force will make the same mass accelerate faster ($F = ma$, where F is the force, m is the mass, and a is the acceleration).

Newton's third law is often stated, "For every action, there is an equal and opposite reaction." The flying of jet planes is only one of the many processes explained by this law.

The Principia also contains the law of gravity; see *A Closer Look 5.3: Newton's Law of Universal Gravitation.* One application of Newton's law of gravity is "weighing" the planets. (See *Figure It Out 5.2: Newton's Version of Kepler's Third Law.*)

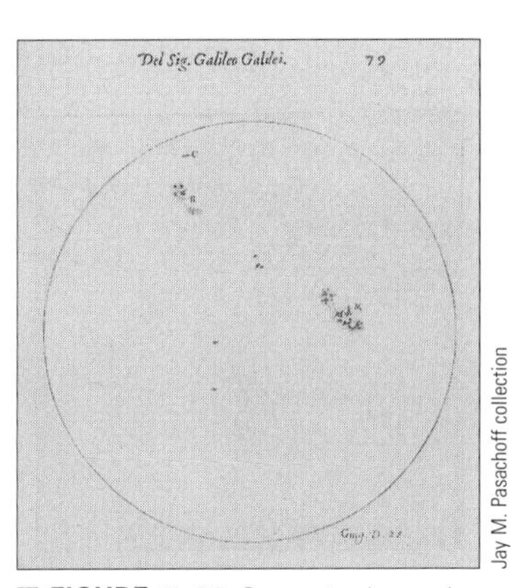

Jay M. Pasachoff collection

■ **FIGURE 5-22** Sunspots observed by Galileo in 1612, as engraved in his book of the following year.

Jay M. Pasachoff

■ **FIGURE 5-23** One of the houses in Florence, Italy, where Galileo lived, with his portrait over the door.

Star Party 5.2 Galileo's Observations

With a small telescope, it is easy to recreate many of Galileo's observations. Try these with the telescope on your campus.

1. Observe Venus over a period of several months. This is most easily done when it is visible in the evening, but at some times of the year the morning is more suitable. You should be able to see it go through different phases, like the Moon does.
2. Observe Jupiter over the course of several nights. You should be able to see its four bright moons, the "Galilean satellites." Their relative positions change daily, or even over the course of a few hours. If you draw a picture each night for a few weeks, you'll even be able to determine their orbital periods and at least roughly verify that they obey Kepler's third law.
3. Observe the Moon. You should see craters and maria ("seas"), as did Galileo. The views will be best during the crescent, quarter, or somewhat gibbous phases, especially near the "terminator" (the border between the sunlit and dark parts of the Moon). During a full moon, the features tend to appear washed out because of the lack of shadows.
4. Point the telescope to a bright region of the Milky Way and notice that many stars are visible. Use the eyepiece that gives the widest field of view for this observation. Do you think Galileo was surprised to find that the Milky Way consists of many faint stars?
5. A small telescope can be used to view sunspots, as did Galileo. But do *not* look directly at the Sun through the telescope. A better technique is to project the Sun's image onto a sheet of cardboard or some other clean surface. (Galileo also used this projection technique.) You can look directly through the telescope *only if* you are certain that you have the right kind of filter at the top end of the telescope, and that it is securely fastened. Never use a filter in the eyepiece of the telescope, since it will get very hot and can break, leading to possible blindness.

One of the most popular tales of science is that an apple fell on Newton's head, leading to his discovery of the concept of gravity. Though no apple actually hit Newton's head, the story that Newton himself told, years later, is that he saw an apple fall and realized that just as the apple fell to Earth, the Moon is falling to Earth, though its forward motion keeps it far from us. (In any short time interval, the distance the Moon travels toward Earth gets compensated by the distance the Moon also travels forward; the result over many such time intervals is a stable orbit, rather than a collision course with Earth.)

A Closer Look 5.3 Newton's Law of Universal Gravitation

When Isaac Newton famously saw an apple fall, he masterfully deduced that the same force that makes the apple fall toward Earth similarly attracts the Moon toward the Earth. He published the result in his *Philosophiae Naturalis Principia Mathematica* in 1687.

Newton realized that gravitation follows an inverse-square law—that is, the strength of the force is directly related to the inverse ("one over") of the square of the distance. The force is also directly proportional to each of the two masses involved, with no preference for one over the other. Thus the formula fits Newton's third law of motion, that for every action there is an equal and opposite reaction. Gravitational tugs are mutual—the Earth is pulling on your mass (giving you weight) just as strongly as your mass is pulling on the Earth.

Newton's law of gravitation shows that

$$F \propto \frac{m_1 m_2}{d_2}.$$

For Kepler's third law, the constant is different for each different central object; see *Figure It Out 5.2: Newton's Version of Kepler's Third Law.* But in Newton's law of gravitation, the constant of proportionality (G) is the same in all circumstances throughout the Universe: It is universal. Newton's law of gravitation, then, is

$$F \propto G\frac{m_1 m_2}{d_2}.$$

From this law, we see that if the same body were twice as far away, the force of gravitation on it from another body would be only one-fourth (one-half squared) as great. Furthermore, if the central body doubles in mass, its force of gravitation at a given distance is twice as great (not four times as great).

See *Lives in Science: Isaac Newton* for a brief summary of Newton's life.

FIGURE IT OUT 5.2

Newton's Version of Kepler's Third Law

Modern astronomers often make rough calculations to test whether physical processes under consideration could conceivably be valid. Astronomy has also had a long tradition of exceedingly accurate calculations. Pushing accuracy to one more decimal place sometimes leads to important results.

For example, Kepler's third law, in its original form—the orbital period of a planet squared is proportional to its distance from the Sun cubed (P^2 = constant × R^3)—holds to a reasonably high degree of accuracy and seemed completely accurate when Kepler did his work. But now we have more accurate observations, and they verify Newton's version.

Newton derived the formula for Kepler's third law, taking account of his own formula for the law of universal gravitation. He found that

$$P^2 = \frac{4\pi^2}{G(m_1 + m_2)} R^3,$$

where m_1 and m_2 are the masses of the two bodies. For planetary orbits, if m_1 is the mass of the Sun, then m_2 is the mass of the planet.

Newton's formulation of Kepler's third law shows that the constant of proportionality (the constant number by which you multiply the variable on the right-hand side of the equation to find the result given as the variable on the left-hand side) is determined by the mass of the Sun. Thus, we can use the formula to find out the Sun's mass.

Newton's version of Kepler's third law, as applied to any planet's orbit around the Sun, also shows that the planet's mass contributes to the value of the proportionality constant, but its effect is very small because the Sun is so much more massive. (The effect can be detected even today only for the most massive planets.) Newton therefore actually predicted that the constant in Kepler's third law depends slightly on the mass of the planet, being different for different planets. The observational confirmation of Newton's prediction was a great triumph.

Kepler's third law, and its subsequent generalization by Newton, applies not only to planets orbiting the Sun but also to any bodies orbiting other bodies under the control of gravity. Thus it also applies to satellites orbiting planets. We determine the mass of the Earth by studying the orbit of our Moon, and we determine the mass of Jupiter by studying the orbits of its moons.

For many decades, we were unable to reliably determine the mass of Pluto because we could not observe a moon in orbit around it. The discovery of a moon of Pluto in 1978 finally allowed us to determine Pluto's mass, and we found that our previously best estimates (based on Pluto's now-discredited gravitational effects on Uranus) were way off. The same formula can be applied to binary stars to find their masses.

When he calculated that the acceleration of the Moon fit the same formula as the acceleration of the apple, he knew he had the right method. Newton was the first to realize that gravity was a force that acted in the same way throughout the Universe. Many pieces of the original apple tree have been enshrined (■ Fig. 5–24). There is no knowing which—if any—of the apple trees now growing in front of Newton's house, still standing at Woolsthorpe, actually descended from the famous original.

Newton's most famous literary quotation is, "If I have seen so far, it is because I have stood on the shoulders of giants." In fact, a whole book has been devoted to all the other usages of this phrase, which turns out to have been in wide use before Newton and thereafter. (Actually, Newton's statement was not merely praising his predecessors; one of his scientific rivals was very short.) As of the 21st century, the pace of science has been so fast that it has even been remarked, "Nowadays we are privileged to sit beside the giants on whose shoulders we stand."

Jay M. Pasachoff

■ **FIGURE 5-24** A piece of Newton's apple tree, now in the Royal Astronomical Society, London. The donor's father was present in about 1818 when some logs were sawed from this famous apple tree, which had blown over.

5.8 Clues to the Formation of Our Solar System

The scientists just discussed provided the basis of our current understanding of our Solar System, which in their time was essentially equivalent to the Universe. Studying the orbits of the planets shows that the Solar System has some basic properties. All the

LIVES IN SCIENCE

Isaac Newton

Isaac Newton was born in Woolsthorpe, Lincolnshire, England, on December 25, 1642. Even now, about 360 years later, scientists still refer regularly to "Newton's laws of motion" and to "Newton's law of gravitation."

Newton's most intellectually fertile years were those right after his graduation from college, when he returned home to the country because fear of the plague shut down many cities. There, he developed his ideas about the nature of motion and about gravitation. In order to derive them mathematically, he invented calculus. Newton long withheld publishing his results, possibly out of shyness. Finally Edmond Halley—who became known by applying Newton's law of gravity to show that a series of reports of comets all really referred to this single comet that we now know as Halley's Comet—persuaded Newton to publish his results. A few years later, in 1687, the *Philosophiae Naturalis Principia Mathematica (Mathematical Principles of Natural Philosophy),* known as *The Principia,* was published with Halley's practical and financial assistance. In it, Newton showed that the motions of the planets and comets could all be explained by the same law of gravitation that governed bodies on Earth. In fact, he derived and generalized Kepler's laws on theoretical grounds.

Isaac Newton in 1702. He was knighted three years later, although for his public work rather than his science.

Collection of David Park

Newton was professor of mathematics at Cambridge University and later in life went into government service as Master of the Mint in London. He was knighted in 1705 (for his governmental rather than his scientific contributions) and lived until 1727. His tomb in Westminster Abbey bears the epitaph: "Mortals, congratulate yourselves that so great a man has lived for the honor of the human race."

planets, for example, move in the same direction around the Sun and are more or less in the same plane.

Considering these regularities gives important clues about how the Solar System may have formed, a topic we will discuss more thoroughly in Chapter 9. In studying the other planetary systems now being discovered at a rapid pace, we are eager to determine how many things about our own Solar System are truly fundamental, and how many may be specific to our case. As we will learn, some of the ideas we thought were basic turned out not to apply to many of the other planetary systems we are finding.

As we have seen, the motion of a planet around the Sun in its orbit is its **revolution.** The spinning of a planet on its own axis is its **rotation.** The Earth, for example, revolves around the Sun in one year and rotates on its axis in one day. (These motions define the terms "year" and "day.") According to Kepler's third law, planets far from the Sun have longer periods of revolution (orbital periods) than planets close to the Sun. The distant planets also move more slowly (see *Figure It Out 5.3: Orbital Speeds of Planets*).

The orbits of the planets take up a disk rather than a full sphere (■ Fig. 5–25). The **inclination** of a planet's orbit is the angle that the plane of its orbit makes with the plane of the Earth's orbit.

■ **FIGURE 5–25** The orbits of the planets, with the exception of Pluto, have only small inclinations to the plane of the Earth's orbit. The orbits of all the planets other than Pluto are nearly circular, but look more extended and obviously elliptical here because of the inclined perspective.

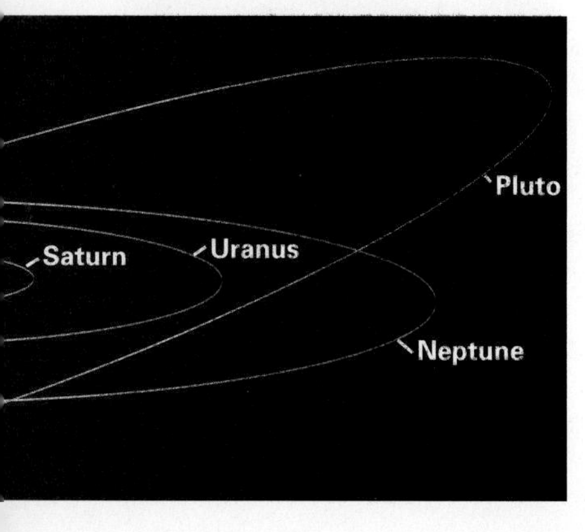

The fact that the planets all orbit the Sun in roughly the same plane is one of the most important generalities about the Solar System. Spinning or rotating objects have a property called **angular momentum.** This angular momentum is larger if an object is more massive, if it is spinning faster, or if mass is farther from the axis of spin.

As will be discussed more extensively in Chapter 9, one important property of angular momentum is that the total angular momentum of a system doesn't change (we say that it is "conserved"), unless the system interacts with something from outside that brings in or takes up angular momentum. (For example, as a spinning ice skater pulls her arms closer to her torso, her rate of spin increases because of the conservation of angular momentum.) Thus, if one body were to spin more and more slowly, or were to begin spinning in the opposite direction, the angular momentum would have to be taken up by another, external body.

FIGURE IT OUT 5.3
Orbital Speed of Planets

The more distant a planet is from the Sun, the slower is its orbital speed. We can derive this from Kepler's third law. Since most of the planets have nearly circular orbits, a reasonable approximation for our purposes here is to assume that the distance travelled by a planet in one full orbit is $2\pi R$, the circumference of a circle of radius R.

If the planet travels at a constant speed (which it would if the orbit were exactly circular), the distance travelled would equal its speed multiplied by time: $d = vt$. Applying this to one full orbit, we find that $2\pi R = vP$, where P is the planet's orbital period. Thus,

$$P = 2\pi R/v, \text{ so } P \propto R/v,$$

where the symbol "$\propto$" means "proportional to."

But Kepler's third law states that $P^2 \propto R^3$. When we substitute the expression $P \propto R/v$ for P into Kepler's law, we get $(R/v)^2 \propto R^3$. Rearranging gives $v^2 \propto 1/R$, and hence

$$v \propto (1/R)^{1/2}.$$

In words, the speed of a planet is inversely proportional to the square root of its orbital radius; distant planets move more slowly than those near the Sun. For example, a planet whose orbital radius is 4 times that of the Earth has an orbital speed that is $(1/4)^{1/2} = 1/2$ as fast as the Earth's orbital speed, which is itself about 30 km/sec. If the planet is 9 times as distant as the Earth, its orbital speed is $(1/9)^{1/2} = 1/3$ that of the Earth.

This "inverse-square-root" relationship is characteristic of systems in which a large central mass dominates over the masses of orbiting particles. Later, we will see that a different result is found for rotating galaxies, where the material is spread out over a large volume.

The planets undoubtedly all revolve around the Sun in the same direction because of how they were formed. They have angular momentum associated with them, and this must have also been the case for the material from which they coalesced. These concepts will be useful when we consider the dozens of new systems of planets around other stars (Chapter 9). After all, it is always better to have more than one example of some concept.

AceAstronomy™ Log into AceAstronomy and select this chapter to see the Astronomy Exercise "Orbital Motion."

CONCEPT REVIEW

The Solar System has been largely explored with spacecraft over the last four decades (Sec. 5.1). Long before that, and even before the telescope was invented, it was known that planets somehow undergo **retrograde motion** (slowly drifting east to west among the stars) besides their generally **prograde motion** (drifting west to east among the stars). Aristotle's cosmological **model,** a description of reality, had the Sun, Moon, and planets orbiting the Earth (Sec. 5.2). Ptolemy enlarged on Aristotle's **geocentric** (Earth-centered) theory and explained retrograde motion by having the planets orbit the Earth on **epicycles,** small circles whose centers move along larger circles called **deferents** at a constant angular rate relative to a point called the **equant** (Sec. 5.2). As a calculational tool, Ptolemy's model could be used to predict the positions of planets with reasonable accuracy.

Copernicus, in 1543, advanced a **heliocentric** (Sun-centered) theory that explained retrograde motion as a projection effect (Sec. 5.3). Copernicus still used circular orbits and a few epicycles, and so had not completely broken with the past, but his idea was nevertheless revolutionary and had lasting consequences. However, Copernicus did not have clear observational evidence favoring his model over that of Ptolemy; instead, the appeal was largely philosophical.

Tycho Brahe made unprecedentedly accurate observations of planetary positions (Sec. 5.4), which Johannes Kepler interpreted quantitatively in terms of **ellipses** (Sec. 5.5a). An ellipse is a closed curve, all of whose points have the same summed distance from two interior points, each of which is called a **focus** (plural **foci**). The **major axis** of an ellipse is the longest line that can be drawn within the ellipse, or the length of that line; it passes through the two foci. The **semimajor axis** is half this length. The **minor axis** is perpendicular to the center of the major axis, or the length of this line; the **semiminor axis** is half of the minor axis. The **eccentricity** of an ellipse, a measure of how far it is out of round, is equal to the distance between the foci divided by the major axis.

Kepler formulated three empirical laws of planetary motion (Sec. 5.5a–c): (1) The orbits are ellipses with the Sun at one **focus.** (2) **(The law of equal areas.)** The line joining the Sun and a planet sweeps out equal areas in equal times. Thus, a planet moves fastest when it is at **perihelion,** the point on the orbit closest to the Sun. (3) The orbital **period** squared is proportional to the semimajor axis cubed. When measured in **Astronomical Units** (the semimajor axis of the Earth's orbit, or Earth's average distance from the Sun) and Earth **years,** Kepler's third law has a particularly simple form.

Galileo first systematically turned a telescope on the sky and discovered features on the Moon, moons of Jupiter, and a full set of phases of Venus (Sec. 5.6). His observational discoveries strongly endorsed the Copernican heliocentric theory and provided evidence against the Ptolemaic geocentric model. Near the end of his life, Galileo studied the motion of falling bodies, providing the foundation for Newton's subsequent work.

Newton developed three basic laws of motion (Sec. 5.7): (1) If no external forces act on an object, its speed and direction of motion stay constant. (2) The force on an object is equal to the object's mass times its acceleration ($F = ma$). (3) When two bodies interact, they exert equal and opposite forces on each other. Newton also discovered the law of gravity, namely that the gravitational force between two objects is proportional to the product of their masses and inversely proportional to the square of the distance between them (Sec. 5.7). He derived Kepler's laws mathematically and invented calculus, which he needed for these studies.

A planet **revolves** around the Sun and **rotates** on its axis (Sec. 5.8). Its orbit is generally **inclined** by only a few degrees to the plane of the Earth's orbit. Spinning or rotating objects have **angular momentum,** roughly a product of an object's mass, physical size, and rotation rate. The total angular momentum of a system can change only if the system interacts with something from the outside. The angular momentum associated with planetary orbits probably came from the material that formed the planets.

QUESTIONS

1. Describe how you can tell in the sky whether a planet is in prograde or retrograde motion.
2. Describe the difference between the Ptolemaic and the Copernican systems in explaining retrograde motion.
3. Can planets interior to the Earth's orbit around the Sun undergo retrograde motion? Explain with a diagram.
4. If Copernicus's heliocentric model did not give significantly more accurate predictions than Ptolemy's geocentric model, why do we now prefer Copernicus's model?
5. Draw the ellipses that represent Neptune's and Pluto's orbits. Show how one can cross the other, even though they are both ellipses with the Sun at one of their foci.
6. The Earth is closest to the Sun in January each year. Use Kepler's second law to describe the Earth's relative speeds in January and July.
7. †Pluto orbits the Sun in about 250 years. From Kepler's third law, calculate its semimajor axis. Show your work. The accuracy you can easily get by estimating the roots by hand rather than with a calculator is sufficient.
8. †Mars's orbit has a semimajor axis of 1.5 A.U. From Kepler's third law, calculate Mars's period. Show your work. The accuracy you can easily get by estimating the roots by hand rather than with a calculator is sufficient.
9. Explain in your own words why the observation of a full set of phases for Venus backs the Copernican system.
10. Do you expect Venus to have a larger angular size in its crescent phase or gibbous phase? Explain.
11. Discuss Galileo's main scientific contributions.
12. Discuss Newton's main scientific contributions.
13. †Explicitly show from Newton's version of Kepler's third law that Kepler's "constant" actually depends on the planet under consideration. (Substitute values for the mass of the Sun and of various planets into the constant.) Does it vary much for different planets?
14. †The discovery of a star orbiting the center of our galaxy with a 15-year period has led to the finding that the mass of the giant black hole at the center of our galaxy is about 3 million times that of the Sun. Using ratios, calculate the semimajor axis of the star's orbit. Convert the answer to light-years.
15. †If a hypothetical planet is 4 times farther from the Sun than Earth is, what is its orbital *speed* (not period) relative to Earth's?
16. †What is the period of a planet orbiting 3 A.U. from a star that is 3 times as massive as the Sun? (*Hint:* Consider Newton's version of Kepler's third law. Use ratios!)
17. Which is greater: The Earth's gravitational force on the Sun or the Sun's gravitational force on the Earth? Explain.
18. †Compare the periods of rotation and revolution of our Moon.
19. If you were standing on a spinning turntable holding barbells close to your chest and you extended your arms, how would your spin be affected? Why?
20. †If you were standing on a turntable and a friend pushed you to give you angular momentum, compare how fast you would spin if you were holding 10-kg vs. 30-kg barbells.
21. **True or false?** An external force is needed to keep an object in motion at a constant speed and a constant direction.
22. **True or false?** In Ptolemy's geocentric theory of the Solar System, Venus could not go through the complete set of phases observed by Galileo.
23. **True or false?** Although the values of the constants may differ, Newton showed that Kepler's laws apply

equally well to the moons orbiting Jupiter, if "moon" and "Jupiter" are substituted for the words "planet" and "Sun," respectively.

24. **True or false?** The orbital period and distance of a planet from the Sun can be used, with Newton's version of Kepler's third law, to estimate the mass of the Sun.
25. **True or false?** Copernicus's greatest contribution to astronomy was that he put forth the notion that retrograde motion can be explained by having the planets move along epicycles.
26. **True or false?** The outer planets orbit the Sun with a smaller physical speed (km/sec) than the inner planets.
27. **Multiple choice:** Which one of the following discoveries was *not* made by Galileo? **(a)** A given planet moves most quickly when it is closest to the Sun. **(b)** Jupiter's four large moons. **(c)** Large numbers of stars in the Milky Way. **(d)** The full set of phases of Venus. **(e)** Craters and mountains on the Moon.
28. †**Multiple choice:** Two planets orbit a star that has the same mass as the Sun. Planet Zaphod orbits at a distance of 2 A.U. while Planet Arthur orbits at a distance of 1 A.U. What is Planet Zaphod's orbital period compared to Planet Arthur's orbital period? **(a)** 8 times longer. **(b)** $\sqrt{8}$ times longer. **(c)** 4 times longer. **(d)** 2 times longer. **(e)** The two periods are equal, since the mass of the star is the same in both cases.
29. †**Multiple choice:** Suppose you discover a comet whose orbital period around the Sun is 10^6 years. What is the semimajor axis of the comet's orbit, in A.U.? **(a)** 10^2. **(b)** 10^3. **(c)** 10^4. **(d)** 10^6. **(e)** 10^9.
30. **Multiple choice:** The retrograde motion of Venus across the sky **(a)** is caused by the "backward" rotation of Venus about its own axis (it spins in a direction opposite that of most planets); **(b)** is caused by the gravitational tug of other planets on Venus; **(c)** is caused by the change in perspective as Venus catches up with, and passes, Earth while both planets orbit the Sun; **(d)** is caused by the motion of Venus along an epicycle whose center orbits the Sun (and hence Earth as well, from our perspective); **(e)** was used by Galileo to explain the complete set of phases of Venus that he observed through his telescope.
31. †**Multiple choice:** Which one of the following statements is *true*? **(a)** A given force accelerates a massive object more than a lower mass object. **(b)** If the distance between two objects is doubled, the gravitational force between them decreases to half its previous value. **(c)** The force on an object due to gravity does *not* depend on the object's mass. **(d)** Neglecting air resistance, a light tennis ball falls just as fast as a heavy lead ball dropped from the same height above the ground. **(e)** If the Sun's gravitational attraction were suddenly cut off, at that instant the Earth would start flying directly away from the Sun—that is, along a path radially away from the Sun.
32. **Fill in the blank:** Although Johannes Kepler developed three laws of planetary motion, __________ developed a physical understanding of these laws.
33. **Fill in the blank:** According to Kepler's first law, the orbits of planets are ellipses with the Sun at one ________.
34. **Fill in the blank:** The line that extends from one edge of an ellipse to the other, passing through both foci of the eclipse, is called the ___________.
35. **Fill in the blank:** In size, the gravitational force pulling an apple toward the Earth is ________ the gravitational force pulling the Earth toward the apple.
36. **Fill in the blank:** If a spinning body expands, its rate of spin decreases, providing an example of the conservation of ________________.
37. **Fill in the blank:** In the model of the Universe developed by Aristotle, the outermost sphere was that of the _________.

†This question requires a numerical solution.

TOPICS FOR DISCUSSION

1. Had you been alive 2000 years ago, do you think you would have believed in a geocentric model of the Universe, or a heliocentric model as did Aristarchus of Samos? Explain.
2. Why does a satellite speed up as it spirals toward the Earth due to friction with the outer atmosphere? Naively, it seems that the friction would cause it to slow down.
3. Referring back to the scientific method discussed in Chapter 1, why was it important to verify Newton's prediction that the "constant" in Kepler's third law actually depends on the mass of the planet under consideration?

MEDIA

AceAstronomy™ Log into AceAstronomy at **http://astronomy.brookscole.com/cosmos3** to access quizzes and animations that will help you assess your understanding of this chapter's topics.

Log into the Student Companion Web Site at **http://astronomy.brookscole.com/cosmos3** for more resources for this chapter including a list of common misconceptions, news and updates, flashcards, and more.

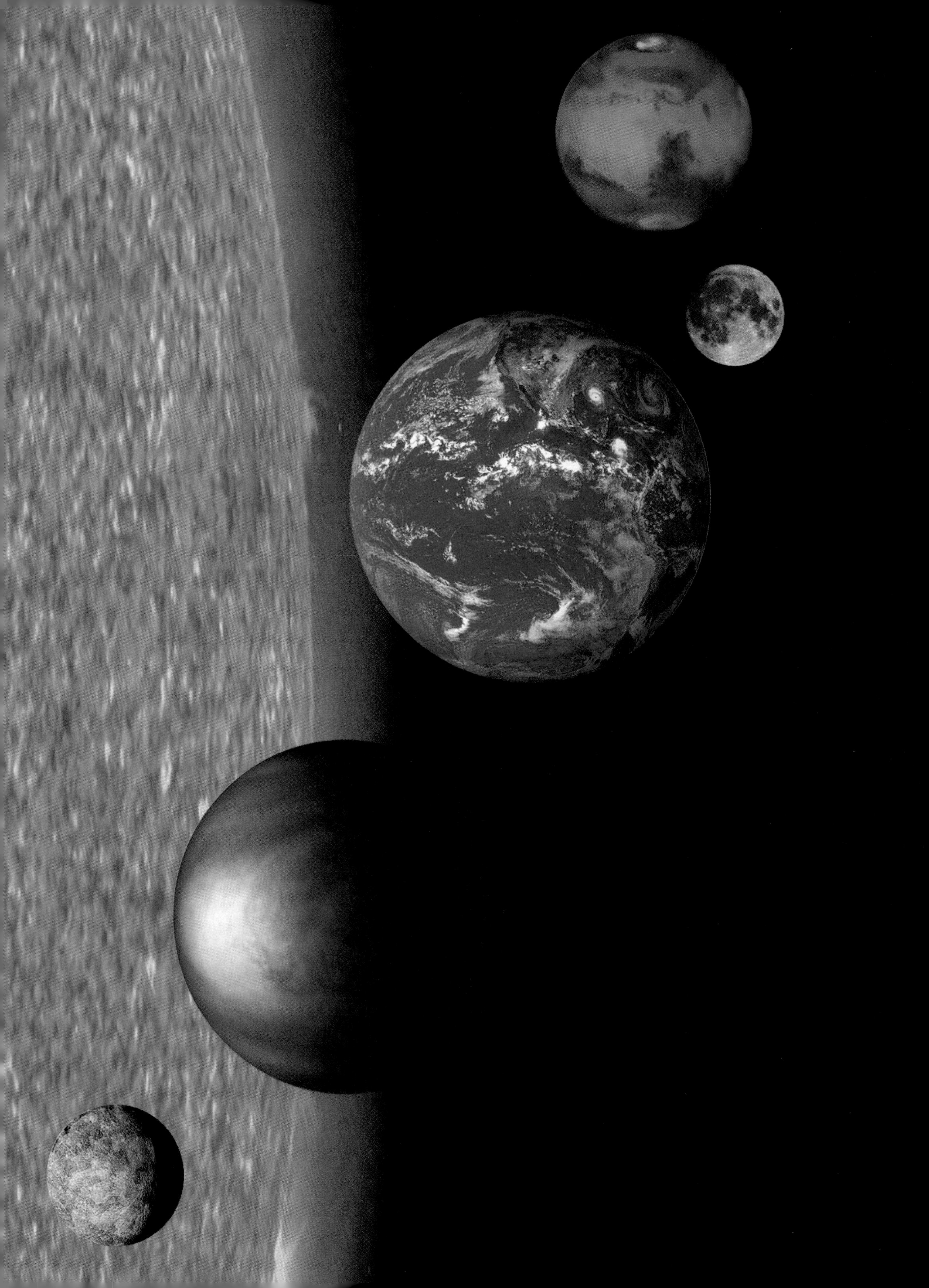

C H A P T E R 6

The Terrestrial Planets: Earth, Moon, and Their Relatives

Mercury, Venus, Earth, and Mars share many similar features. Small compared with the huge planets beyond them, these inner planets also have rocky surfaces surrounded by relatively thin and transparent atmospheres, in contrast with the larger, gaseous/liquid planets. Together, we call these four the **terrestrial planets** (from the Latin "terra," meaning earth), which indicates their significance to us in our attempts to understand our own Earth. In this chapter, we discuss each of these rocky bodies, as well as their moons. (See *A Closer Look 6.1: Comparative Data for the Terrestrial Planets and Their Moons.*)

Venus and the Earth are often thought of as "sister planets," in that their sizes, masses, and densities are about the same (■ Fig. 6–1). But in many respects they are as different from each other as the wicked stepsisters were from Cinderella. The Earth is lush; it has oceans of water, an atmosphere containing oxygen, and life. On the other hand, Venus is a hot, foreboding planet with temperatures constantly over 750 K (900°F), a planet on which life seems unlikely to develop. Why is Venus like that? How did these harsh conditions come about? Can it happen to us here on Earth?

The one Solar-System body other than Earth that humans have visited is our Moon. It is so large relative to Earth that it joins us as a type of a "double-planet system." We will see how space exploration has revealed many of its secrets.

Mars is only 53 per cent the diameter of Earth and has 10 per cent of Earth's mass. Its atmosphere is much thinner than Earth's, too thin for visitors from Earth to rely on to breathe. But Mars has long been attractive as a site for exploration. We remain interested in Mars as a place where we may yet find signs of life or, indeed, where we might encourage life to grow. Mars's two tiny moons are but chunks of rock in orbit.

Mercury, the innermost planet, is more like our Moon than like our own Earth. Its atmosphere is negligible and its surface is seared by solar radiation. An American spacecraft is en route there. A European/Japanese spacecraft is to be launched to Mercury in 2011 or 2012.

ORIGINS

The small, rocky, terrestrial planets and the Moon show many similarities and differences among their interiors, surfaces, and atmospheres. By studying these objects, we can begin to understand the formation and evolution of the Earth, whose properties seem uniquely suitable for the development of complex life.

AIMS

1. Describe our Earth, our Moon, the Moon's origin, and its tidal influence on Earth (Sections 6.1 and 6.2).
2. Learn about the inner planets, four rocky worlds including the Earth (Sections 6.3 to 6.5).
3. Examine the similarities and differences among the terrestrial planets (Sections 6.3 to 6.5).

AceAstronomy™ The AceAstronomy icon throughout this text indicates an opportunity for you to test yourself on key concepts, and to explore animations and interactions of the AceAstronomy website at **http://astronomy.brookscole.com/cosmos3**

NASA's Ames Research Center

■ **FIGURE 6–1** Surface topography on Earth and Venus at the same resolution, imaged with radar.

A montage, in the correct relative scale, of NASA space views of the inner planets, plus the Earth's Moon, with the Sun's edge in the background.

NSSDC at NASA's Goddard Space Flight Center, courtesy of Jay Friedlander.

A Closer Look 6.1 COMPARATIVE DATA FOR THE TERRESTRIAL PLANETS AND THEIR MOONS

	Semimajor Axis of Orbit	Orbital Period	Equatorial Radius (Earth = 100%)	Mass (Earth = 100%)
PLANETS				
Mercury	0.39 A.U.	88 days	38%	5.5%
Venus	0.72 A.U.	225 days	95%	82%
Earth	1.00 A.U.	365.25 days	100%	100%
Mars	1.52 A.U.	687 days	53%	11%
EARTH'S SATELLITE				
The Moon	384,000 km	$27\frac{1}{3}$ days	27% (1738 km)	1.2%
MARS'S SATELLITES				
Phobos	9,378 km	7 hr 39 min	$14 \times 11 \times 9$ km^3	1 millionth %
Deimos	23,459 km	1 d 6 hr 18 min	$8 \times 6 \times 6$ km^3	0.3 millionth %

■ **FIGURE 6–2** A composite of the Earth and Moon, viewed from the Galileo spacecraft as it passed by our region of the Solar System on its mission to Jupiter. Structure on the Earth's surface, especially in South America and Central America, shows clearly.

6.1 Earth: There's No Place Like Home

On the first trip that astronauts ever took to the Moon, they looked back and saw for the first time the Earth floating in space. Nowadays we see that space view every day from weather satellites, so the views from the Jupiter-bound Galileo spacecraft and the Saturn-bound Cassini spacecraft as they passed allowed us to test the instruments on known objects, the Earth and Moon (■ Fig. 6–2).

The realization that Earth is an oasis in space helped inspire our present concern for our environment. Until fairly recently, we studied the Earth only in geology courses and the other planets only in astronomy courses, but now the lines are very blurred. Not only have we learned more about the interior, surface, and atmosphere of the Earth but we have also seen the planets in enough detail to be able to make meaningful comparisons with Earth. The study of **comparative planetology** is helping us to understand weather, earthquakes, and other topics. This expanded knowledge will help us improve life on our own planet.

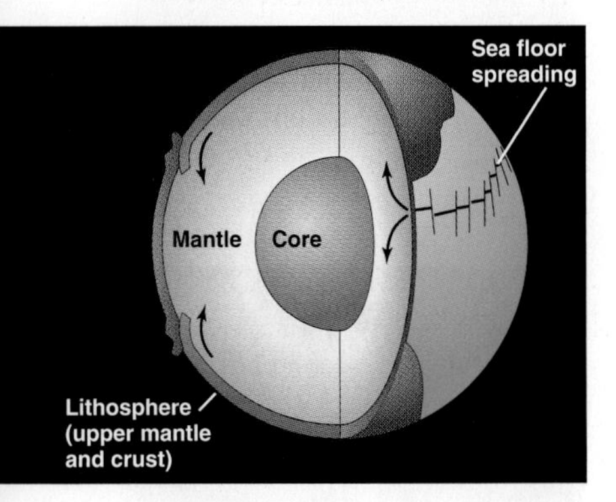

■ **FIGURE 6–3** The interior structure of the Earth. (Adapted from Raymond Siever, "The Earth," *Scientific American*, September 1975, and *The Solar System* [New York: W. H. Freeman and Co., 1975]. ©1975 by Scientific American, Inc. All rights reserved; drawn by Tom Prentiss.)

6.1a The Earth's Interior

The study of the Earth's interior and surface (■ Fig. 6–3) is called **geology.** Geologists study, among other things, how the Earth vibrates as a result of large shocks, such as earthquakes. Much of our knowledge of the structure of the Earth's interior comes from **seismology,** the study of these vibrations. The vibrations travel through different types of material at different speeds.

From seismology and other studies, geologists have been able to develop a picture of the Earth's interior. The Earth's innermost region, the **core,** consists primarily of iron and nickel. Outside the core is the **mantle,** and on top of the mantle is the thin outer layer called the **crust.** The upper mantle and crust are rigid and contain a lot of silicates, while the lower mantle is partially melted.

Such a layered structure must have developed when the Earth was young and molten; the denser materials (like iron) sank deeper than the less-dense ones, as discussed below. But from where did Earth get sufficient heat to become molten?

The Earth, along with the Sun and the other planets, was probably formed from a cloud of gas and dust. Some of the original energy, though not enough to melt the

Earth's interior, came from gravitational energy released as particles came together to form the Earth; such energy is released from gravity between objects when the objects move closer together and collide. The water at the base of a waterfall, for example, is slightly (unnoticeably) hotter than the water at the top; part of the falling water's energy of motion, gained by the pull of gravity, is converted to heat by the collision on the rocks or water at the waterfall's base.

Also, the young Earth was subject to constant bombardment from the remaining debris (dust and rocks), which carried much energy of motion. This bombardment heated the surface to the point where it began to melt, producing lava.

However, scientists have concluded that the major source of energy in the interior, both at early times and now, is the natural radioactivity within the Earth. Certain forms of atoms are unstable—that is, they spontaneously change into more stable forms. In the process, they give off energetic particles that collide with the atoms in the rock and give some of their energy to these atoms. The rock heats up.

The Earth's interior became so hot that the iron melted and sank to the center since it was denser, forming the core. Eventually other materials also melted. As the Earth cooled, various materials, because of their different densities (density is mass divided by volume; see *A Closer Look 6.2: Density*) and freezing points (the temperature at which they change from liquid to solid), solidified at different distances from the center. This process, called "differentiation," is responsible for the present layered structure of the Earth.

Geologists have known for decades that the Earth's iron center consists of a solid inner core surrounded by a liquid outer core. The inner core is solid, in spite of its high (5000°C) temperature, because of the great pressure on it. A new study of 30 years worth of earthquake waves that passed through the Earth's core revealed in 2002 that the inner core has a different inner region, like the pit in a peach. This inner region is less than 10 per cent of the diameter of the "inner core," to continue to use that technical term.

A Closer Look 6.2 DENSITY

The density of an object is its mass divided by its volume. The scale of mass in the metric system was set up so that 1 cubic centimeter of water would have a mass of 1 gram (symbol: g). Though the density of water varies slightly with the water's temperature and pressure, the density of water is always about 1 g/cm^3. Density is always expressed in units of mass (grams, kilograms, and so on) per volume (cubic centimeters, and so on), where the slash means "per." In the units in general use in the United States (but not elsewhere in the world), the density of water is about 62 pounds/cubic foot (though, technically, pounds is a measure of weight—or gravitational force—rather than of mass). Always be careful to note if you are using metric or pound/foot-type units. A spacecraft was lost just as it reached Mars in 1999 because one organization getting it there used metric units while another used pound/foot-type units.

To measure the density of any amount of matter, astronomers must independently measure its mass and its size. They are often able to identify the materials present in an object on the basis of density. For example, since the density of iron is about 8 g/cm^3, Saturn's density of only 0.7 g/cm^3 shows that Saturn is not made largely of iron. On the other hand, the measurement that the Earth's density is about 5 g/cm^3 shows that the Earth has a substantial content of dense elements, of which iron is a likely candidate.

Astronomical objects cover the extremes of density. In the space between the stars, the density is incredibly tiny fractions of a gram in each cubic centimeter (about 10^{-24} g/cm^3), corresponding to just one hydrogen atom. And in neutron stars, the density exceeds a billion tons per cubic centimeter, comparable to that of an atomic nucleus.

Some sample values of the densities of materials are as follows:

water	1 g/cm^3
aluminum	3 g/cm^3
typical rock	3 g/cm^3
iron	8 g/cm^3
lead	11 g/cm^3

Jay M. Pasachoff

■ **FIGURE 6–4** A geyser and its surrounding steam, at Rotorua, New Zealand.

Why does this "peach pit" innermost core make earthquake waves act differently than they do in the surrounding inner core? It could be because this innermost core is a remnant of the original ball of material from which the Earth formed 4.6 billion years ago. Less exciting is the possibility that iron crystals deposited on it had a different orientation after the innermost core reached a critical size. Perhaps the temperature and pressure in that innermost region pack iron crystals differently.

The rotation of the Earth's metallic core helps generate a magnetic field on Earth. (The discovery in 2005 that the Earth's inner core spins 0.009 seconds per year faster than the rest of our planet, giving it an extra full revolution in about 900 years, may affect models of how the magnetic field is generated.) The magnetic field has a north magnetic pole and a south magnetic pole that are not quite where the regular north and south geographic poles are. The Earth's magnetic north pole is in the Arctic Ocean north of Canada. The location of the magnetic poles wanders across the Earth's surface over time. The north magnetic pole is currently moving northward at an average speed of 15 km/year.

AceAstronomy™ Log into AceAstronomy and select this chapter to see the Active Figure called "Seismic Waves."

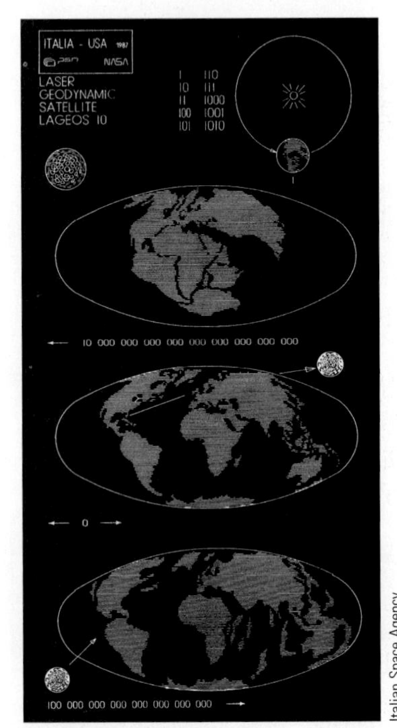

Italian Space Agency

■ **FIGURE 6–5** LAGEOS I and LAGEOS II (**La**ser **Geo**dynamic **S**atellites) bear plaques showing the continents in their positions 270 million years in the past (*top*), at present (*middle*), and 8 million years in the future (*bottom*), according to our knowledge of continental drift. (The dates are given in the binary system.) Data on continental drift are gathered by reflecting laser beams from telescopes on Earth off the 426 retroreflectors that cover the satellites' surface. (Each retroreflector is a cube whose interior reflects incident light back in the direction from which it came.) Measuring the time until the beams return can be interpreted to show the accurate position of the ground station. We can now measure the position to an accuracy of about 1 cm.

6.1b Continental Drift

Some geologically active areas exist in which heat flows from beneath the surface at a rate much higher than average (■ Fig. 6–4). The outflowing **geothermal energy,** sometimes tapped as an energy source, signals what is below. The Earth's rigid outer layer is segmented into **plates,** each thousands of kilometers in extent but only about 50 km thick. Because of the internal heating, the top layers float on an underlying hot layer (the mantle) where the rock is soft, though it is not hot enough to melt completely.

The mantle beneath the rigid plates of the surface churns very slowly, thereby carrying the plates around. This theory, called **plate tectonics,** explains the observed **continental drift**—the drifting of the continents, over eons, from their original positions, at the rate of a few centimeters per year—about the speed your fingernails grow. ("Tectonics" comes from the Greek word meaning "to build.")

Although the notion of continental drift originally seemed unreasonable, it is now generally accepted. The continents were once connected as two super-continents, which may themselves have separated from a single super-continent called Pangaea ("all lands"). Over the past two hundred million years or so, the continents have moved apart as plates have separated. We can see from their shapes how they once fit together (■ Fig. 6–5). We even find similar fossils and rock types along two opposite coastlines that were once adjacent but are now widely separated. Remnants of the magnetic field as measured in rocks laid down when the Earth's magnetic poles had flipped, north and south magnetic poles interchanging (as they do occasionally), are also among the strongest evidence.

Pangaea itself probably formed from the collision of previous generations of continents. The coming together and breaking apart of continents may have had many cycles in Earth's history.

In the future, we expect part of California to separate from the rest of the United States, Australia to be linked to Asia, and the Italian "boot" to disappear. The boundaries between the plates are geologically active areas (■ Fig. 6–6).

Therefore, these boundaries are traced out by the regions where earthquakes and most of the volcanoes occur. The boundaries where two plates are moving apart mark regions where molten material is being pushed up from the hotter interior to the surface, such as the mid-Atlantic ridge (■ Fig. 6–7). Molten material is being forced up through the center of the ridge and is being deposited as lava flows on either side, producing new seafloor.

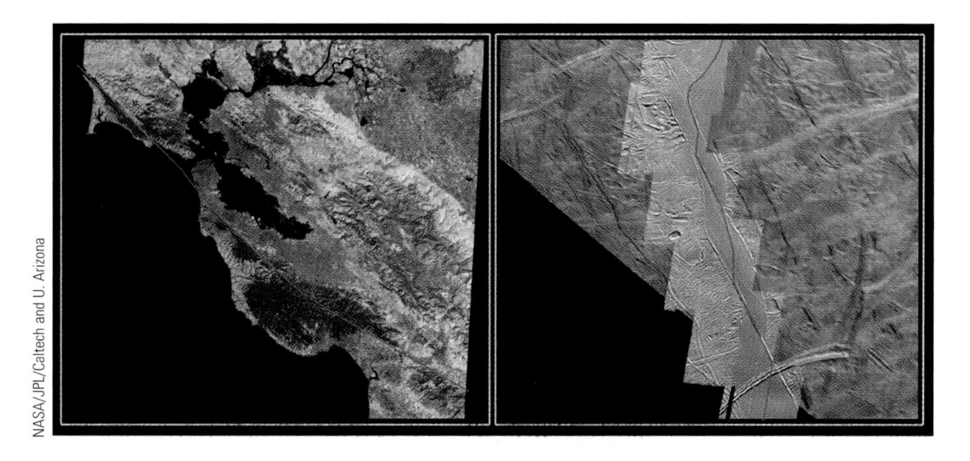
NASA/JPL/Caltech and U. Arizona

■ **FIGURE 6–6** The San Andreas fault on Earth marks the boundary between the North American and Pacific plates. It is a strike-slip fault, in which two blocks of Earth move past each other. This view is paired with a Galileo-spacecraft image of a similar strike-slip fault on Jupiter's moon Europa, shown at the same scale and resolution.

The motion of the plates relative to each other is also responsible for the formation of most great mountain ranges. When two plates come together, one may be forced under the other and the other rises. The great Himalayan mountain chain, for example, was produced by the collision of India with the rest of Asia. The "ring of fire" volcanoes around the Pacific Ocean (including Mt. St. Helens in Washington) were formed when molten material made its way through gaps or weak points between plates.

AceAstronomy™ Log into AceAstronomy and select this chapter to see Astronomy Exercise "Convection and Magnetic Fields."

AceAstronomy™ Log into AceAstronomy and select this chapter to see Astronomy Exercise "Convection and Plate Tectonics."

AceAstronomy™ Log into AceAstronomy and select this chapter to see the Active Figure called "Hotspot Volcanoes."

■ **FIGURE 6–7** The Earth's ocean floor, mapped with microwaves bounced off the ocean's surface from spacecraft as well as with soundings from ships. Features on the seafloor with excess gravity attract the water enough to cause structure on the ocean's surface that can be mapped, though a 2000-meter-high seamount produces a bump on the surface only 2 meters high. High gravity is shown in yellow, orange, and red; low gravity is shown in blue and violet. The feature running from north to south in the middle of the Atlantic Ocean is the *mid-Atlantic ridge*. It marks the boundary between plates that are moving apart. Continents drift at about the rate that fingernails grow. (M. F. Corrin, L. M. Gahagan, and L. A. Lawver, PLATES project, University of Texas Institute for Geophysics)

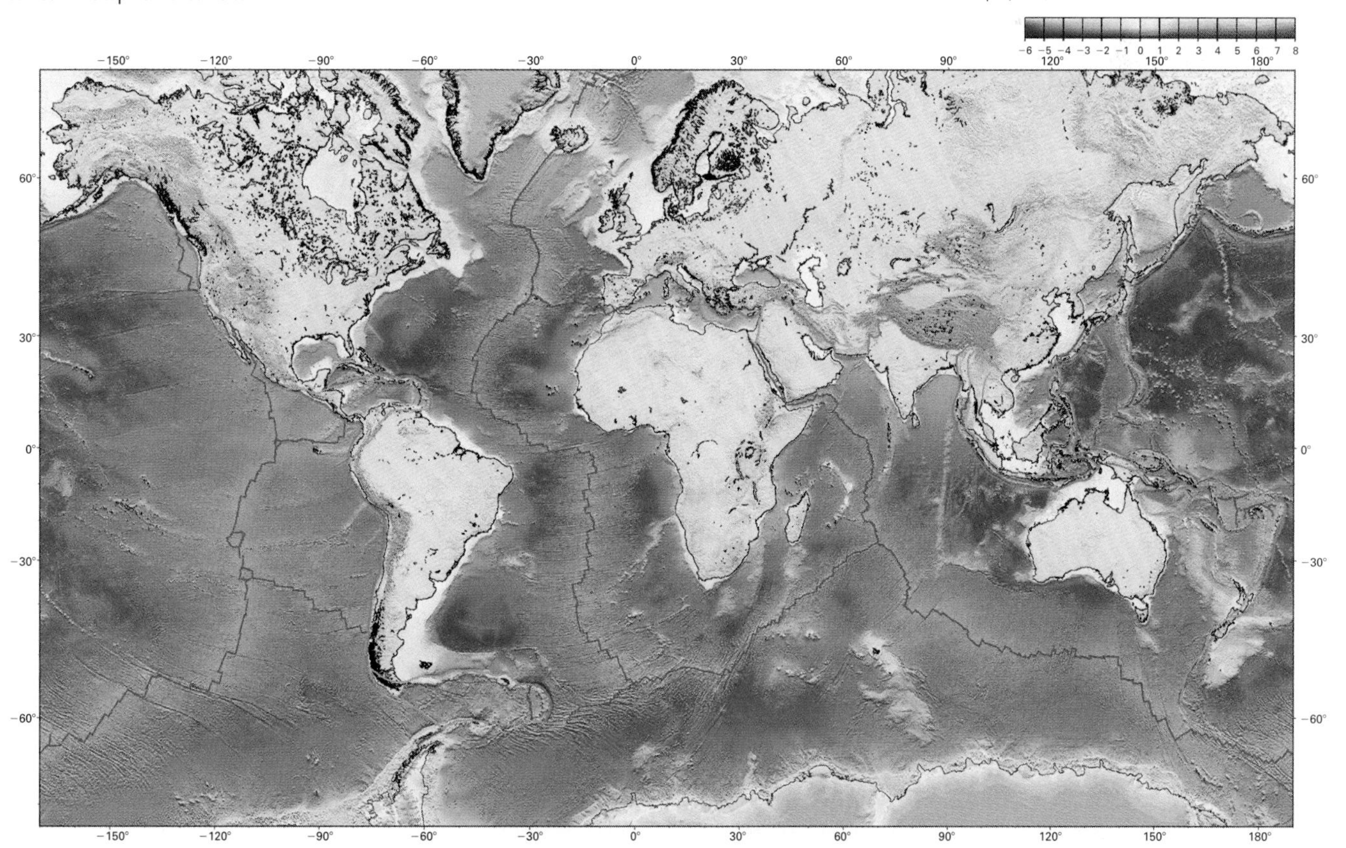

6.1c Tides

It has long been accepted that **tides** are most directly associated with the Moon and to a lesser extent with the Sun. We know of their association with the Moon because the tides—like the Moon's passage across your meridian (the imaginary line in the sky passing from north to south through the point overhead)—occur about an hour later each day.

Tides result from the fact that the force of gravity exerted by the Moon (or any other body) gets weaker as you get farther away from it. Tides depend on the difference between the gravitational attraction of a massive body at different points on another body.

To explain the tides in Earth's oceans, suppose, for simplicity, that the Earth is completely covered with water. We might first say that the water closest to the Moon is attracted toward the Moon with the greatest force and so is the location of high tide as the Earth rotates. If this were the whole story, high tides would occur about once a day. However, two high tides occur daily, separated by roughly $12\frac{1}{2}$ hours.

To see why we get two high tides a day, consider three points, *A*, *B*, and *C*, where *B* represents the solid Earth (which moves all together as a single object and is marked by a point at its center), *A* is the ocean nearest the Moon, and *C* is the ocean farthest from the Moon (■ Fig. 6–8). Since the Moon's gravity weakens with distance, it is greater at point *A* than at *B*, and greater at *B* than at *C*. If the Earth and Moon were not in orbit around each other, all these points would fall toward the Moon, moving apart as they fell because of the difference in force. Thus the high tide on the side of the Earth that is near the Moon is a result of the water being pulled away from the

■ **FIGURE 6–8**

(a) A schematic representation of the tidal effects caused by the Moon (not to scale). The arrows represent the gravitational pull produced by the Moon (exaggerated in the drawing). The water at point A is pulled toward the Moon more than the Earth's center (point B) is; since the Earth is solid, the whole Earth moves with its center. (Tides in the solid Earth exist, but are much smaller than tides in the oceans.) Similarly, the solid Earth is pulled away from the water at point C. Water tends to collect at points A and C (high tide) and to be depleted in regions halfway between them on the Earth's surface, such as on the surface between point B and us (low tide).

Note that although the tidal force results from gravity, it is the *difference* between the gravitational forces at two places and is therefore not the same as the "gravitational force." Tidal forces are important in many astrophysical situations, including interactions between closely spaced galaxies, and rings around planets.

(b) The tidal range at Kennebunkport, Maine, where the shape of the ocean floor leads to an especially high tidal range.

Note that the Moon and the Earth are drawn to a different scale from their separation.

Earth. The high tide on the opposite side of the Earth results from the Earth being pulled away from the water. In between the locations of the high tides the water has rushed elsewhere (to the regions of high tides), so we have low tides.

Since the Moon is moving in its orbit around the Earth, a point on the Earth's surface has to rotate longer than 24 hours to return to a spot nearest to the Moon. Thus a pair of tides repeats about every 25 hours, making $12^{1}/_{2}$ hours between high tides.

The Sun's effect on the Earth's tides is only about half as much as the Moon's effect. Though the Sun exerts a greater gravitational force on the Earth than does the Moon, the Sun is so far away that its force does not change very much from one side of the Earth to the other. And it is only the *difference* in force from one place to another that counts for tides.

Nonetheless, the Sun does matter. We tend to have very high and very low tides when the Sun, Earth, and Moon are aligned (as is the case near the time of full moon or of new moon), because their effects reinforce each other. Conversely, tides are less extreme when the Sun, Earth, and Moon form a right angle (as near the time of a first-quarter or third-quarter moon).

The effect of the tides on the Earth–Moon system slows down the Earth's rotation slightly, by about 1 second per 100,000 years, as we can verify from the timing of solar eclipses that took place thousands of years ago. Also, the interaction is leading to a gradual spiraling away of the Moon from the Earth, though the rate is only centimeters per year. Thus, far in the future (about a billion years from now), it will not be possible to witness a total solar eclipse from Earth; the Moon's angular diameter will be less than that of the Sun's photosphere.

6.1d The Earth's Atmosphere

We name layers of our atmosphere (■ Fig. 6–9) according to the composition and the physical processes that determine their temperatures. The atmosphere contains about 20 per cent oxygen, the gas that our bodies use when we breathe; almost all of the rest is nitrogen. Later, we will see how the small amounts of carbon monoxide, of carbon dioxide, of methane, and of other gases are affecting our climate. When we find a planet around another star (Chapter 9) whose spectrum shows such a high percentage of oxygen, we will infer the presence there of life-forms making the oxygen.

The Earth's weather is confined to the very thin **troposphere.** The ground is a major source of heat for the troposphere, so the temperature of the troposphere decreases as altitude increases. The rest of the Earth's atmosphere, as well as the Earth's surface, is heated mainly by solar energy from above.

A higher layer of the Earth's atmosphere is the "thermosphere." It is also known as the **ionosphere,** since many of the atoms there are ionized—that is, stripped of some of the electrons they normally contain. Most of the ionization is caused by x-ray and

ASIDE 6.1: Avoiding confusion

The greenhouse effect: global warming from gases like carbon dioxide trapping sunlight. The ozone hole: a decrease of atmospheric ozone over Antarctica and surrounding areas each southern-hemisphere springtime, caused by chlorofluorocarbons.

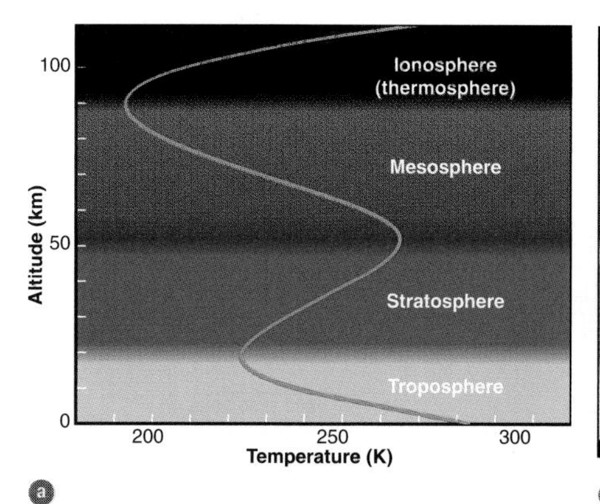

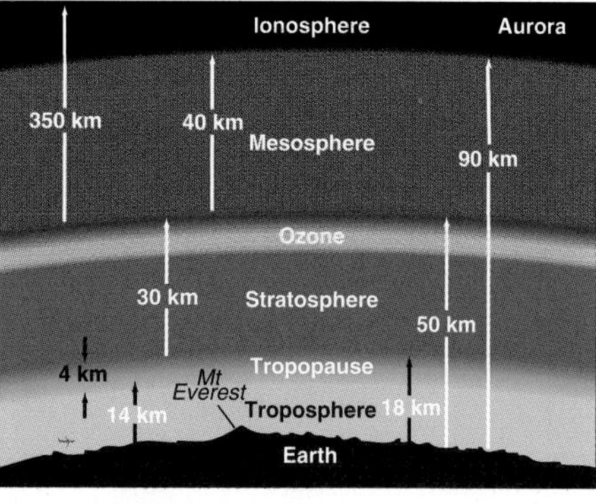

■ **FIGURE 6–9** (a) Temperature in the Earth's atmosphere as a function of altitude. In the troposphere, the energy source is the ground, so the temperature decreases as you go up in altitude. Higher temperatures in other layers result from ultraviolet radiation and x-rays from the Sun being absorbed. (b) The location of the ozone layer, which is depleted over Antarctica in its springtime.

ultraviolet radiation from the Sun, as well as from solar particles. Thus the ionosphere forms during the daytime and diminishes at night. The free electrons in the ionosphere reflect very-long-wavelength radio signals. When the conditions are right, radio waves bounce off the ionosphere, which allows us to tune in distant radio stations.

Observations from high-altitude balloons and satellites have greatly enhanced our knowledge of Earth's atmosphere. Scientists carry out calculations using the most powerful supercomputers to interpret the global data and to predict how the atmosphere will behave. The equations are essentially the same as those for the internal temperature and structure of stars, except that the sources of energy are different, with stars heated from below and the Earth's atmosphere mostly heated from above.

Winds are caused partly by uneven heating of different regions of Earth. The rotation of the Earth also has a very important effect in determining how the winds blow. Comparison of the circulation of winds on the Earth (which rotates in 1 Earth day), on slowly rotating Venus (which rotates in 243 Earth days), and on rapidly rotating Jupiter and Saturn (each of which rotates in about 10 Earth hours) helps us understand the weather on Earth. Our improved understanding allows forecasters of weather (day to day) and climate (long term) to be more accurate.

Comparison with the planet Venus (Section 6.4d) led to our realization that the Earth's atmosphere traps some of the radiation from the Sun, and that we are steadily increasing the amount of trapping. (The word "trapping" is used here in a figurative rather than a literal sense; the energy is actually transformed from one kind to another when air molecules absorb it, and no particular light photons are physically "trapped.") The process by which light is trapped, resulting in the extra heating of Earth's atmosphere and surface, is similar to the process that is generally (though incorrectly) thought to occur in terrestrial greenhouses; it is thus called the **greenhouse effect.** It is caused largely by the carbon dioxide in Earth's atmosphere (■ Fig. 6–10), the amount of which is growing each year because of our use of fossil fuels.

Section 6.4d (below) more fully describes the greenhouse effect, which greatly affects the temperature of Venus. In brief, Earth's atmosphere is warmed both from above by solar radiation, and from below by radiation from the Earth's surface, which is itself warmed by solar radiation. Most solar radiation is in the visible, corresponding to the peak of the Sun's black-body radiation (recall the discussion of this radiation in Chapter 2). The radiation from the Earth's surface is in the form of infrared, which corresponds to the peak wavelength of the black-body curve at Earth's temperature. Some of this infrared radiation is absorbed in the atmosphere by carbon dioxide, water, and, to a lesser extent, by other "greenhouse gases" such as methane.

The greenhouse effect itself is good; it warms us by about 33°C, bringing Earth's atmospheric temperature to the livable range it is now. The important question is

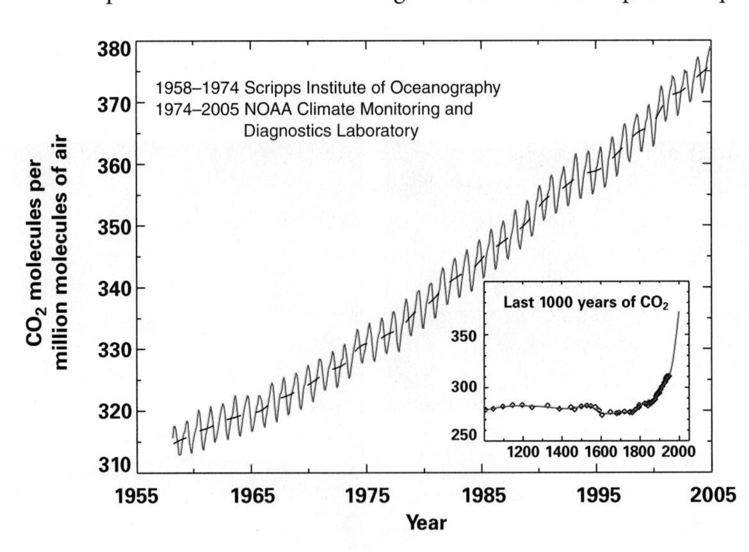

■ **FIGURE 6–10** The concentration of atmospheric carbon dioxide in parts per million (ppm) of dry air as it changed with time, observed at the Mauna Loa Observatory in Hawaii. In addition to the yearly cycle, there is an overall increase. Carbon dioxide is an important contributor to the greenhouse effect. (Pieter Tans, Climate Monitoring and Diagnostics Laboratory, National Oceanic and Atmospheric Administration, Carbon Cycle Group, Boulder, CO, and Charles D. Keeling, Scripps Institute of Oceanography, LaJolla, CA)

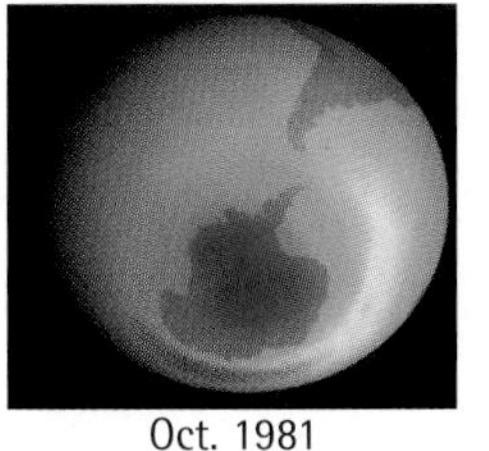

Oct. 1986

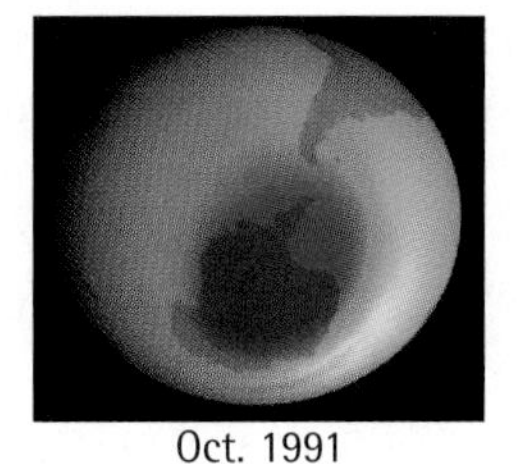

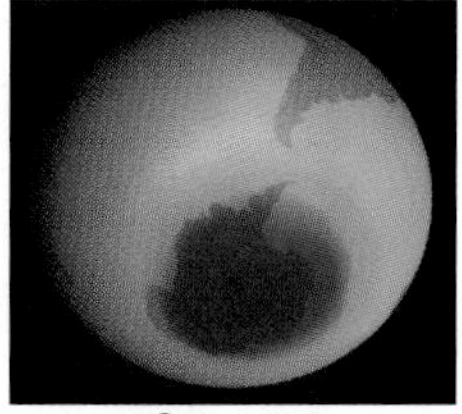

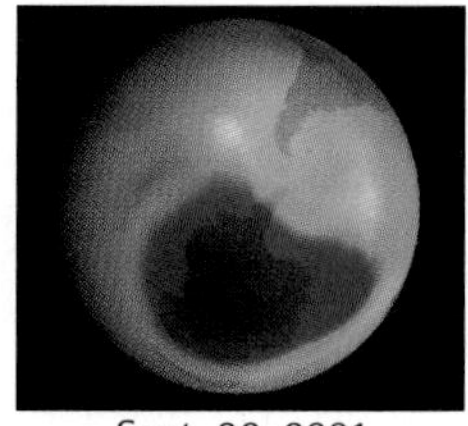

a

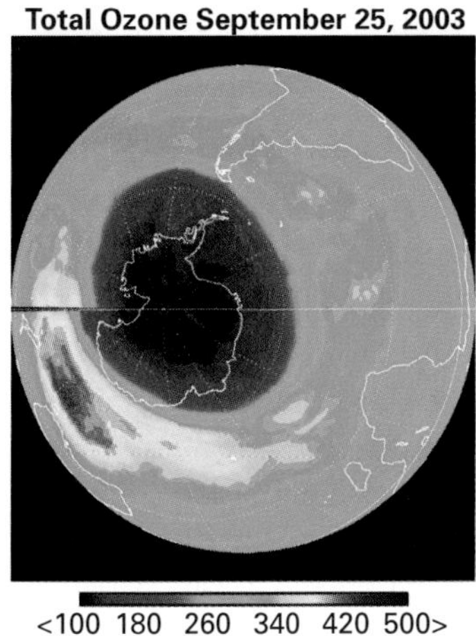

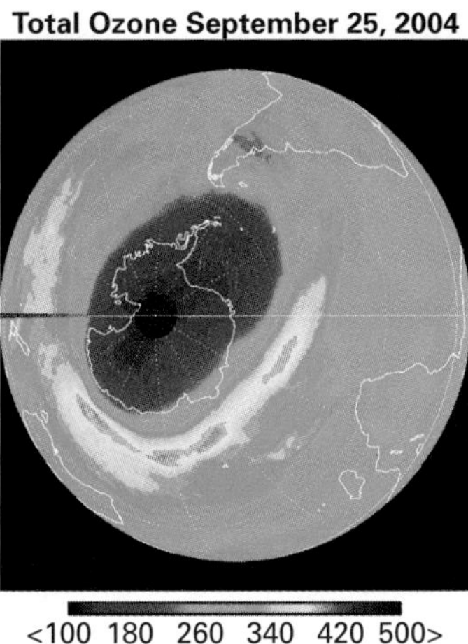

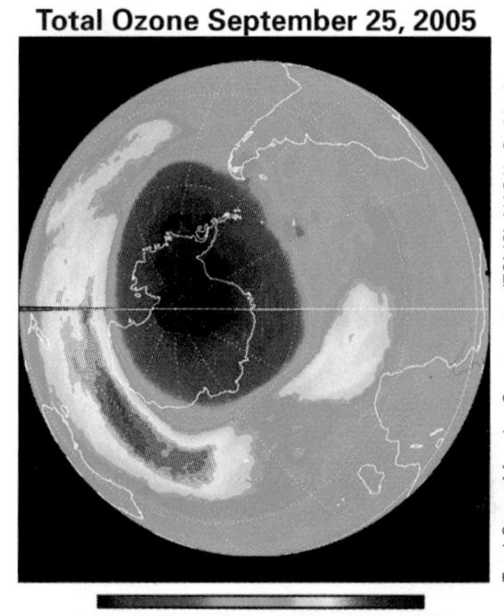

Total Ozone Mapping Spectrometer (TOMS), NASA's Goddard Space Flight Center

b

FIGURE 6–11 The amount of ozone (O_3) over Antarctica in the spring (September and early October for the Earth's southern hemisphere) is much lower (*blue and purple*) than the amount in other seasons. Note the outline of the Antarctic continent (*also shown in blue*) in this display of data from NASA's TOMS (Total Ozone Monitoring Spectrometer). (a) Almost two decades of seasonal decline in the total amount of ozone, shown as the spreading of the area in deep blue, the "ozone hole." (b) Images of the southern region of the Earth, showing the appearance of the ozone hole in recent years during southern-hemisphere spring. The ozone layer is thicker at other times of the year, including the lengthy period of darkness over the south polar region.

whether the amount of greenhouse warming is increasing, progressively raising the Earth's temperature. Such a phenomenon is known as "global warming."

Quite a separate problem is a discovery about the ozone (O_3) in our upper atmosphere. This ozone becomes thinner over Antarctica each Antarctic spring, a phenomenon known as the "ozone hole." The ozone hole apparently started forming only in the mid-1980s, but its maximum size has been larger almost every year since then (Fig. 6–11).

The ozone hole is caused by the interaction in the cold upper atmosphere of sunlight with certain gases we give off near the ground, such as chlorofluorocarbons used in air conditioners and refrigerators. International governmental meetings have arranged cutbacks in the use of these harmful gases. Some success has been achieved, but we have far to go to protect our future atmosphere.

AceAstronomy™ Log into AceAstronomy and select this chapter to see Astronomy Exercise "Primary Atmospheres."

AceAstronomy™ Log into AceAstronomy and select this chapter to see the Active Figure called "Planetary Atmospheres."

AceAstronomy™ Log into AceAstronomy and select this chapter to see Astronomy Exercise "The Greenhouse Effect."

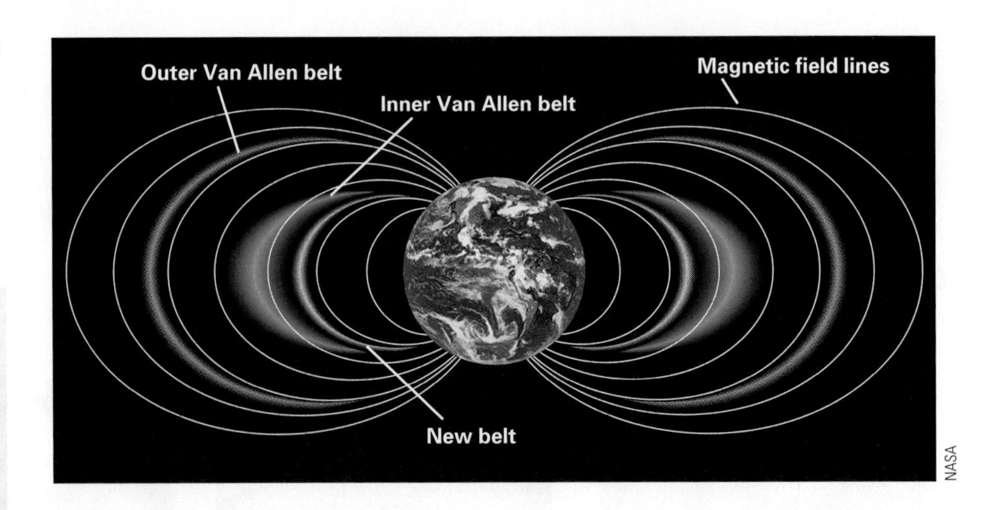

■ **FIGURE 6–12** The outer Van Allen belt contains mainly electrons; the traditional "inner belt" contains mainly protons; and the more recently discovered belt, which is inside the so-called "inner belt," contains mainly ions of oxygen, nitrogen, and neon, for which the distribution of isotopes indicates that they are from interstellar space.

ASIDE 6.2: Planetary magnetic fields

Every planet with a magnetic field—most notably Jupiter—has regions filled with charged particles. Earth's Van Allen belts were the first to be studied.

6.1e The Van Allen Belts

In 1958, the first American space satellite carried aloft, among other things, a device to search for particles carrying electric charge that might be orbiting the Earth. This device, under the direction of James A. Van Allen of the University of Iowa, detected a region filled with charged particles having high energies. Two such regions—the **Van Allen belts**—were found to surround the Earth, like a small and a large doughnut, containing protons and electrons (■ Fig. 6–12). They start a few hundred kilometers above the Earth's surface and extend outward to about 8 times the Earth's radius. A more recently discovered third, innermost belt contains mainly ions of heavier elements from interstellar space.

The particles in the Van Allen belts are trapped by the Earth's magnetic field. Charged particles preferentially move in the direction of magnetic-field lines, and not across the field lines. These particles, often from solar magnetic storms, are guided by the Earth's magnetic field toward the Earth's magnetic poles. When they interact with air molecules, they cause our atmosphere to glow, which we see as the beautiful northern and southern lights—the **aurora borealis** and **aurora australis,** respectively (■ Fig. 6–13).

AceAstronomy™ Log into AceAstronomy and select this chapter to see Astronomy Exercise "Auroras."

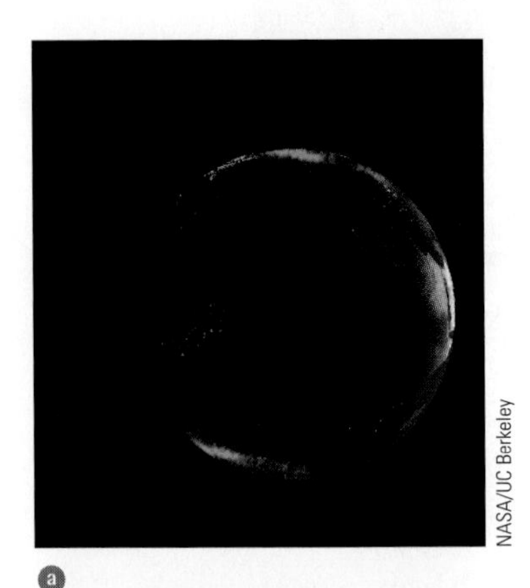

a

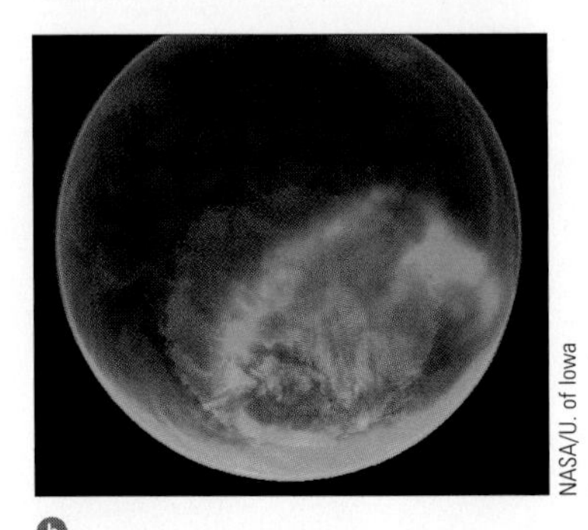

b

c

■ **FIGURE 6–13** a The aurora looking down from NASA's IMAGE (Imager for Magnetopause to Aurora Global Exploration) spacecraft on July 26, 2004. b The aurora borealis and the aurora australis (northern and southern lights) looking down from NASA's Polar spacecraft on July 27, 2004. These auroras followed a coronal mass ejection (see Chapter 10) from the Sun, which provided particles that travelled for a day and a half until they were trapped in Earth's magnetic field. c A ground-based view of an aurora with a meteor from the Perseid shower.

6.2 The Moon

The Earth's nearest celestial neighbor—the Moon—is only 380,000 km (238,000 miles) away from us, on the average. At this distance, it appears sufficiently large and bright to dominate our nighttime sky. The Moon's stark beauty has captured our attention since the beginning of history. Now we can study the Moon not only as an individual object but also as an example of a small planet or a large planetary satellite, since spacecraft observations have told us that there may be little difference between small planets and large moons.

■ **FIGURE 6–14** A waning moon (just past third quarter) showing dark lunar seas (maria) and lighter highlands.

6.2a The Moon's Appearance

Even binoculars reveal that the Moon's surface is pockmarked with craters. Other areas, called **maria** (pronounced mar'ee-a; singular **mare,** pronounced mar'eyh), are relatively smooth and dark. Indeed, the name comes from the Latin word for sea (■ Fig. 6–14). But there are no ships sailing on the lunar seas and no water in them; the Moon is a dry, airless, barren place.

The gravity at the Moon's surface is only one-sixth that of the Earth. Typically you would weigh only 20 or 30 pounds there if you stepped on a scale! The gravity is so weak that any atmosphere and any water that may once have been present would long since have escaped into space.

The Moon rotates on its axis at the same rate that it revolves around the Earth, thereby always keeping the same face in our direction. (To understand this idea, put a quarter on your desk and then slide a dime around it, keeping both flat on the desk and keeping the top of the head on the dime always on the side that is away from you. Notice that though the dime isn't rotating as seen from above, a viewer on the quarter would see the dime at different angles. Then move the dime around the quarter so that the same point on the dime always faces the quarter. Notice that as seen from above the dime rotates as it revolves around the quarter.) Over time, the Earth's gravity locked the Moon in this pattern, pulling on a bulge in the distribution of the lunar mass to prevent the Moon from rotating freely. As a result of this interlock (known as "synchronous rotation") we always see essentially the same side of the Moon from our vantage point on Earth.

ASIDE 6.3: The Moon's rotation rate

The Moon used to rotate about its axis more rapidly, but tides in the solid Moon raised by the Earth's gravity caused a gradual loss of rotation energy, until synchronous rotation was reached.

When the Moon is full, it is bright enough to cast shadows or even to read by. But a full moon is a bad time to try to observe lunar surface structure, for any shadows we see are short, and lunar features appear washed out. When the Moon is a crescent or even a quarter moon, however, the sunlighted part of the Moon facing us is covered with long shadows. The lunar features then stand out in bold relief. Shadows are longest near the **terminator,** the line that separates day from night. Note that nature photographers on Earth, concluding that views with shadows are more dramatic, generally take their best photos when the Sun is low.

Six teams of astronauts in NASA's Apollo program landed on the Moon in 1969–1972. (See *A Closer Look 6.3: The First People on the Moon.*) In some sense, before this period of exploration, we knew more about bright stars than we did about the Moon. As a relatively cold, solid body, the Moon reflects the spectrum of sunlight rather than emitting its own optical spectrum, so we were hard pressed to determine even the composition or the physical properties of the Moon's surface (such as whether you would sink into it!).

6.2b The Lunar Surface

The kilometers of film exposed by the astronauts, the 382 kg of rock brought back to Earth, the lunar seismograph data recorded on tape, meteorites from the Moon that have been found on Earth, and other sources of data have been studied by hundreds of

A Closer Look 6.3 The First People on the Moon

On July 20, 1969, Neil Armstrong and Buzz Aldrin left Michael Collins orbiting the Moon and descended to the lunar surface. After Armstrong stepped out of the spacecraft first and took that first step on the Moon, he, the first person on the Moon, made the famous statement, "That's one small step for man, one giant leap for mankind."

At the time of the 30th anniversary of the Moon landing, looking back, Aldrin said, "I was full of goose bumps when I stepped down onto the surface. I felt a bit disoriented because of the nearness of the horizon. On Earth when you look at the horizon it appears flat, but on the Moon, so much smaller and without hills, the horizon in all directions curved visibly down away from us.

"There was about every type of rock imaginable, all covered with a very fine powder. The rocks themselves actually had no color until you looked closely at the crystals. The thought briefly occurred to me that these rocks had been sitting there for hundreds of millions or billions of years, and that we were the first living beings to see them. But we were too busy to be philosophical for long or to study them closely, and we just grabbed what looked like an interesting assortment. I felt them crunch beneath my feet as we walked around." (*Griffith Observer* interview, July 1999)

ASIDE 6.4: Relative dating

The superposition of one crater on another—when one crater lies wholly or partially over the other—indicates that the underlying crater formed first and is therefore older.

ASIDE 6.5: Radioactive dating

This concept, known as "radioactive dating," is widely used, not only for dating rocks on the Moon, but also for dating rocks on Earth and for finding the ages of archaeological sites, of ancient pottery, and so on. It was even used to date the age of the anthrax that infected people in 2001.

scientists from all over the world. (Meteorites are rocks from space that have landed on Earth; see Chapter 8 for more details.) The data have led to new views of several basic questions, and have raised many new questions about the Moon and the Solar System.

The rocks that were encountered on the Moon are types that are familiar to terrestrial geologists (■ Fig. 6–15). Almost all the rocks are the kind that were formed by the cooling of lava, known as igneous rocks. Basalts are one example.

The Moon and the Earth seem to be similar chemically, though significant differences in overall composition do exist. Some elements that are rare on Earth—such as uranium and thorium—are found in greater abundances on the Moon. (Will we be mining on the Moon one day?) None of the lunar rocks contain any trace of water bound inside their minerals.

Meteoroids, interplanetary rocks that we will discuss in Chapter 8, hit the Moon with such high speeds that huge amounts of energy are released at the impact. The effect is that of an explosion, as though TNT or an H-bomb had exploded. As a result of the Apollo missions, we know that almost all the craters on the Moon come from such impacts. One way of dating the surface of a moon or planet is to count the number of craters in a given area, a method that was used before Apollo. Surely those locations with the greatest number of craters must be the oldest. Relatively smooth areas—like maria—must have been covered over with molten volcanic material at some relatively recent time (which is still billions of years ago, though).

a

b

c

■ **FIGURE 6–15** Different types of lunar rocks brought back to Earth by the Apollo 15 astronauts. A 1-cm block appears for scale. (a) A rock that is a compound of other types of rock. It weighs 1.5 kg. (b) The dark part is dust that covered another rock to which this rock was attached. (c) A 4.5-kg compound rock.

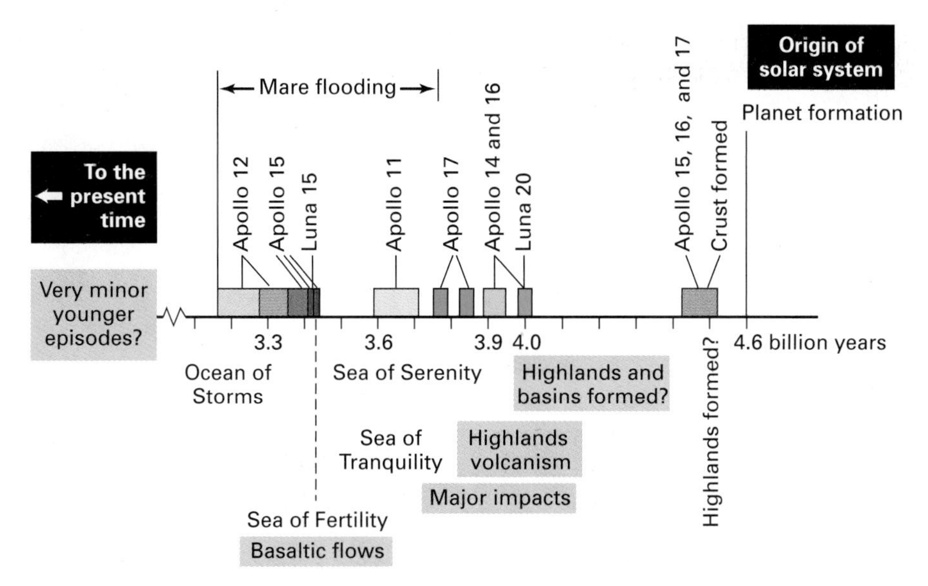

■ **FIGURE 6–16** The chronology of the formation of the lunar surface. The ages of rocks found in 8 missions are shown. Note how many crewed American Apollo missions and uncrewed Soviet Luna missions it took to get a sampling of many different ages on the lunar surface. (Gerald Wasserburg, Caltech)

Obvious rays of lighter-colored matter splattered outward during the impacts that formed a few of the craters. Since these rays extend over other craters, the craters with rays must have formed more recently. The youngest rayed craters may be very young indeed—perhaps only a few hundred million years. The rays darken with time, so rays that may have once existed near other craters are now indistinguishable from the rest of the surface.

Crater counts and the superposition of one crater on another give only relative ages. We found the absolute ages only when rocks from the Moon were physically returned to Earth. Scientists worked out the dates by comparing the current ratio of radioactive forms of atoms to nonradioactive forms present in the rocks with the ratio that they would have had when they were formed. (Varieties of the chemical elements having different numbers of neutrons are known as "isotopes," and radioactive isotopes are those that decay spontaneously; that is, they change into other isotopes even when left alone. Stable isotopes remain unchanged. For certain pairs of isotopes—one radioactive and one stable—we know the proportion of the two when the rock was formed. Since we know the rate at which the radioactive one is decaying, we can calculate how long it has been decaying from a measurement of what fraction is left.) The oldest rocks that were found on the Moon solidified 4.4 billion years ago. The youngest rocks ever found solidified 3.1 billion years ago.

The observations can be explained on the basis of the following general sequence (■ Fig. 6–16): The Moon formed about 4.6 billion years ago. From the oldest rocks, we know that at least the surface of the Moon was molten about 200 million years later. Then the surface cooled. From 4.2 to 3.9 billion years ago, bombardment by interplanetary rocks caused most of the craters we see today. About 3.8 billion years ago, the interior of the Moon heated up sufficiently (from radioactive elements inside) that volcanism began. Lava flowed onto the lunar surface and filled the largest basins that resulted from the earlier bombardment, thus forming the maria (■ Fig. 6–17). By 3.1 billion years ago, the era of volcanism was over. The Moon has been geologically pretty quiet since then.

Up to this time, the Earth and the Moon shared similar histories. But active lunar history stopped about 3 billion years ago, while the Earth continued to be geologically active. Almost all the rocks on the Earth are younger than 3 billion years of age; the oldest single rock ever discovered on Earth has an age of 4.5 billion years, but few

ASIDE 6.6: Comparative planetology: atmospheres

Venus = 90 × Earth
Mars = Earth/160
Mercury = negligible

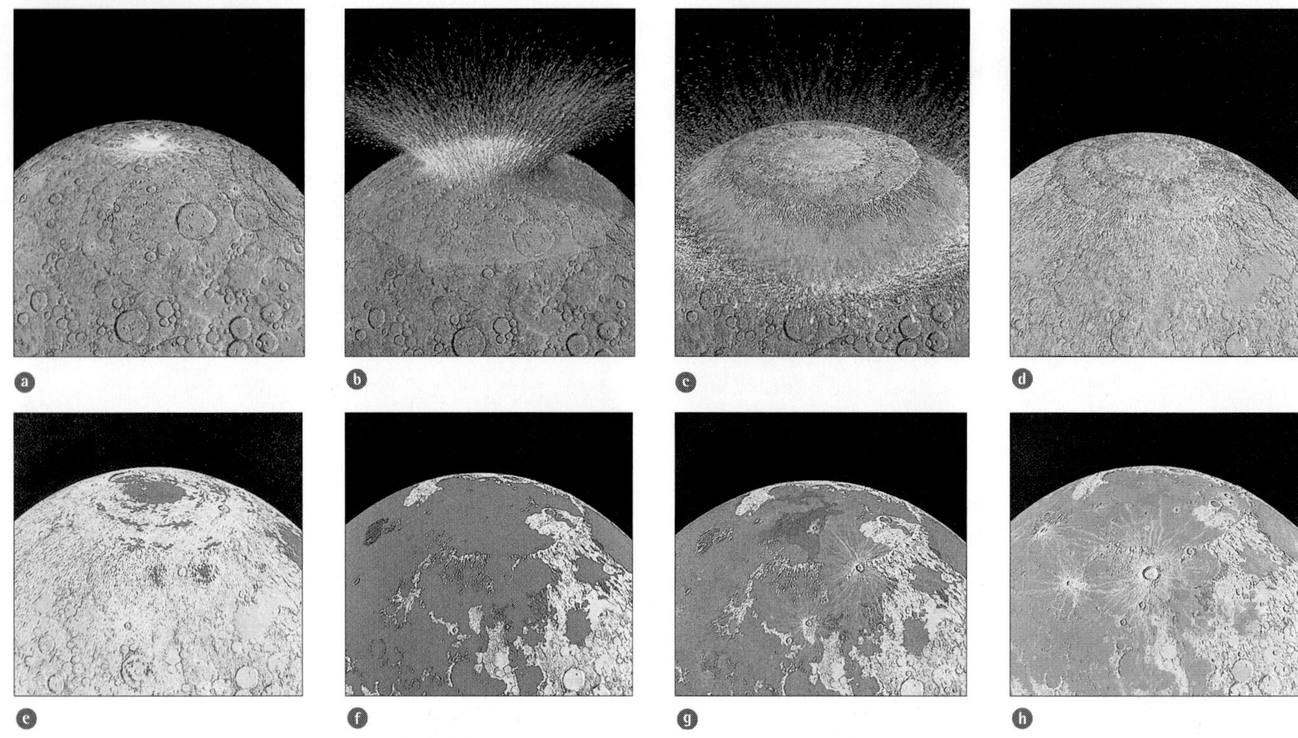

■ **FIGURE 6–17** An artist's view of the formation of the Mare Imbrium region of the lunar surface. (a) An asteroid impact on the Moon, sometime between 3.85 and 4.0 billion years ago. (b) The shock of the asteroid impact began the Imbrium crater. (c) As the dust and heat subsided, the 1300-km Imbrium crater was left. (d) The molten rock (lava) flowed over outlying craters and cooled, leaving lunar mountains. (e) Lava welled up from inside the Moon 3.8 billion years ago. It filled the basin. (f) By 3.3 billion years ago, the lava flooding was nearly complete. (g) The final flow of thick lava came 2.5–3.0 billion years ago. (h) Subsequent cratering has left the Mare Imbrium of today's Moon. (Drawings by Donald E. Davis under the guidance of Don E. Wilhelms of the U.S. Geological Survey; reproduced with the assistance of the Hansen Planetarium)

U.S. Naval Research Laboratory and NASA

U.S. Naval Research Laboratory and NASA

■ **FIGURE 6–18** (a) The far side of the Moon. (b) An oblique view of part of the lunar far side, from Clementine, shows how rough it looks.

such old rocks have been found. Erosion and the remolding of the continents as they move slowly over the Earth's surface have taken their toll. So we must look to extraterrestrial bodies—the Moon or meteorites—that have not suffered the effects of plate tectonics or erosion (which occurs in the presence of water or an atmosphere) to study the first billion years of the Solar System.

Not until the 1990s did spacecraft revisit the Moon. The Clementine spacecraft (named after the prospector's daughter in the old song, since the spacecraft was looking for minerals) took photographs and other measurements. Photographs of the far side of the Moon (■ Fig. 6–18) have shown us that the near and far hemispheres are quite different in overall appearance. The maria, which are so conspicuous on the near side, are almost absent from the far side, which is cratered all over. We shall see in the next section that the difference probably results from the different thicknesses of the lunar crust on the sides of the Moon nearest Earth and farthest from Earth. The difference was first seen in the fuzzy photographs of the far side that were taken by the Soviet Lunik 3 Spacecraft in 1959.

In the 1990s, NASA's Lunar Prospector and Clementine spacecraft mapped the Moon with a variety of instruments (■ Fig. 6–19). Lunar Prospector confirmed indications from the Clementine spacecraft that there is likely to be water ice on the Moon, by detecting more neutrons coming from the Moon's polar regions than elsewhere. Clementine and Lunar Prospector scientists think that these neutrons are given off in interactions of particles coming from the Sun with hydrogen in water ice in craters near the lunar poles, where they are shaded from the Sun's rays.

But the detection is not of water directly, and Apollo 17 astronaut Harrison Schmitt, the only geologist ever to have walked on the Moon, told one of the authors

(J.M.P.) in 1999 that the neutrons may instead have come from the solar wind (■ Fig. 6–20), though as of 2005 most scientists do not agree. He would love to find water there, because it would increase the chance that a crewed Moon base could be supplied on the Moon itself, much easier than bringing everything from Earth. In his estimation, it would take about 10 years to set up such a base, once basic funding is available on Earth. An attempt to test the idea of water in the crater was made by crashing Lunar Prospector into the most likely spot in 1999, but none of the spectrographs on Earth viewing the event detected any signal from the crash.

The European Space Agency has a spacecraft in orbit around the moon. Their SMART-1 (Small Mission for Advanced Research and Technology) spacecraft is basically meant to test new technology, but they may as well test it in lunar orbit. It uses, for the first time, an electric engine, in which a weak but steady puff of ions ejected out the back of the spacecraft accelerated the spacecraft at a slow but steady rate, eventually bringing it into its final lunar orbit in 2005. It is sending back images that include perpetually shaded polar regions where water ice may be found (■ Fig. 6–21).

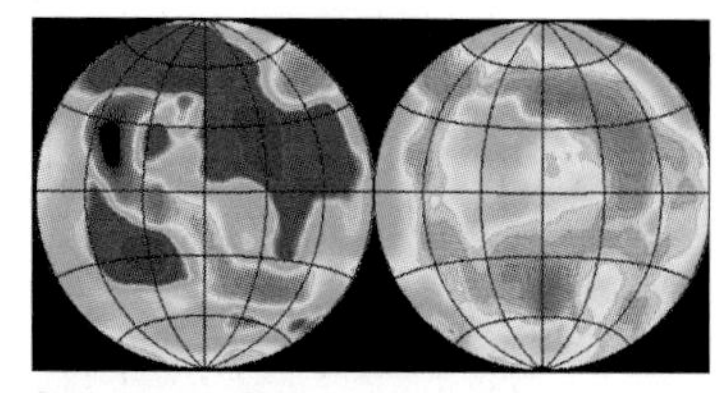

NASA

a

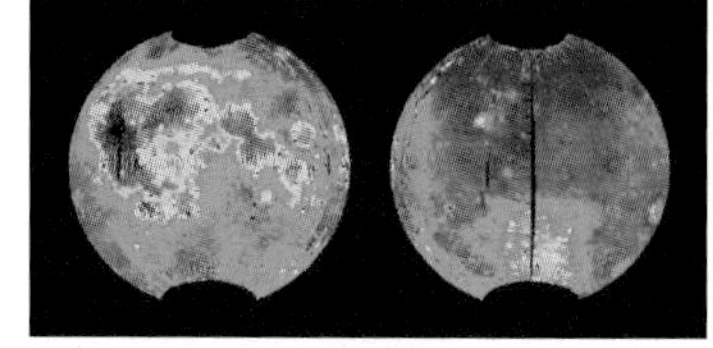

U.S. Naval Research Laboratory and NASA

b

■ **FIGURE 6–19** **a** A Lunar Prospector map of the distribution of gravity on the Moon, showing hemispheres of the near side and the far side. **b** A Clementine map of the distribution of iron on the Moon's surface.

a

b

ESA/SMART-1/SPACE-X Space Exploration Institute

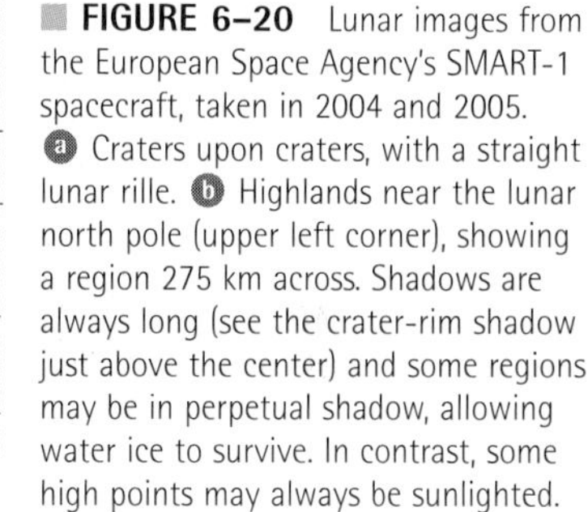

■ **FIGURE 6–20** Lunar images from the European Space Agency's SMART-1 spacecraft, taken in 2004 and 2005. **a** Craters upon craters, with a straight lunar rille. **b** Highlands near the lunar north pole (upper left corner), showing a region 275 km across. Shadows are always long (see the crater-rim shadow just above the center) and some regions may be in perpetual shadow, allowing water ice to survive. In contrast, some high points may always be sunlighted.

NASA

■ **FIGURE 6–21** Scientist-astronaut Harrison Schmitt, the only scientist to have stood on the Moon, during the Apollo 17 mission. The presence of the giant rock indicates that the later Apollo missions went to less-smooth sites than the earlier ones.

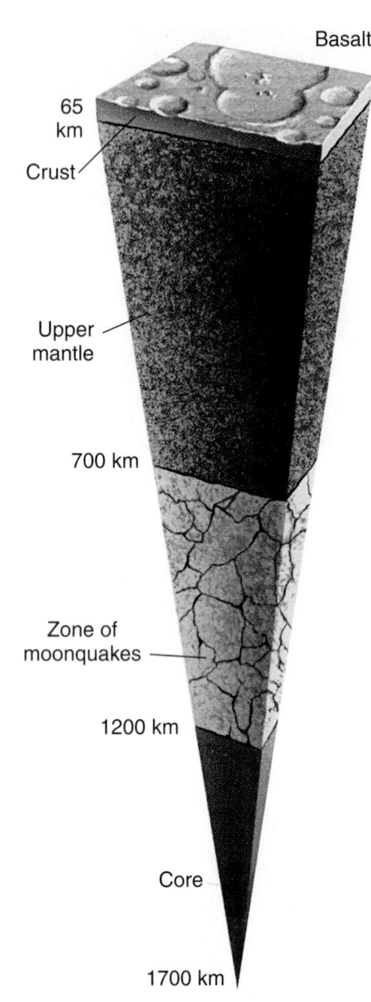

■ **FIGURE 6–22** The Moon's interior. The depth of solidified lava is greater under maria, which are largely on the side of the Moon nearest the Earth.

ASIDE 6.7: Comparative planetology: greenhouse effect

Earth: 33°C extra
Venus: 375°C extra
Mercury: negligible
Mars: negligible

AceAstronomy™ Log into AceAstronomy and select this chapter to see Astronomy Exercise "Cratering."

AceAstronomy™ Log into AceAstronomy and select this chapter to see the Active Figure called "The Moon's Craters."

6.2c The Lunar Interior

Before the Moon landings, it was widely thought that the Moon was a simple body, with the same composition throughout. But we now know it to be differentiated (■ Fig. 6–22), like the planets. Most experts believe that the Moon's core is molten, but the evidence is not conclusive. The lunar crust is perhaps 65 km thick on the near side and twice as thick on the far side. This asymmetry may explain the different appearances of the sides, because lava would be less likely to flow through the far side's thicker crust.

The Apollo astronauts brought seismic equipment to the Moon (■ Fig. 6–23). One type of earthquake wave moves material to the side. Only if the Moon's core were solid would the material move and then return to where it started, continuing the wave. Since that type of wave doesn't come through the Moon, scientists deduce that the Moon's core is probably molten. New computer analysis methods were applied in 2004 to the old data, and thousands of additional moonquakes were discovered.

Tracking the orbits of the Apollo Command Modules and more recently the Clementine and Lunar Prospector spacecraft that orbited the Moon told us about the lunar interior. If the Moon were a perfect, uniform sphere, the spacecraft orbits would have been perfect ellipses. But they weren't. One of the major surprises of the lunar missions of the 1960s, refined more recently, was the discovery in this way of **mascons,** regions of **mass con**centrations near and under most maria. The mascons may be lava that is denser than the surrounding matter, providing a stronger gravitational force on satellites passing overhead.

We also find out about the lunar interior by bouncing powerful laser beams off the Moon's surface. This "laser ranging" (where "ranging" means finding distances) uses several sets of reflectors left on the Moon by the Apollo astronauts. The laser-ranging programs find the distance to the Moon to within a few centimeters—pretty good for an object about 400,000 km away! Variations in the distance result in part from conditions in the lunar interior.

■ **FIGURE 6–23** Buzz Aldrin and the experiments he deployed on Apollo 11. In the foreground, we see the seismic experiment. The laser-ranging retroreflector is behind it.

6.2d The Origin of the Moon

The leading models for the origin of the Moon that were considered at the time of the Apollo missions were as follows.

1. *Fission.* The Moon was separated from the material that formed the Earth; the Earth spun up and the Moon somehow spun off;
2. *Capture.* The Moon was formed far from the Earth in another part of the Solar System, and was later captured by the Earth's gravity; and
3. *Condensation.* The Moon was formed near to (and simultaneously with) the Earth in the Solar System.

But work over the last three decades has all but ruled out the first two of these and has made the third seem less likely. The model now strongly favored, especially because of computer simulations, is

4. *Ejection of a Gaseous Ring.* A planet-like body perhaps twice the size of Mars hit the young Earth, ejecting matter in gaseous (and perhaps some in liquid or solid) form (■ Fig. 6–24). Although some of the matter fell back to Earth, and part escaped entirely, a significant fraction started orbiting the Earth, probably in the same direction as the initial incoming body. The orbiting material eventually coalesced to form the Moon.

Comparing the chemical composition of the lunar surface with the composition of the Earth's surface has been important in narrowing down the possibilities. The mean lunar density of 3.3 grams/cm^3 is close to the average density of the Earth's major upper region (the mantle), and the Moon seems especially deficient in iron. This fact favors the fission hypothesis; had the Moon condensed from the same material as Earth, as in the condensation scenario, it would contain much more iron.

However, detailed examination of the lunar rocks and soils indicates that the abundances of elements on the Moon and Earth's mantle are sufficiently different from each other to indicate that the Moon did not form directly from the Earth. In the collision hypothesis, on the other hand, such differences are expected because of contaminant material from the impactor.

Calculations considering angular momentum (recall the discussion of this concept in Chapter 5) strongly suggest that the fission mechanism doesn't work. Plate-tectonic theory now explains the formation of the Pacific Ocean basin. Before the ejection-of-a-ring theory was considered most probable, it seemed possible that the Pacific Ocean could be the scar left behind when the Moon was ripped from the Earth according to the fission hypothesis.

We are obtaining additional evidence about the capture model by studying the moons of Jupiter and Saturn. The outermost moon of Saturn, for example, is apparently a captured asteroid (small bodies orbiting the Sun, mostly between Mars and Jupiter; see Chapter 8). However, the similarities in composition between the Moon's

ASIDE 6.8: Comparative planetology: temperature at surface

Earth: about 0°C to 40°C (just right)

Venus: about 475°C (much too hot)

Mercury: about −175°C to 425°C (much too cold to much too hot)

Mars: about −125°C to 25°C (too cold to just right)

ASIDE 6.9: Spacecraft to the Moon

NASA is planning a robotic Lunar Reconnaissance Orbiter for 2008 and a robotic lunar lander for 2010. The lander is to search for water ice in the permanently shady parts of craters near the Moon's poles. India is planning to launch their Chandrayan-1 into a lunar polar orbit in 2007 or 2008. It will map the Moon using at least the set of ESA instruments identical to those now orbiting on SMART-1. A 2007 Chinese orbiter is planned.

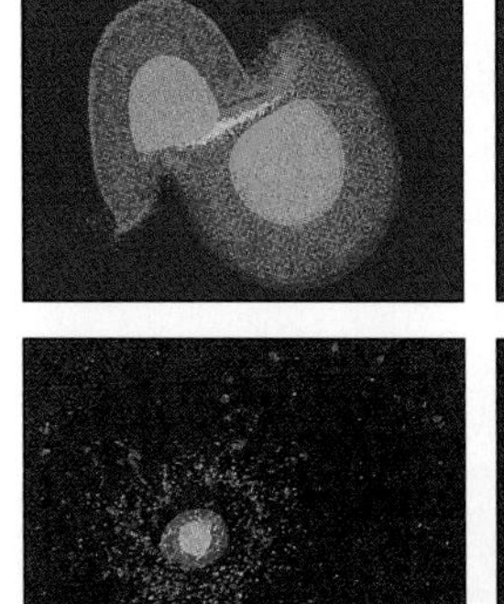
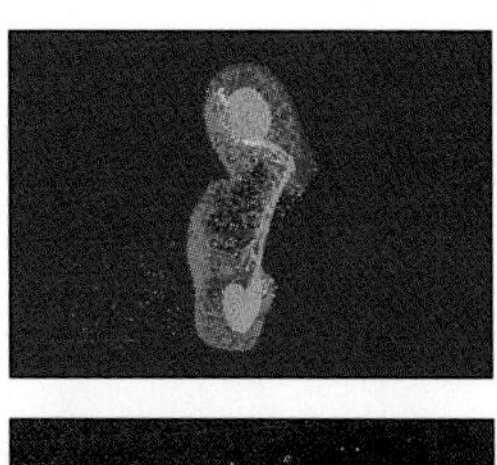
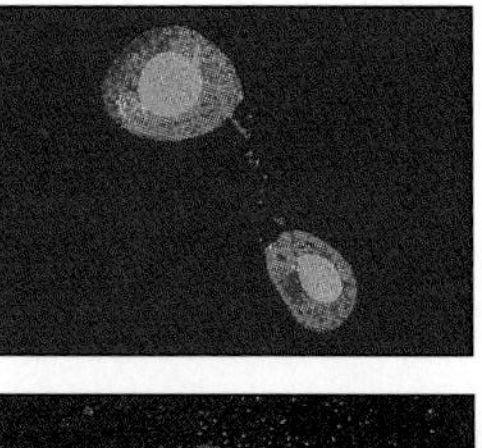
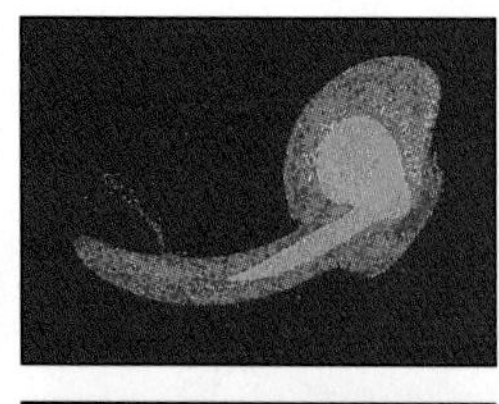
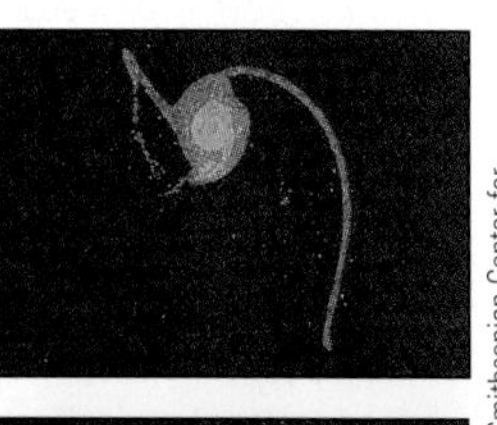
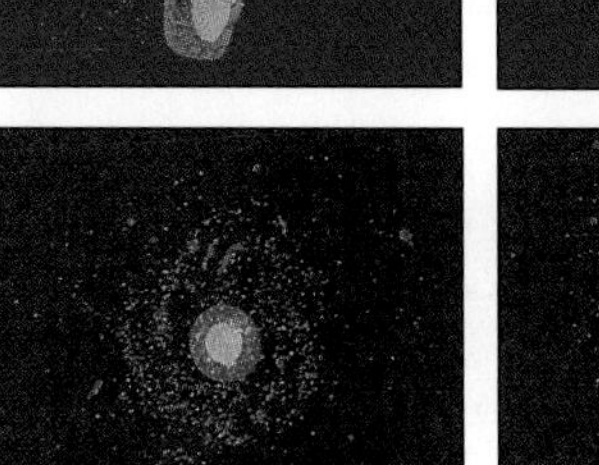
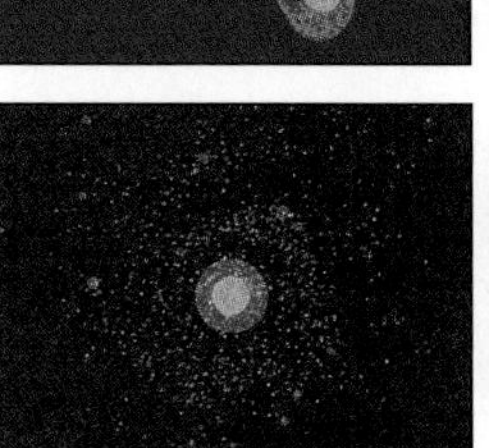
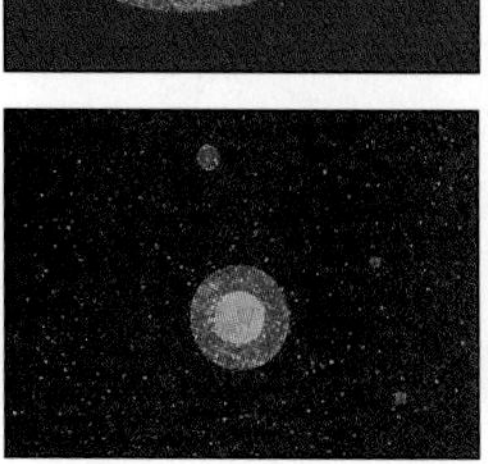

A. G. W. Cameron, Harvard-Smithsonian Center for Astrophysics

■ **FIGURE 6–24** A computer simulation of a collision between the young Earth and an impactor. Each is composed of an iron core and a rock mantle. The internal energy in each increases with both temperature and pressure. Increasing internal energy is shown as red, brown, and yellow for rock, and blue and light green for iron. In the first images, the impactor hits the young Earth, separates, and then falls in again, pulling out a tail of matter. At the end of this sequence, which takes six days, several large clumps and many smaller clumps are left. This material then clumps together to form the Moon.

NASA/Johnson Space Center

■ **FIGURE 6–25** This rock is one of many meteorites found in the Earth's continent of Antarctica. Under a microscope, it seems like a sample from the lunar highlands and quite unlike any terrestrial rock. Eighteen of the known meteorites are thought to have come from the Moon.

and the Earth's mantles argue against the capture hypothesis for the Earth–Moon system. Thus as of now, ejection of a gaseous ring is the most accepted model.

6.2e Rocks from the Moon

A handful of meteorites found in Antarctica, Australia, and Africa have been identified by their chemical composition as having come from the Moon (■ Fig. 6–25). They presumably were ejected from the Moon when craters formed. So we are still getting new moon rocks to study! A few other meteorites have even been found to come from Mars, as we shall discuss later in this chapter.

6.3 Mercury

Mercury is the innermost of our Sun's nine planets. Its average distance from the Sun is $^4/_{10}$ of the Earth's average distance, or 0.4 A.U. Except for distant Pluto, its elliptical orbit around the Sun is the most elongated (eccentric).

Since we on the Earth are outside Mercury's orbit looking in at it, Mercury always appears close to the Sun in the sky (■ Fig. 6–26). At times Mercury rises just before sunrise, and at times it sets just after sunset, but it is never up when the sky is really dark. The Sun always rises or sets within an hour or so of Mercury's rising or setting. As a result, whenever Mercury is visible, its light has to pass obliquely through the Earth's atmosphere. This long path through turbulent air leads to blurred images. Thus astronomers have never had a really clear view of Mercury from the Earth, even with the largest telescopes. Even the best photographs taken from the Earth show Mercury as only a fuzzy ball with faint, indistinct markings. Most people have never seen it at all.

On rare occasions, Mercury goes into **transit** across the Sun; that is, we see it as a black dot crossing the Sun. Transits of Mercury occurred in 1999 (■ Fig. 6–27) and 2003. The next transit of Mercury will occur on November 28, 2006. The entire transit will be visible from the U.S.'s west coast, and the Sun will set during the transit for observers on the east coast. Understanding what we see as Mercury transits helps us understand the much rarer transits of Venus, which we will discuss later in this chapter.

6.3a The Rotation of Mercury

From studies of ground-based drawings and photographs, astronomers did as well as they could to describe Mercury's surface. A few features seemed to be barely distinguished, and the astronomers watched to see how long those features took to rotate around the planet. From these observations they decided that Mercury rotates in the

■ **FIGURE 6–26** Since Mercury's orbit is much closer to the Sun than ours ⓐ, Mercury is never seen against a really dark sky. As a result, even Copernicus apparently never saw it. A view from the Earth appears in ⓑ, showing Mercury and Venus at their greatest respective distances from the Sun. ⓒ Venus, with Mercury slightly to its left, both diagonally below the crescent Moon soon after sunset, in a photograph taken in 2005.

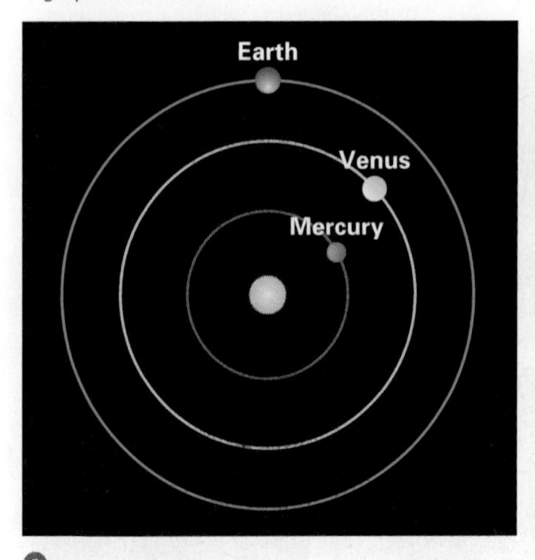

ⓐ

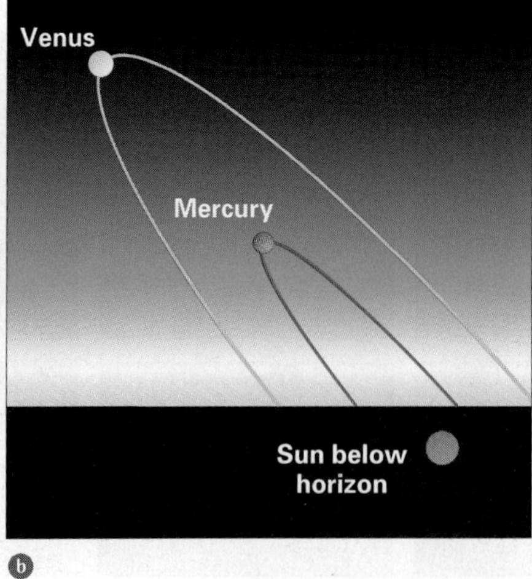

ⓑ

Jay M. Pasachoff

ⓒ

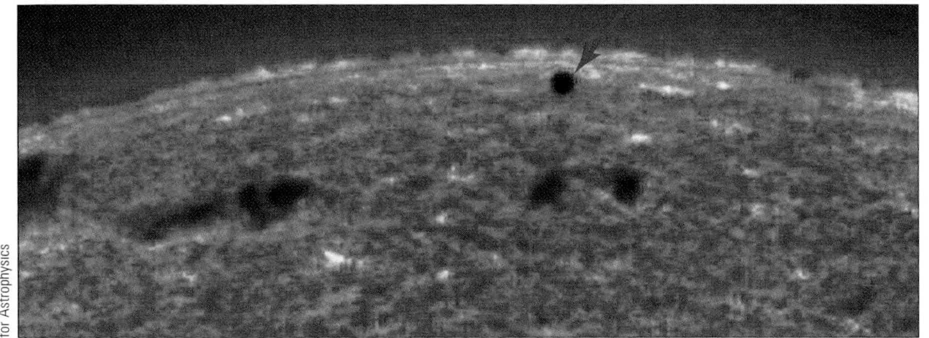

■ **FIGURE 6–27** Mercury silhouetted against the Sun during its 1999 transit, observed from the solar spacecraft known as TRACE. (Other dark spots are sunspots; see Chapter 10.) One of the authors (J.M.P.) participated in analysis of these observations in order to prepare for the much rarer transit of Venus that occurred in 2004 (Section 6.4a).

same amount of time that it takes to revolve around the Sun in its orbit, 88 Earth days. Thus they thought that one side always faces the Sun and the other side always faces away from the Sun. (Recall that one side of the Moon always faces the Earth, a similar phenomenon.) This discovery led to the fascinating conclusion that Mercury could be both the hottest planet (on the Sun-facing side) and the coldest planet (on the other side) in the Solar System.

But when the first measurements were made of Mercury's radio radiation, the planet turned out to be giving off more energy than had been expected. This meant that it was hotter than expected. The dark side of Mercury was too hot for a surface that was always in the shade. (The visible light we see is merely sunlight reflected by Mercury's surface and doesn't tell us the surface's temperature. The radio waves are actually being emitted by the surface, as part of its "thermal" or nearly black-body radiation; see the discussion in Chapter 2.)

Later, we became able to transmit radar from Earth to Mercury. (Radar—**ra**dio **d**etection **a**nd **r**anging—is sending out radio waves so that they bounce off another object, allowing you to study their reflection.) Since one edge of the visible face of Mercury is rotating toward Earth, while the other edge is rotating away from Earth, the reflected radio waves were slightly smeared in wavelength according to the Doppler effect (recall the description of this effect in Chapter 2). This measurement allowed astronomers to determine Mercury's rotation speed, similarly to the way that radar is used by the police to tell if a car is breaking the speed limit. Knowing the rotation speed and Mercury's radius, we could determine the rotation period.

The results were a surprise: it actually rotates every 59 days, not 88 days. Mercury's 59-day period of rotation with respect to the stars is exactly $^2/_3$ of the 88-day orbital period, so the planet rotates three times for every two times it revolves around the Sun.

Mercury's rotation and revolution combine to give a value for the rotation of Mercury relative to the Sun (that is, a mercurian solar day) that is neither 59 nor 88 days long (■ Fig. 6–28). If we lived on Mercury we would measure each "day" (that is,

■ **FIGURE 6–28** Follow, from A to G, the arrow that starts facing to the right, toward the Sun in A. Mercury rotates once with respect to the stars in 59 days, when Mercury has moved only $^2/_3$ of the way around the Sun (E). Note that after one full revolution of Mercury around the Sun (G), the arrow faces away from the Sun. It takes another full revolution, a second 88 days, for the arrow to again face the Sun. Thus Mercury's rotation period with respect to the Sun is twice 88, or 176, days.

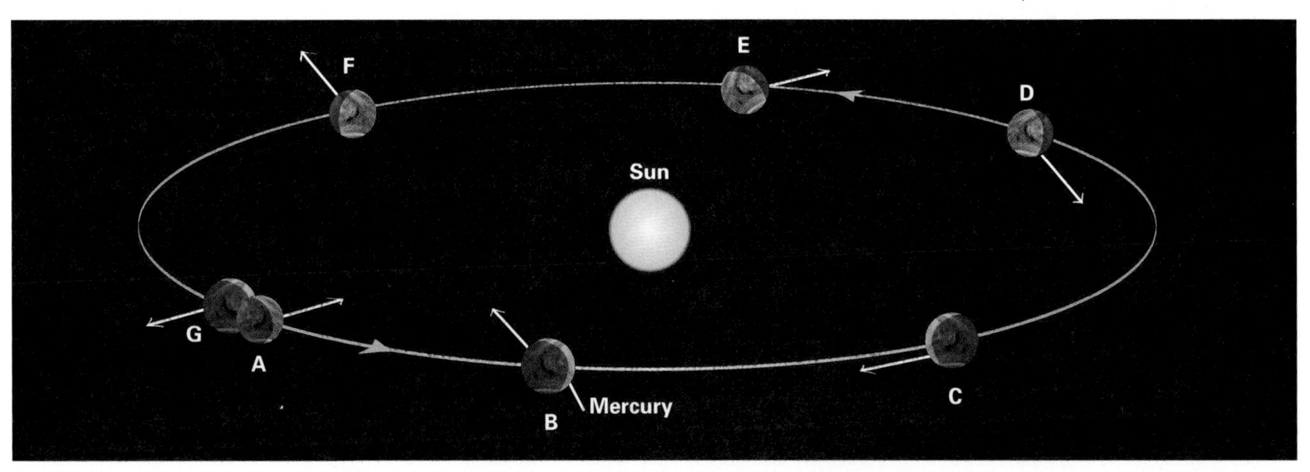

ASIDE 6.10: Comparative planetology: moons

Mercury: none

Venus: none

Mars: 2, only about 20 km diameter each

Earth: 1, of 3476 km diameter

each day/night cycle) to be 176 Earth days long. We would alternately be fried for 88 Earth days and then frozen for 88 Earth days. Since each point on Mercury faces the Sun at some time, the heat doesn't build up forever at the place under the Sun, nor does the coldest point cool down as much as it would if it never received sunlight. The hottest temperature is about 700 K (800°F). The minimum temperature is about 100 K (−280°F).

No harm was done by the scientists' original misconception of Mercury's rotational period, but the story teaches all of us a lesson: we should not be too sure of so-called "facts." Don't believe everything you read in this book, either. It should be fun for you to look back in 20 years and see how much of what we now think we know about astronomy actually turned out to be wrong. After all, science is a dynamic process.

6.3b Mercury Observed from the Earth

Even though the details of the surface of Mercury can't be seen very well from the Earth, other properties of the planet can be better studied. For example, we can measure Mercury's **albedo,** the fraction of the sunlight hitting Mercury that is reflected (■ Fig. 6–29). We can measure the albedo because we know how much sunlight hits Mercury (we know the brightness of the Sun and the distance of Mercury from the Sun). Then we can easily calculate at any given time how much light Mercury reflects, knowing how bright Mercury looks to us and its distance from the Earth. Comparison with the albedoes of materials on the Earth and on the Moon can teach us something of what the surface of Mercury is like.

Let us consider some examples of albedo. An ideal mirror reflects all the light that hits it; its albedo is thus 100 per cent. (The very best real mirrors have albedoes of as much as 96 per cent.) A black cloth reflects essentially none of the light that hits it; its albedo (in the visible part of the spectrum, anyway) is almost 0 per cent. Mercury's overall albedo is only about 10 per cent. Its surface, therefore, must be made of a dark (that is, poorly reflecting) material (though a few regions are very reflective, with albedoes of up to 45 per cent). The albedoes of the Moon's maria are similarly low, 6 to 10 per cent. In fact, Mercury (or the Moon) appears bright to us only because it is contrasted against a relatively dark sky; if it were silhouetted against a white bedsheet, it would look relatively dark.

6.3c Mercury from Mariner 10

In 1974, we learned most of what we know about Mercury in a brief time. We flew right by it. (See *A Closer Look 6.4: Naming the Features of Mercury.*) The tenth in the series of Mariner spacecraft launched by the United States went to Mercury with a variety of instruments on board. First the 475-kg spacecraft passed by Venus and had

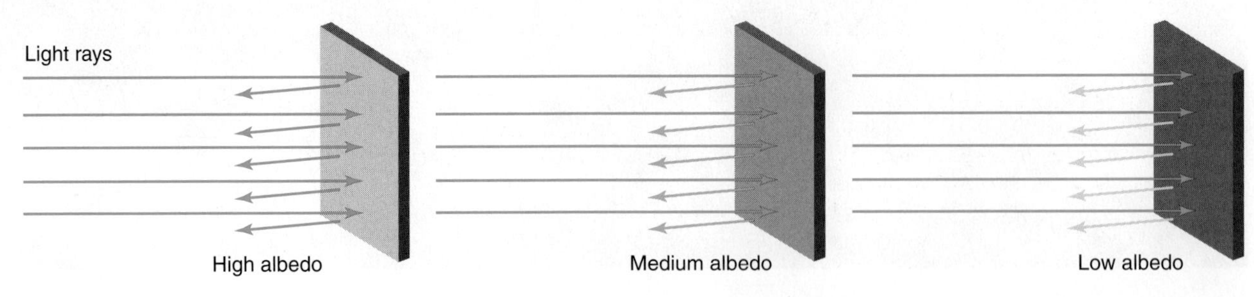

■ **FIGURE 6–29** Albedo is the fraction of radiation reflected. A surface of low albedo looks dark.

A Closer Look 6.4 NAMING THE FEATURES OF MERCURY

Seeing features on the surface of Mercury for the first time led to an immediate need for names. The scarps were named for historical ships of discovery and exploration, such as Endeavour (Captain Cook's ship, with which he went to Tahiti to see the transit of Venus and wound up also exploring what we now call Australia and New Zealand), Santa Maria (Columbus's ship), and Victoria (the first ship to sail around the world, which it did from 1519 to 1522 under Magellan and his successors). Some plains were given the names of the gods equivalent to the Greek's Mercury, such as Tir (in ancient Persian), Odin (an ancient Norse god), and Suisei (Japanese). Craters were named for nonscientific authors, composers, and artists, in order to complement the lunar naming system, which honors scientists.

its orbit changed by Venus's gravity to direct it to Mercury. Tracking its orbit improved our measurements of the gravity of these planets and thus of their masses.

The most striking overall impression is that Mercury is heavily cratered (■ Fig. 6–30). At first glance, it looks like the Moon! But there are several basic differences between the features on the surface of Mercury and those on the lunar surface. Mercury's craters seem flatter than those on the Moon, and they have thinner rims (■ Fig. 6–31). Mercury's higher gravity at its surface may have caused the rims to slump more. Also, Mercury's surface may have been softer, more plastic-like, when most of the cratering occurred. The craters may have been eroded by any of a number of methods, such as the impacts of meteorites or micrometeorites (large or small bits of interplanetary rock). Alternatively, erosion may have occurred during a much earlier period when Mercury may have had an atmosphere, undergone internal activity, or been flooded by lava.

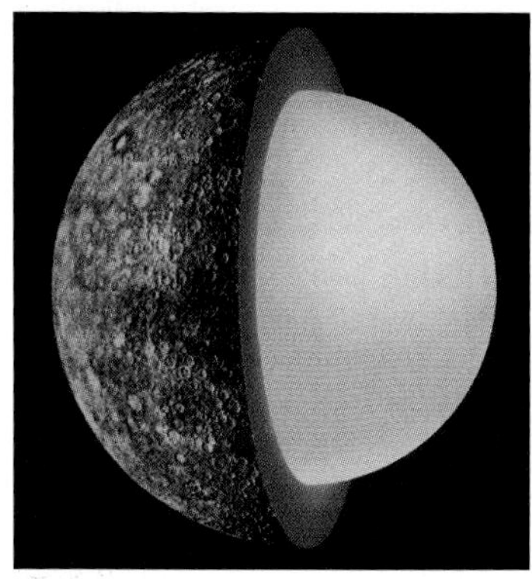
Courtesy of Robert G. Strom from his *Mercury: The Elusive Planet* (Smithsonian Institution Press), artwork by Karen Denomy

■ **FIGURE 6–30** A mosaic of photographs of Mercury from Mariner 10, showing its cratered surface. An artist has added a conception of Mercury's thick core (yellow) and thin crust/upper mantle (red).

Most of the craters seem to have been formed by impacts of meteorites. The Caloris Basin, in particular, is the site of a major impact. The secondary craters, caused by material ejected as primary craters were formed, are closer to the primaries than on the Moon, presumably because of Mercury's higher surface gravity. In many areas, the craters appear superimposed on relatively smooth plains. The plains are so extensive that they are probably volcanic. Their age is estimated to be 4.2 billion years, the oldest features on Mercury.

Smaller, brighter craters are sometimes, in turn, superimposed on the larger craters and thus must have been made afterward. Some craters have rays of higher

a

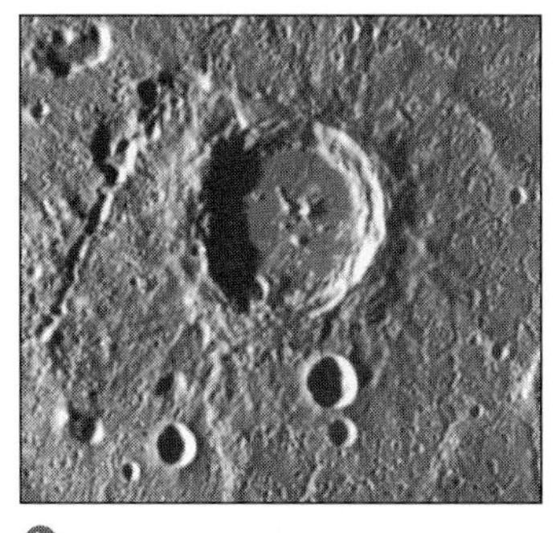
b

c

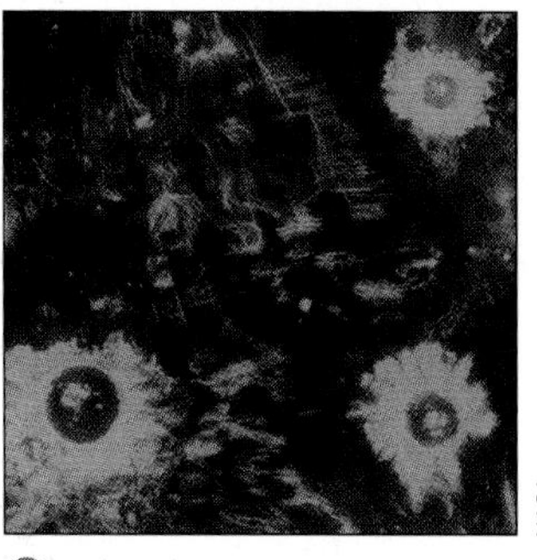
d

NASA

■ **FIGURE 6–31** Comparative planetology: A comparison of (a) craters on Mercury, (b) lunar craters, (c) craters on Mars, and (d) craters on Venus. Though it is not noticeable on these images, material has been thrown out less far on Mercury than on the Moon because of Mercury's higher gravity. Craters on Mars and Venus show flows across their surfaces, possibly resulting from the melting of a buried ice-rich layer and lava flows, respectively. The photographs are dominated by Copernicus on the Moon, 93 km in diameter; a similarly sized flat-bottomed crater on Mercury; Cerulli on Mars, 115 km in diameter; and craters on Venus, 37 to 50 km in diameter.

NASA/JPL/Caltech

■ **FIGURE 6–32** A field of rays on Mercury radiating from a pair of craters, each of which is about 40 km across.

albedo emanating from them (■ Fig. 6–32), just as some lunar craters do. The ray material represents relatively recent crater formation (that is, within the last hundred million years). The ray material must have been tossed out in the impact that formed the crater.

Lines of cliffs hundreds of miles long are visible on Mercury; on Mercury, as on Earth, such lines of cliffs are called **scarps.** The scarps are particularly apparent in the region of Mercury's south pole (■ Fig. 6–33). Unlike fault lines on the Earth, such as California's famous San Andreas fault (responsible for the 1906 San Francisco earthquake), on Mercury there are no signs of geologic tensions like rifts or fissures nearby. These scarps are global in scale, not just isolated. The scarps may actually be wrinkles in Mercury's crust. Perhaps Mercury was once molten, and shrank by 1 or 2 km as it cooled. This shrinking would have caused the crust to buckle, creating the scarps in the quantity that we now observe.

Judging by the fact that Mercury's average density is about the same as the Earth's, its core is probably iron and takes up perhaps 50 per cent of the volume, or 70 per cent of the mass, a much greater proportion than in the case of Earth's core.

Data from Mariner's infrared radiometer indicate that the surface of Mercury is covered with fine dust, as is the surface of the Moon, to a depth of at least several centimeters. Astronauts sent to Mercury, whenever they go, will leave footprints behind. Part of Mercury's surface is very jumbled, probably from the energy released by an impact (■ Fig. 6–34).

The biggest surprise of the mission was the detection of a magnetic field in space near Mercury. The field is weak, only about 1 per cent of the Earth's surface field. It had been thought that magnetic fields were generated by the rapid rotation of molten iron cores in planets, but Mercury is so small that its core should have quickly solidified after forming. So the magnetic field is probably not now being generated. Perhaps the magnetic field has been frozen into Mercury since the time when its core was molten.

6.3d Mercury Research Rejuvenated

After decades with no additional images of Mercury, in 2000 two teams of scientists released their composite images. Each is better than any previous ground-based image (■ Fig. 6–35).

A dozen years after Mariner 10 sent back data about Mercury, an important new discovery about Mercury was made with a telescope on Earth: Mercury has an atmosphere! The atmosphere is very thin, but is still easily detectable in spectra. It was a surprise that Mercury has an atmosphere because it is so hot, since it is relatively close to

NASA/JPL/Caltech

■ **FIGURE 6–34** The fractured and ridged planes of Mercury's Caloris Basin.

NASA

■ **FIGURE 6–33** A prominent scarp near Mercury's limb.

the Sun. "Hot" means that the individual atoms or molecules are moving rapidly in random directions, and to keep an atmosphere a planet must have enough gravity to hold in the particles moving in its atmosphere. Since Mercury has relatively little mass but is hot, the atmospheric particles escape easily and few are left—none from long ago.

Mercury's atmosphere contains more sodium than any other element—150,000 atoms per cubic centimeter compared with 4500 of helium and smaller amounts of oxygen, potassium, and hydrogen. At first, it appeared that the sodium was ejected into Mercury's atmosphere when particles from the Sun or from meteorites hit Mercury's surface. Newer evidence that the potassium and sodium are enhanced when the Caloris Basin is in view indicates, instead, that Mercury's atmosphere may have diffused up through Mercury's crust; the crust is thinner than average in the Caloris Basin.

Mercury's surface features can be mapped from Earth with radar. The radar observations provide altitudes and the roughness of the surface. The radar features show, in part, the half of Mercury not imaged by Mariner 10. It, too, is dominated by intercrater plains, though its overall appearance is different. The craters, with their floors flatter than the Moon's craters, show clearly on the radar maps. The scarps are obvious as well. The highest-resolution images (■ Fig. 6–36) reveal probable water-ice deposits near Mercury's poles. They have been shielded from sunlight for so long that ice (deposited by comets) could have persisted.

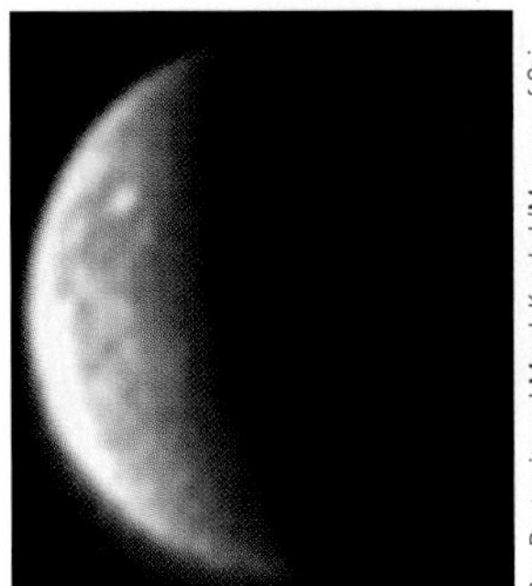

Ron Dantowitz and Marek Kozubal (Museum of Science, Boston) and Scott W. Teare (U. Illinois), with the permission of the *Astronomical Journal*; similar processing was carried out by Jeffrey Baumgardner, Michael Mendillo, and Jody K. Wilson (Boston U.)

■ **FIGURE 6–35** The best visible-light image of Mercury ever taken from Earth. It is a combination of several dozen very short exposures, the result of an effort to find moments when the Earth's atmosphere was the most steady as we looked through it (that is, when the "seeing" was best).

6.3e Mercury's History

Mercury may be the fragment of a giant early collision that nearly stripped it to its core. The stripping would account for the large proportion of iron in Mercury relative to Earth. Still early on, the core heated up and the planet's crust expanded. The expansion opened paths for molten rock to flow outward from the interior, producing the intercrater plains. As the core cooled, the crust contracted, and scarps resulted. Solar tides (tides induced by the Sun) slowed Mercury's original rotation until its rotational period became $^2/_3$ of the orbital period.

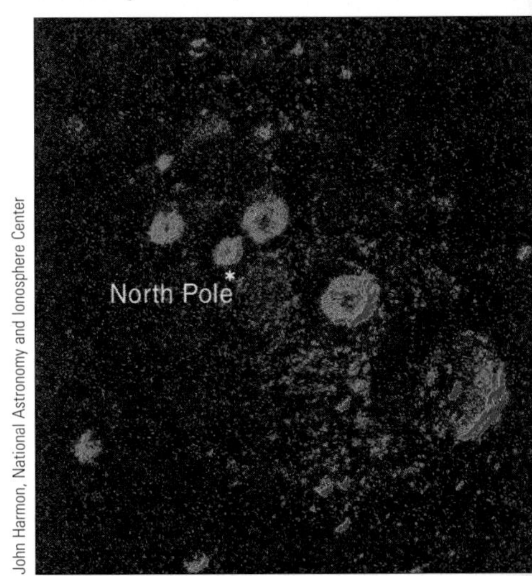

John Harmon, National Astronomy and Ionosphere Center

■ **FIGURE 6–36** Mercury's north pole, imaged with the radar capabilities of the giant radio telescope at Arecibo, Puerto Rico. The image measures 450 km on a side and shows detail down to 1.5 km. The bright features are most likely water-ice deposits located in the permanently shaded floors of craters.

6.3f Back to Mercury, at Last

After decades of neglect, NASA has sent a spacecraft back to Mercury. The **ME**rcury **S**urface, **S**pace **EN**vironment, **GE**ochemistry and **R**anging (or MESSENGER) spacecraft was launched in 2004 and will arrive at Mercury in 2008. MESSENGER will fly by Mercury twice in 2008 and then go into a year-long orbit around Mercury beginning in 2009.

ESA and Japan are planning to launch BepiColombo to Mercury in 2011 or 2012. (Bepi—a nickname—Colombo was a scientist who studied Mercury and spacecraft trajectories.) It will carry two orbiters, one to map the planet in a variety of wavelengths and the other to study Mercury's magnetosphere.

6.4 Venus

Venus orbits the Sun at a distance of 0.7 A.U. Although it comes closer to us than any other planet, we did not know much about it until recently because it is always shrouded in highly reflective clouds (■ Fig. 6–37).

Jay M. Pasachoff

■ **FIGURE 6–37** A crescent Venus, observed with a large telescope on Earth. In visible-light images like this one, we see only a layer of clouds.

6.4a Transits of Venus

Venus was in transit (crossing the Sun's disk) in 2004 and it will do so again in 2012, for the first times since the pair of transits in 1874 and 1882. Previous transits of

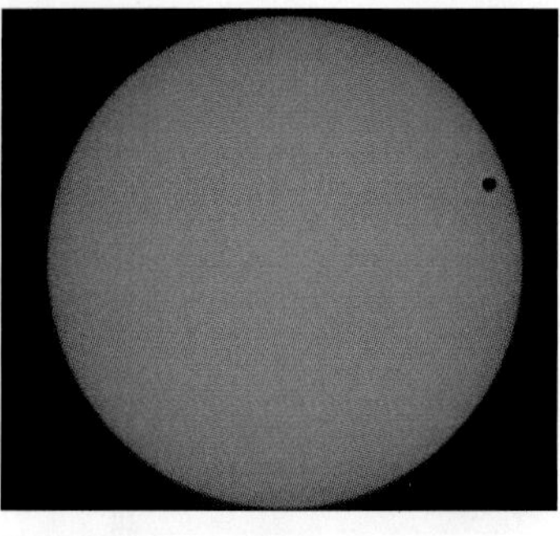
Jay M. Pasachoff, David L. Butts, Joseph W. Gangestad, and Owen W. Westbrook/Williams College Expedition

■ **FIGURE 6–38** The transit of Venus of June 8, 2004.

Venus were important for determining the scale of the Solar System, which we can now measure better by other means. Still, the matter is of historical interest.

The accuracy of observations of transits of Venus in 1761, 1769, 1874, and 1882 (the only ones ever scientifically observed) was impeded by the dreaded "black-drop effect," in which a dark apparent connection between the silhouette of Venus and the edge of the Sun remained for about a minute before Venus was clearly distinctly separated from the edge. (Captain Cook went to Tahiti to observe the 1769 transit; his subsequent exploration of Australia and New Zealand can be considered a bonus of this astronomical expedition.) Timing the transit accurately determined how well the distances between the planets could be known.

One of the authors (J.M.P.) participated in the analysis of the 1999 transit of Mercury observed with a spacecraft aloft to study the Sun. Since this spacecraft was outside the Earth's atmosphere, and since Mercury definitely does not have a substantial atmosphere, the discovery of a black-drop effect in this Mercury transit proved that neither planetary atmosphere was necessary to make such an effect. For Mercury, and presumably for Venus, the effect is caused by a combination of blurring in the telescope and the fact that the brightness of the Sun's disk diminishes near its edge. Though Venus has a substantial atmosphere, it is much too small in height to make the black-drop effect observed during Venus transits.

We used modern instrumentation on the ground and in space to observe the June 8, 2004, Venus transit (■ Fig. 6–38). It was striking to see the black silhouette of Venus against the Sun, something that could be seen looking through a suitable filter even without a telescope. It looked as though a hole had been drilled through the Sun. With a spacecraft in orbit and from the ground, we observed the black drop, a small effect compared with its reports from old records.

We were even able to use a different spacecraft to detect how the amount of light that the Earth received from the Sun dimmed by the 0.1 per cent that corresponded to the fraction of the Sun that was covered by Venus. Even more interesting, the Sun dimmed gradually, as Venus covered parts of the Sun that had different brightnesses. We were able to see up close the causes of transit effects that astronomers are now studying with exoplanets (Chapter 9), planets orbiting around distant stars. Our work should help scientists understand the exoplanet observations better. The 2004 transit of Venus therefore provides a link between historical astronomy and one of the most exciting astronomical investigations of the present and near future.

Don't miss seeing the 2012 transit!

6.4b The Atmosphere of Venus

The clouds on Venus are primarily composed of droplets of sulfuric acid, H_2SO_4, with water droplets mixed in. Sulfuric acid may seem like a peculiar constituent of a cloud, but the Earth, too, has a significant layer of sulfuric acid droplets in its stratosphere, a higher layer of the atmosphere. However, the water in the lower layers of the Earth's atmosphere, circulating because of weather, washes the sulfur compounds out of these lower layers. Venus has sulfur compounds in the lower layers of its atmosphere in addition to those in its clouds.

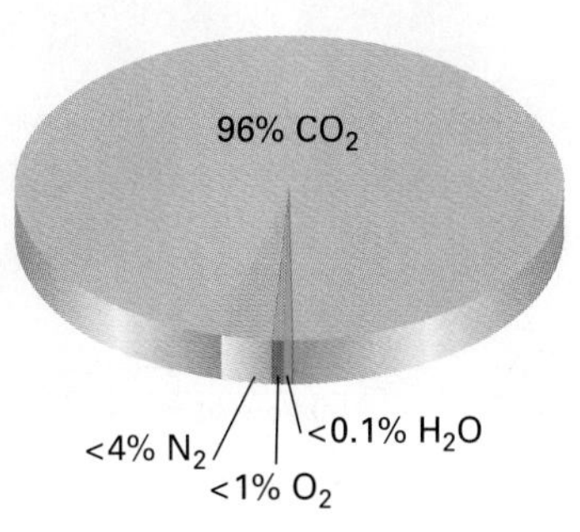

■ **FIGURE 6–39** The composition of Venus's atmosphere.

Observations from Earth show a high concentration of carbon dioxide in the thick atmosphere of Venus. In fact, carbon dioxide makes up over 96 per cent of the mass of Venus's atmosphere (■ Fig. 6–39). The Earth's atmosphere, for comparison, is mainly nitrogen (79 per cent), with a fair amount of oxygen (20 per cent) as well. Carbon dioxide makes up less than 1 per cent of the terrestrial atmosphere.

Because of the large amount of carbon dioxide in its atmosphere, which leads to the atmosphere being so massive, Venus's surface pressure is 90 times higher than the pressure of Earth's atmosphere. Carbon dioxide on Earth dissolves in seawater and rain, eventually forming some types of terrestrial rocks, often with the help of life-forms. (Limestone, for example, has formed from deceased marine life.) If this carbon dioxide were released from the Earth's rocks, along with other carbon dioxide trapped in seawater, our atmosphere would become as dense and have as high a pressure as that of Venus. Venus, slightly closer to the Sun than Earth and thus hotter, has no oceans or rain in which the carbon dioxide can dissolve to help take it up from the atmosphere. Thus the carbon dioxide remains in Venus's atmosphere.

6.4c The Rotation of Venus

In 1961, the radio waves used in radar astronomy penetrated Venus's clouds, allowing us to determine accurately how fast Venus rotates. Venus rotates in 243 days with respect to the stars, in the direction opposite that of the other planets. Venus revolves around the Sun in 225 Earth days. Venus's periods of rotation and revolution combine so that a solar day/night cycle on Venus corresponds to 117 Earth days; that is, the Sun returns to the same position in the sky every 117 days.

The notion that Venus rotates backward used to seem very strange to astronomers, since all the planets revolve around the Sun in the forward direction, and most of the other planets (except Uranus and Pluto, which are on their sides) and most satellites also rotate in that forward direction. Because the laws of physics do not allow the amount of spin (angular momentum; recall the discussion in Chapter 5) to change on its own, and since the original material from which the planets coalesced was undoubtedly rotating, we had expected all the planets to revolve and rotate in the same sense.

Nobody knows definitely why Venus rotates "the wrong way." One possibility is that in the chaos that reigned in the early Solar System, when Venus was forming, a large clump of material struck it at an angle that caused the merged resulting planet to rotate backward. Recent studies of chaotic changes in Venus's angle of rotation show alternative ways that Venus could have wound up rotating backward, perhaps depending on the fact that Venus's atmosphere is so dense. The Sun's gravity raises tidal bulges in Venus's dense atmosphere. Gravitational tugs on these tidal bulges from other planets contribute to making the system chaotic, and eventually the axis could flip over.

The slow rotation of Venus's solid surface contrasts with the rapid rotation of its clouds. The tops of the clouds rotate in the same sense as the surface rotates but about 60 times more rapidly, once every 4 days. Lower parts of the atmosphere, however, rotate very slowly.

6.4d Why Is Venus So Incredibly Hot?

We can determine the temperature of Venus's surface by studying its radio emission, since radio waves emitted by the surface penetrate the clouds. The surface is very hot, about 750 K (900°F), even on the night side. In addition to measuring the temperature on Venus, scientists theoretically calculate what the temperature would be if Venus's atmosphere were transparent to radiation of all wavelengths, both coming in and going out. This value—less than 375 K (215°F)—is much lower than the measured values. The high temperatures derived from radio measurements indicate that Venus traps much of the solar energy that hits it.

The process by which the surface of Venus is heated so much is called the greenhouse effect (■ Fig. 6–40), already discussed briefly in Section 6.1d. The fraction of sunlight not reflected outward by the high clouds and molecules passes through the venusian atmosphere in the form of radiation in the visible part of the spectrum. The sunlight is absorbed by the surface of Venus, which heats up. At the resulting temperatures, the thermal (nearly black-body) radiation that the surface gives off is mostly in the infrared. But the carbon dioxide and other constituents of Venus's atmosphere are largely opaque to infrared radiation; thus, much of the energy is absorbed. Since it

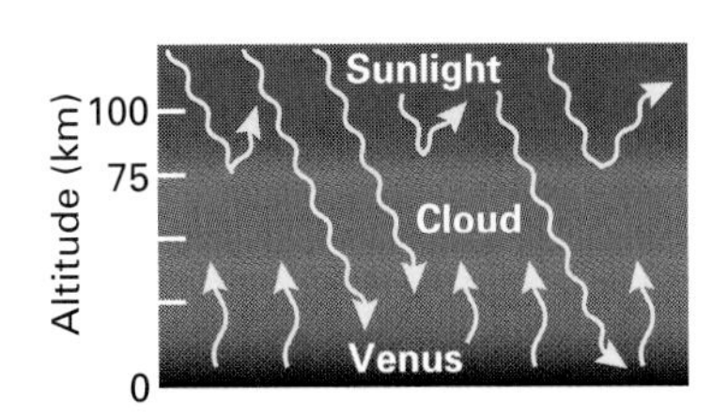

■ **FIGURE 6–40** Though most sunlight is reflected from Venus's clouds, some sunlight penetrates them, so the surface is illuminated with radiation in the visible part of the spectrum. Venus's own radiation is mostly in the infrared, because it is much cooler than the Sun, pushing the peak of its black-body curve to longer wavelengths. This infrared radiation does not escape as readily, leading to a warming of the atmosphere and hence of the surface, a phenomenon known as the greenhouse effect.

doesn't get out, we commonly say that it is "trapped," though not literally of course (the radiation no longer exists after being absorbed). The atmosphere becomes hotter, thereby heating the planet's surface as well.

Thus Venus heats up far above the temperature it would reach if the atmosphere were transparent to infrared radiation. The surface radiates more and more energy as the planet heats up. Finally, a balance is struck between the rate at which energy flows in from the Sun and the rate at which it trickles out (as infrared) through the atmosphere. The situation is so extreme on Venus that we say a "runaway greenhouse effect" has taken place there. Understanding such processes involving the transfer of energy is but one of the practical results of the study of astronomy.

Greenhouses on Earth don't work quite this way. In actual greenhouses, closed glass on Earth prevents the mixing of air inside them (which is mainly heated by conduction from the warmed ground and by radiation from the ground and the greenhouse glass) with cooler outside air. The trapping of solar energy by the "greenhouse effect"—the inability of infrared radiation, once formed, to get out as readily, since it is absorbed by atmospheric carbon dioxide and water—is a less important process in an actual greenhouse or in a car when it is left in the sunlight.

As mentioned in Section 6.1d, the greenhouse effect can be beneficial; indeed, it keeps Earth's temperature at a comfortable level. Earth would be quite chilly without it—about 33°C (60°F) colder, on average.

Earth manages to achieve a comfortable greenhouse effect by recycling its greenhouse gases, primarily CO_2 and H_2O (water). As the oceans evaporate, water vapor builds up in the atmosphere, but then the rain comes down. The rain dissolves atmospheric CO_2, producing carbonic acid. This mild "acid rain" dissolves rocks and forms carbonates, and there are other forms of weathering as well. The carbonate solution eventually reaches the ocean. Various sea creatures then use the carbonates in their shells. When these organisms die, their carbonate shells form sediments on ocean bottoms. Because of plate tectonics, the sediments can later dive down under other plates, when they become part of the magma inside the Earth. The CO_2 is subsequently outgassed by volcanism, and the process repeats.

How did Venus reach its hellish state? It may have begun its existence with a pleasant climate, and possibly even with oceans. Perhaps rain and other processes were not quick enough in removing the atmospheric CO_2 outgassed by volcanoes. As CO_2 accumulated, the greenhouse effect increased, and temperatures rose; oceans evaporated faster, and a runaway greenhouse effect was in progress. The water vapor was gradually broken apart by ultraviolet radiation from the Sun, and the hydrogen (being so light) escaped from Venus. Eventually there were no oceans and no rain—but continued outgassing of CO_2 led to more greenhouse warming and extremely high temperatures.

It is also possible that Venus was essentially born in a hot state, or reached it very quickly. Without oceans, it is difficult for atmospheric CO_2 to become incorporated into rocks as it did on Earth. Extensive greenhouse heating is therefore maintained.

Studies of Venus are important for an overall understanding of Earth's climate and atmosphere. There is a scientific consensus that currently there is human-caused global warming of Earth; the politics and what to do about it remain controversial. Measurements show that the terrestrial atmospheric CO_2 concentration is increasing (Fig. 6–10). Almost all scientists in the field have concluded that the human-induced increase in CO_2 (from the burning of fossil fuels) is one of the main causes of global warming.

Questions have been raised whether a "positive feedback" mechanism that leads to sustained global warming will necessarily occur; there are many variables, and their effects are not yet completely understood. For example, an increase in global temperatures leads to more evaporation of the oceans, but this can increase the cloud cover and hence Earth's reflectivity (albedo), thereby producing a decrease in the amount of visible sunlight reaching Earth's surface. This is a "negative feedback" mechanism that tends to stabilize temperatures. On the other hand, recent computer models suggest that the higher water-vapor content of the atmosphere actually increases greenhouse heating by an amount that more than compensates for the greater reflectivity of the Earth. Earth's atmosphere is a very complicated system, but scientists are gaining confidence in the accuracy of their computer models.

It is very unlikely that the Earth will experience a *runaway* greenhouse effect that makes the planet unfit for *all* forms of life. In the past, there have been times when Earth was much warmer than now, yet a runaway did not occur. Moreover, life on Earth recovered from collisions with large asteroids and comets, some of which dumped enormous quantities of greenhouse gases into the atmosphere.

On the other hand, even a rise of a few degrees in Earth's global average temperature over a timescale of a few decades, which is now generally predicted, would have terrible consequences for humans, including flooding of coastal cities and destruction of many agricultural areas. Also, since we cannot accurately predict what will happen under various circumstances, it is dangerous to act in ways that might disrupt the balance in our atmosphere and adversely affect life on Earth. Venus has shown us the importance of being very careful about how we treat the only home we have.

We need a major push in research, even on a 50-year timescale, to develop large-scale energy supplies from nuclear power, wind power, and other sources that do not give off greenhouse gases and so do not lead to an increased greenhouse effect. Whether such research funds will be available is a different question. Some people are now putting substantial emphasis on starting to adapt to the effects of greenhouse warming in place of (or in addition to) limiting the greenhouse warming itself, especially given the likelihood of using coal for economic growth in the developing world. The "Kyoto Protocol," a multination agreement to limit the emission of greenhouse-causing gases by industrialized nations, went into effect in 2005, when enough countries ratified it. The United States had signed it originally but never ratified it, so the U.S. is not part of the agreement.

6.4e Spacecraft Observations of Venus's Atmosphere

Venus was an early target of both American and Soviet space missions. During the 1960s, American spacecraft flew by Venus, and Soviet spacecraft dropped through its atmosphere. In 1970, the Soviet Venera 7 spacecraft radioed 23 minutes of data back from the surface of Venus before it succumbed to the high temperature and pressure. Two years later, the lander from Venera 8 survived on the surface of Venus for 50 minutes. Both landers confirmed the Earth-based results of high temperatures, high pressures, and high carbon dioxide content.

Several United States spacecraft have observed Venus's clouds and followed the changes in them. The most recent views of changes in Venus's clouds came from the Galileo spacecraft (■ Fig. 6–41) and from the Hubble Space Telescope. Structure in the clouds shows only when viewed in ultraviolet light. The clouds appear as long, delicate streaks, looking like terrestrial cirrus clouds.

Such studies of Venus have practical value. The principles that govern weather on Venus are similar to those that govern weather on Earth, though there are major specific differences, such as the absence of oceans on Venus. The better we understand the interaction of solar heating, planetary rotation, and chemical composition in setting up an atmospheric circulation, the better we will understand our Earth's atmosphere. We then may be better able to predict the weather. The potential financial return from such knowledge is enormous: it would be many times the investment we have made in planetary exploration.

Not one of the spacecraft to Venus has detected a magnetic field. The absence of a magnetic field may indicate either that Venus does not have a liquid core or does not rotate fast enough.

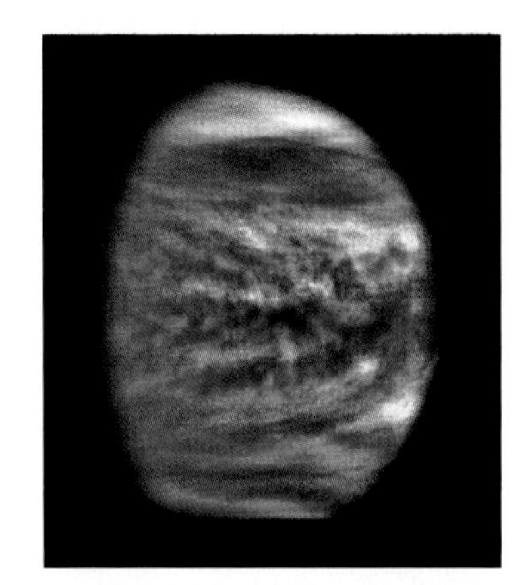

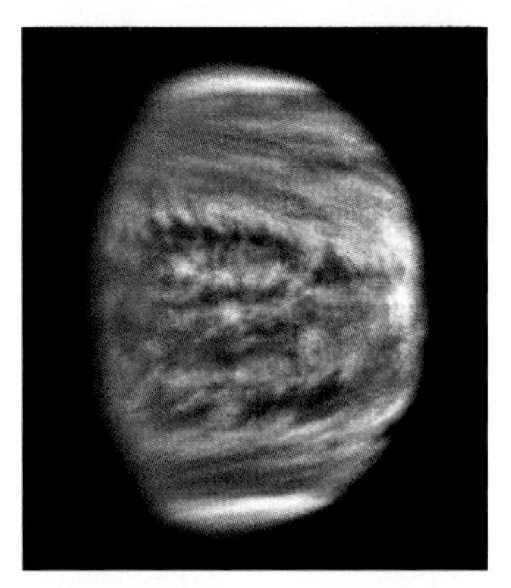

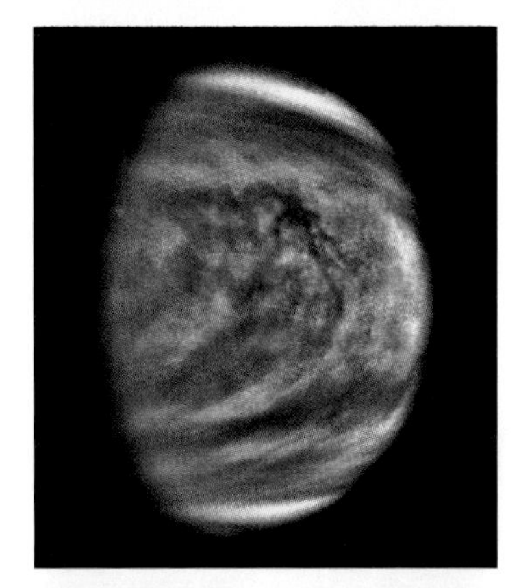

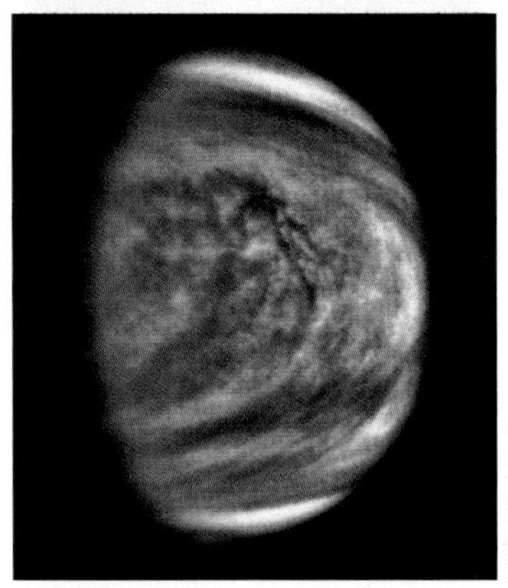

NASA/JPL/Caltech

■ **FIGURE 6–41** Changes in Venus's clouds over several hours, observed from the Galileo spacecraft.

Spacecraft have provided evidence that volcanoes may be active on Venus. The abundance of sulfur dioxide they found varied at different times by a factor of 10. The effect could come from eruptions at least ten times greater than that of even the largest recent terrestrial volcanic eruptions. Lightning that was detected is further evidence that those regions are sites of active volcanoes. On Earth, it is common for the dust and ash ejected by volcanoes to rub together, generating static electricity that produces lightning.

Probes that penetrated the atmosphere and went down to the surface found that high-speed winds at the upper levels are coupled to other high-speed winds at lower altitudes. The lowest part of Venus's atmosphere, however, is relatively stagnant. The probes detected three distinct layers of venusian clouds, separated from each other by regions of relatively high transparency.

One of the probes measured that only about 2 per cent of the sunlight reaching the outer atmosphere of Venus filters down to the surface, making it like a dim terrestrial twilight. Thus most of the Sun's energy is absorbed in or reflected by the clouds, unlike the situation with Earth, and Venus's clouds are more important than Earth's for controlling weather. Most of the light that reaches the surface is orange, so photographs taken on the surface have an orange cast.

6.4f Radar Observations of Venus's Surface

From Venus's size and from the fact that its mean density is similar to that of the Earth, we conclude that its interior is also probably similar to that of the Earth. This means that we expect to find volcanoes and mountains on Venus.

We can study the surface of Venus by using radar to penetrate Venus's clouds. Radars using huge Earth-based radio telescopes, such as the giant 1000-ft (305-m) dish at Arecibo, Puerto Rico, have mapped a small amount of Venus's surface with a resolution of up to about 20 km. (The **resolution** means the size of the finest details that could be detected, so we mean that only features larger than about 20 km can be detected.) Regions that reflect a large percentage of the radar beam back at Earth show up as bright, and other regions as dark.

NASA's Pioneer Venus, Venera 15/16, and, in 1990 to 1993, Magellan were in orbit around Venus, allowing them to make observations over a lengthy time period. They carried radars to study the topography of Venus's surface, mapping a much wider area than could be observed from the Earth.

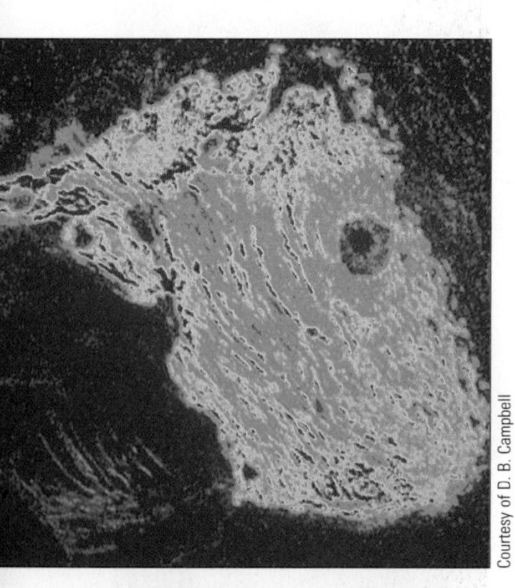
Courtesy of D. B. Campbell

■ **FIGURE 6–42** The Arecibo Observatory's high-resolution radar image of Maxwell Montes, with a resolution of about 2 km. Brighter areas show regions where more power bounces back to us from Venus, which usually corresponds to rougher terrain than darker areas.

From the radar maps (look back at Figure 6–1), we now know that 60 per cent of Venus's surface is covered by a rolling plain, flat to within plus or minus 1 km. Only about 16 per cent of Venus's surface lies below this plain, a much smaller fraction than the two thirds of the Earth covered by ocean floor.

Two large features, the size of small Earth continents, extend several km above the mean elevation. A northern "continent," Terra Ishtar, is about the size of the continental United States. The giant chain of mountains on it known as Maxwell Montes (■ Fig. 6–42) is 11 km high, which is 2 km taller than Earth's Mt. Everest stands above terrestrial sea level. It had formerly been known only as a bright spot on Earth-based radar images. Ishtar's western part is a broad plateau, about as high as the highest plateau on Earth (the Tibetan plateau) but twice as large. An equatorial "continent," Aphrodite Terra, is about twice as large as the northern one and is much rougher.

From the radar maps, it appears that Venus, unlike Earth, is made of only one continental plate. We observe nothing on Venus equivalent to Earth's mid-ocean ridges, at which new crust is carried upward. In particular, the high-resolution Magellan observations (■ Fig. 6–43) show no signs of crust spreading laterally on Venus. Venus may well have such a thick crust that any plate tectonics that existed in the distant past was choked off. Thus "venusquakes" are probably much less common than earthquakes.

NASA/JPL/Caltech

■ **FIGURE 6–43** Maat Mons in a simulated perspective view based on radar data from Magellan and arbitrarily given the orange color that comes through Venus's clouds. Our viewpoint is 560 km north of Maat Mons at an elevation of 1.7 km. We see lava flows extending hundreds of kilometers across the foreground fractured plains. Maat Mons is 8 km high; its height here is exaggerated by a factor of about 20.

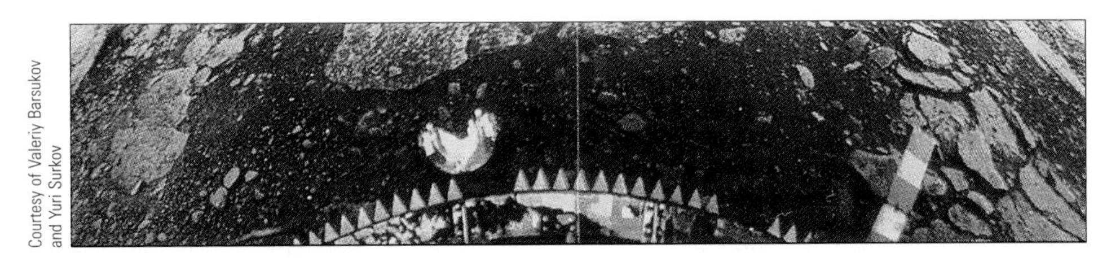

Courtesy of Valeriy Barsukov and Yuri Surkov

■ **FIGURE 6–44** The view from a Soviet spacecraft on Venus's surface, showing a variety of sizes and textures of rocks. It survived on Venus for over 2 hours before it failed because of the high temperature and pressure. The camera first looked off to one side, and then scanned downward as though you were looking down toward your feet. Then it scanned up to the other side. As a result, the left horizon is visible as a slanted boundary at *upper left* and the right horizon is at *upper right*. The base of the spacecraft, Venera 13, and a lens cap are at *bottom center*.

Perhaps giant hot spots, like those that created the Hawaiian Islands on Earth, force mountains to form on Venus in addition to causing volcanic eruptions. The rising lava may also make the broad, circular domes that are seen. Venus got rid of its internal energy through widespread volcanism. Apparently, virtually the whole planet was resurfaced with lava about 500 million years ago. Thus the surface we now study is relatively young, and the history of Venus's surface has been obliterated.

In the 1970s and 1980s, a series of Soviet spacecraft landed on Venus and sent back photographs (■ Fig. 6–44). They also found that the soil resembles basalt in chemical composition and density, in common with the Earth, the Moon, and Mars. They measured temperatures of about 750 K and pressures over 90 times that of the Earth's atmosphere, confirming earlier, ground-based measurements.

Our recent space results, coupled with our ground-based knowledge, show us that Venus is even more different from the Earth than had previously been imagined. Among the differences are Venus's slow rotation, its one-plate surface, the complete resurfacing that occurred relatively recently, the absence of a satellite, the extreme weakness or absence of a magnetic field, the lack of water in its atmosphere, and its high surface temperature and pressure.

6.4g Back to Venus, at Last

NASA's MESSENGER mission to Mercury (Section 6.3f) will fly close to Venus twice, in October 2006 and in June 2007. BepiColombo will also pass by.

The European Space Agency launched a major spacecraft to Venus (■ Fig. 6–45) in November 2005 from Kazakhstan. It arrived at Venus in April 2006. If its plans succeed, it would study its atmosphere and use an infrared window to map its surface from an exaggerated elliptical orbit over 2 Venusian days, which is equivalent to about 500 Earth days. Japan's Venus Climate Orbiter is scheduled for a 2008 launch.

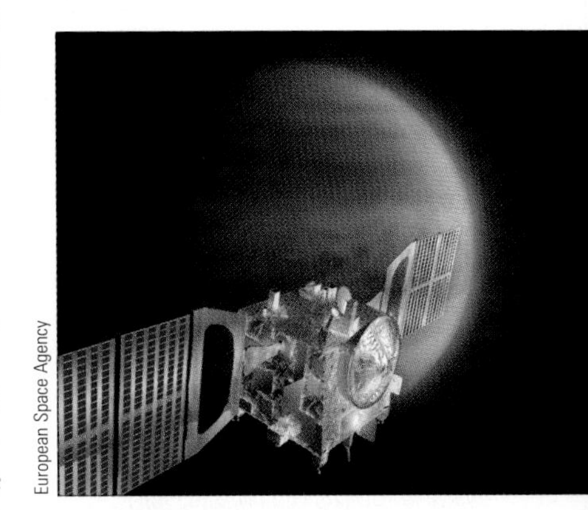

European Space Agency

■ **FIGURE 6–45** An artist's conception of Venus Express.

6.5 Mars

Mars has long been the planet of greatest interest to scientists and nonscientists alike. Its unusual appearance as a reddish object in the night sky with some color changes with its seasons, coupled with some past scientific studies, have made Mars the first place we have looked for extraterrestrial life.

In 1877, the Italian astronomer Giovanni Schiaparelli published the results of a long series of telescopic observations he had made of Mars. He reported that he had seen "canali" on the surface. When this Italian word for "channels" was improperly translated into "canals," which seemed to connote that they were dug by intelligent life, public interest in Mars increased. Percival Lowell, in particular, was fascinated by the prospect of an ancient civilization on Mars. In the late 19th century, he drew detailed maps of Mars that showed an elaborate system of canals, presumably built to bring water to arid regions.

Over the next decades, there were endless debates over just what had been seen. We now know that the channels or canals Schiaparelli and other observers reported are not present on Mars. They were an illusion. In fact, positions of the "canali" do not even always overlap the spots and markings that are actually on the martian surface (■ Fig. 6–46). But hope of finding life elsewhere in the Solar System springs eter-

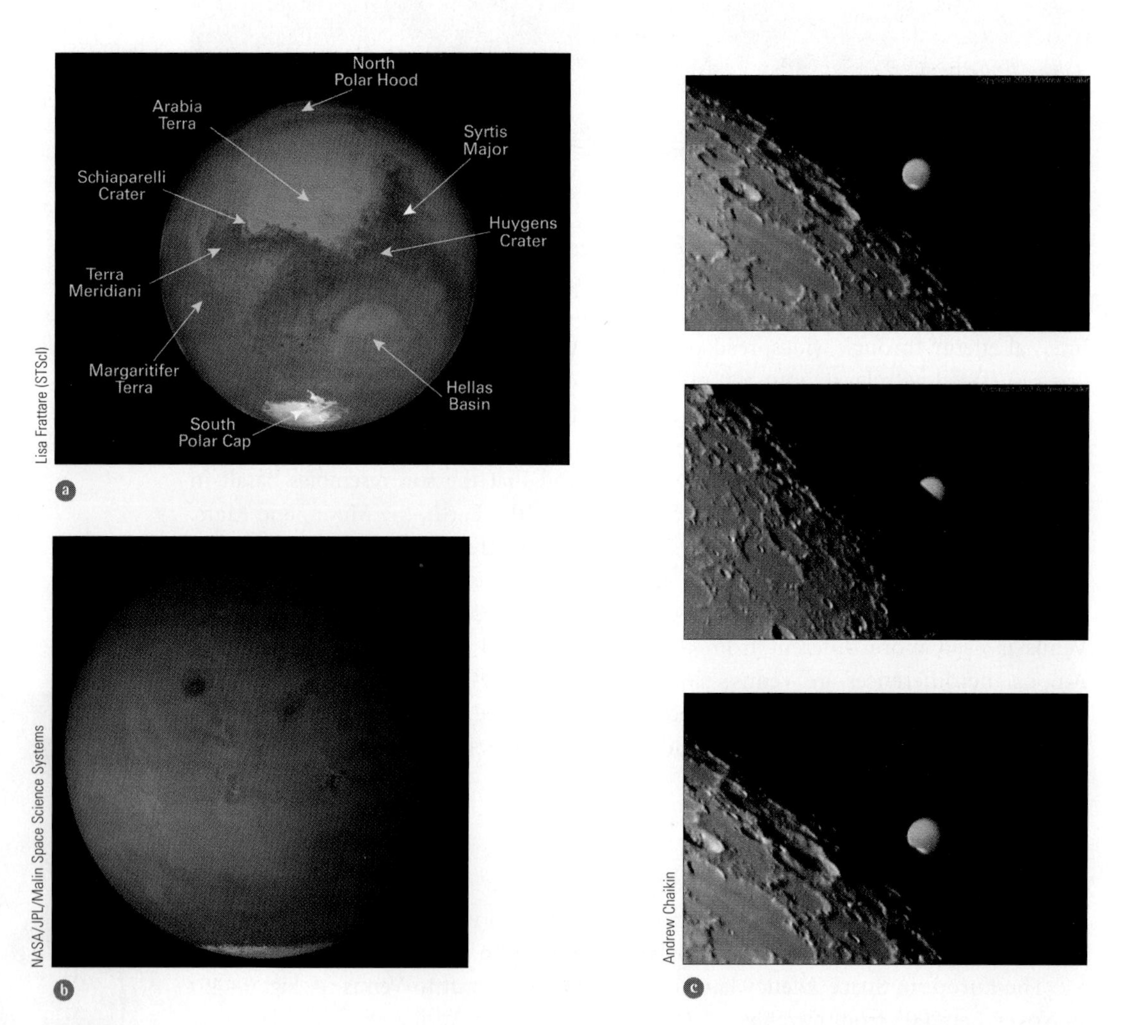

■ **FIGURE 6–46** ⓐ Mars viewed from the Hubble Space Telescope. Dust storms show at the north polar cap and at lower right. ⓑ A composite view from Mars Global Surveyor in 2005, showing Mars's Tharsis plain and set of giant volcanoes. The images in the composite were taken at the same Mars longitude daily over a year, showing northern-hemisphere autumn and southern-hemisphere spring. ⓒ Views from Earth of Mars going close to, and being partly occulted by, the Moon.

nal, and the latest studies have indicated the presence of considerable quantities of liquid water in Mars's past, a fact that leads many astronomers to suggest that life could have formed during those periods.

■ **FIGURE 6–47** Mars's moons Deimos (*lower left*) and Phobos (*lower right*), with the asteroid Gaspra (*top*) for comparison, shown to the same scale. Phobos is about 27 km across. Deimos measures about 15 km × 12 km × 11 km. Its surface is heavily cratered and thus presumably very old. The surface seems smoother than that of Phobos, though, because soil 10 m thick covers Deimos. The spectra of Phobos and Deimos are similar to those of one type of dark asteroid. The Phobos and Deimos images are from Viking; the Gaspra image is from Galileo.

6.5a Characteristics of Mars

Mars is a small planet, 6792 km across, which is only about half the diameter and one-eighth the volume of Earth or Venus, although somewhat larger than Mercury. It has two tiny (about 20 km in diameter), irregularly shaped moons named Phobos and Deimos (■ Fig. 6–47). Mars rotates on its axis in about 24½ Earth hours, meaning that its day/night cycle is similar to that of our own planet.

Mars's atmosphere is very thin, only 1 per cent of Earth's, but it might be sufficient for certain kinds of life. Unlike the orbits of Mercury or Venus, the orbit of Mars is outside the Earth's, so we can observe Mars in the late-night sky.

Mars revolves around the Sun in 1.9 Earth years. The axis of its rotation is tipped at a 25° angle from perpendicular to the plane of its orbit, nearly the same as the Earth's 23½° tilt. Because the tilt of the axis causes the seasons, we know that Mars goes through its year with four seasons just as the Earth does.

We have watched the effects of the seasons on Mars over the last century. In the martian winter, in a given hemisphere, there is a polar cap. As the martian spring comes to the northern hemisphere, the north polar cap shrinks and material at more temperate zones darkens.

The surface of Mars appears mainly reddish-orange when seen from Earth against dark space, probably due to the presence of iron oxide (rust), with darker gray areas that appear blue-green for physiological reasons, as the human eye and brain misjudge the color contrast. The changes in color over time apparently result from dust that is either covering surface rock or is blown off by winds that can have speeds of hundreds of kilometers per hour. Sometimes, global dust storms occur, and we can later watch the color change as the dust blows off the places where it first lands, exposing the dark areas underneath.

From Mars's mass and radius we can easily calculate that it has an average density about that of the Moon, substantially less than the densities of Mercury, Venus, and Earth. The difference indicates that Mars's overall composition must be fundamentally different from that of these other planets. Mars probably has a smaller iron-rich core and a thicker crust than does Earth.

6.5b Mars's Surface

Mars has been the target of several series of spacecraft, most notably some U.S. spacecraft, including Mariner 9 in 1971; a pair of Vikings that started orbiting in 1976 and dropped landers onto the martian surface; and Mars Pathfinder, which landed on July 4, 1997, and placed the Sojourner rover on Mars's surface. NASA's Spirit and Opportunity landers and rovers, which arrived in 2004, have sent back detailed views (■ Fig. 6–48) and have found many signs that water flowed on Mars in the past. (See *A Closer Look 6.5: Mars Exploration Rovers.*) NASA's Mars Global Surveyor and Mars Odyssey are currently orbiting the planet, as is the European Space Agency's Mars Express.

Surface temperatures measured on Mars from orbiters and landers range from a low of about 150 K (−190°F) to over 300 K (80°F). The temperature can vary each day by 35 to 50°C (60 to 90°F).

The surface of Mars can be divided into four major types: volcanic regions, canyon areas, expanses of craters, and terraced areas near the poles. A chief surprise was the discovery of extensive areas of volcanism. The largest volcano—which corresponds in

■ **FIGURE 6–48** A 360° panorama from Spirit, acquired on its 410th to 413th sols (martian days), February 27 to March 2, 2005. The summit near the center is known as Husband Hill, and is about 200 meters away and 45 meters high. We look past an outcrop known as the Cumberland Ridge, with the Tennessee Valley to the left. Some of Spirit's tracks show at right (p. 143).

position to the surface marking long known as Nix Olympica, "the Snow of Olympus"—is named Olympus Mons, "Mount Olympus." It is a huge volcano, 600 km at its base and about 21 km high (■ Fig. 6–49). It is crowned with a crater 65 km wide; Manhattan Island could be easily dropped inside. (The tallest volcano on Earth is Mauna Kea in the Hawaiian Islands, if we measure its height from its base deep below the ocean. Mauna Kea is slightly taller than Mount Everest, though still only 9 km high.)

Perhaps the volcanic features on Mars can get so huge because continental drift is absent there, as on Venus. If molten rock flowing upward causes volcanoes to form, then on Mars the features just get bigger and bigger for hundreds of millions of years, since the volcanoes stay over the sources and do not drift away. (Each of the Hawaiian Islands was formed over a single "hot spot," but gradually drifted away from it. The Big Island of Hawaii is still over the hot spot, which explains why it has active volcanoes and why it is still growing.) Mars's surface gravity is lower than Earth's, adding to the ability of Olympus Mons to get so big.

Another surprise on Mars was the discovery of systems of canyons (■ Fig. 6–50). One tremendous system of canyons—about 5000 kilometers long—is as big as the continental United States and comparable in size to the Rift Valley in Africa, the longest geological fault on Earth. Known as Valles Marineris, it does not appear to have been produced by water, unlike some of the other features discussed below.

Perhaps the most amazing discovery on Mars was the presence of sinuous channels. These are on a smaller scale than the "canali" that Schiaparelli and Lowell believed they had seen, and are entirely different phenomena. Some of the channels show the same characteristic features as streambeds on Earth. Even though liquid water cannot exist on the surface of Mars under today's conditions, it is difficult to think of ways to explain the channels satisfactorily other than to say that they were cut by running water in the past.

Data from the Mars Pathfinder mission strengthen this conclusion: rocks are scattered over a large area in a manner reminiscent of flood plains. The high-resolution Mars Global Surveyor shows what look like sedimentary layers of rock (■ Fig. 6–51). Mars Global Surveyor and the European Space Agency's Mars Express, which is also now in orbit, found additional signs, such as terracing and gullies, showing that water may have

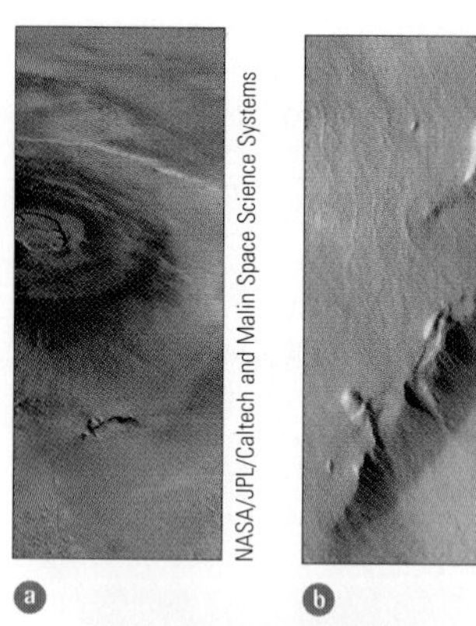

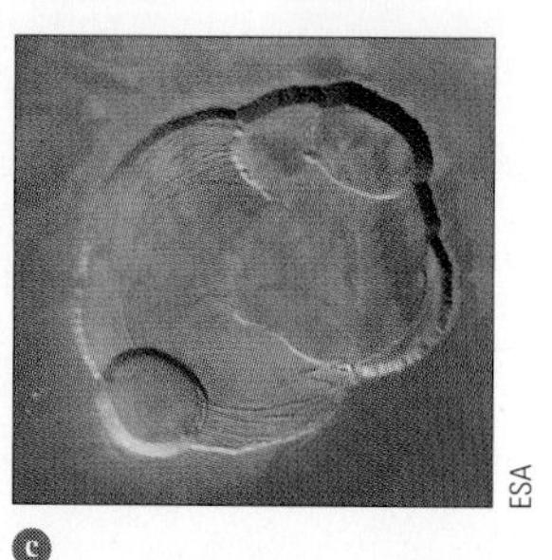

■ **FIGURE 6–49** (a) Olympus Mons, from Mars Global Surveyor. (b) A high-resolution view of a strip across a tiny piece of Olympus Mons, from 2001 Mars Odyssey. This enlargement of a scarp shows lava that flowed in one direction, down the volcano's flank. The low contrast indicates that the lava flows are covered with dust. Few craters show in the lava flows, making them younger than about 500 million years. (c) The caldera of Olympus Mons, from the European Space Agency's Mars Express.

NASA/JPL/Cornell

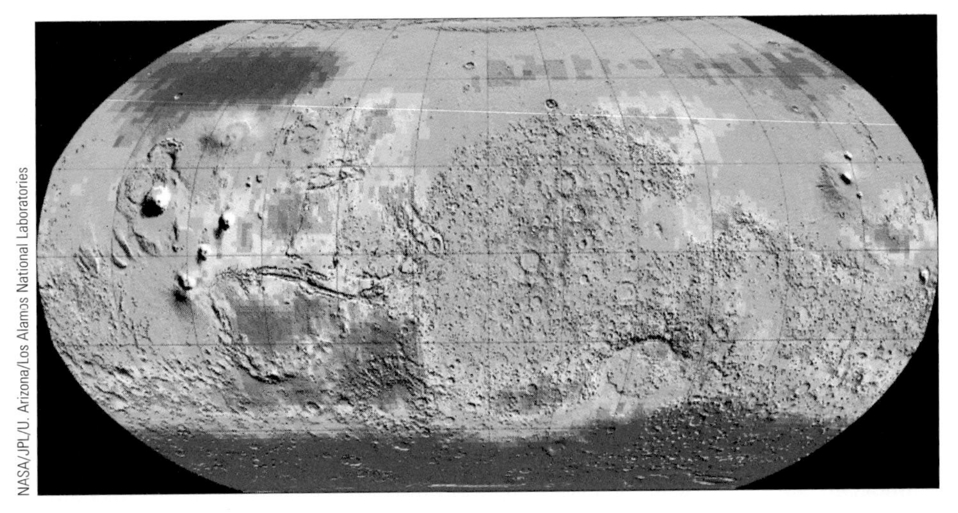

NASA/JPL/U. Arizona/Los Alamos National Laboratories

■ **FIGURE 6–50** Mars, an image made from earlier spacecraft mapping with colors superimposed from observations of mid-energy neutrons detected by the 2001 Mars Odyssey spacecraft. The valleys Valles Marineris (which were not carved by water, but rather are gashes in Mars's crust) appear horizontally across the center; together, they are as long as the United States is wide. At left we see the giant volcanoes. Colors indicate the relative abundances of hydrogen (and thus of water), with decreasing amounts shown ranging from deep blue through green to yellow to red. The hydrogen in the far north is hidden beneath a layer of carbon-dioxide frost (dry ice).

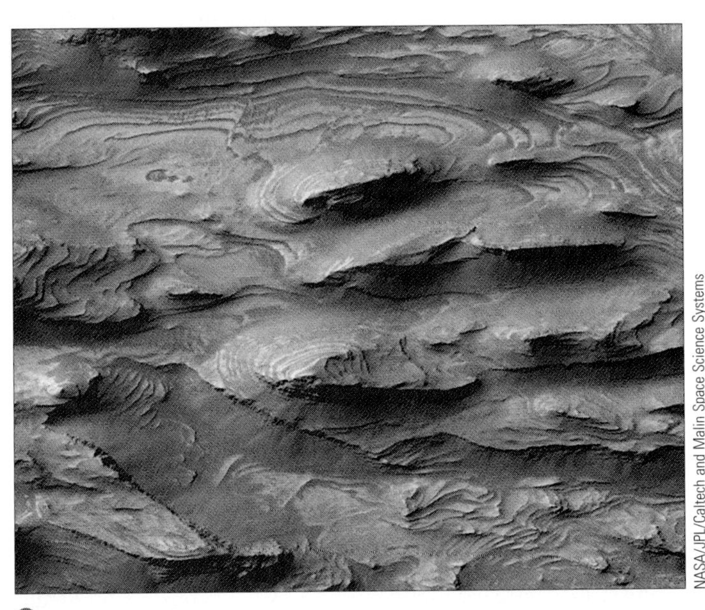

NASA/JPL/Caltech and Malin Space Science Systems

a

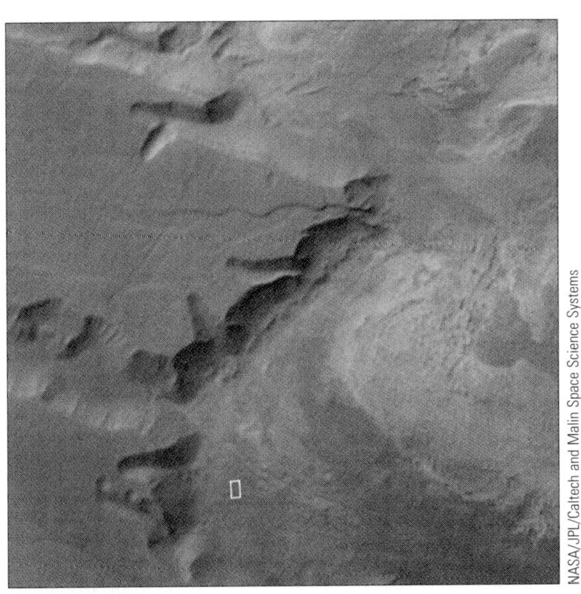

NASA/JPL/Caltech and Malin Space Science Systems

b

■ **FIGURE 6–51** (a) Layered rock, apparently sedimentary, imaged with Mars Global Surveyor, an enlargement of the box shown in (b).

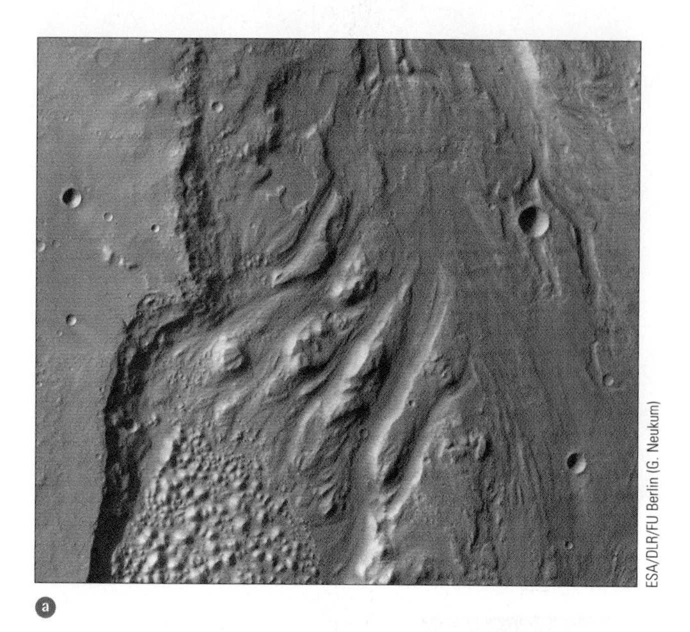

ESA/DLR/FU Berlin (G. Neukum)

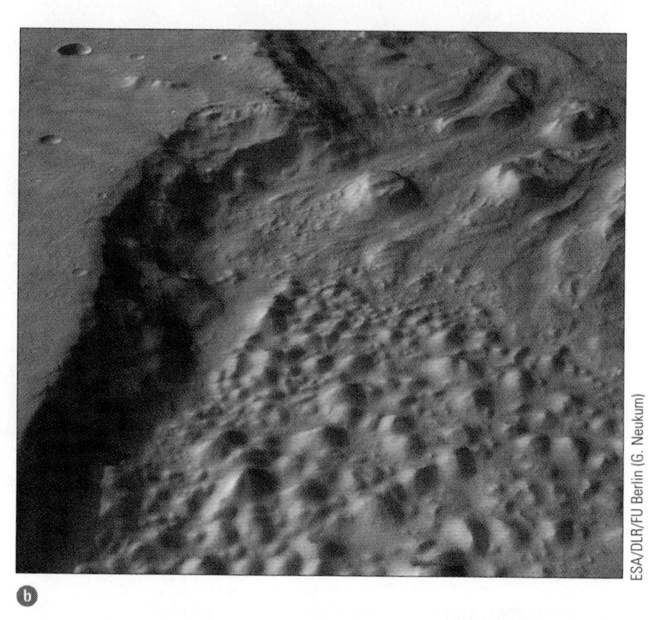

ESA/DLR/FU Berlin (G. Neukum)

■ **FIGURE 6–52** The past presence of water on Mars is revealed by features like the ones shown here, imaged with the European Space Agency's Mars Express. ⓐ Erosion in Ares Vallis, no doubt caused by flowing water. ⓑ The results of outflow in Ares Vallis, reimaged to give a perspective view.

flowed, perhaps fairly recently in geological times (■ Fig. 6–52). A leading model suggests that the source of this water is melting snow on the undersides of snowpacks.

This indication that water most likely flowed on Mars is particularly interesting because biologists feel that water is necessary for the formation and evolution of life as we know it. The presence of water on Mars, even in the past, may therefore indicate that life could have formed and may even have survived.

Where has all the water gone? Most of the water is probably in a permafrost layer—permanently frozen subsoil—beneath middle latitudes and polar regions, although a substantial fraction may have escaped from Mars in gaseous form. Some of the water is bound in the polar caps (■ Fig. 6–53). The large polar caps that extend to latitude 50° during the winter consist of carbon dioxide. But when a cap shrinks during its hemisphere's summer, a residual polar cap of water ice remains in the north, while the south has a residual polar cap of carbon dioxide ice, also with water ice below.

The European Space Agency's Mars Express joined NASA's Mars Global Surveyor in orbit around Mars in late 2003. Since entering its long-term orbit on 28 January 2004, it has been performing studies and global mapping of Mars's atmosphere and surface, analyzing their chemical composition, and sending back images of martian landscapes.

NASA/JPL/Caltech and Malin Space Science Systems

■ **FIGURE 6–53** The martian south polar cap, observed from Mars Global Surveyor just before the start of southern spring on Mars. We see the south polar cap as it retreats; wisps of a dust storm, caused by colder air blowing off the cap into warmer regions, appear just above it. The volcano Arsia Mons is at upper left.

6.5c Mars's Atmosphere

We have found that the martian atmosphere is composed of 95 per cent carbon dioxide with small amounts of carbon monoxide, oxygen, and water. The surface pressure is less than 1 per cent of that near Earth's surface. The current atmosphere is too thin to significantly affect the surface temperature, in contrast to the huge effects that the atmospheres of Venus and the Earth have on climate.

But long ago, up to about a billion years after it formed, Mars had a thicker atmosphere and a hospitable climate. Perhaps partial loss of the atmosphere, as it escaped into space, led to colder overall temperatures and the freezing of CO_2 and water. There would consequently be less greenhouse heating and hence even colder temperatures. This mechanism may have led to the inverse of what happened on Venus.

Observations from the Viking orbiters showed Mars's atmosphere in some detail. The lengthy period of observation led to the discovery of weather patterns on Mars.

A Closer Look 6.5 MARS EXPLORATION ROVERS

A half-dozen years after NASA's tiny Pathfinder rover won the hearts of Internet downloaders around the world, NASA landed a pair of Mars Exploration Rovers in early 2004. Named by a 9-year-old schoolgirl in a contest run by the Planetary Society and LEGO, Spirit and Opportunity have been roaming across Mars's surface for far longer than the few months of the original plan. (Some of the scientists at Cornell who are operating the spacecraft are keeping Mars time in their personal lives; with days that are 24½ Earth days long, they have unusual sleeping habits from the point of view of more ordinary earthlings.)

Spirit has entered a shallow crater and examined its walls, sending back images that reveal the past presence of water. Small, round pebbles, called "blueberries" by the scientists, are universally considered to have formed in water. And closeup examination of several rocks has shown long-term distortions that, in Earth equivalents, would only have formed if the rocks were bathed in an ocean. Layering is yet another sign that the material from which the rocks formed was laid down under water. A small amount of water may have flowed as recently as half a million years ago, though most disappeared within Mars's first billion years.

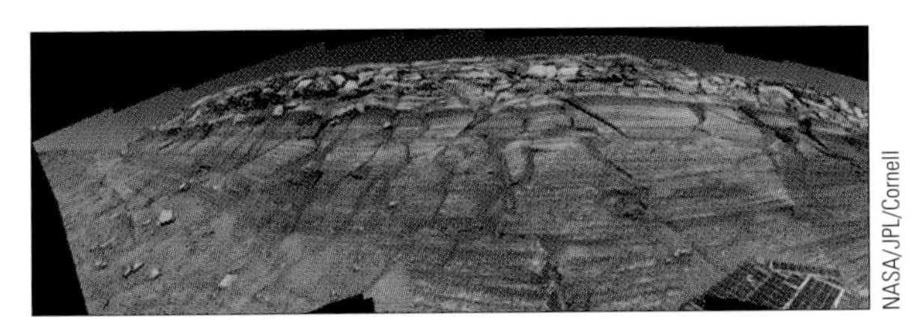

A mosaic of images from the Mars Exploration Rover named Opportunity made this wide-angle (180°) view of the face of Burns Cliff.

Sand dunes on Endurance Crater, imaged by Opportunity and displayed in false color. The chance of getting stuck in the dunes was too high to risk the rover's trying to traverse them.

The fine, parallel laminations in part of this rock, named El Capitan after a peak in a national park in Texas, indicate that it formed in water. Small spherical objects, nicknamed "blueberries," were found in the same region as the laminations. A spectrometer revealed the presence of jarosite, a mineral that contains water. The image is from Opportunity.

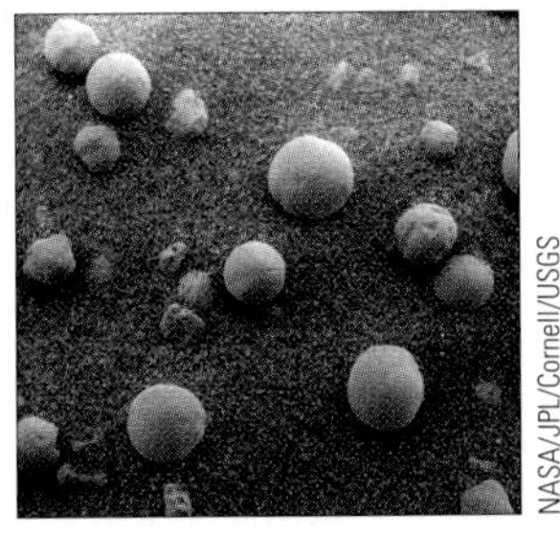

These spherical beads of hematite, 1 to 6 mm across, were remarked to resemble blueberries embedded in a muffin of martian soil. The name stuck. This microscopic image was taken with Opportunity. On Earth, hematite usually forms in the presence of liquid water, making scientists think that Opportunity's landing site had once been soaked in water.

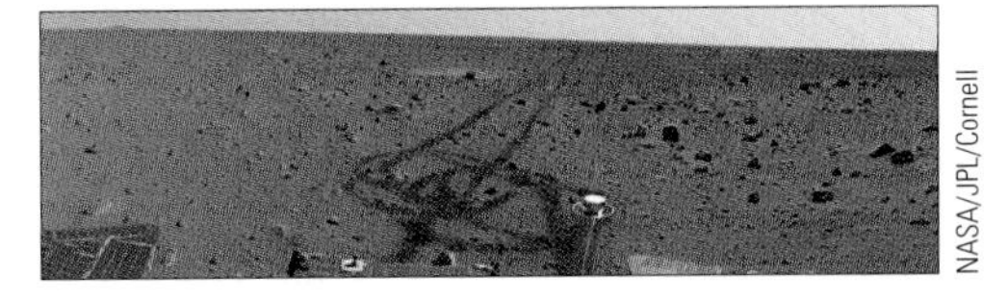

Spirit's tracks show in this view taken en route to Mars's Columbia Hills.

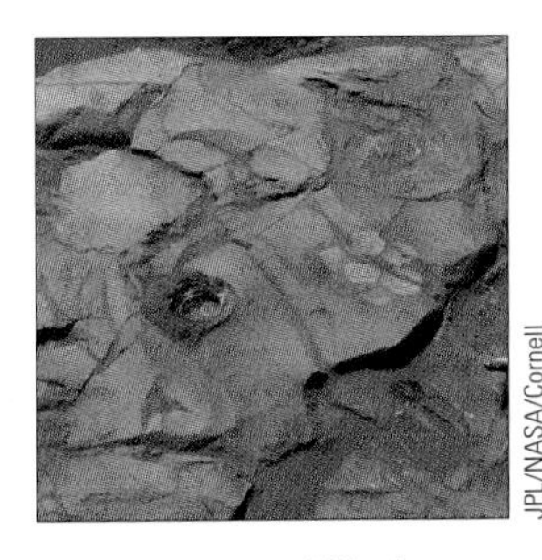

The rovers carry drills that make holes, as at left, or scrub the rock, as at right, to avoid surface dust as they prepare to measure the rock's chemical composition. This image was taken by Spirit.

The rovers' wheels have dug trenches in the dirt to explain the soil and what is just under it. We see an image from Spirit.

The current spacecraft orbiting Mars, NASA's Mars Global Surveyor and 2001 Mars Odyssey as well as ESA's Mars Express, map the weather there even better. The 2001 Mars Odyssey records narrow strips of Mars's surface at exceedingly high resolution (■ Fig. 6–54). Additional information continues to come from the Hubble Space Telescope. With Mars's rotation period similar to that of Earth, some features of its weather are similar to our own. Studies of Mars's weather have already helped us better understand windstorms in Africa that affect weather as far away as North America.

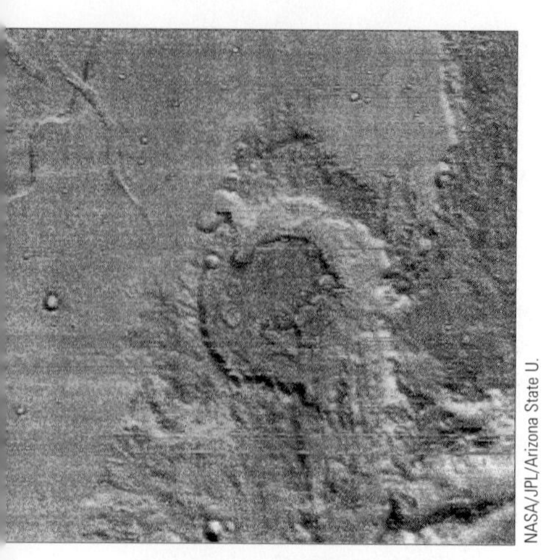
NASA/JPL/Arizona State U.

■ **FIGURE 6–54** Terra Sirenum, in a mosaic of several 2001 Mars Odyssey strips, using images through three infrared filters translated into visible-light colors. The false-color differences correspond to differences in the composition of the surface materials. The field of view is about 30 km by 30 km.

Studies of the effect of martian dust storms on the planet led to the idea that the explosion of many nuclear bombs on Earth might lead to a "nuclear winter." Dust thrown into the air would shield the Earth's surface from sunlight for a lengthy period, with dire consequences for life on Earth. Improvement of computer models for the circulation of atmospheres is contributing to the investigations. The effect, at present, seems smaller than first feared, so it is perhaps better described as "nuclear fall," which would still be something to avoid. A dust-caused winter would have resulted from the collision of one or more asteroids with Earth, as at least one did 65 million years ago; we shall discuss such collisions in Chapter 8 when commenting on the cause of the extinction of the dinosaurs at that time.

These models, and observations of Mars, also help us to understand the effects of smaller amounts of matter we are putting into the atmosphere from factories and fossil-fuel power plants and by the burning of forests. Of course, the absence of oceans on Mars (and their associated effects) complicates direct comparisons with Earth.

ASIDE 6.11: Why water?

Conditions on Mars's early surface were suitable for liquid water, but not for liquid carbon dioxide, methane, or other substances.

6.5d The Search for Life on Mars

On July 20, 1976, exactly seven years after the first crewed landing on the Moon, Viking 1's lander descended safely onto a martian plain called Chryse. The views showed rocks of several kinds, covered with yellowish-brown material that is probably an iron oxide (rust) compound. (Many of the initial photographs incorrectly showed an orange-red color, but subsequent careful color balancing of the images revealed the true yellowish-brown.) Some sand dunes were visible. The sky on Mars also turned out to be yellowish-brown, almost pinkish (■ Fig. 6–55); the color is formed as sunlight is scattered by dust suspended in the air as a result of Mars's frequent dust storms.

A series of experiments aboard the lander was designed to search for signs of life. A long arm was deployed and a shovel at its end dug up a bit of the martian surface. The soil was dumped into three experiments that searched for such signs of life as respiration and metabolism. The results were astonishing at first: The experiments sent back signals that seemed similar to those that would be caused on Earth by biological processes. But later results were less spectacular, and nonbiological chemical explanations seem more likely in all cases.

One important experiment gave much more negative results for the chance that there is life on Mars. It analyzed the soil and looked for traces of organic compounds. On Earth, many organic compounds left over from dead forms of life remain in the soil; living organisms themselves are only a tiny fraction of the organic material. Yet these experiments found no trace of organic material.

Unfortunately, any ancient organic material on the surface would have been destroyed by solar ultraviolet radiation, so its absence isn't proof that Mars never had any life. Still, it is a strong argument against the presence of recent life on the surface of Mars. But even if the life signs detected by Viking come from chemical rather than biological processes, as seems likely, we have still learned of fascinating new chemistry going on.

NASA/JPL/Caltech

■ **FIGURE 6–55** A Viking 2 view of Mars. The spacecraft landed on a rock and so was at an angle. The boom that supports Viking's weather station cuts through the center of the picture. ("Chance of precipitation," the local newscaster would say, "is 0 per cent.")

In 1996, study of a martian meteorite found in the Earth's Antarctic revealed possible evidence for ancient primitive (microbial) organisms. The chunk of rock, estimated to be 4.5 billion years old, was blasted from the surface of Mars about 15 million years ago by a collision. There is little doubt that the rock is from Mars, based on analysis of gases trapped within it; also, calculations show the likelihood that rocks blasted off Mars can reach Earth. The meteorite landed in Antarctica about 13,000 years ago. It was the first meteorite from the Allan Hills analyzed in 1984 at the Johnson Space Center, and so is called ALH 84001.

The meteorite contained carbonate globules (■ Fig. 6–56), which generally form in liquid water. It is thought that these globules were produced when water flowed through the rock, which was still on Mars about 3.6 billion years ago. Within the globules, certain types of organic compounds were found. Although these types of compounds can be formed in a number of ways other than by life, and are sometimes seen in normal meteorites, these particular ones were relatively unusual. A few other substances typical of life (such as magnetite, a mineral produced by some types of bacteria) were also seen. But photographs of tube-like structures resembling the tiniest Earth bacteria received the most publicity (■ Fig. 6–57).

Although none of these findings definitively implied the presence of life, taken together they were certainly intriguing (yet still not compelling). Several researchers challenged the conclusions on a number of grounds. For example, they cited evidence for contamination of the meteorite by substances on Earth. The original research team responded to these criticisms with reasonable counterarguments, and stand by their original announcement. Moreover, in 2002 they published a report stating that they had found signs of magnetism of a type formed by life. Still, the results are highly controversial, and additional tests are being carried out; most scientists remain skeptical.

Regardless of whether the discovery of ancient microbial life is correct, it raises an interesting possibility: Earth life may have originated on Mars, and subsequently migrated to Earth in a meteorite. If so, we are the descendants of Martians!

In 2003, the European Space Agency's Mars Express reached Mars. It carried a last-minute, poorly funded but intelligently designed add-on search-for-life laboratory named Beagle 2. (The original Beagle was Charles Darwin's ship in the Galápagos islands.) The last communication from Beagle 2 occurred when it was released from the mother ship, but then it was never heard from again. Its rocket may have misfired, or something might have broken as it bounced around off giant balloons as it landed on Christmas day. Its failure was obviously a big disappointment, though the report of the subsequent investigation revealed that it was inadequately planned and funded from its start.

NASA

■ **FIGURE 6–56** A microscopic view of a small (2.3 mm across), thin section from one of the Mars meteorites. It shows globules of carbonate materials.

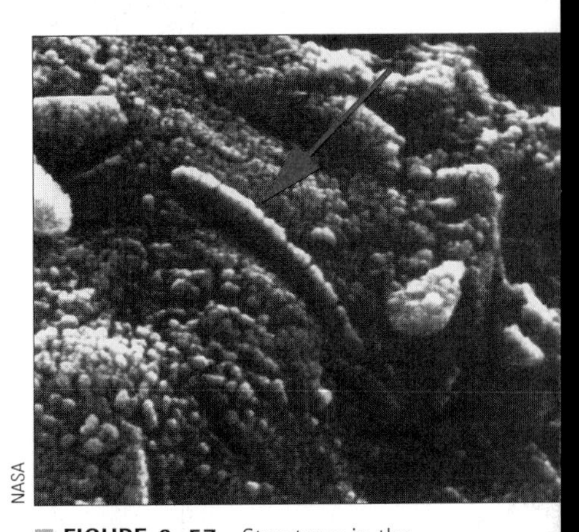

NASA

■ **FIGURE 6–57** Structures in the "Mars meteorite," seen with a special kind of microscope, are thought by a few scientists to be remnants of primitive life-forms but by most other scientists to be natural formations, too small to be fossils of life-forms.

6.5e Mars in Our Future

Because of the tremendous interest in Mars and the possibility of finding out about the origin of life by studying it, space agencies in the United States and elsewhere are planning a whole set of spacecraft to Mars in the near future. Eventually, martian rock and soil will be brought to Earth, though the goal of doing so by about 2015 has been postponed for budgetary reasons. Many scientists feel that only with this material to study using sophisticated Earth-based microscopes and other instruments can we really understand what Mars is like.

The American government has made it a goal for NASA to send astronauts to Mars, and they are redoing many of NASA's priorities to make that happen, even though the trip would be perhaps 30 years in the future. (Astronomers are worrying that this priority is distorting NASA's mission and that it will lead to cutbacks in other important astronomical space science.) Note that though astronauts reached the Moon in only four days, Mars is so much farther away that a trip there would take a couple of years. Because of the cost and other problems, no such astronaut mission is likely for decades.

Mars Reconnaissance Orbiter was launched in August 2005 for arrival in March 2006, carrying cameras and a radar. In the four years or more scheduled (it has enough fuel to last until 2014), it will send back many times more data than all previous spacecraft combined, searching for signs of past water and providing images that will help scientists and engineers choose the landing sites for the next missions. Mars Reconnaissance Orbiter carries a 50-cm-diameter telescope mirror, the largest ever sent to another planet, so it can resolve surface features as small as a desktop. It will also survey Mars's weather and climate, and it will act as a high-speed relay for communications from other spacecraft in orbit around Mars or on Mars's surface.

Those next missions include the Phoenix Mars Scout (■ Fig. 6–58), scheduled for launch in mid-2007, to land in order to search for organic chemicals, and the Mars Science Laboratory, a rover scheduled for launch in late-2009.

In the meantime, we watch Mars from a variety of spacecraft on and around it. From Earth, it is only a tiny spot in the sky (■ Fig. 6–59).

Courtesy of the Phoenix Mission

■ **FIGURE 6–58** This 2005 photo shows the lander of the Phoenix Mission to Mars suspended in the hot-fire test bed at Lockheed Martin Space Systems in Denver.

Moon image: T. A. Rector, I. P. Dell'Antonio NOAO/AURA/NSF; All other images: Hubble Space Telescope STScI/AURA/NASA; Designed graphics: J. R. Gott (Princeton) and F. J. Summers (STScI)

FIGURE 6–59 Sizes of celestial objects as seen from Earth. Shown here are the Moon, six planets, and some images from the Hubble Space Telescope at the correct relative size as they would appear in the sky. Note how small a field of view Hubble has when compared to the full moon. Hubble observes less than one-millionth of the sky at a time. In these images, the pillars in the Eagle Nebula and the Hubble Deep Field were observed with the Wide Field and Planetary Camera 2, whereas the Tadpole galaxy and the cluster of galaxies were observed with the Advanced Camera for Surveys.

CONCEPT REVIEW

The **terrestrial planets,** the four inner planets, are rocky worlds. Venus and the Earth have similar sizes, masses, and densities, but are otherwise very different. Mars is about half the diameter of the Earth and has an atmosphere $^1/_{100}$ of the pressure of the Earth's. Compared with the giant planets we will discuss in the next chapter, the terrestrial planets are smaller, rocky, and dense, and they have fewer moons and no rings. **Comparative planetology** has applications for understanding weather, earthquakes, and other topics of use to us on Earth (Sec. 6.1).

Geology is the study of the Earth's interior and surface, and often uses **seismology,** the study of waves that pass through the Earth, to study the Earth's interior (Sec. 6.1a). The Earth has a layered structure, with a **core, mantle,** and **crust.** Radioactivity heats the interior, providing **geothermal energy** (Sec. 6.1b). Continental **plates** move around the surface, as explained by **plate tectonics,** the current theory that elaborates on the earlier notion of **continental drift.**

Tides are a differential force caused by the fact that the near side of the Earth is closer to the Moon than the Earth's center, and so is subject to higher gravity, while the far side is farther than the center and is subject to lower gravity (Sec. 6.1c). The Earth's weather is confined in its atmosphere to the **troposphere** (Sec. 6.1d). The atmosphere and surface of the Earth are substantially warmer than they would have been without the **greenhouse effect.** In the **ionosphere,** atoms are ionized and the temperature is higher. The **Van Allen belts** of charged particles were a space-age discovery (Sec. 6.1e). Such particles lead to the **aurora borealis** and **aurora australis,** the northern and southern lights.

Our own Moon is midway in size between Pluto and Mercury. Its volcanic surface shows smooth **maria** (singular: **mare**) and cratered highlands (Sec. 6.2a). Shadows reveal relief best along the **terminator,** the day-night line. Most of the craters are from meteoritic impacts (Sec. 6.2b). The oldest rocks are from 4.4 billion years ago; on the Earth, erosion and plate tectonics have erased most of the oldest rocks though a few equally old terrestrial rocks have been found. **Mascons** are mass concentrations of high density under the Moon's surface (Sec. 6.2c). The current leading model for the Moon's formation is that a body perhaps twice the size of Mars hit the proto-Earth, ejecting matter into a ring that coalesced (Sec. 6.2d). A handful of meteorites found on Earth have been identified by their chemical composition as having come from the Moon (Sec. 6.2e).

Mercury always appears close to the Sun in the sky; rarely it even goes into **transit** across the Sun's disk (Sec. 6.3). The correct rotation period of Mercury was first measured using radar, and proved to be linked to its orbital period (Sec. 6.3a). Like the Moon, Mercury has a dark (low **albedo**) surface (Sec. 6.3b). Mercury is heavily cratered, and shows **scarps** that resulted from a planetwide shrinkage (Sec. 6.3c). It has a weak magnetic field, which is a surprise because we think Mercury's core has solidified. It is too close to the Sun to retain a thick atmosphere, but a very thin atmosphere does exist (Sec. 6.3d). Mercury may be the fragment of a giant early collision that nearly stripped it to its core (Sec. 6.3e). New spacecraft are on their way to Mercury, or are now being planned (Sec. 6.3f).

In 2004, Venus transited the Sun, for the first time in over 100 years (Sec. 6.4a). The highly reflective clouds that shroud Venus, giving it a high albedo, are mainly droplets of sulfuric acid, though the atmosphere consists primarily of carbon dioxide (Sec. 6.4b). Venus's atmospheric surface pressure is 90 times higher than Earth's. Radar penetrated Venus's clouds to show that Venus is rotating slowly backward and to map its surface (Sec. 6.4c).

Venus's surface temperature is very high, about 750 K, heated by the greenhouse effect (Sec. 6.4d). Earth escaped such a fate, but we should be careful about disrupting the balance in our atmosphere, lest conditions change too quickly. The atmosphere of Venus has been studied with various spacecraft (Sec. 6.4e). Radar is used to map the surface of Venus (Sec. 6.4f). Higher-**resolution** radar reveals smaller surface features. New spacecraft are on their way to Venus, or are now being planned (Sec. 6.4g).

Stories about life on Mars have long inspired study of this planet. We now know the seasonal surface changes to be the direct result of winds that arise as the sunlight hits the planet at different angles over a martian year (Sec. 6.5a). Mars's weather and dust storms help us understand our own weather. Mars's surface appears reddish-orange because of rusty dust; careful calibration of photos taken on Mars shows that it is actually yellowish-brown.

Mars boasts of giant volcanoes and a canyon longer than the width of the continental United States (Sec. 6.5b). There is strong evidence that Mars had liquid water flowing on its surface long ago. Mars's current atmosphere is very thin, with a surface pressure less than 1 per cent of that near Earth's surface (Sec. 6.5c). But long ago, up to about a billion years after it formed, Mars had a thicker atmosphere and a hospitable climate.

In the 1970s, the Viking landers found no evidence for life on Mars, despite initially promising signs (Sec. 6.5d). Studies of a martian meteorite provided suggestive, but not convincing, evidence for the presence of ancient, primitive life on Mars. A series of spacecraft to Mars should eventually culminate in a sample return, in part to help settle this very controversial claim (Sec. 6.5e).

QUESTIONS

1. How did the layers of the Earth arise? Where did the energy that is flowing as heat come from?
2. What carries the continental plates around over the Earth's surface?
3. **(a)** Explain the origin of tides. **(b)** If the Moon were farther away from the Earth than it actually is, how would tides be affected?

4. Draw a diagram showing the positions of the Earth, Moon, and Sun at a time when there is the least difference between high and low tides.
†5. Calculate your weight if you were standing on the Moon.
6. Look at a globe and make a list or sketches of which pieces of the various continents probably lined up with each other before the continents drifted apart.
7. To what locations, relative to the Earth-Sun line, does the Earth's terminator correspond?
8. What does cratering tell you about the age of the surface of the Moon, compared to that of the Earth's surface?
9. Why are we more likely to learn about the early history of the Earth by studying the rocks from the Moon than those on the Earth?
10. Why may the near side and far side of the Moon look different?
11. Discuss one of the proposed theories to describe the origin of the Moon. List points both pro and con.
12. How can we get lunar material to study on Earth?
13. Assume that on a given day, Mercury sets after the Sun. Draw a diagram, or a few diagrams, to show that the height of Mercury above the horizon depends on the angle that the Sun's path in the sky makes with the horizon as the Sun sets. Discuss how this depends on the latitude or longitude of the observer.
14. If Mercury did always keep the same side toward the Sun, would that mean that the night side would always face the same stars? Draw a diagram to illustrate your answer.
15. Explain why a day/night cycle on Mercury is 176 Earth day/night cycles long.
16. What did radar tell us about Mercury? How did it do so?
17. If ice has an albedo of 70–80%, and volcanic rocks typically have albedoes of 5–20%, what can you say about the surface of Mercury based on its measured albedo?
18. If you increased the albedo of Mercury, would its surface temperature increase or decrease? Explain.
19. How would you distinguish an old crater from a new one?
20. What evidence is there for erosion on Mercury? Does this mean there must have been water on the surface?
21. List three major findings of Mariner 10.
22. Make a table displaying the major similarities and differences between the Earth and Venus.
23. Why does Venus have more carbon dioxide in its atmosphere than does the Earth?
24. Why do we think that there have been significant external effects on the rotation of Venus?
25. Suppose a planet had an atmosphere that was opaque in the visible but transparent in the infrared. Describe how the effect of this type of atmosphere on the planet's temperature differs from the greenhouse effect.
26. Why do radar observations of Venus provide more data about the surface structure than a flyby with close-up optical cameras?
27. Why do we say that Venus is the Earth's "sister planet"?
28. Describe the most current radar observations of Venus.
29. What are some signs of volcanism on Venus?
30. List three of the features of Mars that made scientists think that it was a good place to search for life.
†31. Compare the tallest volcanoes on Earth and Mars relative to the diameters of the planets. Give the ratios.
32. What evidence exists that there is, or has been, water on Mars?
33. Describe the composition of Mars's polar caps.
34. Consult an atlas and compare the sizes of the Grand Canyon in Arizona and the Rift Valley in Africa. How do they compare in size with the giant canyon on Mars?
35. Compare the temperature ranges on Venus, Earth, and Mars.
36. List the evidence from Mars landers for and against the existence of life on Mars.
37. Plan a set of experiments or observations that you, as a martian scientist, would have an uncrewed spacecraft carry out on Earth to find out if life existed here. What data would your spacecraft radio back if it landed in a cornfield? In the Sahara? In the Antarctic? In New York's Times Square?
38. **True or false?** All four terrestrial planets show evidence of water flowing on their surfaces, either now or in the distant past.
39. **True or false?** Earth's Moon has a composition similar to that of the Earth's crust, consistent with the hypothesis that the Moon formed after a large object collided with Earth.
40. **True or false?** At any given coastal location on Earth, high tide occurs only once per day, when the Moon is overhead.
41. **True or false?** The terrestrial planets have iron cores and rocky outer parts; the iron sank when they were young and molten.
42. **True or false?** Of the four terrestrial planets, Earth is the only one that now has obvious plate tectonics.
43. **Multiple choice:** Which one of the following statements about the greenhouse effect, Venus, and Earth is *false*? **(a)** An extreme, possibly runaway, greenhouse effect occurred on Venus, making its planetary surface the hottest in the Solar System. **(b)** Venus's atmosphere is much thicker than that of Earth—but some of Earth's gases are trapped in rocks and oceans. **(c)** If we dump much more carbon dioxide into Earth's atmosphere, Earth might become significantly hotter due to the greenhouse effect. **(d)** The greenhouse effect occurs when an atmosphere is transparent to optical (visible)

light but opaque to infrared light. (**e**) No greenhouse effect currently occurs on Earth, and this is a good thing for humans.

†44. **Multiple choice:** If the Earth's radius suddenly shrank by a factor of 2 but the Earth's mass remained unchanged, how much would you weigh while standing on the new (smaller) surface of the Earth? (**a**) 16 times as much. (**b**) 4 times as much. (**c**) Twice as much. (**d**) Half as much. (**e**) Your weight would remain unchanged.

45. **Multiple choice:** Which one of the following statements about the Earth-Moon system is *true*? (**a**) The same half of the Moon's surface is perpetually dark (craters on that side never see sunlight), leading us to call it the "dark side of the Moon." (**b**) At a given location on Earth, there are two high tides each day—one caused by the gravitational pull of the Sun, and the other by the gravitational pull of the Moon. (**c**) High tide occurs on the side of the Earth nearest to the Moon, while low tide occurs on the opposite side of the Earth. (**d**) By observing the Moon long enough from Earth, night after night, we are able to draw a map of its entire surface. (**e**) The orbital period of the Moon around the Earth and the rotation period of the Moon around its axis are equal.

46. **Multiple choice:** Why does Mars appear reddish-orange from Earth? (**a**) Its surface temperature is lower than that of the Earth, which appears blue, and according to Wien's law Mars is therefore redder. (**b**) Its thick atmosphere consists of reddish-orange clouds. (**c**) Its rocks have suffered "rusting" and contain reddish-orange iron oxides. (**d**) It is moving away from us rapidly and hence is Doppler redshifted. (**e**) There is a large population of Martians wandering around, and they tend to have a reddish-orange skin color.

47. **Multiple choice:** The surface of Venus is best observed using (**a**) ultraviolet satellites; (**b**) radar; (**c**) large optical telescopes; (**d**) large radio telescopes; or (**e**) infrared telescopes.

48. **Fill in the blank:** Earth's ________ occur in regions where Earth's magnetic field lines intersect its atmosphere, as a result of collisions between charged particles and air molecules.

49. **Fill in the blank:** The lunar ______ must have formed more recently than the cratered highlands, covering up older craters.

50. **Fill in the blank:** _______ is often seen as the brightest evening or morning "star" in the sky, in part because it is shrouded in highly reflective clouds.

51. **Fill in the blank:** A planet is said to be in _______ when it appears to move across the face of the Sun.

52. **Fill in the blank:** The process of forming layers within a planet, because of differences in density between materials, is called ________.

†This question requires a numerical solution.

TOPICS FOR DISCUSSION

1. It is sometimes said that the U.S. mission to the Moon was entirely motivated by the Soviet Union's launch of the Sputnik satellite in 1957. Do you think the scientific benefits of lunar landings would have been sufficient reason to take the risks and spend the funds?

2. Do you think the evidence for global warming of Earth is strong? Will it be too late to reverse this trend, if and when the effect becomes so large that its presence is unambiguous?

3. How likely do you think it is that humans will eventually "terraform" Mars so that its climate becomes suitable for humans?

MEDIA

Virtual Laboratories

- Planetary Geology
- Tides and Tidal Forces in Astronomy
- Planetary Atmospheres and Their Retention

AceAstronomy™ Log into AceAstronomy at **http://astronomy.brookscole.com/cosmos3** to access quizzes and animations that will help you assess your understanding of this chapter's topics.

Log into the Student Companion Web Site at **http://astronomy.brookscole.com/cosmos3** for more resources for this chapter including a list of common misconceptions, news and updates, flashcards, and more.

C H A P T E R 7

The Jovian Planets: Windswept Giants

ORIGINS

We find that the properties of giant planets differ from those of Earth because they formed much farther out from the center of the solar nebula. Comparison of features on these planets and their moons leads to a greater understanding of the origin of Earth's properties.

AIMS

1. Understand the giant planets of our Solar System, their moons, and their rings (Sections 7.1 to 7.4).
2. Describe similarities and differences among Earth, the giant planets, and their moons (Sections 7.1 to 7.4).
3. See how spacecraft have vastly improved our understanding of the giant planets, their moons, and their rings (Sections 7.1 to 7.4).

Jupiter, Saturn, Uranus, and Neptune are **giant planets**; they are also called the **jovian planets.** They are much bigger, more massive, and less dense than the inner, terrestrial planets; see *A Closer Look 7.1: Comparative Data for the Major Worlds.* Their internal structure is entirely different from that of the four inner planets. In this chapter, we also discuss a set of moons of these giant planets, some of which range in diameter between 1/2 and 1/4 the size of the Earth, as large as Mercury or Pluto. Close-up space observations have shown that these moons are themselves interesting objects for study.

Jupiter is the largest planet in our Solar System (■ Fig. 7–1). Some of its very numerous moons are close in size to the terrestrial planets and show fascinating surface structure.

Saturn has long been famous for its beautiful rings. We now know, however, that each of the other giant planets also has rings. When seen close-up, as on the opposite page, the astonishing detail in the rings is very beautiful.

Uranus and Neptune were known for a long time to us as mere points in the sky. Spacecraft views have transformed them into objects with more character.

The age of first exploration of the giant planets, with spacecraft that simply flew by the planets, is over. We now are in the stage of space missions to orbit the planets, with a Jupiter orbiter having recently completed its mission and a Saturn orbiter that started collecting data in 2004. These missions study the planets, their rings, their moons, and their magnetic fields in a much more detailed manner than before.

7.1 Jupiter

Jupiter, the largest and most massive planet, dominates the Sun's planetary system. It alone contains two-thirds of the mass in the Solar System outside of the Sun, 318 times as much mass as the Earth (but only 0.001 times the Sun's mass). Jupiter has at least 52 moons of its own and so is a miniature "planetary system" (that is, several planet-like objects orbiting a central object) in itself. It is often seen as a bright object in our night sky, and observations with even a small telescope reveal bands of clouds

NASA/JPL/U. Arizona

■ **FIGURE 7–1** Crescent Jupiter, in a farewell view as the Cassini spacecraft left the Jupiter system in 2001.

AceAstronomy™ The AceAstronomy icon throughout this text indicates an opportunity for you to test yourself on key concepts, and to explore animations and interactions of the AceAstronomy website at **http://astronomy.brookscole.com/cosmos3**

This close-up view of Saturn and its rings was taken in 2005 from the Cassini spacecraft. Mimas is seen in front of the rings' shadows falling on Saturn's disk, just below part of the disk illuminated by light passing through the rings' Cassini Division. A section of Saturn's rings, so thin that we can see through it, appears at the bottom.
NASA/JPL/Space Science Institute

A Closer Look 7.1 Comparative Data for the Major Worlds

Planet	Semimajor Axis of Orbit	Orbital Period	Equatorial Radius ÷ Earth's	Mass ÷ Earth's
Jupiter	5.2 A.U.	12 years	11.2	318
Saturn	9.5 A.U.	29 years	9.4	95
Uranus	19.2 A.U.	84 years	4.0	15
Neptune	30.1 A.U.	164 years	3.9	17

More detailed information appears in Appendices 3 and 4.

across its surface and show four of its moons, the Galilean satellites. (See *Star Party 7.1: Observing the Giant Planets.*)

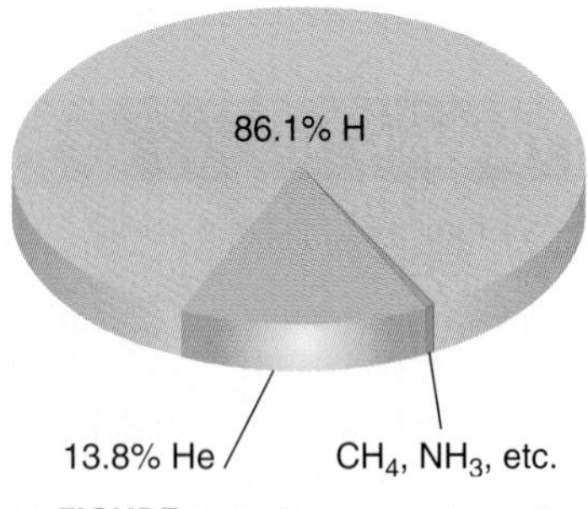

FIGURE 7–2 The composition of Jupiter.

Jupiter is more than 11 times greater in diameter than the Earth (see *Figure It Out 7.1: The Size of Jupiter*). From its mass and volume, we calculate its density to be 1.3 g/cm^3, not much greater than the 1 g/cm^3 density of water. This low density tells us that any core of heavy elements (such as iron) makes up only a small fraction of Jupiter's mass. Jupiter, rather, is mainly composed of the light elements hydrogen and helium. Jupiter's chemical composition is closer to that of the Sun and stars than it is to that of the Earth (■ Fig. 7–2), so its origin can be traced directly back to the solar nebula with much less modification than the terrestrial planets underwent.

Jupiter has no crust. At deeper and deeper levels, its gas just gets denser and denser, turning mushy and eventually liquefying about 20,000 km (15 per cent of the way) down. Jupiter's core, inaccessible to direct study, is calculated to be made of heavy elements and to be larger and perhaps 10 times more massive than Earth.

Jupiter's "surface" (actually, the top of the clouds that we see) rotates in about 10 hours, though different latitudes rotate at slightly different speeds. Regions with different speeds correspond to different bands; Jupiter has a half-dozen jet streams while Earth has only one in each hemisphere. Jupiter's clouds are in constant turmoil; the shapes and distribution of bands can change within days. The bright bands are called "zones" and the dark bands are called "belts," but the strongest winds appear on the boundaries between them. The zones seem to be covered by a uniformly high cloud deck. The belts have both towering convective clouds and lightning, as well as clear spaces that allow glimpses of the deeper atmosphere.

Star Party 7.1 Observing the Giant Planets

Use a telescope, perhaps the one on your campus, to view the giant planets. Jupiter and Saturn are among the brightest objects in the sky, and are easy to locate, but Uranus and especially Neptune are more challenging. (To get current positions of planets, consult *Sky & Telescope* or *Astronomy* magazines, the *Field Guide to the Stars and Planets*, or planetarium software.)

You should be able to see a few of the main bands of Jupiter, and maybe even the Great Red Spot, in addition to the bright Galilean satellites. Sometimes one or more of these moons is behind Jupiter or casts a shadow on its disk. Their relative positions change from night to night; the innermost moon, Io, takes only 1.8 days to complete an orbit.

Saturn's bands are less distinct that those of Jupiter, but the rings look magnificent. You may be able to see the dark gap in the rings, known as Cassini's division, as well as the shadow of the rings on the disk of the planet. The rings appear edge-on to our line of sight twice during Saturn's 29-year orbital period, making them difficult to detect at those times. They will be edge-on next in 2009 and 2016. Only one moon, Titan, is readily visible through a small telescope; it looks like a faint star.

Uranus and Neptune are too small and too far away to reveal detail, but you may be able to tell that they have resolved disks, unlike the much more distant, point-like stars. Try to decide whether Uranus and Neptune look greenish to your eyes.

FIGURE IT OUT 7.1
The Size of Jupiter

It is easy and instructive to estimate the physical diameter of Jupiter from its angular size. Jupiter's disk appears to be about 50 arc seconds across when it is closest to the Earth. (You can actually measure this yourself, by looking through a telescope with an eyepiece whose field of view you have already calibrated—perhaps by looking at the Moon, which is 30 arc minutes in diameter.) Since Jupiter is roughly 5 A.U. from the Sun and the Earth is 1 A.U. from the Sun, Jupiter must be about 4 A.U. from Earth when at its closest.

Now we need to convert an angular diameter into a physical one, measured in kilometers. If the angular size (ϕ, the Greek letter "phi") of an object is measured in radians, and if the object appears small in the sky, then its physical size (s) is given by its distance (d) multiplied by its angular size. This relation, very useful in astronomy and known as the *small-angle formula*, is expressed as

$$s = d\phi.$$

One radian is equal to 206,265 arc seconds, because in a full circle 2π radians correspond to 360°. Converting Jupiter's angular diameter of 50 arc seconds into radians, we have $\phi = 50/206{,}265 = 0.00024$ radians. Since 1 A.U. $= 1.5 \times 10^8$ km, the distance of Jupiter is 6.0×10^8 km. According to the small-angle formula, then, Jupiter's physical diameter must be $s = d\phi = (6.0 \times 10^8 \text{ km}) \times (0.00024) = 1.4 \times 10^5$ km. This size is about 11 times larger than Earth's diameter of 12,800 km.

The most prominent feature of Jupiter's surface is a large reddish oval known as the **Great Red Spot.** It is two to three times larger in diameter than the Earth. Other, smaller spots are also present.

Jupiter emits radio waves, which indicates that it has a strong magnetic field and strong "radiation belts." Actually, these are belts of magnetic fields filled with trapped energetic particles—large-scale versions of the Van Allen belts of Earth (see the discussion in Chapter 6).

AceAstronomy™ Log into AceAstronomy and select this chapter to see the Active Figure called "Small-Angle Formula."

7.1a Spacecraft to Jupiter

Our understanding of Jupiter was revolutionized in the 1970s, when first Pioneer 10 (1973) and Pioneer 11 (1974) and then Voyager 1 and Voyager 2 (both in 1979) flew past it. The Galileo spacecraft arrived at Jupiter in 1995, when it dropped a probe into Jupiter's atmosphere and went into orbit in the Jupiter system. The spacecraft plunged into Jupiter's atmosphere on September 21, 2003, ending a tremendously successful mission. (It was sent on that course in large part to avoid the possibility that it could eventually hit and contaminate the Galilean satellite Europa, on which some scientists speculate that life may exist, as we will discuss in Section 7.1g(ii).) The Cassini spacecraft flew by Jupiter, en route to Saturn, in 2000–2001 (■ Fig. 7–3). Each spacecraft

NASA/JPL/U. Arizona
a

NASA/JPL/U. Arizona
b

■ **FIGURE 7–3** Jupiter from the Cassini spacecraft in 2000. (a) Jupiter's bands and Great Red Spot are visible; Jupiter's moon Europa casts its shadow. (b) Jupiter's moon Ganymede is visible alongside Jupiter. Ganymede is larger than the planet Mercury.

carried many types of instruments to measure various properties of Jupiter, its satellites, and the space around them.

ASIDE 7.1: Comparative planetology: rings

Jupiter: One relatively narrow ring, discovered by Voyager.

Saturn: Glorious ring system with many rings and divisions.

Uranus: Eleven narrow rings, discovered during a stellar occultation.

Neptune: Several rings discovered during a stellar occultation; one ring has several clumps.

7.1b The Great Red Spot

The Great Red Spot is a gaseous "island" a few times larger across than the Earth (■ Fig. 7–4). It is the vortex of a violent, long-lasting storm, similar to large storms on Earth, and drifts about slowly with respect to the clouds as the planet rotates. From the sense of its rotation (counterclockwise rather than clockwise in the southern hemisphere), measured from time-lapse photographs, we can tell that it is a pressure high rather than a low. We also see how it interacts with surrounding clouds and smaller spots. The Great Red Spot has been visible for at least 150 years, and maybe even 300 years. Sometimes it is relatively prominent and colorful, and at other times the color may even disappear for a few years.

Why has the Great Red Spot lasted this long? Heat, energy flowing into the storm from below it, partly maintains its energy supply. The storm also contains more mass than hurricanes on Earth, which makes it more stable. Furthermore, unlike Earth, Jupiter has no continents or other structure to break up the storm. Also, we do not know how much energy the Spot gains from the circulation of Jupiter's upper atmosphere and eddies (rotating regions) in it. Until we can sample lower levels of Jupiter's atmosphere, we will not be able to decide definitively.

Studying the eddies (swirls) in Jupiter's atmosphere helps us interpret features on Earth. For example, one hypothesis to explain Jupiter's spots holds that they are similar to circulating rings that break off from the Gulf Stream in the Atlantic Ocean.

7.1c Jupiter's Atmosphere

Heat emanating from Jupiter's interior churns the atmosphere. (In the Earth's atmosphere, on the other hand, most of the energy comes from the outside—from the Sun.) Pockets of gas rise and fall, through the process of convection, as described for the Sun in Chapter 10. The bright bands ("zones") and dark bands ("belts") on Jupiter represent different cloud layers (■ Fig. 7–5).

Wind velocities show that each hemisphere of Jupiter has a half-dozen currents blowing eastward or westward. The Earth, in contrast, has only one westward current at low latitudes (the trade winds) and one eastward current at middle latitudes (the jet stream).

■ **FIGURE 7–4** ⓐ The Great Red Spot, in an image from the Galileo spacecraft. It rotates in the anticyclonic (counterclockwise) sense. The Great Red Spot is now about 26,000 km by 134,000 km and has been gradually shrinking for 50 years. It goes 20 to 40 km deep. ⓑ This Galileo image reveals clouds of ammonia ice, shown in light blue to the upper left of the Great Red Spot, which appears in false color. The ammonia-ice clouds are produced by powerful updrafts of ammonia-laden air caused by turbulence in Jupiter's atmosphere that results from the Great Red Spot. Red-orange shows high-level clouds, yellow shows mid-level clouds, and green shows low-level clouds.

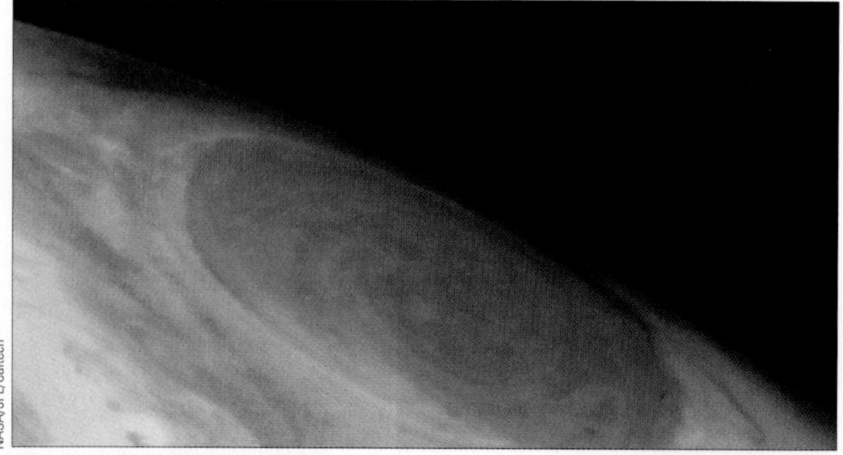

NASA/JPL/Caltech

ⓐ

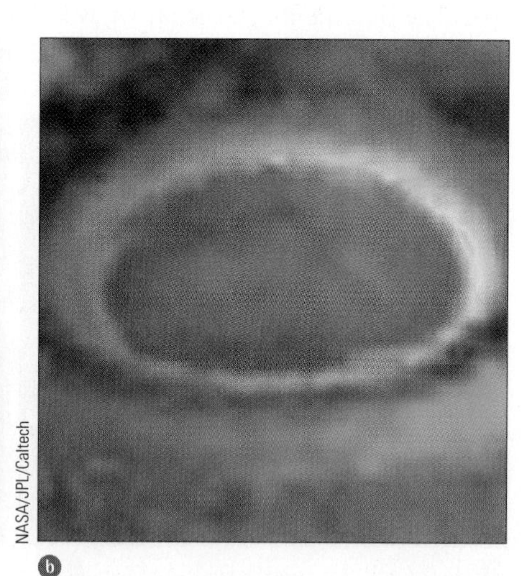

NASA/JPL/Caltech

ⓑ

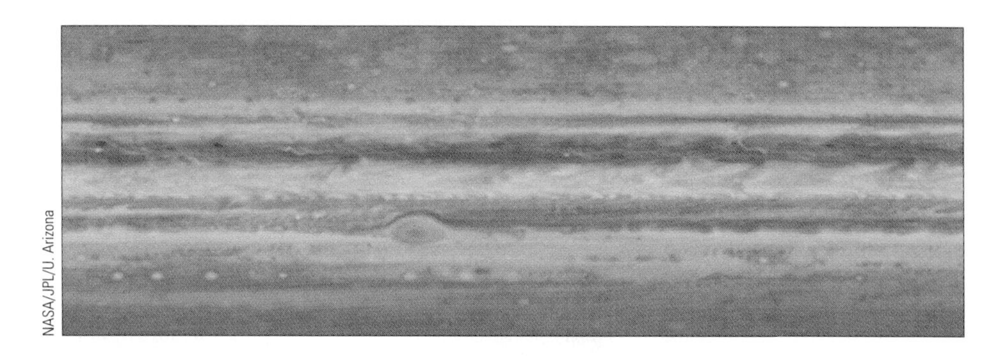

NASA/JPL/U. Arizona

■ **FIGURE 7–5** Bright and dark bands on Jupiter, as seen from the Cassini spacecraft. Computer processing has unrolled the surface.

On December 7, 1995, a probe dropped from the Galileo spacecraft transmitted data for 57 minutes as it fell through Jupiter's atmosphere. It gave us accurate measurements of Jupiter's composition; the heights of the cloud layers; and the variations of temperature, density, and pressure. It went through about 600 km of Jupiter's atmosphere, only about 1 per cent of Jupiter's radius. The probe found that Jupiter's winds were stronger than expected and increased with depth, which shows that the energy that drives them comes from below.

Extensive lightning storms, including giant-sized lightning strikes called "superbolts," were discovered from the Voyagers. The Galileo spacecraft photographed giant thunderclouds on Jupiter, which indicates that some regions are relatively wet and others relatively dry. The probe found less water vapor than expected, probably because it fell through a dry region.

7.1d Jupiter's Interior

Most of Jupiter's interior is in liquid form. Jupiter's central temperature may be between 13,000 and 35,000 K. The central pressure is 100 million times the pressure of the Earth's atmosphere measured at our sea level due to Jupiter's great mass pressing in. (The Earth's central pressure is 4 million times its atmosphere's pressure, and Earth's central temperatures are several thousand degrees.) Because of this high pressure, Jupiter's interior is probably composed of ultra-compressed hydrogen surrounding a rocky core consisting of perhaps 10 Earth masses of iron and silicates (■ Fig. 7–6).

Jupiter radiates 1.6 times as much heat as it receives from the Sun. It must have an internal energy source—perhaps the energy remaining from its collapse from a primordial gas cloud 20 million km across or from the accretion of matter long ago. Jupiter is undoubtedly still contracting inside and this process also liberates energy. It lacks the mass necessary by a factor of about 75, however, to have heated up enough to become a star, generating energy by nuclear processes (see Chapter 12). It is therefore not "almost a star," contrary to some popular accounts.

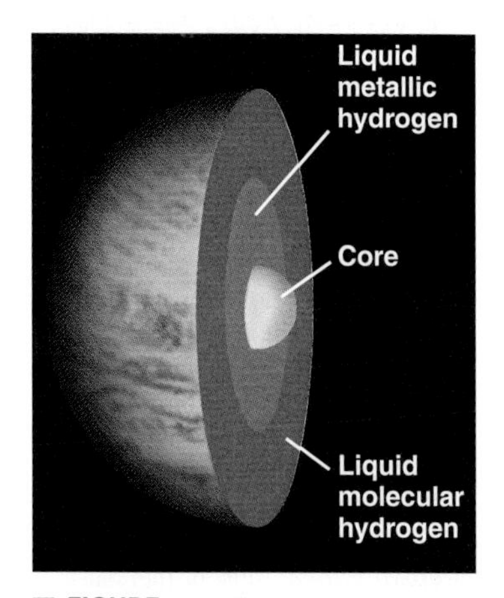

■ **FIGURE 7–6** The current model of Jupiter's interior.

7.1e Jupiter's Magnetic Field

The space missions showed that Jupiter's tremendous magnetic field is even more intense than many scientists had expected (■ Fig. 7–7*a*). At the height of Jupiter's clouds, the magnetic field strength is 10 times that of the Earth, which itself has a rather strong field.

The inner field is shaped like a doughnut, containing several shells of charged particles, like giant versions of the Earth's Van Allen belts (Fig. 7–7*b*). The outer region of Jupiter's magnetic field interacts with the particles flowing outward from the Sun. When this solar wind is strong, Jupiter's outer magnetic field (shaped like a pancake) is pushed in. When the high-energy particles interact with Jupiter's magnetic field, radio emission results. Jupiter's magnetic field leads Jupiter to have giant auroras (Fig. 7–7*c*).

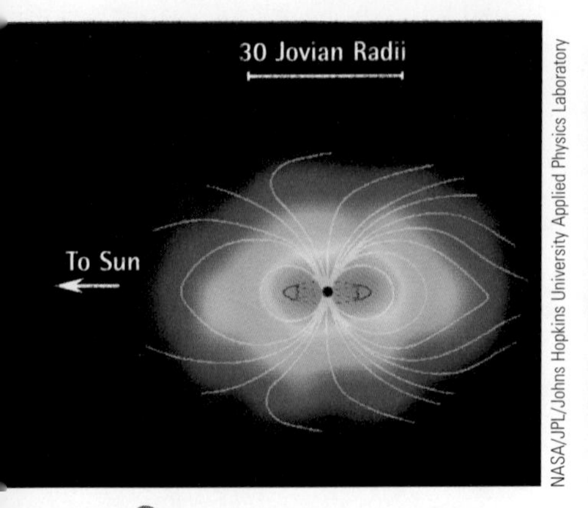

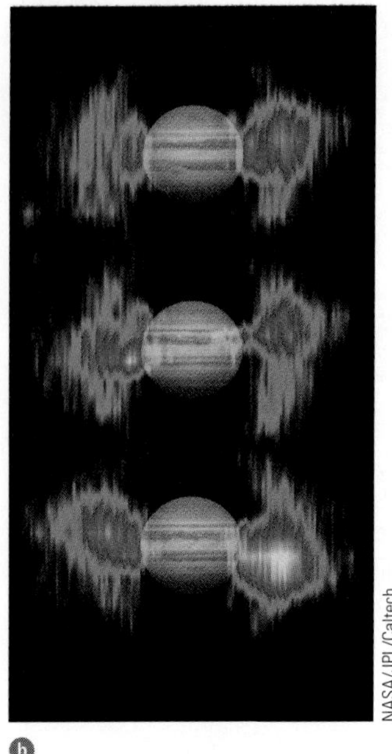

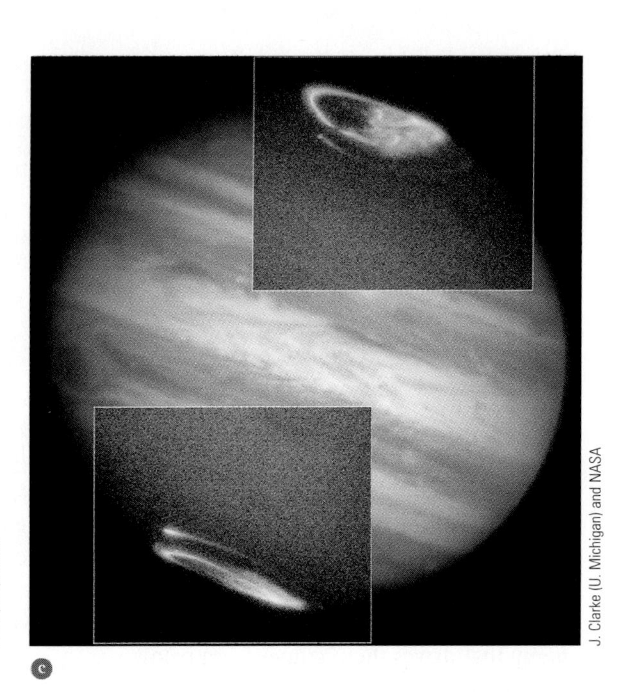

■ **FIGURE 7–7** (a) Cassini could detect neutral atoms expelled from the magnetosphere. A black dot shows Jupiter's size. Lines of Jupiter's magnetic field and a cross section of the doughnut-shaped ring of material from Jupiter's moon Io are sketched. (b) Cassini's radio instrument, operating at a wavelength of 2.2 cm, detected high-energy electrons in the radiation belts. A picture of Jupiter to scale is superimposed. (c) The northern and southern auroral ovals on Jupiter, imaged in the ultraviolet by the Hubble Space Telescope. They are about 500 km above the level in Jupiter's atmosphere where the pressure is the same as Earth's surface pressure.

7.1f Jupiter's Ring

Though Jupiter wasn't expected to have a ring, Voyager 1 was programmed to look for one just in case; Saturn's rings, of course, were well known, and Uranus's rings had been discovered only a few years earlier during ground-based observations. The Voyager 1 photograph indeed showed a wispy ring of material around Jupiter at about 1.8 times Jupiter's radius, inside the orbit of its innermost moon.

As a result, Voyager 2 was targeted to take a series of photographs of the ring. From the far side looking back, the ring appeared unexpectedly bright, probably because small particles in the ring scattered the light toward the spacecraft. Within the main ring, fainter material appears to extend down to Jupiter's cloud tops (■ Fig. 7–8). The ring particles were knocked off Jupiter's inner moons by micrometeorites. Whatever their origin, the individual particles probably remain in the ring only temporarily.

■ **FIGURE 7–8** Jupiter's ring (overexposed), with Jupiter's moon Europa behind it. The Galileo spacecraft was looking at light scattered forward through the ring, and at the side of Europa away from the Sun, illuminated by reflected light off Jupiter itself (the analogue of "earthshine" on the Moon; see Chapter 4).

7.1g Jupiter's Amazing Satellites

Four of the innermost satellites were discovered by Galileo in 1610 when he first looked at Jupiter through his small telescope. These four moons (Io, Europa, Ganymede, and Callisto) are called the **Galilean satellites** (■ Fig. 7–9). One of these moons, Ganymede, 5276 km in diameter, is the largest satellite in the Solar System and is larger than the planet Mercury.

The Galilean satellites have played a very important role in the history of astronomy. The fact that these particular satellites were noticed to be going around another planet, like a solar system in miniature, supported Copernicus's Sun-centered model of the Solar System. Not everything revolved around the Earth! It was fitting to name the Galileo spacecraft after the discoverer of Jupiter's moons.

■ **FIGURE 7–9** Three of the four Galilean satellites of Jupiter alongside Jupiter.

Jupiter also has dozens of other satellites, some known or discovered from Earth and others discovered by the Voyagers. None of these other satellites is even 10 per cent the diameter of the smallest Galilean satellite.

Through first Voyager-spacecraft and then Galileo-spacecraft close-ups (■ Fig. 7–10), the satellites of Jupiter have become known to us as worlds with personalities of their own. The four Galilean satellites, in particular, were formerly known only as dots of light. Not only the Galilean satellites, which range between 0.9 and 1.5 times the size of our own Moon, but also the smaller ones that have been imaged in detail turn out to have interesting surfaces and histories.

■ **FIGURE 7–10** The four Galilean satellites of Jupiter, alongside Jupiter's Great Red Spot, in a montage made from Galileo images.

7.1g(i) Pizza-like Io

Io, the innermost Galilean satellite, provided the biggest surprises. Scientists knew that Io gave off particles as it went around Jupiter, and other scientists had predicted that Io's interior would be heated by its flexing (to be discussed below). Voyager 1 discovered that these particles resulted from active volcanoes on the satellite, a nice confirmation of the earlier ideas. Eight volcanoes were seen actually erupting, many more than erupt on the Earth at any one time. When Voyager 2 went by a few months later, most of the same volcanoes were still erupting.

Though the Galileo spacecraft could not go close to Io for most of its mission, for fear of getting ruined because of Jupiter's strong radiation field in that region, it could obtain high-quality images of Io and its volcanoes (■ Fig. 7–11*a*). Finally, it went within a few hundred kilometers of Io's surface, and found that 100 volcanoes were erupting simultaneously. Io's surface (Fig. 7–11*b* and *c*) has been transformed by the volcanoes, and is by far the youngest surface we have observed in the Solar System.

Why does Io have so many active volcanoes? Gravitational forces from Ganymede and Europa distort Io's orbit slightly, which changes the tidal force on it from Jupiter in a varying fashion. This changing tidal force flexes Io, creating heat from friction that heats the interior and leads to the volcanism.

The surface of Io is covered with sulfur and sulfur compounds, including frozen sulfur dioxide, and the thin atmosphere is full of sulfur dioxide. It certainly wouldn't be a pleasant place to visit! Io's surface, orange in color and covered with strange formations because of the sulfur, led Brad Smith, the head of the Voyager imaging team, to remark that "It's better looking than a lot of pizzas I've seen."

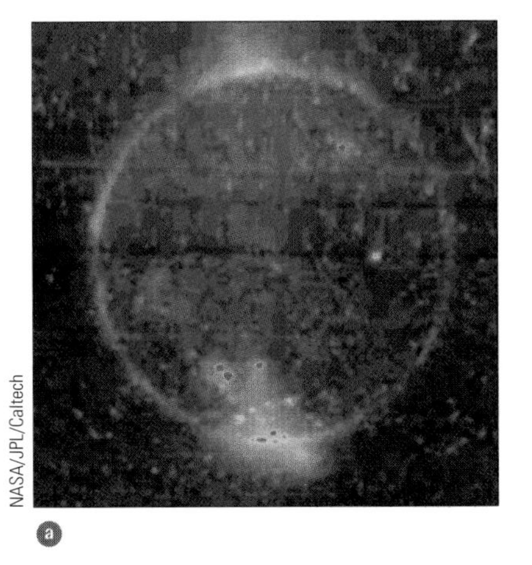

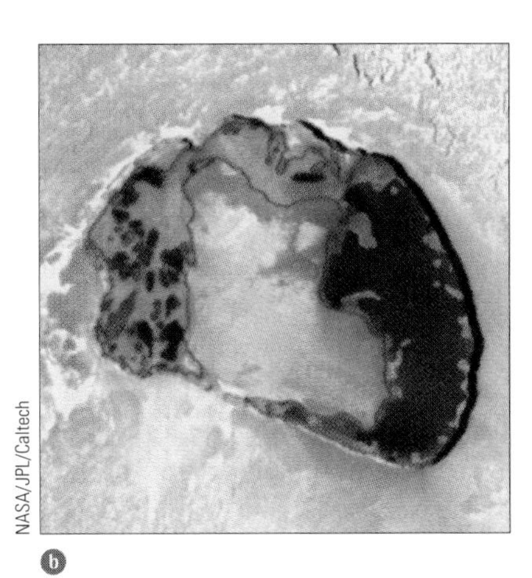

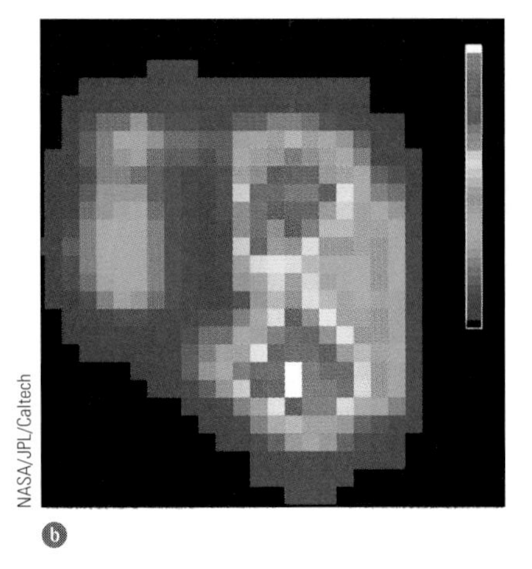

■ **FIGURE 7–11** ⓐ Io, from the Galileo spacecraft, when Io was in Jupiter's shadow. This false-color image uses red to show the most intense features, with the intensity range descending through yellow and green to blue. The diffuse glow on the left limb extends 800 km, 10 times greater in height than the volcano Prometheus below it, which emitted the particles causing the glow. The yellow and red surface spots show erupting magma. Close flybys have revealed over 100 erupting volcanoes. Many of the volcanic plumes active now are different from ones viewed by the Voyagers in 1979. ⓑ A close-up view of Io, from Galileo's last pass. This 2001 optical image shows Tupan Caldera. ⓒ An infrared image of Tupan Caldera, showing the temperature in false color; white is hottest, followed by red and yellow. The hottest regions, probably hot lavas, are darkest in the visible-light image. The cool region in the center of the crater may be an island. Tupan was named for the Brazilian god of thunder.

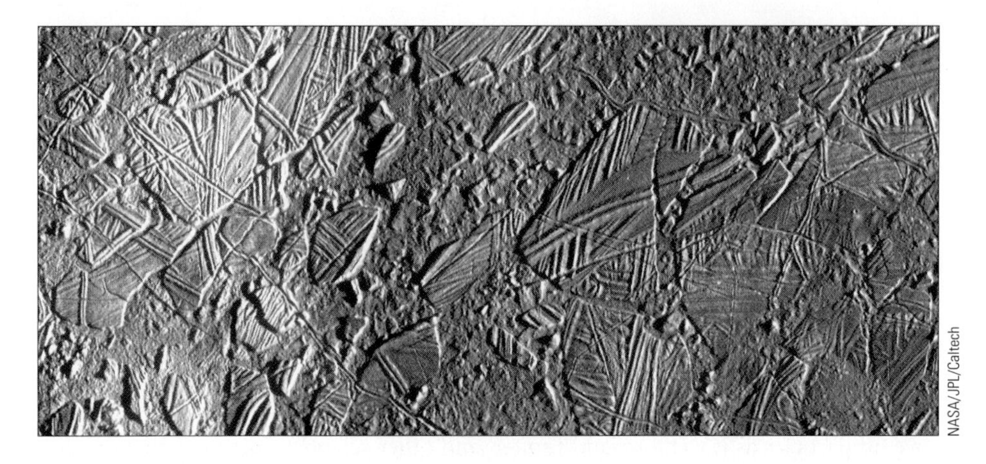

■ **FIGURE 7–12** Close-up views of icy regions on Europa, almost certainly the crust of a global ocean.

Galileo imaging shows many mountains too tall to be supported by sulfur, so stronger types of rock must be involved, with a crust at least 30 km thick above the molten regions. Also, Galileo's infrared observations show that some of the volcanoes are too hot to be sulfur volcanism. The surface changed substantially even as the Galileo spacecraft watched it. By the time Galileo made its sixth, last, and closest pass of Io in 2001, it had raised the total of identified volcanoes to 120.

7.1g(ii) Europa, a Possible Abode for Life

Europa, Jupiter's Galilean satellite with the highest albedo (reflectivity), has a very smooth surface and is covered with narrow, dark stripes. The lack of surface relief, mapped by the Galileo spacecraft to be no more than a couple of hundred meters high, suggests that the surface we see is ice.

The markings may be fracture systems in the ice, like fractures in the large fields of sea ice near the Earth's north pole, as apparently verified in Galileo close-ups (■ Fig. 7–12). Some longer ridges can be traced far across Europa's surface. Few craters are visible, suggesting that the ice was soft enough below the crust to close in the craters. Either internal radioactivity or, more likely, a gravitational tidal heating like that inside Io provides the heat to soften the ice.

■ **FIGURE 7–13** Ganymede, Jupiter's largest satellite. Many impact craters, some with systems of bright rays, are visible. The large, dark region has been named Galileo Regio. Low-albedo features on satellites are named for astronomers.

Because Europa possibly has a liquid-water ocean and extra heating, many scientists consider it a worthy location to check for signs of life. We can only hope that the ice crust, which may be about 10–50 km thick, is thin enough in some locations for us to be able to penetrate it to reach the ocean that may lie below—and see if life has ever existed there.

7.1g(iii) Giant Ganymede

The largest satellite in the Solar System, Ganymede, shows many craters (■ Fig. 7–13) alongside weird, grooved terrain (■ Fig. 7–14). Ganymede is bigger than Mercury but less dense; it contains large amounts of water ice surrounding a rocky core. But an icy surface is as hard as steel in the cold conditions that far from the Sun, so it retains the craters from perhaps 4 billion years ago. The grooved terrain is younger.

Ganymede shows many lateral displacements, where grooves have slid sideways, like those that occur in some places on Earth (for example, the San Andreas fault in California). It is the only place besides the Earth where such faults have been found. Thus, further studies of Ganymede may help our understanding of terrestrial earthquakes. The Galileo spacecraft found a stronger magnetic field for Ganymede than expected, so perhaps Ganymede is more active inside than previously supposed.

7.1g(iv) Pockmarked Callisto

Callisto, the outermost of Jupiter's Galilean satellites, has so many craters (■ Fig. 7–15) that its surface must also be the oldest. Callisto, like Europa and Ganymede, is

NASA/JPL/Caltech

■ **FIGURE 7–14** A computer-generated view, with exaggerated perspective, of ridges on Ganymede. Ganymede's icy surface has been fractured and broken into many parallel ridges and troughs. Such bright grooved terrain covers over half of Ganymede's surface.

covered with ice. A huge bull's-eye formation, Valhalla, contains about 10 concentric rings, no doubt resulting from an enormous impact. Perhaps ripples spreading from the impact froze into the ice to make Valhalla.

Callisto had been thought to be old and uninteresting, but observations from the Galileo spacecraft have revised the latter idea, by showing changes: There are fewer small craters than expected, so the small craters that must have once been there were probably covered by dust that meteorite impacts eroded from larger craters, or disintegrated by themselves through electrostatic charges.

The Galileo spacecraft's measurements of Callisto's gravity from place to place show that its mass is concentrated more toward its center than had been thought. This concentration indicates that heavier materials inside have sunk, and perhaps even indicates that there is an ocean below Callisto's surface. Callisto's interactions with Jupiter's magnetic field have been interpreted to back up the idea of an internal ocean.

7.1g(v) Other Satellites

Galileo's last pass near a Jupiter satellite occurred in 2002 at Amalthea, a small, inner, potato-shaped satellite. (See *A Closer Look 7.2: Jupiter and Its Satellites in Mythology.*) The cameras weren't used; Jupiter's magnetosphere especially close to the planet was primarily studied.

From tracking the spacecraft, Amalthea's mass was measured from its gravitational attraction. The mass coupled with the observed volume gives the density, which turned out to be unexpectedly low, close to that of water ice. Jet Propulsion Laboratory scientists deduced that Amalthea seems to be a loosely packed pile of rubble. Amalthea is thus probably mostly rock with perhaps a little ice, rather than a mix of rock and iron, which would be denser. Amalthea, and presumably other irregular satellites, seem to have been broken apart, with the pieces subsequently drawn roughly together.

Many much smaller satellites of Jupiter are being discovered from the ground, given the existence of mosaics of sensitive CCDs (electronic detectors; see Chapter 3) that cover larger regions of the sky than previously possible and of computer processing methods to analyze the data. In recent years, dozens of small moons, some only 2 km across, have been discovered around Jupiter, bringing Jupiter's total of moons up to at least 52.

These "irregular" satellites are, no doubt, captured objects that were once in orbit around the Sun. This origin is different from that of the large, regular satellites (like

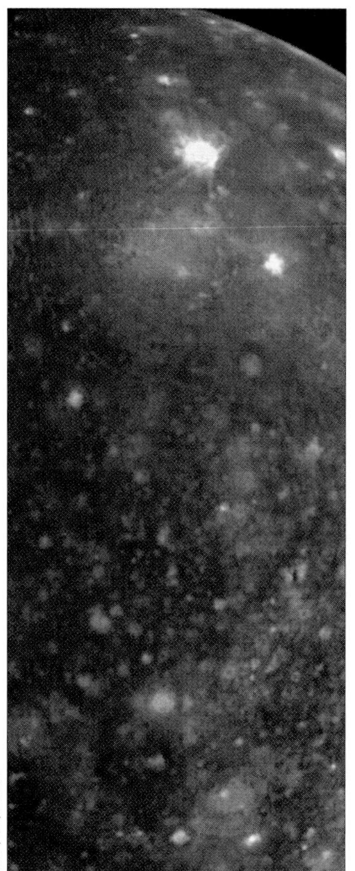
NASA/JPL/Caltech

■ **FIGURE 7–15** The side of Callisto that leads in its orbit around Jupiter, imaged from the Galileo spacecraft. (Callisto and the other Galilean satellites are locked in synchronous orbit, so that their same sides always face Jupiter, similar to the case of Earth's Moon.) The impact region Asgard (*top center*) is surrounded by concentric rings that are up to 1700 km in diameter. The bright icy materials excavated by the younger craters contrast with the darker and redder coatings on older surfaces. Galileo images show, surprisingly, that Callisto has very few craters smaller than 100 m in diameter.

A Closer Look 7.2 Jupiter and Its Satellites in Mythology

Jupiter was the supreme god in Roman mythology. He was also called Jove, so "jovian" refers to Jupiter—the jovian planets are those that are large like Jupiter, namely Jupiter, Saturn, Uranus, and Neptune.

All of Jupiter's moons except Amalthea are named after lovers of Zeus in Greek mythology, the Greek equivalent of Jupiter. Amalthea, a goat-nymph, was Zeus's nurse, and out of gratitude he made her into the constellation Capricorn, a goat. Zeus changed Io into a heifer to hide her from his jealous wife, Hera; in honor of Io, Earth's crescent moon has horns, as do bulls.

Ganymede was a Trojan youth carried off by an eagle to be Jupiter's cup bearer (the constellation Aquarius). Callisto was changed into a bear as punishment for her affair with Zeus. She was then slain by mistake, and rescued by Zeus by being transformed into the Big Bear in the sky, the constellation Ursa Major. Jealous Hera persuaded the sea god to forbid Callisto to ever bathe in the sea, which is why Ursa Major never sinks below the horizon as seen from the latitude of Greece.

Europa was carried off to Crete on the back of Zeus, who took the form of a white bull. She became King Minos's mother. Pasiphae (also the name of one of Jupiter's moons) was the wife of Minos and the mother of the Minotaur.

the Galilean moons), and from that of the set of smaller satellites in close orbits around Jupiter that apparently are remnants of collisions. Both these latter types of moons are thought to have formed from a disk of gas and dust about the planet. The other jovian planets also have these three types of satellites.

Studies of Jupiter's moons tell us about the formation of the Jupiter system, and help us better understand the early stages of the entire Solar System. NASA's next New Horizons mission, in a series of small spacecraft, is to visit Jupiter and its moons in the middle of the next decade. NASA has approved a preliminary phase of the Juno mission, to be launched in 2010 to study Jupiter's interior and atmosphere from polar orbit. The spacecraft is to find out, from gravity studies, if the planet has an ice-rock core and to study how much water and ammonia Jupiter's atmosphere holds.

AceAstronomy™ Log into AceAstronomy and select this chapter to see the Active Figure called "Explorable Jupiter."

7.2 Saturn

Saturn, like Jupiter, Uranus, and Neptune, is a giant planet. Its diameter, without its rings, is 9 times that of Earth; its mass is 95 Earth masses. It is a truly beautiful object in a telescope of any size. The glory of its system of rings makes it stand out even in small telescopes (see *Star Party 7.1: Observing the Giant Planets*). The view from the Cassini spacecraft, now in orbit around it, is breathtaking (■ Fig. 7–16).

NASA/JPL/Space Science Institute

■ **FIGURE 7–16** A high-resolution view of Saturn, imaged with NASA's Cassini spacecraft. The original digital image can be blown up to wall size without showing noticeable fuzziness.

The giant planets have low densities. Saturn's is only 0.7 g/cm^3, 70 per cent the density of water (■ Fig. 7–17). The bulk of Saturn is hydrogen molecules and helium, reflecting Saturn's formation directly from the solar nebula. Saturn is thought to have a core of heavy elements, including rocky material, making up about the inner 20 per cent of its diameter.

Voyagers 1 and 2 flew by Saturn in 1980 and 1981, respectively. Cassini, a joint NASA/European Space Agency mission, arrived at Saturn in 2004. It is orbiting the Saturn system, going up close in turn to various of Saturn's dozens of moons. We will be discussing Cassini observations throughout the following sections.

■ **FIGURE 7–17** Since Saturn's density is lower than that of water, it would float, like Ivory soap, if we could find a big enough bathtub. But it would leave a ring!

7.2a Saturn's Rings

The rings extend far out in Saturn's equatorial plane, and are inclined to the planet's orbit. Over a 30-year period, we sometimes see them from above their northern side, sometimes from below their southern side, and at intermediate angles in between (■ Fig. 7–18). When seen edge-on, they are almost invisible.

The rings of Saturn consist of material that was torn apart by Saturn's gravity or material that failed to collect into a moon at the time when the planet and its moons were forming. However, the rings may have formed fairly recently, within the past few hundred million years.

Every massive object has a sphere, called its **Roche limit,** inside of which blobs of matter do not hold together by their mutual gravity. The forces that tend to tear the blobs apart from each other are tidal forces. They arise, like the Earth's tides, because some parts of an object are closer to the planet than others and are thus subject to higher gravity. The difference between the gravity force farther in and the gravity force farther out is the tidal force.

The radius of the Roche limit is usually 2½ to 3 times the radius of the larger body, closer to the latter for the relative densities of Saturn and its moons. The Sun also has a Roche limit, but all the planets lie outside it. The natural moons of the various planets lie outside their respective Roche limits. Saturn's rings lie inside Saturn's

ASIDE 7.2: Comparative planetology: spots

Great Red Spot on Jupiter: Lasts hundreds of years.

Great Dark Spot on Neptune: There when Voyager flew by (1989) and gone a few years later as shown by Hubble images.

NASA and The Hubble Heritage Team (STScI/AURA); R. G. French (Wellesley College), J. Cuzzi (NASA/Ames), L. Dones (SwRI), and J. Lissauer (NASA/Ames)

ⓐ

NASA/ESA/J. Clarke (Boston U.) and Z. Levay (STScI)

ⓑ

■ **FIGURE 7–18** ⓐ Saturn, from the Hubble Space Telescope, in a series of views over a five-year period. ⓑ These auroras on Saturn, seen in Hubble's visible-light images, show significant changes over the course of a few days.

Roche limit, so it is not surprising that the material in the rings is spread out rather than collected into a single orbiting satellite.

Artificial satellites that we send up to orbit the Earth are constructed of sufficiently rigid materials that they do not break up even though they are within the Earth's Roche limit; they are held together by forces much stronger than gravity.

Saturn has several concentric major rings visible from Earth. The brightest ring is separated from a fainter broad outer ring by an apparent gap called **Cassini's division.** (The 17th-century astronomer Jean-Dominique Cassini (1625–1712), who moved from Italy to France in 1671, discovered several of Saturn's moons as well as the division in the rings, the latter in 1675. It was very appropriate to name not only the ring division but also NASA's spacecraft to Saturn after him.) Another ring is inside the brightest ring. We know that the rings are not solid objects, because the rotation speed of the outer rings is slower than that of rings closer to Saturn.

Radar waves bounced off the rings show that the particles in the rings are at least a few centimeters, and possibly a meter, across. Infrared studies show that at least their outer parts consist of ice.

The images from the Voyagers revolutionized our view of Saturn, its rings, and its moons. The Cassini mission is providing even more detailed views, as it orbits for years instead of merely flying by (■ Fig. 7–19). Only from spacecraft can we see the rings from a vantage point different from the one we have on Earth. Backlighted views showed that Cassini's division, visible as a dark (and thus apparently empty) band from Earth, appeared bright, so it must contain some particles.

The rings are thin, for when they pass in front of stars, the starlight easily shines through. Studies of the changes in the radio signals from the Voyagers when they went

■ **FIGURE 7–19** Saturn and many of its moons, photographed from the Cassini spacecraft (except for a few Voyager or generic images, whose labels are marked with *). In the lower half, the positions of the rings and moons are shown to scale.

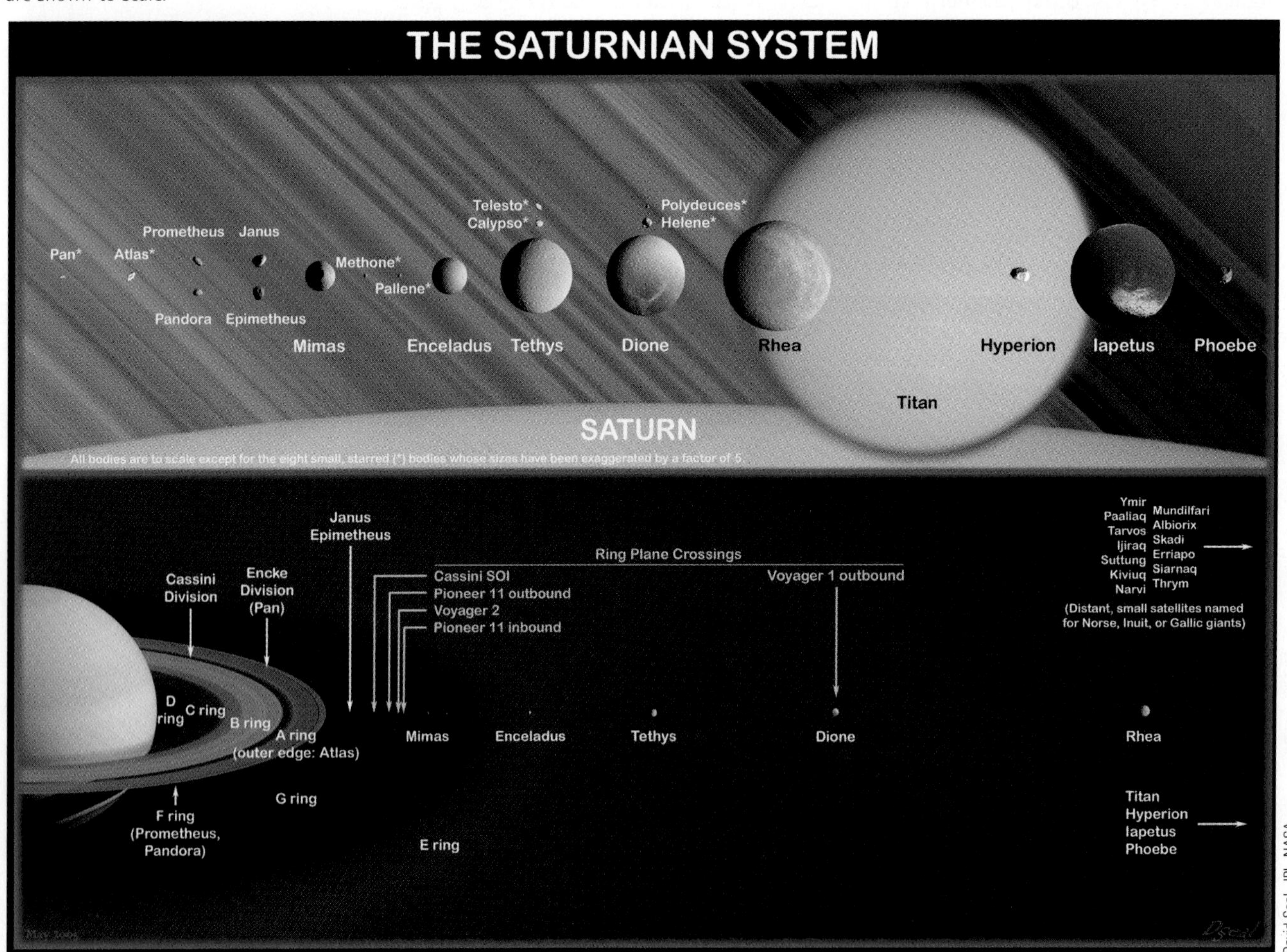

NASA/JPL/Space Science Institute

■ **FIGURE 7–20** The several rings we see from Earth break up into thousands upon thousands of narrow ringlets when viewed up close with the Cassini mission's cameras.

NASA/JPL/Space Science Institute

■ **FIGURE 7–21** Cassini imaged these density waves in Saturn's rings, caused by the gravity of Saturn's satellites.

behind the rings showed that the rings are only about 20 m thick. Relative to the diameter of the rings, this is equivalent to a CD (compact disc) that is 30 km across though still its normal thickness!

The closer the spacecraft got to the rings, the more individual rings became apparent. Each of the known rings was actually divided into many thinner rings. The number of these rings (sometimes called "ringlets") is in the hundreds of thousands. The images from the Cassini spacecraft (■ Fig. 7–20) surpassed even Voyager's views of ringlets.

The outer major ring turns out to be kept in place by a tiny satellite orbiting just outside it. At least some of the rings are kept narrow by "shepherding" satellites that gravitationally affect the ring material, a concept that we can apply to rings of other planets. Density waves (■ Fig. 7–21) were seen in the ring.

A post-Voyager theory said that many of the narrowest gaps may be swept clean by a variety of small moons. These objects would be embedded in the rings in addition to the icy snowballs that make up most of the ring material. The tiny moon Pan has been observed clearing out the Encke Gap in just this way. Theorists suppose that smaller moons probably clear out several other gaps in Saturn's rings, and Cassini is finding some of these moons.

AceAstronomy™ Log into AceAstronomy and select this chapter to see the Active Figure called "Roche Limit."

7.2b Saturn's Atmosphere

Like Jupiter, Saturn rotates quickly on its axis; a complete period is only 10 hours, in spite of Saturn's diameter being over 9 times greater than Earth's. The rapid rotation causes Saturn to be larger across the equator than from pole to pole. This equatorial bulging makes Saturn look slightly "flattened." Jupiter also looks flattened, or oblate, for this reason.

The structure in Saturn's clouds is of much lower contrast than that in Jupiter's clouds. It is not a surprise that the chemical reactions can be different; after all, Saturn is colder than Jupiter. Saturn has extremely high winds, up to 1800 km/hr, 4 times faster than the winds on Jupiter. Cassini is tracking the winds with higher precision

ASIDE 7.3: Comparative planetology: atmospheres

All the giant planets have atmospheres that are almost all hydrogen and helium. The small percentage of heavier elements is greater in Uranus and Neptune than in Jupiter and Saturn.

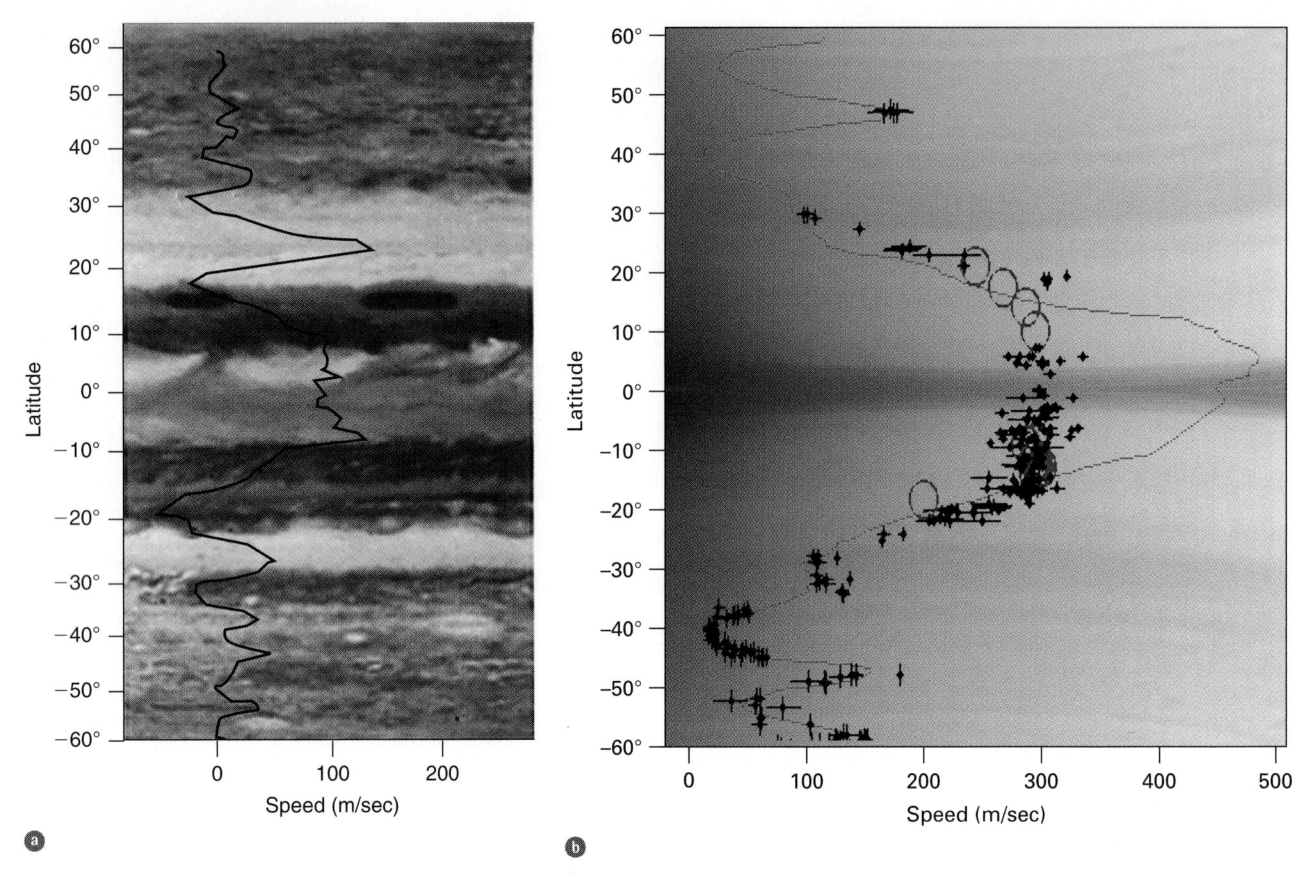

■ **FIGURE 7–22** ⓐ The winds of Jupiter, a graph of their speed on a background that shows Jupiter's surface and how it varies with latitude. (Redrawn from NASA/JPL/Caltech.) ⓑ The winds of Saturn. The solid line shows the Voyager data and the points show newer data from the Hubble Space Telescope during 1996 to 2004. Note how the winds appear much less strong at the equator in the newer data. The situation is under study to see if the change is global or if a wind-shear effect is showing. (Courtesy of A. Sánchez-Lavega; background courtesy of NASA/JPL/Space Science Institute)

than was previously possible. On Saturn, the variations in wind speed do not seem to correlate with the positions of bright and dark bands, unlike the case with Jupiter (■ Fig. 7–22). As on Jupiter, but unlike the case for Earth, the winds seem to be driven by rotating eddies, which in turn get most of their energy from the planet's interior. Such differences provide a better understanding of storm systems in Earth's atmosphere.

7.2c Saturn's Interior and Magnetic Field

Saturn radiates about twice as much energy as it absorbs from the Sun, a greater factor than for Jupiter. One interpretation is that only 2/3 of Saturn's internal energy remains from its formation and from its continuing contraction under gravity. The rest would be generated by the gravitational energy released by helium sinking through the liquid hydrogen in Saturn's interior. The helium that sinks has condensed because Saturn, unlike Jupiter, is cold enough.

Saturn gives off radio signals, as does Jupiter, a pre-Voyager indication to earthbound astronomers that Saturn also has a magnetic field. The Voyagers found that the magnetic field at Saturn's equator is only 2/3 of the field present at the Earth's equator. Remember, though, that Saturn is much larger than the Earth and so its equator is much farther from its center.

Saturn's magnetic field contains belts of charged particles (analogous to Van Allen belts), which are larger than Earth's but smaller than Jupiter's. (Saturn's surface magnetic field is 20 times weaker than Jupiter's.) These particles interact with the atmosphere near the poles and produce auroras (■ Fig. 7–23).

J. P. Trauger (JPL) and NASA

■ **FIGURE 7–23** Saturn's auroras, showing the effect of the planet's magnetic field. This ultraviolet image was taken with the Hubble Space Telescope.

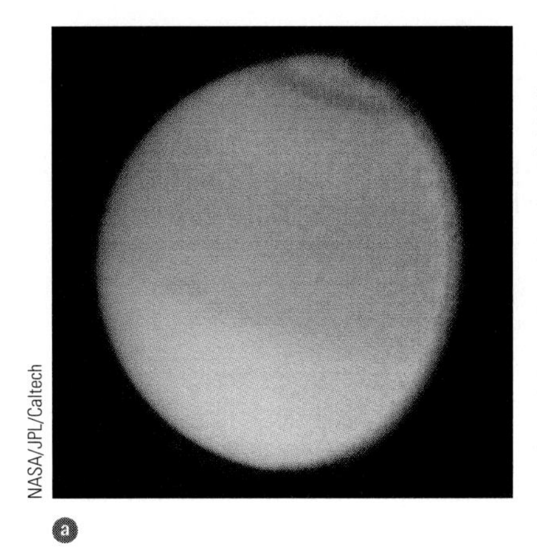

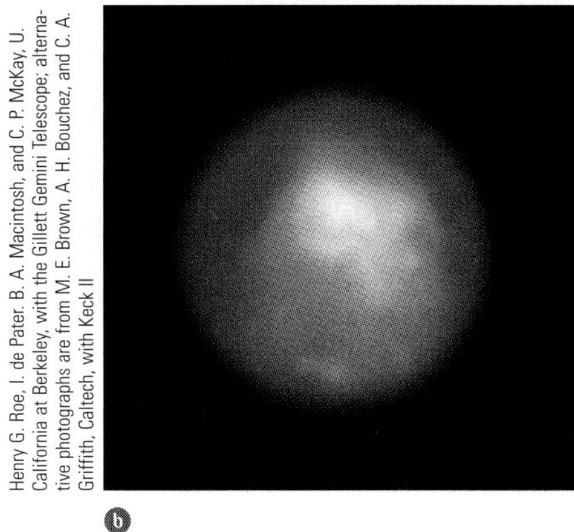

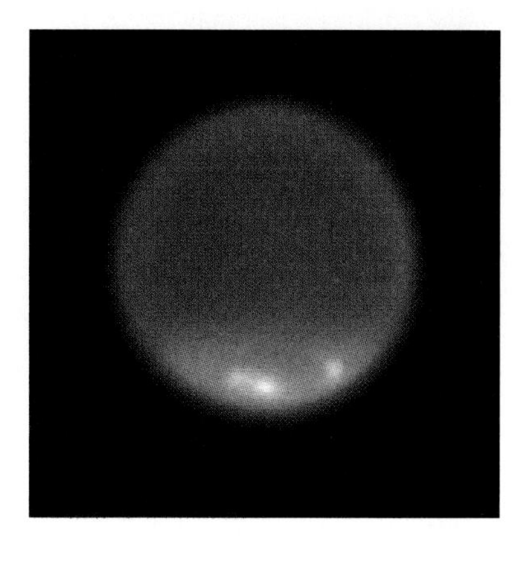

■ **FIGURE 7–24** ⓐ Titan was disappointingly featureless even to Voyager 1's cameras because of its thick, smoggy atmosphere. Its northern polar region was relatively dark. The southern hemisphere was lighter than the northern. ⓑ New imaging with adaptive optics on the Keck II and Gemini North telescopes showed a bright continent on Titan (*left*) and methane clouds near its south pole (*right*).

7.2d Saturn's Moon Titan

At 40 per cent the diameter of the Earth, Titan (see *A Closer Look 7.3: Saturn's Satellites in Mythology*) is an intriguing body for a number of reasons. Titan has an atmosphere that was detected from Earth pre-Voyager. Studies of how the radio signals faded when Voyager 1 went behind Titan showed that Titan's atmosphere is denser than Earth's. The surface pressure on Titan is $1^1/_2$ times that on Earth.

Titan's atmosphere is opaque, apparently because of the action of sunlight on chemicals in it, forming a sort of "smog" and giving it its reddish tint. Smog on Earth forms in a similar way. The Voyagers showed (■ Fig. 7–24) several layers of haze. They detected nitrogen, which makes up the bulk of Titan's atmosphere, as it does Earth's. Methane is a minor constituent, perhaps 1 per cent. A greenhouse effect is present, making some scientists wonder whether Titan's surface may have been warmed enough for life to have evolved there. There is also an "anti-greenhouse effect" caused by the smog.

The temperature near the surface, deduced from measurements made with Voyager's infrared radiometer, is only about −180°C (93 K), somewhat warmed (12°C) by a combination of the greenhouse effect and the anti-greenhouse effect. This temperature is near that of methane's "triple point," at which it can be in any of the three physical states—solid, liquid, or gas. So methane may play the role on Titan that water does on Earth.

Scientists wanted to know whether Titan is covered with lakes or oceans of methane mixed with ethane, and whether other parts are covered with methane ice or snow. Below, we will see how a recent spacecraft gave some answers. Also, using filters in the near-infrared, first the Hubble Space Telescope and then ground-based telescopes have been able to penetrate Titan's haze to reveal some structure on its surface (Fig. 7–24), though no unambiguous lakes were found.

Some of the organic molecules formed in Titan's atmosphere from lightning storms and other processes may rain down on its surface. Thus the surface, largely hidden from our view, may be covered with an organic crust about a kilometer thick, perhaps partly dissolved in liquid methane. These chemicals are similar to those from which we think life evolved on the primitive Earth. But it is probably too cold on Titan for life to begin.

A Closer Look 7.3 SATURN'S SATELLITES IN MYTHOLOGY

Saturn's moons are named after the Titans. In Greek mythology, the Titans were the children and grandchildren of the early supreme god Uranus and of Gaea, the goddess of the Earth, who had been fertilized by a drop of Uranus's blood. The Olympic gods like Zeus (Roman: Jupiter) followed the Titans.

Titan is so intriguing, and potentially so important, that the lander of the Cassini mission was sent to plunge through its atmosphere. So while the lander of Jupiter's Galileo mission went into the planet itself, this lander, known as Huygens, penetrated the clouds around Saturn's largest moon when it arrived on January 14, 2005. (The lander is named after Christiaan Huygens, who first realized that Saturn was surrounded by a ring and who also, in 1655, discovered Titan.)

At higher levels, the Huygens probe was buffeted by strong winds that ranged up to 400 km/h, though the winds calmed near the surface. Radio telescopes on Earth measured the winds by following the Doppler shifts in the probe's signals, since an onboard Doppler experiment failed.

As the Huygens lander drifted downward on its parachute for $2^1/2$ hours, it measured the chemical composition of Titan's atmosphere. It then imaged the surface below it when it got sufficiently far below the haze. The shape of a shoreline is visible (■ Fig. 7–25*a*). No liquid substances were seen (there was never a glint of reflection of the type that occurs off a shiny body, for example), even though methane or ethane had been expected. The dark material is thought to be tar-like and has probably settled out of the atmosphere. Some "islands" in it may be raised material that diverted the flow.

The branching of systems that were visible clearly show that some liquid flowed in the past. Though few clouds were seen, perhaps methane or ethane rainstorms occurred intensely though rarely. A metal rod attached to the bottom of the Huygens lander penetrated 15 cm into the surface, but it apparently had to break through a hard crust to do so. The crust might have formed from a recent inundation. A device on the lander measured a puff of methane, so perhaps there had been methane frost on the surface before Huygens landed there.

A final triumph came when the probe survived for 90 minutes on the surface, sending back pictures of ice blocks (Fig. 7–25*b*). These pieces of ice are rounded, apparently also revealing the past presence of flowing liquid.

Later analysis of infrared images seems to show a 30-km-wide structure that may be a volcanic dome. Such an ice volcano could be caused by the energy generated by tidal stresses within Titan caused by its elliptical orbit. It would release methane to the atmosphere. So the methane long measured in Titan's atmosphere might not be from a methane-rich hydrocarbon ocean after all.

7.2e Saturn's Other Satellites

So many of Saturn's other moons proved to have interesting surfaces when seen close up from Cassini that we show a variety of images in *A Closer Look 7.4: Saturn's Rings and Moons from Cassini.* The spacecraft continues to fly by a variety of Saturn's moons,

NASA/JPL/ESA/U. Arizona

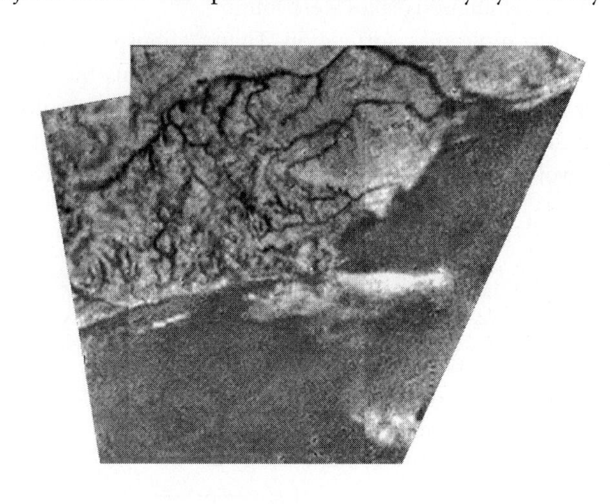

■ **FIGURE 7–25** ⓐ Eroded blocks of ice on Titan's surface, imaged during the 90 minutes that Huygens survived on Titan's surface. ⓑ A shoreline and a branching riverbed were imaged by the Huygens probe's camera as it drifted downward through Titan's atmosphere on January 14, 2005.

A Closer Look 7.4 Saturn's Rings and Moons from Cassini

The Cassini mission is orbiting Saturn, passing close to various of its dozens of moons.

1. Giant Titan, the largest moon (2575 km in radius), seems hazy from afar; elsewhere, we show the views from Cassini's Huygens lander. (a) A true-color image from August 21, 2005. (b) Infrared penetrates the haze, as seen in this August 22, 2005, image. (c) Radar on the Cassini orbiter penetrated the smog to make this image, about 175 km across, with half-kilometer resolution, better than the typical resolution of the optical cameras.

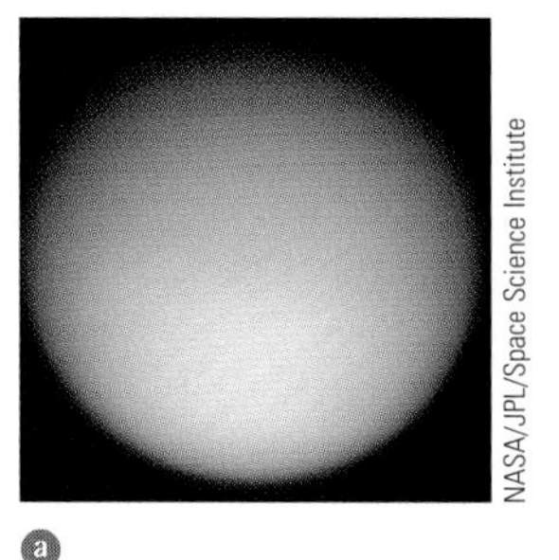
NASA/JPL/Space Science Institute

(a)

NASA/JPL/Space Science Institute

(b)

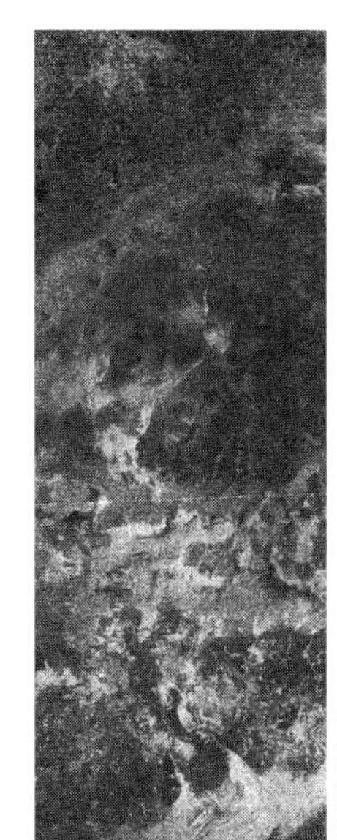
NASA/JPL/Caltech

(c)

2. (a) Enceladus (about 250 km in radius, though not quite round) shows smooth regions, indicating that its surface had been melted, a sign of internal heating. It also has regions covered with impact craters. Linear grooves may be geologic faults. Its south polar region is venting water vapor and icy particles, giving it a tenuous atmosphere. (b) Spray ejected from Enceladus's south polar region.

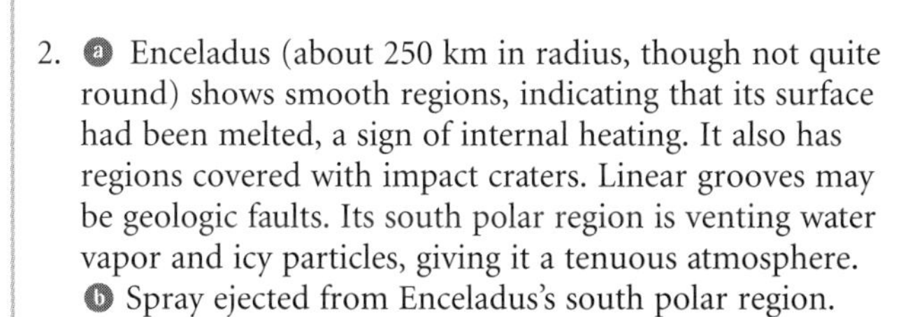

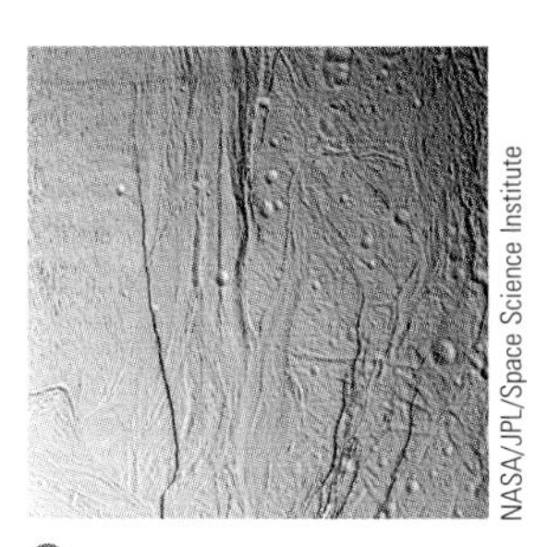
NASA/JPL/Space Science Institute

(a)

NASA/JPL/Space Science Institute

(b)

3. Mimas (about 200 km in radius, though not quite round) has an impact crater named Herschel, after the discoverer of Uranus. The crater, which resembles *Star Wars'* Death Star, is 130 km across.

NASA/JPL/Space Science Institute

4. Rhea (765 km in radius) has craters as large as 300 km across.

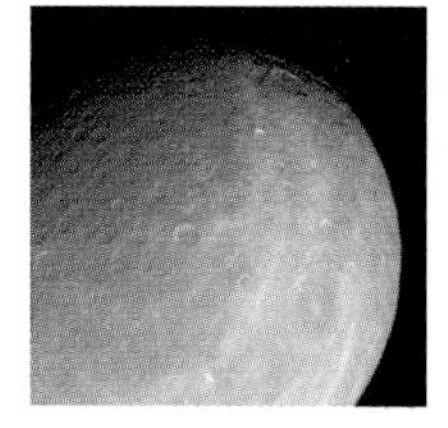
NASA/JPL/Space Science Institute

5. Iapetus (720 km in radius), strangely, has one very dark side and one bright side. The dark side faces forward as Iapetus orbits. The low albedo may result from sunlight reacting with methane that wells up, forming hydrocarbons.

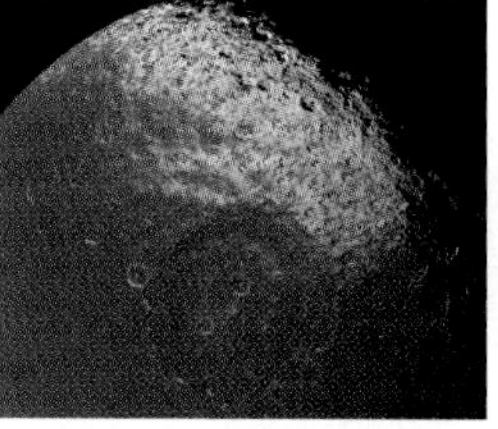
NASA/JPL/Space Science Institute

6. Saturn's rings take on unusual structure when affected by the orbiting moons. Here we see the wake of tiny Prometheus (70 km × 50 km × 40 km in radius).

NASA/JPL/Space Science Institute

making images as well as other measurements. Also, as was the case for Jupiter, dozens of small, irregular moons continue to be found from space images and from Earth-based telescopes.

7.3 Uranus

The two other giant planets beyond Saturn—Uranus (pronounced "U′ran-us"; see *A Closer Look 7.5: Uranus and Neptune in Mythology*) and Neptune—are each about 4 times the diameter of (and about 15 times more massive than) the Earth. They reflect most of the sunlight that hits them, which indicates that they are covered with clouds.

Like Jupiter and Saturn, Uranus and Neptune don't have solid surfaces. Their atmospheres are also mostly hydrogen and helium, but they have a higher proportion of heavier elements. Some of the hydrogen may be in a liquid mantle of water, methane, and ammonia. At the planets' centers, a rocky core contains mostly silicon and iron, probably surrounded by ices. From their average densities, we have deduced that the cores of Uranus and Neptune make up substantial parts of those planets, differing from the relatively more minor cores of Jupiter and Saturn.

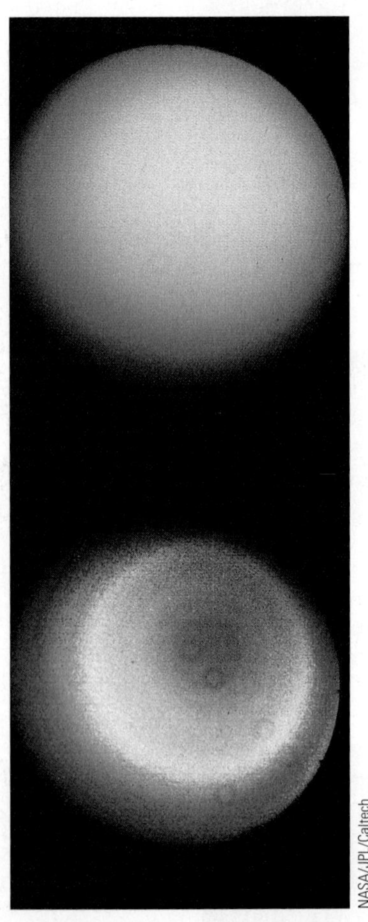

■ **FIGURE 7–26** Uranus, as the Voyager 2 spacecraft approached. The picture at top shows Uranus as the human eye would see it. At bottom, false colors bring out slight differences in contrast. The "donut" shapes are out-of-focus dust spots on the camera.

Uranus was the first planet to be discovered that had not been known to the ancients. The English astronomer and musician William Herschel reported the discovery in 1781. Actually, Uranus had been plotted as a star on several sky maps during the hundred years prior to Herschel's discovery, but had not been singled out as anything other than an ordinary star.

Uranus revolves around the Sun in 84 years at an average distance of more than 19 A.U. Uranus appears so tiny that it is not much bigger than the resolution we are allowed by Earth's atmosphere. Uranus is apparently surrounded by thick clouds of methane ice crystals (■ Fig. 7–26), with a clear atmosphere of molecular hydrogen above them. The trace of methane gas mixed in with the hydrogen makes Uranus look greenish.

Uranus is so far from the Sun that its outer layers are very cold. Studies of its infrared radiation give a temperature of −215°C (58 K). There is no evidence for an internal heat source, unlike the case for Jupiter, Saturn, and Neptune.

The other planets rotate such that their axes of rotation are very roughly parallel to their axes of revolution around the Sun. Uranus is different (as is Pluto), for its axis of rotation is roughly perpendicular to the other planetary axes, lying only 8° from the plane of its orbit (■ Fig. 7–27). Sometimes one of Uranus's poles faces the Earth, 21 years later its equator crosses our field of view, and then another 21 years later the other pole faces the Earth. Polar regions remain alternately in sunlight and in darkness for decades.

The strange seasonal effects that result on Uranus became obvious in 1999, when a series of Hubble Space Telescope views taken over the preceding years revealed the activity in the clouds of Uranus's once-every-84-years springtime. When we understand just how the seasonal changes in heating affect the clouds, we will be closer to understanding our own Earth's climate.

A Closer Look 7.5 URANUS AND NEPTUNE IN MYTHOLOGY

In Greek mythology, Uranus was the personification of Heaven and ruler of the world, the son and husband of Gaea, the Earth. Neptune, in Roman mythology, was the god of the sea, and the planet Neptune's trident symbol reflects that origin.

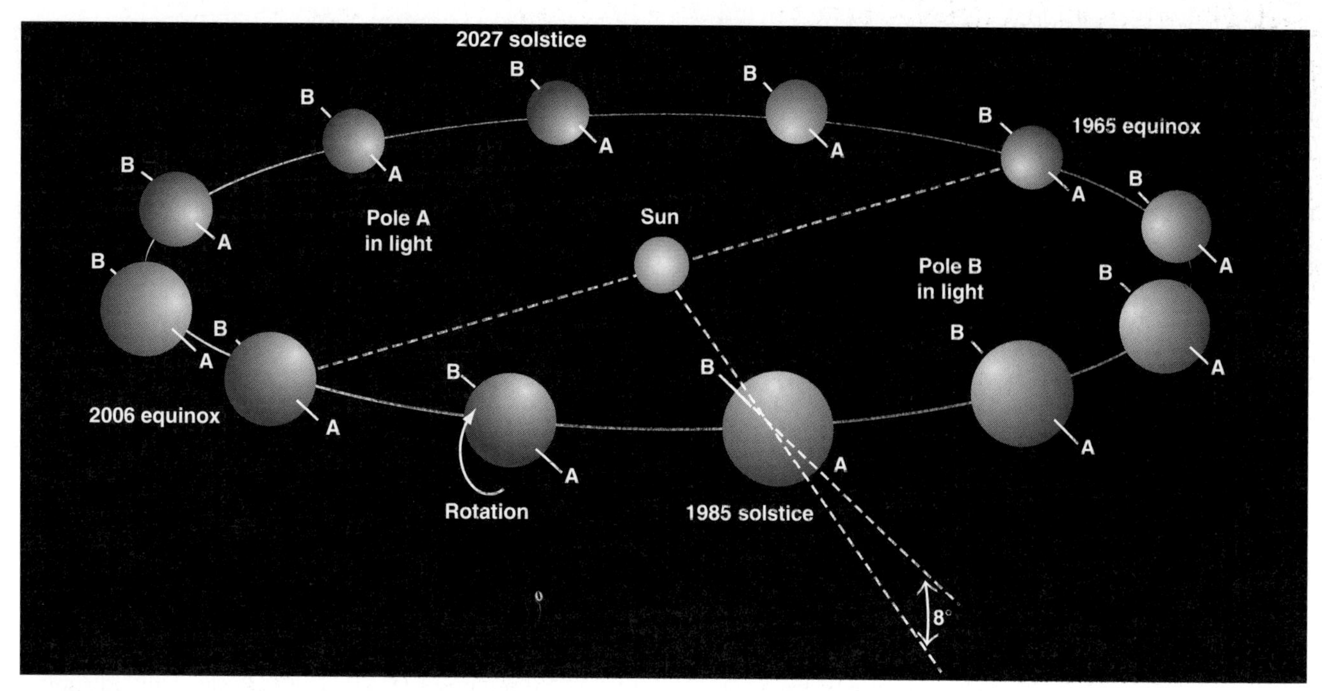

■ **FIGURE 7–27** Uranus's axis of rotation lies roughly in the plane of its orbit. Notice how the planet's poles come within 8° of pointing toward the Sun, while 1/4 of an orbit before or afterward, the Sun is almost over Uranus's equator.

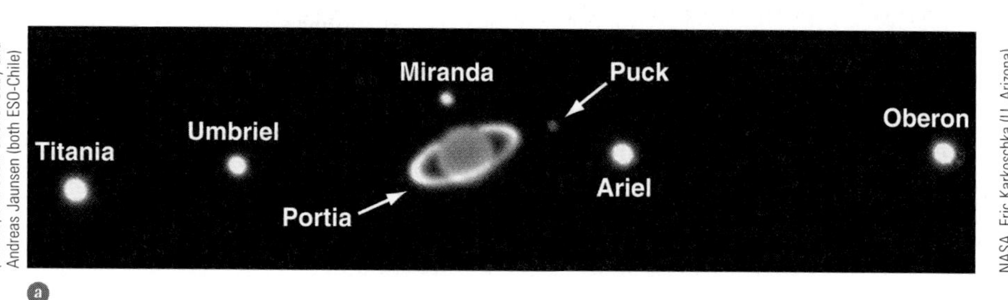

ESO/Emmanuel Lellouch and Therese Encrenaz of the Observatoire de Paris (France) and Jean-Gabriel Cuby and Andreas Jaunsen (both ESO-Chile)

NASA, Eric Karkoschka (U. Arizona), Heidi Hammel (Space Science Institute, Boulder), and STScI

■ **FIGURE 7–28** ⓐ Uranus and its rings, imaged in the near-infrared with the "Antu" Unit Telescope of the European Southern Observatory's Very Large Telescope in Chile. Using a filter at the wavelength of the methane absorption from Uranus's disk diminishes the intensity of the disk. Since we see the rings by reflected sunlight, which does not have a diminished intensity at the methane wavelength, the rings appear relatively bright. We also see several of Uranus's moons. ⓑ Infrared views of Uranus, taken on July 9, 2004, with the adaptive optics system at the Keck telescope. The rings and moons show better at 2.2 micrometers, because the methane absorption of the atmosphere makes the planet's disk appear dark. The atmospheric clouds show better at 1.6 micrometers, while the rings are barely visible.

Voyager 2 reached Uranus in 1986. It revealed most of our current understanding of Uranus, its rings, and its moons. (■ Fig. 7–28 shows a more recent view.) Its moon Miranda, for example, though relatively small, has a surface that is extremely varied and interesting (■ Fig. 7–29).

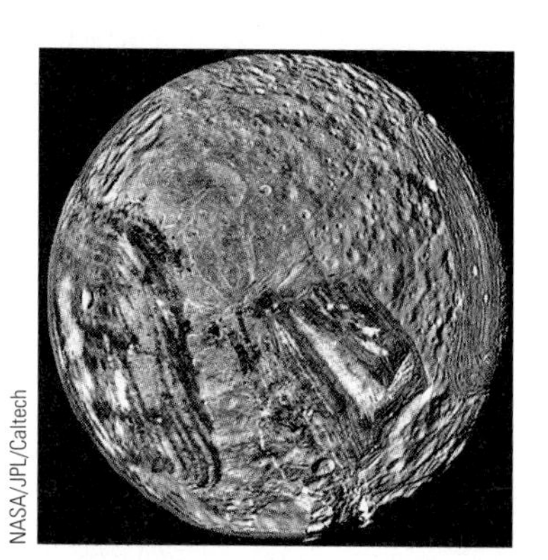

NASA/JPL/Caltech

■ **FIGURE 7–29** Uranus's moon Miranda, apparently broken up by impacts and solidified in new configurations, providing the obviously jumbled surface.

7.3a Uranus's Atmosphere

Even though Voyager 2 came very close to Uranus's surface, as close as 107,000 km (a quarter of the distance from the Earth to the Moon), it saw very little detail on it. Thus Uranus's surface is very bland. Apparently, chemical reactions are more limited than on Jupiter and Saturn because it is colder. Uranus's clouds form relatively deep in the atmosphere.

A dark polar cap was seen on Uranus, perhaps a result of a high-level photochemical haze added to the effect of sunlight scattered by hydrogen molecules and helium atoms. At lower levels, the abundance of methane gas (CH_4) increases. It is this gas that absorbs the orange and red wavelengths from the sunlight that hits Uranus. Thus most of the light that is reflected back at us is blue-green.

Tracking some of the ten clouds that were detected by Voyager 2 revealed the rotation period of their levels in Uranus's atmosphere. The larger cloud, at 35° latitude, rotated in 16.3 hours. The smaller, fainter cloud, at 27° latitude, rotated in 16.9 hours. Observations through color filters give evidence that these clouds are higher than their surroundings by 1.3 km and 2.3 km, respectively.

It was a surprise to find that both of Uranus's poles, even the one out of sunlight, are about the same temperature. The equator is nearly as warm. Comparing such a strange atmosphere with our own will help us understand Earth's weather and climate better.

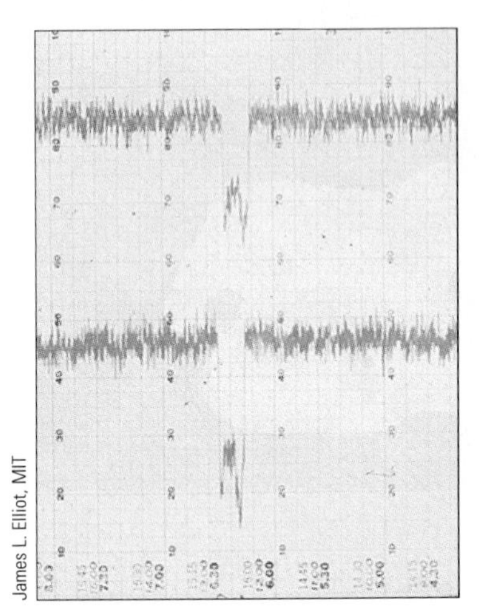

James L. Elliot, MIT

■ **FIGURE 7–30** A graph of the intensity of starlight during an occultation of a star by one of Uranus's rings. Each color image, recorded through a different filter, is shown in a different color ink.

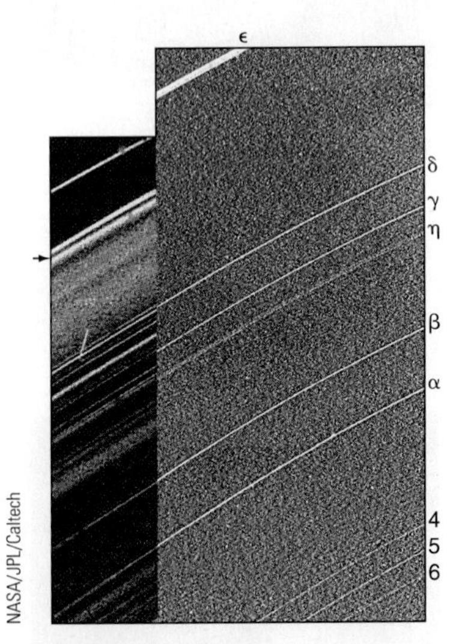

NASA/JPL/Caltech

■ **FIGURE 7–31** When the Voyager spacecraft looked through the rings back toward the Sun, the backlighted view (*left*) revealed new dust lanes between the known rings, which also show in this view. The streaks are stars. The forward-lighted view (*right*) is provided for comparison. One of the new rings discovered from Voyager is marked with an arrow.

7.3b Uranus's Rings

In 1977, astronomers on Earth watched as Uranus occulted (passed in front of) a faint star. Predictions showed that the occultation would be visible only from the Indian Ocean southwest of Australia. The scientists who went to study the occultation from an instrumented airplane turned on their equipment early, to be sure they caught the event. Surprisingly, about half an hour before the predicted time of occultation, they detected a few slight dips in the star's brightness (■ Fig. 7–30). They recorded similar dips, in the reverse order, about half an hour after the occultation.

The dips indicated that Uranus is surrounded by several rings, some of which have since been photographed by Voyager 2 and with the Hubble Space Telescope. Each time a ring went between us and the distant star, the ring blocked some of the starlight, making a dip. Eleven rings are now known. They are quite dark, reflecting only about 2 per cent of the sunlight that hits them.

The rings have radii 1.7 to 2.1 times the radius of the planet. They are very narrow from side to side; some are only a few km wide. How can narrow rings exist, when we know that colliding particles tend to spread out? The discovery of Uranus's narrow rings led to the suggestion that a small unseen satellite (a "shepherd moon") in each ring keeps the particles together. As we saw, this model turned out to be applicable to at least some of the narrow ringlets of Saturn later discovered by the Voyagers.

Voyager provided detailed ring images (■ Fig. 7–31). Interpreting the small color differences is important for understanding the composition of the ring material. Quite significant was the single long-exposure, backlighted view taken by Voyager. Study of these data has shown that less of the dust in Uranus's rings is very small particles compared with the dust in the rings of Saturn and Jupiter.

The rings of Uranus are apparently younger than 100 million years of age, since the satellites that hold them in place are too small to hold them longer. Thus there must have been a more recent source of dust. That source may have been a small moon destroyed by a meteoroid or comet. The larger particles seen only when the rings were backlighted may have come from a different source; perhaps they came from the surfaces of Uranus's current moons. We are now realizing that ring systems are younger and change more over time than had been thought.

AceAstronomy™ Log into AceAstronomy and select this chapter to see the Active Figure called "Uranus's Ring Detector."

7.3c Uranus's Interior and Magnetic Field

Voyager 2 detected Uranus's magnetic field. It is intrinsically about 50 times stronger than Earth's. Surprisingly, it is tipped 60° with respect to Uranus's axis of rotation.

Even more surprising, it is centered on a point offset from Uranus's center by 1/3 its radius. Our own Earth's magnetic field is nothing like that! Since Uranus's field is so tilted, it winds up like a corkscrew as Uranus rotates. Uranus's magnetosphere contains belts of protons and electrons, similar to Earth's Van Allen belts.

Voyager also detected radio bursts from Uranus every 17.24 hours. These bursts apparently come from locations carried in Uranus's interior by the magnetic field as the planet rotates. Thus Uranus's interior rotates slightly more slowly than its atmosphere. Since Uranus's and Neptune's interiors are similar, we will defer additional comments about the former for a few paragraphs.

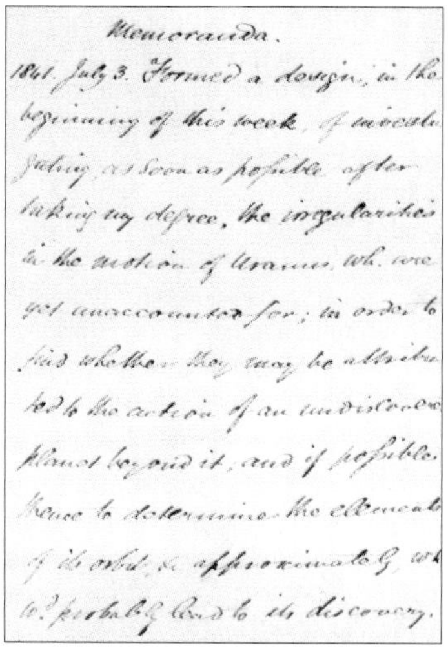

Master and Fellows of St. Johns College of Cambridge

■ **FIGURE 7–32** From John Couch Adams's diary, kept while he was in college. "1841, July 3. Formed a design in the beginning of this week, of investigating, as soon as possible after taking my degree, the irregularities in the motion of Uranus which are yet unaccounted for."

7.4 Neptune

Neptune is even farther from the Sun than Uranus, 30 A.U. compared to about 19 A.U. Neptune takes 164 years to orbit the Sun. Its discovery was a triumph of the modern era of Newtonian astronomy. Mathematicians analyzed the amount that Uranus (then the outermost known planet) deviated from the orbit it would follow if gravity from only the Sun and the other known planets were acting on it. The small deviations could have been caused by gravitational interaction with another, as yet unknown, planet.

The first to work on the problem successfully was John C. Adams in England. In 1845, soon after he graduated from Cambridge University (■ Fig. 7–32), he predicted positions for the new planet. But neither of the two main astronomers in England made and analyzed observations to test this prediction quickly enough.

A year later, the French astronomer Urbain Leverrier independently worked out the position of the undetected planet. The French astronomers didn't test his prediction right away either. Leverrier sent his predictions to an acquaintance at the observatory in Berlin, where a star atlas had recently been completed. The Berlin observer, Johann Galle, discovered Neptune within hours by comparing the sky against the new atlas.

Neptune has not yet made a full orbit since it was located in 1846. But it now seems that Galileo inadvertently observed Neptune in 1613 and recorded its position, which more than doubles the period of time over which it has been studied. Galileo's observing records from January 1613 (when calculations indicate that Neptune had passed near Jupiter) show stars that were very close to Jupiter (■ Fig. 7–33), yet modern catalogues do not contain one of them. Galileo even once noted that one of the "stars" actually seemed to have moved from night to night, as a planet would. The object that Galileo saw was very close to but not quite exactly where our calculations of Neptune's orbit show that Neptune would have been at that time. Presumably, Galileo saw Neptune. We have used the positions he measured to improve our knowledge of Neptune's orbit!

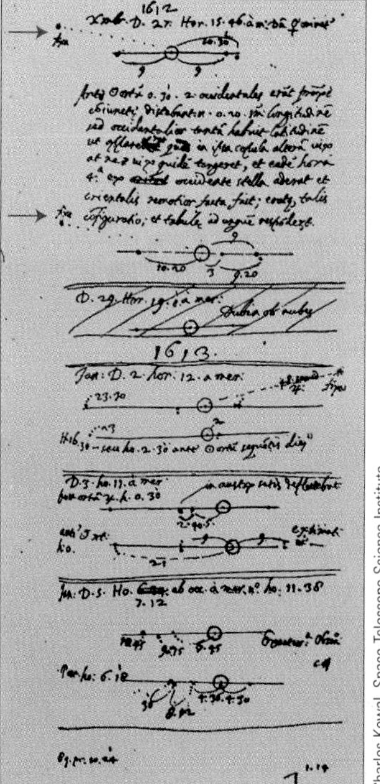

Charles Kowal, Space Telescope Science Institute

■ **FIGURE 7–33** Galileo's notebook from late December 1612, showing a * marking an apparently fixed star to the side of Jupiter and its moons. The "star" was evidently Neptune. The episode shows the importance of keeping good lab notebooks, a lesson we should all take to heart.

7.4a Neptune's Atmosphere

Neptune, like Uranus, appears greenish in a telescope because of its atmospheric methane (■ Fig. 7–34). Some faint markings could be detected on Neptune even before adaptive optics systems became available on Earth (■ Fig. 7–35). It was thus known before Voyager that Neptune's surface was more interesting than Uranus's. Still, given its position in the cold outer Solar System, nobody was prepared for the amount of activity that Voyager discovered.

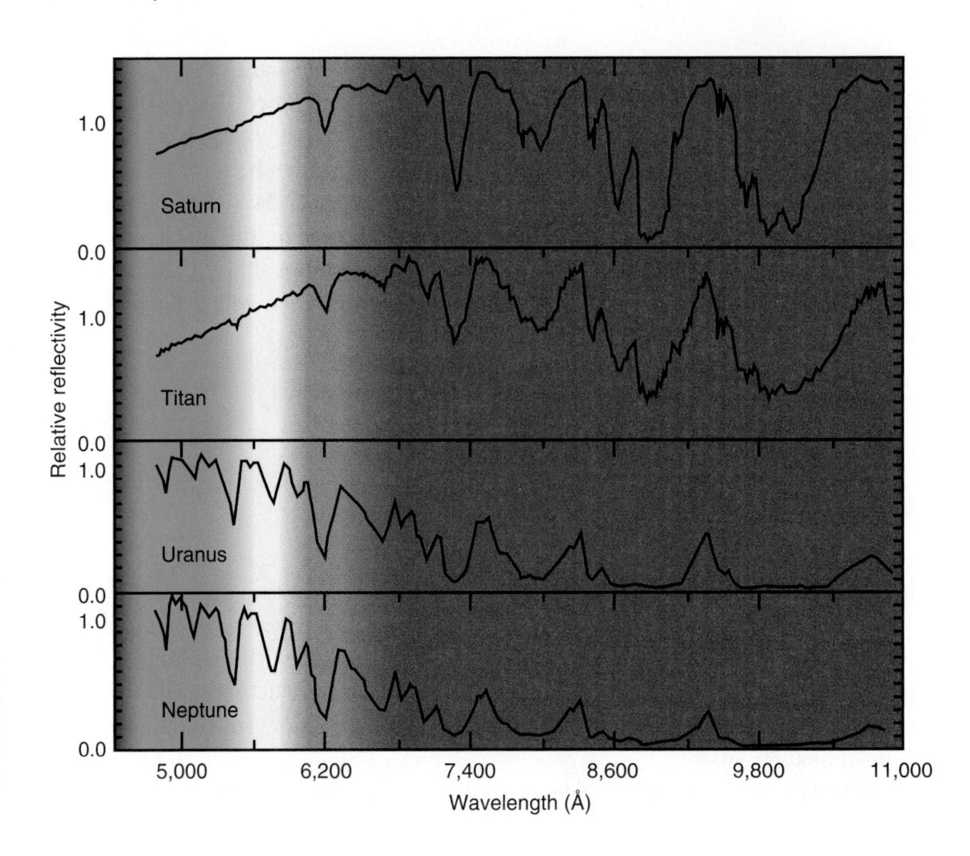

■ **FIGURE 7–34** Each of the four graphs shows a spectrum divided by the Sun's spectrum. The spectra dip at wavelengths where there is a lot of absorption. The dark, broad troughs are from methane absorption; these wavelengths (the right side of the graphs) are in the infrared. Notice how strong the methane absorption is for Uranus and Neptune, and how it leaves mainly blue and green radiation for us to see. (Tobias Owen, Institute for Astronomy, U. Hawaii, after R. Danhy)

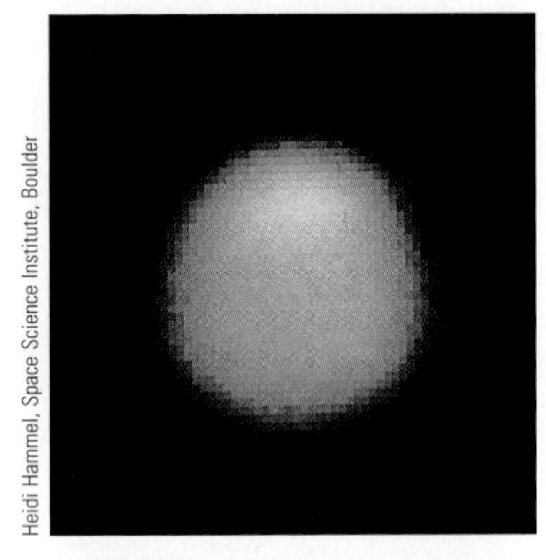
Heidi Hammel, Space Science Institute, Boulder

■ **FIGURE 7–35** Neptune, imaged from Earth in the near-infrared light (8900 Å) that is strongly absorbed by methane gas in Neptune's atmosphere, making most of the planet appear dark. The bright regions are clouds of methane ice crystals above the methane gas, reflecting sunlight. The northern hemisphere appeared brighter than the southern, a reversal of the earlier situation. Neptune's disk was 2.3 arc seconds across; the angular resolution observable was less than 1 arc second.

As Voyager approached Neptune, active weather systems became apparent (■ Fig. 7–36). An Earth-sized region that was soon called the **Great Dark Spot** (■ Fig. 7–37) became apparent. Though colorless, the Great Dark Spot seemed analogous to Jupiter's Great Red Spot in several ways. For example, it was about the same size relative to its planet, and it was in the same general position in its planet's southern hemisphere (■ Fig. 7–38). Putting together a series of observations into a movie, scientists discovered that it rotated counterclockwise, as does the Great Red Spot. Thus it was anti-cyclonic, which made it a high-pressure region. Clouds of ice crystals, similar to Earth's cirrus but made of methane, form at the edge of the Great Dark Spot as the high pressure forces methane-rich gas upward. But this Great Dark Spot had disap-

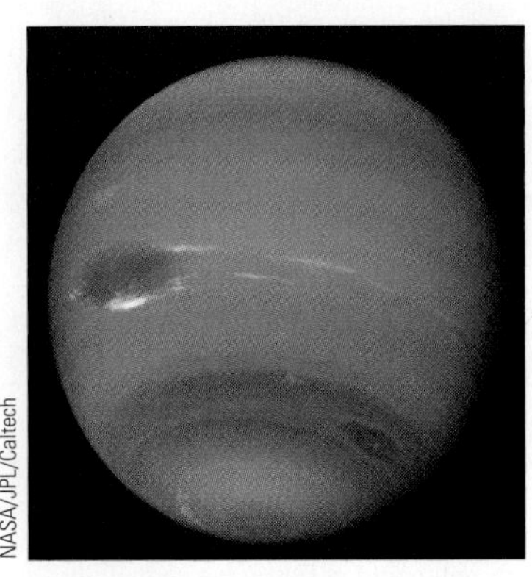
NASA/JPL/Caltech

■ **FIGURE 7–36** Neptune, from Voyager 2, with its Great Dark Spot, a giant storm in Neptune's atmosphere.

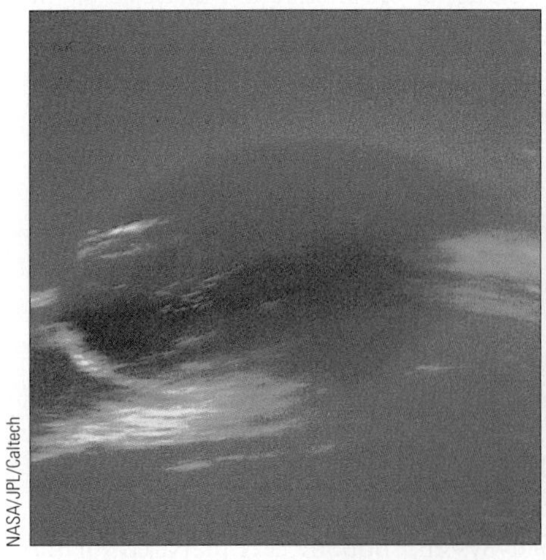
NASA/JPL/Caltech

■ **FIGURE 7–37** The Great Dark Spot, as seen by Voyager in 1989. It was about the size of the Earth, but has since disappeared.

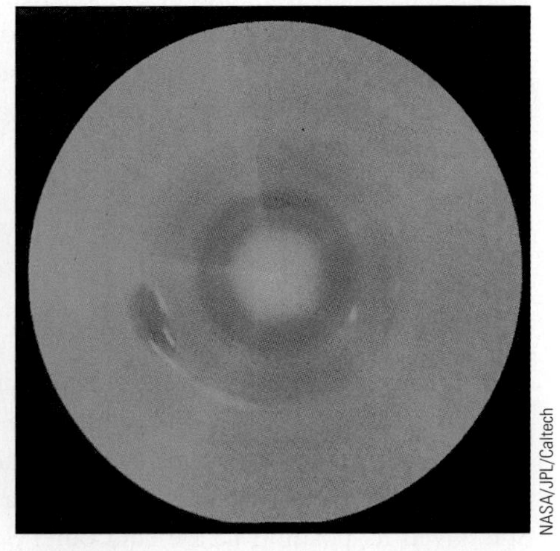
NASA/JPL/Caltech

■ **FIGURE 7–38** Cloud systems in Neptune's southern hemisphere. Neptune's south pole is at the center of this polar projection.

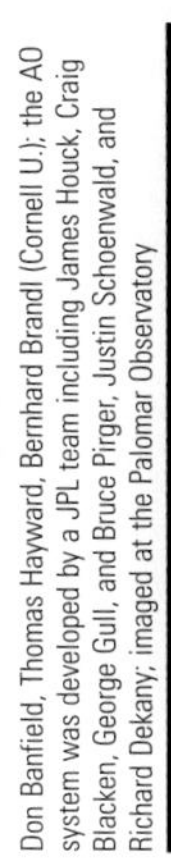

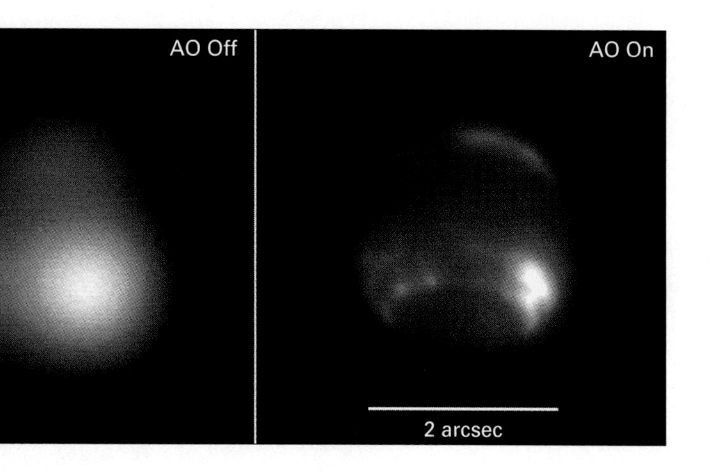

W. M. Keck Observatory

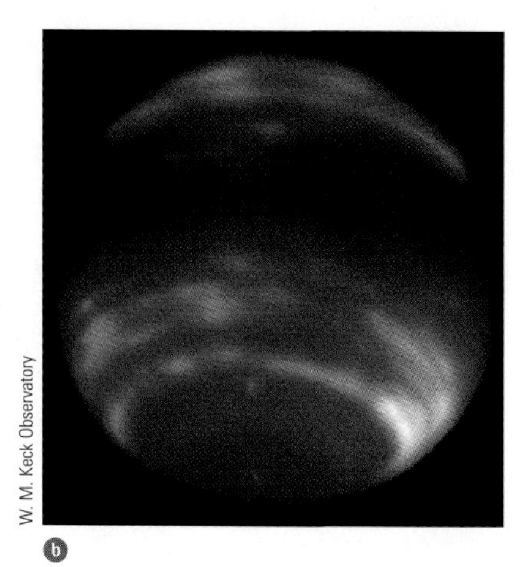

■ **FIGURE 7–39** Neptune, (a) observed with the 5-m Palomar Hale telescope and with the adaptive optics (AO) system turned off (*left*) and on (*right*). (b) Neptune, observed with adaptive optics on the Keck II telescope. A storm is prominent at lower right.

peared when the Hubble Space Telescope photographed Neptune a few years later, so it was much less long lived than Jupiter's Great Red Spot.

Several other cloud systems were also seen by Voyager 2. Clouds can, in addition, be viewed with the Hubble Space Telescope and can now even be viewed with ground-based telescopes using adaptive-optics systems (■ Fig. 7–39).

7.4b Neptune's Interior and Magnetic Field

Voyager 2 measured Neptune's average temperature: 59 K, that is, 59°C above absolute zero (−214°C). This temperature, though low, is higher than would be expected on the basis of solar radiation alone. Neptune gives off about 2.7 times as much energy as it absorbs from the Sun. Thus there is an internal source of heating, unlike the case of Uranus, which otherwise seems like a similar planet.

Why is the average density of Uranus and Neptune higher than that of Jupiter and Saturn? It may reflect slight differences in the origins of those planets. Their densities show that Uranus and Neptune have a higher percentage of heavy elements than Jupiter and Saturn. Perhaps the rocky cores they built up from the solar nebula were smaller, giving them less gravity and thus attracting less hydrogen and helium. Or, perhaps the positions of Uranus and Neptune farther out in the solar nebula put them in a region where either the solar nebula was less dense, or their slower orbital motion moved them through less gas. Perhaps they formed after the solar wind cleared out most of the hydrogen and helium.

Voyager detected radio bursts from Neptune every 16.11 hours. Thus Neptune's interior must rotate with this rate. On Neptune and Uranus, as on the Earth, equatorial winds blow more slowly than the interior rotates. By contrast, equatorial winds on Venus, Jupiter, Saturn, and the Sun blow more rapidly than the interior rotates. We now have quite a variety of planetary atmospheres to help us understand the basic causes of circulation.

Voyager discovered and measured Neptune's magnetic field. The field, as for Uranus, turned out to be both greatly tipped and offset from Neptune's center (■ Fig. 7–40). Thus the tentative explanation for Uranus that its field was tilted by a collision is not plausible; such a rare event would not be expected to happen twice. Astronomers favor the explanation that the magnetic field is formed in an electrically conducting shell outside the planets' cores. The fields of Earth and Jupiter, in contrast, are thought to be formed deep within the core. The tilted and offset magnetic fields are some of the biggest surprises found by Voyager 2.

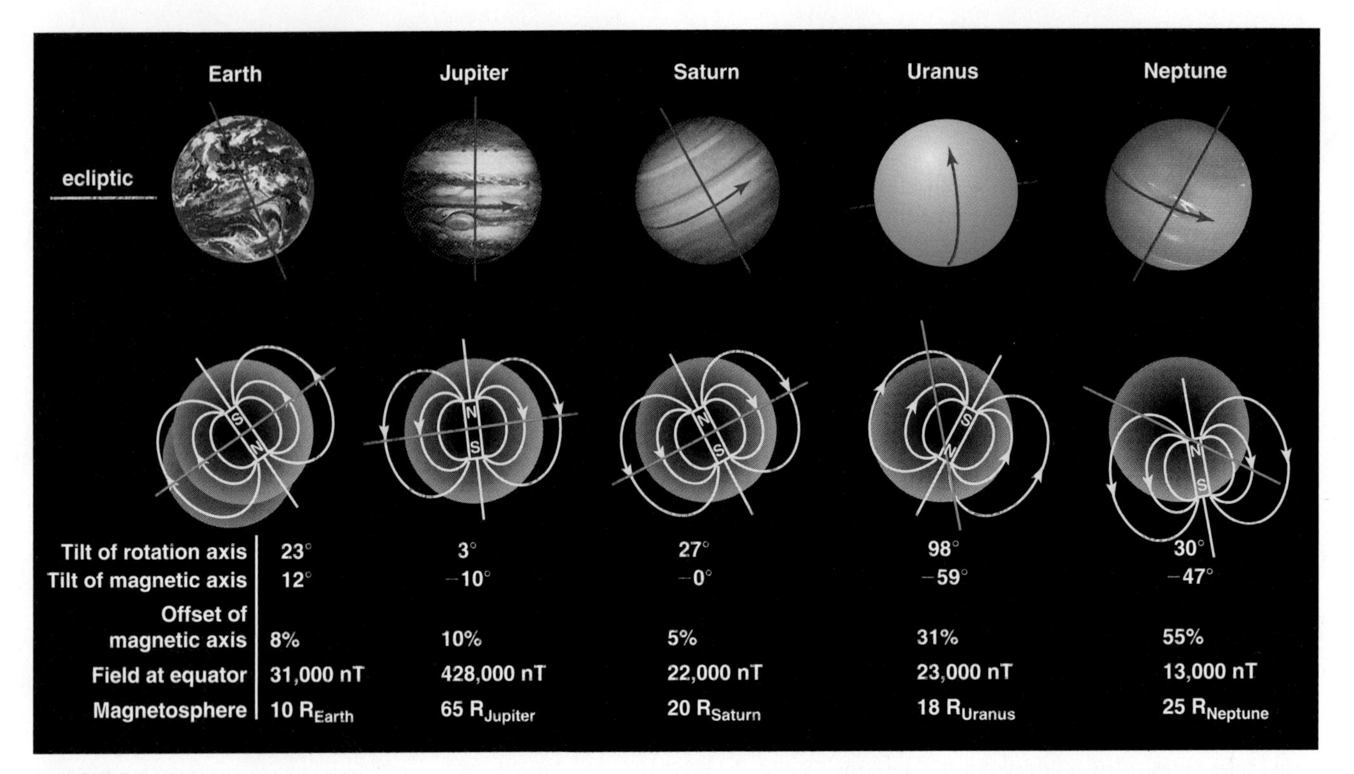

■ **FIGURE 7–40** The magnetic fields of the planets. Mercury, whose magnetic field is only 1% that of Earth's, is not shown. The fields at the planets' equators are given in nanoteslas (nT). 10,000 nT = 1 gauss, 1/100 the strength of a toy magnet. (Information from Norman Ness; style from *Sky & Telescope*)

7.4c Neptune's Rings

Before Voyager's arrival, astronomers wanted to know whether Neptune has rings, like the other giant planets. There was no obvious reason why it shouldn't. Some observations from Earth detected dips during occultations of stars, similar to those produced by Uranus's rings, while others didn't. It was thought that perhaps Neptune had incomplete rings, "ring arcs."

As Voyager came close to Neptune, it radioed back images that showed conclusively that Neptune has narrow rings. (See *A Closer Look 7.6: Naming the Rings of Neptune.*) Further, it showed the rings going all the way around Neptune. The material in the densest of Neptune's rings is very clumpy. The clumps had led to the incorrect idea of "ring arcs." The clumpy parts of the ring had blocked starlight, while the other parts and the other rings were too thin to do so (■ Figs. 7–41 and 7–42). The rings can now be studied from the Hubble Space Telescope and with ground-based telescopes having adaptive optics (see discussion in Chapter 3).

The fact that Neptune's rings are so much brighter when seen backlighted tells us about the sizes of particles in them. The most detectable parts of the rings have at least a hundred times more dust-sized grains than most of the rings of Uranus and Saturn. Since dust particles settle out of the rings, new sources must continually be active. Probably moonlets collide and are destroyed. Though much less dusty, Saturn's outer ring and Uranus's rings are similarly narrow.

A big surprise came in 2005: Neptune's arcs are fading and one of the arcs, *Liberté*, is much weaker. At this rate, it would disappear within 100 years. Neptune's clumpy rings are a much more transitory system than had been realized.

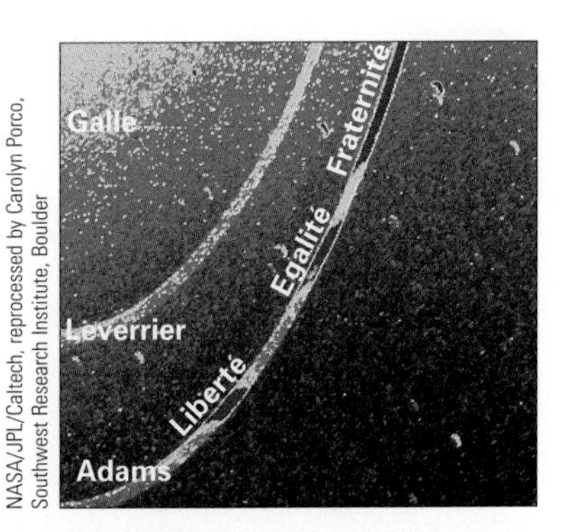

NASA/JPL/Caltech, reprocessed by Carolyn Porco, Southwest Research Institute, Boulder

■ **FIGURE 7–41** Three of Neptune's rings, showing the clumpy structure in one of them that led ground-based scientists to conclude from occultation data that Neptune had ring arcs. The Leverrier and Adams rings are 53,000 and 63,000 km, respectively, from Neptune's center. These photographs were taken as Voyager 2 left Neptune, 1.1 million km later, and so are backlighted. The main clumps are 6° to 8° long.

7.4d Neptune's Moon Triton

Neptune's largest moon, Triton, is a little larger than our Moon and has a retrograde (backward) orbit. It is named after a sea god who was a son of Poseidon. It is massive enough to have a melted interior. Its density is 2.07 grams/cm^3, so it is probably about

A Closer Look 7.6 Naming the Rings of Neptune

The rings of Neptune were named by a committee of the International Astronomical Union. The major rings are named after the two theoretical predictors and the actual observational discoverer of Neptune. The three subdivisions of the outer ring, at the suggestion of the French member of the committee, were named *Liberté, Egalité,* and *Fraternité* after the slogan of the French Revolution (1789). A fourth clump, named later, was called *Courage* (pronounced in the French way), in part because that is another worthwhile value that should be used to defend the other values.

70 per cent rock and 30 per cent water ice. It is denser than any jovian-planet satellite except Io and Europa.

Even before Voyager 2 visited, it was known that Triton has an atmosphere. Since Triton was Voyager 2's last objective among the planets and their moons, the spacecraft could be sent very close to Triton without fear of its destruction. The scientists waited eagerly to learn if Triton's atmosphere would be transparent enough to see the surface. It was. The atmosphere is mostly nitrogen gas, like Earth's.

Triton's surface is incredibly varied. Much of the region Voyager 2 imaged was near Triton's south polar cap (■ Fig. 7–43). The ice appeared slightly reddish. The color probably shows the presence of organic material formed by the action of solar ultraviolet light and particles from Neptune's magnetosphere hitting methane in Triton's atmosphere and surface. Nearer to Triton's equator, nitrogen frost was seen.

Many craters and cliffs were seen. They could not survive if they were made of only methane ice, so water ice (which is stronger) must be the major component. Since Neptune's gravity captures many comets in that part of the Solar System, most of the craters are thought to result from collisions with comets.

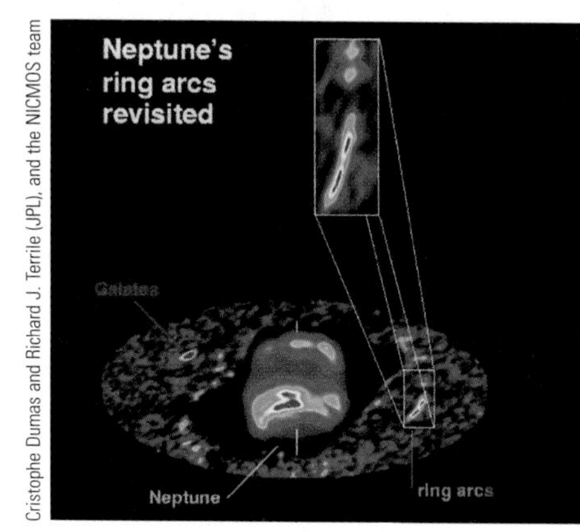

Cristophe Dumas and Richard J. Terrile (JPL), and the NICMOS team

■ **FIGURE 7–42** Neptune, its moon Galatea, and some of its apparent ring arcs, imaged with the Hubble Space Telescope.

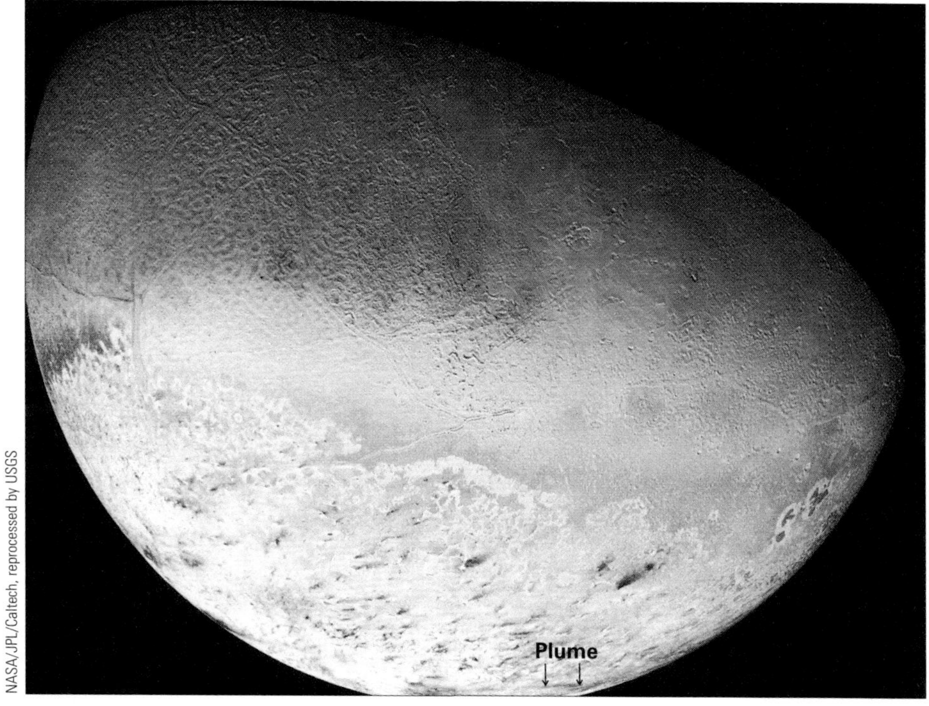

NASA/JPL/Caltech, reprocessed by USGS

■ **FIGURE 7–43** Triton's southern region, including its south polar ice cap. Most of the cap may be nitrogen ice. The pinkish color may come from the action of ultraviolet light and magnetospheric radiation upon methane in the atmosphere and surface. Most of the dark streaks represent material spread downwind from eruptions. One erupting plume is the long, thin, dark streak marked with arrows. Though the streaks are dark relative to other features on Triton, they are still ten times more reflective than the surface of our own Moon. The most detailed information on the photograph was taken in black and white through a clear filter; color information was added from lower-resolution images.

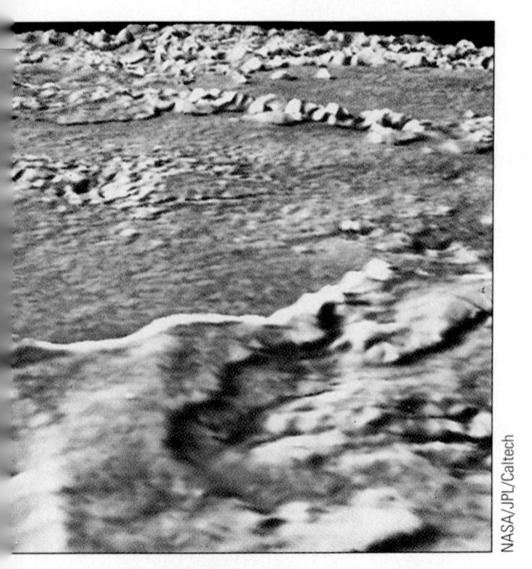

■ **FIGURE 7–44** A computer-generated perspective view of one of Triton's caldera-like depressions.

Triton's surface showed about 50 dark streaks parallel to each other as is readily visible near the bottom of Figure 7–43. They are apparently dark material vented from below. The material is spread out by winds. A leading model is that the Sun heats darkened methane ice on Triton's surface. This heating vaporizes the underlying nitrogen ice, which escapes through vents in the surface. A couple of these "ice volcanoes" were erupting when Voyager 2 went by. Since the streaks are on top of seasonal ice, they are all probably less than 100 years old.

Much of Triton is so puckered that it is called the "cantaloupe terrain." It contains depressions 30 km in diameter, crisscrossed by ridges (■ Fig. 7–44).

Triton has obviously been very geologically active. Since it is in a retrograde (backward) orbit around Neptune, it was probably born elsewhere in the Solar System and later captured by Neptune. Tidal forces from Neptune would have kept Triton molten until its orbit became circular. While it was molten, the heavier rocky material would have settled to form a 2000-km-diameter core.

One of the authors (J.M.P.) was fortunate to participate in a series of Earth-based expeditions to observe occultations of stars by Triton. Only rarely does Triton go in front of a star bright enough to study, but James Elliot, the discoverer of Uranus's rings during an occultation, has organized a few such expeditions. In 1997, several collaborating expeditions of astronomers went first to Australia and later to Hawaii to try to get in Triton's shadow cast onto the Earth.

It turns out that on the second occasion the group got the best observations with the Hubble Space Telescope, which happened to pass the right place at the right time. When Triton blocked out the star, which was about 6 times brighter, the brightness of the merged image dropped. From the way in which it dropped, we could analyze what Triton's atmosphere was like. The temperature turned out to be a few degrees warmer than it was when Voyager flew by, so our joint publication, in the journal *Nature*, was called "Global Warming on Triton."

CONCEPT REVIEW

Jupiter (318 times more massive than Earth), Saturn (95 times more massive), Uranus (15 times more massive), and Neptune (17 times more massive) are **giant planets,** also known as **jovian planets** (Introductory section). They have extensive gaseous atmospheres, liquid interiors, and rocky cores. Each of them has many moons and at least one ring. All were visited by at least one Voyager spacecraft.

Jupiter's diameter is 11 times that of Earth (Sec. 7.1). We learned a tremendous amount about the planet and its moons from the two Voyager (both in 1979) and Galileo (1995–2003) spacecraft (Sec. 7.1a). Jupiter's **Great Red Spot** (Sec. 7.1b), a giant circulating storm, is a few times larger than Earth. Jupiter's atmosphere shows bright and dark bands, which are different cloud layers (Sec. 7.1c). A probe dropped from the Galileo spacecraft measured the atmosphere's composition, density, and other properties. Jupiter has an internal source of energy, perhaps the energy remaining from its contraction (Sec. 7.1d). Its magnetic field is very strong (Sec. 7.1e), and it has a thin ring (Sec. 7.1f).

Jupiter's **Galilean satellites** (Sec. 7.1g) are comparable in size to our Moon. They were studied in detail from the Voyagers and the Galileo spacecraft. Io has over 100 volcanoes, many of which are erupting at any given time (Sec. 7.1g(i)). Europa is covered with smooth ice, perhaps over an ocean (Sec. 7.1g(ii)). Ganymede (Sec. 7.1g(iii)) and Callisto (Sec. 7.1g(iv)) are heavily cratered. Jupiter has dozens of additional moons (Sec. 7.1g(v)), making it look like a miniature planetary system of its own.

Saturn has a very low density, lower even than water (Sec. 7.2). Saturn's rings were produced by tidal forces that prevented the formation of a moon, and are within Saturn's **Roche limit** (Sec. 7.2a). The rings are so thin that stars can be seen through them. **Cassini's division** is the major break between inner and outer rings. Saturn's atmosphere shows fewer features than Jupiter's, probably because it is colder (Sec. 7.2b). Saturn has an internal source of heat, perhaps both energy from its collapse and energy from sinking helium (Sec. 7.2c). Saturn's moon Titan has a thick, smoggy atmosphere (Sec. 7.2d). The Cassini spacecraft, in 2004, arrived at the Saturn system, which it is still studying in detail. It dropped the Huygens probe down to Titan's surface. On the way down, it photographed channels that were cut by running liquid, undoubtedly methane or ethane. Saturn has many additional satellites, some of which were imaged by the Cassini spacecraft (Sec. 7.2e).

Uranus was the first planet to be discovered that had not been known to the ancients (Sec. 7.3). It consists largely of hydrogen and helium, although its core makes up a more substantial part of the planet than in the case of Jupiter or

Saturn. There is no internal energy source. Uranus rotates on its side, so it is heated by the Sun in a strange way over an 84-year period of revolution.

Its atmosphere appears very bland (Sec. 7.3a); chemical reactions are more limited than on Jupiter and Saturn because it is colder. Uranus's extensive clouds form relatively deep in the atmosphere. Methane is a major constituent of Uranus's atmosphere, including the clouds, and it accounts for the planet's greenish color. Uranus has several thin rings, discovered by an occultation of a star observed from Earth (Sec. 7.3b). The magnetic field of Uranus is greatly tipped and offset from the planet's center (Sec. 7.3c).

Neptune was discovered in 1846, after mathematicians predicted its existence based on discrepancies between the observed and expected positions of Uranus in the sky (Sec. 7.4). Apparently, Galileo saw it in 1613, but did not recognize it as a new planet. Like Uranus, Neptune appears greenish in a telescope because of its atmospheric methane (Sec. 7.4a). Methane clouds also cover Neptune. When Voyager 2 flew by in 1989, Neptune's surface showed a **Great Dark Spot,** a high-pressure region analogous to Jupiter's Great Red Spot. It disappeared between the Voyager visit and a later Hubble Space Telescope view. Neptune has a substantial core, as does Uranus; unlike the latter, however, it has an internal energy source (Sec. 7.4b). Neptune's magnetic field is strange, resembling that of Uranus. These magnetic fields are thus probably formed in electrically conducting shells of gas. Neptune's rings are clumpy (Sec. 7.4c). The clumps are dynamic, fading faster than had been expected.

Neptune's moon Triton is very large (Sec. 7.4d). Voyager 2 took high-resolution images that showed a varied surface with plumes from ice volcanoes. Triton has been very geologically active. Occultation studies show that it is gradually warming.

QUESTIONS

1. Why does Jupiter appear brighter than Mars despite Jupiter's greater distance from the Earth?
2. Even though Jupiter's atmosphere is very active, the Great Red Spot has persisted for a long time. How is this possible?
3. What advantages over the 5-m Palomar telescope on Earth did Voyagers 1 and 2 have for making images of the outer planets?
4. Other than photography, what are two types of observations made from the Voyagers?
5. How does the interior of Jupiter differ from the interior of the Earth?
6. †Given that Jupiter's diameter is 11.2 times that of Earth, about how many Earths could fit inside of Jupiter? (Neglect the empty spaces that exist below closely packed spheres.)
7. Why did we expect Io to show considerable volcanism?
8. Contrast the volcanoes of Io with those of Earth.
9. Compare the surfaces of Callisto, Io, and the Earth's Moon. Explain what this comparison tells us about the ages of features on their surfaces.
10. What are the similarities between Jupiter and Saturn?
11. What is the Roche limit and how does it apply to Saturn's rings?
12. How did we know, even before Voyager, that Saturn's rings are very thin?
13. What did the Voyagers reveal about Cassini's division?
14. How are narrow rings thought to be kept from spreading out?
15. What is strange about the direction of rotation of Uranus?
16. Which of the giant planets are known to have internal heat sources?
17. †What fraction of its orbit has Neptune traversed since it was discovered? Since it was first seen by Galileo?
18. Compare Neptune's Great Dark Spot with Jupiter's Great Red Spot and with the size of the Earth.
19. Compare the rings of Jupiter, Saturn, Uranus, and Neptune.
20. Describe the major developments in our understanding of Jupiter, from ground-based observations to the Voyagers to the Galileo and Cassini spacecraft.
21. Describe the major developments in our understanding of Saturn, from ground-based observations to the Voyagers and Cassini.
22. Describe the major developments in our understanding of Uranus, from ground-based observations to the Voyagers.
23. Discuss Miranda's appearance and explain why it is so surprising.
24. Describe the major developments in our understanding of Neptune, from ground-based observations to the Voyagers.
25. Compare Triton in size and surface with other major moons, like Ganymede and Titan.
26. **True or false?** Through a telescope on Earth, we can sometimes see Jupiter and Saturn in their crescent phases, just as the Moon is sometimes a crescent.
27. **True or false?** The rings of Uranus are kept narrow by the presence of adjacent small moons that gravitationally confine the particles to the rings.
28. **True or false?** Uranus has extreme seasons because its rotation axis is nearly in its orbital plane.
29. **True or false?** Tidal forces from Jupiter heat its large moons Io and Europa, leading to extensive volcanism on Io and an icy slush beneath Europa's surface.
30. **True or false?** The existence of Neptune was correctly predicted after astronomers noted that Uranus's observed and expected positions in the sky differed.
31. **Multiple choice:** The low average densities of Jupiter and Saturn compared with Earth suggest that **(a)** they are hollow; **(b)** their gravitational attraction has

squeezed material out of their cores; (**c**) they consist mostly of water; (**d**) they contain large quantities of light elements, such as hydrogen and helium; or (**e**) volcanic eruptions have ejected all the iron that was originally in their cores.

32. **Multiple choice:** Jovian planets have rings because (**a**) their thick gaseous atmospheres would disintegrate any small rock that enters them; (**b**) there is too much material to have fit into the ball of each planet; (**c**) tidal forces prevent the material in rings from forming into moons; (**d**) Jovian planets rotate very rapidly, and some material near the equator of these planets was flung outward, forming the rings; or (**e**) tidal forces cause volcanic eruptions on some moons and part of this material subsequently escaped the gravity of the moons, forming the rings.
33. **Multiple choice:** Which one of the following statements about the jovian planets is *false*? (**a**) They generally have the thickest atmospheres of all planets in the Solar System. (**b**) They are the most massive of all the planets in the Solar System since they are mainly composed of heavy elements like iron. (**c**) Despite their size, they rotate about their axis very rapidly, in fewer than 24 hours. (**d**) They have many moons, probably due to their large gravitational fields. (**e**) They are the largest of all planets in the Solar System, with diameters up to nearly 1/10 the diameter of the Sun.
34. **Multiple choice:** The rings of Uranus were discovered by accident, during observations of its (**a**) phases, (**b**) occultation of a star, (**c**) rotation period, (**d**) transit across the Sun's disk, or (**e**) moons.
35. **Multiple choice:** Jupiter and Saturn radiate more energy than they receive from the Sun. The most likely source of this excess energy is (**a**) gravitational contraction or settling, (**b**) hydrogen fusion, (**c**) chemical reactions, (**d**) radioactive decay, or (**e**) the magnetic fields of these planets.
36. **Fill in the blank:** The distance from a planet within which it is not possible to form a moon through gravitational attraction between particles is called ________.
37. **Fill in the blank:** The _______ spacecraft dropped a probe into Titan's atmosphere in late 2004.
38. **Fill in the blank:** Perhaps the most likely moon of the Solar System on which life could have developed is _______, given its warm interior and partially melted water ice.
39. **Fill in the blank:** The ________ of Uranus and Saturn are strange, with tilted axes offset from the planet centers.
40. **Fill in the blank:** Small moons that keep the material in narrow rings from spreading apart are called _________.

†This question requires a numerical solution.

TOPICS FOR DISCUSSION

1. Why do you think the chemical compositions of Jupiter and Saturn are so close to that of the Sun, but those of the terrestrial planets are not?
2. Given that we have already learned so much about the outer planets with uncrewed spacecraft, should we next send humans to them, or do you think using uncrewed spacecraft and robots (like the Spirit and Opportunity rovers on Mars) makes more sense?

MEDIA

Virtual Laboratories

Planetary Atmospheres and Their Retention

AceAstronomy™ Log into AceAstronomy at **http://astronomy.brookscole.com/cosmos3** to access quizzes and animations that will help you assess your understanding of this chapter's topics.

Log into the Student Companion Web Site at **http://astronomy.brookscole.com/cosmos3** for more resources for this chapter including a list of common misconceptions, news and updates, flashcards, and more.

People in Astronomy

Carolyn Porco

Jay M. Pasachoff

Carolyn Porco is a Senior Research Scientist at the Space Science Institute (SSI) in Boulder, Colorado; an Adjunct Professor in the Department of Planetary Sciences at the University of Arizona; and an Adjunct Professor at the University of Colorado in Boulder. Since 1990, she has been the leader of the Imaging Team for the Cassini mission currently in orbit around Saturn. She is also the Director of the Cassini Imaging Central Laboratory for Operations (CICLOPS) at SSI in Boulder, is the creator/editor of the CICLOPS website (ciclops.org), and writes the site's home page opening greeting to the public.

Dr. Porco grew up in the Bronx, New York, and attended the State University of New York at Stony Brook. Wanting to specialize in planetary studies, she did her graduate work in the Division of Geological and Planetary Sciences at the California Institute of Technology in Pasadena. In 1983, armed with her Ph.D., she joined Professor Bradford Smith, head of the Voyager imaging team, at the Lunar and Planetary Laboratory at the University of Arizona in Tucson, became a member of the Voyager imaging team, and headed the team's working group studying the rings during the Voyager Neptune encounter. In 1990, she was selected as team leader for the imaging experiment on Cassini.

Cassini's cameras are working so well. How does that make you feel?

Fabulous! And very, very relieved. Not only are the cameras functioning beautifully, but the software, databases, and processes that we have built at CICLOPS are also working very well. To run a flight experiment, like the Cassini imaging investigation, is an enormously complicated affair. For any sequence of images we take, the spacecraft pointing "dwell and slew" profile [that is, point in one direction and then move rapidly across the sky to another direction] has to be carefully designed, and the shuttering of the cameras has to be perfectly synchronized with that motion or else you get smeared images. Of course, the images also have to be properly exposed. And doing all this from a platform whose position is constantly changing relative to the bodies we want to image, and accounting for the variations in solar illumination, viewing geometry, and the different intrinsic brightnesses of the bodies around Saturn—all of which affect the exposure time—is quite a challenge. Then, once images hit the ground, we at CICLOPS process the best ones for release to the public, and also ensure that each image and the auxiliary information that describes the circumstances under which it was taken get accurately recorded and archived.

And **all** of it, I'm proud to say, is working—efficiently and smoothly. So, I'm tremendously gratified that the many years of obsessive devotion to this project—in which I had to clear the decks of everything (including any semblance of a normal life)—have paid off.

And now all of us can wake up each day to a new and dazzling sight from an alien "planetary system" that is vastly distant from us. It's all quite remarkable and sometimes even seems unreal. Only it's not!

. . . ALL OF US CAN WAKE UP EACH DAY TO A NEW AND DAZZLING SIGHT FROM AN ALIEN "PLANETARY SYSTEM" THAT IS VASTLY DISTANT FROM US.

It's worth mentioning that the same is true for the other scientific instruments: they are also working very well. And this is the case despite the fact that we scientists are operating on shoestring budgets. Cassini is often thought of as a big, expensive (and therefore well-funded) mission. Not so. It's true that, after it's over, it will have cost more than $3 billion and that's a lot of money. But this funding is spread over more than 18 years, and it is a mission of **tremendous** scope involving hundreds of scientists. And I can tell you that they are all working far above and beyond their compensation.

Unlike missions like Voyager and Galileo, Cassini is very "modern" in that the operations of the science instruments are distributed to the home institutions of the leaders of the science teams. This means that the conduct of each experiment is in the hands of the scientists: i.e., the people who know and care the most about what is required for a scientifically fruitful investigation. But it also means that the scientists are working furiously to run the experiments and, simultaneously, examine and publish their results, and they are all inadequately funded to do these jobs.

At the end of the day, we owe our tremendous success to the engineers who did a remarkable job in the first 7 years of the mission building the spacecraft and the instruments, and to the dedication and hard work of the scientists who believe that the pursuit of truth and knowledge is the noblest of causes and worth whatever effort it requires. But it has not been easy.

What was it like to work in a group on Cassini for so many years?

This project has been ongoing now for 15 years and over that time, things have changed. The challenges we have faced have morphed over time—including the political and budgetary ones—and even the complexion of the project has changed. And it hasn't always been nice or pleasant. So, it's just like real life.

I often say: Cassini is not so much a mission as it is a way of life.

CASSINI IS NOT SO MUCH A MISSION

AS IT IS A WAY OF LIFE.

What is your biggest surprise (so far) with the Cassini results?

The sheer clarity and beauty of our images are the biggest surprises to me. I've said this so much I sound like a broken record, but it's true. I just didn't expect the kind of details that we are able to see. Remember, I'm a former Voyager imaging scientist. For 24 years, I carried around with me the impressions of Saturn and its rings and moons that were left by all those Voyager images. If someone mentioned Enceladus, I'd immediately call to mind one or two "best" Voyager images since that's all we had. The same for Rhea, Titan, Cassini's division in the rings, etc. But now, all those sights pale in comparison to what we can see with Cassini.

It's like going for Lasik [laser eye corneal-shaping] surgery and, overnight, being able to see.

What surprised you about the data that Huygens sent back?

The most stunning information returned by Huygens was that image, taken by the Descent Imager at altitudes of about 8 to 16 km, of the dendritic drainage pattern and shoreline on the surface. And when I say stunning, I mean it literally. I found myself, in Darmstadt, Germany, walking around in a daze after those images came back, as if someone had hit me over the head with a frying pan. I felt transformed. And what was stunning was the unmistakable nature of the pattern: it **had** to be carved by some kind of fluid flowing across the surface and carving channels of the type we see here on Earth.

Our imaging experiment, from the orbiter, had returned images of scenes on Titan that were not easy to interpret. Remember, we are in the game of pattern recognition when studying Titan since there are no shadows to tell us up from down. And our images did not show anything that looked unambiguous. But with that one Huygens image, we had something that we were all hoping for: a pattern that had an unambiguous cause. It is a cause that is Earthlike and therefore familiar. It is that meeting of the familiar with the alien that makes the heart leap.

How and when did you get interested in space science?

I've been interested in the study of the Universe and the exploration of space since I was a young teenager. As a youth, I was a seeker, a thinker. I first became very interested in philosophy and religion and the "big questions": What am I? What are we? Where are we? What is out there? These inquiries ultimately drove me to think about the structure of the Universe, of our cosmic neighborhood. My first sighting of Saturn through a friend's telescope on a roof in the Bronx during this time was the ultimate hook.

What might your cameras show when they reach Pluto on New Horizons?

If history is any precedent, anything I say will be wrong! But I'm putting my bets on a surface that is modified by all the same processes that we see working on Triton, the largest moon of Neptune.

Do you have any advice for undergraduates taking science courses but without a professional interest in going on in science?

Yes! Science is nothing more than a method of asking questions and getting answers. It is, in this regard, the art and practice of critical thinking. Knowing how to separate truth from fiction is a vital skill for living daily life and making important decisions, whether these decisions concern the selection of someone for public office or investment in the stock market. The fact that we can use the methods of scientific inquiry for understanding the natural world around us is a tremendous bonus—a reward—

that comes with the effort. To know is to be empowered, and with science comes knowledge. Don't retreat from it; embrace it!

TO KNOW IS TO BE EMPOWERED,

AND WITH SCIENCE COMES KNOWLEDGE.

And remember: life is good, but a life of knowledge is even better.

What's next for you?

Good question, and one that is on my mind a lot these days. First and foremost, I'm still working on Cassini and will be until it is over. And that will be mid-2008, unless we are given the funding to execute an extended mission beyond that. But right now, I'm writing a book called "*The Captain's Log*" about the journey of Cassini, from start to finish, that will also tell the tale of my own personal journey. Not an autobiography, mind you, but a tale of exploration from both the scientific and spiritual points of view. And, as you might guess, it will be a story told in images, printed on high quality glossy paper. I'm very jazzed about this.

As for the post-Cassini period, I'm thinking broadly. If my book is successful, I may continue to write. Of course, I will have hundreds of thousands of Cassini images and will want to reap the fruits of my labor and immerse myself in the study of Saturn's rings and moons. I have my own company, Diamond Sky Productions, that is geared toward combining the scientific, as well as artful, use of planetary images and computer graphics for the presentation of science to the public. We contributed some lovely computer graphics depicting the orbits of the Voyagers over some 40,000 years to the A&E TV documentary on Voyager's 25th anniversary called *Cosmic Journey*. So, I may be doing more of that in the future. Then again, sometimes I think I may do something completely different and establish a foundation to help women in Africa. Who knows?

If I've learned anything in my life, I've learned that just about anything is possible if you're willing to devote yourself to it. So, what I'll be doing in 5 or 10 years is not completely clear right now, but I'm sure it will be something creative, exciting, and rewarding. Otherwise, I wouldn't be doing it!

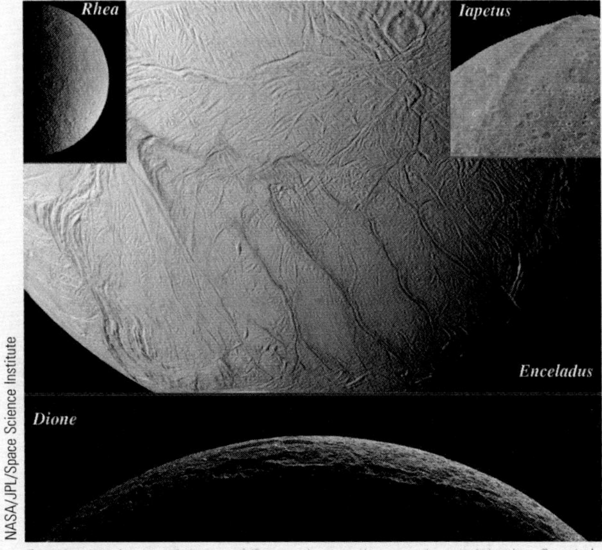

Dramatic views of four of Saturn's satellites, taken with the Cassini satellite, for which Carolyn Porco heads the research team.

CHAPTER 8

Pluto, Comets, and Space Debris

We have learned about the Solar System's giant planets, which range in size from about 4 to about 11 times the diameter of the Earth. We have seen that our Solar System has a set of terrestrial planets, which range in size from the Earth down to 40 per cent the diameter of the Earth. This size range includes the four inner planets as well as seven planetary satellites.

The remaining object that has long had the name "planet," Pluto, is only 20 per cent the diameter of Earth but is still over 2300 km across, so there is much room on it for interesting surface features. Recently, additional objects like it, but smaller, have been found in the outer reaches of the Solar System. We shall see how we determined Pluto's odd properties, and what the other, similar objects are.

Besides the planets and their moons, many other objects are in the family of the Sun. The most spectacular, as seen from Earth, are comets (■ Fig. 8–1). Bright comets have been noted throughout history, instilling great awe of the heavens. Comets have long been seen as omens, usually bad ones. As Shakespeare wrote in *Julius Caesar,* "When beggars die, there are no comets seen; The heavens themselves blaze forth the death of princes."

Asteroids, which are minor planets, and chunks of rock known as meteoroids, are other residents of our Solar System. We shall see how they and the comets are storehouses of information about the Solar System's origin.

Asteroids, meteoroids, and comets are suddenly in the news as astronomers are finding out that some come relatively close to the Earth. We are realizing more and more that collisions of these objects with the Earth can be devastating for life on Earth. Every few hundred thousand years, one large enough to do very serious damage should hit, and every few tens of millions of years, an enormous collision can produce a mass extinction of life on Earth. Apparently, a comet or an asteroid caused the dinosaurs to become extinct some 65 million years ago.

Should we be worrying about asteroid, meteoroid, or comet collisions? Should we be monitoring the sky around us better? Should we be planning ways of diverting an oncoming object if we were to find one?

AceAstronomy™ The AceAstronomy icon throughout this text indicates an opportunity for you to test yourself on key concepts, and to explore animations and interactions of the AceAstronomy website at **http://astronomy.brookscole.com/cosmos3**

A copper projectile from NASA's Deep Impact spacecraft rammed into Comet Tempel 1 on July 4, 2005, making a football-field-sized crater and exposing the top layers of the comet's nucleus. We see here a false-color view of the plume of ejecta and the nucleus's silhouette. Following the ejecta's progression revealed the comet's mass and thus, given its measured size, its density. Since the density is only about one-tenth that of Earth's rocks, the comet's materials must be very loosely packed. *A Closer Look 8.2: Deep Impact* shows more about the event. NASA/JPL/UMD

ORIGINS

We see how objects in the outer part of the Solar System, including comets, are left over from the origin of our Solar System and so can be studied to learn about conditions in our Solar System's early years. We find that asteroids, meteoroids, and comets may lead to our demise, just as they produced episodes of major extinction in the past when they collided with Earth.

AIMS

1. Explore the properties of distant Pluto, whose status as a planet is now questioned (Section 8.1).
2. Learn about the space debris that makes up the outer part of our Solar System, including Kuiper-belt objects and comets (Sections 8.2 and 8.3).
3. Discuss some other minor constituents of our Solar System, such as asteroids and meteoroids (Sections 8.4 and 8.5).
4. See how collisions between space debris and the Earth can lead to profound changes in life on Earth (Section 8.5).

Jay M. Pasachoff

■ **FIGURE 8–1** Comet Hale-Bopp was so bright it was even visible over Central Park in the middle of New York City.

8.1 Pluto

Pluto, the outermost known planet, is a deviant. Its elliptical orbit is the most out of round (eccentric) and is inclined by the greatest angle with respect to the Earth's orbital plane (the "ecliptic" plane, defined in Chapter 4), near which the other planets revolve.

Pluto's elliptical orbit is so eccentric that part lies inside the orbit of Neptune. Pluto was closest to the Sun in 1989 and moved farther away from the Sun than Neptune in 1999. So Pluto is still relatively near its closest approach to the Sun out of its 248-year period, and it appears about as bright as it ever does to viewers on Earth. It hasn't been as bright for over 200 years. It is barely visible through a medium-sized telescope under dark-sky conditions.

The discovery of Pluto was the result of a long search for an additional planet that, together with Neptune, was believed to be slightly distorting the orbit of Uranus. Finally, in 1930, Clyde Tombaugh, hired at age 23 to search for a new planet because of his experience as an amateur astronomer, found the dot of light that is Pluto (■ Fig. 8–2). It took him a year of diligent study of the photographic plates he obtained at the Lowell Observatory in Arizona. From its slow motion with respect to the stars over the course of many nights, he identified Pluto as a new planet.

8.1a Pluto's Mass and Size

Even such basics as the mass and diameter of Pluto are very difficult to determine. It had been hard to deduce the mass of Pluto because to do so was, at first, thought to require measuring Pluto's effect on Uranus, a far more massive body. (The orbit of Neptune, known for less than a hundred years at the time Pluto was discovered, was too poorly known to be of much use.) Moreover, Pluto has made less than one revolution around the Sun since its discovery, thus providing little of its path for detailed study. As recently as 1968, it was mistakenly concluded that Pluto had 91 per cent the mass of the Earth, instead of the correct value of 0.2 per cent.

The situation changed drastically in 1978 with the surprise discovery (■ Fig. 8–3) that Pluto has a satellite. The moon was named Charon, after the boatman who rowed passengers across the River Styx to the realm of Pluto, god of the underworld in Greek mythology. (Its name is informally pronounced "Shar´on," similarly to the name of the discoverer's wife, Charlene, by astronomers working in the field.) The presence of a satellite allows us to deduce the mass of the planet by applying Newton's form of Kepler's third law (Chapter 5). Charon is 12 per cent of Pluto's mass, and Pluto is

Lowell Observatory photographs

Lowell Observatory photographs

■ **FIGURE 8–2** Small sections of the photographic plates on which Clyde Tombaugh discovered Pluto. On February 18, 1930, Tombaugh noticed that one dot among many had moved between January 23, 1930 *(left)*, and January 29, 1930 *(right)*. He didn't have the arrows on the plates, of course, to aid his search!

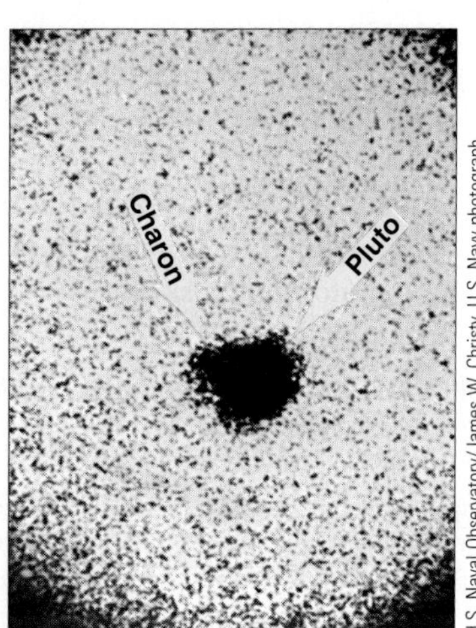

U.S. Naval Observatory/James W. Christy, U.S. Navy photograph

■ **FIGURE 8–3** The image of Pluto and Charon on which Charon was discovered.

only 1/500 the mass of the Earth, ten times less than had been suspected just before the discovery of Charon.

Pluto's rotation axis is nearly in the ecliptic, like that of Uranus. This is also the axis about which Charon orbits Pluto every 6.4 days. Consequently, there are two five-year intervals during Pluto's 248-year orbit when the two objects pass in front of (that is, occult) each other every 3.2 days, as seen from Earth. Such mutual occultations were the case from 1985 through 1990. When we measured their apparent brightness, we received light from both Pluto and Charon together (they are so close together that they appeared as a single point in the sky). Their blocking each other led to dips in the total brightness we received.

From the duration of fading, we deduced how large they are. Pluto is 2300 km in diameter, smaller than expected, and Charon is 1200 km in diameter. Charon is thus half the size of Pluto. Further, it is separated from Pluto by only about 8 Pluto diameters, compared with the 30 Earth diameters that separate the Earth and the Moon. So Pluto/Charon is almost a "double-planet" system.

The rate at which the light from Pluto/Charon faded also gave us information that revealed the reflectivities (albedoes) of their surfaces, since part of the surface of the blocked object remained visible most of the time. The surfaces of both vary in brightness (■ Fig. 8–4). Pluto seems to have a dark band near its equator, some markings on that band, and bright polar caps.

In 1990, the Hubble Space Telescope took an image that showed Pluto and Charon as distinct and separated objects for the first time, and they can now be viewed individually by telescopes on Mauna Kea in Hawaii (■ Fig. 8–5) and elsewhere where the "seeing" is exceptional. The latest Hubble views show that Pluto has a dozen areas of bright and dark, the finest detail ever seen on Pluto, whose diameter is smaller than that of the United States (■ Fig. 8–6). But we don't know whether the bright areas are

John R. Spencer, then Institute for Astronomy, U. Hawaii, now SWRI

■ **FIGURE 8–4** An image drawn to explain the observed variation of the total light as Pluto and Charon hide (occult) each other. The relative sizes, colors, and albedoes are approximately correct.

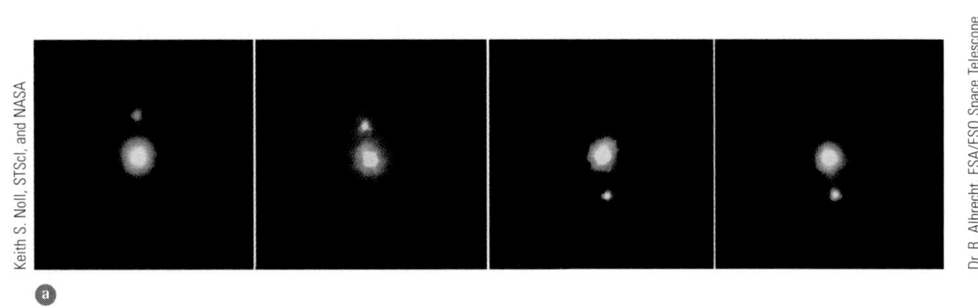

Keith S. Noll, STScI, and NASA

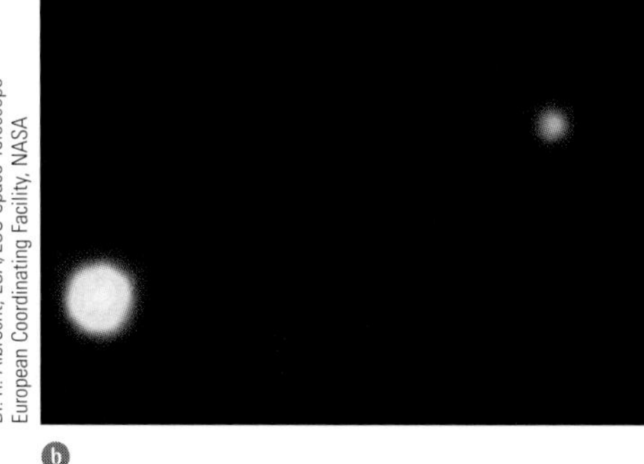

Dr. R. Albrecht, ESA/ESO Space Telescope European Coordinating Facility, NASA

■ **FIGURE 8–5** ⓐ This series of ground-based views of Pluto and Charon, showing them orbiting around each other, was taken in 1999 at the highest possible resolution at near-infrared wavelengths, with adaptive optics (mirror-shape distortion in real time, to compensate for atmospheric turbulence; see discussion in Chapter 3) using the new Gemini North telescope on Mauna Kea in Hawaii. ⓑ Pluto and Charon, clearly distinguished with the Hubble Space Telescope.

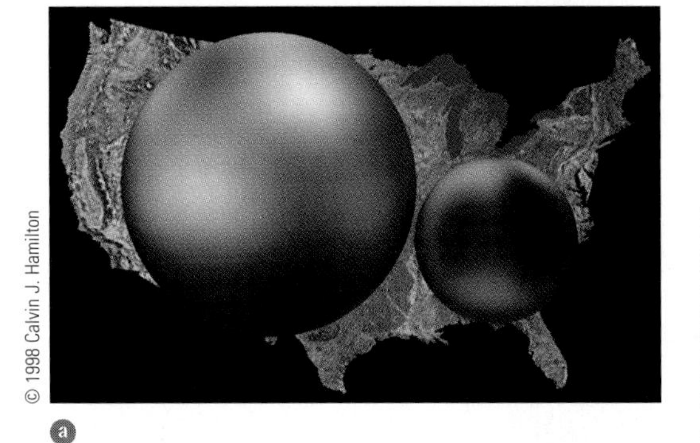

© 1998 Calvin J. Hamilton

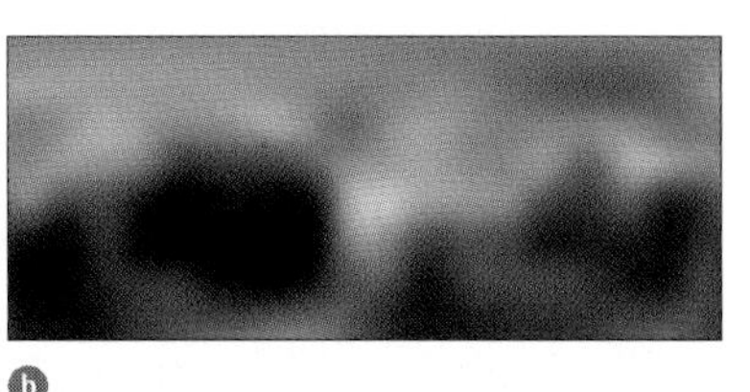

■ **FIGURE 8–6** ⓐ Pluto and Charon, showing their sizes relative to each other and to the United States. The surface features correspond to Hubble Space Telescope observations, the most detailed ever made of Pluto's surface. The spatial resolution of the Hubble image of Pluto is as though you were standing in Los Angeles and looking at the spots on a soccer ball in San Francisco. ⓑ A Hubble map released in 2005. Dark areas may be dirty water-ice and brighter areas are frosts of various molecules. Methane-ice is shown as red. The map extends from the north pole down to 60° south latitude.

bright because they are high clouds near mountains or low haze and frost. We merely know that there are extreme contrasts on Pluto's surface.

If we were standing on Pluto, the Sun would appear over a thousand times fainter than it does to us on Earth. Consequently, Pluto is very cold; infrared measurements show that its temperature is less than 60 K. From Pluto, we would need a telescope to see the solar disk, which would be about the same size that Jupiter appears from Earth.

8.1b Pluto's Atmosphere

Pluto occulted—passed in front of and hid—a star on one night in 1988. Astronomers observed this occultation to learn about Pluto's atmosphere. If Pluto had no atmosphere, the starlight would wink out abruptly. Any atmosphere would make the starlight diminish more gradually. The observations showed that the starlight diminished gradually and unevenly. Thus Pluto's atmosphere has layers in it. Another such occultation wasn't observed until 2002, when (again) Pluto was seen to make the star wink out for a minute or so on two separate occasions.

From the 1988 occultation, astronomers were also able to conclude that the bulk of Pluto's atmosphere is nitrogen. A trace of methane must also be present, since the methane ice on Pluto's surface, detected from its spectrum, must be evaporating.

Still, Pluto's atmospheric pressure is very low, only 1/100,000 of Earth's. The data from the first occultation seemed to show a change at a certain height in Pluto's atmosphere, leading to the deduction that either the atmosphere had a temperature inversion or that the lower atmosphere contained a lot of dust. The lone high-quality scan obtained in July 2002 showed no such change at a certain height in the rate at which the star's light was dimming as it passed through Pluto's atmosphere.

Then, in August 2002, a group of scientists, of which one of the authors (J.M.P.) was a member, succeeded in observing an occultation of a star by Pluto on ten different telescopes, several of them on Mauna Kea (■ Fig. 8–7). J.M.P.'s team from Williams College obtained a thousand data points in a 5-minute interval of the occultation, part

■ **FIGURE 8–7** (a) A series showing the images as Pluto occulted a star in August 2002. Earth's atmospheric conditions were so good that Charon was easily visible alongside Pluto (though the images are not distinct). We used the name P131.1 for the star, giving its place on a list of possible stars to be occulted. (b) A light curve from the Williams College observations made with the 2.2-m University of Hawaii telescope on Mauna Kea. The five-minute drop in total intensity, as Pluto occulted the background star, is obvious. (Jay M. Pasachoff, Bryce A. Babcock, David R. Ticehurst, and Steven P. Souza (Williams College)) (c) A set of false-color images from the 2005 occultation of a star by Charon: (*left*) all objects visible; (*center*) Charon hiding the star, dropping the total intensity; (*right*) all objects visible again. The occultation took less than a minute.

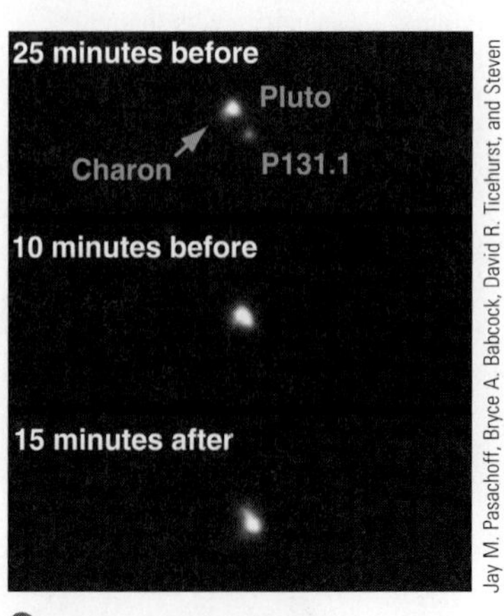

(a)

Jay M. Pasachoff, Bryce A. Babcock, David R. Ticehurst, and Steven P. Souza (Williams College)

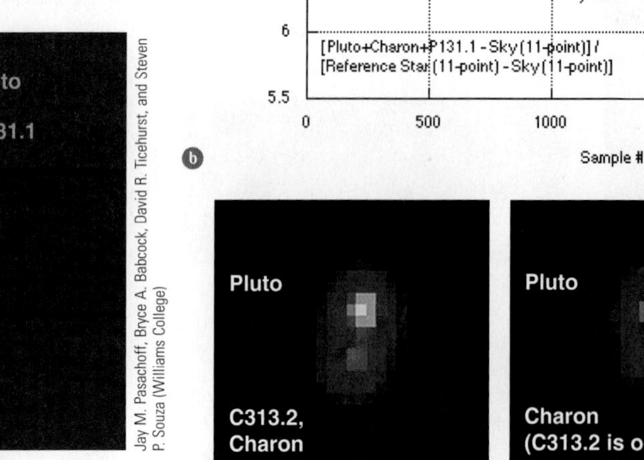

(b)

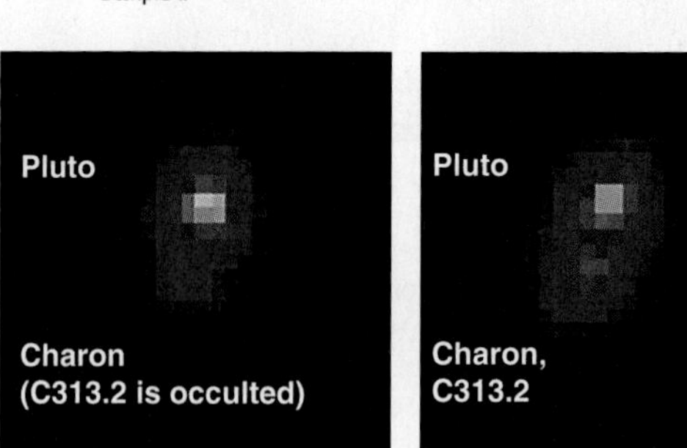

(c)

A. A. S. Gulbis, J. L. Elliot, M. J. Person, E. R. Adams, S. D. Kern, and E. A. Kramer (MIT), B. A. Babcock, J. W. Gangestad, J. M. Pasachoff, and S. P. Souza (Williams College), D. J. Osip (OCIW), M. Emilio (Brazil), and T. Tuvikene (Chile/Estonia) of a 20-minute data run. Further work in 2005 on a similar occultation of a star but this time by Pluto's moon, Charon, gave the MIT-Williams College consortium success on all but one of the five telescopes in South America they used.

Our Pluto results showed an expansion of its atmosphere, which would result from a global warming since 1988. Perhaps some contribution to that warming comes from the changing orientation of Pluto's darker spots with respect to incoming solar radiation. We also saw some bright spikes in the light curve, which could be signs of waves or turbulence in Pluto's atmosphere. Further, observations from several telescopes showed that Pluto's atmosphere is not quite round, undoubtedly resulting from strong winds.

Our Charon results pinned down its size, and therefore density, better than ever before, but even the high-time-resolution observations did not show an atmosphere.

As Pluto goes farther from the Sun, as it is now doing, its atmosphere is generally predicted to freeze out and snow onto the surface. Though some calculations indicate that this might not be so, it is still possible that if we want to find out about the atmosphere, we had better get a spacecraft there within a decade or two, or we'll have to wait another 200 years for the atmosphere to form again. NASA's New Horizons mission, after a period of on-again, off-again for funding reasons, is a small satellite that at the time of this writing is scheduled to be launched in 2006 and to reach Pluto a decade later. Its investigators used Hubble to find two additional, small (under 100-km) moons of Pluto.

8.1c What Is Pluto?

From Pluto's mass and radius, we calculate its density. It turns out to be about 2 g/cm^3, twice the density of water and less than half the density of Earth. Since ices have even lower densities than Pluto, Pluto must be made of a mixture of ices and rock. Its composition is more similar to that of the satellites of the giant planets, especially Neptune's large moon Triton, than to that of Earth or the other inner planets.

Ironically, now that we know Pluto's mass, we calculate that it is far too small to cause the deviations in Uranus's orbit that originally led to Pluto's discovery. The discrepancy probably wasn't real: The wrong mass had been assumed for Neptune when

A Closer Look 8.1 Is Pluto a Planet?

When Pluto was discovered, it was announced as the ninth planet. Its discovery resulted from a specific search for a ninth planet, though we have subsequently found that this did not validly come from the predictions. Since 1930, Pluto has had the status of planet. But after we found that Pluto has only 1/500 the mass of Earth, some scientists began to think that Pluto should be demoted. Demoted to what? That was the question.

The discovery of a set of Kuiper-belt objects with orbits not very different from that of Pluto gives a possible home for categorizing Pluto. Indeed, Pluto is probably just one of the largest Kuiper-belt objects. But, given many decades of having nine planets, are we really prepared to go back to eight? (Many astronomers argue vehemently that the case is very strong.) If you remember the names of the planets as the initial letters of "**M**y **V**ery **E**ducated **M**other **J**ust **S**ent **U**s **N**ine **P**izzas," are you prepared to change to "**M**y **V**ery **E**ducated **M**other **J**ust **S**ent **U**s **N**othing" or "**N**achos"?

Word got around in 1999 that Pluto might get numbered in other systems of astronomical nomenclature. For example, as a Trans-Neptunian Object, it might be made TNO-1, or even, to single it out, TNO-0. Also, the asteroid numbers were at that time getting up close to 10,000, and Pluto could have been given that special, round number in the asteroid numbering scheme.

These leaks disturbed so many people that the official committee on nomenclature of the International Astronomical Union issued a statement that Pluto would not be demoted, at least not yet. A committee of the Division of Planetary Sciences of the American Astronomical Society also made a statement that Pluto would remain designated as a planet for the time being. Scientists attending a debate on the subject at the International Astronomical Union's General Assembly in 2000 voted that Pluto was indeed a planet.

Given the history, the authors of this book are in favor of retaining for now Pluto's planethood, even though it is *also* a big Kuiper-belt object. But as new searches turn up Kuiper-belt objects, including at least one larger than Pluto, Pluto's status as a full-fledged planet has become harder to maintain, unless we count 10+ planets.

predicting the orbit of Uranus. The discovery of Pluto was purely the reward of Clyde Tombaugh's hard work in conducting a thorough search in a zone of the sky near the ecliptic.

Pluto, with its moon and its atmosphere, has some similarities to the more familiar planets. Pluto remains strange in that it is so small next to the giants, and that its orbit is so eccentric and so highly inclined to the ecliptic. Increasingly, Pluto is being identified with a newly discovered set of objects in the outer Solar System, which we will now study.

Is Pluto even a planet? It is so small, so low in mass, and in such an inclined orbit with respect to the eight inner planets that perhaps it should only be called an asteroid, a "Kuiper-belt object," or a "Trans-Neptunian Object." As we will see in the next section, another such object even bigger than Pluto turned up in 2005. Should both be called planets, leaving the possibility that we may soon know of even more? Or should Pluto be demoted to asteroid or the mere status of a Trans-Neptunian Object? As of this writing, the matter is undecided; see *A Closer Look 8.1: Is Pluto a Planet?*

8.2 Kuiper-belt Objects

Beyond the orbit of Neptune, a population of icy objects with diameters of a few tens or hundreds of kilometers is increasingly being found. The planetary astronomer Gerard Kuiper (pronounced koy′per) suggested a few decades ago that these objects would exist and should be the source of many of the comets that we see. As a result, these objects are now known as the **Kuiper-belt objects,** or, less often, **Trans-Neptunian Objects.**

The Kuiper belt is probably about 10 A.U. thick and extends from the orbit of Neptune about twice as far out (■ Fig. 8–8*a*). About 1000 Kuiper-belt objects have been found so far, and tens of thousands larger than 100 km across are thought to exist. The objects may be left over from the formation of the Solar System.

They are generally very dark, with albedoes of only about 4 per cent. Pluto, by contrast, has an albedo of about 60 per cent. Still, Pluto is one of the largest of the Kuiper-belt objects, so much larger than most of the others that it is covered with frost. Triton may have initially been a similar object, subsequently captured by Neptune.

A Kuiper-belt object larger than Pluto's moon Charon was found in 2001, about half of Pluto's diameter. One that may be even somewhat larger was found in 2002, though the uncertainty limits of these two Kuiper-belt objects overlap. The newer one, tentatively and unofficially named Quaoar (pronounced "kwa-whar") after the Indian tribe that inhabited today's Los Angeles, was even imaged with the Hubble Space Telescope, so we have a firmer grasp of its diameter, 1300 km, slightly over half that of Pluto. The size, in turn, gives us the albedo (12 per cent), which is larger than had been assumed for Kuiper-belt objects.

David Jewitt of the University of Hawaii and Jane Luu, now at MIT's Lincoln Lab, have been the discoverers of most of the known Kuiper-belt objects. They found the first one in 1992 and they and several other astronomers are looking for more.

Michael Brown of Caltech and his colleagues stunned the world in July 2005, as this book was going to press, with their discovery of an outer-Solar-System object even larger than Pluto (■ Fig. 8–8*b*). Initially named 2003 UB313, it was first sighted in 2003 but not confirmed until 2005. The object is now 97 A.U. out from the Sun, more than twice as far out as Pluto. It takes over 500 years to orbit the Sun. Its orbit is tilted an incredible 44°, taking it so high out of the ecliptic that no previous planet hunter found it. Undoubtedly, it was thrown into that highly inclined orbit after a close gravitational encounter with Neptune. Is it a 10th planet? That is really a matter of semantics, but words can count. Keep in touch with this book's website or with other sources to find out the latest on it.

ASIDE 8.1: Xena and Gabrielle

2003 UB313, arguably our Solar System's 10th planet, has been unofficially nicknamed Xena by its discoverer, who later found a much smaller, fainter satellite of the warrior princess Xena that he nicknamed Gabrielle, the name of a faithful companion to Xena on a syndicated television series. More classical, official names should be forthcoming.

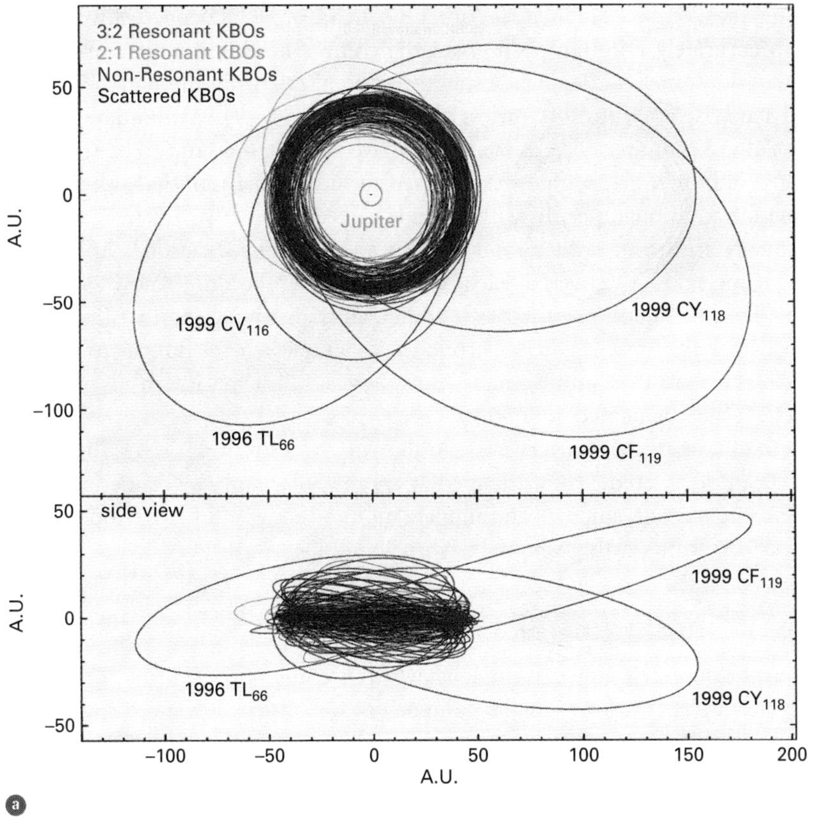

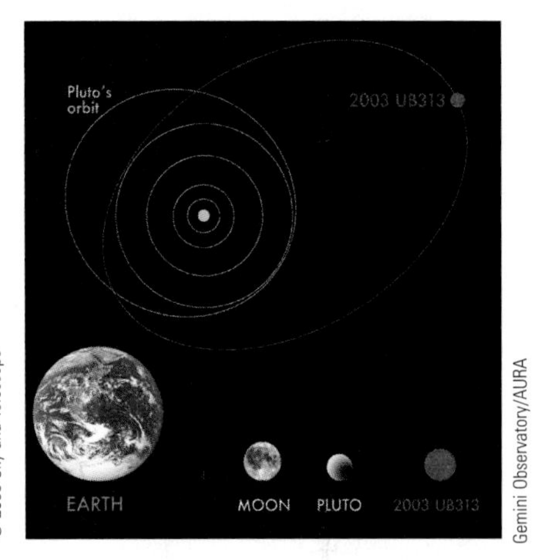

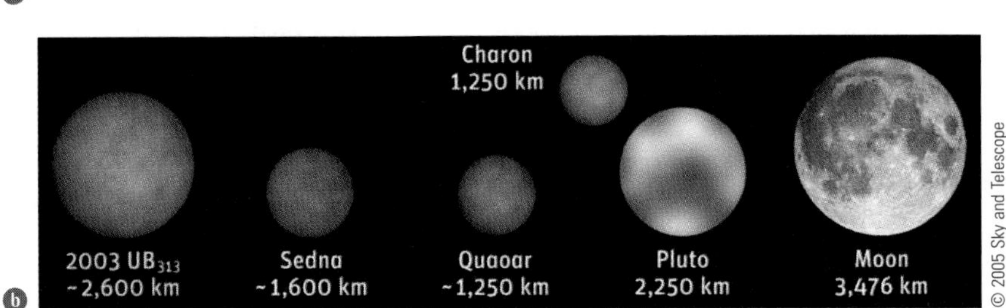

■ **FIGURE 8–8** (a) Orbits of Jupiter (*small green circle*) and the Kuiper-belt objects. (b) A comparison of the sizes of our Moon with Pluto and its moon, Charon, and with recently discovered Kuiper-belt objects in the outer Solar System.

A few objects may once have been Kuiper-belt objects but now come somewhat closer to the Sun, crossing the orbits of the outer planets. About 100 of these "centaur" objects a few hundred kilometers across may exist. Since they are larger and come closer to the Earth and Sun than most Kuiper-belt objects, we can study them better. On at least one, a coma (typical of comets, as we will soon see) was seen, so these centaurs are intermediate between comets and asteroids.

NASA's New Horizons mission is to go to some Kuiper-belt objects after it visits Pluto. Our MIT-Williams consortium certainly hopes to pick up an occultation of a star by one or more of these Kuiper-belt objects, which would accurately determine its diameter and albedo.

8.3 Comets

Nearly every decade, a bright **comet** appears in our sky. From a small, bright area called the **head,** a **tail** may extend gracefully over one-sixth (30°) or more of the sky. The tail of a comet is always directed roughly away from the Sun, even when the comet is moving outward through the Solar System.

Although the tail may give an impression of motion because it extends out only to one side, the comet does not move noticeably with respect to the stars as we casually

Raymond Pojman

■ **FIGURE 8–9** Comet Hyakutake as it moved with the stars; note the star trails.

watch during the course of a night. With binoculars or a telescope, however, an observer can accurately note the position of the comet's head and after a few hours can detect that the comet is moving at a slightly different rate from the stars. Still, both comets and stars rise and set more or less together (■ Fig. 8–9). Within days, weeks, or (even less often) months, a bright comet will have become too faint to be seen with the naked eye, although it can often be followed for additional months with binoculars and then for additional months with telescopes.

Most comets are much fainter than the one we have just described. About two dozen new comets are discovered each year, and most become known only to astronomers. If you should ever discover a comet, and are among the first three people to report it to the International Astronomical Union Central Bureau for Astronomical Telegrams at the Smithsonian Astrophysical Observatory in Cambridge, Massachusetts, it will be named after you.

Hundreds of comets that go very close to the Sun or even hit it, destroying themselves, have been discovered by (and named after) the Solar and Heliospheric Observatory (SOHO) spacecraft, since it can uniquely monitor a region of space too close to the Sun to be seen from Earth given our daytime blue skies.

8.3a The Composition of Comets

At the center of a comet's head is its **nucleus,** which is composed of chunks of matter. The most widely accepted theory of the composition of comets, advanced in 1950 by Fred L. Whipple of the Harvard and Smithsonian Observatories, is that the nucleus is like a "dirty snowball." It may be made of ices of such molecules as water (H_2O), carbon dioxide (CO_2), ammonia (NH_3), and methane (CH_4), with dust mixed in.

The nucleus itself is so small that we cannot observe it directly from Earth. Radar observations have verified in several cases that it is a few kilometers across. The rest of the head is the **coma** (pronounced coh´ma), which may grow to be as large as 100,000 km or so across (■ Fig. 8–10). The coma shines partly because its gas and dust are reflecting sunlight toward us and partly because gases liberated from the nucleus get enough energy from sunlight to radiate.

The tail can extend 1 A.U. (150,000,000 km), so comets can be the largest objects in the Solar System. But the amount of matter in the tail is very small—the tail is a much better vacuum than we can make in laboratories on Earth.

Many comets actually have two tails (■ Fig. 8–11). The **dust tail** is caused by dust particles released from the ices of the nucleus when they are vaporized. The dust particles are left behind in the comet's orbit, blown slightly away from the Sun by the pressure of sunlight hitting the particles. As a result of the comet's orbital motion, the dust tail usually curves smoothly behind the comet. The **gas tail** is composed of gas blown outward from the comet, at high speed, by the "solar wind" of particles emitted by the Sun (see our discussion in Chapter 10). It follows the interplanetary magnetic field. As puffs of gas are blown out and as the solar wind varies, the gas tail takes on a structured appearance. Each puff of matter can be seen.

A comet—head and tail together—contains less than a billionth of the mass of the Earth. It has jokingly been said that comets are as close as something can come to being nothing.

Akira Fujii

■ **FIGURE 8–10** The coma and inner tail of Halley's Comet in 1986.

8.3b The Origin and Evolution of Comets

It is now generally accepted that trillions of tail-less comets surround the Solar System in a sphere perhaps 50,000 A.U. (that is, 50,000 times the distance from the Sun to the Earth, or almost 1 light-year) in radius. This sphere, far outside Pluto's orbit, is the **Oort comet cloud** (named after the Dutch scientist Jan Oort). The total mass of matter in the cloud may be only 1 to 10 times the mass of the Earth. In current models, most of the Oort cloud's mass is in the inner 1000 to 10,000 A.U.

Akira Fujii

■ **FIGURE 8–11** Comet Hale-Bopp, technically C/1995 O1, was the brightest comet in decades at its peak in March and April 1997. Its 40-km-across nucleus gave off large quantities of gas and dust that made its long tails. The broad, yellowish dust tail curved behind the comet's head. We see it by the sunlight that it reflects toward us. The much fainter, blue tail is made of ions (an ion is an atom that has lost an electron) given off by the comet's nucleus and is blown in the direction opposite to the Sun by the solar wind. We detect this "gas tail" by its own emission, though it was much less visible to the eye than it was to the camera. Comet Hale-Bopp was so bright that scientists discovered a third type of tail: a faint one consisting of neutral sodium, not visible in this image.

Occasionally one of these comets leaves the comet cloud. In the early years of the Oort model, it was thought that sometimes the gravity of a nearby star tugged an incipient comet out of place. Currently, astronomers tend to think that gravity from the disk of our Milky Way Galaxy does most of the tugging. In any case, the comet generally gets directly ejected from the Solar System, but in some cases the comet can approach the Sun. The comet's orbit may be altered, sometimes into an elliptical orbit, if it passes near a giant planet, most frequently Jupiter. Because the comet cloud is spherical, comets are not limited to the plane of the ecliptic, which explains why one major class of comets comes in randomly from all directions.

Another group of comets has orbits that are much more limited to the plane of the Solar System (Earth's orbital plane). They probably come from the Kuiper belt beyond the orbit of Neptune, a flatter distribution of objects ranging from about 25 to 50 A.U.

We seem to discover more of these Kuiper-belt-origin comets than we expect compared with Oort-cloud-origin comets. Perhaps the discrepancy has to do with the way comets die. New calculations show that since so few dormant comets are found, the comets must mainly break up and disappear. Maybe Oort-cloud comets, coming from so far out in the Solar System, change temperature regimes so much more quickly than Kuiper-belt comets that they are preferentially disrupted.

Until recently, astronomers tended to say that the long-period comets, those with orbital periods longer than 200 years, came from the Oort cloud while comets with periods shorter than 200 years came from the Kuiper belt (■ Fig. 8–12*a*). Part of the reason for this division was merely that we had observed comet orbits reliably for only about 200 years. Most of the long-period comets have semimajor axes close to 20,000 A.U.,

ASIDE 8.2: Chiron, the Centaur

Chiron was named after the wisest of the Centaurs in Greek mythology; Chiron was the tutor of the heroes Achilles and Hercules. With its 51-year period, Chiron's semimajor axis is between those of Saturn and Uranus. Astronomers think it escaped from the Kuiper belt. When it was discovered it was classified as asteroid number 2060, but when it was seen to be active by showing a coma, it was reclassified as periodic comet 95 (that is, 95P).

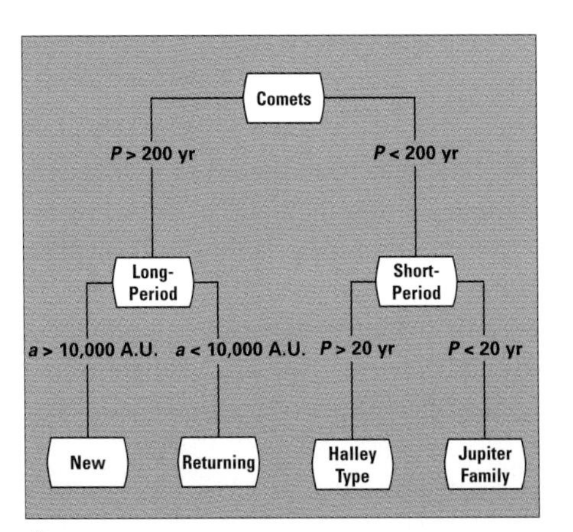

a

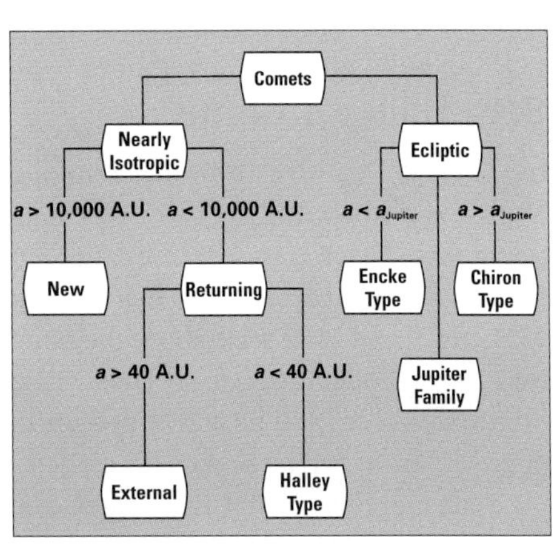

b

■ **FIGURE 8–12** (a) The former classification of comets, now superseded. (b) The current classification of comets. The symbol *a* stands for the semimajor axis of the comet's orbit. Encke's comet has a small orbit and has many returns. Chiron is in the outer parts of the Solar System. (After Alan Stern, SwRI)

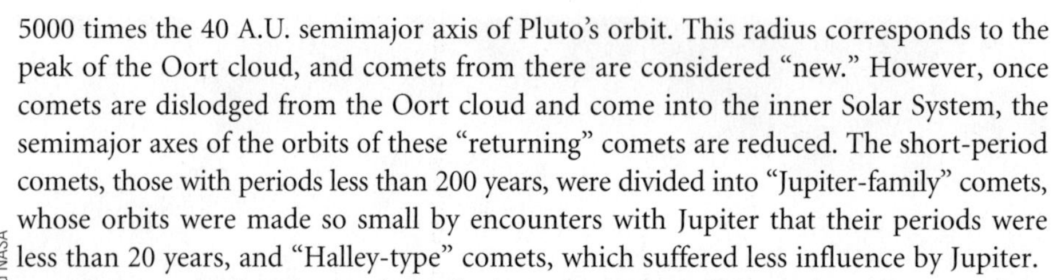

Naval Research Laboratory, LASCO on the Solar and Heliospheric Observatory: SOHO is a project of international cooperation between ESA and NASA

■ **FIGURE 8–13** The outer solar corona extends outward near the solar equator in this view from the Solar and Heliospheric Observatory (SOHO) spacecraft. A comet plunging into the Sun is also seen. A disk that blocks the bright everyday Sun and the finger that holds the disk in place appear in silhouette; the size of the ordinary solar disk is drawn on. In the background, we see stars and the Milky Way. Hundreds of comets have been discovered on SOHO images.

5000 times the 40 A.U. semimajor axis of Pluto's orbit. This radius corresponds to the peak of the Oort cloud, and comets from there are considered "new." However, once comets are dislodged from the Oort cloud and come into the inner Solar System, the semimajor axes of the orbits of these "returning" comets are reduced. The short-period comets, those with periods less than 200 years, were divided into "Jupiter-family" comets, whose orbits were made so small by encounters with Jupiter that their periods were less than 20 years, and "Halley-type" comets, which suffered less influence by Jupiter.

A new comet classification basically depends on the influence of Jupiter. One of the two major classes consists of those that come from all directions. Almost all of these come from the Oort cloud. Comets in the other major class are called "ecliptic," since the comet orbits are aligned close to the plane of the Solar System, the ecliptic plane (Fig. 8–12*b*), rather than being highly tilted. Almost all of these ecliptic comets come from the Kuiper belt. In the new scheme, fewer comets change their classifications over time.

Notice that comets on highly eccentric orbits spend most of their time far away from the Sun, an excellent example of Kepler's second law (Chapter 5).

As a comet gets closer to the Sun than those distant regions, the solar radiation begins to vaporize the ice in the nucleus. The tail forms, and grows longer as more of the nucleus is vaporized. Even though the tail can be millions of kilometers long, it is still so tenuous that only 1/500 of the mass of the nucleus may be lost each time it visits the solar neighborhood. Thus a comet may last for many passages around the Sun. But some comets hit the Sun and are destroyed (■ Fig. 8–13).

We shall see in the following section that meteoroids can be left in the orbit of a disintegrated comet. Some of the asteroids, particularly those that cross the Earth's orbit, may be dead comet nuclei. In recent years, a handful of asteroids—notably Chiron in the outer Solar System—have shown comas or tails, making them comets; conversely, a few comets have died out and seem like asteroids. So we may have misidentified some of each in the past.

How did comets get where they are? We will say more about the formation of the Solar System in Chapter 9. There, we will see that there were many small particles that clumped together in the early eras. Some of these clumps interacted gravitationally with other clumps and even with Jupiter and other planets as they were formed. Many of these clumps were ejected from the region of their formation, often where the asteroid belt now is between Mars and Jupiter, and wound up forming the Oort comet cloud. Other clumps were already beyond the orbit of Neptune, where fewer interactions took place. Those clumps formed the Kuiper belt.

Because new comets come from the places in the Solar System that are farthest from the Sun and thus coldest, they probably contain matter that is unchanged since the formation of the Solar System. So the study of comets is important for understanding the birth of the Solar System. Moreover, some astronomers have concluded that early in Earth's history, the oceans formed when an onslaught of water-bearing comets collided with Earth, although this view is still controversial.

National Portrait Gallery, London

■ **FIGURE 8–14** Edmond Halley.

8.3c Halley's Comet

In 1705, the English astronomer Edmond Halley (Halley is pronounced to rhyme with "Sally," and not with "say′lee") (■ Fig. 8–14) applied a new method developed by his friend Isaac Newton to determine the orbits of comets from observations of their positions in the sky. He reported that the orbits of the bright comets that had appeared in 1531, 1607, and 1682 were about the same. Moreover, the intervals between appearances were approximately equal, so Halley suggested that we were observing a single comet orbiting the Sun, and he accounted for the slightly different periods with Newton's law of gravity from interactions with planets.

Halley predicted that this bright comet would again return in 1758. Its reappearance on Christmas night of that year, 16 years after Halley's death, was the proof of

■ **FIGURE 8–15** Halley's Comet following its 1986 closest approach to the Sun. Both the dust tail and the gas tail are visible. The color image was composed from blue-, green-, and red-sensitive photographs taken at slightly different times, so the comet's motion makes each star appear three different colors.

Halley's hypothesis (and Newton's method). The comet has thereafter been known as Halley's Comet (■ Fig. 8–15). Since it was the first known "periodic comet" (i.e., the first comet found to repeatedly visit the inner parts of the Solar System), it is officially called 1P, number 1 in the list of periodic (P) comets.

It seems probable that the bright comets reported every 74 to 79 years since 240 B.C. were earlier appearances of Halley's Comet. The fact that it has been observed dozens of times endorses the calculations that show that less than 1 per cent of a cometary nucleus's mass is lost at each passage near the Sun.

Halley's Comet came especially close to the Earth during its 1910 return, and the Earth actually passed through its tail. Many people had been frightened that the tail would somehow damage the Earth or its atmosphere, but the tail had no noticeable effect. Even then, most scientists knew that the gas and dust in the tail were too tenuous to harm our environment.

The most recent close approach of Halley's Comet was in 1986. It was not as spectacular from the ground in 1986 as it was in 1910, for this time the Earth and comet were on opposite sides of the Sun when the comet was brightest. Since we knew long in advance that the comet would be available for viewing, special observations were planned for optical, infrared, and radio telescopes. For example, spectroscopy showed many previously undetected ions in the coma and tail.

When Halley's Comet passed through the plane of the Earth's orbit, it was met by an armada of spacecraft. The best was the European Space Agency's spacecraft Giotto (named after the 14th-century Italian artist who included Halley's Comet in a painting), which went right up close to Halley. Giotto's several instruments also studied Halley's gas, dust, and magnetic field from as close as 600 km from the nucleus.

The most astounding observations were undoubtedly the photographs showing the nucleus itself (■ Fig. 8–16), which turns out to be potato-shaped (■ Fig. 8–17). It

■ **FIGURE 8–17** This potato looks remarkably like the nucleus of Halley's Comet, except it is 100,000 times shorter.

■ **FIGURE 8–16** The arrow points to the nucleus of Halley's Comet from the Giotto spacecraft. The dark, potato-shaped nucleus is visible in silhouette. It is about 16 km by 8 km in size; the frame is 30 km across. Two bright jets of gas and dust reflecting sunlight are visible to the left of Halley's nucleus. The small, bright region in the center of the nucleus is probably a raised region of its surface that is struck by sunlight.

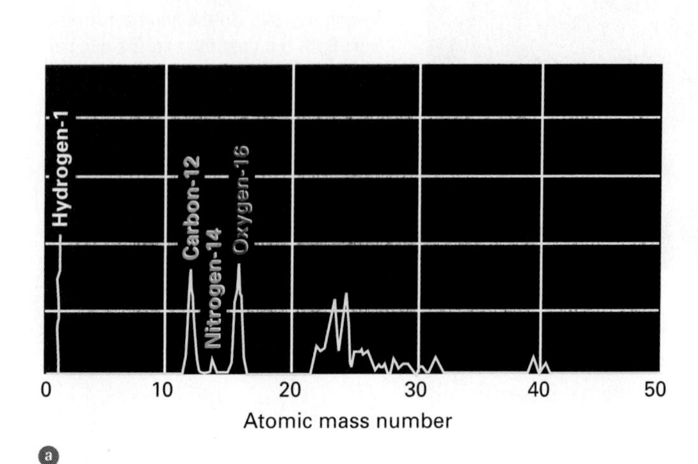

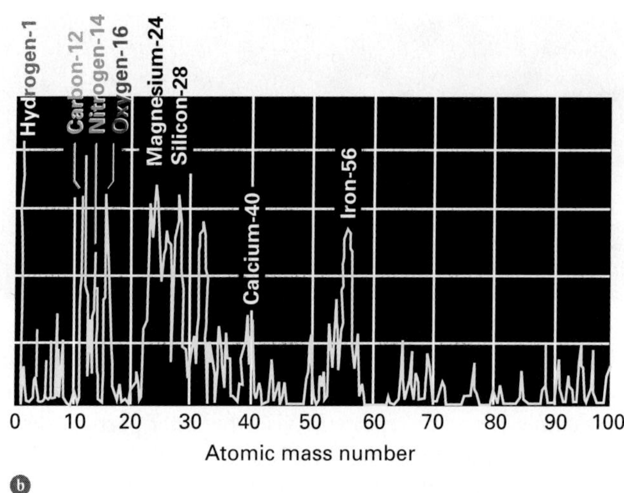

■ **FIGURE 8–18** The mass spectrometer aboard the Vega spacecraft showed the number of nuclear particles (the atomic mass number) of each element in Halley's dust particles. (a) A CHON dust particle, composed only of carbon, hydrogen, oxygen, and nitrogen. (b) Other dust particles contain other elements as well. (Soviet VEGA team)

is about 16 km in its longest dimension, half the size of Manhattan Island. The "dirty snowball" theory of comets was confirmed in general, but the snowball is darker than expected. It is as black as velvet, with an albedo of only about 3 per cent. Further, the evaporating gas and dust is localized into jets that are stronger than expected. They come out of fissures in the dark crust. We now realize that comets may shut off not when they have lost all their material but rather when the fissures in their crusts close.

Giotto carried 10 instruments in addition to its camera. Among them were mass spectrometers to measure the types of particles present, detectors for dust, equipment to listen for radio signals that revealed the densities of gas and dust in the coma, detectors for ions, and a magnetometer to measure the magnetic field.

About 30 per cent of Halley's dust particles are made only of hydrogen, carbon, nitrogen, and oxygen (■ Fig. 8–18). This simple composition resembles that of the oldest type of meteorite. It thus indicates that these particles may be from the earliest years of the Solar System.

Many valuable observations were also obtained from the Earth. For example, radio telescopes were used to study molecules. Water vapor is the most prevalent gas, but carbon monoxide and carbon dioxide were also detected. The comet was bright enough that many telescopes obtained spectra (■ Fig. 8–19).

The next appearance of Halley's Comet, in 2061, again won't be spectacular. Not until the one after that, in 2134, will the comet show a long tail to earthbound observers. Fortunately, though Halley's Comet is predictably interesting, a more spectacular comet appears every 10 years or so. When you read in the newspaper that a bright comet is here, don't wait to see it another time. Some bright comets are at their best for only a few days or a week.

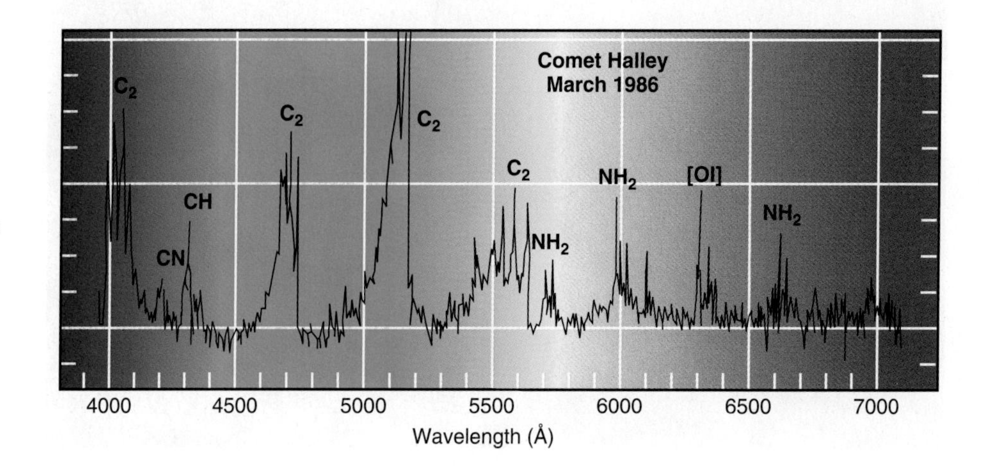

■ **FIGURE 8–19** The visible spectrum of Halley's Comet on March 14, 1986, when it was at its brightest. We see several spectral lines and bands from molecules. Their relative strengths are approximately the same as they are in other comets, except for NH_2, which was slightly stronger in Halley. (Susan Wyckoff, Arizona State U., Tempe)

NASA/STScI

■ **FIGURE 8–20** Comet Shoemaker-Levy 9 viewed with the Hubble Space Telescope. Over twenty fragments, each with its own tail, appear in this image taken six months before their collisions with Jupiter.

8.3d Comet [Shoemaker-]Shoemaker-Levy 9

A very unusual comet gave thrills to people around the world. In 1993, Eugene Shoemaker, Carolyn Shoemaker, and David Levy discovered their ninth comet in a search with a wide-field telescope at the Palomar Observatory. (The authors of this book like to give each Shoemaker individual credit for the discovery, as in the chapter subheading, though the comet is generally and formally called Shoemaker-Levy 9.)

This comet looked weird—it seemed squashed. Higher-resolution images taken with other telescopes, including the Hubble Space Telescope (■ Fig. 8–20), showed that the comet had broken into bits, forming a chain that resembled beads on a string. Even stranger, the comet was in orbit not around the Sun but around Jupiter, and would hit Jupiter a year later. Apparently, several decades earlier the comet was captured in a highly eccentric orbit around Jupiter, and in 1992, during its previous close approach, it was torn apart into more than 20 pieces by Jupiter's tidal forces.

Telescopes all around the world and in space were trained on Jupiter when the first bit of comet hit. The site was slightly around the back side of Jupiter, but rotated to where we could see it from Earth after about 15 minutes. Even before then, scientists were enthralled by a plume rising above Jupiter's edge. When they could view Jupiter's surface, they saw a dark ring (■ Fig. 8–21). Infrared telescopes detected a tremendous amount of radiation from the heated gas. Over a period of almost a week, one bit of the comet after another hit Jupiter, leaving a series of Earth-sized rings and spots as Jupiter rotated. The largest dark spots could be seen for a few months even with small backyard telescopes. (On one of the April 2005 solar eclipse cruises, David Levy sometimes wore a T-shirt that said "My comet crashed.")

The dark material showed us the hydrocarbons and other constituents of the comet. Spectra showed sulfur and other elements, presumably dredged up from lower levels of Jupiter's atmosphere than we normally see.

ASIDE 8.3: Shoemaker's craters

Eugene Shoemaker was the first to show convincingly that the Barringer crater is of meteoritic, rather than volcanic, origin. He also provided compelling evidence that most lunar craters were produced by similar impacts. Tragically, Shoemaker died in a freak car accident in 1997, while hunting in Australia for additional impact craters.

■ **FIGURE 8–21** ⓐ Jupiter, with scars from the July 1994 crash of Comet Shoemaker-Levy 9. The image was taken with the Hubble Space Telescope. ⓑ The shape of the ring left on Jupiter's surface showed that the fragment of Comet Shoemaker-Levy 9 that caused it entered at an angle. The image was taken with the Hubble Space Telescope. ⓒ This infrared image, taken with one of the Keck telescopes, shows bright plumes of hot gas from the collisions of comet fragments with Jupiter.

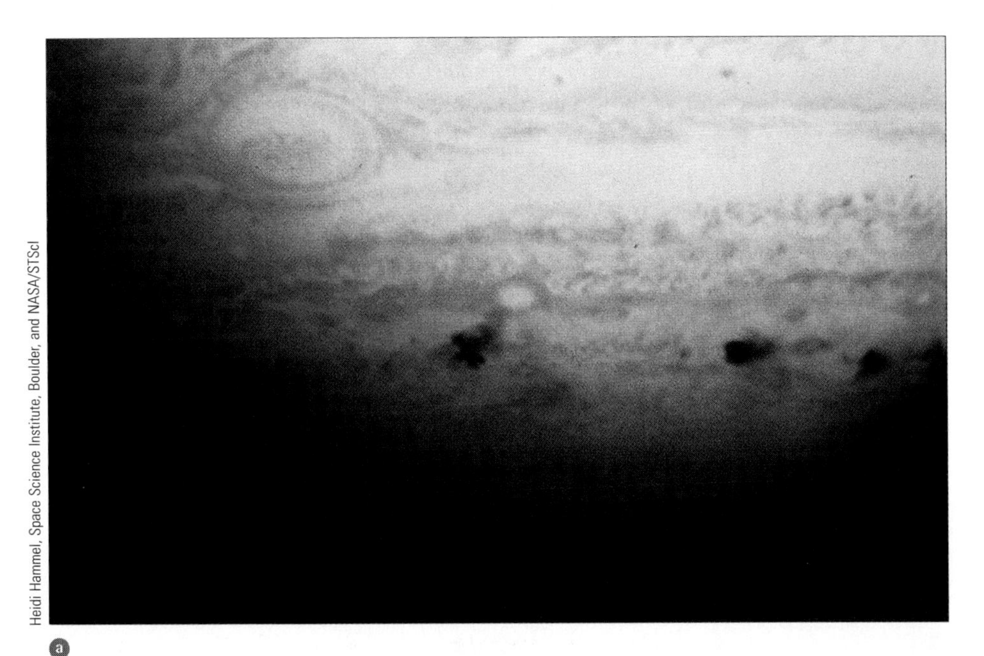
ⓐ
Heidi Hammel, Space Science Institute, Boulder, and NASA/STScI

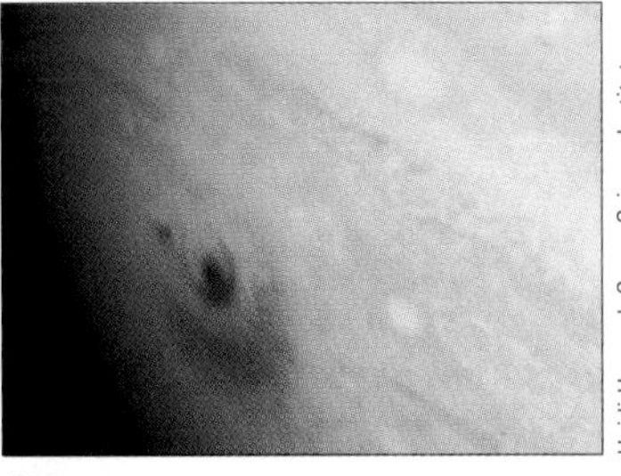
ⓑ
Heidi Hammel, Space Science Institute, Boulder, and NASA/STScI

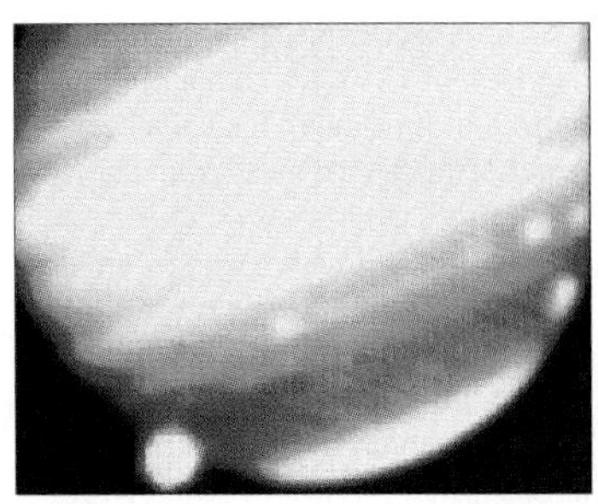
ⓒ
W. M. Keck Observatory/Imke de Pater, James Graham, Garrett Jernigan (U. California, Berkeley)

■ **FIGURE 8–22** ⓐ Comet Hale-Bopp (C/1995 O1) in the sky, showing both a whitish dust tail and a bluish gas tail. ⓑ Comet Ikeya-Zhang passing the Andromeda Galaxy in 2001.

The biggest comet chunk released the equivalent of 6 million megatons of TNT—100,000 times more than the largest hydrogen bomb. Had any of the fragments hit Earth, they would have made a crater as large as Rhode Island, with dust thrown up to much greater distances. Had the entire comet (whose nucleus was 10 km across) hit Earth at one time, much of life could have been destroyed. So Comet Shoemaker-Levy 9 made us even more wary about what may be coming at us from space.

8.3e Recently Observed Comets

In 1995, Alan Hale and Thomas Bopp independently found a faint comet, which was soon discovered to be quite far out in the Solar System. Its orbit was to bring it into the inner Solar System, and it was already bright enough that it was likely to be spectacular when it came close to Earth in 1997. It lived up to its advance billing (■ Fig. 8–22*a*).

Telescopes of all kinds were trained on Comet Hale-Bopp, and hundreds of millions of people were thrilled to step outside at night and see a comet just by looking up. Modern powerful radio telescopes were able to detect many kinds of molecules that had not previously been recorded in a comet.

Occasionally, other bright comets, such as C/2002 C1, Comet Ikeya-Zhang (Fig. 8–22*b*), turn up and are fun to watch.

8.3f Spacecraft to Comets

NASA's Deep Space 1 mission flew close to Comet 19P/Borrelly in 2001. It obtained more detailed images of the bowling-pin-shaped nucleus (■ Fig. 8–23) than even Giotto's views of Halley's nucleus. This comet's surface, and therefore probably the surfaces of comet nuclei in general, was rougher and more dramatic than expected. Deep Space 1 found smooth, rolling plains that seem to be the source of the dust jets, which are more concentrated than Halley's. Darkened material, perhaps extruded from underneath, covers some regions and accentuates grooves and faults. Borrelly's albedo in these places is less than 1 per cent, while Borrelly's overall albedo is only 4 per cent.

Borrelly is thought to have originated in the Kuiper belt, in contrast to Halley's Comet's origin in the Oort cloud. This difference would explain why Halley's Comet gives off many carbon compounds while Borrelly gives off more water and ammonia than carbon. Still, compared with Halley, Borrelly gives off relatively little water, perhaps because so much of its surface is inactive.

Scientists have yet to explain why the solar wind is deflected around Borrelly's nucleus in an asymmetric fashion. The center of the plasma in Borrelly's coma is some

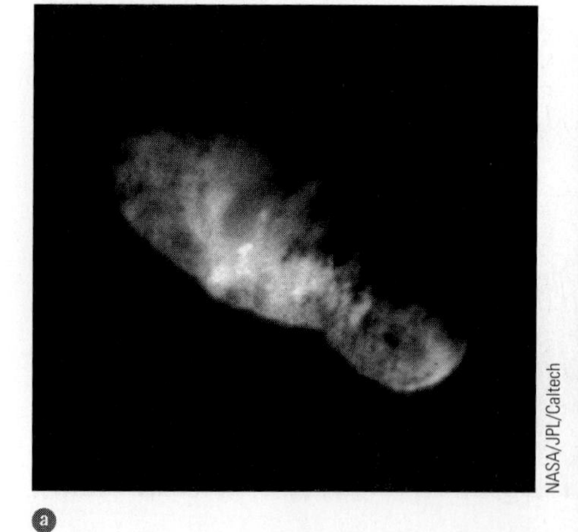

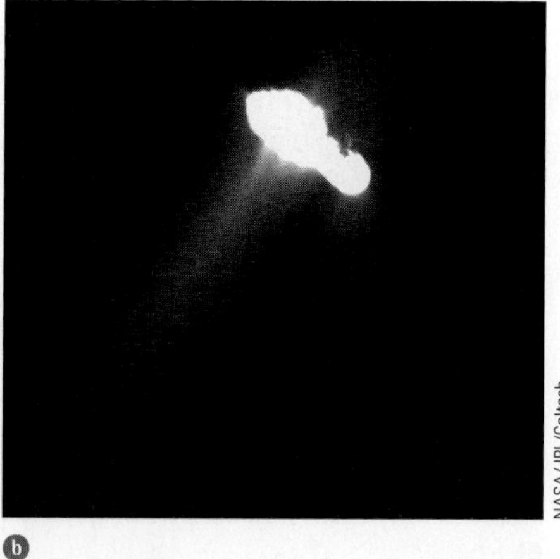

■ **FIGURE 8–23** Comet Borrelly (19P), 8 km long, viewed by NASA's Deep Space 1 probe. ⓐ Seen from a distance of only 3500 km, a variety of terrains is visible, including mountains and faults. Sunlight is coming from the bottom of the image; the comet's narrower end is tipped away from us. We see features as small as 50 m across. ⓑ Dust jets are seen coming from a region at least 3 km wide on the nucleus, which is overexposed in this image. The brightest jet is at lower left; material seen on the dark (night) side at upper right may have been swept around the nucleus from the day side.

2000 km off to the side, as strange as if a supersonic jet's shock wave were displaced far to the airplane's side.

NASA's Stardust mission, launched in 1999, went to Comet Wild 2 (pronounced Vilt-too), a periodic comet with a six-year orbit. When it got there in 2004, it not only photographed the comet but also gathered some of its dust. It carries an extremely lightweight material called aerogel (■ Fig. 8–24), and flew through the comet with the aerogel exposed so that the comet dust could stick in it. Stardust's orbit will bring it back near Earth in January 2006, when it will parachute the aerogel down to the Utah desert. (A parachute that didn't open in a 2004 mission to gather solar wind particles, Genesis, makes everybody worried.)

Jay M. Pasachoff

■ **FIGURE 8–24** Aerogel, a frothy, extremely light substance. Even a large amount weighs very little, so a rocket can carry more of it into space than it could of a denser material.

A major European Space Agency spacecraft, Rosetta, was launched in 2004 to orbit with a comet for some years and to land a probe on the comet's nucleus in 2014. It is heading for Comet 67P/Churyumov-Gerasimenko. It will use three gravity assists from Earth and one from Mars to reach the comet, passing asteroids (2867) Steins in 2008 and (21) Lutetia in 2010, both in the asteroid belt, on the way. (Asteroids are discussed in Section 8.5.) Rosetta will drop a lander, Philae, onto the comet's nucleus. Just as the Rosetta Stone, now in the British Museum, enabled Egyptian hieroglyphics to be deciphered by having the same text in three scripts (hieroglyphics, Demotic, and Greek), scientists hope that the Rosetta spacecraft will prove to be the key to deciphering comets. (Philae was an island in the Nile on which an obelisk was found that helped to decipher the hieroglyphics of the Rosetta Stone.)

Rosetta is to orbit the comet at an altitude of only a few kilometers, mapping its surface and making other measurements, for 18 months, including the comet's closest approach to the Sun and therefore, it is hoped, its increasing activity. The lander is to work for some weeks, taking photographs and drilling into the surface.

NASA's Deep Impact spacecraft crashed a 370-kg projectile into Comet Tempel 1 in 2005. The remainder of the spacecraft studied the impact, which should have formed a football-field-sized crater some 7 stories deep. Astronomers were at telescopes all around the Earth, and were using telescopes in space like Hubble, to record the impact (■ Fig. 8–25); see *A Closer Look 8.2: Deep Impact.*

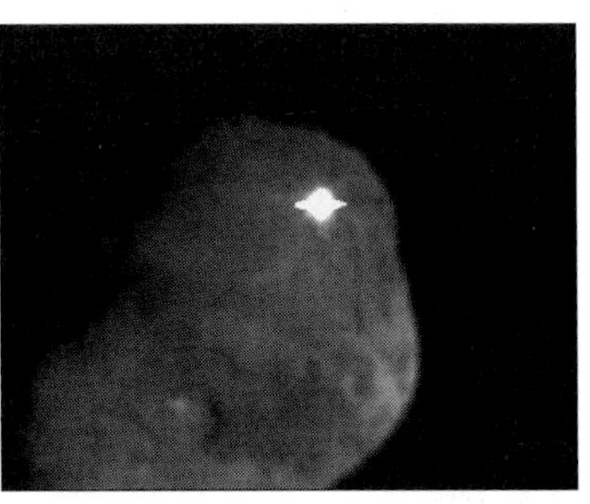

NASA/JPL/UMD

■ **FIGURE 8–25** NASA's Deep Impact probe crashed into a comet in July 2005.

AceAstronomy™ Log into AceAstronomy and select this chapter to see the Active Figure called "Build a Comet" and to see the Astronomy Exercise "Comets."

8.4 Meteoroids

There are many small chunks of matter orbiting in the Solar System, ranging up to tens of meters across and sometimes even larger. When these chunks are in space, they are called **meteoroids.** When one hits the Earth's atmosphere, friction and the compression of air in front of it heat it up—usually at a height of about 100 km—until all or most of it is vaporized. Such events result in streaks of light in the sky (■ Fig. 8–26), which we call **meteors** (popularly, and incorrectly, known as **shooting stars** or **falling stars**).

When a fragment of a meteoroid survives its passage through the Earth's atmosphere, the remnant that we find on Earth is called a **meteorite.** Counting even tiny meteorites, whose masses are typically a milligram, some 10,000 tons of this interplanetary matter land on Earth's surface each year.

8.4a Types and Sizes of Meteorites

Space is full of meteoroids of all sizes, with the smallest being most abundant. Most of the small particles, less than 1 mm across, may come from comets. The large particles, more than 1 cm across, may generally come from collisions of asteroids in the asteroid belt (see Section 8.5).

A Closer Look 8.2 DEEP IMPACT

On July 4, 2005, NASA's Deep Impact spacecraft went right up to Comet Tempel 1, also known as 9P, the ninth periodic comet on astronomy's official list. The comet is 14 km by 4 km in size. The day before, the spacecraft had ejected a 370-kg copper ball, and as cameras on the fly-by probe and on the impactor watched, the ball smashed into the comet's surface. Debris flew, forming a crater.

The Hubble Space Telescope, the Chandra X-ray Observatory, and many other telescopes on the ground and in space photographed the event. When the dust cleared, the fact that a crater was visible disproved the idea that the impactor might merely disappear into a pile of rubble. The mission resulted in scientists gaining a much improved understanding of the internal composition of this comet.

As the dust from Deep Space's impact settled, Mike A'Hearn of the University of Maryland and colleagues watched the edge of the ejecta broadening day by day as it fell back on the comet's surface. On the assumption that we are seeing "ballistic trajectories"—that is, paths following only Newton's laws of gravity and motion—the scientists were able to measure the comet's gravity and thus its mass. It turned out that its density is so low that 75 per cent of its volume must be empty. The outer tens of meters seem to be dirty snowflakes, an elaboration of Whipple's old, standard dirty-snowball model. The comet's nucleus resembles a handful of fluffy snow that you haven't compressed to make a snowball.

Another surprise is that more organic material—material containing carbon-oxygen molecules—was released than expected, compared with water vapor, carbon monoxide, and carbon dioxide. This result is support for the idea that our Earth's atmosphere and the atmospheres of other planets around our Sun and other stars were seeded with organics, leading to the development of life.

"I've gone from working 60 hours a week to 70 hours a week, but the fun part has gone up from 1% to 50%," said Mike A'Hearn.

NASA/JPL/UMD

The surface of Comet 9P/Tempel 1 before impact, taken from the camera on the impactor.

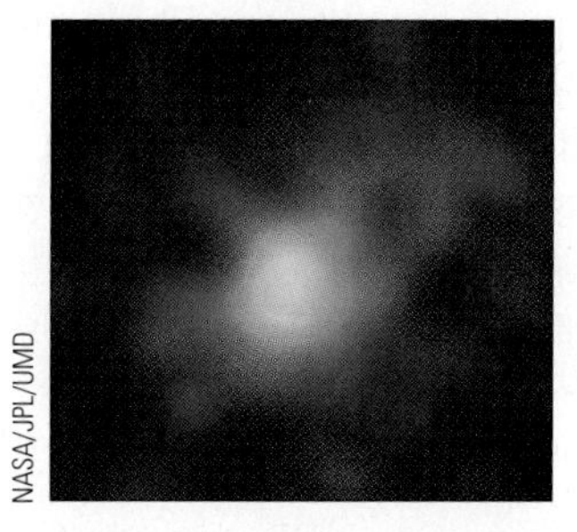
NASA/JPL/UMD

A Chandra X-ray Observatory view from 4 days before the collision, showing that the x-rays come primarily from the interaction between charged oxygen ions in the solar wind and the comet's neutral gases.

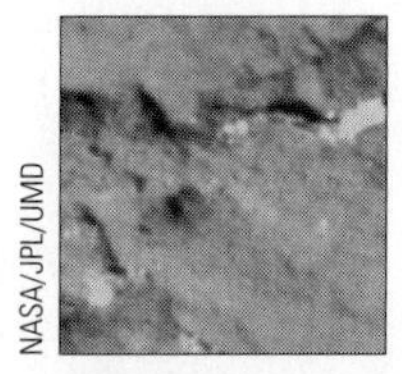
NASA/JPL/UMD

A closeup view from the impactor only 20 seconds before the collision, showing small detail on the comet's surface, including the impact site.

NASA/JPL/UMD

A Hubble Space Telescope series of views. The first is before the impact, showing light reflected off the nucleus. The others, up to an hour and a half after the impact, show the plume that formed, reflecting sunlight toward Earth.

NASA/JPL/UMD

Imaged from the fly-by probe 67 seconds later.

Tiny meteorites less than a millimeter across, **micrometeorites,** are the major cause of erosion (what little there is) on the Moon. Micrometeorites also hit the Earth's upper atmosphere all the time, and remnants can be collected for analysis from balloons or airplanes or from deep-sea sediments. They are often sufficiently slowed down by Earth's atmosphere to avoid being vaporized before they reach the ground.

Some of the meteorites that are found have a very high iron content (about 90 per cent); the rest is nickel. These iron meteorites are thus very dense—that is, they weigh quite a lot for their volume (■ Fig. 8–27*a*).

Most meteorites that hit the Earth are stony in nature. Because they resemble ordinary rocks (■ Fig. 8–27*b*) and disintegrate with weathering, they are not easily discovered unless their fall is observed. That difference explains why most meteorites discovered at random are made of iron. But when a fall is observed, most meteorites

ASIDE 8.4: Meteorites everywhere

Meteorites can land on bodies other than Earth, such as the Moon, Mars, or other asteroids.

Fred Bruenjes

■ **FIGURE 8–26** A compound exposure covering six hours of the Perseid meteor shower, which, weather permitting, is easy to see in the sky every August 11/12. 51 meteors show in this image. The radiant in Perseus is a perspective effect, though the meteors actually travel through space in parallel tracks.

Jay M. Pasachoff

a

Jay M. Pasachoff

b

■ **FIGURE 8–27** a A 15-ton iron meteorite, the Willamette meteorite, at the Rose Center for Earth and Space at the American Museum of Natural History in New York City. Note how dense it has to be to have this high a mass, given its size. b The largest mass of the Cape York meteorite, a 35-ton piece known as "Ahnighito," or "the Tent." This stony meteorite was discovered by Inuits in Greenland in the early 1800s and was brought to New York by William Peary and Matthew Henson in 1897. It is the largest on display in a museum (second in size only to a meteorite still in the ground in Namibia) and is in the American Museum of Natural History in New York City.

■ **FIGURE 8–28** The Barringer "meteor crater" (actually a meteorite crater) in Arizona. (The family pronounces its name with a hard "g.") It is 1.2 km in diameter. Dozens of other terrestrial craters are now known, many from aerial or space photographs. The largest may be a depression over 400 km across under the Antarctic ice pack, comparable with lunar craters. Another very large crater, in Hudson Bay, Canada, is filled with water. Most are either disguised in such ways or have eroded away.

recovered are made of stone. Some meteorites are rich in carbon, and some of these even have complex molecules like amino acids.

A large terrestrial crater that is obviously meteoritic in origin is the Barringer Meteor Crater in Arizona (■ Fig. 8–28). It resulted from what was perhaps the most recent large meteoroid to hit the Earth, for it was formed only about 50,000 years ago.

Every few years a meteorite is discovered on Earth immediately after its fall. The chance of a meteorite landing on someone's house or car is very small, but it has happened (■ Fig. 8–29)! Often the positions in the sky of extremely bright meteors are tracked in the hope of finding fresh meteorite falls. The newly discovered meteorites are rushed to laboratories in order to find out how long they have been in space by studying their radioactive elements.

Over 10,000 meteorites have been found in the Antarctic, where they have been well preserved as they accumulated over the years. Though the Antarctic ice sheets flow, the ice becomes stagnant in some places and disappears, revealing meteorites that had been trapped for over 10,000 years.

Some odd Antarctic meteorites are now known to have come from the Moon or even from Mars. Recall that in Chapter 6 we even discussed controversial evidence for ancient primitive life-forms on Mars, found in one such meteorite. As of mid-2005, the conclusion hasn't been entirely ruled out, but few scientists accept it. As the late Carl Sagan said, "Extraordinary claims require extraordinary evidence," and the evidence from this meteorite is not convincing, at least not yet.

Meteorites that have been examined were formed up to 4.6 billion years ago, the beginning of the Solar System. The relative abundances of the elements in meteorites thus tell us about the solar nebula from which the Solar System formed. In fact, up to the time of the Moon landings, meteorites and cosmic rays (charged particles from outer space) were the only extraterrestrial material we could get our hands on.

8.4b Meteor Showers

Meteors sometimes occur in **showers,** when meteors are seen at a rate far above average. Meteor showers are named after the constellation in which the **radiant,** the point from which the meteors appear to come, is located. The most widely observed—the Perseids, whose radiant is in Perseus—takes place each summer around August 12 and the nights on either side of that date. The best winter show is the Geminids, which takes place around December 14 and whose radiant is in Gemini.

On any clear night a naked-eye observer with a dark sky may see a few **sporadic meteors** an hour—that is, meteors that are not part of a shower. (Just try going out to a field in the country and watching the sky for an hour.) During a shower, on the other hand, you may typically see one every few minutes. Meteor showers generally result from the Earth's passing through the orbits of defunct or disintegrating comets and hitting the meteoroids left behind. (One meteor shower comes from an asteroid orbit.)

■ **FIGURE 8–29** ⓐ A Peekskill, New York, woman heard a noise on October 9, 1992, and went outside to find extensive damage to her car. Under the car, she found a just-fallen meteorite. ⓑ This 1.3-kg meteorite crashed through a roof in New Zealand in 2004 and landed in someone's living room. A chip was taken out of the meteorite as it plunged through the roof of the house and interior ceilings.

ⓐ

ⓑ

A Closer Look 8.3 Meteor Showers

Shower	Date of Maximum	Approximate Duration	Limits	Number Per Hour at Maximum	Parent Object
Quadrantids	Jan. 4	1 day	Jan. 1–6	100	—
Lyrids	Apr. 22	2 days	Apr. 19–24	10	C/1861 G1 (Thatcher)
Eta Aquarids	May 5	3 days	May 1–8	20	1P/Halley
Delta Aquarids	July 27–28	7 days	July 15–Aug. 15	30	—
Perseids	Aug. 12	5 days	July 25–Aug. 18	70	107P/Swift-Tuttle
Orionids	Oct. 21	2 days	Oct. 16–26	30	1P/Halley
Taurids	Nov. 8	10 days	Oct. 20–Nov. 30	10	2P/Encke
Leonids	Nov. 17	1 day	Nov. 15–19	10	55P/Tempel-Tuttle
Geminids	Dec. 14	3 days	Dec. 7–15	60	3200 Phaethon

Though the Perseids and Geminids can be counted on each year, the Leonid meteor shower (whose radiant is in Leo) peaks every 33 years, when the Earth crosses the main clump of debris from Comet Tempel-Tuttle. On November 17/18, 1998, one fireball (a meteor brighter than Venus) was visible each minute for a while (■ Fig. 8–30), and on November 17/18, 1999 through 2001, thousands of meteors were seen in the peak hour. We will now have to wait until about 2031 for the next Leonid peak.

The visibility of meteors in a shower depends in large part on how bright the Moon is; you want as dark a sky as possible. See *Star Party 8.1: Observing a Meteor Shower,* and *A Closer Look 8.3: Meteor Showers* for more details. Meteors are best seen with the naked eye; using a telescope or binoculars merely restricts your field of view.

Jay M. Pasachoff

■ **FIGURE 8–30** A "fireball"—an extremely bright meteor—streaking away from the backward-question-mark shape of the constellation Leo, photographed during the 1999 Leonid meteor shower. Thousands of fainter meteors were seen in the peak hour.

8.5 Asteroids

The nine known planets were not the only bodies to result from the gas and dust cloud that collapsed to form the Solar System 4.6 billion years ago. Thousands of **minor planets,** called **asteroids,** also resulted. We detect them by their small motions in the sky relative to the stars (■ Fig. 8–31). Most of the asteroids have elliptical orbits between the orbits of Mars and Jupiter, in a zone called the **asteroid belt.** It is thought that Jupiter's gravitational tugs perturbed the orbits of asteroids, leading to collisions among them that were too violent to form a planet.

Asteroids are assigned a number in order of discovery and then a name: (1) Ceres, (16) Psyche, and (433) Eros, for example. Often the number is omitted when discussing well-known asteroids. Though the concept of the asteroid belt may seem to imply a lot of asteroids close together, asteroids rarely come within a million

■ **FIGURE 8–31** Asteroid (5100) Pasachoff moves significantly against the background of stars in this series of exposures spanning six minutes. The asteroid appears elongated only because of its motion during the individual exposures.

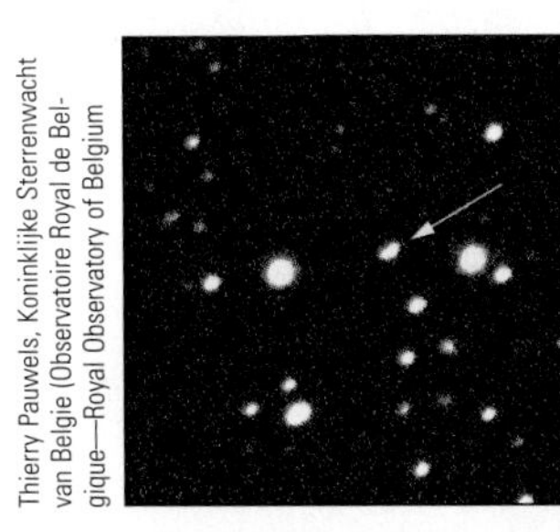

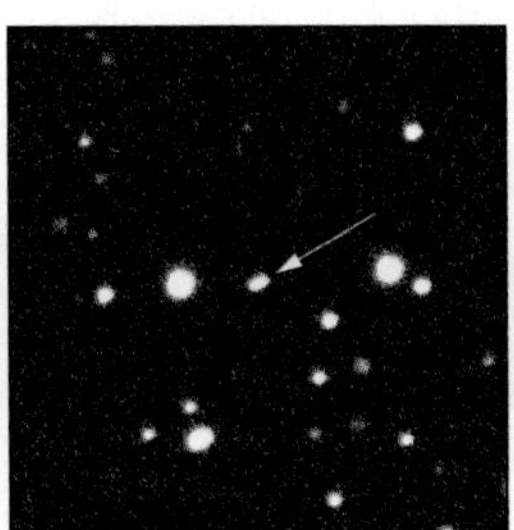

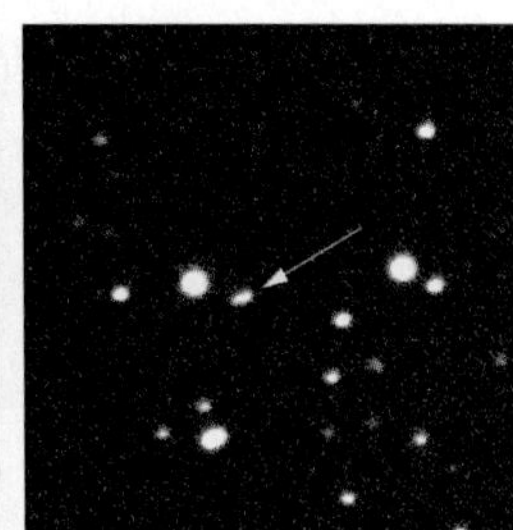
Thierry Pauwels, Koninklijke Sterrenwacht van Belgie (Observatoire Royal de Belgique—Royal Observatory of Belgium

Star Party 8.1 Observing a Meteor Shower

Meteor showers can be fun to watch. You might see a "shooting star" nearly every minute or two. They are best viewed after midnight, when the Earth is rotated so that we are plowing into the stream of pebbles and ice chunks that produces the meteors. It is also very helpful if the Moon is down or at most a thin crescent; a bright, moonlit sky tends to drown out the fainter meteors. A list of annual meteor showers is given in *A Closer Look 8.3: Meteor Showers.*

The Perseids around August 12 is the most commonly observed meteor shower. (For most others, bundle up against the cold.) Bring a reclining chair, blanket, or sleeping bag if possible, and find as dark a site as you can with a wide-open sky. Lie back, relax, and gaze up at the sky for at least half an hour. Any part of the sky is fine, but you will probably see the most meteors in the general direction of the "radiant," (for example Perseus, in the case of the Perseids). Try to see how many meteors you can count. Note any especially bright ones. Sometimes they leave a vapor "trail" (or "train") for a few seconds. Occasionally, colors are seen in the streak of light.

kilometers of each other. Occasionally, collisions do occur, producing the small chips that make meteoroids.

ASIDE 8.5: Rapidly orbiting asteroid

One asteroid, discovered in 2003, is known to have an orbit that is entirely within Earth's.

8.5a General Properties of Asteroids

Only about 6 asteroids are larger than 300 km in diameter. Hundreds are over 100 km across (■ Fig. 8–32), roughly the size of some of the moons of the planets, but most are small, less than 10 km in diameter. Perhaps 100,000 asteroids could be detected with Earth-based telescopes; automated searches are now discovering asteroids at a prodigious rate. Yet all the asteroids together contain less mass than the Moon.

Spacecraft en route to Jupiter and beyond travelled through the asteroid belt for many months and showed that the amount of dust among the asteroids is not much

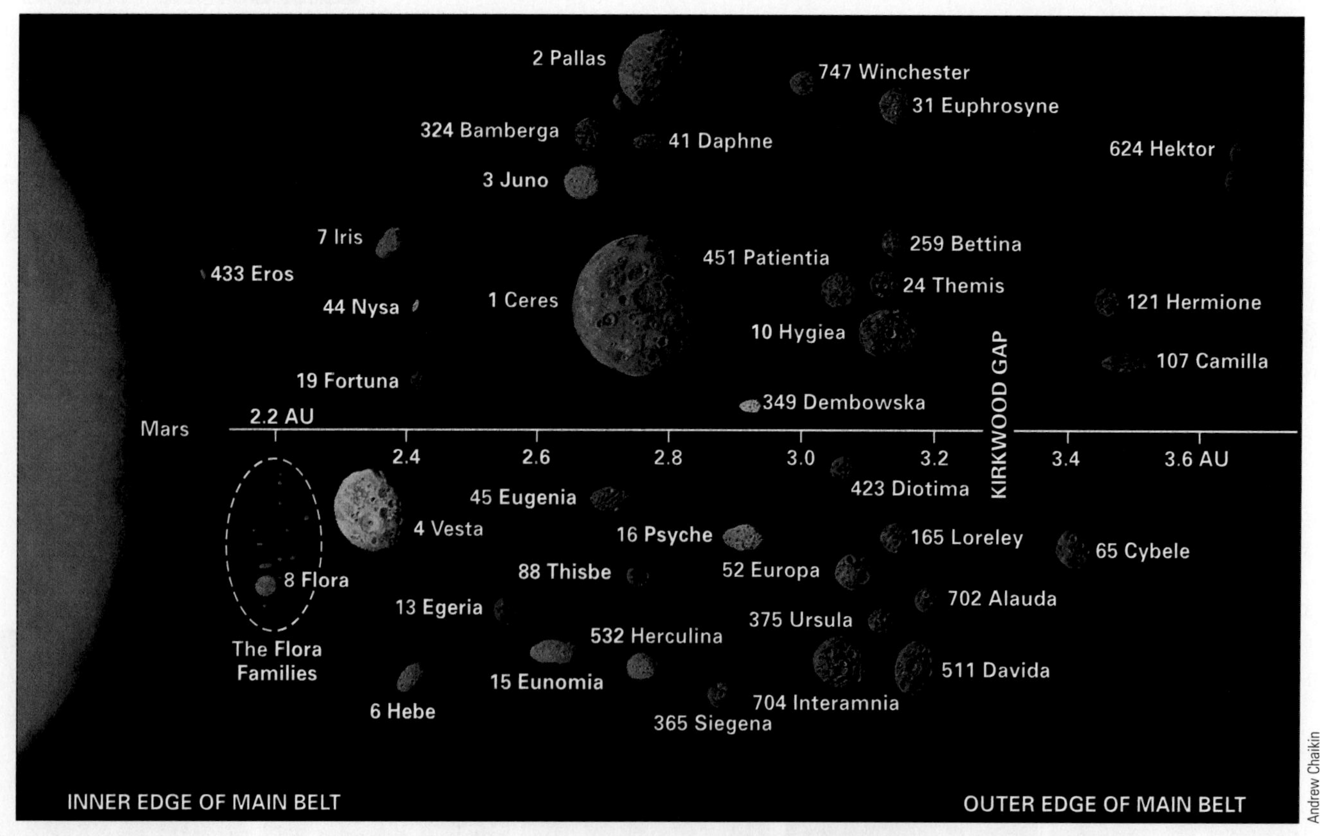

■ **FIGURE 8–32** The sizes of the asteroid belt's larger asteroids and their relative albedoes.

A Closer Look 8.4 The Extinction of the Dinosaurs

On Earth 65 million years ago, long before there were humans, dinosaurs and about 2/3 of the other species died in a catastrophe known as the Cretaceous/Tertiary (K/T) extinction (see the figure below). Evidence has been accumulating that this extinction was sudden and was caused by a Near-Earth Object or a comet hitting the Earth. Among the signs is the fact that the rare element iridium, which is more common in asteroids and comets than in Earth's crust, is widely distributed around the Earth in a thin layer laid down 65 million years ago, presumably after it was thrown up in the impact.

The impact would have raised so much dust into the atmosphere that sunlight could have been shut out for months; plants and animals would not have been able to survive. Moreover, there is evidence for widespread fires 65 million years ago: Charcoal is found in the corresponding strata. The impact would have raised the Earth's temperature enough to start such fires.

A giant buried crater in Mexico that was probably the site of the impact (see the figure below) has been discovered. It is one of the largest craters on Earth, about 200 kilometers across. Its age has been measured to be 65 million years—precisely coinciding with the extinction event. The idea of an asteroid or comet impact therefore seems compelling.

Research is continuing in this exciting area. Revised dating of another giant crater led to the idea that perhaps several asteroids or comets hit the Earth within a few hundred thousand years, and that they provided the "one–two punch" that knocked out the various species over a relatively short but not as abrupt time interval. An even more devastating extinction event took place 250 million years ago, terminating the Permian period and beginning the Triassic.

The dinosaurs and many other species disappeared suddenly 65 million years ago, marking the end of the Cretaceous period, the third part of the Mesozoic era. In the movie *Jurassic Park* (the Jurassic—208 to 146 million years ago—was the second part of the Mesozoic era), dinosaurs were cloned from their DNA, something we can't do—yet. No human has ever been contemporaneous with a dinosaur.

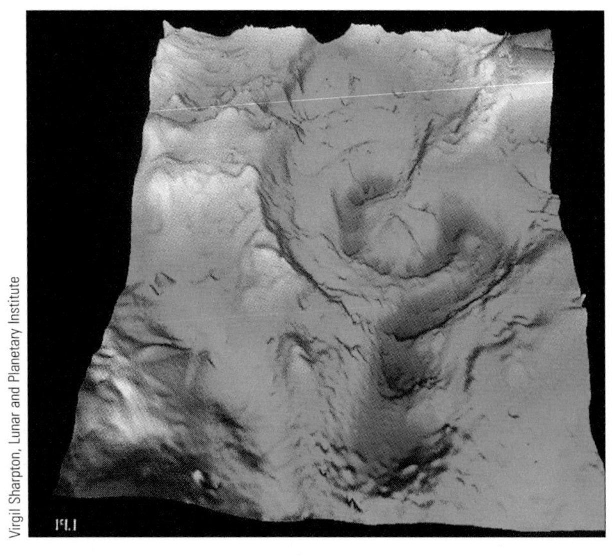

Virgil Sharpton, Lunar and Planetary Institute

The impact structure known as Chicxulub, at the north end of Mexico's Yucatán Peninsula, was formed by an asteroid or comet 12 to 15 km in diameter. In this false-color map, which shows structure under a top layer of a kilometer of sediment, the blue areas are low-density rocks broken up by the impact. The green mound at the center is denser and probably represents a rebound at the point of impact.

greater than the amount of interplanetary dust in the vicinity of the Earth. So the asteroid belt is not a significant hazard for space travel to the outer parts of the Solar System.

Asteroids are made of different materials from each other, and represent the chemical compositions of different regions of space. The asteroids at the inner edge of the asteroid belt are mostly stony in nature, while the ones at the outer edge are darker (because they contain more carbon). Most of the small asteroids that pass near the Earth belong to the stony group. Three of the largest asteroids belong to the high-carbon group. A third group is mostly composed of iron and nickel. The differences may be telling us about conditions in the early Solar System as it was forming and how the conditions varied with distance from the young Sun. Many of the asteroids must have broken off from larger, partly "differentiated" bodies in which dense material sank to the center (as in the case of the terrestrial planets; see our discussion in Chapter 6).

ASIDE 8.6: Peregrine Falcon

A Japanese space mission, Hayabusa (which means "Peregrine Falcon"), formerly MUSES-C, was launched in 2003 in order to rendezvous with a near-Earth Apollo-type asteroid, (25143) Itokawa, about 600 m × 300 m × 260 m in size, in 2005. It is to return a few grams of the asteroid sample to Earth in 2007.

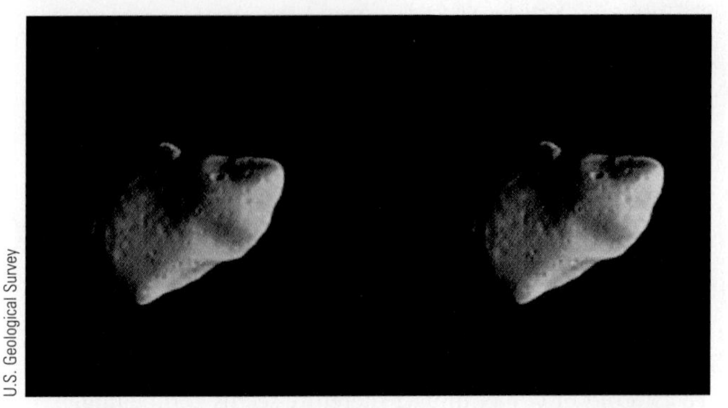

■ **FIGURE 8–33** Gaspra, in this high-resolution image taken by the Galileo spacecraft, looks like a fragment of a larger body. The illuminated part seen here is 16 km by 12 km in size. Galileo discovered that Gaspra rotates every 7 hours. We see a natural color image of Gaspra at *left.* At *right,* the color-enhanced image brings out variations in the composition of Gaspra's surface.

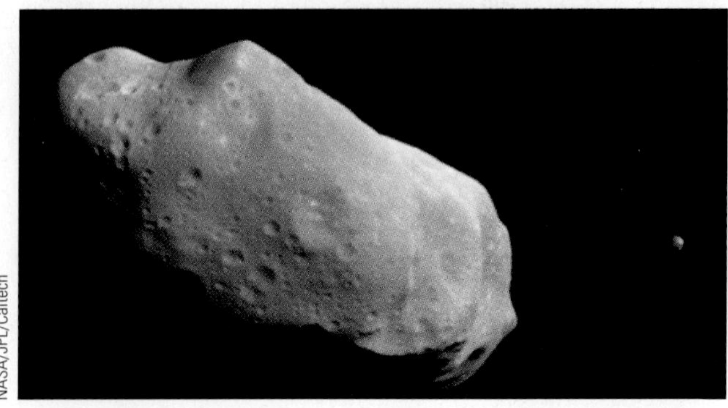

■ **FIGURE 8–34** Ida and its moon (or satellite), Dactyl, photographed from the Galileo spacecraft at a range of 10,500 km. Filters passing light at 4100 Å (violet), 7560 Å (near-infrared), and 9860 Å (infrared) are used, which enhances the color since the CCD is most sensitive in the near-infrared. Subtle color variations show differences in the physical state and composition of the soil. The bluish areas around some of the craters suggest a difference in the abundance or composition of iron-bearing minerals there. Dactyl differs from all areas of Ida in both infrared and violet filters. Asteroid pairs are turning out to be fairly common.

ASIDE 8.7: Ceres, the first asteroid

Analysis of Hubble Space Telescope images of (1) Ceres, matched with occultation results to give its absolute dimensions, have shown that its shape indicates that its interior is differentiated. Ceres, therefore, may have a mantle rich in water ice surrounding a rocky core. NASA's Dawn mission is to orbit Ceres in 2016.

The path of the Galileo spacecraft to Jupiter sent it near the asteroid (951) Gaspra in 1991 (■ Fig. 8–33). It detected a magnetic field from Gaspra, which means that the asteroid is probably made of metal and is magnetized. Galileo passed the asteroid (243) Ida in 1993, and discovered that the asteroid has an even smaller satellite (■ Fig. 8–34), which was then named Dactyl. Other double asteroids have since been discovered, and astronomers newly recognize the frequency of such pairs. For example, ground-based astronomers found a 13-km satellite orbiting 200-km-diameter (45) Eugenia every five days. (Note that Eugenia's low number shows that it was one of the first asteroids discovered.)

8.5b Near-Earth Objects

Some asteroids are far from the asteroid belt; their orbits approach or cross that of Earth. We have observed only a small fraction of these types of **Near-Earth Objects,** bodies that come within 1.3 A.U. of Earth.

The Near Earth Asteroid Rendezvous (NEAR) mission passed and photographed the main-belt asteroid (253) Mathilde in 1997. The existence of big craters that would have torn a solid rock apart, and the asteroid's low density, lead scientists to conclude that Mathilde is a giant "rubble pile," rocks held together by mutual gravity.

NEAR went into orbit around (433) Eros on Valentine's Day, 2000 (■ Figs. 8–35 and 8–36), when it was renamed NEAR Shoemaker after the planetary geologist Eugene Shoemaker. Eros was the first near-Earth asteroid that had been discovered. It is 33 km by 13 km by 13 km in size. NEAR Shoemaker photographed craters, grooves, layers, house-sized boulders, and a 20-km-long surface ridge.

The existence of the craters and ridge, which indicates that Eros must be a solid body, disagrees with the previous suggestions of some scientists that most asteroids are

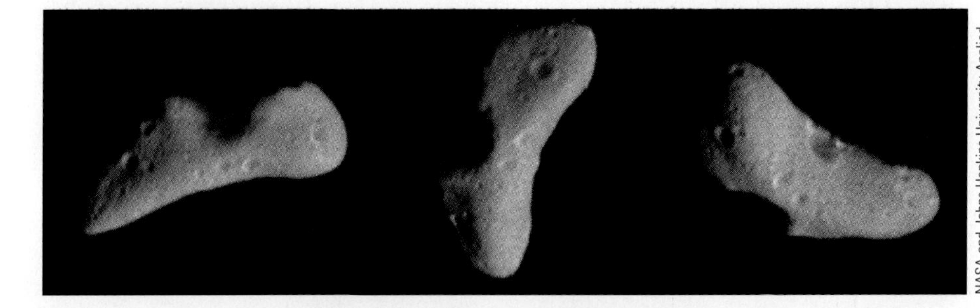

■ **FIGURE 8–35** The rotation of asteroid (433) Eros, imaged from the Near Earth Asteroid Rendezvous mission (renamed NEAR Shoemaker). Eros's surface looks a uniform butterscotch color in visible light, but shows more variation in the infrared.

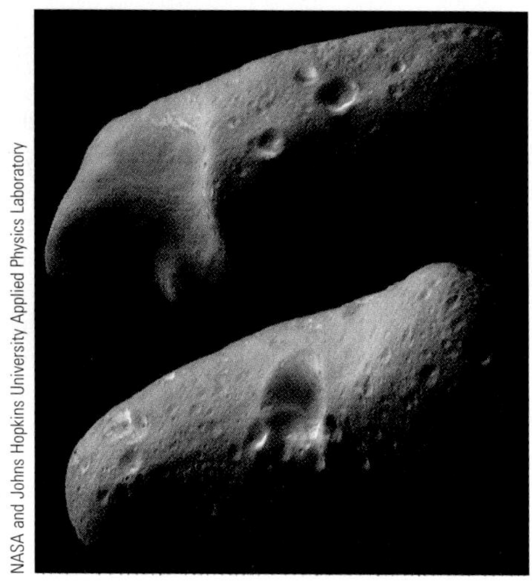

■ **FIGURE 8–36** (a) Mathilde imaged from the Near Earth Asteroid Rendezvous mission in 1997. (b) Eros, from NEAR Shoemaker in orbit only 220 km above its surface. Surface detail down to 35 m is visible. Note how different its opposite hemispheres appear. Eros's surface is really six times brighter than Mathilde's, whose surface brightness has been relatively enhanced in the image. Mathilde typifies a type of asteroid found especially in the outer part of the asteroid belt, while Eros is more typical of asteroids from the inner part of the asteroid belt.

mere rubble piles as Mathilde seems to be. The impact that formed the largest crater, 8 km across and now named Shoemaker, is thought to have formed most of the large boulders found across Eros's surface.

Eros's density, 2.4 g/cm^3, is comparable to that of the Earth's crust, about the same as Ida's, and twice Mathilde's. From orbit, NEAR Shoemaker's infrared, x-ray, and gamma-ray spectrometers measured how the minerals vary from place to place on Eros's surface. The last of these even survived the spacecraft's landing on Eros (■ Fig. 8–37), and radioed back information about the composition of surface rocks.

Scientists analyzing the data have found abundances of elements similar to that of the Sun and of a type of primitive meteorite known as chondrites that are the most common type of meteorite found on Earth. They have concluded that Eros is made of primitive material, unchanged for 4.5 billion years, so we are studying the early eras of the Solar System with it. NEAR Shoemaker's observations show that Eros was probably broken off billions of years ago from a larger asteroid as a uniformly dense fragment. This solidity contrasts with Mathilde's rubble-pile nature.

Besides providing much detailed information, the close-up studies of these objects are allowing us to verify whether the lines of reasoning we use with ground-based asteroid observations give correct results.

Near-Earth asteroids (■ Fig. 8–38) may well be the source of most meteorites, which could be debris of collisions that occurred when these asteroids visit the asteroid belt. Eventually, most Earth-crossing asteroids will probably collide with the Earth. Over 1000 of them are greater than 1 km in diameter, and none are known to be larger than 10 km across. Statistics show that there is a 1 per cent chance of a collision of this tremendous magnitude per millennium. This rate is pretty high on a cosmic scale. Such collisions would have drastic consequences for life on Earth.

Smaller objects are a hundred times more common, with a 1 per cent chance that an asteroid greater than 300 m in diameter would hit the Earth in the next century. Such a collision could kill thousands or millions of people, depending on where it lands.

The question of how much we should worry about Near-Earth Objects hitting us is increasingly discussed, including at a meeting sponsored by the United Nations. Even Hollywood movies have been devoted to the topic, though at present we can't send out astronauts to deflect or break up the objects the way the movies showed.

Several projects are under way to find as many Near-Earth Objects as possible. Current plans are to map 90 per cent of them in the next couple of decades, and the pace of discovery is accelerating. Several projects use CCD detectors, repetitive scanning, and computers to locate asteroids and are discovering thousands each year, some of which are Near-Earth Objects.

■ **FIGURE 8–37** The last two photographs of Eros taken from NEAR Shoemaker as it descended on February 12, 2001. The last photograph, taken from 120 m above the surface, shows a field of view only 6 m across and resolves features as small as 1 cm across! At the bottom of the final image, the vertical streaks show where the transmission of the photo was cut when NEAR Shoemaker impacted Eros.

■ **FIGURE 8–38** The title figure from Antoine de Saint-Exupéry's book *The Little Prince* tending his asteroid.

CONCEPT REVIEW

Pluto, which emerged in 1999 from inside Neptune's orbit, is so small and so far away that we know little about it (Sec. 8.1). The discovery of its moon, Charon, allowed us to calculate that Pluto contains only 1/500 the mass of the Earth (Sec. 8.1a). Mutual occultations of Pluto and Charon revealed sizes and surface structures of each. Pluto has a very thin nitrogen atmosphere, at least when it is closest to the Sun in its orbit (Sec. 8.1b).

For the time being, Pluto is hanging onto its designation as a planet (Sec. 8.1c), but is probably also one of the largest of the **Kuiper-belt objects,** also called **Trans-Neptunian Objects** (Sec. 8.2). About 1000 such objects are now known, some of which are about half the diameter of Pluto, and at least one is larger than Pluto, rejuvenating the discussion of whether Pluto should be called a planet. They appear to be left over from the formation of the Solar System.

In a **comet,** a long **tail** that points away from the Sun may extend from a bright **head** (Sec. 8.3). The head consists of the **nucleus,** which is like a dirty snowball, and the gases of the **coma** (Sec. 8.3a). The **dust tail** contains dust particles released from the dirty ices of the nucleus. The **gas tail** is blown out behind the comet by the solar wind. Comets we see come from a huge spherical **Oort comet cloud** far beyond Pluto's orbit or from a flattened belt of comets (the Kuiper belt) just beyond Neptune's orbit (Sec. 8.3b).

Spacecraft imaged the nucleus of Halley's Comet up close in 1986 and showed it to be an ellipsoidal snowball about 16 km long and 8 km across (Sec. 8.3c). The impact of the tidally disrupted Comet Shoemaker-Levy 9 with Jupiter in 1994 produced temporary Earth-sized scars in Jupiter's atmosphere (Sec. 8.3d). Comet Hale-Bopp in 1997 was not only spectacular for the general public but also allowed many scientific studies (Sec. 8.3e). The Deep Impact mission crashed into a comet in 2005 (Sec. 8.3f), providing information on its composition.

Meteoroids are chunks of rock in space (Sec. 8.4). When one hits the Earth's atmosphere, we see the streak of light from its vaporization as a **meteor,** often called a **shooting star.** A fragment that survives and reaches Earth's surface is a **meteorite.** Bits of space dust are **micrometeorites** (Sec. 8.4a). Most meteorites that are found are made of an iron–nickel alloy, but when meteorites are seen to fall, they are most often stony. The meteorites bring us primordial material to study and occasionally pieces of the Moon or Mars.

Many meteors are seen in **showers,** which occur when the Earth crosses the path of a defunct or disintegrating comet (Sec. 8.4b); these meteors appear to come from a single **radiant** in the sky. **Sporadic meteors** appear at the rate of a few each hour.

Asteroids are **minor planets** (Sec. 8.5). Most asteroids are in the **asteroid belt** between the orbits of Mars and Jupiter. They were prevented from forming a planet by Jupiter's gravitational tugs. Asteroids range up to about 1000 km across (Sec. 8.5a). Most of them are stony, while some contain carbon and others consist largely of iron and nickel. The Galileo and Near Earth Asteroid Rendezvous (renamed NEAR Shoemaker) missions have studied asteroids up close, and NEAR Shoemaker landed on the asteroid Eros (Sec. 8.5b). It found a solid object of primitive composition, in contrast with the rubble pile that is the asteroid Mathilde.

It is plausible that an impact of a large asteroid or comet threw so much dust into the Earth's atmosphere that it led to the extinction of many species, including dinosaurs, 65 million years ago (Sec. 8.5b and *A Closer Look 8.4*). We worry about a **Near-Earth Object** or a comet hitting the Earth, which could lead to the demise of civilization, or of the more likely collision with the Earth of a smaller object, which could cause substantial damage, killing thousands or millions of people.

QUESTIONS

1. What fraction of its orbit has Pluto traversed since it was discovered?
2. What evidence suggests that Pluto is not a "normal" planet?
3. Describe what mutual occultations of Pluto and Charon are and what they have told us.
4. Compare Pluto to the giant planets.
5. Compare Pluto to the terrestrial planets.
6. †If Pluto's surface temperature is 60 K, use Wien's law (Chapter 2) to calculate the wavelength at which it gives off most of its thermal (black-body) radiation. In what part of the electromagnetic spectrum is this?
7. Which of Kepler's laws has enabled the discovery of Charon to help us find the mass of Pluto?
8. Describe the observations of Pluto's atmosphere when Pluto occulted a star.
9. What are Kuiper-belt objects, and how is Pluto thought to be related to them?
10. What creates the tail of a comet? Is something chemically burning?
11. Which part of a comet has the most mass?
12. From where are the comets that we see thought to come?
13. †How far is the Oort comet cloud from the Sun, relative to the distance from the Sun to Pluto?
14. Describe some of the Giotto spacecraft's discoveries about Halley's Comet.
15. †Use Kepler's third law (Chapter 5) and the known period of Halley's Comet (about 76 years) to deduce its semimajor axis. Which planet has a comparable semimajor axis?
16. Discuss the significance of the collision of Comet Shoemaker-Levy 9 with Jupiter.
17. What is the relation of meteoroids and asteroids?
18. What, physically, is a shooting star?
19. Why don't most meteoroids reach the Earth's surface?
20. Explain the process by which a meteor shower occurs.
21. Why might some meteor showers last only a day while others can last several weeks?
22. Why are meteorites important in our study of the Solar System?

23. Compare the sizes and surfaces of asteroids with the moons of the planets.
24. How might the study of the apparent brightness of a star during an occultation by an asteroid tell us whether the asteroid has an atmosphere?
25. Why is the study of certain types of asteroids important for understanding our long-term survival on Earth?
26. What have the Galileo, NEAR Shoemaker, Deep Space 1, and Deep Impact spacecraft found out about comets and asteroids?
27. **True or false?** Comets are made of mostly rocky material from the asteroid belt.
28. **True or false?** The retrograde motion of Pluto can be attributed to the relatively high eccentricity of its orbit around the Sun, as compared to normal planets.
29. **True or false?** Periodic comets generally grow brighter each time they appear, since they gradually accumulate more material from the outer parts of the Solar System.
30. **True or false?** A "shooting star" or "falling star" is not a star at all, but rather a rock from space that enters Earth's atmosphere at a high speed and burns up due to friction.
31. **True or false?** The Giotto spacecraft observations of Comet Halley showed the evaporating gas and dust to come out of fissures in the comet's dark crust, creating jets.
32. **Multiple choice:** Pluto **(a)** is always the ninth planet from the Sun; **(b)** has no moons; **(c)** has a density intermediate between that of the terrestrial planets and the giant planets; **(d)** has a nearly circular orbit around the Sun, due to its interactions with other objects in its vicinity; or **(e)** is massive enough to substantially affect the orbit of Uranus, which led to its discovery.
33. **Multiple choice:** Which one of the following statements about the main asteroid belt is *true*? **(a)** The main asteroid belt is probably debris that would have formed a planet had Jupiter not prevented it. **(b)** The mass of all the asteroids together is roughly that of the Earth. **(c)** The iron/nickel asteroids are the most common kind. **(d)** There is no evidence that the material from which asteroids formed was ever part of one or more larger bodies that could have experienced differentiation. **(e)** All asteroids appear to be quite solid objects, rather than collections of smaller rocks held together by gravity.

†34. **Multiple choice:** Suppose you discover a comet having an orbital period of 1000 years around our Sun. Roughly what is its semimajor axis? **(a)** 10 A.U. **(b)** 32 A.U. **(c)** 10^2 A.U. **(d)** $10^{4.5}$ A.U. **(e)** 10^6 A.U.

35. **Multiple choice:** If an asteroid with a diameter of 30 km were to collide with the Earth, it would probably **(a)** burn up in the atmosphere before reaching the ground; **(b)** only crush a car, if one happened to be at the point of impact; **(c)** devastate an area no larger than 30 km in diameter; **(d)** create a crater more than 300 km in diameter, but have little or no effect on the rest of the Earth and its inhabitants; or **(e)** bring an end to human civilization, or at least destroy much of the human race.
36. **Fill in the blank:** Pluto is probably a large member of the ________, rather than a true planet.
37. **Fill in the blank:** A meteoroid that has landed on the surface of a planet or a moon is called a _________.
38. **Fill in the blank:** When Earth passes through the orbit of an old, disintegrating comet, a ________ can occur.
39. **Fill in the blank:** A thin layer of unusually high levels of the element ______ in the strata of the Earth's crust deposited 65 million years ago provides evidence for the impact theory of the Cretaceous/Tertiary extinction.
40. **Fill in the blank:** It is now generally thought that billions, or even trillions, of tailless comets surround the Solar System in a spherical shell nearly one light-year in radius called the ________.

†This question requires a numerical solution.

TOPICS FOR DISCUSSION

1. Now that many objects, some of them quite large, have been discovered in the Kuiper belt, do you think Pluto should still be called a planet?
2. Do you worry about asteroid or comet collisions with Earth? What could be done to save the Earth if an asteroid were discovered sufficiently far in advance of the collision?
3. Some scientists have suggested that the dinosaurs were about to become extinct anyway, without an asteroid or comet collision, due to change in Earth's climate and other reasons. If so, does this detract from the impact theory, in view of evidence that 2/3 of all species perished quite suddenly and a large crater having the right age has been found?

MEDIA

Virtual Laboratories

Kuiper-belt Objects

AceAstronomy™ Log into AceAstronomy at **http://astronomy.brookscole.com/cosmos3** to access quizzes and animations that will help you assess your understanding of this chapter's topics.

Log into the Student Companion Web Site at **http://astronomy.brookscole.com/cosmos3** for more resources for this chapter including a list of common misconceptions, news and updates, flashcards, and more.

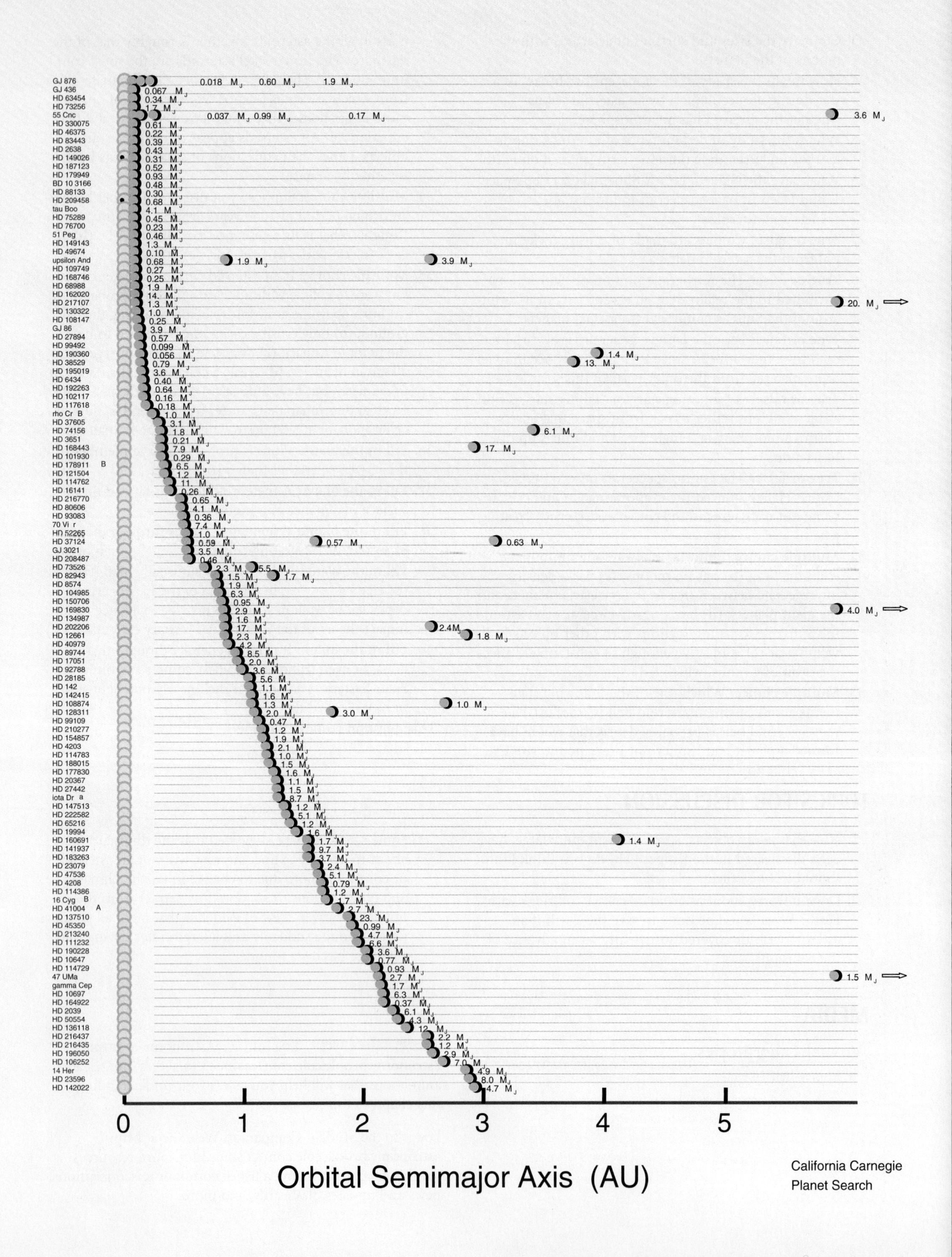
GJ 876 0.018 MJ 0.60 MJ 1.9 MJ
GJ 436 0.067 MJ
HD 63454 0.34 MJ
HD 73256 1.7 MJ
55 Cnc 0.037 MJ 0.99 MJ 0.17 MJ 3.6 MJ
HD 330075 0.61 MJ
HD 46375 0.22 MJ
HD 83443 0.39 MJ
HD 2638 0.43 MJ
HD 149026 0.31 MJ
HD 187123 0.52 MJ
HD 179949 0.93 MJ
BD 10 3166 0.48 MJ
HD 88133 0.30 MJ
HD 209458 0.68 MJ
tau Boo 4.1 MJ
HD 75289 0.45 MJ
HD 76700 0.23 MJ
51 Peg 0.46 MJ
HD 149143 1.3 MJ
HD 49674 0.10 MJ
upsilon And 0.68 MJ 1.9 MJ 3.9 MJ
HD 109749 0.27 MJ
HD 168746 0.25 MJ
HD 68988 1.9 MJ
HD 162020 14. MJ
HD 217107 1.3 MJ 20. MJ
HD 130322 1.0 MJ
HD 108147 0.25 MJ
GJ 86 3.9 MJ
HD 27894 0.57 MJ
HD 99492 0.099 MJ
HD 190360 0.056 MJ 1.4 MJ
HD 38529 0.79 MJ 13. MJ
HD 195019 3.6 MJ
HD 6434 0.40 MJ
HD 192263 0.64 MJ
HD 102117 0.16 MJ
HD 117618 0.18 MJ
rho Cr B 1.0 MJ
HD 37605 3.1 MJ
HD 74156 1.8 MJ 6.1 MJ
HD 3651 0.21 MJ
HD 168443 7.9 MJ 17. MJ
HD 101930 0.29 MJ
HD 178911 B 6.5 MJ
HD 121504 1.2 MJ
HD 114762 11. MJ
HD 16141 0.26 MJ
HD 216770 0.65 MJ
HD 80606 4.1 MJ
HD 93083 0.36 MJ
70 Vir 7.4 MJ
HD 52265 1.0 MJ
HD 37124 0.59 MJ 0.57 MJ 0.63 MJ
GJ 3021 3.5 MJ
HD 208487 0.46 MJ
HD 73526 2.3 MJ 5.5 MJ
HD 82943 1.5 MJ 1.7 MJ
HD 8574 1.9 MJ
HD 104985 6.3 MJ
HD 150706 0.95 MJ
HD 169830 2.9 MJ 4.0 MJ
HD 134987 1.6 MJ
HD 202206 17. MJ 2.4MJ
HD 12661 2.3 MJ 1.8 MJ
HD 40979 4.2 MJ
HD 89744 8.5 MJ
HD 17051 2.0 MJ
HD 92788 3.6 MJ
HD 28185 5.6 MJ
HD 142 1.1 MJ
HD 142415 1.6 MJ
HD 108874 1.3 MJ 1.0 MJ
HD 128311 2.0 MJ 3.0 MJ
HD 99109 0.47 MJ
HD 210277 1.2 MJ
HD 154857 1.9 MJ
HD 4203 2.1 MJ
HD 114783 1.0 MJ
HD 188015 1.5 MJ
HD 177830 1.6 MJ
HD 20367 1.1 MJ
HD 27442 1.5 MJ
iota Dra 8.7 MJ
HD 147513 1.2 MJ
HD 222582 5.1 MJ
HD 65216 1.2 MJ
HD 19994 1.6 MJ
HD 160691 1.7 MJ 1.4 MJ
HD 141937 9.7 MJ
HD 183263 3.7 MJ
HD 23079 2.4 MJ
HD 47536 5.1 MJ
HD 4208 0.79 MJ
HD 114386 1.2 MJ
16 Cyg B 1.7 MJ
HD 41004 A 2.7 MJ
HD 137510 23. MJ
HD 45350 0.99 MJ
HD 213240 4.7 MJ
HD 111232 6.6 MJ
HD 190228 3.6 MJ
HD 10647 0.77 MJ
HD 114729 0.93 MJ
47 UMa 2.7 MJ 1.5 MJ
gamma Cep 1.7 MJ
HD 10697 6.3 MJ
HD 164922 0.37 MJ
HD 2039 6.1 MJ
HD 50554 4.3 MJ
HD 136118 12. MJ
HD 216437 2.2 MJ
HD 216435 1.2 MJ
HD 196050 2.9 MJ
HD 106252 7.0 MJ
14 Her 4.9 MJ
HD 23596 8.0 MJ
HD 142022 4.7 MJ
0
1
2
3
4
5
Orbital Semimajor Axis (AU)
California Carnegie
Planet Search

CHAPTER 9

Our Solar System and Others

ORIGINS

Discoveries of planetary systems around other stars are providing insights into the process by which planets form. The exoplanets are also possible abodes for life outside our own Solar System.

AIMS

1. Discuss the formation of the Solar System (Section 9.1).
2. Learn the methods by which we have discovered over 160 planets orbiting other stars, and consider the properties of these planets (Section 9.2).
3. See how the discovery of planets and protoplanetary disks around other stars affects our thinking about the formation of our own Solar System (Section 9.3).

Astronomers have long speculated about the origin of our Solar System. They have noted regularities in the way the planets orbit the Sun and in the spacing of the planetary orbits. But until recently, astronomers have been limited to studying one planetary system: our own.

In 1600, Giordano Bruno was burned, naked, at the stake for wondering about the plurality of worlds, to use the term of that day. Finally, after hundreds of years of wondering whether they existed, another planetary system was discovered in 1991. And in the last decade, planetary systems galore were discovered around stars like the Sun. While in 1990 we knew of only 9 planets, all around our Sun, as of this writing (late-2005) we know of more than 160 additional planets around other stars (■ Fig. 9–1).

In this chapter, we will first discuss our own Solar System and its formation. Then we will describe how we find other planetary systems and how they may have been formed. This discussion, in turn, should give us insights into how well we understand the origin of our own Solar System.

9.1 The Formation of the Solar System

Many scientists studying the Earth and the planets are particularly interested in an ultimate question: How did the Earth and the rest of the Solar System form? We can accurately date the formation by studying the oldest objects we can find in the Solar System and allowing a little more time. For example, astronauts found rocks on the Moon older than 4.4 billion years. We have concluded that the Solar System formed about 4.6 billion years ago.

9.1a Collapse of a Cloud

Our best current idea is that 4.6 billion years ago, a huge cloud of gas and dust in space collapsed, pulled together by the force of gravity. What triggered the collapse

© Gerry Gropp

■ **FIGURE 9–1** Exoplanet hunters Geoff Marcy and Paul Butler, with the 3-m Shane telescope at Lick Observatory they used for many of their early observations. With their record of success (they are the world's most prolific discoverers of exoplanets, having discovered over 100 of them), they are now granted lots of telescope time at the world's largest telescopes.

AceAstronomy™ The AceAstronomy icon throughout this text indicates an opportunity for you to test yourself on key concepts, and to explore animations and interactions of the AceAstronomy website at **http://astronomy.brookscole.com/cosmos3**

156 of the exoplanets known within 100 parsecs of the Sun, including the 18 multi-planet systems.
California & Carnegie Planet Search

Department of Defense Visual Information Center

■ **FIGURE 9–2** Larger and smaller shock waves formed by an F/A-18 Hornet jet as it broke the sound barrier are visible. The clouds of condensation formed as the shocked air cooled.

Simon Bruty/*Sports Illustrated*

■ **FIGURE 9–3** Sarah Hughes spinning as she wins her gold medal at the 2002 Winter Olympic Games in Salt Lake City, Utah. Note how her arms are held in, to allow her to spin the fastest, as we see from the way her skirt flares outward. Moving more of her mass closer to her axis of spin makes her spin faster, in order to keep her total angular momentum constant.

isn't known. It might have been gravity pulling together a random cloud, or it might have been a shock wave from a nearby supernova (■ Fig. 9–2). As the gas pressure increased, eventually the rate of collapse decreased to a slower contraction.

You have undoubtedly noticed that ice skaters spin faster when they pull their arms in (■ Fig. 9–3). Similarly, the cloud that ultimately formed the Solar System began to spin faster as it collapsed and contracted. The original gas and dust may have had some spin, and this spin is magnified in speed by the collapse, because the total amount of "angular momentum" doesn't change.

It spins faster because the quantity known as angular momentum stays steady in a rotating system, unless the system is acted on by an outside force (assuming the force is not directed at the center of the object). The amount of angular momentum depends on the speed of the spin and on how close each bit of spinning mass is to the axis of spin; see our discussion at the end of Chapter 5.

Objects that are spinning around tend to fly off, and in our case this force eventually became strong enough to counteract the effect of gravity pulling inward. Thus the Solar System stopped contracting in one plane. Perpendicular to this plane, there was no spin to stop the contraction, so the solar nebula ended up as a disk. The central region became hot and dense, eventually becoming hot enough for the gas to undergo nuclear reactions (see Chapter 12), thus forming our Sun.

In the disk of gas and dust, we calculate that the dust began to clump (■ Fig. 9–4). Smaller clumps joined together to make larger ones, and eventually **planetesimals,** bodies that range from about 1 kilometer up to a few hundred kilometers across, were formed. Gravity subsequently pulled many planetesimals together to make **protoplanets** (pre-planets) orbiting a **protosun** (pre-Sun).

The protoplanets then contracted and cooled to make the planets we have today, and the protosun contracted to form the Sun (■ Fig. 9–5). Some of the planetesimals may still be orbiting the Sun; that is why we are so interested in studying small bodies of the Solar System like comets, meteoroids, and asteroids. Most of the unused gas and dust, however, was blown away by a strong solar wind.

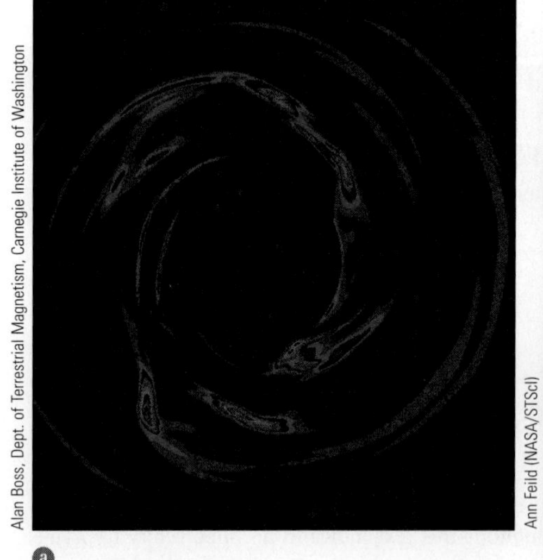
Alan Boss, Dept. of Terrestrial Magnetism, Carnegie Institute of Washington

(a)

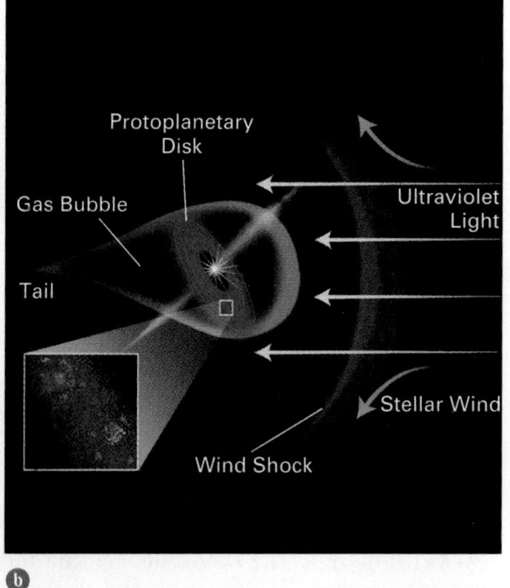

Ann Feild (NASA/STScI)

(b)

■ **FIGURE 9–4** (a) A false-color representation of density clumps in a model of the early solar nebula with a radius of 30 A.U. A cross section through the midplane of the nebula is shown. Red corresponds to the highest density, and black to the lowest density. Five clumps have formed in only 338 years. The model is by Alan Boss. (b) Artist's conception of the effect of strong ultraviolet (UV) light on a young star and its planet-forming disk. The UV light heats the gas so much (to 10,000 K) that it effectively evaporates from the disk and forms a comet-like tail downstream.

FIGURE 9–5 The formation of our Solar System. (a) The four inner, terrestrial planets. The first panel shows the initial disk of dust and gas, where dust grains stick together to become 10-km-diameter planetesimals. In the second panel, planetesimals grow over about 10,000 years to form 2000-km diameter planetary embryos. In the third panel, gas disappears over the next million to 10 million years. In the fourth panel, the gravity of the planetary embryos changes their orbits and brings them together to form the current planets over about 100 million years. Other objects are ejected. (b) The four outer, giant planets. The first panel shows the initial disk of dust and gas. In the second panel, the runaway growth of planetesimals forms very large (10-Earth-mass) cores in a few million to 10 million years. In the third panel, Jupiter and Saturn capture gas from the solar nebula before it dissipates. Uranus and Neptune started out as cores that formed between Jupiter and Saturn but were tossed outward to their present orbits without gathering gas. In the fourth panel, we see some comets as well as bodies like Pluto and Triton forming as Kuiper-belt objects that weren't incorporated in the giant planets.

9.1b Models of Planet Formation

In one of the main models for the formation of the outer planets, the solar nebula first collapsed into several large blobs. These blobs then became the outer planets (as we saw in Figure 9–4). The two outermost blobs lost most of their gaseous atmospheres as a result of strong ultraviolet light from the young Sun that also removed the gas from the nebula at Saturn's orbit and beyond (as we saw in Figure 9–5*b;* see also Figure 9–16). As a result, they became the moderately massive planets, Uranus and Neptune. Because of the strength of their internal gravity, Saturn lost only a portion of its gas, while Jupiter was unaffected, thus becoming the most massive planet.

In another model, a solid core (resembling a terrestrial planet) condensed first for each of the outer planets. The gravity of this core then attracted the gas from its surroundings. For this second model, the relative amounts of elements in the rocks and in the gases would also differ from planet to planet. Spacecraft found that Jupiter, Saturn, Uranus, and Neptune have different relative amounts of some of the elements in their atmospheres. Additionally, the atmospheres of Jupiter and Saturn are very much more massive than the atmospheres of Uranus and Neptune. In this model, the cores of Uranus and Neptune may contain 10 to 15 times the mass of the Earth. Also in this model, Jupiter's core is thought to have only 0 to 3 Earth masses, much less than the cores of Uranus and Neptune, which is an argument against this model.

In the inner Solar System, the terrestrial planets are the accumulation of planetesimals. Our Earth and its neighbor planets were formed out of planetesimals made of rocky material. The rocky material, mainly silicates, condensed at the temperatures of these planets' distances from the Sun and were not blown outward by particles from the forming Sun. These terrestrial planets never became massive enough to accumulate a massive atmosphere the way the giant planets did. And, because the inner planets are closer to the Sun and therefore hotter, gas in their atmospheres moved relatively fast. Thus free hydrogen and helium escaped from the Earth's low gravity, while Jupiter and the other giant planets have huge atmospheres of hydrogen and helium, matching the atmosphere of the Sun.

Early on, the Sun would have been spinning very fast. But much of the excess angular momentum was transported upward by a "bipolar outflow"; see the discussion in Chapter 12.

In 2001, John Chambers and George Wetherill of the Carnegie Institution of Washington proposed a model extending standard ideas of Solar-System formation to explain not only the range of planets we see but also the relative emptiness of the asteroid belt. Many people mistakenly think that the asteroids are the remains of a large planet that exploded, but the asteroid belt actually contains so little material that not even a small moon could have been present there. In their model, planetesimals formed everywhere throughout the solar nebula, including the asteroid belt. Some of the planetesimals coalesced into planets.

Jupiter grew especially rapidly, and its gravity kicked out material in the asteroid-belt region that orbited an integer number of times in the time that Jupiter itself made one orbit (or some other integer number of orbits). This aspect of Jupiter's gravity, and that it formed gaps in the asteroid belt, has long been known.

Chambers and Wetherill have added the idea that objects in this region that are not quite in the places affected so strongly by Jupiter's gravity were pushed into these zones by gravitational encounters among themselves, including some planetesimals and small planets that had formed there. They have checked the idea with computer simulations. The process also made the orbits of the giant planets more circular, matching observations. Further, it sent some objects that would contain volatile substances (those that evaporate easily) like water to crash into the Earth. This process may explain how we got our oceans, something more often attributed to comets.

It is interesting that nothing about our current model of planetary-system formation implies that the Solar System is unique. As we will see next, we are increasingly finding systems of planets around other stars. To our surprise, their properties aren't like those of our Solar System: We can't see if they have small, rocky planets close in to the star, but we know that they don't all have massive ones farther out.

Maybe some of our modelling for our own Solar System has been wrong because we looked for regularity in the distribution of planets while we really had a random distribution.

9.2 Extra-solar Planets (Exoplanets)

People have been looking for planets around other stars for decades. Many times in the last century, astronomers reported the discovery of a planet orbiting another star, but for a long time each of these reports had proven false. Finally, in the 1990s, the discovery of **extra-solar planets** (planets outside, "extra-," the Solar System) seemed valid. They have also become known as **exoplanets.**

9.2a Discovering Exoplanets

Since exoplanets shine only by reflecting a small amount of light from their parent stars (i.e., the stars that they orbit), they are very faint and extremely difficult to see in the glare of their parent stars with current technology. So the search for planets has not concentrated on visible sightings of these planets. Rather, it has depended on watching for motions in the star that would have to be caused by something orbiting it.

9.2a(i) Astrometric Method

The earlier reports, now rejected, were based on tracking the motion of the nearest stars across the sky with respect to other stars. The precise measurement of stellar positions and motions is called **astrometry,** so this method is known as the **astrometric method.**

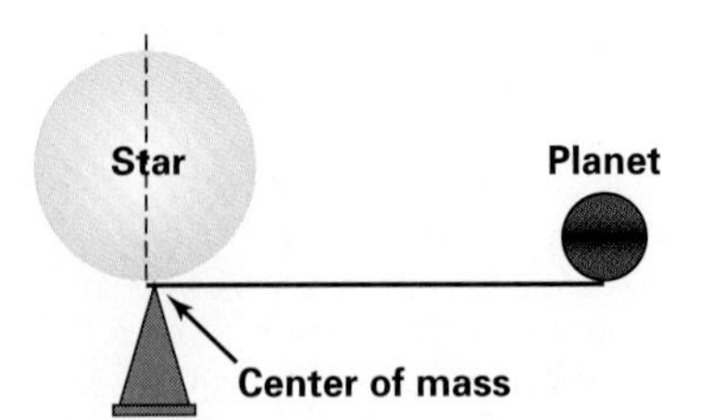

■ **FIGURE 9–6** The center of mass for a two-body system is defined to be the point at which the two objects balance each other, like on a seesaw. In the case of a star-planet system, the center of mass is very close to the massive star, and far from the planet. Both objects orbit their common center of mass.

If a star wobbled from side to side, it would reveal that a planet was wobbling invisibly the other way, so that the star/planet system was moving together in a straight line. (Technically, the "center of mass" of the system has to move in a straight line, unless its motion is distorted by some outside force. The center of mass is illustrated in ■ Fig. 9–6, and is described more fully in Chapter 11, when we discuss binary stars. Both the star and the planet orbit their common center of mass, though the star is much closer to the center of mass than the planet is. Thus, the star moves very slightly, in a kind of "reflex motion" caused by the orbiting planet.)

Astrometric measurements have been made over the last hundred years or so, and a few of the nearby stars whose motions in the sky were followed seemed to show such wobbling. But the effects always turned out to be artifacts of the measuring process. Nevertheless, the astrometric method is still being used by some astronomers, and maybe they will eventually detect an exoplanet with it.

9.2a(ii) Timing of Radio Pulsars

The first extra-solar planet was discovered in 1991 around a pulsar, a weird kind of collapsed star (see our discussion in Chapter 13) that gives off extremely regular pulses of radio waves with a period that is a fraction of a second. The pulses came more frequently for a time and then less frequently in a regular pattern. So the planet around this pulsar must be first causing the pulsar to move in our direction, making the pulses come more rapidly, and then causing it to move in the opposite direction, making the pulses come less often. (As described above, the planet and star are actually orbiting their common center of mass.)

But this planet must have formed after the catastrophic explosion that destroyed most of the star after its inner core collapsed; the planet could not have survived the stellar explosion. Thus, it was not the kind of planet that is born at the same time as the star, like Earth.

Even when the existence of two more planets (and possibly a fourth planet) orbiting that pulsar was established, all with masses comparable to those of the terrestrial planets in our Solar System, the pulsar system seemed too unusual to think much about, except by specialists. It didn't help that another pulsar planet, discovered slightly earlier, turned out to be a mistaken report.

ASIDE 9.1: Measuring an exoplanet's mass

The parent star of one of the planets discovered from its radial-velocity variations (periodic Doppler shifts) has also been observed, with the Hubble Space Telescope, to move to-and-fro-and-to in the sky. These astrometric measurements of the star GJ876 have been coupled with the radial-velocity measurements to deduce the angle at which we are looking at the system. Thus, we can determine the planet's mass pretty accurately: 1.9 to 2.4 times Jupiter's mass. Further work has discovered two other planets around this star, as we describe in the main text.

9.2a(iii) Periodic Doppler Shifts

In the 1990s, techniques were developed using the Doppler effect. Recall that Chapter 2 describes how the Doppler effect is a shift in the wavelengths of light that has been emitted, caused by motion of the source or the receiver along the line of sight (that is, by the "radial velocity").

The planet has a large orbit around the center of mass, moving rapidly. But the parent star, being much more massive than the planet, is much closer to the center of mass (Fig. 9–6) and therefore moves much more slowly in a smaller orbit. With sufficiently good spectrographs and numerous observations, this slight "wobble" can be detected as a periodically changing Doppler shift in the star's spectrum (■ Fig. 9–7); the radial velocity of the star varies in a periodic way.

The breakthrough came because new computer methods were found that measure the changing Doppler shifts very precisely. On a computer, the star's spectrum can be simulated including changes in the wavelengths of light, just as happens in the Doppler effect. These simulated spectra can be compared to the observed spectrum of that star, until an excellent match is found. This method allows very small Doppler shifts to be detected, and the speeds of stars toward or away from us can be measured to a precision of 1 meter/sec, a leisurely walking speed. (This very high precision was only recently achieved, with the Keck-I telescope; at most other sites, the precision has typically been 3 meters/sec or worse.)

The first surprising report came in 1995 from a Swiss astronomer and his student, Michel Mayor and Didier Queloz. They found a planet around a nearby star, 51 Pegasi. One strange thing about the planet is that it seemed to be a giant planet, at least half as massive as Jupiter, but with an orbit far inside what would be Mercury's orbit in our Solar System. The planet orbited 51 Peg in only 4.2 days, much faster than any planet in our Solar System.

Two American astronomers, Geoff Marcy and Paul Butler, then at San Francisco State University and the University of California, Berkeley, had been collecting similar data on dozens of other stars. But they had assumed, reasonably, that a planet like

Doppler shift due to stellar wobble

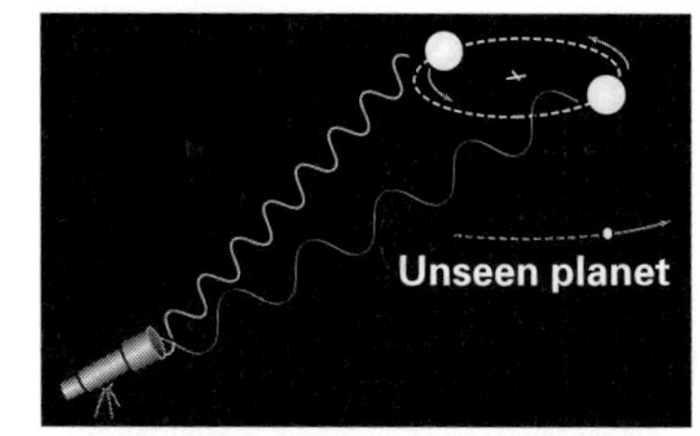

■ **FIGURE 9–7** The technique used to detect planets around Sun-like stars by searching for Doppler shifts in the stars' spectra. The unseen planet and the star orbit their common center of mass (marked with a white cross), with the star having the much smaller (but non-zero) orbital distance because of its much larger mass. As the star comes toward us, the light it emits appears blueshifted. Half an orbit later, when it is going away from us, its light appears redshifted. The cycle then repeats itself. By observing a star over many days, weeks, months, or years, periodic Doppler shifts like these are revealed, signaling the presence of the unseen planet through its gravitational tug; see Figure 9–11 for an example. (After Geoffrey W. Marcy, University of California at Berkeley; and R. Paul Butler, Carnegie Institution of Washington)

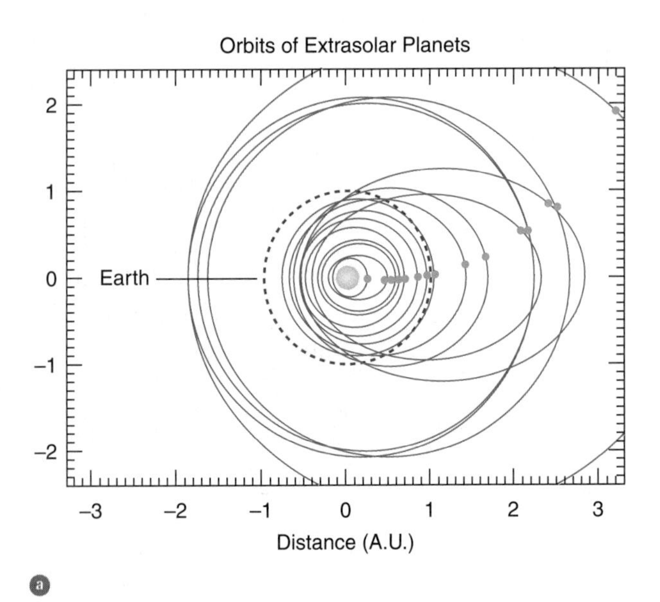

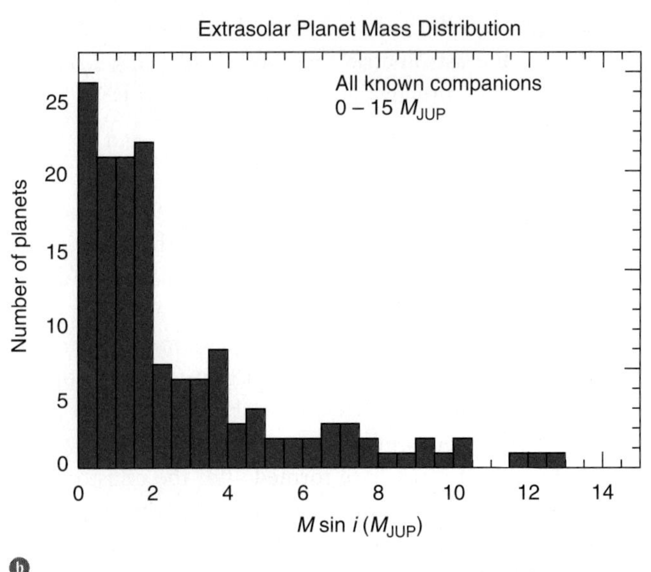

■ **FIGURE 9–8** (a) The orbits of some of the exoplanets overlaid on the same scale. We see "hot Jupiters" close in, eccentric orbits farther out, and circular orbits exceeding the size of Earth's. (b) The distribution of masses of the planets. Actually, we can measure only lower limits (minimum values) to the masses, since we don't know the angle at which we see the systems. The horizontal axis shows the mass times the sine of the orbit's angle of inclination. (Sines are always less than 1, so the actual mass is always greater than the value on this graph.) (After Geoffrey W. Marcy, University of California at Berkeley; and R. Paul Butler, Carnegie Institution of Washington)

Jupiter around another star would take years to orbit, so they were collecting years of data while perfecting their analysis techniques to measure exceedingly small speeds. They hadn't run their spectra through the computer programs they were writing to measure the Doppler shifts. When they heard of the Mayor and Queloz results, they quickly examined their existing data and also observed 51 Peg. They soon verified the planet around 51 Peg and discovered planets around several other stars (see the drawing opening this chapter).

Most of the new objects turned out to be giant planets either in extremely elliptical orbits or in circular orbits very close to the parent stars (■ Fig. 9–8*a*). Most of these planets are orbiting stars within 50 light-years from us, not extremely far away but not the very closest few dozen stars either. Stars this close are bright enough for us to carry out the extremely sensitive spectroscopic measurements.

One limitation of the method is that we generally don't know the angle of the plane in which the planets are orbiting their parent stars. The Doppler-shift method works only for the part of the star's motion that is toward or away from us, and not for the part that is from side to side. So the planets we discover can be more massive than our measurements suggest; we are able to find only a minimum value to their masses (Fig. 9–8*b*).

In the first few years of exoplanet discovery, this problem left the nagging question of whether the objects were really planets or merely low-mass companion stars. They might even be objects called "brown dwarfs," which have between about 10 and 75 times Jupiter's mass, not quite enough to make it to "star status" (see our discussion in Chapter 12, and in Section 9.2c below). Nevertheless, most astronomers believed these objects are planets, because there is a large gap in mass between them and low-mass stars. Very few intermediate-mass objects (10–75 Jupiter masses) had been found, yet they should have been easily detected if they existed. This gap suggested that the new objects are much more numerous than brown dwarfs.

The discovery, in 1999, of a system of three planets around the star Upsilon Andromedae clinched the case that at least most of the objects are planets. Other multiple-exoplanet systems were found thereafter. It seems most unlikely that a system

■ **FIGURE 9–9** An artist's conception of the system of three exoplanets around Gliese 876, a cool, red star that is 15 light-years away from Earth in the direction of the constellation Aquarius. The smallest exoplanet has only 7 to 8 times the Earth's mass and orbits the star every two days. Since it is so close to the star, it probably doesn't have a very thick atmosphere like our Solar System's giant planets; instead, it is probably rocky (with an iron core) like our Sun's terrestrial planets.

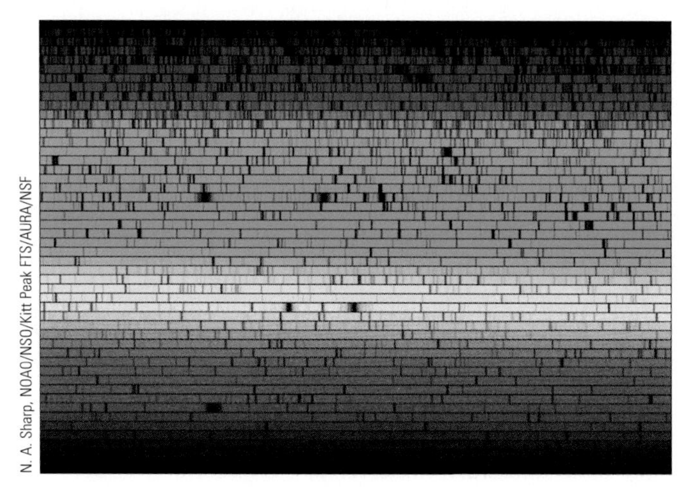

■ **FIGURE 9–10** A stellar spectrum, so spread out (by the telescope's spectrograph) that it appears in strips. The detailed measurements that led to the planet's detection were made possible by recent improvements to the spectrograph on the Keck-I telescope.

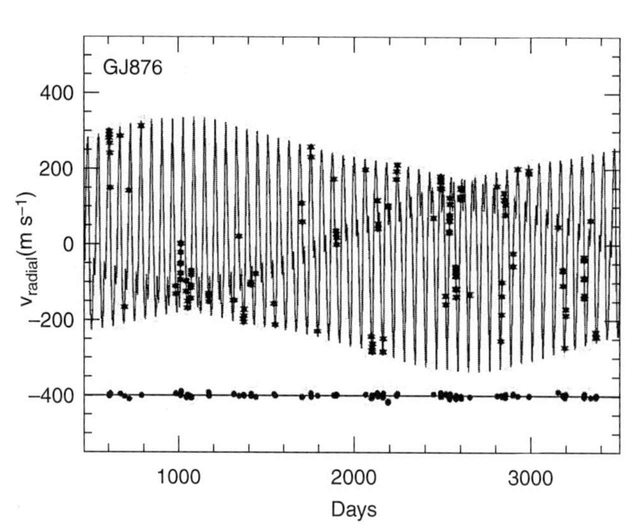

■ **FIGURE 9–11** The actual velocity measurements over time of the star, Gliese 876, whose rocky exoplanet has only about 7 to 8 times the Earth's mass. The measurements also reveal the masses of the two outer planets to be 2 and 0.5 times Jupiter's mass, respectively, where Jupiter's mass is about 318 times Earth's mass. The points are fit by the wobbling line; the newly discovered, low-mass exoplanet causes the small, very frequent wiggles. At the bottom, near velocity -400 km/sec, a straight line shows the tiny deviations of the points from the curve; the fit is nearly perfect. (Geoffrey W. Marcy, R. Paul Butler, Debra Fischer, Steven S. Vogt, Jack J. Lissauer, and Eugenio J. Rivera)

would have formed with four closely spaced stars (or brown dwarfs) in it, while a system with one star and three planets is reasonable. One of Upsilon Andromedae's planets even has an orbit that corresponds to Venus's in our Solar System, not as elliptical as the orbits of the planets around other stars. This planet is in a zone that may be not too hot for life nor too cold for it. Though such a massive planet would be gaseous, and so not have a surface for life to live on, it could have a moon with a solid surface, just as the giant planets in our Solar System have such moons.

The exciting announcements continue. We now know of several systems that each contain a few planets. Our methods are still not sufficiently sensitive to find "minor" bodies like our own Earth. However, a considerable advance was announced in 2005: the existence of a planet whose mass is as low as only 7 to 8 times Earth's (■ Fig. 9–9), by a group of scientists including Geoff Marcy and Paul Butler. This exoplanet orbits a star, known as Gliese 876 or GJ876, that is only 15 light-years from Earth. Astonishingly, the planet orbits its parent star with a period of only two days, meaning that it is only 1/50 of an A.U. out, just 10 times the star's radius. It is so close to its star's surface that its temperature is probably 200°C to 400°C, so it isn't a candidate for bearing life as we know it. Being so hot, yet having relatively low mass and thus moderately weak gravity, it could not have retained a lot of gas. It is therefore apparently the first rocky, terrestrial-type planet ever found orbiting another star.

Finding this planet required improving the Keck telescope's system for detecting small Doppler shifts (■ Fig. 9–10). The detailed observations (■ Fig. 9–11) allowed a computer model to account for the angle at which we are viewing the system, which turns out to be inclined to our line of sight by 40°. That measurement allowed the mass of the orbiting planets themselves to be determined, and not merely lower limits as before. The measurements revealed that the slight discrepancy in the orbits of the two already-measured planets could be resolved by the presence of the third body, the newer exoplanet.

9.2a(iv) Transiting Planets

Since 1999, astronomers have not had to resort only to periodically changing Doppler shifts to detect exoplanets. One planet was discovered to have its orbit aligned so that

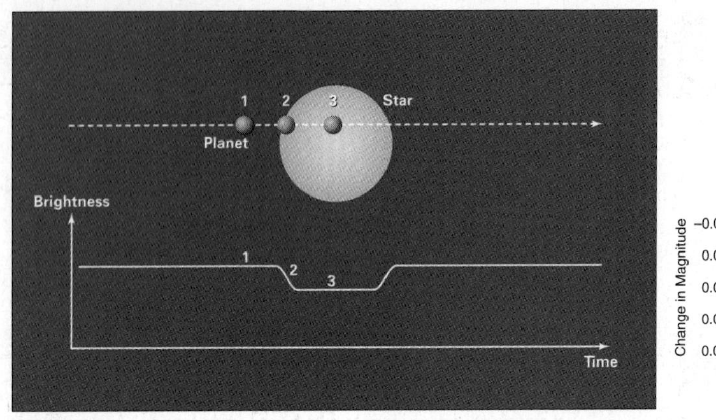

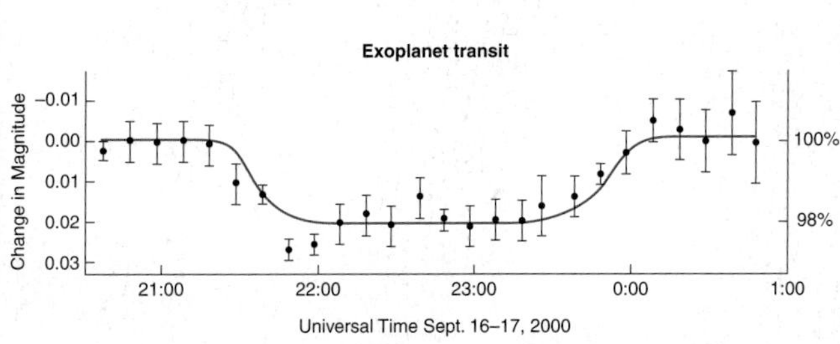

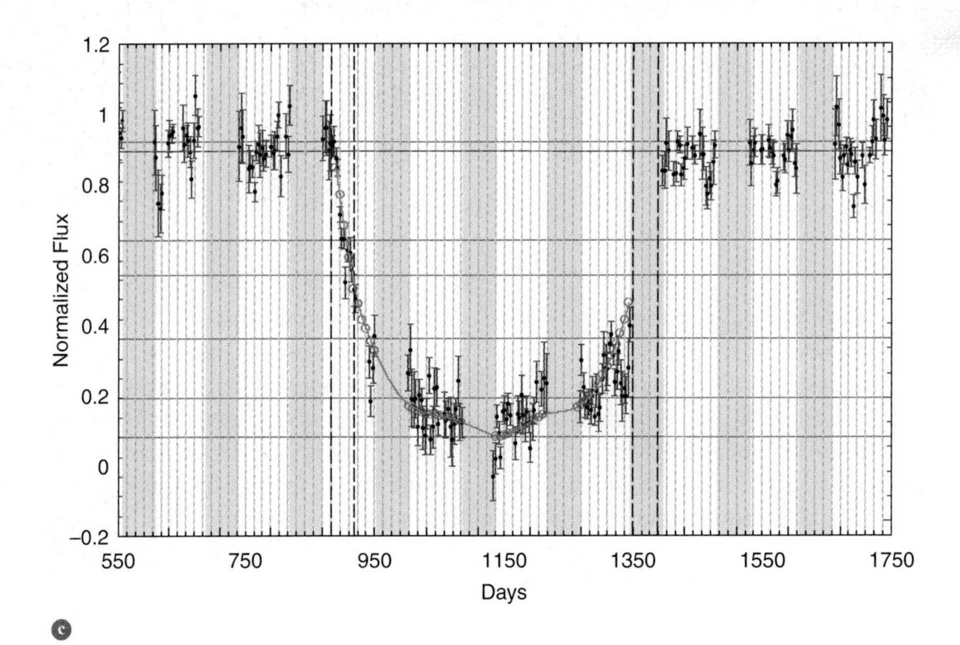

■ **FIGURE 9–12** ⓐ Schematic showing the dips that would result from an extra-solar planet. ⓑ Observations made by amateur astronomers detecting the 2-hour-long passage of an extra-solar planet across its parent planet by a 2% dip in the measured brightness. ⓒ The light curve of the transit of Venus on June 8, 2004, measured with NASA's ACRIMsat spacecraft in Earth's orbit. (b: Arto Oksanen, Nyrölä Observatory, Astronomical Association Jyväsklän Sirius, Finland, from *Sky & Telescope*, January 2001; c: Jay M. Pasachoff, Glenn Schneider, and Richard Willson (NASA))

the planet went in front of the star each time around—that is, it underwent a **transit.** The dip in the star's brightness of a few per cent can be measured not only by professional but also by amateur astronomers (■ Fig. 9–12).

Since the planet's orbital plane is along our line of sight, its mass can be accurately determined, and this turns out to be 63 per cent of Jupiter's mass. This result confirms our conclusion that at least some (and probably most) of the "exoplanets" really are planets rather than more massive brown dwarfs.

The transit method has also revealed, through spectroscopy, sodium in the exoplanet's atmosphere. The atmosphere produced some absorption lines in the starlight passing through it, and these were detected in very high-quality spectra. In the future, it is possible that astronomers will detect evidence for life on other planets by analyzing the composition of their atmosphere using the transit method.

During the past few years, several additional examples of transiting planets have been found. Many more exoplanets will be discovered in this way from the ground and from space. This method is analogous to observing the transit of Venus across the disk of our Sun (Fig. 9–12*c*).

Progress in the search for exoplanets is so rapid that you should keep up by looking at the Extrasolar Planets Encyclopaedia at http://www.obspm.fr/planets and the Marcy site at http://exoplanets.org, both linked through this book's web pages.

9.2b The Nature of Exoplanet Systems

We have discovered enough exoplanets to be able to study their statistics. About 1 per cent of nearby solar-type stars have jovian planets in circular orbits that take between 3 and 5 Earth days. These are sometimes called "hot Jupiters," since they are so close to their parent stars (within 1/10 A.U.) that temperatures are very high. Another 7 per cent of these nearby stars have jovian planets whose orbits are very eccentric. As we observe for longer and longer periods of time, we have better chances of discovering planets with lower masses or with larger orbits.

The discovery of several planetary systems instead of just our own will obviously change the models for how planetary systems are formed. Since giant planets couldn't form in the torrid conditions close in to the parent star, theorists work from the idea that the exoplanets formed far out, as Jupiter did, and migrated inward. Thus it follows that these planets may be jostled loose from their orbits and put in orbits that bring them closer to their parent stars. Maybe their orbits shrink as the planets encounter debris in the dusty disk from which they formed, or shrink along with an overall swirling inward of the whole disk of orbiting material.

Perhaps the highly elliptical orbits didn't start out that way, but were produced by gravitational interactions that completely ejected some planets from the system. The gravitational interactions between two planets can lead to the ejection of one planet, leaving the other in a very eccentric orbit. Another idea is that interactions between a planet and the protoplanetary disk can cause high eccentricities. Once a planet has part of its orbit close to its parent star, tidal forces can circularize the orbit. Other planets can spiral all the way into the star and be destroyed.

The most accepted explanation for the hot Jupiters, which orbit so close to their stars, is that they were formed farther out and migrated in. But the 2005 discovery of a hot Jupiter in a triple star system complicates matters, since the two farther-out stars would have disrupted any protoplanetary disk. So this discovery is being interpreted as evidence against the migration model. Perhaps that model was partly based on residual prejudice that our own Solar System's outer giant planets are normal. Perhaps the triple system had a very different type of protoplanetary disk than the one with which we are familiar from our own system, or the planet was captured.

ASIDE 9.2: Comparative planetology

Our Solar System: Massive planets in almost circular orbits beyond 5 A.U., with orbital periods of many years.

Other Planetary Systems: Massive planets often in very eccentric orbits much smaller than 1 A.U. and with orbital periods of months, or in circular orbits with periods of days.

9.2c Brown Dwarfs

As mentioned in Section 9.2a(iii), some of the objects found with the Doppler shift technique might actually be too massive to be true "planets" (more than 13 Jupiter masses). If so, they are **brown dwarfs,** which are in some ways "failed stars."

Each of the brown dwarfs has less than 75 Jupiter masses (or 7.5 per cent the mass of the Sun), which is not enough for them to become normal stars, shining through sustained nuclear fusion of *ordinary* hydrogen, as we shall discuss in Chapter 12. Their central temperatures and pressures are just not high enough for that. (However, they do fuse a heavy form of hydrogen known as "deuterium," so they are not complete failures as stars. We discuss the origin of deuterium, all of which was formed in the first few minutes after the Big Bang, in Chapter 19.) Brown dwarfs can be thought of as the previously "missing links" between normal stars and planets.

Though no detection of a brown dwarf was accepted until 1995, hundreds have now been observed. Many of these are in the Orion Nebula, while others are alone in space. The current best model for brown dwarfs is that they are formed similarly to the way that normal stars are: in contracting clouds of gas and dust. A disk forms, perhaps even with planets in it, though the material in it contracts onto the not-quite-star.

This idea is backed up by observations with the European Southern Observatory's Very Large Telescope, which has detected an excess of near-infrared radiation from many brown dwarfs. The scientists involved interpret their observations as showing

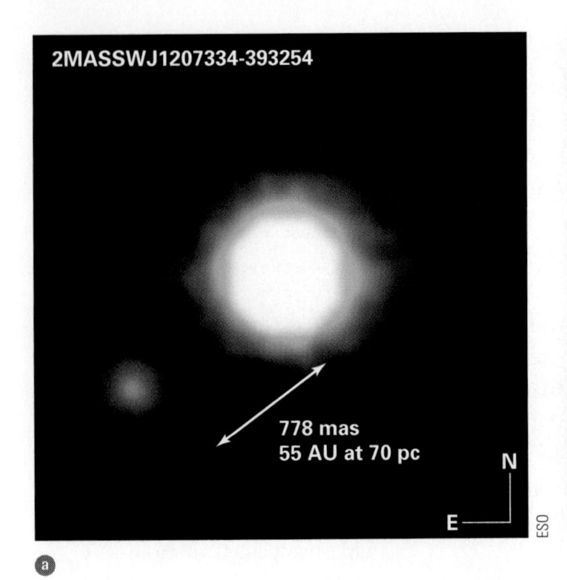

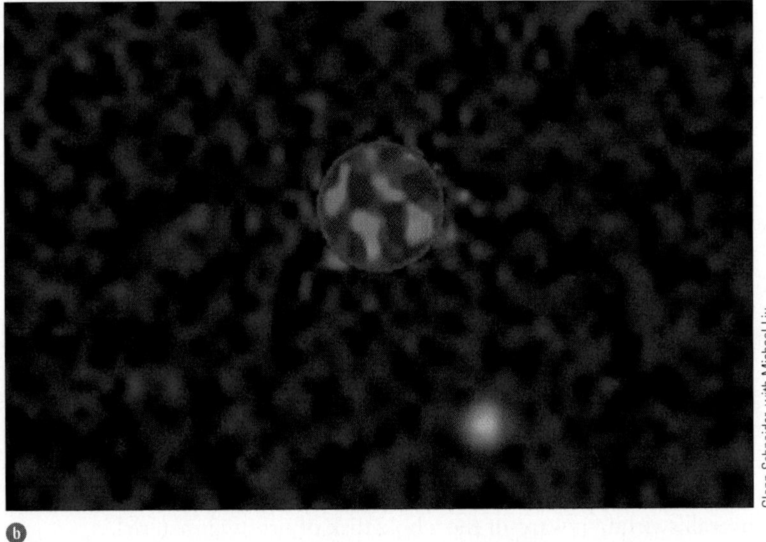

■ **FIGURE 9–13** The first images ever of an exoplanet. This giant object orbits a brown dwarf at a distance of 55 A.U. (a) An image taken with the Very Large Telescope in Chile. (b) A Hubble Space Telescope view, with the central circle showing where the star's relatively bright image is hidden.

that the radiation is from dusty disks. Further, they conclude that since both regular stars and brown dwarfs have such disks, they must also form in similar ways.

Most exciting, a planet next to a brown dwarf (presumably orbiting it) has been imaged (■ Fig. 9–13), the first planet to be imaged around any star. The star with its planet is 200 light-years away from us; observations from the ground and from space have shown that the two objects are moving through space together. The planet is 55 A.U. away from its parent star.

9.2d Future Discovery Methods

NASA's Kepler mission, planned for a 2008 launch, is to carry a 1-m telescope to detect transits of planets across stars. To do so, it will continuously monitor the brightness of 100,000 stars. The 2004 transit of Venus served as an analogue to the type of thing Kepler will study: One of this book's authors (J.M.P.) and colleagues reported on the dip in the total amount of sunlight reaching Earth because Venus was blocking about 0.1 per cent of the Sun's disk (Fig. 9–12*c*).

No telescope now in space has enough resolution to directly detect (image) a planet closely orbiting a normal star (rather than a brown dwarf as described in the preceding section). It will take an interferometry system, with two widely separated telescopes working together, to make such a detection. Installation of interferometric equipment at the two 10-meter Keck telescopes in Hawaii and at the four 8-meter units of the Very Large Telescope in Chile is nearing completion, which should lead to direct detection of some young, giant planets that are luminous at infrared wavelengths.

Though we have mainly detected giant planets, we are now beginning to detect small planets in those systems. To do even better, NASA's Space Interferometry Mission, now called SIM PlanetQuest (■ Fig. 9–14*a*), is on the drawing board, but it will not be launched until at least 2011. NASA's Terrestrial Planet Finder is to be able to image small planets and is slated for launch in the third decade of this century. In 2005, both were delayed by an unfortunate shift in NASA's priorities.

For these missions, NASA is currently examining two approaches to high-contrast imaging. The first would use a huge telescope for direct imaging, though blocking the

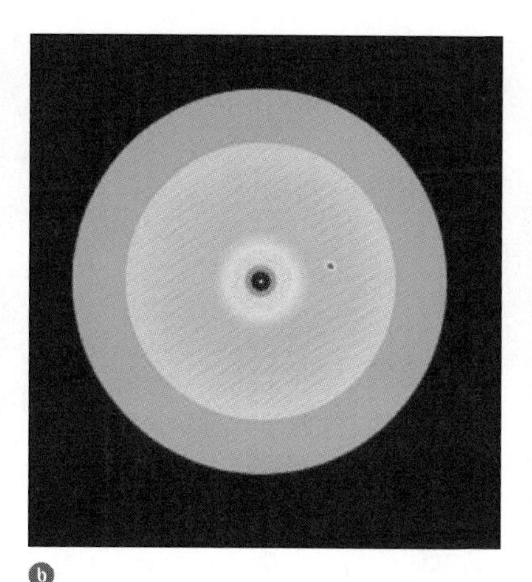
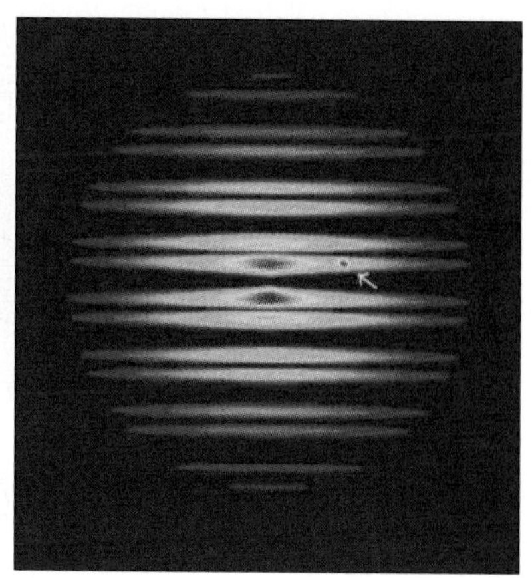

NASA/JPL/Caltech; courtesy of Chris Lindensmith, Stephen Unwin, Jim Marr, Gene Serabyn, and Peter Kahn

■ **FIGURE 9–14** ⓐ An artist's conception of SIM PlanetQuest, formerly called the Space Interferometry Mission. ⓑ In a technique called "nulling," an interferometer makes images in which part is sensitive and part is insensitive. The star itself *(modelled at left)* is always placed on an insensitive part, reducing its brightness in the image *(as at right)* from being so overwhelming. In this simulated image, the star is at the center, with the peak of its brightness not showing because of the central dark band. Only bits of the star's brightness are visible, on the bright bands just above and below the central dark band. The model exoplanet appears to the right of the star *(arrow)*. This technique should make the existence of planets more obvious.

starlight itself to reveal accompanying planets. The second would use several infrared telescopes flying in formation and coupled to form a "nulling interferometer," in which the response at the stellar position is minimized (Fig. 9–14*b*). NASA is considering an earlier, smaller, less expensive mission in its Discovery class of spacecraft to try out the technologies.

The European Space Agency (ESA) also plans spacecraft to find exoplanets. But again, funding reasons have delayed or abandoned their plans. In 2004, they cancelled their Eddington mission to search for Earth-like systems by looking for their transits. Their Gaia mission is to measure positions for a billion stars, and it may discover 10,000 planets! It should be launched by 2012. ESA's Darwin mission is to analyze the atmospheres of Earth-like planets to search for signs of life, but not until at least 2015. It is to consist of a flotilla of three 3-m telescopes on spacecraft in formation.

Also, thousands of stars are being monitored to see if their gravity focuses and brightens the light from other stars behind them, a process called "gravitational lensing" that we will discuss in Chapters 16 and 17. The hope is that not only a background star will brighten equally in all colors over a year or more but also, during that time, a planet will cause an increase in brightness for a day or so. Worldwide groups are searching for such "microlensing" events.

ASIDE 9.3: Exoplanet microlensing

Some candidate events have been found, including one detection of a planet with only about five Earth masses, but more work is necessary before they will be widely accepted as true evidence of exoplanets.

AceAstronomy™ Log into AceAstronomy and select this chapter to see the Astronomy Exercise "Extra-solar Planets."

9.3 Planetary Systems in Formation

We are increasingly finding signs that planetary systems are forming around other stars. One of the first signs was the discovery of an apparent disk of material around a southern star in the constellation Pictor (■ Fig. 9–15). The best observations of it were made with the Hubble Space Telescope, and may show signs of orbiting planets. An even nearer planetary disk has been found, enabling observations with higher resolution (■ Fig. 9–16).

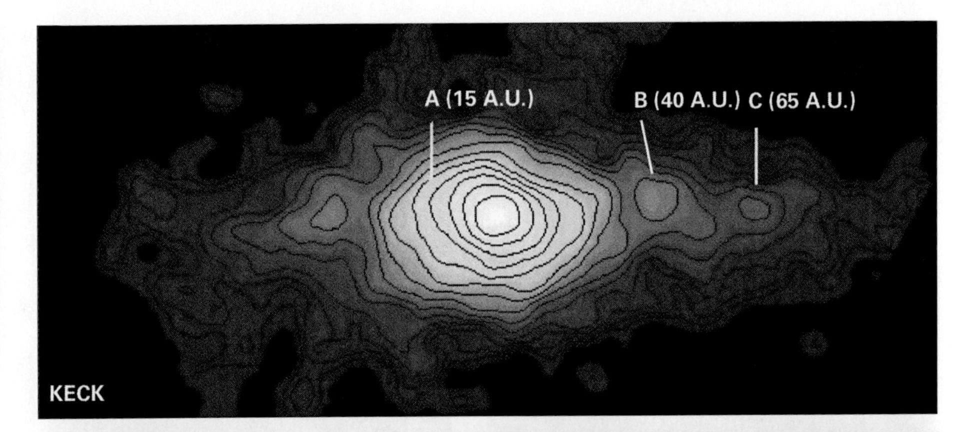

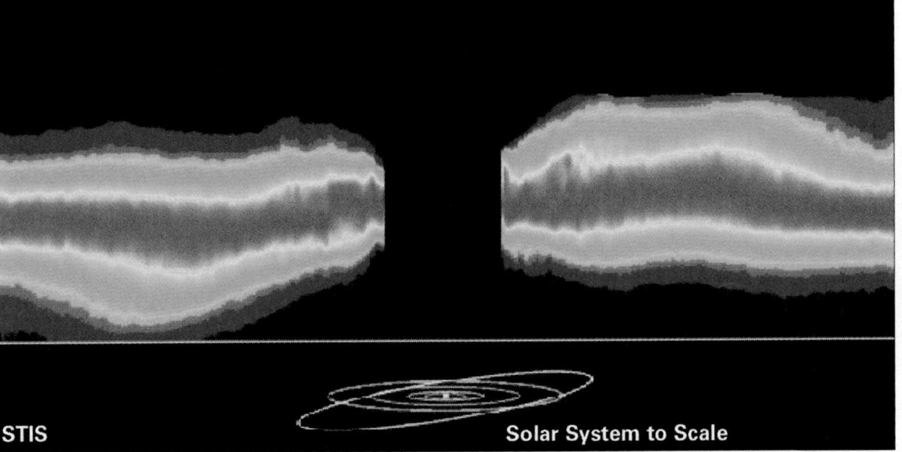

■ **FIGURE 9–15** ⓐ This picture of what may be a planetary system in formation was taken with a ground-based telescope. A circular "occulting disk" blocked out the star β (beta) Pictoris; shielding its brightness and observing in the infrared revealed the material that surrounds it. ⓑ The region around the southern star β Pictoris. *(top)* A view of the dust from the Keck-II telescope at the medium infrared wavelength of 18 micrometers. The inner contours are misaligned with respect to the outer disk; this warp is labelled A. Labels B and C may show the position of tilted dust rings. *(middle)* A Hubble view with its Space Telescope Imaging Spectrometer. We apparently see a side view of a disk of dust, thought to be a planetary system in formation. *(bottom)* Our Solar System to the same scale as the β Pictoris images.

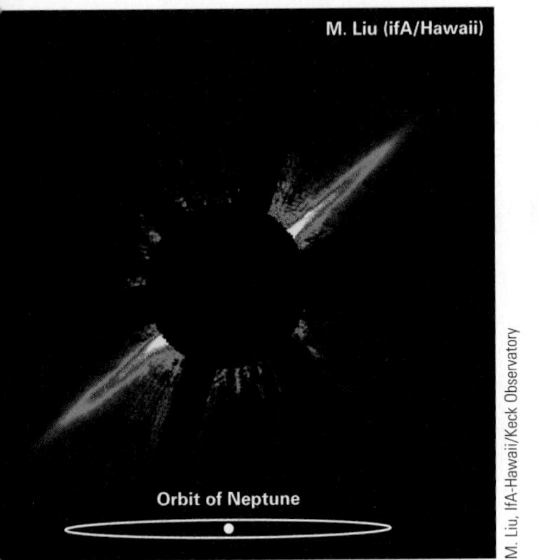

■ **FIGURE 9–16** The star AU Microscopii is closer to Earth than β Pictoris, so even more detail can be observed in the disk that has been discovered around it. This Keck-II image shows the disk's dust particles reflecting the starlight. The image is 100 A.U. wide; the black mask hides the inner 15 A.U. Features as small as 0.5 A.U. are resolved on the image, revealing clumps of dust in the inner disk. These clumps are presumably formed by the gravitational influence of planets, though we cannot see the planets themselves. The clumps are 25 to 40 A.U. from the central star, corresponding for our own Sun to the orbits of Neptune and Pluto. AU Microscopii itself, only 33 light-years away from us, is a dim red star, with half the Sun's mass and 1/10 its energy output. It is 12 million years old, much younger than the Sun's 4.6 billion years, and is apparently in its phase of planetary formation.

Other images of regions in space known as "stellar nurseries" show objects that appear to be protoplanetary disks (■ Fig. 9–17). Observations reported in 2005 show that these objects contain about as much mass as a planetary system, clinching the idea that they are locations where planets are forming. Locations where there are planets in formation glow in the infrared (■ Fig. 9–18), because of the wavelength of the peak of the black-body curves for the temperature of warm dust (see our discussion of black bodies in Chapter 2); thus, the Spitzer Space Telescope and, much later, the Webb Space Telescope, should give many insights into planetary formation. The Hubble Space Telescope has imaged such a ring of dust around the nearby star Fomalhaut (■ Fig. 9–19), recording signs that a planet is tugging on it gravitationally.

A Wesleyan University team has found a young Sun-like star that winks on and off, apparently being eclipsed by dust grains, rocks, or asteroids that are orbiting it in a clumpy disk. The star is near the Cone Nebula, a prolific nursery of young stars. It fades drastically over 2.4 days to only 4 per cent of its maximum brightness, stays dim for another 18 days, and then brightens for another 2.4 days out of every 48.3 days. No single object could provide such a long eclipse, so only an orbiting collection of smaller objects seems to match the observations. The actual eclipse could be from a wave of gas and dust triggered by the masses of these objects. The system, only 3 million years old, is changing from month to month.

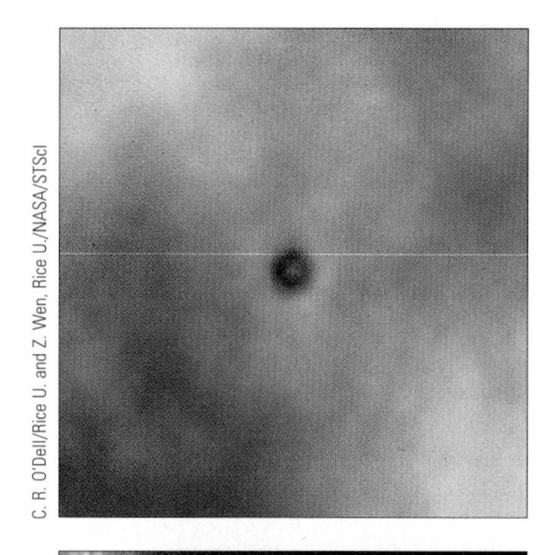

C. R. O'Dell/Rice U. and Z. Wen, Rice U./NASA/STScI

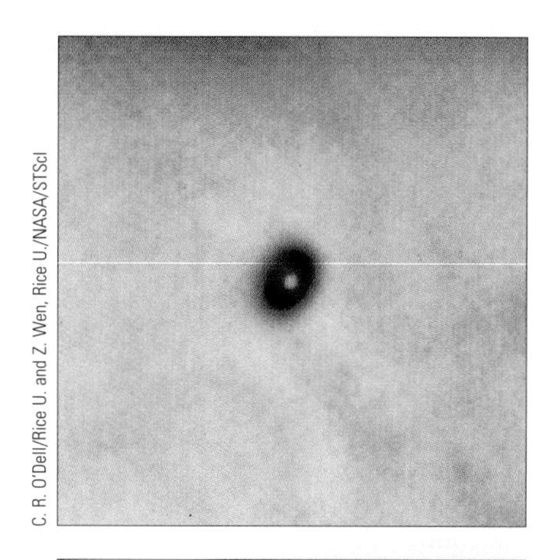

C. R. O'Dell/Rice U. and Z. Wen, Rice U./NASA/STScI

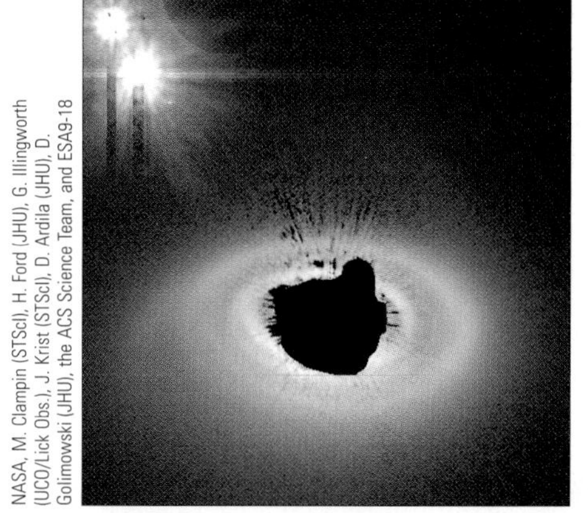

NASA, M. Clampin (STScI), H. Ford (JHU), G. Illingworth (UCO/Lick Obs.), J. Krist (STScI), D. Ardila (JHU), D. Golimowski (JHU), the ACS Science Team, and ESA9-18

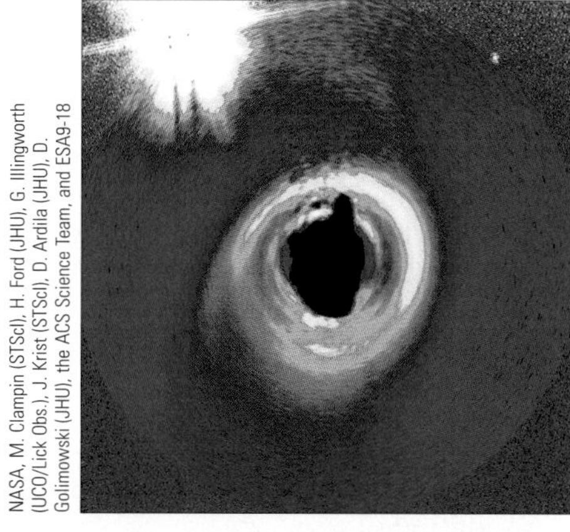

NASA, M. Clampin (STScI), H. Ford (JHU), G. Illingworth (UCO/Lick Obs.), J. Krist (STScI), D. Ardila (JHU), D. Golimowski (JHU), the ACS Science Team, and ESA9-18

■ **FIGURE 9–17** *(top)* The Hubble Space Telescope has revealed protoplanetary disks around several stars, such as these in the Orion Nebula in the constellation Orion. *(bottom)* Detail in the circumstellar dust disk around a young star, seen in black and white and in false color. The star, only 5 million years old, is 320 light-years from us and is in the constellation Libra. It has long been known to be a candidate for having a disk because of its excess infrared emission. The Hubble Space Telescope's Advanced Camera for Surveys, using its coronagraph that has a dark spot to block out the bright central region, shows that the disk is a tightly wound spiral. One of the spiral arms seems to be associated with a nearby double star *(upper left edge of frame)*, so interaction with the double star may be causing the structure. It had previously been known that the inner 3 billion km in radius around the star, about 20 A.U. (roughly the orbit of Uranus, for our Solar System), is relatively clear of dust. Perhaps that region was swept clear by a planet or planets.

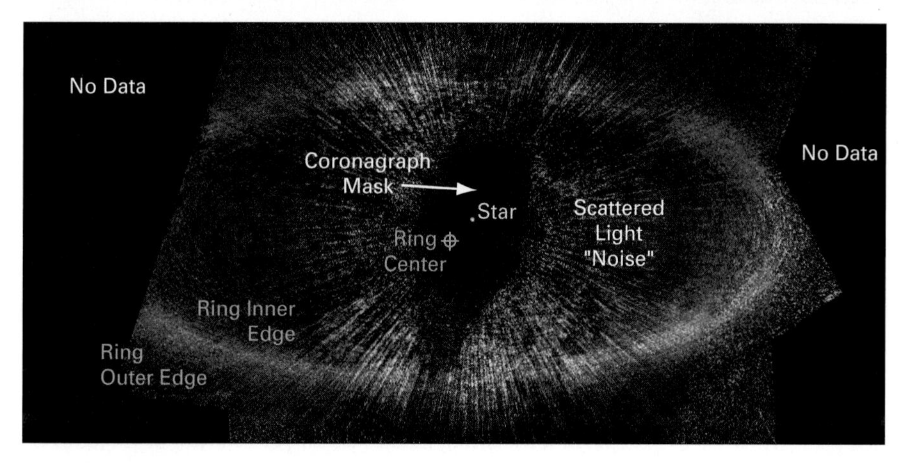

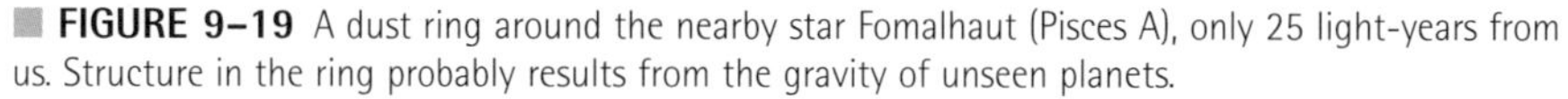

■ **FIGURE 9–19** A dust ring around the nearby star Fomalhaut (Pisces A), only 25 light-years from us. Structure in the ring probably results from the gravity of unseen planets.

Michael Liu and Glenn Schneider

■ **FIGURE 9–18** An almost edge-on dust ring around the star HR4796A. It is thought that planets are forming, or will soon be forming, in this disk.

CONCEPT REVIEW

The conservation of angular momentum explains why planetary systems in formation contract into a disk (Sec. 9.1a). We think that small clumps of dust joined to make **planetesimals,** and planetesimals combined to make **protoplanets** orbiting the **protosun.** In one of the main models for the formation of the Solar System's outer planets, the solar nebula first collapsed into several large blobs, which then became the outer planets (Sec. 9.1b). In another model, a solid core (resembling a terrestrial planet) condensed first for each of the outer planets. The gravity of this core then attracted the gas from its surroundings.

The search for planets around stars other than the Sun, known as **extra-solar planets** or **exoplanets,** has been going on for many decades (Sec. 9.2). They are extremely difficult to see in the glare of their "parent star" (the star they orbit) with current technology (Sec. 9.2a). Instead, we watch for motions in the star that are caused by something orbiting it.

The **astrometric method** depends on **astrometry,** the precise measurement of stellar positions and motions; the presence of an unseen planet is deduced from the star's observed wobble in the sky (Sec. 9.2a(i)). However, so far no exoplanets have been discovered this way. Some planets were finally found in 1991 by timing the radio pulses from a "pulsar," a weird kind of collapsed star to be discussed in Chapter 13, but these are very unusual planets that have little to do with normal planets orbiting Sun-like stars (Sec. 9.2a(ii)).

Most exoplanets have been discovered since 1995 using the Doppler shift technique (Sec. 9.2a(iii)), in which periodic changes in the radial velocity of a star (the subtle "reflex motion" caused by the orbiting planet) are measured with high-quality spectrographs. One limitation of the method is that we generally don't know the angle of the plane in which the planets are orbiting their parent stars; thus, only the minimum mass of each planet can be derived.

This problem left the nagging question of whether the objects were really planets or merely low-mass companion stars. The discovery of a three-planet system helped dispel these concerns, as did the large observed gap in mass between the purported exoplanets and the lowest-mass stars. Now several such planetary systems are known, including one in which the planet appears to be rocky (rather than a gas giant) and has a mass of only 7 to 8 Earth masses.

A few planets have been observed when they go in **transit** across the face of their parent stars (Sec. 9.2a(iv)). The total amount of light detected from the star drops by a very small amount during the transit. In one case, the transit method has revealed, through spectroscopy, sodium in the exoplanet's atmosphere.

All the exoplanets discovered so far are giant planets, though terrestrial planets probably remain undiscovered in those planetary systems. The planets are often in quite eccentric (elongated) orbits, perhaps as a result of previous gravitational encounters with other planets or the protoplanetary disk (Sec. 9.2b). Some of the planets are in circular orbits very close to their parent stars, which may mean that they were formed elsewhere in the systems and later drifted inward. It remains to be seen whether our Solar System is typical of others.

Some of the objects found with the Doppler shift technique might actually be **brown dwarfs,** which are sometimes called "failed stars" because they don't fuse ordinary hydrogen into helium (Sec. 9.2c). They can be thought of as links between planets and normal stars. Recently, a planetary companion to one brown dwarf has been imaged at infrared wavelengths.

In the future, we expect many transiting planets to be discovered with the Kepler spacecraft (Sec. 9.2d). Moreover, interferometers on the ground and in space should improve our detection capabilities, especially as the technique of "nulling" is improved.

We are increasingly finding signs that planetary systems are forming around other stars (Sec. 9.3). A probable disk around the star β (beta) Pictoris and apparent protoplanetary disks observed with the Hubble Space Telescope seem to be direct observations of planetary systems in formation.

QUESTIONS

1. What is the difference between a protoplanet and a planet?
2. What role do planetesimals play in the origin of the planets?
3. Discuss the choice between models for the formation of the outer planets.
4. How does the pulsar-planet system differ from the exoplanet systems discovered subsequently?
5. Explain the method by which most planets have been discovered around other stars.
6. Does the Doppler-shift (radial-velocity) method for deducing the existence of a planet orbiting a star depend on the star's distance from Earth? (Assume the star's apparent brightness is independent of distance.)
7. †It turns out that the ratio of the speed of a planet to the speed of its parent star (as they orbit their common center of mass) is equal to the inverse of the ratios of their masses. If Earth orbits the Sun with a speed of 30 km/sec, what is the Sun's corresponding orbital speed (i.e., its "reflex motion," induced by the Earth's motion)?
8. Why may it be that so far only giant planets and not terrestrial planets have been discovered around other stars?
9. Why can we measure only minima for the mass of an exoplanet in most cases?
10. Many of the giant extra-solar planets are quite close to their parent stars. If they initially formed farther out, how could they have ended up at their observed positions?
11. What was the special importance of the 1999 discovery of a system of several (rather than just one) planets orbiting a star?
12. What are two reasons that the discovery of a planet transiting a Sun-like star is desirable?
13. †If a transiting planet has 10 per cent the diameter of its parent star, the observed brightness of the star will dim by what percentage during the transit?
14. Give two reasons that the recent (mid-2005) discovery of a 7 to 8 Earth-mass planet is important.
15. How do brown dwarfs differ from planets and from stars?
16. Why is β (beta) Pictoris an interesting object?
17. †If the temperature of a dust ring around a star is 30 K, at what wavelength does its black-body spectrum peak? In what spectral region is this wavelength?
18. Discuss how the Kuiper belt in our Solar System may be related to some of the observed phenomena found around other stars.
19. Why do infrared observations help us learn about planetary systems in formation?

20. Describe the basic idea of "nulling interferometry" and its applications to extra-solar planets.
21. **True or false?** Many of the known extra-solar planets are peculiar compared with planets in our own Solar System; these exoplanets have very eccentric orbits, or they orbit very close to the star.
22. **True or false?** Exoplanets can sometimes be detected by monitoring the brightness of a star, and seeing it decrease as the planet blocks part of the star's light.
23. **True or false?** The apparent observed absence of low-mass, Earth-like exoplanets orbiting normal stars is surprising; with current technology, we should have been able to detect them, if they exist.
24. **True or false?** Many exoplanets have already been found with the astrometric method, in which a star's measured position in the sky moves back and forth very slightly.
25. **True or false?** The periodic Doppler-shift method is biased toward finding planets whose orbital plane is along the line of sight, rather than closer to the plane of the sky.
26. **Multiple choice:** In most cases, astronomers currently find massive planets around distant stars by (**a**) looking at the massive planet's Doppler-shifted light as it orbits the star; (**b**) taking spectra of the star to look for contamination from the light of a massive planet; (**c**) looking for a periodic Doppler shift in the star's spectrum as a massive planet causes the star to wobble slightly; (**d**) pointing the Hubble Space Telescope to a spot about 5 to 10 A.U. away from the star, knowing that massive planets can only orbit near that distance; or (**e**) measuring the Doppler shift of the star, and seeing whether it keeps on increasing with time, as would be expected if a massive planet were pulling on it.
27. **Multiple choice:** The giant (jovian) planets are large compared with the terrestrial planets because the giant planets (**a**) have higher densities; (**b**) sweep up material from a much longer orbital path; (**c**) have many moons; (**d**) obtained and retained more gas and ice due to their large distance from the Sun; or (**e**) did not suffer as many major, destructive collisions as the terrestrial planets (like Earth, when the Moon formed).
28. **Multiple choice:** Jupiter's chemical composition is most similar to that of (**a**) Earth; (**b**) the Sun; (**c**) Pluto; (**d**) Venus; or (**e**) meteorites.
29. **Multiple choice:** Since 1995, at least 160 extra-solar planets around normal main-sequence stars have been detected with ground-based optical telescopes. Which one of the following statements about these planets is *true*? (**a**) All of them are less massive than Jupiter. (**b**) In some cases, the planet's mass is as small as Earth's mass. (**c**) The measured masses of the planets are generally only upper limits; that is, their true masses might be smaller. (**d**) In some cases, they are so close to their parent stars that they complete a full orbit in only a few days. (**e**) They all have circular or nearly circular orbits.
30. **Fill in the blank:** Objects having low masses compared with normal stars, but higher masses than planets, are known as _______, and are sometimes called "failed stars."
31. **Fill in the blank:** Theories of the formation of the Solar System suggest that planets formed through the accumulation of _______, much smaller bodies.
32. **Fill in the blank:** A planet and its parent star orbit their common _________, though it is far closer to the star than to the planet.
33. **Fill in the blank:** In 2004, astronomers measured a slight dimming of the Sun when ______ transited across it.
34. **Fill in the blank:** A spinning, collapsing cloud of gas and dust forms a disk because of the conservation of ___________.

†This question requires a numerical solution.

TOPICS FOR DISCUSSION

1. Consider the properties of the various types of exoplanets found thus far. Do you think life as we know it would be common or rare in these different cases? Might life be present on some of the moons of these exoplanets?
2. In your opinion, how important is it to spend Federal funds on the development of spacecraft that could detect Earth-like planets around other stars?

MEDIA

Virtual Laboratories

Extra-solar Planets

AceAstronomy™ Log into AceAstronomy at **http://astronomy.brookscole.com/cosmos3** to access quizzes and animations that will help you assess your understanding of this chapter's topics.

Log into the Student Companion Web Site at **http://astronomy.brookscole.com/cosmos3** for more resources for this chapter including a list of common misconceptions, news and updates, flashcards, and more.

CHAPTER 10

Our Star: The Sun

Not all stars are far away; one is close at hand. By studying the Sun, we not only learn about the properties of a particular star but also can study processes that undoubtedly take place in more distant stars as well. We will first discuss the **quiet Sun,** the solar phenomena that typically appear every day. Afterward, we will discuss the **active Sun,** solar phenomena that appear nonuniformly on the Sun and vary over time.

We study the Sun from Earth's surface not only from telescopes permanently set up on mountaintops and elsewhere but also from temporary observation sites set up to observe total solar eclipses (■ Fig. 10–1). Further, several important spacecraft send back varied and high-resolution images of the Sun from outside our atmosphere, allowing us to study aspects of the Sun and parts of its spectrum that were previously unavailable to us.

In this chapter, we discuss mainly the outer layers of the Sun, which are visible to us. In Chapter 12, we will discuss the deep-down source of energy of the Sun and of the other stars—nuclear fusion in the core, where temperatures and pressures are extremely high.

ORIGINS

The Sun, the source of light and heat on Earth, is critical to the existence of humans. Its properties are typical of stars.

AIMS

1. Learn about the structure of our Sun, the only star we can study in detail (Sections 10.1 and 10.2).
2. Describe sunspots, flares, prominences, and other solar phenomena that vary with the 11-year solar-activity cycle (Section 10.2).
3. See how Einstein's general theory of relativity, first verified at a total solar eclipse, is a theory of gravity (Section 10.3).

10.1 What Is the Sun's Basic Structure?

We think of the Sun as the bright ball of gas that appears to travel across our sky every day. We are seeing only one layer of the Sun, part of its atmosphere. The layer that we see is called the **photosphere** (■ Fig. 10–2), which simply means the sphere from which the light comes (from the Greek *photos,* "light"); it is typically called the Sun's "surface," though any object placed there would fall through it, of course, if it didn't burn.

AceAstronomy™ The AceAstronomy icon throughout this text indicates an opportunity for you to test yourself on key concepts, and to explore animations and interactions of the AceAstronomy website at http://astronomy.brookscole.com/cosmos3

ASIDE 10.1: Why the Sun shines

In this chapter, we discuss what the Sun looks like and how we study it. In Chapter 12, we discuss why the Sun shines.

The Sun in various wavelengths. The central image is a composite of 3 ultraviolet wavelengths, high-resolution images from the Transition Region and Coronal Explorer (TRACE) spacecraft. The filters used show us only the hottest gas, at millions of degrees. We clearly see how the coronal gas is shaped into loops, which are kept in that form by magnetic fields. The smaller images are, clockwise from the top, a magnetic map, a visible-light image, five individual images with different TRACE filters, and an x-ray image.

Images from TRACE (Lockheed Martin Solar and Astrophysics Lab and NASA): 2 UV- and 3 EUV-wavelength mosaics and central mosaic, Yohkoh (Montana State University, ISAS and NASA): x-ray image, SOI/MDI on SOHO (Stanford Lockheed Institute for Space Research, NASA and ESA): white light and magnetogram. The Transition Region and Coronal Explorer, TRACE, is a mission of the Stanford-Lockheed Institute for Space Research, and part of the NASA Small Explorer program.

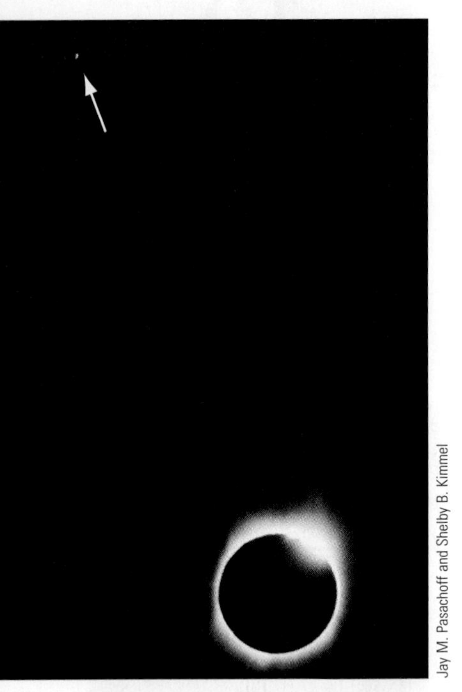

Jay M. Pasachoff and Shelby B. Kimmel

■ **FIGURE 10–1** The April 8, 2005, total solar eclipse, observed from the middle of the Pacific Ocean. We see the solar corona in the sky, surrounding the dark disk of the Moon. Venus is marked with an arrow.

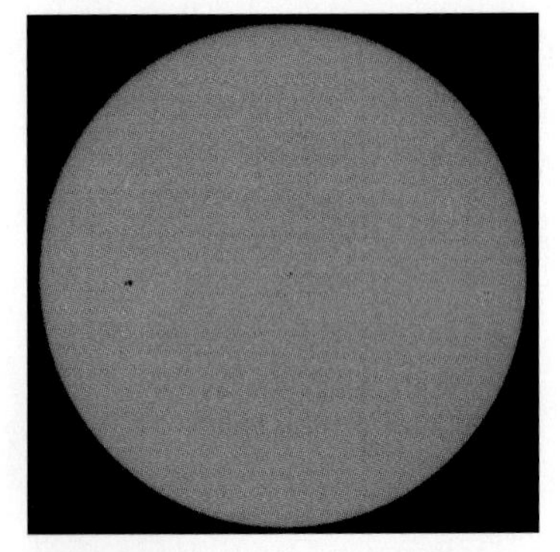

■ **FIGURE 10–2** The solar photosphere in an image taken on April 8, 2005, the day of a total eclipse.
MDI on SOHO/Stanford Lockheed Institute for Space Research, NASA, and ESA

> **ASIDE 10.2: Einstein about the Sun**
>
> Albert Einstein wrote to a sixth-grade class this answer to their question about what would happen if the Sun "burned out": "Without sunlight there is: no wheat, no bread, no grass, no cattle, no meat, no milk, and everything would be frozen. No LIFE."

The photosphere is about 110 Earth diameters across, so over a million Earths (110 cubed) could fit inside the Sun! We find the size of the Sun by knowing its distance and measuring its angular diameter. (We can now find the distance by bouncing radio waves off the Sun's outer atmosphere directly, but historically, the scale of the Solar System was first found accurately by measuring transits of Venus—when Venus is silhouetted against the Sun. Such events take place very rarely; we discussed the transit of June 2004 in Chapters 6 and 9.)

As is typical of many stars, about 92 per cent of the atoms and nuclei in the outer parts are hydrogen, just under 8 per cent are helium, and a mixture of all the other elements makes up the remaining approximately two tenths of one per cent (see *A Closer Look 10.1: The Most Common Elements in the Sun's Photosphere*). The overall composition of the Sun's interior is not very different. We basically find out about the chemical composition of the Sun and stars by studying their spectra. Recently, the hard-to-find abundance of helium has been determined most accurately with the technique of solar seismology that we discuss below (Section 10.1a).

The Sun is sometimes considered a typical star, in the sense that stars much hotter and much cooler, and stars intrinsically much brighter and much fainter, exist. (Actually, there are many more stars that are fainter and cooler than there are stars like our Sun or hotter.) Radiation from the photosphere peaks (is brightest) in the middle of the visible spectrum; after all, our eyes evolved over time to be sensitive to that region of the spectrum because the greatest amount of the solar radiation is emitted there. If we lived on a planet orbiting an object that emitted mostly x-rays, we, like Superman, might have x-ray vision. (Of course, unlike Superman, we would be only passively receiving x-rays, not sending them out.)

Note that the Sun is white, not yellow, though it is often thought of as being yellow. It only appears yellow, or even orange or red, when it is close to the horizon: The blue and green light is selectively absorbed and scattered by Earth's atmosphere. (The scattered blue light produces the color of the daytime sky, as we discussed in Chapter 4.) Also, on many photographs, the filter used to cut down the solar brightness to a safe level favors the yellow.

Beneath the photosphere is the solar **interior** (■ Fig. 10–3). All the solar energy is generated there at the solar **core,** which is about 10 per cent of the solar diameter at this stage of the Sun's life. The temperature there is about 15,000,000 K and the density is sufficiently high to allow nuclear fusion to take place, as we will discuss in Chapter 12.

The photosphere is the lowest level of the **solar atmosphere.** Though the Sun is gaseous throughout, with no solid parts, we still use the term "atmosphere" for the upper part of the solar material because it is relatively transparent.

Just above the photosphere is a jagged, spiky layer about 10,000 km thick, only about 1.5 per cent of the solar radius. This layer glows colorfully pinkish when seen during an eclipse (■ Fig. 10–4*a*), when the photosphere is hidden, and is thus called the **chromosphere** (from the Greek *chromos,* "color").

Above the chromosphere, a ghostly white halo called the **corona** (from the Latin, "crown") extends tens of millions of kilometers into space (Fig. 10–4*b*). The corona is continually expanding into interplanetary space and in this form is called the **solar wind.** The Earth is bathed in the solar wind, a stream of particles having many different speeds.

10.1a The Photosphere

The Sun is a normal star with a surface (photospheric) temperature of about 5800 K, neither the hottest nor the coolest star. (One way we know the temperature is that the Sun's visible spectrum closely resembles the black-body curve, discussed in Chapter 2, for that temperature.) The Sun is the only star sufficiently nearby to allow us to study

A Closer Look 10.1 THE MOST COMMON ELEMENTS IN THE SUN'S PHOTOSPHERE

FOR EACH			Symbol	Atomic Number	Atomic Weight
1,000,000	atoms of	hydrogen, there are	H	1	1
85,000	atoms of	helium	He	2	4
850	atoms of	oxygen	O	8	16
400	atoms of	carbon	C	6	12
120	atoms of	neon	Ne	10	20
100	atoms of	nitrogen	N	7	14
47	atoms of	iron	Fe	26	56

Note: This table gives the relative ***numbers*** *of atoms. To calculate the relative masses, multiply each relative number by the atomic weight of that element before summing the column. One result is that 25% of the mass of the Sun is helium.*

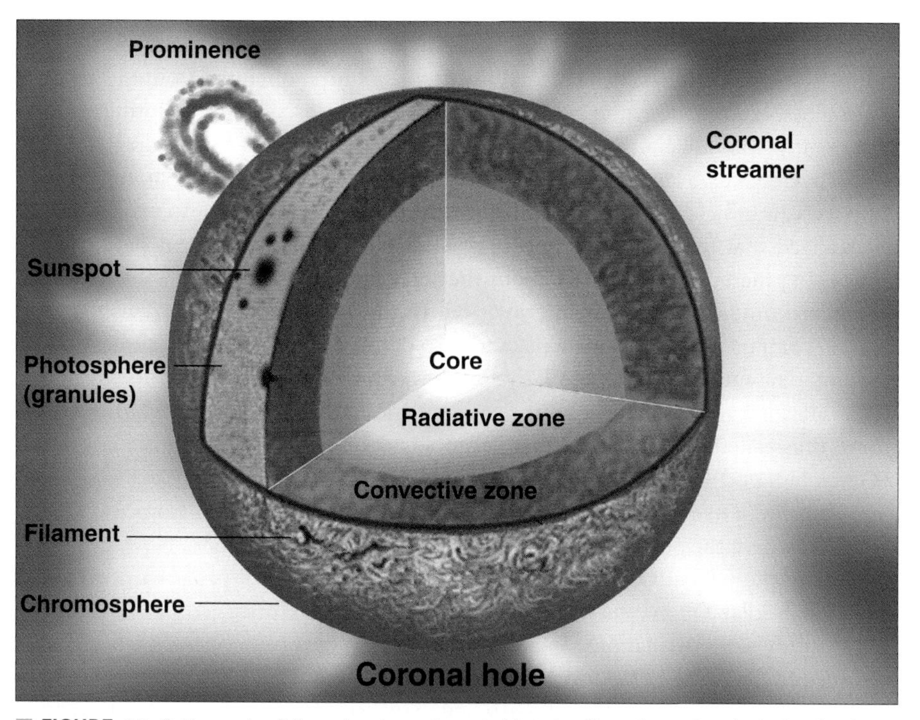

■ **FIGURE 10–3** The parts of the solar atmosphere and interior. The solar surface is depicted as it appears through a hydrogen filter; thus, here we actually see the chromosphere, except in the narrow sliver labeled "photosphere (granules)." The photosphere is the visible surface of the Sun.

The interior has a core, where the energy is generated, surrounded by a zone in which the dominant method of energy transfer is radiation and a zone in which convection (a "boiling" effect) dominates. On the surface, the photosphere, we find sunspots as well as loops of gas that are dark filaments when seen in projection against the disk and that are bright prominences when seen off the Sun's edge. The chromosphere, a thin layer above the photosphere, shows best in hydrogen-light and is composed of small spikes known as spicules. The white corona above it extends far into space. Its shape, including its huge streamers, is determined by the Sun's magnetic field. (Courtesy of Encyclopaedia Britannica, Inc.; illustration by Anne Hoyer Becker; from "A New Understanding of Our Sun," by Jay M. Pasachoff, *1989 Britannica Yearbook of Science and the Future.*)

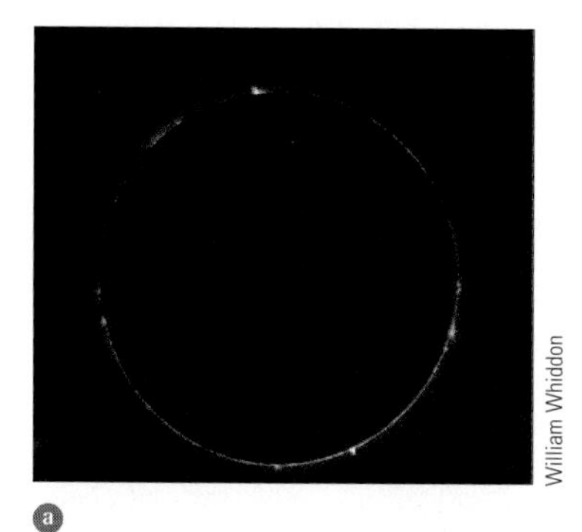

a

b

■ **FIGURE 10–4** (a) The chromosphere of the Sun, photographed during the total solar eclipse of April 8, 2005. Notice how pink it looks. (b) The inner solar corona surrounding the dark silhouette of the Moon at the same eclipse.

its surface in detail. We can resolve parts of the surface about 700 km across, roughly the distance from Boston to Washington, D.C.

When we study the solar surface in **white light**—all the visible radiation taken together—with typical good resolution (about 1 arc second), we see a salt-and-pepper texture called **granulation** (■ Fig. 10–5). The effect, caused by **convection,** is similar to that seen in boiling liquids on Earth: Hot pockets (cells) of gas are more buoyant than

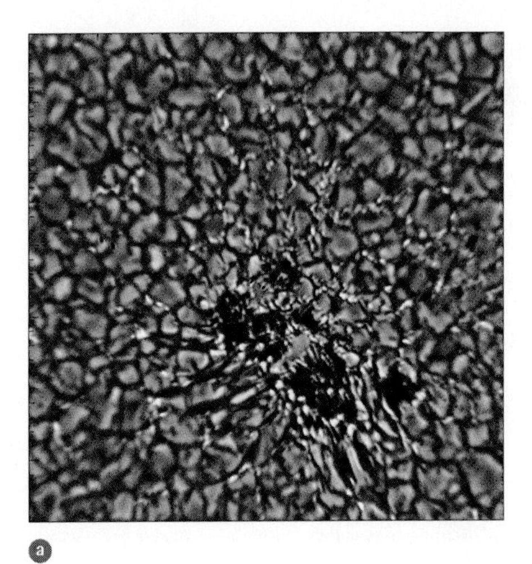

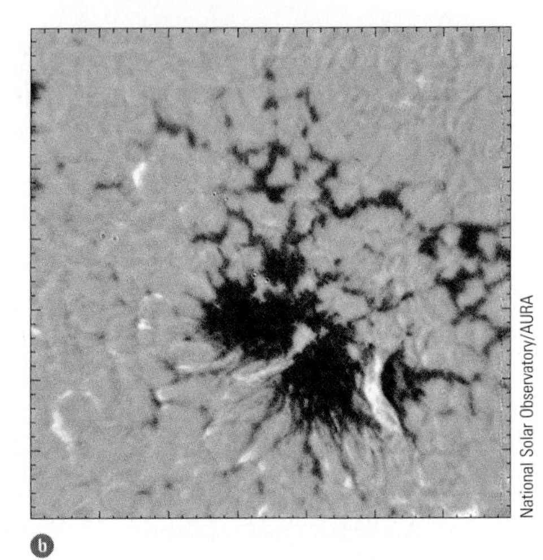

National Solar Observatory/AURA

■ **FIGURE 10–5** ⓐ A visible-light view of a tiny part of the solar surface, showing the salt-and-pepper granulation at the highest detail available from observatories on Earth. ⓑ A magnetic-field map of the same region.

ASIDE 10.3: The solar wind

The "slow solar wind" is from a general expansion of the corona. Particles in it usually take about 20 days to reach Earth. The "fast solar wind" particles travel at twice that speed or greater. They come from coronal holes. A solar flare in 2005 ejected particles moving so rapidly that they reached Earth in only 15 minutes, even faster than the usual hour from a flare, giving even less time for astronauts in space to protect themselves.

surrounding regions, so they rise and deposit their energy at the surface. This causes them to become denser, and so they subsequently sink. But the granules, or convective cells of gas, are about the same size as the limit of our resolution, so they are difficult to study. No spacecraft has yet exceeded the best resolution obtained from the ground.

On close examination, the photosphere as a whole oscillates—vibrating up and down slightly—as can be studied using the Doppler effect. The first period of vibration discovered was 5 minutes long, and for many years astronomers thought that 5 minutes was a basic duration for oscillation on the Sun. However, astronomers have since realized that the Sun simultaneously vibrates with many different periods, and that studying these periods tells us about the solar interior. Indeed, we are able to test the standard model of the solar interior (that is, the way temperatures, densities, and chemical compositions vary with distance from the center) through studies of the vibrations. The method works similarly to the way terrestrial geologists investigate the Earth's interior by measuring seismic waves on the Earth's surface; the studies of the Sun are thus called, by analogy, "solar seismology" or **helioseismology** (after Helios, the Greek Sun god).

To find the longest periods, astronomers have observed the Sun from the Earth's south pole, where the Sun stays above the horizon for months on end. Now, even better, they use the Global Oscillation Network Group (GONG), a program centered at the National Solar Observatory that has erected a network of telescopes around the world to study solar oscillations. In addition, a NASA/European Space Agency spacecraft called SOHO (the Solar and Heliospheric Observatory) is stationed in space at a location from which it continuously views the Sun, and has also assembled long runs of observations lasting many months and covering over a decade.

Studies of solar vibrations thus far have told us about the temperature and density at various levels in the solar interior, and about how fast the interior rotates (■ Fig. 10–6). We have used these studies to test the standard model of the Sun's interior to unprecedented accuracy. Since other stars probably behave like the Sun, we are learning about the interiors of stars in general. Helioseismology can even be used to image, though not with high resolution, what the *back* side of the Sun looks like—a good trick, since we are essentially seeing through over a million kilometers of opaque gas.

The spectrum of the solar photosphere, like that of almost all stars, is a continuous spectrum crossed by absorption lines (■ Fig. 10–7; see also the discussion in Chapter 2). Hundreds of thousands of these absorption lines, which are also called Fraunhofer lines, have been photographed and catalogued. They represent sets of spectral lines from most of the chemical elements. Iron has many lines in the spectrum. The hydrogen lines are strong but few in number. From the spectral lines, we can figure out the relative abundances (the percentages) of the elements. Not only the Sun but also all other ordinary stars have only absorption lines in their spectra.

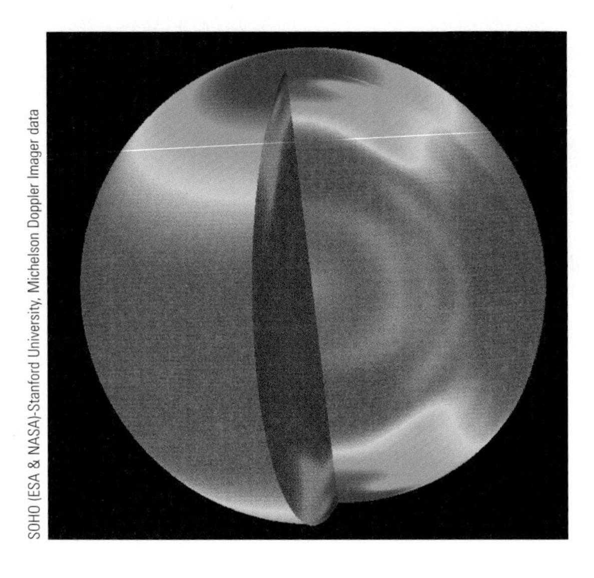

SOHO (ESA & NASA)-Stanford University, Michelson Doppler Imager data

■ **FIGURE 10–6** The rotation periods of different layers and regions of the Sun, based on helioseismology results, shown in false color. Measurements reveal that the surface rotation periods, which vary from 26 days at the equator to 36 days at the poles, persist inward through the solar convection zone. Further toward the center, the data show that the Sun rotates as a solid body with a 27-day period.

10.1b The Chromosphere

When we look at the Sun through a filter that passes only the red light from hydrogen gas, the chromosphere is opaque. Thus through a hydrogen-light filter (■ Fig. 10–8*a*), our view is of the chromosphere. The view has brighter and darker areas, with the brighter areas in the same regions as "sunspots" (see Section 10.2a), but the chromosphere looks different from the photosphere below it.

An important theme valid throughout much of astronomy is that when you look at astronomical objects in detail, you see that they may vary in properties quite a lot from the average you see with the image blurred out. For example, under high resolution, we see that the chromosphere is not a spherical shell around the Sun but rather is composed of small "spicules." These jets of gas rise and fall, and have been compared in appearance to blades of grass or burning prairies (Fig. 10–8*b, c*).

Spicules are more or less cylinders about 700 km across and 7000 km tall. They seem to have lifetimes of about 5 to 15 minutes, and there may be approximately half a million of them on the surface of the Sun at any given moment. They are about the same size as the granules, and are matter ejected into the chromosphere, presumably from the boiling effect that also makes the granules.

Chromospheric matter appears to be at a temperature of 7000 to 15,000 K, somewhat higher than the temperature of the photosphere. The extra energy comes in part from the mechanical motions of convective granules, and in part from magnetic fields, although the relative proportion of these contributions is not yet known. Ultraviolet spectra of stars recorded by spacecraft have shown unmistakable signs of chromospheres in Sun-like stars. Thus by studying the solar chromosphere, we are also learning what the chromospheres of other stars are like.

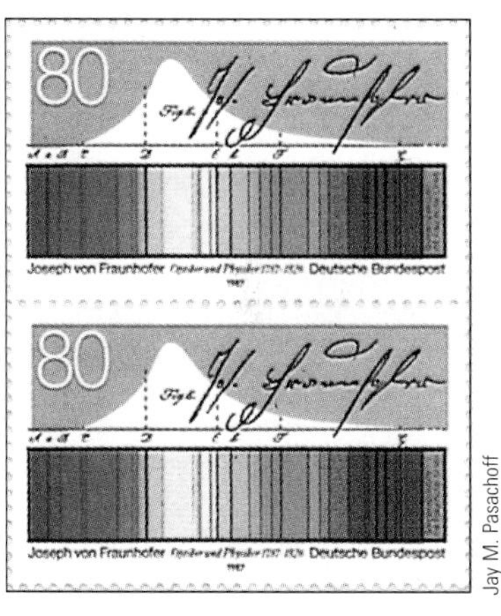

Jay M. Pasachoff

■ **FIGURE 10–7** Fraunhofer's original spectrum, shown on a German postage stamp. We see a continuous spectrum, from red on the left to violet on the right, crossed from top to bottom by the dark lines that Fraunhofer discovered, now known as "Fraunhofer lines" or "absorption lines."

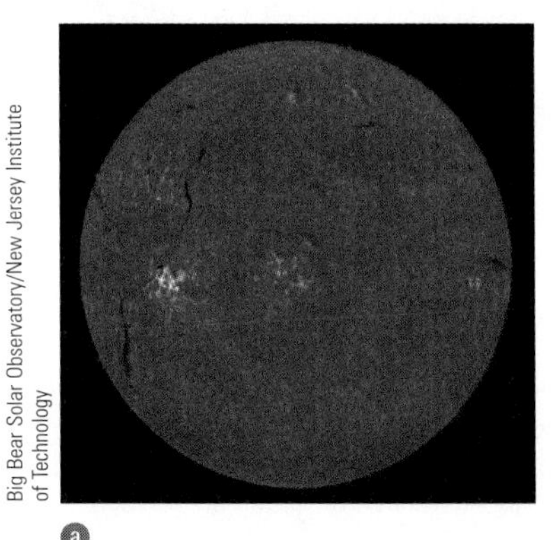

Big Bear Solar Observatory/New Jersey Institute of Technology

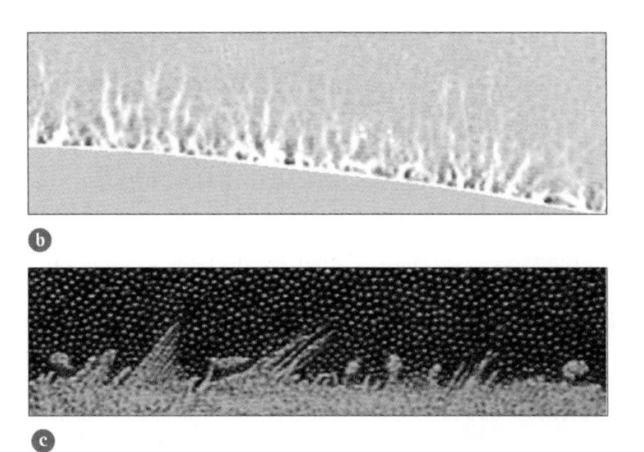

■ **FIGURE 10–8** (a) The Sun, photographed in hydrogen radiation, shows bright areas called plages and dark filaments around the many sunspot regions. We see an image from April 8, 2005, the day of a total eclipse. (b) A high-resolution image of the Sun, showing spicules, taken by one of the authors (J.M.P.) and his students with the Swedish Solar Telescope in the Canary Islands. (c) An original, 19th-century drawing showing how the spicules look like burning prairies.

b: Jay M. Pasachoff, David L. Butts, Joseph W. Gangestad, Owen W. Westbrook, and Jennifer Yee, with NASA support; c: From Fr. Angelo Secchi, *Die Sonne*, courtesy of Jay M. Pasachoff

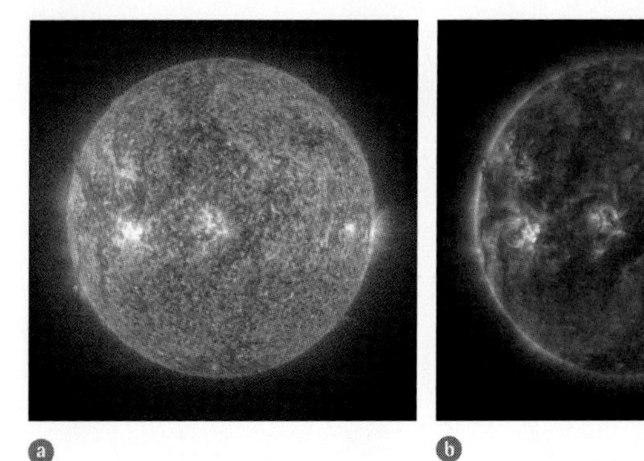
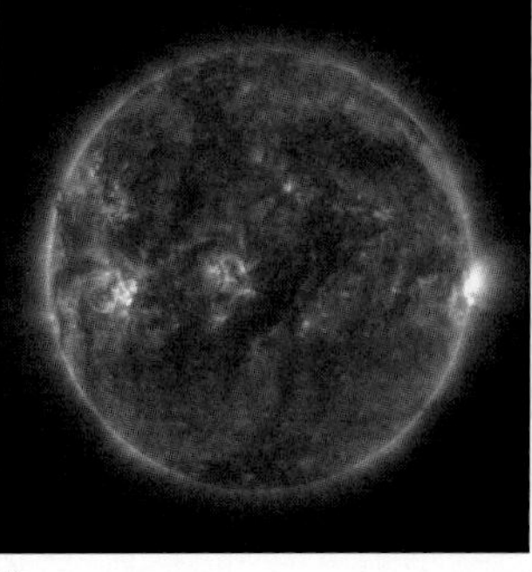
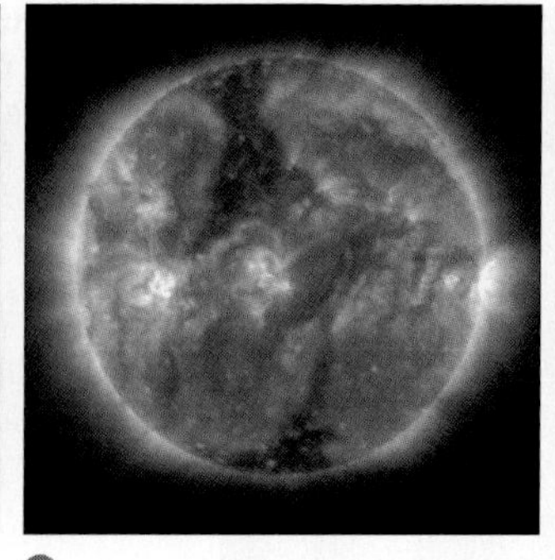
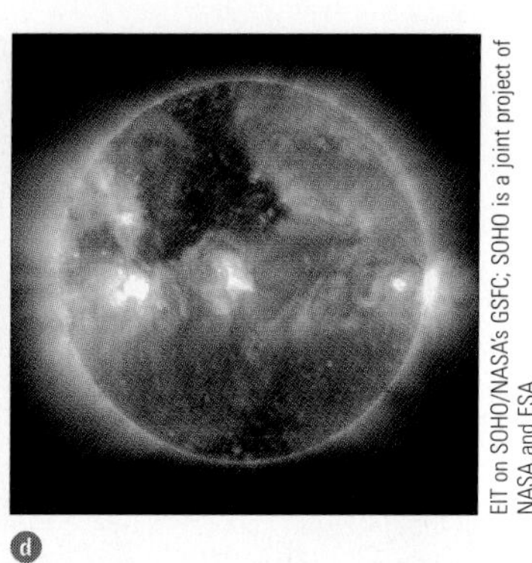

EIT on SOHO/NASA's GSFC; SOHO is a joint project of NASA and ESA

■ **FIGURE 10–9** A set of images taken almost simultaneously through the four filters of the Extreme-ultraviolet Imaging Telescope (EIT) on the Solar and Heliospheric Observatory (SOHO) in space. These images were taken at the same time a total solar eclipse was observed, at about Universal Time 2100 on April 8, 2005. (a) Ionized helium gas at a temperature of about 60,000 K, with a small contribution from ionized silicon. This wavelength shows the chromosphere. (b) Ionized iron at a temperature of about 1,000,000 K. (c) Ionized iron at a temperature of about 1,500,000 K. (d) Ionized iron at a temperature of about 2,000,000 K to 2,500,000 K. The last three wavelengths show the corona. Images like this are taken hourly every day of the year. We choose to display images from April 8, 2005, to allow comparison with the eclipse images of that day.

Studying the solar chromosphere once led to a major discovery: that of helium. A yellow spectral line was seen in the chromosphere at a nineteenth-century total solar eclipse, and didn't quite match the known yellow lines from sodium. The gas was called "helium," after the Greek Sun god, Helios, since it was known only on the Sun. It took decades before helium was isolated on Earth by chemists. The Sun's photosphere isn't hot enough for helium absorption lines to show in its spectrum.

10.1c The Corona

During total solar eclipses, when first the photosphere and then the chromosphere are completely hidden from view, a faint white halo around the Sun becomes visible. This corona is the outermost part of the solar atmosphere, and technically extends throughout the Solar System. At the lowest levels, the corona's temperature is about 2,000,000 K. The heating mechanism of the corona is magnetic, though the details are hotly debated.

The Extreme-ultraviolet Imaging Telescope (EIT) on the SOHO spacecraft images the corona every few minutes. By looking through filters that pass only light given off by gas at a very high temperature, the spacecraft can make images of the corona even in the center of the Sun's disk (■ Fig. 10–9). On the occasional dates of solar eclipses, these images can be lined up with the eclipse images, to allow astronomers to trace many of the coronal streamers seen during an eclipse back to their roots on the Sun's surface (■ Fig. 10–10).

Even though the temperature of the corona is so high, the actual amount of energy in the solar corona is not large. The temperature quoted is actually a measure

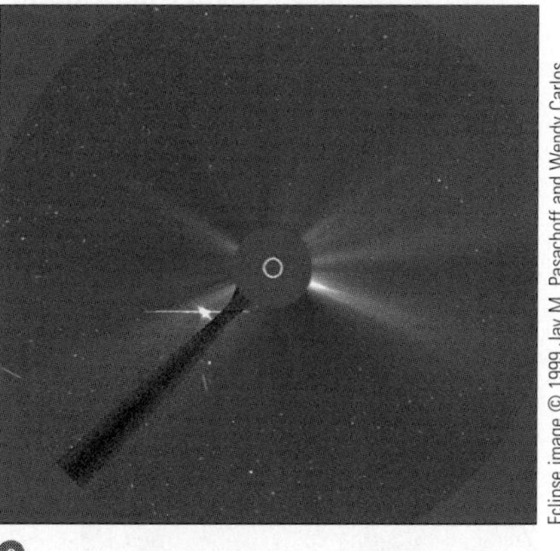

Eclipse image © 1999 Jay M. Pasachoff and Wendy Carlos, merged with a SOHO/EIT image from NASA/Goddard Space Flight Center, courtesy of Joe Gurman, NASA's GSFC

■ **FIGURE 10–10** (a) The Sun at the time of the total solar eclipse of February 26, 1998. A composite image made from photographs of the solar corona taken during the eclipse by one of the authors (J.M.P.) surrounds the solar disk. It shows the equatorial streamers and polar tufts in the corona over a wide range of intensity. Pasted over the dark lunar disk at the center of the image is a false-color image from the EIT on the SOHO spacecraft (see Fig. 10–9). This image shows the coronal temperature based on observations of the corona in the ultraviolet taken at about the same time as the eclipse. (b) A coronagraph on the SOHO spacecraft takes images of the corona every few minutes, but it has to block out the whole photosphere plus the inner corona to do so. Here is an image from eclipse day 2005, matching the images in the previous figure. (c) Another coronagraph on SOHO hides more of the inner and middle corona but shows the outer corona, again here from eclipse day 2005.

of how fast individual particles (electrons, in particular) are moving. There aren't very many coronal particles, even though each particle has a high speed. The corona has less than one-billionth the density of the Earth's atmosphere, and would be considered a very good vacuum in a laboratory on Earth.

For this reason, the corona serves as a unique and valuable celestial laboratory in which we may study gaseous "plasmas" in a near-vacuum. Plasmas are gases consisting of positively and negatively charged particles and can be shaped by magnetic fields. We are trying to learn how to use magnetic fields on Earth to control plasmas, retaining charged particles within a small volume, in order to provide energy through nuclear fusion.

Photographs of the corona show that it is very irregular in form. Beautiful long **streamers** extend away from the Sun in the equatorial regions. The shape of the corona varies continuously and is thus different at each successive eclipse. The structure of the corona is maintained by the magnetic field of the Sun.

NASA's TRACE (Transition Region and Coronal Explorer) spacecraft makes extremely high-resolution observations of the solar corona by observing in the ultraviolet. It shows clearly how the corona is made up of loops of gas, which are held in their shapes by the Sun's magnetic field (■ Fig. 10–11). A successor in providing high resolution, a joint Japanese-American-British spacecraft known (until launch) as Solar-B, will go aloft in 2006 (and at that time, its name will be changed). A still more versatile spacecraft for high-resolution imaging will be NASA's Solar Dynamics Observatory about two years later. It is part of NASA's "Living with a Star" program.

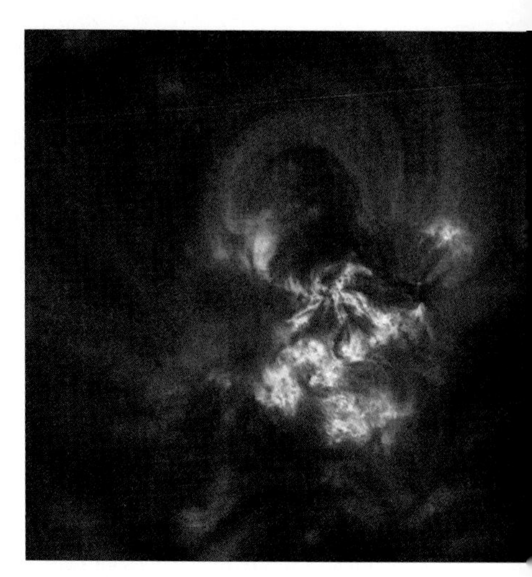

■ **FIGURE 10–11** Loops of hot coronal gas from the TRACE spacecraft, imaged in the ultraviolet. See also the photo opening this chapter.

Alan Title, Lockheed Martin Advanced Technology Center, and Leon Golub, Harvard-Smithsonian Center for Astrophysics

The corona is normally too faint to be seen except during an eclipse of the Sun because it is fainter than the everyday blue sky. But at certain locations on mountain peaks on the surface of the Earth, the sky is especially clear and dust-free, and the innermost part of the corona can be seen with special telescopes. The limited extent of the corona seen in this way shows how valuable the observations at eclipses and from space are.

Several spacecraft, both crewed and robotic, have used devices that made a sort of artificial eclipse to photograph the corona hour by hour in visible light. One of the instruments aboard SOHO makes such observations very well at present. SOHO studies the corona to much greater distances from the solar surface than can be studied with coronagraphs on Earth. (SOHO's coronagraphs cannot see the innermost part, which we can still study best at eclipses.)

Among the major conclusions of this research is that the corona is much more dynamic than we had thought. For example, many blobs of matter were seen to be ejected from the corona into interplanetary space, one per day or so. These "coronal mass ejections" (see Sec. 10.2d) sometimes even impact the Earth, causing surges in power lines and zapping—even occasionally destroying the capabilities of—satellites that bring you television or telephone calls. SOHO and other spacecraft far above the Earth in the direction of the Sun give us early warning when solar particles pass them. NASA has a major program called the Sun–Earth Connection, studying what we increasingly call "space weather" and unifying the study of the Sun, the Earth, and the space between them.

The visible region of the coronal spectrum, when observed at eclipses, shows continuous radiation, absorption lines, and emission lines (■ Fig. 10–12). The presence

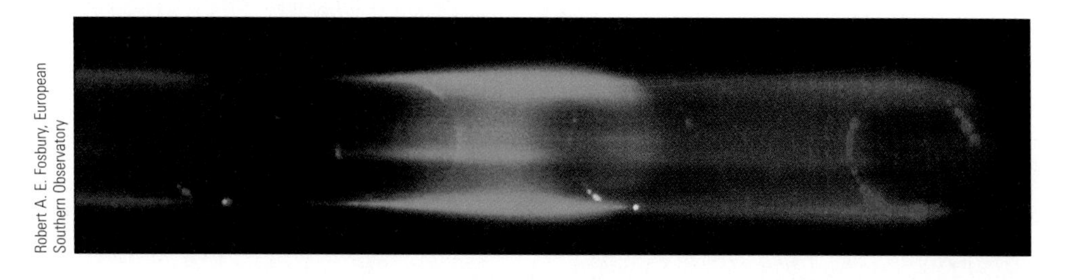

Robert A. E. Fosbury, European Southern Observatory

■ **FIGURE 10–12** A spectrum of the prominences and corona at the 1999 total solar eclipse. The last traces of photospheric spectrum show as the band of color. The bright points are the emission lines from the chromosphere and prominences. The yellow helium D_3 line was the emission line from which helium was first identified over a hundred years ago. The element was named helium because at that time it was found only in the Sun (*helios* in Greek). Faintly visible in the green and in the red are complete circles that are emission lines in the spectrum of the corona. The continuum color with absorption lines, which was visible in Figure 10-7, has been hidden by the Moon.

■ **FIGURE 10–13** From a few mountain sites, the innermost corona can be photographed without need for an eclipse on many days of the year. The corona shows up best in its green emission lines from thirteen-times ionized iron, iron that has lost thirteen electrons. Another technique, used at Mauna Loa, Hawaii, detects the "polarization" of the corona (like the effect seen with polarizing sunglasses) and so distinguishes it from the sky.

Sacramento Peak Observatory, National Solar Observatory/AURA

of emission lines, which we see in silhouette against the dark sky, is in contrast to the photosphere's continuum and absorption lines (but never emission lines), which appear in the everyday spectrum of the Sun.

The coronal emission lines do not correspond in wavelength to any normal spectral lines known in laboratories on Earth or in other stars, and for decades in the late 19th- and early 20th-century, their identification was one of the major problems in solar astronomy. The lines were even given the name of an element: "coronium." (After all, the element helium was first discovered on the Sun.) In the late 1930s, it was discovered that the emission lines arose in atoms that had lost about a dozen electrons each. This was the major indication that the corona was very hot (and that coronium doesn't exist). The corona must be very hot indeed, millions of degrees, to have enough energy to strip that many electrons off atoms. This very hot corona also reflects the photospheric spectrum to us, but the Doppler shift produced by the very rapidly moving plasma particles broadens the absorption lines so they are no longer visible, thus leaving only a continuum.

While the coronal emission lines (■ Fig. 10–13) tell us about the coronal gas, the visible coronal absorption lines are mere reflections of the absorption lines in the spectrum of the Sun's photosphere. To provide these absorption lines, the photospheric spectrum is reflected toward us by dust in interplanetary space far closer to the Earth than to the Sun.

The gas in the corona is so hot that it emits mainly x-rays, photons of high energy. The photosphere, on the other hand, is too cool to emit x-rays. As a result, when photographs of the Sun are taken in the x-ray region of the spectrum (from satellites, since x-rays cannot pass through the Earth's atmosphere), they show the corona and its structure rather than the photosphere.

The x-ray images also reveal very dark areas at the Sun's poles and extending downward across the center of the solar disk. These dark locations are **coronal holes,** regions of the corona that are particularly cool and quiet (■ Fig. 10–14). The density of gas in those areas is lower than the density in adjacent areas. There is usually a coronal hole at one or both of the solar poles. Less often, we find additional coronal holes at lower solar latitudes. The regions of the coronal holes seem very different from other parts of the Sun. The fast streams in the solar wind flow to Earth mainly out of the coronal holes, so it is important to study the coronal holes to understand our environment in space.

The most detailed x-ray images support the more recent ultraviolet high-resolution images in showing that most, if not all, the radiation appears in the form of loops of gas joining separate points on the solar surface (■ Fig. 10–15). We must understand the physics of coronal loops in order to understand how the corona is heated. It is not

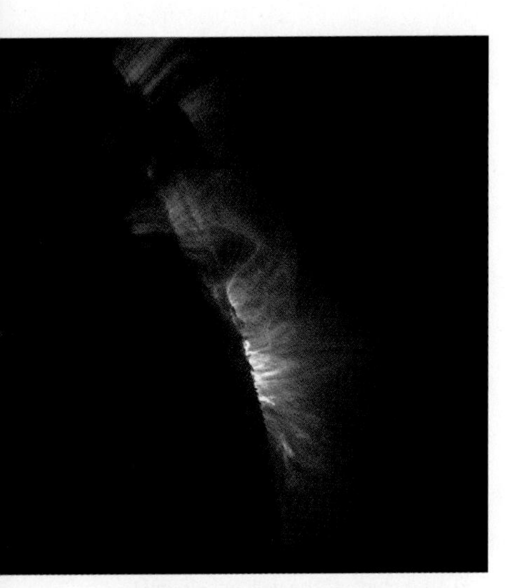

■ **FIGURE 10–15** An ultraviolet image, taken from space on the day of the April 8, 2005, total solar eclipse, shows small coronal loops. The image was made with the TRACE spacecraft. The expeditions of one of the authors (J.M.P.) observed such loops from the ground during the eclipse, with lower spatial resolution but higher time resolution.

Alan Title, Lockheed Martin Advanced Technology Center, and Leon Golub, Harvard-Smithsonian Center for Astrophysics

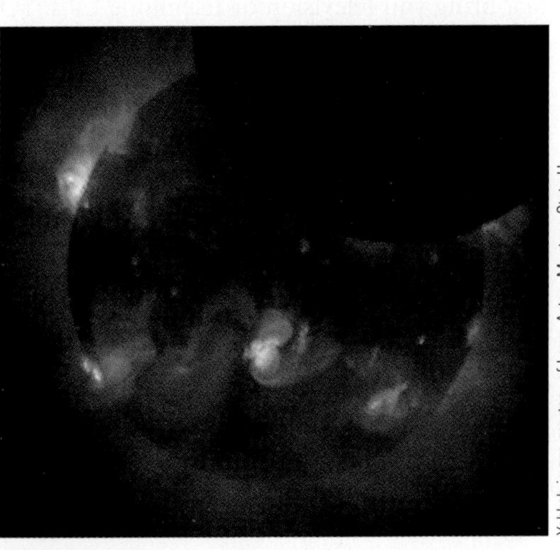

■ **FIGURE 10–14** The corona in relatively low-energy ("soft") x-rays observed from the Yohkoh spacecraft. As the Sun was totally eclipsed on the ground on August 11, 1999, Yohkoh saw a partial eclipse. Unfortunately, during an annular eclipse on December 15, 2001, almost at its 10th anniversary, the spacecraft got confused, lost its bearings, and died. It reentered Earth's atmosphere and burned up in 2005. Formerly called Solar-A, it is to be succeeded in 2006 by the Solar-B mission, a joint effort of Japan, the United States, and Great Britain.

Yohkoh image courtesy of Loren Acton, Montana State U.

sufficient to think—as was done for decades, to simplify calculations—in terms of a uniform corona, since the corona is obviously not uniform.

10.1d The Scientific Value of Eclipses

We discussed solar eclipses and eclipse expeditions in Chapter 4. In these days of orbiting satellites, why is it worth making an expedition to observe a total solar eclipse for scientific purposes?

There is much to be said for the benefits of eclipse observing. Eclipse observations are a relatively inexpensive way, compared to space research, of observing the outer layers of the Sun. Artificial eclipses made by spacecraft hide not only the photosphere but also the inner corona. And for some kinds of observations, space techniques have not yet matched ground-based eclipse capabilities. Some solar scientists will again be on the ground for the total solar eclipses visible from China in 2008 and 2009, coordinating with non-eclipse and spacecraft observations.

AceAstronomy™ Log into AceAstronomy and select this chapter to see the Active Figure called "The Sun" and to see the Astronomy Exercise "Convection and Magnetic Fields."

10.2 Sunspots and Other Solar Activity

The Sun's photosphere and its immediate surroundings exhibit a variety of phenomena including sunspots, flares, coronal mass ejections, filaments, plages, and prominences, as we describe in this section. Their frequency appears to vary with an 11-year cycle.

10.2a What Are Those Blemishes on the Sun?

If you examine the Sun through a properly filtered telescope, you may notice some **sunspots** (■ Fig. 10–16), which appear relatively dark when seen in white light. Sunspots were discovered in 1610, independently by Galileo and by others shortly after the

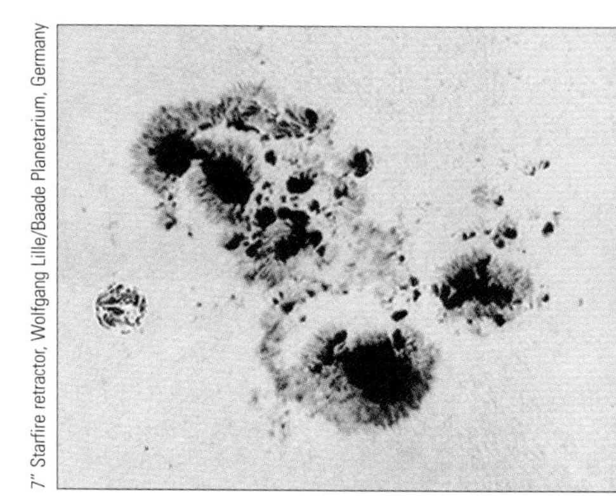

7" Starfire retractor, Wolfgang Lille/Baade Planetarium, Germany

■ **FIGURE 10–16** A sunspot group, showing the dark umbra surrounded by the lighter penumbra for different spots in the group. A photo of the Earth is superimposed to show its relative size.

Star Party 10.1 Observing Sunspots

Sunspots seem dark in contrast with the rest of the Sun, but would be as bright as the Moon if they could be in the sky by themselves. A couple of types of sunspot telescopes were invented in the last decade that are available for only about $300. They can be taken outdoors and set up in a few seconds so that anyone can see the day's sunspots on a screen a few inches across.

Any telescope can be used to project an image of the Sun, including the sunspots, on a piece of paper or on a screen. (Don't look up through the telescope at the Sun, however, as emphasized in Chapter 4.) Furthermore, inexpensive but safe solar filters are available that can be put over the front of a telescope, cutting down the Sun's intensity by a factor of about 100,000, to allow you to look through the telescope and filter at the Sun. It is interesting to see how the sunspots change from day to day and as the Sun rotates.

Jay M. Pasachoff and Steven P. Souza

A small sunspot telescope, with the beam folded by reflection of several mirrors to make the system compact and easy to set up.

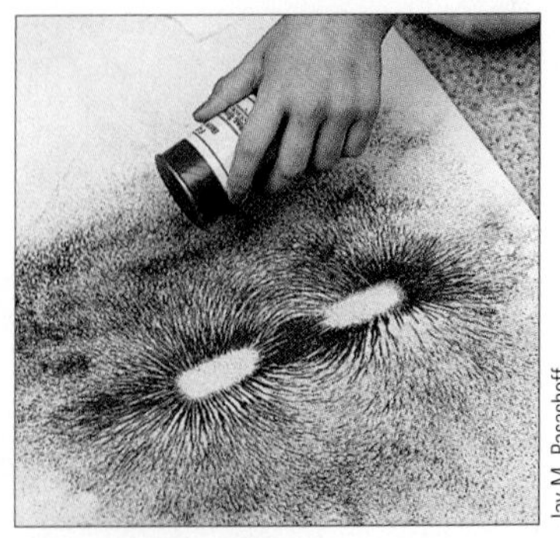

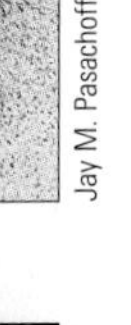

a

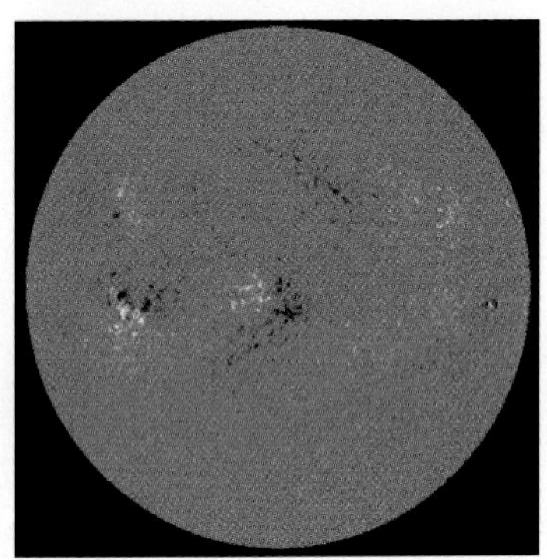

b

■ **FIGURE 10–17** (a) Lines of force from a bar magnet are outlined by iron filings. One end of the magnet is called a north pole and the other is called a south pole. Similar poles ("like poles," where "like" means "similar")—a pair of norths or a pair of souths—repel each other, and unlike (1 north and 1 south) attract each other. Lines of force go between opposite poles. (b) The magnetic field of the Sun on April 8, 2005, the day of an eclipse, to match earlier solar images in this chapter. Black and white, respectively, show opposite polarities. Compare with the sunspot diagram of Figure 10-2.

telescope was first put to astronomical use; an occasional sunspot had been seen with the naked eye previously, and Kepler even saw one in 1607 with a pinhole projection.

Sunspots look dark because they are giving off less radiation per unit area than the photosphere that surrounds them. Thus they are relatively cool (about 2000 K cooler than the photosphere), since cooler gas radiates less than hotter gas (recall our discussion in Chapter 2). Actually, if we could somehow remove a sunspot from the solar surface and put it off in space, it would appear bright against the dark sky; a large one would give off as much light as the full moon seen from Earth. A sunspot includes an apparently dark central region, called the **umbra** (from the Latin for "shadow"; plural: **umbrae**). The umbra is surrounded by a **penumbra** (plural: **penumbrae**), which is not as dark. Some large sunspots are quite long-lived, lasting for over a month. If you follow them day after day, you can see how the Sun rotates at their latitudes.

To explain the origin of sunspots, we must understand magnetic fields. When iron filings are put near a simple bar magnet, the filings show a pattern (■ Fig. 10–17*a*). The magnet is said to have a north pole and a south pole, and the magnetic field linking them is characterized by what we call **magnetic-field lines** (after all, the iron filings are spread out in what look like lines). The Earth (as well as some other planets) has a magnetic field that has many characteristics in common with that of a bar magnet. The structure seen in the solar corona, including the streamers, shows matter being held by the Sun's magnetic field.

The strength of the solar magnetic field is revealed in spectra. George Ellery Hale showed, in 1908, that the sunspots are regions of very high magnetic-field strength on the Sun, thousands of times more powerful than the Earth's magnetic field (Fig. 10-17*b*). Sunspots usually occur in pairs, and often these pairs are part of larger groups. In each pair, one sunspot will be typical of a north magnetic pole and the other will be typical of a south magnetic pole.

The strongest magnetic fields in the Sun occur in sunspots. The magnetic fields in sunspots are able to restrain charged matter; they keep hot gas from being carried upward to the surface. As a result, sunspots are cool and dark.

10.2b The Solar-Activity Cycle

In about 1850, it was realized that the number of sunspots varies with an 11-year cycle (■ Fig. 10–18), the **sunspot cycle,** which is one aspect of the more general **solar-activity cycle.** Every 11-year cycle, the north magnetic pole and south magnetic pole on the Sun reverse; what had been a north magnetic pole is then a south magnetic pole and vice versa. So it is 22 years before the Sun returns to its original state, making the real period of the solar-activity cycle 22 years.

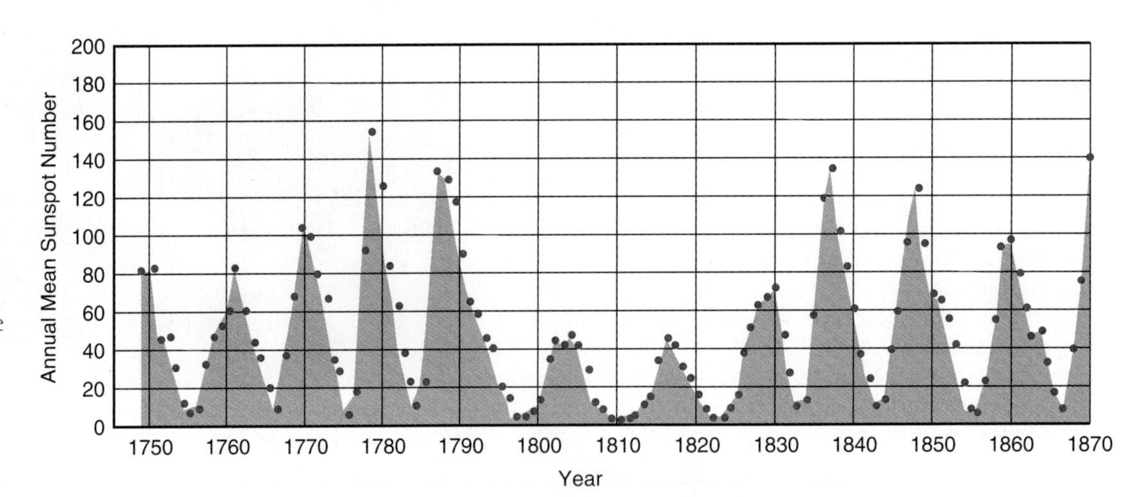

■ **FIGURE 10–18** The 11-year sunspot cycle is but one manifestation of the solar-activity cycle. (Data from Solar Influences Data Analysis Center, Royal Observatory of Belgium)

We passed through the maximum of the sunspot cycle—the time when there is the greatest number of sunspots—in 2001–2002. The number of sunspots has generally been decreasing, and is expecting to continue decreasing through about 2006 or 2007. Thereafter, it will generally rise, reaching its maximum in 2012 or 2013.

Careful studies of the solar-activity cycle are now increasing our understanding of how the Sun affects the Earth, part of the Sun-Earth connection. Although for many years, scientists were skeptical of the idea that solar activity could have a direct effect on the Earth's weather, today's scientists currently seem to be accepting more and more the possibility of such a relationship.

An extreme test of the interaction may be provided by the interesting probability that there were essentially no sunspots on the Sun from 1645 to 1715! This period, called the **Maunder minimum,** was largely forgotten until the late 20th century. An important conclusion is that the solar-activity cycle may be considerably less regular than we had thought.

Much of the evidence for the Maunder minimum is indirect, and has been challenged, as has the specific link of the Maunder minimum with colder climate during that period. It would be reasonable for several mechanisms to affect the Earth's climate on this timescale, rather than only one.

Precise measurements made from spacecraft have shown that the total amount of energy flowing out of the Sun varies slightly, by up to plus or minus 0.2 per cent. On a short timescale, the dips in energy seem to correspond to the existence of large sunspots. Astronomers are now trying to figure out what happens to the blocked energy. On a longer timescale, the effect goes the other way. Spacecraft observations have shown that as sunspot minimum is reached, the Sun becomes overall slightly fainter. As the last maximum was approached, the Sun brightened back up, suggesting the correlation.

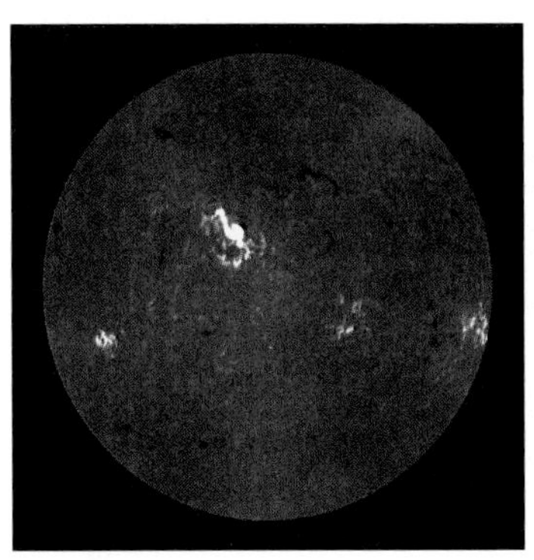

Big Bear Solar Observatory/New Jersey Institute of Technology

■ **FIGURE 10–19** The brightening of regions of the Sun, seen in the hydrogen-alpha line, marking the position of a solar flare. To make a flare, magnetic-field lines change the way their north and south magnetic poles are connected. This "reconnection" releases high-energy particles, which follow the magnetic-field lines back down to lower levels of the solar atmosphere. We see these regions brightening drastically on this Hα image of the May 13, 2005, solar flare.

10.2c Fireworks on the Sun, and Space Weather

Violent activity sometimes occurs in the regions around sunspots. Tremendous eruptions called solar **flares** (■ Fig. 10–19) can eject particles and emit radiation from all parts of the spectrum into space. These solar storms begin abruptly, over a few seconds, and then can last up to four hours. Temperatures in the flare can reach 5 million kelvins, even hotter than the quiet corona.

Until a decade or so ago, scientists thought that most of the particles affecting the Earth came from solar flares. But spacecraft revealed that the corona was ejecting puffs of mass with a frequency of once every day or so. These **coronal mass ejections** cause many of the magnetic storms on Earth. At first, it was thought that flares caused coronal mass

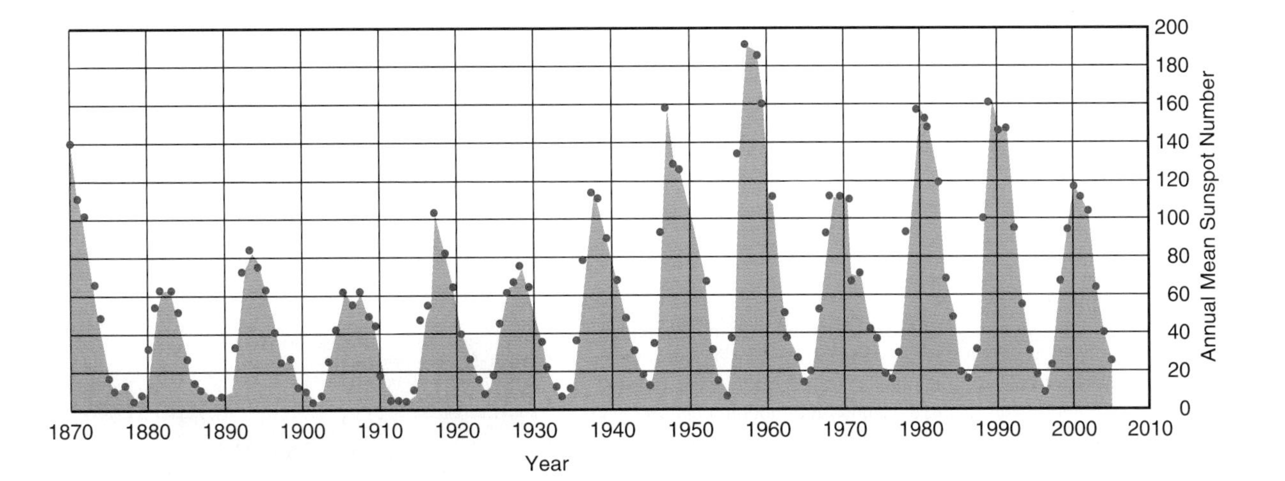

ASIDE 10.4: Freezing in the hot corona

Although the corona is very hot, meaning that individual particles move very rapidly in random directions, its density is so low that you would freeze to death if you were immersed in the corona (assuming the photosphere were somehow blocked from you). The hot particles would hit you too rarely to replenish the energy your body radiates into space.

ejections, but that is now known not to be the case. The link between coronal mass ejections and solar flares is being debated; they often but not always occur together.

Ultraviolet radiation and x-rays that are given off by flares reach Earth at the speed of light in 8 minutes and can disrupt radio communications, because they ionize the upper part of Earth's atmosphere. Flare particles that are ejected reach the Earth in a few hours or days and can cause the auroras (see the discussion in Chapter 6) and even surges on power lines that lead to blackouts of electricity. Particles from coronal mass ejections usually take a day or so to reach Earth, but in some cases, these particles can reach Earth in less than an hour. So astronauts of the future are now thought to have less time to seek safety in their interplanetary spacecraft than had been assumed.

Because of the solar-terrestrial relationships, high priority is placed on understanding solar activity and being able to predict it. The study of space weather—the effect of the Sun on interplanetary space and the Earth—has increasing importance, as robotic and crewed spacecraft spend more time outside the protection of the Earth's atmosphere. The website at http://www.spaceweather.com can keep you up-to-date on a daily basis.

No specific theory for explaining the eruption of solar flares is generally accepted. But it is agreed that a tremendous amount of energy is stored in the solar magnetic fields in sunspot regions. Something unknown triggers the release of energy.

10.2d Filaments and Prominences

Studies of the solar atmosphere through filters that pass only hydrogen radiation also reveal other types of solar activity. Dark **filaments** are seen threading their way for up to 100,000 km across the Sun in the vicinity of sunspots. They mark the locations of zero magnetic field that separate regions of magnetic field pointing in opposite directions. There are also bright areas, called "plages," often associated with sunspots.

When filaments happen to be on the Sun's visible edge, they project into space, often in beautiful shapes; these are called **prominences** (■ Fig. 10–20). Prominences can be seen with the eye during solar eclipses and glow pinkish at that time because of their emission in hydrogen and a few other spectral lines (■ Fig. 10–21). They can be observed from the ground even without an eclipse, if a filter that passes only light emitted by hydrogen gas is used.

Prominences appear to be composed of matter in a condition of temperature and density similar to matter in the quiet chromosphere, somewhat hotter than the photosphere. Sometimes prominences can hover above the Sun, supported by magnetic fields, for weeks or months. Other prominences change rapidly.

AceAstronomy™ Log into AceAstronomy and select this chapter to see the Astronomy Exercises "Sunspot Cycle I" and "Sunspot Cycle II."

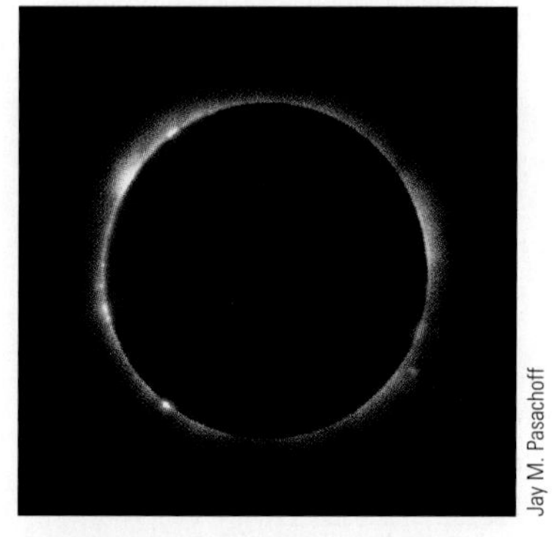

Jay M. Pasachoff

■ **FIGURE 10–21** Prominences and the chromosphere appear pinkish in this image from the 1999 total solar eclipse. Their color comes mainly from the red Hα emission line that they radiate, but with some other emission lines mixed in. Note that the spectral lines are in emission only because we see them off the edge of the Sun silhouetted against the dark sky; spectral lines in the bright photosphere, on the other hand, are absorption lines. Prominences (as well as the corona) change with time, as can be seen by comparing this image with that in Fig. 10–4*a*.

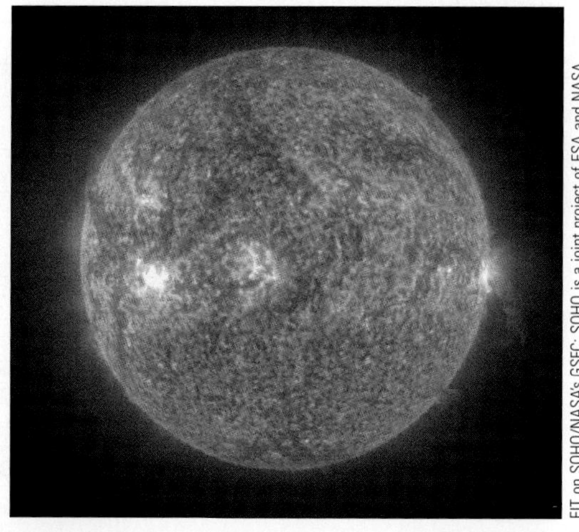

EIT on SOHO/NASA's GSFC; SOHO is a joint project of ESA and NASA

■ **FIGURE 10–20** A giant solar prominence sticks off the Sun. This is a view in the fundamental ultraviolet line of ionized helium. This eruption took place several hours earlier than the image shown in Figure 10-9.

10.3 The Sun and the Theory of Relativity

The intuitive notion we have of gravity corresponds to the theory of gravity advanced by Isaac Newton in 1687. We now know, however, that Newton's theory and our intuitive ideas are not sufficient to explain the Universe in detail. Theories advanced by Albert Einstein in the first decades of the 20th century provide us with a more accurate understanding (see *Lives in Science 10.1: Albert Einstein*).

Einstein was a young clerk in the patent office in Switzerland when, in 1905, he published five scientific papers that revolutionized three major subfields of physics. One

ASIDE 10.5: The Einstein Year of 2005

Worldwide celebrations and symposia were held to mark the 100th anniversary of the "miraculous" year 1905 when Albert Einstein proved the existence of atoms, explained the photoelectric effect, and proposed the special theory of relativity.

LIVES IN SCIENCE

10.1 Albert Einstein

Albert Einstein was born in Ulm, Germany, in 1879. He graduated from a university in Switzerland. Unable to get an academic position, he went to work in the Patent Office as an examiner. Out of this lemon, he made lemonade: He later stated that this job with its set working hours left him all the rest of his time to do physics unconstrained.

In 1905, Einstein published papers on five major ideas of physics—a record never matched in the history of the subject. First, he explained the mechanism by which light falling on a metal surface causes electrons to be given off. This work turned out to be basic to our current ideas of atoms and how they interact with light. (Later, it was the official reason for his being awarded the Nobel Prize in Physics.) It showed that particles of light exist, the particles we now call photons (see our discussion in Chapter 2).

Next, for his doctoral thesis, he found a way to estimate the dimensions of molecules by statistically studying their motion in liquids. In his third, somewhat related paper, he explained the jiggling motion (known as "Brownian motion") of tiny particles that a careful observer can see moving around inside liquids. Einstein said that the particles are hit by atoms over and over again, and are displaced from their original positions by distances proportional to the square root of time.

In his fourth paper, he advanced his special theory of relativity, which described motion and gave a central role to the speed of light. He showed that time, space, and motion are all relative to the observer; absolute measurements of these quantities cannot be made. This work culminated in a fifth paper, in which Einstein presented $E = mc^2$, probably the most famous equation in all of physics. (We will discuss it in Chapter 12.)

Courtesy of the Archives, California Institute of Technology and the Albert Einstein Archives, courtesy of the Hebrew Univ. and Library

Einstein's work made him well known in scientific circles, and he was offered university professorships. Over the next decade, he worked incessantly on incorporating gravity into his theories, and by 1914 had a prediction of the angle by which light passing near the Sun would be bent. He was then a professor in Berlin. German scientists went to Russia to try to observe this effect during a total solar eclipse, but were interned during the war and could not make their observations. The fact that they did not observe the eclipse in 1914 turned out to be a blessing, for Einstein had not yet found an adequate form of the theory, and his prediction was too low by a factor of 2. By 1916, Einstein had revised his theory, and his prediction was the value that we now know to be correct. When the British scientist Arthur Eddington verified the prediction at the 1919 eclipse, Einstein triumphed.

The coming to power of the Nazis in the 1930s forced Einstein to renounce his German citizenship. He was persecuted as a Jew and his work was attacked as "Jewish physics," as though one's religion had some bearing on scientific truth. He accepted an offer to come to America to be the first professor at the Institute for Advanced Study, which was being set up in Princeton, New Jersey, not far from Princeton University.

During his years in Princeton, Einstein continued to work on scientific problems. However, his ideas were far from the mainstream and, partly as a result, his work on unifying the fundamental forces of nature did not succeed. Of course, Einstein had long worked far from the mainstream, but this time he was avoiding the basic physics of atoms—quantum mechanics—which he felt was an incomplete representation of nature.

Einstein was a celebrity in spite of his modesty; he was so much photographed that he once gave his profession as "artist's model." He devoted himself to pacifist and Zionist causes and was very influential.

Einstein was a strong backer of the State of Israel and was even once asked to be its president. His understanding of the Nazi peril forced him to put aside his pacifism in that context. On one important occasion, scientists wishing to warn President Roosevelt that atomic energy could lead to a Nazi bomb enlisted Einstein to help them reach the president's ear.

Einstein died in 1955. A project is now well under way to publish all his papers in a uniform set of books. The volume containing his childhood documents has given insights into the formation of his thought. For his achievements and their influence on modern science and technology, Einstein was named "Person of the Century" in the December 31, 1999, issue of *TIME* magazine.

■ **FIGURE 10–22** Just as light is bent by space, a putted golf ball follows the warp of the green. Here Tiger Woods celebrates his successful accounting for the curvature of space.

of the papers dealt with the effect of light on metals, two dealt with molecules and the irregular motions of small particles in liquids, and the final two dealt with motion.

Einstein's work on motion became known as his "special theory of relativity." It is "special" in the sense that it does not include the effect of gravity and is therefore not "general." "Special" thus means "limited" in this case. Einstein's special theory of relativity assumes that the speed of light—300,000 km/sec—is an important constant that cannot be exceeded by real objects. (More precisely, no information can go through space faster than the speed of light.) This theory must be used to explain the motion of objects moving very fast—near the speed of light—and has been thoroughly tested and verified.

In 1916, Einstein came up with his **general theory of relativity,** which explains gravity and also deals with accelerations. In this theory, objects simply move freely in a curved (warped) space, making it look like they are pulled by gravity. In fact, Einstein showed that time is a dimension, almost but not quite like the three spatial dimensions, and it can also be warped. Thus Einstein worked in four-dimensional space–time.

Picture a ball rolling on a golf green. If the green is warped, the ball will seem to curve (■ Fig. 10–22). If the surface could be flattened out without distortions, though, or if we could view it from a perspective in which the surface were flat, we would see that the ball is rolling in a straight line (defined to be the shortest path between two points). This analogy shows the effect of a two-dimensional space (a surface) curved into an extra dimension. Einstein's general theory of relativity treats mathematically what happens as a consequence of the curvature (warping) of space–time.

In Einstein's mathematical theory, the presence of mass or energy curves the space, just as your bed's surface ceases to be flat when you put your weight on it. Light travelling on Einstein's curved space tends to fall into the dents, just as a ball rolling on your bed would fall into its dents. Einstein, in particular, predicted that light travelling near the Sun would be slightly bent because the Sun's gravity must warp space (■ Fig. 10–23).

The Sun, as the nearest star to the Earth, has been very important for testing some of the predictions of Einstein's general theory of relativity. The theory could be checked by three observational tests that depended on the presence of a large mass like the Sun for experimental verification.

First, Einstein's theory showed that the closest point to the Sun (the "perihelion") of Mercury's elliptical orbit would move slightly around the sky over centuries—that is, it would gradually "precess," or shift. Precession is produced mainly by gravitational interactions with the other planets, but even after all of them had been taken into account, a small amount of the observed effect remained unexplained (only 43 arc seconds per century, out of the total of 570 arc seconds per century). Calculations with Einstein's theory accounted precisely for the amount of this residual precession of the perihelion

ASIDE 10.6: How fast is gravity?

Though it is commonly accepted that gravity travels at the speed of light, a report in 2003 that this agreement was confirmed observationally–using a new theoretically derived method–has not been widely accepted.

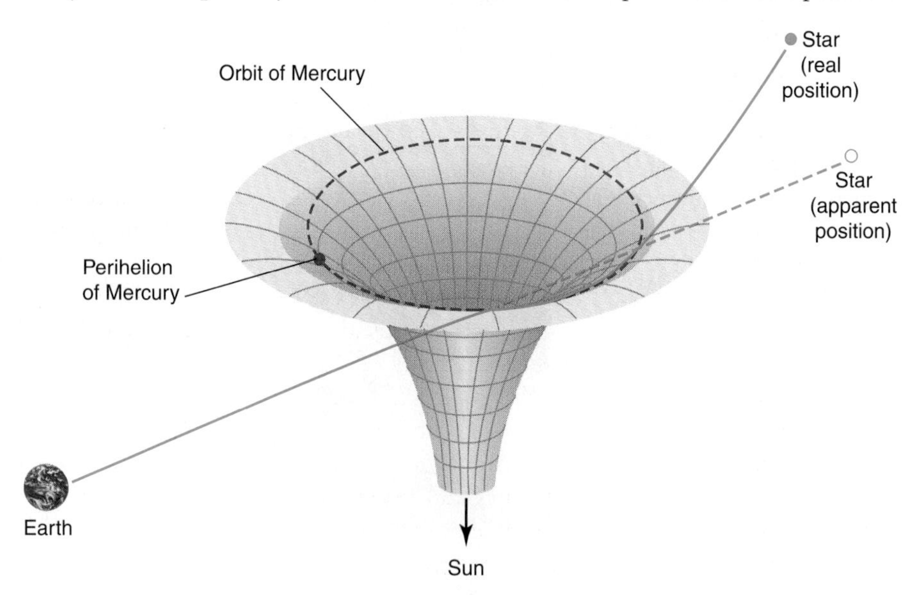

■ **FIGURE 10–23** According to Einstein's general theory of relativity, the presence of a massive body warps the space nearby, here illustrated with the curved grid. The solid line shows the actual path of light. We on Earth, though, trace back the light in the direction from which it came. At any given instant, we have no direct knowledge that the light's path has curved, so we assume that the light has always been travelling straight. Also shown is the orbit of Mercury and its perihelion, or point of closest approach to the Sun. The location of the perihelion slowly shifts with time due to several effects, including one that was unexplained without Einstein's general theory of relativity.

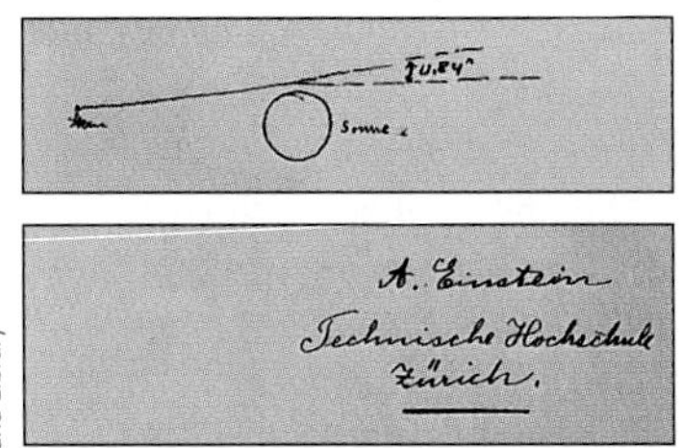

■ **FIGURE 10–24** The prediction, in Einstein's own handwriting, of the deflection of starlight by the Sun. In this early version of his theory, Einstein predicted half the value he later calculated. The number he gives is slightly further wrong, because of his arithmetic error. (There is a story that one of his fellow chamber-music players once said, when Einstein missed a beat, "Einstein, can't you count?") We also see his signature with the name of his university.

of Mercury. It was a plus for Einstein to have explained it, but the true test of a scientific theory is really whether it predicts new things rather than simply explaining old ones.

The second test arose from a major new prediction of Einstein's theory: Light from a star would act as though it were bent toward the Sun by a very small amount (■ Fig. 10–24), though twice the amount that would be predicted by pre-Einsteinian theory. We on Earth, looking back past the Sun, would see the star from which the light was emitted as though it were shifted slightly away from the Sun. Only a star whose radiation grazed the observed edge ("limb") of the Sun would seem to undergo the full deflection; the effect diminishes as one considers stars farther away from the solar limb.

To see the effect, one has to look near the Sun at a time when the stars are visible, and this could be done only during a total solar eclipse. The effect in the amount predicted by Einstein was verified at the total solar eclipse of 1919. Scientists hailed this confirmation of Einstein's theory, and from the moment of its official announcement, Einstein was recognized by scientists and the general public alike as the world's greatest scientist (■ Fig. 10–25).

Similar observations have been made at subsequent eclipses, though they are very difficult to conduct. Fortunately, the effect is constant throughout the electromagnetic spectrum, and the test can now be performed more accurately by observing how the Sun bends radiation from radio sources, especially quasars (which are discussed in Chapter 17). The results agree with Einstein's theory to within 1 per cent, enough to make the competing theories very unlikely. As we shall see later on, the effect is now very well tested farther out in the Universe.

The third traditional test of general relativity was to verify the prediction of Einstein's theory that gravity would cause the spectrum to be redshifted, an effect known as "gravitational redshifting." This effect is very slight for the Sun, but has been detected. It has best been verified for extremely dense stars known as white dwarfs (see Chapter 13), in which mass is very tightly packed together.

As a general rule, scientists try to find theories that not only explain the data at hand, but also make predictions that can be tested. This is an important part of the scientific method. Because the bending of radiation by a certain amount was a prediction of the general theory of relativity that had not been measured before the prediction, its verification provided more convincing evidence of the theory's validity than the theory's ability to explain the previously known shifting (precession) of Mercury's orbit. Now, decades later, the general theory of relativity is a standard part of the arsenal of tools of a theoretical astronomer. We will meet it in several places later in this book.

In particular, in Chapter 13 we will introduce the concept of gravitational waves, one of Einstein's own predictions based on the general theory of relativity. Astronomers have detected the effects of gravitational waves by studying two unusual pulsar systems. Major facilities have recently been constructed in several places around the globe to try to detect gravitational waves directly. In Chapter 14, we will see how the general theory of relativity is integral for understanding black holes.

In Chapter 16, we will meet gravitational lensing, a phenomenon in which the mass of a nearer object, such as a cluster of galaxies, affects light from farther objects, sometimes smearing it into beautiful arcs. In Chapter 17, we will learn about the giant black holes thought to be at the centers of quasars. And in Chapters 18 and 19, we will discuss cosmology, which is based on Einstein's general theory of relativity.

ECLIPSE SHOWED GRAVITY VARIATION

Diversion of Light Rays Accepted as Affecting Newton's Principles.

HAILED AS EPOCHMAKING

British Scientist Calls the Discovery One of the Greatest of Human Achievements.

Special Cable to THE NEW YORK TIMES.

LONDON, Nov. 8.—What Sir Joseph Thomson, President of the Royal Society, declared was "one of the greatest—perhaps the greatest—of achievements in the history of human thought" was discussed at a joint meeting of the Royal Society and the Royal Astronomical Society in London yesterday, when the results of the British observations of the total solar eclipse of May 29 were made known.

There was a large attendance of astronomers and physicists, and it was generally accepted that the observations were decisive in verifying the prediction of Dr. Einstein, Professor of Physics in the University of Prague, that rays of light from stars, passing close to the sun on their way to the earth, would suffer twice the deflection for which the principles enunciated by Sir Isaac Newton accounted. But there was a difference of opinion as to whether science had to face merely a new and unexplained fact or to reckon with a theory that would completely revolutionize the accepted fundamentals of physics.

The discussion was opened by the Astronomer Royal, Sir Frank Dyson, who described the work of the expeditions sent respectively to Sobral, in Northern Brazil, and the Island of Principe, off the west coast of Africa. At each of these places, if the weather were propitious on the day of the eclipse, it would be possible to take during totality a set of photographs of the obscured sun and a number of bright stars which happened to be in its immediate vicinity.

The desired object was to ascertain whether the light from these stars as it passed by the sun came as directly toward the earth as if the sun were not there, or if there was a deflection due to its presence. And if the deflection did occur the stars would appear on the

■ **FIGURE 10–25** The first report of the eclipse results, from *The New York Times* in 1919, showing how the eclipse captured the public's attention.

CONCEPT REVIEW

The everyday layer ("surface") of the **quiet Sun** we see is the **photosphere;** phenomena in it that appear nonuniformly and change with time are signs of the **active Sun.** Beneath it is the solar **interior,** with the energy generated at the solar **core,** whose temperature is 15 million kelvins (Sec. 10.1). The **solar atmosphere** above the photosphere contains the pinkish **chromosphere** and the white **corona.** The corona expands into space as the **solar wind,** a stream of particles that bathes the Earth.

The Sun's photosphere, observed in all the visible light together (which is known as **white light**), is covered with tiny **granulation** (Sec. 10.1a). Each granule is a pocket of rising hot gas or falling cool gas driven by the process of **convection,** which is similar to water boiling. Oscillations (vibrations) of the surface reveal conditions in the solar interior; the study of the Sun in this manner is called **helioseismology** by analogy with Earth seismology. The spectrum of the photosphere shows millions of Fraunhofer (absorption) lines; as with other normal stars, it does not produce emission lines.

The Sun's chromosphere is a very thin layer, somewhat hotter than the photosphere below it (Sec. 10.1b). It consists of many spicules, which are jets of gas that rise and fall. Its spectrum consists of emission lines.

The corona, best seen at total solar eclipses, contains gas with a temperature of 2 million kelvins (Sec. 10.1c); thus, it is actually a plasma, consisting of positively and negatively charged particles, and it produces emission lines. The magnetic field holds the coronal gas into **streamers.** Regions where the corona is less dense and cooler than average are **coronal holes.** The slow solar wind is a general coronal expansion, and the fast solar wind comes from coronal holes. Although instruments on spacecraft can be used to block the photosphere, some chromospheric and coronal phenomena are still best observed from the ground during solar eclipses (Sec. 10.1d).

Sunspots are regions in the photosphere with strong magnetic field and are cooler than the surrounding photosphere; the magnetic fields inhibit the rise of hot gases from beneath the sunspot (Sec. 10.2a). Each sunspot has a dark **umbra** surrounded by a lighter **penumbra.** Sunspots usually appear in pairs having a north pole, a south pole, and **magnetic-field lines** linking them; these pairs are often in larger groups.

The **solar-activity cycle,** including the **sunspot cycle,** lasts about 11 years (or about 22 years, if magnetic polarity is included); the average number of sunspots periodically rises and falls (Sec. 10.2b). The **Maunder minimum** was a 17th- and 18th-century period when sunspots were essentially absent.

Solar **flares** and **coronal mass ejections** are eruptions of tremendous amounts of energy (Sec. 10.2c). Electromagnetic radiation and particles from flares can disrupt radio communications or produce electrical surges, affecting satellites and leading to blackouts of electricity. **Prominences,** which appear pinkish during total solar eclipses, are **filaments** seen in silhouette off the edge of the Sun (Sec. 10.2d).

In 1905, Albert Einstein published five very important papers, including two on the "special theory of relativity," which assumes that the speed of light is an important constant that cannot be exceeded by real objects moving through space (Sec. 10.3). In 1916, Einstein published his new theory of gravity, the **general theory of relativity;** mass and energy are assumed to warp (curve) space and time, and objects move freely within curved space–time. Some basic tests of this theory involved the large mass of the Sun. Specifically, it was verified that the Sun's mass affects the orbit of Mercury; the perihelion (point of closest approach to the Sun) shifts, or precesses, slowly with time. Observational verification that the Sun bends starlight passing near it, as predicted by Einstein, instantly made him world famous.

QUESTIONS

1. Sketch the Sun, labelling the interior, the photosphere, the chromosphere, the corona, sunspots, and prominences.
2. Draw a graph showing the Sun's approximate temperatures, starting with the core and going upward through the corona.
3. Define and contrast a prominence and a filament.
4. Why are we on Earth particularly interested in coronal holes?
5. List three phenomena that vary with the solar-activity cycle.
6. Discuss what can be learned from studies of the vibration (oscillation) of the Sun's atmosphere.
7. Why can't we observe the corona every day from the Earth's surface?
8. How do we know that the corona is hot?
9. If the corona is so much hotter than the photosphere, why isn't it much brighter than the photosphere, per unit area?
10. Describe relative advantages of ground-based eclipse studies and of satellite studies of the corona.
11. What is the process of convection? Give an example in everyday life.
12. Describe the sunspot cycle.
13. Why do we say that the true solar-activity cycle actually has a period of 22 years, rather than 11 years?

†14. Large groups of sunspots can be relatively long-lived (a few months), and they remain essentially fixed at the same physical location on the photosphere throughout their lives. Suppose a sunspot group appears to move from the center to the edge of the Sun's disk in 7 days. What is the approximate rotation period of the Sun? (A rough estimate will suffice; don't worry about not being able to see the sunspot if it is exactly at the edge of the Sun.)

†15. **(a)** From the table in *A Closer Look 10.1: The Most Common Elements in the Sun's Photosphere,* calculate the percentage (by number) of helium atoms in the

Sun and the percentage of iron atoms. (**b**) Calculate the percentage of *mass* taken up by helium atoms and by iron atoms.

16. What are solar flares and coronal mass ejections?
17. What feature in the corona shows the magnetic field there?
18. What are two ways that the Sun's corona can be studied from the Solar and Heliospheric Observatory (SOHO)?
19. What is the difference between the special theory of relativity and the general theory of relativity?
20. Why was the Sun useful for checking the general theory of relativity?
21. Explain in your own words how Einstein's general theory of relativity accounts for the Sun's gravity.
22. **True or false?** The hottest region of the entire Sun is the core, even though the corona emits profusely at x-ray wavelengths.
23. **True or false?** During times of sunspot maximum, an unusually large number of prominences, solar flares, and coronal mass ejections also occur on the Sun.
24. **True or false?** It is safe and easy to view sunspots with the naked eye, without filters, during a total solar eclipse.
25. **True or false?** According to Einstein's general theory of relativity, mass causes the surrounding space–time to curve.
26. **True or false?** If a dark sunspot could be viewed alone, without the glare of the surrounding photosphere, it would still appear quite dark because it emits very little light.
27. **Multiple choice:** Sunspots appear dark because (**a**) they are patches of the photosphere that occasionally burn up, creating soot; (**b**) the changing magnetic polarity of the Sun causes gas in the sunspot to cool down substantially; (**c**) they are regions in which strong magnetic fields make it difficult for fresh supplies of hot gas to reach the photosphere; (**d**) they are much hotter than the surrounding area, so their emission peaks at ultraviolet wavelengths, which our eyes cannot see; or (**e**) they are holes in the photosphere through which the cooler interior of the Sun is visible.
28. **Multiple choice:** As the solar atmosphere expands outward from the Sun, into interplanetary space, it becomes the (**a**) chromosphere; (**b**) corona; (**c**) spicules; (**d**) prominences; or (**e**) solar wind.

†29. **Multiple choice:** Suppose the temperature of a sunspot is 4000 K, and that of the surrounding photosphere is about 6000 K. What is the ratio (sunspot to photosphere) of the wavelengths at which the roughly black-body spectra are brightest? *(Hint: see Chapter 2.)* (**a**) $^2/_3$. (**b**) $^3/_2$. (**c**) $(^2/_3)^2$. (**d**) 3. (**e**) 2000.

†30. **Multiple choice:** Suppose the temperature of a sunspot is 4000 K, and that of the surrounding photosphere is about 6000 K. Per unit area, about how much energy does the sunspot emit, compared with the photosphere? *(Hint: see Chapter 2.)* (**a**) $^2/_3$. (**b**) $^3/_2$. (**c**) 2000. (**d**) $(^2/_3)^2$. (**e**) $(^2/_3)^4$.

†31. **Multiple choice:** If you observe a sunspot whose diameter is about 1/20 that of the Sun, it is roughly _____ times the diameter of the Earth. (**a**) $^1/_{20}$. (**b**) $^1/_5$. (**c**) 1. (**d**) 5. (**e**) 20.

32. **Fill in the blank:** During a total solar eclipse, stars near the Sun's edge are observed to be a bit ________ the Sun than they would have been had the Sun not been present, in the amount predicted by Einstein's general theory of relativity.
33. **Fill in the blank:** In the few days after a powerful solar flare, we might expect to see _______ at night in Earth's atmosphere.
34. **Fill in the blank:** By far the most abundant element in the Sun is _________.
35. **Fill in the blank:** The Sun's color, when it is high in the sky and there isn't much pollution, is _______.
36. **Fill in the blank:** The gradual ______ of Mercury's orbit provided one test of Einstein's general theory of relativity.

†This question requires a numerical solution.

TOPICS FOR DISCUSSION

1. Try to put yourself in the positions of the ancients. What would you think if someone suggested that the stars are simply very distant Suns?
2. Do you think it is worthwhile to spend Federal funding on the solar-terrestrial connection? Consider both short-term effects, such as electricity blackouts caused by particles from solar flares, and long-term effects, such as global cooling produced by events similar to the Maunder minimum.

MEDIA

Virtual Laboratories

Helioseismology

AceAstronomy™ Log into AceAstronomy at **http://astronomy.brookscole.com/cosmos3** to access quizzes and animations that will help you assess your understanding of this chapter's topics.

Log into the Student Companion Web Site at **http://astronomy.brookscole.com/cosmos3** for more resources for this chapter including a list of common misconceptions, news and updates, flashcards, and more.

C H A P T E R 11

Stars: Distant Suns

ORIGINS

The stars are simply distant versions of our Sun. By studying their properties, such as surface temperature and intrinsic brightness, we can achieve a deeper understanding of the Sun and how its light and other forms of energy originate.

The properties of all stars, including the Sun, depend mostly on their mass, which in special cases can be determined from studies of binary stars. Cepheid variable stars give celestial distances, which we will later see are important for studying the age and evolution of our Universe. Star clusters can be used to determine the ages of stars, and to study the evolution of typical stars like our Sun.

AIMS

1. Learn how the colors and spectral types of stars tell us their surface temperatures (Section 11.1).
2. Discuss the most fundamental way of measuring the distances of stars, and see how their intrinsic brightnesses (luminosities) are determined (Sections 11.2 to 11.3).
3. See how stars can be classified into several different groups when their surface temperatures and luminosities are plotted on a graph (Section 11.4).
4. Explore the motions of stars along and perpendicular to our line of sight (Section 11.5).
5. Investigate binary stars, and understand how they are important for measuring stellar masses (Section 11.6).
6. Learn about stars whose brightness changes with time, some of which are used to measure distances (Section 11.7).
7. Discuss star clusters and how we determine their ages (Section 11.8).

The thousands of stars in the sky that we see with our eyes, and the millions more that telescopes reveal, are glowing balls of gas. Their bright surfaces send us the light that we see. Though we learn a lot about a star from studying its surface, we can never see through to a star's interior, where the important action goes on.

In this chapter, we will discuss the surfaces of stars and what they tell us. First we explain how we tell the temperatures of stars and what we observe to study them. We also explain how stars move, and how we determine their distances. Then we will learn about stars that come with friends: other stars or groups of stars. Only when we finish these useful studies will we go on to discuss the stellar interiors, in Chapter 12.

11.1 Colors, Temperatures, and Spectra of Stars

11.1a Taking a Star's Temperature

Let us continue with the discussion of spectra that we began in Chapter 2. When you heat an iron poker in a fire, it begins to glow and then becomes red hot. If we could make it hotter still, it would become white hot, and eventually bluish-white. To understand the temperature of the poker or of a hot gas, we measure its spectrum (■ Fig. 11–1). A dense, opaque gas or a solid gives off a continuous spectrum—that is, light changing smoothly in intensity (brightness) from one color to the next.

Recall from Chapter 2 that a "black body" is a perfect emitter: Its spectrum depends only on its temperature, not on chemical composition or other factors. We can approximate the spectrum of the visible radiation from the outer layer of a star as a **black-body curve.** The black-body curve is also called the Planck curve, in honor of the physicist Max Planck. Its derivation about 100 years ago was a triumph in the early development of quantum physics.

AceAstronomy™ The AceAstronomy icon throughout this text indicates an opportunity for you to test yourself on key concepts, and to explore animations and interactions of the AceAstronomy website at **http://astronomy.brookscole.com/cosmos3**

Stars are forming in this region of our Galaxy, in the direction of the constellation Cygnus; we see glowing dust in a false-color image made with the Spitzer Space Telescope. The image shown is approximately half the Moon's diameter. The data were taken in three infrared wavelengths between 3 micrometers and 8 micrometers (for comparison, yellow light is about 0.6 micrometers).
NASA/JPL-Caltech/S. Carey (Caltech)

■ **FIGURE 11–1** When a narrow beam of light is dispersed (spread out) by a prism, we see a spectrum. Incoming light is usually passed through an open slit to provide the narrow beam.

A different black-body curve corresponds to each temperature (■ Fig. 11–2). Note that as the temperature increases, the gas gives off more energy at every wavelength. Indeed, per unit of surface area, a hot black body emits much more energy per second than a cold one.

Moreover, the wavelength at which most energy is given off is farther and farther toward the blue as the temperature increases. The wavelength of this peak energy is shown with a dashed line. At temperatures of 4000 K, 5000 K, 6000 K, and 7000 K, the peak of the black-body curve is at wavelengths of 7200 Å (red), 5800 Å (yellow), 4800 Å (blue), and 4100 Å (violet), respectively. Thus, the hottest stars look blue and the coolest ones are red. Those of intermediate temperature (like the Sun) appear white, despite having spectra that peak around yellow or green wavelengths, because of the physiological response of our eyes.

Astronomers rely heavily on the quantitative expression of the two radiation laws noted in the preceding paragraphs to determine various aspects of stars (see Chapter 2, *Figure It Out 2.2: Black-Body Radiation and Wien's Law; Figure It Out 2.3: Black-Body Radiation and the Stefan-Boltzmann Law*). For example, by simply measuring the

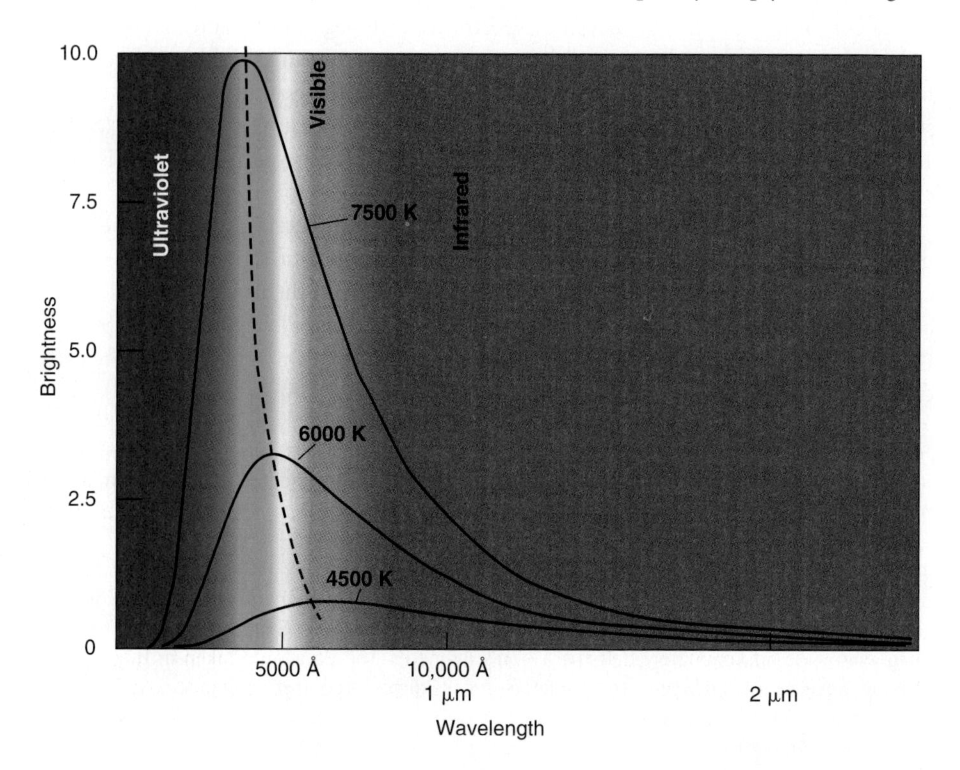

■ **FIGURE 11–2** The brightness (intensity) of radiation at different wavelengths for different temperatures. Black bodies—ideally radiating matter—give off radiation that follows these curves. Stars follow these curves fairly closely.

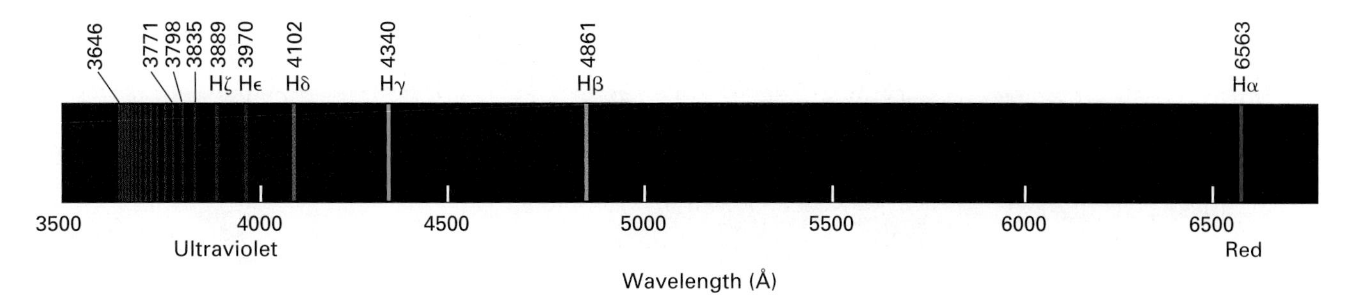

FIGURE 11–3 The series of spectral lines of hydrogen that appears in the visible part of the spectrum. The strongest line in this series, Hα (H-alpha), is in the red. This "Balmer series" is labelled with the first letters of the Greek alphabet. Note how the series converges toward the ultraviolet end of the spectrum. That location is known as the "Balmer limit."

wavelength of the peak brightness of a star's spectrum, astronomers can take the star's temperature (though not with great accuracy, because the exact wavelength of the peak can be difficult to measure and the star isn't a perfect black body).

11.1b How Do We Classify Stars?

All the spectral lines of normal stars are absorption lines, also called dark lines. The absorption lines cause a star's spectrum to deviate from that of a black body, but not by much. Studying the absorption lines has been especially fruitful in understanding the stars and their composition.

We can duplicate on Earth many of the contributors to the spectra of the stars. Since hydrogen has only one electron, hydrogen's spectrum (Chapter 2) is particularly simple (■ Fig. 11–3). Atoms with more electrons have more possible energy levels, resulting in more choices for jumps between energy levels. Consequently, elements other than hydrogen have more complicated sets of spectral lines.

When we look at a variety of stars, we see many different sets of spectral lines, usually from many elements mixed together in the star's outer layers. Hydrogen is usually prominent, though often iron, magnesium, sodium, or calcium lines are also present. The different elements can be distinguished by looking for their distinct *patterns* of spectral lines.

Early in the 20th century, Annie Jump Cannon at the Harvard Observatory classified hundreds of thousands of stars by their visible-light spectra (■ Fig. 11–4). She first classified them by the strength of their hydrogen lines, defining stars with the strongest lines as "spectral type A," stars of slightly weaker hydrogen lines as "spectral type B," and so on. (Many of the letters wound up not being used.)

It was soon realized that hydrogen lines were strongest at some particular temperature, and were weaker at hotter temperatures (because the electrons escaped completely from the atom) or at cooler temperatures (because the electrons were only in the lowest possible energy levels, which do not allow the visible-light hydrogen lines to form).

Rearranging the spectral types in order of temperature—from hottest to coolest—gave O B A F G K M. Recently, even cooler objects—stars and brown dwarfs—have been found, and are designated L-type and (still more recently) T-type.

Harvard College Observatory

FIGURE 11–4 Part of the "computing staff" of the Harvard College Observatory in 1917, when the word "computer" meant a person and not a machine. Annie Jump Cannon, 5th from the right, classified over 500,000 spectra in the decades following 1896. Later in this chapter, we discuss the work of Henrietta Leavitt, 5th from the left, who studied variable stars. Our basic ideas of cosmology are based on one of her discoveries.

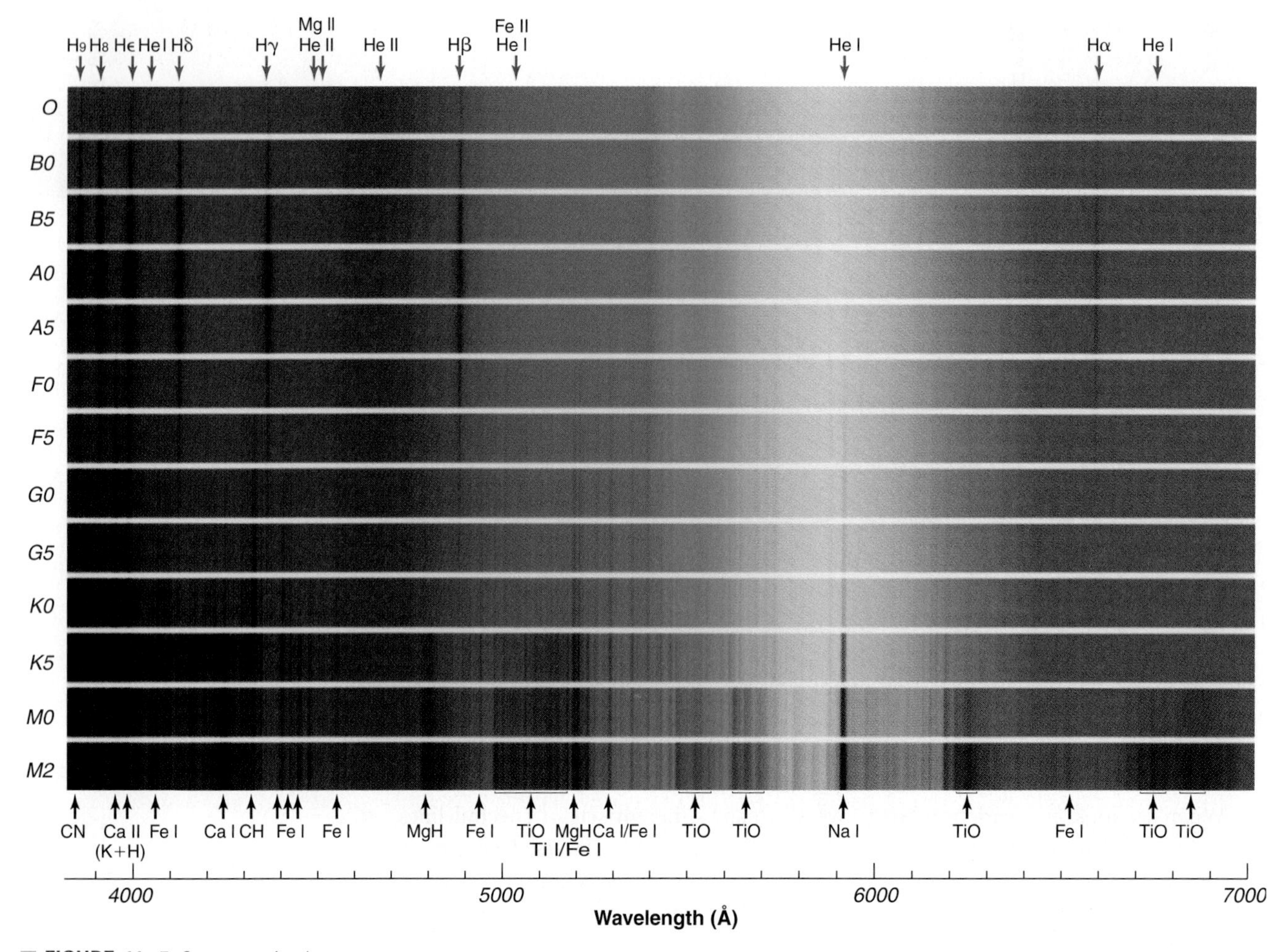

■ **FIGURE 11–5** Computer simulation of the spectra of stars from a wide range of spectral types. The hottest stars are at the top and the coolest at the bottom. Notice how the hydrogen lines decline in darkness for types either above or below A-type stars. (An astronomer would simply say "A stars.") The original lettering was in order of the strength of the hydrogen lines. Only subsequently was the temperature ordering discovered. (Roger Bell, University of Maryland, and Michael Briley, U. Wisconsin at Oshkosh)

(However, most of the T-type objects and some of the L-type objects are actually brown dwarfs, and hence, in a sense, aren't fully stars; see the discussion in Chapter 12.) Generations of astronomy students have memorized the order using the mnemonic "**O**h, **b**e **a** **f**ine **g**irl [or **g**uy], **k**iss **m**e." The authors invite you to think of your own mnemonic for O B A F G K M L T, and we will be glad to receive your entries at this book's website.

Looking at a set of stellar spectra in order (■ Fig. 11–5) shows how the hydrogen lines are strongest at spectral type A, which corresponds to a surface temperature of roughly 10,000 K.

A pair of spectral lines from calcium becomes strong in spectral type G (a type that includes the Sun), at about 6000 K (■ Fig. 11–6). In very cool stars, those of spectral type M, the temperatures are so low (only 3000 K) that molecules can survive, and we see complicated sets of spectral lines from them. At the other extreme, the hottest stars, of spectral type O, can reach 50,000 K. Some stars of type O have shells of hot gas around them and give off emission lines, though stars generally show only absorption lines in their spectra.

Astronomers subdivide each spectral type into ten subtypes ranging from hottest (0) to coolest (9); thus, for example, we have A0 through A9, and then F0 through F9. One of the first things most astronomers do when studying a star is to determine its spectral type and thus its surface temperature. The Sun is a type G2 star, corresponding to a temperature of 5800 K.

We have known that stars are made mainly of hydrogen and helium only since the 1920s. When Cecilia Payne (later Cecilia Payne-Gaposchkin) at Harvard first suggested the idea that stars consist largely of hydrogen, it seemed impossible. But this was what

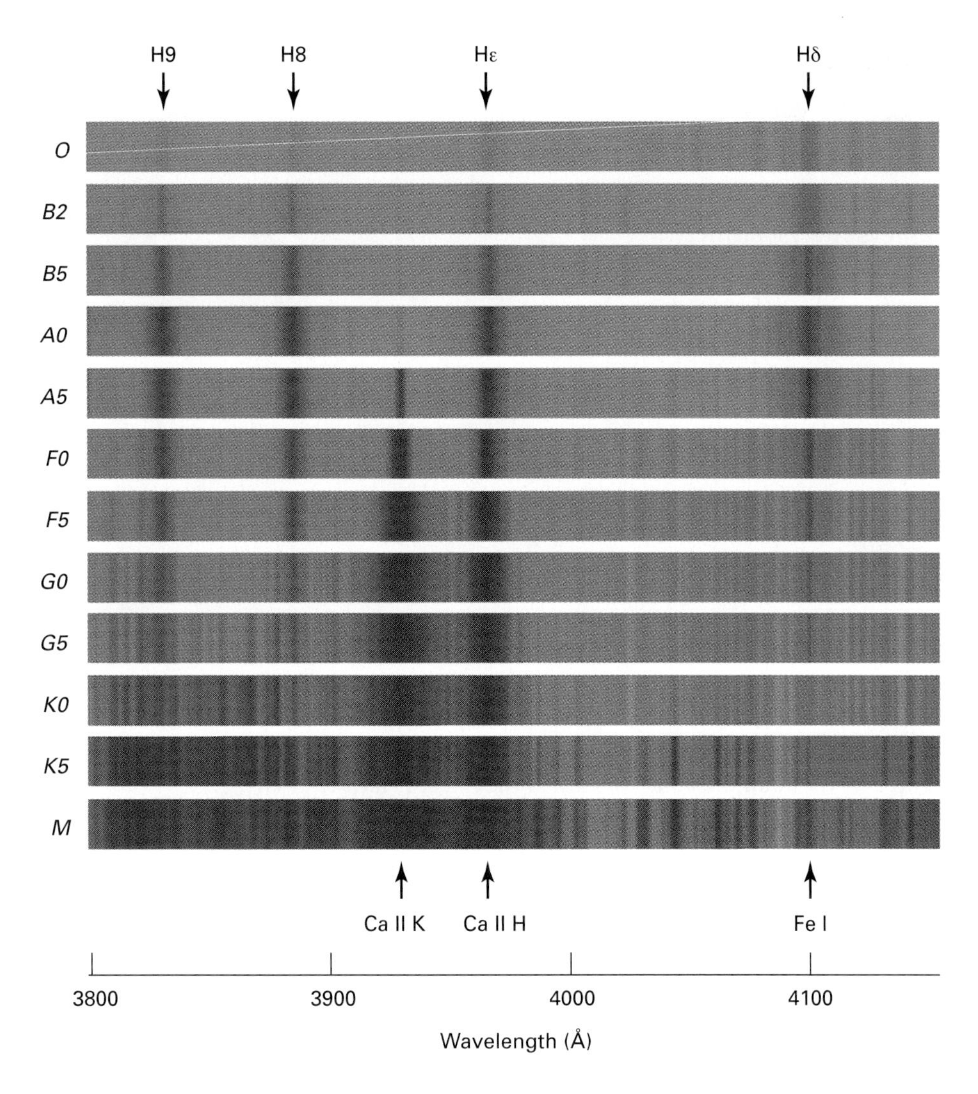

■ **FIGURE 11–6** The violet and blue part of the previous image, displayed to show more clearly the change in visibility of the H and K lines of ionized calcium (labelled Ca II, spoken "calcium 2") versus the Hε (H-epsilon) line of hydrogen as spectral types go from cooler to hotter. (Roger Bell, University of Maryland, and Michael Briley, U. Wisconsin at Oshkosh)

her analyses of stellar spectra implied, and she was right. It was years before other astronomers accepted this conclusion.

AceAstronomy™ Log into AceAstronomy and select this chapter to see the Astronomy Exercises "Emission and Absorption Spectra" and "Stellar Atomic Absorption Lines."

11.2 How Distant Are the Stars?

When you spot your friends on campus, you might use your binocular vision to judge how far away they are. Your brain interprets the slightly different images from your two eyes to give such nearby objects some three-dimensionality and to assess their distance. Also, we often unconsciously judge how far away an object is by assessing its apparent size compared with the sizes of other objects.

But the stars are so far away that they appear as points, so we cannot judge their size. And our eyes are much too close together to give us binocular vision at that great distance. Only for the nearest stars can we reliably measure their distance fairly directly even with our best methods.

The distances of nearby stars can be determined by **triangulation** (also known as "trigonometric parallax")—a sort of binocular vision obtained by taking advantage of our location on a moving platform (the Earth). The basic idea is that the position of a nearby object shifts relative to distant objects when viewed from different lines of sight (■ Fig. 11–7).

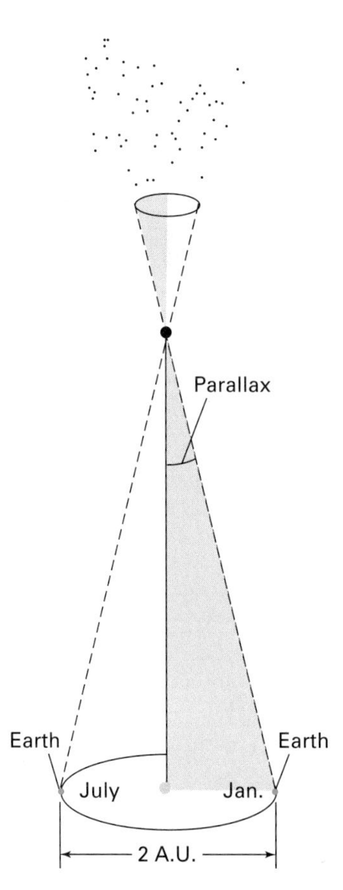

■ **FIGURE 11–7** The nearer stars seem to be slightly displaced with respect to the farther stars when viewed from different locations in the Earth's orbit.

FIGURE IT OUT 11.1
Stellar Triangulation

The technique of measuring the distance of a nearby star by triangulation is also known as "trigonometric parallax." We photograph the star twice, about 6 months apart. We then measure the angular shift of the star relative to galaxies or very distant stars, whose positions are stationary or nearly so.

The "parallax" *(p)* of a star is defined to be *half* of the angular shift produced over a 6-month baseline (2 A.U., the diameter of Earth's orbit—see Figure 11–7).

The distance of a star whose parallax is 1" (1 second of arc) is called 1 parsec, abbreviated 1 pc. (This comes from a *par*allax of one *sec*ond of arc. See Chapter 3, *Figure It Out 3.3: Angular Resolution of a Telescope*, for a review of angular measure.) In the more familiar units of light-years (lt yr), 1 pc is about 3.26 lt yr.

It turns out that with this definition of a parsec, the relationship between distance and parallax is very simple. A star's distance d (in parsecs) is simply the inverse (reciprocal) of its parallax p (in seconds of arc):

$$d = 1/p.$$

For example, a star whose measured parallax is 0.5" has a distance of 1/0.5 = 2 pc, or about 6.5 lt yr. The nearest star, Proxima Centauri, has a parallax of 0.77" (its full angular shift over a 6-month period is twice this value, 1.54"), and hence a distance of 1/0.77 = 1.3 pc (or 4.2 lt yr).

For example, if you put a finger in front of your eyes, and close one eye, the finger will appear to be projected against a particular background object. If you close this eye and open the other one, the position of the finger will shift to a different background object. The amount of shift is smaller if the finger is farther away.

Here's how we apply this method to nearby stars (see *Figure It Out 11.1: Stellar Triangulation* for details). At six-month intervals, the Earth carries us entirely across its orbit, halfway around (■ Fig. 11–8*a*). Since the average radius of the Earth's orbit is 1 astronomical unit (A.U.), we move by 2 A.U. This distance is enough to give us a somewhat different perspective on the nearest stars. These stars appear to shift very slightly against the background of more distant stars. Clearly, the more distant the star, the smaller is its angular shift (■ Fig. 11–8*b*). The nearest star—known as Proxima Centauri—appears to shift by only the diameter of a dime at a distance of 2.4 km! It turns out to be about 4.2 light-years away. (It is in the southern constellation of Centaurus, and is not visible from most of the United States.)

By calculating the length of the long side of a giant triangle—by "triangulating"—we can measure distances in this way out to a few hundred light-years. But there are only a few thousand stars that are so close to us, and at the farthest distances the results are not very accurate.

The European Space Agency lofted a satellite, Hipparcos, in 1989 to measure the positions of stars. Measuring the positions and motions of stars is known as **astrometry.** Based on its data, in 1997 the scientists involved released the Hipparcos catalogue, containing distances for 118,000 stars, relatively accurate to about 300 light-years away. A secondary list, the Tycho-2 catalogue, contains less-accurate distances for a

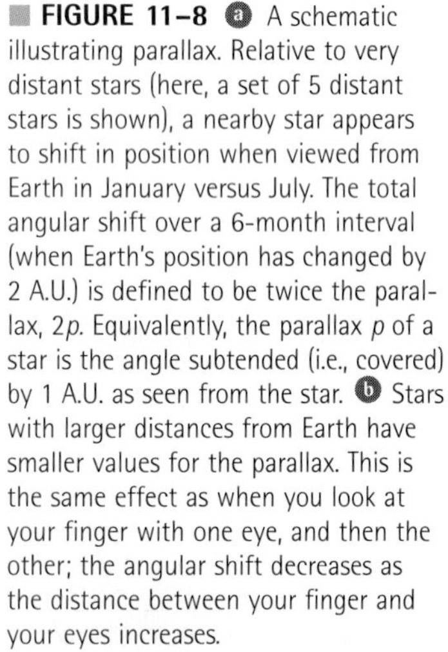

■ **FIGURE 11–8** ⓐ A schematic illustrating parallax. Relative to very distant stars (here, a set of 5 distant stars is shown), a nearby star appears to shift in position when viewed from Earth in January versus July. The total angular shift over a 6-month interval (when Earth's position has changed by 2 A.U.) is defined to be twice the parallax, 2*p*. Equivalently, the parallax *p* of a star is the angle subtended (i.e., covered) by 1 A.U. as seen from the star. ⓑ Stars with larger distances from Earth have smaller values for the parallax. This is the same effect as when you look at your finger with one eye, and then the other; the angular shift decreases as the distance between your finger and your eyes increases.

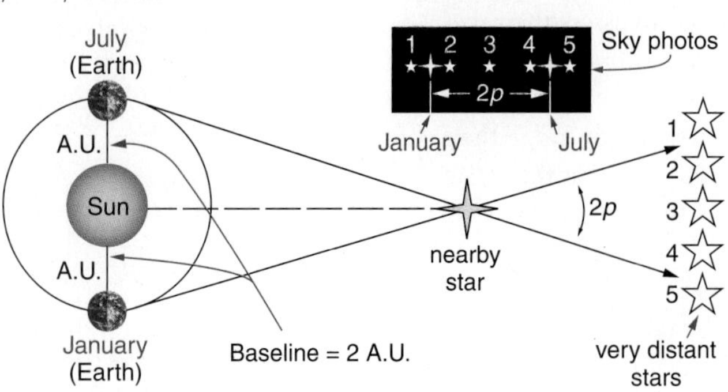

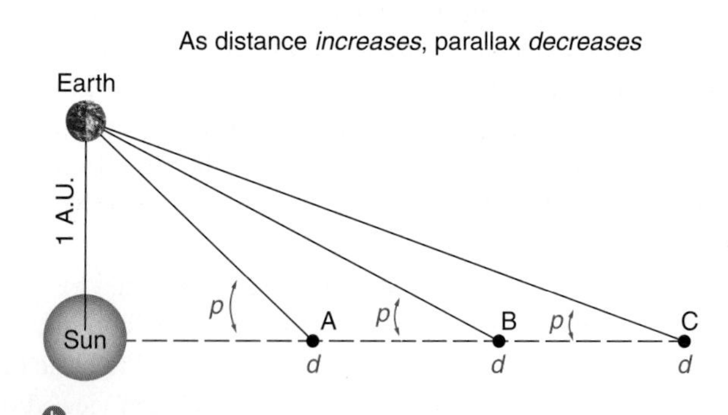

million stars but provides accurate "proper motions" (motion across the sky, which we will discuss in Section 11.5) for a total of two and a half million stars.

These catalogues greatly improve our fundamental knowledge of the distances of nearby stars. Moreover, our understanding of distant stars and even galaxies is often based on the distances to these closer stars (see the discussion in Chapter 16), so the improvement is important for much of astronomy.

Our Milky Way Galaxy is perhaps 100,000 light-years across, so even the 600 light-years (diameter) spanned by the Hipparcos catalogue takes up less than 1 per cent of the diameter of our Galaxy. A successor European Space Agency spacecraft (Gaia) is still on the books for launch in 2011, or beyond. Also, NASA's Space Interferometry Mission (recently renamed SIM PlanetQuest), mentioned in Chapter 9, is also planned for a 2011 launch; it will provide measurements of the positions of stars in our Galaxy with unprecedented accuracy, thereby leading to much improved distances.

ASIDE 11.1: Gaia's wide-field viewing

The electronic camera for Gaia is already being constructed. It is to contain 1.5 gigapixels—that is, 1500 megapixels—about 500 times more pixels than ordinary digital cameras in common use for amateur photography.

AceAstronomy™ Log into AceAstronomy and select this chapter to see the Astronomy Exercises "Parallax" and "Parallax 2."

11.3 How Powerful Are the Stars?

Automobile headlights appear faint when they are far away but can almost blind us when they are up close. Similarly, stars that are intrinsically the same brightness appear to be of different brightnesses to us, depending on their distances. A star's intrinsic brightness is its **luminosity,** or "power." If we use units that show the amount of energy a star gives off in a given time, the Sun's luminosity, for example, is 4×10^{26} joules/second, where a joule (symbol J) is a unit of energy. (Sometimes you will see energy given in ergs, a smaller unit of energy. One joule is 10 million ergs.)

We have previously described (in Chapter 4) how we give stars' apparent brightnesses in "apparent magnitude." To tell how inherently bright stars are, astronomers often use a modification of the magnitude scale known as **absolute magnitude;** see *A Closer Look 11.1: Using Absolute Magnitudes.*

To do so, they have set a specific distance at which to compare stars. Basically, if something is intrinsically fainter, it has a higher absolute magnitude. We label our

A Closer Look 11.1 Using Absolute Magnitudes

To tell how intrinsically bright stars are, astronomers have set a specific distance at which to compare stars. The distance, which is a round number (10 parsecs) in the special units astronomers use for triangulation (parsecs; see *Figure It Out 11.1: Stellar Triangulation*), comes out to be about 32.6 light-years. (Ten parsecs corresponds to a parallax of 0.1", the parallax being what is actually measured at a telescope.)

We calculate how bright a star would appear on the magnitude scale if it were 32.6 light-years away from us. In this way, we are comparing stars as though they were all at that distance.

The apparent brightness that a star would have at this standard distance of 32.6 light-years is its "absolute magnitude." For a star that is actually at the standard distance, its absolute magnitude and apparent magnitude are the same.

For a star that is farther from us than that standard distance, we would have to move it closer to us to get it to that standard distance. This would make its magnitude at the standard distance (its absolute magnitude) brighter than the apparent magnitude we saw for it before it moved. (A brighter star has a lower numerical value, 4th magnitude instead of 6th magnitude, for example.) Then by comparing its absolute magnitude and its apparent magnitude (which can easily be measured at a telescope), astronomers calculate how far away the star must be.

The method links absolute magnitude, apparent magnitude, and distance. Thus, it is particularly valuable, since astronomers have several ways of finding a star's absolute magnitude.

FIGURE IT OUT 11.2

The Inverse-Square Law

The magnitude scale is intuitive but from another era; still, astronomers often use it, and it appears in amateur astronomy magazines. It has the advantage of using small numbers that are relatively easy to remember.

A different approach is to use the apparent brightness itself, b, instead of apparent magnitude. Its unit of measure is energy received per unit area, per unit time. For example, b can be measured in joules/m^2/sec, where a joule is a small unit of energy (formally, one kg-m^2/sec^2). Similarly, instead of absolute magnitude, we can use luminosity L to denote a star's intrinsic brightness or power (the amount of energy it emits per unit time, joules/sec).

A star's apparent brightness depends on its distance, because the light spreads out uniformly in all directions over a sphere whose radius is the distance. The larger the distance, the greater the surface area of the sphere, and so the smaller the amount of light passing through any small section of the sphere (see Figure 11–9). Since the surface area of a sphere is just 4π times the square of the radius, and in this case the radius is the star's distance d, the surface area is $4\pi d^2$. So, we find that the star's apparent brightness b is related to its luminosity L and distance d as follows:

$$b = L/(4\pi d^2).$$

(Note that the units work out correctly: b is in joules/m^2/sec, L is in joules/sec, and d is in meters.) In other words, each unit area of the sphere (for example, 1 m^2) intercepts only a fraction, $1/(4\pi d^2)$, of the total L.

This equation is just the well-known inverse-square law of light. For example, if a star's distance is doubled, it appears four times fainter, because a given amount of light has spread over a surface area four times larger. As an analogy, recall that if you are spraying water out of a hose by putting your thumb over part of the opening, you can either get many people slightly wet (by being relatively far from them) or one person very wet (by being close).

We now have the tools needed to determine the luminosity of a nearby star. First, measure its apparent brightness, b (for example, with a CCD, the type of electronic detector now in universal use by astronomers); this procedure is called "photometry." Also measure its parallax, p (units: seconds of arc), and determine its distance in parsecs from the equation $d = 1/p$ (recall *Figure It Out 11.1: Stellar Triangulation*). Then convert parsecs to meters. (The conversion factor appears in an Appendix.) Finally, use the inverse-square law to compute $L = 4\pi d^2 b$.

It is often convenient to express the luminosity in terms of the Sun's luminosity, $L_{Sun} = 3.83 \times 10^{26}$ joules/sec. For example, a star whose luminosity is 7.66×10^{26} joules/sec is said to have $L = 2L_{\odot}$. (In astronomy, the symbol $\odot$ denotes the Sun. Thus, the Sun's luminosity is often written as an L with this symbol as a subscript, $L_{\odot}$.) The Sun's luminosity is equivalent to an absolute magnitude of 4.8—that is, the Sun would just barely be visible to the naked eye if it were at a distance of about 32.6 light-years (10 pc).

graphs in this book in both absolute magnitude and in luminosity. Some people find one easier to use; others prefer the other.

You can determine how far away a star is by comparing how bright it looks with its intrinsic brightness (its luminosity, whether measured in joules/second or as absolute magnitude). The method works because we understand how a star's energy spreads out with distance (■ Fig. 11–9).

The energy received from a star changes with the square of the distance: the energy decreases as the distance increases. Since one value goes up as the other goes down, it is an inverse relationship. We call it the **inverse-square law.** Using this law, there is a simple way to express the luminosity of a star in terms of its apparent brightness and distance (see *Figure It Out 11.2: The Inverse-Square Law*).

AceAstronomy™ Log into AceAstronomy and select this chapter to see the Active Figure called "Brightness and Distance" and the Astronomy Exercise "Apparent Brightness, Interstellar Matter."

11.4 Temperature–Luminosity Diagrams

If you plot a graph that has the surface temperature of stars on the horizontal axis (x-axis) and the luminosity (intrinsic brightness) of the stars on the vertical axis (y-axis),

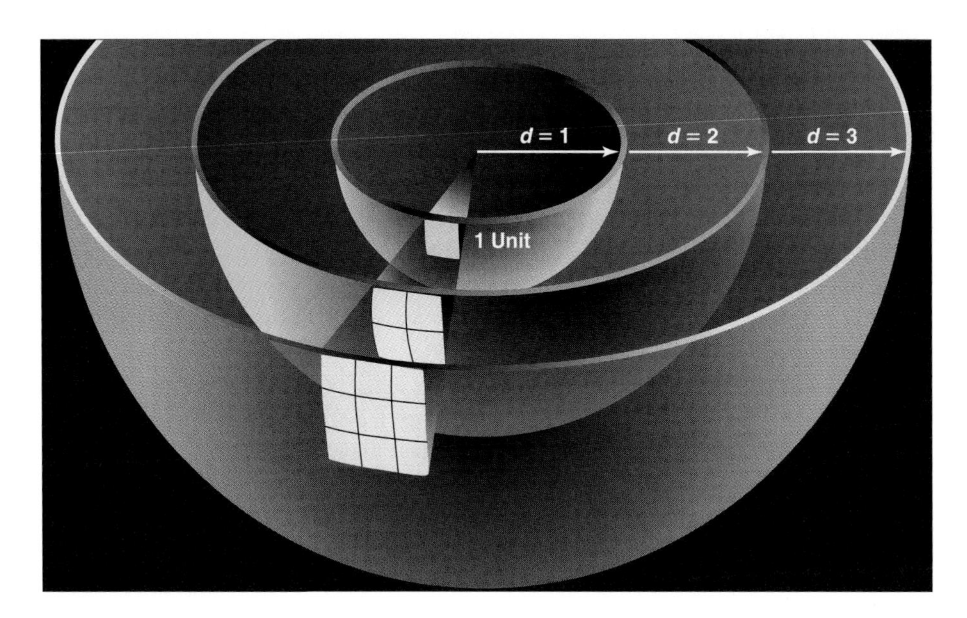

■ **FIGURE 11–9** The inverse-square law of apparent brightness. Radiation passing through a sphere twice as far away as another sphere from a central point (radius, or distance, $d = 2$ units instead of $d = 1$ unit) has spread out so that it covers $2^2 = 4$ times the area. The central point thus appears $1/4$ as bright to an observer located on the second sphere. Similarly, radiation passing through a sphere of radius (distance) $d = 3$ units has spread out so that it covers $3^2 = 9$ times the area, so the central point appears $1/9$ as bright to an observer located on the third sphere. In general, if it were n times farther away, it would cover n^2 times the area, and appear $1/n^2$ as bright.

the result is called a **temperature-luminosity diagram.** The luminosity (vertical axis) can also be expressed in absolute magnitudes, so the term "temperature-magnitude diagram" is synonymous. Sometimes the horizontal axis is labeled with spectral type, since temperature is often measured from spectral type or from a star's color (Section 11.1).

If we plot such a diagram for the nearest stars (■ Fig. 11–10), we find that most of them fall on a narrow band that extends downward (fainter) from the Sun. If we plot such a diagram for the brightest stars we see in the sky, we find that the stars are more scattered, but that all are intrinsically brighter than the Sun.

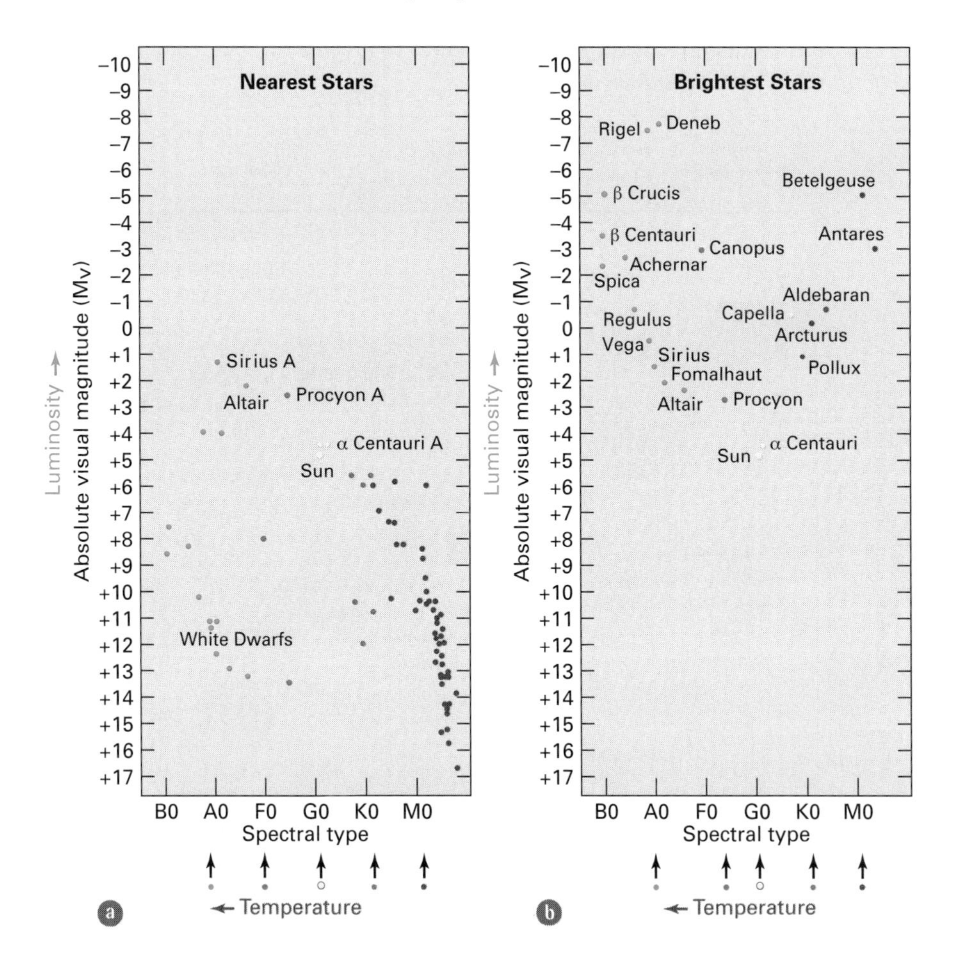

■ **FIGURE 11–10** Temperature-luminosity (or temperature-magnitude) diagrams (a) for the nearest stars in the sky and (b) for the brightest stars in the sky. The brightness scale is given in absolute magnitude—the apparent magnitude a star would have at a standard distance of about 32.6 light-years. The upward arrow indicates increasing luminosity (intrinsic brightness). Because the effect of distance has been removed, the intrinsic properties of the stars can be compared directly on such a diagram.

Note that there are no stars near the top of the nearest-stars diagram (*a*). Thus none of the nearest stars is intrinsically very luminous. Note further that there are no stars near the bottom of the brightest-stars diagram (*b*). Thus none of the stars that appear brightest to us is intrinsically very faint.

American Institute of Physics, Emilio Segrè Visual Archives, Margaret Russell Edmondson Collection

■ **FIGURE 11–11** Henry Norris Russell and his family, circa 1917.

Dorrit Hoffleit—Yale U. Observatory; courtesy of American Institute of Physics, Emilio Segrè Visual Archives

■ **FIGURE 11–12** Ejnar Hertzsprung in the 1930s.

The idea of such plots came to two astronomers in the early years of the 20th century. Henry Norris Russell (■ Fig. 11–11), at Princeton University in the United States, plotted the absolute magnitudes (equivalent to luminosities) as his measure of brightness. Ejnar Hertzsprung (■ Fig. 11–12), in Denmark, plotted apparent magnitudes but did them for a group of stars that were all at the same distance. He could do this by considering a cluster of stars (Section 11.8), since all the stars in the cluster are essentially the same distance away from us. The two methods came to the same thing, in that the magnitudes found for different stars could be directly compared. Temperature-luminosity (temperature-magnitude) diagrams are often known as "Hertzsprung–Russell diagrams" or as "H–R diagrams."

When we graph quite a lot of stars, or put together both nearby and farther stars (■ Fig. 11–13), we can see clearly that most stars fall in a narrow band across the temperature-luminosity diagram. This band is called the **main sequence.** Normal stars in the longest-lasting phase of their lifetime are on the main sequence.

The position of a star on the main sequence turns out to be determined by its mass: Massive stars are hotter and more powerful than low-mass stars. For reasons we will discuss in Chapter 12, the position of a star does not change much while it is on the main sequence: The Sun stays at more or less the same position on the main sequence for about 10 billion years. Stars on the main sequence are called **dwarfs,** so the Sun is a dwarf.

A few stars are more luminous (intrinsically brighter) than main-sequence stars of the same surface temperature. Since the same surface area of gas at the same temperature gives off the same amount of energy per second, these stars must be bigger than the main-sequence stars (see *Figure It Out 11.3: A Star's Luminosity*). They are thus called **giants** or even **supergiants.** The reddish star Betelgeuse in the shoulder of the constellation Orion is a supergiant. Scientists are increasingly able to use interferometers in the infrared and even in the visible to measure the sizes and shapes of very large or nearby stars directly (■ Figs. 11–14 and 11–15); see *A Closer Look 11.2: Proxima Centauri: The Nearest Star Beyond the Sun.*

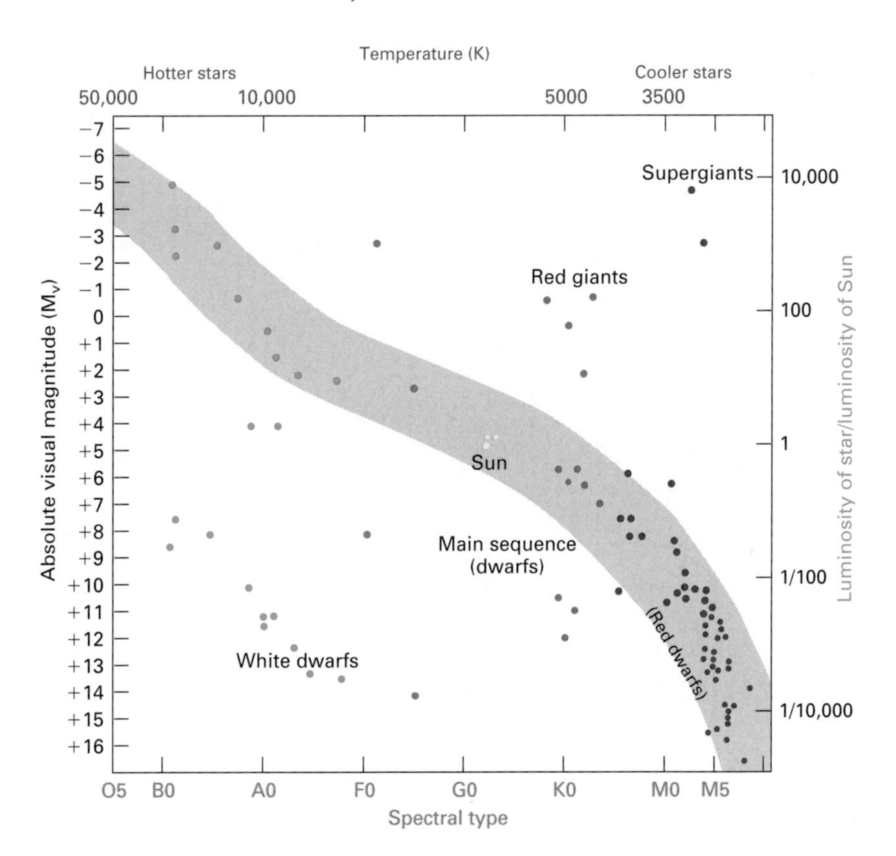

■ **FIGURE 11–13** The temperature-luminosity (temperature-magnitude) diagram with both nearest and brightest stars included. The spectral-type axis *(bottom)* is equivalent to the surface-temperature axis *(top)*. The absolute-magnitude axis *(left)* is equivalent to the luminosity axis *(right)*, which shows the visual luminosity compared with that of the Sun.

The stars at the *upper left* of the main sequence (the gray diagonal band) are more luminous than those at the *lower right*. They are both hotter and larger. We mostly determine the sizes of stars indirectly: We get the luminosity of each bit of surface from their temperatures and then figure out how much surface they must have to appear as luminous as they are.

FIGURE IT OUT 11.3
A Star's Luminosity

We can easily relate the luminosity (intrinsic brightness, or power), radius, and surface temperature of a star. Recall from Chapter 2 (*Figure It Out 2.3: Black-Body Radiation and the Stefan-Boltzmann Law*) the Stefan-Boltzmann law, $E = \sigma T^4$, where E is the energy emitted per unit area (for example, cm^2) per second, T is the temperature, and σ (the Greek letter "sigma") is a constant. To get the luminosity (joules/sec) of a star, we must multiply E by the surface area, $4\pi R^2$ for a sphere of radius R, or

$$L = 4\pi R^2 \sigma T^4.$$

Thus, if we know the luminosity and surface temperature of a star, we can derive its radius R. (The luminosity may have been derived with the inverse-square law by using the apparent brightness and distance, as in *Figure It Out 11.2: The Inverse-Square Law.* The surface temperature can be found from Wien's law, Chapter 2, *Figure It Out 2.2: Black-Body Radiation and Wien's Law.*)

Suppose instead we have two stars, labeled by subscripts 1 and 2, with $L_1 = 4\pi R_1^2 \sigma T_1^4$ and $L_2 = 4\pi R_2^2 \sigma T_2^4$. If we divide one equation by the other, constants such as 4π and σ cancel, and we get

$$(L_1/L_2) = (R_1/R_2)^2 (T_1/T_2)^4.$$

So, for example, if Star 1 has twice the radius of Star 2, but only half the surface temperature of Star 2, then the luminosity of Star 1 will be $^1/_4$ that of Star 2: $(L_1/L_2) = (2)^2(^1/_2)^4 = (4)(^1/_{16}) = {}^1/_4$. This example illustrates the advantage of using *ratios* when solving certain kinds of problems: Nowhere do we have to actually use the values of 4π and σ.

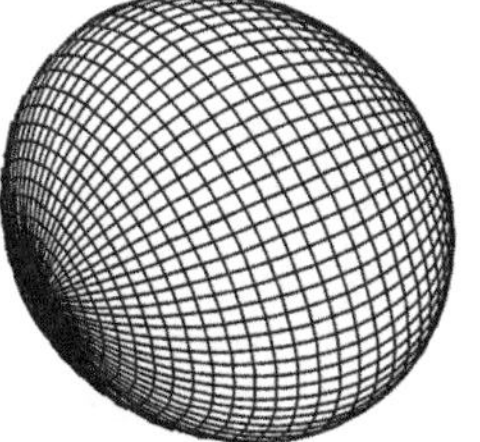
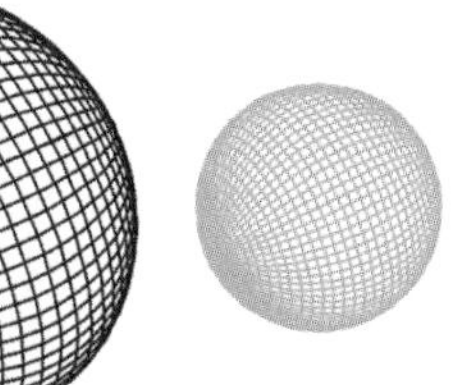

FIGURE 11–14 We can now measure the diameters of the nearest and brightest stars directly, using a technique known as interferometry. For example, measurements made with an interferometer on Palomar Mountain show that Altair, in the constellation Aquila, the 12th brightest star in the sky, is substantially out of round ("aspherical"). The size of the Sun is given for comparison. Altair is 18 times the Sun's mass, yet it spins every 10 hours, about 60 times faster than our Sun. The spin makes Altair's equatorial diameter 14% larger than its polar diameter. (Gerard van Belle, JPL/Caltech)

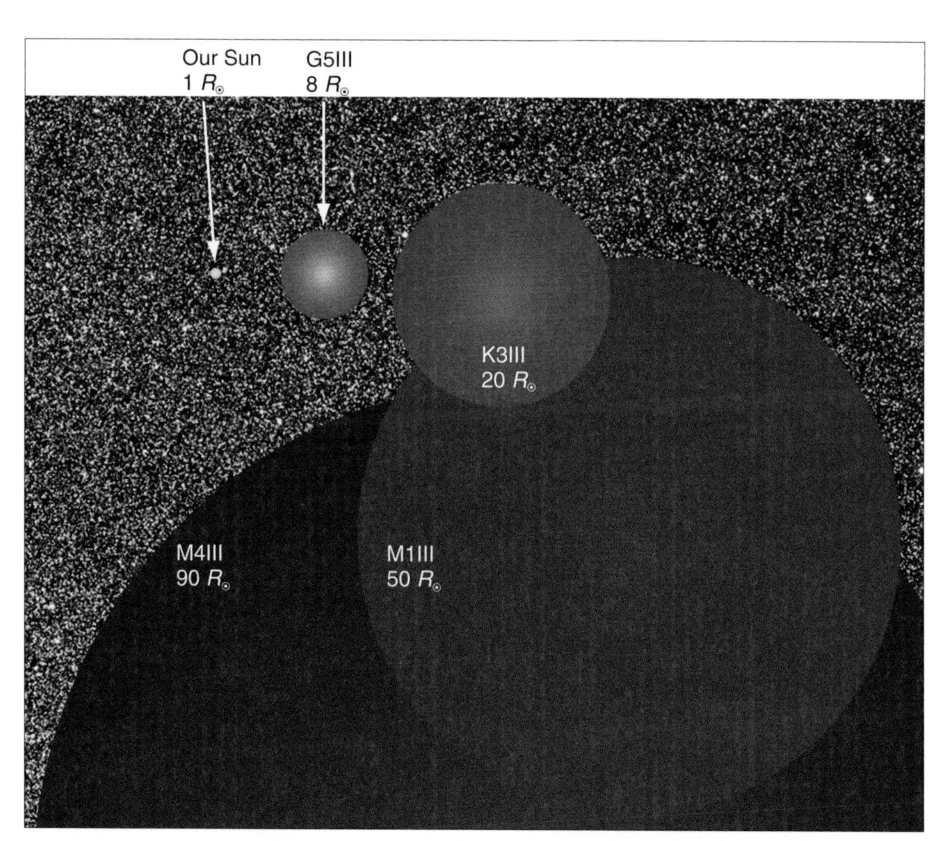

FIGURE 11–15 Palomar interferometer measurements of a variety of giant and supergiant stars show their differences in size directly. The Sun, a dwarf star (Roman numeral V) of spectral type G2, is much smaller than the giant stars (Roman numeral III) that are also shown, and over 300 times smaller than some supergiant stars (Roman numeral I). Our ability to measure the sizes of many stars, instead of merely deducing sizes from their luminosities and surface temperatures, is increasing rapidly because of improvements in interferometry. (Gerard van Belle, JPL/Caltech)

A Closer Look 11.2 PROXIMA CENTAURI: THE NEAREST STAR BEYOND THE SUN

In 2002, scientists using a pair of the Very Large Telescope array's four 8.2-m telescopes succeeded in making interferometric measurements of Proxima Centauri, the nearest star to Earth other than the Sun. They measured its diameter to be 1.02 ± 0.08 thousandths of a second of arc, or about the size of an astronaut on the surface of the Moon as seen from the Earth (or the head of a pin on the surface of the Earth as seen from the International Space Station). Proxima is only one-seventh the size and one-seventh the mass of the Sun. It is on the extreme low-mass end of M stars; if it had only half its mass, it would be a brown dwarf (see Chapter 12) instead of a normal star.

Jack Schmidling

a

b

© 2002, Anglo-Australian Observatory/Royal Observatory, Edinburgh; photograph by David Malin from Digitized Sky Survey data

■ **FIGURE 11–16** a The proper motion of Barnard's star across the sky, noticeable even over an astronomically short time interval of only months or years. It was 50 seconds of arc over the 7-year interval shown, and it would cover the diameter of the Moon in 180 years. (The difference in dot size is an artifact.) b Here we see the angular motion of α (the Greek letter "alpha") Centauri, shown in an overlay of three images taken over a 17-year period, each through a filter of a different color (pure blue, red, and green). When the images of all the other stars are aligned, α Centauri's motion stands out. It would cover the diameter of the Moon in about 500 years.

A few stars are intrinsically fainter than dwarfs (main-sequence stars) of the same color. These less luminous stars are called **white dwarfs.** Do not confuse white dwarfs with normal dwarfs (main-sequence stars); white dwarfs are smaller than normal dwarfs. The Sun is 1.4 million kilometers across, while the white dwarf Sirius B (a companion of the bright star Sirius) is only about 10 thousand kilometers across, roughly the size of the Earth. We shall see in Chapter 13 that giants, supergiants, and white dwarfs are later stages of life for stars.

If the spectral type (that is, surface temperature) of a given star is measured, and if a detailed analysis of the star's spectrum reveals that it is a main-sequence star (this can be determined by examining subtle aspects of the spectrum), then we know its luminosity. Comparison with its observed apparent brightness then yields its distance (see *Figure It Out 11.2: The Inverse-Square Law*). This method for finding distances is known as "spectroscopic parallax." Although it works best for main-sequence stars, it can also be used with white dwarfs, giants, and supergiants, as long as the star can be confidently placed in one of these categories through detailed spectral analysis.

AceAstronomy™ Log into AceAstronomy and select this chapter to see the Astronomy Exercise "H-R/Mass-Luminosity 3D Graph" and the Active Figure called "Animated HR Diagram."

11.5 How Do Stars Move?

11.5a Proper Motions of Stars

The stars are so far away that they hardly move across the sky relative to each other. Don't confuse this relative motion of stars with the nightly motion of stars across the sky, which is merely a reflection of Earth's rotation about its axis (as discussed in Chapter 4), or with the seasonal changes of constellations, which is a reflection of Earth's orbit around the Sun (as discussed in Chapter 1).

Only for the nearest stars can we easily detect any such relative motion among the stars, which is called **proper motion** (■ Fig. 11–16). However, for the most precise work astronomers have to take the effect of the proper motions (over decades) into account. The Hipparcos satellite has improved our measurements of many stars' proper motions, and the future Space Interferometry Mission (SIM PlanetQuest) will provide enormous amounts of additional data.

Note that the proper motion is an *angular* movement of stars across the sky, relative to each other; the units are typically seconds of arc per century. To get the *physical* velocity (speed, in km/sec) of each star in the plane of the sky, perpendicular to our line of sight, we need to also know its distance. After all, a given angular speed could be the result of a nearby star whose physical speed is low, or of a more distant star whose physical speed is higher. For a given physical speed, the angular speed (proper motion) decreases with increasing distance. A good example of this relationship is an airplane taking off from the airport: when it is close, it appears to zoom by, whereas when it is far away, it just crawls across the sky.

11.5b Radial Velocities of Stars

Astronomers can actually measure motions toward and away from us (that is, along our line of sight) much better than they can measure motions from side to side, in the plane of the sky (perpendicular to our line of sight). Recall (Chapters 2 and 9) that a motion toward or away from us on a line joining us and a star (a radius) is called a **radial velocity.** A radial velocity shows up as a Doppler shift, a change of the spectrum so that anything that originally appeared at a given wavelength now appears at another wavelength. We briefly discussed the Doppler shift in Chapter 2, but here we will give a more thorough explanation.

Doppler shifts in sound are more familiar to us than Doppler shifts in light. You can easily hear the pitch of a car's engine drop as the car passes us. We are hearing the wavelength of its sound waves increase as the object passes us and begins to recede. A similar effect takes place with light when we observe light that was emitted by an object that is moving toward or away from us. The effect, though, is not obvious to the eye; sensitive devices are necessary to detect the Doppler shift in light.

To explain the Doppler shift in light or sound, consider an object that emits waves of radiation (■ Fig. 11–17). The waves can be represented by spheres that show the peaks (crests) of the wave. Each sphere is centered on the object and is expanding. Consider an emitter moving in the direction shown by the arrow. In part (*a*), we see that the peak emitted when the emitter was at point 1 becomes a sphere (labeled S_1) around point 1, though the emitter has since moved toward the left. In part (*b*), some time later, sphere S_1 has continued to grow around point 1 (even though the emitter is no longer there). We also see sphere S_2, which shows the position of the peaks emitted when the emitter had moved to point 2. Sphere S_2 is thus centered on point 2, even though the emitter has continued to move on. In (*c*), still later, yet a third peak of the wave has been emitted, sphere S_3, while spheres S_1 and S_2 have continued to expand.

For the case of the moving emitter, observers who are being approached by the emitting source (observers on the left side of *c*) see the three peaks coming past them bunched together. They pass at shorter intervals of time, as though the wavelengths were shorter. This situation corresponds to a color for each wavelength farther to the blue than it started out, so is called a **blueshift.** Observers from whom the emitter is receding (observers on the right side of *c*) see the three peaks at increased intervals of time, as though the wavelengths were longer. This situation corresponds to a color for each wavelength farther to the red than it started out, a **redshift.**

By contrast, for the stationary emitter, all the peaks are centered on the same point. No redshifts or blueshifts arise. So whenever an object is moving away from us, its spectrum is shifted slightly to longer wavelengths. We say that the object's light is redshifted. When an object is moving toward us, its spectrum is blueshifted, that is, shifted toward shorter wavelengths. Even when an object's radiation is already beyond the red (in the infrared or radio regions), we still say it is redshifted whenever the object is receding. Similarly, we say that any approaching object is blueshifted, even for spectral lines beyond the blue.

The fraction of its wavelength by which light is redshifted or blueshifted is the same as the fraction of the speed of light that the object is moving. (That is, the wavelength shift is proportional to the speed of the object; see *Figure It Out 11.4: Doppler Shifts.*) So astronomers can measure an object's radial velocity by measuring the wavelength of

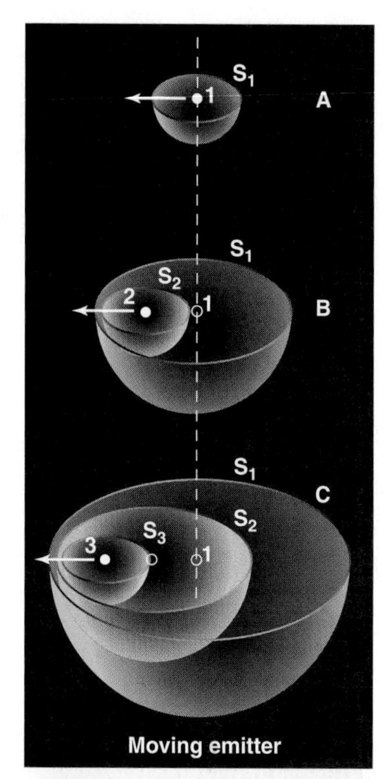

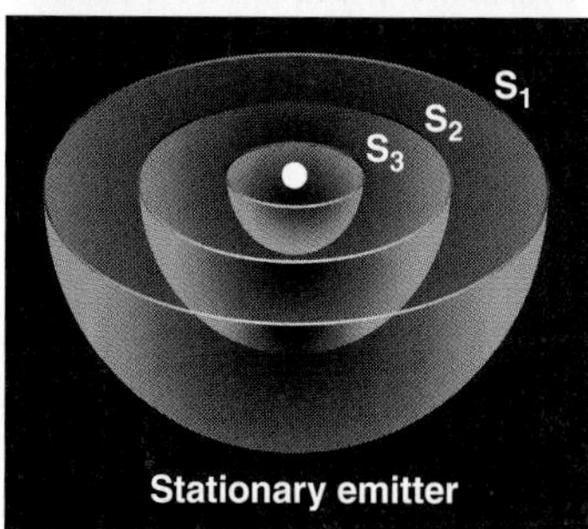

■ **FIGURE 11–17** How Doppler shifts arise. When the emitter is moving toward us, the wave peaks arrive more frequently; we say that the radiation is blueshifted. When the emitter is moving away from us, the wave peaks arrive less frequently; we say that the radiation is redshifted. Radiation from a stationary emitter is not shifted in wavelength. Similarly, there is no wavelength shift in light from an emitter moving transversely (that is, perpendicular) to the line of sight. (There is actually a slight shift with transverse motions that are a significant fraction of the speed of light, or "relativistic," but here we are dealing only with low, "non-relativistic" speeds, for which the transverse shift is negligible.) Note that Doppler shifts also arise if the receiver is moving; only the *relative* radial motion matters.

FIGURE IT OUT 11.4
Doppler Shifts

The wavelength emitted by a source of light that is not moving relative to the observer (it is "at rest") is known as its "rest wavelength" or "laboratory wavelength," λ_0 (pronounced "lambda nought"). If the source is moving relative to the observer, the observed wavelength λ will differ from this. For the Doppler shift of a moving source, we can write

$$\frac{\text{Change in wavelength}}{\text{Rest wavelength}} = \frac{\text{Speed of emitter}}{\text{Speed of light}}, \text{ or } \frac{\Delta\lambda}{\lambda_0} = \frac{v}{c},$$

where v is the speed of the source toward or away from us, c is the speed of light (3×10^5 km/sec), and $\Delta\lambda = \lambda - \lambda_0$ is the change in wavelength. (The uppercase Greek letter "delta," Δ, usually stands for "the change in" or "the difference in.") This equation is valid only if v is much less than c; an expression from the special theory of relativity must be used at high speeds (say, $v > 0.2c$).

Note that when λ is larger than λ_0 (that is, $\lambda > \lambda_0$), the light is redshifted, so the source and observer are receding from each other. Conversely, when $\lambda < \lambda_0$, the light is blueshifted, so the source and observer are approaching each other.

The procedure to use with a star, for example, is to obtain its spectrum, recognize a familiar pattern of lines (such as the Balmer series, which marks the presence of hydrogen), measure the observed wavelength of an absorption or emission line, and compare it with the known wavelength at rest to get $\Delta\lambda$.

For example, if the Hα (H-alpha) absorption line is observed at $\lambda = 6565$ Å, and its rest wavelength is known to be $\lambda_0 = 6563$ Å, we find that

$$\Delta\lambda = \lambda - \lambda_0 = 2 \text{ Å, so } \Delta\lambda/\lambda_0 = 2 \text{ Å}/6563 \text{ Å} = 3 \times 10^{-4}.$$

But we know that this result must be equal to v/c, so

$$v = (3 \times 10^{-4})c = (3 \times 10^{-4})(3 \times 10^5 \text{ km/sec}) = 90 \text{ km/sec}.$$

Thus, we and the star are moving away from each other at about 90 km/sec. (Note that the *relative* radial motion determines the Doppler shift; it doesn't matter whether the emitter or receiver is actually moving.)

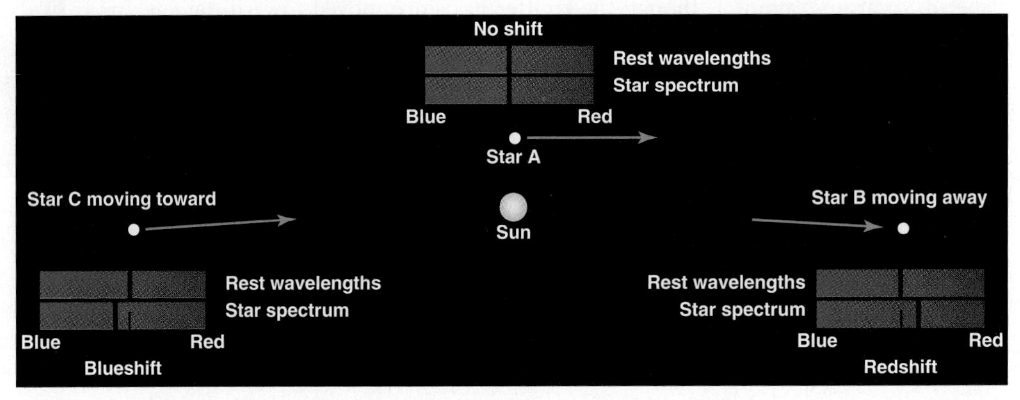

FIGURE 11–18 The Doppler effect in stellar spectra. In each pair of spectra, the position of a spectral line in the laboratory on Earth is shown at top, and the position observed in the spectrum of a star is shown below it. Lines from approaching stars appear blueshifted and lines from receding stars appear redshifted. Lines from stars that are not moving toward or away from us, even if they are moving transversely, are not shifted. A short, dashed, vertical line marks the unshifted position on the spectrum showing shifts.

a spectral line and comparing the wavelength to a similar spectral line measured from a stationary source on Earth (■ Fig. 11–18).

Stars in our Galaxy have only small Doppler shifts, which shows that they are travelling at less than 1 per cent of the speed of light. Though they may be redshifted or blueshifted, the shifts are not enough to change the color that we see when we look at these stars. We shall find, when we discuss galaxies, that the situation is very different for the Universe as a whole (Chapter 16). The observed redshifts of galaxies turn out to be the key to our understanding of the past and future of our Universe (Chapter 18).

AceAstronomy™ Log into AceAstronomy and select this chapter to see the Astronomy Exercise "Doppler Shift."

11.6 "Social Stars": Binaries

Though the Sun is an isolated star in space, most stars are members of pairs or groups. One could say that they are "social," although of course we don't mean this literally, because stars aren't alive and don't make decisions! If our planet were part of such a system, we might see two or more Suns rising and setting (as was the case in the opening scene of the original—Episode 4—*Star Wars* movie and in the closing scene of the latest—Episode 3, The Revenge of the Sith—*Star Wars* movie, an image from which appears as the opener to Chapter 20 on Life in the Universe). And if our Sun were part of a cluster of stars, the nighttime sky would be ablaze with very bright points of light.

By discussing largely observational qualities of stars, and groups of stars, we will continue to prepare ourselves for the theoretical understanding of stars' interiors that takes up the following chapter.

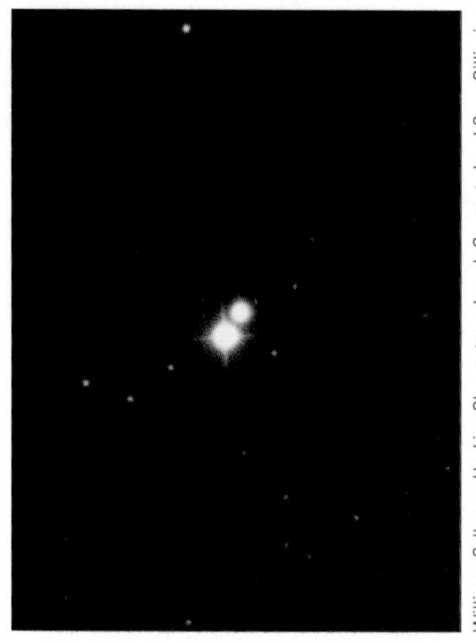

■ **FIGURE 11–19** The double star Albireo, also known as β (the Greek letter "beta") Cygni, contains a B star and a K star. They make a particularly beautiful pair even for amateur observers using small telescopes because of their brightness and contrasting colors (blue and gold, which unfortunately do not appear clearly in most photographs, including this one; the effect is much more striking when viewed visually through a telescope). Albireo is high overhead on summer evenings.

11.6a Pairs of Stars and Their Uses

Sometimes, a star that looks double appears so merely because two unconnected stars happen to appear in essentially the same direction. They are thus examples of **optical doubles,** chance apparent associations. For example, if you look up at the Big Dipper, you might be able to see, even with your naked eye, that the middle star in the handle (Mizar) has a fainter companion (Alcor). Native Americans called these stars a horse and rider.

However, over 50 per cent of the apparently single "stars" you see in the sky are actually multiple systems bound by gravity and orbiting each other. (More precisely, the stars orbit their mutual, or common, center of mass; see *Figure It Out 11.5: Binary Stars.*) Astronomers take advantage of such **binary stars** (or simply "binaries") to find out how much mass stars contain.

11.6a(i) Visual Binaries

When you look at Mizar through a telescope, even after separating it from Alcor you can see that it is split in two. In fact, Mizar was the first double star to have been discovered telescopically (in 1650). These two stars are revolving around each other, and are thus known as a **visual binary.** We see visual binaries best when the stars are relatively far away from each other in their mutual orbits (■ Fig. 11–19). Because of Kepler's third law, these relatively large orbits correspond to long orbital periods—even hundreds of years or longer. We can detect visual binaries better when the star system is relatively near to us compared with other star systems.

11.6a(ii) Spectroscopic Binaries

The two components of Mizar are known as Mizar A and Mizar B. If we look at the spectrum of either one (■ Fig. 11–20), we can see that over a period of days, the spectral lines seem to split and come together again. The spectrum shows that two stars are actually present in Mizar A, and another two are present in Mizar B.

They are **spectroscopic binaries,** since they are distinguished by their spectra. Even when only the spectrum of one star is detectable (■ Fig. 11–21), the other star being

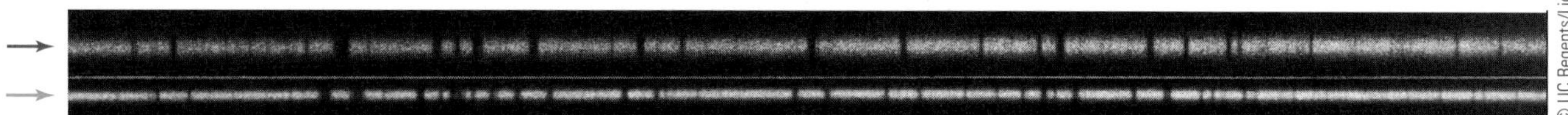

■ **FIGURE 11–20** Two spectra of this star taken 2 days apart show that it is a spectroscopic binary. The lines of both stars are superimposed in the upper stellar absorption spectrum (red arrow). They are separated in the lower spectrum (blue arrow) by an amount that shows that the stars were then moving 140 km/sec with respect to each other.

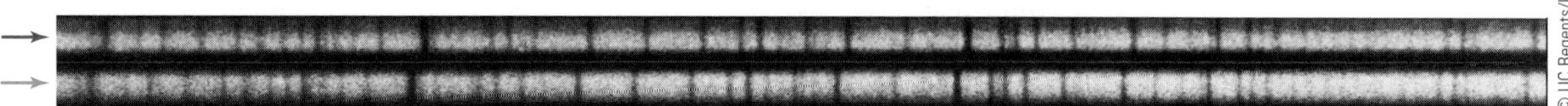

■ **FIGURE 11–21** Two spectra of Castor B, taken at different times. The star's spectrum we see shows a change in the amount of Doppler shift caused by its changing speed toward us. Thus the star must be moving around another star. We conclude that it is a spectroscopic binary, even though only one of the components can be seen. Castor and Pollux are Gemini, the Twins in Greek mythology and a northern constellation.

FIGURE IT OUT 11.5
Binary Stars

The two stars in a binary system generally have elliptical orbits. For simplicity, however, consider the orbits to be circular (see the figure). One can define the center of mass of the system with the equation

$$m_1 r_1 = m_2 r_2,$$

where r_1 and r_2 are the distances of Stars 1 and 2 (respectively) from the center of mass. Note that the center of mass must be along a line joining the two stars.

This behavior is analogous to two weights on opposite ends of a long plank that balances on a point, like a child's seesaw. The larger weight must be closer to the balance point than the smaller weight.

If the two stars have equal mass, they are the same distance from the center of mass. Their orbits have equal size, the stars move with equal speed, and their orbital periods are equal.

If $m_1 > m_2$, then Star 1 is closer to the center of mass than Star 2 is. The orbit of Star 1 is smaller than that of Star 2, and Star 1 moves around the center of mass at a lower speed than Star 2. Their orbital periods remain equal.

If m_1 is much, much larger than m_2 (that is, $m_1 >> m_2$), then Star 1 is very close to the center of mass. The orbit of Star 1 is tiny, and Star 1 moves very slowly around the center of mass. Their orbital periods remain equal. A good example is the Sun–Earth system, ignoring the other planets. The Sun is nearly (but not quite) stationary. (This slight movement of a star makes it possible to detect planets indirectly, as discussed in Chapters 2 and 9.) We often say that the "Earth orbits the Sun," but a formally more correct statement is that the Sun and Earth each orbit their common center of mass.

Most of these basic principles remain valid even when we consider elliptical orbits. The main difference is that the orbital speed of each star varies with position in the orbit; recall Kepler's second law (equal areas in equal times; Chapter 5).

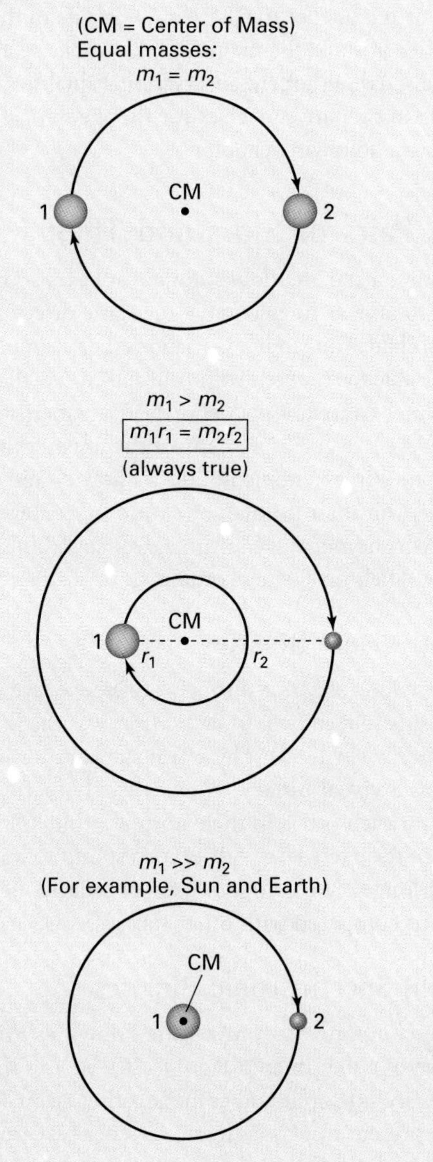

too faint, we can still tell that it is part of a spectroscopic binary if the spectral lines shift back and forth in wavelength over time. These shifts arise from the stars' motions; they are Doppler shifts. The radial velocity (that is, the speed along our line of sight) of each star at any given time can be determined from the Doppler formula.

We give an example of the use of this method in Chapter 14, when we discuss how we find the masses of objects in binary systems that are thought to contain black holes. Unfortunately, since we don't usually know the angle at which we are seeing these orbiting systems, we get only lower limits to the stars' masses rather than actual values for the masses. Since faster-moving stars have greater Doppler shifts, we tend to detect spectroscopic binaries when the stars in them are close to each other. Following Kepler's third law, such large velocities correspond to small orbits.

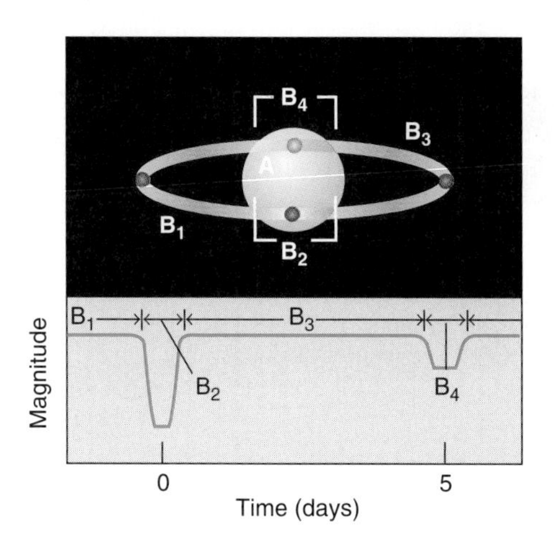

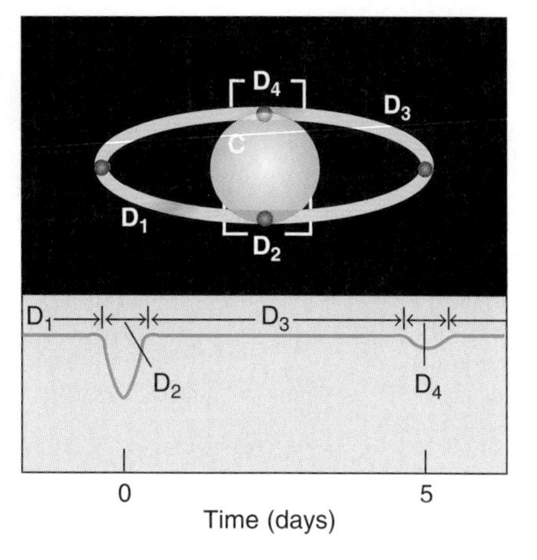

■ **FIGURE 11–22** The shape of the light curve of an eclipsing binary depends on the sizes of the components, the surface temperature, and the angle from which we view them. At *lower left,* we see the light curve that would result for Star B orbiting Star A, as pictured above it. When Star B is in the positions shown at *top* with subscripts, the regions of the light curve marked with the same subscripts result. The eclipse at B_4 is total. At *right,* we see the appearance of the orbit and the light curve for Star D orbiting Star C, with the orbit inclined at a greater angle than at left. The eclipse at D_4 is partial.

11.6a(iii) Eclipsing Binaries

Yet another type of binary star is detectable when the light from a star periodically dims because one of the components passes in front of (eclipses) the other (■ Fig. 11–22). From Earth, we graph the "light curves" of these **eclipsing binaries** by plotting their brightnesses over time.

Eclipsing binaries are detected by us only because they happen to be aligned so that one star passes between the Earth and the other star. Thus all binaries would be eclipsing if we could see them at the proper angle. (At least a few planets around distant stars have been detected in a similar fashion, by a dimming of the total light when it passes in front of, or transits, its parent star, as discussed in Chapter 9.) Since one star can eclipse another only when we are in the plane of its orbit, we know that we are seeing the orbit edge-on and can calculate the masses of the component stars. We can accurately measure stellar masses only for eclipsing binaries.

The same binary system might be seen as different types depending on the angle at which we happen to view the system (■ Fig. 11–23).

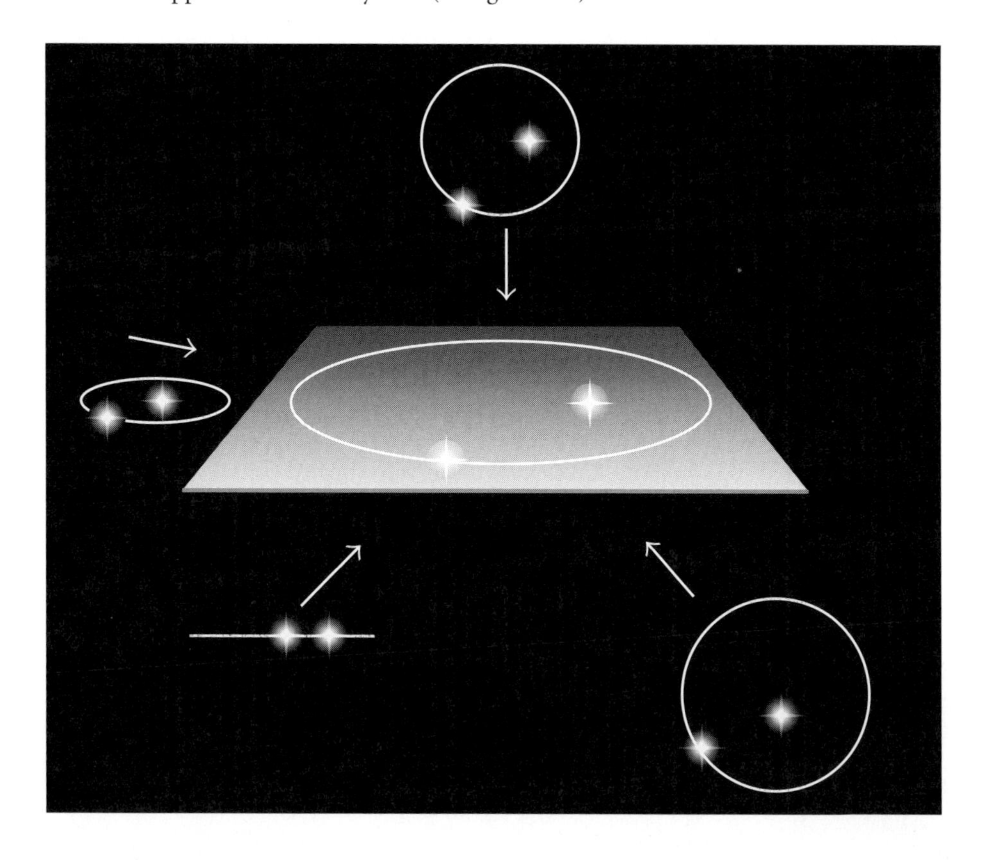

■ **FIGURE 11–23** The appearance of the spectrum of a binary star and whether the lines shift in wavelength over time depend on the angle from which we view it. From far above or below the plane of the orbit, we might see a visual binary, as at *top* and at *lower right.* From close to the plane of the orbit, even if the stars are so close to each other in angle in the sky (closer than shown here at *left*) that they can't be seen as separate stars, we might see only a spectroscopic binary, as at *left.* Similarly, for stars exactly in the plane of the orbit but too close in angle to see as separate objects, we could still see the system as an eclipsing binary, as at *lower left.*

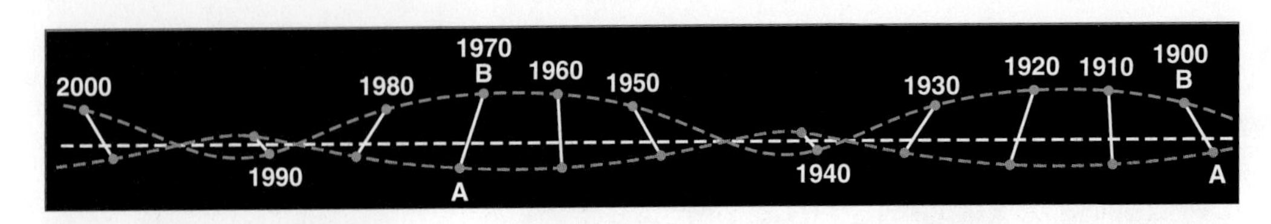

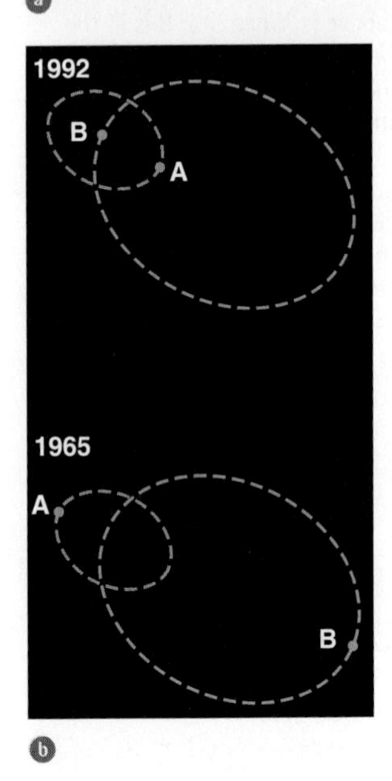

■ **FIGURE 11–24** ⓐ Sirius A and B, an astrometric binary. From studying the motion of Sirius A (usually called simply Sirius), astronomers deduced the presence of Sirius B before it was seen directly. ⓑ The relative positions of Sirius A and B at two extremes of their orbits, when they are closest together *(top)* and farthest apart *(bottom)*.

11.6a(iv) Astrometric Binaries

One more type of double star can be detected when we observe a wobble in the path of a star as its proper motion, its motion with respect to other stars in our Galaxy, takes it slowly across the sky (■ Fig. 11–24). The laws of physics hold that a moving mass must move in a straight line unless affected by a force. The wobble away from a straight line tells us that the visible star must be orbiting an invisible object, pulling the visible star from side to side. Since measurement of the positions and motions of stars is known as astrometry, these stars are called **astrometric binaries.**

A similar method was used to try to detect planets around distant stars, though it has been unsuccessful. The Hipparcos spacecraft, for example, was designed to measure such small motions of stars, and even more precise telescopes are being planned. However, the new planets that have been located around other stars (Chapter 9) were found by using the Doppler effect (and transits, in a few cases) rather than by astrometry. The planets are not massive enough to cause side-to-side shifts that are measurable at the distances to their parent stars.

11.6b How Do We Weigh Stars?

A primary use of binary stars is the measurement of stellar masses (informally, "weighing" stars). We can sometimes determine enough about the stars' orbits around each other that we can calculate how much mass the stars must have to stay in those orbits (see *A Closer Look 11.3: A Sense of Mass: Weighing Stars*).

The most accurate masses are derived from spectroscopic binaries in which the absorption lines of both stars are visible, and which are also eclipsing binaries. For these cases, we can also describe the radii of the two stars reliably. Such binary systems are relatively rare, so we know accurately only about two hundred stellar masses.

11.6c The Mass-Luminosity Relation

It turns out that the mass of a star is its single most important characteristic: Essentially everything else about its properties and evolution depends on mass. In *Figure It Out 11.6: The Mass-Luminosity Relation,* you can learn how the mass is connected to how powerfully the star shines. In particular, massive main-sequence stars are far more luminous and have much shorter lives than low-mass main-sequence stars. Though massive stars start with more fuel, they consume it at a huge rate.

When we discuss the evolution of stars, we will see that matter often flows from one member of a binary system to another. This interchange of matter can drastically alter a star's evolution. And the flowing matter can heat up by friction to the very high temperatures at which x-rays are emitted. In recent years, our satellite observatories above the Earth's atmosphere have detected many such sources of celestial x-rays, and the Chandra X-ray Observatory and the XMM-Newton spacecraft, both launched in 1999, are now the most powerful and sensitive x-ray telescopes ever available.

A Closer Look 11.3 A Sense of Mass: Weighing Stars

The amount of matter in a body is its mass. The standard kilogram resides at the International Bureau of Weights and Measures near Paris and is the ultimate standard of mass. To determine the mass of an object on Earth, we compare it, in some ultimate sense, with the standard kilogram or, more commonly, with secondary standard kilograms held by the various countries.

Kepler's third law, as modified by Newton (Chapter 5), gives us a way of determining the mass of an astronomical body from studying an object orbiting it. But this determination requires knowledge of the universal constant of gravitation, usually called G, that appears in Newton's law of gravity. This constant is measured by laboratory experiments, which limits the resulting accuracy. The first determination of the constant nearly two hundred years ago appeared in a paper entitled "On Weighing the Earth."

Note that weight is actually different from mass. An object's weight on Earth is the gravitational force of the Earth's mass on the object's mass. The object would have a different weight on another planet, but the same mass.

The Earth's mass seems so large to us that we think we can jump up and down without moving the Earth. Actually when we jump up, the Earth moves away from us, according to Newton's third law of motion ("For every action, there is an equal and opposite reaction"). The Earth's mass is about 10^{23} times the mass of a person, though, so it doesn't move much!

The Earth and the rest of the Solar System may be important to us, but they are only minor companions to the Sun. In "Captain Stormfield's Visit to Heaven," by Mark Twain, the Captain races with a comet and gets off course. He comes into heaven by a wrong gate, and finds that nobody there has heard of "the world." ("The world! H'm! there's billions of them!" says a gatekeeper.) Finally, someone goes up in a balloon to try to detect "the world" on a huge map. The balloonist rises into clouds and after a day or two of searching comes back to report that he has found it: an unimportant planet named "the Wart."

Indeed, the Earth is inconsequential in terms of mass next to Jupiter, which is 318 times more massive. And the Sun, an ordinary star, is about 300,000 times more massive than the Earth. Some stars are up to 100 times more massive than the Sun, but others can be as little as $^1/_{12}$ as massive. We shall see that a star's mass is the key factor that determines how the star will evolve and how it will end its life.

Hundreds of billions of stars bound together by gravity make up a galaxy, the fundamental unit of the Universe as a whole. Our own Milky Way Galaxy contains about a trillion (a million million) times as much mass as the Sun.

Since there are billions of galaxies in the Universe, if not an infinite number, the total mass itself in the Universe is not a meaningful number. More often, we discuss the *density* of a region of the Universe—the region's mass divided by its volume. The Universe is so spread out, with so much almost empty space between the galaxies, that the Universe's average density is very low. In Chapter 18 we will see that determining this average density is important for knowing the future of the Universe, whether it will eventually collapse or expand forever. This choice, in turn, dictates whether time will ever end. Thus, mass is among the most important and intriguing quantities to know.

AceAstronomy™ Log into AceAstronomy and select this chapter to see the Active Figure called "Center of Mass" and the Astronomy Exercises "Spectroscopic Binaries," "Eclipsing Binaries," and "Mass/Star Luminosity Relation."

Figure It Out 11.6
The Mass-Luminosity Relation

For *main-sequence stars* (that is, not giants, supergiants, or white dwarfs), we find that the luminosity L (intrinsic brightness) is roughly proportional to the fourth power of mass M, or

$$L \propto M^4,$$

the "mass-luminosity relation." This proportionality ($\propto$) is only approximate; the exponent of M actually varies to some extent along the main sequence. However, it is sufficient for our purposes. Thus, for example, a main-sequence star with twice the Sun's mass is a factor of $2^4 = 16$ times more luminous (powerful) than the Sun.

The mass-luminosity relation for main-sequence stars implies that massive stars have much shorter lives than low-mass stars. Massive stars have more fuel, but they consume it at a disproportionately rapid rate. For example, a main-sequence star with twice the Sun's mass has twice as much fuel, so if all else were equal it would live twice as long. But it actually uses its fuel 16 times more rapidly! So, it can sustain itself for only $^2/_{16}$ (which is equal to $^1/_8$) as long as the Sun.

■ **FIGURE 11–25** The constellation Cetus, the Whale (or Sea Monster). It is the home of the long-period variable star Mira, also known as *o* (the Greek letter "omicron") Ceti.

11.7 Stars That Don't Shine Steadily

Some stars vary in brightness over hours, days, or months. Eclipsing binaries are one example of such **variable stars,** but other types of variable stars are actually individually changing in brightness. Many professional and amateur astronomers follow the "light curves" of such stars, the graphs of how their brightness changes over time.

A very common type of variable star changes slowly in brightness with a period of months or up to a couple of years. The first of these **long-period variables** to be discovered was the star Mira in the constellation Cetus, the Whale (■ Fig. 11–25). At its maximum brightness, it is 3rd magnitude, quite noticeable to the naked eye. At its minimum brightness, it is far below naked-eye visibility (■ Fig. 11–26). Such stars are giants whose outer layers actually change in size and temperature.

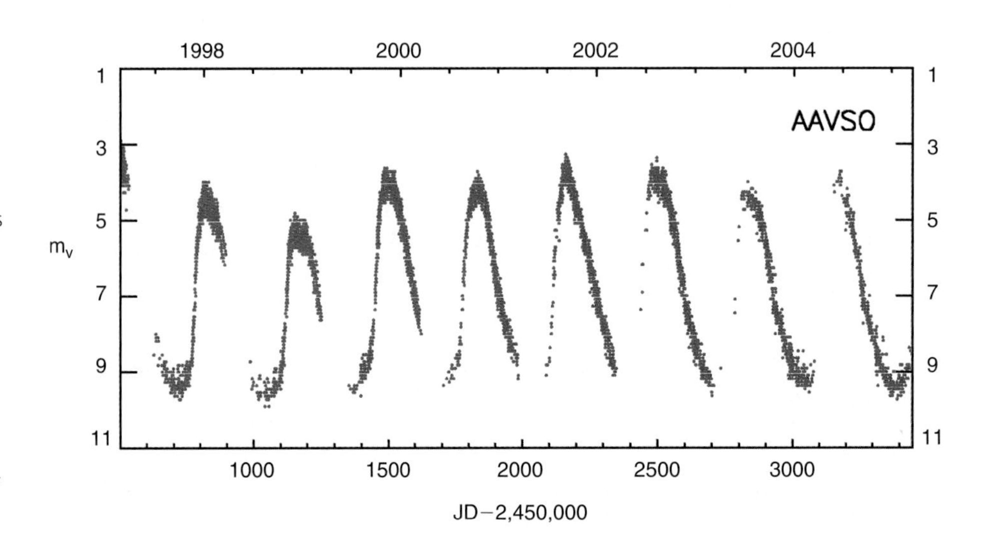

■ **FIGURE 11–26** The light curve for Mira, the first known example of a long-period variable. The horizontal axis shows time and the vertical axis shows apparent magnitude (related to apparent brightness). Notice that both the light-curve's shape and the maximum brightness vary. The Sun may become a Mira-like star in about 5 billion years. The data were gathered and plotted by the American Association of Variable Star Observers (AAVSO). They often use Julian Days (JD), a consecutive count of days, to keep track of time. (American Association of Variable Star Observers)

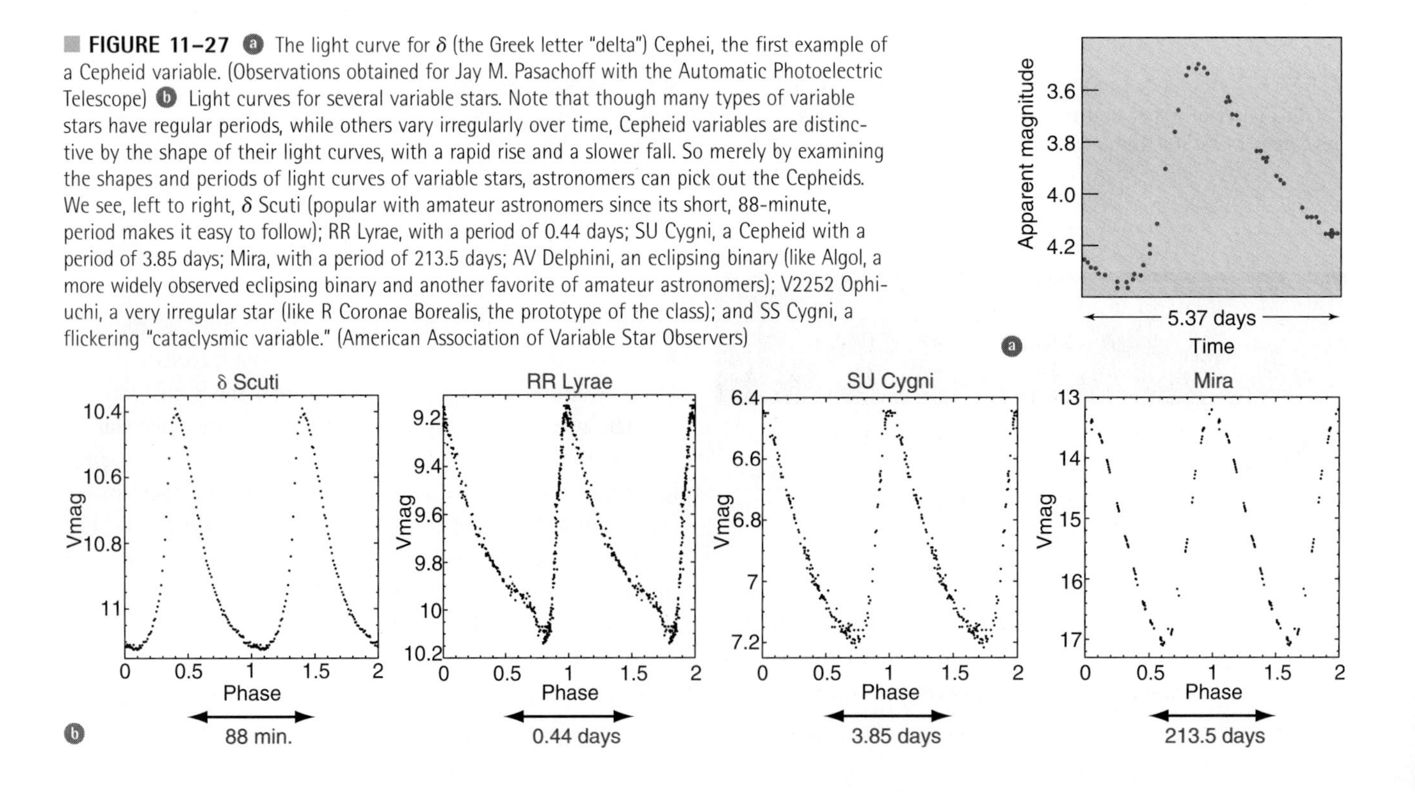

■ **FIGURE 11–27** ⓐ The light curve for δ (the Greek letter "delta") Cephei, the first example of a Cepheid variable. (Observations obtained for Jay M. Pasachoff with the Automatic Photoelectric Telescope) ⓑ Light curves for several variable stars. Note that though many types of variable stars have regular periods, while others vary irregularly over time, Cepheid variables are distinctive by the shape of their light curves, with a rapid rise and a slower fall. So merely by examining the shapes and periods of light curves of variable stars, astronomers can pick out the Cepheids. We see, left to right, δ Scuti (popular with amateur astronomers since its short, 88-minute, period makes it easy to follow); RR Lyrae, with a period of 0.44 days; SU Cygni, a Cepheid with a period of 3.85 days; Mira, with a period of 213.5 days; AV Delphini, an eclipsing binary (like Algol, a more widely observed eclipsing binary and another favorite of amateur astronomers); V2252 Ophiuchi, a very irregular star (like R Coronae Borealis, the prototype of the class); and SS Cygni, a flickering "cataclysmic variable." (American Association of Variable Star Observers)

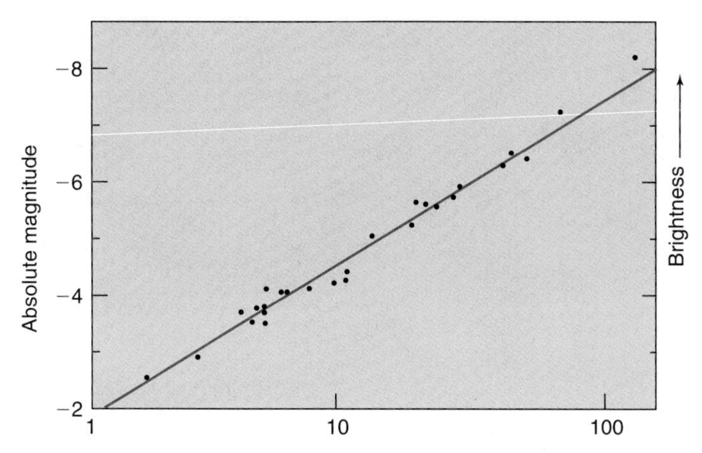

■ **FIGURE 11–28** The period-luminosity relation for Cepheid variables. The stars' average intrinsic brightnesses (average luminosities) are shown on the vertical axis. The upward arrow shows increasing brightness. Note that the Cepheids were found because of the shape of their light curves; only after they were found in this way could Henrietta Leavitt plot them on a graph of period versus brightness to discover the period-luminosity relation, or "Leavitt's law." (J. D. Fernie and R. McGonegal, reprinted from *Astrophys. J.*, **275,** 735, with permission of the U. Chicago Press.)

A particularly important type of variable star for astronomers is the **Cepheid variable,** of which the star δ (the Greek letter "delta") Cephei is a prime example (■ Fig. 11–27). A strict relation exists linking the period over which the star varies with the star's average luminosity (average intrinsic brightness). This "period-luminosity relation" (■ Fig. 11–28) is the key to why Cepheid-variable stars are so phenomenally important, possibly the most important type of star known from the point of view of cosmology. (To mark this fact, we are coining the term "Leavitt's law," to parallel what it led to: "Hubble's law," which we will discuss in Chapter 16.)

After all, we can observe a star's light curve easily, with just a telescope, so we know how long its period is. From the period of a newly observed Cepheid, we know its intrinsic brightness (its luminosity). By comparing its average luminosity with how bright it appears on average, we can tell how far away the star is. We can do so because a given star would appear fainter if it were farther away, according to the inverse-square law of light (see *Figure It Out 11.2: The Inverse-Square Law*). In essence, we are using the Cepheids as **standard candles,** a modern-day use of an old term for a source of known brightness.

Cepheids are supergiant stars, powerful enough that we can detect them in some of the nearby galaxies. Finding the distances to these Cepheids gives us the distances to the galaxies they are in. This chain of reasoning, linking the period of variation of Cepheid variables with their average luminosities, is perhaps the most accurate method we have of finding the distances to nearby galaxies (see our discussion in Chapter 16), and so is at the basis of most of our measurements of the size of the Universe.

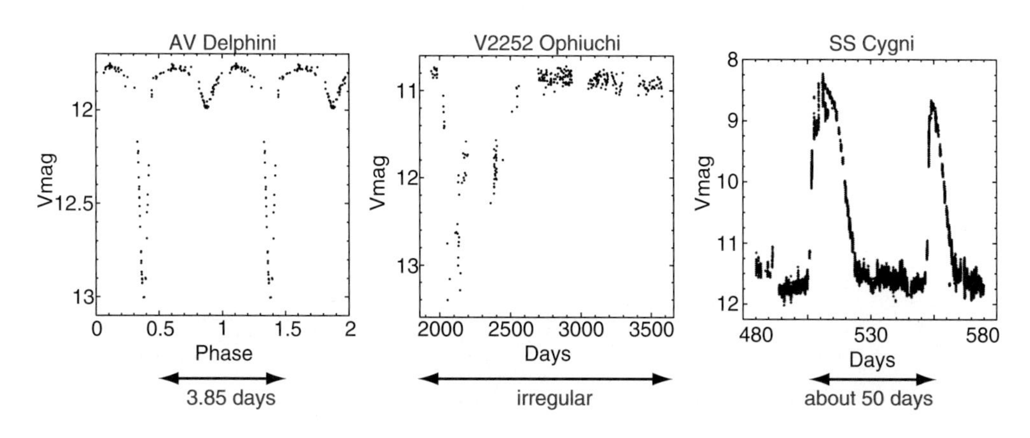

■ **FIGURE 11–29** From the southern hemisphere, the Magellanic Clouds are sometimes high in the sky. They are not quite this obvious to the naked eye compared with a photographic image. Since all stars in a given Magellanic Cloud are essentially the same distance from us, their intrinsic brightnesses (luminosities) are related to each other in the same way as their relative apparent brightnesses when we observe them from Earth. Here we see a time exposure of the Magellanic Clouds with some of the Leonid meteors flashing by during the 2001 meteor storm.

The period-luminosity relation was discovered in the early years of the 20th century by Henrietta Leavitt in the course of her study of stars in the Magellanic Clouds (■ Fig. 11–29). She knew only that all the stars in each of the Magellanic Clouds were about the same distance from us, so she could compare their apparent brightnesses without worrying about distance effects. The Magellanic Clouds turn out to be among the nearest galaxies to us.

The Hubble Space Telescope has been used to detect Cepheids in galaxies farther away than those in which Cepheids could be detected with ground-based telescopes (■ Fig. 11–30). The high resolution of Hubble has enabled Cepheids to be picked out and to have their brightnesses followed over time. The results, as we shall see in the cosmic distance scale discussion (Chapter 18), are at the basis of our knowledge of the size and age of the Universe.

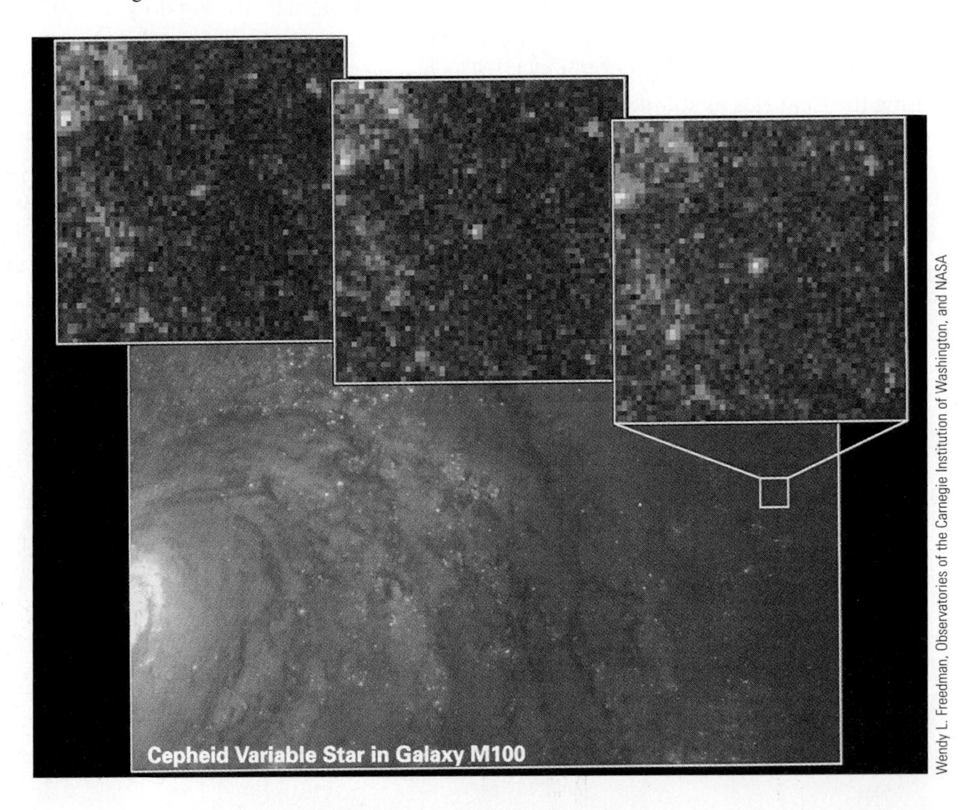

■ **FIGURE 11–30** Brightness variations over time (images are shown for three different times) of a Cepheid variable in a distant galaxy, imaged with the Hubble Space Telescope.

But we need a good way, now lacking, to measure the distances to a few Cepheids reliably in a manner independent of the period-luminosity relation. A cancelled NASA spacecraft was, following its anticipated 2004 launch, supposed to be able to measure accurate, independent distances for 200 Cepheids. We now have to await other astrometric spacecraft, most likely the NASA's Space Interferometry Mission (SIM PlanetQuest) and the European Space Agency's Gaia, both hoped for in 2011 or beyond.

Cepheid variables are rather massive stars, more massive than the Sun, at an advanced and unstable stage of their lives. They are supergiant stars, not on the main sequence. Cepheid variables are changing in size, which leads to their variations in brightness. They contract a bit, which makes them hotter, and because of a peculiarity in the structure of the star, less energy escapes. Thus the pressure rises, and they expand. They overshoot, expanding too far, and the expansion cools them. The energy easily escapes, the pressure decreases, and they respond by contracting again. A cycle is thus set up.

Stars of a related type, **RR Lyrae variables,** have shorter periods—only several hours long—than regular Cepheids. RR Lyrae variables are often (but not exclusively) found in clusters of stars, groupings of stars we will discuss next. Since all RR Lyrae stars have the same average luminosity (average intrinsic brightness), observing an RR Lyrae star and comparing its average apparent brightness with its average luminosity enables us to tell how far away the cluster is.

AceAstronomy™ Log into AceAstronomy and select this chapter to see the Astronomy Exercises "Cepheid Variable" and "Cepheid Variables I and II."

11.8 Clusters of Stars

Clusters, or physical groups of stars in the Milky Way Galaxy, come in two basic varieties, as summarized in *A Closer Look 11.4: Star Clusters in Our Galaxy.* Clusters in other galaxies have similar properties, though there may be a few differences in detail (e.g., some globular clusters in other galaxies may be young, unlike the case in our Galaxy).

11.8a Open and Globular Star Clusters

The face of Taurus, the Bull, is outlined by a "V" of stars, which are close together out in space. On Taurus's back rides another group of stars, the Pleiades (pronounced Plee'a-deez). Both are examples of **open clusters,** groupings of dozens or a few hundred

A Closer Look 11.4 Star Clusters in Our Galaxy

Open Clusters	Globular Clusters
No regular shape	Shaped like a ball, stars more closely packed toward center
Many are young	All are old
Temperature-luminosity diagrams have long main sequences	Temperature-luminosity diagrams have short main sequences
Where stars leave the main sequence tells the cluster's age	All clusters have nearly the same temperature-luminosity diagram and thus nearly the same age
Stars have similar composition to the Sun	Stars have much lower abundances of heavy elements than the Sun
Hundreds of stars per cluster	10,000–1,000,000 stars per cluster
Found in Galactic plane	Found in Galactic halo

stars. We can often see 6 of the Pleiades stars with the naked eye, but binoculars reveal dozens more and telescopes show the rest.

Open clusters (■ Figs. 11–31 and 11–32) are irregular in shape. About 1000 open clusters are known in our Galaxy. They are found near the band of light called the Milky Way (see Chapter 15), which means that they are in the plane of our Galaxy. Specifically, they are generally in or near our Galaxy's spiral arms. (They are often known as "galactic clusters," not to be confused with "clusters of galaxies" which we shall study in Chapter 16.) They are only loosely bound by gravity, gradually dissipating ("evaporating") away as individual stars escape.

In some places in the sky, we see clusters of stars with spherical symmetry. Each such **globular cluster** looks like a faint, fuzzy ball when seen through a small telescope. Larger telescopes reveal many individual stars. Globular clusters (■ Fig. 11–33) contain tens of thousands or hundreds of thousands of stars, many more than typical open clusters. They are more strongly bound than open clusters, although their stars do occasionally get ejected through gravitational encounters with other stars.

Globular clusters usually appear far above and below the Milky Way. About 180 of them are known, and they form a spherical **halo** around the center of our Galaxy. We find their distances from Earth by measuring the apparent brightness of RR Lyrae stars in them.

Star clusters are important because each is a grouping of stars that were all formed at the same time and out of the same material. A further advantage is that all those stars are at the same distance from us. Thus by comparing members of a star cluster, we can often find out what factors make stars evolve differently or at different rates. Below, we will see how using a star cluster enables us to tell the age of its members.

When we study the spectra of stars in open clusters to determine what elements are in them, we find that all the stars (including the Sun; see *A Closer Look 10.1: The Most Common Elements in the Sun's Photosphere*) are about 92 per cent hydrogen (that is, 92 per cent of the number of atoms are hydrogen; by mass, this corresponds to about 73 per cent). Almost all the rest is helium (about 8 per cent by number, or 25 per cent by mass, since each helium atom has about four times the mass of a hydrogen atom). Generally, elements heavier than helium constitute under 0.2 per cent of the number of atoms (or 2 per cent by mass). But the abundances of these "heavy elements" in the stars in globular clusters are ten times lower.

We now think that the globular clusters formed about 11 to 13 billion years ago, early in the life of our Galaxy, from the original gas in the Galaxy. This gas had a very low abundance of elements heavier than hydrogen and helium. As time passed, stars in our Galaxy "cooked" lighter elements into heavy elements through fusion in their interiors, died, and spewed off these heavy elements into space (see our discussion in

■ **FIGURE 11–31** An open cluster of stars in the constellation Carina.

■ **FIGURE 11–32** An open cluster, surrounded by the Rosette Nebula.

■ **FIGURE 11–33** The globular cluster M15 in the constellation Pegasus.

Chapters 12 and 13). The stars in open clusters formed later on from chemically enriched gas. Thus the open clusters have higher abundances of these heavier elements.

X-ray observations have detected bursts of x-rays coming from a few regions of sky, including a half-dozen globular clusters, and the Chandra X-ray Observatory and XMM-Newton are studying more. We think that among the many stars in those clusters, some have lived their lives, have died, and are attracting material from nearby companions to flow over to them. We are seeing x-rays from this material as it "drips" onto the dead star's surface and heats up. So globular clusters still have action going on inside them, in spite of their great age.

11.8b How Old Are Star Clusters?

Large star clusters contain many stars with a wide range of masses. We can determine the ages of such star clusters by looking at their temperature-luminosity (temperature-magnitude) diagrams. For the youngest open clusters, almost all the stars are on the main sequence. But the stars at the upper-left part of the main sequence—which are relatively hot, luminous, and massive—live on the main sequence for much shorter times than the cooler stars, which are fainter and less massive. The points representing their temperatures and magnitudes begin to lie off the main sequence (■ Fig. 11–34).

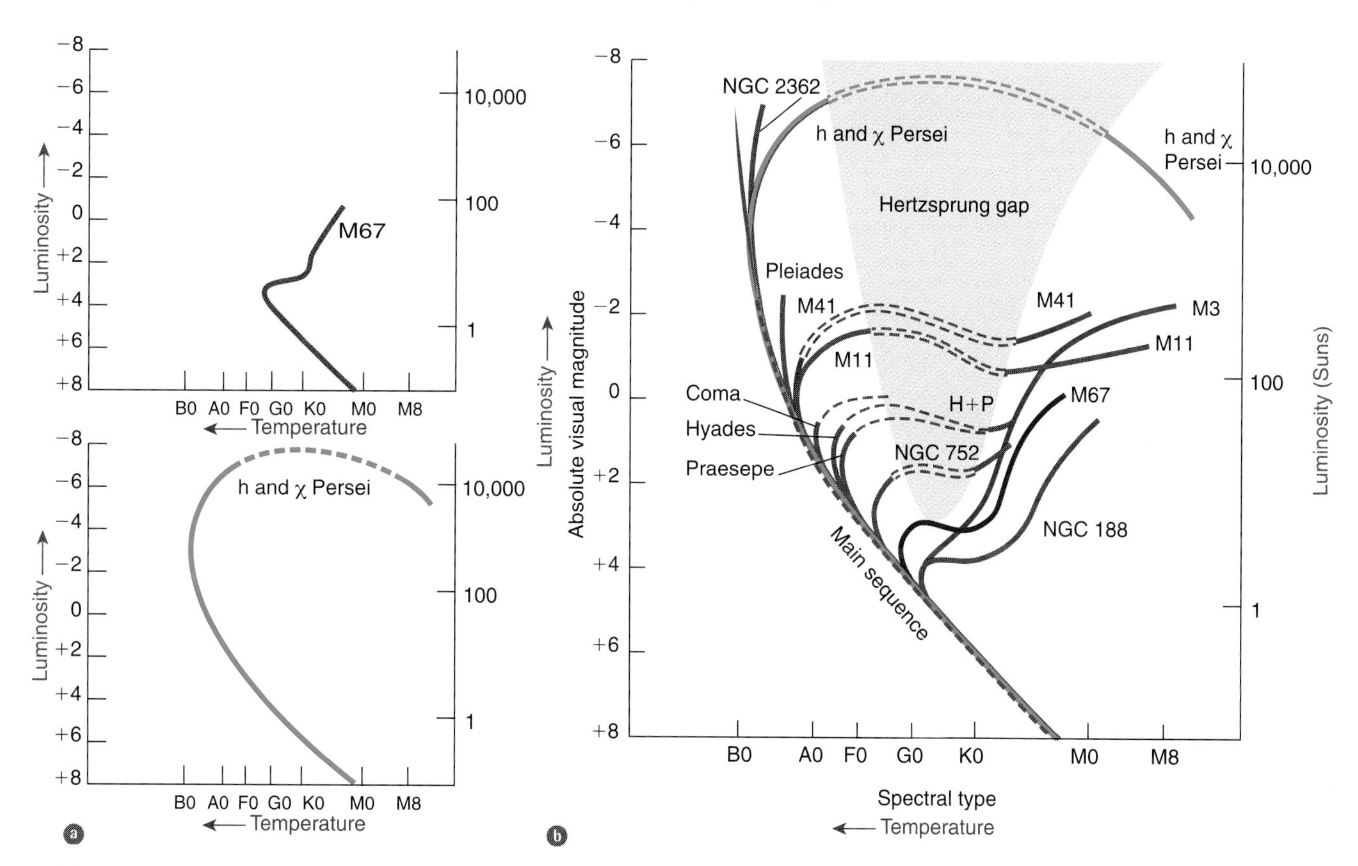

■ **FIGURE 11–34** Temperature-luminosity (temperature-magnitude) diagrams of several open clusters. **a** Individual temperature-luminosity curves showing the temperature and luminosity points on the graph for all the stars in an old open cluster, M67 *(top left)* and for a pair of young open clusters, h and χ (the Greek letter "chi") Persei *(bottom left)*. **b** The temperature-luminosity points of all the stars in one cluster make a curve, and several curves are superimposed here. The two individual curves at *left* are included. Note how the lower parts of the main sequence coincide. Few stars appear in the region marked as the "Hertzsprung gap," because stars with these combinations of luminosity and temperature are apparently unstable, so do not stay in this region of the diagram for long.

If all stars recently reached the main sequence, they would all lie on the curve that extends highest and to the left at the top. The older the stars are, the more likely they are to have evolved to the right. Thus the clusters with more stars matching the original main sequence, like h and χ Persei, are younger. The curve for M3, an old globular cluster, is also shown. (Reprinted by permission of the publisher from *The Milky Way* by B. J. Bok and P. Bok, Cambridge, Mass.: Harvard University Press, Copyright © 1941, 1945, 1957 by the President and Fellows of Harvard College)

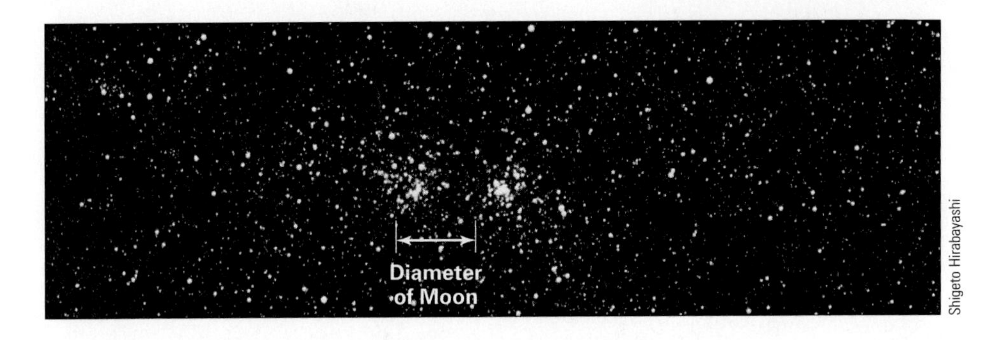

■ **FIGURE 11–35** The double cluster in Perseus, h and χ Persei, a pair of open clusters that is readily visible in a small telescope. Perseus is a northern constellation that is most prominent in the winter sky. In Greek mythology, Perseus slew the Gorgon Medusa and saved Andromeda from a sea monster. See Figure 1–3b for a Chandra X-ray Observatory image of h Persei, part of the research of one of the authors (J.M.P.). Colors represent brightness in various x-ray energy ranges, corresponding to different x-ray wavelengths. Many of the x-rays come from the hot outer layers (coronas) of cool stars like the Sun.

Thus the length of the main sequence on the temperature-luminosity diagram of a cluster gets progressively shorter: It is like the wick of a candle burning down. Since we can calculate the main-sequence lifetimes of stars of different masses, we can tell the age of an open cluster by which stars do not appear on the main sequence. This calculation is similar to estimating how long a candle has been burning if the candle's original length and burn rate are known.

ASIDE 11.2: Stars of all masses

Even in a cluster whose main sequence appears short, consisting only of low-mass stars, we know that massive main-sequence stars were originally present: We can detect their descendants such as red giants and white dwarfs (see Chapter 13).

For example, suppose a cluster does not have any O and B main-sequence stars, but does have A-, F-, G-, K-, M-, and L-type main-sequence stars. We deduce that the cluster is about 100 million years old, because that's how long it takes B-type stars to leave the main sequence. The cluster cannot be a billion years old, because A-type stars would have left the main sequence by that stage. As time passes, the end of the main sequence will move to F stars, then G, and so on—just as the wick of a candle gradually burns down.

The ages of open clusters range from "only" a few million years up to billions of years. The two clusters in the closely spaced pair of clusters known as h and χ (the Greek letter "chi") Persei are among the youngest known (■ Fig. 11–35).

Though different open clusters have different-looking temperature-luminosity diagrams because of their range of ages, all globular clusters have very similar temperature-luminosity diagrams (■ Fig. 11–36). Thus all the globular clusters in our Galaxy are of roughly the same age, about 12 billion years based on the absence of main-sequence stars more massive than the Sun. Recent research, however, is finding an uncertainty in age of two billion years or so.

The ages of globular clusters set a minimum value for the age of the Universe: They must be younger than the Universe itself! The distances set from measurements made with the Hipparcos spacecraft have led to a revision in the calculated ages of some of the globular clusters, and the resulting calculation of their ages is a controversial part of the study of the age of the Universe. For a time around the mid-1990s, the ages determined for globular clusters seemed too old for the age of the Universe deduced otherwise—after all, the Universe can't be younger than something in it. That discrepancy, however, has now been cleared up with further measurements.

The ages of globular clusters are also considered by measuring the brightness of RR Lyrae stars in them. There have long been known to be two types of clusters distinguished by the average period of their RR Lyrae stars—13.2 hours and 15.6 hours, respectively. An interpretation gaining favor is that one of the groups includes globular clusters that were formed in another galaxy and that were captured by our own Galaxy. In this model, these captured clusters are about a billion years younger than

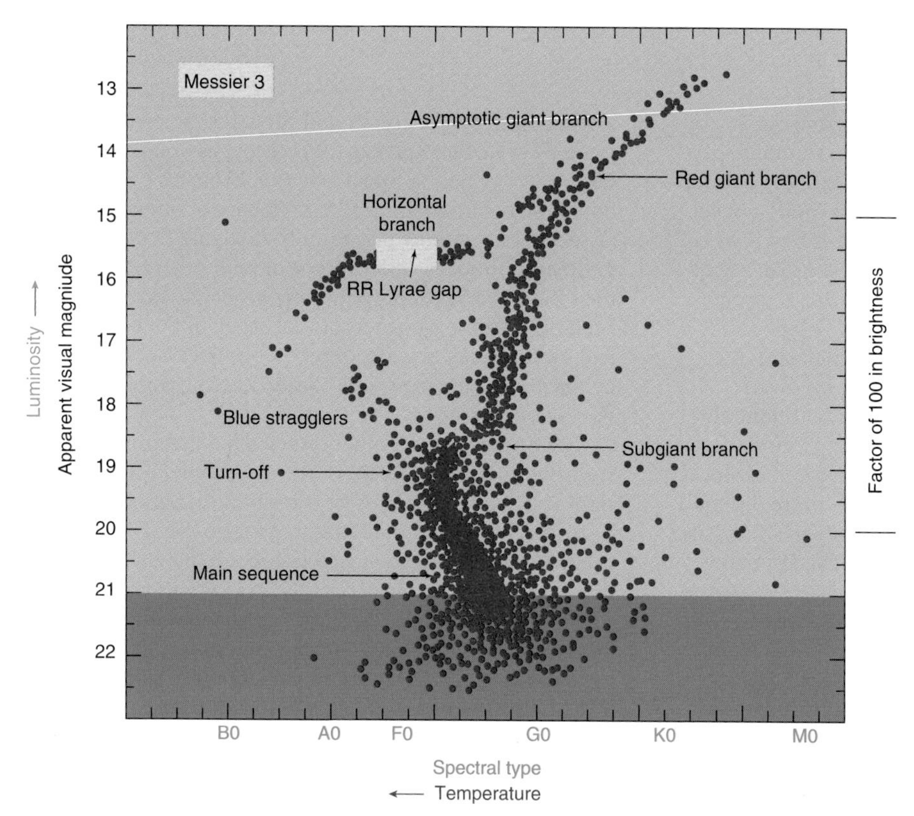

FIGURE 11–36 The temperature-luminosity diagram for the globular cluster M3. The main sequence is very short and stubby, so most of the stars have evolved off it. Thus, M3 and other globular clusters (whose temperature-luminosity diagrams look similar) are very old. (R. Buonanno, A. Buzzoni, C. E. Corsi, F. Fusi Pecci, and A. Sandage, in *Globular Cluster Systems in Galaxies*, ed. J. E. Grindlay and A. G. D. Philip [Dordrecht: Reidel, 1987])

the globular clusters that were formed in our Galaxy, but no clear observational differences in age have been measured. We hope for improvements in instruments, such as the Space Interferometry Mission (SIM PlanetQuest) and Gaia, that will allow the distances to globular clusters to be measured directly by their trigonometric parallaxes; this will lead to more accurate derived ages.

In *A Closer Look 11.5: How We Measure Basic Stellar Parameters,* we summarize many of the concepts introduced in this chapter.

A Closer Look 11.5 How We Measure Basic Stellar Parameters

Here is a summary of some vital types of data for stars more distant than the Sun and how we measure them.

Mass—Usually from studies of the orbits of binary stars.

Size—Usually from the star's apparent brightness, given its surface temperature and distance. (In some cases, the size is determined directly with the technique of interferometry.)

Luminosity—From the star's apparent brightness, given its distance.

Rotation rate—From the Doppler shift, which broadens the star's spectral lines, as one side of the star approaches us relative to the other. (In some cases, the rotation rate is determined from brightness variations as a hot spot or a dark spot rotates to the side of the star facing us, but this technique was not discussed in the text.)

Surface temperature—From the star's spectral lines or from its black-body curve, which can be found accurately enough by measuring the brightness through filters of different colors.

Chemical composition—From detailed analysis of the star's spectral lines.

CONCEPT REVIEW

Heating dense, opaque matter causes it to grow much brighter and to have the maximum in its continuous spectrum shift to shorter wavelengths (Sec. 11.1a). The continuous radiation approximately follows a **black-body curve.** Classifying stars by their spectra led to O B A F G K M L T, with spectral-type A stars having the strongest hydrogen lines (Sec. 11.1b). The spectral sequence is one of progressively decreasing surface temperature, with O-type stars being the hottest. Some L-type objects, and most of the even cooler T-type objects, are actually brown dwarfs.

We find the distance to the nearest stars by **triangulation,** or "trigonometric parallax" (Sec. 11.2), a kind of binocular vision obtained by taking advantage of our location on a moving platform (the Earth). The more distant a star, the smaller is its parallax. Measuring the positions and motions of stars is known as **astrometry.** The Hipparcos spacecraft has greatly improved the accuracy of the astrometry for a hundred thousand stars, and with a lesser improvement for about a million more, but only within about 1 per cent of the diameter of our Galaxy.

A star's intrinsic brightness, or power, is its **luminosity** (Sec. 11.3). **Absolute magnitudes** are the magnitudes that stars would appear to have if they were at a standard distance equivalent to about 32.6 light-years from us. The apparent brightness of a star decreases with distance from it following the **inverse-square law.**

Plotting temperature (often as spectral type) on the horizontal axis and luminosity (intrinsic brightness) on the vertical axis gives a **temperature-luminosity diagram** (Sec. 11.4), also known as a temperature-magnitude diagram or a Hertzsprung-Russell diagram. Most stars appear in a band known as the **main sequence,** which goes from hot, luminous stars to cool, dim stars. Stars on the main sequence are known as **dwarfs.** Some brighter stars exist and are **giants** or even **supergiants.** Less luminous stars than the main sequence for a given temperature are **white dwarfs.**

Stars have small motions across the sky known as **proper motions** (Sec. 11.5a). Proper motion is an *angular* movement of stars across the sky, relative to each other; to get the *physical* velocity (in km/sec) of each star in the plane of the sky, perpendicular to our line of sight, we need to also know its distance. Studies of their Doppler shifts show their **radial velocities** (motions toward or away from us along the line of sight) as **blueshifts** or **redshifts** (Sec. 11.5b). The larger the blueshift or redshift, the proportionally larger is the object's speed toward or away from you.

Optical doubles appear close together by chance (Sec. 11.6a). Most stars are in multiple-star systems, **binary stars,** bound by gravity. **Visual binaries** are revolving around each other and can be detected as double from Earth (Sec. 11.6a(i)). **Spectroscopic binaries** show their double status from spectra (Sec. 11.6a(ii)); one or two sets of spectral lines show periodically changing Doppler shifts. **Eclipsing binaries** pass in front of each other, as shown in their plots of brightness over time, or "light curves" (Sec. 11.6a(iii)). Since astrometry is the study of the positions and motions of stars, **astrometric binaries** can be detected by their wobbling from side to side in their paths across the sky (Sec. 11.6a(iv)).

The primary use of binary stars is the determination of stellar masses (Sec. 11.6b). In the case of *main-sequence stars,* it is found that massive stars are much more luminous (intrinsically bright) than low-mass stars (Sec. 11.6c). Thus, massive stars use up their fuel faster, and have shorter lives, than low-mass stars.

Among **variable stars,** stars that change considerably in brightness over time, **long-period variables,** like Mira, are common (Sec. 11.7). **Cepheid variables** are rare but valuable; the period of a Cepheid variable's variation is linked to its average intrinsic brightness (luminosity), so studying its variation shows its luminosity. Measuring luminosity allows stars to be used as **standard candles.** Comparing intrinsic with apparent brightness gives stars' distances, and they are valuable tools for determining distances to other galaxies. **RR Lyrae variables** also can be used to find distances, particularly to globular clusters in our Milky Way Galaxy.

Clusters, or physical groups of stars, come in two basic varieties. **Open clusters** contain up to a few thousand stars irregularly spread in a small region of sky, generally in the plane of our Galaxy (Sec. 11.8a). They are loosely bound by gravity, but gradually dissipate away as individual stars escape. **Globular clusters,** consisting of 10,000 to a million stars more strongly bound by gravity than in open clusters, have spherical shapes and form a **halo** around the center of our Galaxy. They contain low abundances of heavy elements, so they are very old.

The temperature-luminosity diagrams for open clusters show the ages of the clusters by the length of the cluster's main sequence; massive, short-lived stars leave the main sequence faster than the low-mass, long-lived stars (Sec. 11.8b). The temperature-luminosity diagrams for globular clusters in our Galaxy are all very similar, with short main sequences, indicating that the globular clusters are all about the same age—very old, providing a minimum value for the age of our Universe.

QUESTIONS

1. The Sun's spectrum reaches its maximum brightness at a wavelength of 5000 Å. Would the spectrum of a star whose surface temperature is higher than that of the Sun peak at a longer or a shorter wavelength?

†2. The Sun's surface temperature is about 5800 K and its spectrum peaks at 5000 Å. An O star's temperature may be 40,000 K. **(a)** According to Wien's law, at what wavelength does its spectrum peak? **(b)** In what part of

the spectrum might that peak be? (**c**) Can the peak be observed with the Keck telescopes? Explain.

3. One black body peaks at 2000 Å. Another, of the same surface area, peaks at 10,000 Å. (**a**) Which gives out more radiation at 2000 Å? (**b**) Which gives out more radiation at 10,000 Å?

†4. If a star has a surface temperature four times higher than that of the Sun, but the same surface area, by what factor is the star more luminous than the Sun?

5. Star Albert appears to have the same brightness through red and blue filters. Star Bohr appears brighter in the red than in the blue. Star Curie appears brighter in the blue than in the red. Rank these stars in order of increasing surface temperature.

6. (**a**) What is the difference between continuous radiation and an absorption line? (**b**) What is the difference between continuous radiation and an emission line? (**c**) Graph a spectrum that shows both continuous radiation and absorption lines. (**d**) Can you draw absorption lines without continuous radiation? (**e**) Can you draw emission lines without continuous radiation? Explain.

7. Make up your own mnemonic for spectral types, through at least type L (but perhaps even T, if you feel especially creative).

†8. (**a**) What is the distance of a star whose parallax is 0.2 seconds of arc? (**b**) What is the parallax of a star whose distance is 100 pc?

9. If a star that is 100 light-years from us appears to be 10th magnitude, would its absolute magnitude be a larger or a smaller number? Explain.

†10. If the Sun were 10 times farther from us than it is, how many times less light would we get from it?

11. Sketch a temperature-luminosity diagram, and distinguish among giants, dwarfs, and white dwarfs. Be sure to label the axes.

12. Compare the surface temperatures of white dwarfs with those of dwarfs, and explain their relative luminosities.

†13. If a red giant has half the Sun's surface temperature but 100 times its radius, what is the giant's luminosity relative to that of the Sun?

14. Two stars the same distance from us are the same temperature, but one is a giant while the other is a dwarf. Which appears brighter?

15. Does measuring Doppler shifts depend on how far away an object is? Explain.

16. Consider two stars of the same spectral type and subtype, but not necessarily both on the main sequence. Describe whether this information is sufficient to tell you how the stars differ in (**a**) surface temperature, (**b**) size, and (**c**) distance.

17. Suppose you find an object that has a Doppler shift corresponding to 20% of the speed of light. Is the object likely to be a star in our Galaxy? Why or why not?

18. Use the explanation of Doppler shifts to show what happens to the sound from a fire truck as the truck passes you.

19. (**a**) Sketch a star's spectrum that contains two spectral lines. (**b**) Then sketch the spectrum of the same star if the star is moving toward us. (**c**) Finally, sketch the spectrum if the star is moving away from us.

†20. Suppose a star is moving toward you with a speed of 200 km/sec. (**a**) Will its spectrum appear blueshifted or redshifted? (**b**) At what wavelength will you observe the Hα (H-alpha) line (rest wavelength 6563 Å) in the star's spectrum?

21. Sketch the orbit of a double star that is simultaneously a visual, an eclipsing, and a spectroscopic binary.

22. If the light in a binary star system is greatly dominated by one of the stars (the other star is very faint), how can you tell from its spectrum that the star is a binary?

23. Assume that an eclipsing binary contains two identical stars, but visually to us looks like a single star. (**a**) Sketch the apparent brightness of light received as a function of time. (**b**) Sketch to the same scale another light curve to show the result if both stars were much larger (though still appearing to us as a single star) while the orbit stayed the same.

†24. Suppose you find a binary system in which one star is 6 times as massive as the other star. Which one is closer to the center of mass, and by what factor?

25. Sketch the path in the sky of the visible (with a solid curve) and invisible (with a dotted curve) components of an astrometric binary.

26. (**a**) Explain how astrometric methods were used to search for planets around other stars. (**b**) How were such planets finally found?

27. Explain why massive main-sequence stars have shorter lives than low-mass main-sequence stars.

†28. (**a**) If one main-sequence star is 3 times as massive as another one, what is the ratio of their luminosities? (**b**) What is the ratio of their main-sequence lifetimes? Be sure to state which one is most luminous and which one lives longest.

29. (**a**) Explain briefly how observations of a Cepheid variable in a distant galaxy can be used to find the distance to the galaxy. (**b**) Why can't we use triangulation instead?

†30. If you find a Cepheid variable star in a star cluster, and the Cepheid appears 100 times fainter than another Cepheid with the same period but whose distance is known to be 400 light-years, what is the distance of the star cluster?

31. What ability of the Hubble Space Telescope is enabling astronomers to study Cepheid variables farther out in space than ever before?

32. (**a**) What are the Magellanic Clouds? (**b**) How did we find out the distances to them?

33. What are two distinctions between open and globular clusters?

34. Why is it more useful to study the temperature-luminosity diagram of a cluster of stars instead of one for stars in a field chosen at random?
35. Suppose you find two clusters, one whose main sequence doesn't have O, B, and A stars, and the other whose main sequence doesn't have any O, B, A, and F stars. Which is older? Why?
36. Which are generally older: open clusters or globular clusters? Explain.
37. **True or false?** Thoroughly studying certain types of binary stars is the main way in which astronomers directly measure stellar masses.
38. **True or false?** A star of spectral type K must be less luminous than a star of spectral type F.
39. **True or false?** It would be easier to accurately measure the parallax of a given star from Jupiter than from Earth (though it would take longer).
40. **True or false?** A star cluster whose main sequence contains only F, G, K, M, and L stars is probably older than a star cluster whose main sequence contains only K, M, and L stars.
41. **True or false?** "Open" star clusters in the Milky Way Galaxy generally have hundreds of thousands of densely packed stars.
42. **True or false?** We usually determine the chemical composition of stars by examining absorption lines produced in their relatively cool outer layers.
43. **True or false?** A star having a small proper motion must be moving in the plane of the sky at a physically low speed.
44. **Multiple choice:** Albireo, the "Cal Star" (that is, the official star of the University of California, Berkeley, because of its colors), is a physical binary system of a bright, yellow ("gold") star and a fainter, blue star. Which one of the following statements is *true*? **(a)** The gold star is significantly larger than the blue star. **(b)** The blue star is significantly farther away from Earth than the gold star. **(c)** The blue star is significantly more massive than the gold star. **(d)** The blue star is significantly younger than the gold star. **(e)** The gold star orbits the center of the blue star.

†45. **Multiple choice:** Treating the Sun as a "black body," suppose its surface temperature were doubled from 5800 K to 11,600 K. It would have the same luminosity if, at the same time, its radius **(a)** doubles; **(b)** remains the same; **(c)** decreases to one-half its former size; **(d)** decreases to one-quarter its former size; or **(e)** decreases to one-sixteenth its former size.

†46. **Multiple choice:** If Stars Noelle and Kyle have the same apparent brightness, but Star Noelle is 5 times farther away from us than Star Kyle, then Star Noelle's luminosity (power) is _______ times the luminosity of Star Kyle. **(a)** 25. **(b)** 5. **(c)** 1. **(d)** 1/5. **(e)** 1/25.

47. **Multiple choice:** A binary star that varies in apparent brightness as one member of the binary passes in front of the other is **(a)** an astrometric binary; **(b)** a visual binary; **(c)** an eclipsing binary; **(d)** a spectroscopic binary; or **(e)** impossible, unless at least one of the two stars is itself intrinsically variable.
48. **Multiple choice:** One can determine the radius of a nearby star (that is not necessarily on the main sequence) knowing only its **(a)** chemical composition, distance, and surface temperature; **(b)** radial velocity and luminosity; **(c)** apparent brightness and distance; **(d)** apparent brightness, parallax, and surface temperature; or **(e)** apparent brightness, parallax, and luminosity.

†49. **Multiple choice:** If the amount of nuclear fuel available during the main-sequence stage of a star's life is proportional to the star's mass (M), and the luminosity of a main-sequence star is proportional to the fourth power of the star's mass (M^4), then to what is a star's main-sequence lifetime proportional? **(a)** $1/M^3$. **(b)** M^3. **(c)** M. **(d)** $1/M^5$. **(e)** Insufficient information is given.

50. **Multiple choice:** Which one of the following is a *true* statement for a globular cluster and a *false* statement for an open cluster? **(a)** All stars in the cluster formed at about the same time. **(b)** There are usually about a few hundred stars loosely held together by gravity. **(c)** All stars in the cluster had approximately the same initial chemical composition. **(d)** They are usually found away from the disk and arms of a spiral galaxy. **(e)** A temperature-luminosity plot can reveal the cluster's age.

†51. **Fill in the blank:** Suppose you measure an absorption line from Star Hal to be at $\lambda = 5005$ Å. If you know from laboratory experiments that the same line from a stationary gas is at $\lambda_0 = 5000$ Å, the radial velocity of Star Hal must be _______ km/sec, in the direction _______ us.

†52. **Fill in the blank:** A star whose parallax is 0.04 second of arc is _____ as distant as a star whose parallax is 0.08 second of arc.

53. **Fill in the blank:** _______ are stars that vary in brightness in a regular way, with the period of variation of the most luminous cases being the longest.
54. **Fill in the blank:** Most stars within a few hundred light-years from Earth are in the broad category of __________, though this is not generally true of the brightest visible stars.

†55. **Fill in the blank:** Star Luke is 8 times farther away than Star Obi-Wan but 16 times as luminous. The apparent brightness of Star Luke compared with Star Obi-Wan (Luke/Obi-Wan) is ________.

56. **Fill in the blank:** The luminosities of massive main-sequence stars are much ______ than those of low-mass main-sequence stars.
57. **Fill in the blank:** In a binary star system, the two stars actually orbit their ______ rather than each other.

†This question requires a numerical solution.

TOPICS FOR DISCUSSION

1. Do you find it surprising that we can learn so many fundamental characteristics of stars, despite their enormous distances?
2. Why do you think massive main-sequence stars use up their fuel much faster than low-mass stars, despite having more fuel?
3. Do you think that planetary systems (containing at least several planets) are more or less likely to form around binary stars than around single stars?

MEDIA

Virtual Laboratories

- The Spectral Sequence and H-R Diagram
- Binary Stars, Accretion Disks, and Kelper's Laws

AceAstronomy™ Log into AceAstronomy at **http://astronomy.brookscole.com/cosmos3** to access quizzes and animations that will help you assess your understanding of this chapter's topics.

Log into the Student Companion Web Site at **http://astronomy.brookscole.com/cosmos3** for more resources for this chapter including a list of common misconceptions, news and updates, flashcards, and more.

CHAPTER 12

How the Stars Shine: Cosmic Furnaces

ORIGINS

The formation of stars like the Sun is critical to our existence. With its large supply of fuel, the Sun exists for a long time as it produces energy through nuclear fusion, making possible the development of life on Earth.

AIMS

1. Understand how stars form out of gigantic clouds of gas and dust (Section 12.1).
2. Study the way in which stars generate their energy through nuclear fusion deep in their cores (Sections 12.2 to 12.5).
3. Discuss brown dwarfs, which do not fuse ordinary hydrogen to helium (Section 12.6).
4. Learn how neutrino detection experiments have moved from a test of our most fundamental understanding of how stars shine to a test of the fundamental physics of small particles (Section 12.7).
5. Briefly introduce the end-points of stellar evolution (Section 12.8).

Even though individual stars shine for a relatively long time, they are not eternal. Stars are born out of the gas and dust that exist within a galaxy; they then begin to shine brightly on their own. Eventually, they die. Though we can directly observe only the outer layers of stars, we can deduce that the temperatures at their centers must be millions of kelvins. We can even figure out what it is deep down inside that makes the stars shine.

To determine the probable life history of a typical star, we observe stars having many different ages and assume that they evolve in a similar manner. However, we must take into account the different masses of stars; some aspects of their evolution depend critically on mass.

We start this chapter by discussing the birth of stars. We see how new capabilities of observing in the infrared in addition to the visible are helping us understand star formation (■ Fig. 12–1). We then consider the processes that go on inside a star during its life on the main sequence. Finally, we begin the story of the evolution of stars when they finish the main-sequence stage of their lives. Chapters 13 and 14 will continue the story of what is called "stellar evolution," all the way to the deaths of stars.

Near the end of this chapter we will see that the most important experiment to test whether we understand how stars shine is the search for elusive particles, called neutrinos, from the Sun. Over the past decades a search for them has been made, but only about a third to half of those expected had been found. Recent experiments have provided better ways of detecting neutrinos than we previously had, and they were there all along, though transformed and thus hidden! The results indicate that we did

AceAstronomy™ The AceAstronomy icon throughout this text indicates an opportunity for you to test yourself on key concepts, and to explore animations and interactions of the AceAstronomy website at http://astronomy.brookscole.com/cosmos3

A star-formation region in our Milky Way Galaxy, shown as glowing dust (red) and hot gas (green) in this infrared false-color image made with the Spitzer Space Telescope. We see embryonic stars in the Carina Nebula, 200 light-years across and 10,000 light-years from us. A hot star, Eta Carinae, is too bright for Spitzer to handle, so it was placed off the top of the image. Radiation and stellar winds from Eta Carinae are tearing apart the existing cloud of gas and dust, with shock waves compressing and heating the matter, forming many new stars. NASA/JPL-Caltech/N. Smith (U. Colorado at Boulder)

NOAO/AURA/NSF

■ **FIGURE 12–1** For comparison, a visible-light image of the same region of the Carina Nebula as on the infrared image opening the chapter.

not understand neutrinos as well as we had thought. These astronomical results therefore have added important knowledge about fundamental physics in addition to our understanding of the stars.

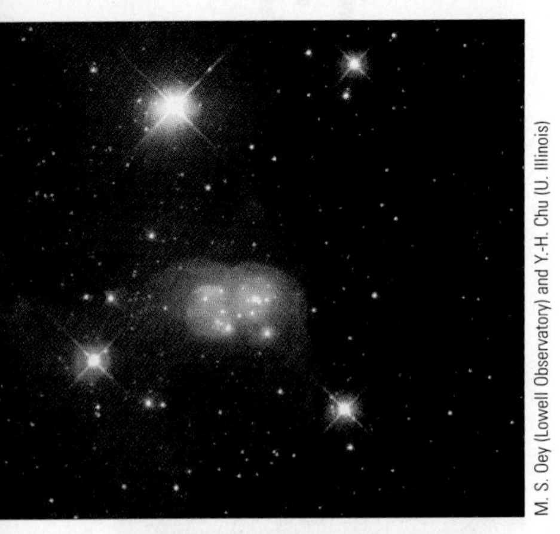
M. S. Oey (Lowell Observatory) and Y.-H. Chu (U. Illinois)

■ **FIGURE 12–2** A Hubble Space Telescope image of gas and dust around a cluster of young, hot stars. The image shows a region about 60 light-years across.

12.1 Starbirth

The birth of a star begins with a nebula—a large region of gas and dust (■ Figs. 12–2 and 12–3, as well as the earlier Chapter Opener and Fig. 12–1). The dust (tiny solid particles) may have escaped from the outer atmospheres of giant stars. The regions of gas and dust (often called clouds, or "giant molecular clouds") from which stars are forming are best observed in the infrared and radio regions of the spectrum, because most other forms of radiation (such as optical and ultraviolet) cannot penetrate them. We discuss the infrared observations largely here, including the new capabilities of NASA's Spitzer Space Telescope, and we leave the radio observations to Chapter 15 on the Milky Way Galaxy.

Stars forming at the present time incorporate this previously cycled gas and dust, which gives them their relatively high abundances of elements heavier than helium in the periodic table (still totaling less than a per cent). In contrast, the oldest stars we see formed long ago when only primordial hydrogen and helium were present, and therefore have lower abundances of the heavier elements, as we discussed near the end of the previous chapter.

12.1a Collapse of a Cloud

Consider a region that reaches a higher density than its surroundings, perhaps from a random fluctuation in density or—in a leading theory of why galaxies have spiral arms—because a wave of compression passes by. Still another possibility is that a nearby star explodes (a "supernova"; see Chapter 13), sending out a shock wave that compresses the gas and dust. In any case, once the cloud gains a higher-than-average density, the gas and dust continue to collapse due to gravity. Energy is released, and the material accelerates inward. Magnetic fields may resist the infalling gas, slowing the infall, though the role of magnetic fields is not well understood in detail.

Eventually, dense cores, each with a mass comparable to that of a star, form and grow like tiny seeds within the vast cloud. These **protostars** (from the prefix of Greek origin meaning "primitive"), which will collectively form a star cluster, continue to collapse, almost unopposed by internal pressure.

But when a protostar becomes sufficiently dense, frequent collisions occur among its particles; hence, part of the gravitational energy released during subsequent collapse goes into heating the gas, increasing its internal pressure. (In general, compression heats a gas; for example, a bicycle tire feels warm after air is vigorously pumped into it.) The rising internal pressure, which is highest in the protostar's center and decreases outward, slows down the collapse until it becomes very gradual and more accurately described as contraction. The object is now called a **pre-main-sequence star** (■ Fig. 12–4).

By this time, the object has contracted by a huge fraction, a factor of 10 million, from about 10 trillion km across to about a million km across—that is, something initially larger than the whole Solar System collapses until most of its mass is in the form of a single star.

During the contraction phase, a disk tends to form because the original nebula was rotating slightly. We discussed this process when considering the "nebular hypothesis" for the formation of our Solar System (Chapter 9). Dusty disks have been found around young stars and pre-main-sequence stars in nebulae such as the Orion Nebula (■ Fig. 12–5). These are sometimes called protoplanetary disks ("proplyds"), and they support the theoretical expectation that planetary systems are common. Jets of gas are

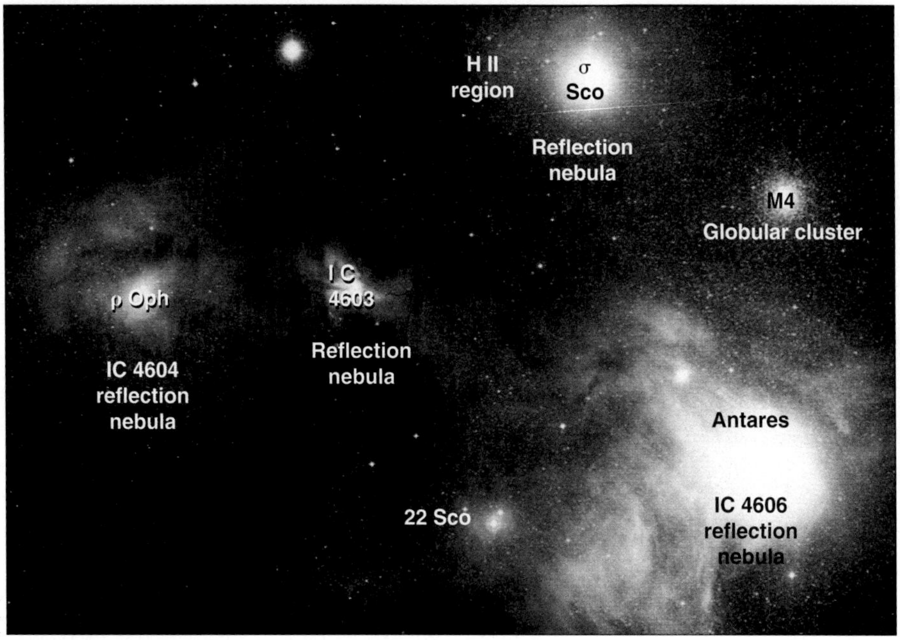

■ **FIGURE 12–3** The clouds of gas and dust around ρ (the Greek letter "rho") Ophiuchi (in the blue reflection nebula). The bright star Antares is surrounded by yellow and red nebulosity. The complex is 500 light-years away. M4, the nearest globular cluster to us, is also seen.

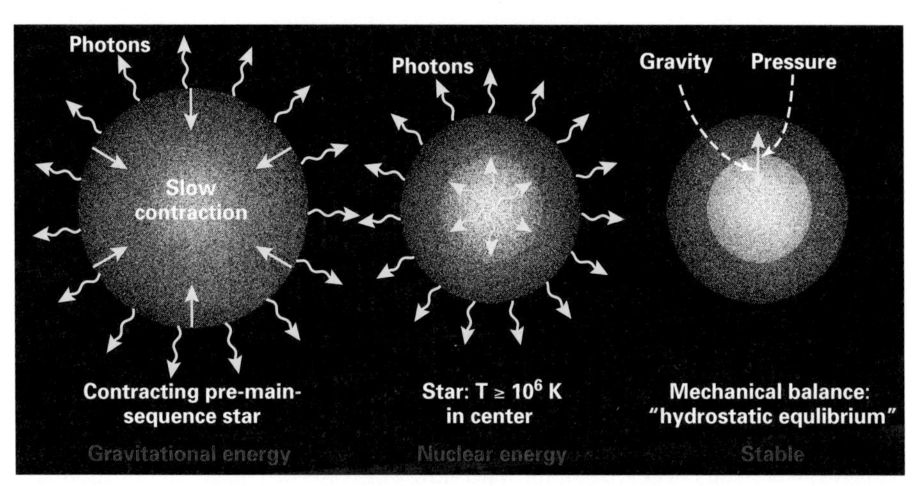

■ **FIGURE 12–4** Stars form out of collapsing gas and dust. The contraction of the pre-main-sequence star *(left)* supplies energy that eventually allows nuclear fusion *(center)* to begin. When the outward pressure of the hot gas matches the inward force of gravity *(right)*, the star is in "hydrostatic equilibrium."

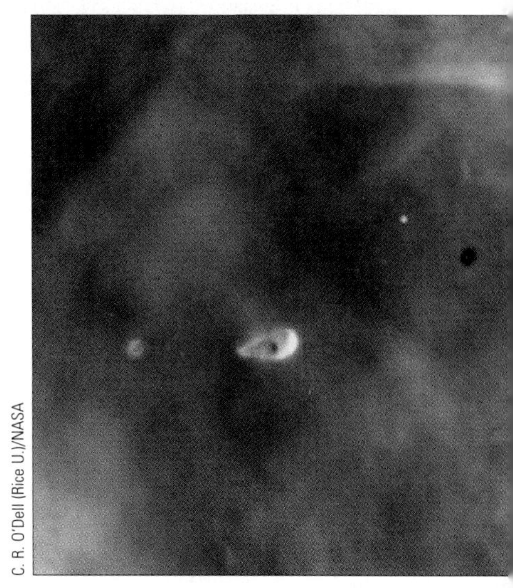

■ **FIGURE 12–5** The Hubble Space Telescope has revealed protoplanetary disks around several stars, as we saw in Section 9.1. Here we view such "proplyds" in a portion of the Orion Nebula only 0.1 light-year across.

commonly ejected in opposite directions out the poles of the rotating pre-main-sequence star (■ Fig. 12–6).

As energy is radiated from the surface of the pre-main-sequence star, its internal pressure decreases, and it gradually contracts. This release of gravitational energy heats the interior, thereby increasing the internal temperature and pressure. It is also the source of the radiated energy. Gravitational energy was released in this way in the early Solar System. As the temperature in the interior rises, the outward force resulting from the outwardly decreasing pressure increases, and eventually it balances the inward force of gravity, a condition known as "hydrostatic equilibrium." As we shall discuss later, this mechanical balance is the key to understanding stable stars; see Figure 12–4.

Theoretical analysis shows that the dust surrounding the stellar embryo we call a pre-main-sequence star should absorb much of the radiation that the object emits.

■ **FIGURE 12–6** An artist's conception of the formation of a protoplanetary disk with a jet of gas being given off along the axis of rotation of the protostar.

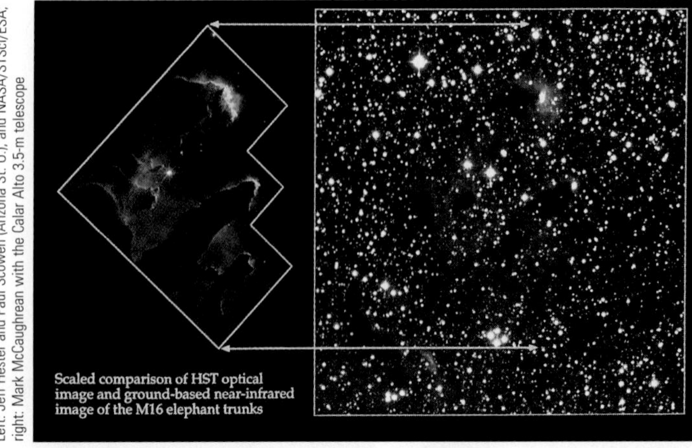

Left: Jeff Hester and Paul Scowen (Arizona St. U.), and NASA/STScI/ESA; right: Mark McCaughrean with the Calar Alto 3.5-m telescope

NASA, ESA, and The Hubble Heritage Team (STScI/AURA)

■ **FIGURE 12–7** ⓐ The orientation of the Hubble Space Telescope visible-light image of the Eagle Nebula on a near-infrared image taken with a ground-based telescope. ⓑ A Hubble detail of one of the pillars of gas and dust.

The radiation from the pre-main-sequence star should heat the dust to temperatures that cause it to radiate primarily in the infrared. Infrared astronomers have found many objects that are especially bright in the infrared but that have no known optical counterparts. These objects seem to be located in regions where the presence of a lot of dust, gas, and young stars indicates that star formation might still be going on.

Imaging in the visible (with the Hubble Space Telescope) and in the infrared (not only with previous infrared space telescopes but now especially with the Spitzer Space Telescope) has shown how young stars are born inside giant pillars of gas and dust inside certain nebulae. The Eagle Nebula is the most famous example because of the beautiful Hubble image showing exquisite detail, with false colors assigned to different filters (■ Fig. 12–7). As hot stars form, their intense radiation evaporates the gas and dust around them, freeing them from the cocoons of gas and dust in which they were born. We see this "evaporation" taking place at the tops of the Eagle Nebula's "pillars." The stars are destroying their birthplaces as they become independent and more visible from afar.

12.1b The Birth Cries of Stars

To their surprise, astronomers have discovered that young stars send matter out in oppositely directed beams, while they had expected to find only evidence of infall. This "bipolar ejection" (■ Fig. 12–8) may imply that a disk of matter orbits such pre-main-sequence stars, blocking an outward flow of gas in the equatorial direction and later coalescing into planets. Thus the flow of gas is channeled toward the poles.

Sometimes clumps of gas appear, but only recently have they been identified with ejections from stars in the process of collapsing. The clumps seem like spinning bullets, though what makes them spin is uncertain (perhaps connected with the magnetic

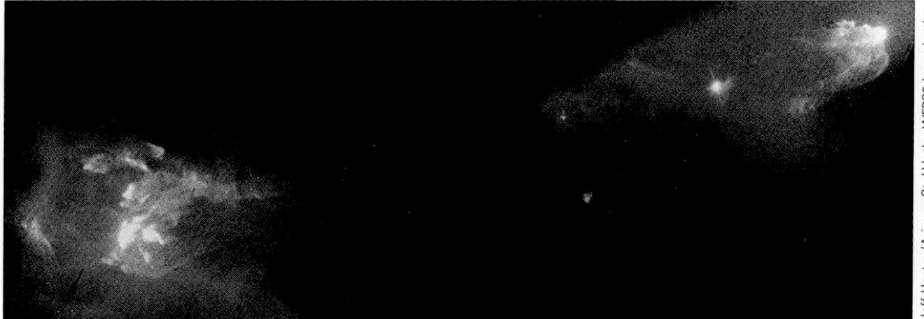

Jeff Hester (Arizona St. U.), the WFPC2 Investigation Definition Team, and NASA

■ **FIGURE 12–8** Hubble Space Telescope observations of Herbig-Haro objects #1 and #2 (HH-1 and HH-2).

NASA/JPL-Caltech/A. Noriega-Crespo (SSC/Caltech), Digital Sky Survey

■ **FIGURE 12–9** An outflow of gas in the Herbig-Haro objects HH–46/47, imaged in the infrared with the Spitzer Space Telescope and displayed in false color. At these infrared wavelengths, we see through the dark cloud (a "Bok globule") shown in the visible-light view in the inset to see molecular gas flowing outward from a bright newborn star. This low-mass star is ejecting supersonic jets of gas in opposite directions. It is 1140 light-years away from Earth in the direction of the southern constellation Vela.

Spitzer's 3.6 μm image is shown as blue, 4.5 μm and 5.8 μm as green, and 8 μm, which shows organic molecules made of carbon and hydrogen, as red. The organic molecules are glowing in a shape that matches that of the dark cloud.

field). The ejection of these spinning clumps helps slow the star's rate of spin, since they carry away angular momentum from what had been a rapidly spinning pre-main-sequence star. At the same time, some gas with low angular momentum is falling in toward the star. Hidden here is perhaps the main unsolved problem in star formation at the moment: How do stars figure out what their final mass will be?

The bipolar ejection appears as "Herbig-Haro objects," clouds of interstellar gas heated by shock waves from jets of high-speed gas. The jets are being ejected from a pre-main-sequence star, a star in the process of being born. Since the pre-main-sequence star is hidden in visible light by a dusty cocoon of gas, infrared observations of Herbig-Haro objects most clearly reveal what is going on (■ Fig. 12–9).

The jets of gas were formed as the pre-main-sequence star contracted under the force of its own gravity. Because a thick disk of cool gas and dust surrounds the pre-main-sequence star, the gas squirts outward along the pre-main-sequence star's axis of rotation at speeds of perhaps 1 million km/hr.

HH–1 and HH–2 (Fig. 12–8) are more irregular in shape than many other Herbig-Haro objects, perhaps because the bow shock wave we are seeing (a shock wave like those formed by the bow of a boat plowing through the water) has broken up. These objects are about 1500 light-years from us, in a star-forming region of the constellation Orion. The smallest features resolved are about the size of our Solar System, and the whole image is only about 1 light-year across.

Several classes of stars that vary erratically in brightness have been found. One of these classes, called T Tauri, contains pre-main-sequence stars as massive as or less massive than the Sun. Presumably, these stars are so young that they have not quite settled down to a steady and reliable existence on the main sequence. (T Tauri stars always have the word "stars" in their name though technically they haven't reached the main sequence, so they are not yet fully formed stars.)

In astronomical teaching, we have the question of whether to first consider the formation of stars in the star section of the book, as here, or in the section about the gas and dust between the stars from which the stars form. We choose to do some of each, and will continue our discussion of stars in formation in that latter location, Chapter 15 on the Milky Way Galaxy.

ASIDE 12.1: Herbig-Haro objects

Herbig-Haro objects are named after George Herbig, formerly of the Lick Observatory and now of the University of Hawaii, and Guillermo Haro of the Mexican National Observatory.

AceAstronomy™ Log into AceAstronomy and select this chapter to see the Active Figure called "Jet Deflection."

12.2 Where Stars Get Their Energy

If the Sun got all of its energy from gravitational contraction, it could have shined for only about 30 million years, not very long on an astronomical timescale. Yet we know that rocks about 4 billion years old have been found on Earth, and up to 4.4 billion years old on the Moon, so the Sun and the Solar System have been around at least that long. Moreover, fossil records of planets and animals, which presumably used the Sun's light and heat, date back billions of years. Some other source of energy must hold the Sun and other stars up against their own gravitational pull.

Actually, a pre-main-sequence star will heat up until its central portions become hot enough (at least one million kelvins) for **nuclear fusion** to take place, at which time it reaches the main sequence of the temperature-luminosity (temperature-magnitude, or Hertzsprung-Russell; see Chapter 11) diagram. Using this process, which we will soon discuss in detail, the star can generate enough energy to support it during its entire lifetime on the main sequence. A star's luminosity and temperature change little while it is on the main sequence; nuclear reactions provide the stability.

ASIDE 12.2: Fusion on Earth

In 2005, an agreement was reached to site a huge international project to show that it is feasible to produce fusion energy on Earth at ITER (International Thermonuclear Experimental Reactor) in France, with a related facility in Japan. The United States and the European Union are important parts of the consortium.

The energy makes the particles in the star move around rapidly. Such rapid, random motions in a gas are the definition of high temperature. The **thermal pressure,** the force from these moving particles pushing on each area of gas, is also high. The varying pressure, which decreases outward from the center, produces a force that pushes outward on any given pocket of gas. This outward force balances gravity's inward pull on the pocket ("hydrostatic equilibrium," which we illustrated in Figure 12–4).

The basic fusion process in main-sequence stars fuses four hydrogen nuclei into one helium nucleus. In the process, tremendous amounts of energy are released. (Hydrogen bombs on Earth fuse hydrogen nuclei into helium, but use different fusion sequences. The fusion sequences that occur in stars are far too slow for bombs.)

A hydrogen nucleus is but a single **proton.** A helium nucleus is more complex; it consists of two protons and two **neutrons** (■ Fig. 12–10). The mass of the helium nucleus that is the final product of the fusion process is slightly less than the sum of the masses of the four hydrogen nuclei (protons) that went into it. A small amount of the mass, m, "disappears" in the process: 0.007 (0.7 per cent) of the mass of the four protons.

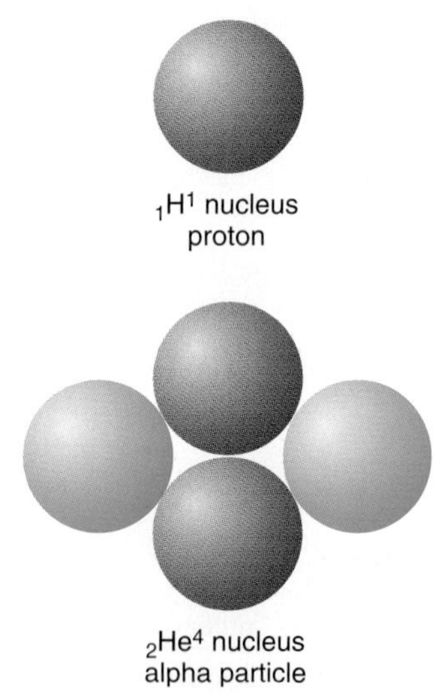

■ **FIGURE 12–10** The nucleus of hydrogen's most common form (isotope) is a single proton, while the nucleus of helium's most common isotope consists of two protons and two neutrons. Protons and neutrons are made of smaller particles called quarks (Chapter 19).

The mass difference does not really disappear, but rather is converted into energy, E, according to Albert Einstein's famous formula

$$E = mc^2,$$

where c is the speed of light; see *Figure It Out 12.1: Energy Generation in the Sun.* Even though m is only a small fraction of the original mass, the amount of energy released is prodigious; in the formula, c is a very large number. This energy is known as the "binding energy" of the nucleus, here specifically that of helium.

The loss of only 0.7 per cent of the central part of the Sun, for example, is enough to allow the Sun to radiate as much as it does at its present rate for a period of about ten billion (10^{10}) years; see *Figure It Out 12.1: Energy Generation in the Sun.* This fact, not realized until 1920 and worked out in more detail in the 1930s, solved the long-standing problem of where the Sun and the other stars get their energy.

All the main-sequence stars are approximately 90 per cent hydrogen (that is, 90 per cent of the atoms are hydrogen), so there is a lot of raw material to fuel the nuclear "fires." We speak colloquially of "nuclear burning," although, of course, the processes are quite different from the chemical processes that are involved in the "burning" of logs or of autumn leaves. In order to be able to discuss these processes, we must first review the general structure of nuclei and atoms.

AceAstronomy™ Log into Ace Astronomy and select this chapter to see the Astronomy Exercises "Nuclear Fusion" and "Hydrostatic Equilibrium."

FIGURE IT OUT 12.1

Energy Generation in the Sun

We can easily calculate the amount of hydrogen needed each second for the nuclear fusion in the Sun to account for its observed luminosity (power). The energy produced by the Sun comes from the following net reaction:

$$4\,{}_1H^1 \rightarrow {}_2He^4 + \text{energy.}$$

The energy is given by Einstein's famous formula, $E = mc^2$, where m is the mass difference between the final helium nucleus and the original four protons—that is, the "binding energy" of the helium nucleus. (In the nuclear fusion process, two protons are converted into two neutrons, but we will ignore the mass difference between neutrons and protons, which is very small.)

The mass of a proton is 1.6726×10^{-27} kg; thus, the mass of the original four protons is 6.6904×10^{-27} kg. The mass of a helium nucleus is 6.6448×10^{-27} kg, which is less by 0.0456×10^{-27} kg (roughly 0.7% of the original mass). Multiplying this value, m, by c^2, we have

$$E = mc^2 = (0.0456 \times 10^{-27}\ \text{kg})(3.00 \times 10^8\ \text{m/sec})^2$$
$$= 4.10 \times 10^{-12}\ \text{kg·m}^2/\text{sec}^2 = 4.10 \times 10^{-12}\ \text{J},$$

since 1 J (1 joule) = 1 kg·m²/sec². So, 4.10×10^{-12} J of energy is liberated every time 4 protons fuse to form a helium nucleus (call this "one reaction").

The luminosity of the Sun is 3.83×10^{26} J/sec. To account for this, we need

$$(3.83 \times 10^{26}\ \text{J/sec})/(4.10 \times 10^{-12}\ \text{J/reaction})$$
$$= 9.33 \times 10^{37}\ \text{reactions/sec.}$$

But each reaction used 4 protons, whose mass is about 6.69×10^{-27} kg. Thus, the total mass of protons used per second is

$$(9.33 \times 10^{37}\ \text{reactions/sec})(6.69 \times 10^{-27}\ \text{kg/reaction})$$
$$= 6.24 \times 10^{11}\ \text{kg/sec.}$$

Let's put this into more familiar units. 1 kg = 2.2 pounds, so about 1.37×10^{12} pounds of protons fuse each second. But 1 ton = 2000 lbs, so this is equivalent to 6.86×10^8 tons per second. The Sun fuses nearly 700 million tons of fuel each second!

Roughly how long will the Sun continue to fuse hydrogen in its core? (This will be its total main-sequence lifetime.) The Sun's mass is 2.0×10^{30} kg, of which about 70% is hydrogen (protons), or 1.4×10^{30} kg. Only about the central 14% will be used for fusion, so the amount of fuel is

$$(0.14)(1.4 \times 10^{30}\ \text{kg}) \approx 2.0 \times 10^{29}\ \text{kg.}$$

Above, we found that the Sun uses about 6.2×10^{11} kg/sec, and at this rate it could shine on the main sequence for

$$(2.0 \times 10^{29}\ \text{kg})/(6.2 \times 10^{11}\ \text{kg/sec}) \approx 3.2 \times 10^{17}\ \text{sec.}$$

But there are about 3.2×10^7 sec/year, so the main-sequence lifetime of the Sun should be about $(3.2 \times 10^{17}\ \text{sec})/(3.2 \times 10^7\ \text{sec/year}) \approx 10^{10}$ years, about 10 billion years! The Sun is currently about half way through its main-sequence life, so it is a middle-aged star.

12.3 Atoms and Nuclei

As we mentioned in Chapter 2, an atom consists of a small nucleus surrounded by electrons. Most of the mass of the atom is in the nucleus, which takes up a very small volume in the center of the atom. The effective size of the atom, the chemical interactions of atoms to form molecules, and the nature of spectra are all determined by the electrons.

12.3a Subatomic Particles

The nuclear particles with which we need to be most familiar are the proton and neutron. Both of these particles have nearly the same mass, 1836 times greater than the mass of an electron, though still tiny. The neutron has no electric charge and the proton has one unit of positive electric charge. The electrons, which surround the nucleus, have one unit each of negative electric charge. When an atom loses an electron, it has a net positive charge of 1 unit for each electron lost. The atom is now a form of **ion** (■ Fig. 12–11).

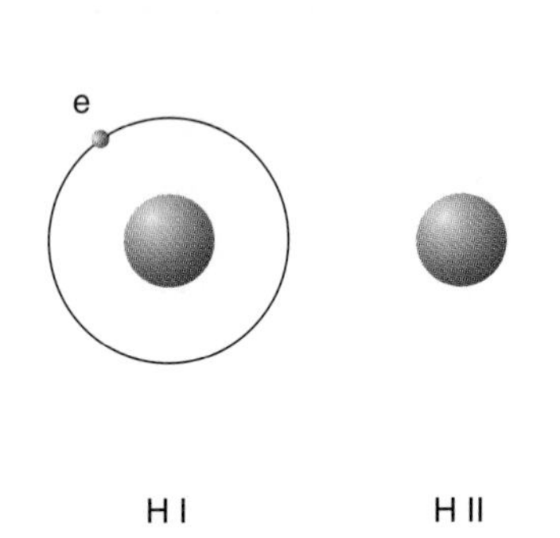

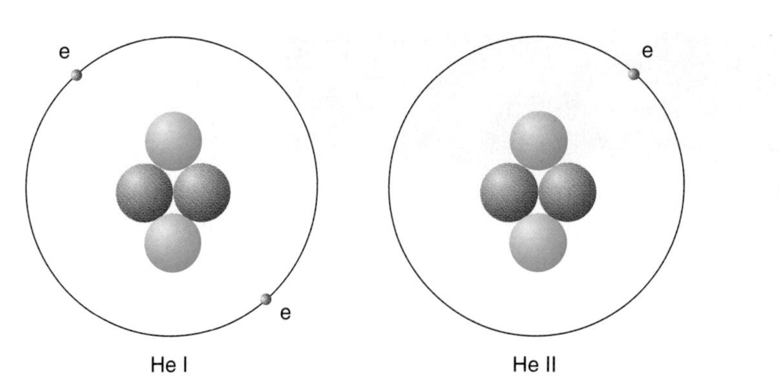

■ **FIGURE 12–11** Hydrogen and helium ions. The sizes of the nuclei are greatly exaggerated with respect to the sizes of the "orbits" of the electrons.

Since the number of protons in the nucleus determines the charge of the nucleus, it also dictates the quota of electrons that the neutral state of the atom must have. To be neutral, after all, there must be equal numbers of positive and negative charges. Each **element** (sometimes called "chemical element") is defined by the specific number of protons in its nucleus. The element with one proton is hydrogen, that with two protons is helium, that with three protons is lithium, and so on.

12.3b Isotopes

Though a given element always has the same number of protons in a nucleus, it can have several different numbers of neutrons. (The number of neutrons is usually somewhere between 1 and 2 times the number of protons. The most common form of hydrogen, just a single proton, is the main exception to this rule.) The possible forms of the same element having different numbers of neutrons are called **isotopes.**

For example, the nucleus of ordinary hydrogen contains one proton and no neutrons. An isotope of hydrogen (■ Fig. 12–12) called deuterium (and sometimes "heavy hydrogen") has one proton and one neutron. Another isotope of hydrogen called tritium has one proton and two neutrons.

> **ASIDE 12.3: Deuteronomy**
>
> There is a joke that the study of deuterons, and the cosmological importance of studying their formation in the first few minutes after the big bang, is known as "deuteronomy."

Most isotopes do not have specific names, and we keep track of the numbers of protons and neutrons with a system of superscripts and subscripts. The subscript before the symbol denoting the element is the number of protons (called the atomic number), and a superscript after the symbol is the total number of protons and neutrons together (called the mass number, or atomic mass).

For example, ${}_1H^2$ is deuterium, since deuterium has one proton, which gives the subscript, and an atomic mass of 2, which gives the superscript. (Note that ${}^2_1H$ is also correct notation.) Deuterium has atomic number equal to 1 and mass number equal to 2. Similarly, ${}_{92}U^{238}$ is an isotope of uranium with 92 protons (atomic number = 92) and mass number of 238, which is divided into 92 protons and 238 − 92 = 146 neutrons.

Each element has only certain isotopes. For example, most of the naturally occurring helium is in the form ${}_2He^4$, with a much lesser amount as ${}_2He^3$.

12.3c Radioactivity and Neutrinos

Sometimes an isotope is not stable, in that after a time it will spontaneously change into another isotope or element; we say that such an isotope is **radioactive.** The most mas-

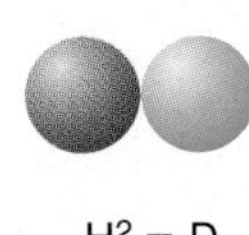

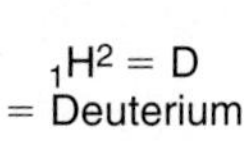

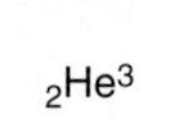

■ **FIGURE 12–12** Isotopes of hydrogen and helium. ${}_1H^2$ (deuterium) and ${}_1H^3$ (tritium) are much rarer than the normal isotope ${}_1H^1$. Similarly, ${}_2He^3$ is much rarer than ${}_2He^4$.

sive elements, those past uranium, are all radioactive, and have average lifetimes that are very short.

It has been theoretically predicted that around element 114, elements should begin being somewhat more stable again. The handful of atoms of element 114 and 116 discovered in 1998 and 1999 are more stable than those of slightly lower mass numbers—lasting even about 5 seconds instead of a small fraction of a second. (A claim that element 118 was also discovered has been withdrawn.)

During certain types of radioactive decay, as well as when a free proton and electron combine to form a neutron, a particle called a **neutrino** is given off. A neutrino is a neutral particle (its name comes from the Italian for "little neutral one").

Neutrinos have a very useful property for the purpose of astronomy: They rarely interact at all with matter. Thus when a neutrino is formed deep inside a star, it can usually escape to the outside without interacting with any of the matter in the star. A photon of electromagnetic radiation, on the other hand, can travel only about 0.5 mm (on average) in a stellar interior before it is absorbed, and it takes about a hundred thousand years for a photon to zig and zag its way to the surface.

The elusiveness of the neutrino not only makes it a valuable messenger—indeed, the only possible direct messenger—carrying news of the conditions inside the Sun at the present time, but also makes it very difficult for us to detect on Earth. A careful experiment carried out over many years has found only about $^1/_3$ the expected number of neutrinos, as we shall soon see.

AceAstronomy™ Log into AceAstronomy and select this chapter to see the Active Figure called "Radioactive Decay."

12.4 Stars Shining Brightly

Let us now use our knowledge of atomic nuclei to explain how stars shine. For a pre-main-sequence star, the energy from the gravitational contraction goes into giving the individual particles greater speeds; that is, the gas temperature rises. When atoms collide at high temperature, electrons get knocked away from their nuclei, and the atoms become fully ionized. The electrons and nuclei can move freely and separately in this "plasma."

For nuclear fusion to begin, atomic nuclei must get close enough to each other so that the force that holds nuclei together, the "strong nuclear force" (to be discussed in Chapter 19), can play its part. But all nuclei have positive charges, because they are composed of protons (which bear positive charges) and neutrons (which are neutral). The positive charges on any two nuclei cause an electrical repulsion between them, which tends to prevent fusion from taking place.

However, at the high temperatures (millions of kelvins) typical of a stellar interior, some nuclei occasionally have enough energy to overcome this electrical repulsion. They come sufficiently close to each other that they essentially collide, and the strong nuclear force takes over. Fusion on the main sequence proceeds in one of two ways, as will be discussed below.

Once nuclear fusion begins, enough energy is generated to maintain the pressure and prevent further contraction. The pressure provides a force that pushes outward strongly enough to balance gravity's inward pull.

In the center of a star, the fusion process is self-regulating. The star finds a balance between thermal pressure pushing out and gravity pushing in. It thus achieves stability on the main sequence (at a constant temperature and luminosity). When we learn how to control fusion in power-generating stations on Earth, which currently seems decades off (and has long seemed so), our energy crisis will be over, since deuterium, the potential "fuel," is so abundant in Earth's oceans.

Don F. Figer, Mark Morris, Ian F. McLean (UCLA), Gene Serabyn (Caltech) and R. Michael Rich (Columbia)

■ **FIGURE 12–13** The Pistol Nebula, gas glowing because of radiation from a massive, highly luminous, main-sequence star. The star, perhaps the most massive known, shines 10 million times more powerfully than our Sun and would fill the Earth's orbit. Hidden behind the dust that lies between us and the center of our Galaxy, the star and nebula are revealed here by the infrared camera aboard the Hubble Space Telescope.

The greater a star's mass, the hotter its core becomes before it generates enough pressure to counteract gravity. The hotter core gives off more energy, so the star becomes brighter (■ Fig. 12–13), explaining why main-sequence stars of large mass have high luminosity. In fact, it turns out that more massive stars use their nuclear fuel at a very much higher rate than less massive stars. Even though the more massive stars have more fuel to burn, they go through it relatively quickly and live shorter lives than low-mass stars, as we discussed in Chapter 11. The next two chapters examine the ultimate fates of stars, with the fates differing depending on the masses of the stars.

AceAstronomy™ Log into AceAstronomy and select this chapter to see the Active Figure called "Gravity vs. Pressure."

12.5 Why Stars Shine

Several chains of nuclear reactions have been proposed to account for the fusion of four hydrogen nuclei into a single helium nucleus. Hans Bethe of Cornell University suggested some of these procedures during the 1930s. The different chain reactions prevail at different temperatures, so chains that are dominant in very hot stars may be different from the ones in cooler stars.

When the temperature of the center of a main-sequence star is less than about 20 million kelvins, the **proton-proton chain** (■ Fig. 12–14) dominates. This sequence uses six hydrogen nuclei (protons), and winds up with one helium nucleus plus two protons. The net transformation is four hydrogen nuclei into one helium nucleus. (Though two of the protons turn into neutrons, here this isn't the main point.)

But the original six protons contained more mass than do the final single helium nucleus plus two protons. The small fraction of mass that disappears is converted into an amount of energy that we can calculate with the formula $E = mc^2$; see *Figure It Out 12.1: Energy Generation in the Sun.* According to Einstein's special theory of relativity, mass and energy are equivalent and interchangeable, linked by this equation.

For stellar interiors significantly hotter than that of the Sun, the **carbon-nitrogen-oxygen (CNO) cycle** dominates. This cycle begins with the fusion of a hydrogen nucleus (proton) with a carbon nucleus. After many steps, and the insertion of four protons, we are left with one helium nucleus plus a carbon nucleus. Thus, as much carbon remains at the end as there was at the beginning, and the carbon can start the cycle again.

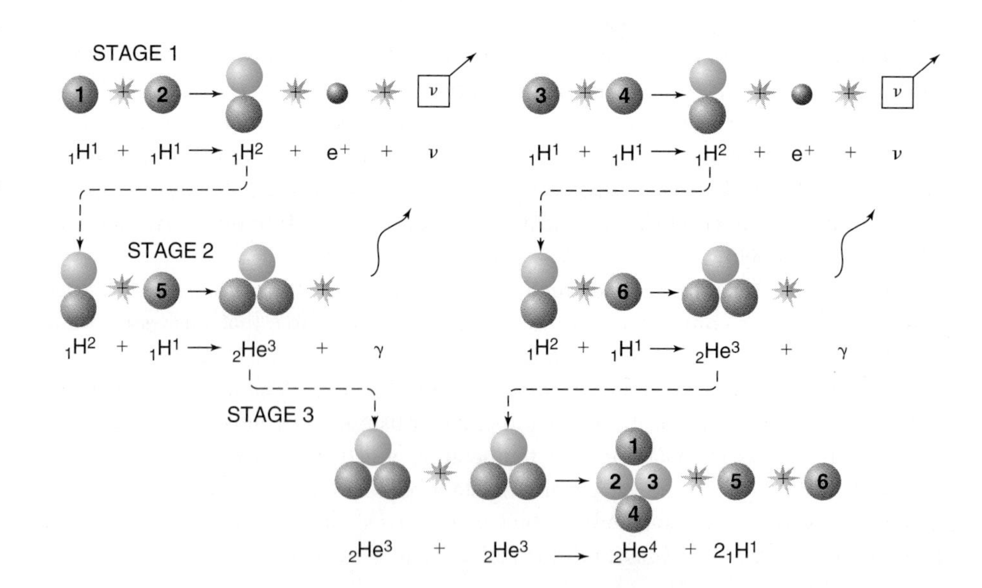

■ **FIGURE 12–14** The proton-proton chain; e^+ stands for a positron (the equivalent of an electron, but with a positive charge; an "antielectron"), ν (the Greek letter "nu") is a neutrino, and γ (the Greek letter "gamma") is electromagnetic radiation at a very short wavelength. (Gamma rays are similar to x-rays but correspond to photons of even higher energies.)

In the first stage, two nuclei of ordinary hydrogen (protons) fuse to become a deuterium (heavy hydrogen) nucleus, a positron, and a neutrino. The neutrino immediately escapes from the star, but the positron soon collides with an electron. They annihilate each other, forming gamma rays.

Next (stage 2), the deuterium nucleus fuses with yet another proton to become an isotope of helium with two protons and one neutron. More gamma rays are released.

Finally (stage 3) two of these helium isotopes fuse to make one nucleus of ordinary helium plus two protons. The protons are numbered to help you keep track of them.

As in the proton-proton chain, four hydrogen nuclei have been converted into one helium nucleus during the CNO cycle, 0.7 per cent of the mass has been transformed, and an equivalent amount of energy has been released according to $E = mc^2$. Main-sequence stars more massive than about 1.1 times the Sun are dominated by the CNO cycle.

Later in their lives, when they are no longer on the main sequence, stars can have even higher interior temperatures, above 10^8 K. They then fuse helium nuclei to make carbon nuclei. The nucleus of a helium atom is called an "alpha particle" for historical reasons. Since three helium nuclei ($_2He^4$) go into making a single carbon nucleus ($_6C^{12}$), the procedure is known as the **triple-alpha process.**

A series of other processes can build still heavier elements inside very massive stars. These processes, and other element-building methods, are called **nucleosynthesis** (new′clee-oh-sin′tha-sis). The theory of nucleosynthesis in stars can account for the abundances (proportions) we observe of the elements heavier than helium. Currently, we think that the synthesis of isotopes of the lightest elements (hydrogen, helium, and lithium) took place in the first few minutes after the origin of the Universe (Chapter 19), though some of the observed helium was produced later by stars.

12.6 Brown Dwarfs

When a pre-main-sequence star has at least 7.5 per cent of the Sun's mass (that is, it has about 75 Jupiter masses), nuclear reactions begin and continue, and it becomes a normal star. But if the mass is less than 7.5 per cent of the Sun's mass, the central temperature does not become hot enough for nuclear reactions using ordinary hydrogen (protons) to be sustained. (Masses of this size do, however, fuse deuterium into helium, but this phase of nuclear fusion doesn't last long because there is so little deuterium in the Universe relative to ordinary hydrogen.)

These objects shine dimly, shrinking and dimming as they age. They came to be called **brown dwarfs,** mainly because "brown" is a mixture of many colors and people didn't agree how such supposedly "failed stars" would look, and also because they emit very little light (■ Fig. 12–15). When old, they have all shrunk to the same radius, about that of the planet Jupiter. We have met them already in Section 9.2c.

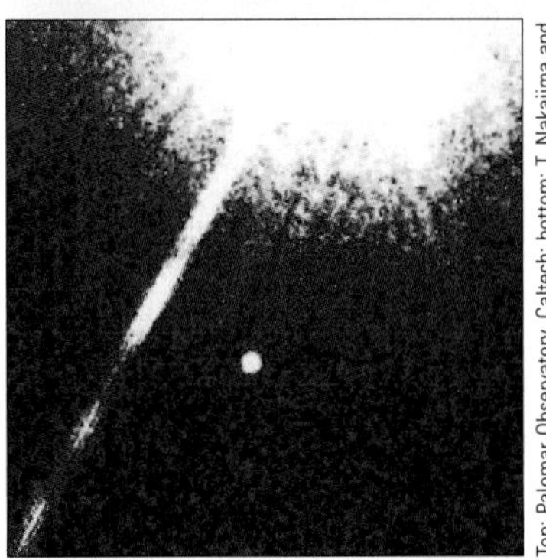

Top: Palomar Observatory, Caltech; bottom: T. Nakajima and S. Kulkarni (Caltech), S. Durrance and F. Golimowski (Johns Hopkins U.), and NASA

■ **FIGURE 12–15** The first brown dwarf to be unambiguously discovered, Gliese 229B. A ground-based image *(top)* barely shows the brown dwarf orbiting the larger star, because the atmosphere blurs the larger star's image, but the Hubble Space Telescope shows the brown dwarf clearly *(bottom)*. (The straight streak is an artifact in the imaging.)

For decades, there was a debate as to whether brown dwarfs exist, but finally some were found in 1995. We now know of about 1000, because of the advances in astronomical imaging and in spectroscopy, not only in the visible but also in the infrared. The coolest ones, of spectral type T, show methane and water in their spectra, like giant planets but unlike normal stars.

It is difficult to tell the difference between a brown dwarf and a small, cool, ordinary star, unless the brown dwarf is exceptionally cool. One way is to see whether an object has lithium in its spectrum. Lithium is a very fragile element, and undergoes fusion in ordinary stars, which converts it to other things. So if you detect lithium in the spectrum of a dim star, it is probably a brown dwarf (which isn't sustaining nuclear fusion using protons) rather than a cool, ordinary dwarf star of spectral class M or L, which are the coolest stars on the main sequence (and thus have begun to sustain their nuclear fusion). A complication is that very young M and L stars might not be old enough to have burned all their lithium, leading to potential confusion with brown dwarfs.

How do we tell the difference between brown dwarfs and giant planets in cases where they are orbiting a more normal star? Some astronomers would like to distinguish between them by the way that they form: While planets form in disks of dust and gas as the central star is born, brown dwarfs form like the central star, out of the collapse of a cloud of gas and dust. But we can't see the history of an object when we look at it, so it is hard to translate the distinction into something observable.

All of the proposed tests are difficult to make. So, currently, for lack of definitive methods, the distinction is usually made on the basis of mass: Any orbiting object with a mass less than 13 times Jupiter's is called a planet, while the range 13 to 75 Jupiter masses corresponds to brown dwarfs. (Objects less massive than 13 Jupiter masses not orbiting stars are sometimes called "free-floating planets" since they are not planets in the conventional sense of the word.)

The rationale for using 13 Jupiter masses as the dividing line between planets and brown dwarfs is that above this mass, fusion of deuterium occurs for a short time, whereas below this mass, no fusion ever occurs. Thus, although brown dwarfs are not normal stars, they *do* fuse nuclei for a short time, and hence aren't completely "failed stars" as many people call them.

Brown dwarfs are being increasingly studied, especially in the infrared. Hubble Space Telescope images show that one of the nearby ones is double, with the components separated by 5 A.U. By watching it over a few years, we should be able to measure its orbit and derive the masses of the components.

12.7 The Solar Neutrino Experiment

Astronomers can apply the equations that govern matter and energy in a star, and make a model of the star's interior in a computer. Though the resulting model can look quite nice, nonetheless it would be good to confirm it observationally.

However, the interiors of stars lie under opaque layers of gas. Thus we cannot directly observe electromagnetic radiation from stellar interiors. Only neutrinos escape directly from stellar cores. Neutrinos interact so weakly with matter that they are hardly affected by the presence of the rest of the Sun's mass. Once formed, they zip right out into space, at (or almost at) the speed of light. Thus they reach us on Earth about 8 minutes after their birth.

Neutrinos should be produced in large quantities by the proton-proton chain in the Sun, as a consequence of protons turning into neutrons, positrons, and neutrinos; see Figure 12–14. (A positron is an "antielectron," an example of antimatter. Whenever a particle and its antiparticle meet, they annihilate each other.)

Brookhaven National Laboratory

FIGURE 12–16 The original neutrino telescope, deep underground in the Homestake Gold Mine in South Dakota to shield it from other types of particles from space. The telescope is mainly a tank containing 400,000 liters of perchloroethylene (dry-cleaning solution).

12.7a Initial Measurements

For over three decades, astrochemist Raymond Davis has carried out an experiment to search for neutrinos from the solar core, set up in consultation with the theorist John Bahcall, whose calculations long drove the theory. Davis set up a tank containing 400,000 liters of a chlorine-containing chemical (Fig. 12–16). One isotope of chlorine can, on rare occasions, interact with one of the passing neutrinos from the Sun. It turns into a radioactive form of argon, which Davis and his colleagues at the University of Pennsylvania can detect. He needs such a large tank because the interactions are so rare for a given chlorine atom. In fact, he detects fewer than 1 argon atom formed per day, despite the huge size of the tank.

Over the years, Davis and his colleagues detected only about 1/3 the number of interactions predicted by theorists. Where is the problem? Is it that astronomers don't understand the temperature and density inside the Sun well enough to make proper predictions? Or is it that the physicists don't completely understand what happens to neutrinos after they are released?

The latest thinking is that neutrinos actually change after they are released. According to a theoretical model, the neutrinos of the specific type produced inside the Sun change, before they reach Earth, into all three types (called "flavors") of neutrinos that are known. But Davis's experiment is sensitive only to the specific flavor

Y. Totsuka, U. Tokyo, Inst. for Cosmic Ray Research

a

Y. Totsuka, U. Tokyo, Inst. for Cosmic Ray Research

b

■ **FIGURE 12–17** (a) The Super-Kamiokande neutrino detector in Japan. We see the empty stainless steel vessel, 40 m across, and some of the 13,000 photomultiplier tubes, each 50 cm across, surrounding it. Neutrinos interacting within molecules of the 50,000 tons of highly purified water (note the boat, photographed as the detector was partly filled) lead to the emission of blue flashes of light that are being detected. (b) The Kamiokande and Super-Kamiokande experiments detected neutrinos from the Sun and provided a test not only of how the stars shine but also of whether our basic understanding of fundamental physics is correct. Here we see thousands of Super-Kamiokande's light-sensitive "photomultiplier" tubes. In one of the most dramatic and expensive mistakes ever made in science, most of these tubes "popped" and were destroyed in 2001, when the tank was being refilled after being emptied for the first time. Apparently, one defective tube broke, and sent out a shock wave that destroyed thousands of other tubes.

("electron neutrinos") released by the Sun. Thus only $^1/_3$ the original prediction is expected, and that is what we detect.

The chlorine experiment, though it has run the longest by far, is no longer the only way to detect solar neutrinos. An experiment in Japan (■ Fig. 12–17), led by Masatoshi Koshiba, first set up to study protons and whether they decay, has used a huge tank of purified water to verify that the number of neutrinos is less than expected. For their pioneering work in the detection of astrophysical neutrinos, Davis and Koshiba were together awarded half the 2002 Nobel Prize in Physics.

Other sets of experiments in Italy and in Russia that use gallium, an element that is much more sensitive to neutrinos than chlorine, also show that up to half of the expected neutrinos are missing. The chlorine was sensitive only to neutrinos of very high energy, which come out of only a small fraction of the nuclear reactions in the Sun and not out of the basic proton-proton chain. Gallium is sensitive to a much wider range of interactions, including the most basic ones.

Neutrinos were first thought of theoretically as particles that have no rest mass—that is, particles that would have no mass if they weren't moving ("at rest"). The Japanese experiment has shown that neutrinos probably have a tiny rest mass after all. Theoretically, neutrinos can oscillate from one type to another only if they have some rest mass. So this result fits with the current ideas that we are seeing neutrino oscillations from one type to another.

12.7b The Sudbury Neutrino Observatory

A U.S.–Canadian experiment in Sudbury, Ontario, Canada, began collecting data in 1999 (■ Fig. 12–18). It has even more sensitive detection capability than earlier experiments. It uses a large quantity of "heavy water," water whose molecules contain deuterium instead of the more normal hydrogen isotope. This experiment, like those in Japan, looks for light given off when neutrinos hit the water.

The Sudbury Neutrino Observatory (SNO) has apparently resolved the remaining mysteries about the missing solar neutrinos. They aren't missing after all, since SNO has been able to detect the correct rate in one of its configurations, which is sensitive to all flavors of neutrinos. In another mode, sensitive only to the electron neutrinos that the Sun gives off, it confirms the deficit found by other detectors. So SNO has confirmed that most of the neutrinos change in flavor en route from the Sun to the Earth. The solar-neutrino problem was indeed in the neutrino physics rather than in our understanding of the temperature of the Sun's core.

Lawrence Berkeley National Laboratory and Sudbury Neutrino Observatory

■ **FIGURE 12–18** The outside of the tank of deuterated water ("heavy water") that is the heart of the Sudbury Neutrino Observatory.

SNO will provide even more data over the coming years. In particular, it should be able to determine the effect of the Earth's mass on neutrinos by comparing what it detects during daylight hours with what it detects during nighttime hours, when the neutrinos from the Sun have to pass through the Earth to reach the detector.

The various neutrino experiments may cause a revolution in fundamental ideas of physics. Similarly, there are also important repercussions for physics from a set of astronomical observations we will discuss later (Section 18.5), showing that measurements of distant supernovae (exploding stars) indicate that the expansion of the Universe is accelerating. The relation between physics and astronomy is close, though the point of view that physicists and astronomers have in tackling problems may be different.

12.8 The End States of Stars

The next two chapters discuss the various end states of stellar evolution. The mass of an isolated, single star determines its fate and we provide a figure here as a type of "coming attractions" (■ Fig. 12–19). This brief section and diagram thus set the stage for what is coming next, which is so interesting that this brief introduction leads to two separate chapters.

In Chapter 13, we discuss low-mass stars, like the Sun, which wind up as white dwarfs. We also discuss the more massive stars, which use up their nuclear fuel much more quickly. These high-mass stars are eventually blown to smithereens in supernova explosions, becoming neutron stars or even black holes! Chapter 14 is devoted entirely to exotic black holes and their properties.

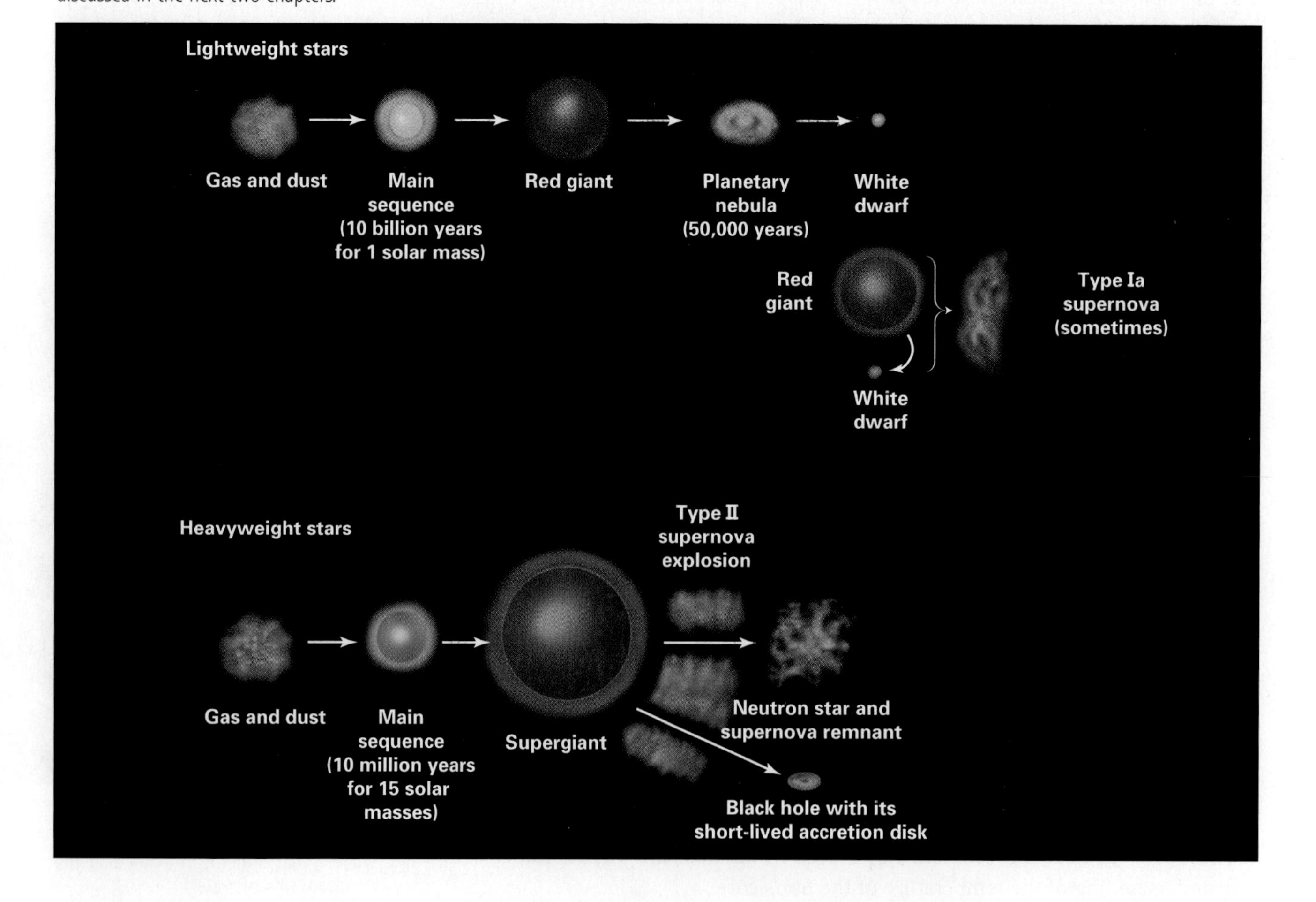

■ **FIGURE 12–19** A summary of the stages of stellar evolution for stars of different masses, using the boxing terms "lightweight" and "heavyweight" to mean low mass and high mass, respectively. The various types of objects shown will be defined and discussed in the next two chapters.

CONCEPT REVIEW

Stars are formed from dense regions in nebulae, "clouds" of gas and dust that are best observed at infrared and radio wavelengths (Sec. 12.1). The collapsing cloud becomes a cluster of **protostars** and subsequently **pre-main-sequence stars,** which are powered by gradual gravitational contraction (Sec. 12.1a). Eventually, the outward force from the pressure balances the inward force of gravity, a condition known as "hydrostatic equilibrium." The dust heats up and becomes visible in the infrared. Some planetary systems in formation may have been discovered around young stars (Sec. 12.1b).

Stars get their energy from **nuclear fusion** (Sec. 12.2). The resulting **thermal pressure** balances gravity. The basic fusion process in the Sun and other main-sequence stars is a merger of four hydrogen nuclei (**protons**) into one helium nucleus (two protons and two **neutrons**), with the difference in mass (the "binding energy" of the helium nucleus) transformed to energy according to Einstein's famous equation, $E = mc^2$.

Atoms that have lost electrons are **ions** (Sec. 12.3a). Each **element** is defined by the number of protons in its nucleus. Forms of the same element with different numbers of neutrons are **isotopes** (Sec. 12.3b). Isotopes that decay spontaneously are **radioactive** (Sec. 12.3c). **Neutrinos** are light, neutral particles given off in some radioactive decays as well as during nuclear fusion in stars. They are so elusive that they escape from a star immediately.

In the hot cores of stars, collisions have stripped electrons away from the atomic nuclei, leaving the atoms fully ionized (Sec. 12.4). Nuclear fusion requires very high temperatures (millions of kelvins); only at high speeds can some of the protons overcome their electrical repulsion and get sufficiently close to each other for the "strong nuclear force" (which holds protons and neutrons together in nuclei) to be effective. The greater a star's mass, the hotter its core becomes before it generates enough pressure to counteract gravity, so the star is more luminous.

The **proton-proton chain** is the basic fusion process in the Sun, while the **carbon-nitrogen-oxygen cycle** dominates in hotter, more massive stars (Sec. 12.5). At still higher temperatures, when stars have left the main sequence, the **triple-alpha process** is dominant, where an alpha particle is a helium nucleus. Element-building processes are what is meant by **nucleosynthesis.**

Brown dwarfs are sometimes called "failed stars," with insufficient mass to begin nuclear fusion of protons, ordinary hydrogen (Sec. 12.6). However, they do fuse deuterium (a rare, heavy form of hydrogen) to helium for a short time, so in a sense they do act like stars. Brown dwarfs have a mass range of 13 to 75 Jupiter masses (which is 1.3 to 7.5 per cent of the Sun's mass). Being very cool and dim, brown dwarfs were not discovered until 1995, but new infrared surveys of the sky have revealed large numbers of them.

Theorists predict a certain number of neutrinos that should result from fusion in the Sun. Careful experiments have generally found only about one-third to one-half that number (Sec. 12.7a). To explain this deficit, physicists have suggested that most of the neutrinos change from one type to several different types on their journey from the Sun to the Earth. Recently, measurements at the Sudbury Neutrino Observatory showed that the predicted number of neutrinos is indeed being produced by the Sun, and that they change type as they travel (Sec. 12.7b).

Stars have several different endpoints to their evolution after the main sequence (Sec. 12.8). These will be discussed in the next two chapters.

QUESTIONS

1. Since individual stars can live for millions or billions of years, how can observations taken at the current time tell us about stellar evolution?
2. **(a)** Why are regions of star formation difficult to study at optical and ultraviolet wavelengths? **(b)** Which wavelengths are better for this purpose, and why?
3. What is the source of energy in a pre-main-sequence star?
4. At what point does a pre-main-sequence star become a normal star?
5. Arrange the following in order of development: pre-main-sequence star, nebula, protostar, Sun.
6. What are T Tauri stars and Herbig-Haro objects?
7. †Give the number of protons, the number of neutrons, and the number of electrons in ordinary hydrogen ($_1H^1$), lithium ($_3Li^7$), and iron ($_{26}Fe^{56}$) atoms.
8. **(a)** If you remove one neutron from helium, the remainder is what element? **(b)** If you remove one proton from helium, what element is left?
9. Why is four-times ionized helium never observed?
10. Explain why nuclear fusion takes place only in the center of stars rather than on their surface.
11. What is the major fusion process that takes place in the Sun?
12. What is meant by the term "nucleosynthesis"?
13. What forces are in mechanical balance for a star to be on the main sequence?

14. (**a**) Qualitatively, how do the main-sequence lifetimes of massive stars compare with those of low-mass stars? (**b**) Explain why this is so.

†15. Use the speed of light and the distance between the Earth and the Sun to verify that it takes neutrinos 8.3 minutes to reach us from the Sun.

16. What does it mean for the temperature of a gas to be higher?

†17. If 0.7% of the mass of four protons gets emitted as energy when they fuse to form a helium nucleus, how many joules of energy are produced? Show your work. (Note that a joule is a $kg \cdot m^2/sec^2$.)

18. Why do neutrinos give us different information about the Sun than light does?

19. What is a brown dwarf?

20. Why are the results of the solar neutrino experiment so important?

21. Summarize the current status of our understanding of solar neutrinos.

22. **True or false?** Nuclear reactions occur throughout almost the entire Sun, interior to the photosphere, because of the high temperature of the gas everywhere within the Sun.

†23. **True or false?** If a helium nucleus has 0.7% less mass than four free protons (each of mass m_p), then the equation $E = (0.007)(4m_p)c^2$ represents the amount of energy generated by the Sun each time a new helium nucleus forms through nuclear fusion.

24. **True or false?** As a main-sequence star ages, its surface becomes much hotter and it moves to a new location on the main sequence in a temperature-luminosity diagram.

25. **True or false?** Brown dwarfs are formed by the same initial process that forms main-sequence stars, but they are less massive.

26. **True or false?** Stars generate their energy through nuclear fission, the breaking apart of unstable heavy elements into lighter ones.

27. **Multiple choice:** As a protostar is formed from a cloud of gas and becomes a pre-main-sequence star, which one of the following does *not* occur? (**a**) The cloud contracts due to gravity. (**b**) The density of the gas increases. (**c**) The pressure of the gas increases. (**d**) The temperature of the gas increases. (**e**) The protons begin to fuse.

28. **Multiple choice:** In a stable star, the gravitational forces are balanced by (**a**) gravitational forces acting in the opposite direction; (**b**) thermal pressure; (**c**) the strong nuclear force; (**d**) chemical reactions; or (**e**) electromagnetic forces.

†29. **Multiple choice:** The number of protons, neutrons, and electrons in a neutral atom of $_{26}Fe^{56}$ is, respectively, (**a**) 26, 30, 26; (**b**) 30, 26, 30; (**c**) 26, 30, 56; (**d**) 30, 26, 26; or (**e**) 26, 56, 26.

30. **Multiple choice:** The amount of energy released in a nuclear fusion reaction is directly proportional to the (**a**) mass of the initial components; (**b**) mass of the final products; (**c**) mass difference between the initial components and final products; (**d**) mass of the proton; or (**e**) mass of the photon.

31. **Multiple choice:** Which one of the following statements regarding stars is *false*? (**a**) Stars form from large clouds of gas and dust that fragment as they collapse. (**b**) Even though they are gravitationally contracting, pre-main-sequence stars do not emit electromagnetic radiation; they have not yet started fusing hydrogen to helium. (**c**) Brown dwarfs, sometimes called "failed stars," undergo fusion for only a short time compared with M-type main-sequence stars. (**d**) Once the core of a sufficiently massive contracting star gets hot enough, hydrogen starts fusing to helium and the star settles on the main sequence. (**e**) The light (gamma rays) emitted by the nuclear reactions in the core of a main-sequence star does not immediately escape the star because the ionized gases are quite opaque.

32. **Fill in the blank:** Brown dwarfs emit electromagnetic radiation primarily at ________ wavelengths.

33. **Fill in the blank:** Forms of the same element having different numbers of neutrons are called ________.

34. **Fill in the blank:** Pre-main-sequence stars generate energy predominantly through the process of ________.

35. **Fill in the blank:** Theorists predicted that the Sun should be producing a large number of ________, but for a long time the measure rate on Earth was substantially less.

†36. **Fill in the blank:** The number of neutrons in a nucleus of $_7N^{15}$ is ______.

†This question requires a numerical solution.

TOPICS FOR DISCUSSION

1. Can you convince yourself that radio waves are not much affected by gas and small particles? Consider whether you can hear a radio on a foggy or smoggy day.
2. Do you think astronomers were overly bold in predicting that the solar neutrino problem would be resolved by changes in our theories of fundamental particles, rather than by abandoning the standard model of solar energy production?
3. A solitary star that is still forming must get rid of its angular momentum through the ejection of spinning jets (and, to some extent, by storing it in planets that form in the disk). Is there as much of a problem for a cloud from which a binary star forms?

MEDIA

AceAstronomy™ Log into AceAstronomy at **http://astronomy.brookscole.com/cosmos3** to access quizzes and animations that will help you assess your understanding of this chapter's topics.

Log into the Student Companion Web Site at **http://astronomy.brookscole.com/cosmos3** for more resources for this chapter including a list of common misconceptions, news and updates, flashcards, and more.

People in Astronomy

Gibor Basri

Gibor Basri

Gibor Basri is Professor of Astronomy at the University of California, Berkeley. He grew up in Colorado and went to Stanford University for his B.Sc. Then he returned to Colorado for graduate school in the department then called Astrogeophysics, where he wrote his thesis about the chromospheres of supergiant stars. More recently, he was one of the astronomers who finally proved that brown dwarfs really exist. Professor Basri is a high-resolution spectroscopist and also does stellar model atmospheres on the computer. He works in the optical, ultraviolet, and infrared regions of the spectrum. He recently was awarded a Miller Research Professorship to carry out a year's work on his current interest in brown dwarfs. He has summarized the latest in that exciting field in the *Annual Reviews of Astronomy and Astrophysics*, *Scientific American*, and *Sky and Telescope*.

What are you most interested in now?

I am interested in the question of the formation of substellar objects, namely giant planets and brown dwarfs. More generally, this involves the formation of stars. It's all part and parcel of the same process. This has also gotten me involved in the debate on what a "planet" is: Is Pluto a planet, and where is the line between brown dwarfs and planets? That has been a lot of fun, and gotten me involved with the public and the media in an entertaining way. I mostly use it as a way to inform people about all the advances that have occurred in our understanding.

I am now also involved in the search for Earth-like planets around other stars, through NASA's Kepler mission. I truly believe that this will be one of the breakthroughs of the next decade, with enormous public interest in it. Unfortunately, it will be close to 2010 before we have definitive results.

Were you very surprised that all of a sudden we have dozens of new planets outside our Solar System?

The discovery of brown dwarfs and extra-solar planets at the same time made 1995 very exciting. This is one of those convergences in science where people are looking and looking and being frustrated, when all of a sudden the dam breaks and it all falls out. It was very exciting to be part of that process. I wasn't surprised that they exist, but the particular discoveries were surprising and thrilling.

Why do you think it took so long to find brown dwarfs?

It was partly a matter of technology development. They are very faint and people weren't looking quite hard enough, basically. And it was also a matter of developing the right tests to be sure that we have brown dwarfs. When Gliese 229B was found, it was a no-brainer since it was so cool that it couldn't possibly be a star. Just prior to that, there were some brown dwarfs that we identified because they had lithium in them, which was more subtle but also convincing. Once people realized that you could find them, everybody went after them, and they just started dropping out of the sky, so to speak. Also, it happens that the all-sky infrared surveys started up around then, and that has been a major source of new brown dwarfs, too.

What is your personal method for finding brown dwarfs?

I got into this game mostly because of the advent of the Keck telescopes. I was privileged to be a Keck user and wanted to find something exciting to do with the new world's largest telescopes. The idea of the new lithium test for brown dwarfs had already been suggested by astronomers at the Canary Islands, but they were finding that their telescope was not quite up to the task. So I got involved in that and we were lucky enough to make the first discoveries and confirm that brown dwarfs exist.

The basic idea is that stars will destroy lithium when they start their hydrogen fusion. Most brown dwarfs will never get hot enough to destroy lithium, so they will retain it. So you can do a simple spectroscopic test to see if lithium is still present or not. If a star is a very faint red object and it shows lithium, then it is probably a brown dwarf.

Along the way, in applying that test carefully, we discovered that the age scale for young open clusters was off. It turns out they are all about 50 per cent older than we had thought. This is because the normal way for finding the ages of those clusters involves high-mass stars turning off the main sequence, and those stars have convective nuclear burning cores. Convective overshoot is a poorly understood process, but the cores of those stars can basically grab extra hydrogen and live longer. The stellar evolution people were aware of this potential problem, but they didn't have a way of calibrating how much hydrogen would be grabbed. We know how long it takes stars to destroy lithium. You look down the pre-main sequence until you see that lithium is not yet destroyed; you see how bright those objects are and that gives you the age (since they get fainter with time in a known way).

When did you get interested in astronomy?

I think I was about 8 years old. I came to it through science fiction. When I started reading science fiction, I thought it was really cool stuff, and I started learning about space. My father was a physicist and he encouraged me, and I never lost my interest after that. However, I did do a career report in the 8th grade and decided that astronomy was too small and esoteric a field to be a realistic career. Later, I was majoring in physics at Stanford, and realized that I only wanted to do it if I could do it as astrophysics. So I just thought I'd go to grad school, and thought I'd see how long my astronomical career lasted.

You teach both elementary and advanced classes. How do you contrast them?

I have everything from basic astronomy students to grad students. I find that almost everybody has some interest in astronomy. Of course, the grad students want to do it as a career so they have a deeper interest, but I never have trouble conversing about astronomy with a student at any level.

How exciting has astronomy been for you since you started your career?

It turned out to be a particularly good time to get into astronomy, just as I became a grad student. We had just opened up space astronomy. I was able, my last student year, to work with the IUE [the International Ultraviolet Explorer] to do hands-on, real-time observations in space. Also, computers have gotten better and better and better. And detectors have become far more sensitive and versatile. It has been a particularly exciting time for the last 20 years or so, and I think it will go on for the next 20 years.

What excites you most about new planets being found: astronomical reasons or the idea that they are a possible location for life?

In the end, it is the question of life that is exciting, but I am also excited about the fact that we now have techniques that allow us to attack this question empirically. The radial velocity method has been very successful for giant planets, but we will discover large numbers of Earth-sized planets by transits, which is what will be done by the Kepler mission. At the end of that mission, we should know the rough numbers of terrestrial planets in our Galaxy. This is a watershed in humans' understanding of our context in the Universe.

What do you think are some of the major discoveries that will be made in the next 20 years?

As I said, I think that we will discover Earth-like planets. We will also make the first measurements of atmospheres on other planets. I think we will make major progress toward the question of life on Mars and Europa (and perhaps other Solar System locations like Titan). Cosmology will resolve the question of the Hubble constant, the cosmological constant, and the fate of the Universe. We will understand the formation of stars and planets in great detail, and the formation of galaxies in much better detail than we have now. Ground-based telescopes will remove atmospheric blurring well, and from space we will measure parallaxes and motions of stars across our Galaxy, giving us a precise distance scale and a real understanding of dynamics in our Galaxy and beyond. Vast amounts of data at all wavelengths will be available over the web for many to analyze in new and different ways. And computer models of very complicated systems with very great resolution and physical detail will provide a new means of observing the cosmos.

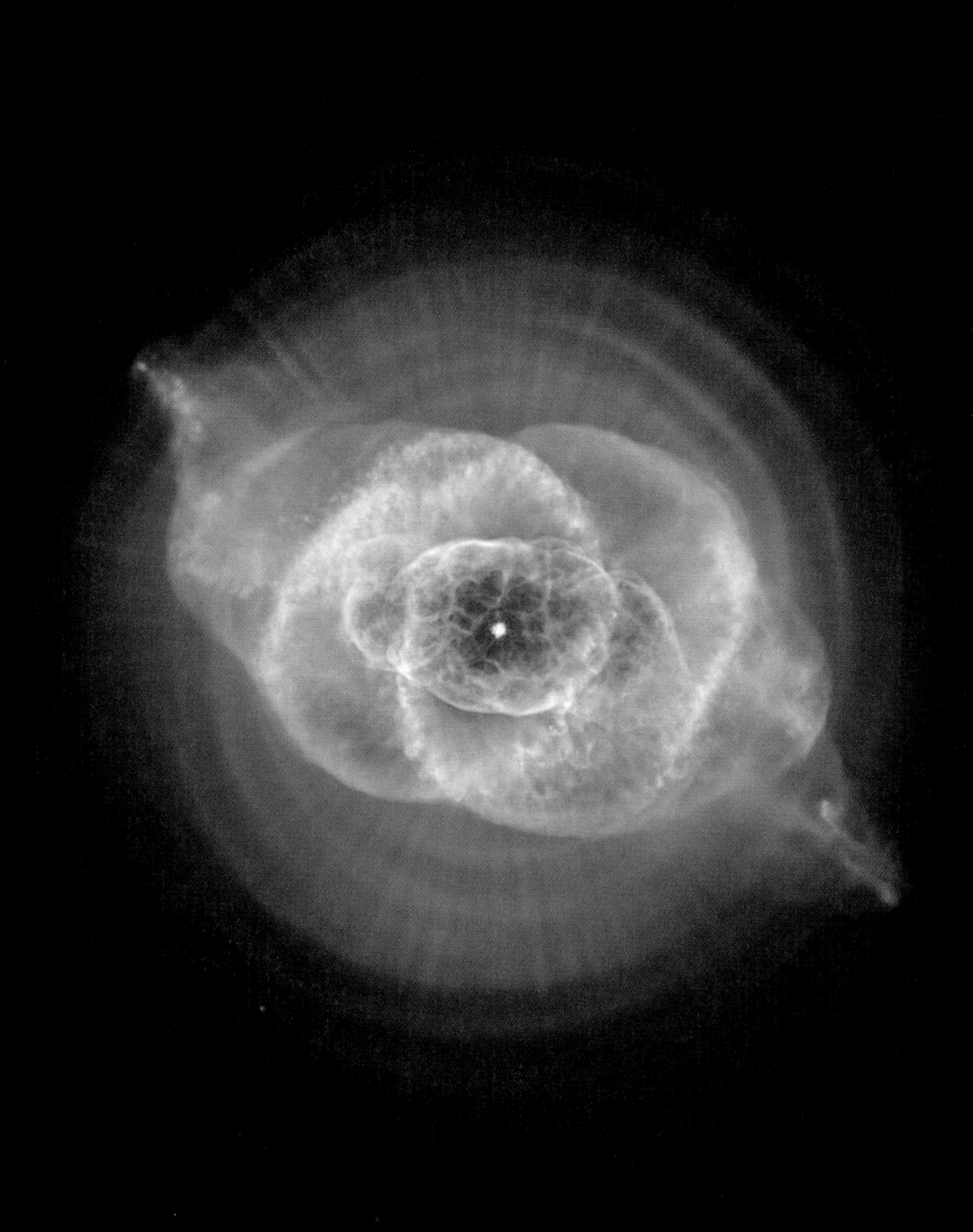

CHAPTER 13

The Death of Stars: Recycling

The more massive a star is, the shorter its stay on the main sequence. The most massive stars may be there for only a few million years. A star like the Sun, on the other hand, is not especially massive and will live on the main sequence for about ten billion years. Since it has taken over four billion years for humans to evolve, it is a good thing that some stars can be stable for such long times.

In this chapter, we will first discuss what will happen when the Sun dies: It will follow the same path as other single **lightweight stars,** stars born with up to about 10 (but possibly as low as 8) times the mass of the Sun. They will go through planetary nebula (■ Fig. 13–1) and white-dwarf stages. Then we will discuss the death of more massive stars, greater than about 8 or 10 times the Sun's mass, which we can call **heavyweight stars.** They go through spectacular stages. Some wind up in such a strange final state—a black hole—that we devote the entire next chapter to it.

ORIGINS

We discover how the heavy elements in the Earth, so necessary for life itself, came into existence. We discuss the ultimate fate of our Sun, billions of years into the future. And, we learn about the origins of interesting classes of stars, including red giants, white dwarfs, and neutron stars, and about the origins of exploding stars.

AIMS

1. Follow the evolution of low-mass stars like the Sun after they live their main-sequence core hydrogen-burning lives (Section 13.1).
2. Learn about red giants, planetary nebulae, white dwarfs, and novae (Section 13.1).
3. Understand supernovae, the catastrophic explosions of certain types of stars, and how they eject heavy elements into the cosmos (Section 13.2).
4. Describe the origin, properties, and utility of neutron stars, often visible as pulsars (Section 13.3).

13.1 The Death of the Sun

Though the details can vary, most solitary (that is, single) stars containing less than about 8–10 times the Sun's mass will have the same fate. (Later we will see that stars in tightly bound binary systems can end their lives in a different manner.) Since the Sun is a typical star in this mass range, we focus our attention on stars of one solar mass when describing the various evolutionary stages of stars. We quote only approximate timescales for each stage; more massive stars will evolve faster, and less massive stars will evolve more slowly.

13.1a Red Giants

As fusion exhausts the hydrogen in the center of a star (after about 10 billion years on the main sequence, for the Sun), its core's internal pressure diminishes because temperatures are not yet sufficiently high to fuse helium into heavier elements. Gravity pulls the core in, heating it up again. Hydrogen begins to "burn" more vigorously in

AceAstronomy™ The AceAstronomy icon throughout this text indicates an opportunity for you to test yourself on key concepts, and to explore animations and interactions of the AceAstronomy website at **http://astronomy.brookscole.com/cosmos3**

ASIDE 13.1: Terminology

We colloquially use the boxing terms "lightweight" and "heavyweight," though we are referring to mass rather than weight.

The Cat's Eye Nebula, a planetary nebula, the late stage of evolution of a lightweight star like the Sun. It is imaged here in the Hubble Heritage Project.
NASA, ESA, HEIC, and the Hubble Heritage Team (STScI/AURA)

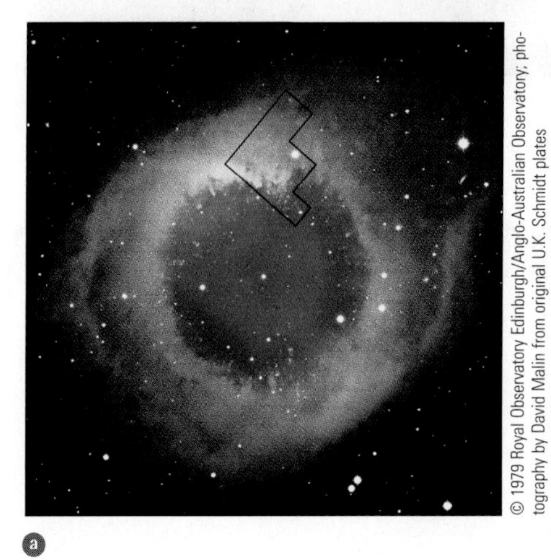

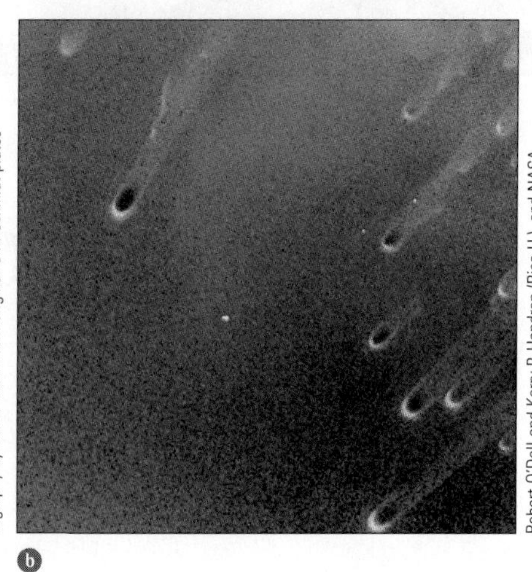

■ **FIGURE 13–1** ⓐ A ground-based image of the Helix Nebula, a very nearby planetary nebula. ⓑ A Hubble Space Telescope close-up of part of the Helix Nebula shows comet-shaped features formed by the outward flow of stellar wind from the central star. (The full Hubble field of view is outlined in part (*a*).) This wind pushes on nebular knots or even breaks up the shell of gas. In the Hubble image, the heads of the knots are 100 A.U. in diameter, roughly the size of our Solar System (out to the orbit of Pluto), and the tails are perhaps 1000 A.U. long. New observations from the Hubble Space Telescope and ground-based observatories have shown that the Helix Nebula isn't a helix after all. It consists of a 6,000-year-old inner disk and a 6,600-year-old outer ring that are almost at right angles to each other. One way that such a complicated shape could form is in a binary system.

the now hotter shell around the core. (The process is once again nuclear fusion, not the chemical burning we have on Earth.)

The new energy causes the outer layers of the star to swell by a factor of 10 or more. They become very large, so large that when the Sun reaches this stage, its diameter will be roughly 30 per cent the size of the Earth's orbit, about 60 times its current diameter. The solar surface will be relatively cool for a star, only about 3000 K, so it will appear reddish. Such a star is called a **red giant.** Red giants appear at the upper right of temperature-luminosity (Hertzsprung-Russell) diagrams. The Sun will be in this stage, or on the way to it after the main sequence, for about a billion years, only 10 per cent of its lifetime on the main sequence.

Red giants are so luminous that we can see them at quite a distance, and a few are among the brightest stars in the sky. Arcturus in Boötes and Aldebaran in Taurus are both red giants.

The contracting core eventually becomes so hot that helium will start fusing into carbon (via the triple-alpha process; see our discussion in Chapter 12) and oxygen nuclei, but this stage will last only a brief time (for a star), perhaps 100 million years. During this time, the star becomes slightly cooler and fainter. We end up with a star whose core is carbon and oxygen, and is surrounded by shells of helium and hydrogen that are undergoing fusion.

Not being hot enough to fuse carbon nuclei, the core once again contracts and heats up, generating energy and causing the surrounding shells of helium and hydrogen to fuse even faster than before. This input of energy makes the outer layers expand again, and the star becomes an even larger red giant. (Note that stars less massive than 0.45 times the mass of the Sun don't ever produce a carbon-oxygen core; their fusion process creates only helium.)

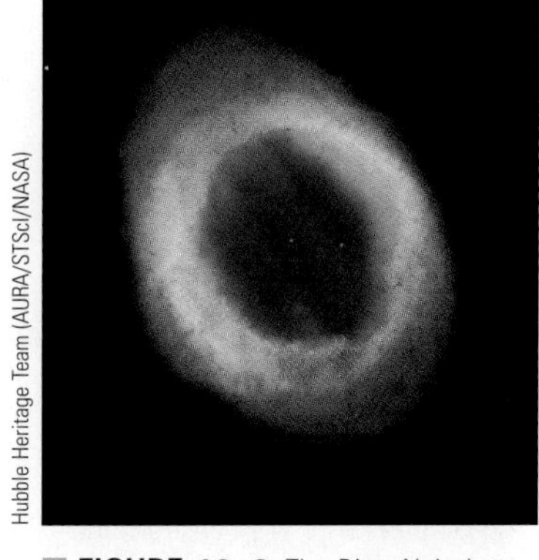

■ **FIGURE 13–2** The Ring Nebula, a planetary nebula, imaged here in the Hubble Heritage Project. Radiation from the hot central star provides the energy for the nebula to shine.

13.1b Planetary Nebulae

As the star grows still larger during the second red-giant phase, the loosely bound outer layers can continue to drift outward until they leave the star. Perhaps the outer layers escape as a shell of gas, in a relatively gentle ejection that we can think of as a "cosmic burp." Or perhaps they drift off gradually (in the form of a "wind"), and a second round of gas sometimes comes off at a more rapid pace. This second round of gas plows into the first round, creating a visible shell (■ Fig. 13–2). Each of these two models has its proponents, and observations are being carried out to discover which is valid in most cases as red giants continue to evolve.

In any case, we know of about a thousand such glowing shells of gas in our Milky Way Galaxy. Each shell contains roughly 20 per cent of the Sun's mass. They are exceptionally beautiful. In the small telescopes of a hundred years ago, though, they appeared as faint greenish objects, similar to the planet Uranus. These objects were thus named **planetary nebulae.**

The remaining part of the star in the center is the star's exposed hot core, which reaches temperatures of 100,000 K and so appears bluish. It is known as the "central star of a planetary nebula." It is on its way to becoming a white dwarf (see next section). Ultraviolet radiation from this hot star partly ionizes gas in the planetary nebula, causing it to glow at optical wavelengths: electrons cascade down various energy levels after recombining with the positive ions. Also, collisions kick the ions into higher energy levels, and photons are emitted as the electrons in the ions jump back down to lower energy levels.

We now know that planetary nebulae generally look greenish because the gas in them emits mainly a few strong spectral emission lines that include greenish ones, specifically lines of doubly ionized oxygen (■ Fig. 13–3). Uranus seems green for an entirely different reason (principally the molecule methane). But the name "planetary nebulae" remains. (Note: Planetary nebulae are almost never called just "nebulae"; always use the adjective "planetary.")

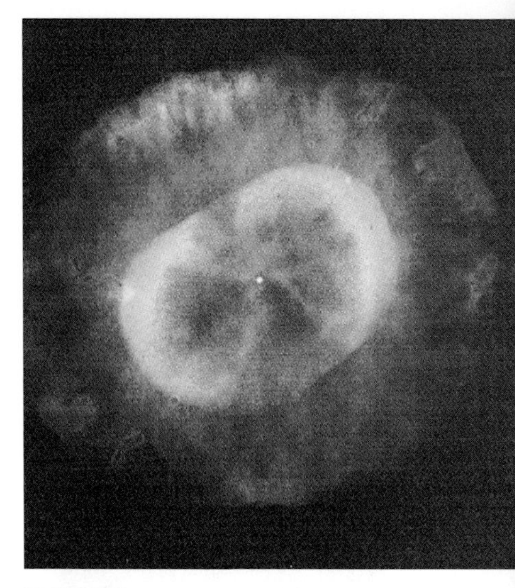

■ **FIGURE 13–3** A Hubble Space Telescope view of the planetary nebula NGC 7662 shows the basic parts of elliptical planetary nebulae. We see the central cavity and shell caused by the fast wind and, around it, the material given off earlier. Colors show degrees of ionization and thus the energy of photons: singly ionized (red), doubly ionized (green), and triply or more ionized (purple).

U. Washington—U.S. Naval Observatory—Cornell U.—Arcetri Obs. (Florence, Italy) collaboration, courtesy of Bruce Balick, U. Washington

Astronomers are very interested in "mass loss" from stars, and planetary nebulae certainly provide many of the prettiest examples, even if hotter and more massive stars give off mass at a more prodigious rate.

The best-known planetary nebula is the Ring Nebula in the constellation Lyra (Fig. 13–2). It is visible in even a medium-sized telescope as a tiny apparent smoke ring in the sky. Only photographs reveal the vivid colors. The Dumbbell Nebula is another famous example. The Helix Nebula (Fig. 13–1) is so close to us that it covers about half the apparent diameter in the sky as the full moon, though it is much fainter.

The Hubble Space Telescope has viewed planetary nebulae with a resolution about 10 times better than most images from the ground, and has revealed new glories in them. Its infrared camera provided views of different aspects of some of the planetary nebulae (■ Fig. 13–4). To our surprise, planetary nebulae turn out to be less round than previously thought. The stars aren't losing their mass symmetrically in all directions. In some cases, the star shedding mass is only one member of a binary system, and the presence of the secondary star affects the direction of mass loss.

Each planetary-nebula stage in the life of a Sun-like star lasts only about 50,000 years; after that time, the nebula spreads out and fades too much to be seen at a distance. But there can be many such stages, with the total amount of time spent at this phase of stellar evolution being perhaps a hundred thousand to a million years.

Raghvendra Sahai and John Trauger (JPL), the WFPC2 Science Team, and NASA

(a)

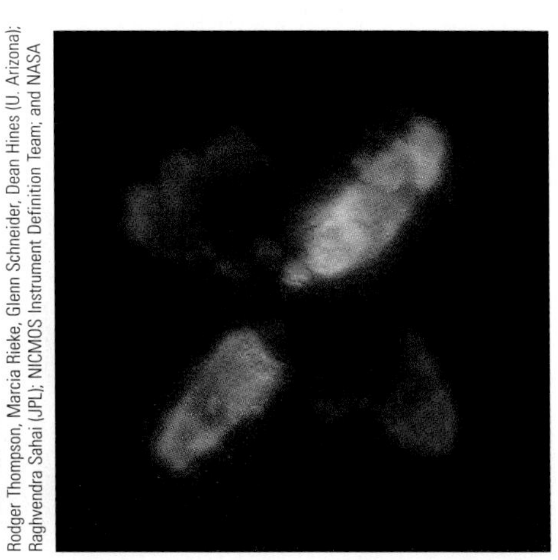

Rodger Thompson, Marcia Rieke, Glenn Schneider, Dean Hines (U. Arizona); Raghvendra Sahai (JPL); NICMOS Instrument Definition Team; and NASA

(b)

■ **FIGURE 13–4** A Hubble Space Telescope view of a planetary nebula being born, the Egg Nebula. (a) In this visible-light view, the central star is hidden by a dense lane of dust. We are seeing light from the central star scattered toward us by dust farther away. Some of the light escapes through relatively clear places, so we see beams coming out of the polar regions. The circular arcs are presumably the shells of gas and dust that were irregularly ejected from the central star. (b) The false-color view from Hubble's Near Infrared Camera and Multi-Object Spectrometer (NICMOS) shows starlight reflected by dust particles (blue) and radiation from hot molecular hydrogen (red). The collision between material ejected rapidly along a preferred axis and the slower, outflowing shells causes the molecular hydrogen to glow.

■ **FIGURE 13–5** The sizes of white dwarfs are not very different from that of the Earth. A white dwarf contains about 300,000 times more mass than the Earth, however. The Sun's curvature is exaggerated; see page 110 for the correct scale.

13.1c White Dwarfs

Through a series of winds and planetary-nebula ejections, all stars that are initially up to 8 (or perhaps even 10) times the Sun's mass manage to lose most of their mass. The remaining stellar core is less than 1.4 times the Sun's mass. (The Sun itself will have only 0.6 of its current mass at that time, in about 6 billion years.)

When this contracting core reaches about the size of the Earth, 100 times smaller in diameter than it had been on the main sequence, a new type of pressure succeeds in counterbalancing gravity so that the contraction stops. This "electron degeneracy pressure" is the consequence of processes that can be understood only with quantum physics. It comes from the resistance of electrons to being packed too closely together; they become "degenerate" (indistinguishable from each other in certain respects), and end up differing mainly in their energy levels. This resistance results in a type of star called a **white dwarf** (■ Fig. 13–5).

The Sun is 1.4 million km (nearly a million miles) across. When its remaining mass (about 0.6 solar masses) is compressed into a volume 100 times smaller across, which is a million times smaller in volume, the density of matter goes up incredibly. A single teaspoonful of a white dwarf would weigh about 5 tons! Such a high density may have been momentarily achieved in a recent terrestrial laboratory experiment.

A white dwarf's mass cannot exceed 1.4 times the Sun's mass; it would become unstable and either collapse or explode. This theoretical maximum was worked out by an Indian university student, S. Chandrasekhar (usually pronounced "chan dra sek′ har" in the United States), en route to England in 1930. It is called the "Chandrasekhar limit." In a long career in the United States, Chandrasekhar became one of the most distinguished astronomers in history, and shared the 1983 Nobel Prize in Physics with William A. Fowler for this early research. A major NASA spacecraft, the Chandra X-ray Observatory, is named after him.

Because they are so small, white dwarfs are very faint and therefore hard to detect. Only a few single ones are known. We find most of them as members of binary systems. Even the brightest star, Sirius (the Dog Star), has a white-dwarf companion, which is named Sirius B and sometimes called "The Pup" (■ Fig. 13–6).

A very odd and interesting system of white dwarfs was discovered with the Chandra X-ray Observatory. From the periodicity of the x-ray observations (■ Fig. 13–7*a*), the system is thought to contain two white dwarfs orbiting each other so closely that the orbit has a period of only five minutes (■ Fig. 13–7*b*)! That makes the stars' separation only 80,000 km, roughly a fifth of the distance between Earth and the Moon! The white dwarfs are thought to be about 1600 light-years from us, though the distance is poorly known. They are in the direction of the constellation Cancer. We will see in Section 13.3f that there are testable consequences from such an extreme, violent wrenching around of these objects.

White-dwarf stars have all the energy they will ever have. Over billions of years, they will gradually radiate the energy stored in the motions of their hot positive ions

ASIDE 13.2: Intermediate-mass stars

Stars whose initial mass is in the range of 8 to 10 solar masses can fuse carbon to magnesium and neon; they subsequently become oxygen-neon-magnesium white dwarfs. However, some of them might explode as core-collapse supernovae before losing their hydrogen envelope; the theory and observations are still uncertain.

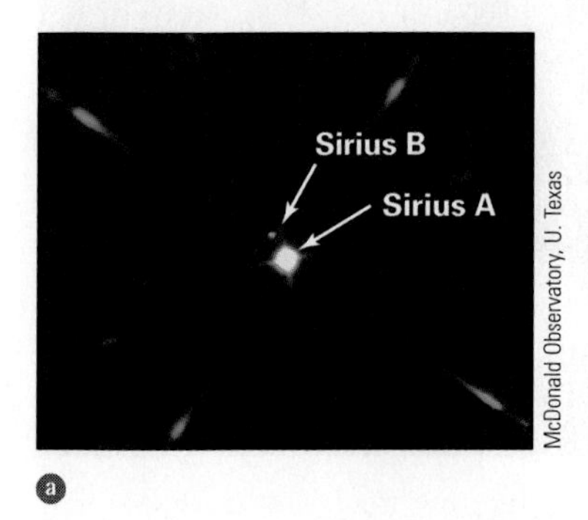

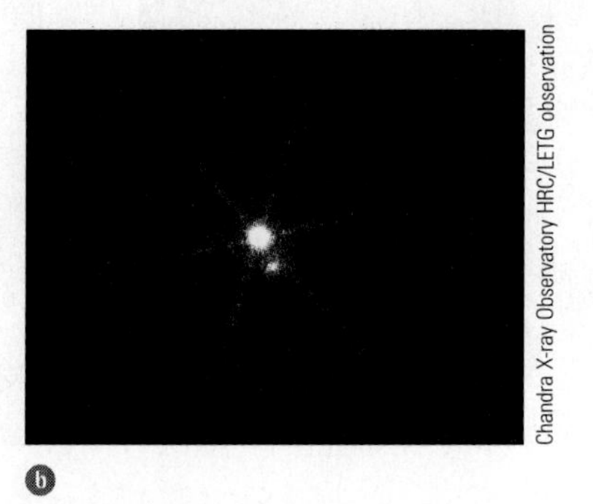

■ **FIGURE 13–6** (a) Sirius A with its companion white dwarf Sirius B appearing as a faint dot nearby. (Ignore the four spikes, which are an artifact.) Sirius A and B have been moving apart from each other since their closest approach in 1993; they are now about 5 seconds of arc apart and the separation will increase to 10 arc sec in 2043. (b) A Chandra X-ray Observatory image of Sirius A and B. Interestingly, in x-rays, Sirius B is the brighter of the pair.

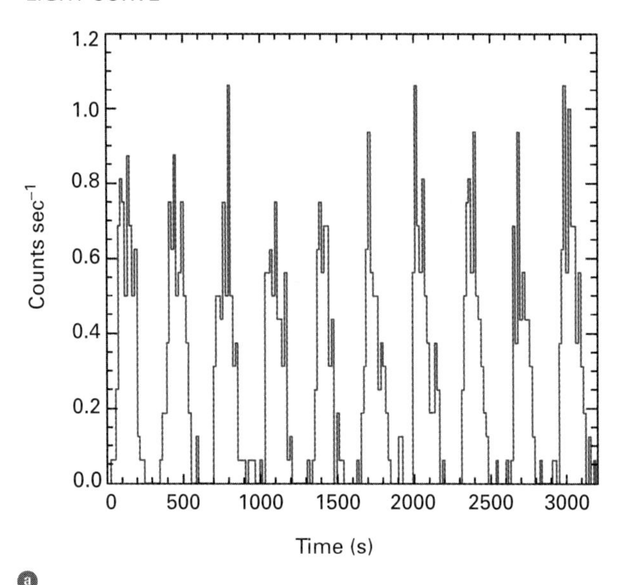

■ **FIGURE 13–7** ⓐ The x-ray light curve (x-ray brightness vs. time) of a binary white-dwarf system, measured with the Chandra X-ray Observatory. (NASA/CXC/GSFC/T. Strohmayer) ⓑ An artist's impression of the two orbiting white dwarfs, whose separation is much less than that of the Earth-Moon system. Gravitational waves are emitted, making the two white dwarfs gradually spiral in toward each other. Eventually, they will merge.

(carbon and oxygen nuclei, which are not degenerate). We can call them "retired stars," since they are spending their life savings of stored energy; they are not undergoing nuclear fusion as normal stars do. White dwarfs eventually become so cool and dim that they can no longer be seen, at least not easily. Some astronomers refer to such objects as "black dwarfs," but there is no clear boundary between a white dwarf and a black dwarf. The Sun will wind up in this state.

From the above discussion, we see that the main-sequence stage is the longest of a star's active lifetime. All subsequent stages are much shorter, except for the white-dwarf stage. But white dwarfs do not create nuclear energy; they simply use their life savings of stored energy. So, they are not considered an "active" phase of a star's life.

13.1d Summary of the Sun's Evolution

The entire "post-main-sequence" evolution of the Sun, a representative solitary low-mass star, can be tracked in a temperature-luminosity diagram (■ Fig. 13–8), or Hertzsprung-Russell diagram (recall our discussion in Chapter 11). We will say informally that the Sun "moves" through the diagram, but of course we really mean that the combination of luminosity and surface temperature changes with time, and that this changing set of values is reflected by a "trajectory" in the diagram.

Referring to Figure 13–8, the basic features are as follows (we will ignore small wiggles that represent details in the evolution). Initially the Sun will become a red giant, growing much more luminous (moving up in the diagram) and a bit cooler (moving slightly to the right), as its helium core contracts while surrounded by a hydrogen-burning shell. This stage will take about a billion years.

When the core becomes sufficiently hot to ignite helium, producing carbon (through the triple-alpha process) and oxygen nuclei, the Sun will briefly (for only about 100 million years) become a bit less luminous and hotter, thus accounting for the slight wiggle seen in the temperature-luminosity diagram. It will subsequently become an even larger red giant, growing even more luminous (moving up in the diagram) and

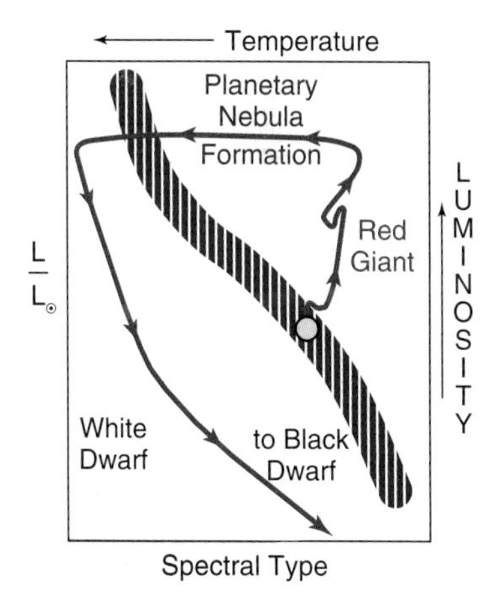

■ **FIGURE 13–8** The post-main-sequence evolution of the Sun, roughly sketched on a temperature-luminosity diagram (Hertzsprung-Russell diagram); see text for details. The broad hatched area represents the main sequence; massive stars are at the upper left, while low-mass stars are at the bottom right. The evolutionary tracks corresponding to different masses are not shown, but in general low-mass stars follow curves similar to that of the Sun, while stars more massive than 8 to 10 Suns become red supergiants before exploding as supernovae.

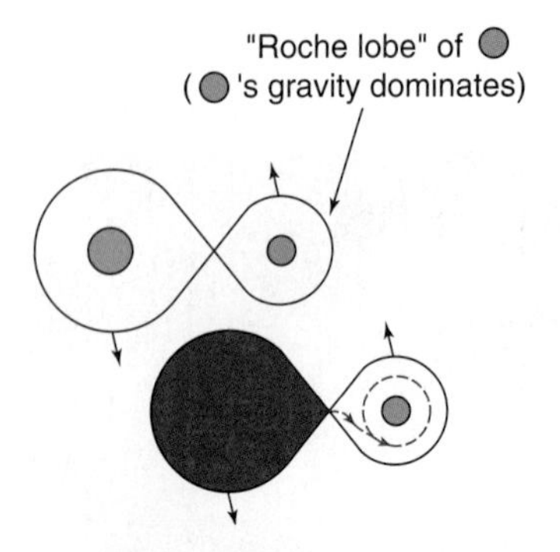

■ **FIGURE 13–9** ⓐ The Roche lobe of each star in a physical binary system is the region where the gravity of that star dominates. As the two stars orbit each other, the "figure-8" pattern also rotates. ⓑ When the initially more massive star evolves to fill its Roche lobe, its outer gases can be transferred to the other lower-mass star, thereby increasing its mass and making it evolve faster.

■ **FIGURE 13–10** A binary system in which a normal star fills its Roche lobe and transfers material to a white-dwarf companion. The material forms an "accretion disk" around the white dwarf. Such systems can produce novae, which are "surface explosions" of white dwarfs: Either the material on the surface undergoes a nuclear runaway, or a clump of material falls onto the surface from the accretion disk.

slightly cooler (moving a bit to the right) as the carbon-oxygen core contracts while surrounded by helium-burning and hydrogen-burning shells.

The Sun will then lose its outer envelope of gases through winds and gentle ejections, producing planetary nebulae and revealing a hotter surface (since those layers used to be in the hot interior). The luminosity, however, will stay roughly the same, as the burning shells of hydrogen and helium generate the energy. Thus, the Sun will quickly (over the course of 10^5 to 10^6 years) move to the left in the temperature-luminosity diagram.

As the supply of hydrogen and helium dwindles, the Sun will contract, but at the same time its temperature will decrease; thus, it will move steeply down toward the lower right of the diagram.

When the contracting Sun reaches roughly the size of the Earth, pressure from electron degeneracy will dominate completely, thwarting further contraction. However, the Sun will continue to cool, so its luminosity will continue to drop, though not as rapidly as when it was contracting. Thus, the Sun will move less steeply toward the lower right of the diagram. This stage will last tens of billions of years, eventually producing a very dim white dwarf (a "black dwarf," according to some astronomers).

13.1e Binary Stars and Novae

Single stars evolve in a simple manner. In particular, their main-sequence lifetime depends primarily on their mass. Most stars, however, are in binary systems, and the stars can exchange matter.

Surrounding each star is a region known as the **Roche lobe,** in which its gravity dominates over that of the other star (■ Fig. 13–9*a*). The Roche lobes of the two stars join at a point between them, forming a "figure-8" shape. (Édouard Roche was a 19th-century French mathematician.)

Consider two main-sequence stars. As the more massive star evolves to the red giant phase, it fills its Roche lobe, and gas can flow from this "donor" star toward the lower-mass companion (■ Fig. 13–9*b*). The recipient star can gain considerable mass, and it subsequently evolves faster than it would have as a single star. Note that the flowing matter forms an **accretion disk** around the recipient star because of the rotation of the system (■ Fig. 13–10).

If one star is already a white dwarf and the companion (donor) fills its Roche lobe (for example, on its way to the red giant phase), a **nova** can result (■ Fig. 13–11). For millennia, apparently new stars (**novae,** pronounced "no´vee" or "no´vay," the plural

■ **FIGURE 13–11** Nova Cygni 1975. ⓐ During its outburst, in which it brightened by over 50,000 times and easily reached naked-eye visibility, and ⓑ after it faded (marked with an arrow). The star that became the nova was very faint, and was not noticeable except in deep images.

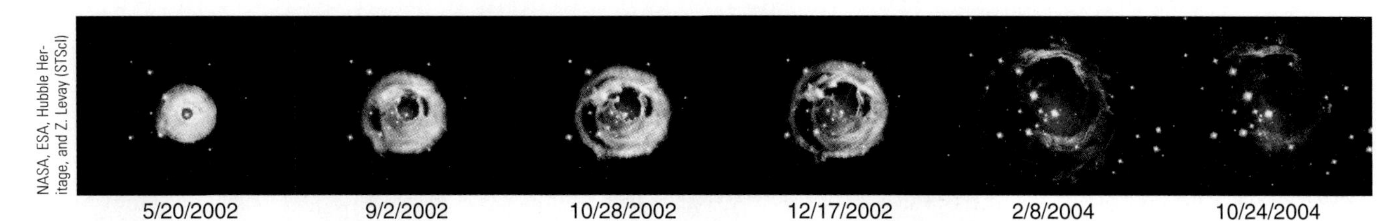

■ **FIGURE 13–12** V838 Monocerotis, a peculiar nova surrounded by shells of gas that were "lit up" by the nova outburst. As shells at progressively larger distances from the nova were illuminated, we saw the "light echoes" evolve with time. The images shown here were taken between May 2002 and October 2004 with the Hubble Space Telescope's Advanced Camera for Surveys, using blue, visible, and infrared filters. The exposure times are equivalent, as shown by the background stars, so the fading of V838 Mon itself is real.

of nova) have occasionally become visible in the sky. Actually, however, a nova is an old star that brightens by a factor of a hundred to a million (corresponding to 5 to 15 magnitudes) in a few days or weeks. It then fades over the course of weeks, months, or years. The ejected gas may eventually become visible as an expanding shell.

The brightening of a nova can occur in at least two ways. First, as gas from the donor accumulates in the accretion disk, the disk becomes unstable and suddenly dumps material onto the white dwarf. The gravitational energy is converted to emitted light. Alternatively, matter can accumulate on the white dwarf's surface and suddenly undergo nuclear fusion, when it gets hot and dense enough. This mechanism can brighten the system far more than the release of gravitational energy. The fusion produces a few of the elements heavier than helium (such as carbon and oxygen, and very rarely up to silicon and sulfur) in the Periodic Table of the Elements. This process involves only about 1/10,000 of a solar mass, so it can happen many times. We do indeed see some novae repeat their outbursts after an interval of years or decades.

An especially peculiar, and still poorly understood, nova was V838 Monocerotis. This object brightened by a large amount, but probably for a different physical reason than normal novae. It was surrounded by many shells of gas that were "lit up" by the nova outburst (■ Fig. 13–12). These "light echoes" evolved with time, as shells at different distances from the nova were successively illuminated.

AceAstronomy™ Log into AceAstronomy and select this chapter to see the Active Figure called "Future of the Sun."

AceAstronomy™ Log into AceAstronomy and select this chapter to see the Active Figure called "Stellar Evolution of High and Low Mass Stars."

AceAstronomy™ Log into Ace Astronomy and select this chapter to see the Astronomy Exercise "Mass/Star-Lifetime Relation."

13.2 Supernovae: Stellar Fireworks!

Though most stars with about the mass of the Sun gradually puff off faint planetary nebulae, ending their lives as white dwarfs, some of them join more massive stars in going off with a spectacular bang. Let us consider these celestial fireworks.

13.2a Core-Collapse Supernovae

Stars that are more than 8–10 times as massive as the Sun whip through their main-sequence lifetimes at a rapid pace. These prodigal stars use up their store of hydrogen very quickly. A star containing 15 times as much mass as the Sun may take only 10 million years from the time it reaches the main sequence until it fully uses up the hydrogen in its core. This timescale is 1000 times faster than that of the Sun.

ASIDE 13.3: Rebound!

For a stunning analogy demonstrating the mechanical "rebound" mechanism in a core-collapse supernova, hold a tennis ball on top of a basketball several feet above the ground, and drop both of them simultaneously. The tennis ball will rebound to a great height. (Note that in reality, however, an extra push from neutrinos is probably required for a successful explosion.)

Andrea Dupree (Harvard-Smithsonian Center for Astrophysics)

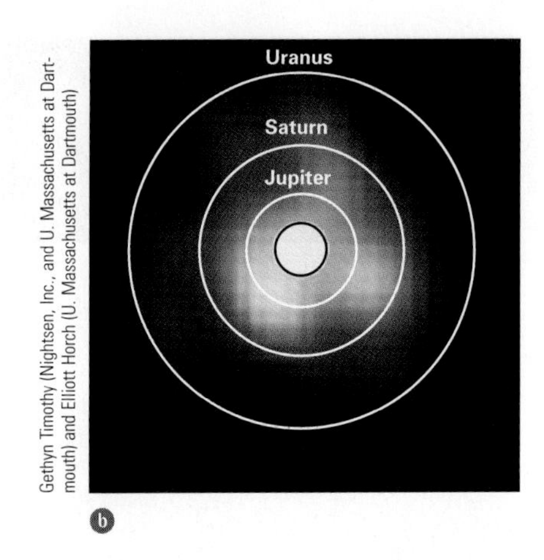

Gethyn Timothy (Nightsen, Inc., and U. Massachusetts at Dartmouth) and Elliott Horch (U. Massachusetts at Dartmouth)

■ **FIGURE 13–13** ⓐ The red supergiant star Betelgeuse, α [alpha] Orionis, is revealed by an ultraviolet image with the Hubble Space Telescope to have an atmosphere the size of Jupiter's orbit. A huge, hot, bright spot is visible on its surface. It is 2000 K hotter than surrounding gas. ⓑ When viewed farther in the ultraviolet with Hubble's imaging spectrograph, Betelgeuse is even larger, by a factor of 1.5. (The images are not reproduced at the same scale. The scale of this image is shown by an overlay of the orbits of planets in our Solar System.) Betelgeuse's atmosphere shows more structure at this far-ultraviolet wavelength. The central white disk shows the size of Betelgeuse's image in visible light.

For these massive stars, the outer layers expand as the helium core contracts. The star has become so large that we call it a **red supergiant.** Betelgeuse, the star that marks the shoulder of Orion, is the best-known example (■ Fig. 13–13).

Eventually, the core temperature reaches 100 million degrees, and the triple-alpha process begins to transform helium into carbon. Some of the carbon nuclei then fuse with a helium nucleus (alpha particle) to form oxygen. The carbon-oxygen core of a supergiant contracts, heats up, and begins fusing into still heavier elements. The ashes of one set of nuclear reactions become the fuel for the next set. Each stage of fusion gives off energy.

Finally, even iron builds up. Layers of elements of progressively lower mass surround the iron core, somewhat resembling the shells of an onion. But when iron fuses into heavier elements, it takes up energy instead of giving it off. No new energy is released to make enough pressure to hold up the star against the force of gravity pulling in; thus, the iron doesn't fuse.

Instead, the mass of the iron core increases as nuclear fusion of lighter elements takes place, and its temperature increases. Eventually the temperature becomes so high that the iron begins to break down (disintegrate) into smaller units like helium nuclei. This breakdown soaks up energy and reduces the pressure. The core can no longer counterbalance gravity, and it collapses. The core's density becomes so high that electrons are squeezed into the nuclei. They react with the protons there to produce neutrons and neutrinos. Additional neutrinos are emitted spontaneously at the exceedingly high temperature (10 to 100 billion kelvins) of the collapsing core. All of these neutrinos escape within a few seconds, carrying large amounts of energy.

The collapsing core of neutrons overshoots its equilibrium size and rebounds outward, like someone jumping on a trampoline. The rebounding core collides with the inward-falling surrounding layers and propels them outward, greatly assisted by the plentiful neutrinos (only a very tiny fraction of which actually interact with the gas).

The star explodes, achieving within one day a stupendous optical luminosity rivaling the brightness of a billion normal stars. It has become a **supernova** (■ Fig. 13–14), and it will continue to shine for several years, gradually fading away. So much energy is available that very heavy elements, including those heavier than iron, form in the ejected layers. The core remains as a compact sphere of neutrons called a **neutron star** (to be discussed in more detail in Section 13.3a). There is even some evidence that occasionally, the neutron star further collapses to form a black hole (Chapter 14).

Such **supernovae** (the plural of supernova), known as **Type II** (they show hydrogen lines in their spectra, unlike Type I supernovae), mark the violent death of heavyweight stars that have retained at least part of their outer layer of hydrogen. Since the fundamental physical mechanism is the collapse of the iron core, they are also one type of **core-collapse supernova.**

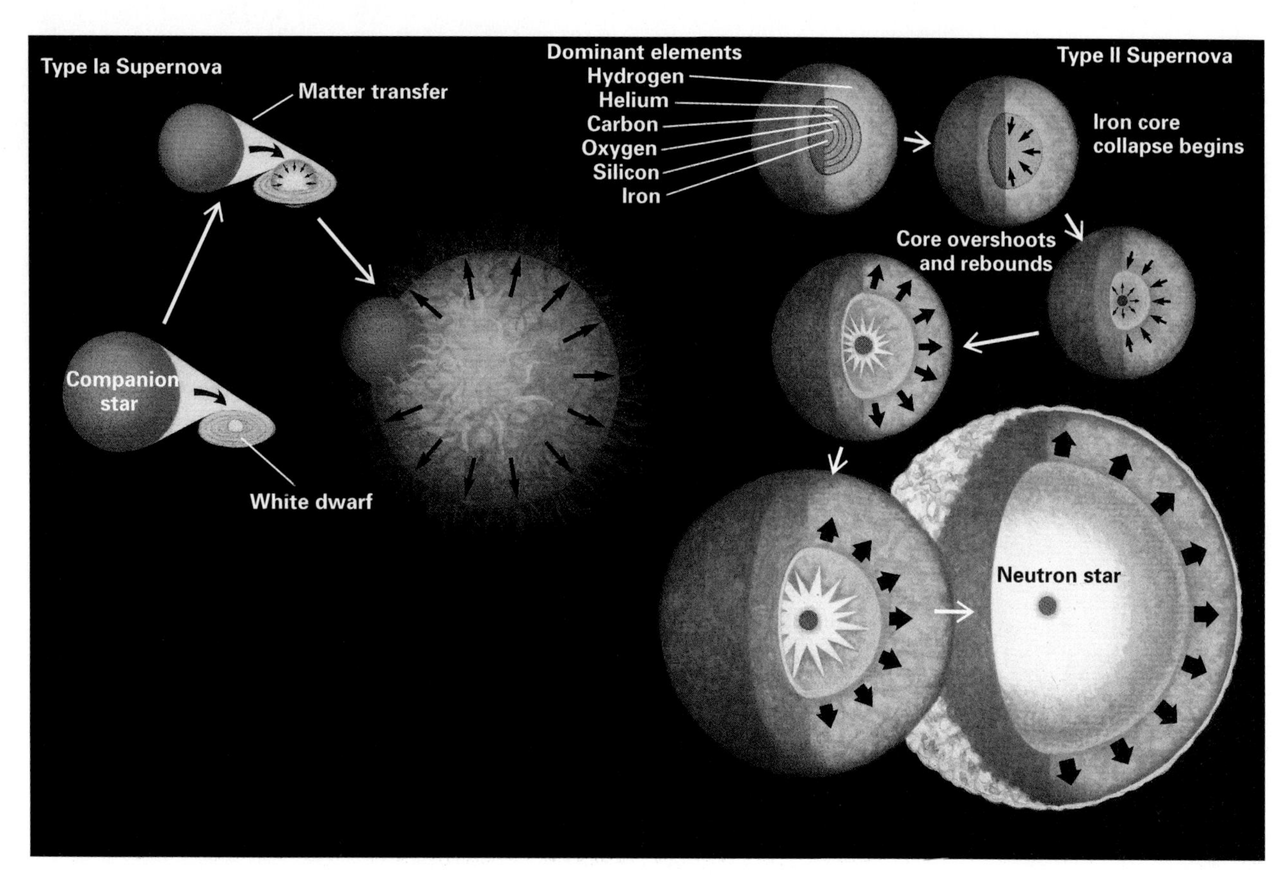

■ **FIGURE 13–14** Type Ia supernovae *(left),* which show no hydrogen in their spectra, come from the incineration of a white dwarf that is gaining matter from a neighboring star and reaches the Chandrasekhar limit, 1.4 times the Sun's mass. Type II supernovae *(right),* whose spectra exhibit lines of hydrogen, are the explosions of massive stars, usually from the supergiant phase. When iron forms at the center of the onion-like layers of heavy elements, the iron core of the star collapses. In this model of the collapse, the core overshoots its final density and rebounds. Neutrinos also push outward. The shock wave from the rebound and the neutrinos blast off the star's outer layers. (Courtesy of Encyclopaedia Britannica, Inc.; from the *1989 Britannica Yearbook of Science and the Future;* illustration by Jane Meredith)

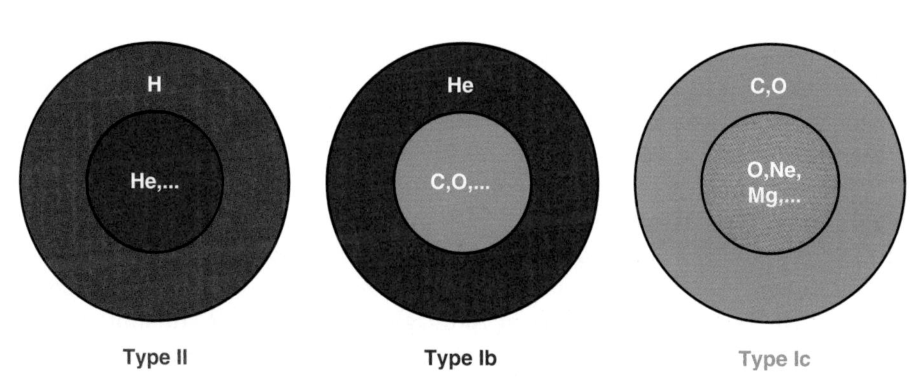

■ **FIGURE 13–15** Massive stars usually have an extensive hydrogen envelope *(left),* but they can sometimes lose their outer envelopes of hydrogen *(middle)* and even helium *(right),* through winds or transfer to a binary companion. Yet their iron cores implode in the manner shown in Fig. 13–14 for Type II supernovae; a neutron star (or black hole) is formed, and the outer layers are ejected. Thus, these are "core-collapse" supernovae, as are Type II supernovae, but their spectra lack hydrogen. "Type Ib" supernovae show helium in their spectra, while "Type Ic" supernovae lack helium. They come from stars like those shown at *middle* and *right,* respectively.

Two physically related kinds of supernovae are known as Type Ib and Type Ic. Both lack hydrogen lines (hence the "Type I" classification), but those of Type Ib show helium lines whereas Type Ic do not. We think they arise from massive stars undergoing iron core collapse (and the subsequent rebound). Prior to this, however, they got rid of their outer atmosphere of gases (■ Fig. 13–15), either through winds and small eruptions or by transferring material to a companion star. We shall see in Chapter 14 that some Type Ic supernovae are associated with stupendous bursts of gamma rays.

13.2b White-Dwarf Supernovae (Type Ia)

Another type of supernova, **Type Ia,** comes from carbon-oxygen white dwarfs in closely spaced (tightly bound) binary systems (Fig. 13–14). (We mentioned Types Ib and Ic, variations of explosions of heavyweight stars, in the previous section.) Gas can be transferred from the companion to the white dwarf, via an accretion disk, similarly to what was discussed for novae (Section 13.1e) and illustrated in Figure 13–10. However, for some reason the white dwarf avoids surface explosions (novae), which tend to eject all of the material that had settled on the surface; in so doing, the white dwarf's mass can grow.

If the companion adds too much matter to the white dwarf, causing the white dwarf to reach the Chandrasekhar limit of 1.4 solar masses, it becomes unstable and undergoes a runaway chain of nuclear-fusion reactions. Heavy elements are synthesized from carbon and oxygen, thereby liberating a huge amount of energy. This energy makes the white dwarf literally explode, leaving no compact remnant (unlike the case in a core-collapse supernova, which produces a neutron star or a black hole). After exploding, the supernova takes about three weeks to brighten, but then fades to obscurity over the course of a year. Because they come from white dwarfs, Type Ia supernovae are also sometimes called **white-dwarf supernovae** or "thermonuclear supernovae" (meaning that nuclear reactions occur in a very hot gas of atomic nuclei).

Such Type Ia supernovae become as powerful as 10 billion Suns, the energy coming from the decay of radioactive heavy elements produced by the incineration. Because they are so luminous, they are detectable far into space and can be used to measure distances of galaxies billions of light-years away. This distant visibility allows astronomers to determine the expansion history of the Universe, its age, and possibly even its fate, as we will see in Chapters 16 and 18. One of the most surprising discoveries in quite a long time is that the Universe appears to be expanding at an accelerating rate, propelled by a mysterious form of "dark energy." Einstein postulated "cosmic antigravity" of this type in 1917; ironically, he later retracted the idea, believing it to be nonsense (Chapter 18). He may well have been too hasty in his retraction.

Theoretical models can now account fairly well for the spectrum and the amount of light from a Type Ia supernova. This basic physical picture for the explosion seems fairly secure, but the exact nature of the companion is unclear. One problem is that if the companion were a normal star donating hydrogen to the white dwarf, why do we not see any hydrogen in the spectrum? Although the hydrogen on the white dwarf's surface could have fused into helium prior to the explosion, the hydrogen blown away from the atmosphere of the companion star might be visible. Perhaps there is just too little hydrogen to be seen; the companion is able to retain the vast majority of its atmosphere.

But another possibility is that the companion star is itself a white dwarf. If the two stars are sufficiently close together, they will gradually spiral toward each other as they emit gravitational radiation (see Section 13.3f). Eventually, the less massive white dwarf will be ripped apart by tidal forces, forming a disk of material around the more massive white dwarf. As material in the disk falls onto the white dwarf, its mass grows, and the white dwarf explodes when the mass reaches the Chandrasekhar limit.

Some evidence for the first hypothesis (slow transfer of gas from a relatively normal star to a white dwarf) was recently found by one of the authors (A.F.) and his collaborators in the case of Tycho's supernova of 1572 (discussed further in the next section), thought to be of Type Ia. By analyzing the radial velocities and proper motions (see our explanation of these terms in Chapter 11) of stars along the general line of sight to the supernova, they found one otherwise ordinary star hurtling through space at an unusually high speed. This runaway star may have been the donor companion of the white dwarf, released from its orbit when the white dwarf exploded.

13.2c Observing Supernovae

Only in the 1920s was it realized that some of the "novae"—apparently new stars—that had been seen in other galaxies (■ Fig. 13–16) were really much more luminous than ordinary novae seen in our own Milky Way Galaxy. These supernovae are very different kinds of objects. Whereas novae are small eruptions involving only a tiny fraction of a star's mass, supernovae involve entire stars. A supernova may appear about as bright as the entire galaxy it is in.

Unfortunately, we have seen very few supernovae in our own Galaxy, and none since the invention of the telescope. The most recent ones definitely noticed were observed by Kepler in 1604 and Tycho in 1572. A relatively nearby supernova might appear as bright as the full moon, and be visible night and day. Since studies in other large galaxies show that supernovae erupt every 30 to 50 years on the average, we appear to be due, although a few supernovae have probably occurred in distant, obscured parts of our Galaxy. Maybe the light from a nearby supernova will reach us tonight. Meanwhile, scientists must remain content with studying supernovae in other galaxies (see *A Closer Look 13.1: Searching for Supernovae*).

Photography of the sky has revealed some two dozen regions of gas in our Galaxy that are **supernova remnants,** the gas spread out by the explosion of a supernova (Fig. 13–11). The most studied supernova remnant is the Crab Nebula in the constellation Taurus (■ Fig. 13–17*a*). The explosion was noticed widely in China, Japan, and Korea in A.D. 1054; there is still debate as to why Europeans did not see it.

If we compare photographs of the Crab taken decades apart, we can measure the speed at which its filaments are expanding. Tracing them back shows that they were together at one point, at about the time the bright "guest star" was seen in the sky by the observers in Asia, confirming the identification. The rapid speed of expansion—thousands of kilometers per second—also confirms that the Crab Nebula comes from an explosive event. Moreover, spectra of the expanding gases reveal the presence of

■ **FIGURE 13–16** Supernova 2004dj is visible near the edge of this Hubble Space Telescope image of the galaxy NGC 2403, only 11 million light-years away. A Type II supernova (that is, showing hydrogen in its spectrum), it is thought to have been produced by a star having an initial mass of 15 Suns.

NASA, ESA, A. V. Filippenko (U. California, Berkeley), P. Challis (Harvard-Smithsonian Center for Astrophysics)

ASIDE 13.4: Prescient predictions

Fritz Zwicky and Walter Baade coined the word "supernova" in 1931, although several other astronomers had previously pointed out the concept of an exceptionally luminous class of nova. A few years later, Zwicky and Baade correctly suggested that supernovae represent the collapse of ordinary stars into neutron stars, releasing tremendous amounts of energy and producing cosmic rays.

ASIDE 13.5: Supernova designations

Supernovae are named in order of discovery in a given calendar year. Thus, the first supernova discovered in 2005 was SN 2005A, the second was SN 2005B, and so on through SN 2005Z. Subsequent supernovae are named SN 2005aa, SN 2005ab, and so on through SN 2005az, followed by SN 2005ba, SN 2005bb, etc.

David F. Malin (Anglo-Australian Observatory) and Jay M. Pasachoff, from Palomar Observatory plates made available by the California Institute of Technology, courtesy of Robert Brucato

(a)

N. A. Sharp, REU program/NOAO/AURA/NSF

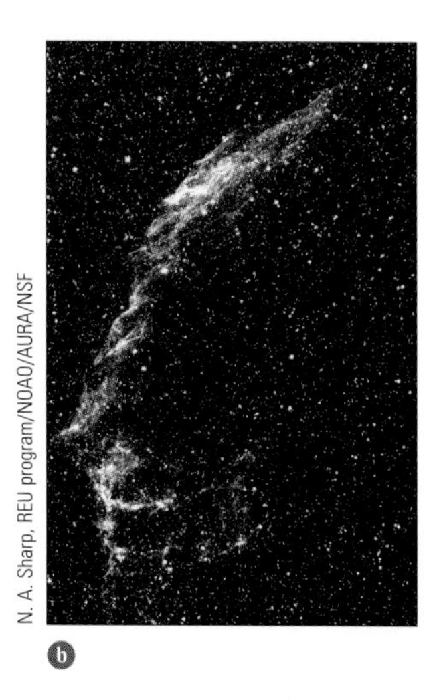

(b)

■ **FIGURE 13–17** (a) The Crab Nebula, a prominent supernova remnant, the result of the great supernova of A.D. 1054. It is 6 light-years across. See also Figure 13-31. (b) The supernova optical remnant in Cygnus known as the Veil Nebula. The supernova that formed it exploded over 20,000 years ago, and the remnant covers over 100 light-years.

A Closer Look 13.1 Searching for Supernovae

A group led by one of the authors (A.F.) is currently running the world's most successful search for nearby supernovae (up to a few hundred million light-years away), with a 0.76-m robotic telescope (KAIT, the Katzman Automatic Imaging Telescope; see the lower-left figure) at Lick Observatory in California. KAIT has been used to discover over 500 supernovae since 1998, including a record high of 95 in 2003, and the first supernova of the new millennium regardless of one's definition of the new millennium (SN 2000A, SN 2001A). An example of a KAIT discovery is shown in the lower-right figure: SN 1998dh in the galaxy "NGC 7541."

The telescope automatically obtains CCD images of over 1000 galaxies per night (on average), cycling through about 7000 galaxies in one week. It then repeats the list of observations. Over the course of a whole year, about 14,000 galaxies are monitored; the available galaxies change with the seasons, along with the changing constellations (recall our discussion in Chapter 1, *Star Party 1.1: Using the Sky Maps*). The computer software processes the images, and then compares new images of galaxies with old images of the same galaxies, performing a digital subtraction.

Most of the time, nothing new is seen in the most recent images, but in about 50 to 100 images each night there is a new spot. These are supernova candidates—but they could also be poorly subtracted stars, pixels affected by cosmic rays, or asteroids that happened to be in the field of view. Thus, undergraduate students scan the supernova candidates (each student is responsible for one particular night of the week), determining which are the best supernova candidates and requesting that KAIT obtain confirmation images the next night. When they are correct, they are officially credited with the supernova discovery.

With the assurance that some new supernovae will be discovered each month, time on larger telescopes can be booked to follow these supernovae in some detail. In particular, a spectrum is obtained of each supernova to determine its type. Sometimes a whole series of spectra is obtained, over many months, to study in detail the expanding gases. One goal is to determine which chemical elements are produced by different kinds of explosions. Another is to understand the supernovae sufficiently well to be able to use them for cosmological distance determinations, as we will discuss in Chapters 16 and 18.

Laurie Hatch, Lick Observatory

The Katzman Automatic Imaging Telescope (KAIT), a 0.76-m robotic telescope operated by one of the authors (A.F.) and his team at Lick Observatory, a 2-hour drive from San Francisco, California. KAIT automatically searches for supernovae in about 14,000 galaxies, obtaining CCD images of more than 1000 galaxies on a typical night.

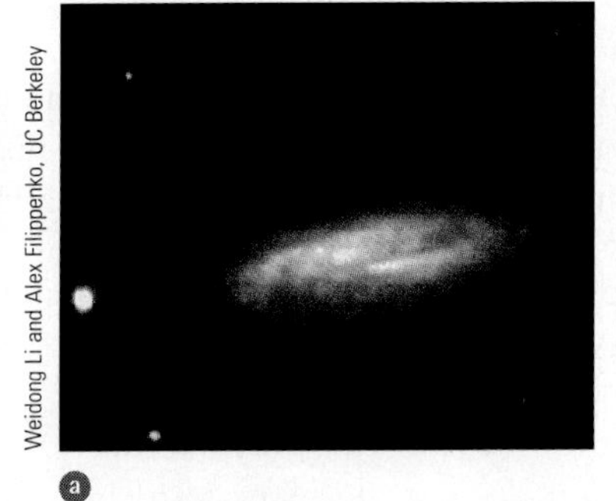

a

b

Weidong Li and Alex Filippenko, UC Berkeley

A supernova found by KAIT in a spiral galaxy, about 130 million light-years away from Earth. (a) An image of the galaxy NGC 7541 before the supernova went off. (b) Another image, clearly showing the supernova (SN 1998dh). Though only a single star, it is comparable in brightness to the entire galaxy consisting of billions of stars. It appears as bright as foreground stars (at left) in our own Milky Way Galaxy, which are only about 1000 light-years away from us.

unusually high abundances (proportions) of heavy elements, as expected if such elements are indeed produced by the explosion.

The Chandra X-ray Observatory is giving us high-resolution x-ray images of supernova remnants (■ Fig. 13–18). The x-rays reveal exceptionally hot gas produced by the collision of the supernova with gas surrounding it.

13.2d Supernovae and Us

The heavy elements that are formed and thrown out by both core-collapse supernovae and white-dwarf supernovae are necessary for life as we know it. Directly or indirectly, supernovae are the only known source of most heavy elements, especially those past

NASA, ESA, CXO, and P. Ruiz-Lapuente (U. Barcelona)

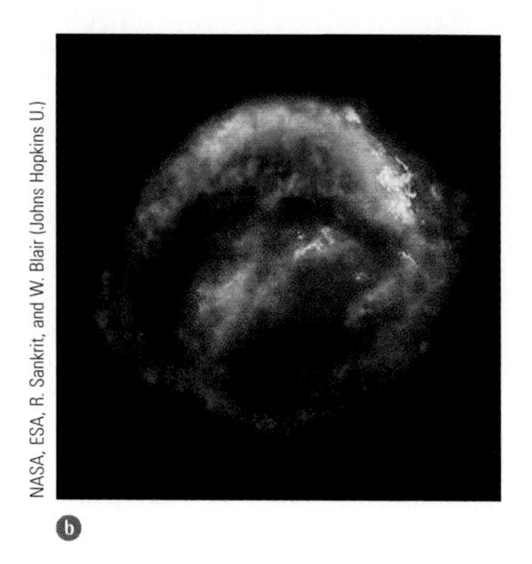

NASA, ESA, R. Sankrit, and W. Blair (Johns Hopkins U.)

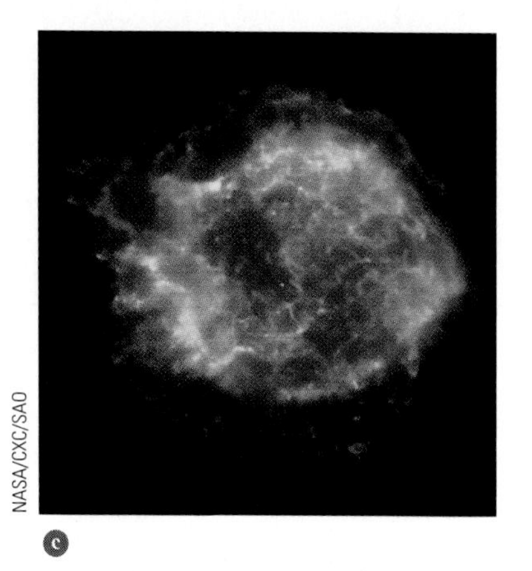

NASA/CXC/SAO

FIGURE 13–18 (a) A Chandra X-ray Observatory view of Tycho's supernova remnant, expanding since 1572, superimposed on an optical image showing stars along the line of sight. The extreme bottom portion of the x-ray remnant fell off the edge of the Chandra detector. (b) A composite image of Kepler's supernova remnant, expanding since 1604, made with data from the Chandra (x-ray), Hubble (optical), and Spitzer (infrared) observatories. (c) A Chandra view of the Cassiopeia A supernova remnant, expanding since about 1680.

iron (Fe) on the Periodic Table of the Elements. They are spread through space and are incorporated in stars and planets that form later on.

Specifically, the Sun and our Solar System were made from the debris of many previous generations of stars. So we humans, who depend on heavy elements for our existence, are here because of supernovae and this process of recycling material.

Think about it: The carbon in your cells, the oxygen that you breathe, the calcium in your bones, and the iron in your blood are just a few examples of the elements produced long ago by stars and their explosions. (Other examples are the silver, gold, and platinum in jewelry—but these are not vital for the existence of life!) Thus, as the late Carl Sagan was fond of saying, "We are made of star stuff [or stardust]."

13.2e Supernova 1987A!

An astronomer's delight, a supernova quite bright but at a safe distance, appeared in 1987. On February 24 of that year, Ian Shelton, then of the University of Toronto, was photographing the Large Magellanic Cloud, a small galaxy 170,000 light-years away, with a telescope in Chile. Fortunately, he chose to develop his photograph that night.

When he looked at the photograph still in the darkroom, he saw a bright star where no such star belonged (■ Fig. 13–19). He went outside, looked up, and again saw the star in the Large Magellanic Cloud, this time with his naked eye. He had discovered the nearest supernova to Earth seen since Kepler saw one in 1604. By the next night, the news was all over the world, and all the telescopes that could see Supernova 1987A (the first supernova found in 1987) were focused on it. Some of these telescopes, as

FIGURE 13–19 The region of the Tarantula Nebula *(lower right)* in the Large Magellanic Cloud. We see the region (a) before and (b) after February 23, 1987. Supernova 1987A shows clearly at upper left in *(b)*. It was a Type II supernova, having hydrogen in its spectrum.

well as the Hubble Space Telescope, continue to observe the supernova on a regular basis to this day.

Hubble's high resolution shows clear views of an inner ring of material produced prior to the supernova, slowly expanding around the supernova (■ Fig. 13–20). The supernova debris is in the process of meeting up with the ring, and the collision should cause the supernova debris and the ring to brighten substantially over the next few years. Already, we see many individual "hot spots" brightening in the ring of material (■ Fig. 13–21). Two outer rings are also visible.

One exciting thing about such a close supernova is that we even know which star had erupted! Pre-explosion photographs showed that a blue supergiant star had been where the supernova now is (■ Fig. 13–22*a*). It had been thought that Type II supernovae always erupt from red supergiants, not blue ones. Most do come from red supergiants (as shown by observations of a few other relatively nearby supernovae), but perhaps this particular star was a blue supergiant because the Large Magellanic Cloud is deficient in heavy elements relative to large galaxies like the Milky Way, affecting the star's structure. Another idea is that the star had merged with a binary companion prior to exploding, making it turn into a blue supergiant.

Hubble Heritage Team

■ **FIGURE 13–20** The inner and outer rings of ejected gas around Supernova 1987A. This gas was lost from the star prior to its explosion.

In the leading current model, the progenitor star for the supernova formed about 10 million years ago. A million years ago, a slow stellar wind blew off its outer layers that formed a large cloud of cool gas, baring a hot stellar surface. Before the star exploded, a high-speed stellar wind from the hot surface carved out a cavity in the cool gas, though leaving some fingers of cooler gas extending inward. We now see the edge of this cavity as the ring imaged by the Hubble Space Telescope, which lit up from the strong ultraviolet light given off as the supernova exploded. The shock wave from the supernova reached the ring in 2005, and the x-ray strength increased greatly. The Chandra X-ray Observatory image (Fig. 13–22*b*) shows hot gas, millions of kelvins, matching the optical bright spots. The optical and x-ray spots result from a collision of the shock waves with the fingers of cool gas. Scientists expect the ring to brighten still more.

Supernova 1987A did not brighten as much as expected, and the fact that it was from a blue supergiant may explain why. Blue supergiants are smaller than red ones, so they are not able to radiate as much light in the early stages of the explosion, and more of the explosion energy is used up in expanding the star. Theoretical models coupled with the observations indicate that the star had once contained 20 times the mass of the Sun, with 6 times the mass of the Sun in the form of a helium core. Through winds and gentle ejections, it had already lost 4 solar masses by the time it exploded.

The rate at which the supernova faded matched the rate of decay of radioactive cobalt into stable iron. Its brightness corresponded to the 0.07 solar mass of cobalt that theory predicted would be formed (initially as radioactive nickel, which decayed into cobalt). Moreover, gamma-ray telescopes observing Supernova 1987A detected photons having the specific energies emitted only by a radioactive, short-lived isotope of cobalt. The radioactive cobalt must have been formed by the supernova; it could not have been present in the 10-million-year-old star prior to the explosion.

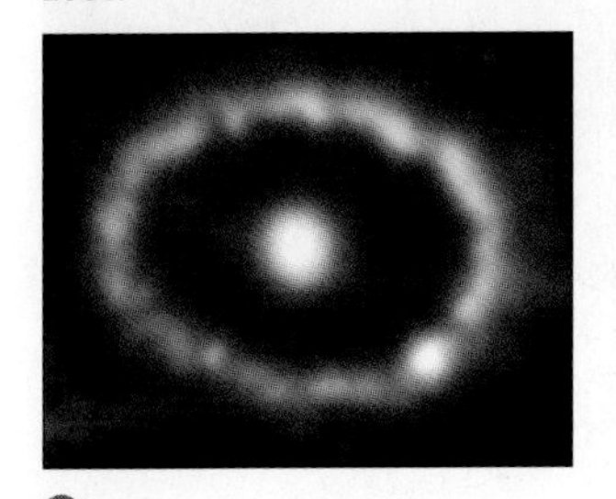
(a)

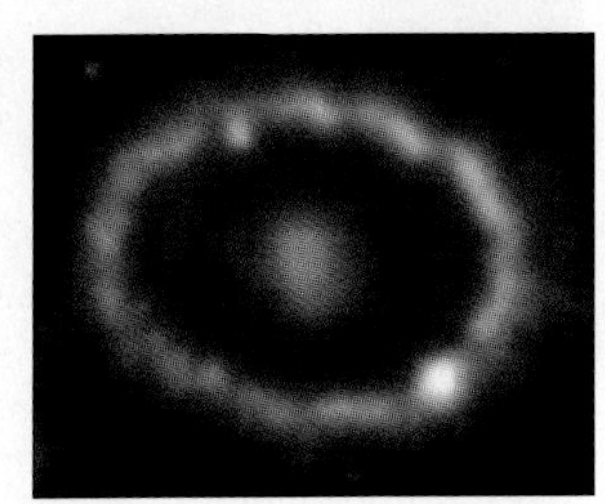
(b)

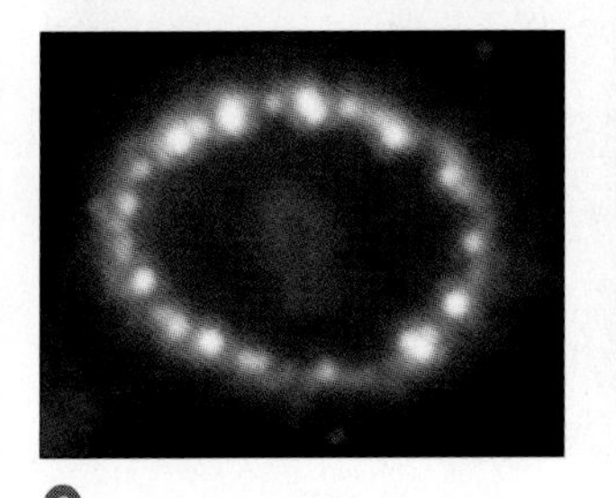
(c)

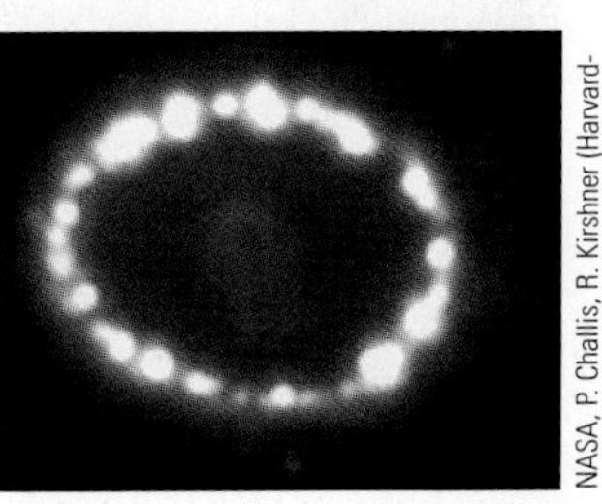
(d)

NASA, P. Challis, R. Kirshner (Harvard-Smithsonian Center for Astrophysics) and B. Sugerman (STScI)

■ **FIGURE 13–21** A ring of ejected gas so close to Supernova 1987A that it is clearly visible only from the Hubble Space Telescope. The ring is made of gas thrown off by the star prior to the explosion itself. Debris expanding from the explosion is reaching the ring, making spots on it brighten. (a) From 1995. (b) From 1998. (c) From 2003. (d) From 2005.

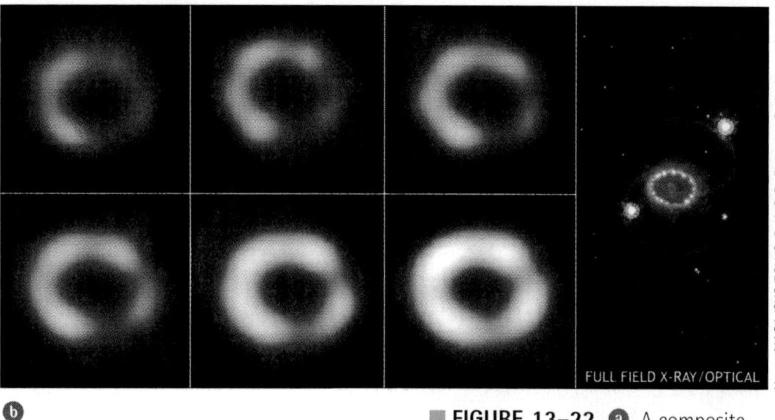

FIGURE 13–22 (a) A composite photo in which a pre-explosion photograph of the blue supergiant star that exploded is superimposed as a negative (thus appearing black) on an image of Supernova 1987A. Several stars appear as black dots in the negative. (b) Chandra's images of x-rays from the ring around Supernova 1987A brightening over several years, with a Hubble optical image from 2005. Chandra's x-ray spectra show that the hot spots result from the shock wave hitting dense fingers of gas protruding inward from the ring.

When the core of a massive star collapses, creating a neutron star and a supernova, theorists tell us that many neutrinos are produced. In fact, 99 per cent of the energy is carried by neutrinos. The solar neutrino experiment using chlorine in liquid form (Chapter 12) was not sensitive enough to the energy range of neutrinos emitted by Supernova 1987A. Fortunately, at least two other experiments that had been set up for other purposes were operating during this event. Both experiments contained large volumes of extremely pure water surrounded by sensitive phototubes to measure any light given off as a result of interactions in the water.

One of the detectors, in a salt mine in Ohio, reported that 8 neutrinos had arrived and interacted within a 6-second period on February 23, 1987 (■ Fig. 13–23). Normally, an interaction of some kind with a neutrino was seen only about once every five days. Another detector, in a zinc mine in Japan, detected a burst of 11 neutrinos. (The detectors were in mines to shield them from other types of particles.) These few neutrinos marked the emergence of a new observational field of astronomy: extra-solar neutrino astronomy. The 2002 Nobel Prize in Physics was awarded in part for the discovery of neutrinos from Supernova 1987A.

The fact that the neutrinos arrived three hours before we saw the optical burst matches our theoretical ideas about how a star collapses and rebounds to make a supernova. (The neutrinos escape from the stellar core at essentially the speed of light, whereas the surface of the exploding star doesn't brighten until a few hours later, when the shock wave from the rebounding material reaches the star's surface and heats it.) Also, the amount of energy carried by neutrinos, taking account of the tiny fraction that we detect, matches that of theoretical predictions. Our basic ideas of how Type II supernovae occur were validated: A neutron star had indeed been created, at least temporarily. We are now awaiting the emergence of definite signs of a pulsar, the type of star we shall discuss in the next section.

The observations have also given us important basic knowledge about neutrinos themselves. If neutrinos have mass, they would travel at different speeds depending on how much energy they were given. The fact that the neutrinos arrived so closely spaced in time placed a sensitive limit on how much mass they could have. So though the current observations of neutrinos show that they probably have some mass, we also know that they don't have much mass. Understanding the mass of neutrinos is important for understanding how much matter there is in the Universe, as we shall see in our discussion of galaxies and cosmology (Chapters 16 and 18).

ASIDE 13.6: Ghostly neutrinos

If neutrinos were massless, like particles of light (photons), in empty space they would all travel at the same speed, the speed of light. Since they have been shown to have a slight mass, they must be travelling slightly slower than the speed of light.

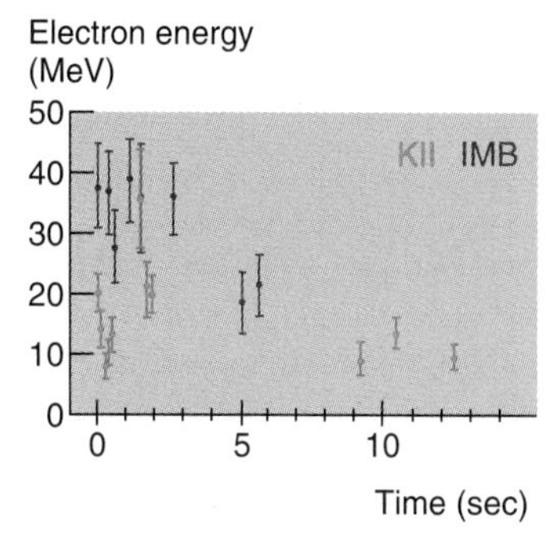

FIGURE 13–23 A graph showing the detection of neutrinos from two detectors. A short "burst" of neutrinos, substantially above the background "noise" (some of which is produced by other particle interactions), is visible.

13.2f Cosmic Rays

So far, our study of the Universe in this book has relied on information that we get by observing electromagnetic radiation—including not only visible light but also gamma

ASIDE 13.7: Cosmic rays and weather

Cosmic rays may even affect the Earth's weather, by encouraging clouds to form in the Earth's atmosphere. The flux of cosmic rays that reach Earth varies over the solar-activity 11-year cycle.

rays, x-rays, ultraviolet, infrared, and radio waves. Moon rocks and meteorites have given important insights as well. But we also receive a few high-energy particles from space.

These **cosmic rays** (misnamed historically; we now know that they aren't rays at all) are nuclei of atoms moving at tremendous speeds. Some of the weaker cosmic rays come from the Sun while other cosmic rays come from farther away. Cosmic rays provide about 1/5 of the radiation environment of Earth's surface and of the people on it. (Almost all the radiation we are exposed to comes from cosmic rays, from naturally occurring radioactive elements in the Earth or in our bodies, and from medical x-rays.)

For a long time, scientists have debated the origin of the non-solar "primary cosmic rays," the ones that actually hit the Earth's atmosphere as opposed to cosmic rays that hit the Earth's surface. (Our atmosphere filters out most of the primary cosmic rays. When they hit the Earth's atmosphere, the collisions with air molecules generate "secondary cosmic rays.")

Because cosmic rays are charged particles—we receive mostly protons and also some nuclei of atoms heavier than hydrogen, and many fewer electrons—our Galaxy's magnetic field bends them. Thus we cannot trace back the paths of cosmic rays we detect to find their origin. It seems that most middle-energy cosmic rays were accelerated to their high speeds in supernova explosions.

Primary cosmic rays can be captured with high-altitude balloons or satellites. Many images taken with CCDs, whether those on the Hubble Space Telescope, on the Solar and Heliospheric Observatory, or on Earth, show streaks from cosmic rays (■ Fig. 13–24). When studying your CCD data, you have to eliminate the pixels that show these "cosmic-ray hits."

a

b

Chien Peng (U. Arizona) and Alex Filippenko (UC Berkeley)

■ **FIGURE 13–24** The central part of the galaxy M77, made by combining Hubble Space Telescope images obtained through red, green, and blue filters. In (a), many cosmic-ray hits can be seen as streaks and specks. (See arrows; the shape of each one depends on the angle at which the cosmic ray hits the CCD.) In (b), the cosmic-ray hits are gone. They were removed by comparing pairs of images obtained through each filter: Pixels that were much brighter in one image than the other were assumed to have been hit by cosmic rays, and were replaced by the unaffected pixels.

Stacks of suitable plastics (the observations were formerly made with thick photographic emulsions) show the damaging effects of cosmic rays passing through them. Scientists are now worried about cosmic rays damaging computer chips vital for navigation in airplanes as well as in spacecraft, and, for safety, engineers are providing chips that can work even when slightly damaged in this way.

When primary cosmic rays hit the Earth's atmosphere, they cause flashes of light that can be detected with telescopes on Earth. A project for observing secondary cosmic rays by studying light they generate as they plow through a cubic kilometer of clear natural ice is going into operation underground near the South Pole. A similar project, using clear Mediterranean water, is also progressing. A huge cosmic-ray-detector project covering hundreds of kilometers of ground is underway in Argentina and Utah, with the former site to be built first. A smaller project, already active in Japan, hasn't detected the super-energetic cosmic rays that have been occasionally measured, but the larger project could resolve the current debate about their existence.

13.3 Pulsars: Stellar Beacons

We have examined the fate of the outer layers of a massive star that explodes as a core-collapse supernova. But what about the core itself? Let us now discuss cores that wind up as superdense stars. In the next chapter, we will see what happens when the core is too massive to ever stop collapsing.

13.3a Neutron Stars

The collapsed core in a core-collapse supernova is a very compact object, generally a neutron star consisting mainly of neutrons, as already mentioned in Section 13.2a. They are like a single, giant atomic nucleus, without the protons (■ Fig. 13–25). Neutron stars have measured masses of about 1.4 Suns, but some might exist with up to 2 or 3 solar masses; astronomers don't understand the limit for neutron stars as accurately as they know that 1.4 solar masses is the limit for white dwarfs.

Neutron stars are only about 20 or 30 km across (Fig. 13–25); a teaspoonful would weigh a billion tons. At such high densities, the neutrons resist being further compressed; they become "degenerate," as we can explain using laws of quantum physics, acting similarly to the way electrons behave in a white dwarf (see our discussion in Section 13.1c). A pressure ("neutron-degeneracy pressure") is created, which counterbalances the inward force of gravity.

When an object contracts, its magnetic field is compressed. As the magnetic-field lines come together, the field gets stronger. A neutron star is so much smaller than the Sun that its magnetic field should be about a trillion times stronger.

When neutron stars were first discussed theoretically in the 1930s, the chances of observing one seemed hopeless. But we currently can detect signs of them in several independent and surprising ways, as we now discuss.

AceAstronomy™ Log into AceAstronomy and select this chapter to see the Active Figure called "Neutron Star."

■ **FIGURE 13–25** A neutron star is the size of a city, even though it may contain a solar mass or more. Here we see the ghost of a neutron star superimposed on a photograph of New York City. A neutron star might have a solid, crystalline crust about a hundred meters thick. Above these outer layers, its atmosphere probably takes up only another few centimeters. Since the crust is crystalline, there may be irregular structures like mountains, which would only poke up a few centimeters through the atmosphere.

13.3b The Discovery of Pulsars

Recall that the light from stars twinkles in the sky because the stars are point-like objects, with the Earth's atmospheric turbulence bending the light rays. Similarly, point-like radio sources (radio sources that are so small or so far away that they have no apparent length or breadth) fluctuate in brightness on timescales of a second because of variations in the density of electrons in interplanetary space. In 1967, a special radio telescope was built to study this radio twinkling; previously, radio astronomers had mostly ignored and blurred out the effect to study the objects themselves.

In 1967 Jocelyn Bell (now Jocelyn Bell Burnell) was a graduate student working with Professor Antony Hewish's special radio telescope (■ Fig. 13–26). As the sky swept over the telescope, which pointed in a fixed direction, she noticed that the signal occasionally wavered a lot in the middle of the night, when radio twinkling was usually low.

Her observations eventually showed that the position of the source of the signals remained fixed with respect to the stars rather than constant in terrestrial time (for example, always occurring at exactly midnight). This timing implied that the phenomenon was celestial rather than terrestrial or solar.

Bell and Hewish found that the signal, when spread out, was a set of regularly spaced pulses, with one pulse every 1.3373011 seconds (■ Fig. 13–27). The source was briefly called LGM, for "**L**ittle **G**reen **M**en," because such a signal might come from an extraterrestrial civilization!

But soon Bell located three other sources, pulsing with regular periods of 0.253065, 1.187911, and 1.2737635 seconds, respectively. Though they could be LGM2, LGM3, and LGM4, it seemed unlikely that extraterrestrials would have put out four such beacons at widely spaced locations in our Galaxy. The objects were named **pulsars**—to

■ **FIGURE 13–26** Jocelyn Bell Burnell, the discoverer of pulsars. She did so with a radio telescope—actually a field of aerials—at Cambridge, England.

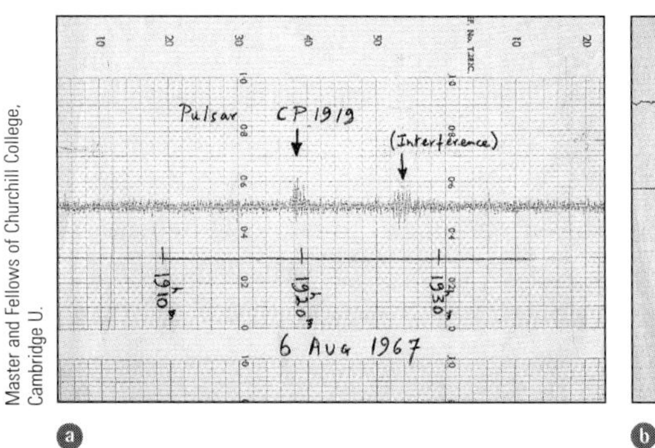

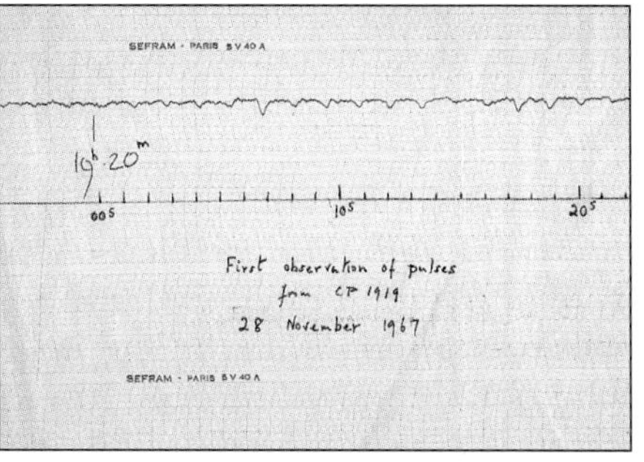

■ **FIGURE 13–27** ⓐ The up-and-down variations on this chart led Jocelyn Bell Burnell to suspect that something interesting was going on. ⓑ The chart record showing the discovery of the individual pulses from CP 1919. Here downward blips are actually increases in brightness.

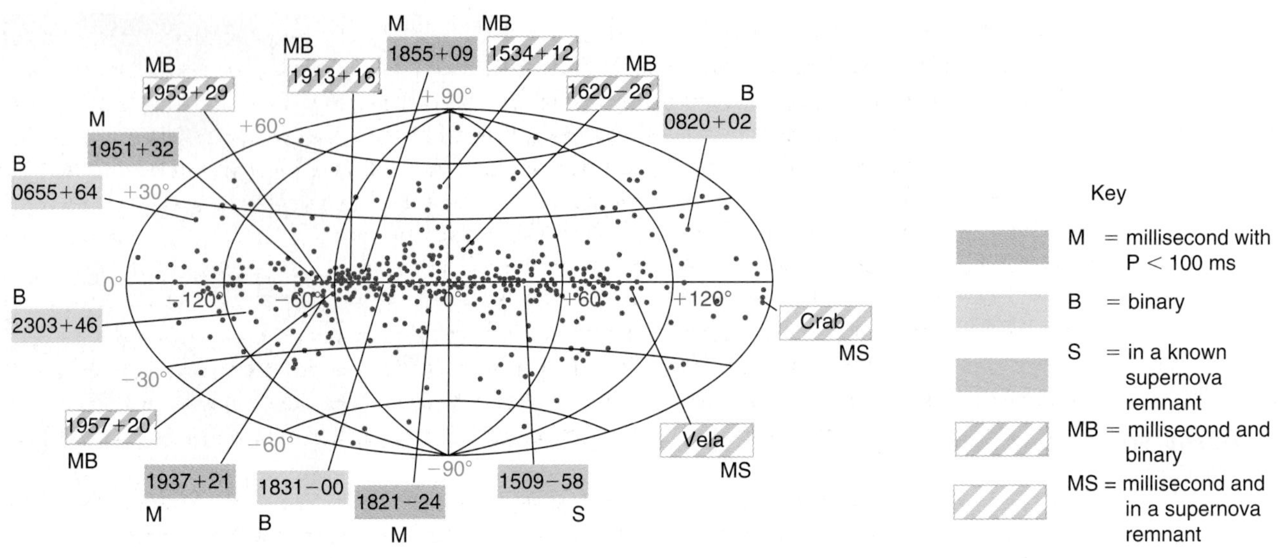

■ **FIGURE 13–28** The distribution of over 500 of the 2000 known pulsars on a map that shows the entire sky, with the plane of the Milky Way along a horizontal line at center. From the concentration of pulsars along the plane of our Galaxy, we can conclude that the pulsars are members of our Galaxy; otherwise, we would have expected to see as many near the poles of the map. The concentration of pulsars near −60° galactic longitude on this map merely represents the fact that this section of the sky has been especially carefully searched. (Joseph H. Taylor, Jr., at the Princeton Pulsar Physics Laboratory)

indicate that they gave out pulses of radio waves—and announced to an astonished world. It was immediately apparent that they were an important discovery, but what were they?

13.3c What Are Pulsars?

Other observatories set to work searching for pulsars, and dozens were soon found. They were all characterized by very regular periods, with the pulse itself taking up only a small fraction of a period. When the positions of many of the roughly 1400 known pulsars are plotted on a map of the whole sky (■ Fig. 13–28), we easily see that they are concentrated along the relatively flat plane of our Galaxy. Thus they must be in our Galaxy; if they were located outside our Galaxy we would see them distributed uniformly around the sky or even partly obscured near our Galaxy's plane where the Milky Way might block something behind it.

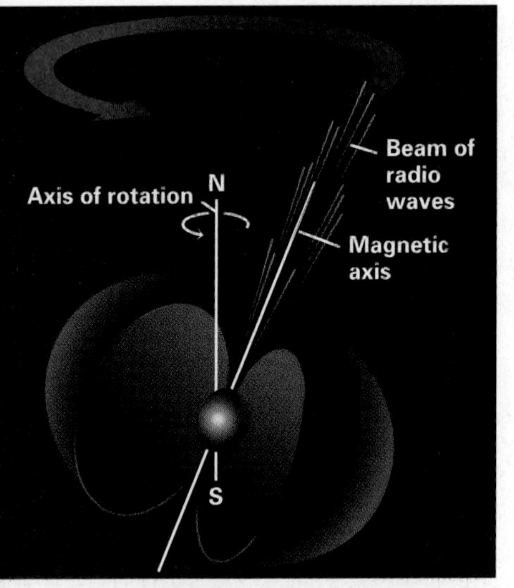

■ **FIGURE 13–29** In the lighthouse model for pulsars, which is now commonly accepted, a beam of radiation flashes by us once every pulsar period. (A beam pointing in the opposite direction gives an additional pulse, if the magnetic axis is perpendicular to the rotation axis.) Similarly, a lighthouse beam appears to flash by a ship at sea.

The question of what a pulsar is can be divided into two parts. First, we want to know why the pulses are so regular—that is, what the "clock" is. Second, we want to know where the energy comes from.

We can get pulses from a star in two ways: If the star oscillates in size ("vibrates") or if it rotates. (The only other possibility—collapsed stars orbiting each other—would result in progressively decreasing pulse periods as the objects release energy and spiral inward.) The theory worked out for ordinary variable stars had shown that the speed with which a star oscillates depends on its average density. Ordinary stars would oscillate much too slowly to be pulsars, and even white dwarfs would oscillate somewhat too slowly. Further, neutron stars would oscillate too rapidly to be pulsars. So oscillations were excluded.

That left only rotation as a possibility. And it can be calculated that a white dwarf is too large to rotate fast enough to cause pulsations as rapid as those that occur in a pulsar; it would be torn apart. So, the only remaining possibility is the rotation of a neutron star. We have solved the problem, or at least come to a plausible hypothesis, by the process of elimination.

Over the years, more and more evidence has accumulated to back the **lighthouse model** for pulsars (■ Fig. 13–29), and there is general agreement on it. Just as a lighthouse seems to flash light every time its beam points toward you, a pulsar is a rotating neutron star.

How is the energy generated? There is much less agreement about that, and the matter remains unsettled. Remember that the magnetic field of a neutron star is

Neutron stars are only about 20 or 30 km across (Fig. 13–25); a teaspoonful would weigh a billion tons. At such high densities, the neutrons resist being further compressed; they become "degenerate," as we can explain using laws of quantum physics, acting similarly to the way electrons behave in a white dwarf (see our discussion in Section 13.1c). A pressure ("neutron-degeneracy pressure") is created, which counterbalances the inward force of gravity.

When an object contracts, its magnetic field is compressed. As the magnetic-field lines come together, the field gets stronger. A neutron star is so much smaller than the Sun that its magnetic field should be about a trillion times stronger.

When neutron stars were first discussed theoretically in the 1930s, the chances of observing one seemed hopeless. But we currently can detect signs of them in several independent and surprising ways, as we now discuss.

AceAstronomy™ Log into AceAstronomy and select this chapter to see the Active Figure called "Neutron Star."

Background: NASA's Johnson Space Center

■ **FIGURE 13–25** A neutron star is the size of a city, even though it may contain a solar mass or more. Here we see the ghost of a neutron star superimposed on a photograph of New York City. A neutron star might have a solid, crystalline crust about a hundred meters thick. Above these outer layers, its atmosphere probably takes up only another few centimeters. Since the crust is crystalline, there may be irregular structures like mountains, which would only poke up a few centimeters through the atmosphere.

13.3b The Discovery of Pulsars

Recall that the light from stars twinkles in the sky because the stars are point-like objects, with the Earth's atmospheric turbulence bending the light rays. Similarly, point-like radio sources (radio sources that are so small or so far away that they have no apparent length or breadth) fluctuate in brightness on timescales of a second because of variations in the density of electrons in interplanetary space. In 1967, a special radio telescope was built to study this radio twinkling; previously, radio astronomers had mostly ignored and blurred out the effect to study the objects themselves.

In 1967 Jocelyn Bell (now Jocelyn Bell Burnell) was a graduate student working with Professor Antony Hewish's special radio telescope (■ Fig. 13–26). As the sky swept over the telescope, which pointed in a fixed direction, she noticed that the signal occasionally wavered a lot in the middle of the night, when radio twinkling was usually low.

Her observations eventually showed that the position of the source of the signals remained fixed with respect to the stars rather than constant in terrestrial time (for example, always occurring at exactly midnight). This timing implied that the phenomenon was celestial rather than terrestrial or solar.

Bell and Hewish found that the signal, when spread out, was a set of regularly spaced pulses, with one pulse every 1.3373011 seconds (■ Fig. 13–27). The source was briefly called LGM, for "**L**ittle **G**reen **M**en," because such a signal might come from an extraterrestrial civilization!

But soon Bell located three other sources, pulsing with regular periods of 0.253065, 1.187911, and 1.2737635 seconds, respectively. Though they could be LGM2, LGM3, and LGM4, it seemed unlikely that extraterrestrials would have put out four such beacons at widely spaced locations in our Galaxy. The objects were named **pulsars**—to

© Royal Observatory Edinburgh, Brian Hadley

■ **FIGURE 13–26** Jocelyn Bell Burnell, the discoverer of pulsars. She did so with a radio telescope—actually a field of aerials—at Cambridge, England.

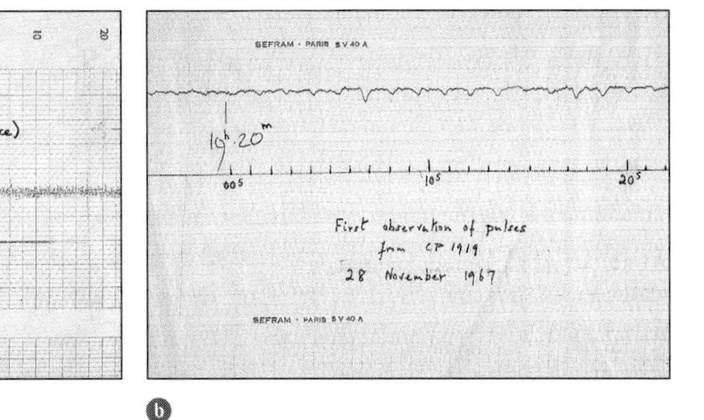

Master and Fellows of Churchill College, Cambridge U.

(a) (b)

■ **FIGURE 13–27** (a) The up-and-down variations on this chart led Jocelyn Bell Burnell to suspect that something interesting was going on. (b) The chart record showing the discovery of the individual pulses from CP 1919. Here downward blips are actually increases in brightness.

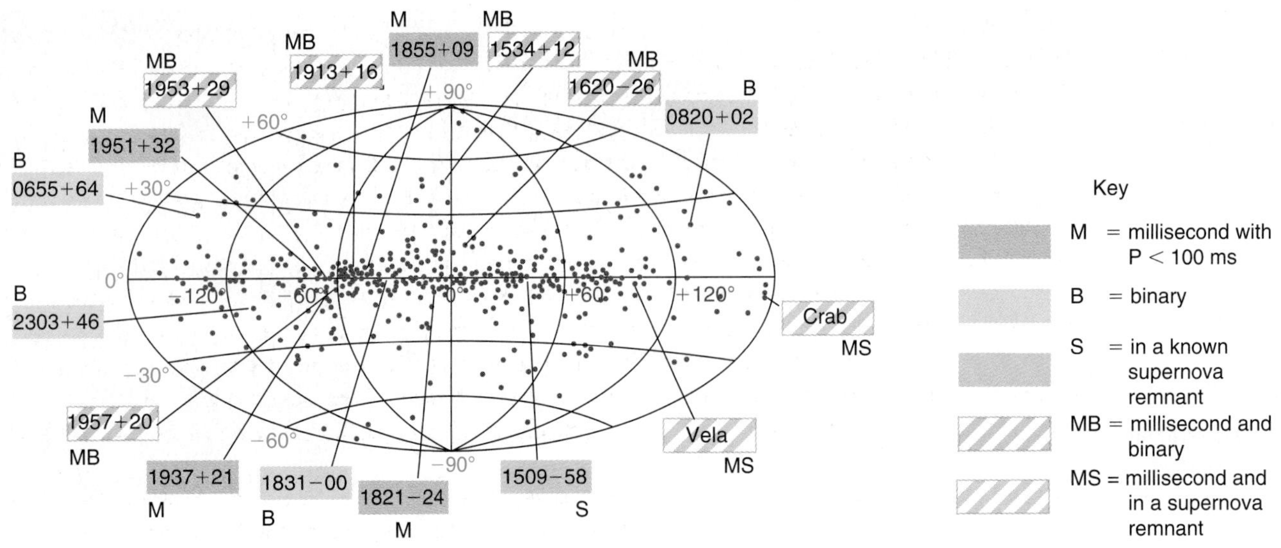

■ **FIGURE 13–28** The distribution of over 500 of the 2000 known pulsars on a map that shows the entire sky, with the plane of the Milky Way along a horizontal line at center. From the concentration of pulsars along the plane of our Galaxy, we can conclude that the pulsars are members of our Galaxy; otherwise, we would have expected to see as many near the poles of the map. The concentration of pulsars near −60° galactic longitude on this map merely represents the fact that this section of the sky has been especially carefully searched. (Joseph H. Taylor, Jr., at the Princeton Pulsar Physics Laboratory)

indicate that they gave out pulses of radio waves—and announced to an astonished world. It was immediately apparent that they were an important discovery, but what were they?

13.3c What Are Pulsars?

Other observatories set to work searching for pulsars, and dozens were soon found. They were all characterized by very regular periods, with the pulse itself taking up only a small fraction of a period. When the positions of many of the roughly 1400 known pulsars are plotted on a map of the whole sky (■ Fig. 13–28), we easily see that they are concentrated along the relatively flat plane of our Galaxy. Thus they must be in our Galaxy; if they were located outside our Galaxy we would see them distributed uniformly around the sky or even partly obscured near our Galaxy's plane where the Milky Way might block something behind it.

The question of what a pulsar is can be divided into two parts. First, we want to know why the pulses are so regular—that is, what the "clock" is. Second, we want to know where the energy comes from.

We can get pulses from a star in two ways: If the star oscillates in size ("vibrates") or if it rotates. (The only other possibility—collapsed stars orbiting each other—would result in progressively decreasing pulse periods as the objects release energy and spiral inward.) The theory worked out for ordinary variable stars had shown that the speed with which a star oscillates depends on its average density. Ordinary stars would oscillate much too slowly to be pulsars, and even white dwarfs would oscillate somewhat too slowly. Further, neutron stars would oscillate too rapidly to be pulsars. So oscillations were excluded.

That left only rotation as a possibility. And it can be calculated that a white dwarf is too large to rotate fast enough to cause pulsations as rapid as those that occur in a pulsar; it would be torn apart. So, the only remaining possibility is the rotation of a neutron star. We have solved the problem, or at least come to a plausible hypothesis, by the process of elimination.

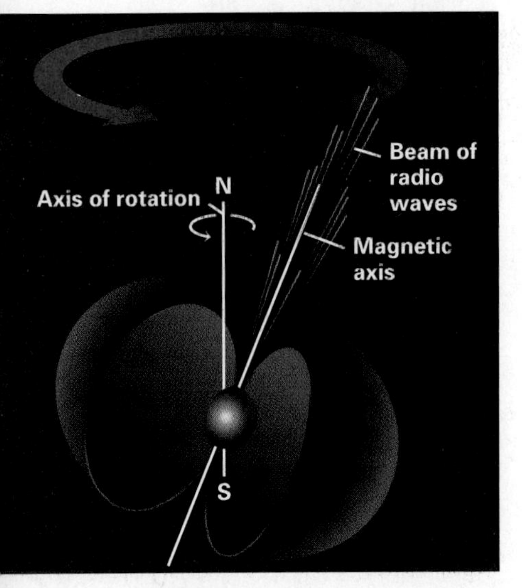

■ **FIGURE 13–29** In the lighthouse model for pulsars, which is now commonly accepted, a beam of radiation flashes by us once every pulsar period. (A beam pointing in the opposite direction gives an additional pulse, if the magnetic axis is perpendicular to the rotation axis.) Similarly, a lighthouse beam appears to flash by a ship at sea.

Over the years, more and more evidence has accumulated to back the **lighthouse model** for pulsars (■ Fig. 13–29), and there is general agreement on it. Just as a lighthouse seems to flash light every time its beam points toward you, a pulsar is a rotating neutron star.

How is the energy generated? There is much less agreement about that, and the matter remains unsettled. Remember that the magnetic field of a neutron star is extremely high. This can lead to a powerful beam of radio waves. If the magnetic axis is tilted with respect to the axis of rotation (which is also true for the Earth, whose magnetic north pole is quite far from true north), the beam from stars oriented in certain ways will flash by us at regular intervals. We wouldn't see other neutron stars if their beams were oriented in other directions.

In 1974, Hewish received the Nobel Prize in Physics, largely for his discovery of pulsars. Given the crucial role played by Jocelyn Bell, it is unfortunate that she was not honored in this way as well. At that time, it was not the custom of the Nobel Prize committees to honor work done by a graduate student while also honoring the advisor, but the custom has changed largely as a result of the omission in this case.

13.3d The Crab, Pulsars, and Supernovae

Several months after the first pulsars had been discovered, strong bursts of radio energy were found to be coming from the direction of the Crab Nebula. Observers detected that the Crab pulsed 30 times per second, almost ten times more rapidly than the fastest other pulsar then known. This very rapid pulsation definitively excluded white dwarfs from the list of possible explanations.

The discovery of a pulsar in the Crab Nebula made the theory that pulsars were neutron stars look more plausible, since neutron stars should exist in supernova remnants like this one. And the case was clinched when it was discovered that the clock in the Crab pulsar was not precise—it was slowing down slightly. The energy given off as the pulsar slowed down was precisely the amount of energy needed to keep the Crab Nebula shining. The source of the Crab Nebula's energy had been discovered!

Astronomers soon found, to their surprise, that an optically visible star in the center of the Crab Nebula could be seen apparently to turn on and off 30 times per second. Actually the star only appears "on" when its beamed light is pointing toward us as it sweeps around. Long photographic exposures had always hidden this fact, though the star had been thought to be the remaining core because of its spectrum, which oddly doesn't show any emission or absorption lines. Later, similar observations of the star's blinking on and off in x-rays were also found (■ Fig. 13–30). Even more recently, the high-resolution observations by the Hubble Space Telescope and Chandra X-ray Observatory of the Crab Nebula revealed interesting structure near its core (■ Fig. 13–31).

Although every pulsar is thought to be a neutron star, not every neutron star will be visible as a pulsar. The spinning beam of radiation (in other words, the lighthouse

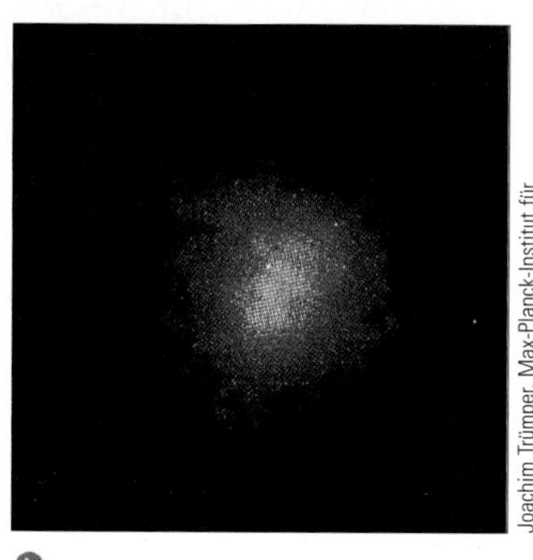

Joachim Trümper, Max-Planck-Institut für Extraterrestrische Physik

■ **FIGURE 13–30** ROSAT views of the pulsar in the Crab Nebula. (a) An x-ray view of the main pulse collected only when the pulsar is on—that is, when the beam of radiation is sweeping by the Earth. (b) The off phase of the pulsar. The satellite imaged x-rays, as indicated by its name, Roentgen satellite. It was named after Wilhelm Roentgen, who discovered x-rays in 1895.

November 26 January 31 April 6

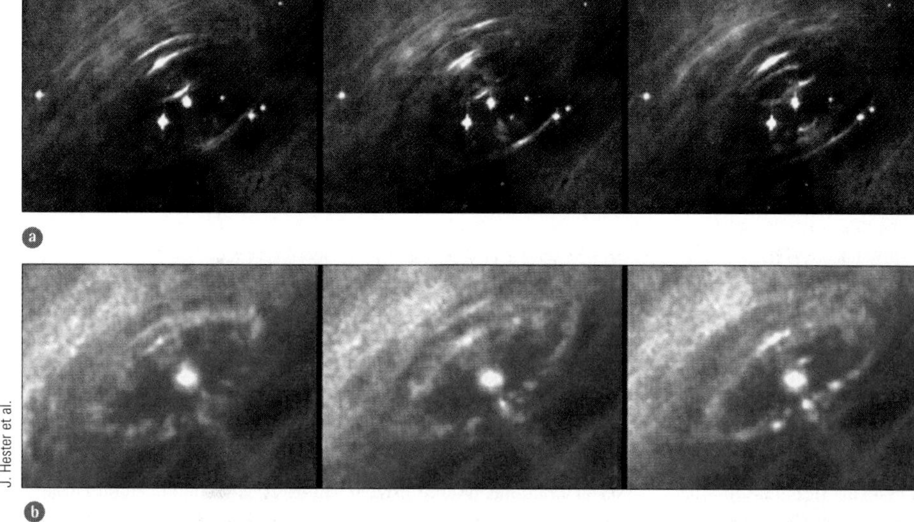

X-ray: NASA/CXC/ASU/J. Hester et al.; Optical: NASA/HST/ASU/J. Hester et al.

■ **FIGURE 13–31** The central region of the Crab Nebula. (a) Images by the Hubble Space Telescope. Wisps of gas take on an apparent whirlpool shape. (b) Images by the Chandra X-ray Observatory, showing an outer ring $^2/_3$ light-year across and an inner ring.

beam) might not intersect our line of sight, and the neutron star might not even have a beam of radiation if it is rotating too slowly or has too weak a magnetic field.

The pulsar with the longest known period was identified in 1999. Its period is over 8 seconds, longer than can be easily accounted for by any of the current models of how the energy for the pulses is generated.

13.3e Slowing Pulsars and Fast Pulsars

The Crab, when discovered, was the most rapidly pulsing known pulsar, and it is slowing down by the greatest amount. But most other pulsars have also been found to be slowing gradually. The theory had been that the younger the pulsar, the faster it was spinning and the faster it was slowing down. After all, the Crab came from a supernova explosion only almost 1000 years ago.

So, in 1982, scientists were surprised to find a pulsar spinning 20 times faster—642 times per second. Even a neutron star rotating at that speed would be on the verge of being torn apart. And this pulsar is hardly slowing down at all; it may be useful as a long-term time standard to test even the atomic clocks that are now the best available to scientists. The object, which is in a binary system, is thought to be old—over a hundred million years old—because of its gradual slowdown rate. Astronomers conclude that its rotation rate has been speeded up in an interaction with its companion.

This pulsar's period is 1.56 milliseconds (0.00156 second), so it became known as "the millisecond pulsar." Dozens more millisecond pulsars have since been discovered. Each pulses rapidly enough to sound like a note in the middle of a piano keyboard (■ Fig. 13–32) if its rotation frequency were converted to an audio signal.

Many of the millisecond pulsars we have detected are in globular star clusters. So many stars are packed together in globular clusters that the former companion star might have been stripped off in most of the cases. But recently, observations with the Chandra X-ray Observatory revealed a closely spaced binary consisting of a millisecond pulsar and a normal star. In this system, known as 47 Tuc W (it is in the constellation Tucana, and after 47 Tuc A, B, and so on, it was the Wth to be discovered), the neutron star makes a complete rotation every 0.00235 second (2.35 milliseconds). It has almost certainly been spun up by accretion of material from its companion star.

13.3f Binary Pulsars and Gravitational Waves

Joseph Taylor and Russell Hulse set out to discover lots of pulsars. One of the millisecond pulsars they found seemed less regular in its pulsing than expected. It turned out that another neutron star in close orbit with the pulsar was rapidly pulling it to and fro, shortening and lengthening the interval between the pulses that reached us on Earth. The system is called "the binary pulsar," though one of its neutron stars is not a pulsar. It has an elliptical orbit that can be traced out by studying small differences in the time of arrival of the pulses. The pulses come a little less often when the pulsar is moving away from us in its orbit, and a little more often when the pulsar is moving toward us.

■ **FIGURE 13–32** Many of the millisecond pulsars have periods fast enough to hear as musical notes when we listen to a signal at the frequency of their pulse rate. The display, of the first set of millisecond pulsars discovered, is in order of celestial longitude. (Walter Brisken (NRAO/VLA), at the Princeton Pulsar Physics Laboratory)

Einstein's general theory of relativity explained the slight change over decades in the orientation of Mercury's orbit around the Sun. The gravity in the binary-pulsar system is much stronger, and the effect is much more pronounced. Calculations show that the orientation of the pulsar's orbit should change by 4° per year (■ Fig. 13–33), which is verified precisely by measurements.

Another prediction of Einstein's general theory of relativity is that **gravitational waves,** wiggles in the curvature of space–time caused by fluctuations of the positions of masses, should travel through space. The process would be similar to the way that electromagnetic radiation, caused by fluctuations in electricity and magnetism, travels through space.

Gravitational waves have never been detected directly. But the motion of the binary pulsar in its orbit is speeding up by precisely the rate that would be expected if the system were giving off gravitational waves (■ Fig. 13–34). (The two stars are getting closer and closer together, so according to Kepler's third law their orbital period is decreasing.) So scientists consider that the existence of gravitational waves has been verified in this way. The Nobel Prize in Physics was awarded in 1993 to Taylor and Hulse for their discovery and analysis of the binary pulsar.

An even better system for testing general relativity was discovered in 2003. While we can detect pulsars from only one of the Taylor-Hulse pulsar members, for the double pulsar in the constellation Puppis pulses are detected from both neutron stars. One of the pulsars spins every 23 milliseconds while the other spins every 2.8 seconds, much more slowly.

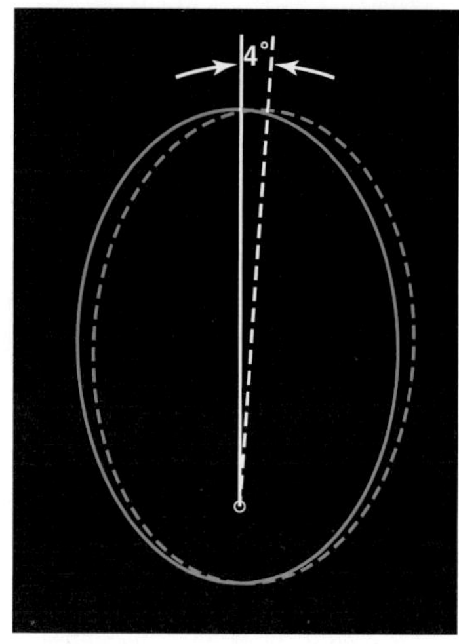

■ **FIGURE 13–33** In the binary pulsar's orbit, the near point to the star it is orbiting moves around by 4° per year. This measurement matches, and thus endorses, the prediction of Einstein's general theory of relativity. For convenience, the diagram shows the farthest point of the orbit rather than the nearest point; it, too, moves by 4° per year.

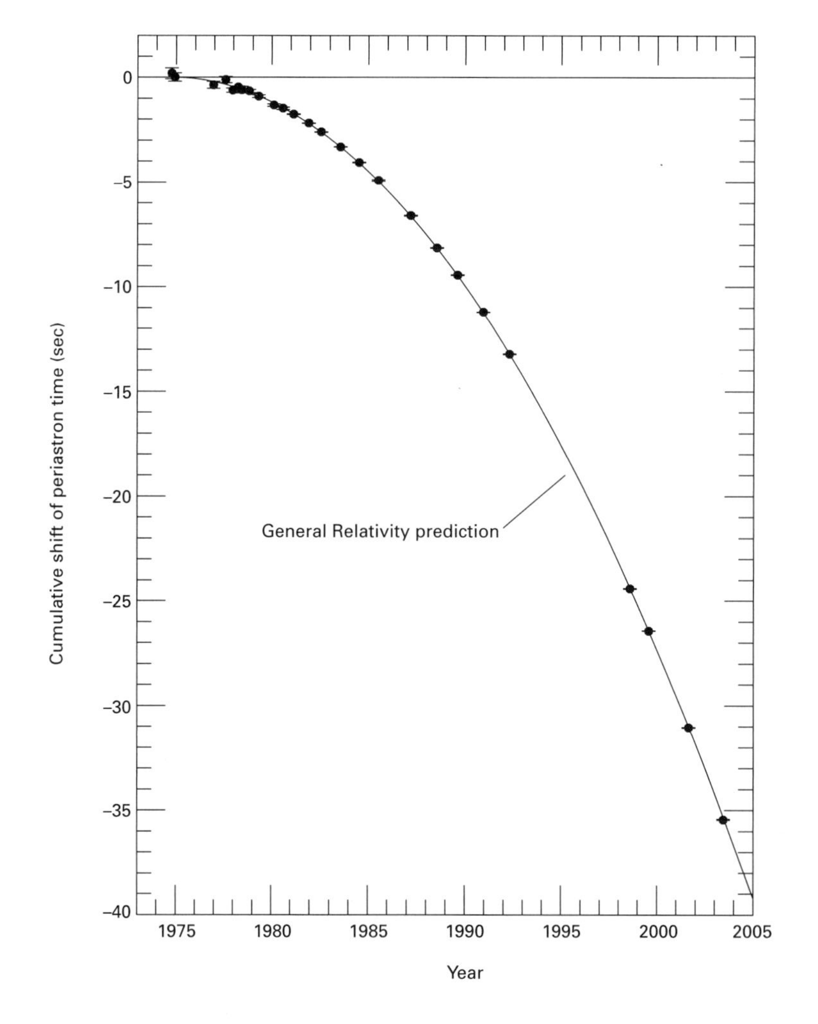

■ **FIGURE 13–34** The rate at which the Taylor-Hulse pair of pulsars is orbiting (black points with error bars) slows in a way that precisely matches the predictions of Einstein's general theory of relativity (solid curve). (Joel Weisberg, Carleton College)

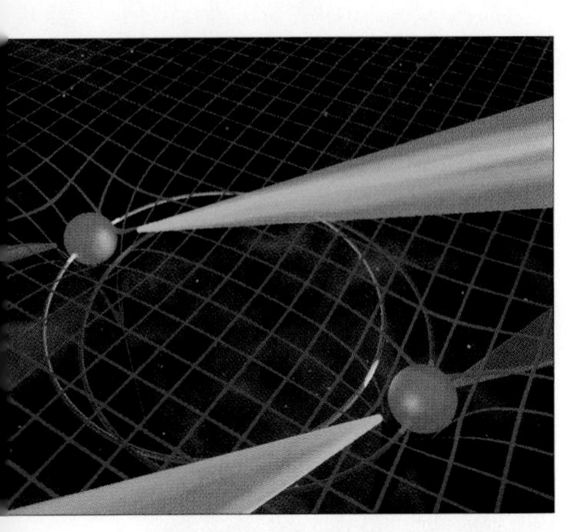

■ **FIGURE 13–35** An artist's conception of the double pulsar in Puppis, whose orbit is shrinking 7 millimeters per day. (Michael Kramer and Jodrell Bank)

The two pulsars orbit each other every 2.4 hours (■ Fig. 13–35), making them much closer together than the Taylor-Hulse neutron stars with their 8-hour orbital period. Thus the Puppis system gives off gravitational waves at an even greater rate. Measurements show that the orbit is shrinking—as gravitational waves carry off energy—by 7 millimeters per day, an incredibly tiny rate to be able to measure. The measurements are in excellent agreement with the predictions of Einstein's theory.

Because both pulsars can be detected, scientists have been able to measure the masses of the individual objects. One of the neutron stars has 1.35 times the Sun's mass and the other has 1.25 solar masses. There is some mass contribution from the energy in the stars' gravitational fields (after all, $E = mc^2$), which puts each near the Chandrasekhar limit, matching theory.

The Taylor-Hulse method of detecting gravitational waves was indirect. Only the speeding up of the orbit caused by the waves, rather than the waves themselves, was measured. Efforts to sense gravitational waves directly by whether they set large metal bars vibrating have failed, because the experiments lacked the necessary sensitivity.

A major pair of observational facilities has been built to try to detect gravitational waves directly in another manner. This Laser Interferometer Gravitational-wave Observatory (LIGO) uses complicated lasers, optics, and electronics to see whether a 100-m length is distorted as a gravitational wave goes by. Nearly identical LIGOs have been set up in Hanford, Washington, and in Livingston, Louisiana (■ Fig. 13–36), to make sure that the system isn't tricked by local effects such as nearby logging or traffic. The Livingston site has perpendicular evacuated tubes 4 km long while the Hanford site has 2-km and 4-km tubes within the same vacuum. Any signal detected must affect all three interferometers to be believed. A similar site is under construction in Europe. Thus far, no clear cases of gravitational waves have been detected.

LIGO is already being upgraded to improve its sensitivity. Scientists are planning for LISA (Laser Interferometer Space Antenna), a NASA/ESA 3-dimensional interferometer to be launched into space in about 2012. (The 3 dimensions are laid out by 3 perpendicular arms coming from pairs of 3 individual spacecraft.) LISA is to be in an Earth-trailing orbit around the Sun, and should be able to search even more sensitively for gravitational waves.

The prime system we know of for LISA to detect is the pair of rapidly orbiting white dwarfs referred to in Section 13.1c. Calculations based on the observed decrease of the x-ray period show that gravitational waves should be given off at such a high

Jay M. Pasachoff

a

b

c

■ **FIGURE 13–36** a, b The perpendicular arms of the Laser Interferometer Gravitational-wave Observatory (LIGO) at Livingston, Louisiana. c The vacuum chamber in which the laser beams are split, sent to mirrors 4 km away, and then recombined.

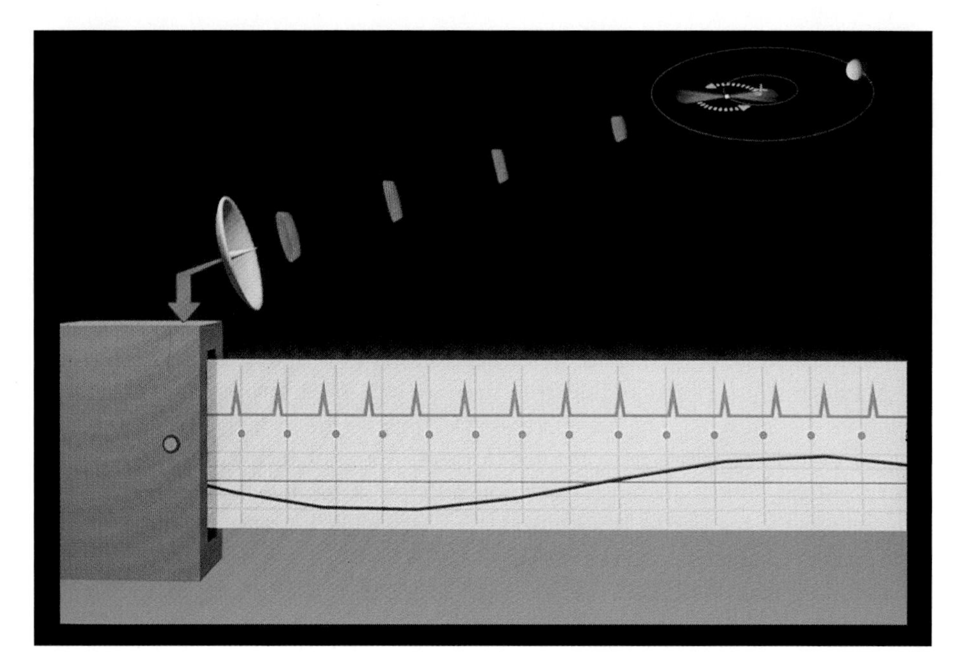

■ **FIGURE 13–37** The pulses from the pulsar, shown in blue schematically on a chart, sometimes arrive slightly before or after regular pulses would (red dots). The average pulse period is 6.2 milliseconds. The difference is graphed at the bottom. It reveals that there must be planets orbiting the pulsar. (After a Cornell U. visualization, imaging by Chris Hildreth and Wayne Lytle; courtesy of Alex Wolszczan, Pennsylvania State U.)

rate that the orbit shrinks by about 1 centimeter per hour! The waves' frequency would be prime for LISA's sensitivity, though not a good frequency for even the improved LIGO.

13.3g A Pulsar with a Planet

Earlier in this book, in Chapters 2 and 9, we described the discovery of planets around other stars. But these extra-solar planets were not the first to be discovered. In 1991, the first ones were discovered by observing a pulsar.

The detections were from observations of a pulsar that pulses very rapidly—162 times each second. The arrival time of the pulsar's radio pulses varied slightly (■ Fig. 13–37), indicating that something is orbiting the pulsar and pulling it slightly back and forth. Alex Wolszczan, now at Penn State, has concluded that the variations in the pulse-arrival time are caused by three planets in orbit around the pulsar. These planets are 0.19, 0.36, and 0.47 A.U. from the pulsar, within about the same distance that Mercury is from the Sun. They revolve in 25.3-, 66.5-, and 98.2-day periods, respectively. The system is 2000 light-years from us, too faint for us to detect optically.

The presence of the planets was conclusively verified when they interacted gravitationally as they passed by each other. The two most massive planets are calculated from the observations to be somewhat larger than Earth, each containing about 4 times its mass. The innermost planet is much less massive than Earth. The existence of a fourth planet farther out in the system is possible but uncertain.

Astronomers think that neutron stars are formed in supernova explosions, so any original planets almost certainly didn't survive the explosion. Most likely, the planets formed after the supernova explosion, from a disk of material in orbit around the neutron star remnant. These pulsar planets are not the ones on which we expect life will have arisen!

13.3h X-Ray Binaries

Neutron stars are now routinely studied in a way other than their existence as radio pulsars. Many neutron stars in binary systems interact with their companions. As gas from the companion forms an accretion disk much like those discussed in Section 13.1e (Fig. 13–10), or is funneled toward the neutron star's poles by a strong magnetic field, the gas heats up (due to friction and compression) and gives off x-rays.

X-ray telescopes in orbit (such as the Chandra X-ray Observatory) detect the x-rays, sometimes in pulses if there is a rotating beam of emission. But in many of these binary systems, unlike the case for normal pulsars, the pulse rate usually speeds up. Gas from the accretion disk spins up the neutron star as it lands on the surface.

Although white dwarfs and neutron stars are indeed peculiar endpoints of stellar evolution, the last remaining possibility—black holes—are so interesting and bizarre that we devote the entire next chapter to them. Many of them are found in x-ray binary systems similar to those harboring neutron stars, but apparently the stellar remnant was too massive to be a neutron star. Meanwhile, however, it would be useful to go back to Chapter 12 and review Figure 12–29, in which we gave a preview of stellar evolution; with the knowledge you have gained in this chapter, you should now understand that diagram better.

AceAstronomy™ Log into AceAstronomy and select this chapter to see the Active Figure called "End States of Stars."

CONCEPT REVIEW

Stars have different fates depending on the mass with which they were born. **Lightweight stars** are those with mass up to about 10 (but perhaps only 8) times the Sun's mass, while **heavyweight stars** have larger masses (Introductory section). When the Sun and other low-mass stars exhaust their central hydrogen, they will swell and become luminous **red giants** as their cores contract and the outer layers expand (Sec. 13.1a). The outer layers subsequently drift off as **planetary nebulae,** which glow because their gases are ionized by radiation from the hot dying star (Sec. 13.1b).

The remaining core continues to contract until electrons won't be compressed further (they become "degenerate"), and the core becomes a **white dwarf** (Sec. 13.1c). The entire post-main-sequence evolution of the Sun can be conveniently traced on a temperature-luminosity (Hertzsprung–Russell) diagram (Sec. 13.1d). The evolution of other solitary (single) stars is similar, but differs in detail.

If a star is in a binary system, its evolution can be sped up by the transfer of matter from a companion star filling its **Roche lobe,** the region in which the companion's gravity dominates (Sec. 13.1e). The flowing gas forms an **accretion disk** around the recipient star due to the rotation of the system. Matter falling onto a white dwarf from a companion star can release gravitational energy suddenly, or even undergo rapid nuclear fusion on the white dwarf's surface, flaring up to form a **nova** (plural **novae**).

Heavyweight stars, having more than 8 to 10 times the Sun's mass, become **red supergiants** (Sec. 13.2a). Heavy elements build up in layers inside these stars. When the innermost core is iron, it collapses and the surrounding layers rebound (explode); the result is a **Type II supernova** (plural **supernovae**), a type of **core-collapse supernova** that has hydrogen in its spectrum. The remaining compact object is a very dense **neutron star,** consisting almost entirely of neutrons. If the evolved massive star had previously lost its outer envelope of gases, it would explode in a similar manner, but hydrogen lines would be absent in its spectrum; these are known as Type Ib (helium present) and Type Ic (no helium) supernovae.

Hydrogen-deficient **Type Ia** supernovae, in contrast, occur when carbon-oxygen white dwarfs in binary systems receive too much mass from their companions to remain in that state, and undergo a nuclear runaway that completely destroys the white dwarf (Sec. 13.2b). This happens at, or near, the Chandrasekhar limit, the maximum mass of a white dwarf (about 1.4 solar masses). Such **white-dwarf supernovae** do not produce a compact remnant (neutron star or black hole), unlike the case in core-collapse supernovae.

The last bright supernovae to have been seen in our Galaxy were in 1572 and 1604 (Sec. 13.2c). Many supernovae are discovered each year in other galaxies, however.

Supernovae in our Galaxy have left detectable **supernova remnants,** the expanding debris of the explosions. The gases ejected by supernovae are enriched in heavy elements produced prior to, and during, the explosions (Sec. 13.2d). We owe our existence to supernovae, being made of their debris.

The Type II Supernova 1987A, in a satellite galaxy known as the Large Magellanic Cloud, was the brightest supernova since 1604 and provided many valuable insights (Sec. 13.2e). The star was a massive, evolved star, as predicted, although its exact nature (blue, instead of red, supergiant) was a surprise. Neutrinos from the supernova, as well as the detection of radioactive heavy elements, show that we had the basic ideas of the theory of Type II supernovae correct. **Cosmic rays** are particles accelerated to high energy, perhaps from supernova explosions (Sec. 13.2f).

The cores of massive stars, after the supernova explosions, consist of degenerate neutrons that cannot be compressed further; they are called neutron stars (Sec. 13.3a). Some give off beams of radiation as they rotate like a lighthouse, and we detect pulses in the radio spectrum. We now know of thousands of these **pulsars** (Sec. 13.3b). They are explained by the **lighthouse model,** in which two oppositely directed beams are seen only when they rotate into our line of sight (Sec. 13.3c). The discovery of a very rapid pulsar in the Crab Nebula, a young supernova remnant, provided strong support for the hypothesis that pulsars are neutron stars (Sec. 13.3d).

Most pulsars slow down as they radiate energy (Sec. 13.3e). Some pulsars, however, spin very rapidly, up to a few hundred times per second. They were probably spun up by accretion of gas from a companion star.

Two binary-star systems consisting of a pair of neutron stars, at least one of which is a pulsar, are proving very useful for testing the general theory of relativity (Sec. 13.3f). Their orbital periods are getting shorter at a rate that agrees with the idea that **gravitational waves** are given off. Observational facilities have been built to directly detect gravitational waves, so far without success.

Strange planets have been discovered around one pulsar, but they must have formed after the supernova explosion (Sec. 13.3g). Neutron stars in binary systems can be studied in a way other than their existence as radio pulsars: They give off x-rays emitted by hot gas in an accretion disk fed by the companion star (Sec. 13.3h).

QUESTIONS

1. What happens to the core and envelope of a star at the end of its main-sequence stage?
2. Why doesn't the helium core of a red giant immediately start fusing to heavier elements?
3. Why does a red giant appear reddish?
4. Sketch a temperature-luminosity diagram, label the axes, and point out the location of red giants.
5. If you compare a photograph of a nearby planetary nebula taken 100 years ago with one taken now, how would you expect them to differ?
6. Why is the surface of a star hotter after the star sheds a planetary nebula?
7. What forces balance to make a white dwarf?
8. What is the difference between the Sun and a 1-solar-mass white dwarf?
9. When the Sun becomes a white dwarf, about how much mass will it contain? Where will the rest of the mass have gone?
10. Which has a higher surface temperature, the Sun or a white dwarf having the same spectral type?
11. What is the relation of novae and white dwarfs?
12. Is a nova really a new star? Explain.
13. If we see a massive main-sequence star (a heavyweight star), what can we conclude about its age, relative to the ages of most stars? Why?
14. Distinguish between what is going on in novae and supernovae.
15. In what way do we distinguish observationally between Type Ia and Type II supernovae?
16. What are the physical differences between Type Ia and Type II supernovae, in terms of the kinds of stars that explode and their explosion mechanisms?
17. What is special about the iron core of a massive star? Why does it collapse and then produce a supernova?
18. †In a supernova explosion of a 15-solar-mass star, about how much material is ejected (blown away)?
19. Explain how a very massive star might explode as a Type I supernova (that is, with no hydrogen in its spectrum), yet still be a core-collapse supernova like the Type II variety.
20. From where did the heavy elements in your body come?
21. Why do we think that the Crab Nebula is a supernova remnant?
22. Would you expect the appearance of the Crab Nebula to change in the next 1000 years? How?
23. What are two reasons why Supernova 1987A was significant?
24. From what event in Supernova 1987A did the neutrino burst come?
25. What is the difference between cosmic rays and x-rays?
26. What keeps a neutron star from collapsing?
27. †Compare the Sun, a white dwarf, and a neutron star in size. Include a sketch.

28. In what part of the spectrum do all pulsars give off the energy that we study? (A few pulsars also emit significant energy at other wavelengths.)
29. How do we know that pulsars are in our Galaxy?
30. Why do we think that the lighthouse model explains pulsars?
31. How did studies of the Crab Nebula pin down the explanation of pulsars?
32. How has the Taylor-Hulse binary pulsar been especially useful?
33. Why is the binary pulsar in Puppis even more useful than the Taylor-Hulse pulsar?
34. Compare the discovery of gravitational waves with the method of detecting them now being worked on.
35. **True or false?** The decreasing orbital period of a neutron star binary system provides astronomers with strong evidence for the validity of Einstein's general theory of relativity.
36. **True or false?** A white dwarf shines because of ongoing nuclear fusion of heavy elements in its core.
37. **True or false?** Neutron stars are commonly seen in star clusters, where they are among the brightest red stars.
38. **True or false?** The Sun will initially shrink in size, becoming a white dwarf, before moving on to the red-giant stage of its evolution.
39. **True or false?** A supernova whose spectrum shows no sign of hydrogen may have been produced by collapse of the core of a very massive star.
40. **Multiple choice:** When a main-sequence star runs out of hydrogen fuel in its core, (**a**) the core expands and thus heats up; (**b**) the core expands and thus cools down; (**c**) the core contracts and thus heats up; (**d**) the core contracts and thus cools down; or (**e**) the core remains about the same size, but heats up as fusion of helium to carbon begins immediately after the hydrogen fuel is gone.

†41. **Multiple choice:** Suppose white dwarf Yoda is the size of the Earth and has a surface temperature of 3000 K. If the Earth's surface temperature is 300 K, how many times more luminous than the Earth is white dwarf Yoda? (**a**) 10^2. (**b**) 10^3. (**c**) 10^4. (**d**) 10^8. (**e**) 3000×300, which is 9×10^5.

42. **Multiple choice:** White dwarf stars (**a**) are the end states only of stars whose initial mass is much greater than that of the Sun; (**b**) in some cases consist largely of carbon and oxygen; (**c**) in some cases consist largely of uranium and other very heavy elements; (**d**) shine only while nuclear reactions continue within them; or (**e**) support themselves against the pull of gravity in the same way as normal stars like the Sun, using the pressure exerted by hot gases within them.
43. **Multiple choice:** Which one of the following statements about Supernova 1987A is *true*? (**a**) Most of the energy from the explosion was released in the form of tiny, uncharged particles called neutrinos. (**b**) A newly formed white dwarf surrounded by a planetary nebula was left over after the explosion. (**c**) The star that exploded was a red giant with about the same mass as our Sun. (**d**) Heavy elements were not detected in the ejected material because they were ripped apart by the explosion. (**e**) The star exploded because of a runaway chain of nuclear reactions in the star's core.
44. **Multiple choice:** Typical novae occur when (**a**) a red giant star ejects a planetary nebula; (**b**) two neutron stars merge, forming a more massive neutron star; (**c**) an extremely massive star collapses, and also ejects its outer atmosphere; (**d**) matter accreted from a companion star unstably ignites on the surface of a white dwarf; or (**e**) a neutron star's magnetic field becomes strong enough to produce two oppositely directed jets of rapidly moving particles.
45. **Multiple choice:** Which one of the following statements about supernovae is *false*? (**a**) At optical wavelengths, a supernova can appear about as bright as the entire galaxy of stars in which it's located. (**b**) A supernova produces an expanding volume of gas that is rich in heavy elements. (**c**) Nobody has seen a bright supernova in our own Milky Way Galaxy during the past century. (**d**) A neutron star is produced when a white dwarf exceeds a certain mass limit and becomes a supernova, experiencing a runaway chain of nuclear fusion reactions. (**e**) One kind of supernova occurs when the iron core of a massive star collapses.
46. **Fill in the blank:** Mass exchange from one star to the other can occur if one star extends into the __________ of its companion.
47. **Fill in the blank:** A white dwarf is supported against gravitational collapse by ________, a weird kind of quantum-mechanical pressure.
48. **Fill in the blank:** A red giant forms a __________ when its outer atmosphere becomes unstable, producing a gentle ejection.
49. **Fill in the blank:** One kind of ________ occurs when a clump of material is suddenly dumped onto the surface of a white dwarf in a binary system.
50. **Fill in the blank:** A rapidly rotating neutron star having a strong magnetic field can sometimes be observed as a _________.

†This question requires a numerical solution.

TOPICS FOR DISCUSSION

1. How can we be so confident in our theory of the Sun's future evolution?
2. Does the fact that you are made of stardust ("star stuff") give you a sense of unity with the cosmos?
3. If only about 10 neutrinos from SN 1987A were detected by each underground tank containing several thousand tons of water, and if a typical human consists of 100 pounds of water, what are the odds that a given human body directly detected a neutrino from SN 1987A?

MEDIA

Virtual Laboratories

- Binary Stars, Accretion Disks, and Kepler's Laws
- White Dwarfs, Novae, and Supernovae
- Neutron Stars and Pulsars
- Cosmic Rays

AceAstronomy™ Log into AceAstronomy at **http://astronomy.brookscole.com/cosmos3** to access quizzes and animations that will help you assess your understanding of this chapter's topics.

Log into the Student Companion Web Site at **http://astronomy.brookscole.com/cosmos3** for more resources for this chapter including a list of common misconceptions, news and updates, flashcards, and more.

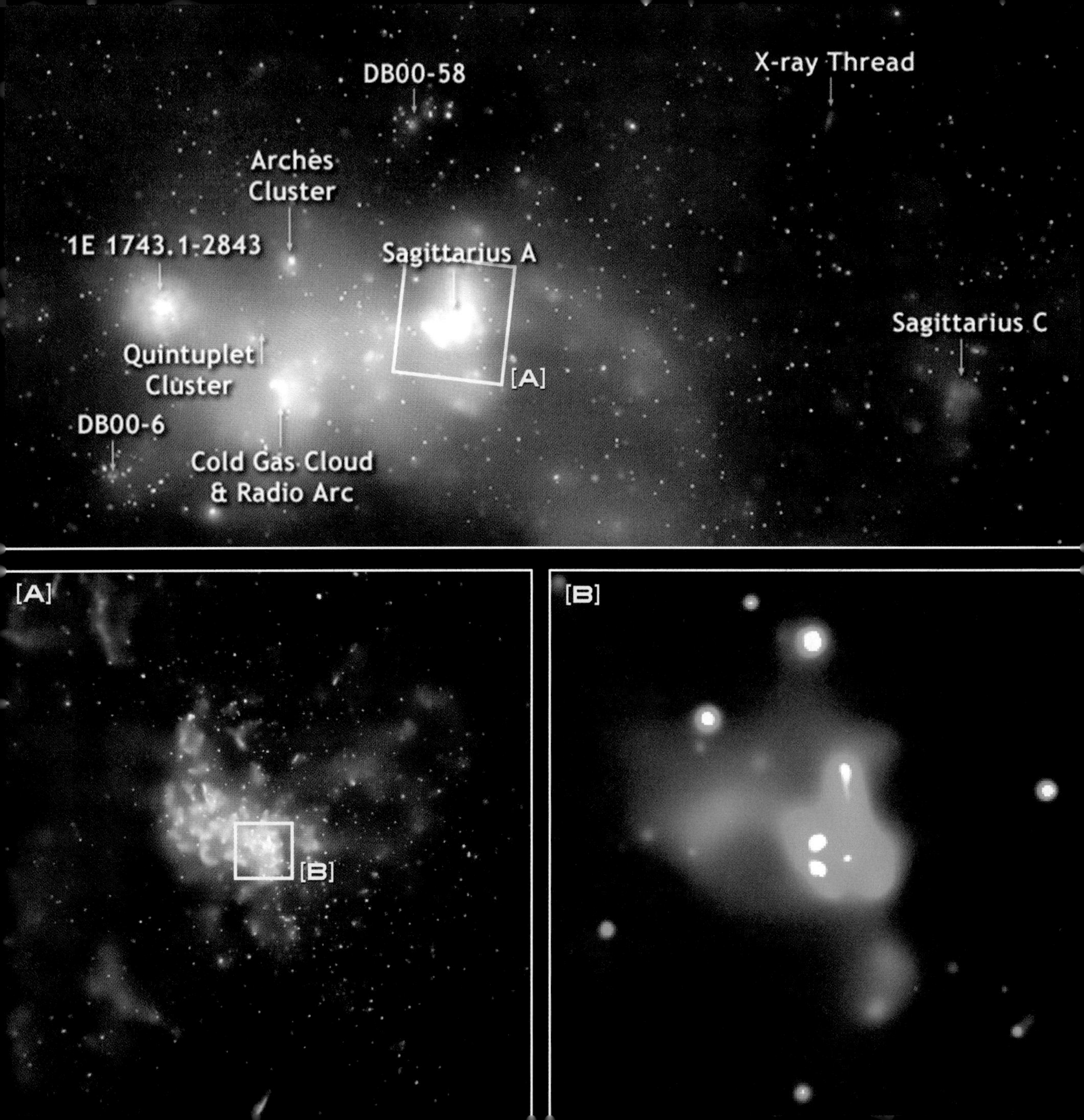

DB00-58
X-ray Thread
Arches Cluster
1E 1743.1-2843
Sagittarius A
Sagittarius C
Quintuplet Cluster
[A]
DB00-6
Cold Gas Cloud & Radio Arc
[A]
[B]
[B]

CHAPTER 14

Black Holes: The End of Space and Time

ORIGINS

Gravity is the dominant force shaping the structure and evolution of the Universe. Black holes provide a laboratory for testing the most extreme predictions of Einstein's general theory of relativity, our best existing description of gravity.

AIMS

1. Understand the properties of black holes, regions in which gravity is so incredibly strong that nothing, not even light, can escape (Sections 14.1 to 14.6).
2. Discuss the observational evidence for stellar-mass black holes in binary systems, and for supermassive black holes in the centers of many galaxies (Sections 14.7 and 14.8).
3. Learn about the new category of intermediate-mass black holes (Section 14.9).
4. Explore enigmatic gamma-ray bursts, the probable birth cries of black holes (Section 14.10).
5. Introduce the concept of mini black holes that evaporate (Section 14.11).

The peculiar forces of electron and neutron degeneracy pressure (an effect of quantum physics) support dying lightweight stars and some heavyweight stars against gravity. The strangest case of all occurs at the death of the most massive stars, which contained much more than 10 and up to about 100 solar masses when they were on the main sequence. After these stars undergo supernova explosions, some may retain cores of over 2 or 3 solar masses. Nothing in the Universe is strong enough to hold up the remaining mass against the force of gravity, so it collapses.

The result is a **black hole,** in which the matter disappears from contact with the rest of the Universe (■ Fig. 14–1). Later, we shall discuss much more massive black holes than those that result from the collapse of a star. Most astronomers now agree that black holes are widespread and very important in our Universe.

14.1 The Formation of a Stellar-Mass Black Hole

Astronomers had long assumed that the most massive stars would somehow lose enough mass to wind up as white dwarfs. When the discovery of pulsars (neutron stars) ended this prejudice, it seemed more reasonable that black holes could exist. If more than 2 or 3 times the mass of the Sun remains after the supernova explosion, the star collapses through the neutron-star stage. We know of no force that can stop the collapse. (In some cases, the supernova explosion itself may fail, so the remaining mass can be much larger than 2 or 3 solar masses.)

AceAstronomy™ The AceAstronomy icon throughout this text indicates an opportunity for you to test yourself on key concepts, and to explore animations and interactions of the AceAstronomy website at http://astronomy.brookscole.com/cosmos3

◀ Many black holes probably exist near the center of our Milky Way Galaxy. (top) A Chandra X-ray Observatory view of a 400-light-year-across region of the center of our Galaxy, in the direction of the constellation Sagittarius. "Sagittarius A" (usually called "Sagittarius A*") marks the location of a supermassive black hole. ⓐ The detail box in the upper portion includes about 2000 individual x-ray sources, many of which are probably black holes several times more massive than our Sun in close orbits with normal stars. ⓑ A small region only 6 light-years across centered on Sagittarius A*. Since even such a small region contains several of the x-ray binaries, a dense swarm of 10,000 black holes and neutron stars may well have formed around Sagittarius A*.

Top: NASA/U Mass/D. Wang et al. Bottom left: NASA/CXC/MIT/F. K. Bagonoff et al. Bottom right: NASA/CXC/UCLA/M. Mumo et al.

Roy Bishop, Acadia University

■ **FIGURE 14–1** Directions to a small, deep cove on the shore of the Bay of Fundy. It got its name (probably a century or two ago) because it appears very dark as seen from the sea.

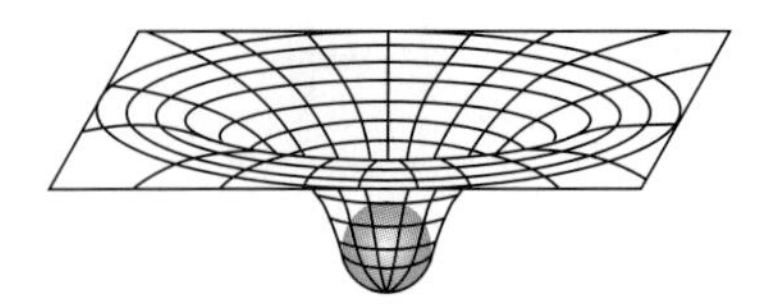

■ **FIGURE 14–2** Mass warps (bends) space and time in its vicinity. Here we show just two of the three warped spatial dimensions, analogous to the effect of a billiard ball on a stretched sheet of rubber. Note that the surface warps around a third, perpendicular dimension, not part of the surface itself.

We may then ask what happens to an evolved 5-, 10-, or 50-solar-mass star as it collapses, if it retains more than 2 or 3 solar masses. As far as we know, it must keep collapsing, getting denser and denser. Einstein's general theory of relativity suggests that the presence of mass or energy warps (bends) space and even time in its vicinity (■ Fig. 14–2); recall our discussion of the Sun's gravity in Section 10.3. The greater the density of the material, the more severe is the warp.

As we explore in more detail below, a strong gravitational field appears to bend light; thus, radiation leaving the star other than perpendicularly to the surface is bent more and more as the star contracts and its surface gravity increases. In addition, according to general relativity, the light we see from a distance will become progressively more redshifted as the star contracts. Eventually, when the mass has been compressed to a certain size, light from the star can no longer escape into space. The star has withdrawn from our observable Universe, in that we can no longer receive radiation from it. We say that the star has become a black hole, specifically a **stellar-mass black hole.**

Why do we call it a black hole? We think of a black surface as a surface that reflects none of the light that hits it. Similarly, any radiation that hits a black hole continues into its interior and is not reflected or transmitted out. In this sense, the object is perfectly black.

ASIDE 14.1: Frozen star

The term "black hole" was coined by John A. Wheeler of Princeton University. Previously, some Soviet astrophysicists had instead suggested "frozen star"; the time dilation effect (see Section 14.4) would make a collapsing star appear to get no smaller than a certain minimum size.

14.2 The Photon Sphere

Let us consider what happens to radiation emitted by the surface of a star as it contracts. Although what we will discuss applies to radiation of all wavelengths, let us simply visualize standing on the surface of the collapsing star while holding a flashlight.

On the surface of a supergiant star, if we shine the beam at any angle, it seems to go straight out into space. As the star collapses, two effects begin to occur. (We will ignore the star's outer layers, which are unimportant here. We also ignore the star's rotation.) Although we on the surface of the star cannot notice the effects ourselves, a friend on a planet revolving around the star could detect them and radio information back to us about them. For one thing, our friend could see that our flashlight beam is redshifted.

Second, our flashlight beam would be bent by the gravitational field of the star (■ Fig. 14–3). If we shine the beam straight up, it would continue to go straight up. But the further we shine it away from the vertical, the more it would be bent from the vertical. When the star reaches a certain size, a horizontal beam of light would not escape (■ Fig. 14–4).

From this time on, only if the flashlight is pointed within a certain angle of the vertical does the light continue outward. This angle forms a cone, with its apex at the flashlight, and is called the **exit cone** (■ Fig. 14–5). As the star grows smaller yet, we find that the flashlight has to be pointed more directly upward in order for its light to escape. The exit cone grows smaller as the star shrinks.

When we shine our flashlight in a direction outside the exit cone, the light is bent sufficiently that it falls back to the surface of the star. When we shine our flashlight exactly along the side of the exit cone, the light goes into orbit around the star, neither escaping nor falling onto the surface.

The sphere around the star in which light can orbit is called the **photon sphere.** Its radius is calculated theoretically to be 4.5 km for each solar mass present. It is thus 45 km in radius for a star of 10 solar masses, for example.

As the star continues to contract, theory shows that the exit cone gets narrower and narrower. Light emitted within the exit cone still escapes. The photon sphere remains at the same height even though the matter inside it has contracted further, since the total amount of matter within has not changed.

■ **FIGURE 14–3** As the star contracts, a light beam emitted other than straight outward, along a radius, will be bent. It will also become gravitationally redshifted—that is, the wavelength of the light will increase as the light recedes from the star.

■ **FIGURE 14–4** Light can be bent so that it circles the star, but in this case there is no gravitational redshifting because the light is at a constant distance from the star's center. If the light is bent so much that it approaches the star, it would become blueshifted.

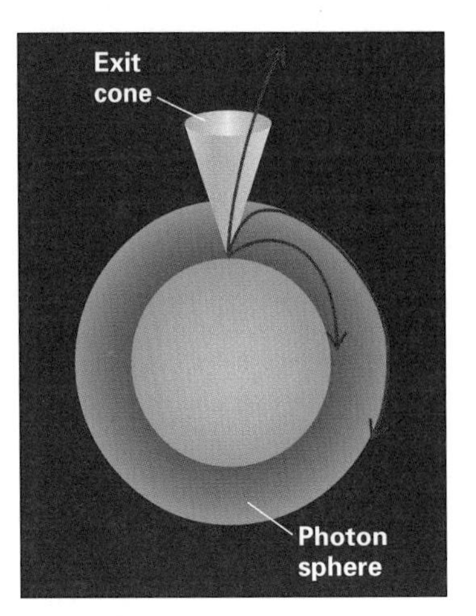

■ **FIGURE 14–5** When the star (shown by the inner sphere) has contracted enough, only light emitted within the exit cone escapes. Light emitted on the edge of the exit cone goes into the photon sphere. The further the star contracts within the photon sphere, the narrower the exit cone becomes.

14.3 The Event Horizon

We might think that the exit cone would simply continue to get narrower as the star shrinks. But Einstein's general theory of relativity predicts that the cone vanishes when the star contracts beyond a certain size. Even light travelling straight up can no longer escape into space, as was worked out in 1916 by Karl Schwarzschild while solving Einstein's equations.

The radius of the star at this time is called the **Schwarzschild radius.** The imaginary surface at that radius is called the **event horizon** (■ Fig. 14–6). (A horizon on Earth, similarly, is the limit to which we can see.) Its radius is exactly $^2/_3$ times that of the photon sphere, 3 km for each solar mass. Formally, the equation for the Schwarzschild radius is $R = 2GM/c^2$, where M is the star's mass, G is Newton's constant of gravitation, and c is the speed of light.

> **ASIDE 14.2: Not quite a black hole?**
>
> It is conceivable, though not probable, that there exists a weird state of matter that allows a collapsed star to have a mass larger than the most massive "normal" neutron star, yet not technically be a black hole. It would have a radius smaller than a normal neutron star, yet somewhat larger than the Schwarzschild radius.

14.3a A Newtonian Argument

We can visualize the event horizon in another way, by considering a classical picture based on the Newtonian theory of gravitation. The picture is essentially that conceived in 1783 by John Michell and in 1796 by Pierre Laplace.

A projectile launched from rest must be given a certain minimum speed, called the **escape velocity,** to escape from the other body, to which it is gravitationally attracted. For example, we would have to launch rockets at 11 km/sec (40,000 km/hr) or faster in order for them to escape from the Earth, if they got all their velocity at launch. (From a more massive body of the same size as Earth, the escape velocity would be higher.)

Note that the escaping object still feels the gravitational pull of the other body; one cannot "cut off" or block gravity. However, the object's speed is sufficiently high that it will never turn around and fall back down.

Now imagine that this body contracts; we are drawn closer to the center of the mass. As this happens, the escape velocity rises. If the body contracts to half of its former

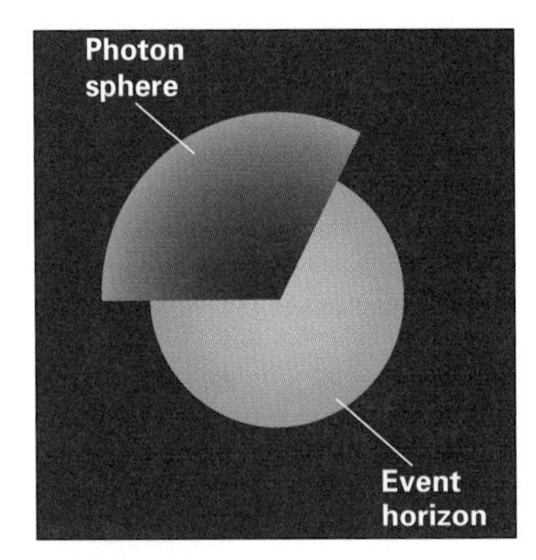

■ **FIGURE 14–6** When the star becomes smaller than its Schwarzschild radius, we can no longer observe it. It has passed its event horizon.

■ **FIGURE 14–7** This drawing is from Charles Addams, long-time *New Yorker* cartoonist.

radius, for instance, the gravitational force on its surface increases by a factor of four, since gravity follows an inverse-square law; the corresponding escape velocity therefore also increases (although not by a factor of four).

When all the mass of the body is within its Schwarzschild radius, the escape velocity becomes equal to the speed of light. Thus, even light cannot escape (■ Fig. 14–7). If we begin to apply the special theory of relativity, which deals with motion at very high speeds, we might then reason that since nothing can go faster than the speed of light, nothing can escape.

14.3b Black Holes in General Relativity

Now let us return to the picture according to the general theory of relativity, which explains gravity and the effects caused by large masses. A black hole curves space (actually, space-time) to such a large degree that light cannot escape; it remains trapped within the event horizon.

The size of the Schwarzschild radius of the event horizon is directly proportional to the amount of mass that is collapsing: $R = 2GM/c^2$. A star of 3 solar masses, for example, would have a Schwarzschild radius of 9 km. A star of 6 solar masses would have a Schwarzschild radius of 18 km.

One can calculate the Schwarzschild radii for less massive stars as well, although the less massive stars would be held up in the white dwarf or neutron-star stages and not collapse to their Schwarzschild radii. The Sun's Schwarzschild radius is 3 km. The Schwarzschild radius for the Earth is only 9 mm; that is, the Earth would have to be compressed to a sphere only 9 mm in radius in order to form an event horizon and be a black hole.

Anyone or anything on the surface of a star as it passes its event horizon would not be able to survive. An observer would be stretched out and torn apart by the tremendous difference in gravity between his head and feet (■ Fig. 14–8). This resulting force is called a **tidal force,** since this kind of difference in gravity also causes tides on Earth (recall our discussion in Chapter 6). The tidal force on an observer would be smaller near a more massive black hole than near a less massive black hole: Just outside the event horizon of a massive black hole, the observer's feet would be only a little closer to the center (in comparison with the Schwarzschild radius) than the head.

If the tidal force could be ignored, the observer on the surface of the star would not notice anything particularly wrong locally as the star passed its event horizon, except that the observer's flashlight signal would never get out. (On the other hand, the view of space outside the event horizon would be highly distorted.)

Once the star passes inside its event horizon, it continues to contract, according to the general theory of relativity. Nothing can ever stop its contraction. In fact, the classical mathematical theory predicts that it will contract to zero radius—it will reach a **singularity.** Quantum effects, however, probably prevent it from reaching exactly zero radius.

Even though the mass that causes the black hole has contracted further, the event horizon doesn't change. It remains at the same radius forever, as long as the amount of mass inside is constant.

Note that if the Sun were turned into a black hole (by unknown forces), Earth's orbit would not be altered: The masses of the Sun and Earth would remain constant, as would the distance between them, so the gravitational force would be unchanged. Indeed, the gravitational field would remain the same everywhere outside the current radius of the Sun. Only at smaller distances would the force be stronger.

AceAstronomy™ Log into AceAstronomy and select this chapter to see the Active Figure called "Schwarzchild Radius of a Black Hole."

AceAstronomy™ Log into AceAstronomy and select this chapter to see the Astronomy Exercise "Escape Velocity."

ASIDE 14.3: Feeling stretched

An observer would not be torn apart outside the event horizon by the tidal force of a supermassive black hole, such as those discussed in Chapter 17; stellar-mass black holes are more dangerous. The stretching process near a stellar-mass black hole is sometimes informally called "spaghettification"!

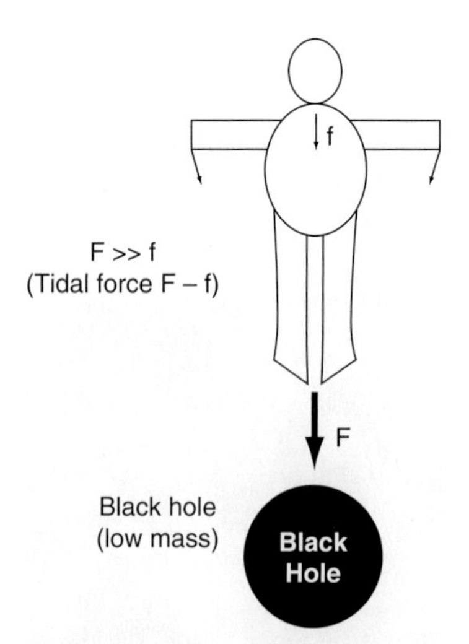

■ **FIGURE 14–8** If you are near a black hole, the force of gravity on your feet (long, bold arrow marked "F") exceeds that on your head (short arrow marked "f"), so you get "tidally" stretched, analogous to the Moon causing the Earth's tides (see our discussion in Chapter 6).

14.4 Time Dilation

According to the general theory of relativity, if you were far from a black hole and watching a space shuttle carrying some of your friends fall into it, your friends' clocks would appear to run progressively slower as they approached the event horizon. (Eventually they would get torn apart by tidal forces, as discussed above, but here we will ignore this complication.) From your perspective, time would be slowing down (or "dilated") for them; indeed, it would take an infinite amount of time (as measured by your clock) for them to reach the horizon. From your friends' perspective, in contrast, no such **time dilation** occurs; it takes a finite amount of time for them to reach and cross the event horizon, and shortly thereafter they hit the singularity.

On the other hand, if your friends were to approach the event horizon and subsequently escape from the vicinity of the black hole, they would have aged less than you did (for example, only 3 months instead of 30 years). This is a method for jumping into the future while aging very little!

However, this strategy doesn't increase longevity—your friends' lives would not be extended. Locally, they would not read more books or see more movies than you would in the same short time interval (3 months in our case). In contrast, while viewing you from the vicinity of the black hole, they would see you read many books and watch many movies, signs that you are aging much more than they are (30 years in our case).

ASIDE 14.4: Approaching the event horizon

Because of time dilation, even a collapsing star never reaches the event horizon, as seen by an outside observer. However, the star very quickly fades from sight: It emits light infrequently because of time dilation, the light becomes progressively more redshifted, and only the photons heading radially outward can eventually escape (that is, the exit cone is very narrow). Thus, everything that has ever fallen into the black hole seems to be in a thin, dark "membrane" slightly outside the event horizon.

14.5 Rotating Black Holes

Once matter is inside a black hole and reaches the singularity, it loses its identity in the sense that from outside a black hole, all we can tell is the mass of the black hole, the rate at which it is spinning (more precisely, its angular momentum), and what total electric charge it has. These three quantities are sufficient to completely describe the black hole, assuming it has reached an equilibrium configuration (that is, matter is not still falling asymmetrically toward the singularity).

Thus, in some respects, black holes are simple objects to describe physically, because we only have to know three numbers to characterize each one. The theorem that summarizes the simplicity of black holes is often colloquially stated by astronomers active in the field as "a black hole has no hair."

The theoretical calculations about black holes we have discussed in previous sections are based on the assumption that black holes do not rotate. But this assumption is only a convenience; we think, in fact, that the rotation of a black hole is one of its important properties. It took decades before Einstein's equations were solved for a black hole that is rotating. (The realization that the solution applied to a rotating black hole came after the solution itself was found.)

In this more general case, an additional special boundary—the **stationary limit**—appears, with somewhat different properties from the original event horizon. Within the stationary limit, no particles can remain at rest even though they are outside the event horizon.

The equator of the stationary limit of a rotating black hole has the same diameter as the event horizon of a nonrotating black hole of the same mass. But a rotating black hole's stationary limit is squashed. The event horizon touches the stationary limit at the poles. Since the event horizon remains a sphere, it is smaller than the event horizon of a nonrotating black hole (■ Fig. 14–9).

The space between the stationary limit and the event horizon is the **ergosphere.** This is the region in which particles cannot be at rest. In principle, we can get energy and matter out of the ergosphere. For example, if one sends an object into the ergosphere along an appropriate trajectory, and part of the object falls into the black hole

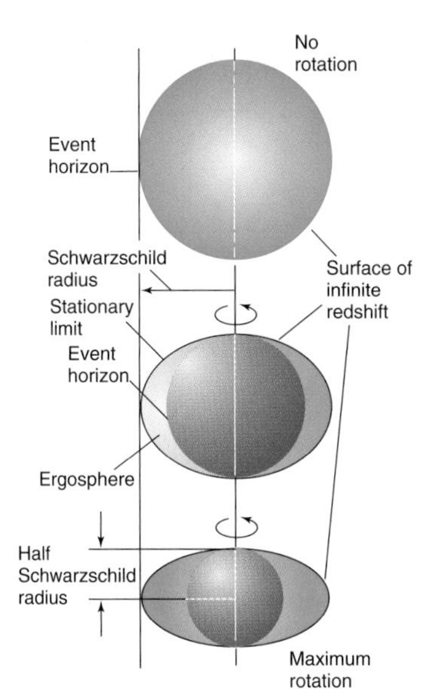

■ **FIGURE 14–9** Side views of black holes of the same mass and different rotation speeds. The region between the stationary limit and the event horizon of a rotating black hole is called the ergosphere (from the Greek word *ergon*, meaning "work") because, in principle, work can be extracted from it.

> **ASIDE 14.5: White holes**
>
> From our perspective, a nearby black hole can be thought of as being connected, through a wormhole, with a "white hole" at another location. A "white hole" is like the inverse of a black hole: Matter and radiation come out from it, but cannot go back in. We have no evidence for the existence of white holes in our Universe; they almost certainly do not exist.

at the right moment, the rest of the object can fly out of the ergosphere with more energy than the object initially had. The extra energy was gained from the rotation of the black hole, causing it to spin more slowly.

A black hole can rotate up to the speed at which a point on the event horizon's equator is travelling at the speed of light. The event horizon's radius is then half the Schwarzschild radius. If a black hole rotated faster than this, its event horizon would vanish. Unlike the case of a nonrotating black hole, for which the singularity is always unreachably hidden within the event horizon, in this case distant observers could receive signals from the singularity.

Such a point is called a **naked singularity.** Most theorists assume the existence of a law of "cosmic censorship," which requires all singularities to be "clothed" in event horizons—that is, not naked. Since so much energy might erupt from a naked singularity, we can conclude from the fact that we do not find signs of them that there are probably none in our Universe. Thus, each black hole cannot exceed its maximum rate of rotation (or a naked singularity could be seen).

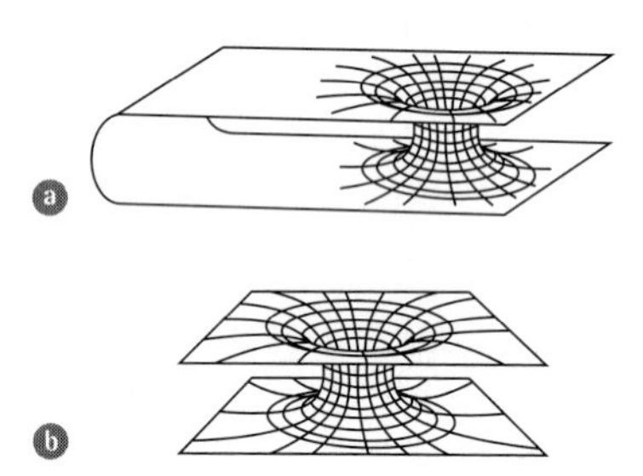

■ **FIGURE 14–10** Wormholes hypothetically connect black holes in distant parts of (a) our Universe, or (b) perhaps even in different universes.

14.6 Passageways to Distant Lands?

In science fiction movies (such as *Contact*), it is often claimed that one can travel through a black hole to a distant part of our Universe in a very short amount of time, or perhaps even travel to other universes. This misconception arises in part from diagrams such as ■ Figure 14–10 for two nonrotating black holes. One black hole is connected to another black hole by a tunnel, or **wormhole** (officially called an "Einstein-Rosen bridge"), and it appears possible to traverse this shortcut.

However, this map is misleading; it does not adequately describe the structure of space–time inside a black hole. In particular, one would need to travel through space faster than the speed of light (which nobody can do) to avoid the singularity and end up in a different region. Thus, nonrotating black holes definitely seem to be excluded as passageways to distant lands.

In the case of a *rotating* black hole, on the other hand, travel through the wormhole at speeds slower than that of light, avoiding the singularity, initially seems feasible. In fact, it appears as though one could travel back to one's starting point in space, possibly arriving at a time prior to departure! This is quite disturbing, since "causality" could then be violated. For example, the traveller could affect history in such a way that he or she would not have been born and could not have made the journey!

More detailed analysis, however, shows that this favorable geometry of a rotating black hole is only valid for an *idealized* black hole into which no material is falling (or has previously fallen). As soon as an object actually tries to traverse the wormhole, the passageway closes! One would need to have a very exotic form of matter with antigravitating properties to keep the wormhole open. There is no evidence for the existence of such matter, at least not in the form where it can be gathered. (In Chapter 18, we will introduce the concept of "dark energy," which causes the expansion of the Universe to accelerate. But this energy is uniformly spread throughout space, and cannot be concentrated into a small volume.)

"Would you tell me, please, which way I ought to go from here?"
"That depends a good deal on where you want to get to," said the Cat.
"I don't much care where—" said Alice.
"Then it doesn't matter which way you go," said the Cat.
"—so long as I get *somewhere*," Alice added as an explanation.
"Oh, you're sure to do that," said the Cat, "if only you walk long enough."

■ **FIGURE 14–11** Lewis Carroll's Cheshire Cat, from *Alice in Wonderland*, shown here in John Tenniel's drawing. The Cheshire Cat is analogous to a black hole in that the cat left its grin behind when it disappeared, just as a black hole leaves its gravity behind when it disappears from view. Alice thought that the Cheshire Cat's persisting grin was "the most curious thing I ever saw in my life!" We might say the same about a black hole and its persisting gravity.

14.7 Detecting a Black Hole

A star collapsing to become a black hole would blink out in a fraction of a second, so the odds are unfavorable that we would actually see the crucial stage of star collapse as it approached the event horizon. And a black hole is too small to see directly. But all hope is not lost for detecting a black hole. Though the black hole disappears, it leaves its gravity behind. It is a bit like the Cheshire Cat from *Alice in Wonderland*, which fades away leaving only its grin behind (■ Fig. 14–11).

14.7a Hot Accretion Disks

Suppose a black hole is orbited by a normal star that fills its "Roche lobe" (see Fig. 13–9 in Chapter 13)—that is, the region in which its gravitational pull dominates over that of the black hole. Gas crossing the intersection point in the "figure 8" will begin to accelerate toward the black hole; like all objects in the Universe, the black hole attracts matter.

Some of the gas will be pulled directly into the black hole, never to be seen again. But other matter will go into orbit around the black hole, and will orbit at a high speed. This added matter forms an **accretion disk** (■ Fig. 14–12*a*); "accretion" is growth in size by the gradual addition of matter. We previously encountered accretion disks in Chapter 13, when we discussed novae (Section 13.1e) and Type Ia supernovae (Section 13.2b).

It seems likely that the gas in orbit near the black hole will be heated to a very high temperature by friction, and by compression into a very small volume. The gas radiates strongly in the x-ray region of the spectrum, giving off bursts of radiation sporadically as clumps develop in the disk. The inner 200 km should reach ten million kelvins. Thus, though we cannot observe the black hole itself, we can hope to observe x-rays from the gas surrounding it.

A NASA satellite known as the Rossi X-ray Timing Explorer is measuring such rapid changes in x-rays, and it detects a number of interesting flickering x-ray sources. The Chandra X-ray Observatory and the XMM-Newton Mission, both launched in 1999, are finding still more black-hole candidates.

In fact, a large number of x-ray sources are binary-star systems. Some may contain black holes, but the majority of them probably have neutron stars. To be certain that you have a black hole, it is not enough to find an x-ray source that gives off sporadic pulses, for both black holes and neutron stars can produce such pulses. We also have to show that a collapsed star of greater than 3 solar masses is present.

14.7b Cygnus X-1: The First Plausible Black Hole

We can determine masses only for certain binary stars. When we search the position of the x-ray sources, we look for a single-lined spectroscopic binary (that is, an apparently single star whose spectrum shows a periodically changing Doppler shift that indicates the presence of an invisible companion; we discussed such binary stars in Chapter 11). Then, if we can show that the companion is too faint to be a normal, main-sequence star, it must be a collapsed star.

If, further, the mass of the unobservable companion is greater than 3 solar masses, it is likely to be a black hole, assuming that the general theory of relativity is the correct theory of gravity and that our present understanding of matter is correct. To definitively show that the black-hole candidate is indeed a black hole, however, additional evidence is needed, as we will discuss below.

The first and most discussed, though no longer the most persuasive, case is named Cygnus X-1 (Fig. 14–12), where X-1 means that it was the first x-ray source to be

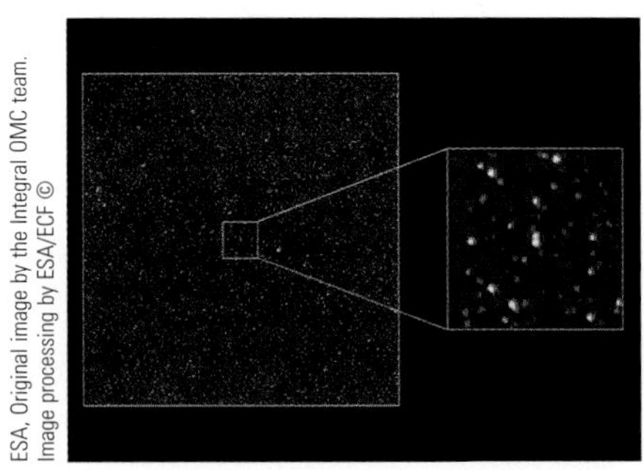

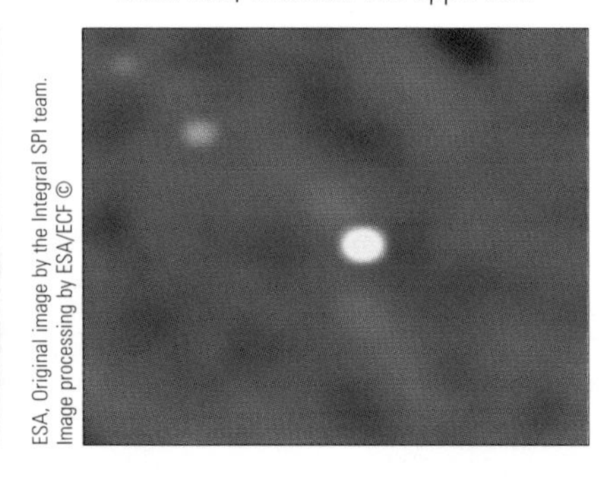

■ **FIGURE 14–12** The black hole candidate Cygnus X-1. ⓐ An artist's conception of the binary system, where gas from the visible, blue supergiant star HDE 226868 is swirling over to and around the black hole. The mutual orbits around each other take 5.6 days. ⓑ The region of the source imaged with the Optical Monitoring Camera on the European Space Agency's Integral satellite, which was launched in 2002 to study gamma rays. ⓒ Gamma rays from the accretion disk surrounding the black hole in Cygnus X-1, viewed with the Spectrometer on Integral (SPI). The binary x-ray source Cygnus X-3, which involves a neutron star rather than a black hole, is toward the upper left.

discovered in the constellation Cygnus. A 9th-magnitude star called HDE 226868 has been found at its location. This star has the spectrum of a blue supergiant; its mass is uncertain but is thought to be about 20 times that of the Sun. Its radial velocity varies with a period of 5.6 days, indicating that the supergiant and the invisible companion are orbiting each other with that period.

From the orbit, it is deduced that the invisible companion must probably have a mass greater than 7 solar masses and less than about 13 solar masses. This range makes it very likely that it is a black hole. But if the visible star is abnormal, and doesn't really have a mass of 20 times that of the Sun, the invisible companion could be less massive, so the case for it being a black hole is not absolutely conclusive.

ASIDE 14.6: A black-hole bubble

Radio astronomers have discovered a rapidly expanding 10-light-year-diameter bubble around Cygnus X-1, and optical astronomers have seen it as well. Its boundary is a shock wave formed where the jet hits the interstellar medium. This discovery confirms that black holes affect the space around them in more ways than with just their gravity.

14.7c Other Black-Hole Candidates

Although Cygnus X-1 provides good evidence for the existence of stellar-mass black holes in our Galaxy, the large (and uncertain) mass of the visible star complicates the measurement of the mass of the dark object. More conclusive evidence for black holes comes from binary systems in which the visible star has a very small mass (for example, a K- or M-type main-sequence star); uncertainties in its mass are nearly irrelevant (see *Figure It Out 14.1: Binary Stars and Kepler's Third Law*).

However, when such systems brighten suddenly at x-ray wavelengths, light from the accretion disk also dominates at other wavelengths, and the low-mass star becomes too difficult to see. Observers must wait until the outburst subsides, and the companion star once again becomes visible, before they are able to measure the orbital parameters.

The first well-studied system of this type is A0620–00, which is in the constellation Monoceros. During an x-ray outburst in 1975, the source was even brighter than Cygnus X-1, but it subsequently faded and the properties of the visible star were measured. The derived minimum mass of the invisible companion was 3.2 Suns, close to but above the limit at which a collapsed star could be a neutron star instead of a black hole.

In 1992, an even more convincing case was found: The dark object in the x-ray binary star V404 Cygni has a mass of at least 6 Suns, but probably closer to 12 solar masses (■ Fig. 14–13). One of the authors (A.F.) has found four additional black-hole candidates with more than 5 times the Sun's mass, as well as two somewhat weaker

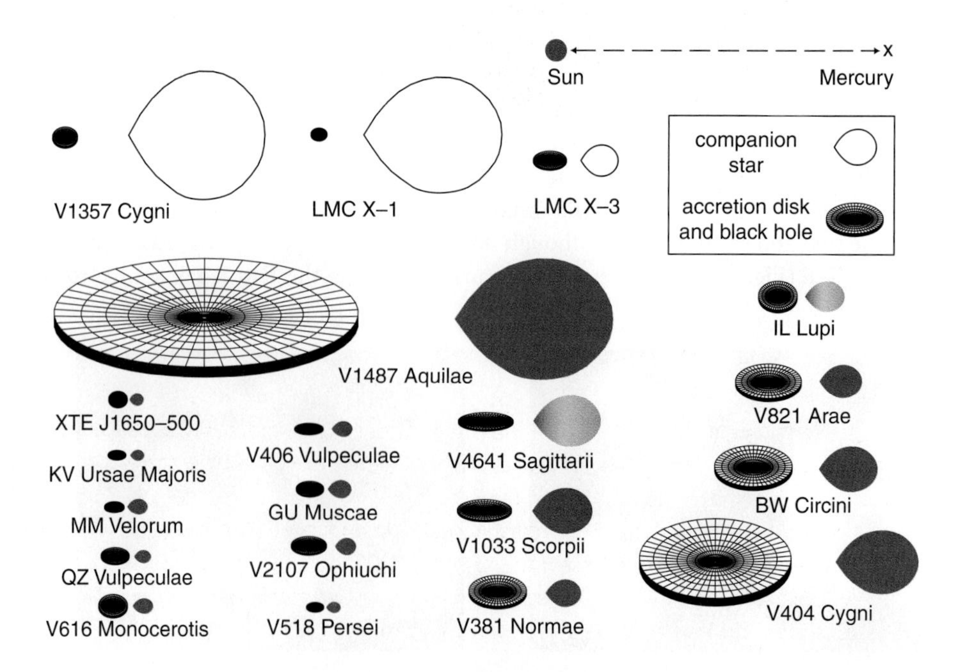

■ **FIGURE 14–13** The sizes of the accretion disks, companions, and their separations for some of the leading black-hole candidates. (Jerome A. Orosz, San Diego State U.)

FIGURE IT OUT 14.1

Binary Stars and Kepler's Third Law

Consider Kepler's third law, as generalized by Newton and discussed in Chapter 5:

$$(m_1 + m_2)P^2 = (4\pi^2/G)R^3.$$

If m_1 (the mass of the invisible object) is much larger than m_2 (the mass of the visible star), then we can safely ignore m_2 in the sum. Thus

$$m_1 P^2 = (4\pi^2/G)R^3, \text{ or } m_1 = (4\pi^2/GP^2)R^3,$$

regardless of the exact value of m_2.

If the visible star has a circular orbit of radius R, it travels the circumference ($2\pi R$) in one orbital period (P) at speed v, so

$$2\pi R = vP \ (\textit{distance} = \textit{speed} \times \textit{time}).$$

Dividing by 2π, we have $R = vP/2\pi$. Substituting this expression for R into the previous equation for m_1, we find

$$m_1 = (4\pi^2/GP^2)(vP/2\pi)^3 = v^3P/2\pi G.$$

If v and P are respectively measured as the amplitude (height) and period of the radial-velocity curve (observed speed versus time) of the visible star, then m_1 is found from this equation. Actually, this value is merely the minimum mass that the invisible object can have, since the measured radial velocity is less than the true orbital velocity if the orbit isn't exactly perpendicular to the line of sight. Thus we aren't actually able to find the true mass without more information.

For example, the figure below shows the radial-velocity curve for the visible star in an x-ray binary system. When the star is moving most directly toward us or away from us, we see the maximum blueshift or redshift of the absorption lines, −520 km/sec and +520 km/sec, respectively. At other times in the circular orbit, a smaller part of the velocity is radial, so intermediate values are measured, but we conclude that v = 520 km/sec for this system. The orbital period is measured to be only 8.3 hours. Using these values in the equation $m_1 = v^3P/2\pi G$ (and being careful with units!), we find that m_1 = 5.0 solar masses. This is just the minimum mass of the dark object; further considerations suggest that the true mass is 8 to 9 Suns. Hence, the visible star's companion is almost certainly a black hole.

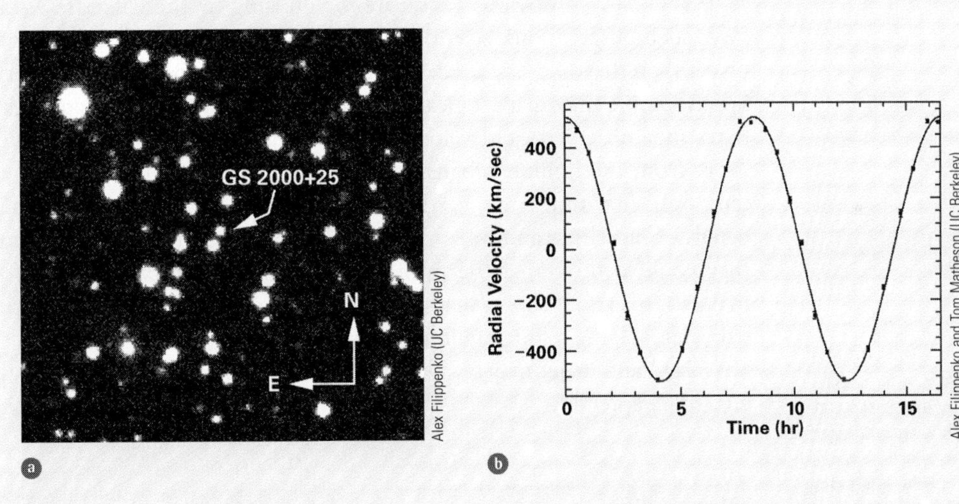

a An image of the visible star in the x-ray binary system known as GS 2000+25, long after it had faded from its x-ray outburst in 1988. b Radial-velocity curve of the visible star, obtained by one of the authors (A.F.) with the Keck-I 10-m telescope.

cases for black holes. By mid-2005, about two dozen good or excellent stellar-mass black-hole candidates in binary systems had been identified and measured in our Galaxy.

Astronomers are now trying to conclusively demonstrate that these systems do indeed contain black holes. After all, the argument thus far has been somewhat indirect: We think black holes are present partly because we don't know of any kind of object denser than a neutron star but not quite a black hole.

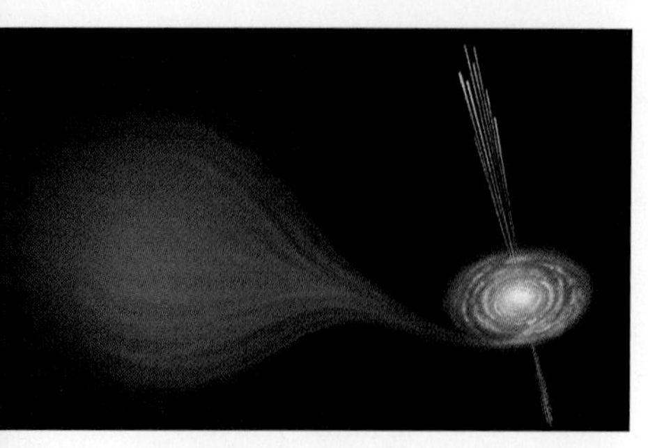

■ **FIGURE 14–14** A model of SS433 in which the radiation emanates from two narrow beams of matter that are given off by the disk of matter orbiting the compact object (a neutron star or a black hole). When we discuss active galaxies in Chapter 17, we will see that high-speed jets are not unusual on the much larger galactic scale.

From a model of Modecai Milgrom, Bruce Margon, Jonathon Katz, and George Abell

One of the most intriguing tests is to see whether the material from the accretion disk lands on a hard stellar surface, releasing its considerable kinetic energy in the form of electromagnetic radiation, or instead passes quietly through an event horizon, never to be seen again. So far, the results seem to favor the presence of an event horizon rather than a stellar surface in several black-hole candidates, but more research is necessary in this area.

Recently, a few solitary black holes (not in binary systems) may have been detected as they gravitationally bent and magnified light from more distant stars along our line of sight. This phenomenon of "gravitational lensing" is discussed in Chapter 16.

14.7d The Strange Case of SS433

One of the oddest x-ray binaries is known as SS433, based on its number in a catalogue. It is in the constellation Aquila, the Eagle, and is about 16,000 light-years away. From measurements of Doppler shifts, we detect gas coming out of this x-ray binary at about 25 per cent of the speed of light, a huge speed for a source in our Galaxy.

Until recently, the most widely accepted model (■ Fig. 14–14) considered SS433 to be a neutron star surrounded by a disk of matter it has taken up from a companion star. Measurements of the Doppler shifts in optical light show us light coming toward us from one jet and going away from us from the other jet at the same time. The disk would wobble like a top (a precession, similar to the one we briefly discussed for the Earth's axis in Chapter 4). As it wobbles, the apparent to-and-fro velocities decrease and increase again, as we see the jets from different angles. We have even detected the jets in radio waves and x-rays (■ Fig. 14–15).

SS433 was an obvious candidate for imaging by the Chandra X-ray Observatory. With such observations, scientists found that the jets originated closer to the compact object than they had expected, only about 0.3 A.U., closer than Mercury is to the Sun. Spectroscopic information gleaned with Chandra showed temperatures (from the isotopes they observed) and velocities in the jets; they measured that the gas in the jet drops from 100 million kelvins to 10 million kelvins within about two million kilometers, only five times the Earth–Moon distance. From this information, and from details of how long and how often the companion star blocks the compact x-ray star, they could deduce that the probable mass of the compact object is about 16 solar masses. This result shows that the object is likely to be a black hole rather than a neutron star, solving a long-standing question.

SS433 and its jet may be a relatively close-by analog to the quasars that we will discuss in Chapter 17.

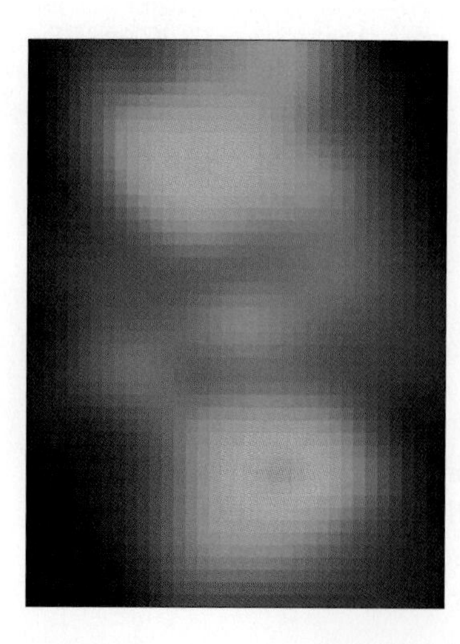

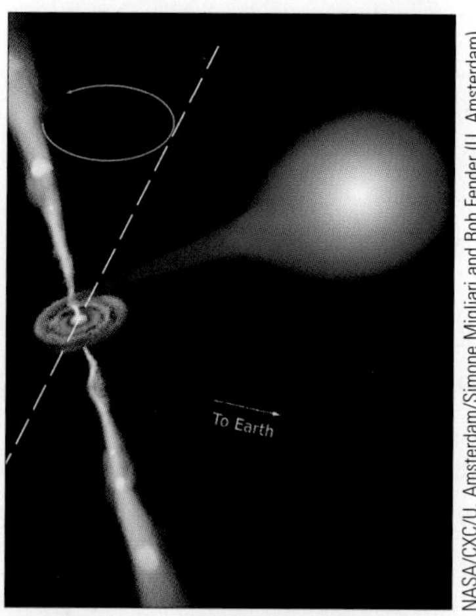

NASA/CXC/U. Amsterdam/Simone Migliari and Rob Fender (U. Amsterdam) and Mariano Mendez (National Institute for Space Research, SRON)

■ **FIGURE 14–15** An x-ray image of SS433 from the Chandra X-ray Observatory, accompanied by an artist's conception of how the lobes are produced. We see a binary system, which is thought to contain a massive star (bloated blue balloon-shaped object) and a black hole with its accretion disk. Jets emitted from the accretion disk produce the lobes. The detection of iron so highly ionized that only one of its electrons remains indicates that the temperature of the gas is 50 million kelvins. The jets wobble *(blue circular arrow)* between the positions shown and the dotted line. Finding the lobes so hot so far out (1/4 light-year) implies that the gas in them has been reheated. Apparently, blobs of gas ejected from near the black hole catch up with earlier ejections after travelling outward for months at about 1/4 the speed of light (as measured by the Doppler effect). The pileup reheats the gas.

14.8 Supermassive Black Holes

Although the stellar-mass black holes discussed above provided the best initial evidence for the existence of black holes, we now have firm observational support for black holes containing millions or billions of solar masses. Let us discuss these **supermassive black holes** briefly in this black-hole chapter, and say more about them in Chapter 17, when we consider quasars and active galaxies.

The more mass involved, the lower the density needed for a black hole to form. For a very massive black hole, one containing hundreds of millions or even billions of solar masses, the density would be so low when the event horizon formed that it would be close to the density of water.

Thus if we were travelling through the Universe in a spaceship, we couldn't count on detecting a black hole by noticing a volume of high density, or by measuring large tidal effects. We could pass through the event horizon of a high-mass black hole without being stretched tidally to oblivion, though visual effects would certainly be bizarre. We would never be able to get out, but it would be over an hour before we would notice that we were being drawn into the center at a rapidly accelerating rate.

Where could such a supermassive black hole be located? The center of our Milky Way Galaxy almost certainly contains a black hole of about 3.7 million solar masses. Though we do not observe radiation from the black hole itself, the gamma rays, x-rays, and infrared radiation we detect come from the gas surrounding the black hole. (See the image opening this chapter.)

Other galaxies and quasars (see Chapter 17) are also probable locations for massive black holes. The Hubble Space Telescope is being used to take images of galaxies with the highest possible resolution, and is finding extremely compact, bright cores at the centers of some of them. Many of these cores probably contain black holes that have a large concentration of stars around them, making their central regions unusually luminous.

The Hubble Space Telescope has also been able to take spectra very close to the centers of certain galaxies, though that particular instrument on the Hubble is currently broken. The Doppler shifts on opposite sides of the galaxies' cores, especially in disks of gas surrounding these cores, show how fast these points are revolving around the centers of the galaxies (■ Fig. 14–16). The speed, in turn, shows how much mass is present in such a small volume. Only a giant black hole can have so much mass in such a small volume.

It now seems that most galaxies have a supermassive black hole at their centers. A majority of these galaxies look relatively normal, but some are "active" (Chapter 17), giving off exceptionally large amounts of electromagnetic radiation from very small volumes. The Chandra X-ray Observatory, the XMM-Newton Mission, and the Hubble Space Telescope, working together, are greatly increasing our knowledge of such supermassive black holes. Chandra has found evidence of a jet in one galaxy (■ Fig. 14–17) and confirmed the presence of a supermassive black hole at the center of the Andromeda Galaxy (■ Fig. 14–18). X-ray spectra it took showed features that very strongly suggested that the objects are black holes.

Radio techniques that provide very high resolution can be used to study the jets of gas emitted from the vicinity of supermassive black holes. The jet from the galaxy M87, the central galaxy in the Virgo Cluster of galaxies, has been imaged in various

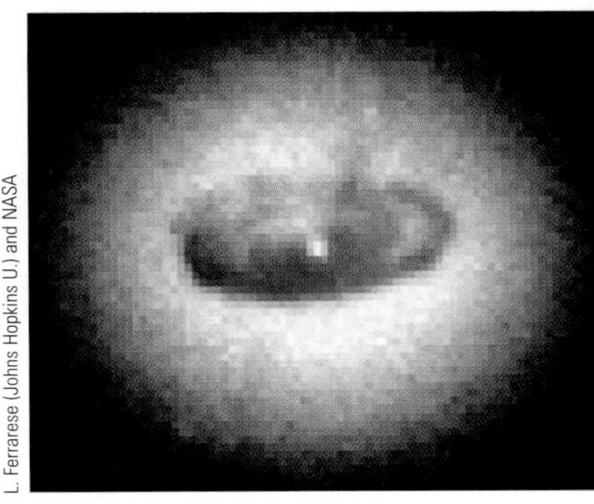
L. Ferrarese (Johns Hopkins U.) and NASA

■ **FIGURE 14–16** A Hubble Space Telescope image of a spiral-shaped disk of dust fueling a massive black hole in the center of the galaxy NGC 4261. The disk shown is 800 light-years wide and is 100 million light-years from us. Astronomers have measured the speed of the gas swirling around the central object. The object must have 1.2 billion times the mass of our Sun in order to have enough gravity to keep the orbiting gas in place. But it is concentrated into a region of space not much larger than our Solar System. Only a black hole easily meets both these requirements.

NASA/Chandra X-ray Center/SAO

■ **FIGURE 14–17** A Chandra X-ray Observatory view of an elongated cloud of gas in the center of the galaxy NGC 4151. A leading theory for the elongated shape is that we are seeing a jet of x-rays emitted from the vicinity of a central supermassive black hole. The colors represent x-ray intensities.

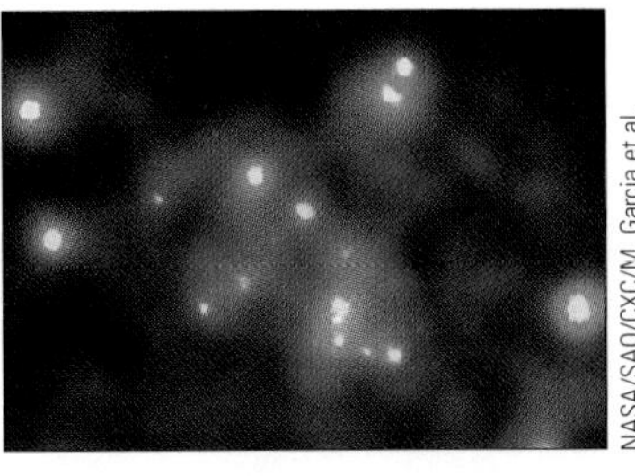
NASA/SAO/CXC/M. Garcia et al.

■ **FIGURE 14–18** A Chandra X-ray Observatory view of a tiny region at the core of the Andromeda Galaxy. A hotter object 10 light-years north of (that is, slightly above) the relatively cool object shown in blue in this false-color image is at the very center of the galaxy and is deduced to be a black hole about 100 million times more massive than the Sun.

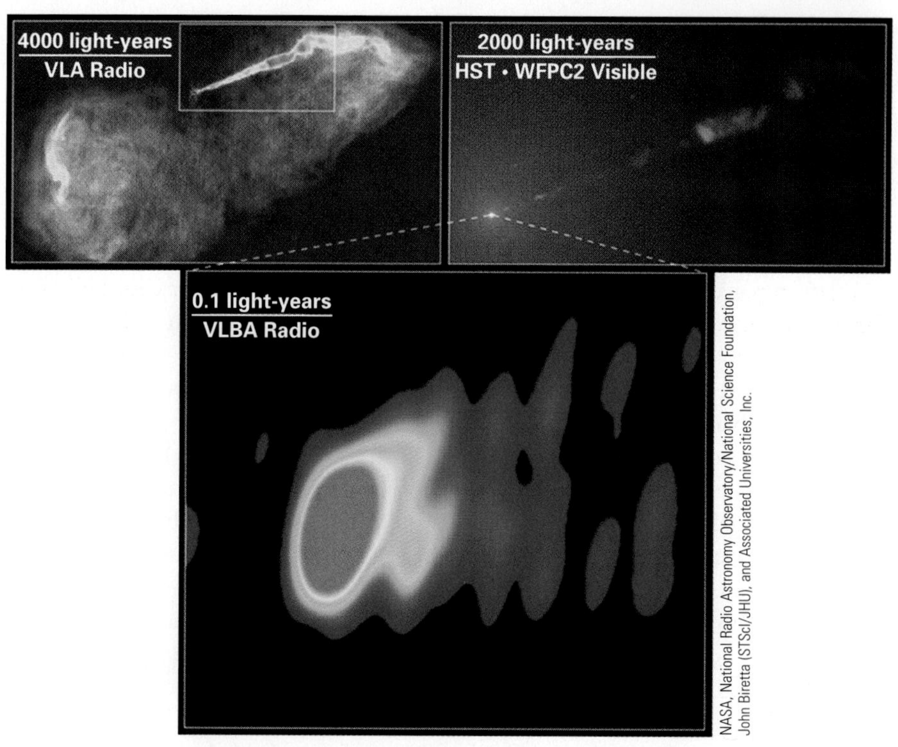

NASA, National Radio Astronomy Observatory/National Science Foundation, John Biretta (STScI/JHU), and Associated Universities, Inc.

■ **FIGURE 14–19** A combination of observations in different parts of the spectrum reveals the region of formation of the jet at the center of the giant elliptical galaxy M87 (an object that we will discuss further in Chapter 17). A radio image made with the Very Large Array (VLA) of radio telescopes in New Mexico shows the jet, which is imaged in visible light with the Hubble Space Telescope. The region of origin of that jet can be imaged in radio waves with the Very Long Baseline Array (VLBA), a network of radio telescopes spanning the globe, giving the resolution of a single radio telescope close to the size of the Earth. The color coding shows the intensity of the radiation recorded.

ways (■ Fig. 14–19). The image has been traced so close to the central object that it shows a widening of the jet. This widening seems to indicate that the jet comes from the accretion disk rather than from the central black hole itself.

14.9 Moderation in All Things

Until recently, many scientists doubted that there were black holes with masses intermediate between those of stars—say, 10 times the Sun's mass—and those in the centers of all or most galaxies—perhaps a million to a billion times the Sun's mass. But one of the latest advances in black-hole astrophysics is the discovery of so-called **intermediate-mass black holes,** with masses of "only" 100–10,000 times the Sun's, bridging this gap. Chandra X-ray images reveal luminous, flickering sources in some galaxies, suggesting that intermediate-mass black holes are accreting material sporadically from their surroundings.

One possible place for such black holes to form is in the centers of dense clusters of stars, perhaps even young globular clusters, which retain the black holes after they form. Indeed, in 2002, measurements of the motions of stars in the central regions of two old globular clusters seemed to imply the presence of intermediate-mass black holes, although the interpretation of the data is still somewhat controversial.

■ **FIGURE 14–20** A visible-light image of the galaxy M74, the 74th object in Charles Messier's catalogue, with the position of an ultraluminous x-ray source (ULX) marked. This source is thought to be a black hole of about 10,000 times the Sun's mass because of its quasi-periodic two-hour variation. The red spots are x-ray sources observed with Chandra.

X-ray: NASA/CXC/U. of Michigan/J. Liu et al.; Optical: NOAO/AURA/NSF/T. Boroson

Chandra observations of one x-ray source in the galaxy M74 (■ Fig. 14–20) discovered strong variations in its x-ray brightness that repeated almost periodically about every two hours. Because they are almost, but not quite, periodic, the phenomenon is known as "quasi-periodic oscillations." The object is one of many known as "ultraluminous x-ray sources," which are 10 to 1000 times stronger in x-rays than neutron stars or black holes of stellar mass. These sources are being observed from both NASA's Chandra and the European Space Agency's XMM-Newton.

The question has been whether this type of source was an intermediate-mass black hole or a stellar-mass black hole for which we are looking straight into the jet, increasing the x-ray intensity. The existence of quasi-periodic oscillations tends to show that the intermediate-mass idea is the actual situation.

Observations of many black-hole x-ray sources of all types have shown that the quasi-period of oscillations is linked to the mass of the black hole. Based on this relation, this source in M74 would have a mass about 10,000 times that of our Sun. Such intermediate-mass black holes, it is thought, would have resulted from a merger of perhaps hundreds of stellar-mass black holes in a star cluster, or are nuclei of small galaxies that are being incorporated in (one might say "eaten" by) a larger galaxy.

14.10 Gamma-Ray Bursts: Birth Cries of Black Holes?

One of the most exciting astronomical fields in the middle of this first decade of the new millennium is the study of **gamma-ray bursts.** These events are powerful bursts of extremely short-wavelength radiation, lasting only about 20 seconds on average. Each gamma-ray photon, to speak in terms of energies instead of wavelength, carries a relatively large amount of energy. Gamma-ray bursts flicker substantially in gamma-ray brightness and each of them has a unique light curve (■ Fig. 14–21). The joke so far has been "If you see one gamma-ray burst, you've seen one gamma-ray burst," a far cry from "you've seen them all."

14.10a How Far Away Are Gamma-Ray Bursts?

Rarely have astronomers had ideas as wrong as the first ones about gamma-ray bursts. Indeed, in previous editions of this textbook, we discussed gamma-ray bursts in the chapter on the Milky Way Galaxy, since that is where people first thought they were coming from. But now we know that most of them are in extremely distant galaxies, typically billions of light-years away. To appear as bright as they do to us (or to our gamma-ray satellites), they must come from exceptionally powerful explosions—and indeed, we now think they represent the formation of black holes during certain kinds of stellar collapse.

The gamma-ray burst story started with the "Vela" military satellites used in the late 1960s to search for gamma rays that might come from surreptitious atomic-bomb tests. Some bursts of gamma rays were indeed detected by the orbiting satellites, but

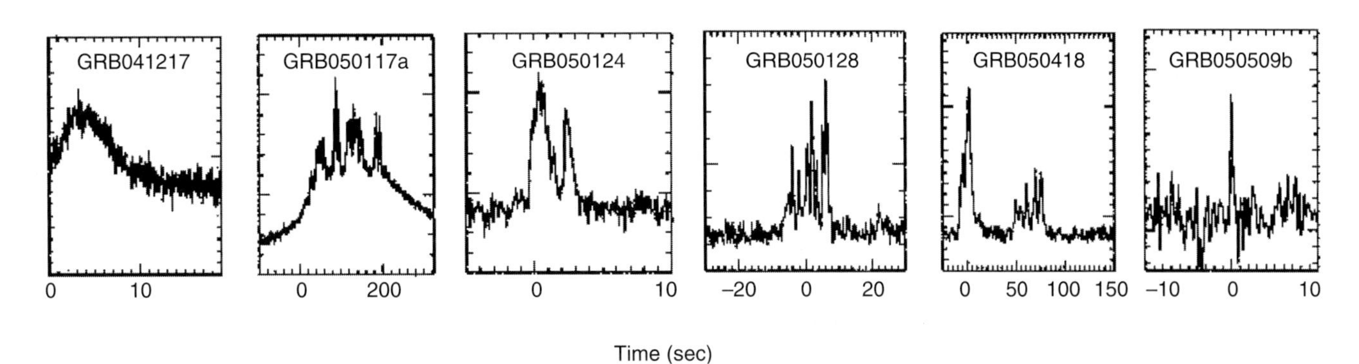

■ **FIGURE 14–21** A montage of six gamma-ray burst light curves from the Swift satellite. The gamma-ray brightness is plotted against time. Each of the light curves differs from the others. (NASA's Swift Mission, courtesy of Lynn Cominsky, Sonoma St. U.)

A. Fruchter and NASA

■ **FIGURE 14–22** A Hubble Space Telescope image of the optical afterglow of gamma-ray burst 990123. At its peak, the optical afterglow was about four million times brighter and could be seen through a good pair of binoculars, despite being in a galaxy over 9 billion light-years away from Earth. The host galaxy shows signs of having recently formed a substantial number of very massive stars.

they came from above, not from below. Moreover, their properties seemed inconsistent with the explosions of nuclear bombs. However, since gamma-ray telescopes don't give accurate positions, astronomers were unable to link the bursts to specific astronomical objects.

As the data gradually accumulated, it became clear that gamma-ray bursts were not associated with the plane of our Milky Way Galaxy, unlike most bright stars or neutron stars (pulsars; see Figure 13–28 for the latter). This meant that gamma-ray bursts come from either a very large, spherical halo around our Galaxy or from far outside our Galaxy, perhaps from extremely distant galaxies; otherwise they would have shown some variation with respect to the plane of the Milky Way.

In the mid-1990s, an Italian satellite named BeppoSax ("Beppo" was the nickname of a distinguished Italian scientist, "Sa" was for satellite, and the "x" was for "x-ray") was able to swiftly reposition itself when it sensed a gamma-ray burst to point an x-ray telescope at that part of the sky. The x-ray telescope on board, in turn, gave a good enough position to allow optical satellites on the ground and in space to look at that region. In some cases, an optical "afterglow" was visible at the same position. Additional studies with the largest telescopes, including the Keck telescopes in Hawaii and the Hubble Space Telescope (■ Fig. 14–22), revealed that these afterglows were coincident with faint galaxies billions of light-years away. Thus, gamma-ray bursts must be incredibly powerful to appear as bright as they do from such large distances.

In 2004, NASA launched a new satellite, Swift, to discover gamma-ray bursts and swiftly turn toward them, obtaining accurate positions with x-ray and ultraviolet/optical telescopes. It has been used together with NASA's High-Energy Transient Explorer 2 (HETE-2) to study both gamma-ray bursts and x-ray flashes, similar phenomena that give off most of their bursts of radiation in their respective parts of the spectrum.

Both HETE-2 and Swift immediately relay the positions of gamma-ray bursts to ground-based observatories via the Internet. A network of small telescopes, the Robotic Optical Transient Experiment (ROTSE), spanning the world so that at least one is always in darkness, can study the associated optical burst, sometimes even while the gamma-ray burst is still going on! Several other telescopes have also been automated to make follow-up observations of gamma-ray bursts. For example, one of the authors (A.F.) operates a robotic telescope at Lick Observatory (see *A Closer Look 13.1: Search for Supernovae*) that is able to slew over to the position of a gamma-ray burst and take relatively deep images within a few tens of seconds after the outburst (■ Fig. 14–23).

a

b

Weidong Li and Alex Filippenko, University of California, Berkeley

■ **FIGURE 14–23** ⓐ An image of the optical afterglow of gamma-ray burst 021211, obtained only a few minutes after the burst with a 0.76-meter robotic telescope at Lick Observatory. ⓑ An image of the same region of the sky, taken several hours later with the same telescope. Note how much the optical afterglow has faded.

14.10b Models of Gamma-Ray Bursts

Analysis of observations of gamma-ray bursts shows convincingly that their radiation is highly beamed, like that of pulsars (Section 13.3c)—it cannot be emitted uniformly over the sky. A leading hypothesis is known as the "fireball model," in which some sort of compact engine releases blobs of material at speeds close to that of light along a narrow "jet." Rapidly moving blows overtake more slowly moving ones, creating internal shocks within the jet that release the gamma rays. When the blobs subsequently collide with surrounding gas, additional shocks are produced and emit the afterglow visible at wavelengths other than gamma rays. The afterglow fades with time as the shock's energy dissipates.

How are these "relativistic jets" formed? A clue comes from the fact that at least some gamma-ray bursts, those with long duration (over 2 seconds) and characterized by lower-energy gamma rays, often appear in galaxies undergoing vigorous bursts of massive-star formation. Because massive stars live only short lives, such stars are also dying in those galaxies. The jet may have something to do with the rotation of the massive collapsing star known as a *collapsar.*

If the massive star is rotating just before its core collapses, the collapsed part will spin even faster than before because of the conservation of angular momentum, a concept we have already encountered several times in this book. A disk will form, as in the formation of planetary systems. Since escaping material has a hard time going through the disk, it flows along the axis of rotation instead. If the star does not have a thick envelope, the material might burst all the way through the envelope (■ Fig. 14–24).

Recall (Section 13.2a) that massive stars without hydrogen envelopes, or even without helium envelopes, are thought to produce Type Ib and Type Ic supernovae, respectively (Fig. 13–15). Thus, perhaps long-duration gamma-ray bursts are essentially these kinds of supernovae, but with the core collapsing all the way to form a rapidly rotating black hole instead of just a neutron star, thereby releasing an even larger amount of gravitational energy. The rotation energy of the black hole and the surrounding disk might also be tapped, as discussed in Section 14.5. If the jet breaks through the star's surface and happens to be pointing at Earth, we will see an incredibly bright flash, especially because the brightness of the relativistic material moving along the line of sight is enhanced.

■ **FIGURE 14–24** Artist's conception of the collapsar model of a gamma-ray burst. A rotating, massive star largely devoid of its hydrogen and helium envelopes collapses at the end of its life, forming a black hole and two oppositely directed jets along the star's rotation axis. The jets break out of the star, and if our line of sight is along one of them, we see a gamma-ray burst.

An expectation of the collapsar model is that a Type Ib or Ic supernova should become visible at the location of the gamma-ray burst when its afterglow has faded sufficiently. During the past few years, compelling evidence has indeed been found for Type Ic supernovae associated with several long-duration gamma-ray bursts. A few of these supernovae show signs of being especially powerful, prompting some astronomers to call them "hypernovae."

There are some other viable models for gamma-ray bursts, especially the short-duration bursts (less than 2 seconds) characterized by higher-energy gamma rays, which have not been studied as extensively as long-duration gamma-ray bursts. These objects are currently thought to come from collisions of collapsed objects, either two neutron stars (forming a black hole) or a neutron star and a black hole (increasing the mass of the black hole). For example, a tight (short separation) binary neutron star system emits gravitational waves (Section 13.3f), with the neutron stars spiraling toward each other and eventually merging to form a black hole. The neutron stars would become torn apart near the final stages of merging, forming a disk with an axis of rotation. A jet of relativistic particles and radiation could then emerge along that axis.

Although currently the best bet is that gamma-ray bursts are associated with the formation of stellar-mass black holes, there is lots of work left to do in this field. Swift and other telescopes will record many more bursts; Swift is finding one every three days or so. Maybe more similarities among different gamma-ray bursts will become clear. Robotic ground-based telescopes will conduct successful follow-up observations at other wavelengths, with progressively shorter time delays after the gamma-ray burst discoveries. Speeding up the rate at which astronomy is done is unusual and difficult. Gamma-ray bursts are but the latest of violent phenomena discovered in the Universe, which is by no means as placid as had been long thought.

ASIDE 14.7: Black-hole evaporation

Hawking's quantum process should apply to any black hole. However, the rate of evaporation is negligible for all but the smallest black holes. Indeed, in most cases it is easily exceeded by the rate at which a black hole swallows (accretes) surrounding gas.

14.11 Mini Black Holes

We have discussed how black holes can form by the collapse of massive stars. But theoretically a black hole should result if a mass of any amount is sufficiently compressed. No object containing less than 2 or 3 solar masses will contract sufficiently under the force of its own gravity in the course of stellar evolution. The density of matter was so high at the time of the origin of the Universe (see Chapter 19) that smaller masses may have been sufficiently compressed to form **mini black holes.**

■ **FIGURE 14–25** Stephen Hawking, who is the Lucasian Professor of Mathematics at Cambridge University, the position once held by Isaac Newton. Among his many theoretical ideas were black-hole radiation and the evaporation of mini black holes.

Stephen Hawking (■ Fig. 14–25), an English astrophysicist, has suggested their existence. Mini black holes the size of pinheads would have masses equivalent to those of asteroids. There is no observational evidence for a mini black hole, but they are

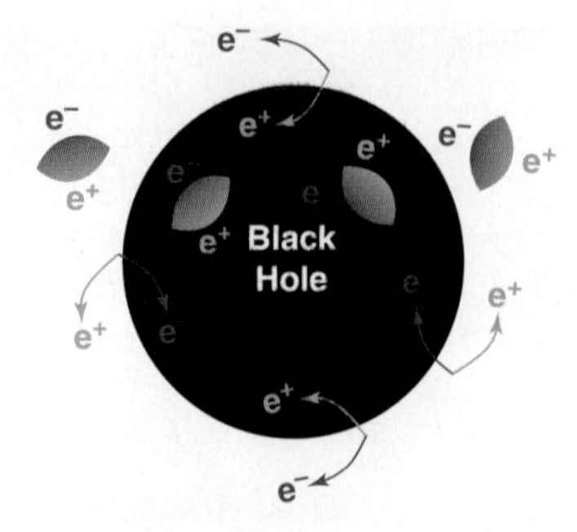

■ **FIGURE 14–26** The formation of "virtual pairs" of particles and antiparticles (here represented as electrons and positrons) in and near a black hole. In some cases, the particle or antiparticle can escape, while its partner goes into the black hole with negative energy and decreases the black hole's mass.

theoretically possible. Hawking has deduced that small black holes can emit energy in the form of elementary particles (electrons, neutrinos, and so forth). The mini black holes would thus evaporate and eventually disappear, with the final stage being an explosion of gamma rays.

Hawking's idea that black holes radiate may seem to be a contradiction to the concept that mass can't escape from a black hole. But when we consider effects of quantum physics, the simple picture of a black hole that we have discussed up to this point is not sufficient. Physicists already know that "virtual pairs" of particles and antiparticles can form simultaneously in empty space, though they disappear by destroying each other a short time later. Hawking suggests that a black hole so affects space near it that the particle or antiparticle disappears into the black hole, allowing its partner to escape (■ Fig. 14–26). Photons, which are their own antiparticles, appear as well.

As the mass of the mini black hole decreases, the evaporation rate increases, and the typical energy of emitted particles and photons increases. The final result is an explosion of gamma rays.

Although "gamma-ray bursts" have indeed been detected in the sky (see the preceding section), their observed properties are not consistent with the explosions of mini black holes. For example, gamma-ray bursts flicker substantially in gamma-ray brightness, with each one being different (Fig. 14–21). Such behavior is not expected of exploding black holes.

Instead, as we have seen, gamma-ray bursts may result from neutron stars merging to form black holes, or from supernovae in which the collapsing core forms a black hole. Gamma-ray bursts are extremely powerful, and probably have something to do with the formation of black holes, but unfortunately for Hawking they are not evidence for exploding mini black holes.

AceAstronomy™ Log into AceAstronomy and select this chapter to see the Astronomy Exercise "Black Hole."

AceAstronomy™ Log into AceAstronomy and select this chapter to see the Astronomy Exercise "Black Hole 2."

CONCEPT REVIEW

Black holes result when too much mass is present in a collapsed star to stop at the neutron-star stage (Introductory section). The strong gravitational field of a **stellar-mass black hole** bends light, according to Einstein's general theory of relativity (Section 14.1). Only light within an **exit cone** escapes (Section 14.2). Light on the edge of the exit cone orbits in the **photon sphere.**

When the radius of the star becomes equal to the **Schwarzschild radius,** it is so compact that the exit cone closes, and no light escapes; the star is within its **event horizon** (Section 14.3), and it is called a black hole. The Schwarzschild radius, only 3 km for a black hole having the mass of the Sun, is directly proportional to the mass. In the Newtonian view, when a star's size decreases to the Schwarzschild radius, the **escape velocity** from the star's surface reaches the speed of light (Section 14.3a).

Near a stellar-mass black hole, or on the surface of a star collapsing to form a black hole, the **tidal force** can be immense, pulling a local observer apart (Section 14.3b). This effect occurs because of the large difference in gravity between his head and feet: the force on the latter is much larger than that on the former. The tidal force on an observer would be smaller near a more massive black hole than near a less massive black hole.

Once a collapsing star passes inside its event horizon, it continues to contract, according to the general theory of relativity. Nothing can ever stop its contraction. In fact, the classical mathematical theory predicts that it will contract to zero radius—it will reach a **singularity** (Section 14.3b). (Quantum effects probably prevent it from reaching exactly zero radius.) The event horizon, however, does not change in size as long as the amount of mass inside is constant.

If you were far from a black hole, watching clocks fall toward it, you would see them run progressively more slowly, an effect known as **time dilation** (Section 14.4). Indeed, it would take an infinite amount of time (as measured by your clock) for them to reach the horizon. From the falling clocks' perspective, in contrast, no such time dilation occurs. On the other hand, if the falling clocks were to approach the event horizon and subsequently escape from the vicinity of the black hole, less time would have passed for them than for you; they would have aged less.

In equilibrium, a black hole "has no hair"—in other words, its only measurable properties are mass, angular momentum (loosely, the rate at which it is spinning), and total electric charge (Section 14.5).

The spherical event horizon of a rotating black hole is smaller than that of a nonrotating black hole of the same mass (Section 14.5). Moreover, rotating black holes have a **stationary limit** within which no particles can remain at rest, even though they are outside the event horizon. The space between the stationary limit and the event horizon is the **ergosphere,** from which energy (work) can be extracted. If a black hole were to rotate faster than a certain value, its event horizon would vanish, revealing a **naked singularity** from which distant observers could receive signals. Most theorists assume that all singularities are clothed by an event horizon.

At first glance, it appears as though black holes provide passageways to distant parts of our Universe, or even to other universes (Section 14.6). Such **wormholes** do not work, even in principle, for nonrotating black holes; one would need to travel faster than the speed of light to traverse them. In the case of a rotating black hole, on the other hand, travel through the wormhole at speeds slower than that of light, avoiding the singularity, initially seems feasible. But actually, the gravity of any object attempting to traverse a wormhole would squeeze it shut, precluding a safe journey.

We look for black holes from collapsed stars in regions where flickering x-rays come from an **accretion disk** (Section 14.7a). We think a black hole is present where, as for Cygnus X-1, we find a visible object that is being yanked gravitationally to and fro by an invisible object that is too massive and too faint to be anything other than a black hole (Section 14.7b). About two dozen probable stellar-mass black holes in our Galaxy have been identified and measured (Section 14.7c). Astronomers are now searching for signs of event horizons in these systems. A peculiar x-ray binary known as SS433, now thought to contain a black hole, shoots out jets of matter reaching 25% the speed of light (Section 14.7d).

Astronomers conclude that **supermassive black holes,** with millions or billions of times the mass of the Sun, exist in the centers of quasars and most galaxies (Section 14.8). Measurements of stars and rotating gas disks in the central regions of these objects reveal very high speeds, necessitating strong gravitational fields in very small volumes. Even our own Milky Way Galaxy contains a central black hole having several million solar masses.

Evidence for **intermediate-mass black holes** (containing about 100 to 10,000 solar masses) has been found (Section 14.9). Such black holes may have resulted from a merger of perhaps hundreds of stellar-mass black holes in a star cluster, or perhaps they are nuclei of small galaxies being consumed by a larger galaxy.

One of the most exciting new astronomical fields is the study of **gamma-ray bursts,** brief flashes of extremely short-wavelength radiation (Section 14.10). Although their distances were for many years highly controversial, most of them are now known to come from galaxies billions of light-years away (Section 14.10a). Their radiation is highly beamed, like that of pulsars; it cannot be emitted uniformly over the sky (Section 14.10b). A leading hypothesis is known as the "fireball model," in which some sort of compact engine releases blobs of material at speeds close to that of light along a narrow "jet."

Gamma-ray bursts may be the birth cries of certain types of stellar-mass black holes (Section 14.10b). Specifically, those of relatively long duration (typically a few tens of seconds) appear to be produced when the core of a very massive star deficient in hydrogen and helium collapses to form a black hole, with the remaining material ejected as a peculiar supernova. Those of short duration (typically less than 2 sec) are less well understood, but may result from either the merging of two neutron stars to form a black hole, or the merging of a neutron star and a black hole.

Mini black holes may have formed in the early Universe, but there is no direct evidence for them (Section 14.11). If they were indeed present, they should gradually evaporate by a quantum process, finally resulting in a violent burst of gamma rays. They are not, however, the origin of observed gamma-ray bursts.

QUESTIONS

1. Why doesn't the pressure from electrons or neutrons prevent a sufficiently massive star from becoming a black hole?
2. Why is a black hole blacker than a black piece of paper?
3. Discuss the escape of light from the surface of a star that is collapsing to form a black hole.
4. Explain the bending of light as a property of a warping of space, as discussed in Chapter 10.

†5. How would the gravitational force at the surface of a star change if the star contracted to one fifth of its previous diameter, without losing any of its mass?

†6. What is the Schwarzschild radius of a nonrotating black hole of 10 solar masses? What is the radius of its photon sphere?

†7. What would be your Schwarzschild radius, if you were a black hole?

†8. What is the relation in size of the photon sphere and the event horizon of a nonrotating black hole?

9. If you were to throw a mortal enemy into a black hole, would you be able to actually see him cross the event horizon of the black hole? Explain.

10. Describe what we mean when we say that time slows down near a black hole.

11. Explain what we mean by the statement "A black hole has no hair."

12. If someone close to a black hole were shining a blue flashlight beam outward, how would the color that you see be affected, if you are far from the black hole?

13. If you were an astronaut in space, could you escape (**a**) from within the photon sphere of a nonrotating black hole, (**b**) from within the ergosphere of a rotating black hole, or (**c**) from within the event horizon of any black hole? Explain each of these cases.

14. If the Sun suddenly became a black hole (it won't!), would the orbits of planets in our Solar System be affected?

15. How could the mass of a black hole that results from a collapsed star increase?

16. How could mini black holes, if they exist, lose mass? What can eventually happen to them?

17. Would we always know when we reached a black hole's event horizon by noticing its high density? Explain.

18. Is it possible to detect a black hole that is not part of a binary system, or is not surrounded by stars and gas in a galaxy? How?

19. Under what circumstances does the presence of an x-ray source associated with a spectroscopic binary suggest to astronomers the presence of a black hole?

20. Why is SS433 so unusual?

21. When and how were mini black holes formed?

22. Where are supermassive black holes thought to exist?

23. What is the link between quasi-periodic oscillations and intermediate-mass black holes?

24. **True or false?** The "ergosphere" is a region outside the event horizon of a rotating black hole from which the rotational energy of the black hole can sometimes be extracted.

25. **True or false?** The radius of the event horizon of a nonrotating black hole is proportional to the square of the black hole's mass.

26. **True or false?** Any object trying to orbit a black hole will be swallowed after orbiting just a few times.

27. **True or false?** Light can orbit in circles around a nonrotating black hole, at a radius of 1.5 times the Schwarzschild radius of the black hole.

28. **True or false?** If we measure the distance and the gamma-ray apparent brightness of a "gamma-ray burst" (GRB), we can accurately determine its actual gamma-ray luminosity by assuming that the gamma rays are emitted uniformly in all directions.

29. **Multiple choice:** Which one of the following statements about "gamma-ray bursts" is *false*? (**a**) They have been observed most often in distant galaxies. (**b**) They may be the result of merging neutron stars. (**c**) They may result from the quantum evaporation of black holes. (**d**) They emit as much, if not more, energy as a normal supernova. (**e**) They may be caused by the collapse of a massive star at the end of its life, forming a black hole.

30. **Multiple choice:** Which one of the following statements about detecting black holes in binary systems is true? (**a**) Absorption lines from both the black hole and the companion star can be seen in the spectrum. (**b**) Kepler's third law must be used to determine the product of the masses (i.e., $m_1 m_2$) of the two components of the binary. (**c**) The black hole periodically passes between us and the companion star, causing the latter to periodically disappear, and allowing us to deduce the black hole's presence. (**d**) Because of the generally unknown inclination of the orbit, spectroscopic observations only give observers an upper limit (i.e., maximum value) for the mass of a black hole. (**e**) The strongest candidates for black holes are generally found in binary systems with low-mass K and M main-sequence stars.

31. **Multiple choice:** The presence of a black hole in a galaxy core can be inferred from (**a**) the total mass of the galaxy; (**b**) the speeds of stars near the core; (**c**) the color of the galaxy; (**d**) the distance of the galaxy from the Milky Way Galaxy; or (**e**) the diminished brightness of starlight in the galaxy core, relative to surrounding areas.

32. **Multiple choice:** Which one of the following statements about black holes is *false*? (**a**) Inside a black hole, matter is thought to consist primarily of iron, the end point of nuclear fusion in massive stars. (**b**) Photons escaping from the vicinity of (but not inside) a black hole lose energy, yet still travel at the speed of light. (**c**) Near the event horizon of a small black hole (mass = a few solar masses), tidal forces stretch objects apart. (**d**) A black hole that has reached an equilibrium configuration can be described entirely by its mass, electric charge, and amount of spin ("angular momentum"). (**e**) A black hole has an "event horizon" from which no light can escape, according to classical (i.e., nonquantum) ideas.

33. **Fill in the blank:** Mass causes the surrounding space–time to _______.

34. **Fill in the blank:** According to Einstein's general theory of relativity, a black hole is completely described by its _____, _______, and ________.

35. **Fill in the blank:** The ________ of a black hole is its boundary, defining the region from within which nothing can escape (ignoring quantum effects).

36. **Fill in the blank:** All of the matter that falls into a black hole becomes concentrated in a very small volume called a _________.

†This question requires a numerical solution.

TOPICS FOR DISCUSSION

1. If you were given the chance, would you travel close to a black hole and then return to Earth, having aged very little compared with your friends and family? Would you, effectively, want to permanently jump to the future (and, if so, by what amount of time)?
2. What sorts of problems could be produced by the violation of causality—that is, if you could travel through a wormhole and return before your departure?
3. If gamma-ray bursts are beamed, are the energy requirements per gamma-ray burst (calculated from the observed brightness of each burst) smaller than if their energy were emitted uniformly in all directions? Is the number of gamma-ray bursts we detect in the sky affected by the beaming?

MEDIA

Virtual Laboratories

General Relativity, Black Holes, and Gravitational Lensing

AceAstronomy™ Log into AceAstronomy at **http://astronomy.brookscole.com/cosmos3** to access quizzes and animations that will help you assess your understanding of this chapter's topics.

Log into the Student Companion Web Site at **http://astronomy.brookscole.com/cosmos3** for more resources for this chapter including a list of common misconceptions, news and updates, flashcards, and more.

CHAPTER 15

The Milky Way: Our Home in the Universe

ORIGINS

Our Sun is part of the Milky Way Galaxy, an enormous collection of billions of stars bound by gravity. New stars forming from dense clouds of gas and dust in spiral arms provide clues to the Sun's birth.

AIMS

1. Understand the nature of the Milky Way, the faint band of light arching across the dark sky (Section 15.1).
2. Learn about our Galaxy and its constituents, and appreciate the position of our Sun and Earth, far from its center (Sections 15.2 to 15.6).
3. Discuss the shape of our Galaxy and why it has spiral arms (Sections 15.7, 15.8).
4. Explore the gas and dust between the stars–interstellar matter from which new stars form (Sections 15.9, 15.13).
5. See how radio observations are critical to studying interstellar matter and mapping our Galaxy (Sections 15.10 to 15.12, 15.14).

We have already described the stars, which are important parts of any galaxy, and how they are born, live, and die. In this chapter, we describe the gas and dust (small particles of matter) that are present to some extent throughout a galaxy. Substantial clouds of this gas and dust are called **nebulae** (pronounced "neb´yu-lee" or "neb´yu-lay"; singular: **nebula**); "nebula" is Latin for "fog" or "mist." New stars are born from such nebulae. We also discuss the overall structure of the Milky Way Galaxy and how, from our location inside it, we detect this structure.

15.1 Our Galaxy: The Milky Way

On the clearest moonless nights, when we are far from city lights, we can see a hazy band of light stretching across the sky (■ Fig. 15–1). This band is the **Milky Way**—the gas, dust, nebulae, and stars that make up the Galaxy in which our Sun is located. All this matter is our celestial neighborhood, typically within a few hundred or a thousand light-years from us. If we look a few thousand light-years in a direction away from that of the Milky Way, we see out of our Galaxy. But it is much, much farther to the other galaxies and beyond.

Don't be confused by the terminology: The Milky Way itself is the band of light that we can see from the Earth, and the **Milky Way Galaxy** is the whole galaxy in which we live. Like other large galaxies, our Milky Way Galaxy is composed of perhaps a few hundred billion stars plus many different types of gas, dust, planets, and so on. In the directions in which we see the Milky Way in the sky, we are looking through the relatively thin, pancake-like disk of matter that forms a major part of our Milky Way

AceAstronomy™ The AceAstronomy icon throughout this text indicates an opportunity for you to test yourself on key concepts, and to explore animations and interactions of the AceAstronomy website at **http://astronomy.brookscole.com/cosmos3**

◀ The Milky Way appears as the red horizontal bar in these full-sky maps released by the Wilkinson Microwave Anisotropy Probe (WMAP). The map projection is such that the Milky Way defines the equator, and the north and south Galactic poles are at top and bottom, respectively. The microwave wavelengths decrease from top to bottom, covering the range from 13 mm to 3 mm (corresponding to frequencies of 23 GHz to 94 GHz, beyond your radio FM dial). The Milky Way is less significant at the shorter microwave wavelengths. Around it, a smoothly varying background has been subtracted and we see primordial fluctuations in the temperature of the Universe's cosmic microwave background that we will discuss in Chapter 19.
NASA's Wilkinson Microwave Anisotropy Probe (WMAP) Science team

ASIDE 15.1: Our Galaxy

When we refer to our Milky Way Galaxy, we say "our Galaxy" or "the Galaxy" with an uppercase "G." When we refer to other galaxies, we use a lowercase "g." This is similar to the convention used for the Solar System (our Solar System), so as not to confuse it with other planetary systems (other solar systems).

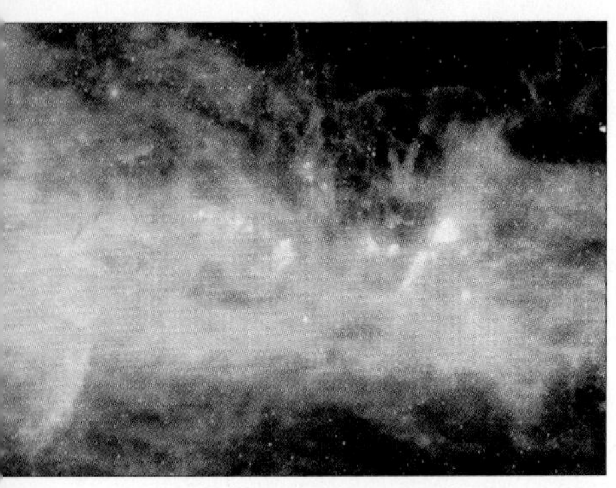

■ **FIGURE 15–1** The Milky Way, shown in this image and the next, illustrating the whole panorama in visible light. This image shows the Perseus Spiral Arm of our Galaxy in types of radiation that are not visible to the eye—radio waves and infrared light—translated into false color. The observations and display are part of the Canadian Galactic Plane Survey.

Jayanne English (CGPS/STScI) using data acquired by the Canadian Galactic Plane Survey (NRC/NSERC) and produced with the support of Russ Taylor (U. Calgary)

Galaxy. This disk is about 90,000 light-years across, an enormous, gravitationally bound system of stars.

The Milky Way appears very irregular when we see it stretched across the sky—there are spurs of luminous material that stick out in one direction or another, and there are dark lanes or patches in which much less can be seen (see *Star Party 15.1: Observing the Milky Way*). This patchiness is due to the splotchy distribution of nebulae and stars.

Here on Earth, we are inside our Galaxy together with all of the matter we see as the Milky Way (■ Fig. 15–2). Because of our position, we see a lot of our own Galaxy's matter when we look along the plane of our Galaxy. On the other hand, when we look "upward" or "downward" out of this plane, our view is not obscured by matter, and we can see past the confines of our Galaxy.

15.2 The Illusion That We Are at the Center

The gas in our Galaxy is more or less transparent to visible light, but the small solid particles that we call "dust" are opaque. So the distance we can see through our Galaxy depends mainly on the amount of dust that is present. This is not surprising: We can't always see far on a foggy day. Similarly, the dust between the stars in our Galaxy dims the starlight by absorbing it or by scattering (reflecting) it in different directions.

The dust in the plane of our Galaxy prevents us from seeing very far toward its center with the unaided eye and small telescopes. With visible light, on average we can see only one tenth of the way in (about 2000 light-years), regardless of the direction we look in the plane of the Milky Way. These direct optical observations fooled astronomers at the beginning of the 20th century into thinking that the Earth was near the center of the Universe (■ Fig. 15–3).

We shall see in this chapter how the American astronomer Harlow Shapley (pronounced to rhyme with "map′lee," as in "road map") realized in 1917 that our Sun is not in the center of the Milky Way. This fundamental idea took humanity one step further away from thinking that we are at the center of the Universe. Copernicus, in 1543, had already made the first step in removing the Earth from the center of the Universe.

■ **FIGURE 15–2** A visible-light view of the Milky Way, which wraps around the sky.

Axel Mellinger, Universität Potsdam

■ **FIGURE 15–3** A cross-sectional view of the structure of the Universe, as perceived by astronomers near the turn of the 20th century. The effects of interstellar gas and dust had not been taken into account, so they mistakenly concluded that the Sun was near the center of the Universe.

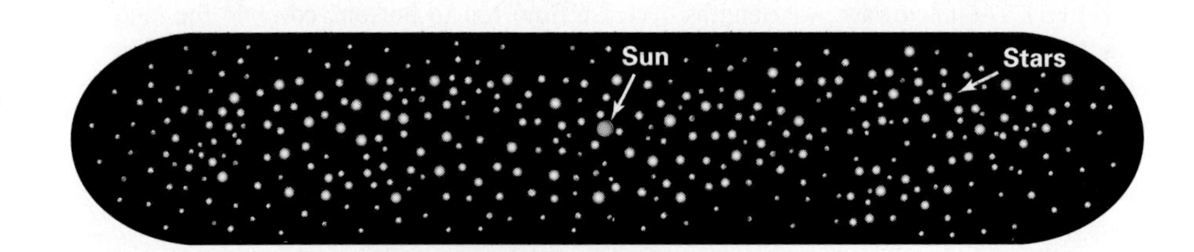

Star Party 15.1 Observing the Milky Way

To see the Milky Way, go far away from city lights on a dark, moonless night and look up in the sky. Depending on the season, you should be able to see a faint band of light somewhere in the sky. It will become more and more apparent as your eyes become adapted to the darkness, a process that usually takes about 15 minutes. The Milky Way is best seen in the summer sky, especially toward the constellations Scorpius and Sagittarius, but the winter Milky Way is also quite bright.

Notice how irregular some parts of the Milky Way appear, due to different amounts of gas, dust, and stars. There is an especially prominent, dark rift through Cygnus (the Swan), visible in the summer and early fall.

After viewing with your unaided eye, try using binoculars; you will probably see many open star clusters and nebulae sprinkled throughout the Milky Way. Refer to the Sky Maps on the inside covers of this book for their names and positions, and also consult Chapter 1 ("Observing the Constellations"). The Orion Nebula, in the sword of Orion, is an impressive emission nebula. The Pleiades (the Seven Sisters, in Taurus) form an obvious open star cluster surrounded by a very faint reflection nebula. If you have a small telescope, you can get reasonably detailed views of several other star clusters and nebulae that are marked on the Sky Maps.

In the 20th century, astronomers began to use wavelengths other than optical ones to study the Milky Way Galaxy. In the 1950s and 1960s especially, radio astronomy gave us a new picture of our Galaxy. In the 1980s and 1990s, we began to benefit from space infrared observations at wavelengths too long to pass through the Earth's atmosphere. The latest infrared telescope, launched by NASA in 2003, is the Spitzer Space Telescope. Infrared and radio radiation can pass through the Galaxy's dust and allow us to see our Galactic center and beyond.

A new generation of telescopes on high mountains enables us to see parts of the infrared and submillimeter spectrum. The Atacama Large Millimeter Array, now being built in Chile (see an artist's concept at the end of this chapter), will give us high-resolution views in the millimeter part of the spectrum. Giant arrays of radio telescopes spanning not only local areas but also continents and the Earth itself enable us to get crisp views of what was formerly hidden from us.

15.3 Nebulae: Interstellar Clouds

The original definition of "nebula" was a cloud of gas and dust that we see in visible light, though we now detect nebulae in a variety of ways. When we see the gas actually glowing in the visible part of the spectrum, we call it an **emission nebula** (■ Fig. 15–4).

Gas is ionized by ultraviolet light from very hot stars within the nebula; it then glows at optical (and other) wavelengths when electrons recombine with ions and cascade down to lower energy levels, releasing photons. Additionally, free electrons can collide with atoms (neutral or ionized) and lose some of their energy of motion, kicking the bound electrons to higher energy levels. Photons are emitted when the excited bound electrons jump down to lower energy levels, so the gas glows even more. The spectrum of an emission nebula therefore consists of emission lines.

Emission nebulae often look red (on long-exposure images; the human eye doesn't see these colors directly), because the red light of hydrogen is strongest in them. Electrons are jumping from the third to the second energy levels of hydrogen, producing the Hα [alpha] emission line in the red part of the spectrum (6563 Å).

Other types of emission nebulae can appear green in photographs, because of green light from doubly ionized oxygen atoms. Additional colors occur as well. Don't be misled by the pretty, false-color images that you often see in the news. In them, color is assigned to some specific type of radiation and need not correspond to colors that the eye would see when viewing the objects through telescopes.

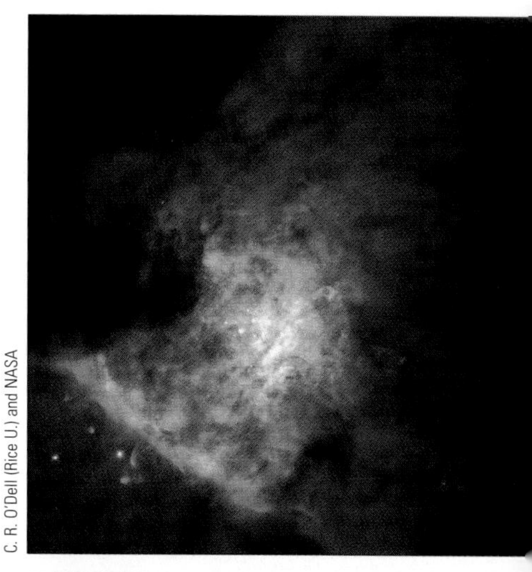

C. R. O'Dell (Rice U.) and NASA

■ **FIGURE 15–4** The central part of the Orion Nebula, a large emission nebula about 1500 light-years away. The gas is ionized by ultraviolet radiation from the cluster of hot stars.

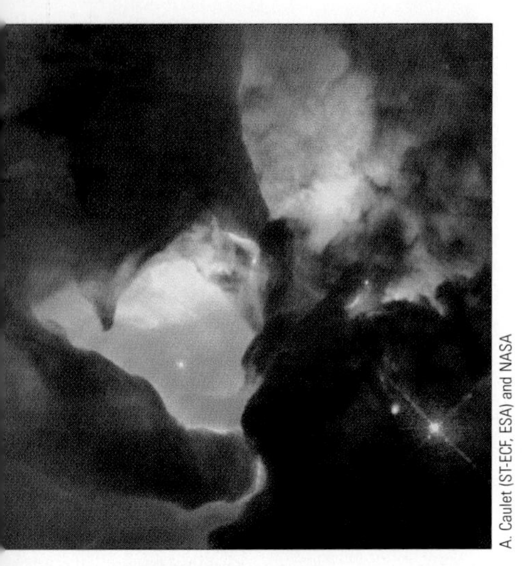

■ **FIGURE 15–5** The center of the Lagoon Nebula, imaged in high detail with the Hubble Space Telescope. Dark, absorbing dust appears as black. Emission from ionized sulfur atoms (red), from doubly ionized oxygen atoms (blue), and from electrons recombining with ionized hydrogen and cascading down to lower energy levels (green) are put together in this false-color view. The hot star at lower right appears as bright white, with red surroundings and reddish spikes that are an artifact caused by the telescope. This star provides the ultraviolet radiation that ionizes the brightest region of the nebula.

Sometimes a cloud of dust obscures our vision in some direction in the sky. When we see the dust appear as a dark silhouette (■ Fig. 15–5), we call it a **dark nebula** (or, often, an **absorption nebula,** since it absorbs visible light from stars behind it).

The Horsehead Nebula (■ Fig. 15–6) is an example of an object that is simultaneously an emission and an absorption nebula. The reddish emission from glowing hydrogen gas spreads across the sky near the leftmost (eastern) star in Orion's belt. A bit of absorbing dust intrudes onto the emitting gas, outlining the shape of a horse's head. We can see in the picture that the horsehead is a continuation of a dark area in which very few stars are visible. In this region, dust is obscuring the stars that lie beyond.

Clouds of dust surrounding relatively hot stars, like some of the stars in the star cluster known as the Pleiades (■ Fig. 15–7), are examples of **reflection nebulae.** They merely reflect the starlight toward us without emitting visible radiation of their own. Reflection nebulae usually look bluish for two reasons: (1) They reflect the light from relatively hot stars, which are bluish, and (2) dust reflects blue light more efficiently than it does red light. (Similar scattering of sunlight in the Earth's atmosphere makes the sky blue; see *A Closer Look 4.1: Colors in the Sky*).

Whereas an emission nebula has its own spectrum, as does a neon sign on Earth, a reflection nebula shows the spectral lines of the star or stars whose light is being reflected. Dust tends to be associated more with young, hot stars than with older stars, since the older stars would have had a chance to wander away from their dusty birthplaces.

The Great Nebula in Orion (Fig. 15–4) is an emission nebula. In the winter sky, we can readily observe it through even a small telescope or binoculars, and sometimes it has a tinge of color. We need long photographic exposures or large telescopes to study its structure in detail. Deep inside the Orion Nebula and the gas and dust alongside it, we see stars being born this very minute; many telescopes are able to observe in the infrared, which penetrates the dust. An example in a different region of the sky is shown in ■ Figure 15–8.

We have already discussed (in Chapter 13) some of the most beautiful emission nebulae in the sky, composed of gas thrown off in the late stages of stellar evolution.

■ **FIGURE 15–6** The Horsehead Nebula in Orion is dark dust superimposed on glowing gas.

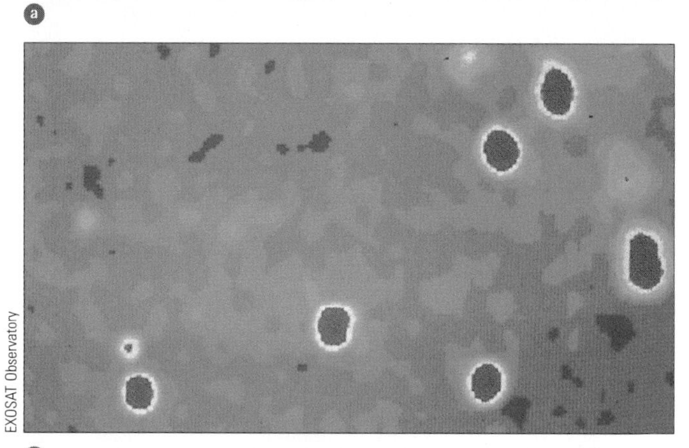

© 1985 Royal Observatory Edinburgh/Anglo-Australian Observatory, photograph from original U.K. Schmidt Plates by David Malin

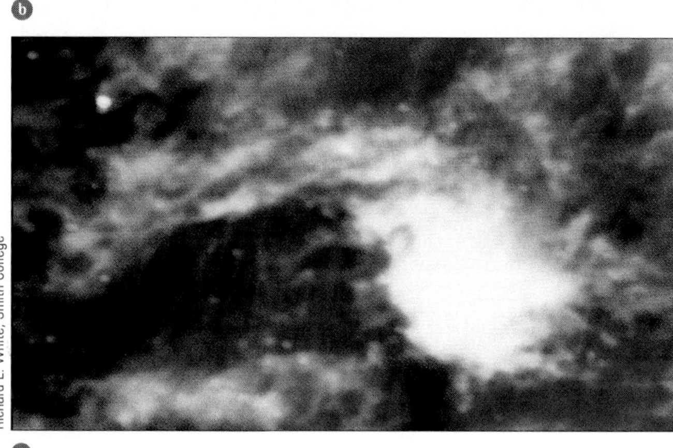

EXOSAT Observatory

Richard E. White, Smith College

■ **FIGURE 15–7** ⓐ The Pleiades in visible light, showing the blue reflection nebula. The spikes from bright stars in the open cluster are caused by the telescope optics. ⓑ This x-ray view of the Pleiades penetrates the dust to show the relatively hot stars. It was made from the EXOSAT in orbit. ⓒ The Pleiades (buried in the large region that appears white) have left a wake (dark region extending leftward) as they moved through the interstellar medium, as we see in this infrared image. Partly as a result of detecting the motion, scientists think that the Pleiades are undergoing a chance passage through the interstellar matter that forms the reflection nebula.

European Southern Observatory

■ **FIGURE 15–8** An image of the Chameleon I region made with the Very Large Telescope in Chile. The three colors used were visual (yellow), red, and infrared. The infrared radiation penetrates the dust especially well.

They include planetary nebulae (■ Fig. 15–9) and supernova remnants. Thus, nebulae are closely associated with both stellar birth and stellar death. The chemically enriched gas blown off by unstable or exploding stars at the end of their lives becomes the raw material from which new stars and planets are born. As we emphasized in Chapter 13, we are made of the ashes of stars!

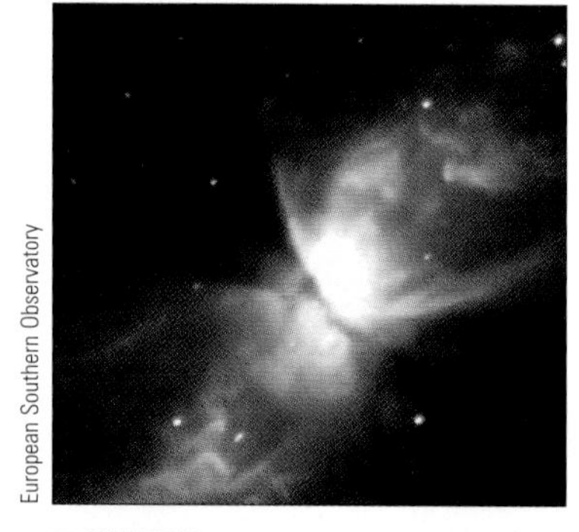

European Southern Observatory

■ **FIGURE 15–9** The Butterfly Nebula, a planetary nebula imaged with the Very Large Telescope of the European Southern Observatory, in Chile.

15.4 The Parts of Our Galaxy

It was not until 1917 that the American astronomer Harlow Shapley realized that we are not in the center of our Milky Way Galaxy. He was studying the distribution of globular clusters and noticed that, as seen from Earth, they are all in the same general area of the sky. They mostly appear above or below the Galactic plane and thus are not heavily obscured by the dust. When he plotted their distances and directions, he

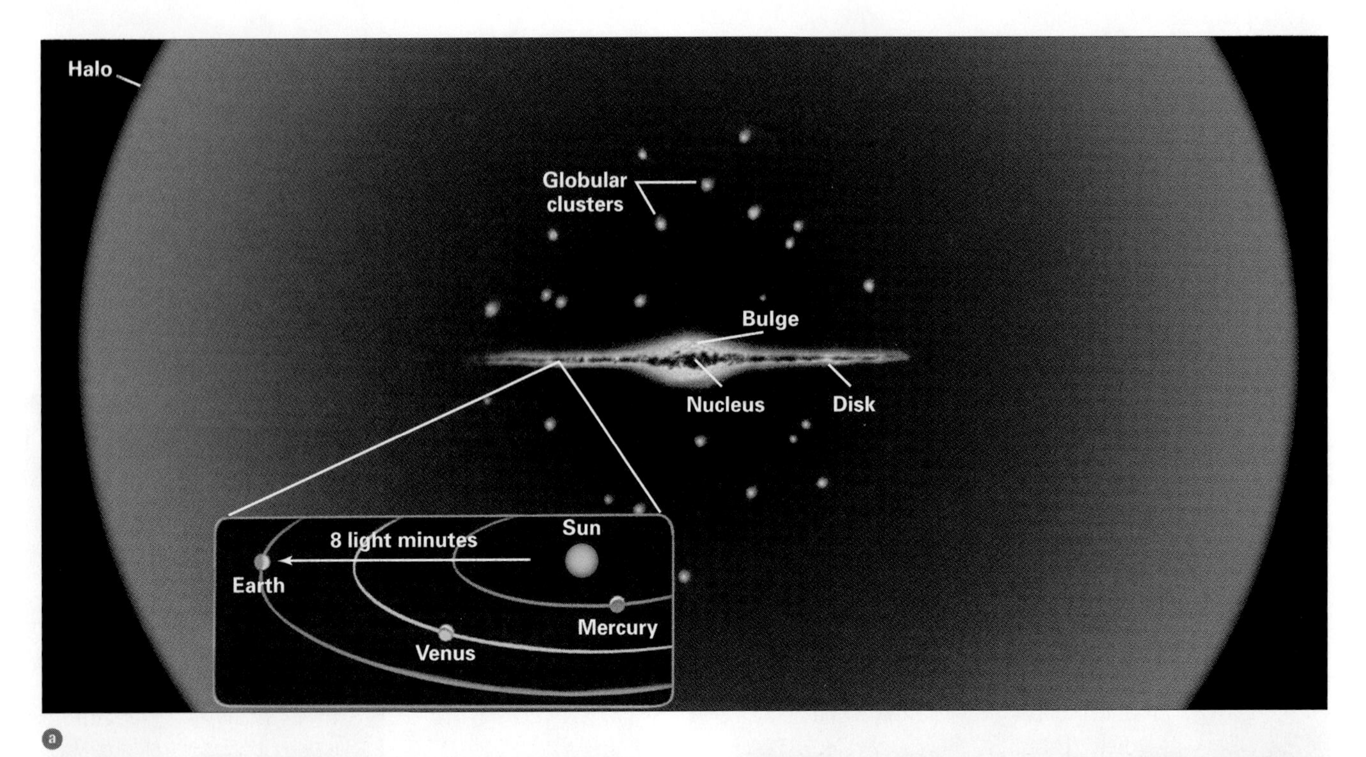

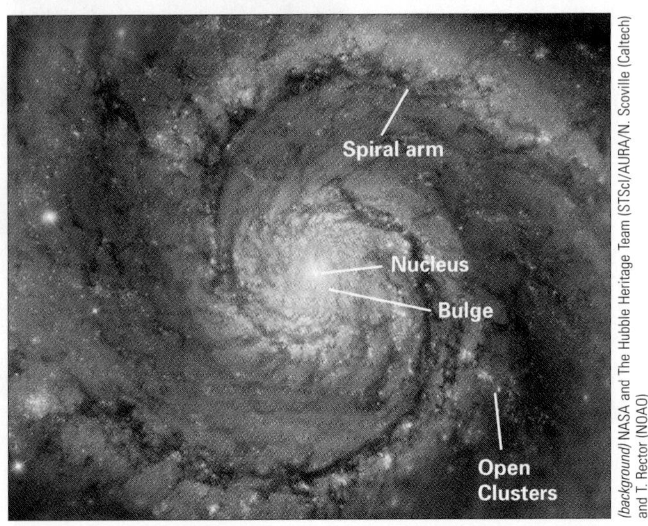

(background) NASA and The Hubble Heritage Team (STScI/AURA/N. Scoville (Caltech) and T. Rector (NOAO)

NASA and The Hubble Heritage Team (STScI/AURA/G. R. Meurer and T. M. Heckman (JHU), C. Leitherer, J. Harris, and D. Calzetti (STScI); and M. Sirianni (JHU)

■ **FIGURE 15–10** (a) The drawing shows our Milky Way Galaxy's nuclear bulge surrounded by the disk, which contains the spiral arms. The globular clusters are part of the halo, which extends above and below the disk. (From Earth, we see fewer clusters near the plane of our Galaxy than are actually there because interstellar dust gets in the way. This drawing was made from the actual observed distribution of globular clusters.) From the fact that most of the clusters appear in less than half of our sky, Harlow Shapley deduced that the Galactic center is in the direction indicated. (b) The center of the Whirlpool Galaxy, M51, imaged with the Hubble Space Telescope. Its spiral arms, with its dust clouds in which stars are being born, show clearly. Our Galaxy probably looks something like this one if seen from a comparable vantage point. (c) Another spiral galaxy, M100 in the constellation Coma Berenices, imaged with the European Southern Observatory's Very Large Telescope.

noticed that they formed a spherical halo around a point thousands of light-years away from us (■ Fig. 15–10*a*).

Shapley's touch of genius was to realize that this point is likely to be the center of our Galaxy. After all, if we are at a party and discover that everyone we see is off to our left, we soon figure out that we aren't at the party's center. Other spiral galaxies are also shown (Figs. 15–10*b* and 15–10*c*) for comparison and to show something of what our Galaxy must look like when seen from high above it.

Though Shapley correctly deduced that the Sun is far from our Galactic center, he actually overestimated the distance. The reason is that dust dims the starlight, making the stars look too far away, and he didn't know about this "interstellar extinction." The amount of dimming can be determined by measuring how much the starlight has been **reddened:** Blue light gets scattered and absorbed more easily than red light, so the star's color becomes redder than it should be for a star of a given spectral type. This is the same reason sunsets tend to look orange or red, not white (again see *A Closer Look 4.1: Colors in the Sky*).

Our Galaxy has several parts:

1. ***The nuclear bulge.*** Our Galaxy has the general shape of a pancake with a bulge at its center that contains millions of stars, primarily old ones. This **nuclear bulge** has the Galactic **nucleus** at its center. The nucleus itself is only about 10 light-years across.
2. ***The disk.*** The part of the pancake outside the bulge is called the Galactic **disk.** It extends 45,000 light-years or so out from the center of our Galaxy. The Sun is located about one half to two thirds of the way out. The disk is very thin—2 per cent of its width—like a phonograph record, CD, or DVD. It contains all the young stars and interstellar gas and dust, as well as some old stars. The disk is slightly warped at its ends, perhaps by interaction with our satellite galaxies, the Magellanic Clouds. Our Galaxy looks a bit like a hat with a turned-down brim.

It is very difficult for us to tell how the material in our Galaxy's disk is arranged, just as it would be difficult to tell how the streets of a city were laid out if we could only stand on one street corner without moving. Still, other galaxies have similar properties to our own, and their disks are filled with great **spiral arms**—regions of dust, gas, and stars in the shape of a pinwheel (Figure 15–10*b*). So, we assume the disk of our Galaxy has spiral arms, too. Though the direct evidence is ambiguous in the visible part of the spectrum, radio observations have better traced the spiral arms.

The disk looks different when viewed in different parts of the spectrum (■ Fig. 15–11). Infrared and radio waves penetrate the dust that blocks our view in visible light, while x-rays show the hot objects best.

ASIDE 15.2: Reddening vs. redshift

Note that this "reddening" is not a Doppler redshift! Interstellar reddening suppresses blue light relative to red, but does not shift the wavelengths of absorption or emission lines, as the Doppler effect does.

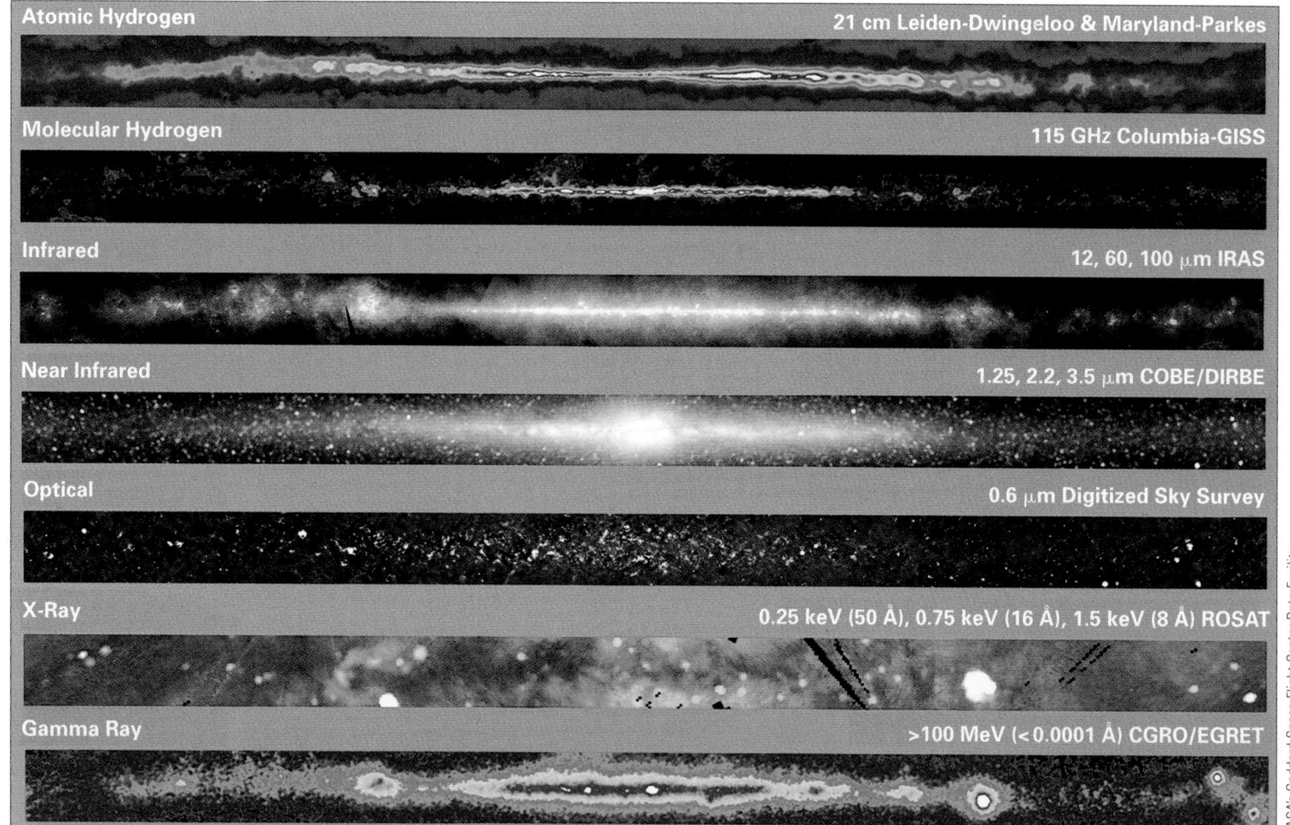

■ **FIGURE 15–11** Views of the Milky Way at a wide variety of wavelengths. In all cases, we see the concentration toward the disk of the Galaxy. In some wavelength/frequency/energy bands, such as the IRAS infrared view, the radiation comes all the way from the Galactic center. We see data in the 21-cm radio-wavelength band from neutral hydrogen, the 115-gigahertz (2.6-millimeter) radio-wavelength band from molecular hydrogen, infrared bands from 100 μm down to 1.25 μm imaged by spacecraft, a visible band at 0.6 μm = 6000 Å, three x-ray bands at approximately 8–50 Å, and a gamma-ray band, at wavelengths shorter than 0.0001 Å, 100,000 times shorter than the x-ray bands.

■ **FIGURE 15–12** A cluster of stars near the center of our Galaxy, imaged with the infrared camera aboard the Hubble Space Telescope.

3. ***The halo.*** Old stars (including the globular clusters) and very dilute interstellar matter form a roughly spherical Galactic **halo** around the disk. The inner part of the halo is at least as large across as the disk, perhaps 60,000 light-years in radius. The gas in the inner halo is hot, 100,000 K, though it contains only about 2 per cent of the mass of the gas in the disk. As we discuss in Chapter 16, the outer part of the halo extends much farther, out to perhaps 200,000 or 300,000 light-years. Believe it or not, this Galactic outer halo apparently contains 5 or 10 times as much mass as the nucleus, disk, and inner halo together—but we don't know what it consists of! We shall see in Section 16.4 that such "dark matter" (invisible, and detectable only through its gravitational properties) is a very important constituent of the Universe.

15.5 The Center of Our Galaxy

We cannot see the center of our Galaxy in the visible part of the spectrum because our view is blocked by interstellar dust. Radio waves and infrared, on the other hand, penetrate the dust. The Hubble Space Telescope, with its superior resolution, has seen isolated stars where before we saw only a blur (■ Fig. 15–12).

In 2003, NASA launched an 0.85-m infrared telescope, the Spitzer Space Telescope (Section 3.8c, Figure 3–32*a*). Its infrared detectors are more sensitive than those on earlier infrared telescopes. Spitzer completes NASA's series of Great Observatories, including the Compton Gamma Ray Observatory (now defunct), the Chandra X-ray Observatory, and the Hubble Space Telescope.

■ **FIGURE 15–13** A wide-field view of the center of our Galaxy, observed with the VLA at a wavelength of 90 cm.

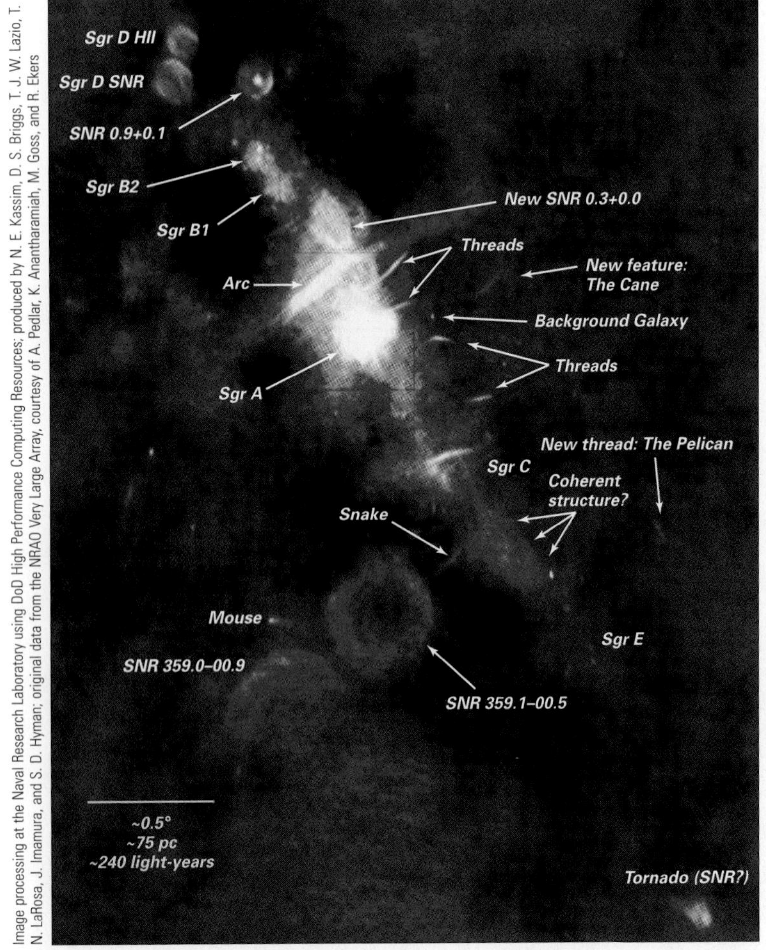

One of the brightest infrared sources in our sky is the nucleus of our Galaxy, only about 10 light-years across. This makes it a very small source for the prodigious amount of energy it emits: as much energy as radiated by 80 million Suns. It is also a radio source and a variable x-ray source.

High-resolution radio maps of our Galactic center (■ Fig. 15–13) show a small bright spot, known as Sgr A* (pronounced "Saj A-star"), in the middle of the bright radio source Sgr A. The radio radiation could well be from gas surrounding a central giant black hole (as shown in the image opening Chapter 14). Extending somewhat farther out, a giant Arc of parallel filaments stretches perpendicularly to the plane of the Galaxy (■ Fig. 15–14).

As we discuss further in Chapter 17, adaptive optics techniques in the near-infrared have allowed very rapid motions of stars to be measured much nearer the Galactic center than was previously possible (■ Fig. 15–15). The orbits measured show the presence of a supermassive black hole that is about 3.7 million times the Sun's mass. One of the stars comes within an astonishing 17 light-hours of Sgr A*.

Observations of the Galactic center with the Chandra X-ray Observatory and the European Space Agency's INTEGRAL gamma-ray spacecraft (■ Fig. 15–16) reveal the presence of hot, x-ray luminous gas and stars there.

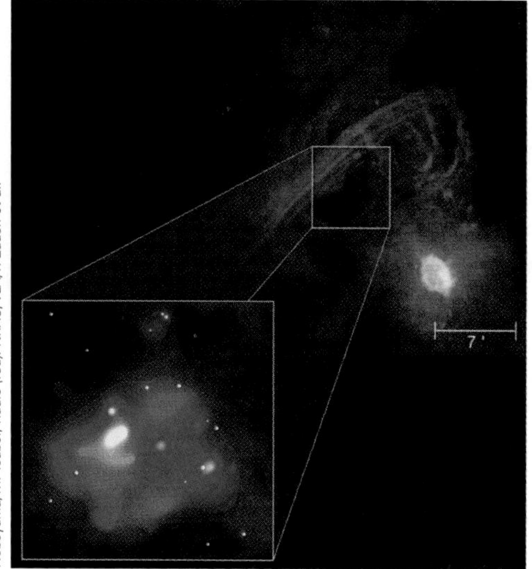

X-ray *(blue)*: NASA/CXC/Northwestern/F. Zadeh et al.; Millimeter Wavelength *(green)*: Nobeyama/M. Tsuboi; Radio *(red)*: NRAO/VLA/F. Zadeh et al.

■ **FIGURE 15–14** The VLA shows (in red) that a vast Arc of parallel filaments stretches over 130 light-years perpendicularly to the plane of our Galaxy, which runs from upper left to lower right. The field of view is about 1° across; the wavelength used was in the continuous emission near 21 cm. The Arc is also visible on Figure 15–13. A Chandra view in x-rays of part of the field is singled out and shown at lower left (blue). The current model for the x-ray emission is that it results from the collision of electrons from the filaments with a cloud containing a million solar masses of cold gas.

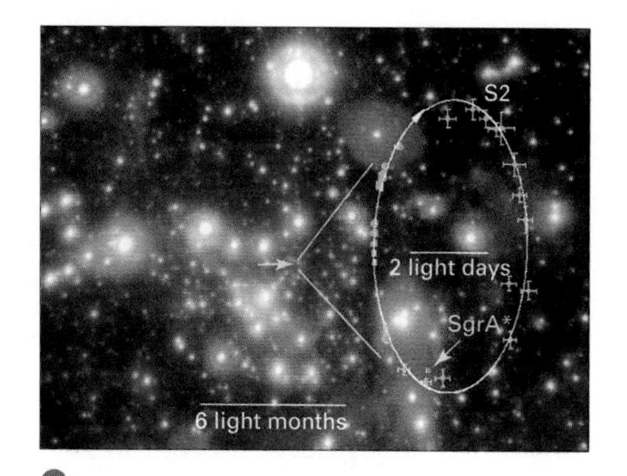

(a)

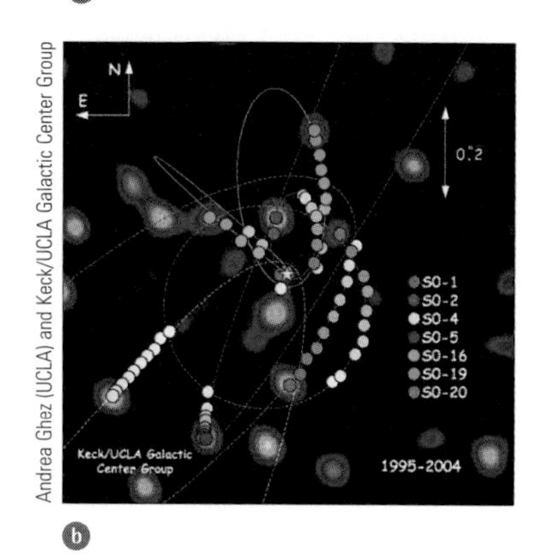

Andrea Ghez (UCLA) and Keck/UCLA Galactic Center Group

(b)

Andrea Ghez (UCLA) and Keck/UCLA Galactic Center Group

(c)

■ **FIGURE 15–15** Using adaptive optics to study stars orbiting close to the center of our Galaxy, scientists have deduced that a giant black hole with 3.7 million times the Sun's mass is lurking there. (a) The center of the Milky Way, with measurements made of the position of a star near it from the Very Large Telescope by a largely German research group. The star comes within only about 17 light-hours of Sgr A*. Following Kepler's second law, the star moves much more quickly when close to Sgr A*. At the 2 μm wavelength used, about 10% of the emitted photons reach us. (b) Images made at the Keck telescope by a research group from the University of California, Los Angeles (UCLA), working at 3.8 μm, showing the orbits of several stars in the central 1 square arcsecond around Sgr A*. The background image was taken in 2004; all the stars on it are seen to move over time. Advancing time is shown with increasing saturation of the colors of the symbols. (c) An image at 3.8 μm made by the UCLA group, using a "laser guide star," a technique having the adaptive optics use a laser spot projected on the sky as a known image to allow calculation of the feedback needed for adjustment of the mirror shape.

a: European Southern Observatory: Rainer Schödel, Thomas Ott, Reinhard Genzel, Reiner Hofmann, and Matt Lehnert (Max-Planck-Institut für extraterrestrische Physik, Garching, Germany); Andreas Eckart and Nelly Mouawad (Physikalisches Institut, Universität zu Köln, Cologne, Germany); Tal Alexander (The Weizmann Institute of Science, Rehovot, Israel); Mark J. Reid (Harvard-Smithsonian Center for Astrophysics, Cambridge, MA, USA); Rainer Lenzen and Markus Hartung (Max-Planck-Institut für Astronomie, Heidelberg, Germany); François Lacombe, Daniel Rouan, Eric Gendron, and Gérard Rousset (Observatoire de Paris—Section de Meudon, France); Anne-Marie Lagrange (Laboratoire d'Astrophysique, Observatoire de Grenoble, France); Wolfgang Brandner, Nancy Ageorges, Chris Lidman, Alan F.M. Moorwood, Jason Spyromilio, and Norbert Hubin (ESO); and Karl M. Menten (Max-Planck-Institut für Radioastronomie, Bonn, Germany)

■ **FIGURE 15–16** (a) A close-up of the region of Sgr A* from Chandra. (b) A mosaic from the Chandra X-ray Observatory showing x-ray emission from the Milky Way. (c) Emission from the galactic center imaged with INTEGRAL at energies near the boundary between x-rays and gamma rays. The grid marks are separated by 0.5°. The image shows an extended source around Sgr A*.

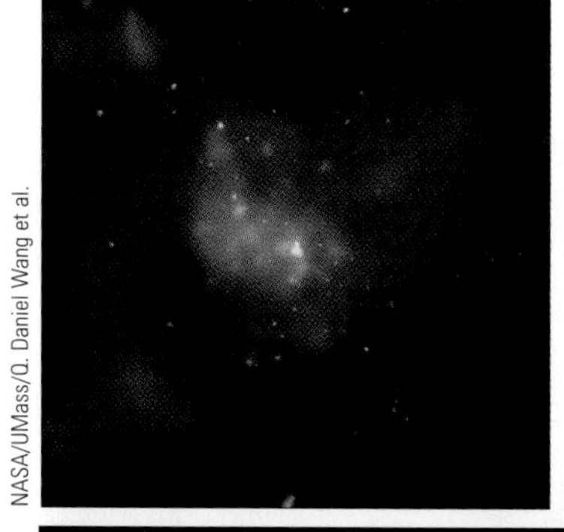

NASA/UMass/Q. Daniel Wang et al.

(a)

NASA/MIT/Frederick K. Baganoff et al.

(b)

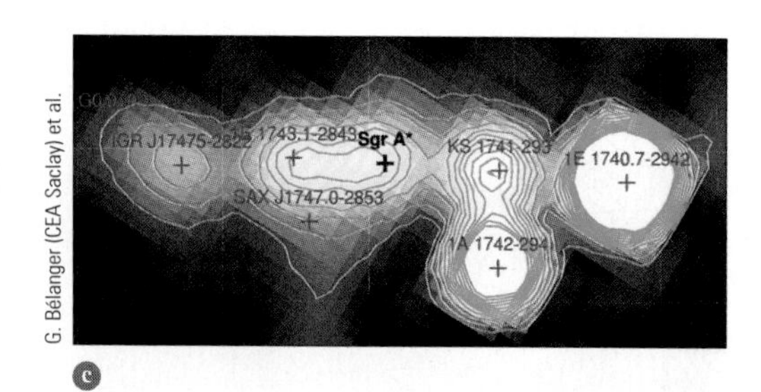

G. Bélanger (CEA Saclay) et al.

(c)

15.6 All-Sky Maps of Our Galaxy

The study of our Galaxy provides us with a wide range of types of sources to study. Many of these have been known for decades from optical studies (■ Fig. 15–17, and also the fifth panel from the top in Fig. 15–11). The infrared sky looks quite different (Fig. 15–11, third and fourth panels from the top), with its appearance depending strongly on wavelength. The radio sky provides still different pictures, depending on the wavelength used (Fig. 15–11, top two panels).

Maps of our Galaxy in the x-ray region of the spectrum (Fig. 15–11, second panel from the bottom) show the hottest individual sources (such as x-ray binary stars) and diffuse gas that was heated to temperatures of a million degrees by supernova explosions. The Compton Gamma Ray Observatory produced maps of the steady gamma rays (Fig. 15–11, bottom panel), most of which come from collisions between cosmic rays (see our discussion in Section 13.2f) and atomic nuclei in clouds of gas.

A different instrument on the Compton Gamma Ray Observatory detected bursts of gamma rays that last only a few seconds or minutes (■ Fig. 15–18). These **gamma-ray bursts,** which were seen at random places in the sky roughly once per day, are especially intriguing. NASA's Swift satellite, mentioned in Sections 3.7a and 14.10a, was sent aloft in 2004 specifically to study them in detail.

Though some models suggested that the gamma-ray bursts were produced within our Galaxy (either very close to us or in a very extended halo), more recent observations have conclusively shown that most of them are actually in galaxies billions of light-years away. As we discussed in Chapter 14, these distant gamma-ray bursts may be produced when extremely massive stars collapse to form black holes, or when a neutron star merges with another neutron star or with a black hole.

The Chandra X-ray Observatory is producing more detailed images of x-ray sources than had ever before been available. Studies of the highest-energy electromagnetic radiation like x-rays and gamma rays, and of rapidly moving cosmic-ray particles (Section 13.2f) guided to some extent by the Galaxy's magnetic field, are part of the field of **high-energy astrophysics.** Riccardo Giacconi received a share of the 2002 Nobel Prize in Physics for his role in founding this field.

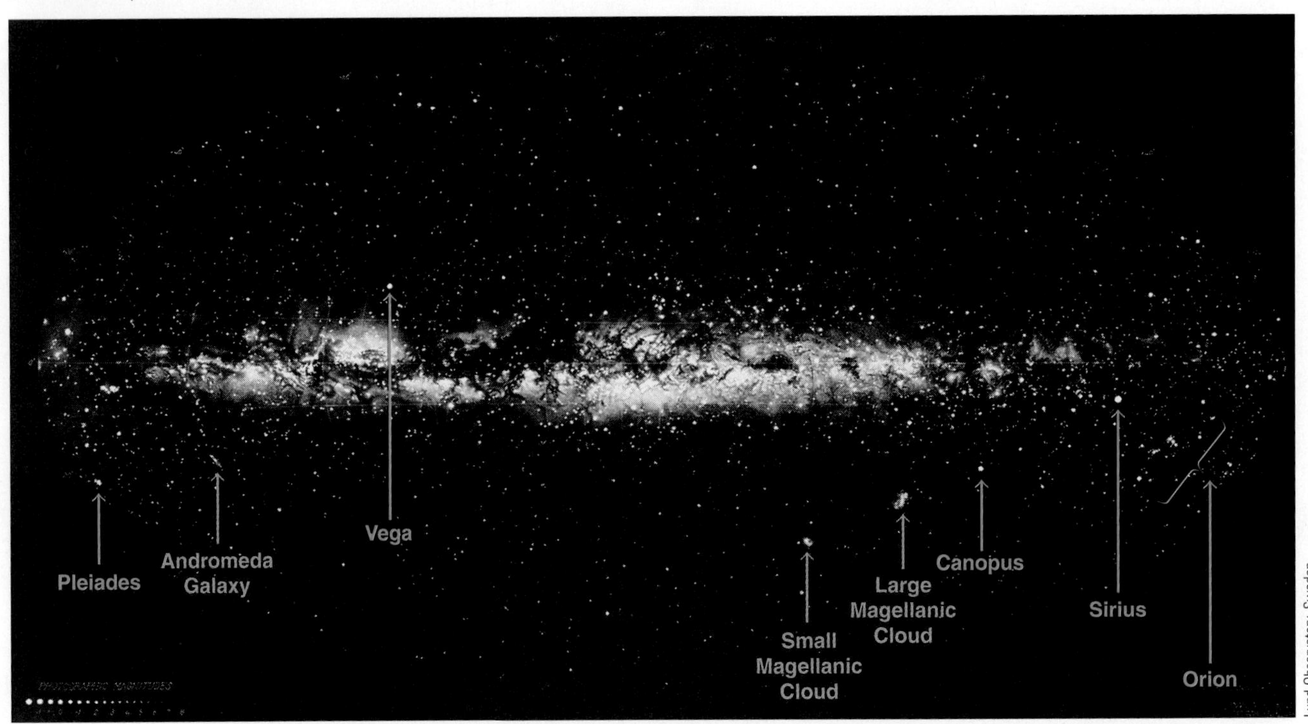

■ **FIGURE 15–17** A drawing of the Milky Way as it appears to the eye, in visible light. Seven thousand stars plus the diffuse glow of the Milky Way are shown in this panorama.

Lund Observatory, Sweden

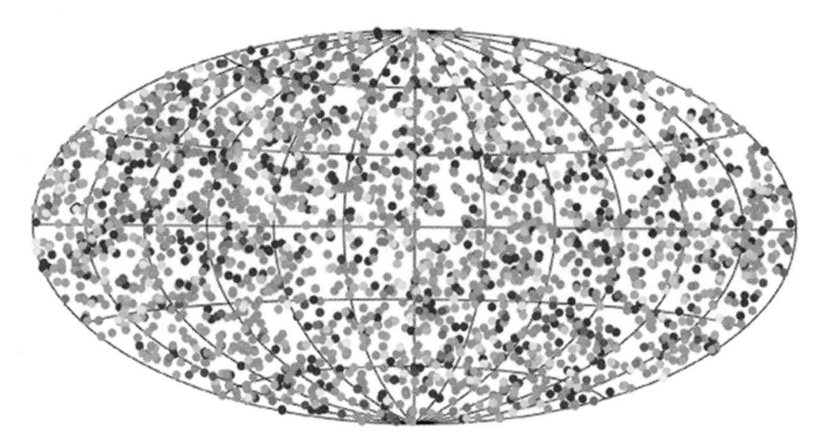

BATSE, Compton Gamma Ray Observatory Science Support Center, NASA's Goddard Space Flight Center

■ **FIGURE 15–18** The distribution of the 2704 gamma-ray bursts recorded with the Burst and Transient Source Experiment (BATSE) on the Compton Gamma Ray Observatory before the spacecraft's demise. The total energy received from the burst is color coded. Surprisingly, the sources are randomly distributed, so they seemed either very close to us or far outside our Galaxy. Studies with x-ray telescopes, the Hubble Space Telescope, ground-based telescopes, and now the Swift satellite have revealed that most of the gamma-ray bursts are very distant, and hence are incredibly powerful.

15.7 Our Pinwheel Galaxy

It is always difficult to tell the shape of a system from a position inside it. Think, for example, of being somewhere inside a maze of tall hedges; we would find it difficult to trace out the pattern. If we could fly overhead in a helicopter, though, the pattern would become very easy to see (■ Fig. 15–19). Similarly, we have difficulty tracing out the spiral pattern in our own Galaxy, even though the pattern would presumably be apparent from outside the Galaxy. Still, by noting the distances and directions to objects of various types, we can determine the Milky Way's spiral structure.

Young open clusters are good objects to use for this purpose, for they are always located in spiral arms. We think that they formed there and that they have not yet had time to move away (■ Fig. 15–20). We know their ages from the length of their main sequences on the temperature-luminosity diagram (Chapter 11). Also useful are main-sequence O and B stars; the lives of such stars are so short we know they can't be old. But since our methods of determining the distances to open clusters, as well as to O and B stars, from their optical spectra and apparent brightnesses are uncertain to 10 per cent, they give a fuzzy picture of the distant parts of our Galaxy. Parallaxes measured from the Hipparcos spacecraft do not go far enough out into space to help in mapping our Galaxy. We need new astrometric satellites.

Other signs of young stars are the presence of emission nebulae. We know from studies of other galaxies that emission nebulae are preferentially located in spiral arms. In mapping the locations of emission nebulae, we are really again studying the locations of the O stars and the hottest of the B stars, since it is ultraviolet radiation from these hot stars that provides the energy for the nebulae to glow.

It is interesting to plot the directions to and distances of the open clusters, the O and B stars, and the clouds of ionized hydrogen known as H II (pronounced "H two") regions as seen from Earth. When we do so, they appear to trace out bits of three spiral arms, which are relatively nearby.

Interstellar dust prevents us from using this technique to study parts of our Galaxy farther away from the Sun. However, another valuable method of mapping the spiral structure in our Galaxy involves spectral lines of hydrogen and of carbon monoxide in the radio part of the spectrum. Radio waves penetrate the interstellar dust, allowing us to study the distribution of matter throughout our Galaxy, though getting the third dimension (distance) that allows us to trace out spiral arms remains difficult. We will discuss the method later in this chapter.

Jay M. Pasachoff

a

Jay M. Pasachoff

b

■ **FIGURE 15–19** The view from the level of a maze (a) makes it hard to figure out the pattern, which is readily visible from above (b).

© 1981 NOAO/CTIO, photo by Gabriel Martin

■ **FIGURE 15–20** The Jewel Box, an open cluster of stars. The orange star is κ (the Greek letter "kappa") Crucis, a red supergiant (a cool star), while most of the other stars are bluish and therefore hot.

ASIDE 15.3: The Sun's "Galactic age"

With a distance of about 26,000 light-years from the center of our Galaxy, and a speed of 200 km/sec, the Sun takes about 250 million years to complete one orbit. So, the Sun has orbited the Galaxy about 18 times since its birth 4.6 billion years ago. If we think of "Galactic years" being analogous to Earth's years, the Sun is a young adult. (But, of course, the Sun is about halfway through its main-sequence life, and thus is a middle-aged star in those terms.)

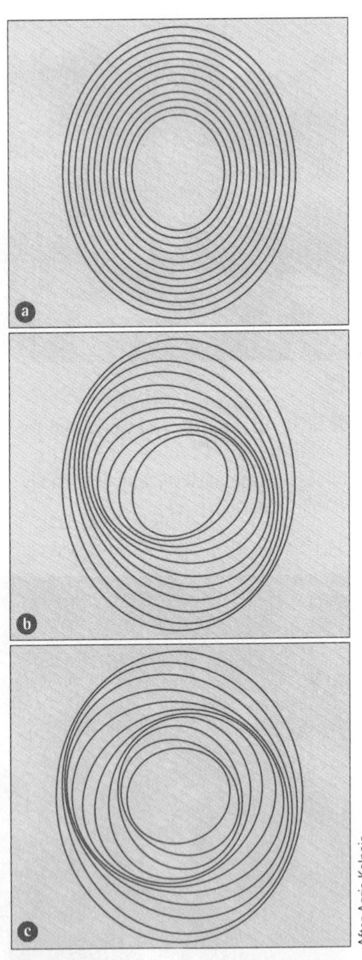

■ **FIGURE 15–21** Each part of the figure includes the same set of ellipses; the only difference is the relative alignment of their axes. Consider that the axes are rotating slowly and at different rates. The places where their orbits are close together take a spiral form, even though no actual spiral of physically connected points exists. The spiral structure of a galaxy may arise from an analogous effect.

15.8 Why Does Our Galaxy Have Spiral Arms?

The Sun revolves around the center of our Galaxy at a speed of approximately 200 kilometers per second. At this rate, it takes the Sun about 250 million years to travel once around the center, only 2 per cent of the Galaxy's current age. (Our Galaxy, after all, must be older than its globular clusters, whose age we discussed in Chapter 11.) But stars at different distances from the center of our Galaxy revolve around its center in different lengths of time. (As we will see in Chapter 16, the Galaxy does not rotate like a solid disk.) For example, stars closer to the center revolve much more quickly than does the Sun. Thus the question arises: Why haven't the arms wound up very tightly, like the cream in a cup of coffee swirling as you stir it?

The leading current solution to this conundrum says, in effect, that the spiral arms we now see do not consist of the same stars that would previously have been visible in those arms. The spiral-arm pattern is caused by a **spiral density wave,** a wave of increased density that moves through the gas in the Galaxy. This density wave is a wave of compression, not of matter being transported. It rotates more slowly than the actual material and causes the density of passing material to build up. Stars are born at those locations and appear to form a spiral pattern (■ Fig. 15–21), but the stars then move away from the compression wave.

Think of the analogy of a crew of workers fixing potholes in two lanes of a four-lane highway. A bottleneck occurs at the location of the workers; if we were in a traffic helicopter, we would see an increase in the number of cars at that place. As the workers continue slowly down the road, fixing potholes in new sections, we would see what seemed to be the bottleneck moving slowly down the road. Cars merging from four lanes into the two open lanes need not slow down if the traffic is light, but they are compressed more than in other (fully open) sections of the highway. Thus the speed with which the bottleneck advances is much smaller than that of individual cars.

Similarly, in our Galaxy, we might be viewing only some galactic bottleneck at the spiral arms. The new, massive stars would heat the interstellar gas so that it becomes visible. In fact, we do see young, hot stars and glowing gas outlining the spiral arms, providing a check of this prediction of the density-wave theory. This mechanism may work especially well in galaxies with a companion that gravitationally perturbs them (as seen in the opening image in Chapter 16).

15.9 Matter Between the Stars

The gas and dust between the stars is known as the **interstellar medium** or "interstellar matter." The nebulae represent regions of the interstellar medium in which the density of gas and dust is higher than average.

For many purposes, we may consider interstellar space as being filled with hydrogen at an average density of about 1 atom per cubic centimeter. (Individual regions may have densities departing greatly from this average.) Regions of higher density in which the atoms of hydrogen are predominantly neutral are called **H I regions** (pronounced "H one regions"; the Roman numeral "I" refers to the neutral, basic state). Where the density of an H I region is high enough, pairs of hydrogen atoms combine to form molecules (H_2). The densest part of the gas associated with the Orion Nebula might have a million or more hydrogen molecules per cubic centimeter. So hydrogen molecules (H_2) are often found in H I clouds.

A region of ionized hydrogen, with one electron missing, is known as an **H II region** (from "H two," the second state—neutral is the first state and once ionized is the second). Since hydrogen, which makes up the overwhelming proportion of interstellar gas, contains only one proton and one electron, a gas of ionized hydrogen contains

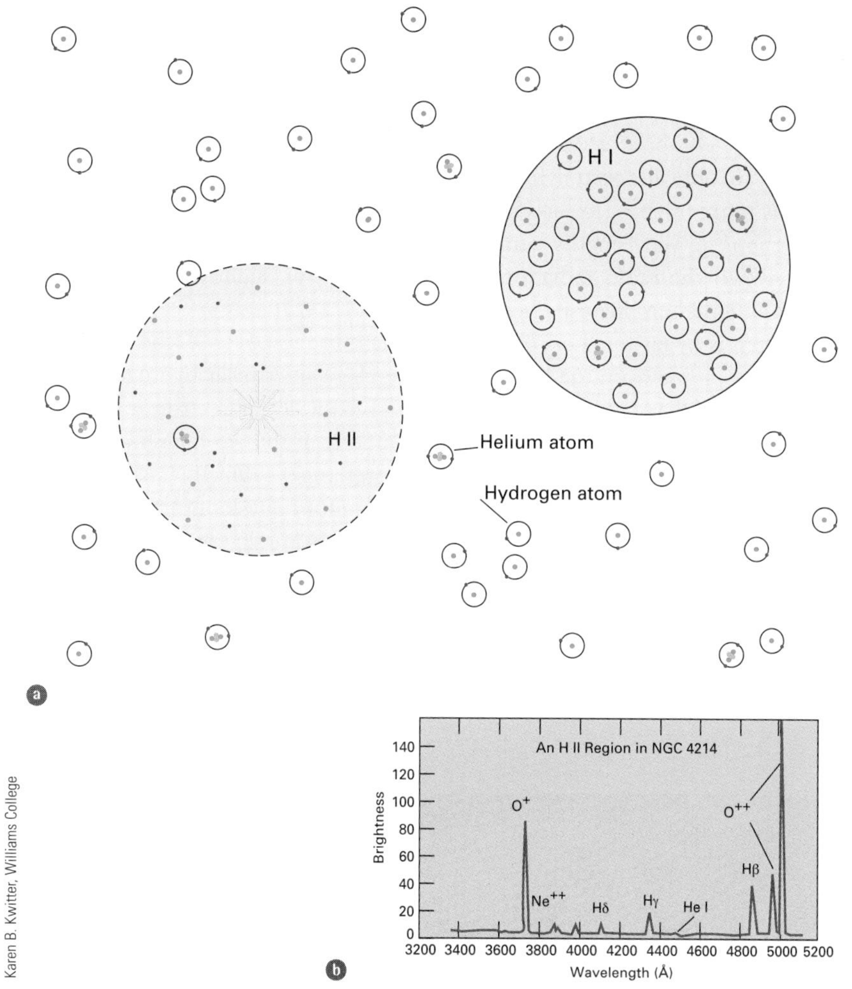

■ **FIGURE 15–22** (a) H I clouds are regions of neutral hydrogen of higher density than average, and H II regions are regions of ionized hydrogen. The protons and electrons that result from the ionization of hydrogen by ultraviolet radiation from a hot star, and the neutral hydrogen atoms, are shown schematically. The larger dots represent protons or neutrons, and the smaller dots represent electrons. The star that provides the energy for the H II region is shown. (b) The optical spectrum of an H II region in the galaxy NGC 4214. Emission lines are marked.

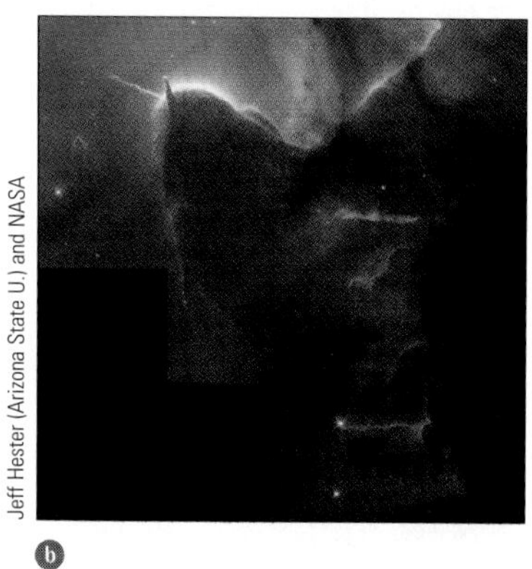

■ **FIGURE 15–23** (a) The Trifid Nebula, M20, in Sagittarius, is the red H II region divided into (at least) three visible segments by absorbing dust lanes. The red Hα emission line is diluted with blue light scattered from the hot central stars. The blue reflection nebula at the top is unconnected to the Trifid. In the dark regions, the extinction (obscuration) is so high that we cannot see stars behind the dust causing the absorption. (b) A high-resolution Hubble Space Telescope view of about 1/4 the diameter of the red nebula shown in part *(a)*.

individual protons and electrons. Wherever a hot star provides enough energy to ionize hydrogen, an H II region (emission nebula) results (■ Fig. 15–22 and ■ Fig. 15–23).

Studying the optical and radio spectra of H II regions and planetary nebulae tells us the abundances (proportions) of several of the chemical elements (especially helium, nitrogen, and oxygen). How these abundances vary from place to place in our Galaxy and in other galaxies helps us choose between models of element formation and of galaxy evolution.

Tiny grains of solid particles are given off by the outer layers of red giants. They spread through interstellar space, and dim the light from distant stars. This "dust" never gets very hot, so most of its radiation is in the infrared. The radiation from dust scattered among the stars is faint and very difficult to detect, but the radiation coming from clouds of dust surrounding newly formed stars is easily observed from ground-based telescopes and from infrared spacecraft. They found infrared radiation from so many stars in our Galaxy that we think that about one star forms in our Galaxy each year.

Since the interstellar gas is often "invisible" in the visible part of the spectrum (except at the wavelengths of certain weak emission lines), different techniques are needed to observe the gas in addition to observing the dust. Radio astronomy is the most widely used technique, so we will now discuss its use for mapping our Galaxy.

ASIDE 15.4: Ionized gases

Though all hydrogen is ionized in an H II region, only some of the helium is ionized. Heavier elements have sometimes lost 2 or even 3 electrons.

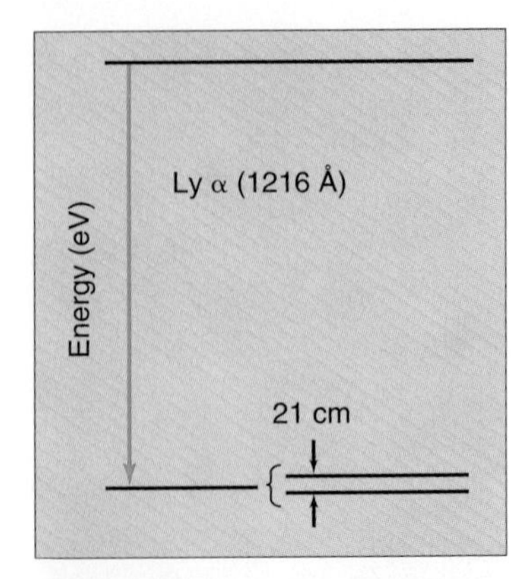

■ **FIGURE 15–24** An energy-level diagram showing that 21-cm radiation results from an energy difference between two sublevels in the lowest principal energy state of hydrogen. The energy difference is much smaller than the energy difference that leads to Lyman α (the Greek letter "alpha"), which was described in Chapter 2. So, because $E = h\nu = hc/\lambda$, the wavelength is much longer.

15.10 Radio Observations of Our Galaxy

The first radio astronomy observations were of continuous radiation; no spectral lines were known. If a radio spectral line is known, Doppler-shift measurements can be made, and we can tell about motions in our Galaxy. What is a radio spectral line? Remember that an optical spectral line corresponds to a wavelength of the optical spectrum that is more intense (for an emission line) or less intense (for an absorption line) than neighboring wavelengths. Similarly, a radio spectral line corresponds to a wavelength at which the radio radiation is slightly more, or slightly less, intense. A radio station is an emission line on a home radio.

Since hydrogen is by far the most abundant element in the Universe, the most-used radio spectral line is a line from the lowest energy levels of interstellar hydrogen atoms. This line has a wavelength of 21 cm. A hydrogen atom is basically an electron "orbiting" a proton. Both the electron and the proton have the property of spin, as if each were spinning on its axis. The spin of the electron can be either in the same direction as the spin of the proton or in the opposite direction. The rules of quantum physics prohibit intermediate orientations. The energies of the two allowed conditions are slightly different.

If an atom is sitting alone in space in the upper of these two energy states, with its electron and proton spins aligned in the same direction, there is a certain small probability that the spinning electron will spontaneously flip over to the lower energy state and emit a bundle of energy—a photon (■ Fig. 15–24). We thus call this a **spin-flip transition** (■ Fig. 15–25). The photon of hydrogen's spin-flip transition corresponds

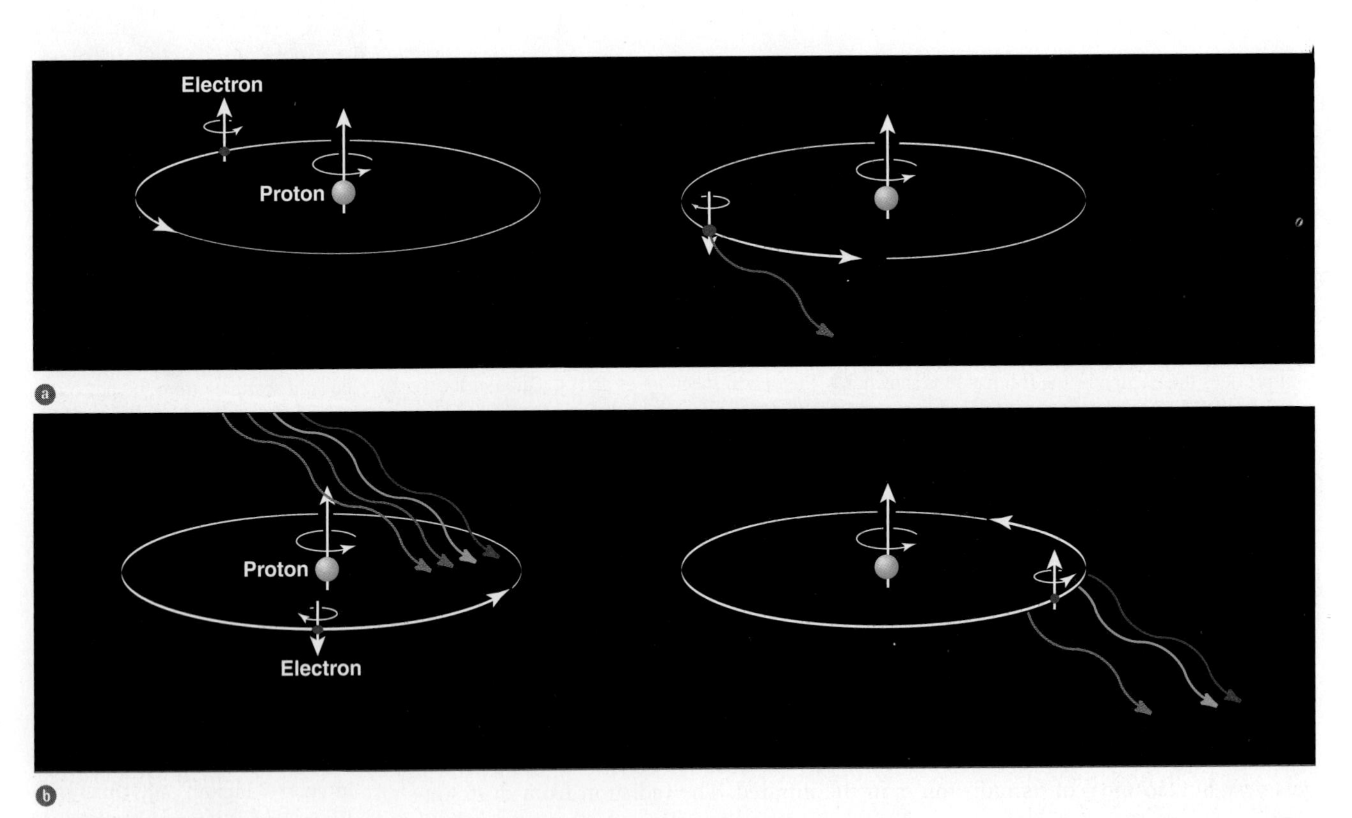

■ **FIGURE 15–25** In contrast to the energy-level diagram of the preceding figure, here we see what a spin-flip would look like if we could observe on such a submicroscopic level. ⓐ When the electron in a hydrogen atom flips over so that it is spinning in the opposite direction from the spin of the proton, an emission line at a wavelength of 21 cm results. To illustrate this effect, we show no rays coming in and the one ray emitted by the spontaneous spin-flip going outward (toward lower right). Note from the arrows the change in the direction of the electron's spin. ⓑ When an electron takes energy from a passing beam of radiation, causing it to flip from spinning in the direction opposite to that of the proton's spin to spinning in the same direction, then a 21-cm line in absorption results. We illustrate this effect schematically by showing four rays coming in from the upper left, but since only the one of them at the wavelength that matches the spontaneous spin-flip shown at the top causes the spin to flip back, only three rays go out. Actually, the spin-flip takes place only for a wavelength of 21 cm, so the ray has no actual color perceptible to our eyes; we arbitrarily use blue to show it. Physicists often show spin by wrapping their right hands around the spinning object in the direction of spin with their thumbs up. The direction of the thumb is called the direction (orientation) of the spin axis. We show the spin axes defined in this way as vertical lines with arrows.

to radiation at a wavelength of 21 cm—**the 21-cm line.** If the electron flips from the higher to the lower energy state, we have an emission line. If it absorbs energy from passing continuous radiation, it can flip from the lower to the higher energy state and we have an absorption line.

If we were to watch any particular group of hydrogen atoms in the slightly higher energy state, we would find that it would take 11 million years before half of the electrons had undergone spin-flips; we say that the "half-life" is 11 million years for this transition. Thus, hydrogen atoms are generally quite content to sit in the upper state! But there are so many hydrogen atoms in space that enough 21-cm radiation is given off to be detected. The existence of the line was predicted in 1944 and discovered in 1951, marking the birth of spectral-line radio astronomy.

15.11 Mapping Our Galaxy

The 21-cm hydrogen line has proven to be a very important tool for studying our Galaxy (■ Fig. 15–26) because this radiation passes unimpeded through the dust that prevents optical observations very far into the plane of our Galaxy. It can even reach us from the opposite side of our Galaxy, whereas light waves penetrate the dust clouds in the Galactic plane only about 10 per cent of the way to the Galactic center, on average.

Astronomers have ingeniously been able to find out how far it is to the clouds of gas that emit the 21-cm radiation. They use the fact that gas closer to the center of our Galaxy rotates with a shorter period than the gas farther away from the center. Though there are substantial uncertainties in interpreting the Doppler shifts in terms of distance from the Galaxy's center, astronomers have succeeded in making some maps. These maps show many narrow arms but no clear pattern of a few broad spiral arms like those we see in other galaxies (Chapter 16). The question emerged: Is our Galaxy really a spiral at all? With the additional information from studies of molecules in space that we describe in the next section, we finally made further progress.

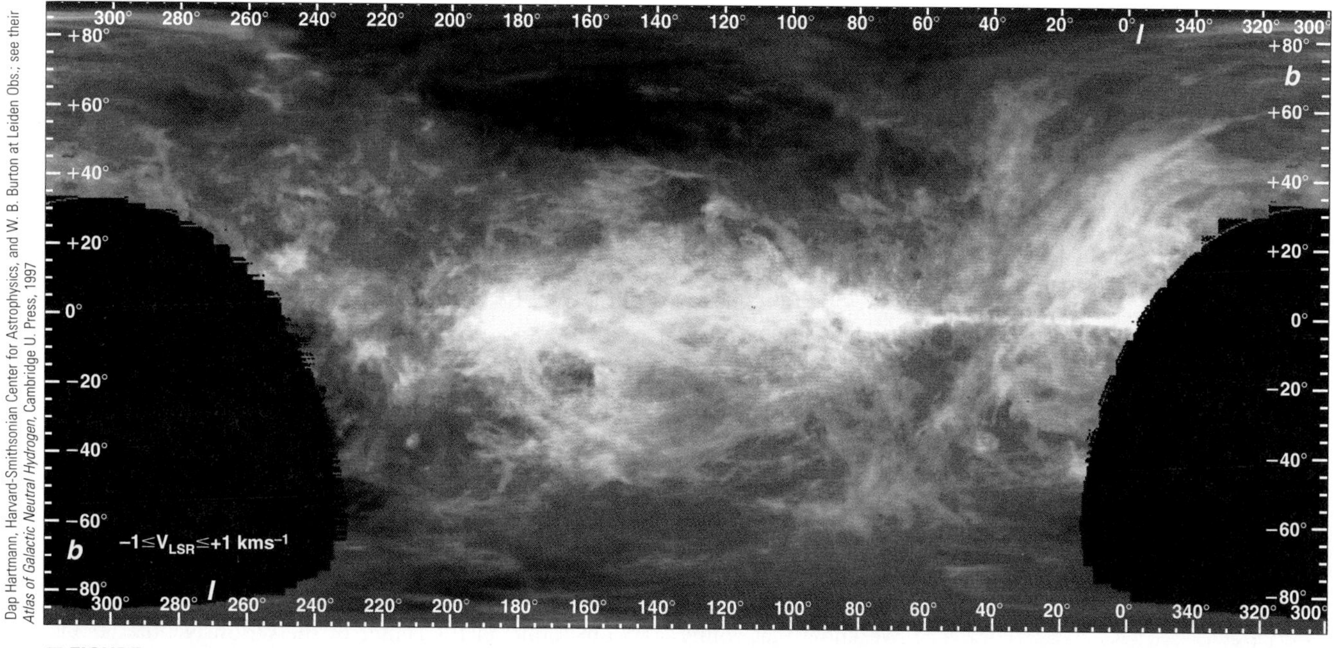

■ **FIGURE 15–26** A map of our Galaxy's 21-cm radiation, which comes from interstellar neutral hydrogen, in a Mercator projection.

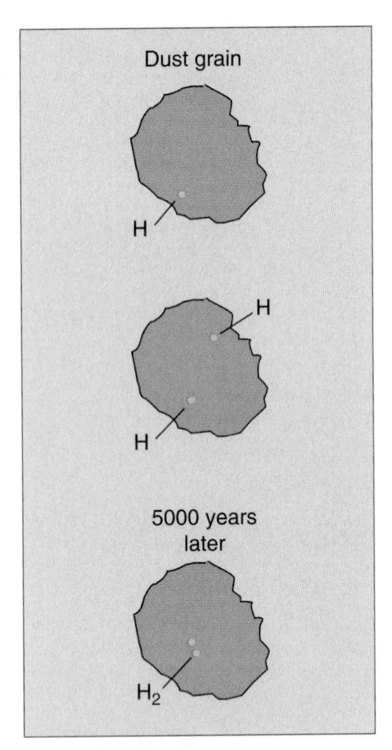

■ **FIGURE 15–27** Hydrogen molecules are formed in space with the aid of dust grains at an intermediate stage. Terrestrial laboratory research reported in 2005 showed that irregularities on the dust grains appear to be needed to form the molecules.

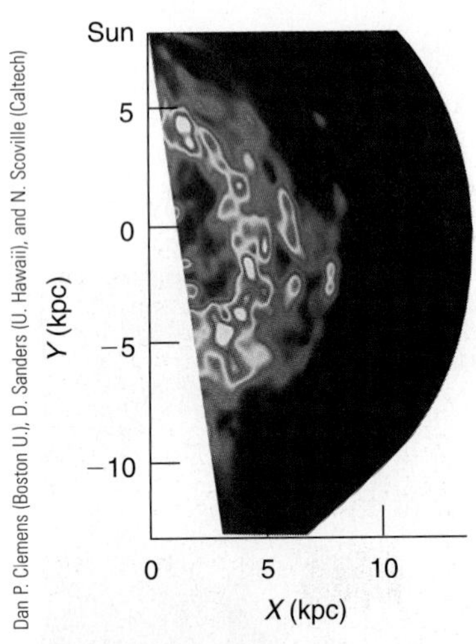

■ **FIGURE 15–28** A map based on CO data does not show clear spiral arms. The point at (0, 0) on the graph marks the center of our Galaxy. The regions 4–6 kpc and near 7 kpc are more ring-like than spiral. Studies of other galaxies are showing that clear spiral arms are common in the outer regions of galaxies but not necessarily in the inner regions, which may be the case with our Galaxy as well.

15.12 Radio Spectral Lines from Molecules

Radio astronomers had only the hydrogen 21-cm spectral line to study for a dozen years, and then only the addition of one other group of lines for another five years. Then radio spectral lines of water (H_2O) and ammonia (NH_3) were found. The spectral lines of these molecules proved surprisingly strong, and were easily detected once they were looked for. Over 100 additional types of molecules have since been found.

The earlier notion that it would be difficult to form molecules in space was wrong. In some cases, atoms apparently stick to interstellar dust grains, perhaps for thousands of years, and molecules build up (■ Fig. 15–27). Though hydrogen molecules form on dust grains, most of the other molecules may be formed in the interstellar gas, or in the atmospheres of stars, without need for grains.

Studying the spectral lines provides information about physical conditions—temperature, densities, and motion, for example—in the gas clouds that emit the lines. Studies of molecular spectral lines have been used together with 21-cm line observations to improve the maps of the spiral structure of our Galaxy (■ Fig. 15–28). Observations of carbon monoxide (CO), in particular, have provided better information about the parts of our Galaxy farther out from the Galaxy's center than our Sun. We use the carbon monoxide as a tracer of the more abundant hydrogen molecular gas, since the carbon monoxide produces a far stronger spectral line and is much easier to observe; molecular hydrogen emits extremely little.

15.13 The Formation of Stars

We have already discussed (in Chapter 12) some of the youngest stars known and how stars form. Here we will discuss star formation in terms of the gas and dust from which stars come. Astronomers have found that **giant molecular clouds** are fundamental building blocks of our Galaxy. Giant molecular clouds are 150 to 300 light-years across. There are a few thousand of them in our Galaxy. The largest giant molecular clouds contain about 100,000 to 1,000,000 times the mass of the Sun. Since giant molecular clouds break up to form stars, they only last 10 million to 100 million years.

Most radio spectral lines seem to come only from the molecular clouds. (Carbon monoxide is the major exception, for it is widely distributed across the sky.) Infrared and radio observations together have provided us with an understanding of how stars are formed from these dense regions of gas and dust. Carbon-monoxide observations reveal the giant molecular clouds, but it is molecular hydrogen (H_2) rather than carbon monoxide that contains a vast majority of the mass.

Many radio spectral lines have been detected only in a particular cloud of gas, the Orion Molecular Cloud. It is located close to a visible part, which we call the Orion Nebula, of a larger cloud of gas and dust. The Orion Molecular Cloud contains about 500 thousand times the mass of the Sun. It is relatively accessible to our study because it is only about 1500 light-years away. Even though less than 1 per cent of the Cloud's mass is dust, that is still a sufficient amount of dust to prevent ultraviolet light from nearby stars from entering and breaking the molecules apart. Thus molecules can accumulate.

The properties of the molecular cloud can be deduced by comparing the radiation from its various molecules and by studying the radiation from each molecule individually. The average density is a few hundred to a thousand particles per cubic centimeter, but the cloud center may have up to a million particles per cubic centimeter. This central region is still billions of times less dense than our Earth's atmosphere, though it is much denser than the typical interstellar density of about 1 particle per cubic centimeter.

We know that young stars are found in the center of the Orion Nebula (■ Fig. 15–29). The Trapezium (■ Fig. 15–30), a group of four hot stars readily visible in a

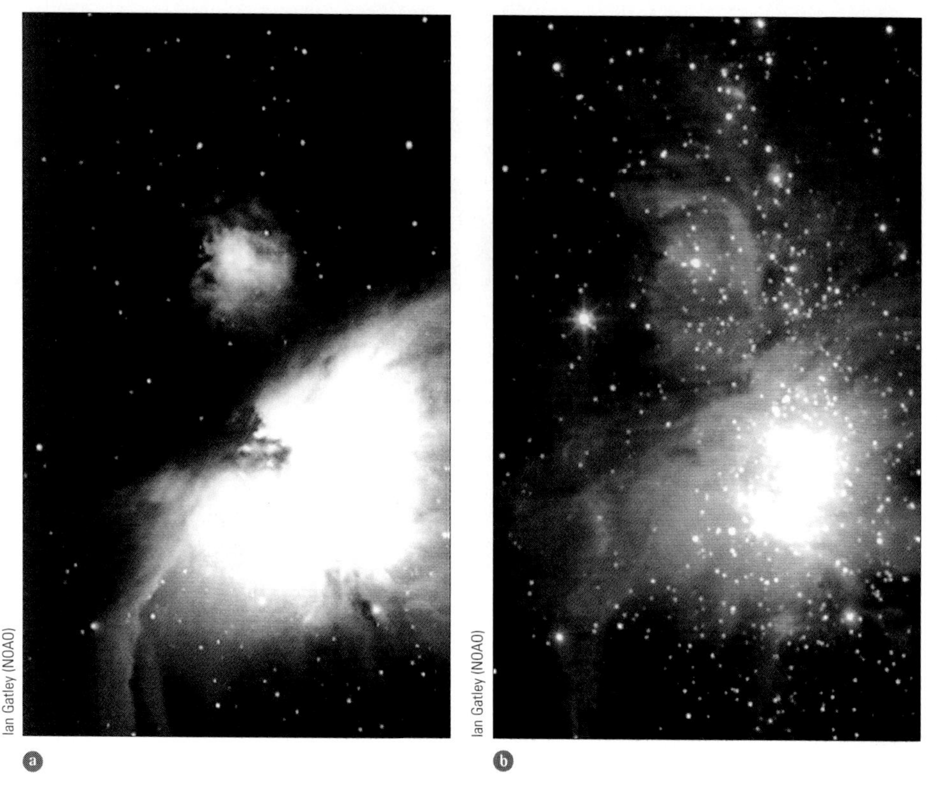

■ **FIGURE 15–29** ⓐ A visible-light image of the Orion Nebula, an H II region on the side of a molecular cloud. ⓑ A view of Orion taken with an infrared array. Three infrared wavelengths known as J, H, and K (1.25, 1.6, and 2.2 micrometers, respectively) were used to make this false-color view.

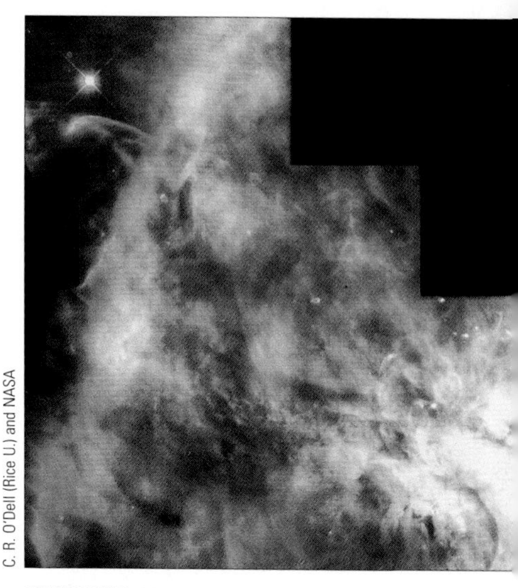

■ **FIGURE 15–30** Part of the Orion Nebula only 1.6 light-years across diagonally, imaged with the Hubble Space Telescope with much higher detail than possible from the ground. One of the Trapezium stars that provides energy for the nebula to glow is at top left. In the image, nitrogen emission is shown as red, hydrogen emission is shown as green, and oxygen emission is shown as blue. Astrometric measurements show the brightest object (Fig. 15-32) and two radio sources in the Trapezium moving away from the location where they were 500 years ago. They and other "runaway objects" may be young stars emerging from a multiple-star system that was disrupted by internal gravitational interactions.

small telescope, is the source of ionization and energy for the Orion Nebula. The Trapezium stars are relatively young, about 100,000 years old. The Orion Nebula, though prominent at visible wavelengths, is but an H II region located along the near side of the much more extensive molecular cloud (■ Fig. 15–31).

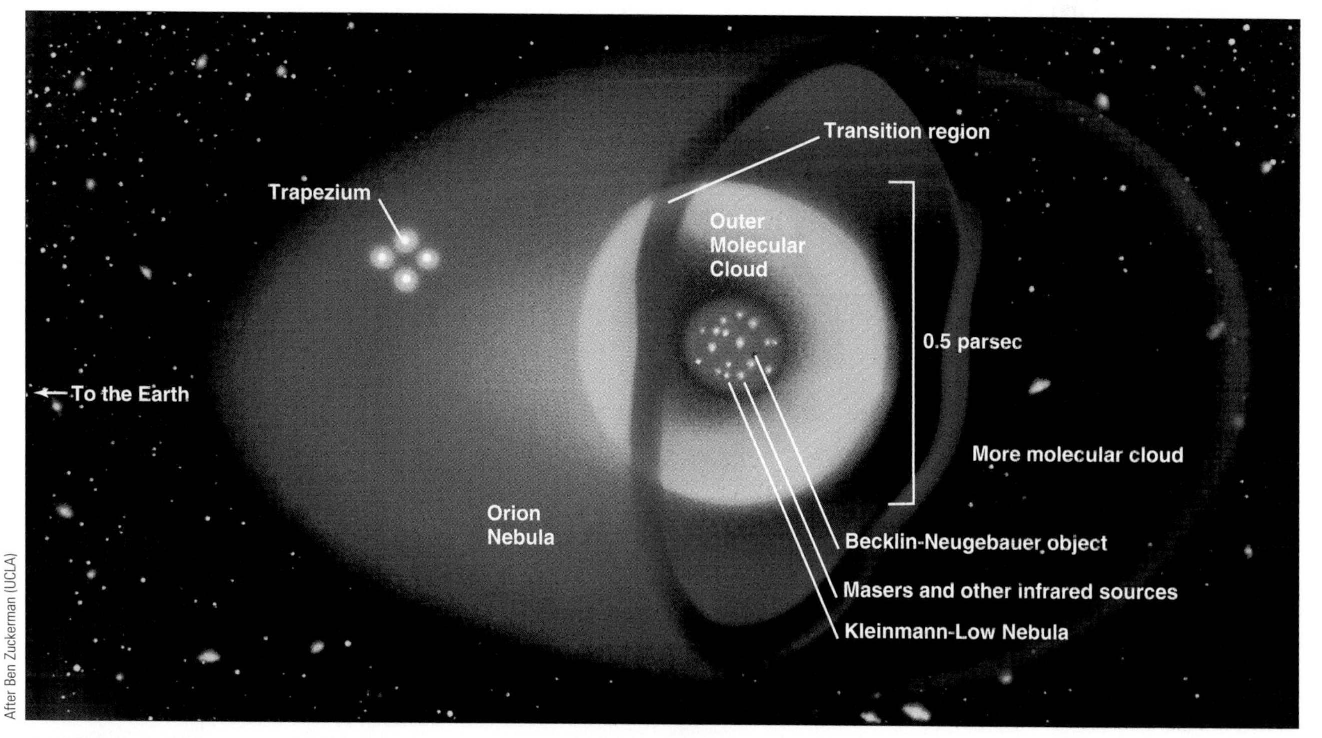

■ **FIGURE 15–31** A model of the structure of the Orion Nebula and the Orion Molecular Cloud.

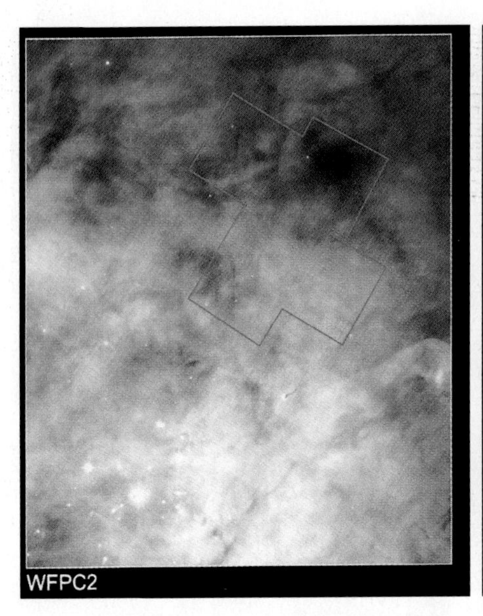

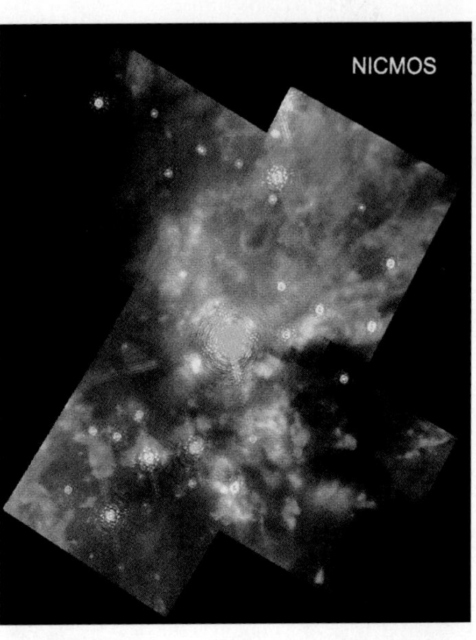

FIGURE 15–32 (a) Optical and (b) infrared views of the heart of the Orion Molecular Cloud, obtained with instruments (WFPC2 and NICMOS, respectively) on the Hubble Space Telescope. Most of the stars revealed in the infrared image were hidden from optical view; they are in a chaotic, active region of star birth. Stars and dust that has been heated by the intense starlight are shown as yellow-orange. Emission from excited hydrogen molecules is shown in blue. The brightest region is the "Becklin-Neugebauer object" (BN), a massive, young star. Outflowing gas from BN causes the bow shock above it. The view is about 0.4 light-years across the diagonal; details as small as our Solar System are visible.

WFPC2 image: C. Robert O'Dell and Shui Kwan Wong (Rice U.) and NASA; NICMOS image: Rodger Thompson, Susan Stolovy, and Marcia Rieke (U. Arizona), Glenn Schneider (Steward Obs., U. Arizona); Edwin Erickson (SETI Institute/NASA's Ames Research Center); David Axon (STScI); and NASA

The Near-Infrared Camera and Multi-Object Spectrometer (NICMOS) on the Hubble Space Telescope is able to record infrared light that had penetrated the dust, bringing us images of newly formed stars within the Orion Molecular Cloud (■ Fig. 15–32).

15.14 At a Radio Observatory

What is it like to go observing at a radio telescope? First, you decide just what you want to observe, and why. You have probably worked in the field before, and your reasons might tie in with other investigations underway. Then you decide with which telescope you want to observe, usually the most suitable one accessible to you; let us say it is the Very Large Array (VLA) of the National Radio Astronomy Observatory. You send in a written proposal describing what you want to observe and why.

Your proposal is read by a panel of scientists. If the proposal is approved, it is placed in a queue to wait for observing time. You might be scheduled to observe for a five-day period to begin six months after you submitted your proposal.

At the same time, you might apply (usually to the National Science Foundation) for financial support to carry out the research. Your proposal possibly contains requests for some salary for yourself during the summer, and salary for a student or students to work on the project with you. You are not charged directly for the use of the telescope itself—that cost is covered in the observatory's overall budget.

You carry out your observing at the VLA headquarters at Socorro, New Mexico. A trained telescope operator runs the mechanical aspects of the telescope. You give the telescope operator a computer program that includes the coordinates of the points in the sky that you want to observe and how long to dwell at each location. The telescopes (■ Fig. 15–33) operate around the clock—one doesn't want to waste any observing time.

The electronics systems that are used to treat the incoming signals collected by the radio dishes are particularly advanced. Computers combine the output from the 27 telescopes and show you a color-coded image, with each color corresponding to a different brightness level (■ Fig. 15–34). Standard image-processing packages of programs are available for you to use back home, with the radio community generally using a different package from that used in the optical community.

You are expected to publish the results as soon as possible in one of the scientific journals, often after you have given a presentation about the results at a professional meeting, such as one of those held twice yearly by the American Astronomical Society.

Jay M. Pasachoff

■ **FIGURE 15–33** The Very Large Array (VLA), near Socorro, New Mexico, with its dishes in its most compact configuration. An Extended VLA (EVLA), with additional antennas and other improvements, is under construction. A similar Giant Metrewave Radio Telescope, a set of radio telescopes spaced over many kilometers, is in operation near Pune, India.

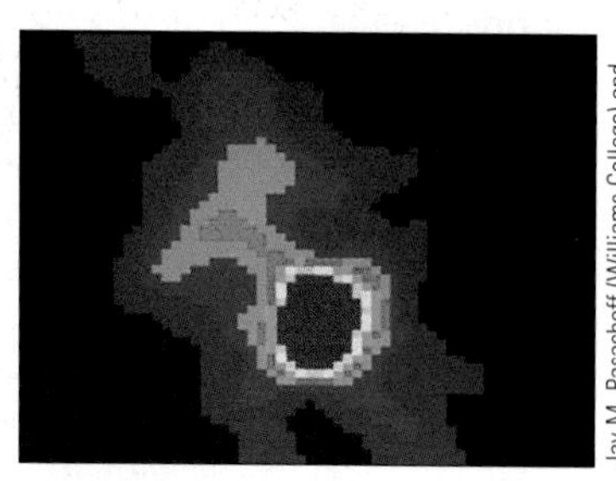

Jay M. Pasachoff (Williams College) and Donald A. Lubowich (Hofstra U.)

■ **FIGURE 15–34** The sum of all our partial maps showed the Arc near the center of our Galaxy. (See it also, at a shorter wavelength, in Figure 15–14.)

© European Southern Observatory

■ **FIGURE 15–35** A set of at least 50 linked telescopes, the Atacama Large Millimeter Array (ALMA), is to be set up on a high, dry plain in Chile. It is a major international collaboration.

Astronomy has become a very collaborative science. Many consortia of individual scientists, such as those studying distant supernovae, have dozens of members. Telescope projects have also become so huge that collaboration is necessary. The Atacama Large Millimeter Array (ALMA), to be built in Chile on a high plain where it hasn't rained in decades (■ Fig. 15–35), will use at least 50 high-precision radio telescopes as an interferometer to examine our Galaxy and other celestial objects with high resolution. It is a joint project of the United States' National Science Foundation, the European Space Agency, and Chile.

AceAstronomy™ Log into AceAstronomy and select this chapter to see the Active Figure called "Explorable Milky Way."

AceAstronomy™ Log into AceAstronomy and select this chapter to see the Astronomy Exercise "Milky Way 3D."

CONCEPT REVIEW

Gas and dust (small particles of matter) are present to some extent throughout a galaxy, between the stars; a **nebula** (plural **nebulae**) is a substantial cloud of such gas and dust (Introductory paragraph). At night, from a good observing location, we see a band of stars, nebulae, gas, and dust—the **Milky Way**—stretch across the sky (Section 15.1). It is our internal, edge-on view of our Galaxy, the **Milky Way Galaxy.** With the unaided eye, we cannot see more than a few thousand light-years through the Milky Way because of the large amount of dust; thus, it appears that we are close to the center of our Galaxy, but this is only an illusion (Section 15.2).

Emission nebulae glow because ultraviolet radiation from hot stars ionizes the gas, and electrons jumping down to lower energy levels emit light (Section 15.3). **Absorption nebulae** (**dark nebulae**) block radiation that comes from behind them. **Reflection nebulae,** like those near the stars of the Pleiades, reflect radiation; they often look blue, for the same reason that the sky is blue.

By studying the distribution of globular star clusters in the sky, Harlow Shapley deduced that we are not at the center of our Galaxy (Section 15.4). However, he overestimated our distance from the center because he didn't know about interstellar extinction, the dimming of starlight by dust. We now can measure this effect by noticing that it also **reddens** starlight: Dust preferentially scatters (reflects) or absorbs the violet and blue light, while the longer wavelengths pass through more easily.

Our Galaxy has a **nuclear bulge** centered on the **nucleus,** and surrounded by a flat **disk** that contains **spiral arms** (Section 15.4). A spherical **halo** includes the globular clusters, and has a much greater diameter than the disk. We can detect the very center of our Galaxy in infrared light, radio waves, x-rays, or gamma rays that penetrate the dust between us and it (Section 15.5). The Galactic center is a bright infrared source and almost certainly contains a very massive black hole.

Although mysterious **gamma-ray bursts** were at one time considered to be in our Galaxy, we now know that many of them originate billions of light-years away, perhaps from the formation of black holes (Section 15.6). NASA has sent the Swift satellite aloft to study them. The gamma rays and x-rays throughout our Galaxy and elsewhere are studied as part of **high-energy astrophysics.**

Our Galaxy is a pinwheel-shaped spiral galaxy, with spiral arms marked by massive stars, open clusters, and nebulae (Section 15.7). However, from our vantage point inside the Galaxy, it is difficult to accurately trace out the arms. These spiral arms appear to be caused by a slowly rotating **spiral density wave** (Section 15.8).

The matter between the stars, the **interstellar medium,** is mainly hydrogen gas (Section 15.9). Emission nebulae are mostly **H II regions,** regions of ionized hydrogen. Clouds of higher density in which the atoms of hydrogen are predominantly neutral are called **H I regions.** Hydrogen molecules (H_2) are very difficult to detect, but we think they are plentiful in regions where they are protected by dust from ultraviolet radiation.

Observations of our Galaxy at radio wavelengths have been very important (Section 15.10). Specifically, the observed **21-cm line** comes from the **spin-flip transition** of hydrogen atoms, when the spin of the electron changes relative to that of the proton. Studies of the 21-cm line have enabled us to map our Galaxy by finding the distances to H I regions (Section 15.11). Observations of interstellar molecules, primarily carbon monoxide, have also been valuable in this regard (Section 15.12).

Giant molecular clouds, containing a hundred thousand to a million times the mass of the Sun, are fundamental building blocks of our Galaxy; they are the locations at which new stars form (Section 15.13). Infrared satellites and radio telescopes have permitted mapping of the Orion Molecular Cloud and others. The Atacama Large Millimeter Array is a major international project to provide high-resolution imaging (Section 15.14).

QUESTIONS

1. Why do we think our Galaxy is a spiral?
2. How would the Milky Way appear if the Sun were closer to the edge of our Galaxy?
3. Compare **(a)** absorption (dark) nebulae, **(b)** reflection nebulae, and **(c)** emission nebulae.
4. How can something be both an emission and an absorption nebula? Explain and give an example.
5. If you see a red nebula surrounding a blue star, is it an emission or a reflection nebula? Explain.
6. Discuss the key observations that led to the discovery of the Sun's location relative to the center of our Galaxy.
7. †If the Sun is 8 kpc from the center of our Galaxy and it orbits with a speed of 200 km/sec, show that the Sun's orbital period is about 250 million years. (Assume the orbit is circular.)
8. Why may some infrared observations be made from mountain observatories while all x-ray observations must be made from space?
9. Describe infrared and radio results about the center of our Galaxy.
10. Discuss the possible explanations for gamma-ray bursts in the sky.
11. Why does the density-wave theory of spiral arms lead to the formation of stars?
12. **(a)** What are three tracers that we use for the spiral structure of our Galaxy? **(b)** What are two reasons why we expect them to trace spiral structure?
13. Discuss how infrared observations from space have added to our knowledge of our Galaxy.
14. While driving at night, you almost instinctively judge the distance of an oncoming car by looking at the apparent brightness of its headlights. Is your estimate correct if you don't account for fog along the way? Discuss your answer.
15. Describe the relation of hot stars to H I (neutral hydrogen) and H II (ionized hydrogen) regions.
16. Briefly define and distinguish between the redshift of a gas cloud and the reddening of that cloud.
17. What determines whether the 21-cm lines will be observed in emission or absorption?
18. Describe how a spin-flip transition can lead to a spectral line, using hydrogen as an example.
19. †Suppose the rest wavelength of the 21-cm line of hydrogen were exactly 21.0000 cm. **(a)** If this line from a particular cloud is observed at a wavelength of 21.0021 cm, is the cloud moving toward us or away from us? **(b)** How fast?
20. Why are dust grains important for the formation of some types of interstellar molecules?
21. Describe the relation of the Orion Nebula and the Orion Molecular Cloud.
22. Optical astronomers can observe only at night. In what time period can radio astronomers observe? Explain any difference.

23. **True or false?** By measuring the wavelength of the peak of the reflected light received from a reflection nebula, we can determine the temperature of the nebula using Wien's law.
24. **True or false?** A dark (absorption) nebula blocks the light from background stars, and is sometimes so dense that new stars are forming within it.
25. **True or false?** When the relative spin directions of the proton and electron in a hydrogen atom change, a photon having a wavelength of about 21 cm is either emitted or absorbed.
26. **True or false?** The spiral arms in a galaxy such as the Milky Way consist of the same groups of stars throughout the entire lives of these arms.
27. **True or false?** Interstellar dust and gas tend to absorb and scatter blue light more than red light, causing stars to appear redder than their true colors.
28. **Multiple choice:** Gaseous emission nebulae in the Milky Way Galaxy look red because **(a)** they are moving away from us, so that the light is redshifted; **(b)** many electrons are jumping from the third to the second energy levels of hydrogen, producing Hα emission; **(c)** they absorb red light from their surroundings; **(d)** they have temperatures of only about 100 K, and Wien's law tells us that the light they emit is therefore red; or **(e)** they are made mostly of iron compounds, like rust.
29. **Multiple choice:** In 1917, our perception of the Milky Way Galaxy changed when Harlow Shapley noticed **(a)** a high concentration of neutron stars in the halo of our Galaxy; **(b)** pulsars concentrated around the region of our Galaxy surrounding the Sun; **(c)** open star clusters located around the center of our Galaxy; **(d)** a concentration of other planetary systems near the edges of our Galaxy; or **(e)** globular star clusters centered around a point far from the Sun.
30. **Multiple choice:** At the present time, stars in our Galaxy tend to form most readily in **(a)** giant molecular clouds in spiral arms; **(b)** the Galactic halo; **(c)** the central supermassive black hole; **(d)** the Galactic bulge; or **(e)** globular clusters.
31. **Multiple choice:** The 21-cm line observed by radio astronomers comes from **(a)** electrons in hydrogen atoms jumping from the third to the second energy levels; **(b)** the rotation of hydrogen molecules; **(c)** the atomic hydrogen spin-flip transition; **(d)** dust grains in molecular clouds; or **(e)** carbon monoxide (CO) molecules.
32. **Multiple choice:** Which one of the following statements concerning the Milky Way Galaxy is *false*? **(a)** It is flattened out into a disk because of its rotation. **(b)** It is a spiral galaxy, probably of the "barred" variety. **(c)** Old globular star clusters are found primarily in its halo. **(d)** Open star clusters and emission nebulae are found primarily in its spiral arms. **(e)** Most of it can be seen from Earth at optical wavelengths with a small telescope.
33. **Fill in the blank:** To form some types of molecules in interstellar space, grains of _____ appear to be necessary.
34. **Fill in the blank:** A(n) ______ nebula can form when gas is ionized by a nearby young star, usually of spectral type O.
35. **Fill in the blank:** The most distant parts of our Galaxy are most easily seen at _____ wavelengths.
36. **Fill in the blank:** A(n) ________ nebula reflects (scatters) light from stars that are near the gas.
37. **Fill in the blank:** Studies of the motions of stars near the center of our Galaxy suggest that a massive _______ is present there.

†This question requires a numerical solution.

TOPICS FOR DISCUSSION

1. Can Harlow Shapley's conclusion regarding the Sun's location in our Galaxy be considered an extension of the Copernican revolution?
2. If the band of light called the Milky Way stretched only halfway around the sky, forming a semicircle rather than a circle, what would you conclude about the Sun's location in our Galaxy?

MEDIA

AceAstronomy™ Log into AceAstronomy at **http://astronomy.brookscole.com/cosmos3** to access quizzes and animations that will help you assess your understanding of this chapter's topics.

Log into the Student Companion Web Site at **http://astronomy.brookscole.com/cosmos3** for more resources for this chapter including a list of common misconceptions, news and updates, flashcards, and more.

CHAPTER 16

A Universe of Galaxies

At the beginning of the 20th century, the nature of the faint, fuzzy "spiral nebulae" was unknown. In the mid-1920s, Edwin Hubble showed that they are distant galaxies like our own Milky Way Galaxy, and that the Universe is far larger than previously thought. Galaxies are the fundamental units of the Universe, just as stars are the basic units of galaxies.

Like stars, many galaxies are found in clusters, and there are also superclusters separated by enormous voids. By looking back in time at very distant galaxies and clusters, we can study how they formed and evolved. Surprisingly, we now know that all these enormous structures consist largely of "dark matter" that emits little or no electromagnetic radiation.

ORIGINS

We find that galaxies, of which many billions are known to exist, are the fundamental units of the Universe. Essentially all stars, including our Sun, are within them. Understanding the birth and lives of galaxies is necessary for a complete picture of our existence.

AIMS

1. Learn about the discovery of galaxies, and become familiar with the different types of galaxies and their habitats (Sections 16.1 to 16.3).
2. Investigate dark matter and how pervasive it is (Section 16.4).
3. Discover how gravitational lensing can tell us about both luminous and dark matter (Section 16.5).
4. Understand how astronomers can study the birth and lives of galaxies by looking far back in time (Section 16.6).
5. Introduce the expansion of the Universe (Section 16.7).
6. Explore the evolution of galaxies and large-scale structure (Sections 16.8 to 16.10).

16.1 The Discovery of Galaxies

In the 1770s, the French astronomer Charles Messier was interested in discovering comets. To do so, he had to be able to recognize whenever a new fuzzy object (a candidate comet) appeared in the sky. To minimize possible confusion, he thus compiled a list of about 100 diffuse objects that could always be seen, as long as the appropriate constellation was above the horizon. Some of them are nebulae, and others are star clusters, which can appear fuzzy through a small telescope such as that used by Messier. To this day, the objects in Messier's list are commonly known by their

AceAstronomy™ The AceAstronomy icon throughout this text indicates an opportunity for you to test yourself on key concepts, and to explore animations and interactions of the AceAstronomy website at **http://astronomy.brookscole.com/cosmos3**

An infrared image of the Whirlpool Galaxy, M51, made with the Spitzer Space Telescope. (See also an optical image as Fig. 15–10*b* and an x-ray image as Fig. 16–1.) It uses false color to show radiation at four different infrared wavelengths: 3.6 μm (reproduced blue), 4.5 μm (reproduced green), 5.8 μm (reproduced orange), and 8.0 μm (reproduced red). The first two of these wavelengths show mainly stars, while the second two show mainly interstellar dust. Most of the dust is made of carbon-based organic molecules, and marks locations of future star formation. The thin spoke-like filaments linking spiral arms in the infrared image were unexpected and are unexplained.

The Whirlpool consists of a main spiral galaxy and a smaller companion with which the spiral is gradually merging. Note that the companion has little infrared emission, showing that its stars are older. The ongoing collision is thought to have caused (or influenced) the spiral structure and to be triggering star formation.

This pair of galaxies is in the constellation Canes Venatici, the Hunting Dogs, and is 31 million light-years away from Earth.

NASA/JPL-Caltech/R. Kennicutt (Univ. of Arizona)/DSS

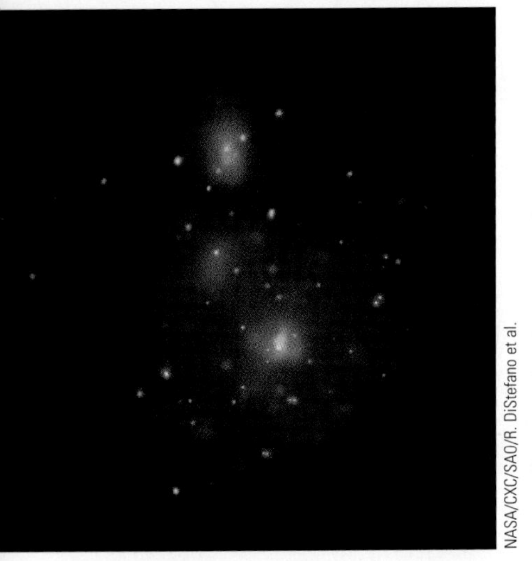

NASA/CXC/SAO/R. DiStefano et al.

■ **FIGURE 16–1** A Chandra X-ray Observatory view of M51, the galaxy seen in visible light and infrared in the chapter opener. This false-color view shows the more energetic x-rays as blue, medium energy x-rays as green, and lower energy x-rays as red. A number of x-ray sources are visible, with temperatures between 1 and 4 million kelvins.

Jack Newton

a

b

■ **FIGURE 16–2** a M81 and M82 in the constellation Ursa Major, two nearby galaxies in Charles Messier's catalogue of fuzzy, comet-like objects. b A close-up of M82 from the 8-meter Japanese national telescope, Subaru, on Mauna Kea. Extending outward from the galaxy's center are filaments of gas, seen in the red Hα light of hydrogen, tracing hydrogen that is forced outward by a superwind from numerous supernovae. The horizontal streak extending from the bright star is the result of an overloaded pixel in the CCD camera used to make the image.

Messier numbers (■ Figs. 16–1 and 16–2). They are among the brightest and most beautiful objects in the sky visible from mid-northern latitudes. A set of photographs of all the Messier objects appears as an Appendix.

Other astronomers subsequently compiled additional lists of nebulae and star clusters. By the early part of the 20th century, several thousand nebulae and clusters were known. The nebulae were especially intriguing: Although some of them, such as the Orion Nebula (■ Fig. 16–3), seemed clearly associated with bright stars in our Milky Way Galaxy, the nature of others was more controversial. When examined with the largest telescopes then available, many of them showed traces of spiral structure, like pinwheels, but no obvious stars (■ Fig. 16–4).

Photo © UC Regents/Lick Observatory

■ **FIGURE 16–3** The central region of the Orion Nebula in our Milky Way Galaxy, which is powered by the cluster of hot, massive stars. Green shows ionized oxygen.

NASA and The Hubble Heritage Team (STScI/AURA); G. R. Meurer and T. M. Heckman (JHU), C. Leitherer, J. Harris, and D. Calzetti (STScI), and M. Sirianni (JHU)

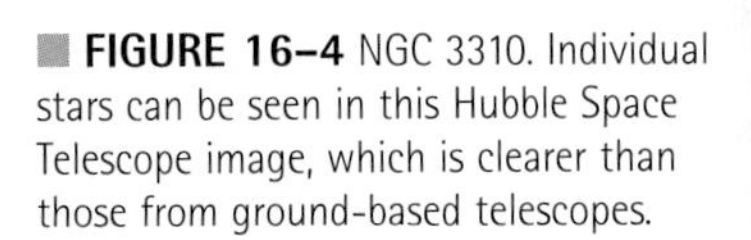

■ **FIGURE 16–4** NGC 3310. Individual stars can be seen in this Hubble Space Telescope image, which is clearer than those from ground-based telescopes.

16.1a The Shapley-Curtis Debate

Some astronomers thought that these so-called **spiral nebulae** were merely in our own Galaxy, while others suggested that they were very far away—"island universes" in their own right, so distant that the individual stars appeared blurred together. The distance and nature of the spiral nebulae was the subject of the well-publicized "**Shapley-Curtis debate,**" held in 1920 between the astronomers Harlow Shapley and Heber Curtis. Shapley argued that the Milky Way Galaxy is larger than had been thought, and could contain such spiral-shaped clouds of gas. Curtis, in contrast, believed that they are separate entities, far beyond the outskirts of our Galaxy. This famous debate is an interesting example of the scientific process at work.

Shapley's wrong conclusion was based on rather sound reasoning but erroneous measurements and assumptions. For example, one distinguished astronomer thought he had detected the slight angular rotation of a spiral nebula, and Shapley correctly argued that this change would require a preposterously high physical rotation speed if the nebula were very distant. It turns out, however, that the measurement was faulty. Shapley also argued that a bright nova that had appeared in 1885 in the Andromeda Nebula, M31, the largest spiral nebula (■ Fig. 16–5), would be far more powerful than any previously known nova if it were very distant. Unfortunately, the existence of supernovae (this object turned out to be one), which are indeed more powerful than any known nova, was not yet known.

a

b

■ **FIGURE 16–5** The Andromeda Galaxy, also known as M31 and NGC 224, is the nearest large spiral galaxy to the Milky Way. Its Hubble type is Sb, and it is accompanied by many much smaller galaxies. Two of these are the elliptical galaxies M32 *(left of middle)* and NGC 205 *(lower middle)*. These galaxies are only 2.4 million light-years from Earth. (a) In this visible-light image, the older red and orange stars give the central regions of M31 a yellowish cast, in contrast to the blueness of the spiral arms from the younger stars there. (b) Spitzer Space Telescope views of Andromeda, by observing dust at different temperatures with its different wavelength bands, show the spiral arms continuing from the outside to their base inside the previously known star-forming ring. This ring is revealed to be separate from the arms, and to have a hole in it where the satellite galaxy M32 punched through.

■ **FIGURE 16–6** Edwin Hubble, who did his groundbreaking work at the Mt. Wilson Observatory above Los Angeles, California. Here he is shown at the Palomar Observatory, on Palomar Mountain near San Diego, California.

On the other hand, Curtis's conclusion that the spiral nebulae were external to our Galaxy was based largely on an incorrect notion of our Galaxy's size; his preferred value was much too small. Moreover, he treated the nova in Andromeda as an anomaly.

16.1b Galaxies: "Island Universes"

The matter was dramatically settled in the mid-1920s, when observations made by **Edwin Hubble** (■ Fig. 16–6) at the Mount Wilson Observatory in California proved that the spiral nebulae were indeed "island universes" (now called galaxies) well outside the Milky Way Galaxy. Using the 100-inch (2.5-m) telescope, Hubble discovered very faint Cepheid variable stars in several objects, including the Andromeda Nebula.

As we saw in Chapter 11, Cepheids are named after their prototype, the variable star δ (the Greek letter "delta") Cephei. Their light curves (brightness vs. time) have a distinctive, easily recognized shape. Cepheids are intrinsically very luminous stars, 500 to 10,000 times as powerful as the Sun, so they can be seen at large distances, out to a few million light-years, with the 100-inch telescope used by Hubble.

Cepheids are very special to astronomers because measuring the period of a Cepheid's brightness variation (using what we are calling Leavitt's law, after Henrietta Leavitt) gives you its average luminosity. And comparing its average luminosity with its average apparent brightness tells you its distance, using the inverse-square law of light. The Cepheids in the spiral nebulae observed by Hubble turned out to be exceedingly distant. The Andromeda Nebula, for example, was found to be over 1 million light-years away (the value is now known to be about 2.4 million light-years)—far beyond the measured distance of any known stars in the Milky Way Galaxy.

From the distance and the measured angular size of the Andromeda Nebula, its physical size was found to be enormous. Clearly, the "spiral nebulae" were actually huge, gravitationally bound stellar systems like our own Milky Way Galaxy, not relatively small clouds of gas like the Orion Nebula (and so the Andromeda Nebula was renamed the Andromeda Galaxy).

The effective size of the Universe, as perceived by humans, increased enormously with this realization. In essence, Hubble brought the Copernican revolution to a new level; not only is the Earth just one planet orbiting a typical star among over 100 billion stars in the Milky Way Galaxy, but also ours is just one galaxy among the myriads in the Universe! Indeed, it is humbling to consider that the Milky Way is one of roughly 50 to 100 billion galaxies within the grasp of the world's best telescopes such as the Keck twins and the Hubble Space Telescope.

ASIDE 16.1: Island universes

The name "island universes" had originated with the philosopher Immanuel Kant in 1755, though at that time there was no evidence for the existence of galaxies outside our own.

16.2 Types of Galaxies

Galaxies come in a variety of shapes. In 1925, Edwin Hubble set up a system of classification of galaxies, and we still use a modified form of it.

16.2a Spiral Galaxies

There are two main "Hubble types" of galaxies. We are already familiar with the first kind—the **spiral galaxies.** The Milky Way Galaxy and its near-twin, the Andromeda Galaxy (M31; look back at Figure 16–5), are relatively large examples containing several hundred billion stars. (Most spiral galaxies contain a billion to a trillion stars.) Another near-twin is NGC 7331 (■ Fig. 16–7).

Spiral galaxies consist of a bulge in the center, a halo around it, and a thin rotating disk with embedded spiral arms. There are usually two main arms, with considerable structure such as smaller appendages. Doppler shifts indicate that spiral galaxies rotate in the sense that the arms trail.

NASA/JPL-Caltech/M. Regan (STScI), and the SINGS Team

■ **FIGURE 16–7** NGC 7331, a spiral galaxy that is thought to resemble our own Milky Way Galaxy so closely that it has been called our twin. It is 50 million light-years away, far beyond the stars in the constellation Pegasus. It was discovered in 1784 by William Herschel, who also discovered not only Uranus but also infrared light. This false-color infrared view, obtained with the Spitzer Space Telescope, reproduces wavelengths of 3.6 μm in blue, 4.5 μm in green, 5.8 μm in yellow, and 8.0 μm in red. At the shorter wavelengths, most of the radiation comes from stars, especially old, cool ones. At longer wavelengths, the starlight is less significant, and we see mainly radiation from glowing clouds of interstellar dust.

The longer wavelengths of this Spitzer image show a ring of active star formation about 20,000 light-years from the center of the galaxy. There is enough gas in the ring to produce about four billion Suns.

The three background galaxies in the image are about 10 times farther away than NGC 7331. The red dots are even more-distant galaxies. The blue dots are foreground stars in the Milky Way.

ASIDE 16.2: Our home is a barred spiral galaxy

For many years, it had been suspected that our own Milky Way Galaxy is a barred spiral, probably of the SBbc variety. New observations with the Spitzer Space Telescope provide strong evidence that our Galaxy contains a bar, possibly up to 27,000 light-years long.

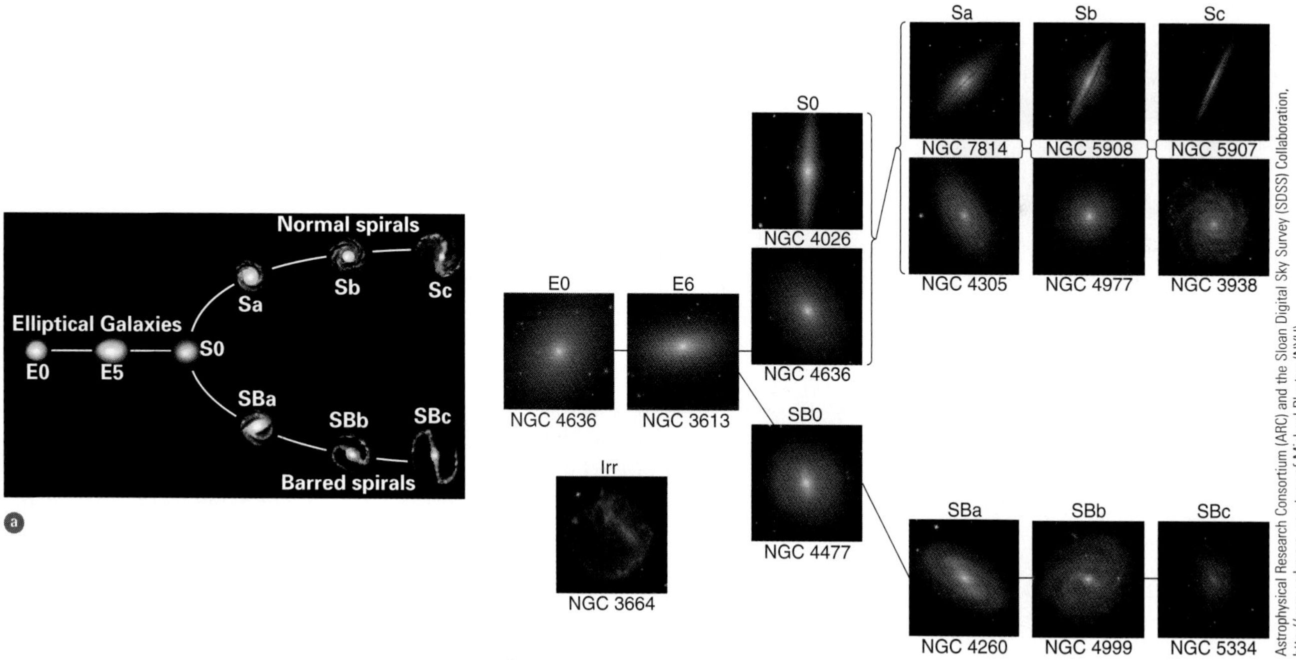

Astrophysical Research Consortium (ARC) and the Sloan Digital Sky Survey (SDSS) Collaboration, http://www.sdss.org, courtesy of Michael Blanton (NYU)

■ **FIGURE 16–8** (a) The Hubble "tuning fork" classification of galaxies. It is not meant to imply an evolutionary sequence among galaxies. (b) Images of all the Hubble types of galaxies, taken during the Sloan Digital Sky Survey. At top right we see a row of galaxies edge-on and a row face-on. The irregular (Irr) galaxy is not part of the tuning fork.

■ Figure 16–8, which is based on a very similar "tuning-fork diagram" drawn by Hubble, shows different types of spiral galaxies. Relative to the disk, the bulge is large in some spiral galaxies (known as "Sa"), which also tend to have more tightly wound spiral arms. The bulge is progressively smaller (relative to the disk) in spirals known as "Sb," "Sc," and "Sd," which also tend to have more loosely wound spiral arms. Moreover, spirals with smaller bulge-to-disk ratios generally have more gas and dust, and larger amounts of active star formation within the arms at the present time. Spiral

■ **FIGURE 16–9** ⓐ An edge-on view of NGC 4565, a type Sb spiral galaxy in the constellation Coma Berenices. The dust lane in the plane of the galaxy can be seen. ⓑ The Sombrero galaxy (M104), named because it resembles a Mexican hat, is a type Sa spiral galaxy in the constellation Virgo. It is also seen close to edge-on, with its dust lane showing prominently.

■ **FIGURE 16–10** A barred spiral galaxy (SBbc), NGC 1365, in the constellation Eridanus.

■ **FIGURE 16–11** A combination infrared/visible view of the Sombrero Galaxy, M104, already introduced in Figure 16–9*b*. The visible-light view was taken with the Hubble Space Telescope's Advanced Camera for Surveys. The infrared view (in false color), taken with the Spitzer Space Telescope, shows how the dust is concentrated in a ring.

galaxies viewed along, or nearly along, the plane of the disk (that is, "edge-on") often exhibit a dark dust lane that appears to divide the disk into two halves (■ Fig. 16–9).

In nearly one half of all spirals, the arms unwind not from the nucleus, but rather from a relatively straight bar of stars, gas, and dust that extends to both sides of the nucleus (■ Fig. 16–10). These "barred spirals" are similarly classified in the Hubble scheme from "SBa" to "SBd" (the "B" stands for "barred"), in order of decreasing size of the bulge and increasing openness of the arms (as we saw in Figure 16–8). In many cases, the distinction between a barred and nonbarred spiral is subtle. Studies show that our own Milky Way Galaxy is a barred spiral, probably of type SBbc (that is, intermediate between SBb and SBc).

Because they contain many massive young stars, the spiral arms appear bluish in color photographs. Between the spiral arms, the whitish-yellow disks of spiral galaxies contain both old and relatively young stars, but not the hot, massive, blue main-sequence stars, which have already died. Very old stars dominate the bulge, and especially the faint halo (which is difficult to see), and so the bulge is somewhat yellow/orange or even reddish in photographs (as we saw in Figure 16–5).

Since young, massive stars heat the dusty clouds from which they formed, resulting in the emission of much infrared radiation, the current rate of star formation in a galaxy can be estimated by measuring its infrared power. Space telescopes such as the Infrared Astronomical Satellite (IRAS, in the mid-1980s) and the Infrared Space Observatory (ISO, in the mid-1990s) were very useful for this kind of work, and it is being continued with the infrared camera on the Hubble Space Telescope and, at even longer infrared wavelengths, with the Spitzer Space Telescope.

About half of the energy emitted by our own Milky Way Galaxy is in the infrared, indicating that a lot of stars are being formed. But we don't know why the Andromeda Galaxy, which in optical radiation resembles the Milky Way, emits only 3 per cent of its energy in the infrared. This galaxy and the Sombrero Galaxy emit infrared mostly in a ring rather than in spiral arms (■ Fig. 16–11).

16.2b Elliptical Galaxies

Hubble recognized a second major galactic category: **elliptical galaxies** (■ Fig. 16–12). These objects have no disk and no arms, and generally very little gas and dust.

Unlike spiral galaxies, they do not rotate very much. At the present time, nearly all of them consist almost entirely of old stars, so they appear yellow/orange or even reddish in true-color photographs. The dearth of gas and dust is consistent with this composition: There is insufficient raw material from which new stars can form. In many ways, then, an elliptical galaxy resembles the bulge of a spiral galaxy.

Elliptical galaxies can be roughly circular in shape (which Hubble called type E0), but are usually elongated (from E1 to E7, in order of increasing elongation). Since the classification depends on the observed appearance, rather than on the intrinsic shape, some E0 galaxies must actually be elongated, but are seen end-on (like a cigar viewed from one end).

Most ellipticals are dwarfs, like the two main companions of the Andromeda Galaxy (look back at Figures 16–5 and 16–12*b*), containing only a few million solar masses—a few per cent of the mass of our Milky Way Galaxy. Some, however, are enormous, consisting of a few trillion stars in a volume several hundred thousand light-years in diameter (Fig. 16–12*a*). Many ellipticals may have resulted from two or more spiral galaxies colliding and merging, as we will discuss later.

a

b

FIGURE 16–12 a M87 (NGC 4486), a giant elliptical galaxy in the constellation Virgo. Two more-distant galaxies appear to the lower right. b The central region of the small elliptical galaxy M32 (see also Figure 16–5), imaged with the Hubble Space Telescope.

16.2c Other Galaxy Types

"Lenticular" galaxies (also known as "S0" [pronounced "ess-zero"] galaxies) have a shape resembling an optical lens; they combine some of the features of spiral and elliptical galaxies. They have a disk, like spiral galaxies. On the other hand, they lack spiral arms, and generally contain very little gas and dust, like elliptical galaxies. Hubble put them at the intersection between spiral and elliptical galaxies in his classification diagram (which we showed as Figure 16–8). Though sometimes called "transition galaxies," this designation should not be taken literally: The diagram is not meant to imply that spiral galaxies evolve with time into ellipticals (or vice versa) in a simple manner, contrary to the belief of some astronomers decades ago.

A few per cent of galaxies at the present time in the Universe show no clear regularity. Examples of these "irregular galaxies" include the Small and Large Magellanic Clouds, small satellite galaxies that orbit the much larger Milky Way Galaxy (■ Fig. 16–13). Sometimes traces of regularity—perhaps a bar—can be seen. Irregular galaxies generally have lots of gas and dust, and are rapidly forming stars. Indeed, some of them emit 10 to 100 times as much infrared as optical energy, probably because the rate of star formation is greatly elevated.

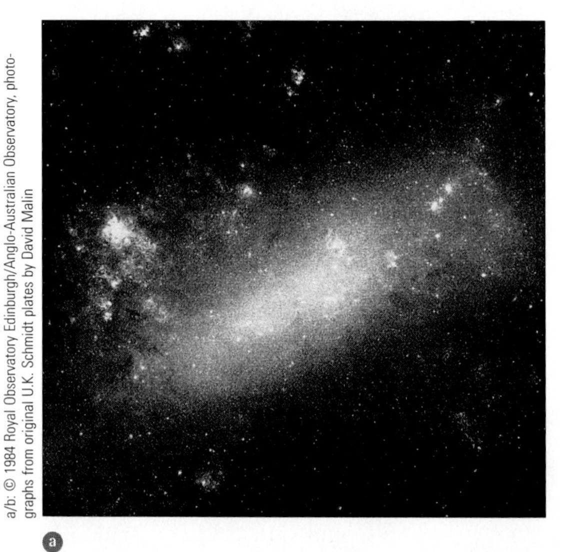

a

b

FIGURE 16–13 a The Large Magellanic Cloud and b the Small Magellanic Cloud–small, irregular galaxies that orbit around our Milky Way Galaxy. Their respective distances from us are about 170,000 light-years and 210,000 light-years. The Clouds were first reported to Europe by 16th-century Portuguese navigators who had travelled to the Cape of Good Hope at the southern tip of Africa. They were named a few decades later in honor of Ferdinand Magellan, who was then circumnavigating the globe by that route (although Magellan himself died before completing the journey).

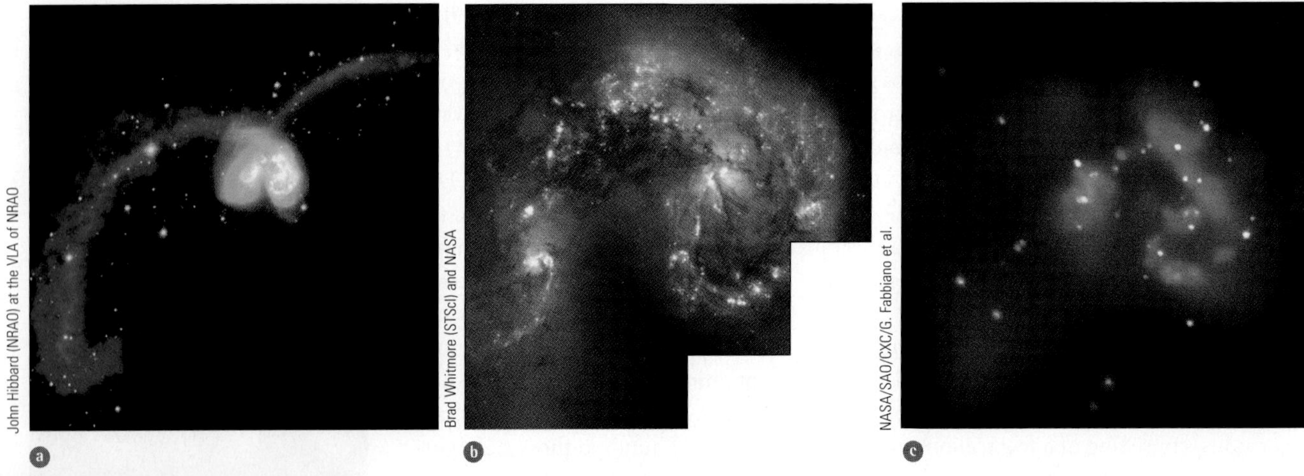

John Hibbard (NRAO) at the VLA of NRAO

Brad Whitmore (STScI) and NASA

NASA/SAO/CXC/G. Fabbiano et al.

NASA/JPL-Caltech/Z. Wang (Harvard-Smithsonian CfA)

■ **FIGURE 16–14** ⓐ The peculiar, interacting spiral galaxies NGC 4038 and NGC 4039, known as The Antennae. False color (optical: green and white; radio: blue) is used to highlight the tidal tails in this view. ⓑ A Hubble Space Telescope close-up of The Antennae that shows details of gas and dust in the two central regions. ⓒ A Chandra X-ray Observatory view. The dozens of bright points are neutron stars or black holes pulling gas off nearby stars. The bright patches come from the accumulated power of thousands of supernovae. ⓓ A Spitzer Space Telescope view, showing glowing dust.

Some galaxies are called "peculiar." These often look roughly like spiral or elliptical galaxies, but have one or more abnormalities. For example, some peculiar galaxies look like interacting spirals (■ Fig. 16–14), or like spirals without a nucleus (that is, like rings, ■ Fig. 16–15), or like ellipticals with a dark lane of dust and gas. The ring galaxies are the result of collisions of galaxies.

16.3 Habitats of Galaxies

Most galaxies are not solitary; instead, they are generally found in gravitationally bound binary pairs, small groups, or larger **clusters of galaxies.** Binary and multiple galaxies consist of several members. An example is the Milky Way Galaxy with its two main companions (the Magellanic Clouds, ■ Fig. 16–16), or Andromeda and its two main satellites (as we saw in Figure 16–5). Both Andromeda and the Milky Way have several even smaller companions.

Curt Struck and Philip Appleton (Iowa State U.), Kirk Borne (Hughes STX Corp.), Ray Lucas (STScI), and NASA

■ **FIGURE 16–15** A Hubble Space Telescope image of the Cartwheel Galaxy, in the constellation Sculptor, a ring galaxy probably caused by one of its satellite galaxies passing through it.

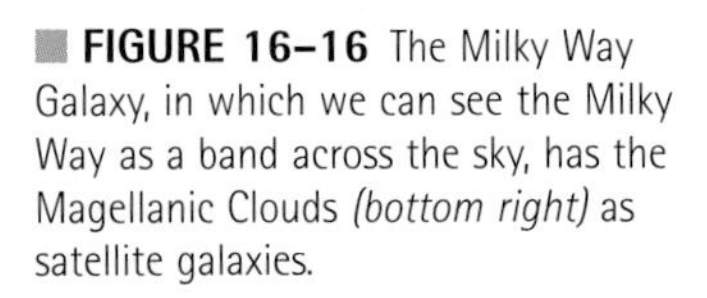

Axel Mellinger, Universität Potsdam

■ **FIGURE 16–16** The Milky Way Galaxy, in which we can see the Milky Way as a band across the sky, has the Magellanic Clouds *(bottom right)* as satellite galaxies.

Galaxies and clusters of galaxies all over the Universe are studied with the Hubble Space Telescope in the ultraviolet, visible, and near-infrared, the Spitzer Space Telescope in the infrared, and the Chandra X-ray Observatory in x-rays. NASA's Galaxy Evolution Explorer (GALEX), launched in 2003, is a small satellite that is studying galaxies and surveying the sky in the ultraviolet.

16.3a Clusters of Galaxies

The Local Group is a small cluster of about 30 galaxies, some of which are binary or multiple galaxies. Its two dominant members are the Andromeda (M31) and Milky Way Galaxies. M33, the Triangulum Galaxy (■ Fig. 16–17), is a smaller spiral. M31 and M33, at respective distances of 2.4 and 2.6 million light-years, are the farthest objects you can see with your unaided eye (see *Star Party 16.1: Observing Galaxies*). The Local Group also contains four irregular galaxies, at least a dozen dwarf irregulars (■ Fig. 16–18), four regular ellipticals, and the rest are dwarf ellipticals or the related "dwarf spheroidals." The diameter of the Local Group is about 3 million light-years.

The Virgo Cluster (in the direction of the constellation Virgo, but far beyond the stars that make up the constellation), at a distance of about 50 million light-years, is the largest relatively nearby cluster (■ Fig. 16–19). It consists of at least 2000 galaxies spanning the full range of Hubble types, covering a region in the sky over 15° in diameter—about 15 million light-years. The Coma Cluster of galaxies (in the direction of the constellation Coma Berenices) is very rich, consisting of over 10,000 galaxies at a distance of about 300 million light-years (■ Fig. 16–20).

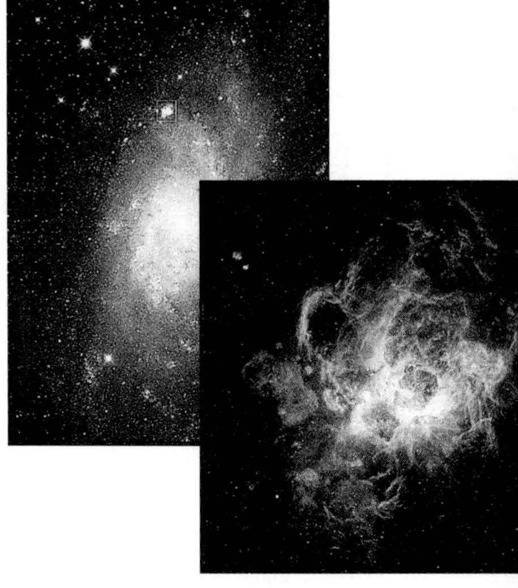

■ **FIGURE 16–17** The Triangulum Galaxy, M33, a member of the Local Group. It is about 2.6 million light-years away, in nearly the same direction as the Andromeda Galaxy but considerably smaller. At lower right we see a nebula in M33, imaged with the Hubble Space Telescope.

Black-and-white image from Palomar Observatory, Caltech; color HST image from Hui Yang (U. Illinois) and NASA

© 1981 Royal Observatory Edinburgh/Anglo-Australian Observatory, photograph from original U.K. Schmidt plates by David Malin

■ **FIGURE 16–19** A visible-light image of the central part of the extensive Virgo Cluster of galaxies (in the direction of the constellation Virgo); with a distance of about 50 million light-years, it is the nearest large cluster to the Milky Way Galaxy. Fewer than a dozen of the roughly 2000 cluster galaxies are shown here.

■ **FIGURE 16–18** This Hubble view of the Sagittarius Dwarf Irregular Galaxy shows its hot, bluish stars at a distance of 3.5 million light years. Hubble's Advanced Camera for Surveys is able to resolve individual stars in such nearby members of the Local Group of galaxies. Studying them reveals how stars are still forming out of the gas that is present. Background objects, usually reddish/brown, often show spiral arms; they are galaxies at much larger distances.

This program marks an international collaboration between Italian (Yazan Momany, Enrico V. Held, and Marco Gullieuszik), E.S.O. (Ivo Saviane and Luigi Bedin), U.S.A. (Michael Rich, Luca Rizzi), and Dutch (Konrad Kuijken) institutions. This image was produced by The Hubble Heritage Team (STScI/AURA). Credit: NASA, ESA, and The Hubble Heritage Team (STScI/AURA). Acknowledgment: Y. Momany (University of Padua)

NOAO/Nigel Sharp

■ **FIGURE 16–20** The extremely rich Coma Cluster (in the direction of the constellation Coma Berenices), about 300 million light-years away and containing over 10,000 galaxies, only a fraction of which can be seen in this visible-light image. There are two dominant, large elliptical galaxies.

Star Party 16.1 Observing Galaxies

The stars that make up the outlines of constellations, and essentially all other stars that you see in the night sky (even faint ones that require binoculars or a small telescope), are part of our Milky Way Galaxy. But a separate "island universe," the Andromeda Galaxy (M31; see Figure 16–5), is visible to the unaided eye if you know where to look.

On a clear, dark, moonless night you can see it as a faint, fuzzy patch of light between the brightest stars of the constellations Andromeda and Cassiopeia (consult the Sky Maps inside the back cover of this text). Through binoculars, it can be traced over an arc length of several degrees. (The width of your thumb at the end of your outstretched arm covers an arc length of about 2 degrees.) Ponder that you are seeing light that left the Andromeda Galaxy about 2.4 million years ago, around the time when early hominids were developing on Earth!

The smaller and fainter galaxy M33 (Fig. 16–17), 2.6 million light-years away, is also sometimes visible in the constellation Triangulum near Andromeda, but it is considerably more difficult to see than M31.

With a good pair of binoculars, it is possible to just barely see the galaxies M81 and M82 (Fig. 16–2). They are about 12 million light-years away, in the direction of the constellation Ursa Major. M81 is a type Sb spiral galaxy, while M82 is an irregular galaxy with a very large amount of gas and dust. Both galaxies are much easier to see with your college telescope, if one is available.

If you have access to a telescope, try looking for galaxies in the Virgo Cluster. In clear, dark skies, many galaxies will be visible in a relatively small area of the sky. The light that you see actually left those galaxies about 50 million years ago, not long after the time when dinosaurs became extinct on Earth and long before people walked the Earth.

A majority of the galaxies in rich clusters are ellipticals, not spirals. There is often a single, very large central elliptical galaxy (sometimes two) that is cannibalizing other galaxies in its vicinity, growing bigger with time (■ Fig. 16–21).

X-ray observations of rich clusters reveal a hot intergalactic gas (10 million to 100 million K) within them, containing as much (or more) mass as the galaxies themselves (■ Fig. 16–22). The gas is clumped in some clusters, while in others it is spread out more smoothly with a concentration near the center. This may be an evolutionary effect; the clumps occur in clusters that only recently formed from the gravitational attraction of their constituent galaxies and groups of galaxies. As clusters age, the gas within them becomes more smoothly distributed and partly settles toward the center.

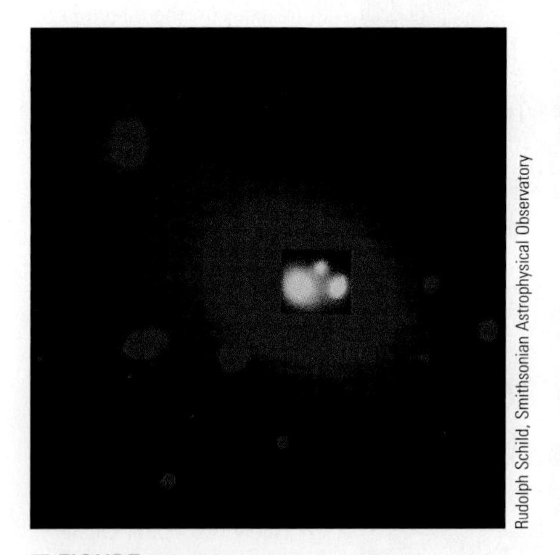

Rudolph Schild, Smithsonian Astrophysical Observatory

■ **FIGURE 16–21** An example of galactic cannibalism. Computer processing displays the central region of the large galaxy in yellow, and we also see two smaller galaxies in the process of being swallowed. The large galaxy, NGC 6166, is the central one in the rich cluster Abell 2199.

NASA/Chandra X-ray Observatory Center/SAO

■ **FIGURE 16–22** A false-color x-ray image of 3C 295, a cluster of galaxies, taken with the Chandra X-ray Observatory. We see more irregularities than had been expected from earlier observations of clusters of galaxies, which didn't show such faint detail. It appears that the cluster has been built up from smaller elements like the constituent galaxies or small groups of galaxies.

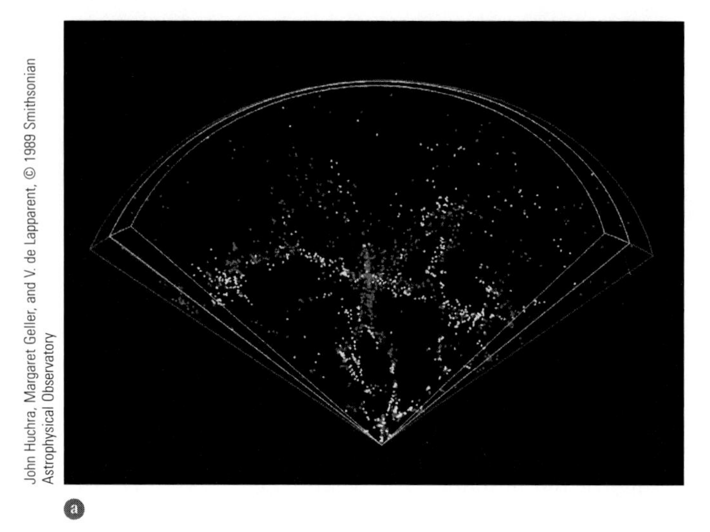

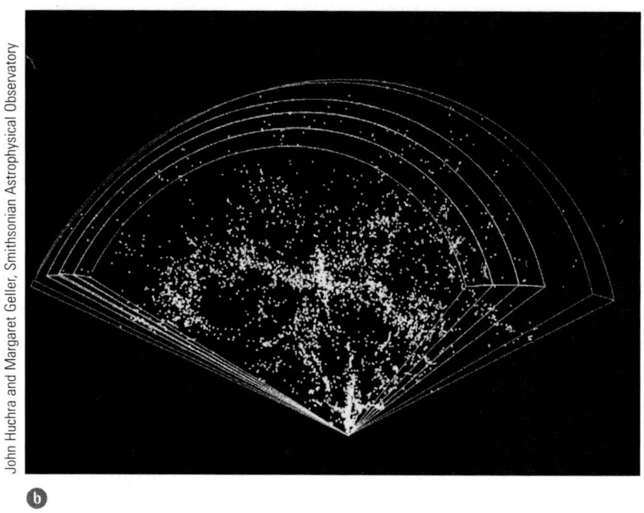

FIGURE 16–23 "Slices of the Universe," illustrating the distribution of galaxies, a pioneering study that later led to the large-scale mapping by 2dF and Sloan that we will discuss later in this chapter. Each slice shows distance from the Earth (going along straight lines outward from the Earth, which is at the bottom point) versus position on the sky (in right ascension) for all galaxies in a strip of declination 6° in size. (Right ascension corresponds to longitude and declination to latitude on the sky; see Chapter 4.) Each wedge extends out to a distance of about one billion light-years. (a) Distance vs. position of galaxies in two adjacent slices centered on declinations of 29.5° (pink dots) and 35.5° (white dots). (b) Positions of galaxies in four adjacent slices. Note how the structures apparent in one slice continue to the next, indicating that the galaxies define giant bubbles, like the suds in a kitchen sink, or perhaps long, sponge-like structures in space. The Coma Cluster of galaxies is in the center, apparently at the intersection of several bubbles.

16.3b Superclusters of Galaxies

Clusters are seen to vast distances, in a few cases up to 8 billion light-years away. When we survey their spatial distribution, we find that they form clusters of clusters of galaxies, appropriately called **superclusters.** These vary in size, but a typical diameter is about 100 million light-years. The Local Group, dozens of similar groupings nearby, and the Virgo Cluster form the Local Supercluster.

Superclusters often appear to be elongated and flattened. The thickness of the Local Supercluster, for example, is only about 10 million light-years, or one tenth of its diameter. Superclusters tend to form a network of bubbles, like the suds in a kitchen sink (Fig. 16–23). Large concentrations of galaxies (that is, several adjacent superclusters) surround relatively empty regions of the Universe, called **voids,** that have typical diameters of about 100 million light-years (but sometimes up to 300 million light-years). They formed as a consequence of matter gravitationally accumulating into superclusters; the regions surrounding the superclusters were left with little matter.

Does the clustering continue in scope? Are there clusters of clusters of clusters, and so on? The present evidence suggests that this is not so. Surveys of the Universe to very large distances do not reveal many obvious super-superclusters. There are, however, a few giant structures such as the "Great Wall" that crosses the center of the slices shown in Figure 16–23. We will discuss in Chapter 19 how the "seeds" from which these objects formed were visible within 400,000 years after the birth of the Universe.

ASIDE 16.3: The distribution of galaxies

Galaxies rarely live alone. A useful analogy is to consider isolated houses (galaxies), small villages (loose groups like the Local Group), towns and cities (small and large clusters), and metropolitan areas. Even isolated houses (say, in the countryside) are usually accompanied by smaller buildings such as barns and tool sheds, just as large galaxies often have smaller companions.

16.4 The Dark Side of Matter

There are now strong indications that much of the matter in the Universe does not emit any detectable electromagnetic radiation, but nevertheless has a gravitational influence on its surroundings. One of the first clues was provided by the flat (nearly constant) **rotation curves** of spiral galaxies.

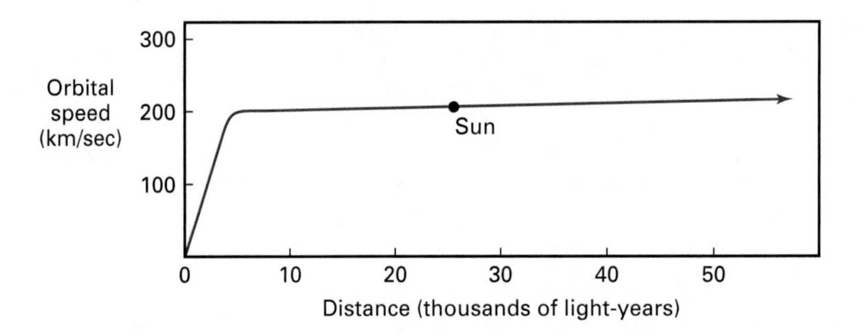

FIGURE 16–24 The rotation curve of our Milky Way Galaxy. The vertical axis shows the speed with which stars travel as they orbit the center, and the horizontal axis is their distance from the center. Note how constant ("flat") the speed is beyond distances of a few thousand light-years; it certainly doesn't decline, and might even rise a little.

16.4a The Rotation Curve of the Milky Way Galaxy

The rotation curve of any spinning galaxy is a plot of its orbital speed as a function of distance from its center. For example, the rotation curve of the Milky Way Galaxy has been determined through studies of the motions of stars and clouds of gas (Fig. 16–24). It rises from zero in the center, to a value of about 200 km/sec at a radial distance of about 5000 light-years. The rotation curve farther out is rather "flat"; the orbital speed stays constant, all the way out to distances well beyond that of the Sun.

The speed of a star at any given distance from the center is determined by the gravitational field of matter enclosed within the orbit of that star—that is, by the matter closer to the center. (It can be shown that matter at larger distances, outside the star's orbit, does not affect the star's speed as long as the galaxy's disk has a smooth, symmetric distribution of matter.) So, we can use the rotation curve to map out the distribution of mass within our Galaxy. The speeds and distances of stars near our Galaxy's edge, for example, are used to measure the mass in the entire Galaxy.

Specifically, Kepler's third law can be manipulated to give an expression for the mass *(M)* enclosed within an orbit of radius *R* from the center (see *Figure It Out 16.1: Calculating the Mass from the Rotation Curve*). A similar method is used to find the amount of mass in the Sun by studying the orbits of the planets, or the amount of a planet's mass by observing the orbits of its moons. In the 17th century, Newton developed this technique as part of his derivation and elaboration of Kepler's third law of orbital motion (see Chapter 5).

If we insert the Sun's distance from the Galactic center (26,000 light-years) and the Sun's orbital speed (200 km/sec) into the formula, we find that the matter within the Sun's orbit has a mass of about 100 billion (10^{11}) solar masses! But the mass of a typical star is about half that of the Sun. Thus, if most of the matter in our Galaxy is in the form of stars (rather than interstellar gas and dust, black holes, etc.), we conclude that there are about *200 billion stars* within the Sun's orbit, closer to the Galaxy's center than the Sun is.

The next thing to notice is that the flat rotation curve of the Milky Way Galaxy (Fig. 16–24) is quite different from the rotation curve of the Solar System. As discussed in Chapter 5, the orbital speeds of distant planets are slower than those of planets near the Sun (see *Figure It Out 5.3: Orbital Speed of Planets*), instead of being roughly independent of distance. In the Solar System, the Sun's mass greatly dominates all other masses; the masses of the planets are essentially negligible in comparison with the Sun. But in the Milky Way Galaxy, the flat rotation curve implies that except in the central region (where the rotation curve isn't flat), the mass grows with increasing distance from the center.

The growth in mass of our Galaxy continues to large distances beyond the Sun's orbit. (The rotation curve way out from the center is usually determined from the measured speeds of clouds of hydrogen gas, which can be easily seen at radio wavelengths.) This is very puzzling because few stars are found in those regions: The number of stars falls far short of accounting for the derived mass.

For example, at a distance of 130,000 light-years from the center, the enclosed mass is about 5×10^{11} solar masses, and the corresponding number of typical stars

FIGURE IT OUT 16.1

Calculating the Mass from the Rotation Curve

The rotation curve, a plot of orbital speed versus distance from the orbit's center, can be used to calculate the mass enclosed within an object's orbit. We shall first apply it to our Solar System; see also *Figure It Out 5.3: Orbital Speed of Planets.*

Recall from Chapter 5 that Newton's generalized version of Kepler's third law is

$$P^2 = \frac{4\pi^2}{G(m_1 + m_2)} R^3,$$

where m_1 and m_2 are the masses of the two bodies (1 and 2). Suppose m_2 (a planet) is negligible relative to m_1 (the Sun). Then we can ignore m_2 in the sum $(m_1 + m_2)$, and Kepler's third law becomes

$$P^2 = \frac{4\pi^2}{Gm_1} R^3.$$

If the planet's orbit is circular (which is roughly true), then the circumference of the orbit ($2\pi R$) must equal the planet's speed multiplied by the period: $2\pi R = vP$. This is merely an application of *distance* = *speed* × *time,* which is correct for constant speed. Thus, $P = 2\pi R/v$. If we now substitute this into the above equation for P^2 and rearrange terms, we find that

$$m_1 = v^2R/G.$$

So, knowing Earth's distance from the Sun ($R = 1.5 \times 10^8$ km) and its orbital speed ($v = 30$ km/sec), we can deduce that the Sun's mass is $m_1 = 2 \times 10^{33}$ grams. This is about 330,000 times the Earth's mass, thereby justifying our assumption that Earth's mass (m_2) is negligible.

The formula $m_1 = v^2R/G$ also turns out to give the mass of a galaxy (m_1) within a radius R from its center. Let's use $R = 26{,}000$ light-years and $v = 200$ km/sec, as is the case for the Sun in the Milky Way Galaxy. Being careful to keep track of units, we find that the mass within the Sun's orbit is

$$\begin{aligned} m_1 &= (200\ \text{km/sec})^2(10^5\ \text{cm/km})^2(26{,}000\ \text{lt yr}) \\ &\quad \times (9.5 \times 10^{17}\ \text{cm/lt yr})/(6.673 \times 10^{-8}\ \text{cm}^3/\text{g/sec}^2) \\ &= 1.5 \times 10^{44}\ \text{g}. \end{aligned}$$

But the mass of the Sun is about 2×10^{33} g, so $m_1 = (1.5 \times 10^{44}\ \text{g})/(2 \times 10^{33}\ \text{g/solar mass}) \approx 7.4 \times 10^{10}$ solar masses, or roughly 10^{11} solar masses, as stated in the text.

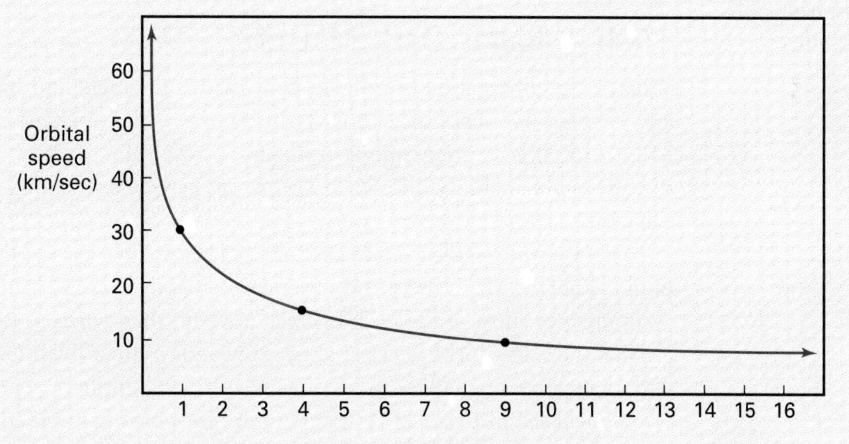

The rotation curve of the Solar System, showing the orbital speeds for planets (or particles) at different distances from the Sun. Earth, with a distance of 1 A.U., has an orbital speed of 30 km/sec, as marked. Particles at distances of 4 A.U. and 9 A.U. would have orbital speeds of 15 km/sec and 10 km/sec, respectively. See also *Figure It Out 5.3: Orbital Speed of Planets.*

(each having half a solar mass) would be about a trillion—yet there are too few stars visible, by a large margin. Indeed, studies of the outer parts of the Milky Way Galaxy throughout the electromagnetic spectrum do not reveal sufficient quantities of material to account for the derived mass.

16.4b Dark Matter Everywhere

We conclude that the Milky Way Galaxy contains large quantities of **"dark matter"**—material that exerts a gravitational force, but is invisible or at least very difficult to see!

■ **FIGURE 16–25** Vera Rubin, who was the first to show that many spiral galaxies have flat rotation curves and thus contain substantial dark matter.

This material has sometimes been called the "missing mass," especially in older texts, but the term is not appropriate because the *mass* is present. Instead, it is the *light* that's missing.

Many other spiral galaxies also have flat (speed roughly constant) rotation curves, as was first shown by Vera Rubin (■ Fig. 16–25). Estimates suggest that 80 to 90 per cent of the mass of a typical spiral galaxy consists of dark matter. However, it has been shown that the amount of matter in the disk cannot exceed what is visible by more than a factor of 2. Instead, the dark matter is probably concentrated in an extended, spherical, outer halo of material that extends to perhaps 200,000 light-years from the galactic center.

Similar studies show that elliptical galaxies also contain large amounts of dark matter. Gas and stars are moving so quickly that they would escape if the visible matter alone produced the gravitational field. There must be another, more dominant, contribution to gravity in these galaxies.

The orbital speeds of galaxies in binary pairs, groups, and clusters can be used to determine the masses of these systems. Astronomers find that in essentially all cases, the amount of mass required to produce the observed orbital speeds is larger than that estimated from the visible light (which is assumed to come from stars and gas). A related technique is to measure the typical speeds of particles of gas bound to a cluster of galaxies—and once again, the particles could not be gravitationally bound to the cluster if its mass consisted only of that provided by the visible matter.

Decades ago, the Caltech astronomer Fritz Zwicky was the first to point out that clusters of galaxies could not remain gravitationally bound if they contain only visible matter. He postulated the existence of some form of dark matter. However, this idea was largely ignored or dismissed—it was too far ahead of its time.

ASIDE 16.4: Some distant stars

Some galaxies, such as NGC 300, show quite a few faint stars at enormous distances from their centers. To see these stars, one must search very carefully. Nevertheless, dark matter dominates the outer regions of most galaxies.

16.4c What Is Dark Matter?

What is the physical nature of the dark matter in single and binary galaxies, groups, and clusters? We just don't know—this is one of the outstanding unsolved problems in astrophysics. A tremendous number of very faint normal stars (or even brown dwarfs) is a possibility, though it seems unlikely, extrapolating from the numbers of the faintest stars that we can study.

There is some evidence (see Section 16.5 below) that part of the dark matter consists of old white dwarfs. If these and other corpses of dead stars (neutron stars, black holes) accounted for most of the dark matter, however, then where is the chemically enriched gas that the stars must have ejected near the ends of their lives? Other candidates for the dark matter are small black holes, massive planets ("Jupiters"), and neutrinos.

We will see in Chapter 19 that certain kinds of measurements indicate that only a small fraction of the dark matter can consist of "normal" particles such as protons, neutrons, and electrons; the rest must be exotic particles. Most of the normal dark matter consists of tenuous, million-degree gas in galactic halos. This gas was recently detected by the absorption spectra it produced in the radiation from background objects, and also from its emission at relatively long x-ray wavelengths. Though no longer technically "dark" (because we have seen it!), such matter is still generally considered to be part of the "dark matter" that pervades the Universe; it is difficult to detect.

Probably the most likely candidate for a majority of the dark matter (the "abnormal" part) is undiscovered subatomic particles with unusual properties, left over from the big bang, such as **WIMPs**—"weakly interacting massive particles." Physicists studying the fundamental forces of nature suggest that many WIMPs exist, though it is disconcerting that none has been unambiguously detected in a laboratory experiment.

If it is unsatisfactory to you that most of the mass in our Galaxy (and indeed, most of the mass in the Universe!) is in some unknown form, you may feel better by knowing that astronomers also find the situation unsatisfactory. But all we can do is go out and conduct our research, and try to find out more. Clever new techniques, such as one described below, may provide the crucial clues that we seek.

16.5 Gravitational Lensing

The phenomenon of **gravitational lensing** of light provides a powerful probe of the amount (and in some cases the nature) of dark matter. About one hundred cases of gravitational lensing have been found thus far.

If the light from a distant object passes through a gravitational field, the light is bent—that is, it follows a curved path through the warped space–time. This is analogous to (but differs in detail from) the effect that a simple glass lens has on light. It is, perhaps, more akin to the warping of images we get from a fun-house mirror, but the terms "lens" and "lensing" have caught on.

We have already encountered lensing in Chapter 10: Recall that Einstein's general theory of relativity predicts that the Sun should bend the light of stars beyond it by an amount twice that predicted with Newtonian theory, and that this effect was first measured in 1919 by Arthur Eddington and others during a total solar eclipse.

If an observer, a galaxy acting as a gravitational lens, and a more distant, very compact object are nearly perfectly aligned (colinear), the distant object will look like a circle (known as an "Einstein ring") centered on the lens (■ Fig. 16–26). More usually, deviations from symmetry (for example, slight misalignment of the lens) lead to the formation of several discrete, well-separated images (■ Fig. 16–27). The apparent brightness of the object is magnified, in some cases by large amounts.

ASIDE 16.5: Gravitational mirages

The Einstein ring, or the separation between different images of a gravitationally lensed object, is often so small that the object appears spatially unresolved (that is, point-like) even through a telescope. The object does, however, look brighter than it would have without being lensed.

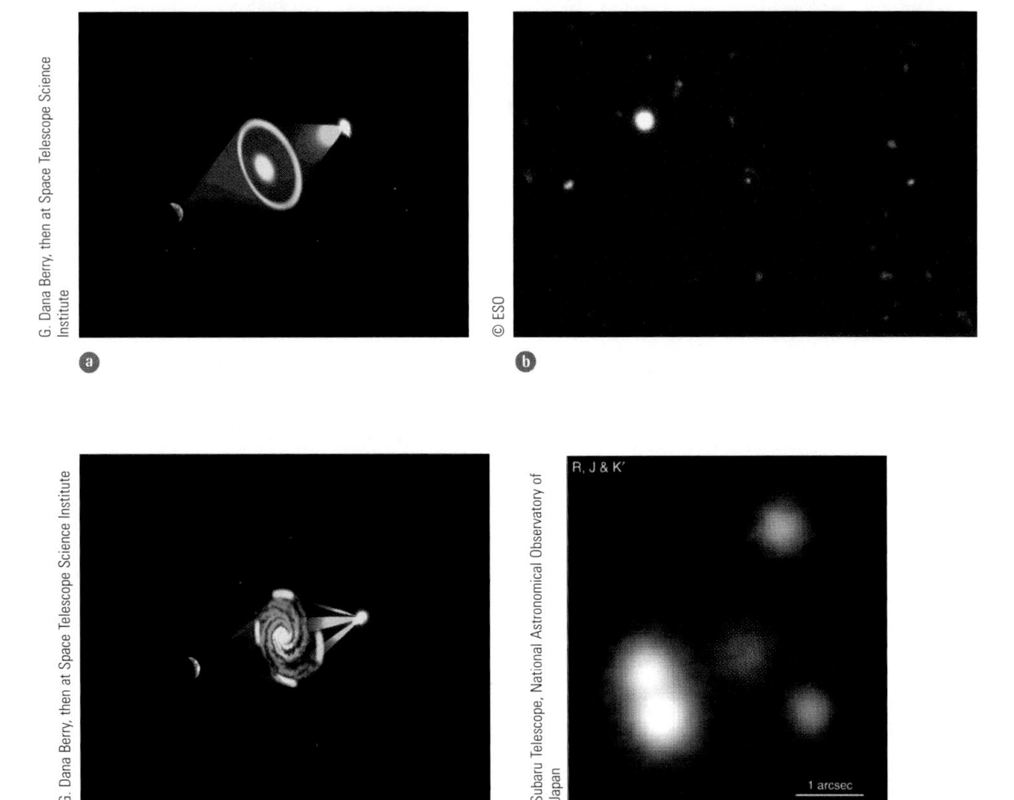

■ **FIGURE 16–26** ⓐ Simulation of the formation of an "Einstein ring" from the gravitational lensing of a compact object by a lensing object, a galaxy or cluster of galaxies, that is along a straight line between the observer and the compact object. The apparent brightness of the lensed object is increased due to the magnification of the gravitational lens. Einstein would be very pleased with the actual discovery of such objects; he had thought that the effect–predicted not only by himself but also by one earlier theorist–would never be observed. ⓑ This cosmic mirage, most of a circular Einstein ring, popped into view when observed with one of the 8.2-m telescopes of the Very Large Telescope Array in Chile. The distances of the lensing galaxy and the lensed galaxy are nearly 8 and 12 billion light-years, respectively. The magnification factor of at least 13 helps us see the lensed galaxy at a time when the Universe was only about 12% of its current age.

■ **FIGURE 16–27** ⓐ A slight misalignment of the lensing object leads instead to the formation of several discrete images of the compact lensed object. ⓑ A visible-light image of a lensed object (a quasar, as described in the next chapter), PG1115+080, that was not directly in line with the lensing object.

■ **FIGURE 16–28** If the lensed object is an extended (rather than a compact) galaxy, the lensed images often appear as arcs. (a) Distant galaxies gravitationally lensed by the cluster Abell 2218 appear as arcs centered on the cluster and imaged with the Hubble Space Telescope. The bluish color of many of the lensed galaxies indicates that their rate of star formation is high. (b) A set of rings, arcs, and other gravitationally lensed patterns detected by the Hubble Space Telescope in various sky directions.

When a massive cluster of galaxies lenses many distant galaxies, a collection of arcs tends to be seen (■ Fig. 16–28). The number of arcs, their magnification factors, and their distorted pattern depend on the mass of the cluster, as well as on the distribution of mass within the cluster. This method measures the total mass (visible and dark) in the cluster, and gives results consistent with those obtained from other techniques. Again, we conclude that dark matter dominates most clusters of galaxies.

Projects in which the apparent brightness of millions of stars in the Large and Small Magellanic Clouds are systematically monitored have revealed that occasionally, a star brightens and fades over the course of a few weeks. (These galaxies provide a nice background field of stars that are out of the Galactic plane.) The light curve has exactly the shape expected if a compact lens were to pass between the star and us, temporarily focusing the star's light toward us (■ Fig. 16–29). Moreover, the shape and height (amplitude) of the light curve is independent of the filter through which it was obtained, as predicted for gravitational lensing and unlike the case for intrinsically variable stars.

Some of these searches for lensed stars in the Magellanic Clouds were motivated by the opportunity of finding so-called "massive compact halo objects," or **MACHOs.** In principle, such objects would be too massive to be brown dwarfs, yet could not be normal main-sequence stars because we would see them. White dwarfs are a possible candidate, but in this case the population of stars that produced them must have been devoid of very low-mass stars, since we do not see enough stars in the halo, where astronomers expect that MACHOs would be found.

The searches have already detected hundreds of lensed stars in the Magellanic Clouds. There is considerable controversy about the interpretation of these detections.

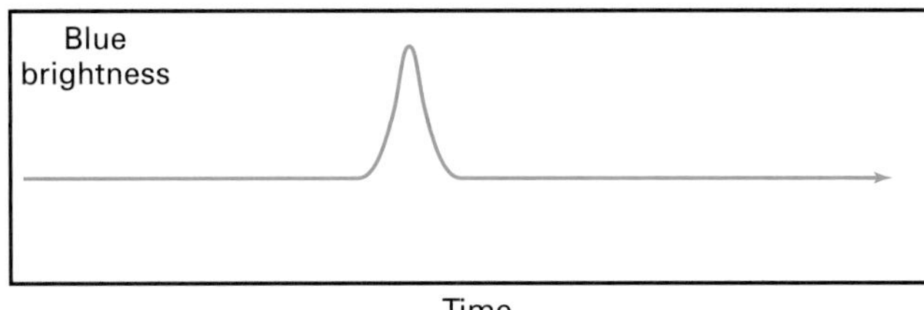

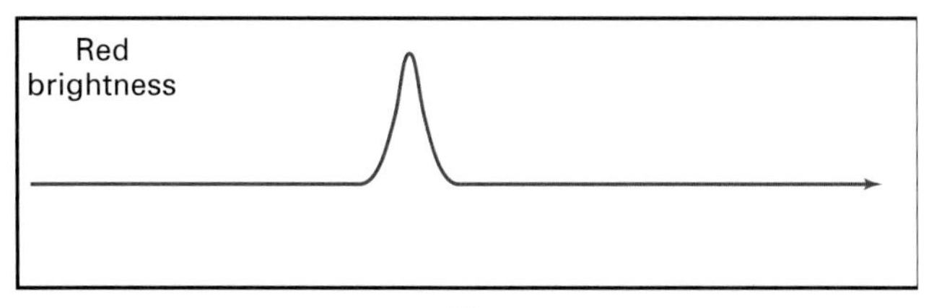

FIGURE 16–29 Light curves, through two different filters, of a star that has been gravitationally lensed by a dark, compact object. Although several discrete images are presumably present, even through a telescope the star appears point-like and its brightness can be measured. The star's light is temporarily focused toward us, with the same magnification at all wavelengths. (The important point is that, as we show here, the shape and height of the light curves through the different color filters are identical.) Unlike the case for variable stars, which brighten and fade periodically, a star is generally lensed only once while we are watching, since the chances of a lens lining up perfectly are exceedingly small.

In a few especially favorable cases, the distance of the lens has been determined. In two cases, the lens turns out to be in the Large Magellanic Cloud itself, rather than in the halo of the Milky Way Galaxy. Another Hubble discovery of a lens in the direction of the Large Magellanic Cloud turns out to be an ordinary star in the disk of the Milky Way Galaxy. So far, the searches have not revealed any definitive lenses in the halo of our Galaxy.

If most of the lenses are not in our Galaxy's halo, then the evidence for dark, compact objects greatly decreases. Indeed, the most recent estimates (late 2005) suggest that no more than about 10 or 20 per cent of the halo's mass consists of MACHOs. Thus, it appears that much of the dark matter in the halo may consist of subatomic particles like the WIMPs mentioned above. It is exciting to think that astronomical observations may end up providing the crucial evidence for the existence of tiny, otherwise undetectable particles.

Gravitational lensing also helps us find out about dark matter by revealing how mass is distributed in galaxies. Just how centralized mass is in galaxy cores can, in principle, be revealed by the study of gravitational lenses with multiple images.

16.6 The Birth and Life of Galaxies

It is difficult or impossible to determine what any given nearby galaxy (or our own Milky Way Galaxy) used to look like, since we can't view it as it was long ago. However, as we discussed in Chapter 1, the finite speed of light effectively allows us to view the past history of the Universe: We see different objects at different times in the past, depending on how long the light has been travelling toward us. At least in a statistical manner we can explore galactic evolution by examining galaxies at progressively larger distances or **lookback times,** and hence progressively farther back in the past.

An important but likely valid assumption is that we live in a typical part of the Universe, so that nearby galaxies are representative of galaxies everywhere. Hence, galaxies several billion light-years away, viewed as they were billions of years ago, probably resemble what today's nearby galaxies used to look like. The refurbished Hubble Space Telescope has led to the most progress in this field, since it provides detailed images of faint, distant galaxies. Also, the Chandra X-ray Observatory has revealed objects that might be very primitive, distant galaxies; the seemingly uniform x-ray glow that previous x-ray telescopes had detected is actually produced by many individual discrete sources.

It is crucial to know the distances of the very distant galaxies, but they are so far away that no Cepheid variables or other normal stars can be seen and compared with nearby examples. So, an indirect technique is used: Hubble's law, as described below.

16.7 The Expanding Universe

Early in the 20th century, Vesto Slipher of the Lowell Observatory in Arizona noticed that the optical spectra of "spiral nebulae" (later recognized by Edwin Hubble to be separate galaxies) almost always show a **redshift** (recall our discussion of redshifts and blueshifts in Chapter 11). The absorption or emission lines seen in the spectra have the same *patterns* as in the spectra of normal stars or emission nebulae, but these patterns

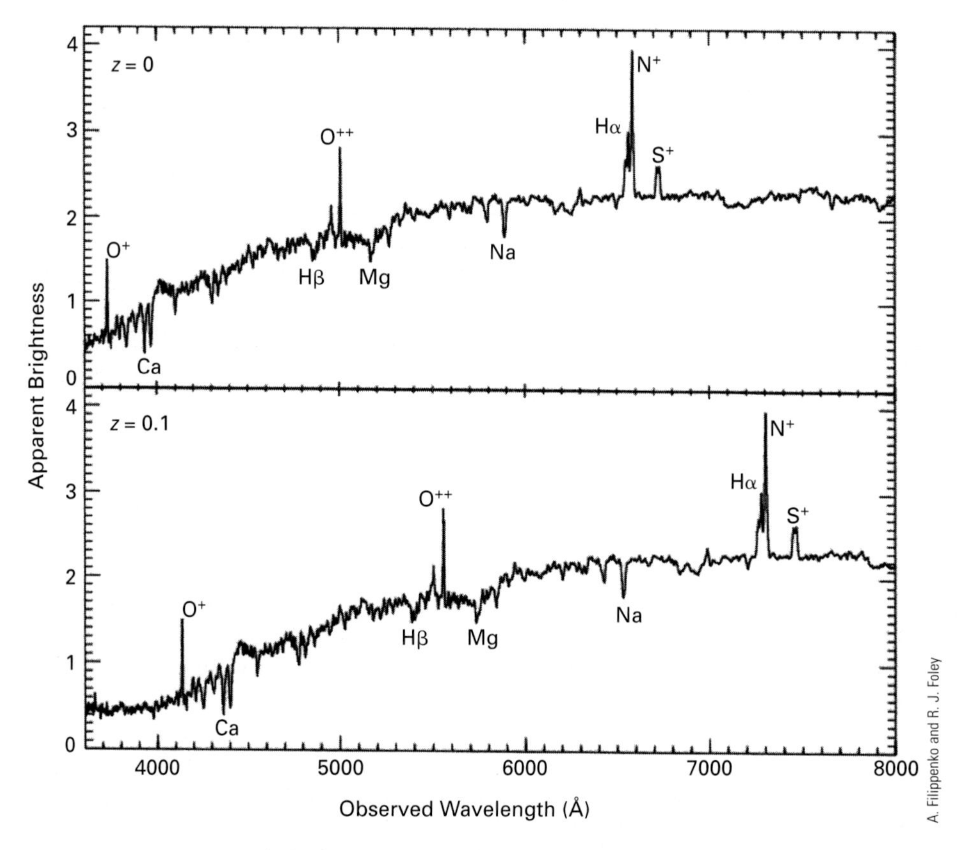

■ **FIGURE 16–30** *(top)* The optical spectrum of a very nearby galaxy, having essentially zero redshift. Various emission and absorption lines are marked. *(bottom)* The spectrum of the same galaxy, if it were at a redshift $z = 0.1$ (in other words, a recession speed of about 30,000 km/sec; see *Figure It Out 16.2: Redshifts and Hubble's Law*). The same patterns of emission and absorption lines are visible, but their wavelengths are shifted toward the red by 10 per cent.

are displaced (that is, shifted) to longer (redder) wavelengths (■ Fig. 16–30). Under the assumption that the redshift results from the Doppler effect, we can conclude that most galaxies are moving away from us, regardless of their direction in the sky. (In Chapter 18, we will see that the redshift is actually caused by the stretching of space, but the equation is the same as that for the Doppler effect, at least at low redshifts.)

In 1929, using newly derived distances to some of these galaxies (from Cepheid variable stars), Hubble discovered that the displacement of a given line (that is, the redshift) is *proportional* to the galaxy's distance. (In other words, when the redshift we observe is greater by a certain factor, the distance is greater by the same factor.)

Thus, under the Doppler-shift interpretation, the recession speed, v, of a given galaxy must be proportional to its current distance, d (■ Fig. 16–31*a*). This relation is

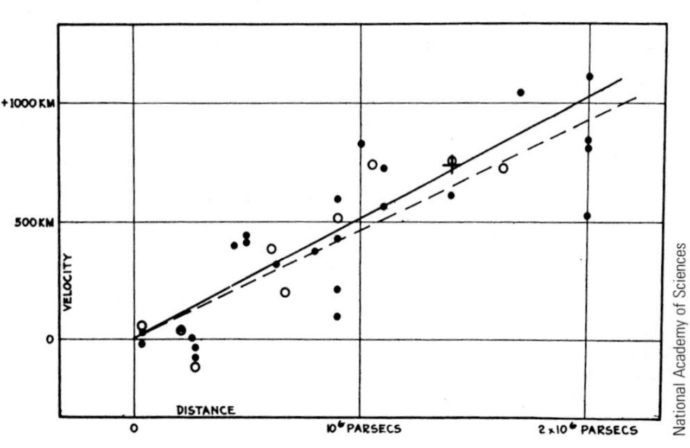

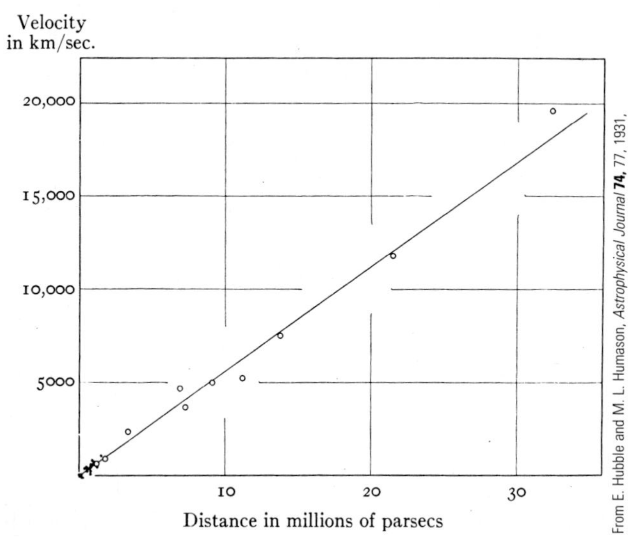

■ **FIGURE 16–31** ⓐ The speed-distance relation, in Hubble's original diagram from 1929. Note that the units of velocity (speed) on the vertical axis should be km/sec, not km; Hubble, or the artist who labeled the graph, made a mistake in the units. One million parsecs equals 3.26 million light-years. Dots are individual galaxies; open circles represent groups of galaxies. The scatter to one side of the line or the other is substantial. ⓑ By 1931, Hubble and Humason had extended the measurements to greater distances, and Hubble's law was well established. All the points shown in the 1929 work appear bunched in the lower-left corner of this graph. The relation $v = H_0 d$ represents a straight line of slope H_0. These graphs use older distance measurements than we now use, and so give different values for H_0 than we now derive.

known as **Hubble's law,** $v = H_0 d$ (see *Figure It Out 16.2: Redshifts and Hubble's Law*), where H_0 (pronounced "H naught") is the *present-day* value of the constant of proportionality, H (the factor by which you multiply d to get v). H_0 is known as **Hubble's constant.**

For various reasons, Edwin Hubble's original data were suggestive but not conclusive. Subsequently, Hubble's assistant and disciple Milton Humason joined Hubble in very convincingly showing the relationship (Fig. 16–31*b*). (Interestingly, Humason had first come to Mt. Wilson as a mule-team driver, helping to bring telescope parts up the mountain. He worked his way up within the organization.)

This behavior is similar to that produced by an explosion: Bits of shrapnel are given a wide range of speeds, and those that are moving fastest travel the largest distance in a given amount of time. Although Edwin Hubble himself initially resisted this idea, the implication of Hubble's law is that the Universe is *expanding*! As we shall see in Chapter 18, however, there is no unique center to the expansion, so in this way it is not like an explosion. Moreover, the expansion of the Universe marks the creation of *space itself,* unlike the explosion of a bomb in a preexisting space.

One of the greatest debates in 20th century astronomy has been over the value of Hubble's constant. In Chapter 18 we will explain in detail how it is determined. Measurements of the recession speeds of galaxies at known distances showed that $H_0 = 50$ to 80 km/sec/Mpc; as of late 2005, the value is known to be 71 km/sec/Mpc to within

ASIDE 16.6: Not a short stroll

An "Mpc" is a megaparsec, a million parsecs. Parsecs are units of distance used by astronomers; 1 parsec (1 pc) is equal to 3.26 light-years (see Chapter 11). A capital M is the prefix meaning a million.

FIGURE IT OUT 16.2
Redshifts and Hubble's Law

For a given absorption or emission line in a spectrum of a galaxy, define $\Delta\lambda$ (the Greek letters uppercase "delta" and lowercase "lambda") to be the displacement (shift) in the wavelength—that is, the observed wavelength (λ) minus the wavelength measured at rest in a laboratory gas (λ_0). Although different lines have different displacements $\Delta\lambda$, the ratio of the displacement to the rest wavelength ($\Delta\lambda/\lambda_0$) is the same for all lines in a galaxy's spectrum.

This quantity, $z = \Delta\lambda/\lambda_0$, is defined as the redshift (z) of the galaxy. When multiplied by 100 per cent, the redshift z can be considered to be the percentage by which lines are shifted toward longer wavelengths. For example, if $z = 0.5$, then a line with $\lambda_0 = 6000$ Å is shifted by $\Delta\lambda = 3000$ Å (50 per cent of 6000 Å) to an observed wavelength of $\lambda = 9000$ Å.

Empirically, Hubble found that the redshift is larger in the spectra of distant galaxies than in the spectra of nearby galaxies. Thus, redshift is proportional to distance (or, $z \propto d$). If this redshift z is produced by motion away from us, then we can use the Doppler formula to derive v, the speed of recession. Since $z = \Delta\lambda/\lambda_0 \approx v/c$ (the Doppler formula; see Chapter 11), we have

$$z \approx v/c, \text{ and therefore } v \approx cz,$$

where c is the speed of light (3×10^5 km/sec). The approximation $v \approx cz$ is accurate as long as the redshift z is not larger than about 0.2 (but the definition $z = \Delta\lambda/\lambda_0$ is exact for all redshifts). For example, if $z = 0.1$, then $0.1 = \Delta\lambda/\lambda_0 \approx v/c$, and so $v \approx 0.1c = 0.1 \times (3 \times 10^5$ km/sec$) = 3 \times 10^4$ km/sec.

Since speed v is proportional to redshift z, which Hubble found to be proportional to distance d, a galaxy's speed of recession must be proportional to its distance. This relation is known as Hubble's law,

$$v = H_0 d,$$

where H_0 is the constant of proportionality, Hubble's constant. Thus, a galaxy twice as far away from us as another galaxy recedes twice as fast; if 10 times as distant, it recedes 10 times as fast.

Note that $v = H_0 d$ does *not* imply that the speed of a given galaxy increases with time as its distance increases. We have defined H_0 to be the present-day value of Hubble's constant H, but the value of H actually decreases with time. (Indeed, in most simple models of the Universe, $H \propto 1/t$.) If no mysterious long-range repulsive effects exist in the Universe, then the speed of a given galaxy is at best constant with time, and in fact should gradually decrease due to the attractive gravitational force of all other matter in the Universe.

Since "Hubble's constant" changes with time, some astronomers prefer to call it the "Hubble parameter." It is a constant only in *space,* having the same value throughout the Universe at any given time.

FIGURE IT OUT 16.3

Using Hubble's Law to Determine Distances

Suppose you find a galaxy that is so faint and far away that conventional techniques for measuring distances, such as the use of Cepheid variable stars, cannot be used, or give inaccurate results. You can instead do the following. (1) Obtain a spectrum of the galaxy. (2) Measure the redshift, $z = \Delta\lambda/\lambda_0$, of an absorption or emission line that you know has rest wavelength λ_0. (3) Compute the approximate recession speed v from $z \approx v/c$—that is, solve for the recession speed $v \approx cz$. (4) Compute the distance d from Hubble's law, $v = H_0 d$, using your preferred value for Hubble's constant H_0—in other words, solve for the distance $d = v/H_0$.

For example, suppose an absorption line with a known rest wavelength $\lambda_0 = 5000$ Å is at an observed wavelength $\lambda = 5100$ Å in the spectrum of a galaxy. Then

$$\Delta\lambda = \lambda - \lambda_0 = 5100 \text{ Å} - 5000 \text{ Å} = 100 \text{ Å, so}$$
$$z = \Delta\lambda/\lambda_0 = (100 \text{ Å})/(5000 \text{ Å}) = 0.02.$$

Therefore the recession speed is $v \approx cz = (3 \times 10^5$ km/sec) $\times$ 0.02 = 6000 km/sec. If we assume that H_0 = 71 km/sec/Mpc, we find that the galaxy's distance is

$$d = v/H_0 = (6000 \text{ km/sec})/(71 \text{ km/sec/Mpc}) \approx 85 \text{ Mpc},$$

or about 280 million light-years, since 1 Mpc = 3.26 million light-years.

Several large "redshift surveys" of galaxies have each produced many thousands of distances, allowing study of the large-scale, three-dimensional distribution of galaxies. These are the studies that led to the conclusion that galaxies tend to be found in superclusters separated by large voids (see Figures 16–23 and 16–43).

One of the largest published sets of distances (there are 221,414 of them) is the "Two Degree Field Galaxy Redshift Survey" conducted with the Anglo-Australian 4-m telescope in Australia and named because the field of view is unusually wide for a telescope that size: 2°. The overall project is known as 2dF. (A 6dF project, with a field three times wider, is under way.) The Sloan Digital Sky Survey, which uses a telescope in New Mexico, surpassed that record.

10 per cent. (We will discuss these definitive measurements, from NASA's Wilkinson Microwave Anisotropy Probe spacecraft, in Chapter 19.) For example, a galaxy 10 Mpc (32.6 million light-years) away recedes from us with a speed of about 710 km/sec, and a galaxy 20 Mpc away recedes with a speed of about 1420 km/sec.

Hubble's constant is always quoted in the strange units of km/sec/Mpc. The value H_0 = 71 km/sec/Mpc simply means that for each megaparsec (3.26 million light-years) of distance, galaxies are receding 71 km/sec faster.

The expansion of the Universe will be the central theme in Chapters 18 and 19; we will discuss its implications and associated phenomena. For now, however, let us simply use Hubble's law to determine the distances of very distant galaxies and other objects. Knowing the value of H_0, a measurement of a galaxy's recession speed v then gives the distance d, since $d = v/H_0$; see *Figure It Out 16.3: Using Hubble's Law to Determine Distances* and *Figure It Out 16.4: Relativistic Effects.*

Note that Hubble's law *cannot* be used to find the distances of stars in our own Galaxy, or of galaxies in our Local Group; these objects are gravitationally bound to us, and hence the expansion of the intervening space is overcome. Moreover, Hubble's law does *not* imply that objects in the Solar System or in our Galaxy are themselves expanding; they are bound together by forces strong enough to overcome the tendency for empty space to expand. Hubble's law applies to distant galaxies and clusters of galaxies; the space between us and them is expanding.

16.8 The Search for the Most Distant Galaxies

With the Hubble Space Telescope, we obtained relatively clear images of faint galaxies that are suspected to be very distant. The main imaging camera of the time (the Wide

FIGURE IT OUT 16.4
Relativistic Effects

The conversion between redshift and recession speed ($z \approx v/c$) used in *Figure It Out 16.2* and *16.3* is not accurate when z is larger than about 0.2, and it fails badly at redshifts approaching or exceeding 1. Instead, the relativistic equation

$$z = \frac{\sqrt{(1 + v/c)}}{\sqrt{(1 - v/c)}} - 1$$

provides a better approximation at high redshifts (but see the caveat below).

When the derived "distance" of an object is very large (say, billions of light-years), it is better to think of it as a "lookback time"—that is, how long the light that we are currently receiving has been travelling from the object. Since the Universe expanded during the light's long journey, it had to cover a greater distance than if the Universe were static.

Thus, the derived "distance" corresponds neither to the object's distance when the light was emitted (which was smaller), nor to the object's distance now (which is larger), but rather to the distance at some intermediate time. This is much more confusing than simply saying that the light was emitted a certain time ago—the lookback time. Such lookback times can be compared with the age of the Universe for various models; see the example in Table 16–1. So, remember—if a newspaper article says that astronomers found a galaxy 8 billion light-years away, what it really means is that the light we are seeing was emitted 8 billion years ago!

It is interesting to note that even the relativistic equation given above is not exactly right (though in this text it will suffice), since the redshift is actually due to the expansion of space rather than motion through space. An object at high redshift is *currently* at a greater distance than when it emitted the light you now see, because at each moment the space that light had already traversed subsequently expanded while the light was still on its way to us. Hence, the high-redshift object's formally calculated speed (as defined by the total distance divided by the lookback time) can actually exceed the speed of light! But this is not a violation of Einstein's special theory of relativity; the object did not move *through* space at a "superluminal" speed, and no information was transmitted at $v > c$.

TABLE 16–1

Redshift	Lookback time
0	0 years
0.1	1.3 billion years
0.2	2.4 billion years
0.3	3.4 billion years
0.4	4.3 billion years
0.6	5.7 billion years
0.8	6.8 billion years
1.0	7.7 billion years
1.5	9.3 billion years
2.0	10.3 billion years
2.5	11.0 billion years
3.0	11.5 billion years
4.0	12.1 billion years
5.0	12.5 billion years
6.0	12.7 billion years
7.0	12.9 billion years
∞	13.7 billion years

Note: Here we assume that the Universe's matter density is 27% of the critical density, and that the energy density associated with the cosmological constant is 73% of the critical density; see Chapter 18 for definitions and details. We also adopt a Hubble constant of 71 km/sec/Mpc. The Universe is 13.7 billion years old in this model.

Field/Planetary Camera 2) exposed on a small area of the northern sky for 10 days in December 1995. Though it covers only about one 30-millionth of the area of the sky (roughly the apparent size of a grain of sand held at arm's length), this **Hubble Deep Field** contains several thousand extremely faint galaxies (■ Fig. 16–32). If we could photograph the entire sky with such depth and clarity, we would see about 50 to 100 billion galaxies, each of which contains billions of stars!

■ **FIGURE 16–32** Part of the original Hubble Deep Field, a tiny section of the sky in the northern celestial hemisphere, near the Big Dipper. About a thousand galaxies at a range of distances are visible in this exposure. In fact, only a few foreground stars in our own Milky Way Galaxy can be seen.
Robert Williams and the Hubble Deep Field Team (STScI), and NASA

NASA, ESA, S. Beckwith (STScI) and the HUDF Team

■ **FIGURE 16–33** Part of the Hubble Ultra Deep Field, a region in the constellation Fornax that appears almost blank to ground-based telescopes. Overall, it looks similar to the original Hubble Deep Field but with finer resolution. This exposure, a million seconds (about 11 days) long, shows even fainter galaxies and was made more quickly, thanks to the efficiency of Hubble's Advanced Camera for Surveys. In this image, blue and green are their true colors but red shows an infrared image. The data were collected over 400 Hubble orbits.

Three years later, the Hubble Space Telescope got very deep images of another region, this time in the southern celestial hemisphere: the Hubble Deep Field—South. It looks similar to the northern field, even though it is nearly in the opposite direction in the sky, providing some justification for our assumption that the Universe is reasonably uniform over large scales. Later, after the Advanced Camera for Surveys was installed on Hubble, it was used to make a **Hubble Ultra Deep Field** (■ Fig. 16–33). These regions of the deep fields and ultra deep field have since been observed by many other telescopes on the ground and in space, notably the Chandra X-ray Observatory.

Other deep surveys further strengthen our conclusion that we live in a rather typical place in the Universe. Spectra obtained with large telescopes, especially the two Keck telescopes in Hawaii, confirm that many galaxies in the Hubble Deep/Ultra Deep Fields and other deep surveys have large redshifts and hence are very far away. Though a few of the galaxies have relatively low redshifts, typical redshifts of the faintest objects are between 1 and 4 (■ Fig. 16–34). Light that we observe at visible wavelengths actually corresponds to ultraviolet radiation emitted by the galaxy, but shifted redward by 100 per cent to 400 per cent! If we convert these redshifts to "distances" (or, more precisely, lookback times—see Table 16–1), we find that the galaxies are billions of light-years away. We see them as they were billions of years in the past, when the Universe was much younger than it is now. Note that when astronomers say that light from a high-redshift galaxy comes from "the distant universe," what they really mean is "distant parts of our Universe." We do not receive light from other universes, even if they exist!

In the past few years, many galaxies with redshifts exceeding 5 (that is, all the spectral lines are shifted by over 500 per cent, putting them at more than 600 per cent of their original values) were discovered (■ Fig. 16–35), and there are quite a few with redshifts over 6. As of late 2005, at least one galaxy has been reported with a redshift of about 7. Several objects are suspected of having even higher redshifts, though the spectra are not yet good enough to be certain. The lookback time corresponding to redshift 6 is about 12.7 billion years (Table 16–1); we are seeing denizens of an era shortly (a billion years) after the birth of the Universe. They are probably newly formed galaxies.

ASIDE 16.7: Better eyes for the Hubble

The Advanced Camera for Surveys, installed on the Hubble Space Telescope in 2002, has the capability of observing areas larger than the Hubble Deep Fields (imaged with the Wide Field/Planetary Camera 2) with smaller exposure times. This ability makes this new camera much more efficient for surveys, as its name implies.

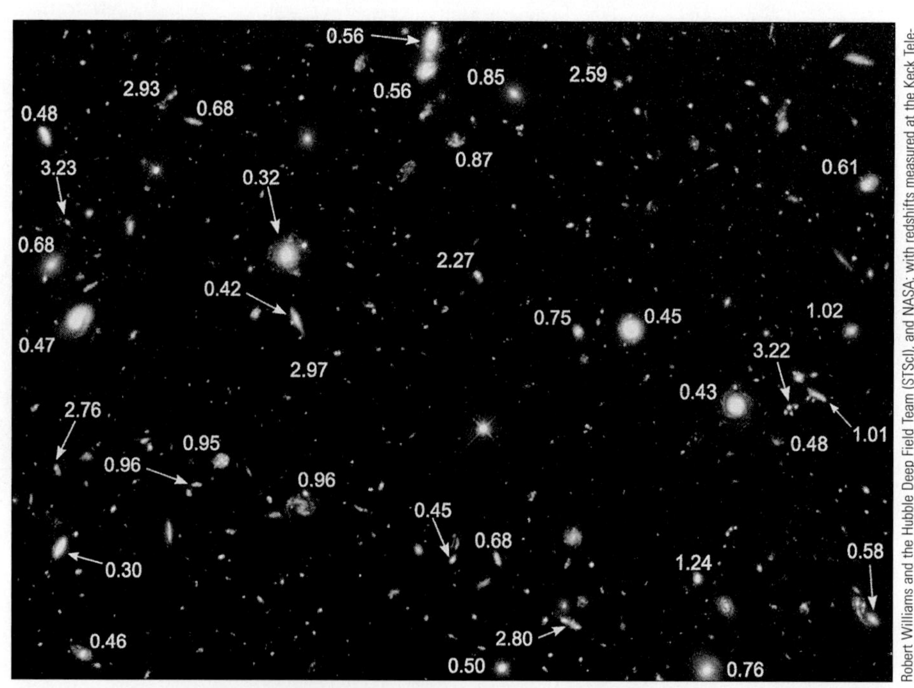

Robert Williams and the Hubble Deep Field Team (STScI) and NASA; with redshifts measured at the Keck Telescope, compiled by M. Dickinson and Z. Levay

■ **FIGURE 16–34** Part of the original Hubble Deep Field, with measured redshifts (mostly from the Keck telescopes) indicated for many of the galaxies.

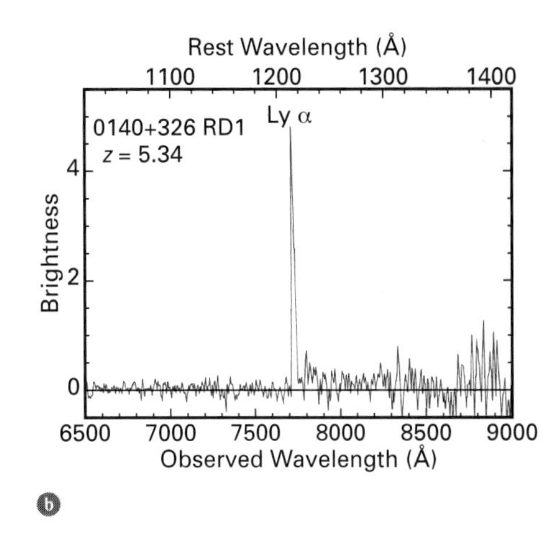

FIGURE 16–35 (a) The image and (b) the spectrum of a distant galaxy, obtained with the Keck telescopes. The measured redshift is 5.34, until recently one of the highest known. The spectrum shows ultraviolet light that was redshifted to optical wavelengths.

Out of the thousands of galaxies found in images such as the Hubble Deep and Ultra Deep Fields, how do we go about choosing those that are likely to be at high redshift? (After all, with limited time for spectroscopy using large telescopes such as Keck, it is important to improve the odds if the goal is to find the highest redshifts.) One very effective technique is to first measure the "color" of each galaxy (Fig. 16–36). For two reasons, those that are likely to be very far away look very red, and have little if any of the blue or ultraviolet light normally emitted by stars. First, their redshift moves light to redder (longer) wavelengths. Second, there is often another galaxy or large cloud of gas along the way, and the hydrogen gas within it completely absorbs the ultraviolet light.

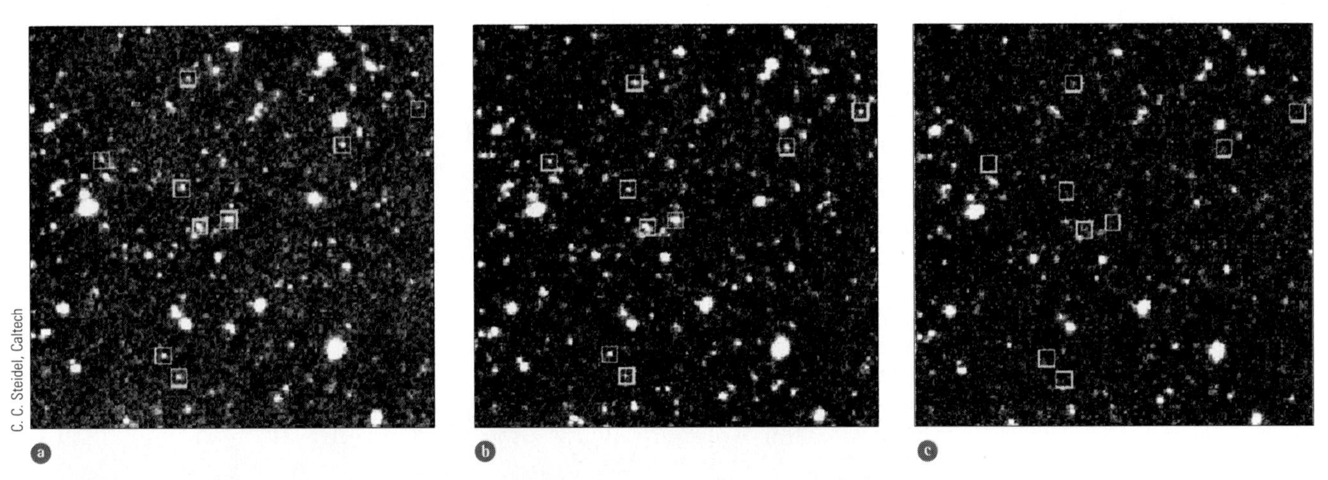

FIGURE 16–36 Deep images through several different filters, obtained at the Keck Observatory, showing faint galaxies: (a) red filter, (b) green filter, (c) ultraviolet filter. Galaxies marked with boxes are visible in the red and green images, but are extremely faint or invisible in the ultraviolet image, suggesting that they are extremely far away; their intrinsic ultraviolet light has been redshifted, as well as absorbed by intervening clouds of gas.

NASA and The Hubble Heritage Team (STScI/AURA), and Rogier Windhorts (ASU)

■ **FIGURE 16–37** Galaxies can appear dramatically different at different wavelengths. The galaxy NGC 6782 looks very different in ultraviolet and visible light, as shown in these images obtained with the Hubble Space Telescope. (a) In the ultraviolet, an inner, ring-shaped region in which intense bursts of star formation have occurred is obvious. An outer ring is faintly visible. The nucleus consists mostly of old stars, and is relatively faint in the ultraviolet. (b) The visible-light/infrared image, on the other hand, shows the pinwheel shape that is typical of spiral galaxies. The image reveals that the ring of star formation appears at the outer edge of the galaxy's bar. The image is made from three exposures: one in the blue, one in the yellow, and one in the near-infrared.

16.9 The Evolution of Galaxies

Comparisons of the appearance of distant galaxies and nearby galaxies provide clues to the way in which galaxies evolve. These comparisons must be done carefully to avoid wrong conclusions. For example, a visible-light image of a high-redshift galaxy actually corresponds to ultraviolet radiation emitted by the galaxy and shifted into the visible band, so it would not be fair to compare the image with the visible-light appearance of a nearby (and hence essentially unshifted) galaxy.

One way around this problem is to compare the visible-light images of high-redshift galaxies with ultraviolet images of nearby galaxies, as obtained with the Hubble Space Telescope. These images emphasize regions containing hot, massive, young stars that glow brightly in the ultraviolet (■ Fig. 16–37).

Another technique is to obtain infrared images of the high-redshift galaxies, and compare them with the visible-light images of nearby galaxies. Such images tend to be dominated by light from older, less massive stars that more accurately reflect the overall shape of the galaxy rather than pockets of recent, intense star formation (■ Fig. 16–38). So far, the clearest infrared images have been made with the Hubble Space Telescope. (Its infrared camera ran out of solid nitrogen coolant in 1998, sooner than anticipated. A new method of cooling the camera was used in equipment installed during the servicing mission in 2002, and it is now working even better than before.)

These data, together with various types of analysis such as computer simulations (for example, of what happens when two galaxies collide and merge), provide many interesting results. One spectacular conclusion is that most spiral galaxies used to look quite peculiar; there were essentially no large galaxies with distinct, well-formed spiral arms beyond redshift 2. By redshift 1 there were quite a few of them, but many took on their current, mature shapes more recently, in the past 5 billion years (that is, at redshifts below about 0.5).

Another conclusion is that there used to be a large number of small, blue, irregular galaxies that formed stars at an unusually high rate (■ Fig. 16–39). Their strange shape might be partly caused by an irregular distribution of young star clusters. However, they appear peculiar even at infrared wavelengths, which are more sensitive to older stars, so they must be structurally disturbed. Some of them probably later merged together to form larger galaxies, including disturbed spirals. Perhaps others faded and are now difficult to find because they are so dim.

It appears that most elliptical galaxies formed early in the Universe, beyond redshift 2 (lookback time 10 billion years); there are many old-looking, well-formed ellip-

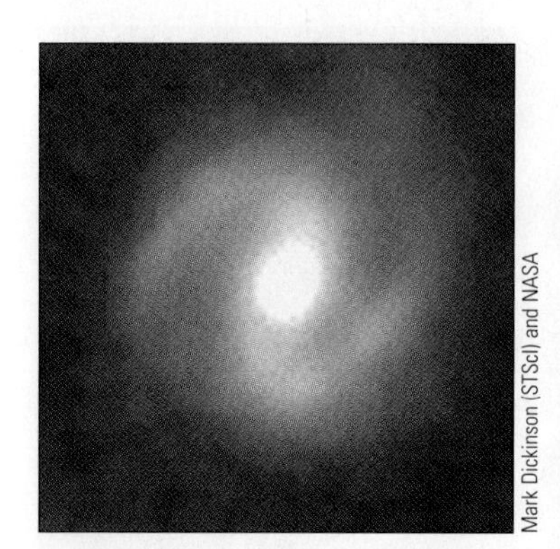

Mark Dickinson (STScI) and NASA

■ **FIGURE 16–38** An infrared image of a high-redshift galaxy ($z = 1$), obtained with the Hubble Space Telescope. This can be compared with visible-light (optical) images of nearby galaxies that have only a small redshift, since the intrinsic wavelengths viewed are then the same in both cases.

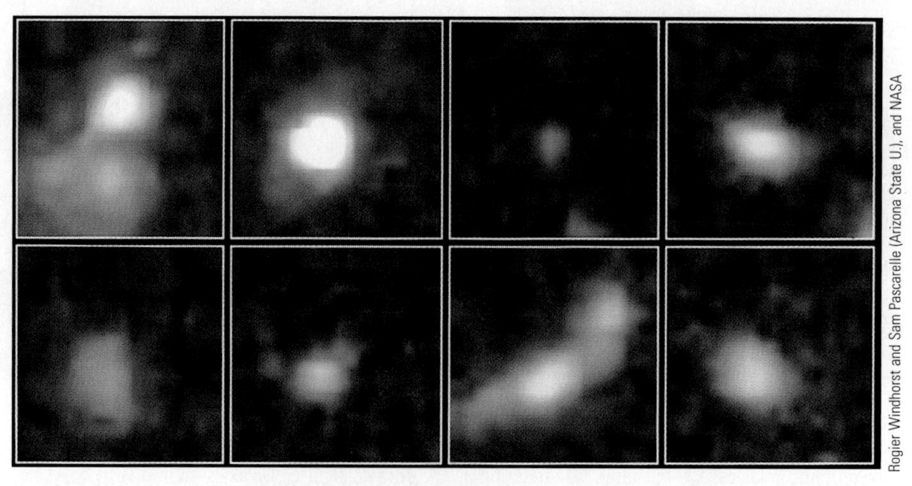

Rogier Windhorst and Sam Pascarelle (Arizona State U.), and NASA

■ **FIGURE 16–39** Small, generally blue, irregular galaxies that have unusually high rates of star formation, imaged with the Hubble Space Telescope. These objects have lookback times of about 10 billion years. Some of them probably merged to form larger galaxies; others may have faded away with time.

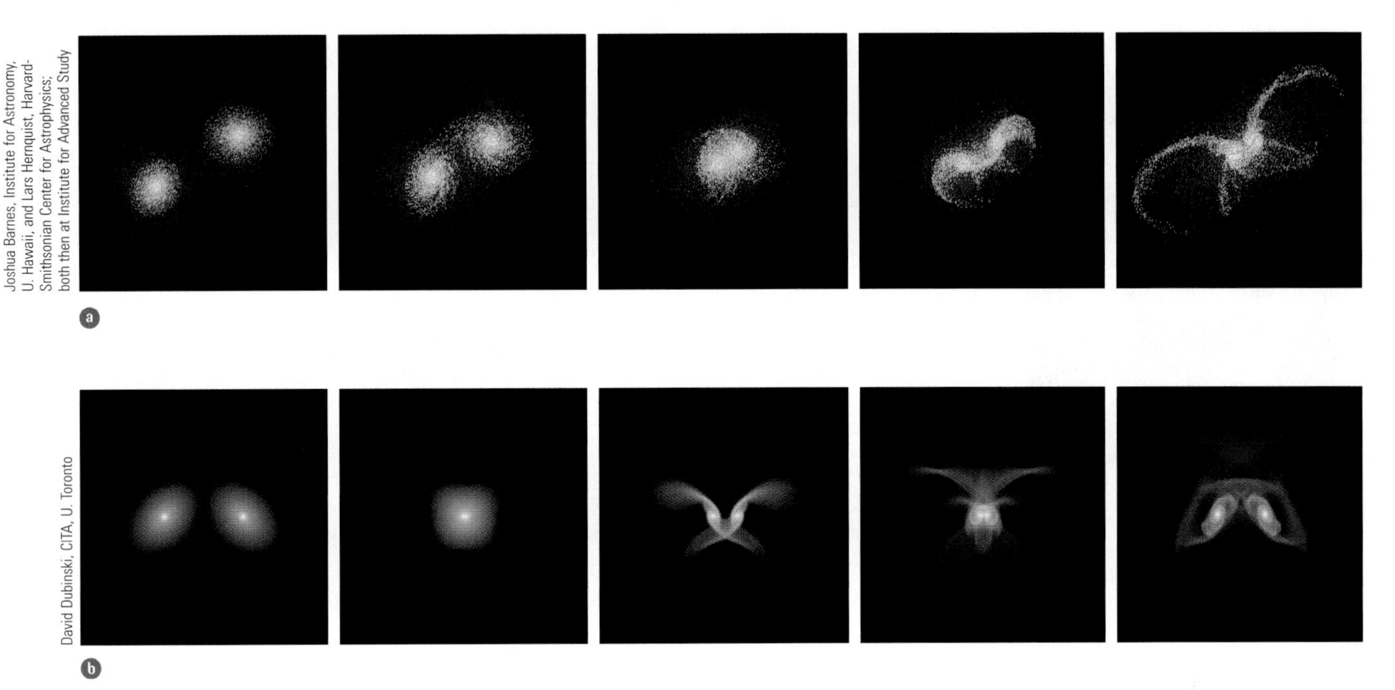

■ **FIGURE 16–40** Calculations made with supercomputers, showing time sequences of the very close encounter of two identical model galaxies. (a) A time interval of 250 million years separates successive frames. About 10,000 particles represent each galaxy. (b) Advances in computing power allow even more particles to be considered. The formation of tidal tails is obvious. Extension of such calculations to even later times shows the formation of an elliptical galaxy from the merged pair. The Antennae (which we saw in Figure 16–14) are thought to have taken their shape in the manner shown above.

ticals between redshifts of 1 and 2. On the other hand, we also think that an elliptical galaxy can be produced by the collision and **merging** of two spiral galaxies. Many new stars are created from the interstellar gas in the spirals. Computer models of the merging process also show that long "tails" of material are sometimes temporarily formed (■ Fig. 16–40), just as we see in nearby examples of interacting galaxies (■ Fig. 16–41; also see the example we showed in Figure 16–14). Later these tails disappear, leaving a more normal looking elliptical galaxy, but with a population of stars younger than in the really ancient ellipticals. In fact, the Milky Way Galaxy and the Andromeda Galaxy are approaching each other and may collide (or barely miss each other) in about 5 or 6 billion years. Subsequently, they are likely to merge and become an elliptical galaxy over the course of billions of years.

The total rate at which stars currently form in the Universe is rather small compared with what it was billions of years ago. We see that the star formation rate has decreased since a redshift of 1 (8 billion years ago) to the present time. The rate may have been constant at still larger redshifts, up to about 4 or 5 (12 billion years ago), though we are unsure because many high-redshift galaxies are cloaked with dust. The dust seems to have been produced by the first few generations of stars, making it difficult to detect high-redshift galaxies at visible wavelengths. But infrared and submillimeter telescopes are finding them in progressively larger numbers, so we can expect a more accurate census in the near future.

As galaxies age, they evolve chemically, primarily because of supernovae that create many of the heavy elements through nuclear reactions and disperse them into the cosmos. Large, massive galaxies, whose gravitational fields don't allow much of the gas to escape, tend to become more chemically enriched than small galaxies that are not able to retain the hot gas. So, we don't expect to find many rocky, Earth-like planets in small (dwarf) galaxies like the Magellanic Clouds. The formation of massive galaxies like the Milky Way seems to be a critical step for the existence of humans.

■ **FIGURE 16–41** The Mice, NGC 4676, with long tails of stars and gas drawn out as two galaxies collided, imaged with the Advanced Camera for Surveys on the Hubble Space Telescope. Supercomputer simulations such as those in Figure 16–40 show that we are seeing two spiral galaxies approximately 160 million years after their closest encounter.

NASA, H. Ford (JHU), G. Illingworth (USCS/LO), M. Clampin (STScI), G. Hartig (STScI), the ACS Science Team, and ESA; the ACS Science Team: H. Ford, G. Illingworth, M. Clampin, G. Hartig, T. Allen, K. Anderson, F. Bartko, N. Benitez, J. Blakeslee, R. Bouwens, T. Broadhurst, R. Brown, C. Burrows, D. Campbell, E. Cheng, N. Cross, P. Feldman, M. Franx, D. Golimowski, C. Gronwall, R. Kimble, J. Krist, M. Lesser, D. Magee, A. Martel, W. J. McCann, G. Meurer, G. Miley, M. Postman, P. Rosati, M. Sirianni, W. Sparks, P. Sullivan, H. Tran, Z. Tsvetanov, R. White, and R. Woodruff

■ **FIGURE 16–42** The two slices of space viewed by the Two Degree Field (2dF) project, in an artist's conception. The vertex of each wedge is the Earth, and the wedges were chosen to show some galaxies in each celestial hemisphere. The observations were made with a telescope in Australia.

16.10 Evolution of Large-Scale Structure

We have also studied the evolution of **large-scale structure** in the Universe. By getting the redshifts of hundreds of thousands of galaxies over large regions of the sky (■ Fig. 16–42), the growth of clusters, superclusters, and voids can be traced (■ Fig. 16–43). As mentioned earlier, it appears that superclusters and giant walls of galaxies are the largest structures in the Universe; we have no clear evidence for super-superclusters of galaxies.

Galaxies generally preceded clusters, and then gravitationally assembled themselves into clusters. Many clusters formed relatively recently, within the past 5 billion years (that is, at redshifts below 0.5), and in fact are still growing now. However, cluster formation did begin earlier. Some very large, well-formed clusters have been found at redshift 1 (8 billion light-years away), and evidence exists for substantial concentrations of matter (which later formed clusters) at a redshift of 4, corresponding to a lookback time of about 12 billion years.

The observed distribution of superclusters and voids (as the 2dF mapping project showed in Figure 16–43) can be compared with the predictions of various theoretical models using computer simulations (■ Fig. 16–44). One important conclusion is that dark matter pervades the Universe; otherwise, it is difficult to produce very large structures. Galaxies and clusters seem to form at unusually dense regions ("peaks") in the dark matter distribution, like snow on the peaks of mountains.

Agreement with observations seems best when most of the dark matter used in the simulations is "cold"—that is, moving relatively slowly compared with the speed of light. Work is in progress to determine what specific type of **cold dark matter** is likely to account for most of the material. Simulations that use primarily **hot dark matter**

■ **FIGURE 16–44** This supercomputer calculation of the formation of clusters of galaxies shows the existence of filaments and voids. The details of the results depend on whether cold dark matter, hot dark matter, or a mixture of the two fills the Universe. The ability of astrophysicists to model the formation of galaxies in the Universe increases as computing power increases.

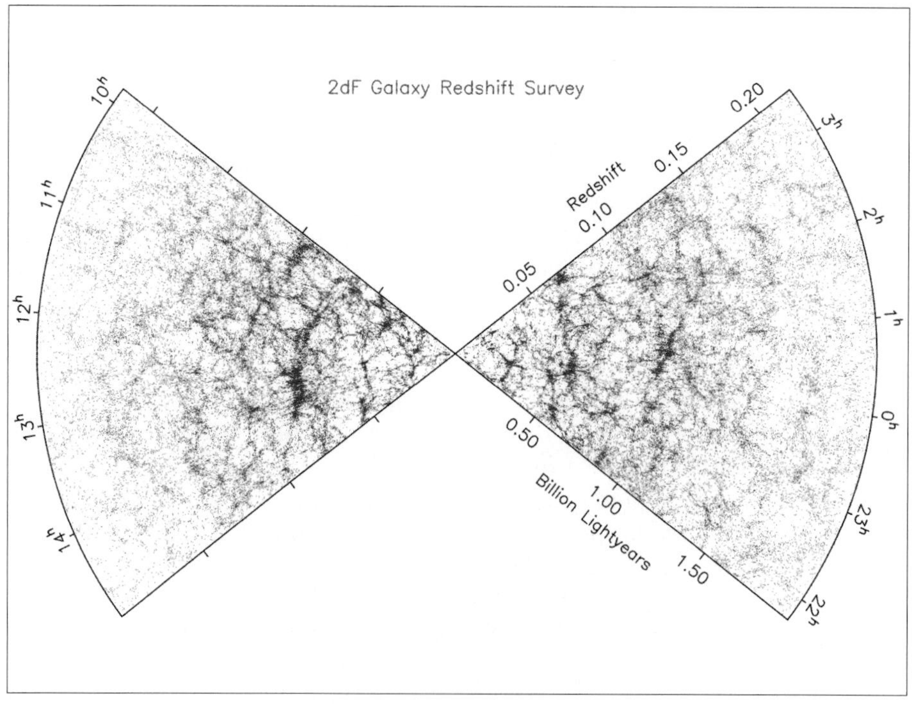

■ **FIGURE 16–43** A map of space, made by the Two Degree Field Galaxy Redshift Survey (2dF GRS), showing a slice whose vertex is at the Earth, as in Figures 16–23 and 16–42. The distances come from the measured redshifts. We see over two hundred thousand galaxies and the structure they reveal in the Universe. Many clusters of galaxies, superclusters, filaments, and voids can be traced. This plot shows the completed survey.

(such as neutrinos, with speeds close to that of light) do not produce galaxy distributions that resemble those observed. Specifically, hot dark matter has a hard time clustering on small scales, like those of individual galaxies and small clusters.

Results announced in 2003 from NASA's Wilkinson Microwave Anisotropy Probe (WMAP; see Chapter 19) show that about 15 per cent of the matter in the Universe consists of protons, neutrons, and electrons—although only about one-quarter of this normal matter resides in stars and other visible objects, the rest being in the form of relatively hot gas, MACHOs, and other constituents. The remaining 85 per cent is matter that does *not* consist of normal particles; moreover, it is dark. This is the cold dark matter discussed above. It may be mostly WIMPs (see Section 16.4c), but we don't yet know this for sure, and no actual WIMPs have ever been directly detected in a laboratory. Again, we emphasize that the nature of dark matter is one of the outstanding mysteries of modern astrophysics!

ASIDE 16.8: The composition of the Universe

When we say that 15 per cent of the *matter* in the Universe consists of protons, neutrons, and electrons, we are really talking about only 4 to 4.5 per cent of the *total* (matter plus energy). This is because matter (most of which is dark) constitutes only about 30 per cent of the Universe; the rest, as we shall see in Chapter 18, is *dark energy,* which is very different from dark (or luminous) matter.

CONCEPT REVIEW

Around the turn of the 20th century, some astronomers believed that the **spiral nebulae** were clouds of gas within our Milky Way star system, while others thought they might be huge, completely separate systems of stars bound together by gravity (Section 16.1a). Their distance and nature was the subject of the famous **Shapley-Curtis debate.** In the mid-1920s, **Edwin Hubble** recognized Cepheid variable stars within a few of these objects, and their observed faintness implied an enormous distance (Section 16.1b). Thus, the true nature of the spiral nebulae was finally unveiled: They are distant galaxies like our own Milky Way Galaxy, and the Universe is far larger than previously thought. Galaxies are the fundamental units of the Universe; each one contains billions of stars spread over a diameter exceeding 10,000 light-years.

There are various types of galaxies. **Spiral galaxies** contain a thin, rotating disk of stars and gas; new stars tend to form in the spiral arms (Section 16.2a). They also have a spherical bulge of old stars. **Elliptical galaxies** have no disk or arms, and generally very little gas and dust; they consist almost entirely of old stars (Section 16.2b). They are roughly spherical or elongated in shape, and don't rotate much. Lenticular (S0) galaxies are a cross between spirals and ellipticals (Section 16.2c). The shapes of irregular and peculiar galaxies are strange. Galaxies frequently occur in **clusters of galaxies** (Section 16.3a), and these in turn congregate in **superclusters** separated by large **voids** (Section 16.3b).

We now have strong indications that much of the matter in the Universe emits little if any detectable electromagnetic radiation, but it has a gravitational influence on its surroundings. One of the first clues was provided by the flat (nearly constant) **rotation curves** of spiral galaxies, including our own Milky Way Galaxy (Section 16.4a). These are in stark contrast with the orbital speeds of planets in the Solar System, which decrease with increasing distance from the Sun. There is also evidence for such **dark matter** in pairs, groups, and clusters of galaxies (Section 16.4b).

The exact composition of the dark matter is unknown (Section 16.4c); possibilities include white dwarfs, neutron stars, black holes, brown dwarfs, large "planets," and neutrinos. Most of the dark matter is probably exotic subatomic particles such as **WIMPs** (weakly interacting massive particles). Much of what we call the "normal" dark matter (that is, consisting of protons, neutrons, and electrons) is gas at a temperature of about a million degrees; it has now barely been detected, and therefore isn't really "dark." Studies of **gravitational lensing** can help reveal the presence of dark matter (Section 16.5). They suggest that a small fraction of the dark matter in our Milky Way Galaxy may consist of **MACHOs** (massive compact halo objects), though the vast majority of it seems to be WIMPs.

Clues to the formation and evolution of galaxies are found from observations of very distant galaxies, at large **lookback times;** we are viewing them as they were when the Universe was far younger than it is today (Section 16.6). Their distances are obtained from the **redshifts** measured in their spectra and interpreted according to the Doppler formula and **Hubble's law** (Section 16.7). The Universe is expanding, with the speeds of different galaxies being proportional to their current separations from us, and the constant of proportionality is called **Hubble's constant.**

Galaxies at a variety of lookback times are seen in the **Hubble Deep Fields** and other long-exposure photographs of the Universe (Section 16.8). Recently, galaxies with redshifts exceeding 5, 6, and even 7 have been found; we see them as they were only one billion years after the birth of the Universe. Hubble's Advanced Camera for Surveys is more efficient than the earlier cameras.

Though most elliptical galaxies formed very long ago, some were produced relatively recently by the **merging** of two spiral galaxies (Section 16.9). Early in the history of the Universe, there were many small, irregular galaxies that may have coalesced to form large spirals. Massive galaxies can retain the heavy elements synthesized and blown out by supernovae, while small galaxies tend not to become as chemically enriched.

Computer models of the **large-scale structure** of the Universe suggest that clusters and superclusters of galaxies formed at the peaks of a more uniform distribution of dark matter (Section 16.10). **Cold dark matter,** which moves slowly compared with light, provides a better match to the data than does rapidly moving **hot dark matter.** Cold dark matter constitutes most (about 85 percent) of the matter in the Universe, but it cannot consist of normal protons, neutrons, and electrons; though it may be mostly WIMPs, its exact nature is one of the outstanding mysteries of modern astrophysics. Matter itself constitutes only about 30% of the Universe; the rest (70%) is gravitationally repulsive "dark energy," to be discussed in Chapter 18.

QUESTIONS

1. Summarize the topic (and its significance) of the Shapley-Curtis debate.
2. What was the physical principle used by Edwin Hubble to demonstrate that the "spiral nebulae" are very distant and large?
3. Explain why Cepheid variable stars are so useful for determining the distances of galaxies.
4. What are the main morphological types of galaxies, and their most important physical characteristics?
5. In which types of structures are galaxies found?
6. Why can't we determine the distances to galaxies using the geometric method of trigonometric parallax (triangulation), as we do for stars?
7. How is the discovery of other galaxies an extension of the Copernican revolution?
8. Describe the rotation curve of the Milky Way Galaxy and contrast its shape with that of the Solar System.
9. †Show that the mass of the Milky Way Galaxy within a radius of 10,000 light-years from the center is about 3×10^{10} solar masses. (Assume a rotation speed of 200 km/sec at that radius.)
10. How does the shape of the Milky Way Galaxy's rotation curve imply the presence of dark matter, given the distribution of visible light in the Galaxy?
11. The sense of rotation of galaxies is determined spectroscopically. How might this be done, in practice?
12. What is the evidence for dark matter in larger structures such as clusters of galaxies?
13. Describe the phenomenon of gravitational lensing.
14. Define MACHOs, state the tentative evidence for their presence, and discuss their possible nature.
15. How can we statistically study the evolution of galaxies many billions of years ago?
16. What if the speed of light were infinite? Would there still be a way for astronomers to look back at galaxies as they appeared long ago?
17. What is meant by the redshift of a galaxy?
18. State Hubble's law, explain its meaning, and argue that the Universe is expanding.
19. Is your body expanding according to Hubble's law? Is the Earth expanding?
20. Can Hubble's law be used to determine the distance of **(a)** α Centauri and other stars relatively near the Sun, **(b)** the center of our Galaxy, **(c)** the Large Magellanic Cloud, a satellite of our Galaxy, and **(d)** a cluster of galaxies very far from the Local Group?
21. †What is the recession speed of a galaxy having a distance of 50 Mpc, if Hubble's constant is 71 km/sec/Mpc?
22. †At what speed is a galaxy 100 million light-years away receding from us, if Hubble's constant is 71 km/sec/Mpc? (Be careful with units!)
23. †What is the redshift z of a galaxy whose Hα absorption line (rest wavelength 6563 Å) is measured to be at 6707 Å?
24. †Calculate the distance of a galaxy with a measured redshift of $z = 0.03$, using 71 km/sec/Mpc for Hubble's constant.
25. Of what importance are spectra and clear images of distant galaxies, in studies of galactic evolution?
26. Summarize some of the tentative conclusions drawn about the formation and evolution of galaxies.
27. **True or false?** Spiral galaxies contain only relatively young stars; there are no ancient stars anywhere within them.
28. **True or false?** The nearly "flat" rotation curve of our Milky Way Galaxy implies that the mass of our Galaxy interior to a given radius grows linearly with increasing radius (that is, $M \propto r$).
29. **True or false?** Giant "voids" in the Universe are the result of enormous explosions that blew matter out of those regions.
30. **True or false?** Some elliptical galaxies appear to have formed as a result of the merging of two spiral galaxies.
31. **True or false?** If you measure the redshift of a certain galaxy to be 4 right now, then an observer living in that galaxy would measure the Milky Way Galaxy to have a redshift of 4 right now.
32. **Multiple choice:** Which one of the following has never been used by astronomers measuring distances? **(a)** Cepheid variable stars. **(b)** Trigonometric parallax of nearby galaxies. **(c)** Hubble's law. **(d)** Main-sequence stars. **(e)** Inverse-square law of light.
33. **Multiple choice:** Which one of the following has *not* been used to suggest that there is "dark matter" in the Universe? **(a)** The orbital speeds of planets around the Sun. **(b)** The orbital speeds of stars in our Milky Way Galaxy and in other spiral galaxies. **(c)** The orbital speeds of galaxies in binary pairs. **(d)** The orbital speeds of galaxies in clusters. **(e)** The gravitational lensing of background galaxies by foreground clusters of galaxies.

34. **Multiple choice:** How do we try to detect MACHOs? (**a**) By looking at a dimming of stars when MACHOs pass in front of them. (**b**) By observing the Large and Small Magellanic Clouds night after night. (**c**) Through their interactions with WIMPs. (**d**) By looking at the Doppler shift of stars. (**e**) From their occasional emission of x-rays.

†35. **Multiple choice:** If an absorption line of calcium is normally found at a wavelength of about 4000 Å in a laboratory gas, and you see it at 4400 Å in the spectrum of a galaxy, what is the approximate distance to the galaxy? (You may assume that Hubble's constant is 50 km/sec/Mpc.) (**a**) 600 Mpc. (**b**) 2000 light-years. (**c**) 6 Mpc. (**d**) 6 million light-years. (**e**) 6000 Mpc.

36. **Multiple choice:** Spiral arms are usually the most prominent features in the disk of a spiral galaxy. Which one of these statements about spiral arms is *false*? (**a**) Clouds of gas and dust are mostly found in spiral arms. (**b**) Emission nebulae are mostly found in spiral arms. (**c**) Spirals arms contain most of the hot, young, massive stars. (**d**) Spiral arms contain most of the mass in the galaxy. (**e**) Spiral arms consist mostly of dark matter.

37. **Fill in the blank:** High-redshift arcs around clusters of galaxies at lower redshifts are now thought to be examples of _________.

38. **Fill in the blank:** Most of the dark matter is thought to consist of _____; it is not the "normal" dark matter composed of neutrons, protons, and electrons.

39. **Fill in the blank:** Edwin Hubble realized that some of the "spiral nebulae" visible in photographs must be outside our own Milky Way Galaxy because he saw very faint ______ within them, yet such stars are known to be intrinsically luminous.

40. **Fill in the blank:** Distant galaxies have large ______ times, allowing us to study the evolution of galaxies.

†41. **Fill in the blank:** If Hubble's constant is 100 km/sec/Mpc, then a galaxy whose redshift $z = 0.05$ has a distance of ______.

†This question requires a numerical solution.

TOPICS FOR DISCUSSION

1. How is the discovery of other galaxies an extension of the Copernican revolution?
2. Are you bothered by the notion that most of the matter in the Universe might be dark, detectable only through its gravitational influence? Can you think of any alternative explanations for the data?
3. Can you think of possible explanations for the redshifts of galaxies that do not involve the expansion of the Universe? How could you test these hypotheses?

MEDIA

Virtual Laboratories

- General Relativity, Black Holes, and Gravitational Lensing
- The Astronomical Distance Scale
- Dark Matter
- Large-Scale Structure

AceAstronomy™ Log into AceAstronomy at **http://astronomy.brookscole.com/cosmos3** to access quizzes and animations that will help you assess your understanding of this chapter's topics.

Log into the Student Companion Web Site at **http://astronomy.brookscole.com/cosmos3** for more resources for this chapter including a list of common misconceptions, news and updates, flashcards, and more.

People in Astronomy

Sandra Faber

Photo with Hubble Deep Field, courtesy R. R. Jones, Hubble Deep Field Team, NASA

Sandra Faber is Professor of Astrophysics and Astronomy at the Lick Observatory, University of California at Santa Cruz. She was born in Boston, grew up in Cleveland and Pittsburgh, and attended Swarthmore College, Swarthmore, Pennsylvania. After graduate school at Harvard's Department of Astronomy, she accepted a position at the Lick Observatory, "the only job [she has] ever had."

Professor Faber has always wondered about the large-scale features in the Universe: why there are galaxies, why they look as they do, and how the Universe began. Her work takes her regularly to the telescope, and she was for a time Co-Chair of the Scientific Steering Committee of the Keck Observatory as they planned and built their 10-m telescopes.

Among the prizes she has received are the Bart J. Bok Prize from Harvard University and the Dannie Heineman Prize of the American Astronomical Society. She is a Trustee of the Carnegie Institution of Washington and a member of the National Academy of Sciences. In 1996 she was elevated to University Professor at the University of California. Recently she served on the blue-ribbon panel that advised NASA that the Hubble Space Telescope should be serviced once more, by astronauts using the Space Shuttle.

In 1980, she joined six other scientists in a study that eventually showed a large-scale flow of galaxies at a million miles per hour toward the direction in the sky where the constellation Centaurus is located. Our Milky Way is part of this flow. The flow is caused by the gravitational attraction of a large supercluster of galaxies, one of the largest structures yet seen in the Universe. They nicknamed it the Great Attractor. Its existence implies, yet again, that most of the matter in the Universe is dark and invisible to telescopes, in this case the dark halos that surround the visible galaxies that compose the Great Attractor.

Professor Faber has two grown daughters. Her husband, Andy, is an attorney in San Jose.

What kind of research are you doing with the Keck telescopes?

The Keck telescopes are working as well as anyone had hoped and, in fact, have now been copied around the world. Astronomers are finding that they just cannot study faint objects at the visible edge of the Universe without the huge aperture of a Keck or something similar. I led a team to build a giant spectrograph for Keck-II. This spectrograph is now being used in the DEEP Survey (**D**eep **E**xtragalactic **E**volutionary **P**robe, linking observations with the Keck telescopes and the Hubble Space Telescope) to collect spectra of 50,000 faint galaxies in order to map the Universe as it was billions of years back in time. Most of the data are in, and we are measuring changes in the galaxy population over the last half of the age of the Universe.

How abut the Hubble Space Telescope?

The Space Telescope has been the biggest roller coaster of my scientific career. First there was euphoria just after launch, when all seemed to be going well. Then we discovered the flawed primary mirror (I was part of a three-person group that diagnosed the error and reported it to NASA). The whole project hastily regrouped and replanned to limp along and do some science with the flawed mirror, while hundreds of people conceived of a strategy to fix the telescope and carried it out brilliantly. The telescope has performed beautifully after the repair mission and has delivered more important data than any other telescope in history, and I now find myself with the rewarding assignment of entertaining audiences with slides of gorgeous Hubble images.

The main lesson is, never give up. Pull victory from defeat. The second lesson is that a team of dedicated people can accomplish amazing feats.

What things have you learned with the Hubble Space Telescope?

I've been part of a team searching for supermassive black holes at the centers of galaxies. Hubble can find them by spotting stars very close to the hole that are orbiting super-fast. Some of these black holes are many billions of solar masses in size. Our team has shown that a big black hole lurks at the center of nearly every large galaxy.

I've also been part of another team using the Hubble Space Telescope along with the Keck-II telescope to study the most distant galaxies in the Universe. This is the DEEP Survey mentioned above. Hubble images are crucial because their high resolution shows us distant galaxies in detail, from which we can measure Hubble types and other important quantities. A big discovery from DEEP is that we have actually detected disk galaxies turning into elliptical galaxies, probably by colliding with each other to create disorganized "starpiles." It is very exciting to actually see how galaxies assumed their final forms and how the Hubble sequence was made.

Tell us about your study of the Great Attractor.

Like most of the important things I have done scientifically, the motivation for the project was all wrong. We started out to survey the properties of nearby elliptical galaxies, such things as their brightness, radii, and so on. And we wound up finding a method that could estimate the absolute size of each galaxy and hence tell you how far away each object is. Knowing that, from the Hubble law you could predict the redshift (velocity) of every galaxy. When we compared these predictions with the measured velocities, we found a big discrepancy, and this could be interpreted simply as a streaming motion of all the nearby galaxies toward the center of a hitherto unidentified mass concentration.

This came as a total surprise. We couldn't believe there was such a large supercluster of galaxies so close-by that nobody had noticed before. But fortunately there was a graduate student in Cambridge, England—Ofer Lahav—who had just stored complete galaxy catalogues in the computer, and he was able to make a gigantic map of all the galaxies in that direction in the sky. In this new picture, the Great Attractor appeared for the first time.

Were you surprised at all the interest your results generated?

No. I think we generated some of the interest because we were so surprised ourselves. We were stunned. In graduate school, I was taught that the Hubble expansion was very uniform. The typical streaming motion of galaxies [motion relative to the Hubble expansion] was only supposed to be about 100 km/sec, so it was a total surprise to find peculiar motions 6 times larger than that.

When you were in grad school at Harvard years ago, would you have been surprised at such a discovery?

Totally. I have never had long-range goals as a professional astronomer. I've always been a short-range opportunist, so it always comes as a surprise when something interesting turns up.

What in your view is the relation of observation with theory?

Deep down I feel a little sorry for theoreticians because they see the Universe only through the eyes of the observers. An observer at a telescope with a good project is like an explorer in the New World—the view over each new ridge is new. On the other hand, we would never actually understand anything without theory to back it up. So they fit together like hand in glove. Each is essential.

How did you get interested in astronomy?

I was one of those kids who was deeply interested in science. It really didn't matter too much what kind. I had star charts, but I also had a rock collection and read books on spiders. It was only later, when graduating from high school, that I began to focus on astronomy. The reason was simple: I wanted to know where the Universe came from and why it is how it is.

And are you making progress toward understanding the Universe?

I think so, in broad outline—very broad. Actually, I have come to believe there are many universes—an infinite number perhaps. I'd love to know how different they all are from one another, but I don't know the answer to that. However, I do think that ours is roughly the way it is because we are in it. It takes certain restrictions to create intelligent life. Within those restrictions, it is a matter of chance, but the basic restrictions are set by our existence in this Universe. Our Universe has the properties it does because they are required to make our kind of intelligent life.

Perhaps there is an analogy here. Ancient peoples might have wondered why the Earth is as it is. We know now that there are probably millions of planets, but most of them are like the other eight planets in our Solar System—that is, not hospitable to our kind of life. However, out of those millions, there are probably many planets rather similar to our Earth that would do quite nicely. And within that broad selection, our existence on this particular Earth is a matter of chance.

In the same way, I believe that our Universe is just one of many hospitable universes we could inhabit. Our being in this particular one is of no special interest. The really interesting implication is that there must exist "out there" many more universes of vastly different types, most of them possibly so bizarre that intelligent life would find them quite hostile. Out of all of these, ours has the properties it does because we are in it. Recent breakthroughs in quantum cosmology have even found a plausible way to generate all those universes in a never-ending, infinite cascade of big bangs. This idea is speculation right now, but chasing it down is going to provide a lot of excitement in the years ahead.

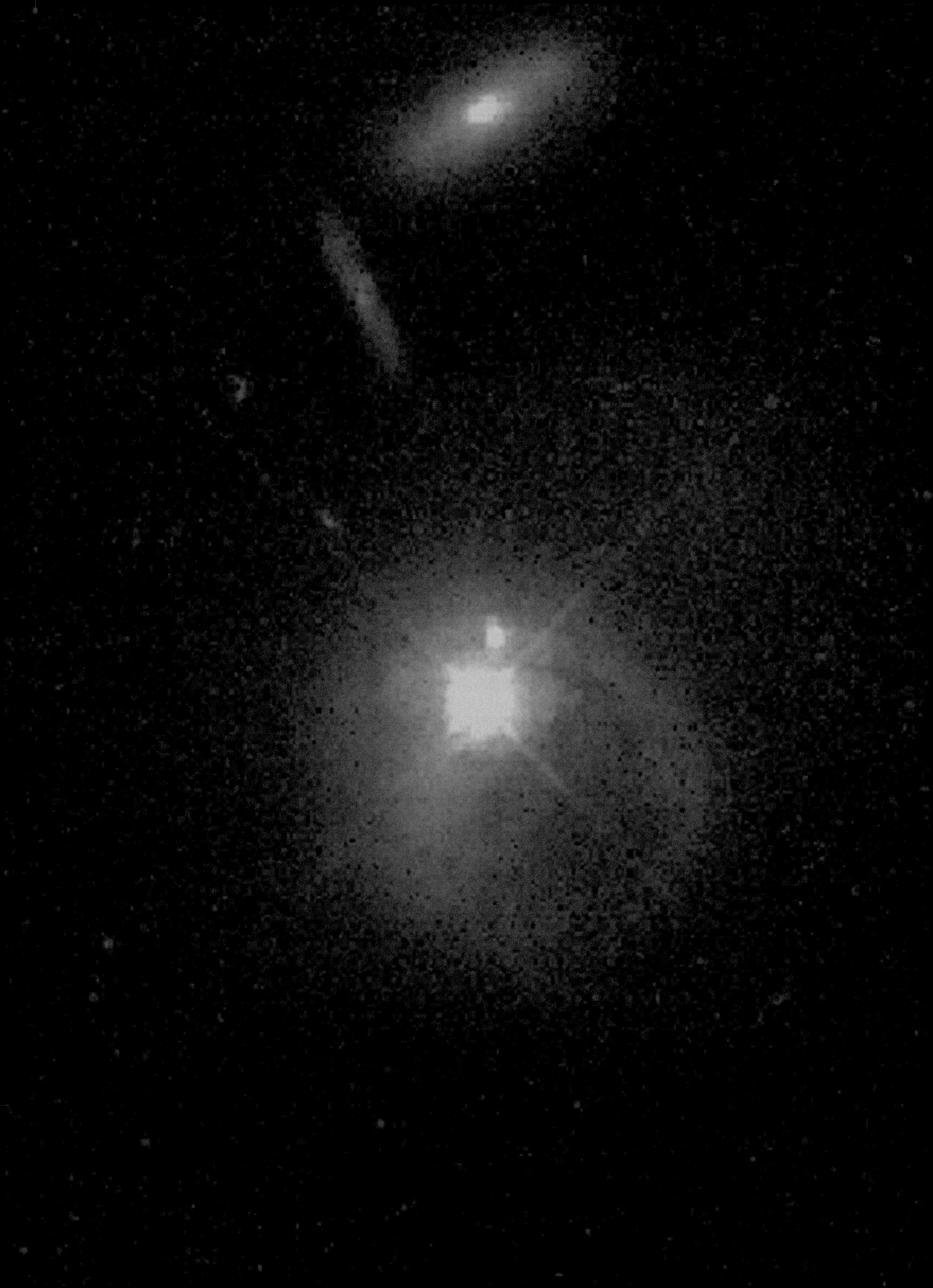

CHAPTER 17

Quasars and Active Galaxies

Quasars, and the way in which they became understood, have been one of the most exciting stories of the last forty years of astronomy. First noticed as seemingly peculiar stars, quasars turned out to be some of the most powerful objects in the Universe, and represent violent forces at work. We think that giant black holes, millions or even billions of times the Sun's mass, lurk at their centers. A quasar shines so brightly because its black hole is pulling in the surrounding gas, causing the gas to glow vividly before being swallowed.

Our interest in quasars is further piqued because many of them are among the most distant objects we have ever detected in the Universe. Since, as we look out, we are seeing light that was emitted farther and farther back in time, observing quasars is like using a time machine that enables us to see the Universe when it was very young. We find that quasars were an early stage in the evolution of large galaxies. As time passed, gas in the central regions was used up, and the quasars faded, becoming less active. Indeed, we see examples of active galaxies relatively near us, and in some of these the presence of a massive black hole has been all but proven.

ORIGINS

We explore the significance of quasars as an early phase in the origin and evolution of large galaxies, such as our Milky Way Galaxy. We find that quasars allow us to study the distribution of matter, both visible and dark, in the vast space between us and them.

AIMS

1. Describe the discovery of quasars and their relationship to galaxies with unusually energetic centers (Sections 17.1 to 17.2).
2. Explore the physical nature of quasars and the source of their stupendous power (Sections 17.3 to 17.5).
3. Discuss the evidence for supermassive black holes in the centers of nearby galaxies (Section 17.6).
4. Understand how beams of particles and radiation from quasars and active galaxies produced certain types of observed phenomena (Section 17.7).
5. Illustrate how quasars allow us to probe the amount and nature of intervening material at high redshifts (Section 17.8).

17.1 Active Galactic Nuclei

The central regions of normal galaxies tend to have large concentrations of stars. For example, at infrared wavelengths we can see through our Milky Way Galaxy's dust and penetrate to the center. When we do so, we see that the bulge of our Galaxy becomes more densely packed with stars as we look closer to the nucleus. With so many stars confined there in a small volume, the nucleus itself is relatively bright. This concentrated brightness appears to be a natural consequence of galaxy formation; gas settles in the central region due to gravity, and subsequently forms stars.

In a minority of galaxies, however, the nucleus is far brighter than usual at optical and infrared wavelengths, when compared with other galaxies at the same distance

AceAstronomy™ The AceAstronomy icon throughout this text indicates an opportunity for you to test yourself on key concepts, and to explore animations and interactions of the AceAstronomy website at **http://astronomy.brookscole.com/cosmos3**

A Hubble Space Telescope image of the quasar PKS 2349–014, whose spectral lines are redshifted by 17.3 per cent—that is $z = 0.173$. The loops of gas around this quasar suggest that it is being fueled by gas from the merging of two galaxies. The galaxies above the quasar are probably interacting with it.

John N. Bahcall and Sofia Kirhakes (Institute for Advanced Study), and Donald P. Schneider (Pennsylvania State U.), and NASA

a/b: Chien Peng (U. Arizona), Alex Filippenko (UC Berkeley), and NASA

■ **FIGURE 17–1** Hubble Space Telescope images of the central regions of (a) the normal galaxy NGC 7626 and (b) the active galaxy NGC 5548. Both galaxies have the same general appearance (Hubble type), but note the bright, star-like nucleus in the active galaxy. The spikes of light from this star-like nucleus are a telescope artifact ("diffraction spikes," like those in images of bright stars).

(■ Fig. 17–1). Indeed, when we compute the optical luminosity (power) of the nucleus from its apparent brightness and distance, we have trouble explaining the result in terms of normal stars: It is difficult to cram so many stars into so small a volume.

Such nuclei are also often very powerful at other wavelengths, such as x-rays, ultraviolet, and radio. These galaxies are called "active" to distinguish them from normal galaxies, and their luminous centers are known as **active galactic nuclei.** Clusters of ordinary stars rarely, if ever, produce so much x-ray and radio radiation.

Active galaxies that are extraordinarily bright at radio wavelengths often exhibit two enormous regions (known as "lobes") of radio emission far from the nucleus, up to a million light-years away. The first "radio galaxy" of this type to be detected, Cygnus A (■ Fig. 17–2), emits about a million times more energy in the radio region of the spectrum than does the Milky Way Galaxy.

Close scrutiny of such radio galaxies sometimes reveals two long, narrow, oppositely directed "jets" joining their nuclei and lobes (■ Fig. 17–3). The jets are thought to consist of charged particles moving at close to the speed of light and emitting radio waves. Sometimes radio galaxies appear rather peculiar when we look at visible wavelengths, and the jet is visible in x-rays, as in the case of Centaurus A (■ Fig. 17–4).

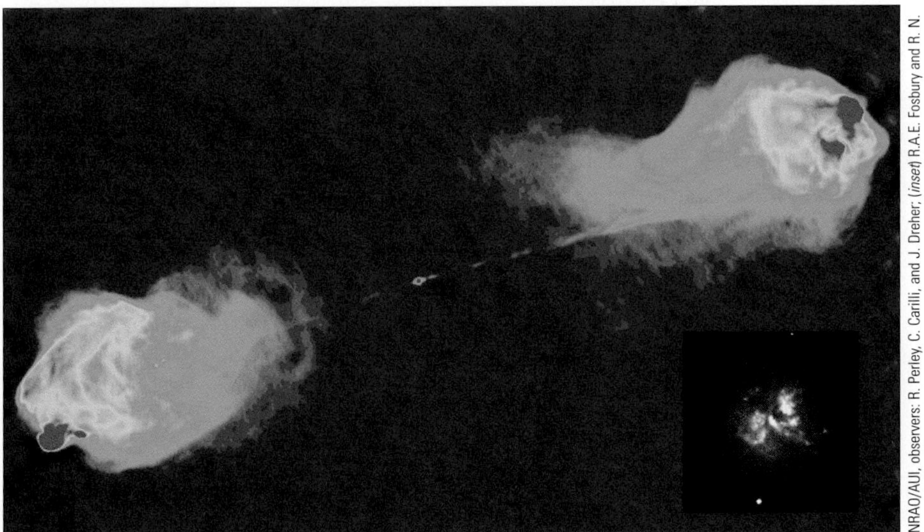

NRAO/AUI, observers: R. Perley, C. Carilli, and J. Dreher; (*inset*) R.A.E. Fosbury and R. N. Hook/NASA, ESA, Keck Obs.

■ **FIGURE 17–2** A radio map of Cygnus A, with shading indicating the intensity of the radio emission in the two giant lobes. A Hubble Space Telescope optical image, with the blue component provided with a Keck telescope, shows detailed structure. Particles feeding the large radio lobes come from the very central part.

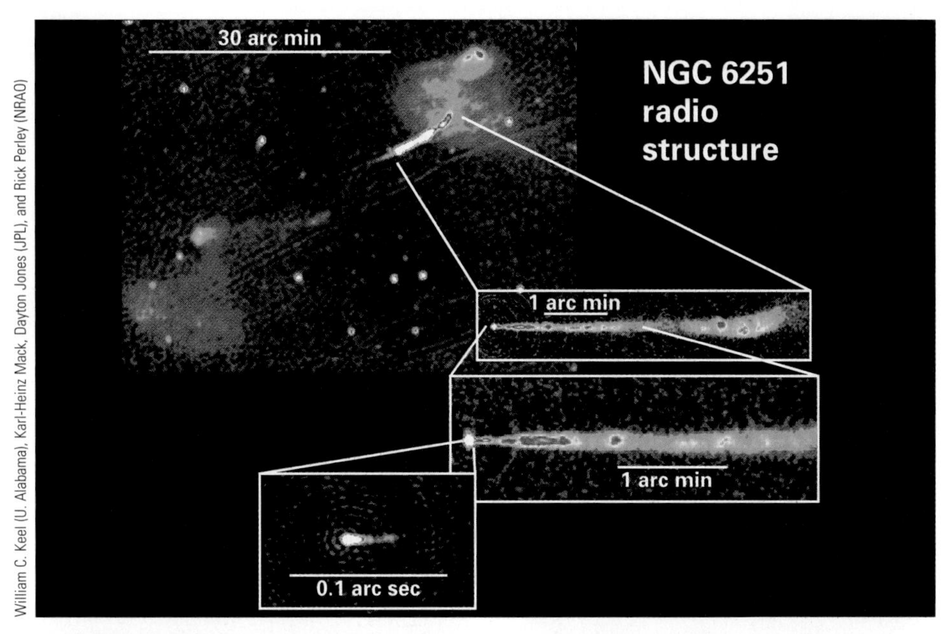

William C. Keel (U. Alabama), Karl-Heinz Mack, Dayton Jones (JPL), and Rick Perley (NRAO)

■ **FIGURE 17–3** Radio maps of the active galaxy NGC 6251 shown on four scales. (At the distance of NGC 6251, 30 arc minutes corresponds to about 3 million light-years, 1 arc minute is about 100,000 light-years, and 0.1 arc second is about 170 light-years.) One of the radio jets is pointing roughly in our direction, and thus appears much brighter than the other jet, which points away from us.

© 1980 Anglo-Australian Observatory

NASA/Chandra X-ray Observatory Center/SAO

■ **FIGURE 17–4** (a) Centaurus A (NGC 5128) looks like an elliptical or S0 galaxy seen through an extensive dust lane. (b) A Chandra Observatory x-ray view of the jet in Centaurus A. The orientation is the same but the image scale is 2.5 times greater (in other words, a closer-up view).

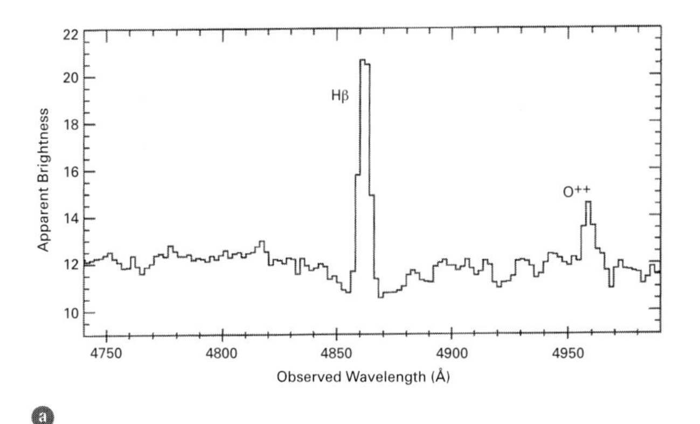

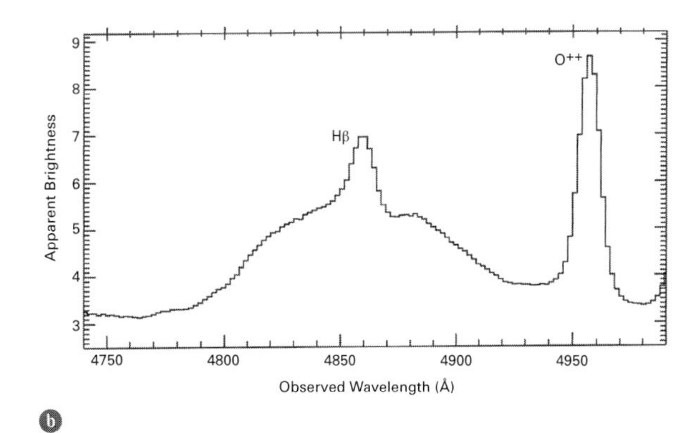

■ **FIGURE 17–5** A very small part of the optical spectra of the nuclei of two galaxies, plotted with the redshifts removed (so that we see light at the emitted wavelengths rather than the observed, redshifted wavelengths). (a) A normal galaxy in which hot, massive stars have recently formed. Note the two narrow emission lines, whose width indicates that gas moves with a speed of only a few hundred km/sec. There is also some Hβ absorption, formed by relatively massive stars. (b) A typical active galactic nucleus, in this case the Seyfert galaxy NGC 5548 (an image of which is shown in Figure 17–1). Note the broad Hβ emission line along with narrower lines (in this case, O^{++} and the narrow component of Hβ). The broad line is formed by gas, close to the center, that is moving very rapidly (thousands of km/sec). The narrow lines come from gas, farther out from the nucleus, having a speed of just a few hundred km/sec. (A. Filippenko and R. J. Foley, UC Berkeley)

Optical spectra of the active nuclei often show the presence of gas moving with speeds in excess of 10,000 km/sec, far higher than in normal galactic nuclei. We measure these speeds from the spectra, which have broad emission lines (■ Fig. 17–5). Atoms that are moving toward us emit photons that are then blueshifted, while those that are moving away from us emit photons that are then redshifted, thereby broadening the line by the Doppler effect. Early in the 20th century, Carl Seyfert was the first to systematically study galaxies with unusually bright optical nuclei and peculiar spectra, and in his honor they are often called "Seyfert galaxies."

Although spectra show that gas has very high speeds in supernovae as well, the overall observed properties of active galactic nuclei generally differ a lot from those of supernovae, making it unlikely that stellar explosions are responsible for such nuclei. Indeed, it is difficult to see how stars of any kind could produce the unusual activity. However, for many years active galaxies were largely ignored, and the nature of their central powerhouse was unknown.

17.2 Quasars: Denizens of the Distant Past

Interest in active galactic nuclei was renewed with the discovery of **quasars** (shortened form of "quasi-stellar radio sources"), the recognition that quasars are similar to active galactic nuclei, and the realization that both kinds of objects must be powered by a strange process that is unrelated to stars.

17.2a The Discovery of Quasars

In the late 1950s, as radio astronomy developed, astronomers found that some celestial objects emit strongly at radio wavelengths. Catalogues of them were compiled, largely at Cambridge University in England, where the method of pinpointing radio sources was developed. For example, the third such Cambridge catalogue is known as "3C," and objects in it are given numerical designations like 3C 48.

Although the precise locations of these objects were difficult to determine with single-dish radio telescopes (since they had poor angular resolution), sometimes within the fuzzy radio image there was an obvious probable optical counterpart such as a supernova remnant or a very peculiar galaxy. More often, there seemed to be only a bunch of stars in the field—yet which of them might be special could not be identified, and in any case there was no known mechanism by which stars could produce so much radio radiation.

Special techniques were developed to pinpoint the source of the radio waves in a few instances. Specifically, the occultation (hiding) of 3C 273 by the Moon provided an unambiguous identification with an optical star-like object. When the radio source winked out, we knew that the Moon had just covered it while moving slowly across the background of stars. Thus, we knew that 3C 273 was somewhere on a curved line marking the front edge of the Moon. When the radio source reappeared, we knew that the Moon had just uncovered it, so it was somewhere on a curved line marking the Moon's trailing edge at that time. These two curves intersected at two points, and hence 3C 273 must be at one of those points. Though one point seemed to show nothing at all, the other point was coincident with a bluish, star-like object about 600 times fainter than the naked-eye limit.

When the positions of other radio sources were determined accurately enough, it was found that they, too, often coincided with faint, bluish-looking stars (■ Fig. 17–6). These objects were dubbed "quasi-stellar radio sources," or "quasars" for short. Optically they looked like stars, but stars were known to be faint at radio wavelengths, so they had to be something else. Object 3C 273 seemed to be especially interesting: A jet-like feature stuck out from it, visible at optical wavelengths (■ Fig. 17–7) and radio wavelengths (■ Fig. 17–8).

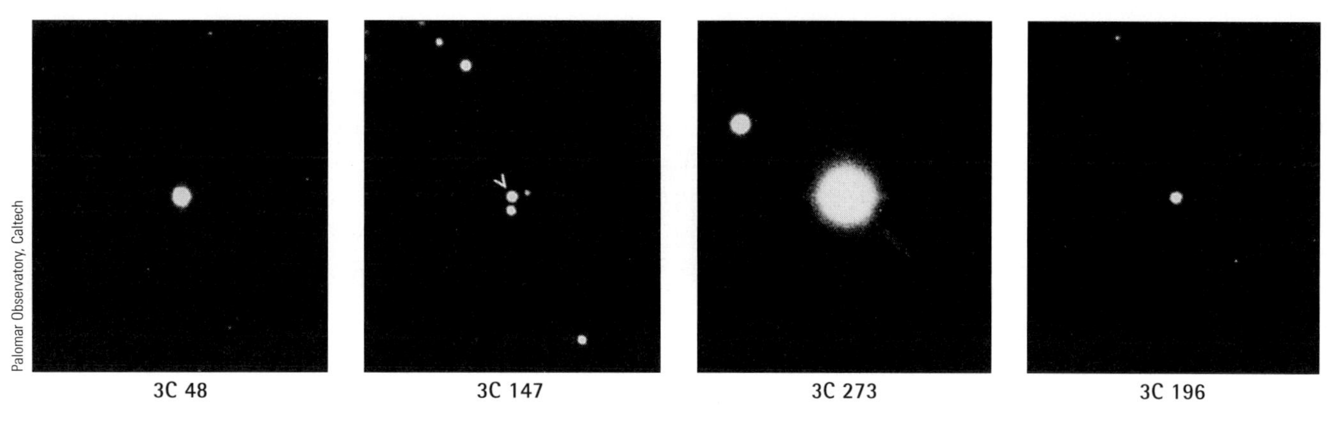

■ **FIGURE 17–6** Original optical photographs of four of the first known quasars, from the photographic era before the current era of making electronic images. The objects appear star-like, but measurements with radio telescopes show that they are much brighter than normal stars at radio wavelengths.

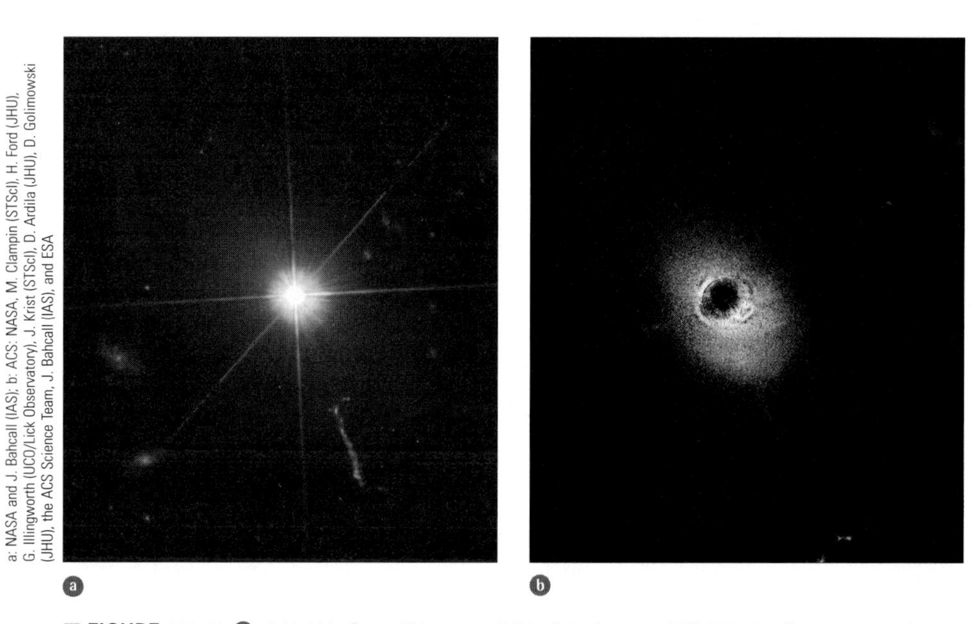

■ **FIGURE 17–7** ⓐ A Hubble Space Telescope visible-light image of 3C 273, the first quasar to be unambiguously identified, shows a star-like central brightness. The diffraction spikes going off in several directions are artifacts formed in the telescope. The image also shows, to the lower right, a faint jet. ⓑ Hubble's Advanced Camera for Surveys includes a dark spot that artificially eclipses the central quasar 3C 273, greatly dimming its light. This provides a much better view of the quasar's host galaxy because it removes the glare from the ultra-bright central source.

17.2b Puzzling Spectra

Several astronomers, including Maarten Schmidt of Caltech, photographed the optical spectra of some quasars with the 5-m (200-inch) Hale telescope at the Palomar Observatory. These spectra turned out to be bizarre, unlike the spectra of normal stars. They showed bright, broad emission lines, at wavelengths that did not correspond to lines emitted by laboratory gases at rest. Moreover, different quasars had emission lines at different wavelengths.

Schmidt made a breakthrough in 1963, when he noticed that several of the emission lines visible in the spectrum of 3C 273 had the pattern of hydrogen—a series of lines with spacing getting closer together toward shorter wavelengths—though not at

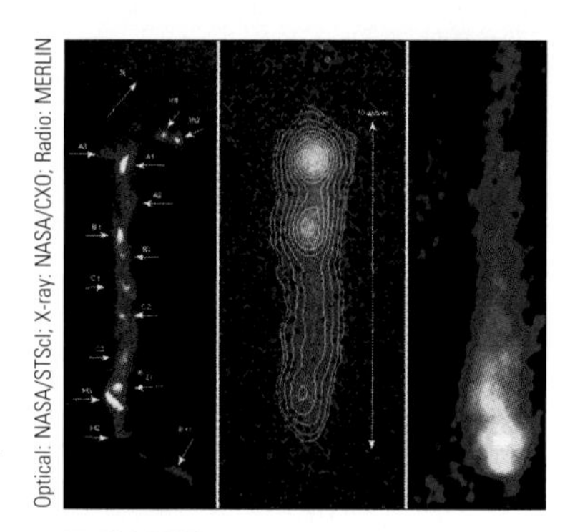

■ **FIGURE 17–8** The jet of the quasar 3C 273 stems from the nucleus. *Left to right,* we see the jet in visible light from Hubble, x-rays from Chandra, and radio waves from MERLIN, a British set of radio telescopes.

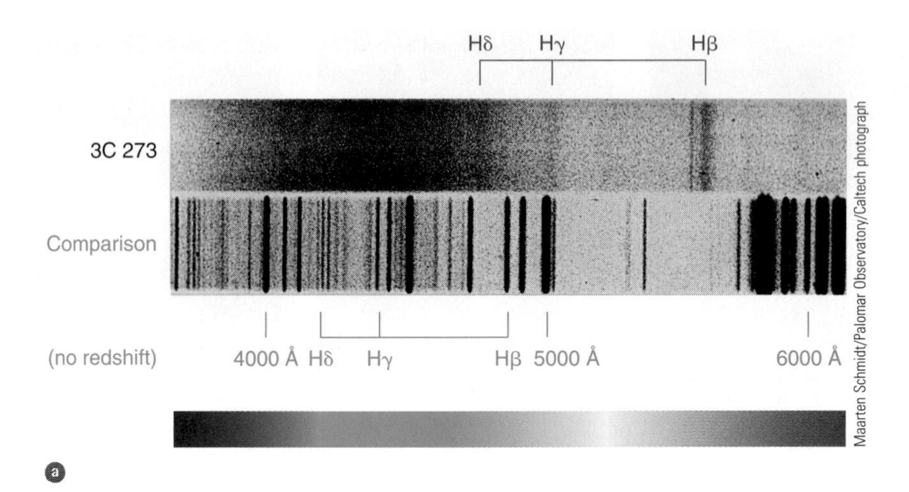

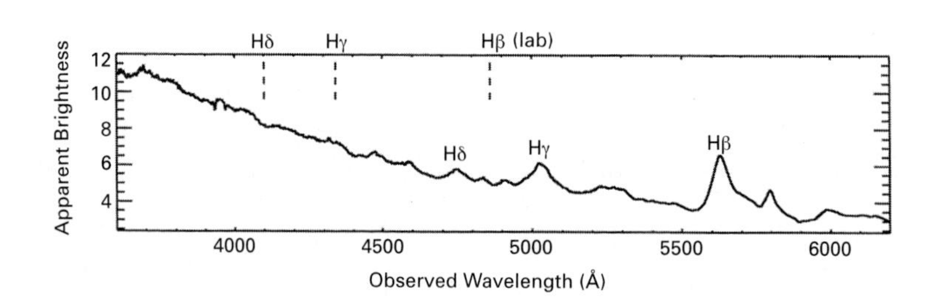

■ **FIGURE 17–9** (a) The upper spectrum is a photographic optical spectrum of the quasar 3C 273. Dark lines represent emission, not absorption. The hydrogen Balmer emission lines (Hβ, Hγ, and Hδ) appear as fuzzy (broad) bands. When properly scanned, the spectrum would look much like the modern, well-calibrated spectrum of 3C 273 shown in (b).

The comparison spectrum in *(a)* is of a hot lamp in the telescope dome; it consists of hydrogen, helium, neon, and other elements that emit lines at known wavelengths. This "comparison spectrum" establishes the wavelength scale. A color bar shows the colors of the different wavelengths of the comparison spectrum.

The hydrogen Balmer lines (Hβ, Hγ, and Hδ) in the quasar *(upper)* spectrum in *(a)* are at longer wavelengths (labels in red) than in the comparison spectrum (labels in green). Similarly, in *(b)*, the observed hydrogen emission lines appear redshifted to wavelengths longer than their laboratory values (indicated at the top). The redshift of 16% corresponds, according to Hubble's law (with H_0 = 71 km/sec/Mpc), to a distance of 2.2 billion light-years. (b: A. Filippenko and R. J. Foley, UC Berkeley)

the normal hydrogen wavelengths (■ Fig. 17–9). He realized that he could simply be observing hot hydrogen gas (with some contaminants to produce the other lines) that was Doppler shifted. The required redshift would be huge, about 16% (that is, $z = \Delta\lambda/\lambda_0 = 0.16$), corresponding to 16% of the speed of light (since $z \approx v/c$, or $v \approx cz$, valid for z less than about 0.2).

This possibility had not been recognized because nobody expected stars to have such large redshifts. Also, the spectral range then available to astronomers, who took spectra on photographic film, did not include the bright Balmer-α line of hydrogen (that is, Hα), which is normally found at 6563 Å but was shifted over to 7600 Å in 3C 273. As soon as Schmidt announced his insight, the spectra of other quasars were interpreted in the same manner. Indeed, one of Schmidt's Caltech colleagues, Jesse Greenstein, immediately realized that the spectrum of quasar 3C 48 looked like that of hydrogen redshifted by an even more astounding amount: 37%.

Subsequent searches for blue stars revealed a class of "radio-quiet" quasars—their optical spectra are similar to those of quasars, yet their radio emission is weak or absent. These are often called QSOs ("**q**uasi-**s**tellar **o**bjects"), and they are about ten times more numerous than "radio-loud" quasars. Consistent with the common practice of using the terms interchangeably, here we will simply use "quasar" to mean either the radio-loud or radio-quiet variety, unless we explicitly mention the radio properties.

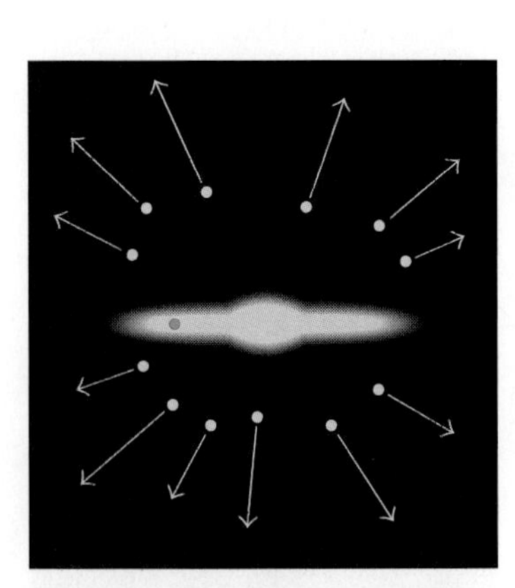

■ **FIGURE 17–10** The idea that quasars were local, and were ejected from our Galaxy, could explain why they all have high redshifts. (The blue dot shows the position of the Sun.) But this hypothesis has a number of problems and was quickly rejected by most astronomers.

17.2c The Nature of the Redshift

How were the high redshifts produced? The Doppler effect is the most obvious possibility. But it seemed implausible that quasars were discrete objects ejected like cannonballs from the center of the Milky Way Galaxy (■ Fig. 17–10); their speeds were very high, and no good ejection mechanism was known. Also, we would then expect some quasars to move slightly across the sky relative to the stars, since the Sun is not at the center of the Galaxy, but such motions were not seen. Even if these problems could be overcome, we would then have to conclude that only the Milky Way Galaxy (and not other galaxies) ejects quasars—otherwise, we would have seen "quasars" with blueshifted spectra, corresponding to those objects emitted toward us from other galaxies.

Similarly, there were solid arguments against a "gravitational redshift" interpretation (recall our discussion of this effect in Chapter 14), one in which a very strong gravitational field causes the emitted light to lose energy on its way out. This possibility was completely ruled out later, as we shall see.

If, instead, the redshifts of quasars are due to the expansion of the Universe (as is the case for normal galaxies), then quasars are receding with enormous speeds and hence must be very distant. Quasar 3C 273, for example, has $z = 0.16$, so $v \approx 0.16c \approx$ 48,000 km/sec. According to Hubble's law, $v = H_0 d$, so if $H_0 = 71$ km/sec/Mpc, then

$$d = v/H_0 \approx (48{,}000 \text{ km/sec})/(71 \text{ km/sec/Mpc}) \approx 680 \text{ Mpc} \approx 2.2 \text{ billion light-years,}$$

a sixth of the way back to the origin of the Universe! A few galaxies with comparably high redshifts (and therefore distances) had previously been found, but they were fainter than 3C 273 by a factor of 10 to 1000, and they looked fuzzy (extended) rather than star-like.

Quasar 3C 273 turns out to be one of the closest quasars. Other quasars found during the 1960s had redshifts of 0.2 to 1, and hence are billions of light-years away.

Note that redshifts greater than 1 do *not* necessarily imply speeds larger than the speed of light, because the approximation $z \approx v/c$ is reasonably accurate only when v/c is less than about 0.2. For higher speeds we may instead use the relativistic Doppler formula to calculate the nominal speed; see *Figure It Out 17.1: The Relativistic Doppler Effect*. However, even calling it a Doppler effect is misleading and, strictly speaking, incorrect: The redshift is produced by the *expansion* of space, not by motion *through* space, and the concept of "speed" then takes on a somewhat different meaning.

Similarly, as discussed in Chapter 16 for galaxies, it makes more sense to refer to the "lookback time" of a given quasar (the time it has taken for light to reach us) than to its distance: $v = H_0 d$ is inaccurate at large redshifts for a number of reasons. The lookback time formula is complicated, but some representative values are given in Table 16–1.

A few dozen quasars with redshifts exceeding 6 have been discovered (■ Fig. 17–11). The highest redshift known for a quasar as of late-2005 is $z = 6.4$, which means that a feature whose laboratory (rest) wavelength is 1000 Å is observed to be at a wavelength 640 per cent larger, or 1000 Å + 6400 Å = 7400 Å. (Recall that $z = \Delta\lambda/\lambda_0$.) The corresponding nominal speed of recession is about 0.96c, and the quasar's lookback time is roughly 12.8 billion years (in a model where the Universe is 13.7 billion years old). We see the quasar as it was when the Universe was about 6.6 per cent of its current age!

How do we detect quasars? Many of them are found by looking for faint objects with unusual colors—that is, the relative amounts of blue, green, and red light differ from those of normal stars. Low-redshift quasars tend to look bluish, because they

ASIDE 17.1: That's really fast!

"Relativistic," to an astronomer or physicist, means that such high speeds are involved that formulas from Einstein's special theory of relativity must be used instead of the simpler formulas that are valid only at low speeds.

ASIDE 17.2: Redshift records

For a long time, quasars held the record of the highest redshifts ever identified for discrete objects (that is, not the cosmic background radiation that we discuss in Chapter 19). However, in the past few years normal galaxies at somewhat higher redshifts have been found.

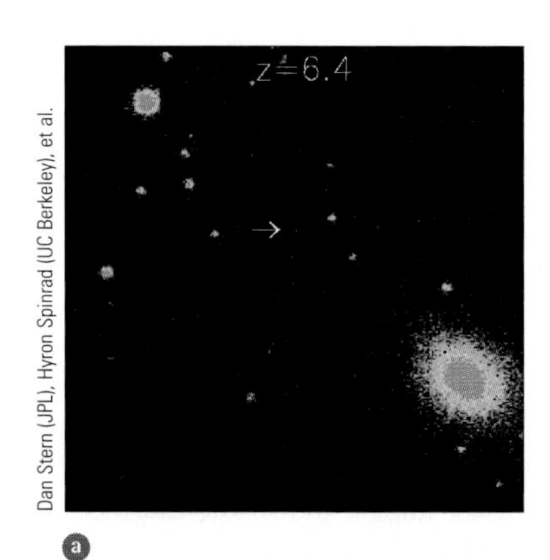

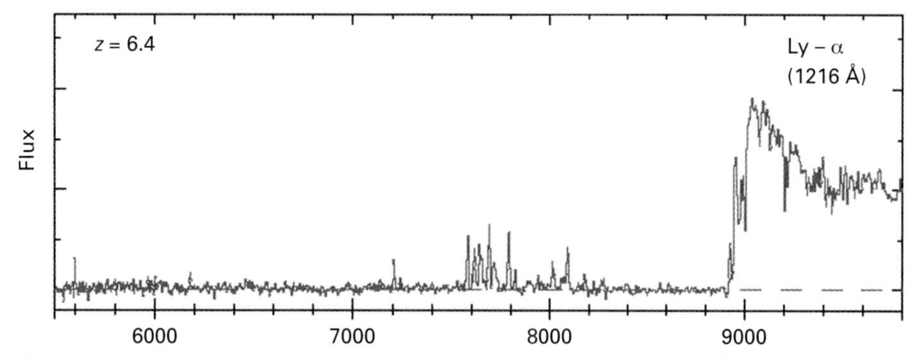

■ **FIGURE 17–11** ⓐ An image of the most distant known quasar, discovered with the Sloan Digital Sky Survey. ⓑ Its spectrum, measured with the Keck telescope, shows that all the emission lines are shifted by 6.4 times their original wavelengths. The high redshift puts the ultraviolet line Lyman-α, normally at 1216 Å, into the near-infrared region of the spectrum. [The original wavelength is shown in parentheses; 1216 Å + (6.4 × 1216 Å) = 9000 Å.] At these redshifts, we are looking back to only about 900 million years after the birth of the Universe.

FIGURE IT OUT 17.1

The Relativistic Doppler Effect

The relativistic Doppler formula is

$$z = [(1 + v/c)/(1 - v/c)]^{1/2} - 1,$$

where z is the redshift and v is the speed of recession. Note that for $v = 0$,

$$z = [(1 + 0)/(1 - 0)]^{1/2} - 1 = 0,$$

the same answer as for the nonrelativistic approximation. But for $v = 0.9c$,

$$z = [(1 + 0.9)/(1 - 0.9)]^{1/2} - 1$$
$$= [1.9/0.1]^{1/2} - 1 = \sqrt{19} - 1 = 3.4,$$

far greater than the nonrelativistic approximation would have given. So even this $z > 1$ corresponds to a speed less than the speed of light.

emit more blue light than typical stars. But the light from high-redshift quasars is shifted so much toward longer wavelengths that these objects appear very red, especially since intergalactic clouds of gas absorb much of the blue light. Quasars have also been found in maps of the sky made with x-ray satellites, and of course with ground-based radio surveys.

After finding a quasar candidate with any technique, however, it is necessary to take a spectrum in order to verify that it is really a quasar and to measure its redshift. As we have seen, the spectra of quasars are quite distinctive, and are rarely confused with other types of objects. Tens of thousands of quasars are now known, and more are being discovered very rapidly, especially by the Sloan Digital Sky Survey.

17.3 How Are Quasars Powered?

Astronomers who conducted early studies of quasars (mid-1960s) recognized that quasars are very powerful, 10 to 1000 times brighter than a galaxy at the same redshift. But while galaxies looked extended in photographs, quasars with redshifts comparable to those of galaxies appeared to be mere points of light, like stars. Their diameters were therefore smaller than those of galaxies, so their energy-production efficiency must have been higher, already making them unusual and intriguing.

17.3a A Big Punch from a Tiny Volume

However, these astronomers were in for a big surprise when they figured out just how compact quasars really are. They noticed that some quasars vary in apparent brightness over short timescales—days, weeks, months, or years (■ Fig. 17–12). This implies that the emitting region is probably smaller than a few light-days, light-weeks, light-months, or light-years in diameter, in all cases a far cry from the tens of thousands of light-years for a typical galaxy.

The argument goes as follows: Suppose we have a glowing, spherical, opaque object that is 1 light-month in radius (■ Fig. 17–13). Even if all parts of the object brightened instantaneously by an intrinsic factor of two, an outside observer would see the object brighten gradually over a timescale of 1 month, because light from the near side of the object would reach the observer 1 month earlier than light from the edge. Thus, the timescale of an observed variation sets an upper limit (that is, a maximum value) to the size of the emitting region: The actual size must be smaller than this upper limit.

Although this conclusion can be violated under certain conditions (such as when different regions of the object brighten in response to light reaching them from other regions, creating a "domino effect"), such models generally seem unnatural. Proper use of Einstein's special theory of relativity (in case the light-emitting material is moving very fast) can also change the derived upper limit to some extent, but the basic conclusion still holds: Quasars are very small, yet they release tremendous amounts of

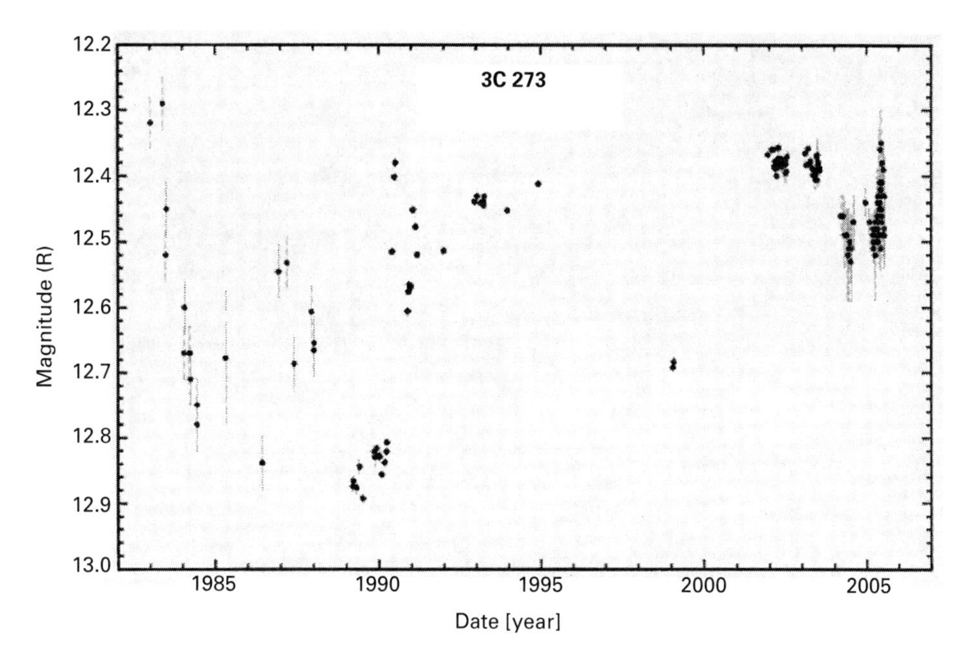

■ **FIGURE 17–12** The apparent brightness of optical radiation from the quasar 3C 273 varies on a timescale of months. Note that a difference from magnitude 12.3 to 12.9, a fading by 0.6 magnitudes, is a fading by a factor of about 2. (Thomas J. Balonek, Terry-Ann Suer, Joseph W. Gangestad, et al., at Colgate U.)

energy. For example, a quasar only 1 light-month across can be 100 times more powerful than an entire galaxy of stars 100,000 light-years in diameter!

17.3b What Is the Energy Source?

The nature of the prodigious (yet physically small) power source of quasars was initially a mystery. How does such a small region give off so much energy? After all, we don't expect huge explosions from tiny firecrackers. There was some indication that these objects might be related to active galactic nuclei: They have similar optical spectra and are bright at radio wavelengths. So, perhaps the same mechanism might be used to explain the unusual properties of both kinds of objects. In fact, maybe active galactic nuclei are just low-power versions of quasars! If so, quasars should be located in the centers of galaxies. Later we will see that this is indeed the case.

The fact that the incredible power source of quasars is very small immediately rules out some possibilities. Such a process of elimination is often useful in astronomy; recall, for instance, how we deduced that pulsars are rapidly spinning neutron stars. It turns out that for quasars, chemical energy is woefully inadequate: They cannot be wood on fire, or even chemical explosives, because the most powerful of these is insufficient to produce so much energy within such a small volume.

Even nuclear energy, which works well for stars, is not possible for the most powerful quasars. They cannot be radiation from otherwise-unknown supermassive stars or chains

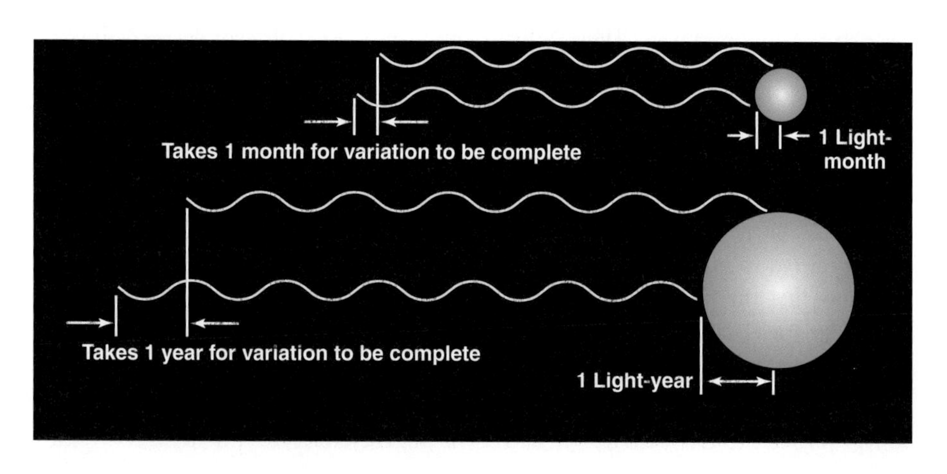

■ **FIGURE 17–13** Why a large object can't appear to fluctuate in brightness as rapidly as a smaller object. Say that each object abruptly brightens at one instant. The wave emitted from the edge of the object takes longer to reach the observer than light from the near side of the object, because it has to travel farther. We don't see the full variation until waves from all parts of the object reach us. The same is true for both opaque and transparent objects. For the latter, light from the far side takes longer to reach us than light from the near side.

of supernovae going off almost all the time, or other more exotic stellar processes, because once again the efficiency of nuclear energy production is not high enough. To produce that much nuclear energy, a larger volume of material would be needed.

The annihilation of matter and antimatter is energetically feasible, since it is 100% efficient. That is, *all* of the mass in a matter–antimatter collision gets turned into photons (radiation), and in principle a very small volume can therefore be tremendously powerful. However, the observed properties of quasars do not support this hypothesis. Specifically, matter–antimatter collisions tend to emit excess amounts of radiation at certain wavelengths, and this is not the case for quasars.

The release of gravitational energy, on the other hand, can in some cases be very efficient, and seemed most promising to several theorists studying quasars in the mid-1960s. We have already discussed how the gravitational contraction of a ball of gas (a protostar), for example, both heats the gas and radiates energy. But to produce the prodigious power of quasars, a very strong gravitational field is needed.

The conclusion was that a quasar is a **supermassive black hole,** perhaps 10 million to a billion times the mass of the Sun, in the process of swallowing ("accreting") gas. The black hole is in the center of a galaxy. The rate at which matter can be swallowed, and hence the power of the quasar, is proportional to the mass of the black hole, but it is typically a few solar masses per year. Although the Schwarzschild radius of, say, a 50 million solar-mass black hole is 150 million km, this is just 1 A.U. (i.e., 8.3 light-minutes, the distance between the Earth and the Sun), and hence is minuscule compared with the radius of a galaxy (many thousands of light-years).

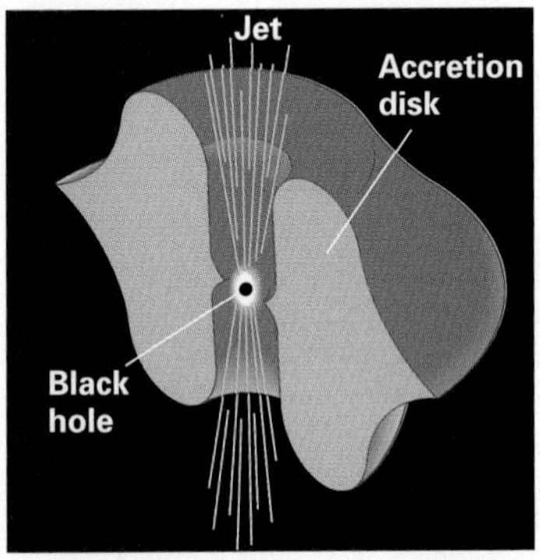

■ **FIGURE 17–14** Cross-sectional view of an accretion disk (doughnut) surrounding a black hole, with high-speed jets of particles emerging from the disk's nozzle. The nozzle points along the black hole's rotation axis.

17.3c Accretion Disks and Jets

The matter generally swirls around the black hole, forming a rotating disk called an **accretion disk** (■ Fig. 17–14), a few hundred to a thousand times larger than the Schwarzschild radius of the black hole (and hence up to a few light-days to a light-week in size). As the matter falls toward the black hole, it gains speed (kinetic energy) at the expense of its gravitational energy, just as a ball falling toward the ground accelerates. Compression of the gas particles in the accretion disk to a small volume, and the resulting friction between the particles, causes them to heat up; thus, they emit electromagnetic radiation, thereby converting part of their kinetic energy into light.

Note that energy is radiated *before* the matter is swallowed by the black hole—nothing escapes from within the black hole itself. This process can convert the equivalent of about 10% of the rest-mass energy of matter into radiation, more than 10 times more efficiently than nuclear energy. (Recall from Chapter 11 that the fusion of hydrogen to helium converts only 0.7% of the mass into energy.)

A spinning, very massive black hole is also consistent with the well-focused "jets" of matter and radiation that emerge from some quasars, typically reaching distances of a few hundred thousand light-years. Again, no material actually comes from *within* the black hole; instead, its origin is the accretion disk. The charged particles in the jets are believed to shoot out in a direction perpendicular to the accretion disk, along the black hole's axis of rotation (Fig. 17–14). They emit radiation as they are accelerated. In addition to the radio radiation, high-energy photons such as x-rays can also be produced (■ Fig. 17–15). The impressive focusing might be provided by a magnetic field, as in the case of pulsars, or by the central cavity in the disk.

NASA/Chandra X-ray Observatory Center/SAO

■ **FIGURE 17–15** An image of the quasar PKS 0637-752 ($z = 0.65$) obtained with the Chandra X-ray Observatory. It is so distant that we see it as it was about 6 billion years ago. Note the "jet" that extends about 200,000 light-years out from the quasar; its power at x-ray energies exceeds that at radio energies, an important constraint that a successful theory of jet formation will have to explain.

Recall that jets are also seen in some types of active galaxies, which appear to be closely related to quasars (Fig. 17–3). As discussed in more detail later in this chapter, we know that the particles move with very high speeds because a jet can sometimes appear to travel faster than the speed of light—an effect that occurs only when an object travels nearly along our line of sight, nearly at the speed of light.

Recently, indirect evidence for accretion disks surrounding a central, supermassive black hole has been found in several active galaxies from observations with various x-ray telescopes (Japan's ASCA, the European Space Agency's XMM-Newton Mission, and NASA's Chandra X-ray Observatory). The specific shape of emission lines from highly ionized iron atoms that must reside very close to the galaxy center resembles that expected if the light is coming from a rotating accretion disk. Moreover, these lines exhibit a "gravitational redshift"—they appear at a somewhat longer wavelength than expected from the recession speed of the galaxy, because the photons lose some energy (and hence get shifted to longer wavelengths) as they climb out of the strong gravitational field near the black hole (see Chapter 14).

Similar emission lines have been seen in x-ray binary systems in which the compact object is likely to be a black hole (see the discussion in Section 14.7). Such lines, in both active galaxies and x-ray binaries, are now being analyzed in detail to detect and study predicted relativistic effects such as the strong bending of light and the "dragging" of space–time around a rotating black hole.

17.4 What Are Quasars?

The idea that quasars are energetic phenomena at the centers of galaxies is now strongly supported by observational evidence. First of all, the observed properties of quasars and active galactic nuclei are strikingly similar. In some cases, the active nucleus of a galaxy is so bright that the rest of the galaxy is difficult to detect because of contrast problems, making the object look like a quasar (■ Fig. 17–16). This is especially true if the galaxy is very distant: We see the bright nucleus as a point-like object, while the spatially extended outer parts (known as "fuzz" in this context) are hard to detect because of their faintness and because of blending with the nucleus.

In the 1970s, a statistical test was carried out with quasars. A selection of quasars, sorted by redshift, was carefully examined. Faint fuzz (presumably a galaxy) was discovered around most of the quasars with the smallest redshifts (the nearest ones), a few of the quasars with intermediate redshifts, and none of the quasars with the largest redshifts (the most distant ones). Astronomers concluded that the extended light was too faint and too close to the nucleus in the distant quasars, as expected. In the 1980s, optical spectra of the fuzz in a few nearby quasars revealed absorption lines due to stars, but the vast majority of objects were too faint for such observations. In any case, the data strongly suggested that quasars could indeed be extreme examples of galaxies with bright nuclei.

More recently, images obtained with the Hubble Space Telescope demonstrate conclusively that quasars live in galaxies, almost always at their centers. With a clear view of the skies above the Earth's atmosphere, and equipped with CCDs, the Hubble Space Telescope easily separates the extended galaxy light from the point-like quasar itself at low redshifts. In some cases the galaxy is obvious (■ Fig. 17–17), but in others

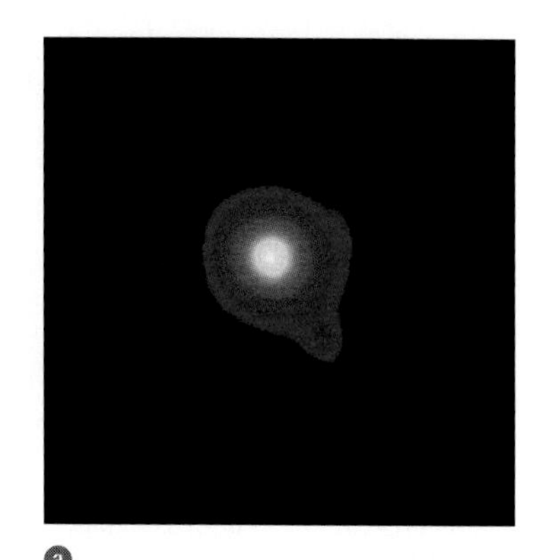

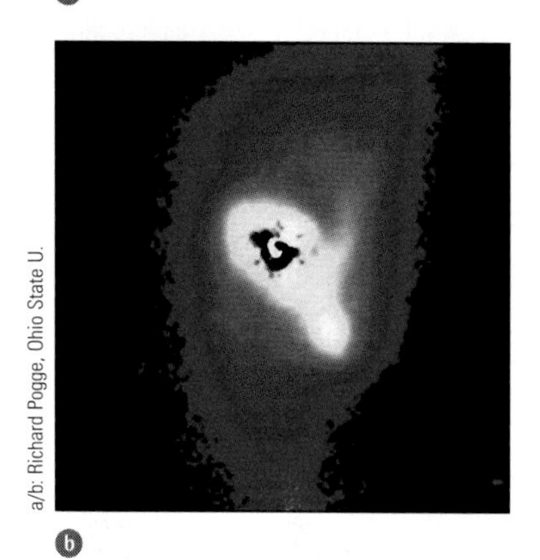

a/b: Richard Pogge, Ohio State U.

■ **FIGURE 17–16** A quasar observed with adaptive optics on the Gemini North telescope on Mauna Kea. (a) A false-color image of the quasar, which has a very bright nucleus (shown in white) surrounded by "fuzz" (blue). The source is so powerful that it emits more than 100 times as much total energy each second as does our entire Milky Way Galaxy. (b) Subtracting the bright peak, and thus decreasing the contrast, more clearly reveals the disk of the host galaxy; the quasar is centered on it. This disk may have been formed during the strong collision and eventual merger of two normal spiral galaxy disks. Perhaps two galaxies merged to trigger the quasar's formation.

a/b: John N. Bahcall and Sofia Kirhakos (Institute for Advanced Study), Donald P. Schneider (Pennsylvania State U.), and NASA

■ **FIGURE 17–17** (a) The quasar PG 0052+251, at redshift $z = 0.155$, is in the center of a normal spiral galaxy. (b) The quasar PHL 909, at redshift $z = 0.171$, is at the core of a normal elliptical galaxy. Both of these images were obtained with the Hubble Space Telescope.

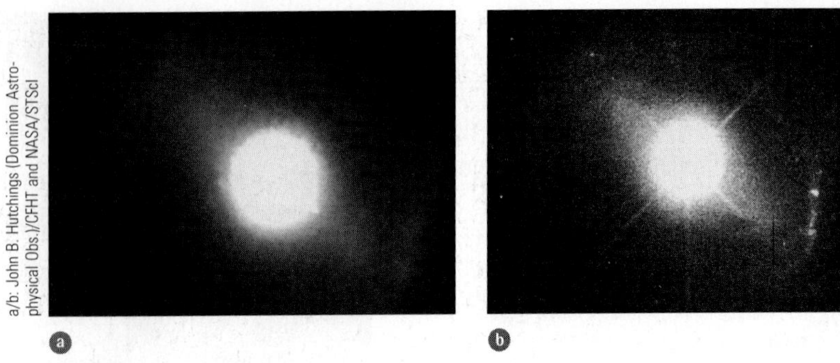

■ **FIGURE 17–18** Quasar fuzz, in the object QSO 1229+204. (a) A ground-based image made with the Canada-France-Hawaii Telescope on Mauna Kea. The quasar is in the core of a spiral galaxy that is colliding with a dwarf galaxy (not seen in this image). The resolution is about 0.5 arc sec. (b) The high resolution of the Hubble Space Telescope reveals structure in the fuzz. On one side of the galaxy we see a string of knots, which are probably massive young star clusters. They may have been formed as a result of the collision. The resolution is about 0.1 arc sec.

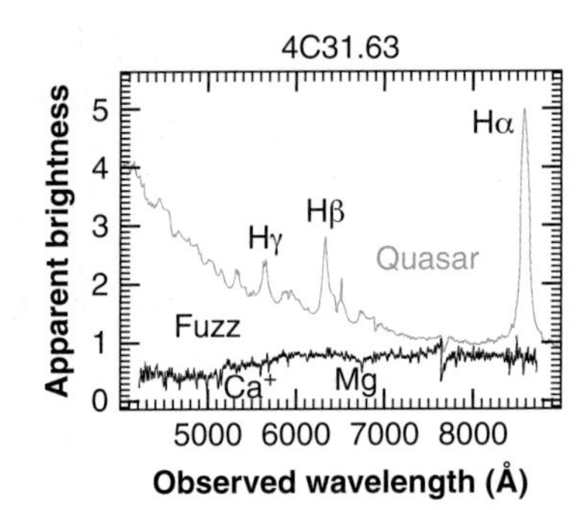

■ **FIGURE 17–19** Keck telescope spectra of *(top)* the quasar 4C 31.63 ($z = 0.296$) and *(below)* of "fuzz." The spectrum of the fuzz shows absorption lines typical of those that are produced by the relatively cool outer parts of normal stars. Thus the fuzz seen around nearby quasars is really starlight from galaxies surrounding the bright central object. In this double graph, the brightness of the quasar was reduced substantially to fit it on the same part of the vertical axis as the fuzz spectrum. (Joe Miller and Andy Sheinis, University of California at Santa Cruz/Lick Observatory)

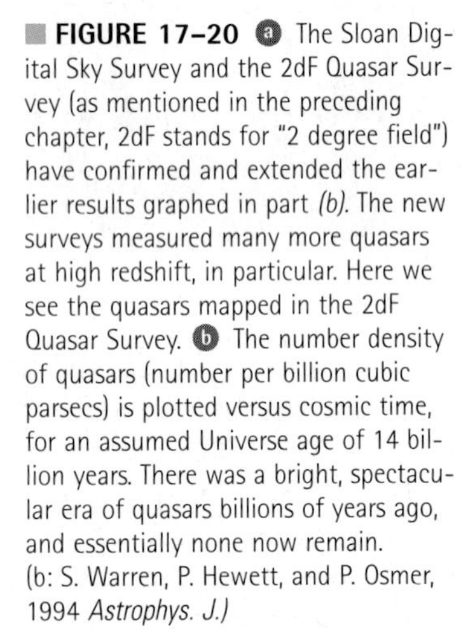

■ **FIGURE 17–20** (a) The Sloan Digital Sky Survey and the 2dF Quasar Survey (as mentioned in the preceding chapter, 2dF stands for "2 degree field") have confirmed and extended the earlier results graphed in part *(b)*. The new surveys measured many more quasars at high redshift, in particular. Here we see the quasars mapped in the 2dF Quasar Survey. (b) The number density of quasars (number per billion cubic parsecs) is plotted versus cosmic time, for an assumed Universe age of 14 billion years. There was a bright, spectacular era of quasars billions of years ago, and essentially none now remain. (b: S. Warren, P. Hewett, and P. Osmer, 1994 *Astrophys. J.*)

it is barely visible, and special techniques are used to reveal it; recall, for example, 3C 273 in Figure 17–7. Further solidifying the association of quasars with galaxies, recent ground-based optical spectra of some relatively nearby quasars ($z = 0.2–0.3$) show unambiguous stellar absorption lines at the same redshift as that given by the quasar emission lines (■ Fig. 17–19).

Quasars exist almost exclusively at high redshifts and hence large distances. The peak of the distribution is at $z \approx 2$ (■ Fig. 17–20), though new studies at x-ray wavelengths suggest that it might be at an even higher redshift. With lookback times of

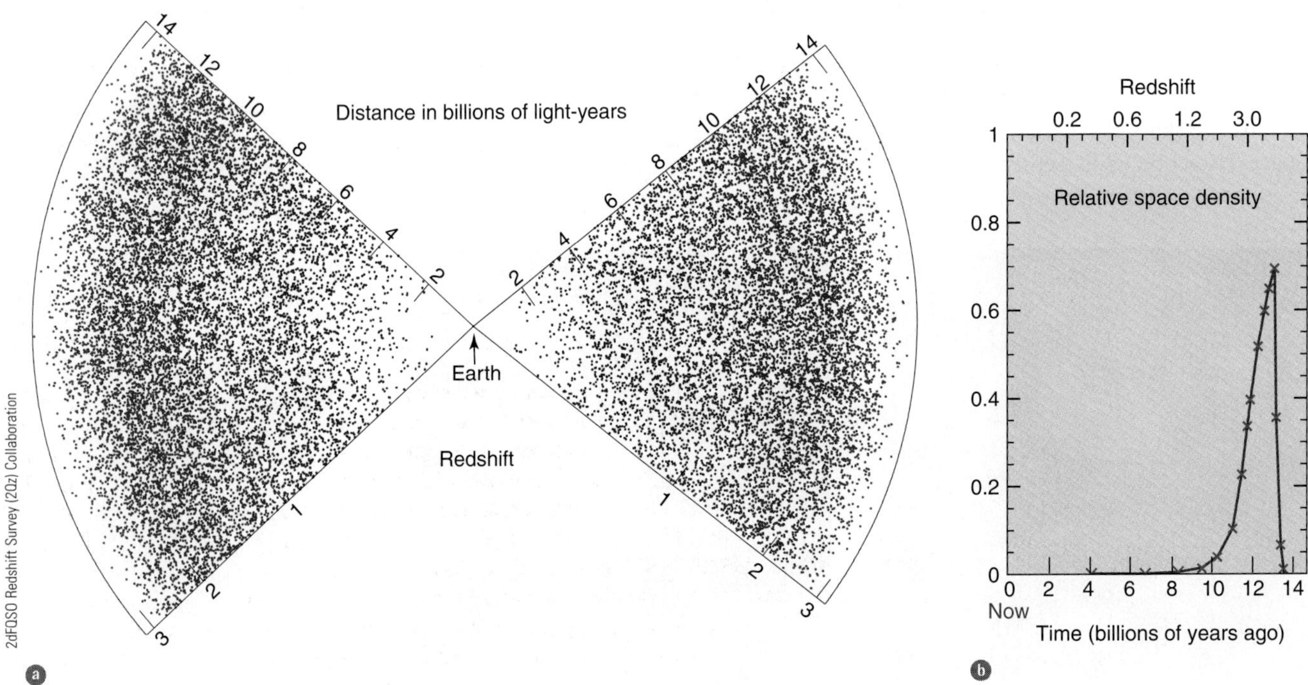

John N. Bahcall and Sofia Kirhakos (Institute for Advanced Study), Donald P. Schneider (Pennsylvania State U.), Mike Disney (U. of Wales), and NASA

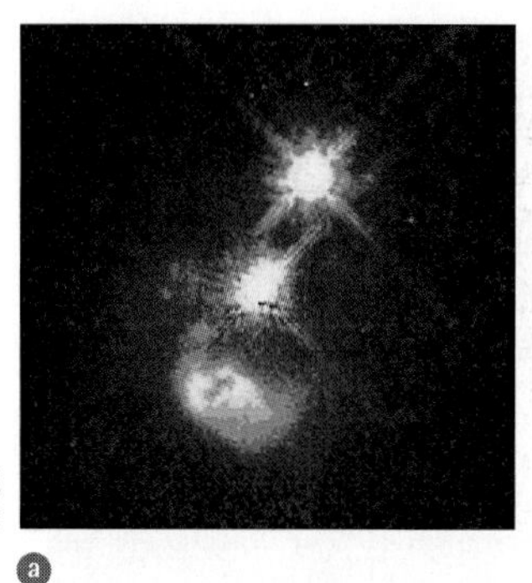

FIGURE 17–21 Quasars in interacting or merging galaxies. (a) A Hubble Space Telescope image showing debris from a collision between two galaxies fueling a quasar. A ring galaxy left by the collision is at bottom; a foreground star in our Galaxy is at top. (b) A Hubble view showing a tidal tail above a quasar, perhaps drawn out by a galaxy that is no longer there. (c) A quasar merging with the bright galaxy that appears just below it. The swirling wisps of dust and gas surrounding them indicate that an interaction is taking place. (d) A pair of merged galaxies have left loops of gas around this quasar.

about 10 billion years, quasars must be denizens of the *young* Universe. What happened to them? Quasars probably faded with time, as the central black hole gobbled up most of the surrounding gas; the quasar shines only while it is pulling in material.

Thus, some of the nearby active and normal galaxies may have been luminous quasars in the distant past, but now exhibit much less activity because of a slower accretion rate. Perhaps even the nucleus of the Milky Way Galaxy, which is only slightly active, was more powerful in the past, when the putative black hole had plenty of material to accrete. Of course, many of the weakly active galaxies we see nearby were probably never luminous enough to be genuine quasars. Either their central black hole wasn't sufficiently massive to pull in much material, or there was little gas available to be swallowed.

Though most quasars are very far away, some have relatively low redshifts (like 0.1). If quasars were formed early in the Universe, how can these quasars still be shining? Why hasn't all of the gas in the central region been used up? High-resolution images (Fig. 17–21) show that in many cases, the galaxy containing the quasar is interacting or merging with another galaxy. This result suggests that gravitational tugs end up directing a fresh supply of gas from the outer part of the galaxy (or from the intruder galaxy) toward its central black hole, thereby fueling the quasar and allowing it to continue radiating so strongly. Some quasars may have even faded for a while, and then the interaction with another galaxy rejuvenated the activity in the nucleus.

Adaptive optics is now allowing high-resolution imaging from mountaintop observatories in addition to the Hubble Space Telescope. An image with adaptive optics on the Gemini North telescope has enabled the central quasar peak of brightness to be subtracted from the overall image. A flat edge-on disk, interpreted to be the host galaxy, was revealed (Fig. 17–16).

17.5 Are We Being Fooled?

A few astronomers have disputed the conclusion that the redshifts of quasars indicate large distances, partly because of the implied enormously high luminosity produced in a small volume. If Hubble's law doesn't apply to quasars, maybe they are actually quite nearby.

■ **FIGURE 17–22** Two galaxies with relatively low redshifts are seen close in the sky to high-redshift quasars. Most astronomers think these are just chance projections along nearly the same lines of sight; the quasars and nearby galaxies are not physically associated with each other.

Specifically, Halton Arp has found some cases where a quasar seems associated with an object of a different, lower redshift (■ Fig. 17–22). However, most astronomers blame the association on chance superposition. There could also be some amplification of the brightnesses of distant quasars, along the line of sight, by the gravitational field of the low-redshift object; this would produce an apparent excess of quasars around such objects.

We now have little reason to doubt the conventional interpretation of quasar redshifts (though of course as scientists we should keep an open mind). Quasars clearly reside in the centers of galaxies having the same redshift. They are simply the more luminous cousins of active galactic nuclei, and a plausible energy source has been found. In addition, gravitational lensing (see below) shows that quasars are indeed very distant.

17.6 Finding Supermassive Black Holes

We argued above, essentially by the process of elimination, that the central engine of a quasar or active galaxy consists of a supermassive black hole swallowing material from its surroundings, generally from an accretion disk. Is there any more direct evidence for this?

Well, the high speed of gas in quasars and active galactic nuclei, as measured from the widths of emission lines, suggests the presence of a supermassive black hole. A strong gravitational field causes the gas particles to move very quickly, and the different emitted photons are Doppler shifted by different amounts, resulting in a broad line. On the other hand, alternative explanations such as supernovae might conceivably be possible; they, too, produce high-speed gas, but without having to use a supermassive black hole.

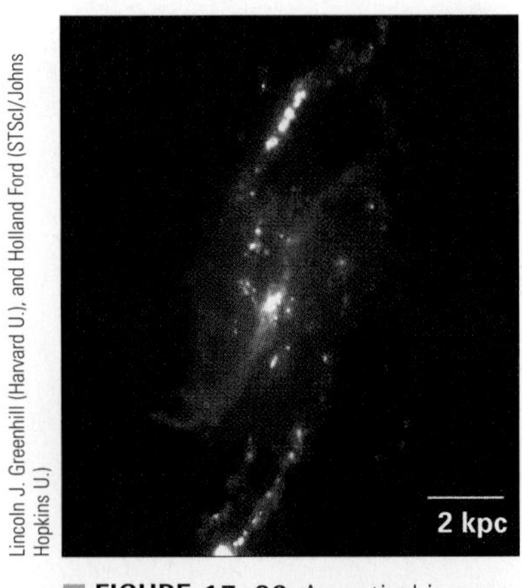

■ **FIGURE 17–23** An optical image of the spiral galaxy NGC 4258 that emphasizes star-forming regions. Radio observations of the nucleus have revealed gas orbiting a central, exceedingly small, very massive object—almost certainly a black hole.

Recently, however, very rapidly rotating disks of gas have been found in the centers of several mildly active galaxies. Their motion is almost certainly produced by the gravitational attraction of a compact central object, because we see the expected decrease of orbital speed with increasing distance from the center, as in Kepler's laws for the Solar System. The galaxy NGC 4258 (■ Fig. 17–23) presents the most convincing case, one in which radio observations were used to obtain very accurate measurements. The typical speed is $v = 1120$ km/sec at a distance of only 0.4 light-year from the center. The data imply a mass of about 3.6×10^7 solar masses in the nucleus (see *Figure It Out 17.2: The Central Mass in a Galaxy*).

The corresponding density is over 100 million solar masses per cubic light-year, a truly astonishing number. If the mass consisted of stars, there would be no way to pack

FIGURE IT OUT 17.2
The Central Mass in a Galaxy

We saw in Chapter 16 that the mass enclosed within a circular orbit of radius R is $M = v^2R/G$, where v is the orbital speed and G is Newton's constant of gravitation. For NGC 4258, we measure $v = 1120$ km/sec at a distance of 0.4 light-year from the center. Properly converting units, we find that $M = 7.1 \times 10^{40}$ g $= 3.6 \times 10^7$ solar masses. Most of this mass must be in a black hole, rather than in stars.

them into such a small volume, at least not for a reasonable amount of time: They would rapidly collide and destroy themselves, or undergo catastrophic collapse. The natural conclusion is that a supermassive black hole lurks in the center. Indeed, this is now regarded as the most conservative explanation for the data: If it's not a black hole, it's something even stranger!

One of the most massive black holes ever found is that of M87, an active galaxy in the Virgo Cluster that sports a bright radio and optical jet (■ Fig. 17–24). Spectra of

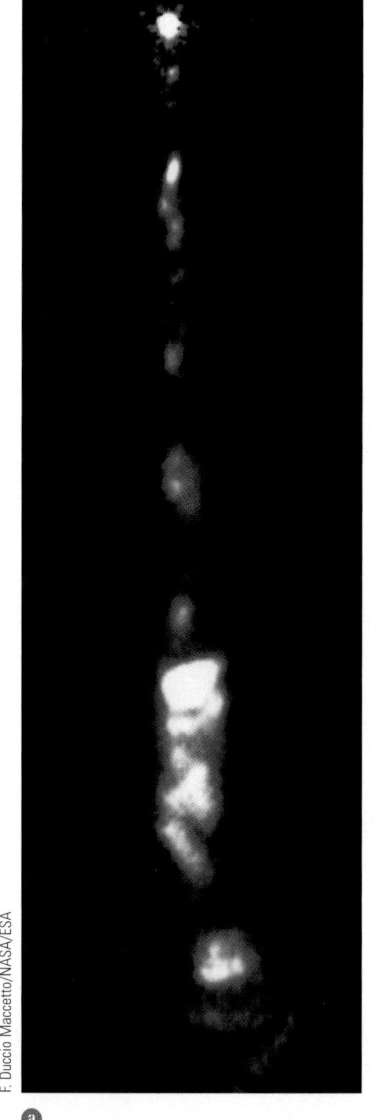

F. Duccio Maccetto/NASA/ESA

■ **FIGURE 17–24** ⓐ The nucleus (bright point at top) and jet of M87, in a computer-processed view from the Faint Object Camera aboard the Hubble Space Telescope. In the jet of high-speed charged particles, which extends 8000 light-years, the image reveals detail as small as 10 light-years across. ⓑ This later Hubble image of M87's nucleus and jet also shows (enlarged) an unusual spiral disk in the galaxy's center. See Figure 17–28 for a radio view of the jet.

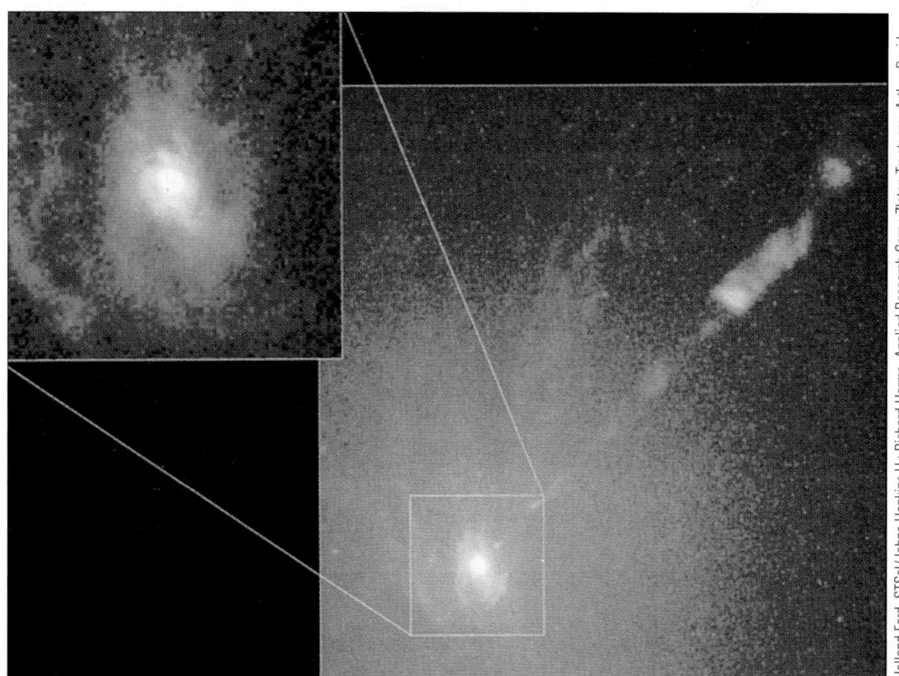

Holland Ford, STScI/Johns Hopkins U.; Richard Harms, Applied Research Corp.; Zlatan Tsvetanov, Arthur Davidsen, and Gerard Kriss, Johns Hopkins U.; Ralph Bohlin and George Hartig, STScI; Linda Dressel and Ajay K. Kochar, Applied Research Corp.; and Bruce Margon, STScI; NASA/ESA/STScI

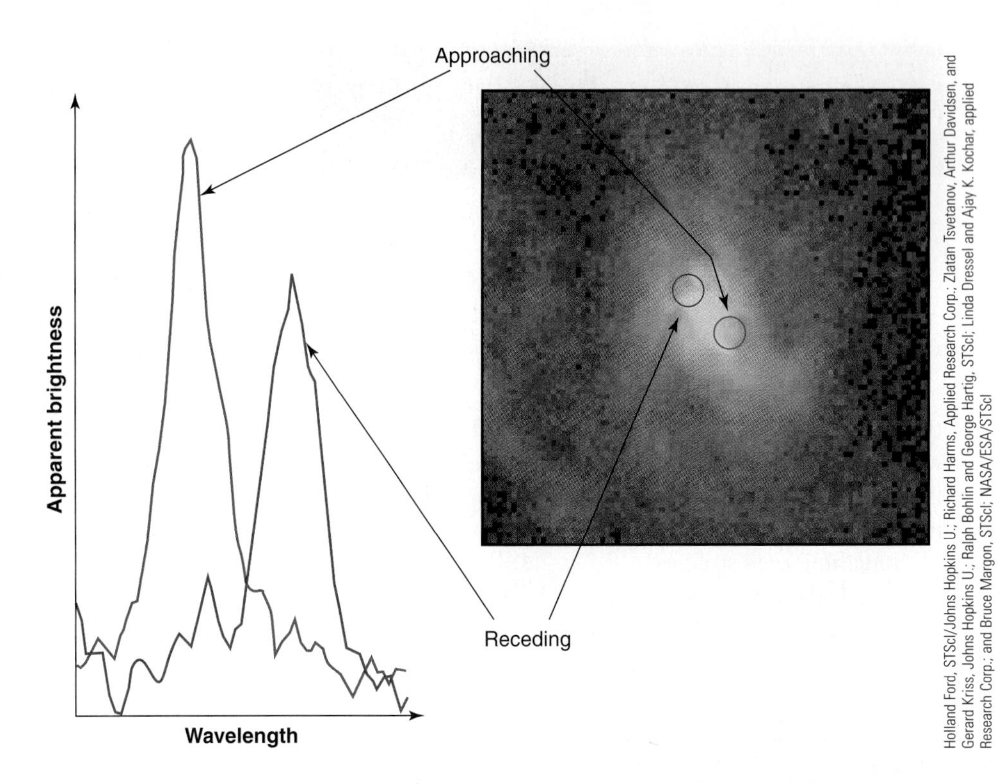

■ **FIGURE 17–25** Spectra of the regions shown on the image of the center of M87, taken with the Faint Object Spectrograph aboard the Hubble Space Telescope, reveal Doppler shifts of the gas. (The single emission line appears at different wavelengths.) The orbital speed of 550 km/sec at this distance of 60 light-years from the nucleus allows astronomers to calculate how much mass must be inside those locations to keep the gas in orbit. The result is about 3 billion solar masses, after various effects like the inclination of the disk are taken into account.

Holland Ford, STScI/Johns Hopkins U.; Richard Harms, Applied Research Corp.; Zlatan Tsvetanov, Arthur Davidsen, and Gerard Kriss, Johns Hopkins U.; Ralph Bohlin and George Hartig, STScI; Linda Dressel and Ajay K. Kochar, applied Research Corp.; and Bruce Margon, STScI; NASA/ESA/STScI

the gas disk surrounding the nucleus were obtained with the Hubble Space Telescope (■ Fig. 17–25), and the derived mass in the nucleus is about 3 billion solar masses.

If some nearby, relatively normal-looking galaxies were luminous quasars in the past, and a significant fraction even show some activity now, we suspect that supermassive black holes are likely to exist in the centers of many large galaxies today. Sure enough, when detailed spectra of the nuclear regions of a few galaxies were obtained (especially with the Hubble Space Telescope), strong evidence was found for rapidly moving stars. The masses derived from Kepler's third law were once again in the range of a million to a billion Suns. By late-2005, the central regions of several dozen galaxies had been observed in this manner, revealing the presence of supermassive black holes.

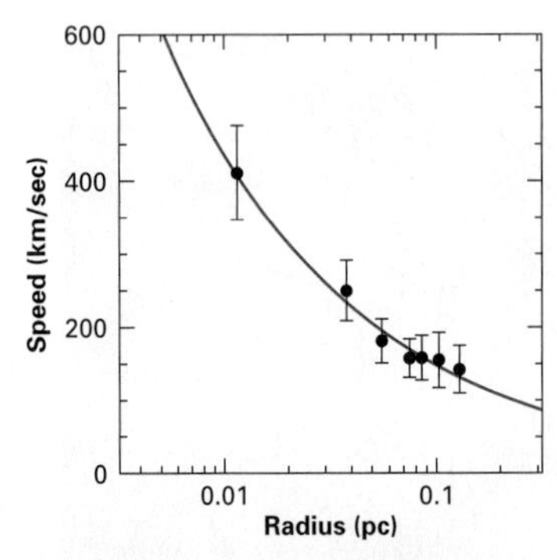

■ **FIGURE 17–26** The observed speeds of stars vs. their distance from the center of our Galaxy, a radio source known as Sgr A* ("Sagittarius A star"). Recent data suggest that the mass is close to 3.7 million solar masses. Note that 1 pc = 3.26 light-years.

See the information about stars orbiting close to our Galaxy's center in Figure 15-15. (Andrea Ghez (UCLA) et al.)

Probably the most impressive and compelling case is our own Milky Way Galaxy. As we discussed in Chapter 15, stars in the highly obscured nucleus were seen from Earth at infrared wavelengths, and their motions were measured over the course of a few years; see Figure 15–15. The data are consistent with stars orbiting a single, massive, central dark object (■ Fig. 17–26). The implied mass of this object is 3.7 million solar masses, and it is confined to a volume only 0.03 light-year in diameter! The only known explanation is a black hole. Thus, our Galaxy could certainly have been more active in the past, though never as powerful as the most luminous quasars, which require a black hole of 10^8 to 10^9 solar masses.

In the past few years, it has been found that the mass of the central black hole is proportional to the mass of the bulge in a spiral galaxy, or to the total mass of an elliptical galaxy (■ Fig. 17–27). But recall from Chapter 16 that the bulges of spiral galaxies are old, as are elliptical galaxies (which resemble the bulges of spiral galaxies). Thus, there is evidence that the formation of the supermassive black hole is related to the earliest stages of formation of galaxies. We don't yet understand this relation, but clearly it offers a clue to physical processes long ago, when most galaxies were being born. Very recent studies show that for a given bulge mass, the more compact the bulge, the more massive the black hole, suggesting an even closer link between bulge formation and black-hole formation.

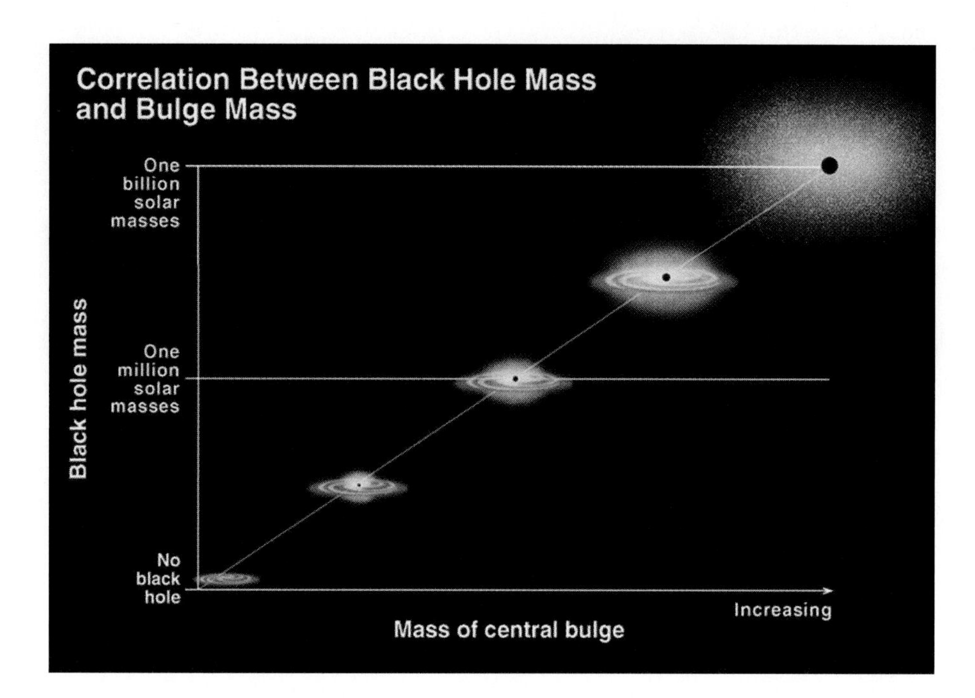

■ **FIGURE 17–27** The mass of the supermassive black hole in the center of a galaxy appears to be larger in spiral galaxies with larger bulges. Similarly, the black-hole mass scales with the mass of the elliptical galaxy in which it resides.

17.7 The Effects of Beaming

Radio observations with extremely high angular resolution, generally obtained with the technique of very-long-baseline interferometry, have shown that some quasars consist of a few small components. In many cases, observations over a few years reveal that the components are apparently separating very fast (■ Fig. 17–28), given the conversion from the angular change in position we measure across the sky to the actual physical speed in km/sec at the distance of the quasar. Indeed, some of the components appear to be separating at **superluminal speeds**—that is, at speeds greater than that of light! But Einstein's special theory of relativity says that no objects can travel through space faster than light, an apparent contradiction.

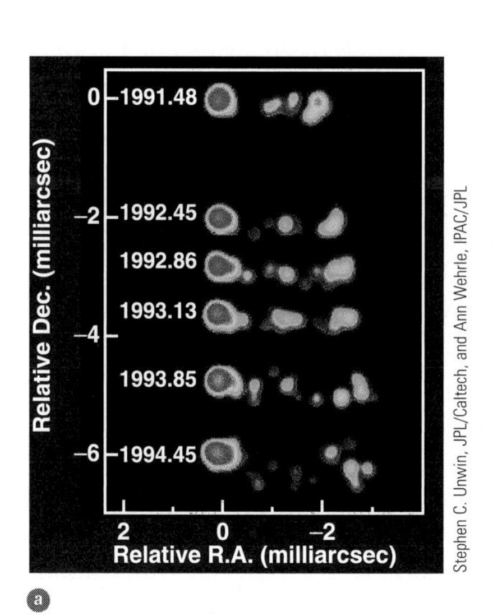

Stephen C. Unwin, JPL/Caltech, and Ann Wehrle, IPAC/JPL

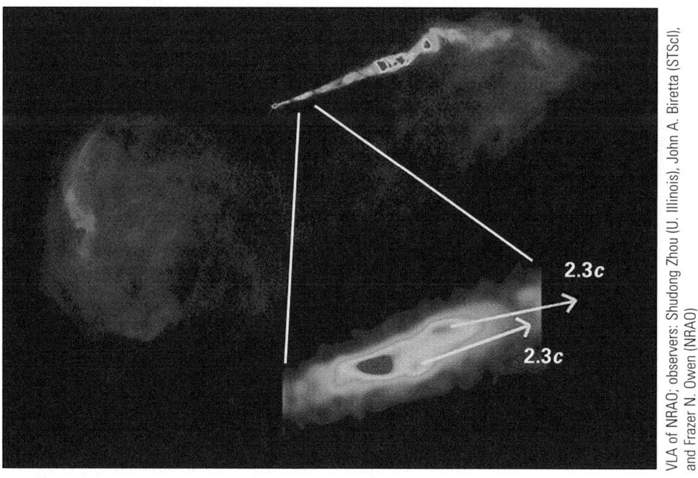

VLA of NRAO; observers: Shudong Zhou (U. Illinois), John A. Biretta (STScI), and Frazer N. Owen (NRAO)

■ **FIGURE 17–28** (a) A series of views of the quasar 3C 279 with radio interferometry at a wavelength of 1.3 cm. The apparent speed translates (for H_0 = 71 km/sec/Mpc) to a speed of 8.5*c*, though such apparent "superluminal speeds" can be explained in conventional terms. (b) Superluminal speeds also appear in the active galaxy M87. The inset shows features that seem to move at 2.3*c*. Red shows the brightest and blue shows the faintest emission.

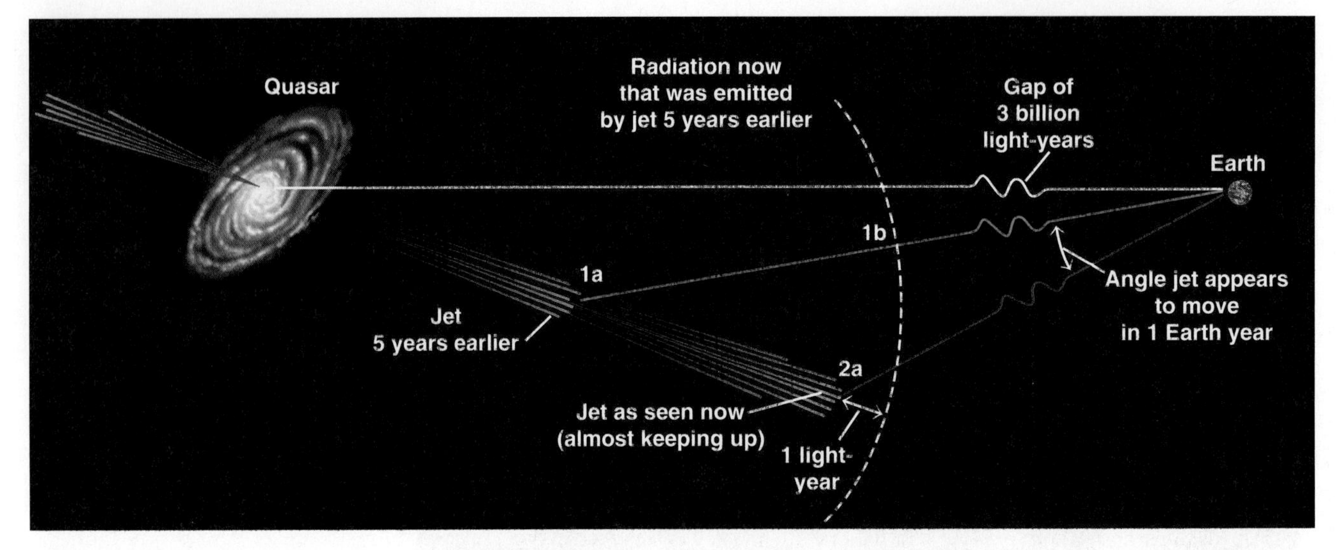

■ **FIGURE 17–29** The leading model for explaining how a jet of gas emitted from a quasar, shown at left with its accretion disk surrounding a central black hole, can appear to exceed the speed of light when seen from Earth *(shown at right)*, in this case 3 billion light-years away. Point 1a marks the position of the jet 5 years earlier than the jet we are seeing now. (It actually takes another 3 billion years for the radiation to reach us, an extra duration that we can ignore for the purposes of this example because we are interested in *differences* in the arrival times of the light from two jet positions.) After an interval of 5 years, the light emitted from point 1a has reached point 1b, while the jet itself has reached point 2a. The jet has been moving so fast that points 2a and 1b are separated by only 1 light-year in this example. The light emitted then from the jet at point 2a is thus only 1 year behind the light emitted 5 years earlier at point 1a. So, within a 1-year interval we receive light from two positions that the jet has taken 5 years to go between. The jet actually therefore had 5 years to move across the sky to make the angular change in position that we see over a 1-year interval. We therefore think that the jet is moving faster than it actually moves.

Astronomers can explain how the components only *appear* to be separating at greater than the speed of light even though they are actually physically moving at allowable speeds (less than that of light). If one of the components is a jet approaching us almost along our line of sight, and nearly at the speed of light, then according to our perspective the jet is nearly keeping up with the radiation it emits (■ Fig. 17–29).

If the jet moves a certain distance in our direction in (say) 5 years, the radiation it emits at the end of that period gets to us sooner than it would have if the jet were not moving toward us. So in fewer than 5 years, we see the jet's motion over 5 full years. In the interval between our observations, the jet had several times longer to move than we would naively think it had. So it could, without exceeding the speed of light, appear to move several times as far.

Whether a given object looks like a quasar or a less-active galaxy with broad emission lines probably depends on the orientation of the jet relative to our line of sight: Jets pointing at us appear far brighter than those that are misaligned. Thus, quasars are probably often beamed roughly toward us, a conclusion supported by the fact that many radio-loud quasars show superluminal motion.

However, if the jet is pointing straight at us, it can greatly outshine the emission lines, and the object's optical spectrum looks rather featureless, unlike that of a normal quasar. It is then called a "BL Lac object," after the prototype in the constellation Lacerta, the Lizard. At the other extreme, if the jet is close to the plane of the sky, dust and gas in a torus (doughnut) surrounding the central region may hide the active nucleus from us (recall Fig. 17–14). The galaxy nucleus itself may then appear relatively normal, although the active nature of the galaxy could still be deduced from the presence of extended radio emission from the jet.

This general idea of beamed, or directed, radiation probably accounts for many of the differences seen among active galactic nuclei. For example, in one type of Seyfert galaxy, the very broad emission lines are not easily visible, despite other evidence that indicates considerable activity in the nucleus. (For example, bright narrow emission lines can be seen.) We think that in some cases, the broad emission lines are present, but simply can't be directly seen because they are being blocked by an obscuring torus of material (■ Fig. 17–30). But light from the broad lines can still escape along the axis of this torus and reflect off of clouds of gas elsewhere in the galaxy. Observations of these clouds then reveal the broad lines, but faintly.

Similarly, some galaxies hardly show any sort of active nucleus directly—it is too heavily blocked from view by gas and dust along our line of sight, in the central torus. However, radiation escaping along the axis of this torus can still light up exposed parts of the galaxy, indirectly revealing the active nucleus (■ Fig. 17–31).

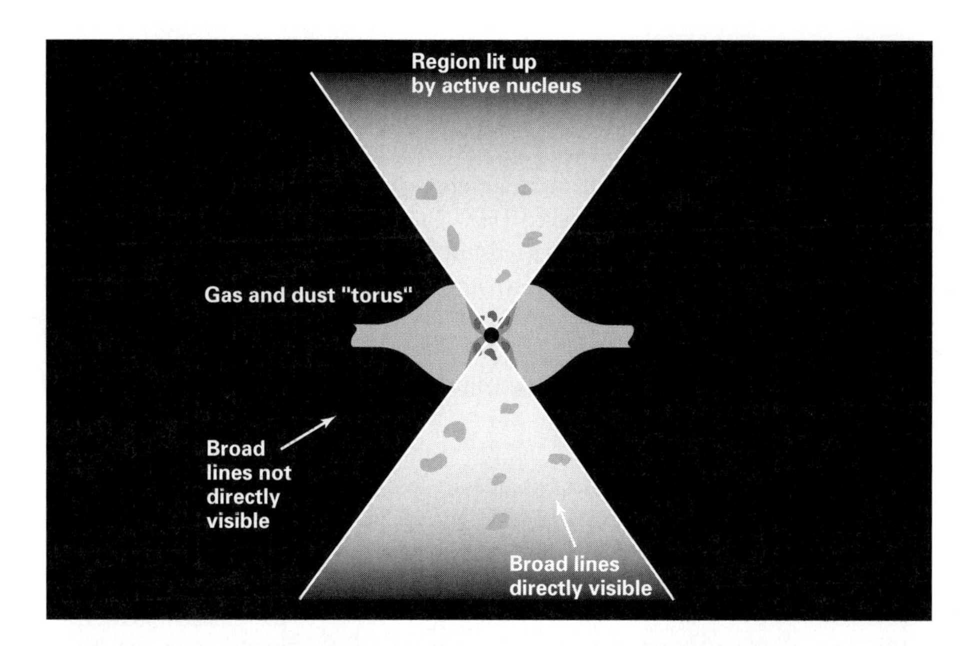

■ **FIGURE 17–30** The appearance from Earth of an active galaxy can depend on the orientation relative to our line of sight of the torus (doughnut) that surrounds it. If we see the nucleus from a direction within the indicated cones, the nucleus appears bright, and we can see the broad emission lines produced by gas (shown as dark clouds) in rapid motion near the supermassive black hole. If instead we view from outside the cones, then the active nucleus and central, rapidly moving clouds can be hidden at optical wavelengths by gas and dust in the torus. Narrow emission lines from gas farther from the nucleus (and within the cones; lighter clouds), however, should still be visible. (Richard Pogge [Ohio State U.] and Alex Filippenko [UC Berkeley])

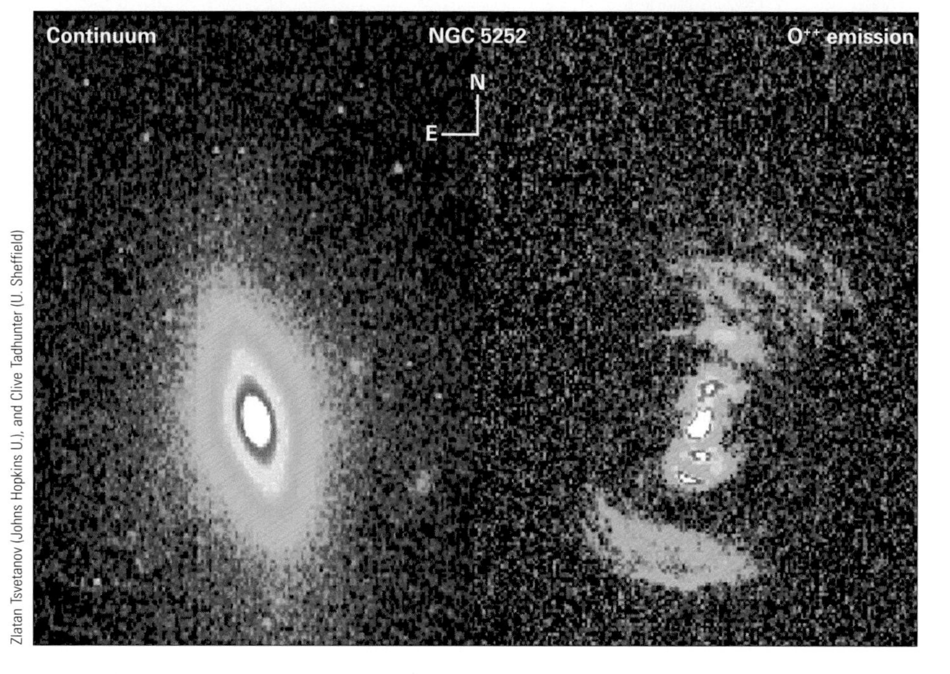

Zlatan Tsvetanov (Johns Hopkins U.), and Clive Tadhunter (U. Sheffield)

■ **FIGURE 17–31** Radiation from the core of this active galaxy, NGC 5252, illuminates matter only within two oppositely directed cones. Presumably the cones are aligned along the axis of a rotating black hole surrounded by a torus of gas and dust that blocks our direct view of the active nucleus, as shown in Figure 17–30.

17.8 Probes of the Universe

Quasars are powerful beacons, allowing us to probe the amount and nature of intervening material at high redshifts. For example, numerous narrow absorption lines are seen in the spectra of high-redshift quasars (■ Fig. 17–32). These spectral lines are produced by clouds of gas at different redshifts between the quasar and us. The lines can be identified with hydrogen, carbon, magnesium, and other elements.

Analysis of the line strengths and redshifts allows us to explore the chemical evolution of galaxies, the distribution and physical properties of intergalactic clouds of gas, and other interesting problems. The lines are produced by objects that are generally too faint to be detected in other ways. One surprising conclusion is that all of the clouds have at least a small quantity of elements heavier than helium. Since stars and supernovae produced these heavy elements, the implication is that an early episode of star formation preceded the formation of galaxies.

ASIDE 17.3: Different kinds of lines

The quasar absorption lines are narrow because they are formed by cold gas in intergalactic space between us and the quasar. Spectral lines formed in the quasar itself, typically emission lines, are broad because the gas moves rapidly in the gravitational field of the supermassive black hole.

Another way in which quasars are probes of the Universe is the phenomenon of **gravitational lensing** of light (Chapter 16). In fact, such lensing was first confirmed through studies of quasars. In 1979, two quasars were discovered close together in the sky, only a few seconds of arc apart (■ Fig. 17–33*a*). They had the same redshift, yet their spectra were essentially identical, arguing against a possible binary quasar. A cluster of galaxies with one main galaxy (Fig. 17–33*b,c*) was subsequently found along the same line of sight, but at a smaller redshift. The most probable explanation is that light from the quasar is bent by the gravity of the cluster (warped space–time), leading to the formation of two distinct images (■ Fig. 17–34). The cluster is acting like a gravitational lens.

Since then, dozens of gravitationally lensed quasars have been found. For a point lens and an exactly aligned object, we can get an image that is a ring centered on the lensing object. Such a case is called an "Einstein ring," and a few are known (■ Fig. 17–35*a*). Some gravitationally lensed quasars have quadruple quasar images that resemble a cross (■ Figs. 17–35*b*, 17–36), or even more complicated configurations (Fig. 17–35*c*). Only gravitational lensing seems to be a reasonable explanation of these objects, the redshifts of whose components are identical.

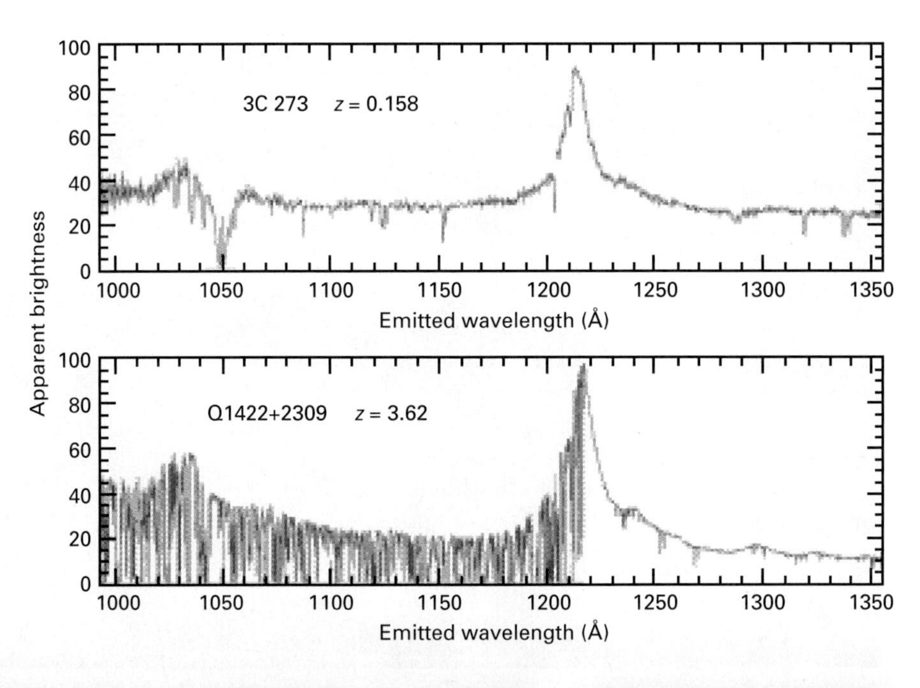

■ **FIGURE 17–32** Spectra of two quasars, plotted with the redshift removed (so that we see light at the emitted wavelengths rather than the observed, redshifted wavelengths). A high-redshift quasar *(bottom)* shows a large number of absorption lines produced by clouds of gas and galaxies along our line of sight to the quasar. Most of these lines, especially the ones in the blue (left) part of the spectrum, correspond to the hydrogen Lyman-α transition produced by intergalactic clouds at many different (and generally large) distances from us. We see the lines at many different wavelengths because of the different redshifts of the clouds. The spectrum of 3C 273 *(top)*, a low-redshift quasar, has far fewer absorption lines, indicating that intergalactic clouds of gas are scarce at the present time in the Universe. (William C. Keel [U. Alabama], Michael Rauch [Caltech], and NASA)

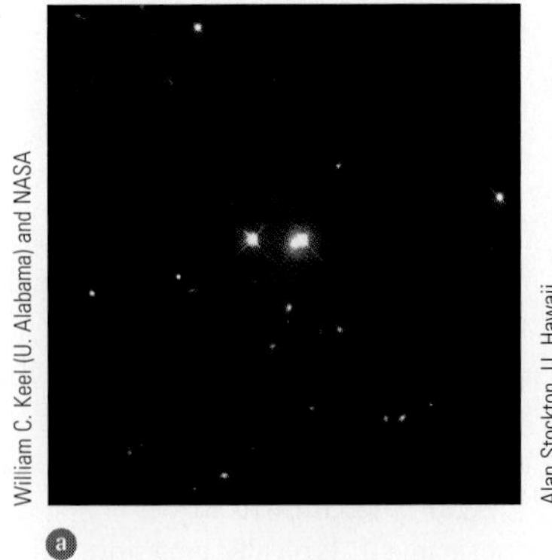

William C. Keel (U. Alabama) and NASA

a

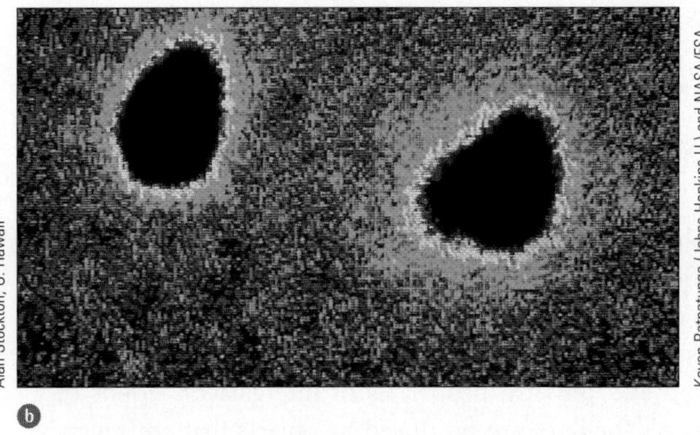

Alan Stockton, U. Hawaii

b

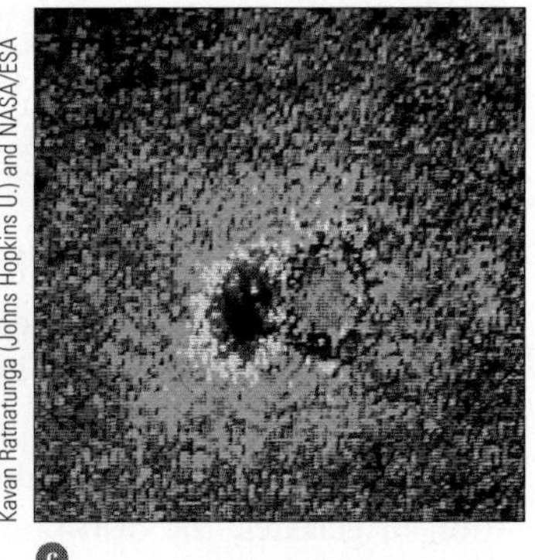

Kavan Ratnatunga (Johns Hopkins U.) and NASA/ESA

c

■ **FIGURE 17–33** (a) A Hubble Space Telescope wide-angle view of two quasars, 0957+561 A and B, that are only about 6 seconds of arc apart in the sky. They have the same redshifts (z = 1.4136) and essentially identical spectra. (b) Narrow-angle view of the two quasars, obtained with a ground-based telescope; they appear slightly oblong because of imperfect tracking by the telescope during the exposure. An intervening galaxy can be seen only 1 second of arc from the quasar on the right, in the direction of the quasar on the left. (c) The image of the quasar on the left has been subtracted from that on the right to reveal just the intervening galaxy.

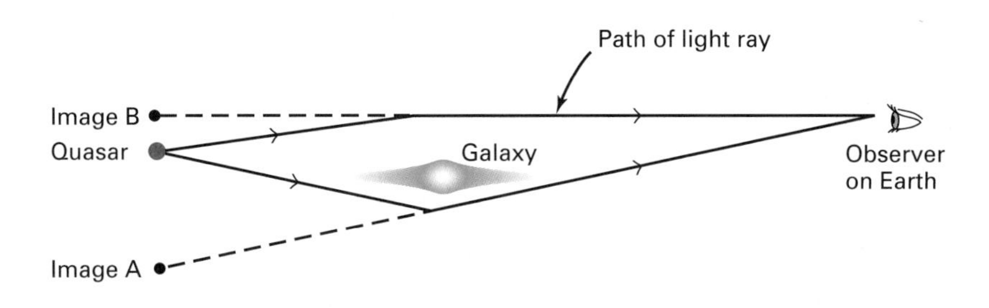

■ **FIGURE 17–34** The geometry of gravitational lensing. Light from a distant quasar can take two different paths on its way to Earth; each is bent because of the gravitational field of the intervening galaxy or cluster of galaxies. The two paths produce two distinct images of the same quasar close together in the sky. Note that the lengths of the two paths generally differ.

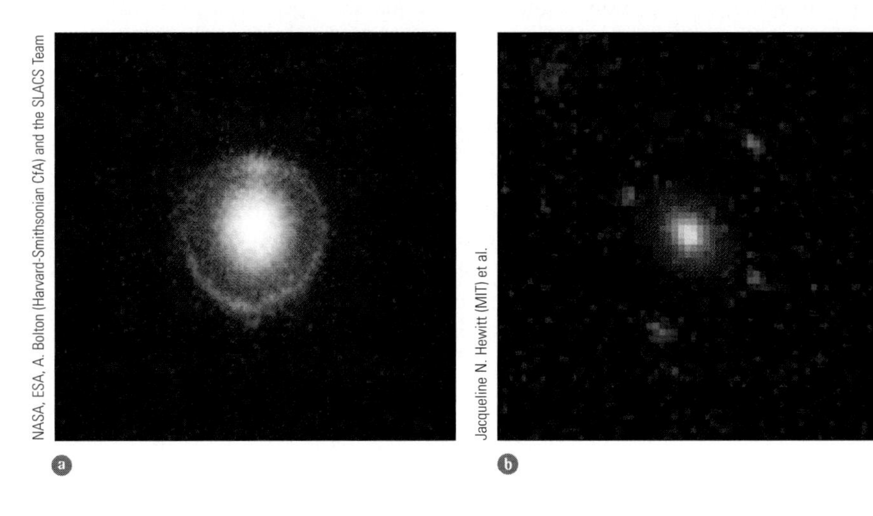

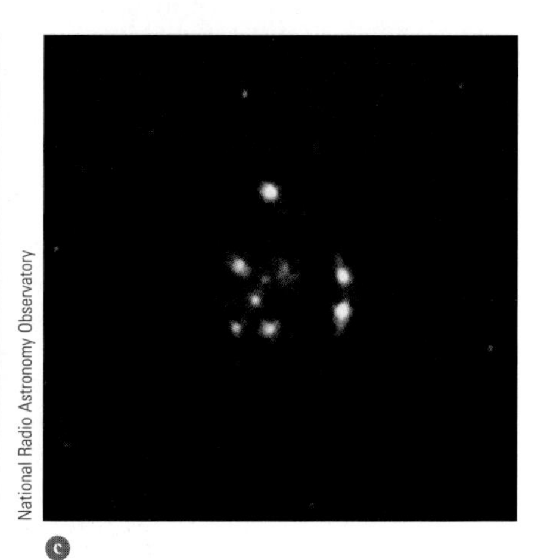

■ **FIGURE 17–35** (a) This particular quasar is so exactly aligned with the intervening gravitational lens that its radio radiation is spread out into an "Einstein ring." Eight newly discovered Einstein rings found by Hubble were reported in 2005, adding to the three previously known. (b) A Hubble Space Telescope image of an "Einstein cross"—a quasar that is gravitationally lensed into four images by a galaxy along the line of sight. In this case, the lensing galaxy is at a much lower redshift than the quasar; the galaxy's reddish nucleus is visible in the center of the image. The four bluish images of comparable brightness are separated by less than 2 seconds of arc and have essentially identical spectra. The rare configuration and identical spectra show that we are indeed seeing gravitational lensing rather than a cluster of quasars. (c) With complicated foreground mass distributions, one can get strange image configurations. In this system there are 6 images of a quasar at redshift 3.235 produced by a compact triple of foreground galaxies.

Moreover, in some cases continual monitoring of the brightness of each quasar image has revealed the same pattern of light variability, but with a time delay between the different quasar images. This delay occurs because the light travels along two different paths of unequal length to form the two quasar images; recall Figure 17–34. The variability pattern is not expected to be identical in two entirely different quasars that happen to be bound in a physical pair.

The multiple imaging of quasars is an exciting verification of a prediction of Einstein's general theory of relativity. The lensing details are sensitive to the total amount and distribution of matter (both visible and dark) in the intervening cluster. Thus, gravitationally lensed quasars provide a powerful way to study dark matter.

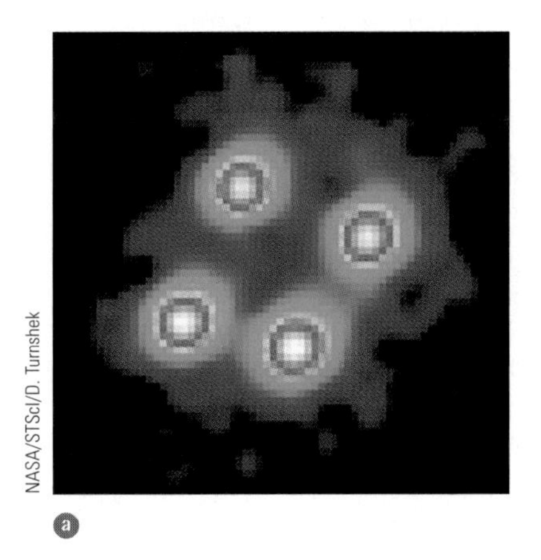

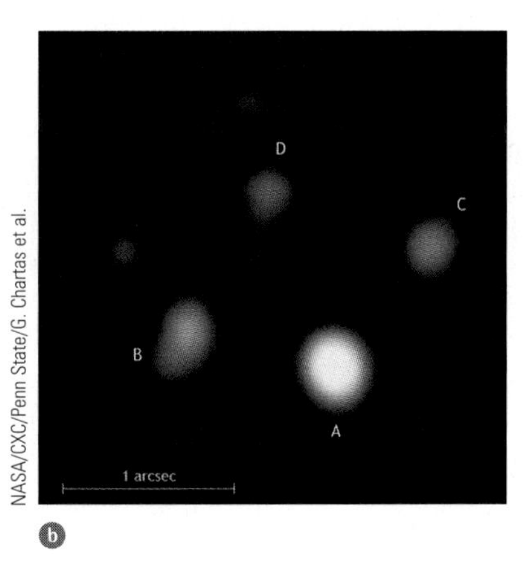

■ **FIGURE 17–36** The "clover leaf" (H1413+117), a quadruply imaged quasar ("Einstein cross") at redshift 2.55. (a) A Hubble Space Telescope view. (b) A Chandra X-ray Observatory view. The four images are only about 1 second of arc apart. Note that a different one of the four images, marked "A," is brighter in the x-ray band. This difference results from "microlensing," when a single or binary star in one of the galaxies between us and the quasar passes right in front of the small, x-ray producing region around the quasar's supermassive black hole. Since the optical light comes from a much larger region, it is not similarly magnified.

CONCEPT REVIEW

The central regions of some galaxies are unusually luminous—they are very powerful emitters of all kinds of electromagnetic radiation, and they sometimes spew out enormous jets of particles (Section 17.1). It is unlikely that normal stellar processes can account for these objects, called **active galactic nuclei.** The mysterious **quasars** are apparently related: They are star-like objects that in some cases shine very brightly at radio wavelengths (Section 17.2a). Their optical spectra were initially difficult to interpret, but eventually astronomers realized that these spectra were substantially redshifted (Section 17.2b). The discovery that quasars have high redshifts implied that they are at vast distances and have astonishingly large luminosities (Section 17.2c).

Rapid variations in the observed brightnesses of quasars were used to deduce that they are small, in some cases less than one light-week across—yet they are more powerful than entire galaxies 100,000 light-years in diameter (Section 17.3a). The energy of a quasar probably comes from matter falling into a **supermassive black hole,** 10 million to a billion times the mass of the Sun, in the central region of a galaxy (Section 17.3b). The matter forms a rotating **accretion disk** around the black hole (Section 17.3c). In some cases, well-focused "jets" of matter and radiation emerge along the disk's axis of rotation.

We now know that quasars are indeed phenomena in the centers of a minority of galaxies (Section 17.4). Hubble Space Telescope images of low-redshift quasars reveal nebulous structures that closely resemble galaxies, and spectra confirm that the extended light comes from stars. Quasars generally lived in the distant past; as the material surrounding the supermassive black holes was used up, the quasars faded, becoming less-active galaxies and finally normal-looking galaxies. Although a few quasars appear close in the sky to galaxies with much lower redshifts, these are thought by most astronomers to be chance projections (Section 17.5).

There is strong evidence for the existence of gigantic black holes in the centers of many relatively normal, nearby galaxies; our own Milky Way Galaxy harbors such a beast (Section 17.6). The more compact and massive the bulge of a spiral galaxy, the more massive is the black hole within it. Some quasars appear to be revived through gravitational interactions and collisions with galaxies, which direct gas toward the central black hole.

Apparent **superluminal speeds** of components seen in some quasars are produced by material moving near (but below) the speed of light, close in direction to our line of sight; they are not violations of Einstein's special theory of relativity (Section 17.7). The spectra of quasars often exhibit absorption lines due to material at intermediate distances, allowing astronomers to study galaxies and clouds of gas that are otherwise too dim to see (Section 17.8). Some quasars look multiple because of **gravitational lensing** by intervening galaxies or clusters.

QUESTIONS

1. Summarize the main observed characteristics of active galactic nuclei.
2. Describe the historical development of the study of quasars, and list their peculiar properties.
3. Why is it useful to find the optical objects that correspond in position with radio sources?
4. Why did the optical spectra of quasars initially seem so mysterious, before the correct interpretation was made?
5. What was the key breakthrough in the interpretation of quasar spectra? What was its significance?
6. Why do you think it was initially difficult for astronomers to entertain the possibility that the spectra of quasars are highly redshifted?
7. It was suggested in the 1960s that quasars might be compact objects ejected at high speeds from the centers of nearby ordinary galaxies. Why does the absence of blueshifted quasars provide strong evidence against this hypothesis?
8. †We observe a quasar with a spectral line at 5000 Å that we know is normally emitted at 4000 Å. **(a)** By what percentage is the line redshifted? **(b)** By approximately what percentage of the speed of light is the quasar receding? (Nonrelativistic formula acceptable.) **(c)** At what speed is the quasar receding in km/sec? **(d)** Using Hubble's law, to what distance does this speed correspond?
9. †Suppose you observe a quasar with a redshift of 3.6—that is, all lines are shifted to longer wavelengths by 360%. At what wavelength would you observe the ultraviolet Lyman-alpha line of hydrogen, whose laboratory (rest) wavelength is 1216 Å? What region of the spectrum is this? (Just give the general name, such as "ultraviolet.")
10. †A certain quasar has redshift $z = 2.5$. At what speed is it receding from us, using the relativistic formula?
11. Explain what is meant by the "lookback time" of a quasar.
12. Outline the argument used to infer that the physical size of quasars is very small.
13. Why does the rapid variability of the apparent brightness of some quasars make "the energy problem" even more difficult to solve using conventional energy sources such as chemical burning or nuclear fusion?
14. What is the most probable physical mechanism that produces a quasar's energy? Is anything actually escaping from within the black hole's event horizon?
15. What are three differences between quasars and pulsars?
16. What is the evidence suggesting that quasars live in the centers of galaxies?
17. Summarize what generally happens to a quasar as it ages.
18. Why are relatively nearby quasars so rare, compared with very distant quasars?
19. Suppose no nearby galaxies exhibited evidence for the presence of supermassive black holes in their centers.

Would this be a problem for the hypothesis of what powers quasars?

20. Why do we think that some relatively nearby quasars are rejuvenated?
21. Explain how we deduce the presence and measure the mass of a supermassive black hole in the center of a galaxy.
22. Qualitatively explain how parts of a quasar could appear to be moving at greater than the speed of light, without violating the special theory of relativity, which states that no physical object can exceed the speed of light.
23. Describe how quasars can be used as probes of matter between them and us.
24. What do we mean by a "gravitationally lensed quasar"?
25. How would the presence or absence of dark matter affect the gravitational lensing of quasars?
26. **True or false?** Most astronomers think that the enormous amount of energy emitted by a quasar is produced by nuclear reactions in gas confined to a radius of a light-year or less.
27. **True or false?** The term "active galaxy" refers to a galaxy undergoing rapid evolutionary change in shape, such as spiral to elliptical.
28. **True or false?** The energy source that produces jets of particles and radiation in the centers of radio galaxies is thought to be associated with a supermassive black hole.
29. **True or false?** Multiple quasars with identical spectra and the same redshift, very near each other in the sky, generally are examples of gravitational lensing of light.
30. **Multiple choice:** After carefully measuring the brightness of an astronomical object for one month, you find that the brightness changes significantly from one night to the next. The safest conclusion is that this object is (**a**) a perfectly normal galaxy; (**b**) no more than one light-month away from the Sun; (**c**) not within our Milky Way Galaxy; (**d**) not larger than about one light-day across; or (**e**) a perfectly normal main-sequence star.

†31. **Multiple choice:** If two quasars have the same apparent brightness, but Quasar R2D2 has a redshift of 0.1 while Quasar C3PO has a redshift of 0.2, the luminosity of Quasar C3PO is ________ that of Quasar R2D2. (**a**) one-quarter; (**b**) one-half; (**c**) equal to; (**d**) twice; or (**e**) four times.

32. **Multiple choice:** Quasars are probably powered by (**a**) extremely hot, young, massive stars, since they are in very young galaxies; (**b**) colliding neutron stars or black holes; (**c**) material falling into a central supermassive black hole; (**d**) supernovae; or (**e**) collisions between matter and antimatter.

†33. **Multiple choice:** Suppose you obtain a spectrum of a quasar. You find that a hydrogen emission line (Lyman-α; rest wavelength $\lambda_0 \approx 1200$ Å) is observed to be at a wavelength of $\lambda \approx 6000$ Å. You also measure two Ly-α absorption lines at 3600 Å and 4800 Å. Which one of the following statements is definitely *true*? (**a**) The quasar is at a redshift of $z = 5$. (**b**) Hydrogen must be present along this line of sight at a redshift of $z = 3$. (**c**) The quasar is lensed into multiple quasar images having redshifts of $z = 2$ and 3. (**d**) There are no hydrogen clouds along this line of sight at a redshift of $z = 6$. (**e**) The absorption lines are produced by hydrogen in the galaxy containing the quasar.

34. **Fill in the blank:** Quasars were first detected in the _____ region of the electromagnetic spectrum.
35. **Fill in the blank:** The "fuzz" seen around some quasars is evidence for a _____ in which the quasar lives.

†36. **Fill in the blank:** Using the relativistic equation, one finds that a quasar receding from us with a speed of 0.6 times the speed of light has a redshift $z =$ _______.

37. **Fill in the blank:** Broad _______ provide evidence that the gas in a quasar is moving around at very high speeds.

†This question requires a numerical solution.

TOPICS FOR DISCUSSION

1. Does it seem paradoxical that black holes, from which nothing can escape, account for the incredible power of quasars?
2. Suppose *no* nearby galaxies exhibited evidence for the presence of supermassive black holes in their centers. Would this be a problem for the hypothesis of what powers quasars?
3. What do you think of the hypothesis, favored by some astronomers, that quasars are ejected from relatively nearby galaxies, and hence are not at the distances implied by their redshifts?

MEDIA

Virtual Laboratories

Active Galactic Nuclei and Quasars

AceAstronomy™ Log into AceAstronomy at **http://astronomy.brookscole.com/cosmos3** to access quizzes and animations that will help you assess your understanding of this chapter's topics.

Log into the Student Companion Web Site at **http://astronomy.brookscole.com/cosmos3** for more resources for this chapter including a list of common misconceptions, news and updates, flashcards, and more.

CHAPTER 18

Cosmology: The Birth and Life of the Cosmos

Cosmology is the study of the structure and evolution of the Universe on its grandest scales. Some of the major issues studied by cosmologists include the Universe's birth, age, size, geometry, and ultimate fate. We are also interested in the birth and evolution of galaxies, topics already discussed in Chapter 16.

As in the rest of astronomy, we are trying to discover the fundamental laws of physics, and use them to understand how the Universe works. We do not claim to determine the purpose of the Universe or why humans, in particular, exist; these questions are more in the domain of theology, philosophy, and metaphysics. However, in the end we will see that some of our conclusions, which are solidly based on the methods of science, nonetheless seem to be untestable with our present knowledge.

18.1 Olbers's Paradox

We begin our exploration of cosmology by considering a deceptively simple question: "Why is the sky dark at night?" The answer seems so obvious ("The sky is dark because the Sun is down, dummy!") that the question may be considered absurd.

Actually, though, it is very profound. If the Universe is static (that is, neither expanding nor contracting), and infinite in size and age, with stars spread throughout it, every line of sight should intersect a shining star (■ Fig. 18–1)—just as in a hypothetical infinite forest, every line of sight eventually intersects a tree. So, the sky should be bright everywhere, even at night. But it clearly isn't, thereby making a paradox—a conflict of a reasonable deduction with our common experience.

One might argue that distant stars appear dim according to the inverse-square law of light, so they won't contribute much to the brightness of the night sky. But the apparent size of a star also decreases with increasing distance. (This can be difficult to comprehend: Stars are so far away that they are usually approximated as points, or as the blur circle produced by turbulence in Earth's atmosphere, but intrinsically they

ORIGINS

Knowledge of the age, geometry, structure, and evolution of the Universe is fundamental to our overall understanding of our place in the Cosmos. The ultimate fate of the Universe is closely linked to its history.

AIMS

1. See how the simple observation that the night sky is dark has profound implications for the origin or extent of the Universe (Section 18.1).
2. Understand the expansion of the Universe, and how it has no center (Section 18.2).
3. Learn how astronomers determine the age of the Universe (Section 18.3).
4. Explore various possibilities for the overall geometry, expansion history, and fate of the Universe (Section 18.4).
5. Discover how the expansion of the Universe now appears to be accelerating, propelled by mysterious "dark energy" (Section 18.5).
6. Discuss major stages in the future of the Universe, should it expand forever (Section 18.6).

ASIDE 18.1: His name was Olbers

The name "Olbers" has an "s" on the end before we make it possessive, so we write of "Olbers's paradox" (or, alternatively, "Olbers' paradox"). It is incorrect to write "Olber's paradox," with an apostrophe before a sole "s," since his name wasn't "Olber."

AceAstronomy™ The AceAstronomy icon throughout this text indicates an opportunity for you to test yourself on key concepts, and to explore animations and interactions of the AceAstronomy website at **http://astronomy.brookscole.com/cosmos3**

The dark night sky with comet Hale-Bopp above the twin Keck telescope domes on Mauna Kea volcano in Hawaii. The puzzle of why the entire sky isn't bright is known as Olbers's paradox.

■ **FIGURE 18–2** The lower halves of trees are sometimes painted white, and mimic seeing a uniform expanse of starlight. Note that the surface brightness of a tree is the same whether the tree is close to us or far away.

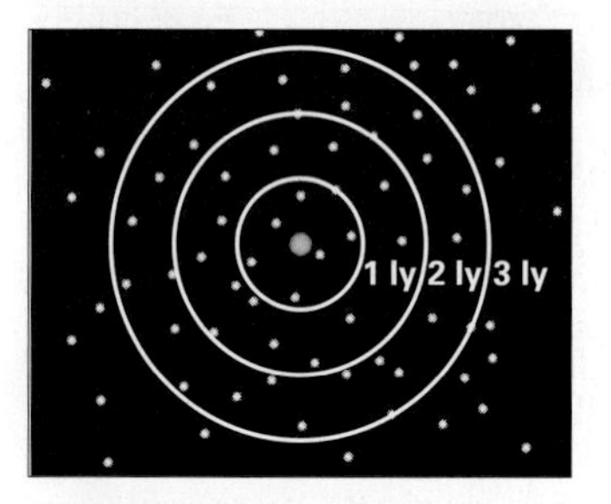

■ **FIGURE 18–3** A model non-expanding universe in which the stars are uniformly distributed in spherical shells, all of the same thickness, centered on the Earth. Individual stars in the distant shells appear dim, but there are more stars (because of the larger volume of progressively more-distant shells). The total brightness contributed by each shell is the same, as long as the universe is infinitely old and stars have always been present. With an infinite number of shells, the sky should be very bright, even at night.

Now suppose the stars were suddenly created everywhere at the same instant. In the first year after creation, the observer would see only those stars within 1 light-year of him or her (that is, all stars within the innermost sphere), and most of the sky would appear not to have stars; thus, it would be dark. With each additional year, he or she would see all stars within another light-year farther away (progressively larger spheres). The total number of visible stars would steadily increase with time, thereby brightening the sky. Stated another way, the size of dark gaps between the stars would steadily decrease as time passed and more stars became visible.

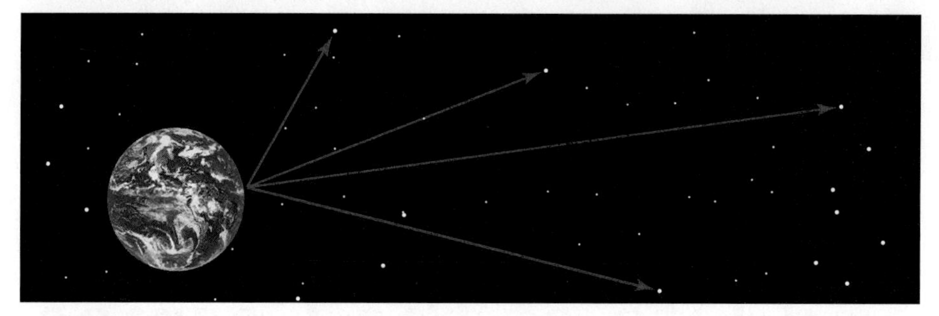

■ **FIGURE 18–1** If we look far enough in any direction in an infinite universe, our line of sight should eventually hit the surface of a star. This leads to Olbers's paradox.

really *do* have a nonzero angular area.) So, a star's brightness per unit area remains the same regardless of distance (■ Fig. 18–2). If there were indeed an infinite number of stars, and if we could see all of them, then every point on the sky would be covered with a star, making the entire sky blazingly bright!

Another way to think about Olbers's paradox is to consider an infinite number of spherical shells centered on Earth (■ Fig. 18–3). Each shell has the same thickness, but the volume occupied by progressively more distant shells grows. Although individual stars in the distant shells appear dimmer than in the nearby shells, there are more stars in the distant shells (because of the growing volume). These two effects exactly cancel each other, so each shell contributes the same amount to the brightness of the sky. With an infinite number of shells, the sky should be infinitely bright—or at least as bright as the surface of a star (since distant stars will tend to be blocked by closer stars along the same line of sight).

This dark-sky paradox has been debated for hundreds of years. It is known as **Olbers's paradox,** though the 19th-century astronomer Wilhelm Olbers wasn't the first to realize the problem. Kepler and others considered it, but not until the 20th century was it solved.

Actually, there are several conceivable resolutions of Olbers's paradox, each of which has profound implications. For example, the Universe might have finite size. It is as though the whole Universe were a forest, but the forest has an edge—and if there are sufficiently few trees, one can see to the edge along some lines of sight. Or, the Universe might have infinite size, but with few or no stars far away. This is like a forest that stops at some point, or thins out quickly, with an open field (the rest of the Universe) beyond it.

Another possible solution is that the Universe has a finite age, so that light from most of the stars has not yet had time to reach us. If the forest suddenly came into existence, an observer would initially see only the most nearby trees (due to the finite speed of light), and there would be gaps between them along some lines of sight.

There are other possibilities as well. Most of them are easily ruled out by observations or violate the Copernican principle (that we are at a typical, non-special place in the Universe), and some are fundamentally similar to the three main suggestions discussed above. One idea is that dust blocks the light of distant stars, but this doesn't work because, if the Universe were infinitely old, the dust would have time to heat up and either glow brightly or be destroyed.

It turns out that the primary true solution to Olbers's paradox is that the Universe has a *finite age,* about 14 billion years. There has been far too little time for the light from enough stars to reach us to make the sky bright. Effectively, we see "gaps" in the sky where there are no stars, because light from the distant stars still hasn't been detected.

For example, consider the static (non-expanding) universe in Figure 18–3. If stars were suddenly created as shown, then in the first year the observer would see only those stars within 1 light-year of him or her, because the light from more distant stars would still be on its way. After 2 years, the observer would see more stars—all those within 2 light-years; the gaps between stars would be smaller, and the sky would be brighter. After 10 years, the observer would see all stars within 10 light-years, so the sky would be even brighter. But it would be a long time before enough very distant stars became visible and filled the gaps.

Our own Universe, of course, is not so simple—it is expanding. To some degree, the expansion of the Universe also helps solve Olbers's paradox: As a galaxy moves away from us, its light is redshifted from visible wavelengths (which we can see) to longer wavelengths (which we can't see). Indeed, the energy of each photon actually decreases because of the expansion. But the effects of expansion are minor in resolving Olbers's paradox, compared with the finite, relatively short age of the Universe.

Regardless of the actual resolution of Olbers's paradox, the main point is that such a simple observation and such a silly-sounding question lead to incredibly interesting possible conclusions regarding the nature of the Universe. So the next time your friends are in awe of the beauty of the stars, point out the profound implications of the darkness, too!

18.2 An Expanding Universe

To see that the Universe has a finite age, we must consider the expansion of the Universe. In Chapter 16, we described how spectra of galaxies studied by Edwin Hubble led to this amazing concept, one of the pillars of cosmology.

18.2a Hubble's Law

Hubble found that in every direction, all but the closest galaxies have spectra that are shifted to longer wavelengths. Moreover, the measured **redshift,** z, of a galaxy is *proportional* to its distance from us, d (■ Fig. 18–4). If this redshift is produced by

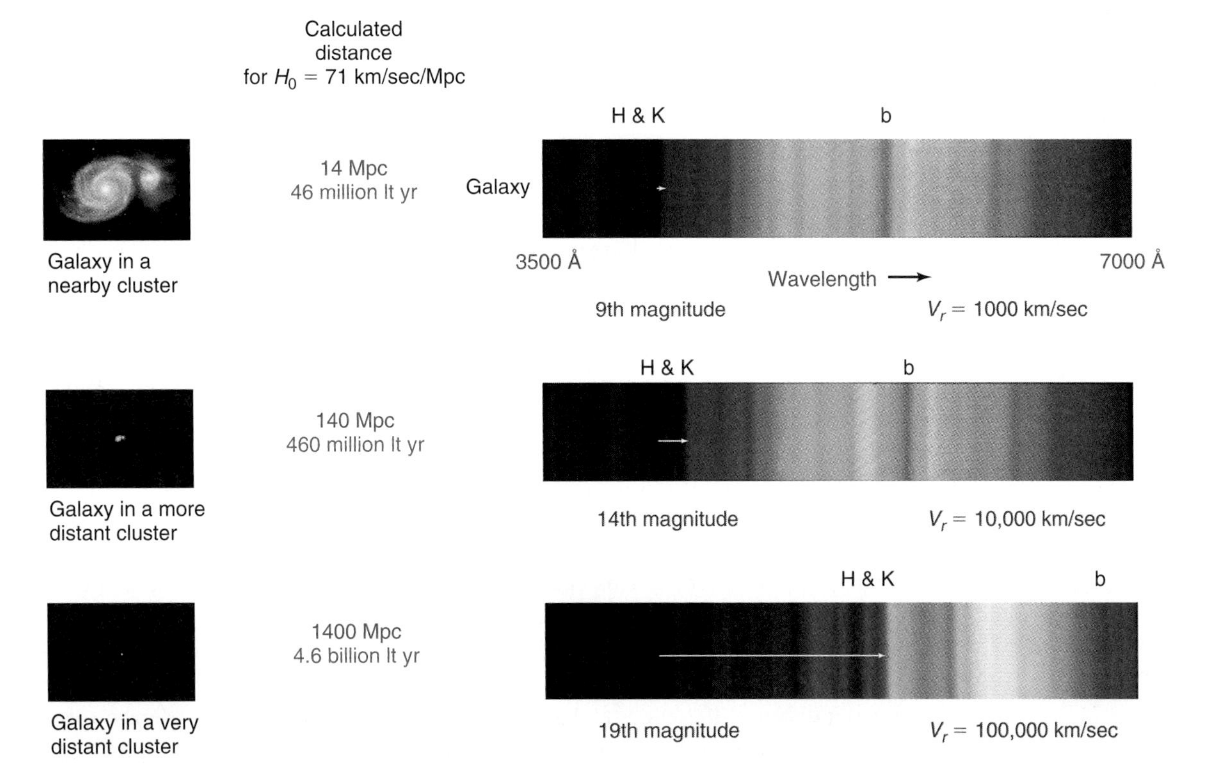

■ **FIGURE 18–4** Spectra are shown at right for the galaxies at left, all reproduced to the same scale. Distances are based on Hubble's constant = 71 km/sec/Mpc. Notice how the farther away a galaxy is, the smaller it looks. The arrow on each horizontal spectrum shows how far the "H and K" lines of singly ionized calcium are redshifted. A redshifted line of magnesium is labeled with Fraunhofer's "b."

(b) Images from N. A. Sharp (NOAO); computer spectra models from Ned Wright (UCLA), from B. Zuckerman and M. Malkan, eds., *The Origin and Evolution of the Universe.*

motion away from us, then we can use the Doppler formula to derive the speed of recession, v.

The recession speed can be plotted against distance for many galaxies in a **Hubble diagram** (■ Fig. 18–5), and a straight line nicely represents the data. The final result is **Hubble's law,** $v = H_0 d$, where H_0 is the *present-day value* of **Hubble's constant,** H. (Note that H is constant throughout the Universe at a given time, but its value decreases with time.)

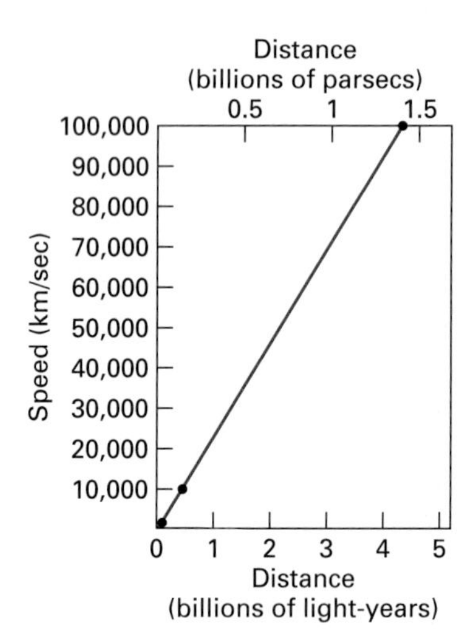

■ **FIGURE 18–5** The Hubble diagram for the galaxies shown in Figure 18–4.

An explosion can give rise to Hubble's law. If we kick a pile of balls, for example, some of them are hit directly and given a large speed, while others fly off more slowly. After a while, we see that the most distant ones are moving fastest, while the ones closest to the original pile are moving slowly (■ Fig. 18–6). The reason is obvious: To have reached a particular distance in a given amount of time, the distant balls must have been moving quickly, while the nearby ones must have been moving slowly. Speed is indeed proportional to distance, which is the same formula as Hubble's law.

Based on Hubble's law, we conclude that the Universe is expanding. Like an expanding gas, its density and temperature are decreasing with time. Extrapolating the expansion backward in time, we can reason that the Universe began at a specific instant when all of the material was in a "singularity" at essentially infinite density and temperature. (This is not the only possible conclusion, but other evidence, to be discussed in Chapter 19, strongly supports it.)

We call this instant the **big bang**—the initial event that set the Universe in motion—though we will see below that it is not like a conventional explosion. "Big bang" is both the technical and popular name for the current class of theories that deal with the birth and evolution of the Universe. In Chapter 19, we will discuss in more detail the reasons astronomers think that the Universe began its life in a hot, dense, expanding state.

ASIDE 18.2: Redshift and speed

Recall from Chapter 16 that the redshift z of an object is defined by $z = \Delta\lambda/\lambda_0$, where $\lambda = \lambda - \lambda_0$ is the observed wavelength of an absorption or emission line in the object's spectrum minus its laboratory (rest) wavelength. If $z \leq 0.2$, a reasonable approximation is the Doppler formula, $z = \lambda/\lambda_0 \approx v/c$, where v is the object's speed of recession.

18.2b Expansion Without a Center

Does the observed motion of galaxies away from us imply that we are the center of expansion, and hence in a very special position? Such a conclusion would be highly anti-Copernican: Looking at the billions of other galaxies, we see no scientifically based reason for considering our Galaxy to be special, in terms of the expansion of the Universe. Historically, too, we have encountered this several times. The Earth is not the only planet, and it isn't the center of the Solar System, just as Copernicus found. The Sun is not the only star, and it isn't the center of the Milky Way Galaxy. The Milky Way Galaxy is not the only galaxy . . . and it probably isn't the center of the Universe.

If our Galaxy *were* the center of expansion, we would expect the number of other galaxies per unit volume to decrease with increasing distance, as shown by the balls in Figure 18–6. In fact, however, galaxies are *not* observed to thin out at large distances—thus providing direct evidence that we are not at the unique center.

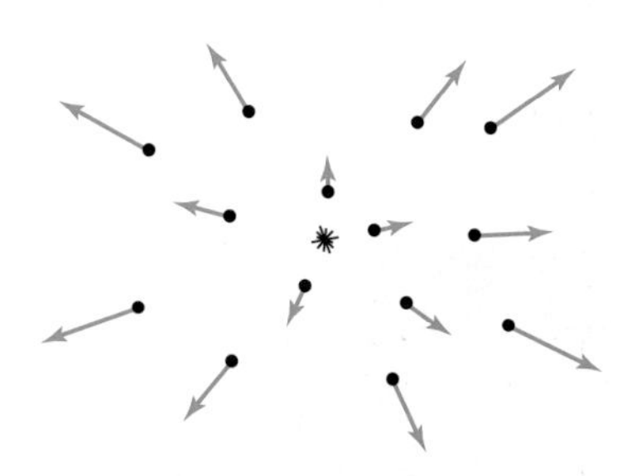

■ **FIGURE 18–6** After receiving a kick, balls go flying away from their common point of origin at different speeds, depending on how directly each one was hit. After any given amount of time, it is clear that the most distant balls are those with the highest speeds; to have travelled farthest, they must be moving fastest.

A different conclusion that is consistent with the data, and also satisfies the Copernican principle, is that there is *no* center—or, alternatively, that all places can claim to be the center. Consider a loaf of raisin bread about to go into the oven. The raisins are spaced at various distances from each other. Then, as the uniformly distributed yeast causes the dough to expand, the raisins start spreading apart from each other (■ Fig. 18–7). If we were able to sit on any one of those raisins, we would see our neighboring raisins move away from us at a certain speed. Note that raisins far away from us recede faster than nearby raisins because there is more dough between them and us, and all of the dough is expanding uniformly.

For example, suppose that after 1 second, each original centimeter of dough occupies 2 cm (Fig. 18–7). From our raisin, we will see that another raisin initially 1 cm away has a distance of 2 cm after 1 second. It therefore moved with an average speed of 1 cm/sec. A different raisin initially 2 cm away has a distance of 4 cm after 1 sec-

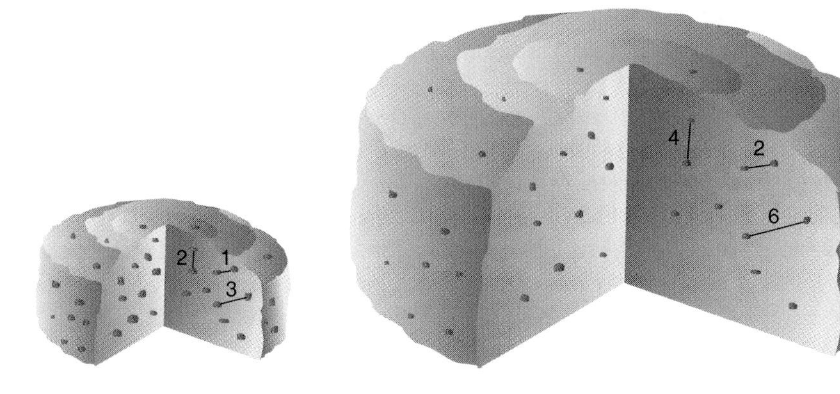

■ **FIGURE 18–7** From every raisin in a rising loaf of raisin bread, every other raisin seems to be moving away from you at a speed that depends on its distance from you. This leads to a relation like Hubble's law between the speed and the distance. Note also that each raisin would be at the center of the expansion measured from its own position, yet the bread is expanding uniformly. The raisins themselves do not expand, just as galaxies do not expand. For a better analogy with the Universe, consider an infinite loaf; unlike the finite analogy pictured here, there is then no overall center to its expansion.

ond, and moved with an average speed of 2 cm/sec. Yet another raisin initially 3 cm away has a distance of 6 cm after 1 second, so its average speed was 3 cm/sec. We see that speed is proportional to distance ($v \propto d$), as in Hubble's law.

It is important to realize that it doesn't make any difference which raisin we sit on; all of the other raisins would seem to be receding, regardless of which one was chosen. (Of course, any real loaf of raisin bread is finite in size, while the Universe may have no limit so that we would never see an edge.) The fact that all the galaxies are receding from us does not put us in a unique spot in the Universe; there is no center to the Universe. Each observer at each location would observe the same effect.

A convenient one-dimensional analogue is an infinitely long rubber band with balls attached to it (■ Fig. 18–8). Imagine that we are on one of the balls. As the rubber band is stretched, we would see all other balls moving away from us, with a speed of recession proportional to distance. But again, it doesn't matter which ball we chose as our home.

Another very useful analogy is an expanding spherical balloon (■ Fig. 18–9). Suppose we define this hypothetical universe to be only the *surface* of the balloon. It has just two spatial dimensions, not three like our real Universe. We can travel forward or backward, left or right, or any combination of these directions—but "up" and "down"

■ **FIGURE 18–8** A one-dimensional representation of the expanding Universe in which balls are attached to an infinite, expanding rubber band. No matter which ball you choose to sit on, all others move away from it. Distant balls recede faster than nearby ones, since each bit of rubber expands uniformly and there is more rubber between you and distant balls than between you and nearby ones.

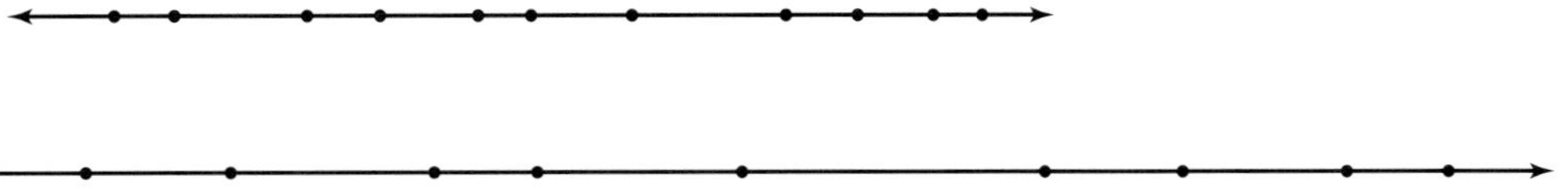

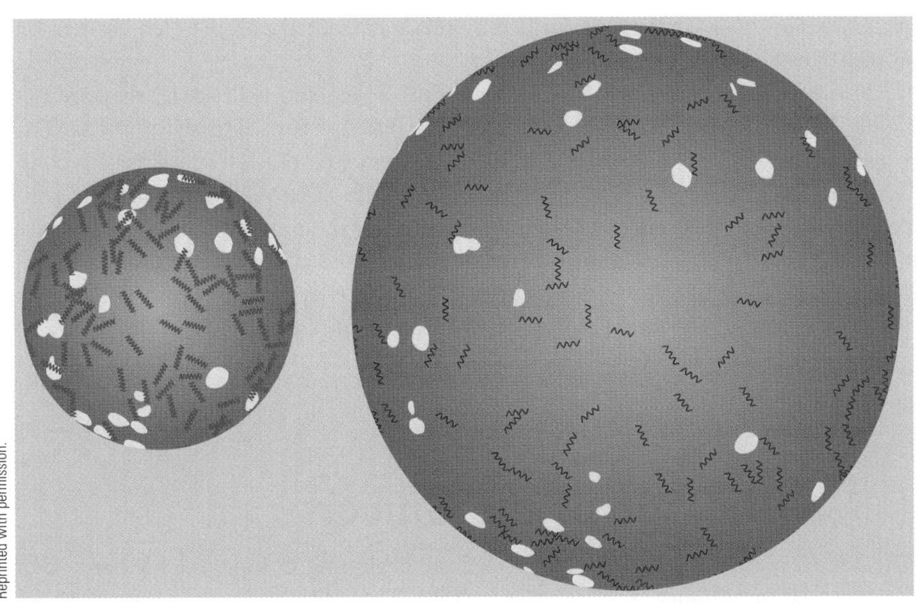

■ **FIGURE 18–9** The surface of a spherical balloon provides a good example of a finite, expanding universe. The space inside and outside the balloon is forbidden in this hypothetical universe. Stickers on the balloon represent galaxies, and their distances from each other increase according to Hubble's law as the balloon expands. The center of expansion is the center of the balloon, but this is not part of the surface itself, and hence is outside this hypothetical universe. Note also that while the stickers don't expand, light waves do expand, since they stretch with the rubber—and this fundamentally gives rise to the redshift. (In this simple drawing, the wavelength is longer at the right, though fewer oscillations are schematically shown in each set of waves drawn.)

(that is, out of, or into, the balloon) are not allowed. All of the laws of physics are constrained to operate along these two directions; even light travels only along the surface of the balloon.

If we put flat stickers on the balloon, they recede from each other as the balloon expands, and flat creatures on them would deduce Hubble's law. A clever observer could also reason that the surface is curved: By walking in one direction, for example, the starting point would eventually be reached. With enough data, the observer might even derive an equation for the surface of the balloon, but it would reveal an *unreachable third spatial dimension.* The observer could conclude that the center of expansion is in this "extra" dimension, which exists only mathematically, not physically!

It is possible that we live in an analogue of such a spherical universe, but with three spatial dimensions that are physically accessible, and an additional, inaccessible spatial dimension around which space mathematically curves. We will discuss this idea in more detail later.

18.2c What Is Actually Expanding?

Besides illustrating Hubble's law and the absence of a unique center, the above analogies accurately reproduce two additional aspects of the Universe. First, according to Einstein's general theory of relativity, which is used to quantitatively study the expansion and geometry of the Universe, it is *space itself* that expands.

The dough or the rubber expands, making the raisins, balls, and stickers recede from each other. They do *not* travel *through* the dough or rubber. Similarly, in our Universe, galaxies do not travel through a preexisting space; instead, space itself is expanding. (We sometimes say that the "fabric of space–time is expanding.") In this way, the expansion of the Universe differs from a conventional explosion, which propels material through a preexisting space.

Second, note that the raisins, balls, and stickers themselves don't expand—only the space *between* them expands. (We intentionally didn't draw ink dots on the balloon, because they would expand, unlike stickers.) Similarly, galaxies and other gravitationally bound objects such as stars and planets do not expand: The gravitational force is strong enough to overcome the tendency of space within them to expand. Nor do people expand, because electrical forces (between atoms and molecules) strongly hold us together.

Strictly speaking, even most clusters of galaxies are sufficiently well bound to resist the expansion. Only the space between the *clusters* expands, and even in these cases the expansion is sometimes diminished by the gravitational pull between clusters (as in superclusters).

However, electromagnetic waves or photons *do* expand with space; they are not tightly bound objects. Thus, for example, blue photons turn into red photons (Fig. 18–9). In fact, this is what actually produces the observed redshift of galaxies. Technically, the redshift is not a Doppler effect, since nothing is moving *through* the Universe, and the Doppler effect was defined in terms of the motion of an object relative to the waves it emits. The Doppler formula remains valid at low speeds, though, and it is convenient to think about the redshift as a Doppler effect, so we will continue to do so—but you should be aware of the deeper meaning of redshifts.

AceAstronomy™ Log into AceAstronomy and select this chapter to see the Active Figure called "Raisin Bread."

18.3 The Age of the Universe

The discovery that our Universe had a definite beginning in time, the big bang, is of fundamental importance. The Universe isn't infinitely old. But humans generally have a fascination with the ages of things, from the Dead Sea scrolls to movie stars. Naturally, then, we would like to know how old the Universe itself is!

■ **FIGURE 18–10** A Hubble image of a globular cluster in the Large Magellanic Cloud, seen as clearly as Earth-bound telescopes image clusters within our own Galaxy. Two bright foreground stars also appear.

Michael Rich, Kenneth Mighell, and James D. Neill Columbia U.); Wendy Freedman (Carnegie Observatories); and NASA

18.3a Finding Out How Old

There are at least two ways in which to determine the age of the Universe. First, the Universe must be at least as old as the oldest objects within it. Thus, we can set a minimum value to the age of the Universe by measuring the ages of progressively older objects within it. For example, the Universe must be at least as old as you—admittedly, not a very meaningful lower limit! More interestingly, it must be at least 200 million years old, since there are dinosaur fossils of that age. Indeed, it must be at least 4.6 billion years old, since Moon rocks and meteorites of that age exist.

The oldest discrete objects whose ages have reliably been determined are globular star clusters in our Milky Way Galaxy (■ Fig. 18–10). The oldest ones are now thought to be 12–13 billion years old, though the exact values are still controversial. (Globular clusters used to be thought to be about 14–17 billion years old.) Theoretically, the formation of globular clusters could have taken place only a few hundred million years after the big bang; if so, the age of the Universe would be about 14 billion years. Since no discrete objects have been found that appear to be much older than the oldest globular clusters, a reasonable conclusion is that the age of the Universe is indeed at least 13 billion years, but not much older than 14 billion years.

A different method for finding the age of the Universe is to determine the time elapsed since its birth, the big bang. At that instant, all the material of which any observed galaxies consist was essentially at the same location. Thus, by measuring the distance between our Milky Way Galaxy and any other galaxy, we can calculate how long the two have been separating from each other if we know the current recession speed of that galaxy.

At this stage, we have made the simplifying assumption that the recession speed has always been *constant*. So, the relevant expression is distance equals speed multiplied by time: $d = vt$. For example, if we measure a friend's car to be approaching us with a speed of 60 miles/hour, and the distance from his home to ours is 180 miles, we calculate that the journey took 3 hours if the speed was always constant and there were no rest breaks. In the case of the Universe, the amount of time since the big bang, assuming a constant speed for any given galaxy, is called the **Hubble time** (see *Figure It Out 18.1: The Hubble Time*).

If, however, the recession speed was *faster* in the past, then the true age is *less* than the Hubble time. Not as much time had to elapse for a galaxy to reach a given distance from us, compared to the time needed with a constant recession speed. Using the previous example, if our friend started his journey with a speed of 90 miles/hour, and gradually slowed down to 60 miles/hour by the end of the trip, then the average speed was clearly higher than 60 miles/hour, and the trip took fewer than 3 hours. (This is why people often break the posted speed limit!)

Astronomers have generally expected such a decrease in speed because all galaxies are gravitationally pulling on all others, thereby presumably slowing down the expansion rate. In fact, many cosmologists have believed that the deceleration in the expansion rate is such that it gives a true age of exactly two-thirds of the Hubble time. This is, in part, a theoretical bias; it rests on an especially pleasing cosmological model. Later in this chapter we will see how attempts have been made to actually measure the expected deceleration.

18.3b The Quest for Hubble's Constant

To determine the Hubble time, we must measure Hubble's constant, H_0. This can be done if we know the distance (d) and recession speed (v) of another galaxy, since

FIGURE IT OUT 18.1

The Hubble Time

There is a simple relationship between the value of Hubble's constant and the time elapsed since the big bang, assuming that the recession speed of any given galaxy has always been constant. For galaxies, we know that Hubble's law holds: $v = H_0 d$. Rearranging this relation, we have

$$d = v/H_0 = vT_0,$$

where we define T_0 (the "Hubble time") to be $1/H_0$, the reciprocal or inverse of Hubble's constant. Note that this equation looks just like the relation $d = vt$ for the distance travelled by an object at constant speed.

Suppose Hubble's constant is 50 km/sec/Mpc. What is the corresponding Hubble time? This is mostly an exercise in keeping track of units. First, 1 pc is approximately equal to 3.26 light-years (see Chapter 11), or 3.09×10^{18} cm, so 1 Mpc = 3.09×10^{24} cm = 3.09×10^{19} km. We then convert H_0 = 50 km/sec/Mpc = 50 km/(Mpc sec) to units of inverse seconds:

$$H_0 = (50\ \text{km}/(\text{Mpc sec}))/(3.09 \times 10^{19}\ \text{km}/\text{Mpc}) = 1.62 \times 10^{-18}/\text{sec}.$$

Thus, $T_0 = 1/H_0 = 6.2 \times 10^{17}$ sec. But one year (that is, 365.25 days at 86,400 seconds per day) is about 3.16×10^7 sec, so

$$T_0 \approx (6.2 \times 10^{17}\ \text{sec})/(3.16 \times 10^7\ \text{sec/year}) = 1.96 \times 10^{10}\ \text{years}.$$

This is roughly 20 billion years! But if H_0 were actually twice as large, 100 km/sec/Mpc, then the corresponding Hubble time would be half as large, about 10 billion years. This illustrates why astronomers want to determine the value of Hubble's constant accurately.

Note that as the Universe ages, T_0 increases, and so H_0 decreases—the value of Hubble's constant decreases with time. However, at a given time, its value is thought to be the same everywhere in the Universe—hence the name "Hubble's constant."

rearrangement of Hubble's law tells us that $H_0 = v/d$. Many galaxies at different distances should be used, so that an average can be taken.

ASIDE 18.3: One enormous distance

Recall that 1 Mpc = 1 megaparsec, a million parsecs, which is 3.26 million light-years.

The recession speed is easy to measure from a spectrum of the galaxy and the Doppler formula. We can't use galaxies within our own Local Group (like the Andromeda galaxy, M31), however, since they are bound to the group by gravity and are not expanding away.

Galaxy distances are notoriously hard to determine, and this leads to large uncertainties in the derived value of Hubble's constant. We can't use triangulation because galaxies are much too far away. In principle, the distance of a galaxy can be determined by measuring the apparent angular size of an object (such as a nebula) within it, and comparing it with an assumed physical size. But this method generally gives only a crude estimate of distance, because the physical sizes of different objects in a given class are not uniform enough.

More frequently, we measure the apparent brightness of a star, and compare this with its intrinsic brightness (luminosity, or power) to determine the distance. This is similar to how we judge the distance of an oncoming car at night: We intuitively use the inverse-square law of light (discussed in Chapter 11) when we see how bright a headlight appears to be. We must be able to recognize that particular type of star in the galaxy, and we assume that all stars of that type are "standard candles" (a term left over from the 19th century, when sets of actual candles were manufactured to a standard brightness)—that is, they all have the same luminosity.

Historically, the best such candidates have been the Cepheid variables, at least in relatively nearby galaxies. Though not all of uniform luminosity, they do obey a period-luminosity relation (■ Fig. 18–11), as shown by Henrietta Leavitt in 1912. Thus, if the variability period of a Cepheid is measured, its average luminosity can be read directly off the graph.

But individual stars are difficult to see in distant galaxies: They merge with other stars when viewed with ground-based telescopes. Other objects that have been used include luminous nebulae, globular star clusters, and novae—though all of them have substantial uncertainties. They aren't excellent standard candles, or are difficult to see in ground-based images, or depend on the assumed distances of some other galaxies.

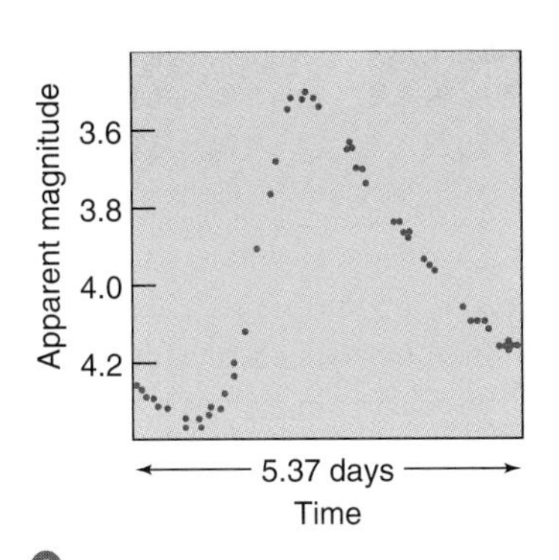

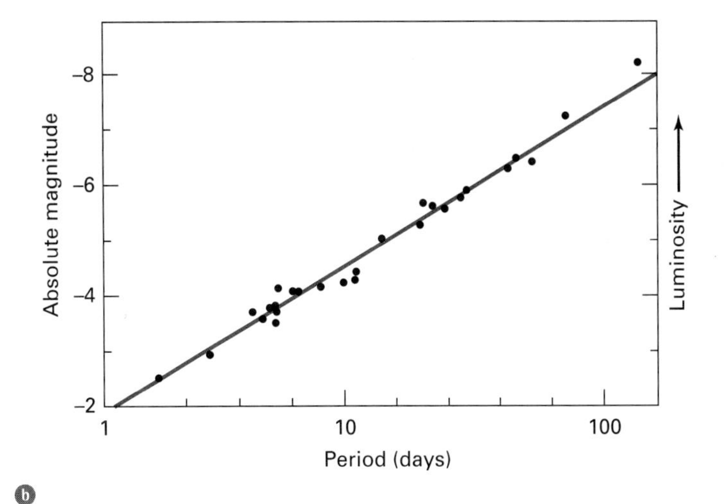

■ **FIGURE 18–11** ⓐ The light curve of a typical Cepheid variable star. ⓑ The period-luminosity relation of Cepheid variable stars, which the authors like to refer to as "Leavitt's law" in honor of Henrietta Leavitt, who discovered it (see our discussion in Chapter 11). Long-period Cepheids have higher average luminosities than short-period Cepheids.

Even entire *galaxies* can be used, if we determine their luminosity from other properties. The luminosity of a spiral galaxy is correlated with how rapidly it rotates, for example. Also, the brightest galaxy in a large cluster has a roughly standard luminosity. Again, however, significant uncertainties are associated with these techniques, or they depend on the proper calibration of a few key galaxies.

Throughout the 20th century, many astronomers have attempted to measure the value of Hubble's constant. Edwin Hubble himself initially came up with 550 km/sec/Mpc, but several effects that were at that time unknown to him conspired to make this much too large. From the 1960s to the early 1990s, the most frequently quoted values were between 70 and 50 km/sec/Mpc, due largely to the painstaking work of Allan Sandage (■ Fig. 18–12), a disciple of Edwin Hubble himself.

Such values yield a Hubble time of about 14–20 billion years, the probable maximum age of the Universe. The true expansion age could therefore be 9–13 billion years, if it were only two-thirds of the Hubble time, as many cosmologists have been prone to believe. These lower numbers may give rise to a discrepancy if the globular clusters are 12–13 billion years old. Despite uncertainties in the measurements, an "age crisis" would exist if the clusters were 14–17 billion years old, as was thought until the mid-1990s.

But some astronomers obtained considerably larger values for H_0, up to 100 km/sec/Mpc. This value gives a Hubble time of about 10 billion years, and two thirds of it is only 6.7 billion years. Even the recently revised ages of globular star clusters are substantially greater (12–13 billion years), leading to a sharp age crisis. The various teams of astronomers who got different answers all claimed to be doing careful work, but there are many potential hidden sources of error, and the assumptions might not be completely accurate. The debate over the value of Hubble's constant has often been heated, and sessions of scientific meetings at which the subject is discussed are well attended.

Note that the value of Hubble's constant also has a broad effect on the perceived size of the observable Universe, not just its age. For example, if Hubble's constant is 71 km/sec/Mpc, then a galaxy whose recession speed is measured to be 7100 km/sec would be at a distance

$$d = v/H_0 = (7100 \text{ km/sec})/(71 \text{ km/sec/Mpc}) = 100 \text{ Mpc}.$$

On the other hand, if Hubble's constant is actually 35.5 km/sec/Mpc, then the same galaxy is twice as distant: $d = (7100 \text{ km/sec})/(35.5 \text{ km/sec/Mpc}) = 200$ Mpc.

Andrew Fraknoi, Astronomical Society of the Pacific

■ **FIGURE 18–12** Allan Sandage *(right)* of the Carnegie Observatories in Pasadena, California, who has devoted much of his career to the measurement of Hubble's constant. He is seen here with Maarten Schmidt of Caltech, the first person to realize that the spectra of quasars are highly redshifted.

18.3c A Key Project of the Hubble Space Telescope

The aptly named Hubble Space Telescope was expected to provide a major breakthrough in the field. It was to obtain distances to many important galaxies, mostly

Wendy L. Freedman, Observatories of the Carnegie Institution of Washington, and NASA

■ **FIGURE 18–13** Cepheid variable stars in the galaxy M100, as observed with the Hubble Space Telescope. They have been used to calibrate the cosmic distance scale through the measurement of Hubble's constant.

through the use of Cepheid variable stars (■ Fig. 18–13). Indeed, very large amounts of telescope time were to be dedicated to this "Key Project" of measuring galaxy distances and deriving Hubble's constant. But astronomers had to wait a long time, even after the launch of the Hubble Space Telescope in 1990, because the primary mirror's spherical aberration (see our discussion in Chapter 4) made it too difficult to detect and reliably measure Cepheids in the chosen galaxies.

Finally, in 1994, the Hubble Key Project team announced their first results, based on Cepheids in only one galaxy (■ Fig. 18–14). Their value of H_0 was about 80 km/sec/Mpc, higher than many astronomers had previously thought. This implied that the Hubble time was 12 billion years; the Universe could be no older, but perhaps significantly younger (down to 8 billion years) if the expansion decelerates with time.

Because these values are less than 14–17 billion years (the ages preferred for globular star clusters at that time), this disagreement brought the age crisis to great prominence among astronomers, who shared it with the public. How could the Universe, as measured with the mighty Hubble Space Telescope, be younger than its oldest contents? There were several dramatic headlines in the news (■ Fig. 18–15).

Admittedly, the Hubble team's quoted value of H_0 had an uncertainty of 17 km/sec/Mpc, meaning that the actual value could be between about 63 and 97 km/sec/Mpc. Thus, the Universe could be as old as 15–16 billion years, especially if there has been little deceleration. The ages of globular clusters were uncertain as well, so it was not entirely clear that the age crisis was severe. But, as is often the case with

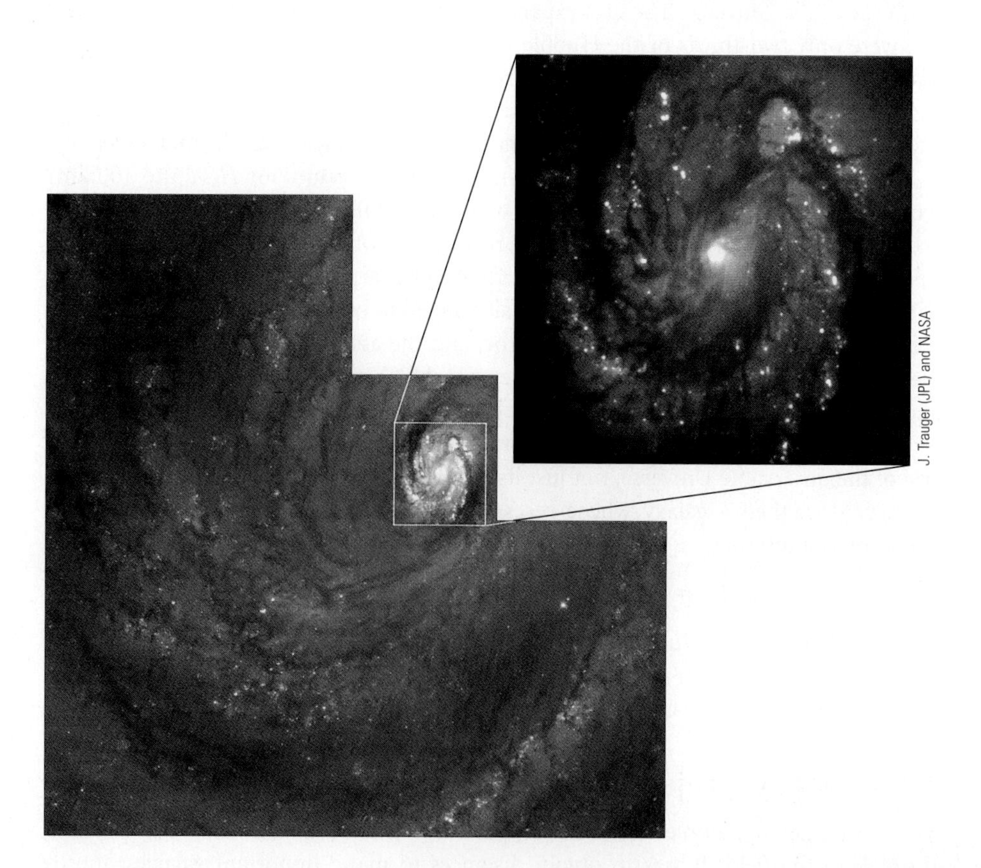

J. Trauger (JPL) and NASA

■ **FIGURE 18–14** A Hubble Space Telescope image of the galaxy M100 in the Virgo Cluster, the first galaxy whose distance was determined by the Hubble Space Telescope's Key Project team that used Cepheid variable stars. The zigzag square on Hubble's Wide Field and Planetary Camera 2 has twice the resolution of the larger squares, so the galaxy's center can be enlarged as at upper right without loss of detail. The Cepheids, though, are in the outer spiral arms.

newspaper and popular magazine articles, these subtleties are ignored or barely mentioned; only the "bottom line" gets reported, especially if it's exciting.

In 2001, the Hubble team announced a final answer, which was based on several methods of finding distances, with Cepheid variables as far out as possible and supernovae pinning down the greatest distances. Their preferred value of H_0 was 72 km/sec/Mpc, with an uncertainty of about 8 km/sec/Mpc (■ Fig. 18–16). But the Hubble team was not the only game in town, and other groups of scientists measured slightly different values. A "best bet" estimate of H_0 = 71 km/sec/Mpc seems reasonable, especially considering the measurements with the Wilkinson Microwave Anisotropy Probe (see our discussion in Chapter 19).

A value of 71 km/sec/Mpc for Hubble's constant means that the Universe has been expanding for 13.9 billion years, if there is no deceleration. By assuming only a small amount of deceleration (not as much as many theorists would have preferred), the Hubble team announced a best-estimate expansion age of 12 billion years for the Universe. Moreover, around 2000, the preferred ages of globular clusters had shifted from 14–17 billion years to only 11–14 billion years, based on accurate new parallaxes of stars from the Hipparcos satellite and on some new theoretical work. This meant that the age discrepancy had subsided to some extent, but did not fully disappear if the globular clusters are actually as old as 13–14 billion years.

But on what basis was the amount of deceleration estimated? We will discuss this more fully in Section 18.5, with the surprising result that the assumed deceleration may have been erroneous. Instead, the expansion rate of the Universe appears to actually be *increasing* with time! This exciting discovery, known as the "accelerating universe," is now accepted by most astronomers and physicists, contrary to the situation when it was initially announced in 1998. As we shall see later in this chapter, recent evidence makes it quite convincing.

■ **FIGURE 18–15** The headline of *Discover* magazine after the 1994 announcement of an expansion age of only 8 to 12 billion years for the Universe. At that time, globular star clusters were thought to be 14 to 17 billion years old.

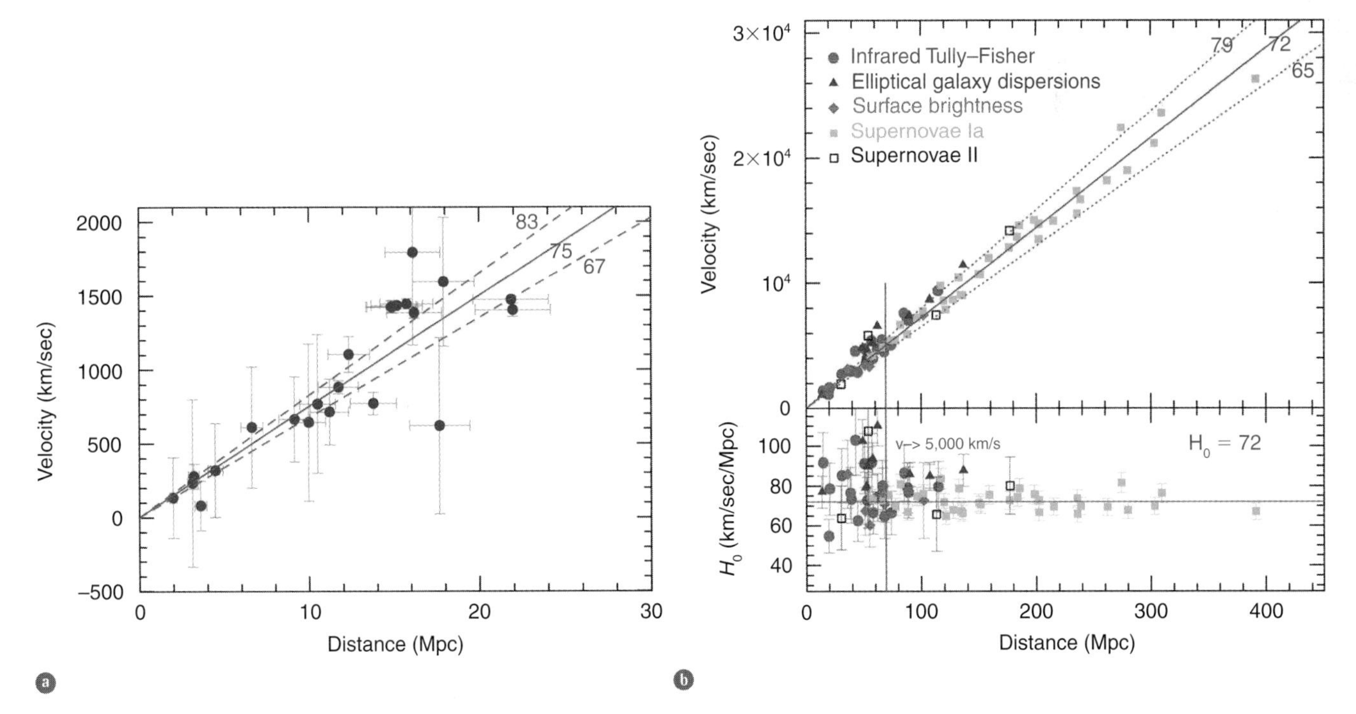

■ **FIGURE 18–16** Hubble's law as determined by the Key Project on the Cosmic Distance Scale of the Hubble Space Telescope. (a) Hubble's law as determined from Cepheid variable stars out to a distance of about 20 Mpc (about 65 million light-years), our "local" region. (b) The Hubble diagram made by evaluating and averaging several distance indicators (which are plotted with different symbols). Including supernovae brings the graph out to 400 Mpc (about 1.3 billion light-years). At the bottom, we see deviations of the individual values of Hubble's constant from the average. In Chapter 19, we will see how the value of Hubble's constant reported from the Wilkinson Microwave Anisotropy Probe (WMAP) is in close agreement with this Hubble Key Project value. (Wendy L. Freedman, Observatories of the Carnegie Institution of Washington, and NASA)

■ **FIGURE 18–17** The central part of the Virgo Cluster of galaxies. This cluster is not receding from the Milky Way Galaxy as quickly as it would if galaxies had no mass and didn't gravitationally attract each other.

The discovery of acceleration implies some very intriguing, but also troubling, new aspects to the nature and evolution of the Universe. If correct, however, it may fully resolve the age crisis: We find that the expansion age of the Universe is 13.7 billion years, consistent with the 12–13 billion year ages of globular clusters estimated most recently.

18.3d Deviations from Uniform Expansion

A major problem with using relatively nearby galaxies for measurements of Hubble's constant is that proper corrections must be made for deviations from the **Hubble flow** (the assumed uniform expansion of the Universe). As we discussed in Chapter 16, there are concentrations of mass (clusters and superclusters) in certain regions, and large voids in others, so a specific galaxy may feel a greater pull in one direction than in another direction. It will therefore be pulled *through* space (relative to the Hubble flow), and its apparent recession speed may be affected. Though the galaxy's recession speed is easy to measure from a spectrum, it might not represent the true expansion of space.

For example, the Virgo Cluster of galaxies (■ Fig. 18–17) is receding from us more slowly than it would if it had no mass: The Milky Way Galaxy is "falling" toward the Virgo Cluster, thereby counteracting part of the expansion of space. Such gravitationally induced **peculiar motions** are typically a few hundred kilometers per second, but can reach as high as 1000 km/sec. Their exact size is difficult to determine without detailed knowledge of the distribution of matter in the Universe. In the case of the Virgo Cluster, the average observed recession speed is about 1100 km/sec, and the peculiar motion is thought to be about 300 km/sec, but this is uncertain. Errors in the adopted "true" recession speed directly affect the derived value of Hubble's constant.

A surprising discovery was that even the Virgo Cluster is moving with respect to the average expansion of the Universe. Some otherwise unseen "Great Attractor" is pulling the Local Group, the Virgo Cluster, and even the much larger Hydra-Centaurus Supercluster toward it. Redshift measurements by a team of astronomers informally known as the "Seven Samurai" showed the location of the giant mass that must be involved. (See the interview in this book with Sandra Faber, its head.) It is about three times farther from us than the Virgo Cluster, and includes tens of thousands of galaxies or their equivalent mass.

Measurements of still more distant galaxies avoid the problem of peculiar motions when trying to determine Hubble's constant. For example, compared with galaxies having recession speeds of 15,000–30,000 km/sec, the peculiar motions are negligible. So, measurements of their distances, when combined with their recession speeds, can yield an accurate value of H_0. The trick is to find their distances—and this can't be done directly with Cepheid variable stars because they aren't intrinsically bright enough.

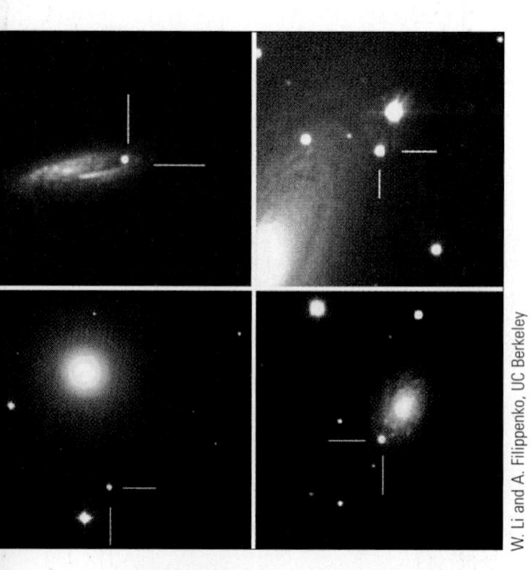

■ **FIGURE 18–18** Type Ia supernovae, marked with cross-hairs, in four different galaxies. Near its peak, a Type Ia supernova can shine as brightly as several billion Suns.

18.3e Type Ia Supernovae as Cosmological Yardsticks

In the 1990s, a remarkably reliable method was developed for measuring the distances of very distant galaxies. It is based on Type Ia supernovae ("white-dwarf supernovae"), which are exploding stars that result from a nuclear runaway in a white dwarf (see our discussion in Chapter 13). When they reach their peak power, these objects shine with the luminosity (intrinsic brightness) of about 10 billion Suns, or about a million times more than Cepheid variables. So, they can be seen at very large distances, 1000 times greater than Cepheid variables (■ Fig. 18–18.).

Most observed Type Ia supernovae are found to have nearly the same peak luminosity, as would be expected since the exploding white dwarf is thought to always have the same mass (the Chandrasekhar limit). Type Ia supernovae are therefore very good "standard candles" for measuring distances. (They do show small variations in peak luminosity, but we have ways of taking this into account—essentially like reading the

ASIDE 18.4: Hubble's constant

The Wilkinson Microwave Anisotropy Probe, to be discussed in Chapter 19, has provided (in conjunction with other data) the most accurate and precise estimate of Hubble's constant, $H_0 = 71 \pm 4$ km/sec/Mpc, in 2003.

wattage label on a light bulb.) By comparing the apparent brightness of a faint Type Ia supernova in a distant galaxy with the supernova's known luminosity, and by using the inverse-square law of light, we obtain the distance of the supernova, and hence of the galaxy in which it exploded (see Fig. 18–18.).

Of course, to apply this method successfully, we need to know the peak luminosity of a Type Ia supernova. But this can be found by measuring the peak apparent brightness of a supernova in a relatively nearby galaxy—one whose distance can be measured by other techniques, such as Cepheid variable stars. So, an important part of the Hubble Key Project was to find the distances of galaxies in which Type Ia supernovae had previously been seen, and in that way to calibrate the peak luminosity of Type Ia supernovae. By 2005, reliable distances to over a dozen such galaxies had been measured. Indeed, our adopted value of H_0 = 71 km/sec/Mpc is partly dependent on this work.

AceAstronomy™ Log into AceAstronomy and select this chapter to see the Active Figure called "Hubble."

AceAstronomy™ Log into AceAstronomy and select this chapter to see the Astronomy Exercise "Hubble Relation."

AceAstronomy™ Log into AceAstronomy and select this chapter to see the Astronomy Exercise "Age of the Universe."

18.4 The Geometry and Fate of the Universe

We have seen that to determine the age of the Universe, its expansion history (in addition to Hubble's constant) must be known. It turns out that, under certain assumptions, the expansion history is closely linked to the eventual fate of the Universe as well as to its overall (large-scale) geometry.

18.4a The Cosmological Principle: Uniformity

Mathematically, we use Einstein's general theory of relativity to study the expansion and overall geometry of the Universe. Since matter produces space–time curvature (as we have seen when studying black holes in Chapter 14), we expect the average density to affect the overall geometry of the Universe. The average density should also affect the way in which the expansion changes with time: High densities are able to slow down the expansion more than low densities, due to the gravitational pull of matter. Thus, the average density appears to be the most important parameter governing the Universe as a whole.

To simplify the equations and achieve reasonable progress, we assume the **cosmological principle:** On the largest size scales, the Universe is very *uniform*—it is **homogeneous** and **isotropic.** Homogeneous means that it has the same average density everywhere at a given time (though the density *can* change with time). Isotropic means that it looks the same in all directions—there is no preferred axis along which most of the galaxies are lined up, for example (■ Fig. 18–19). Note that we can check for isotropy only from our own position in space. However, for even greater simplicity we could suppose that the Universe looks isotropic from *all* points. (In this case of isotropy everywhere, the Universe is also necessarily homogeneous.)

The cosmological principle is basic to most big-bang theories. But it is clearly incorrect on small scales: A human, the Earth, the Solar System, the Milky Way Galaxy, and our Local Group of galaxies have a far higher density than average. Even the supercluster of galaxies to which the Milky Way belongs is somewhat denser than average.

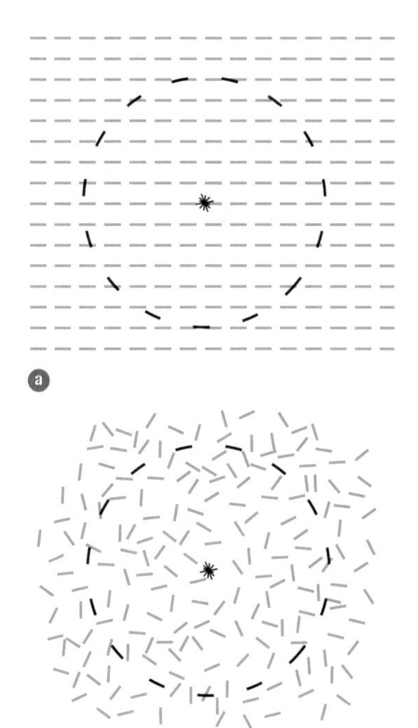

■ **FIGURE 18–19** Illustration of a homogeneous universe that is not isotropic (a), and a universe that is both homogeneous and isotropic (b). The average density of galaxies in a sufficiently large volume is the same in both universes, but in *(a)* the galaxies are all lined up in the same direction. The dashed circle is the observable part of the universe.

However, averaged over volumes about a billion light-years in diameter, the cosmological principle does appear to hold. The largest structures in the Universe seem to be superclusters and huge voids, but these are only a few hundred million light-years in diameter. Moreover, as we will see in Chapter 19, the strongest evidence comes from the "cosmic background radiation" that pervades the Universe: It looks the same in all directions, and it comes to us from a distance of about 14 billion light-years. Thus, over large distances, the Universe is indeed uniform.

18.4b No "Cosmological Constant"?

Another assumption we will make, at least temporarily, is that there are no long-range forces other than gravity, and that only "normal" matter and energy (with an attractive gravitational force) play a significant role—there is no "dark energy" having a repulsive effect.

Prior to Edwin Hubble's discovery that the Universe is expanding, most people thought the Universe is static (neither expanding nor contracting), which in some ways is aesthetically pleasing. Einstein knew that normal gravity should make the Universe contract, so in 1917 he postulated a long-range repulsive force, sort of a "cosmic antigravity," with a specific value that made the Universe static (■ Fig. 18–20). This "fudge factor" became known as the **cosmological constant,** denoted by the Greek capital letter Λ (lambda).

Though not mathematically incorrect, the cosmological constant is aesthetically displeasing, and it implies that the vacuum has a nonzero energy. Einstein was never fond of it, and reluctantly introduced it only because of the existing evidence for a static universe.

In 1929, when Hubble discovered the expansion of the Universe, the entire physical and philosophical motivation for the cosmological constant vanished. The Universe wasn't static, and no forces are needed to make it expand. After all, the Universe could have simply begun its existence in an expanding state, and is still coasting. Einstein renounced the cosmological constant and was unhappy that he had erred; after all, he could have predicted that the Universe is dynamic rather than static.

However, the concept of the cosmological constant itself (or, more generally, repulsive "dark energy"; see Section 18.5d) should perhaps not be considered erroneous. In a sense, it is just a generalization of Einstein's relativistic equations for the Universe. The mistake was in supposing that the cosmological constant has the precise value needed to achieve a static universe—especially since this turns out to be an unstable mathematical solution (slightly perturbing the Universe leads to expansion or collapse).

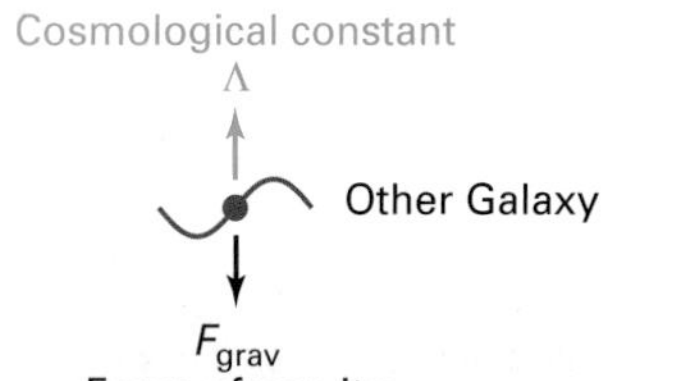

Nevertheless, it isn't clear what could physically produce a nonzero cosmological constant, and the simplest possibility is that the cosmological constant is zero ($\Lambda = 0$). Since there has generally been no strong observational evidence for a nonzero cosmological constant, astronomers have long assumed that its value is indeed zero. This is what we will initially assume here, too—but later in this chapter we will discuss exciting evidence that the cosmological constant (or some kind of "dark energy" that behaves in a similar way) isn't zero after all.

ASIDE 18.5: Stretching the truth?

Einstein is often quoted as saying that he made the "biggest blunder" of his life (or career) with the cosmological constant, but these exact words are not found in any of his written work. They were written only in a memoir by a respected physicist, George Gamow, who was known for exaggerating and joking. So it is possible, and maybe even probable, that Einstein never said this at all.

Cosmological constant
Λ
Other Galaxy
F_{grav}
Force of gravity
Milky Way Galaxy

■ **FIGURE 18–20** Albert Einstein introduced the cosmological constant, Λ (Lambda), in order to produce a static universe. He knew that normal gravity always pulls, so there had to be a counteracting repulsive force to prevent collapse. In this diagram, another galaxy is pulled toward the Milky Way Galaxy by their mutual gravitational attraction, but pushed apart with equal force provided by the cosmological constant.

18.4c Three Kinds of Possible Universes

Given the assumptions of the cosmological principle and no long-range antigravity, and also that no new matter or energy are created after the birth of the Universe, the general theory of relativity allows only three possibilities. These are known as "Friedmann universes" in honor of Alexander Friedmann, who, in the 1920s, was the first to derive them mathematically.

In each case the expansion decelerates with time, but the ultimate fate (that is, whether the expansion ever stops and reverses) depends on the overall *average density*

FIGURE IT OUT 18.2
The Critical Density and Ω_M

The "critical density," ρ_{crit}, is defined to be that density of matter that would allow the Universe to expand forever, but only just barely. Its formula is

$$\rho_{crit} = 3H_0^2/(8\pi G),$$

where H_0 is Hubble's constant and G is Newton's universal constant of gravitation. If the average matter density of the Universe (ρ_{ave}) is greater than, equal to, or less than ρ_{crit}, then Ω_M (Omega of the matter) is greater than, equal to, or less than 1, respectively, since we have defined $\Omega_M = \rho_{ave}/\rho_{crit}$.

For H_0 = 71 km/sec/Mpc, we find that $\rho_{crit} = 9.4 \times 10^{-30}$ g/cm^3, which is the equivalent of only about 5.6 hydrogen atoms per cubic meter of space! Clearly, our local surroundings are much denser, but this isn't relevant. Only the average density of the *Universe as a whole* should be compared with ρ_{crit} to determine the value of Ω_M. In fact, we find that Ω_M is almost certainly less than 1. This means that on large scales the Universe is remarkably empty.

Note that as the Universe ages, the average density of matter in the Universe decreases. However, if the value of Ω_M initially exceeds 1, it does not later drop to 1 or less. This is because the value of H_0 decreases with time, and hence the critical density also decreases. The ratio of the average density to the critical density, Ω_M, remains either above 1 or below 1 (or is exactly equal to 1) forever.

As we will discuss in Sec. 18.5, we now know that in addition to matter, there is mysterious "dark energy" that contributes to the total density of the Universe.

of matter relative to a specific **critical density.** If we define the average matter density divided by the critical density to be Ω_M, where Ω is the Greek capital letter "Omega" and the subscript *M* stands for "matter," then the three possible universes correspond to the cases where this ratio is greater than one, equal to one, and less than one. (Details are explored in *Figure It Out 18.2: The Critical Density and Ω_M.*)

The separation between any two galaxies versus time is shown in ■ Figure 18–21 for the three types of universes. It is best to choose galaxies in different clusters (or even different superclusters, to be absolutely safe), since we don't want them to be bound together by gravity. This galaxy separation is often called the "scale factor" of the Universe; it tells us about the expansion of the Universe itself.

If $\Omega_M > 1$ (that is, the average density is above the critical density), galaxies separate progressively more slowly with time, but they eventually turn around and approach each other (in other words, the recession speed becomes negative), ending in a hot "big crunch." (Some astronomers also jokingly call it a "gnab gib," which is "big bang" backwards!) A good analogy is a ball thrown upward with a speed less than Earth's escape speed; it eventually falls back down. It is conceivable that another big bang subsequently occurs, resulting in an "oscillating universe," but we have little confidence in this hypothesis since the laws of physics as currently stated cannot be traced through the big crunch.

If $\Omega_M = 1$ (that is, the average density is exactly equal to the critical density), galaxies separate more and more slowly with time, but as time approaches infinity, the

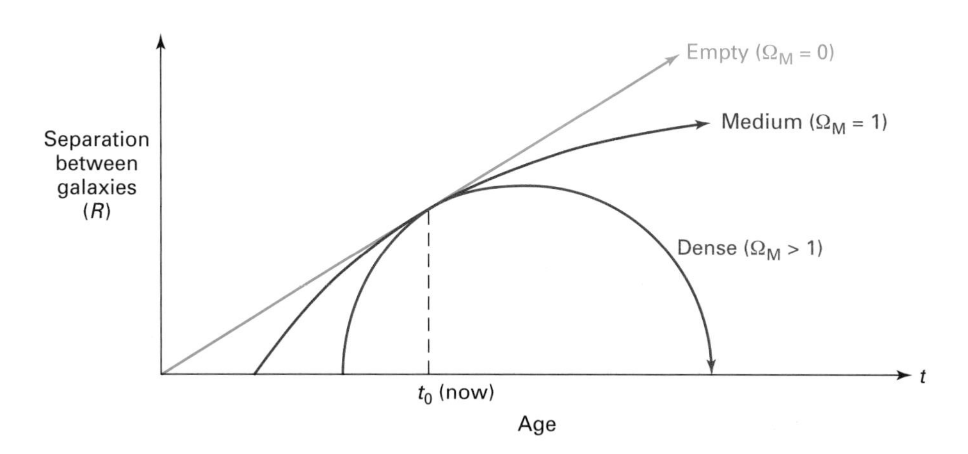

■ **FIGURE 18–21** To trace back the growth of our Universe, we would like to know the rate at which its expansion rate is changing with time. Big-bang models of the Universe are shown; the vertical axis represents the separation between any two galaxies (preferably in different superclusters), and the horizontal axis is time. If the Universe were empty, then the rate of expansion would be constant forever, since there would be no galaxies and hence no gravitational forces. In a more realistic case of a non-empty Universe, gravity tends to decelerate the expansion. However, if the ratio of the average density of matter to the critical density, $\Omega_M < 1$, then the Universe is still able to expand forever. If $\Omega_M > 1$, on the other hand, gravity causes the expansion to eventually halt, and the Universe subsequently recollapses. The dividing line, where the ratio of the matter density is equal to the critical density (that is, $\Omega_M = 1$), corresponds to expansion that continues forever, but just barely: Galaxies recede from each other with a speed that approaches zero as time approaches infinity.

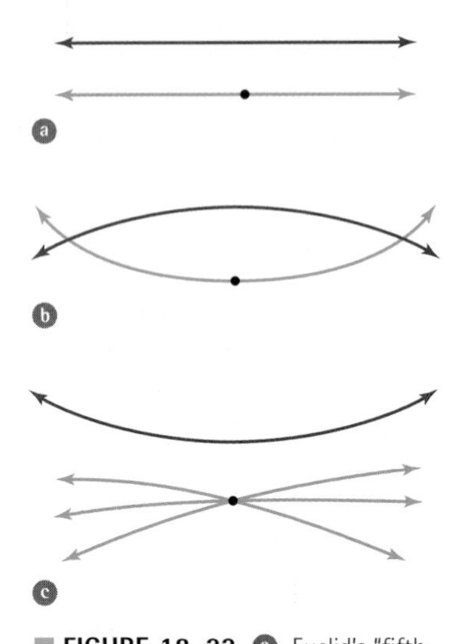

■ **FIGURE 18–22** (a) Euclid's "fifth postulate" for spatially flat geometry states that given a line and a point not on that line, only one unique parallel line can be drawn through the point. (b) In a positively curved geometry, no parallel lines can be drawn through the point. (c) In a negatively curved geometry, many parallel lines can be drawn through the point. (Of course, we cannot accurately draw curved geometries on flat sheets of paper, so these diagrams are only meant to convey the general idea.)

recession speed approaches zero. Thus, the Universe will expand forever, though just barely. The relevant analogy is a ball thrown upward with a speed equal to Earth's escape speed; it continues to recede from Earth ever more slowly, and it stops when time reaches infinity. This turns out to be the type of universe predicted by most "inflation theories" (which we will study in Chapter 19).

If $\Omega_M < 1$ (that is, the average density is below the critical density), galaxies separate more and more slowly with time, but as time approaches infinity, the recession speed (for a given pair of galaxies) approaches a constant, nonzero value. Thus, the Universe will easily expand forever. Once again using our ball analogy, it is like a ball thrown upward with a speed greater than Earth's escape speed; it continues to recede from Earth ever more slowly, but it never stops receding.

These three kinds of universes have different overall geometries. The $\Omega_M = 1$ case is known as a **flat universe** or a **critical universe.** It is described by "Euclidean geometry"—that is, the geometry worked out first by the Greek mathematician Euclid in the third century B.C. In particular, Euclid's "fifth postulate" is satisfied: Given a line and a point not on that line, only one unique parallel line can be drawn through the point (■ Fig. 18–22*a*). Such a universe is spatially flat, formally infinite in volume (but see the caveat at the end of Section 18.4c), and barely expands forever. Its age is exactly two-thirds of the Hubble time, $(2/3)/H_0 = (2/3)T_0$.

In the $\Omega_M > 1$ universe, Euclid's fifth postulate fails in the following way: Given a line and a point not on that line, *no* parallel lines can be drawn through the point (Fig. 18–22*b*). Such a universe has positive spatial curvature, is finite ("closed") in volume, but has no boundaries (edges) like those of a box. Its fate is a hot "big crunch." Generally known as a **closed universe,** it is also sometimes called a "spherical" ("hyperspherical") or "positively curved" universe. Its age is less than two-thirds of the Hubble time.

Finally, in the $\Omega_M < 1$ universe, Euclid's fifth postulate fails in the following way: Given a line and a point not on that line, many (indeed, infinitely many) parallel lines can be drawn through the point (Fig. 18–22*c*). Such a universe has negative spatial curvature, is formally infinite ("open") in volume (but see the caveat at the end of Section 18.4c), and easily expands forever. Generally known as an **open universe,** it is also sometimes called a "hyperbolic" or "negatively curved" universe. Its age is between $(2/3)T_0$ and T_0 (the latter extreme only if $\Omega_M = 0$).

Note that in some texts and magazine articles, the $\Omega_M = 1$ universe is called "closed," but only because it is *almost* closed. It actually represents the *dividing line* between "open" and "closed."

Under certain conditions, flat or negatively curved universes might have exotic shapes with finite volume (see *A Closer Look 18.1: Finite Flat and Hyperbolic Universes*). Even positively curved universes might not be simple hyperspheres. It is difficult, but not impossible, to distinguish such universes from the "standard" ones discussed above, and so far no clear observational evidence for them has been found. Though quite intriguing, in this book we will not further consider this possibility. Keep in mind, though, that convincing support for a finite, strangely shaped universe might be found in the future; we should always be open to potential surprises.

18.4d Two-Dimensional Analogues

It is useful to consider analogues to the above universes, but with only two spatial dimensions (■ Fig. 18–23). The flat universe is like an infinite sheet of paper. One property is that the sum of the interior angles of a triangle is always 180°, regardless of the shape and size of the triangle. Moreover, the area A of a circle of radius R is proportional to R^2 (that is, $A = \pi R^2$). This relation can be measured by scattering dots uniformly (homogeneously) across a sheet of paper, and seeing that the number of dots enclosed by a circle grows in proportion to R^2.

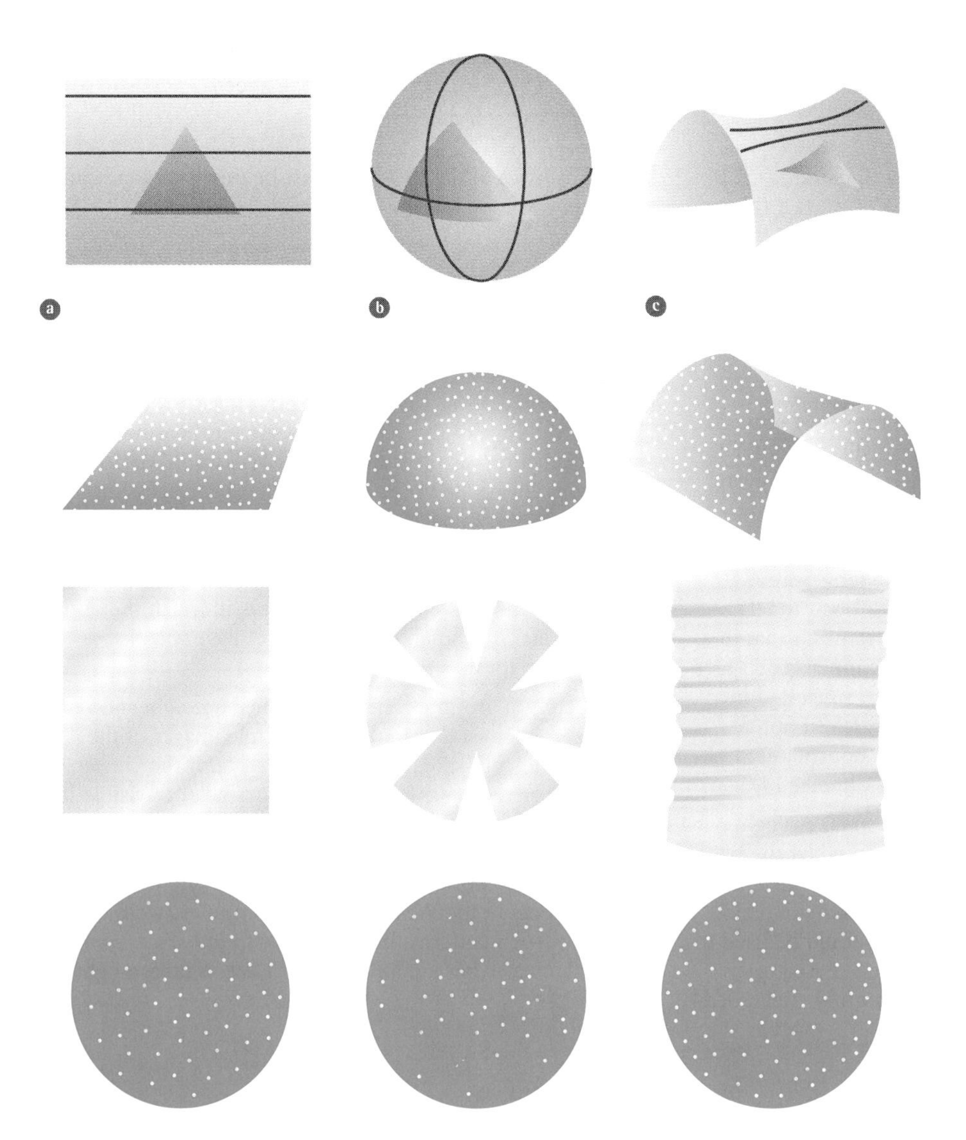

■ **FIGURE 18–23** *(Top row)* Two-dimensional analogues to three-dimensional space. (a) A flat universe is like an infinite sheet of paper; the laws of Euclidean geometry are obeyed. For example, the sum of the interior angles of a triangle is equal to 180°. (b) A positively curved universe is like the surface of a sphere; the sum of the interior angles of a triangle is larger than 180°. (c) A negatively curved universe is like the surface of an infinitely large saddle or potato chip in some respects; the sum of the interior angles of a triangle is smaller than 180°. *(Second row)* Consider dots uniformly distributed on each two-dimensional surface. If we count the total number of dots within circles of larger and larger radius R in flat space, we find that the total number of dots grows exactly in proportion to the area, $A = \pi R^2$. But in positively curved space there are fewer dots within a given radius, and in a negatively curved space there are more dots within a given radius. *(Third row)* It is easiest to visualize these relationships by trying to flatten out the curvature in each case. The positively curved surface, when flattened, gives rise to empty sections, while the flattened negatively curved surface has folds with extra dots. *(Bottom row)* The resulting distribution of dots in the flattened versions shows a relative deficit of dots at large radii in the positively curved surface and a relative excess of dots in the negatively curved surface.

The positively curved universe is like the surface of a sphere. The sum of the interior angles of a triangle is always greater than 180°. For example, a triangle consisting of a segment along the equator of the Earth, and two segments going up to the north pole at right angles from the ends of the equatorial segment, clearly has a sum greater than 180°. Moreover, the area of a circle of radius R falls short of being proportional to R^2. If the sphere is uniformly covered with dots, the number of dots enclosed by a circle grows more slowly with R than in flat space because a flattened version of the sphere contains missing slices.

The negatively curved universe is somewhat like the surface of an infinite horse's saddle or potato chip. These analogies are not perfect because a horse's saddle (or potato chip) embedded in a universe with three spatial dimensions is not isotropic; the saddle point, for example, can be distinguished from other points. The sum of the interior angles of a triangle is always less than 180°. The area of a circle of radius R is more than proportional to R^2. If the saddle is homogeneously covered with dots, the number of dots enclosed by a circle grows more quickly with R than in flat space because a flattened version of the saddle contains extra wrinkles.

With three spatial dimensions, we can generalize to the growth of volumes V with radius R. In a flat universe, the volume of a sphere is proportional to R^3 [that is, $V = (^4/_3)\pi R^3$]. In a positively curved universe, the volume of a sphere is not quite proportional to R^3. In a negatively curved universe, the volume of a sphere is more than proportional to R^3.

A Closer Look 18.1 FINITE FLAT AND HYPERBOLIC UNIVERSES

Mathematicians have pointed out that the local isotropy and homogeneity of a flat or negatively curved (hyperbolic) space might not extend globally, to every part of the universe. (After all, we observe our Universe to be isotropic only from our own vantage point; we don't really know that it must be isotropic everywhere.) If so, one can avoid the conclusion that universes with these geometries are infinite in size.

Indeed, there is an infinite number of examples of finite flat and hyperbolic spaces! The shape might be like that of a torus (doughnut) or a horn, for instance. One could even have unusual positively curved (spherical) spaces that are dodecahedral, or some other exotic shape.

These finite universes evolve in exactly the same way as their infinite counterparts, making them difficult to tell apart observationally. One distinguishing feature is that in a finite universe, light could in principle circumnavigate space one or more times. This possibility was first suggested in 1900 by the German physicist Karl Schwarzschild (after whom the radius of the event horizon of a nonrotating black hole is named; see our discussion in Chapter 14).

Schwarzschild reasoned that if space were small enough, there would be a "hall-of-mirrors" effect caused by the light wrapping around the universe, and we would see multiple images of a single astronomical object; see the figure below. Conversely, the absence of multiple images could be used to determine the minimum size of the universe. Based on observations available at the time, Schwarzschild concluded that the volume of the universe must be much greater than the volume ascribed to the Milky Way.

Over the years, people have used ever-improving observations to extend this minimum size, and most cosmologists now think that the scale over which a spatially flat universe wraps around is probably much larger than the distance to which we can see.

However, some physicists have recently suggested that the image of the edge of the visible Universe, the "cosmic background radiation" as measured by the Wilkinson Microwave Anisotropy Probe (WMAP; see our extensive discussion in Chapter 19), shows deviations from isotropy like those one would expect in a finite universe. Others point to the lack of large-scale features in the WMAP image as another possible sign that the Universe might be finite, in much the same way that water in a small bathtub cannot have waves as large as those found in the open ocean.

Very recently, a group of scientists applied a new version of Schwarzschild's test that looks for multiple images of the same hot and cold spots in the WMAP data. If the Universe were finite on a scale smaller than the diameter of visible space, then we would see the same portion of the early universe when looking in different directions in the sky. The multiple images would give rise to matching circles of hot and cold spots in the WMAP image. Computers were used to look for such patterns, but none was found. This implies that the Universe has a diameter of at least 78 billion light-years.

Neil Cornish and Jeff Weeks

a

© Tee and Charles Addams Foundation

b

Repeating images of the Earth give the illusion that space is infinite in this "torus universe." Actually, though, a single Earth is viewed at many times in the past, as light repeatedly circumnavigates (that is, completes a full path around) the finite universe. The illusion is created by gluing together opposite faces of a cubic cell, the edges of which are indicated with wooden struts.

18.4e What Kind of Universe Do We Live In?

How do we go about determining to which of the above possibilities our Universe corresponds? There are a number of different methods. Perhaps most obvious, we can measure the average density of matter, and compare it with the critical density. The value of Ω_M (again, the ratio of the average matter density to the critical density) is greater than 1 if the Universe is closed, equal to 1 if the Universe is flat (critical), and less than 1 if the Universe is open.

Or, we can measure the expansion rate in the distant past (preferably at several different epochs), compare it with the current expansion rate, and calculate how fast the Universe is decelerating. This can be done by looking at very distant galaxies, which are seen as they were long ago, when the Universe was younger.

We can also examine geometrical properties of the Universe to determine its overall curvature. For example, in principle we can see whether the sum of the interior angles of an enormous triangle is greater than, equal to, or less than 180°. This is not very practical, however, since we cannot draw a sufficiently large triangle. Or, we can see whether "parallel lines" ever meet—but again, this is not practical, since we cannot reach sufficiently large distances.

A better geometrical method is to measure the angular sizes of galaxies as a function of distance. High-redshift galaxies of fixed physical size will appear larger in angular size if space has positive curvature than if it has zero or negative curvature, because light rays diverge more slowly in a closed universe than in a flat universe or in an open universe (■ Fig. 18–24*a*). Or, we could instead look at the apparent brightness of objects as a function of distance. High-redshift objects of fixed luminosity (intrinsic brightness) will appear brighter if space has positive curvature than if it has

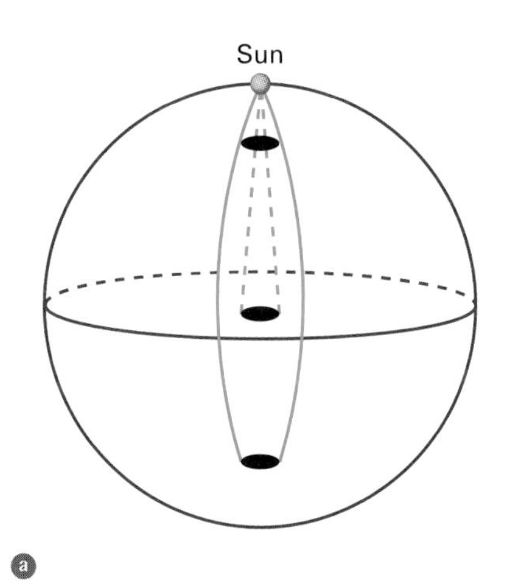

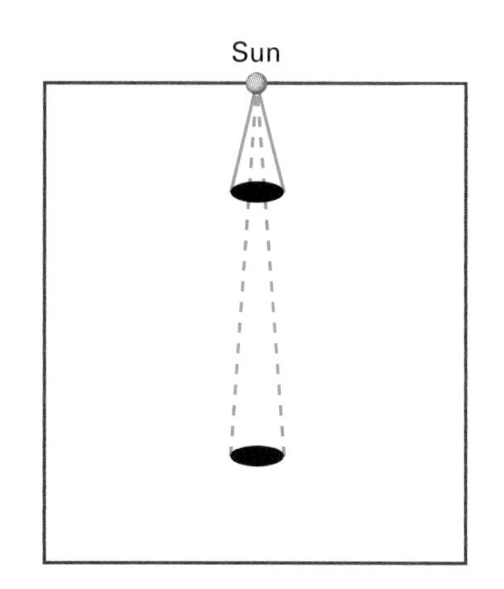

a

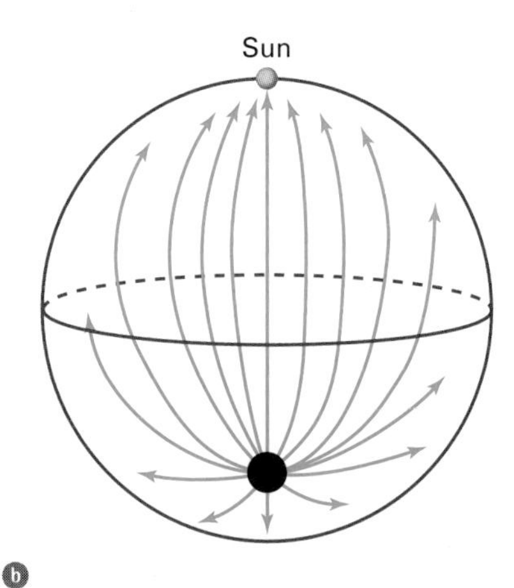

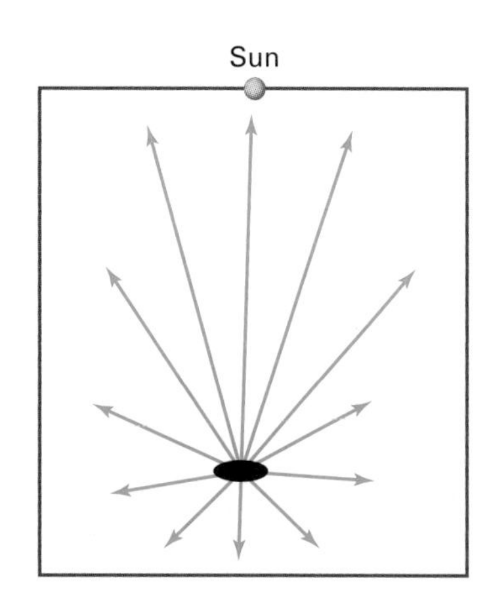

b

■ **FIGURE 18–24** a In flat space *(right)*, distant galaxies look much smaller than a nearby galaxy; in fact, the angular size is inversely proportional to distance. In positively curved space *(left)*, on the other hand, the angular size is somewhat larger than expected when compared with flat space. Extremely distant galaxies can even start looking larger again, due to the curved trajectory of light rays. b In flat space *(right)*, distant galaxies look very faint due to the inverse-square law of light. But in positively curved space *(left)*, the light rays don't diverge as fast, and even begin to converge again at sufficiently large distances; thus, distant objects look brighter than expected.

zero or negative curvature; again, light rays diverge more slowly in a closed universe (Fig. 18–24*b*).

We might also count the number of galaxies as a function of distance to see how volume grows with radius (if galaxies don't evolve with time, something known to be untrue). This is analogous to the measurement of area in two-dimensional universes, as explained in Section 18.4d. If space is flat, volume is exactly proportional to R^3; thus, doubling the surveyed distance (R) should increase the number of galaxies by a factor of 8. On the other hand, if space has positive curvature, the factor will be smaller than 8, while if space has negative curvature, the factor will be larger than 8.

A completely different technique is to measure the relative abundances (proportions) of the lightest elements and their isotopes, which were produced shortly after the big bang. As we will discuss in Chapter 19, these depend on the value of Ω_M.

Some astronomers measure the motions of galaxies and clusters of galaxies through the Universe (that is, relative to the smooth Hubble flow). These are produced by the gravitational tug of other clusters, and hence provide a measure of the mass and distribution of clumped matter (both visible and dark).

There are many other, related techniques. For a number of reasons, all of them (including those listed here) are difficult and uncertain.

18.4f Obstacles Along the Way

One problem is that local, dense objects (planets, stars, galaxies, etc.) produce spatial curvature larger than the gradual, global effect that we seek. Moreover, we know that the Universe is nearly flat, so to detect any slight overall curvature one needs to look very far, and this is difficult. Another problem is that when counting galaxies or determining the average density, how does one know that a representative volume was chosen? After all, there are deviations from uniformity (inhomogeneities) on large scales, such as superclusters of galaxies.

A major difficulty is that galaxies evolve with time. So, we cannot assume that a known type of galaxy has a constant physical size (when measuring angular sizes as a function of distance), and the quantitative evolution of physical size is difficult to predict. We also do not know how the luminosity of a typical galaxy evolves with time, yet galaxies certainly do evolve (■ Fig. 18–25). Counts of galaxies to various distances might be dominated by intrinsic luminosity differences, rather than by the volume of space surveyed.

ASIDE 18.6: The matter density

The Wilkinson Microwave Anisotropy Probe, to be discussed in Chapter 19, has provided (in conjunction with other data) the most accurate and precise estimate of Ω_M, the ratio of the average matter density to the critical density: 0.27 ± 0.02, consistent with expectations. Note that matter consists of visible matter and dark matter.

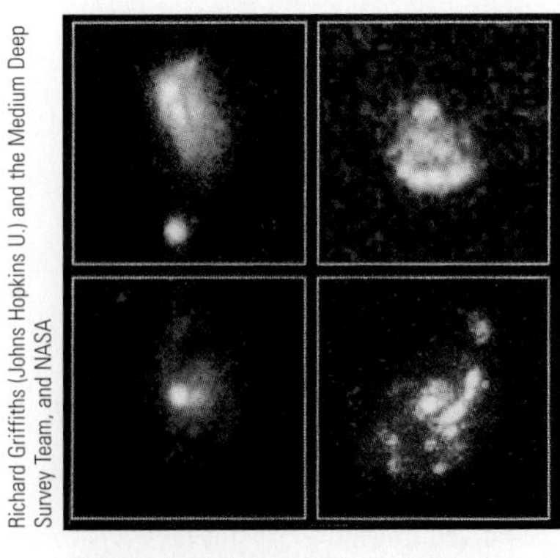

■ **FIGURE 18–25** Typical galaxies photographed by the Hubble Space Telescope, at redshifts between 0.3 and 0.7 (that is, roughly 3.4 to 6.3 billion light-years away). They look peculiar relative to nearby galaxies, suggesting that galaxies evolve over time in shape, size, and luminosity.

Clusters of galaxies also evolve with time, and are therefore subject to similar uncertainties. Determining the total amount of matter in a cluster is difficult; much of the matter is dark, and can be detected only through its gravitational effects. In addition, we do not know how much dark matter is spread somewhat uniformly, rather than clumped in clusters and superclusters, and this affects the calculated average density.

Of course, an observational problem is that distant objects appear small and faint, and are therefore subject to considerable measurement uncertainties.

At the time of writing, in late 2005, there is consensus that $\Omega_M \approx 0.3$ (almost certainly larger than 0.2 but definitely smaller than 0.4). If true, the Universe is spatially infinite (but see the caveat at the end of Section 18.4c) and will expand forever. However, these conclusions are based primarily on studies of clusters of galaxies—their motions, masses, and so on. Uniformly distributed matter and other possible effects, such as the cosmological constant, are not taken into account. There are few tests of deceleration, or of overall geometry.

Yet there are theoretical reasons (described in Chapter 19) for believing that Ω_M (or Ω_{total}, which might include some new kind of energy) is exactly 1, if it is known to be at least relatively close to 1, such as the value of 0.3 currently favored. If it were not *exactly* 1 initially, it should have deviated *very* far from 1 (for example, 10^{-7} or 10^{15}) by the present time. This behavior is like nudging a pencil balanced on its tip: It quickly falls to the surface.

18.5 Measuring the Expected Deceleration

Perhaps the most direct way of determining the expected deceleration of the Universe is to measure the expansion rate as a function of time by looking at very distant objects. As discussed earlier, the separation between two clusters of galaxies varies with time in different ways, depending on the deceleration rate of the Universe (Fig. 18–21). For any of the curves, the slope at a given time is the expansion rate at that time. (Hubble's constant itself is the slope divided by the separation between galaxies at that time.)

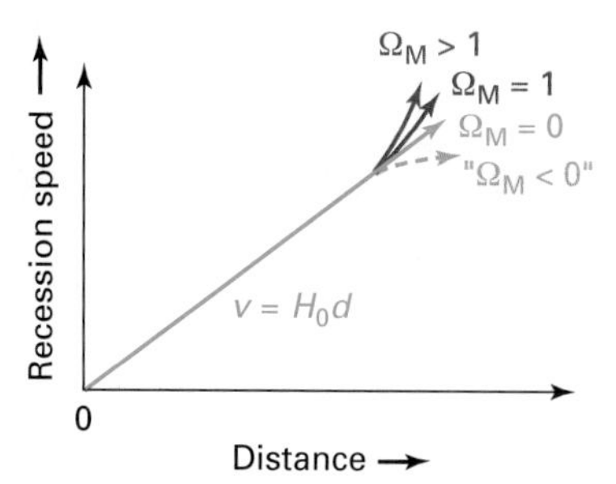

FIGURE 18–26 A plot of the measured recession speeds of objects (as given by their redshifts) vs. their distances. Relatively nearby galaxies all satisfy the same relationship, Hubble's law. But at large recession speeds there are differences that depend on how rapidly the Universe's expansion rate is decelerating, and hence on Ω_M, the ratio of the average density of matter to the critical density. For a given redshift, a galaxy is closer if the Universe is denser than if it is less dense. If the density of matter could be negative (that is, "$\Omega_M < 0$"), distances larger than those in an empty (non-decelerating) universe would be measured.

18.5a The High-Redshift Hubble Diagram

Observationally, we construct the Hubble diagram by plotting the measured recession speed (from the redshift) vs. the distance (from the inverse-square law) for a set of objects (Fig. 18–26). At small distances, Hubble's law holds: Speed is directly proportional to distance ($v = H_0 d$). At very large distances, however, there should be deviations from this. For a given distance, the speed should be higher if Ω_M is greater than 1 than if Ω_M is less than 1, because the Universe used to be expanding much more quickly if Ω_M is greater than 1.

Or, for a given speed, the distance should be larger if Ω_M is less than 1 than if Ω_M is greater than 1, because the expansion didn't slow down as much if Ω_M is less than 1. The distance should be even larger if Ω_M is less than 0 (which is physically impossible, since matter is gravitationally attractive), because the expansion of the Universe will have accelerated, pushing objects even farther away.

What kinds of objects can be seen at sufficiently large distances to accomplish this task? We could try galaxies—but they evolve with time due to mergers, bursts of star formation, and other processes that vary from one galaxy to another and are not well understood. Clusters of galaxies also evolve with time in ways that are difficult to predict accurately, although recent progress in the use of clusters looks promising (see below).

Some astronomers had hoped that quasars would serve well—but they exhibit a tremendous range in luminosity (intrinsic brightness), and they evolve quickly with time. Gamma-ray bursts, being so luminous, are obvious candidates as well, but they too show a wide range in luminosity. On the other hand, very recent attempts to calibrate gamma-ray bursts have shown considerable signs of success; by the time the next edition of this book is written, they may provide important complementary information on the history of the expansion rate.

18.5b Type Ia (White-Dwarf) Supernovae

In the mid- to late-1990s, the most progress on this front was made with Type Ia supernovae, whose utility for finding the Hubble constant has already been discussed (Sec. 18.3e). Type Ia supernovae are nearly ideal objects for such studies. They are very luminous, and although not exactly standard candles, accurate corrections can be made for the differences in luminosity by measuring their light curves. Their intrinsic properties should not depend very much on redshift: Long ago (that is, at high redshift), white dwarfs should have exploded in essentially the way they do now.

Starting in the early 1990s, two teams have found and measured high-redshift ($z = 0.3$ to $z = 0.7$) Type Ia supernovae. The first is led by Saul Perlmutter (Lawrence Berkeley Laboratory), and the second by Brian Schmidt (Australian National University) and Adam Riess (now at the Space Telescope Science Institute). One of us (A.F.) has been fortunate to work with both groups, although his primary association since 1996 has been with the Schmidt/Riess team.

Brian P. Schmidt (Australian National Univ.) and the High-z Supernova Search Team, and NASA

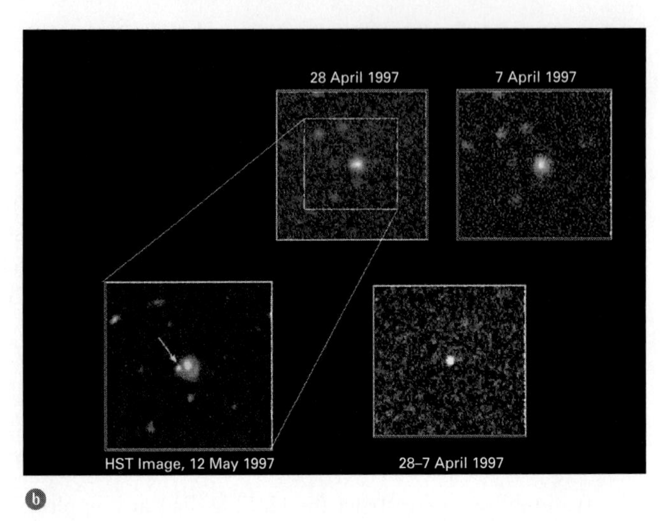

■ **FIGURE 18–27** ⓐ Deep, wide-angle CCD image of the sky showing thousands of faint, distant galaxies. Almost every visible object is another galaxy, not a star in the Milky Way Galaxy. ⓑ Discovery of SN 1997cj (redshift $z = 0.5$). The top two images are small subsets of much larger CCD frames obtained on April 7 and 28, 1997, with a 4-m telescope in Chile. The difference between the two images, at *bottom*, reveals a supernova candidate that was subsequently confirmed to be a Type Ia supernova with a spectrum obtained by one of the Keck 10-m telescopes. At *left* is an image obtained on May 12, 1997, with the Hubble Space Telescope, showing the supernova well separated from the center of the host galaxy.

A Type Ia supernova occurs somewhere in the observable universe every few tens of seconds. Thus, if very deep photographs are made of thousands of distant galaxies (■ Fig. 18–27*a*), the chances of catching a supernova are reasonably good. The same regions of the sky are photographed about a month apart with large telescopes. Comparison of the photographs with sophisticated computer software reveals the faint supernova candidates (Fig. 18–27*b*).

A spectrum of each candidate is obtained, often with the Keck telescopes in Hawaii. This reveals whether the object is a Type Ia supernova, and provides its redshift. Follow-up observations of the Type Ia supernovae with many telescopes, including the Hubble Space Telescope (■ Fig. 18–28), provide their light curves. The peak apparent brightness of each supernova, together with the known (appropriately corrected) luminosity, gives the distance using the inverse-square law.

Incidentally, the light curves of high-redshift supernovae appear broader than those of low-redshift examples; that is, the former take longer to brighten and fade than the latter. This results from the expansion of the Universe: Each successive photon has farther to travel than the previous one. Indeed, this observed "time dilation" effect (Fig. 18–28) currently provides the best evidence that redshifts really are produced by the expansion of the Universe, rather than by some other mechanism (such as light becoming "tired," losing energy during its long journey). The time dilation factor by which the light curves are broader is $1 + z$.

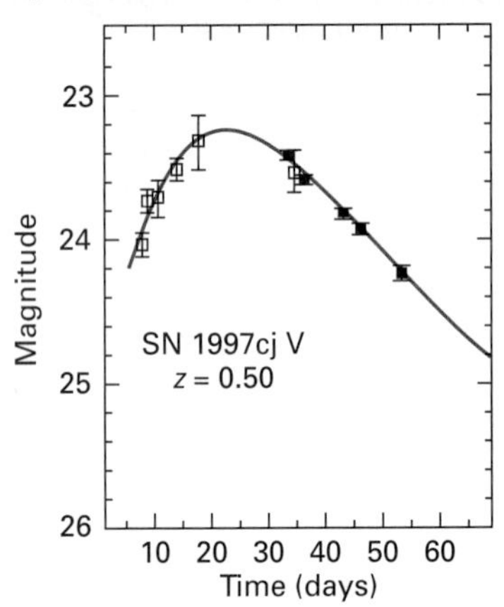

ⓐ

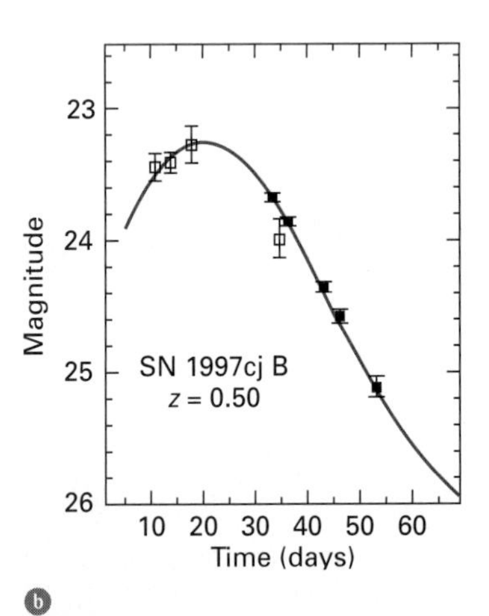

ⓑ

■ **FIGURE 18–28** Light curves of SN 1997cj (redshift $z = 0.5$), obtained with ground-based telescopes and the Hubble Space Telescope. The data were obtained through filters that pass light emitted by the supernova at ⓐ visual and ⓑ blue wavelengths. These light curves are broader than those of low-redshift supernovae because the Universe is expanding, producing a kind of "time dilation." (Peter Garnavich, Notre Dame, the High-z Supernova Search Team, and NASA)

18.5c An Accelerating Universe!

Several dozen supernovae at typical lookback times of 4 to 5 billion years were measured in this manner by early 1998, and many additional ones have been observed by the time of this writing (late 2005). The results are astonishing: Both teams find that the high-redshift supernovae are fainter, and hence more distant, than expected. The data agree best with the "$\Omega_M < 0$" curve in Figure 18–26, yet we know that the matter density is greater than 0 since we exist! Note that the possible presence of antimatter does not resolve this problem: Both matter and antimatter have positive energy, and exert an *attractive* gravitational force.

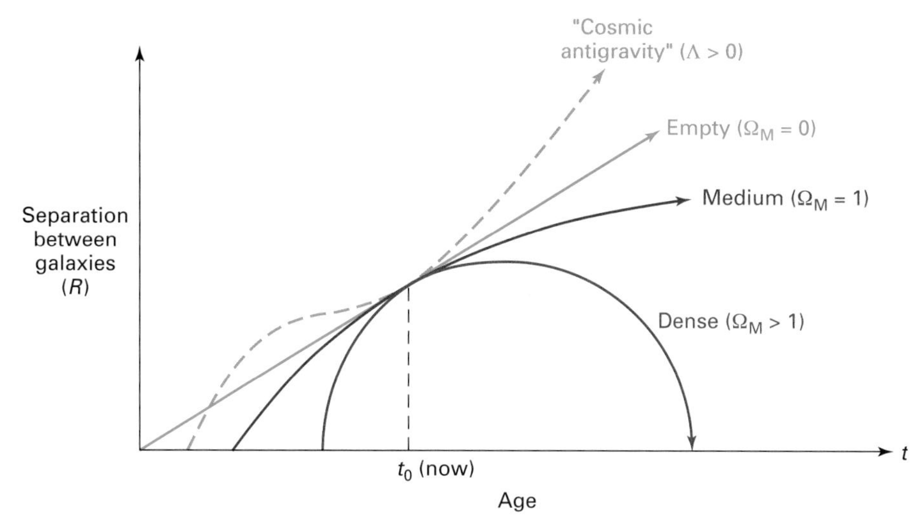

■ **FIGURE 18–29** Similar to Figure 18–21, but with an additional *(dashed)* curve corresponding to what our Universe appears to be doing. The ratio of the average density of matter to the critical density, Ω_M, is about 0.3—but recent data also suggest that the cosmological constant (Λ, Lambda) is positive. If correct, the expansion of the Universe is now accelerating, and will continue to do so forever. The derived current age of the Universe turns out to be about 14 billion years.

We conclude that there is a previously unknown component of the Universe with a long-range *repulsive* effect—essentially a cosmic "antigravity." Its presence is felt only over exceedingly large distances. One cannot use this substance to make "antigravity boots," for example.

The extra stretching of space makes the supernovae more distant than they would have been had the Universe's expansion decelerated throughout its history. Thus, if these results are correct, we now live in an **accelerating universe,** one whose expansion rate is *increasing* rather than decreasing with time (■ Fig. 18–29)! Unless the repulsive effect changes sign and becomes attractive in the future (which it might, since we do not know what causes the effect), the Universe will expand forever, even if it is closed (finite in volume). Distant galaxies are receding from each other faster and faster, so eventually they will fade away, as their light becomes redshifted to essentially zero energy, and as time dilation causes photons to arrive extremely rarely.

The two teams calculate the age of the Universe to be about 13.7 billion years, if $H_0 = 71$ km/sec/Mpc. This is a bit smaller than the value of 13.9 billion years expected if the Universe were empty ($\Omega_M = 0$), but larger than the value of 9.3 billion years expected with $\Omega_M = 1$ (Fig. 18–29). An expansion age of 13.7 billion years is quite consistent with the recently revised ages of globular clusters: 12–13 billion years. The pesky age crisis seems finally to be over!

Although astronomers and physicists were skeptical of the initial results announced by the two teams in February 1998, by the end of that year nobody had found any obvious flaws in their data, analysis methods, or conclusions. Thus, either they were right, or they had been led astray by some subtle effect that was likely to teach us something interesting about the Universe. The discovery gained international prominence (■ Fig. 18–30), and soon the quest for the cause of the acceleration (and the physical nature of the responsible agent) became one of the hottest topics in all of physics.

■ **FIGURE 18–30** The cover of *Science* magazine from December 18, 1998; the discovery of the accelerating expansion of the Universe, made by two teams including one of the authors (A.F.), was named the most important scientific breakthrough of 1998. The caricature of Einstein is surprised because the "bubble universe" that he blew out of his pipe appears to be accelerating in its expansion, rather than decelerating as expected. He is doubly surprised because he predicted (but for the wrong reason) the possibility of large-scale cosmic repulsion in 1917, and later renounced the idea.

18.5d Einstein's Biggest Blunder?

Perhaps the simplest explanation for the observed acceleration is that Einstein's cosmological constant "Lambda" (Λ), the "fudge factor" he introduced to make a static universe, is nonzero. In essence, space appears to have a repulsive aspect to it, a cosmic

ASIDE 18.7: Partners, yet mortal enemies

As we briefly mentioned in Chapters 12 and 14, for every subatomic particle, there is a corresponding antiparticle. This antiparticle has the same mass as the particle, but is opposite in all other properties such as charge and the direction it is spinning. If a particle and its antiparticle meet, they annihilate each other; their total mass is 100 per cent transformed into energy in the amount $E = mc^2$. Antiparticles together make up antimatter.

"antigravity." But instead of exactly negating the attractive force of gravity, the cosmological constant slightly dominates gravity over very large distances, producing a net acceleration.

It is ironic that Einstein introduced the cosmological constant and later completely rejected it, anecdotally calling it his "biggest blunder." The idea itself may not have been wrong; rather, only the exact value that Einstein gave to the cosmological constant was slightly erroneous. Even if the acceleration turns out to be caused by something other than Λ (see below), one could say that Einstein was right after all, because there is indeed a new, previously unanticipated, repulsive effect present in the Universe.

Many theorists find a positive cosmological constant to be very disconcerting. In the context of quantum theories, it suggests that the vacuum has a nonzero, positive energy density (a "vacuum energy") due to **quantum fluctuations**—the spontaneous formation (and then rapid destruction) of virtual pairs of particles and antiparticles. (They are called "virtual" because they form out of nothing and last for only a very short time, unlike "real matter.") For subtle reasons in general relativity having to do with its "negative pressure," a positive energy density of this type causes space to expand faster and faster with time.

Expressed in the same units as Ω_M, essentially as a ratio of densities, the current value of Λ that the two teams measure is $\Omega_\Lambda \approx 0.7$ (making use of the cosmic background radiation as well—see Chapter 19). But theorists generally expected that $\Omega_\Lambda = 0$ due to an exact cancellation of all the quantum fluctuations; otherwise, they predicted the exceedingly large value $\Omega_\Lambda \approx 10^{120}$, or possibly down to "only" 10^{50}, clearly neither of which is actually observed. (We would definitely not exist if the cosmological constant were so large; the Universe would have expanded much too quickly for galaxies and stars to form.) The discrepancy between the observed and expected values of Ω_Λ has been named the greatest error (or embarrassment) ever in theoretical physics!

Although Λ itself is constant with time, the density ratio Ω_M decreases as the Universe ages, while Ω_Λ increases. This, however, leads to another problem: Why should these two density ratios be roughly equal right now? They are measured to be $\Omega_\Lambda \approx 0.7$ and $\Omega_M \approx 0.3$, but they could have been 0.00001 and 0.99999, for example, or any other two numbers (between 0 and 1) whose sum is 1.0000. Do we live in a cosmically "special" time?

18.5e Dark Energy

ASIDE 18.8: Dark energy everywhere

Dark energy is thought to be present throughout space, even here on Earth. However, only on the largest scales (exceeding roughly 100 million light-years) does its density exceed that of luminous and dark matter, making its effects noticeable. Dark energy causes superclusters of galaxies, and especially the space between them, to expand progressively faster. Space does not expand on Earth, in our Solar System, in our Galaxy, or even in our Local Group of galaxies.

Because of these and other problems associated with the cosmological constant, physicists have eagerly sought alternative explanations for the observed acceleration of the Universe. Some of the hypotheses seem rather wild, to say the least. For example, gravity might be "leaking out" of our Universe and into extra dimensions, or perhaps "other universes" in some larger "hyperspace" (see Chapter 19 for more details) are "pulling out" on our Universe!

Most of the alternatives, however, invoke a new kind of particle or field within our Universe, similar to the cosmological constant but having different specific properties. The most general term for the responsible substance (including also the cosmological constant) is **dark energy**—in some ways an unfortunate choice of words because of the possible confusion with "dark matter." Despite Einstein's famous equation, $E = mc^2$, it is important to remember that although "dark energy" has a mass equivalent, it is *not* the same thing as "dark matter." Dark energy causes space (over the largest distances) to expand more and more quickly, whereas dark matter is gravitationally attractive and produces deceleration.

One popular set of dark-energy models, having hundreds of possibilities, is called "quintessence"—named after the Aristotelian "fifth essence" that complements Earth, air, fire, and water. In the quintessence models, the value of "Ω_X" (associated with the new energy "X") decreases with time in a manner similar to that of Ω_M, so it is not surprising that the two values are now roughly comparable. But these hypotheses have their own set of problems. Moreover, detailed observations (including those of Type Ia supernovae) are beginning to rule out entire classes of quintessence models.

It turns out that, regardless of the nature of the dark energy, its value of Ω plus that of matter add up to 1 (that is, $\Omega_X + \Omega_M = \Omega_{total} = 1$). This means that the Universe is *spatially flat* on large scales. Some theorists predicted that the Universe is flat (see our discussion in Chapter 19), but without dark energy. Perhaps the most natural model is one in which the Universe is formally closed (like the three-dimensional version of a sphere), but so incredibly large that it appears flat; this way, we don't need to deal with a formally infinite universe. On the other hand, some theorists have shown that an infinite universe is also a reasonable possibility, strange as it may sound.

ASIDE 18.9: The "big rip"

In one interesting form of quintessence, the amount of dark energy per unit volume actually grows with time. In the future, dark energy could thus become strong enough to rip apart clusters of galaxies, followed successively by the disruption of galaxies, planetary systems, stars, planets, and (regrettably) people. Eventually, even atoms would get ripped apart! Fortunately, this model has substantial theoretical difficulties, and current observations do not appear to strongly support it.

18.5f The Cosmic Jerk

Given how bizarre most of the above dark-energy hypotheses sound, perhaps we should question more carefully the data on which the accelerating-universe conclusion rests. Is it possible that the high-redshift supernovae appear fainter than expected not because of their excessively large distances, but for a different reason? For example, maybe long ago they were intrinsically less luminous than now. Keck spectra of high-redshift supernovae, however, look very similar to those of nearby supernovae (■ Fig. 18–31); there is no clear observational evidence for an intrinsic difference.

Or, perhaps there is dust between us and the high-redshift supernovae, making them look too faint. If this is normal dust, it would redden the light (that is, preferentially absorb and scatter blue photons)—but our procedure already accounts for this kind of dust by measurements of the reddening. If the dust grains are larger than normal, on the other hand, then the reddening is less (that is, blue light is not as preferentially extinguished), making such dust more difficult to detect. Nevertheless, some observable consequences are predicted, yet none has been seen.

If the dust grains are very large, then there would be essentially no difference in the amount by which blue and red light are extinguished, making such dust extremely difficult to detect directly. However, there are theoretical reasons against the formation of such large dust grains, and they would have adverse observable effects on other aspects of cosmology.

A convincing test of the accelerating universe hypothesis is provided by Type Ia supernovae at redshifts exceeding 1. If the unanticipated faintness measured for the $z = 0.3–0.7$ supernovae (discussed above) were caused by evolution of the intrinsic luminosity of supernovae or by the presence of dust, then we would expect supernovae at higher redshifts to experience a still larger effect, and thus to appear even fainter than expected. On the other hand, if the cosmological constant (or some similar effect) were causing the observed faintness of the $z = 0.3–0.7$ supernovae, then we would expect supernovae at higher redshifts to not appear as dim.

The reason for the latter effect is that at redshifts exceeding 1, corresponding to lookback times of at least 8 billion years, the Universe was so young that it should have been decelerating, even though it has been accelerating the past 4 or 5 billion years. Galaxies were closer together back then, so their gravitational attraction for each other was stronger. Moreover, if the density of the dark energy does not decrease very much as the Universe ages (for example, the density stays constant if dark energy is a property of space itself), then its cumulative effect was smaller when galaxies were closer together. For both reasons, attractive gravity should have dominated over antigravity, thereby producing deceleration of the expansion rate when the Universe was young.

In 2004, a team led by Adam Riess (Space Telescope Science Institute), of which one of the authors (A.F.) is a member, announced measurements of about 10 supernovae with redshifts close to or exceeding 1 (■ Fig. 18–32). They were brighter than would

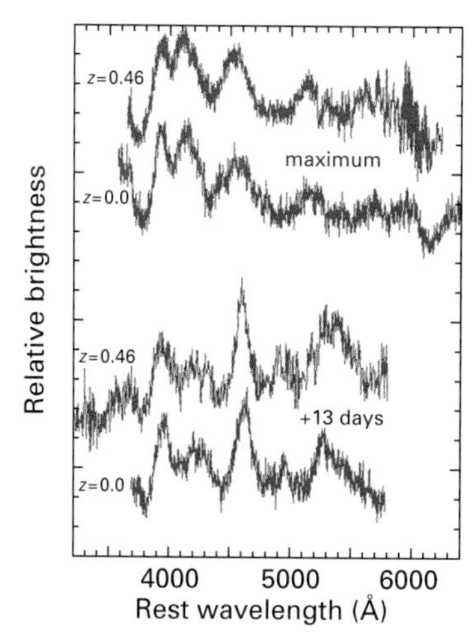

■ **FIGURE 18–31** Spectrum of a Type Ia supernova at redshift $z = 0.46$, compared with that of a nearby (redshift about 0) example, at two different times: *(top)* near peak brightness, and *(bottom)* 13 days after peak brightness. There are no clear differences, suggesting that the two objects are physically very similar. (Alex Filippenko (UC Berkeley), Adam Riess (STScI), and the High-z Supernova Search Team)

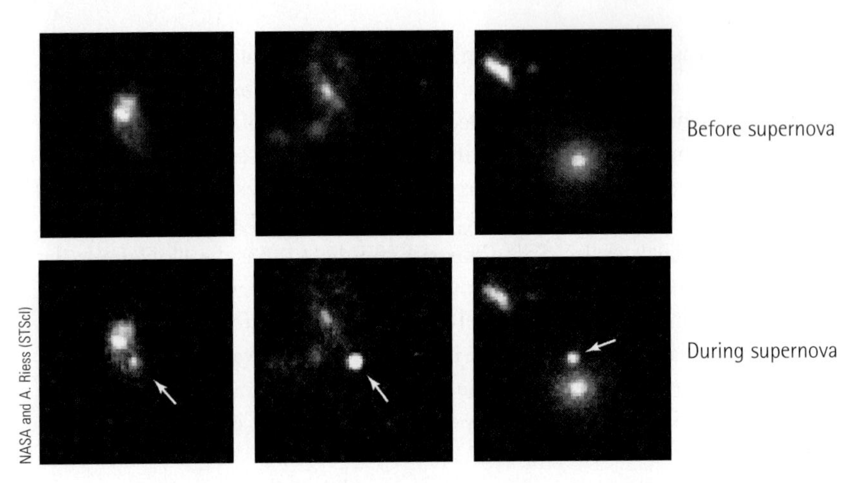

■ **FIGURE 18–32** Three extremely distant Type Ia supernovae, at redshifts close to or exceeding 1, imaged with the Advanced Camera for Surveys on the Hubble Space Telescope. These stars exploded when the Universe was roughly half its present age. Analysis of the data suggests that the expansion of the Universe was decelerating for its first 9 billion years, but then switched to acceleration about 4 or 5 billion years ago.

have been expected if luminosity evolution or dust had affected the results for the previous, redshift 0.3–0.7 supernovae. Instead, the data show that for roughly the first 9 billion years of its existence, the Universe was decelerating. Then, 4 or 5 billion years ago, the expansion rate began to accelerate.

Mathematically, the rate of change of acceleration is called "jerk." (So, to list the relevant terminology, the rate of change of position is called velocity; the rate of change of velocity is called acceleration; and the rate of change of acceleration is called jerk.) A nonzero jerk clearly occurred when the expansion of the Universe went from a state of deceleration to acceleration about 4 to 5 billion years ago. So, it can be said that the Universe experienced a "cosmic jerk" at that time!

Through late-2005, no known effect other than distance has been found to account for the measured brightness of supernovae as a function of redshift. Although we should remain vigilant for subtle problems that could invalidate the conclusions based on supernovae, for the time being they seem reliable.

Moreover, completely independent techniques that give the same results have increased our confidence in the conclusion that the expansion rate of the Universe has been accelerating the past 4–5 billion years. The high-redshift Hubble diagrams constructed from observations of clusters of galaxies and gamma-ray bursts, for example, are consistent with those obtained from Type Ia supernovae, though one should view these new Hubble diagrams with caution because they depend on certain unproven assumptions.

The most convincing support for the existence of dark energy and the accelerating Universe comes from the cosmic background radiation (discussed in Chapter 19), the "afterglow" of the big bang. We will see that the data imply a flat universe, which means that $\Omega_{total} = 1$. Yet, other compelling observations yield $\Omega_M = 0.3$ (that is, the contribution of luminous and dark matter is about 30% of the total density, averaged over large scales). The difference, 0.7, is consistent with the contribution of dark energy suggested by the measurements of Type Ia supernovae together with the large-scale distribution of matter (galaxies, clusters of galaxies) in the Universe.

Detailed studies of the cosmic background radiation (Chapter 19) even confirm a subtle effect expected from dark energy: superclusters of galaxies, being only loosely bound by gravity, expand more quickly than would have been the case without dark energy. We can tell that the superclusters expand this way because cosmic background photons gain some energy from gravity (that is, they become blueshifted) as they travel toward the center of a supercluster, but they lose less energy from gravity (becoming redshifted, but by a smaller amount) as they travel out from the center of the now more extended cluster. The blueshift slightly exceeds the redshift because of the expansion of the supercluster during the time taken for the photons to traverse it. Astronomers recently found a small net blueshift of cosmic background photons in the directions of known superclusters, thus detecting their extra expansion, propelled by dark energy.

18.6 The Future of the Universe

It seems that no matter what kind of universe we live in, the end is somewhat depressing. If the Universe is closed and the "big crunch" occurs in the reasonably near future (say, within 100 billion or a trillion years), some stars will be present throughout much of the time. During the final collapse, however, all matter and energy will be squeezed into a tiny volume, at blazingly high temperatures (since compressed gases heat up). If a rebirth occurs (but there is no good reason to suggest this), it will not contain any traces of complexity from the former universe.

If, on the other hand, the Universe is open (or critical) and expands forever, or if dark energy exists (as seems to be the case) and continues to stretch space in the future, a number of interesting physical stages will be encountered—though the end result will still be dreary. Freeman Dyson (at the Institute for Advanced Study in Princeton) was among the first to discuss these stages, which we will now consider, in turn. The specific timescales mentioned here were calculated in 1998 by Fred Adams (at the University of Michigan) and Greg McLaughlin (now at the University of California, Santa Cruz), who assumed a universe with low matter density and zero dark energy (since its existence was still quite controversial at the time). However, the overall picture remains similar even if the Universe is accelerating and continues to do so.

We now live in the **stelliferous era**—there are lots of stars! The Sun will become a white dwarf in about 6 or 7 billion years. Current M-type main-sequence stars will die as white dwarfs by the time the Universe is about 10^{13} years old. Gas supplies will be exhausted by $t \approx 10^{14}$ years, and the last M-type stars will become white dwarfs shortly thereafter (that is, within another 10^{13} years). L-type main-sequence stars won't last too much longer. Certainly by $t \approx 10^{15}$ years, normal stars will be gone.

The Universe will then enter the **degenerate era** filled with cold brown dwarfs, old white dwarfs ("black dwarfs," since they will be so cold and dim), and neutron stars. There will also be black holes and planets. Most stars and planets will be ejected from galaxies by $t \approx 10^{20}$ years. Black holes will swallow a majority of the remaining objects in galaxies by $t \approx 10^{30}$ years. All remaining objects (that is, outside of galaxies) except black holes will disintegrate by $t \approx 10^{38}$ years, due to proton decay. The timescale for this is very uncertain since the lifetime of a proton has not yet been measured.

Next, the Universe will enter the **black-hole era:** The only discrete objects will be black holes. Stellar-mass black holes will evaporate by $t \approx 10^{65}$ years, because of the Hawking quantum process (see Chapter 14). Supermassive black holes having a mass of a million Suns (such as those commonly thought to be in the centers of today's galaxies) will evaporate by $t \approx 10^{83}$ years. The largest galaxy-mass black holes will evaporate by $t \approx 10^{100}$ years.

Finally, the Universe will enter the **dark era:** There will be only low-energy photons (with a characteristic temperature of nearly absolute zero), neutrinos, and some elementary particles that did not find a partner to annihilate. The electron–positron pairs may form slightly bound "positronium" atoms (■ Fig. 18–33) millions of light-years in size. (They will be disturbed very little, because of the very low space density of objects.) The positronium atoms will decay (combine to form photons) by $t \approx 10^{110}$ years.

If the Universe is open or critical, it seems unlikely that "life" of any sort will be possible beyond $t \approx 10^{14}$ to 10^{15} years, except perhaps in rare cases (such as when brown dwarfs collide and temporarily form a normal star). Certainly beyond $t \approx 10^{20}$ years, the prospects for life appear grim. The Universe will be very cold, and all physical processes (interactions) will be very slow.

On the other hand, Freeman Dyson and others have pointed out that the available timescale for interactions will become very long. Hence, perhaps something akin to life will be possible. This is the "Copernican time principle"—we do not live at a special time, the only one that permits life. Instead, life of unknown forms might be possible far into the future.

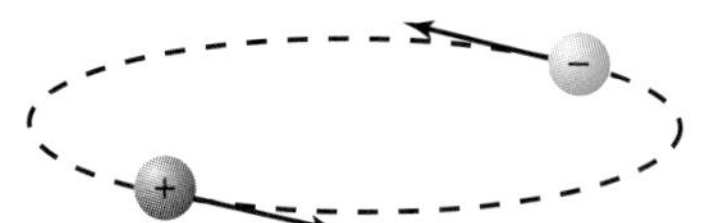

■ **FIGURE 18–33** Schematic of a "positronium atom": an electron bound to an anti-electron (positron), rather than to a proton as in a normal hydrogen atom. Weakly bound pairs of this type, millions of light-years in size, might form during the "dark era" of the Universe, when space is nearly empty of matter.

However, the situation is much bleaker if the Universe is accelerating and continues to do so: It will become exceedingly cold and nearly empty much faster. Within a hundred billion (10^{11}) years, there won't even be any galaxies visible in the sky, because they will have accelerated away to invisible regions of the Universe (too far away to be traversed by light).

In Chapter 19 we will see that other universes might be born from within our Universe, or in other regions of a larger "hyperspace." Thus, in some form the "universe" (more correctly, the "multiverse") might continue to exist, and to support life, essentially forever. Such ideas are intriguing and fun, but perhaps they do not currently belong to the realm of science, for we don't know any observational or experimental ways to test them. Indeed, they are very speculative, much more so than most of the other ideas described in this book.

CONCEPT REVIEW

Cosmology is the study of the Universe on its grandest scale—its birth, evolution, and ultimate fate (Introductory paragraph). **Olbers's paradox**—the darkness of the night sky—has profound implications, and is resolved primarily by the finite age of the Universe (Section 18.1).

Historically, the first piece of evidence for this definite beginning, the **big bang,** was Edwin Hubble's discovery of the expansion of the Universe (Section 18.2a): Most galaxies are moving away from us, and the speed of any galaxy, determined from the **redshift** of its spectrum, is proportional to its distance. The plot of redshift (or recession speed) vs. distance is the **Hubble diagram,** the proportionality is known as **Hubble's law,** and the constant of proportionality is called **Hubble's constant,** H_0. The expansion of the Universe has no definite center within any physically accessible dimensions (Section 18.2b). Note that *space itself* expands; galaxies are not flying apart from each other through a preexisting space (Section 18.2c).

There are at least two ways in which to determine the age of the Universe (Section 18.3a). First, the Universe must be at least as old as the oldest objects within it. Second, the age of the Universe is the elapsed time since the big bang, when all the material that formed our Galaxy and any other was essentially in the same place. The **Hubble time** is the expansion age of the Universe under the assumption that gravity has not been slowing down the expansion; its value depends on Hubble's constant. If the true expansion age of the Universe is only two-thirds of the Hubble time, as expected by some theorists, then its value is smaller than the ages of the oldest globular star clusters, leading to an age crisis.

The quest for an accurate measurement of Hubble's constant has been long and arduous, with many conflicting results (Section 18.3b). Generally, astronomers have used the inverse-square law of light to determine the distances of galaxies from the luminosities and measured brightnesses of certain kinds of stars within them. Cepheid variables have played a very important role in this regard. A major project of the Hubble Space Telescope was to measure the distances of galaxies with Cepheid variables, and hence to determine the Hubble constant (Section 18.3c). One difficulty is that the recession speeds of relatively nearby galaxies can deviate from the **Hubble flow,** the assumed smooth expansion of the Universe (Section 18.3d). Such **peculiar motions** are produced by the gravitational tug of the Local Group and other mass concentrations. Nevertheless, several techniques were combined in the late 1990s and early 2000s to achieve a convincing value for Hubble's constant, $H_0 = 71$ km/sec/Mpc (Section 18.3e).

Mathematically, studies of the expansion history and overall geometry of the Universe are conducted with Einstein's general theory of relativity (Section 18.4a). To simplify the equations and achieve reasonable progress, we assume the **cosmological principle:** On the largest size scales, the Universe is **homogeneous** (has the same average density everywhere) and **isotropic** (looks the same in all directions). Observationally, this assumption appears to be valid. In addition, we initially assume that Einstein's **cosmological constant** (Λ, Lambda), or any other source of long-range repulsion, is zero (Section 18.4b). However, as we discuss later in the chapter, this constant (or something that behaves like it) actually appears to have a nonzero value.

The result is that the ultimate fate of the Universe depends on the ratio of the average matter density of the Universe to a specific **critical density** (Section 18.4c). If the average density exceeds the critical density, then this ratio, known as Ω_M, is greater than 1 and gravity will eventually cause the Universe to collapse to a fiery death informally called the "big crunch." On the other hand, if the average density is low, such that Ω_M is less than or equal to 1, the Universe will expand forever, though more and more slowly with time; indeed, in the case Ω_M is equal to 1, the expansion will halt as time approaches infinity.

The value of Ω_M also determines the overall geometry of the Universe (Section 18.4c). If the average density equals the critical density, we live in a **flat (critical) universe** whose volume is formally infinite, and whose age is two-thirds ($2/3$) of the Hubble time. A **closed universe** has $\Omega_M > 1$; it has positive curvature and finite volume, resembling a sphere (technically, a "hypersphere"). Its age is less than $2/3$ of the Hubble time. The last alternative, an **open uni-**

verse, occurs if $\Omega_M < 1$; it has negative curvature ("hyperbolic"), and its volume is formally infinite. Its age is $^2/_3$ to 1 times the Hubble time. Two-dimensional analogues for such flat, positively curved, and negatively curved universes are an infinite sheet of paper, a sphere, and an infinite horse's saddle or potato chip (Section 18.4d). We can attempt to predict the ultimate fate of the Universe from a number of methods (Section 18.4e), all of which are difficult and have substantial uncertainties (Section 18.4f).

One can determine the expansion history of the Universe by measuring the distance and redshift of objects at a very wide range of distances (Section 18.5a). The most successful recent technique for determining accurate distances at high redshifts has been to measure the apparent brightness of Type Ia (white-dwarf) supernovae (Section 18.5b). The result is that the Universe appears to be expanding faster now than in the past: We live in an **accelerating universe** (Section 18.5c)! This observation resolves the age crisis of the Universe and suggests (but not convincingly) that the Universe will expand forever.

Apparently, a long-range "antigravity" effect exists in the Universe, like the cosmological constant that Einstein had previously postulated with a specific value to obtain a static universe (Section 18.5d). Physically, the cosmological constant may correspond to a vacuum energy consisting of **quantum fluctuations,** the spontaneous formation (followed by rapid destruction) of virtual pairs of particles and antiparticles. However, there are many alternatives to the cosmological constant; we don't know the nature of this **dark energy** that fills the Universe (Section 18.5e). Dark energy is ubiquitous, and it constitutes about 70% of the Universe. All together, there is enough luminous matter, dark matter, and dark energy to make the Universe spatially flat (Euclidean) over the largest scales.

The above conclusions are so astonishing that alternative explanations have been sought, but they have not been compelling (Section 18.5f). One recent triumph is that measurements of supernovae from a time when the Universe was half its present age show that, as expected, the expansion rate was initially decelerating. The change from deceleration to acceleration, a "cosmic jerk," occurred about 4 to 5 billion years ago. Moreover, there are now additional observations, independent of supernovae, which imply the presence of dark energy and cosmic acceleration.

If the Universe lasts forever, many interesting physical processes will occur (Section 18.6). We now live in the **stelliferous** era, filled with stars, but eventually all normal stars will burn out, and the Universe will enter the **degenerate era,** dominated by objects like white dwarfs and neutron stars. After a very long time, these degenerate objects will disintegrate, leaving the Universe in the **black-hole era.** But even black holes decay due to the Hawking evaporation process after exceedingly long times; the Universe will enter its last stage, the **dark era,** as a cold, nearly empty space consisting of extremely low-energy photons and a few elementary particles.

QUESTIONS

1. Describe Olbers's paradox—the darkness of the night sky—and several possible resolutions.
2. Summarize the first observational evidence found for the expansion of the Universe and a possible beginning of time.
3. Why does the recession of galaxies not necessarily imply that the Milky Way Galaxy is at the center of the Universe?
4. Explain how the effective center of expansion can be in an unobservable spatial dimension.
5. **(a)** What do we mean when we say that the Universe is expanding? **(b)** Are galaxies moving through some preexisting space? **(c)** Do humans, planets, stars, and galaxies themselves expand?
6. What are two ways of estimating the age of the Universe?
7. Describe how the current value of Hubble's constant is measured.
8. Why can relatively nearby galaxies give erroneous values for Hubble's constant?
9. †Calculate the Hubble time (that is, the expansion age of the Universe assuming no deceleration) if Hubble's constant is 71 km/sec/Mpc.
10. Suppose we assume that the recession speed of a given galaxy never changes with time. Why is the derived age of the Universe likely to be an overestimate of its true age?
11. If the expansion rate of the Universe were increasing (rather than decreasing) with time, would the true expansion age be less than or greater than that derived by assuming a constant expansion rate?
12. What is the "age crisis" that rocked cosmology until the late 1990s? Discuss two developments that helped alleviate this crisis.
13. Why are exploding white dwarf stars so useful for measuring the distances of galaxies?
14. State the cosmological principle. Why is it a reasonable assumption?
15. Define what we mean by the cosmological constant, Λ.
16. Summarize the possible types of overall geometry for a homogeneous, isotropic universe with a cosmological constant (or dark energy) equal to zero.
17. Describe how the geometry of the Universe is intimately connected to the ultimate fate of the Universe, assuming the cosmological constant (or dark energy) is zero.
18. Explain what we mean by the critical density of the Universe.
19. Discuss why Ω_M, the ratio of the average matter density to the critical density of the Universe, is such an important parameter.

20. **(a)** Summarize different methods for determining the value of Ω_M. **(b)** What are the difficulties in actually implementing these methods?
21. Does the average matter density appear to be large enough to close the Universe?
22. Discuss the main conclusion of the distant supernova studies, and its implications.
23. Explain how Einstein's cosmological constant is relevant to the conclusion in question 22.
24. What do we mean by "dark energy"? How might it affect the expansion of the Universe?
25. If the expansion of the Universe is currently accelerating, can we conclude that the Universe will necessarily expand forever?
26. What is the overall shape of the Universe, as implied by the total amount of luminous matter, dark matter, and dark energy?
27. Why might high-redshift supernovae appear fainter than expected, other than an acceleration of the expansion rate of the Universe?
28. **(a)** Has the expansion rate of the Universe always been accelerating? **(b)** What could cause a change from deceleration to acceleration?
29. What are some of the processes that will eventually occur if the Universe expands forever?
30. **True or false?** The curvature of three-dimensional space cannot be detected because we live within the three spatial dimensions.
31. **True or false?** Current observational evidence suggests that the Universe will eventually collapse; the expansion rate is decelerating rapidly.
32. **True or false?** Most astronomers believe that "Olbers's paradox," the darkness of the night sky, is resolved by the fact that the Universe has a finite age.
33. **True or false?** According to the "cosmological principle," the Universe is homogeneous and isotropic on the largest size scales, but its properties can change with time.
34. **True or false?** According to Einstein's equation, $E = mc^2$, dark energy and dark matter are essentially the same thing.
35. **Multiple choice:** Which one of the following statements about Hubble's constant, H_0, is *false*? **(a)** Hubble's constant changes with time. **(b)** The value of Hubble's constant is best determined with the very nearest galaxies, whose distances can be measured accurately. **(c)** Hubble's constant is believed to be constant throughout the Universe at a given time. **(d)** If the expansion of the Universe has always been constant (neither speeding up nor slowing down), then the age of the Universe is $1/H_0$. **(e)** H_0 is the constant of proportionality between the observed recession speeds (v) and distances (d) of galaxies.
36. **Multiple choice:** Which one of the following statements about the big bang theory is *false*? **(a)** The redshifts of distant galaxies are a consequence of the wavelength of light stretching during its journey to us. **(b)** All, or almost all, clusters of galaxies move away from each other. **(c)** Everything used to be much closer together and hotter than at the present time. **(d)** The matter density of the Universe is decreasing with time. **(e)** There is a unique center within the Universe that coincides with where the Big Bang happened.
37. **Multiple choice:** Observations of what kind of objects were the first to strongly imply that the expansion of the Universe is currently accelerating? **(a)** The cosmic background radiation left over from the big bang. **(b)** Supernovae. **(c)** Gamma-ray bursts. **(d)** Clusters of galaxies. **(e)** Gravitational lenses.
38. **Multiple choice:** Which one of the following distributions *is* isotropic but is *not* homogeneous? **(a)** The distribution of stars in a globular cluster, as measured from its center. **(b)** The distribution of globular clusters in our Galaxy, as measured from Earth. **(c)** The distribution of stars in our Galaxy, as measured from Earth. **(d)** The distribution of galaxy clusters in the Universe averaged over large scales, as measured from Earth. **(e)** The distribution of galaxy clusters in the Universe averaged over large scales, as measured from anywhere.
39. **Multiple choice:** Which one of the following is *not* a possible homogeneous, isotropic, expanding universe according to Einstein's general theory of relativity? (Assume there is no dark energy.) **(a)** A "flat" universe in which the laws of Euclidean geometry are satisfied. **(b)** A "closed" universe in which there is no such thing as parallel lines. **(c)** A "closed" universe that will expand forever. **(d)** An "open" universe whose volume is infinite, and has been infinite since the beginning of time. **(e)** An "open" universe in which the volume of a large sphere of radius R is greater than $(4/3)\pi R^3$.
40. **Fill in the blank:** Two types of stars that have played central roles as cosmological yardsticks are ______ variables and ______ supernovae.
41. **Fill in the blank:** A time far in the future of the Universe, when the density of photons and elementary particles will be very low, is known as the ______.
42. **Fill in the blank:** The expansion of the Universe appears to be accelerating, driven by the repulsive effect of the cosmological constant or, more generally, ______.
43. **Fill in the blank:** A universe with ______ curvature is said to be "hyperbolic," shaped somewhat like a potato chip.
44. †**Fill in the blank:** In a universe where the Hubble constant is 50 Gyr^{-1}, and in which $\Omega_M = 0$ and there is no dark energy, you can determine that the universe has an age of ______.

†This question requires a numerical solution.

TOPICS FOR DISCUSSION

1. Are you convinced that Olbers's paradox really does present a problem whose main possible solutions are profound? Can you think of any other solutions, besides those discussed here?
2. Can you visualize what our Universe might look like, if it's the three-dimensional analogue of the two-dimensional surface of a balloon?
3. Which prospect do you find more depressing: eternal expansion, or the fiery "big crunch"? Or perhaps you find neither one depressing—all things must die, and the Universe is no exception.
4. What do you think Einstein's reaction would be to the discovery that the cosmological constant might actually be nonzero, after he had rejected it (anecdotally as his "biggest blunder")?

MEDIA

Virtual Laboratories

- Cosmology and Cosmic Microwave Radiation
- Large-Scale Structure

AceAstronomy™ Log into AceAstronomy at **http://astronomy.brookscole.com/cosmos3** to access quizzes and animations that will help you assess your understanding of this chapter's topics.

Log into the Student Companion Web Site at **http://astronomy.brookscole.com/cosmos3** for more resources for this chapter including a list of common misconceptions, news and updates, flashcards, and more.

CHAPTER 19

In the Beginning

ORIGINS

The birth of the Universe itself is probably the most fundamental of all questions about origins. The behavior of the Universe at very early times greatly affected its contents, producing the matter of which we are made and the relative proportions of the light elements that dominate all others. It is possible that many universes exist, yet only a small fraction might have conditions suitable for life as we know it.

AIMS

1. Learn about radio radiation that now uniformly permeates the Universe and provides compelling evidence for the big bang, ruling out a competing theory (Sections 19.1 and 19.2).
2. Consider tiny ripples in the cosmic microwave background radiation, and how they tell us many properties of the Universe (Section 19.3).
3. Discuss what happened during the first few minutes after the birth of the Universe (Section 19.4).
4. See how a recently proposed modification of the original big-bang theory in the first wink of the Universe's existence explains some otherwise puzzling observations (Section 19.5).
5. Consider the very speculative possibility of multiple universes, possibly having different properties (Sections 19.6 and 19.7).

By itself, the expansion of the Universe does not prove that there was a big bang; indeed, one could postulate that the Universe had no beginning in time and will have no end. The fatal blow to this "steady-state theory" was the discovery of a faint radio glow that pervades all of space and was produced when the Universe was very young. The existence of a nearly uniform amount of helium and deuterium (heavy hydrogen) throughout the Universe provides additional evidence for a very hot, dense phase in its early history.

But how did the Universe achieve a uniform temperature, and why is space so nearly flat? These troubling issues led to a magnificent but still unproven hypothesis: The early Universe apparently went through a stage of phenomenally rapid expansion, *doubling* its size many times in just a tiny fraction of a second. Such a process may have created the Universe from essentially nothing! And if it happened once, it might happen again; perhaps there exist multiple "universes" physically disconnected from ours.

We emphasize from the start that these last few ideas are quite speculative, much more so than the other material presented in this book. Nevertheless, based on what we now know, they are physically reasonable possibilities.

19.1 The Steady-State Theory

In the previous chapter, we considered the big-bang theory of the Universe, which is motivated in part by the observed recession of galaxies away from our own Milky Way Galaxy (■ Fig. 19–1). We saw that the matter density of the Universe dictates the history, as well as the future, of its expansion rate. Moreover, there is growing evidence for a cosmic "antigravity" effect over large scales that accelerates the expansion of the Universe, causing it to grow larger and less dense at an ever-increasing rate.

AceAstronomy™ The AceAstronomy icon throughout this text indicates an opportunity for you to test yourself on key concepts, and to explore animations and interactions of the AceAstronomy website at **http://astronomy.brookscole.com/cosmos3**

◄ A cluster of galaxies, with faint, blue arcs (a few of which are marked with arrows) representing gravitational lensing of still more-distant galaxies. This image from the Advanced Camera for Surveys on the Hubble Space Telescope shows many more lensed galaxies than were previously recorded in a single image of the cluster.

NASA, N. Benitez (JHU), T. Broadhurst (Racah Institute of Physics/The Hebrew U.), H. Ford (JHU), M. Clampin (STScI), G. Hartig (STScI), G. Illingworth (UCO/Lick Obs.), the ACS Science Team, and ESA

■ **FIGURE 19–1** A Hubble Space Telescope image of galaxies in the direction of the Coma Cluster of galaxies. The smaller, fainter galaxies are generally more distant than the brighter, larger galaxies. Essentially all galaxies are receding from us, with the distant galaxies moving away faster than the nearby ones.

Hubble Space Telescope WFPC Team and NASA, courtesy of William A. Baum, U. Washington

Clemson U. and Donald D. Clayton (Clemson U.)

■ **FIGURE 19–2** Fred Hoyle, one of the main advocates of the steady-state theory of the Universe. Though now known to be incorrect, this theory helped accelerate progress in cosmology because it stimulated new observational and theoretical advances.

Although the expansion of the Universe suggests that there was a definite beginning of time when matter was in a very compressed and hot state, this is not the only logically consistent possibility. Here we consider a reasonable alternative that turned out to be incorrect, but that nevertheless served science well because it forced cosmologists to question critically their assumptions and conclusions.

In 1948, Fred Hoyle of Cambridge University (■ Fig. 19–2) and two of his associates (Hermann Bondi and Thomas Gold) proposed the **steady-state theory** as an alternative to the hot big bang. It is based on a modification of the cosmological principle called the **perfect cosmological principle:** The Universe is homogeneous and isotropic on large scales, and *its average properties never change with time.* Therefore, there was no well-defined beginning (the Universe is infinitely old), and there will be no end; the Universe is simply expanding throughout all time. To keep the average density constant as the Universe expands, new matter must be created continuously, and new galaxies form out of this material.

We might argue against the steady-state theory on the grounds that it requires something to be produced from nothing, which is thought to be impossible over long timescales. That is, we believe in the law of "conservation of energy"—the combination of mass plus energy is neither created nor destroyed.

But this objection turns out to be observationally weak. Only one hydrogen atom must be created per cubic meter, per billion years, to satisfy the requirements of the steady-state theory. This rate is well below the current detection limit of laboratory experiments. In other words, we have not yet verified the law of conservation of energy to this degree of accuracy. Moreover, the original big-bang theory postulates that the Universe was created out of nothing in a single instant ($t = 0$), so it too appears to violate conservation of energy (although we will later see that no significant violation might occur).

Despite its aesthetic appeal to some astronomers, the steady-state theory gradually lost support as observational evidence showed that the Universe has a finite age (about 14 billion years, not much greater than the age of the oldest globular clusters) and that its properties change with time. One of the first arguments for evolution of the Universe was that radio galaxies seem to be more numerous and luminous at large distances.

An even more compelling case was made with quasars (see Chapter 17): They are clearly denizens of the distant past; all have high redshifts, so we see them at large lookback times. There are no nearby quasars, seen as they are in the present-day Universe, but only lower-luminosity active galaxies. In recent years, Hubble Space Telescope images of distant galaxies have shown that they differ from nearby galaxies, providing additional evidence for the evolution of the Universe.

19.2 The Cosmic Microwave Radiation

The fatal blow to the steady-state theory, however, was the discovery of the **cosmic microwave radiation**—a faint afterglow from the Universe's hot past. The steady-state theory has no explanation for it, but big-bang models require it, as we will now discuss.

19.2a A Faint Hiss from All Directions

In 1964 through 1965, two young researchers (Arno A. Penzias and Robert W. Wilson) were observing with a 20-foot horn-shaped radio antenna owned by Bell Labs near Holmdel, New Jersey (■ Fig. 19–3). The antenna had previously been used to detect radio signals from communications satellites, but Penzias and Wilson were allowed to use it for astronomical observations. They initially wanted to accurately measure the radio brightness of Cas A, a prominent supernova remnant that is used to calibrate radio data. Then they wanted to map the 21-cm radio emission in the Milky Way Galaxy.

For the most accurate measurements, they needed to make their microwave antenna as sensitive as possible. After removing all possible sources of noise, they found a very faint but persistent hiss that seemed to defy explanation. It was independent of location in the sky, and also independent of the time of day. If produced by a "black body" (Chapter 2), it corresponded to a temperature of about 3 K, which is just 3°C above absolute zero and is equal to −270°C. They were not aware that it might be the afterglow of the big bang!

Meanwhile, at nearby Princeton University in New Jersey, a team led by Robert Dicke was building a radio telescope with which to detect this afterglow. They had predicted that it should exist if the Universe began in a very hot and compressed state.

Actually, this prediction had already been made in the 1940s, by the Russian astrophysicist George Gamow (who was then working in the United States), but the Princeton group was unaware of it. Gamow and his students, Robert Herman and especially Ralph Alpher, had based their conclusion on the idea that all of the chemical elements were created shortly after the big bang. Although incorrect in detail, this is correct in spirit; only the lightest elements were produced early in the Universe, as we shall see later. The Universe must have been very hot to do this, but expansion would have cooled it, so the radiation should now be that of a cold black body.

Another astronomer alerted the Princeton and Bell Labs teams to each other, and that's how Penzias and Wilson found out what they had discovered. The Princeton group made a measurement at another wavelength, and it agreed with the prediction of a black-body spectrum. Other, subsequent measurements were also consistent with black-body radiation from a gas at $T \approx 3$ K, a very low temperature indeed. In 1990, astronomers announced results from NASA's Cosmic Background Explorer (COBE, pronounced "koh′bee"; ■ Fig. 19–4): The spectrum was that of a perfect, cold black body (■ Fig. 19–5).

Bell Laboratories, Lucent Technologies

■ **FIGURE 19–3** Arno Penzias *(left)* and Robert W. Wilson *(right)* with their horn-shaped antenna in the background. After they removed all possible sources of noise (by fixing faulty connections and loose antenna joints, and by removing "sticky white contributions" from pigeons), a certain amount of radiation remained. It was the 3 K background radiation.

NASA COBE Science Team, courtesy of Nancy Boggess

■ **FIGURE 19–4** The COBE spacecraft (Cosmic Background Explorer), remade smaller to fit on a Delta rocket after the explosion of the space shuttle Challenger in 1986. COBE was launched in 1989 and sent back data for four years. It fantastically improved the accuracy of measurements of the cosmic background radiation.

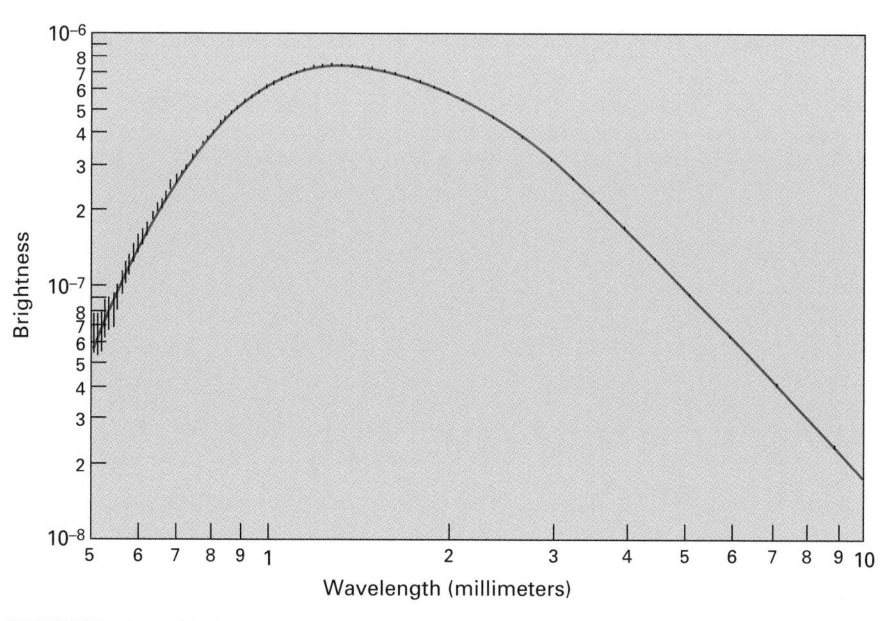

■ **FIGURE 19–5** This first spectrum based on COBE observations drew a standing ovation from astronomers when it was initially shown at an American Astronomical Society meeting. It represents astonishingly good evidence that the cosmic background radiation comes from a black body. The data points and their error bars *(black)* were fit precisely by a black-body curve *(red)* for 2.735 ± 0.06 K, a value that has since been slightly improved, based on the final data set, to 2.728 ± 0.004 K. Current error bars are thinner than the thickness of the curve drawn! (NASA COBE Science Team, special graphing courtesy of E. S. Cheng)

19.2b Origin of the Microwave Radiation

From where, exactly, does a photon of the cosmic microwave background radiation come? The early Universe was very hot and ionized: There were no bound atoms, but rather free electrons and atomic nuclei. Because photons easily scatter (bounce, or reflect) off of free electrons, they cannot travel in a straight line from one object to another, and instead jump randomly from one direction to another. So instead of being transparent, the Universe looked opaque, like thick fog. Even if there were discrete objects at this time, one would not be able to see them. The Universe was filled with photons, however, since it was hot—and they simply scattered around randomly in a sea of hot particles.

As the Universe expanded, it cooled. Eventually, about 400,000 years after the big bang, it reached a temperature of about 3000 K, and electrons were able to combine with protons to form neutral hydrogen atoms. The process is called "recombination"—even though in this case, electrons were combining with protons, forming neutral atoms for the *first* time in the history of the Universe.

Unlike free electrons, electrons bound in atoms are able to interact with photons of only certain specific energies, absorbing them (Chapter 2). They are not effective at scattering photons at other energies. Thus, at this time most photons stopped bouncing around, and became free to travel unhindered. The Universe therefore became transparent to almost all electromagnetic waves. Astronomers say that matter and radiation "decoupled" from each other at $t \approx$ 400,000 years.

As space expanded, the wavelengths of photons increased, just like that of a wavy line drawn on an expanding balloon, so they lost energy. They maintained the spectrum of a black body, but of lower and lower temperature (■ Fig. 19–6). Typical photons went from being optical/infrared to radio (microwave). The cosmic microwave radiation now corresponds to such a low temperature, about 3 K, because the Universe has expanded a great deal over its 14-billion-year life. Since it started off as radiation filling the entire Universe, it is easy to understand why it now comes equally from every direction (that is, it appears isotropic).

In essence, the cosmic background photons come from an opaque "wall" at redshift about 1000 (the ratio of 3000 K to 3 K). We cannot see electromagnetic radiation

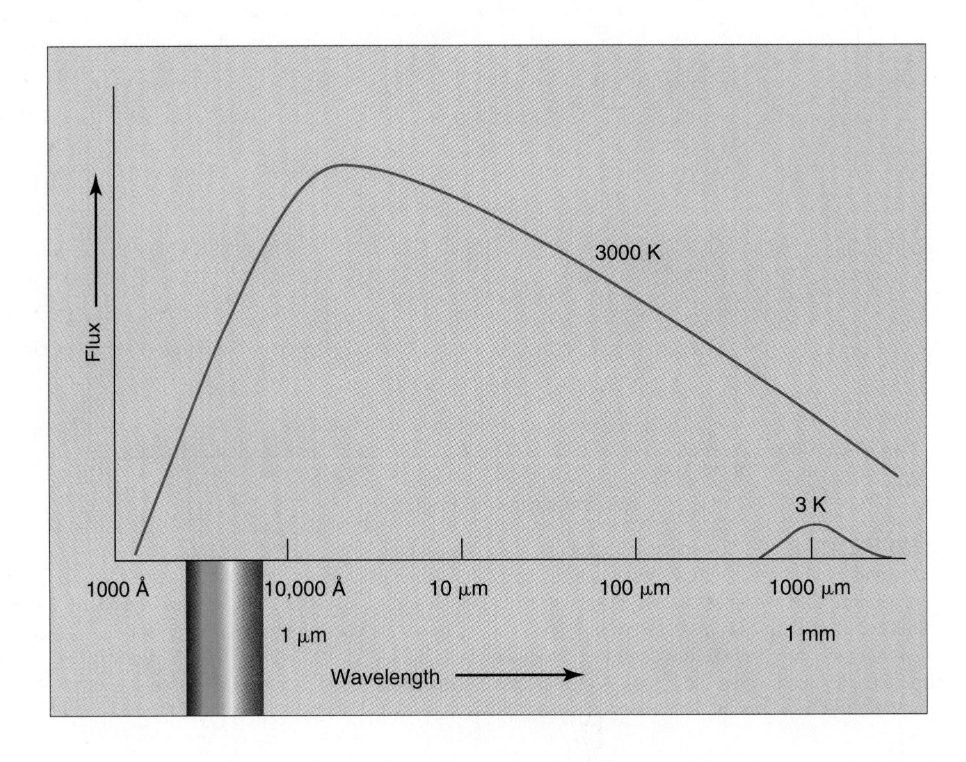

■ **FIGURE 19–6** The cosmic background radiation was set free at a temperature of 3000 K *(blue curve)* when the Universe became transparent at the time of recombination, about 400,000 years after the big bang. As the Universe expanded, the spectrum was transformed into that of a lower temperature gas *(red curve at lower right)*. The rainbow shows the wavelengths of visible light; the cosmic background radiation is detected near its peak at microwave wavelengths and not in visible light or other wavelengths far from its peak.

from beyond this wall because the Universe was opaque. This marks the "boundary" of the observable Universe: We cannot see electromagnetic radiation from times prior to about 400,000 years after the big bang. In principle, we could see neutrinos from $t < 400{,}000$ years, since the Universe was not opaque to them. But unfortunately, neutrinos are very difficult to detect.

The cosmic microwave radiation is left over from when the Universe was very hot, dense, and opaque. Being almost perfectly isotropic, it provides the best-known evidence in support of the cosmological principle that the Universe is homogeneous and isotropic. The steady-state theory has no reasonable explanation for its presence.

Penzias and Wilson shared the 1978 Nobel Prize in Physics for their important discovery: They had been very careful, and had not ignored the faint (almost nonexistent) hiss. Gamow could not receive the Nobel Prize, as he had already died, but it seems somewhat unfair that Alpher was omitted. (It can be shared among a maximum of three people.)

AceAstronomy™ Log into AceAstronomy and select this chapter to see the Active Figure called "Cosmic Microwave Background Radiation."

a 3.3 mm

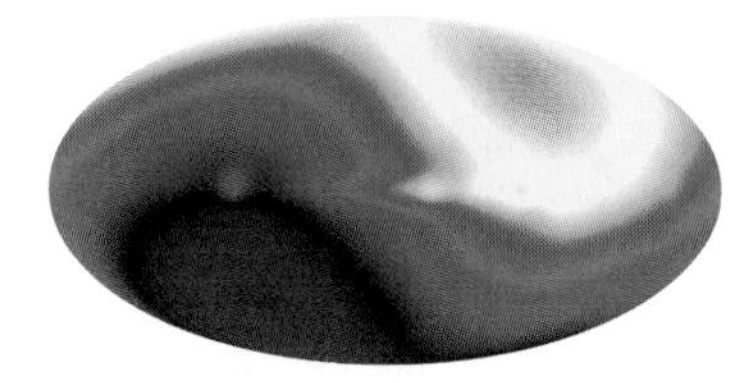

b 5.7 mm

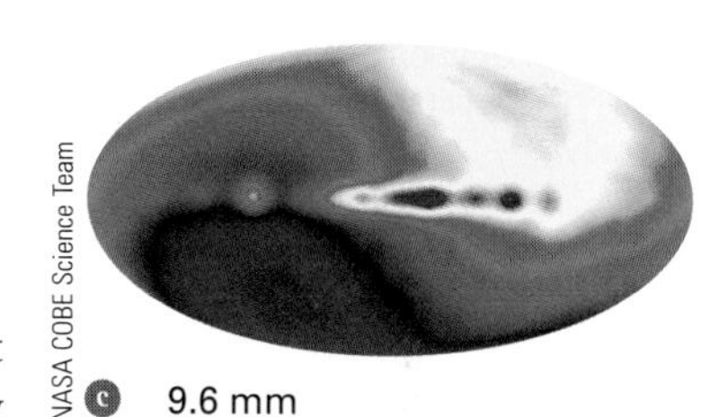

c 9.6 mm

NASA COBE Science Team

FIGURE 19–7 Maps of the sky from all four years of COBE data at radio wavelengths of a 3.3 mm, b 5.7 mm, and c 9.6 mm. At the longest wavelength, we see a discrete source in Cygnus *(green dot at middle left)* and a sign of the Galactic plane *(horizontal red at right)*. The rest of the signal is from the cosmic microwave radiation. The asymmetry from bottom left to top right is produced by the Sun's motion relative to the smoothly expanding Universe. The range shown is 63 millikelvins.

19.3 Deviations from Isotropy

The cosmic microwave background shows a slight deviation from isotropy of about two parts per thousand. It looks slightly hotter in one direction of the sky, and slightly cooler in the opposite direction. Such an "anisotropy" is caused by a combination of motions: the Sun's motion around the center of the Milky Way Galaxy, the Milky Way's motion within the Local Group, and, most important, the Local Group's motion relative to the "Hubble flow" (that is, relative to the smoothly expanding Universe).

These gravitational perturbations are all known. The sum of the motions produces a net motion in a particular direction of space, and the slight temperature anisotropy is attributed to the Doppler effect. The radiation coming from the direction of motion is slightly blueshifted ("hotter"), while that coming from the opposite direction is slightly redshifted ("cooler"). Maps made with COBE very clearly show this anisotropy (■ Fig. 19–7).

19.3a Ripples in the Cosmic Microwave Background

A major puzzle for many years was the apparent *absence* of small variations in the background radiation corresponding to the size scales of clusters and superclusters of galaxies. If these structures formed by the gravitational contraction of matter, there should have been slight ripples in the density of matter early in the Universe. Photons escaping from higher-density clumps would have a slightly different redshift than those coming from regions of average or below-average density.

There are several relevant processes, but perhaps the easiest to understand is that photons lose more energy when they come out of stronger gravitational fields. The different redshifts would translate into slight variations in the associated temperature of the radiation.

No such ripples were found until the early 1990s. Some specific models for the formation of large-scale structure were consequently eliminated from further consideration, because they predicted variations larger than the observed upper limits.

In 1992, a breakthrough was made with the COBE satellite: Tiny temperature variations (about one part per hundred thousand, or 30 microkelvins) were finally found, at least in a statistical sense. (Ripples were present, but no great significance could be given to any particular one.) After a few more years of data collection, individual variations became clearly visible (■ Fig. 19–8). The angular resolution of COBE was low, so even the smallest detected variations correspond to structures billions of light-years across.

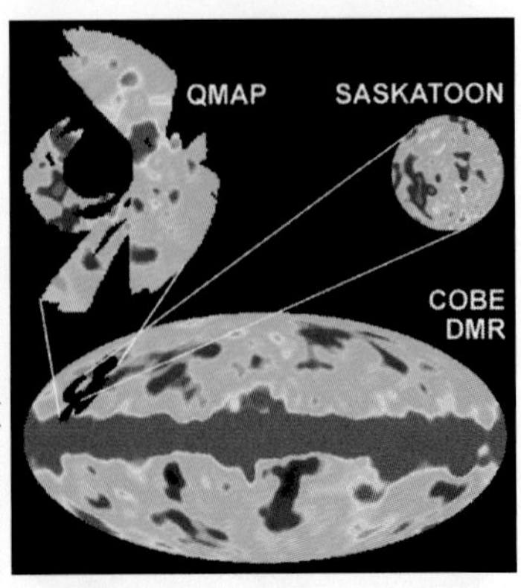

Updated in 2001 from Max Tegmark, Angelica de Oliveira-Costa, Marc Devlin, Barth Netterfield, Lyman Page, and Ed Wollack, in *Astrophys. J.* 474, L77–80, 1996

■ **FIGURE 19–8** A full-sky map of the variations in the 3 K cosmic background radiation, as analyzed from all four years of COBE data, after removal of the Sun's motion relative to the Universe. The temperature variations are typically only 30 microkelvins (0.001 per cent). The red band across the center of this map results from the microwave emission of the Milky Way Galaxy. The variations above and below the Milky Way are real fluctuations in the background radiation. The results are also shown for a balloon experiment known as QMAP and for a ground-based experiment carried out in Saskatoon, Saskatchewan, Canada. The latter observed only a small area around the north celestial pole but obtained higher angular resolution within that region.

Subsequently, several balloon-based experiments found similar variations, but on smaller angular scales (■ Fig. 19–9), corresponding to superclusters of galaxies. These are the "seeds" from which large-scale structure grew in the Universe. They are the imprints of minuscule ripples in the distribution of matter established shortly after the big bang. One of the missions was BOOMERANG: **B**alloon **O**bservations **o**f **M**illimetric **E**xtragalactic **R**adiation **an**d **G**eophysics. Their results and others showed that the Universe is flat (■ Fig. 19–10), as we will now discuss.

19.3b The Overall Geometry of the Universe

By 2001, several missions had shown that the angular size of temperature variations in the cosmic background radiation is typically about one degree (1°; ■ Fig. 19–11*a*). This statement means that the typical angular size of a ripple is around 1°, and it tells us something about the overall geometry of the Universe. In fact, the data are consistent with the presence of considerable amounts of dark matter and other energy in the Universe—enough to make the Universe spatially flat! (Looking at the data showed that the Universe is flat, but analysis of the sizes of the variations gives us much more detail and some explanation as to why and how.)

A spatially flat geometry means that Ω, the ratio of average total density (of luminous and dark matter, normal energy, and other energy) to the critical density, is 1 (see Chapter 18). This "other energy" may well be the "vacuum energy" or "dark energy" associated with the cosmological constant, whose nonzero value was first suggested by observations of distant supernovae (see Chapter 18). Galaxies and clusters of galaxies seem to have formed at peaks in the overall distribution of dark matter.

Results announced in 2003 show the variations at still better resolution (Fig. 19–11*b*). From the results, scientists infer an amount of dark energy matching that determined from supernova observations that led to the accelerating-universe conclusion.

Other observations show that the cosmic microwave background is polarized—that is, radiation coming from a given region of the sky has more of its electric field oscillating in one direction than in the perpendicular direction. Such polarization is typically produced when light scatters off of a surface. For example, light reflected from water or pavement is polarized, and polarizing sunglasses reduce the glare by block-

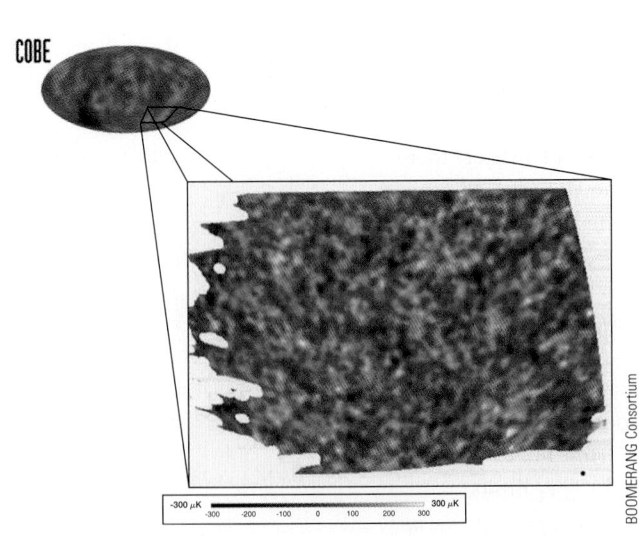

BOOMERANG Consortium

■ **FIGURE 19–9** A map of the tiny variations in the 3 K cosmic background radiation over part of the sky, based on data from the BOOMERANG balloon project. The full moon is shown *(as a black dot at lower right)* for comparison. Note that the temperature differences, as shown in the scale bar, vary from the orange spots to the black spots by only about 600 microkelvins–that is, 0.0006 kelvins.

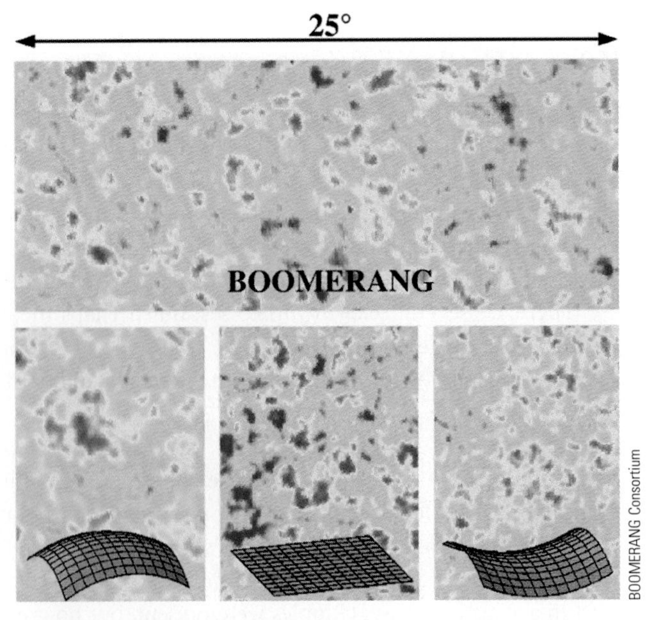

BOOMERANG Consortium

■ **FIGURE 19–10** The BOOMERANG mission's observed variations match the predictions of the flat model for the Universe *(bottom center)* rather than the open or closed models, for which simulations are shown at *bottom left* and *bottom right.*

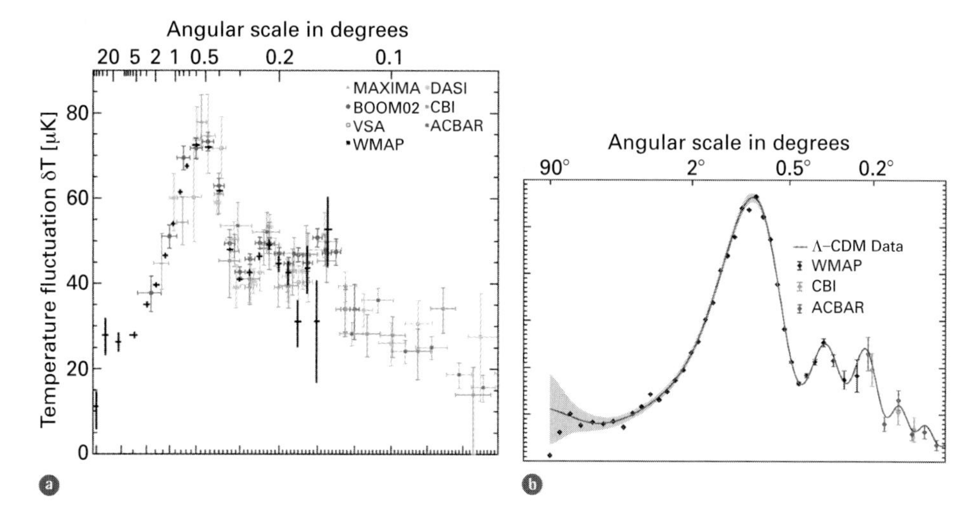

FIGURE 19–11 The relative number and amplitude of temperature variations in the sky versus their angular size. Though variations appear over a wide range of scales, the strongest ones appear to be about 1° in diameter, or twice the full moon. As of 2005, at least three peaks can be seen. The sizes and positions of the peaks in the graph indicate that the Universe is flat and allows the percentage of nuclear particles in the Universe to be calculated. The curves illustrate theoretical fits to the data shown with error bars. Do not confuse such diagrams of variation peaks with Planck curves, which graph intensity versus wavelength for a given temperature. (a) The experimental results, as of the beginning of 2004, showing a selection of the different telescopes devoted to the topic. (b) The experimental results based on the Wilkinson Microwave Anisotropy Probe (WMAP) observations and observations from the CBI and ACBAR telescopes, with a solid line drawn from the post-WMAP best model. (a: Max Tegmark, MIT; b: NASA's Wilkinson Microwave Anistropy Probe (WMAP) Science Team)

ing light having the dominant direction of oscillation. The observed polarization tells us that the cosmic background radiation is indeed coming from a surface of last scattering, a moment before recombination occurred and the Universe became transparent.

19.3c The Wilkinson Microwave Anisotropy Probe (WMAP)

A NASA spacecraft now aloft, the Wilkinson Microwave Anisotropy Probe (WMAP), is covering the whole sky at 30 times the resolution of COBE. It was launched in June 2001 into an orbit a million miles from us, on the side of the Earth opposite to the direction to the Sun. The results of WMAP's first year of data were released in February 2003. The probe is named after David Wilkinson, an important member of the COBE and WMAP teams and a member of the original Princeton team that was among the first to understand the cosmic microwave background.

WMAP's map of the entire sky (Fig. 19–12) is spectacularly detailed. It shows temperature variations measured in microkelvins—millionths of degrees. COBE also

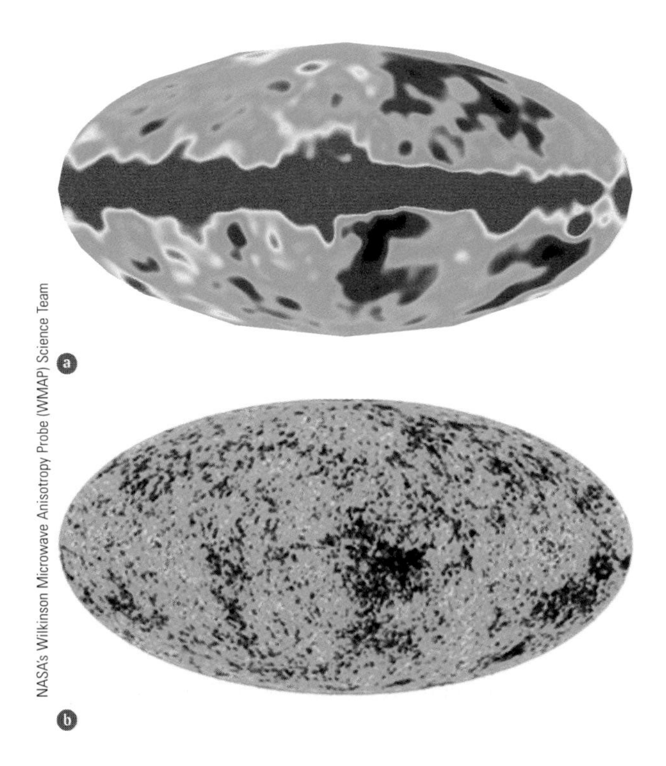

FIGURE 19–12 Temperature variations on the microkelvin level displayed on an all-sky map. (a) The calibration of all four years of COBE data. These variations, about 5 parts per million, show the seeds from which structure in the Universe evolved nearly 14 billion years ago. (b) The higher resolution provided from NASA's Wilkinson Microwave Anisotropy Probe. These data, based on the first year of observations, were released in early 2003. Many cosmological parameters were determined, including Hubble's constant and the percentages of ordinary matter, dark matter, and dark energy. The following three years of data, and eventually the European Space Agency's Planck Mission (to be launched in 2007), should do even better.

See the images opening Chapter 15 for WMAP images at three different microwave wavelengths, which were used to compile the image shown here.

had mapped the whole sky, but at lower resolution, and though several recent experiments had mapped at high resolution, they had covered only tiny portions of the sky. Thus the results from WMAP, taken at five wavelengths between 2.3 mm and 13 mm, represented a major advance.

Comparing the WMAP maps at the different wavelengths allowed the effects of the Milky Way to be subtracted. The results for a variety of cosmological parameters are so precise that they have greatly helped transform cosmology into a precision science.

From studying various features (especially the first peak) in the graph showing the distribution of the sizes of the variations, and making use of additional information from other astronomical studies, we now know the age of the Universe. It is 13.7 billion years, with a formal statistical uncertainty of only 0.2 billion years. (However, various effects tend to increase the uncertainty, perhaps to about 0.5 or even 0.7 billion years.) This value fits with our current understanding of the main-sequence turnoff in globular clusters of 12 billion years and the ages of the oldest white dwarfs in clusters of 12.7 billion years, both with uncertainties of about 1 billion years.

WMAP's data, in conjuction with other studies, show that Hubble's constant is 71 km/sec/Mpc, with an uncertainty of only 6 per cent. This value is in close agreement with the Hubble Key Project's value of 72 km/sec/Mpc. The long battle over the value of Hubble's constant seems finally to be over.

The data, when linked with observations from the Hubble Key Project, the observations of distant supernovae, or maps of the distribution of galaxies in the sky, show that the Universe is flat. The measurements are uncertain to a 66 per cent probability by only 1 per cent, and to a 95 per cent probability by only 3 per cent. The total amount of matter and energy in the Universe, Ω, is thus 1 to a high precision.

WMAP's coverage allows the distribution of the sizes of temperature variations to be measured much more accurately than had previously been done. The distribution of its peaks reveals other cosmological parameters.

In particular, the density of normal types of matter—the kinds of particles that you are made of (technically, "baryons")—comprises only 4.4 per cent of the total. And of that same total, only about 1 per cent is luminous matter, the type that we can detect. Hence, the stuff of which we are made is just a small minority of all that exists; we are like the debris of the Universe. The rest of the 4.4 per cent contributes to the dark matter that scientists have been detecting for some decades by its gravitational effects in the halos of galaxies and in clusters of galaxies. MACHOs are an example of this type of dark matter, but recent results suggest that much of it actually consists of tenuous, million-degree gas in the halos of galaxies.

WMAP shows, further, that another 23 per cent is other types of "cold dark matter." In this context, "cold" means that the matter is moving slowly, not close to the speed of light. We do not yet know what this matter consists of, but the "best bet" is exotic particles left over from the big bang.

We can tell that most of the dark matter cannot be "hot" dark (that is, moving at very high speeds) because if a substantial amount of hot dark matter had been present in the early Universe, star formation would have been suppressed at that time. The WMAP results, in contrast, imply that the first stars formed only about 200 million years after the big bang.

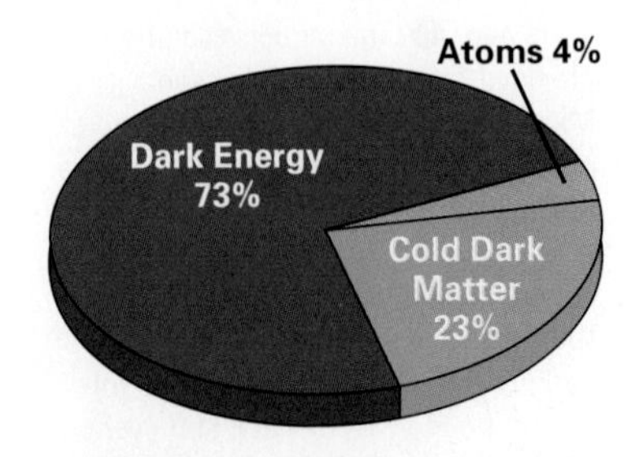

■ **FIGURE 19–13** Pie diagram showing the composition of the Universe. Normal matter (made of protons, neutrons, and electrons) constitutes only about 4% of the total. Cold (slowly moving) dark matter is 23%, and gravitationally repulsive dark energy is 73%.

Thus, normal matter and cold dark matter contribute about 0.27 to the density ratio of the Universe, Ω. Since the total value of Ω is 1.0, that leaves another 73 per cent as "dark energy" (■ Fig. 19–13), which we discussed in Chapter 18. We have even less understanding of dark energy than we do of cold dark matter. We appear to live in a Universe in which we can relate to only 4.4 per cent of its constituents, if even that! As John Bahcall of the Institute for Advanced Study summarized, "We are in an implausibly crazy Universe, but one whose characteristics we know."

The WMAP findings represent strong endorsement of the big-bang model of the Universe. They also endorse some versions of inflationary cosmology ("inflation" will be discussed later in this chapter), though other specific inflationary scenarios are now ruled out. Also, the "quintessence" model that had rivalled Einstein's cosmological constant as a contender for explaining the dark energy now seems somewhat less likely than it had previously.

Spectacular as these results are, they will improve still further as the remaining three years of data are collected and analyzed. (The announcement of findings from Years 2 and 3 is anticipated in 2006.) Yet another improvement should result from the European Space Agency's Planck spacecraft, currently scheduled for launch in 2007.

ASIDE 19.1: The fifth essence

"Quintessence," named after the Aristotelian fifth essence (in addition to earth, air, fire, and water), refers to possible new forces or fields that might be causing the expansion of the Universe to accelerate.

19.4 The Early Universe

The evolution of the Universe can be studied nearly back to the moment of creation, "$t = 0$," with an increasing amount of uncertainty as we go to progressively earlier times. Just how close to $t = 0$ we can get with reasonable accuracy is debated.

19.4a Going Back in Time

Almost all astronomers and physicists agree that we can go quite far back in time because the early Universe was relatively simple: It consisted of uniformly distributed particles and photons in thermal equilibrium (that is, at the same temperature), and we seem to know the few fundamental laws that governed their behavior. Compare this with a complex problem like Earth's weather, which is affected by factors such as local heating, ocean currents, cloud formation, continents, mountain ranges, and so on. It is easier to predict the properties of the early Universe than it is to predict the weather!

Certainly we can reach something like $t \approx 1$ sec, because conditions were similar to those in the central regions of stars, and the nuclear physics is well understood. We also have substantial confidence in our conclusions down to $t \approx 10^{-12}$ sec, since the corresponding temperatures and energy densities have been reproduced in high-energy particle accelerators.

It can also be argued that we can extrapolate all the way to about 10^{-35} sec, or perhaps somewhat earlier; once again, hot gases in equilibrium are relatively easy to understand. However, we do not yet know everything down to such small timescales; our physical theories are incomplete, and no conceivable particle accelerators can achieve temperatures that are high enough to test them. Thus, some of our current conclusions about this era are rather speculative. Turning this problem around, we can use celestial observations to test the physical theories: The Universe is our great "accelerator in the sky."

Although times such as 10^{-35} sec may seem ridiculously short, with no real "action," it is important to understand that *lots* of things could have happened. This is because the interaction timescales and distances were far smaller than they are today.

19.4b A Brief History of the Early Universe

Concerning times before 10^{-43} sec we know essentially nothing, except that the temperature exceeded about 10^{32} K. This is the **Planck time:** It is thought that time itself might be packaged in small units ("quantized") at about this interval, or at least it becomes unpredictable. Some physicists have the notion of "space–time foam," where packets of time and space themselves flit into and out of existence. We must develop a self-consistent quantum theory of gravity to better understand this era.

When the Universe was between 10^{-35} and 10^{-6} sec old, there was equilibrium among particles, antiparticles, and photons. Particle–antiparticle pairs annihilated each other and produced photons. Photons spontaneously formed particle–antiparticle pairs.

ASIDE 19.2: Artificial primordial soup

In 2000, physicists at CERN (the European Laboratory for Particle Physics near Geneva, Switzerland) announced that they may have reproduced a tiny amount of the "primordial soup" of free quarks and gluons by smashing nuclei of lead at very high energies, though this result is still controversial.

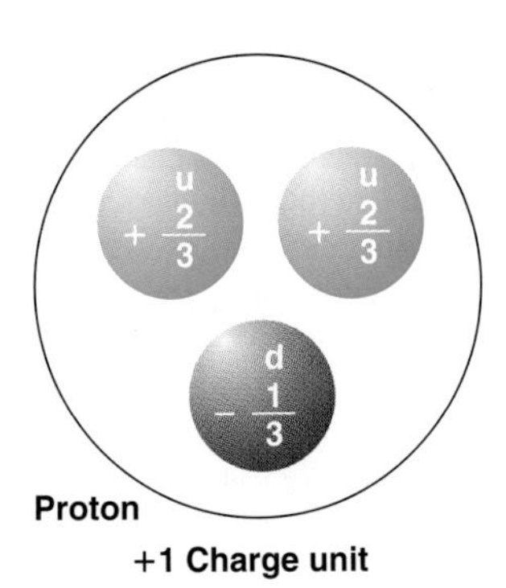

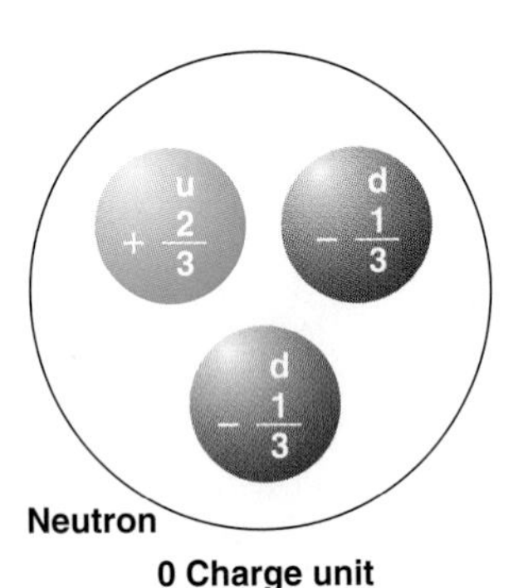

FIGURE 19–14 The most common quarks are *up (u)* and *down (d)*; ordinary matter is made of them. The up quark has an electric charge of $+\frac{2}{3}e$, and the down quark has a charge of $-\frac{1}{3}e$, where *e* is the unit of electric charge. This fractional charge is one of the unusual things about quarks; prior to their invention, it had been thought that all electric charges come in whole numbers. Note how the charge of the proton and of the neutron is the sum of the charges of their respective quarks.

Quarks, particles that are normally bound together by "gluons" to form protons and neutrons, were plentiful and unbound during most of this era.

Quarks come in six types or "flavors": up, down, strange, charmed, top (or truth), and bottom (or beauty). Each flavor also comes in three "colors": blue, green, and red. (These names are whimsical; they don't reflect the real "character" of each quark.) Normal matter consists of the "up" and "down" quarks, together with electrons and electron neutrinos (a type of neutrino associated with electrons). For example, a proton is two "up" quarks (each with a charge of $+\frac{2}{3}e$, where *e* is the unit of electric charge) and one "down" quark (with a charge of $-\frac{1}{3}e$). A neutron is one "up" quark and two "down" quarks (■ Fig. 19–14). Each quark has a corresponding antiquark.

At some stage, probably early in this era of the quark-gluon-photon mixture, a slight imbalance of matter (quarks) over antimatter (antiquarks) was formed. It amounts to about one part per billion. It is not known which specific reactions took place, but laboratory experiments have demonstrated the existence of a matter–antimatter asymmetry for some processes (■ Fig. 19–15). We owe our existence to this tiny asymmetry.

The annihilation of protons, neutrons, and their antiparticles occurred at $t \approx 10^{-6}$ sec, one microsecond, when the temperature was about 10^{13} K, leaving a sea of photons. The annihilation was not complete, due to the slight imbalance of matter over antimatter.

Then there was rapid conversion of protons and neutrons back and forth into each other (■ Fig. 19–16). However, since neutrons are slightly more massive than protons, they are harder to make, and the number of neutrons settled to only about one quarter of that of protons.

By the time the Universe was about 1 sec old, its temperature had dropped to about 10^{10} K (ten billion degrees), and the typical photon energies became too low to spontaneously produce electron–positron pairs. Electrons and positrons therefore annihilated each other for the last time, producing more photons. The remaining (excess) electrons balanced the positive charge of the protons. Moreover, at this time, neutrons began to decay into protons, electrons, and antineutrinos; it was not possible to rapidly replenish the supply of neutrons. This further increased the imbalance between protons and neutrons.

19.4c Primordial Nucleosynthesis

The temperature had dropped to only a billion degrees by an age of about 100 sec, allowing the formation of heavy isotopes of hydrogen, two isotopes of helium, and a little bit of lithium. Before this time, collisions between particles would sometimes produce bound states, but subsequent collisions destroyed them.

At the temperature and density of matter when $t \approx 100$ sec, collisions tended to produce a surplus of bound particles; they were not immediately destroyed (■ Fig. 19–17). Elements heavier than lithium were not formed at this time: The temperature and density dropped too rapidly. By an age of about ten minutes, the Universe had completed its **primordial nucleosynthesis**—the formation of the light elements shortly after the big bang.

Great discoveries sometimes have humble beginnings, especially when the discoverers are not aware that they are making a great discovery. Because of this fact we do not have an attractive picture of the neutral kaon apparatus. I am sorry about this.

Sincerely,

James W Cronin

James W. Cronin

James W. Cronin (U. Chicago)

FIGURE 19–15 James Cronin and Val Fitch received the 1980 Nobel Prize in Physics for their experiment that showed an asymmetry in the decay of a certain kind of elementary particle. This result may point the way to the explanation of why there is now more matter than antimatter in the Universe.

When one of us (J.M.P.) asked Professor Cronin for a picture to illustrate the Nobel-Prize-winning apparatus, he replied with the letter that is excerpted in the figure.

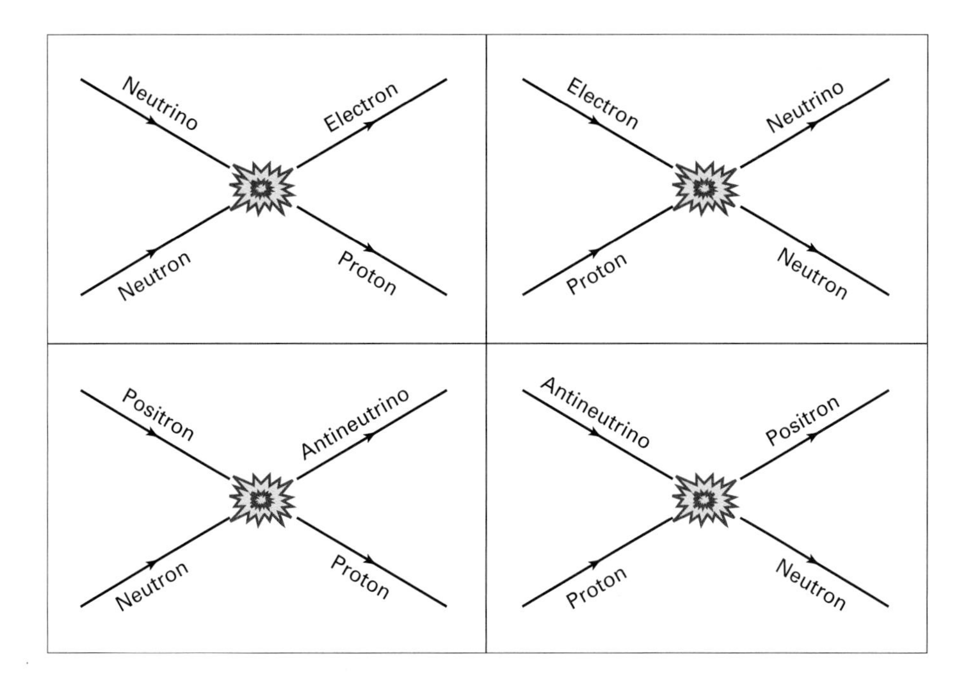

■ **FIGURE 19–16** Methods by which protons and neutrons are converted back and forth into each other, early in the history of the Universe when temperatures were very high. (After Alan Guth (MIT))

Primordial nucleosynthesis is predicted to have happened when the proton to neutron ratio was about 7:1. For every 16 nuclear particles, 2 were neutrons, and these generally combined with 2 protons to form helium, thereby using up 4 of the 16 particles. The other 12 particles remained as protons (simple hydrogen). So, about 25 per cent of the matter (by mass) should have turned into helium, while most of the rest was hydrogen. (A little bit of the hydrogen was in the form of deuterium and tritium, but the latter is unstable and quickly decays.)

The fact that helium is observed to be about 25 per cent by mass throughout the Universe (as far as we can tell) strongly supports the big-bang theory: The Universe had to be hot and dense at early times. The rather uniform relative proportion (abundance) of helium suggests a primordial origin, before stars were born. Heavier elements, on the other hand, exhibit rather large variations in their relative proportions: In some

ASIDE 19.3: Cosmic convergence

The contribution of normal matter to the total energy density of the Universe was found to be 4.4% by the Wilkinson Microwave Anisotropy Probe, in remarkable agreement with the value of 5% derived from primordial nucleosynthesis considerations.

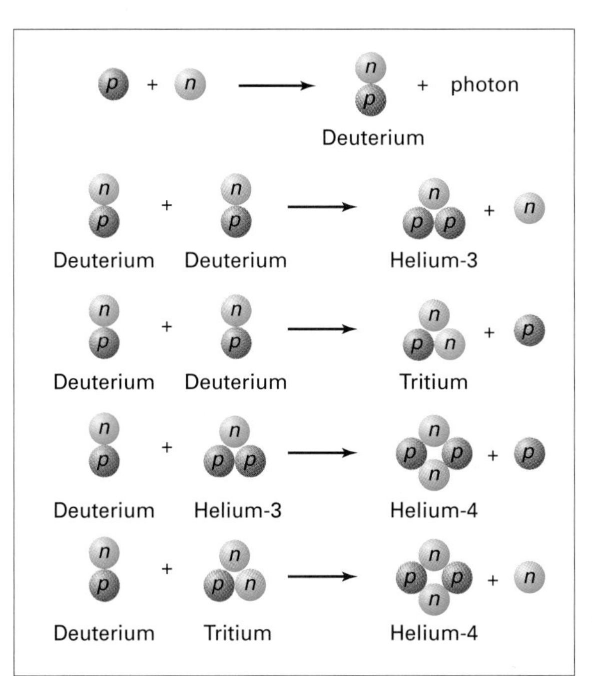

■ **FIGURE 19–17** Reactions during the first few minutes of the Universe's existence that produced various isotopes of the lightest elements, the process of primordial nucleosynthesis. In addition, a very small amount of lithium-7 was produced by other reactions. (After Alan Guth (MIT))

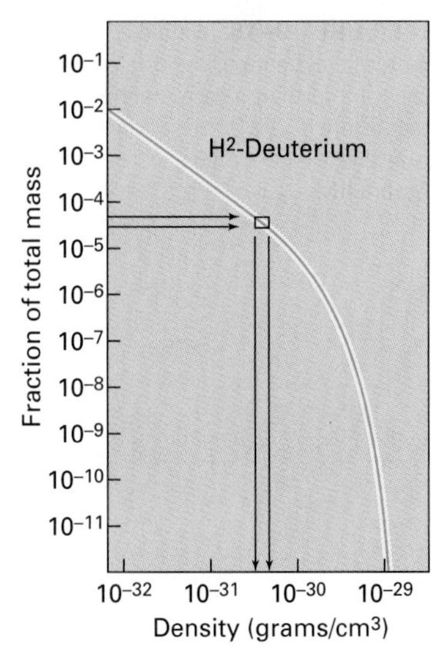

■ **FIGURE 19–18a** The horizontal axis shows the current cosmic density of matter. From our knowledge of the approximate rate of expansion of the Universe, we can deduce what the density was long ago. The abundance of deuterium is very sensitive to the density; thus present-day observations of the relative amount of deuterium tell us the cosmic density, by following the arrow on the graph. (After Robert V. Wagoner (Stanford U.))

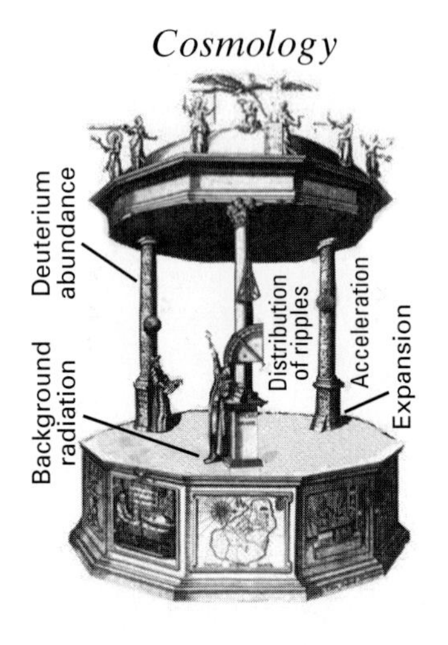

■ **FIGURE 19–18b** Pillars of cosmology: expansion (from Hubble's law), background radiation (and the deductions from studying ripples in it), and deuterium abundance. The evolution of the Universe is an additional pillar. (Based on Kepler's *Rudolphine Tables* [1627], collection of Jay M. Pasachoff)

stars they are rare, and in others they are common, implying an origin that is not primordial or uniform throughout the Universe.

The verification of the predictions of primordial nucleosynthesis is one of the four major pillars on which the big-bang theory rests. (The other three are the observed expansion of the Universe, the existence of the cosmic background radiation, and the observed evolution of the Universe from its beginning about 14 billion years ago.) Being inconsistent with the competing steady-state theory, it is more compelling than the mere expansion of the Universe.

George Gamow and his students had the right idea when they concluded that the Universe must have been hot at an early stage. Their mistake was in thinking that *all* of the elements (instead of just the lightest ones) were synthesized shortly after the big bang. They didn't know about nucleosynthesis in the cores of stars.

The approximate observed abundances of the light elements show that the general ideas of primordial nucleosynthesis are correct. Detailed measurements provide some additional constraints; the exact current proportions reflect conditions in the early Universe. For example, deuterium is rapidly destroyed, forming helium.

Thus, if the density remained high for a long time (the expansion of the Universe was relatively slow), the current deuterium abundance would be very low. The fact that deuterium is reasonably abundant right now (about 10^{-5} of the amount of ordinary hydrogen) tells us that Ω_M, the ratio of the average matter density to the critical density, is only about 0.05 (5 per cent) (■ Fig. 19–18).

In spite of years of trying, nobody succeeded in observing the fundamental spin-flip line of deuterium, the analogue to the 21-cm line of ordinary hydrogen (see our discussion in Chapter 15), until the report of MIT's Alan Rogers and colleagues in 2005. Observing in Massachusetts, they used a field of broadband antennas, linked electronically, to detect deuterium in the direction opposite to that of our Galaxy's center. The background noise in the anticenter direction is so much lower than that toward the center that the deuterium was detectable, in spite of its being weaker. They found that deuterium nuclei are fewer than hydrogen nuclei by a ratio of 2.3×10^{-5}.

The 5 per cent limit measurement of Ω_M only refers to normal matter (protons, neutrons, electrons). There might be additional contributions to Ω_M, but they must be "exotic" matter such as neutrinos and "weakly interacting massive particles" (WIMPs). If we believe that $\Omega_M \approx 0.3$, as suggested by studies of clusters of galaxies and the peculiar motions of galaxies (Chapter 18), then a large fraction of the matter in the Universe is not only dark, but must also be composed of some kind of exotic matter! This conclusion is consistent with the findings of WMAP discussed above (Sec. 19.3c).

19.5 The Inflationary Universe

The original, "standard" hot-big-bang theory is remarkably successful. Its four main observational foundations are the existence of the cosmic microwave background radiation, the relative proportions of various isotopes of the lightest elements (hydrogen, helium, and lithium), the expansion of the Universe, and the evolution of the Universe over a finite amount of time.

However, there are some puzzling aspects of the Universe that the original big-bang theory cannot explain, at least not without imposing "initial conditions" that seem unlikely and unjustified. Here we will discuss two of the main problems. We will also present a brilliant (but still speculative) refinement to the big-bang theory that appears to resolve them by affecting just the first blink-of-an-eye of the Universe's existence.

19.5a Problems with the Original Big-Bang Model

The first apparent difficulty is the great uniformity of the Universe—its observed homogeneity and isotropy. This uniformity is known as the **horizon problem.**

For example, the temperature indicated by the cosmic background radiation is identical (2.728 K) in all directions, excluding the two known effects that produce deviations from isotropy (our Galaxy's motion through the Universe, and early variations in density from which clusters of galaxies formed).

How can the Universe be so uniform? Widely separated regions of the Universe could never have been in thermal equilibrium with each other (that is, at the same temperature): No physical signals, not even those travelling at the speed of light, could have crossed such vast distances over the age of the Universe. Thus, distant regions had no way of "telling" each other which temperature to choose—they are beyond each other's "horizon" (and hence the name, "horizon problem").

For example, assume the Universe is 14 billion years old, and consider two points on opposite sides of the sky. Cosmic background photons are just now reaching us, after travelling almost the entire age of the Universe (minus about 400,000 years when the Universe was still opaque). Clearly, photons from one side have not yet reached the other side; these regions should therefore have different temperatures.

One might suggest that the big bang started as a "point," and hence could have achieved a uniform temperature. But actually, the big bang was never a true "point"; we need a quantum theory of gravity to figure out its initial size. Running the original big-bang equations backward in time, we find that the Universe was *always* larger than the distance light (or any other signal) could have travelled by that age, even if the Universe was always transparent. Indeed, the number of "causally disconnected" regions was larger in the past, making the observed homogeneity and isotropy (uniformity) even more perplexing. For many years, cosmologists simply *assumed,* as an initial condition without physical justification, that the Universe originated with a uniform temperature.

The second major problem with the original big-bang theory is that it has no natural way of explaining why the overall geometry of the Universe is so close to being flat (Euclidean). This is known as the **flatness problem.** Observations tell us that the ratio of matter density to the critical density, Ω_M, is certainly larger than 0.01, just from the luminous matter and a small correction for dark matter in galaxies. But in fact $\Omega_M \approx 0.3$ if the dark matter in clusters of galaxies is included.

Moreover, as discussed in Chapter 18, observations of distant supernovae suggest that there is a significant contribution to the total density from the "dark energy" ("vacuum energy") associated with the cosmological constant, so the total Ω might be 1. And, as we saw above, recent measurements of the cosmic background radiation strongly suggest that the overall geometry of the Universe is flat, meaning that the total Ω is indeed 1.

Why should we be surprised that the Universe is so flat right now ($\Omega \approx 1$)? Well, according to the equations describing the evolution of the Universe in the original big-bang theory, Ω must have been *exceedingly* close to 1 (that is, exceedingly flat) in the distant past (■ Fig. 19–19). Otherwise, as the Universe aged, Ω would have rapidly deviated from 1 by a very large extent, so we should not be measuring $\Omega \approx 1$ at present.

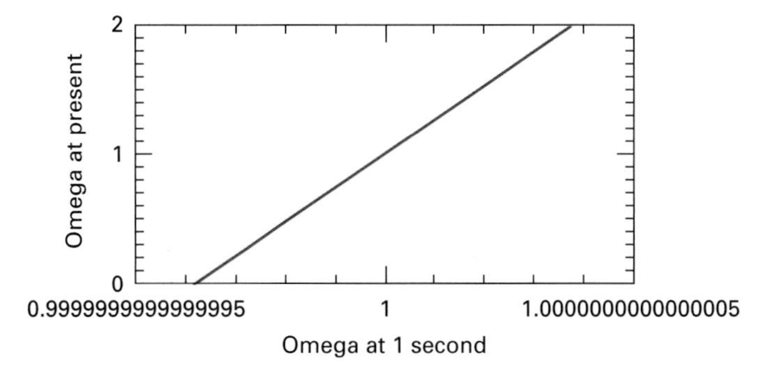

■ **FIGURE 19–19** To achieve a current value close to 1 for the density of the Universe divided by the critical density (that is, $\Omega \approx 1$), the ratio had to be almost exactly 1 to an exceedingly high precision when the Universe was very young (say, 1 second old, as shown here). (After Alan Guth)

Courtesy of Donna Coveney, MIT News Office

Andre Linde (Stanford U.), CERN photo

■ **FIGURE 19–20** ⓐ Alan Guth, one of the inventors of the inflationary model of the Universe, a proposed modification to the original big-bang theory during the first tiny fraction of a second of existence. ⓑ Andrei Linde, who has proposed many different ways in which inflation can occur, including the idea of eternal inflation.

ASIDE 19.4: A googol

"Googol" is the name of the number 10^{100}. The popular search engine "Google" was named after this number, but the creators of the search engine intentionally misspelled it because they believed that most people would naturally spell it this way rather than the correct way.

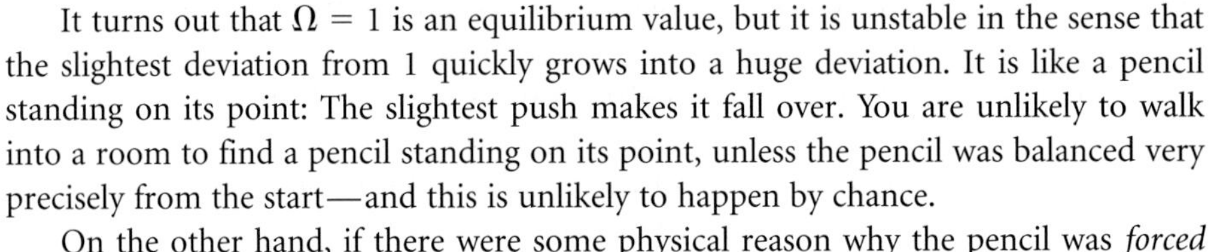

It turns out that $\Omega = 1$ is an equilibrium value, but it is unstable in the sense that the slightest deviation from 1 quickly grows into a huge deviation. It is like a pencil standing on its point: The slightest push makes it fall over. You are unlikely to walk into a room to find a pencil standing on its point, unless the pencil was balanced very precisely from the start—and this is unlikely to happen by chance.

On the other hand, if there were some physical reason why the pencil was *forced* to stand on its point, then you might expect to find it in that position. The original big-bang theory provides no suitable explanation for why the Universe started out with $\Omega = 1$, so for many years cosmologists simply *assumed* this initial condition; they had no reasonable explanation for it.

19.5b Inflation to the Rescue

In 1979, Alan Guth (■ Fig. 19–20*a*), now at the Massachusetts Institute of Technology, came up with a brilliant possible, but still unproven, solution to these problems. His idea, known as **inflation,** is a modification of the original big-bang theory. Andrei Linde (■ Fig. 19–20*b*), then in the Soviet Union but now at Stanford University, made a very similar suggestion at about the same time.

Although Guth's proposal had an important flaw, the basic hypothesis remained very attractive. The flaw was resolved in 1981 by Linde, as well as by Paul Steinhardt (now at Princeton University) working together with Andreas Albrecht (now at the University of California, Davis). Most physicists and astronomers accept some form of inflation, although the proposed details and specific mechanisms currently have much greater variety than (and generally differ from) the theories of a quarter-century ago. But the earliest example, as formulated by Guth, Linde, Steinhardt, and Albrecht, illuminates many of the essential principles, so we discuss it here.

Suppose the Universe started out *much smaller* than the size implied by an extrapolation of the equations of the original big-bang theory. Guth proposed that it might have subsequently "inflated"—expanded *extremely fast* and by an immense factor (■ Fig. 19–21), perhaps 10^{50} or even 10^{100} (a "googol"). The Universe achieved this huge expansion by doubling in size every tiny fraction of a second for a significant

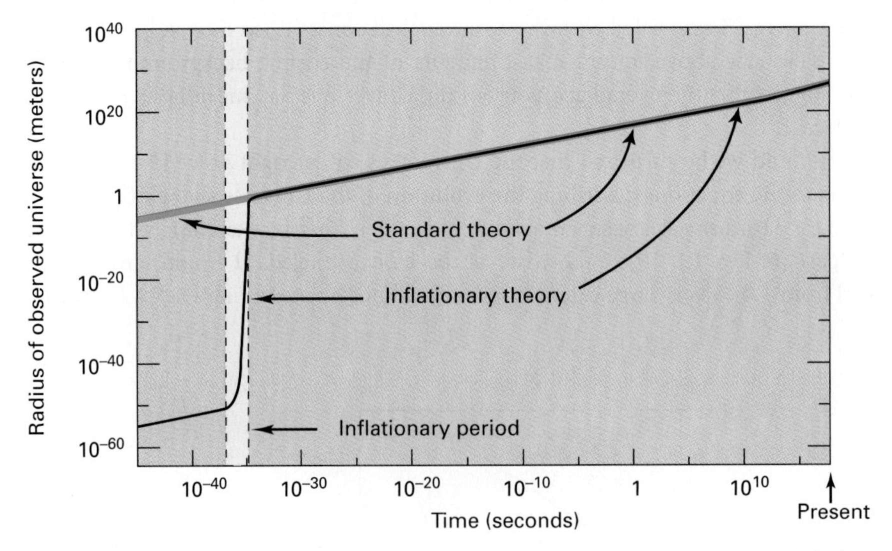

■ **FIGURE 19–21** The rate at which the Universe expands, for both inflationary and noninflationary models. The vertical axis shows the radius of the region that expands to become the presently observable part of the Universe, and the horizontal axis shows time. The inflationary period is marked on the graph as a vertical yellow band. Before inflation, the size of the Universe in the inflationary model is much smaller than in the original big-bang model, allowing the observed Universe to come to a uniform temperature in the time available. (After Alan Guth [MIT])

amount of time, contrary to the constant or decelerating expansion predicted in the original big-bang theory.

An intuitive feel for such "exponential" growth can be obtained by envisioning a chessboard. If there is one grain of wheat on the first square, and the number of grains in each successive square is doubled, the wheat on the 64th square (2^{63} grains) would be the size of a mountain that covers Manhattan!

Notice that before inflating, the tiny initial Universe could have achieved a uniform temperature (thermal equilibrium). Unlike the case in the original big-bang theory, it was sufficiently small that signals could travel across it. Specifically, collisions between particles and photons would have distributed the heat uniformly. But the Universe would have retained its uniformity as it inflated, because there is no physical mechanism by which it could have lost this property. With inflation, we expect the currently observable Universe to be homogeneous and isotropic, eliminating the horizon (uniformity) problem. (For numerical examples, see *Figure It Out 19.1: Inflation of the Early Universe.*)

Moreover, inflation would have flattened out any initial curvature in the Universe, like a balloon expanding to immense size. In effect, the Universe became so large that the entire region we can see with our most powerful telescopes (which is only a minuscule fraction, perhaps 10^{-40} in radius, of what really exists) appears flat or nearly flat, just as the Earth's surface appears flat when viewed over small distances. Inflation therefore resolves the flatness problem.

After the end of inflation, the Universe would have continued expanding at a rate consistent with the original big-bang theory. So, inflation affected only the first brief

FIGURE IT OUT 19.1

Inflation of the Early Universe

Let us consider a numerical example to clarify the concept of inflation in the early Universe. At time $t = 10^{-37}$ sec, the radius of the *currently* observable part of the Universe (at $t \approx 14$ billion years) was about 0.01 m, according to the original big-bang theory. Even if the early Universe were transparent, light could have travelled only

$$d = ct = (3 \times 10^8 \text{ m/sec})(10^{-37} \text{ sec}) = 3 \times 10^{-29} \text{ m}$$

in that amount of time. (The actual distance is somewhat greater, because of the expansion of space itself. Typically it might be a factor of 3 larger—but in this case still only about 10^{-28} m.) This distance is far too small, compared with the radius of the Universe at that time. Thermal equilibrium could not have been established, even with light signals.

Now suppose the radius of the currently observable Universe was actually only 10^{-52} m at $t = 10^{-37}$ sec. This was sufficiently small that the transmission of signals (even slow ones such as particle collisions) *could* have easily produced thermal equilibrium.

If the Universe subsequently inflated by a factor of 10^{51} by $t = 10^{-35}$ sec, then its new radius would be about 0.1 m, in agreement with the original big-bang theory; see Figure 19–21. Thereafter, the radius of the currently observable part of the Universe grew at a rate consistent with the original big-bang theory. Inflation affects *only* the first roughly 10^{-35} sec of the Universe's existence.

Here is another example, this time focusing on the entire Universe (not just the currently observable part). Suppose the Universe had a radius of 10^{-40} m at $t = 10^{-37}$ sec. If it inflated by a factor of 10^{100} by $t = 10^{-35}$ sec, its new radius at the end of inflation would be

$$10^{-40} \text{ m} \times 10^{100} = 10^{60} \text{ m},$$

or about 10^{44} light-years! But the currently observable part of the Universe had a radius of about 0.1 m at $t = 10^{-35}$ sec, and the radius is now about 10^{26} m (or 10^{10} light-years). Both of these radii are much smaller than the total size of the Universe after inflation.

The times discussed here, and especially the initial sizes and the inflation factors, are somewhat arbitrary; they depend on the details of the specific inflationary model under consideration. (In some versions of inflation, the Universe inflates by a factor of 10 to the *trillionth* power!) The important point is that inflation produces a truly *enormous* Universe, far larger than we can see. This means that the geometry of the Universe should appear flat ($\Omega = 1$), as observed.

moments in the life of the Universe. It is not a replacement for the big-bang theory, but rather a modification (or refinement) of it. In fact, inflation can be thought of as the mechanism that "launched" the big bang!

In the inflationary universe, space itself expands far faster than the speed of light. After all, the Universe was initially much smaller than a subatomic particle, yet it quickly inflated to an almost uncountable number of light-years across! Einstein's theory of relativity permits space to expand faster than the speed of light; it only restricts the transmission of *information* to a speed no greater than that of light. But the expansion of space cannot be used to transmit information from one object to another elsewhere. Put another way, relativity states that two objects cannot pass each other at speeds exceeding that of light. But the expansion of space itself does not allow particles to pass each other.

19.5c Forces in the Universe

Why should the Universe have experienced a period of inflationary expansion? Physicists are now considering a variety of possible reasons, some quite speculative and far removed from those advocated by the early pioneers of inflation. But the originally favored reason remains a viable candidate and is instructive to study. It stems from theories that attempt to unify the known fundamental forces of nature.

For much of the 20th century, physicists thought that there are four fundamental forces in nature. The first force, **gravity,** is extremely weak, but it acts over large distances, and all types of matter (and energy) are affected in the same way, regardless of charge or other properties. We are already very familiar with gravity, since it has played a prominent role in many topics discussed in this book.

Electromagnetism, the second force, is about 10^{39} times stronger than gravity, and it acts over large distances, but only charged or magnetized objects are affected. (This particular strength comparison uses an electron and a proton in a hydrogen atom. The exact numbers are different if two protons or two electrons are used.) Since most large objects are electrically neutral and not magnetized, electromagnetic forces are generally not felt over large distances. Electromagnetism is critical, however, to the structure of liquids and solids such as your body, and of course to atoms and molecules themselves.

The third force, known as the **strong nuclear force,** is about 100 times stronger than electromagnetism, but it operates only over an extremely short range. The protons and neutrons in an atomic nucleus are held together by the strong force. Actually, we now know that the "strong force" is just the residue of a more fundamental (and even stronger) "color force" that binds quarks together. However, for historical purposes we will still refer to it as the strong force.

Finally, the **weak nuclear force** is about a million times weaker than electromagnetism, and it operates only over an extremely short range. It is the force that governs certain types of radioactive decay, such as the neutron becoming a proton, an electron, and an antineutrino. Any time neutrinos or antineutrinos interact with matter, the weak force is involved.

In the late 1970s, theorists suggested that the electromagnetic and weak forces are different manifestations of a single more fundamental force, the **electroweak force.** The electromagnetic and weak forces seem different at the low energies characteristic of the Universe today. But at higher energies and temperatures (such as when the Universe was younger than about 10^{-12} sec, and hotter than about 10^{16} K), they behave in a similar manner. We say that the "symmetry is broken" at low energies. In high-energy interactions, the two forces cannot be distinguished and hence appear **symmetric** (unified). A useful analogy is a coin flicked along the surface of a table: When it is spinning rapidly at high energies, it appears symmetric, but we see either "heads" or "tails" when it comes to rest, a "broken symmetry."

ASIDE 19.5: A Brief History of Gravity (A Limerick by Bruce Elliot)

It filled Galileo with mirth
To watch his two rocks fall to Earth.
He gladly proclaimed,
"Their rates are the same,
And quite independent of girth!"
Then Newton announced in due course
His own law of gravity's force:
"It goes, I declare,
As the inverted square
Of the distance from object to source."
But remarkably, Einstein's equation
Succeeds to describe gravitation
As space-time that's curved,
And it's this that will serve
As the planets' unique motivation.
Yet the end of the story's not written;
By a new way of thinking we're smitten.
We twist and we turn,
Attempting to learn
The Superstring Theory of Witten!

High-energy experiments at the CERN particle accelerator in Switzerland, at the Stanford Linear Accelerator Center in California, and elsewhere subsequently confirmed the predictions of the electroweak theory. Specifically, for example, physicists found the expected "W" and "Z" particles in 1983 (■ Fig. 19–22). Thus, we now often say that there are only *three* fundamental forces. James Clerk Maxwell achieved a similar "unification" in the 1860s, when he showed that electricity and magnetism are different manifestations of a more fundamental interaction, electromagnetism.

Physicists are now trying to see whether the three known forces are different low-energy manifestations of only one or two fundamental forces. **Grand unified theories (GUTs)** attempt to unify the electroweak and strong nuclear forces. In principle, these forces behave in a similar (symmetric) manner at temperatures above about 10^{29} K (which characterized the Universe at times less than 10^{-37} sec).

GUTs predict that the big bang produced slightly more matter than antimatter; all of the antimatter subsequently annihilated with matter, leaving behind the matter we see today. GUTs also predict that protons are unstable, though proton decay has not yet been observed (■ Fig. 19–23). The experimental minimum value of about 10^{32} years for the proton lifetime already rules out some candidate GUTs, but many still remain, and we don't know which (if any) is correct.

"Theories of everything" (TOEs), though still rather speculative, attempt to unify the grand unified force and gravity in a single superunified force. Physicists conjecture that these forces are symmetric at temperatures above about 10^{32} K, which characterized the Universe at times less than about 10^{-43} sec (the Planck time).

To understand the physics of this very early era in cosmic history, we need a unification of quantum mechanics and general relativity, the two pillars of 20th century physics—that is, a quantum theory of gravity. No fully self-consistent theory has yet emerged, but many theorists are focusing their attention on **superstring theories** (often called simply string theories). Modern string theory was born with the work of John Schwarz (Caltech) and Michael Green (then of Queen Mary College), and one of the current leaders in this quest is Edward Witten (Institute for Advanced Study, Princeton).

In superstring theories, each fundamental particle (quark, electron, neutrino) is postulated to be a different vibration mode of a tiny, one-dimensional energy packet called a "string" (or perhaps a tiny "membrane," having two or more dimensions). A string is an elongated, one-dimensional analogue to the zero-dimensional points that have often been used in theories of matter. Strings are so small (about 10^{-33} cm) that

Philip Crane

■ **FIGURE 19–22** Carlo Rubbia and Simon van der Meer with the apparatus at CERN with which they and their colleagues discovered the W^+, W^-, and Z^0 particles. These discoveries verified predictions of the electroweak theory. Rubbia and van der Meer received the 1984 Nobel Prize in Physics as a result.

Joe Stancampiano and Karl Luttrell/U. Michigan, © National Geographic Society

■ **FIGURE 19–23** This cavity, deep underground in a salt mine near Cleveland, was filled with 10,000 tons of water to make the Irvine-Michigan-Brookhaven (IMB) detector. To detect flashes of light that result as neutrinos or other elementary particles pass through it, 2400 light-detectors were installed around the sides. Its detection of a few neutrinos from Supernova 1987A confirmed that massive stars can undergo a core collapse to form neutron stars.

■ **FIGURE 19–24** Theorists calculate that superstrings require an 11-dimensional space. We are familiar with only a 4-dimensional space, having 3 spatial dimensions and 1 time dimension.

they seem almost like points. Superstring theories thus far appear to be a promising way to get around theoretical difficulties on the extremely small scale where a quantum theory of gravity is needed.

Recent superstring theories tend to favor an 11-dimensional Universe (revised up from 10 dimensions), including the time dimension, but 7 of the dimensions are too small to see; they are curled up on scales of about 10^{-33} cm, the Planck length (■ Fig. 19–24). A useful analogy is to view a sheet of sandpaper from far above: it looks only two dimensional, but is really three dimensional due to the granularity. Another example is a garden hose: From far away, it appears one dimensional (like a string) instead of three dimensional. Nobody knows why the Universe has only four macroscopically large dimensions (usually called x, y, z, and t).

One prediction of superstring theories is that the Universe should contain "shadow matter." This has a gravitational effect on normal matter, but does not emit any electromagnetic radiation; it is therefore a candidate for dark matter.

ASIDE 19.6: Extra dimensions

Most recently, it has been proposed that at least one of the unseen dimensions is macroscopically large, but very difficult to detect.

19.5d Supercooling the Universe

Right after the big bang, when temperatures were exceedingly high, superstring theories suggest that there should have been just one symmetric, superunified force. Given the uncertainty over what happens before the Planck time (10^{-43} sec), we will instead consider the situation when the Universe was about 10^{-37} sec old, focusing on GUTs. The grand unified (GUT) force and gravity would have existed as distinguishable forces.

At $t \approx 10^{-37}$ sec, the temperature would have dropped to 10^{29} K, below which the GUT force should have started to break symmetry, becoming the strong nuclear force and the electroweak force. This **phase transition** is similar to what happens when liquid water is cooled below a temperature of 0°C: It generally turns to ice (■ Fig. 19–25). When water turns to ice, the freezing process liberates excess energy so that the water molecules can line up to form crystals.

Suppose, however, that the Universe cooled below 10^{29} K *without* breaking symmetry—that is, the strong and electroweak forces remained unified as the GUT force.

ASIDE 19.7: Perfect symmetry

A sphere of liquid water in a vacuum, floating in space, looks the same when viewed from different directions, so physicists say it is "rotationally symmetric." In contrast, an ice crystal has a preferred axis (the axis of symmetry), and hence does not appear the same when viewed from different directions. Physicists say that ice is less symmetric than liquid water when they use the word "symmetry" in this manner.

This **supercooled** Universe would have contained too much energy relative to what it should have had at that temperature. For example, water can sometimes be supercooled below 0°C without turning into ice, if one is very careful not to disturb the water in any way. Its excess energy keeps it in the liquid state. The excess energy is released when the water freezes.

The energy associated with the supercooled, symmetric (GUT) Universe would have caused space to double its size every tiny fraction of a second—that is, the Universe would have entered an epoch of accelerated, inflationary expansion. (This is where the analogy with water breaks down: The excess energy associated with the supercooled phase does *not* make the water expand exponentially!) For reasons that are not intuitively obvious, the Universe's excess energy had a stretching effect on space, like a cosmic "antigravity." Physicists call this excess energy the "energy of the false vacuum" because a conventional vacuum should have zero energy. It is essentially Einstein's cosmological constant Lambda (Λ) (see Chapter 18), but in this case very early in the history of the Universe. The inflationary expansion made the Universe truly *enormous;* it is much, much larger than the regions that we can actually see.

Eventually, perhaps at an age of 10^{-35} sec, the Universe experienced a phase transition, with the electroweak and strong forces separating (that is, the symmetry was broken). At this time all of the excess energy of the false vacuum was released, thereby forming matter, antimatter, and photons. The energy of the false vacuum thus provided all of the energy in the Universe.

With no more "antigravity" force, the inflationary expansion of the Universe subsided, and the regular Hubble expansion began. The Universe started "coasting," but gravity caused the expansion to decelerate. In essence, inflation is what gave the Universe its "big bang"—its initial kick up to a time of about 10^{-35} sec, after which all of the events discussed previously in this chapter unfolded (■ Fig. 19–26).

Jay M. Pasachoff

■ **FIGURE 19–25** Ice (solid water) and liquid water are different states of matter. The phase change between them is, in some ways, analogous to the phase change that must have taken place early in the history of the Universe.

19.5e Successes of Inflation

Despite its abstract and speculative nature, the achievements of the inflationary theory are magnificent. First, it explains the horizon problem—the great uniformity (homogeneity and isotropy) of the Universe. The Universe was sufficiently small to have achieved thermal equilibrium (the same temperature) at very early times, and it maintained this as it inflated.

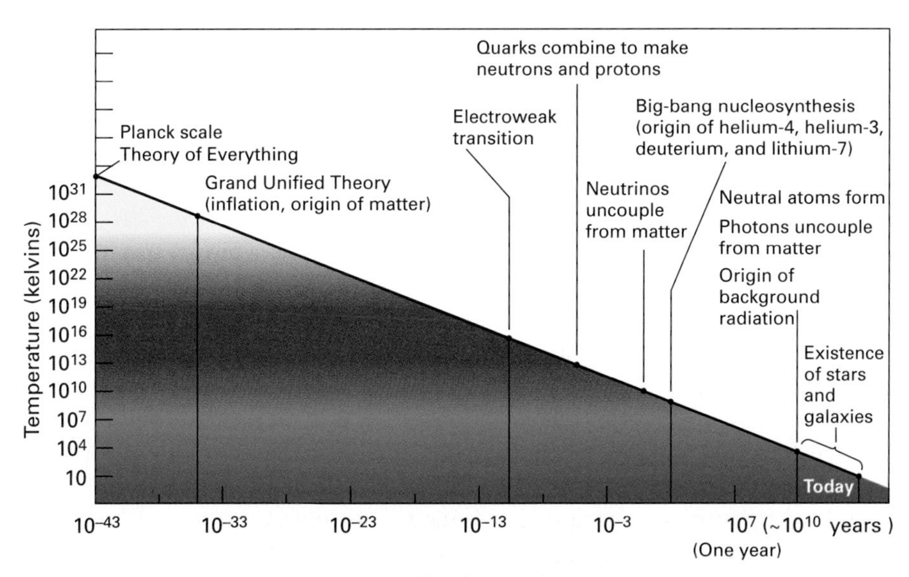

■ **FIGURE 19–26** As the temperature of the Universe cooled, the forces became distinguishable from each other and the Universe evolved. Note that most of the horizontal axis shows the first year of the Universe. (Adapted from "Particle Accelerators Test Cosmological Theory" by David N. Schramm and Gary Steigman. © 1988 by Scientific American, Inc. All rights reserved. Drawn by Andrew Christie.)

ASIDE 19.8: Variable speed of light?

Recently, a few physicists have suggested that a variable speed of light at early times could account for the horizon and flatness problems; inflation would then be unnecessary. However, it is not clear how such a hypothesis would explain other observed aspects of the Universe for which inflation has reasonable solutions. Moreover, the currently observed acceleration of the Universe is reminiscent of the beginning of inflation, suggesting that nature knows how to produce long-range antigravity effects even if we don't yet understand the details.

Second, the Universe is very close to having a spatially flat geometry. After inflation, it was so large that any observable region would appear flat (that is, $\Omega = 1$; the average total density is equal to the critical density).

Third, the observable Universe has few, if any, "magnetic monopoles" (isolated north or south magnetic poles, without the opposite pole being present as well); none has been found despite intensive searches for them. There are also no apparent "cosmic strings"—long, skinny regions that contain huge amounts of mass. Many such "defects" should have been produced by phase transitions soon after the big bang, but the rapid expansion of the Universe during the inflationary era would have whisked them away, thereby accounting for their scarcity.

Fourth, the predictions of inflation are consistent with the observed distribution of matter over various size scales (such as clusters of galaxies and superclusters).

19.5f The Ultimate Free Lunch?

In the inflationary theory, matter, antimatter, and photons were produced by the energy of the false vacuum, which was released following the phase transition. All of these particles consist of positive energy. This energy, however, is exactly balanced by the *negative* gravitational energy of everything pulling on everything else. In other words, the *total* energy of the Universe is zero! It is remarkable that the Universe might consist of essentially nothing, but (fortunately for us!) in positive and negative parts.

You can easily see that gravity is associated with negative energy: If you drop a ball from rest (defined to be a state of zero energy), it gains energy of motion (kinetic energy) as it falls. But this gain is exactly balanced by a larger negative gravitational energy as the ball comes closer to the Earth's center, so the sum of the two energies remains zero.

The idea of a zero-energy Universe, together with inflation, suggests that all one needs is just a tiny bit of energy to get the whole thing started (that is, a tiny volume of energy in which inflation can begin). The Universe then experiences inflationary expansion, but without creating net energy.

What produced the energy before inflation? This is perhaps the ultimate question. As crazy as it might seem, the energy may have come out of *nothing!* The meaning of "nothing" is somewhat ambiguous here. It might be the vacuum in some preexisting space and time, or it could be nothing at all—that is, all concepts of space and time were created with the Universe itself.

Quantum theory provides a natural explanation for how that energy may have come out of nothing (see *Figure It Out 19.2: Heisenberg's Uncertainty Principle*). As mentioned in Chapter 18, it turns out that particles and antiparticles spontaneously form and quickly annihilate each other, microscopically violating the law of energy conservation (■ Fig. 19–27). (Alternatively, if the energy of the particles and antiparticles is "borrowed" from the vacuum, briefly creating a negative-energy "hole," then energy is conserved at all times.)

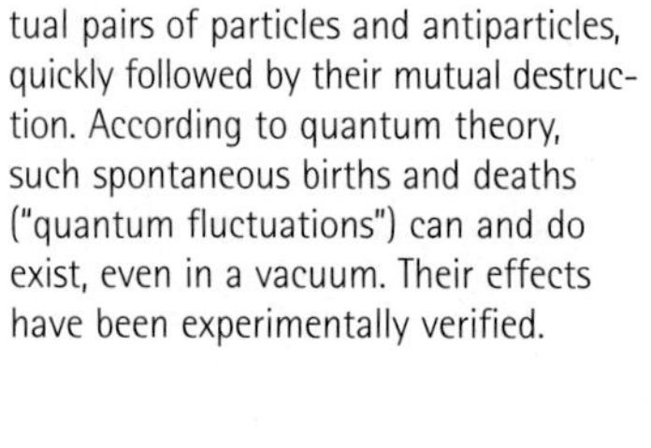

These spontaneous births and deaths of so-called virtual-particle pairs are known as "quantum fluctuations." Indeed, laboratory experiments have proven that quantum fluctuations occur everywhere, all the time. Virtual particle pairs (such as electrons and positrons) directly affect the energy levels of atoms, and the predicted energy levels disagree with the experimentally measured levels unless quantum fluctuations are taken into account.

Perhaps many quantum fluctuations occurred before the birth of our Universe. Most of them quickly disappeared. But one of them lived sufficiently long and had the right conditions for inflation to have been initiated. Thereafter, the original tiny volume inflated by an enormous factor, and our macroscopic Universe was born. The original particle–antiparticle pair (or pairs) may have subsequently annihilated each other, disappearing from the Universe. But even if they didn't, the violation of energy

■ **FIGURE 19–27** Formation of virtual pairs of particles and antiparticles, quickly followed by their mutual destruction. According to quantum theory, such spontaneous births and deaths ("quantum fluctuations") can and do exist, even in a vacuum. Their effects have been experimentally verified.

FIGURE IT OUT 19.2

Heisenberg's Uncertainty Principle

A key feature of quantum theory is the Heisenberg Uncertainty Principle, which implies that one cannot know the exact energy at each point in space in a finite amount of time. The energy at each point must therefore be randomly fluctuating; otherwise, we would know that its value is zero. The creation and annihilation of "virtual" pairs of particles and antiparticles, even in a complete vacuum, is the manifestation of such quantum energy fluctuations.

Quantitatively, virtual particles of energy ΔE spontaneously form and live for a time Δt if the product $\Delta E \Delta t$ is no larger than about $h/2\pi$, where h is Planck's constant:

$$\Delta E \Delta t \leq h/2\pi.$$

In a sense, space itself is uncertain of its energy by an amount up to ΔE, if the timescale over which it is measured is less than or equal to Δt, where the product of the two does not exceed $h/2\pi$.

Thus, the smaller the energy (ΔE) of the virtual particles that are spontaneously created, the larger the time interval (Δt) over which they can exist. For example, a virtual proton–antiproton pair, which has a mass of about 3.3×10^{-24} g and a corresponding energy given by $E = mc^2$, can live up to about 3×10^{-25} sec. But a virtual electron–positron pair can exist much longer, since its mass is nearly 2 thousand times less than that of a virtual proton–antiproton pair.

conservation would be minuscule, not large enough to be measurable (and perhaps even zero, in the case of a pair borrowing energy from the vacuum).

If this admittedly speculative hypothesis is indeed correct, then the answer to the ultimate question is, in the words of Alan Guth, that the Universe is "the ultimate free lunch"! (The Nobel Prize–winning economist Milton Friedman, who asserted that "there is no such thing as a free lunch," would surely be surprised.) The Universe came from nothing, and its total energy is zero, but it nevertheless has incredible structure and complexity. Physicists are attempting to come up with ways to test this idea in principle, but no method has emerged yet; thus, in some respects it is not currently within the realm of science, which requires hypotheses to be experimentally or observationally testable.

ASIDE 19.9: Whose free lunch is it, anyway?

Although Milton Friedman published a book in 1975 called "There's No Such Thing as a Free Lunch," previously the science-fiction author Robert Heinlein had popularized a similar phrase, "There ain't no such thing as a free lunch" (often referred to by the acronym "TANSTAAFL") in his 1966 novel, *The Moon Is a Harsh Mistress.* But apparently, the phrase was in common usage before either Friedman or Heinlein.

19.6 A Universe of Universes

Is our Universe the only universe, or could it be one of many? The entire history of our Universe might be just one episode in the much grander **multiverse** consisting of many (perhaps infinitely many) universes. This idea is the most extreme extension of the Copernican principle. Some scientists, notably the celebrated British cosmologist Lord Martin Rees, find the multiverse idea quite attractive. Others, however, remain skeptical. The concept is certainly quite controversial at this time, and more speculative than even the "free lunch" hypothesis discussed above.

The concept of another "universe" is difficult to define. It could be a disjoint region that occupies dimensions separate from (and physically inaccessible to) those of our Universe. This other universe might, however, be connected with ours via a wormhole—a passage between two black holes (see our discussion in Chapter 14). Another possibility is a region in our physical Universe, or arising from our Universe, in which inflation occurred or ended in a different way.

In some or all cases, the physical constants (or, much more speculatively, perhaps even the laws of physics) in other universes may differ from those in our Universe. By "physical constants" we mean the relative strengths of the various forces, the values of quantities such as the speed of light, and the masses of fundamental particles such as electrons and neutrinos.

ASIDE 19.10: A joke no more

When the authors of this book were children, "Define universe and give three examples" was a popular science joke; it was widely thought that our Universe was all there is.

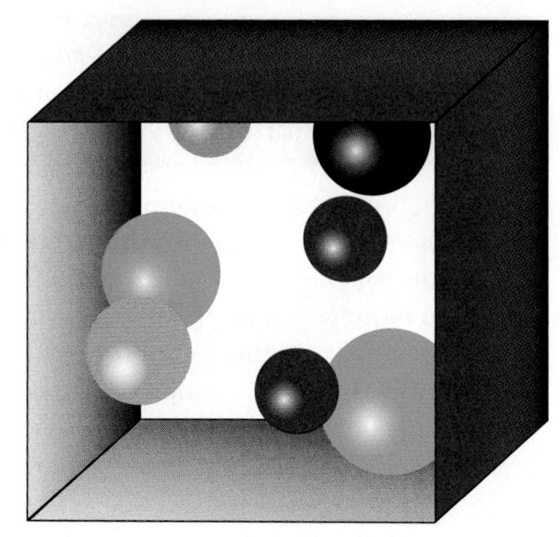

FIGURE 19–28 Many separate universes hypothetically arising from quantum fluctuations in a grander hyperspace, the "multiverse." The different colors of these "bubble universes" are meant to illustrate that the values of their physical constants (or perhaps even their physical laws) could differ.

As summarized below, there are scientifically valid reasons to think that other universes might exist. But by its very nature, this interesting conclusion removes itself from the realm of science, at least temporarily, because we don't know of a way to contact other universes and perform experiments on them. In other words, the hypothesis generally seems to be untestable.

One exception would be a different region in our Universe that eventually becomes observable when photons reach us, far in the future. Also, consider sending light signals through a wormhole: It is conceivable that a future generation will discover matter with antigravity properties (on small scales) and will transport it into the wormhole, in this way perhaps keeping it open. Finally, there may be properties of our Universe that depend on the existence of other universes; thus, our descendants might eventually discover some method of testing for their presence. But currently, at least, we can only speculate about the multiverse.

There seem to be at least three or four ways to produce other universes. For example, distinct quantum fluctuations could arise out of "nothing" (■ Fig. 19–28). This "nothing" might be the vacuum of our Universe, or the "nothing" outside our Universe, in some sort of a larger "hyperspace." If a quantum fluctuation out of "nothing" created our Universe, it seems reasonable that such a process may have occurred many times (thereby making the hyperspace a multiverse), perhaps even infinitely many. We do not yet know whether this is a viable process. To find out, we need a fully self-consistent quantum theory of gravity that unites relativity with quantum physics, and theorists are currently quite excited by the potential of superstring theories.

Other universes could also be regions of different broken symmetry in an inflating volume. As the false vacuum inflates, the phase transition (symmetry breaking) may occur in different ways in different regions (■ Fig. 19–29). This is analogous to water

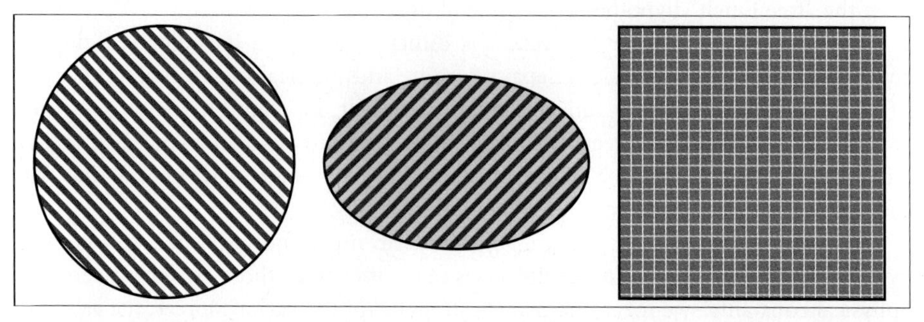

FIGURE 19–29 Regions of different broken symmetry in an inflating volume; each might have a different set of values for the physical constants.

freezing along different axes in a jar: A jumble of ice crystals forms, with various orientations. The values of the physical constants could differ in those regions where symmetry breaking had occurred differently. (More speculatively, the physical laws themselves might differ in them.) These regions could remain separated from each other due to the presence of a still inflating false vacuum between them.

A popular related scenario, promoted largely by Andrei Linde (Fig. 19–20*b*), is that quantum effects in an inflating volume spawn new, distinct, inflating regions within it. Although most of the volume will have undergone a phase transition (symmetry breaking) and stopped inflating, some regions will continue to inflate. These inflating regions will subsequently spawn regions that continue to inflate, even as symmetry is broken throughout most of their volume (■ Fig. 19–30). The process never ends, so it is called "eternal inflation." Again, the values of the physical constants may differ in the different regions of inflation.

Yet another idea is that a new universe can sprout from within any black hole in our Universe. Even more speculatively, the main characteristics of our Universe might even be transferred to its progeny, but with some small mutations. Lee Smolin of the Perimeter Institute for Theoretical Physics in Waterloo, Ontario, Canada, proposed this notion of "universe heredity."

Time

Courtesy of Andre Linde (Stanford U.)

■ **FIGURE 19–30** A series of inflating universes, each giving rise to new regions of inflation. The different colors represent changes in the values of physical constants (or, more speculatively, perhaps even different physical laws) among the universes.

19.7 A Universe Finely Tuned for Life?

We will end this chapter on an intriguing, but very controversial note, one with which some scientists vigorously disagree. It is at the extreme boundary of what can currently be called science, and may be more suitable to theology, philosophy, or metaphysics. But it is worth bringing up for completeness, since it addresses our most basic origins.

Our existence may allow us to draw some interesting conclusions about the Universe—for example, that it contains elements heavier than hydrogen and helium. Such deductions utilize the **anthropic principle:** We exist, hence the Universe must have certain properties or we wouldn't be here to see it.

In a few cases, apparently "mysterious" relationships between numbers were explained with anthropic reasoning. A previously unknown property of the energy levels of the carbon nucleus was actually *predicted* by Fred Hoyle in this manner. He argued that it must exist—otherwise there would not be enough carbon in the Universe.

The values of the physical constants (not to mention the laws of physics) seem to be spectacularly "fine-tuned" for life as we know it—indeed, almost "tailor-made" for humans. In many cases, if things were altered just a tiny amount, the results would be disastrous for life, and even for the production of heavy elements or molecules. The Universe would be "stillborn" in terms of its complexity. Here are a few examples.

If the weak nuclear force were weaker, most neutrons would not have turned into protons shortly after the big bang. The protons and neutrons would have then fused to form helium, leaving little hydrogen. A similar conclusion is reached if the weak nuclear force were much stronger: Protons would quickly combine with each other to form helium nuclei, converting two of the four protons to neutrons.

Now suppose the strong nuclear force were weaker. It turns out that the elements heavier than hydrogen would be unstable. The Periodic Table of the Elements would be very boring, and stars would not produce nuclear energy. Conversely, if the strong nuclear force were stronger, protons would bind together so easily that hydrogen would not exist.

Or, if the mass of a neutron and a proton were more nearly equal (they are the same to within about 0.14 per cent), most protons would be converted to neutrons.

ASIDE 19.11: Possible universes

Ours is not necessarily the *only* interesting Universe, or the *most interesting* possible Universe. But we argue that a far larger number of *uninteresting* possibilities exist.

Without hydrogen, stars would live only a short time, and there would be no hydrogen-rich compounds such as water and hydrocarbons.

Additional examples abound. If gravity were considerably stronger relative to electromagnetism, stars would be more massive, and they would burn out too quickly for very complex life (as we know it) to develop. Also, if certain relationships between the energy levels of different kinds of atomic nuclei were not present, the heavy elements would not have formed; the case of carbon, predicted by Hoyle, was mentioned above. If electrons were not much less massive than protons, complex chemistry would not be possible. Large molecules such as DNA can have well-defined shapes only if the nuclei of atoms are much heavier than electrons.

Some people use these and other "cosmic coincidences" to argue for a divine Creator who specifically designed the Universe in this manner. But this theological conclusion is not testable by the methods of science. Therefore, here we will not consider it further, though perhaps it is true.

Or, one could say that there is only a single Universe, and it has these properties by chance. This conclusion is difficult to accept; the odds are astronomically low (pun intended), if most other possible combinations of physical constants lead to conditions unsuitable for life. However, it may be true.

Finally, one could say that there is an ensemble of universes, perhaps even infinitely many, which span a very wide range of properties (values of the physical constants, particle masses, etc.). We, of course, were dealt a "winning hand"—indeed, a straight flush: We live in a universe that allowed life and sentient beings to develop. Such universes might be exceedingly rare; most random hands are not winners under the rules of poker.

Though an excellent example of the Copernican principle applied to the grandest imaginable scale, the multiverse idea is not directly testable, and is therefore more in the realm of philosophy and theology than of science. It seems that we are still left with the ultimate question: How and why did the Universe (or the hyperspace that contains it) occur? Perhaps there is no way, even in principle, for humans to answer this question.

CONCEPT REVIEW

The expansion of the Universe alone suggests, but does not demand, that there was a big bang (Section 19.1). According to the alternative **steady-state theory,** the Universe had no beginning in time and will have no end. This idea is based on the **perfect cosmological principle:** The average properties of the Universe do not change at all with time.

Historically, there were several observational arguments against the steady-state theory, but the fatal blow was the discovery of the **cosmic microwave radiation,** a faint radio glow that uniformly pervades the Universe (Section 19.2a). It was produced when the Universe was hot, before an age of about 400,000 years, and now corresponds to a very low temperature because of the cooling produced by expansion (Section 19.2b).

In the 1990s, tiny ripples were detected in the cosmic background radiation, corresponding to the seeds from which superclusters and clusters of galaxies were formed (Section 19.3a). The observed angular size of typical variations indicates that the Universe is spatially flat on large scales (Section 19.3b). These ripples were recently mapped in great detail (Section 19.3c). They were used to show that Hubble's constant is 71 km/sec/Mpc, and that normal matter, exotic "cold dark matter," and "dark energy" constitute 4.4%, 23%, and 73% of the total energy density of the Universe, respectively.

One can consider the Universe when it was very young; it consisted of uniformly distributed gas and radiation at the same temperature, governed by just a few fundamental laws (Section 19.4a). Specifically, we discussed the first several minutes after the big bang, starting from an almost inconceivable age of 10^{-43} sec, the **Planck time** (Section 19.4b). Up to an age of about 1 microsecond, the Universe consisted of a primordial soup of **quarks,** gluons (which normally bind quarks together), and photons. A slight dominance of matter over antimatter was established, making possible our existence.

The lightest elements were formed in the first 10 minutes through the process of **primordial nucleosynthesis** (Section 19.4c), and their currently observed proportions show striking agreement with the predictions of the big-bang theory. The relative amounts of different isotopes, combined with other observations, tell us that much of the dark matter in the Universe must consist of exotic particles such as neutrinos.

Some observed aspects of the Universe, such as its incredible uniformity (the **horizon problem**) and nearly flat spatial geometry (the **flatness problem**), are difficult to understand in the context of the original big-bang theory (Section 19.5a). It has been conjectured that the Universe began much smaller than we had thought, was able to achieve a uniform temperature, and subsequently went through an era of extremely rapid expansion known as **inflation** (Section 19.5b). The resulting Universe would then be truly enormous, much larger than what we can see—and our relatively small, observed region would naturally appear to be homogeneous and flat. Inflation affects just the first tiny fraction of a second of the Universe's existence, yet it removes some of the arbitrary assumptions of the original big-bang theory.

Although many alternative possibilities are now being considered, a clue to why inflation occurred may be provided by theories that postulate a high-energy merging of at least three of the four fundamental forces of nature: **gravity, electromagnetism, strong nuclear force,** and **weak nuclear force** (Section 19.5c). Already physicists have shown that electromagnetism and the weak nuclear force are unified at high energies to form the **electroweak force;** the two forces cannot be distinguished and hence appear **symmetric** (unified). At still higher energies, the electroweak and strong nuclear forces might merge, according to **grand unified theories (GUTs).** It is possible that even gravity can be unified with the grand unified force. Perhaps the best candidates for such speculative "theories of everything" (TOEs) are **superstring** (string) **theories,** which postulate that all fundamental particles are different forms of vibration of a tiny string or membrane.

As the Universe expanded and cooled, the various forces should have become apparent as different manifestations of the same force—that is, **phase transitions** should have occurred, like water turning to ice. But if the Universe **supercooled** without undergoing a phase transition, it would have contained excess energy that caused inflation for a short time (Section 19.5d). Inflation solves the horizon and flatness problems; moreover, it explains the absence of "magnetic monopoles," and its predictions are consistent with the observed distribution of matter on various scales (Section 19.5e). It is speculated that before inflation, the Universe could have begun as a quantum fluctuation out of *nothing,* the "ultimate free lunch" (Section 19.5f).

We end with the still more speculative possibility that there are many, perhaps even infinitely many, universes in a **multiverse** (Section 19.6). The values of the physical constants might be different in universes other than our own. According to the **anthropic principle,** our existence allows us to deduce some properties of the Universe (Section 19.7). Many calculations suggest that the consequences of modified constants would be disastrous for life as we know it, or even for the production of heavy elements or molecules. Perhaps our Universe is simply one of few in the multiverse with conditions suitable for the formation of complexity, culminating in life.

QUESTIONS

1. Explain how, in the absence of other information, the expansion of the Universe could be consistent with no beginning and no end in time: a "steady-state universe."
2. Summarize some observational arguments against the steady-state theory.
3. Describe the origin of the cosmic background radiation and the reason it now corresponds to a very low temperature.
4. Would you believe the announcement of the discovery of a galaxy at redshift 10,000? Why or why not?
5. Explain why the cosmic background radiation looks slightly hotter in one direction of the sky, and slightly cooler in the opposite direction.
6. Discuss the significance of tiny inhomogeneities, "ripples," in the cosmic background radiation.
7. Explain why we can consider with some confidence the behavior of the Universe early in its history, even before it was 1 second old.
8. Discuss the significance of the matter–antimatter asymmetry in the Universe.
9. Summarize the main physical events that occurred in the Universe's first 10 minutes of existence.
10. Describe how the currently observed proportions of the lightest chemical elements reflect conditions in the Universe shortly after its birth.
11. What are two observational problems with the original big-bang theory of the Universe?
12. Explain how the main problems with the original big-bang theory can be resolved if, at very early times, the Universe was much smaller than we had thought and subsequently inflated by an enormous factor.
13. Explain how space itself can expand faster than light without violating Einstein's theory of relativity.
14. Describe the fundamental forces of nature and attempts to unify them.
15. Discuss what should have happened to the "grand unified force" as the Universe expanded and cooled at an age of about 10^{-37} sec.
16. Outline how the Universe may have supercooled, leading to inflationary expansion by an almost arbitrarily large factor when the Universe was only 10^{-35} sec old.

†17. Suppose that during inflation, the size-doubling timescale was only 10^{-38} sec. If inflation started at $t = 10^{-37}$ sec and continued until $t = 10^{-35}$ sec, show that about 1000 doublings occurred.

18. Explain how the *total* energy of the Universe might be zero, or nearly zero.

19. Why has it been suggested that the Universe may be the "ultimate free lunch"?

20. Define what is meant by the term "universe."

21. Describe several ways in which there might be other universes besides our own.

22. Outline the basic idea behind superstring theories and the possible existence of dimensions far too small to detect.

23. Discuss how the properties of the Universe depend on the values of the physical constants.

24. Summarize various possibilities for why our Universe seems to be so finely tuned for life.

25. State the anthropic principle and its implications.

26. **True or false?** Sensitive observations have not revealed any inhomogeneities in the temperature of the cosmic microwave background radiation, leading to major conflicts with theories of galaxy formation.

27. **True or false?** The expansion of the Universe is consistent with Hoyle's "steady-state theory" of the Universe.

28. **True or false?** It is quite possible that the Universe started from essentially zero energy and has remained at essentially zero total energy ever since.

29. **True or false?** The most highly redshifted electromagnetic radiation we have detected so far is from quasars or very distant galaxies.

30. **True or false?** The force involved in chemical reactions is the strong nuclear force.

31. **Multiple choice:** There are good reasons to believe that at one time the Universe expanded extremely rapidly. This "inflation" occurred primarily **(a)** within the first 10^{-30} seconds after the Universe was created; **(b)** when the Universe was about 10^{-6} seconds old; **(c)** when the Universe was about 1 second old; **(d)** when the Universe was about 1 million years old; or **(e)** when the Universe was about 10 billion years old.

32. **Multiple choice:** Which one of the following prevented photons from travelling long distances in the early Universe? **(a)** Before inflation the Universe was small enough for the photon to wrap back around to its source. **(b)** Free electrons easily interacted with the photons, absorbing and re-emitting them. **(c)** Neutron decay absorbed most photons. **(d)** The large expansion of the early Universe redshifted the photons to slow speeds. **(e)** The dark energy absorbed the photons.

33. **Multiple choice:** Which one of the following statements about the inflationary theory is *false*? **(a)** The Universe may have inflated because of a strange vacuum energy associated with a "supercooled" state of the Universe. **(b)** The Universe is uniform (i.e., homogeneous and isotropic) on large scales because long ago, before inflation, parts of the Universe that are currently very distant were able to come into equilibrium. **(c)** The Universe started off in a very hot, dense state and has inflated exponentially ever since. **(d)** According to the inflationary theory, the Universe is very much larger than the parts that we currently see. **(e)** During inflation, the speed of the expansion of space can exceed the speed of light.

34. **Multiple choice:** Which one of the following statements about the early universe is *true*? **(a)** There were many forces in the past, which eventually unified, leaving four fundamental forces. **(b)** The Universe started off very large and has remained the same average density due to the random creation of small amounts of matter over time. **(c)** When matter and antimatter combined, a small excess of matter was left over, which later formed stars and galaxies. **(d)** The creation of matter left very little room in the Universe and space was forced to expand as a result. **(e)** Small amounts of heavy elements, including carbon and oxygen, were formed within a few minutes after the big bang.

35. **Multiple choice:** The first helium nuclei in the Universe were synthesized **(a)** during the inflationary epoch; **(b)** by the first generation of stars; **(c)** immediately after the Universe became transparent to radiation; **(d)** when the Universe was roughly 10^{-6} seconds old; or **(e)** when the Universe was only a few minutes old.

36. **Fill in the blank:** According to the ________, we can deduce certain properties of the Universe based on our mere existence.

37. **Fill in the blank:** The cosmic microwave background radiation was released when _____ and ______ combined together for the first time.

38. **Fill in the blank:** The lightest elements formed shortly after the big bang in a fusion process known as ________.

39. **Fill in the blank:** Recent measurements have shown that the ratio of the actual density of the Universe to the critical density is _____, meaning that over very large distances the shape of the Universe is ____.

40. **Fill in the blank:** Rather than being fundamental, protons and neutrons consist of smaller particles called ______, which themselves might be tiny, vibrating packets of energy known as _______.

†This question requires a numerical solution.

TOPICS FOR DISCUSSION

1. Does it even make sense to include times before $t = 0$ in a scientific discussion, given that science is supposed to deal with predictions that are testable in principle?
2. What do you think of the argument that the Universe has a total energy of zero, and that it arose from a quantum fluctuation out of "nothing"?
3. Do you think some forms of "life" could arise in most universes, almost regardless of the specific values of the physical constants?

MEDIA

Virtual Laboratories

Cosmology and Cosmic Microwave Radiation

AceAstronomy™ Log into AceAstronomy at **http://astronomy.brookscole.com/cosmos3** to access quizzes and animations that will help you assess your understanding of this chapter's topics.

Log into the Student Companion Web Site at **http://astronomy.brookscole.com/cosmos3** for more resources for this chapter including a list of common misconceptions, news and updates, flashcards, and more.

CHAPTER 20

Life in the Universe

ORIGINS

One of the most important questions we can pursue in our quest to determine our origin is whether life, and especially intelligent, communicating life, has arisen independently elsewhere in the Universe. Are humans a common phenomenon, or are we alone?

AIMS

1. Understand the nature of life and conditions necessary for its emergence (Section 20.1).
2. Identify the kinds of planets and stars that might be most suitable for the development of life and intelligence (Sections 20.2 and 20.3).
3. Explore ways of searching for, and communicating with, extraterrestrial life (Sections 20.4 and 20.5).
4. Assess the probability that intelligent communicating life exists elsewhere in our Galaxy besides the Earth (Section 20.6).
5. Discuss UFOs and how they relate to the scientific method (Section 20.7).

We have discussed the nine planets and some of the moons in the Solar System, and have found most of them to be places that seem hostile to terrestrial life forms. Yet a few locations besides the Earth—most notably Mars, with its signs of ancient running water, and Europa, with liquid water below its icy crust—have characteristics that suggest life may have existed there in the past, or might even be present now or develop in the future. **Exobiology** is the study of life elsewhere than Earth.

In our first real attempt to search for life on another planet, in the 1970s, NASA's Viking landers carried out biological and chemical experiments with martian soil. The results seemed to show that there is probably no life on Mars (■ Fig. 20–1). A small, inadequately funded British probe, Beagle 2 (the first Beagle having been Darwin's ship), failed as it approached Mars in 2003. (One reason it became so popular among the public may have been that it was supposed to send back tones from the musical group Blur when it landed, though those tones never came.)

NASA continues to explore Mars with robotic spacecraft and one day should have a more sophisticated biology lab landing on it. In the meantime, its Mars Exploration Rovers, Spirit and Opportunity, and the European Space Agency's Mars Express have found clear signs that water flowed on Mars in the distant past, raising hope that life could have formed there at that time and might even have survived.

Although studies of a martian meteorite in 1996 gave some indications of ancient, primitive life on Mars (see Chapter 6), this idea has not been generally accepted, though it is still causing much discussion. Jupiter's moon Europa and Saturn's moon Titan (see Chapter 7) are also intriguing places where scientists think it is possible that life has begun.

Since it seems reasonable that life as we know it would be on planetary bodies, we first discuss the chances of life arising elsewhere in the Solar System, as well as the kinds of stars most likely to have planets suitable for the emergence and development of intelligent life. Next we explore attempts to receive communication signals from intelligent extraterrestrials.

AceAstronomy™ The AceAstronomy icon throughout this text indicates an opportunity for you to test yourself on key concepts, and to explore animations and interactions of the AceAstronomy website at **http://astronomy.brookscole.com/cosmos3**

ASIDE 20.1: A tenth planet?

During the publication process of this book, astronomers announced the discovery of a possible tenth planet in our Solar System. This object, tentatively named 2003 UB313, is likely to be just the largest known Kuiper-Belt Object. Being bigger than Pluto, however, it has greatly fueled the ongoing debate regarding the exact definition of a planet.

◄ Yoda.

NASA/JPL/Caltech

■ **FIGURE 20–1** The surface of Mars from a Viking Lander, which carried a small biology laboratory to search for signs of life. It probably didn't find any. We now still look for life on Mars but are also considering Europa as another possible abode.

■ **FIGURE 20–2** "I'll tell you something else I think. I think there are other bowls somewhere out there with intelligent life just like ours." (Drawing by Frank Modell; © 1987 *The New Yorker Collection 1987*, Frank Modell, from http://cartoonbank.com. All Rights Reserved.)

We also consider a way to estimate the number of communicating civilizations elsewhere in our Milky Way Galaxy, or at least to see which factors most seriously limit our ability to do so. It is possible that humans are the only technologically advanced civilization in our Galaxy, or one of very few. Finally, we explain why most scientists do not consider reported sightings of UFOs to be good evidence for extraterrestrial visitations to Earth.

NASA has formed an institute of "astrobiology" and is making a major push to investigate matters of biology that can be important to understanding the origin of life or to space exploration. The institute is "virtual," in that it has no actual buildings but rather is made of individuals who communicate by e-mail and by occasional meetings.

20.1 The Origin of Life

It would be very helpful if we could state a clear, concise definition of life, but unfortunately that is not yet possible. Biologists state several criteria that are ordinarily satisfied by life forms—reproduction, for example. Still, there exist forms on the fringes of life, such as viruses, that need a host organism in order to reproduce. Scientists cannot always agree whether some of these things are "alive."

In science fiction, authors sometimes conceive of beings that show such signs of life as the capability for intelligent thought (■ Fig. 20–2), even though the being may share few of the other criteria that we ordinarily recognize. In Fred Hoyle's novel *The Black Cloud,* for instance, an interstellar cloud of gas and dust is as alive as (and smarter than!) human beings. But we can make no concrete deductions if we allow such wild possibilities, so exobiologists prefer to limit the definition of life to forms that are more like "life as we know it."

Life on Earth is based on **amino acids**—chains of carbon, in which each carbon atom is bonded to hydrogen and sometimes to oxygen and nitrogen. Chemically, carbon is essentially unique in its ability to form such long chains; indeed, we speak of compounds that contain carbon as being **organic.** But life is selective, incorporating only about 20 of all the possible amino acids.

Similarly, long chains of amino acids form **proteins,** though life utilizes only a small fraction of the multitude of possible combinations of amino acids. The genetic code of any living creature is contained in one extremely long and complex structure: DNA (deoxyribonucleic acid), the famous "double helix."

How hard is it to build up long organic chains? To the surprise of many, an experiment performed in the 1950s showed that making organic molecules is easier than had been supposed. Stanley Miller and Harold Urey, at the University of Chicago, filled a glass jar with simple molecules like water vapor (H_2O), methane (CH_4), and ammonia (NH_3), along with hydrogen gas (H). They exposed it to electric sparks, simulating the vigorous lightning that may have existed in the early stages soon after Earth's formation.

After a few days, long chains of atoms formed in the jar, in some cases complex enough to include simple amino acids, the building blocks of life. Later versions of these experiments created even more complex organic molecules from a wide variety of simple actions on simple molecules (■ Fig. 20–3). Such molecules may have mixed in the oceans to become a "primordial soup" of organic molecules (■ Fig. 20–4).

Since the original experiment of Miller and Urey, most scientists have come to think that the Earth's primitive atmosphere was not made of methane and ammonia, which would have disappeared soon after the Earth's formation. Instead, it may have consisted mostly of carbon dioxide, carbon monoxide, and water, and such a mixture does not generally lead to a large abundance of amino acids.

On the other hand, complex molecules may have formed near geothermal sources of energy under the oceans or on Earth's surface, or perhaps in vents deep under-

Photograph by Andrew Clegg/Cornell University

Louis J. Allamandola, NASA/Ames Research Laboratory

■ **FIGURE 20–3** ⓐ Bishun Khare and Carl Sagan, of Cornell University, in their laboratory several decades ago. The apparatus was used to simulate conditions thought to be present on the primitive Earth, to see whether complex compounds such as amino acids form easily from the gases of which Earth's atmosphere consisted at that time. ⓑ Louis J. Allamandola with equipment used for similar purposes in his Astrochemistry Laboratory at the NASA/Ames Research Center.

NASA'S Ames Research Center

■ **FIGURE 20–4** The simple solution of organic material in the oceans, from which life may have arisen, is informally known as "primordial soup."

ground, where the right raw materials existed. Also, extraterrestrial amino acids have been found in several meteorites that had been long frozen in Antarctic ice, as well as in some other meteorites (■ Fig. 20–5). In any case, some extrasolar planets or moons in our Galaxy may have had primitive atmospheres similar to the mixture used by Miller and Urey, so the results of their experiments are interesting.

Also relevant are laboratory observations with conditions mimicking the low density (strong vacuum) of space. Amino acids have been observed to form.

However, mere amino acids or even DNA molecules are not life itself. A jar containing a mixture of all the atoms that are in a human being is not the same as a human being. This is a vital gap in the chain; astronomers certainly are not yet qualified to say what supplies the "spark" of life.

Still, many astronomers think that since it is not difficult to form complex molecules, primitive life may well have arisen not only on the Earth but also in other locations. The appearance of very simple organisms in Earth rocks that are 3.5 billion years old, and indirect evidence for life as far back as 3.8 billion years (not long after the end of the bombardment suffered by the newly formed Earth), suggests that primitive life arises quite easily. Similarly, the presence of life in what appear to be very harsh environments on Earth (water that is highly acidic or near its boiling point, for example) shows that life can exist in extreme conditions.

The discovery of indigenous life on at least one other planet or moon in the Solar System would provide much support for the hypothesis that simple organisms such as microbes and bacteria form readily. But even if life is not found elsewhere in the Solar System, there are so many other stars in space that it would seem that life might have arisen at some location.

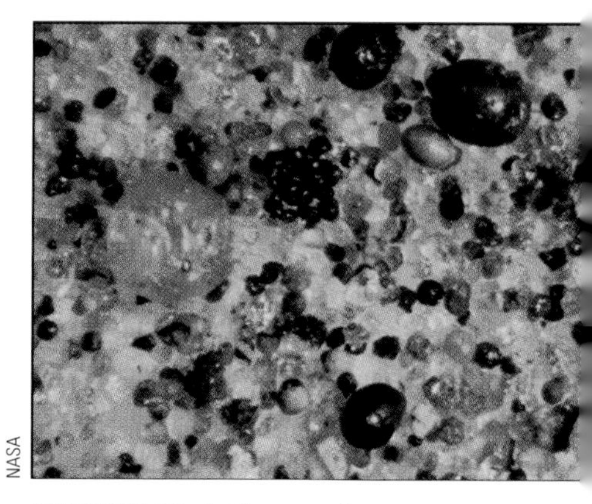
NASA

■ **FIGURE 20–5** Close-up of a part of the Murchison meteorite, in which simple amino acids have been found. Analysis shows that they are truly extraterrestrial and the sample had not simply been contaminated on Earth.

20.2 Life in the Solar System

Life elsewhere in our planetary system, if present at all, is primitive at best (single cells, or perhaps very simple multi-cellular organisms). Among all of the planets and moons, only a few have nonzero odds for life.

NASA's JSC

■ **FIGURE 20–6** The martian meteorite ALH 84001, in which it has been controversially suggested that tiny "nano-fossils" have been found, as shown in Figures 6–56 and 6–57. The conclusion is not widely accepted by other scientists working in this field.

Mars has provided the best evidence thus far, but it is still very controversial (■ Fig. 20–6). Moreover, even if real, life on Mars might not have been independent of Earth—a meteorite from Mars containing simple life may have contaminated the young Earth, "seeding" it with life, though this idea is of course quite speculative.

Europa, one of Jupiter's large moons, looks like a promising environment: Below its icy surface there is almost certainly water slush or even an ocean. A Europa orbiter to study this moon in more detail is being considered by NASA, though its recent "Vision for Space Exploration" of "Moon, Mars, and Beyond" is apparently slowing down all other NASA projects, even the Europa orbiter.

Titan, Saturn's largest moon, has a thick atmosphere of nitrogen molecules. As we saw, in 2004, the Cassini spacecraft sent the Huygens probe into this atmosphere. There is evidence for substantial methane and ethane on Titan, perhaps in the form of lakes; methane can be solid, liquid, or gas at temperatures reasonable for Titan, just as water can have solid, liquid, or gaseous phases on Earth. Huygens, as it approached Titan's surface, imaged what appeared to be a lake shore. The Cassini orbiter's radar has found a reflection that seems lake-like, presumably with the liquid being a tarry substance. The absence of liquid water, however, makes it more difficult for life as we know it to form.

Io, another Galilean satellite, has conditions that may be suitable for life resembling that found near volcanic vents on Earth. Again, however, the apparent absence of water is a major problem.

Note that an intelligent alien who obtains and correctly interprets a spectrum of Earth could deduce the presence of life here. The large amount of free oxygen suggests the continuous production by a process like photosynthesis; otherwise, oxygen would rapidly be depleted because it is so reactive. In addition, methane quickly reacts with oxygen, so the significant amount of methane (largely from cows—"bovine flatulence," to quote the late Carl Sagan) in our atmosphere implies a steady production mechanism—the decay of organic compounds. Such signs have been picked up from spacecraft, such as Galileo and Cassini, while they were near the Earth and headed for the outer planets.

20.3 Suitable Stars for Intelligent Life

If we seek indigenous (that is, originating locally, not from elsewhere), intelligent life on planets orbiting other stars, what kinds of stars have the best odds? Stars that are either near the beginning or end of their lives are not very good bets. For example, intelligent life may take a long time to form, so very young stars are less suitable than older stars. Stars heading to the red giant or white dwarf stages go through rapid changes, making it difficult for complex life to survive. White dwarfs and neutron stars certainly have passed through stages that would have destroyed life.

Some types of main-sequence stars are also not especially suitable. The lives of O-type and B-type stars are probably too short for the development of life of any kind. Planets around A-type stars might have life, but probably not intelligent life. Type-M and type-L stars have a small **ecosphere** (also known as the "habitable zone")—the range of distances in which conditions suitable for life might be found. There is unlikely to be a planet in this narrow region around a low-mass star. Also, such a planet would be in "synchronous rotation": The same hemisphere would always face the star, so one side would be very hot, and the other very cold.

Main-sequence stars of types F, G, and K are the most likely candidates. They live a long time, and models of their ecosphere lead to reasonably large sizes. (In more detail, such models must explain the "Goldilocks effect" in the case of the Solar System: why Venus is too hot, Mars is too cold, but the Earth is just right.)

Single stars, stars in very wide binary systems (with planets orbiting close to one star), or closely spaced binary stars (which planets could orbit at large distances) are most suitable. Planets that move from one star to another in a binary system (for example, like a "figure eight") tend to have unstable orbits and are ejected.

Over 160 extrasolar planets have already been found (see our discussion in Chapter 9). In most cases they orbit F, G, and K stars because the searches specifically targeted those stars. Generally these planets are gas giants, in many cases very close to the star or on highly eccentric orbits and hence probably inhospitable to intelligent life. A few of them, however, appear to have orbital properties potentially suitable for the emergence of life; perhaps at least simple life exists within their atmosphere or on moons orbiting them. The 2005 announcement of a planet, presumably rocky, with only about 7 to 8 times Earth's mass, gives hope of our finding closer analogs to Earth, which encourages those trying to discover signs of life there.

20.4 The Search for Extraterrestrial Intelligence

How should we look for intelligent extraterrestrial life? Perhaps we can find evidence for such life here on Earth, in the form of alien spacecraft that have landed here. After all, Pioneers 10 and 11 and Voyagers 1 and 2 are even now carrying messages out of the Solar System in case an alien interstellar traveller should happen to encounter these spacecraft (■ Fig. 20–7).

The odds, however, seem very small, given the vastness of space. Although spaceships can, in principle, travel between the stars, even rather quickly according to Einstein's special theory of relativity (see *Figure It Out 20.1: Interstellar Travel and Einstein's Relativity*), such journeys are difficult, expensive, dangerous, and probably rare compared to communication with electromagnetic waves.

A potentially more fruitful approach is to search for electromagnetic signals, but some waves are more suitable than others. For example, x-ray and gamma-ray photons have high energy and are therefore expensive to produce, and typical atmospheric gases block them. Ultraviolet photons are absorbed by interstellar dust in the plane of our Milky Way Galaxy. At optical wavelengths, the signal from a planet orbiting a star is very difficult to distinguish from the bright light of the star itself, unless a large

Jay M. Pasachoff
(a)

Jay M. Pasachoff
(b)

■ **FIGURE 20–7** (a) The gold-plated copper record bearing two hours' worth of Earth sounds and a coded form of photographs, carried by Voyagers 1 and 2. The sounds include a car, a steamboat, a train, a rainstorm, a rocket blastoff, a baby crying, animals in the jungle, and greetings in various languages. Musical selections include Bach, Beethoven, rock, jazz, and folk music. (b) The record includes 116 photographs. Among them is this view, which one of the authors (J.M.P.) took when in Australia for a solar eclipse. It shows Heron Island on the Great Barrier Reef in Australia, in order to illustrate an island, an ocean, waves, a beach, and signs of life.

FIGURE IT OUT 20.1

Interstellar Travel and Einstein's Relativity

Interstellar travel is, in principle, possible (but extremely difficult!) for humans to achieve. If the speed is constant, distance is equal to speed multiplied by time,

$$d = vt,$$

and we can calculate that a rocket moving with the Earth's escape velocity (11 km/sec) would reach the bright star Sirius (8.7 light-years away) in 240,000 years. (Recall from Chapter 14 that the escape velocity is the speed needed at liftoff to escape from an object's surface.)

One can imagine a huge spaceship containing a community of explorers that has permanently left Earth. They would have to be self-sufficient (for example, grow their own food), and the journey to Sirius would span roughly 10,000 generations. To be sure, humans would need to overcome enormous obstacles to successfully complete such a trip, but it is not categorically impossible.

But what if you want to make the journey within a single human lifetime? It is possible, according to Einstein's special theory of relativity. The key fact is that time slows down if you move rapidly: A given time interval t_{Earth} (according to a clock on Earth) is reduced to

$$t_{Travel} = t_{Earth} \times [1 - (v/c)^2]^{1/2}$$

for the traveller, where c is the speed of light. For example, at a speed of $v = 0.995c$, the round-trip travel time to Sirius according to an observer on Earth is 17.5 years, but only 1.75 years according to the traveller. (Here we ignore the additional time it takes to accelerate up to that speed, but the general idea still holds.)

At higher speeds, the traveller ages even less: If $v = 0.9999c$, 17.4 years will have passed on Earth, but the journey is only 3 months long according to the traveller's clock! Thus, by moving fast one ages less and effectively "jumps" into the future. However, the actual life span according to the traveller's clock is unchanged: In 3 months of elapsed time, for example, the traveller reads far fewer books than the Earthling does in the corresponding 17.4 years.

How can this be, according to the traveller's frame? He sees the Earth zooming away from him, and Sirius is rapidly approaching. Well, special relativity tells us that lengths along the direction of motion contract. The distance between Earth and Sirius is reduced to

$$L_{Travel} = L_{Earth} \times [1 - (v/c)^2]^{1/2}$$

for the traveller, where L_{Earth} is the length measured by a person on Earth. At $v = 0.995c$, the round-trip distance is only

$$(17.4 \text{ lt yr})(0.1) = 1.74 \text{ lt yr}$$

according to the traveller, and this distance is traversed in only 1.75 years at $v = 0.995c$. At $v = 0.9999c$, the round-trip distance is only 3 light-months, which is traversed in just 3 months. Thus, at high speeds, times dilate and lengths contract. Everything is self-consistent, and rapid interstellar travel is possible in principle.

The overwhelming *practical* problem is that travel close to the speed of light requires astronomically large amounts of energy, if the rocket's mass is non-negligible. According to special relativity, the total energy E of the rocket is given by

$$E = m_0c^2/[1 - (v/c)^2]^{1/2},$$

where m_0 is the rest mass. For $v = 0.995c$, $E \approx 10m_0c^2$—that is, 10 times its rest mass; for $v = 0.9999c$, $E \approx 71m_0c^2$. These are the energies that must be expended to accelerate to such speeds, assuming 100% efficiency! (To give you some feeling for this quantity of energy, the most powerful bombs ever built by humans release the energy equivalent of only about 2 kilograms of material—that is, $E = m_0c^2$, where $m_0 = 2$ kg.) If the efficiency of energy conversion is only 1% (the maximum possible for nuclear energy), then $E \approx 1000m_0c^2$ or $7100m_0c^2$ for the above two examples. These are truly staggering amounts of energy!

When one considers that the rockets sent to the Moon had about 1000 times as much mass as their payload, the energy requirements become even more prohibitive, since the rest mass m_0 is so large. It would be best to gather the fuel along the way, from interstellar gas. Maybe some highly advanced aliens have overcome these and other formidable problems, such as reaching very high speeds in relatively small amounts of time (which requires large accelerations—and these are harmful to life).

amount of energy is concentrated into a narrow range of wavelengths or into a short pulse. And at infrared wavelengths, Earth's warm atmosphere makes the sky bright.

For many years, radio waves have seemed to be the best choice: They are easy and cheap to produce, and are not generally absorbed by interstellar matter. Also, there are few sources of contamination—although the increasing number and strength of radio and television stations is a threat to radio astronomers in the same manner that city lights brighten the sky at optical wavelengths.

At least initially, it seems too overwhelming a task to listen for signals at all radio frequencies in all directions and at all times. One must make some reasonable guesses on how to proceed.

A few frequencies in the radio spectrum seem especially fundamental, such as the 21-cm line of neutral hydrogen (as described in Chapter 15). This wavelength corresponds to 1420 MHz, a frequency over ten times higher than stations at the high end of the normal FM band. We might conclude that creatures on a far-off planet would decide that we would be most likely to listen near this frequency because it is so fundamental. The "water hole," the wavelength range between the radio spectral lines of H and OH, has a minimum of radio noise from background celestial sources, the telescope's receiver, and the Earth's atmosphere, and so is another favored possibility.

Humans have conducted several searches for extraterrestrial radio signals. No unambiguous evidence has been found, but the quest is worthwhile and continues. A telltale signal might be an "unnatural" set of repeating digits, such as the first 100 digits of the irrational number π (the Greek letter "pi"). We must verify that the signal could not be produced by a natural, non-intelligent source (such as a pulsar), and also that it is not of human origin (either unintentionally or intentionally).

Though there is agreement to verify the veracity of the supposition that a signal received is indeed a message or a sign of extraterrestrial life, and to do so before making a public announcement, the discovery of such an extraterrestrial signal would be so momentous that it would surely be released to the public without major delay. A protocol has been accepted by researchers in the field as to how to announce any believable signal from extraterrestrials, so as not to cause fear and panic.

In 1960, Frank Drake conducted the first serious search for extraterrestrial signals. He used a telescope at the National Radio Astronomy Observatory (in West Virginia) for a few months to listen for signals from two of the nearest stars—tau Ceti and epsilon Eridani. He was searching for any abnormal kind of signal—a sharp burst of energy, for example. He called this investigation Project Ozma after the queen of the land of Oz (■ Fig. 20–8) in L. Frank Baum's stories.

A NASA-sponsored group based at the Search for Extraterrestrial Intelligence (SETI) Institute in California began an ambitious effort on October 12, 1992, the 500th anniversary of Columbus's landing in the New World. It made use of sophisticated signal-processing capabilities of powerful computers to search millions of radio channels simultaneously in the microwave region of the spectrum (■ Fig. 20–9). Consisting of both a sky survey and a targeted study of individual stars, in its first fraction of a second it surpassed the entire Project Ozma. But Congress cut off funds anyway, after about a year.

The targeted search of the project, known as Project Phoenix, is continuing, backed by funds contributed by private individuals. Now led by astrophysicist Jill Tarter (■ Fig. 20–10), a real-life model for actress Jodie Foster's character Ellie Arroway in the movie *Contact,* it examines about 1000 stars. No unexplained signals have been found so far, though there have been some exciting false alarms, as in 1997 when an intriguing signal turned out to be from the SOHO spacecraft (which was described in Chapter 10)!

A well-known project is "SETI@home," whose operation is based at the University of California, Berkeley (see http://seti.berkeley.edu). It has established a way for the general public to contribute: During otherwise unused time on your home computer, a special program can automatically analyze data from the giant Arecibo radio telescope in Puerto Rico (■ Fig. 20–11) for signs of extraterrestrial signals (■ Fig. 20–12). Thus, there is a very small but nonzero chance that the first unambiguous evidence of intelligent life elsewhere in the Universe would be found by *your* computer, should you choose to participate!

From *Glinda of Oz,* by L. Frank Baum, illustrated by John R. Neill, copyright 1920

■ **FIGURE 20–8** Dorothy and Ozma climb the magic stairway.

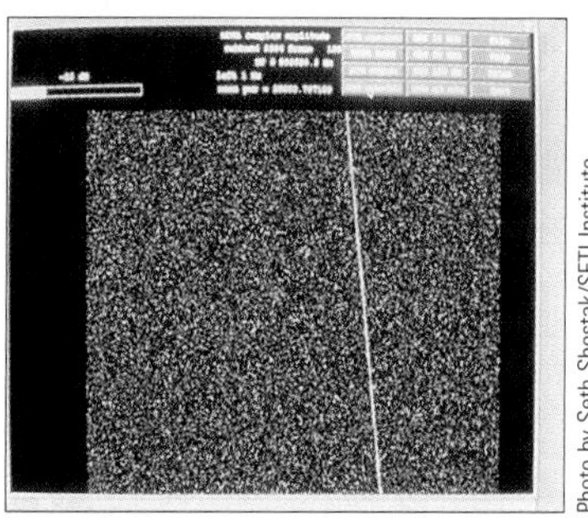

Photo by Seth Shostak/SETI Institute

■ **FIGURE 20–9** This "waterfall" display from Project Phoenix shows the instantaneous power in 462 narrow channels, each only 2 Hz wide. Each dot represents a channel, and the size of the dot is proportional to the strength of the signal. As a new spectrum is plotted in the top row across the screen, the older spectra move downward. Here, the signal changes steadily in frequency, so it slants on the screen. It is readily distinguishable from the background noise.

Photo by Seth Shostak/SETI Institute

■ **FIGURE 20–10** Jill Tarter, Director of Project Phoenix at the SETI Institute.

■ **FIGURE 20–11** The Arecibo radio telescope in Puerto Rico, used once to send a message into space. The telescope is 305 meters across, the largest telescope on Earth; you can see it in the movie *Contact.* It is sensitive enough to pick up a hypothetical cellular telephone signal on Venus.

The SETI@home effort already has nearly 6 million participants in over 200 countries, effectively forming the Earth's largest supercomputer, and the amount of computing time contributed has been over two million computer-years. The scientists conducting the project have taken the most suspicious signals from the widespread data analysis and gone back to Arecibo to observe those sources in detail. Obviously, no confirmation was obtained.

Although most searches for extraterrestrials have been conducted at radio wavelengths, some optical and near-infrared searches are also underway. Because the plane of the Milky Way Galaxy absorbs visible light, we can't expect to survey as many stars as at radio wavelengths, but there are still plenty of them. If we search for short, very intense pulses of light emitted by lasers, we can actually see quite far in the plane of our Galaxy, increasing the number of available stars. More importantly, the laser pulses outshine the light from the star that a planet is orbiting. Finally, such laser pulses are very difficult or impossible to produce by any natural phenomena other than life; detection of them could provide strong evidence for the existence of extraterrestrials.

Only some scientists think that there is a reasonable chance of detecting signals from extraterrestrial beings, at least in the near future. But it would be a shame if there were abundant signals that we missed just because we weren't looking. Of course, now that we have looked and listened for a while, that possibility seems to be reduced. There are other reasons to think that there may not be any extraterrestrials at all in our neighborhood within the Milky Way Galaxy, or perhaps even in our entire Galaxy. For example, the odds are that life would have arisen elsewhere than on Earth first, so why aren't the aliens already here? Maybe they aren't here because they don't exist.

Even if we don't get messages from outer space, there are many scientific spinoffs of the search. Investigating thousands of stars in detail, and mapping the sky in different parts of the spectrum, gives us bits of information that can lead to scientific breakthroughs of other kinds.

ASIDE 20.2: Seeing is believing?

During the past few years, searches for special, narrow optical emission lines have been made in high-resolution spectra of many nearby stars, but without success.

20.5 Communicating with Extraterrestrials

All of the searches described above are passive—astronomers are simply looking for signals from intelligent, communicating extraterrestrials. If such a signal is ever found and confirmed by several cross-checks, we might choose to "reply"—but not until some global consensus has been reached about who will speak for Earth and what they will say. Nevertheless, we have already intentionally sent our own signals toward very distant stars that may or may not be orbited by planets containing intelligent life.

For example, on November 16, 1974, astronomers used the giant Arecibo radio telescope in Puerto Rico to send a coded message about people on Earth (■ Fig. 20–13) toward the globular star cluster M13 (■ Fig. 20–14) in the constellation Hercules. The idea was that the presence of over 200,000 closely packed stars in that loca-

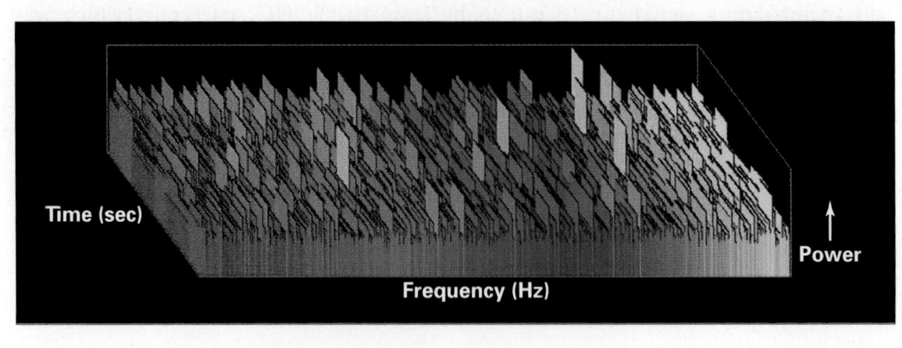

■ **FIGURE 20–12** Output on the screen of a computer that has been used to analyze some of the radio data collected by the "SETI@home" project. There have been no clear signs of intelligent life among the data processed so far.

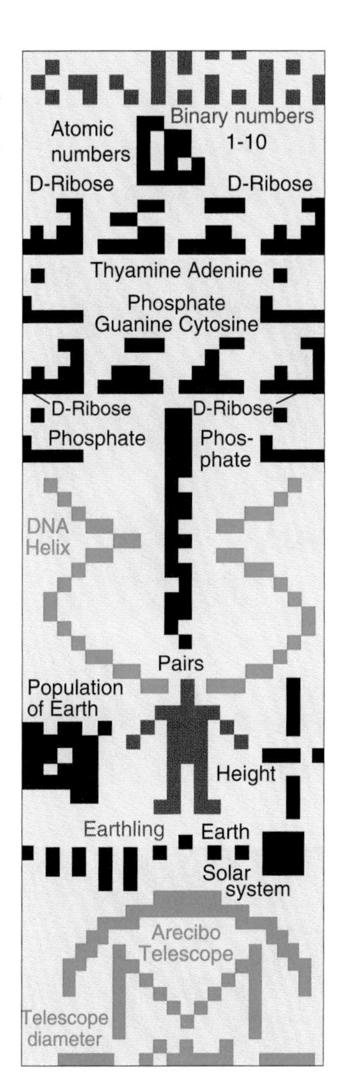

■ **FIGURE 20–13** The message sent to the globular cluster M13, plotted out and with a translation into English added. The basic binary-system count at upper right is provided with a position-marking square below each number. The message was sent as a string of 1679 consecutive characters, in 73 groups of 23 characters each (two prime numbers). There were two kinds of characters (0 and 1), each represented by a frequency; they are shown here as yellow background or other colors, respectively. If interpreted correctly, the diagram provides information about humans (DNA structure, population, etc.).

tion would increase the chances of our signal being received by a civilization in the planetary system of one of them.

But the travel time of the message (at the speed of light) is about 24,000 years to M13, so we certainly cannot expect to have an answer before twice 24,000, or 48,000 years, have passed. If any creature is observing the Sun at the right frequency (2380 MHz) when the signal arrives, the radio brightness of the Sun will increase by 10 million times over a 3-minute period. A similar signal, if received from a distant star, could be the giveaway that there is intelligent life there.

In retrospect, M13 may not have been the best choice as a target, despite its large number of stars. The problem is that the stars in globular clusters, being very old, were formed from gas that did not have a large proportion of heavy elements; it had not gone through many stages of nuclear processing by massive stars and supernovae. Rocky, Earth-like planets are thus not as likely to have formed there as they would have around younger stars.

A quarter century later, in 1999, a new message was sent by Canadian astronomers toward several relatively nearby (50 to 70 light-years away) Sun-like stars, including 16 Cygni B, which is known through Doppler measurements to have at least one planet orbiting it (see our discussion in Chapter 9). The complete message, which is much larger in size, duration, and scope than the one sent in 1974, was transmitted three times over a 3-hour period in the direction of each star. Still later, starting on July 5, 2003, some much more complicated messages were sent out toward five nearby stars (32 to 46 light-years away) from a 70-m radio telescope in Evpatoriya, Ukraine. One would hope to get answers in fewer than 100 years instead of tens of thousands, but nobody is betting on a return message.

Even had we not sent these few directed messages, during much of the 20th century (and so far in the 21st century) we have been unintentionally transmitting radio signals into space on the normal broadcast channels. Aliens within a few tens of light-years could listen to our radio and television programs. A wave bearing the voice of Winston Churchill is expanding into space, and at present is about 60 light-years from Earth. And once a week or so a new episode of *Fear Factor* is carried into the depths of the Universe. Do you think aliens would get a favorable impression of us from most of what they hear?

AceAstronomy™ Log into AceAstronomy and select this chapter to see the Active Figure called "Interstellar Communication."

© Bill and Sally Fletcher

■ **FIGURE 20–14** The globular star cluster M13, toward which a 3-minute message was sent with the Arecibo radio telescope in 1974.

ASIDE 20.3: Personal advertisements in space

When you place a personal advertisement on craigslist.com, you have a choice of using the Deep Space Network to beam it into space in addition to its terrestrial circulation. Tens of thousands of ads have been sent into space in this way.

20.6 The Statistics of Intelligent Extraterrestrial Life

Is it reasonable to expect that there are any signals out there that we can hope to detect with projects such as that at the SETI Institute? What if no intelligent creatures exist elsewhere in our Milky Way Galaxy or even in the observable parts of the Universe?

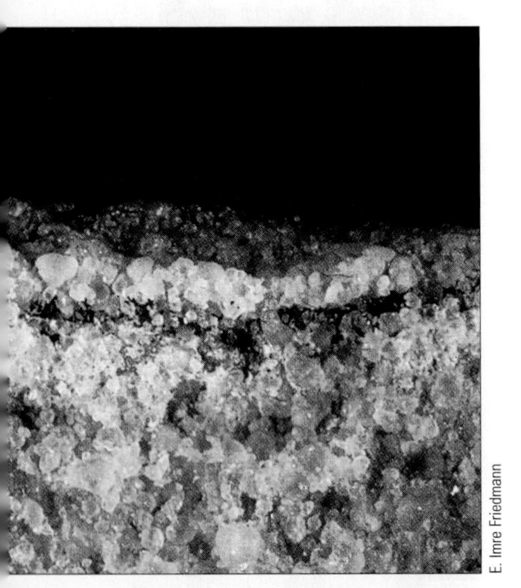

E. Imre Friedmann

a

b

■ **FIGURE 20–15** (a) We see the inside of an Antarctic rock, with a lichen growing safely insulated from the external cold. The image shows the outer 1 cm of the rock. (b) Bacteria that live over 1 km below the Earth's surface and that fluoresce as red patches on this microscope image. The bacteria, discovered in deep wells, survive on hydrogen generated by a chemical reaction between water and iron silicates in the surrounding volcanic rock. Thus they do not need sunlight (or oxygen). Do similar types of microbes exist on Mars?

b: Todd Stevens, photo from Pacific Northwest Regional Laboratory, operated by Batelle Memorial Institute

20.6a The Drake Equation

Instead of phrasing an all-or-nothing question about extraterrestrial life, we can use a procedure developed by Frank Drake, then of Cornell University, and extended by Carl Sagan and Joseph Shklovskii, among others. The overall problem is broken down into a chain of simpler questions, the results of which are multiplied together in what is called the **Drake equation** to give the final answer. The formulation is unusual for an equation because it is really a guide to thought rather than a real way of calculating; there is no right answer or solution to the Drake "equation."

In the standard formulation of the Drake equation, we first estimate the rate at which stars form in our Galaxy. Second, we consider the probability that stars at the centers of planetary systems are suitable to allow intelligent life to evolve. For example, as we have already discussed, the most massive stars evolve rather quickly, remaining stable for too short a time to allow intelligent life a good chance to evolve.

Third, we ask what the probability is that a suitable star has planets. With the new detection of planets orbiting other stars, most astronomers think that the chances are likely to be pretty high.

Fourth, we need planets with suitable conditions for the origin of life, so we multiply by the average number of such planets per suitable star. A planet like Jupiter might be ruled out, for example, because it lacks a solid surface and because its surface gravity is high. (Alternatively, though, one could consider a liquid region, if it were at a suitable temperature, to be as advantageous as the oceans on Earth may have been to the development of life here. Or a moon of the planet could provide the solid surface.) Also, planets probably must be in orbits in which the temperature does not vary too much.

Fifth, we have to consider the fraction of the suitable planets on which life actually begins. This is a very large uncertainty, for if this fraction is zero (with the Earth being the only exception), then we get nowhere with this entire line of reasoning. Still, the discovery that amino acids can be formed in laboratory simulations of at least some kinds of potential primitive atmosphere, and the discovery of complex molecules (such as ammonia, formic acid, and vinegar) in interstellar space, indicate to many astronomers that it is relatively easy to form complicated molecules. As mentioned previously, amino acids, much less complex than DNA but also basic to life as we know it, have even been found in some meteorites.

Moreover, life is found in a wide range of extremes on Earth, including oxygen-free environments near geothermal sulfur vents on the ocean bottom or surface hot springs, under rocks in the Antarctic, and deep underground in some other parts of the Earth (■ Fig. 20–15). These bacteria do not survive in the presence of oxygen; their existence supports the idea that life evolved before oxygen appeared in Earth's atmosphere. So environments on other planets may not be as hostile to life as we had thought, even though they couldn't support the types of animal and plant life with which we are most familiar. It is these hardy examples of life on Earth that give us hope that the possible discovery of primitive life on Mars will someday be confirmed.

Sixth, if we want to hear meaningful signals from aliens, we must have a situation where not just life but intelligent life has evolved. We cannot converse with algae or paramecia, and certainly not with the organic compounds reported on Mars. Furthermore, the life must have developed a technological civilization capable of interstellar communication. So we consider the fraction of stars with life on which intelligence and communication actually develop.

Finally, such a civilization must also live for a fairly long time; otherwise, its existence would be just like a flashbulb going off in an otherwise perpetually dark room. Humans now have the capability of destroying themselves either dramatically in a flurry of hydrogen bombs or more slowly by, for example, altering our climate, lessening our ozone shield (which keeps harmful ultraviolet radiation out), or increasing the level of atmospheric pollution. Natural disasters must also be avoided. That an Earth-crossing asteroid or comet will eventually impact the Earth with major consequences seems statistically guaranteed on a timescale of a few hundred million years, unless we take preventive action.

It is a sobering question to ask whether the typical lifetime of a technological civilization is measured in decades, or whether all the problems that we have—political, environmental, and otherwise—can be overcome, leaving our civilization to last for millions or billions of years. So, to complete our calculation, we must multiply by the average lifetime of a civilization.

We can try to estimate answers for each of these simpler questions within our chain of reasoning, though some of them are actually more like wild guesses. (Consequently, some scientists don't find the exercise useful.) We can then use these answers together to figure out the larger question of the probability that communicating extraterrestrial life exists (see *Figure It Out 20.2: The Drake Equation*).

Fairly reasonable assumptions can lead to the conclusion that there may be thousands, or even tens of thousands, of planets in our Galaxy on which technologically advanced life has evolved. (Handy "Drake equation calculators" can be found at several websites, including that of the SETI Institute, http://www.seti.org). Perhaps overly optimistically, Carl Sagan estimated that there might be a million such planets.

On the other hand, adopting more pessimistic (but possibly more realistic) values for the probabilities of intelligent and technologically advanced life, and for the typical lifetime of such a civilization, gives a much bleaker picture. Indeed, humans may be at the pinnacle of intelligence and technological capability in our Galaxy, with few (if any) comparably advanced civilizations.

20.6b Where Is Everyone?

One interesting argument is as follows: If there are many (say, a million) intelligent, communicating civilizations in our Galaxy, why haven't we detected any of them? Where are they all? Surely most must be far more advanced than we are, and have sent signals if not spacecraft that have reached Earth, yet there is no evidence for them. This reasoning, promoted among others by the late Enrico Fermi of the University of Chicago, suggests that advanced creatures such as us might be very rare, even if primitive life is abundant in our Galaxy. (Note that various surface features claimed by some people to have been built by extraterrestrials are either nonexistent, such as the "face" on Mars—■ Figure 20–16—or have more conventional explanations, like the huge drawings on the Peruvian desert known as the Nazca lines.)

Indeed, a more extreme version of this argument points out that there could be self-sustaining colonies voyaging through space for generations. They need not travel close to the speed of light if families are aboard. Even if colonization of space took place at a rate of only 1 light-year per century, our entire Galaxy would still have been colonized during a span of just a hundred million years, less than one per cent of its lifetime. The fact that we do not find extraterrestrial life here indicates that the Solar System has not been colonized, which in turn may imply that technologically advanced life has not arisen elsewhere in our Galaxy.

Another possibly relevant fact suggesting that life having our capabilities is very rare in our Galaxy is that more than 1 billion species have ever lived on Earth, yet

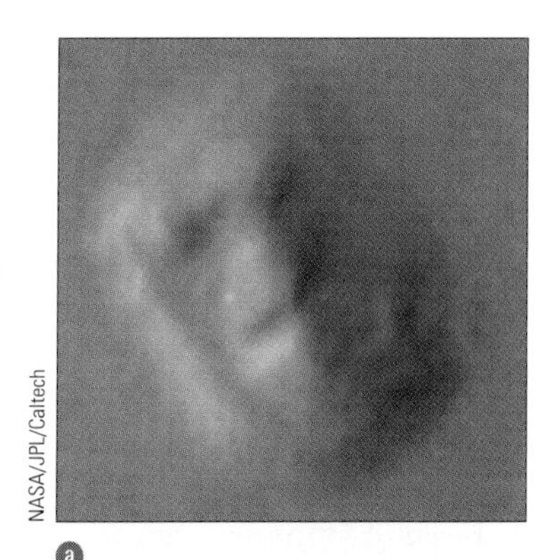

NASA/JPL/Caltech

a

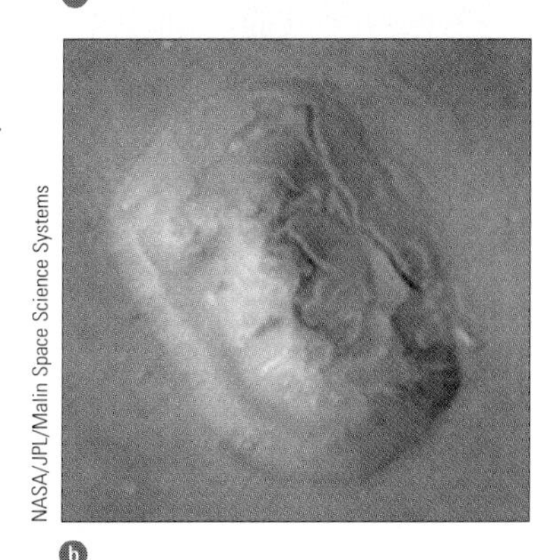

NASA/JPL/Malin Space Science Systems

b

■ **FIGURE 20–16** The "face" on Mars, seen at low resolution from an earlier spacecraft (a), turns out to be an ordinary set of landforms when imaged from spacecraft capable of higher resolution, such as this Mars Global Surveyor image (b).

FIGURE IT OUT 20.2

The Drake Equation

We can make an educated guess regarding the present number of communicating civilizations (those that should be able to contact each other) in our Milky Way Galaxy by using the "Drake equation," first written by Frank Drake in 1961:

$$N = Rf_s f_p n_e f_l f_i f_c L.$$

Let us examine each of the terms in this equation.

R is the star-formation rate in our Galaxy. It is roughly equal to the total number of stars divided by the age of our Galaxy. A reasonable estimate is $R = 10$ stars per year. The true rate is lower now, but was higher in the past.

f_s is the fraction of stars that are suitable. Various criteria can be applied, such as F, G, and K main-sequence stars, and a sufficiently high abundance of heavy elements to make rocky planets possible. Perhaps 0.1 is a good guess.

f_p is the fraction of stars with planetary systems. This fraction could be roughly 0.1—many binary systems might not have planets.

n_e is the number of Earth-like planets or moons per planetary system. We could guess that this number is typically between 0.1 and 1. It might be a little larger than 1, if a planetary system has several suitable moons. For example, in the Solar System, it is not unreasonable to suppose that Mars, Europa, and Titan (besides Earth) had or will have conditions suitable for the formation of life at some time in their existence.

f_l is the fraction of Earth-like planets on which life (even if primitive) arises. This fraction, though quite speculative, might be in the range 10^{-3} to 1.

f_i is the fraction of life-bearing planets on which intelligent life actually arises. Though perhaps even more speculative, it might be 10^{-6} to 1.

f_c is the fraction of intelligent life that has the ability and desire to communicate with aliens. Once again, we can only speculate; perhaps 10^{-3} to 1 is reasonable.

L is the average lifetime of a communicating civilization, or the cumulative lifetime of such civilizations on a given planet (if there are more than one at any given time, or if the annihilation of one is followed at some later time by the emergence of another). Humans have sent radio signals for only about 100 years—and we hope that we have many years ahead of us! However, perhaps advanced civilizations tend to destroy themselves after a short time. The possible range for L is enormous: Let's say 100 to 10^9 years.

The overall result is the product of all these factors, most of which are highly uncertain. The pessimistic estimates give a value of 10^{-12}, suggesting that creatures like humans are exceedingly rare—only one case per 10^{12} galaxies. There are about 10^{11} galaxies in the observable part of the Universe, so we might be the only communicating civilization within this volume!

On the other hand, the most optimistic estimates give a value of 10^9—essentially the lifetime L, since all of the other factors are either about 1 or balance each other out (note that $R \times f_s \approx 1$). This means that one star out of several hundred might now have a communicating civilization. The fraction is even larger if some civilizations colonize more than one star.

Although use of the Drake equation doesn't answer our question very precisely at all, the approach is systematic, and it gives us a range of reasonable values. It also focuses our attention on which factors are highly uncertain and need the most improvement. For example, we can see that our uncertainty in the rate at which stars form is just a minor fraction of the total uncertainty in the final answer; far more important are the uncertainty in the fraction of life with intelligence and the lifetime of a communicating civilization.

apparently we are the only species to have developed space communication or even acquired technology. Similarly, it is sobering to realize that a communicating civilization developed on Earth only in the last century, despite evidence for primitive life on Earth for the past 3.5 to 3.8 billion years.

Recently, astronomers and other scientists have carefully considered the many factors that affect complex life on Earth. For example, the stability of the Earth's axis of rotation depends on the presence of our rather massive Moon. If we had no moon, or only a small one, then Earth's axis would undergo rather rapid, random changes in its orientation, causing large variations in the seasons and climate, and presumably making it more difficult for complex life to develop. Similarly, the presence of the very massive planet Jupiter in the Solar System, yet fairly far from Earth and having a nearly circular orbit, is a blessing: Jupiter's gravitational tugs have cleared out the Solar System, making collisions between Earth and large "killer meteoroids" infrequent.

There are many other relevant factors. The presence of plate tectonics on Earth is very important, for instance, because it allows carbon to be recycled in a steady manner through Earth's atmosphere, oceans, interior, and surface. This "carbon cycle" is important for maintaining global climate stability and for certain aspects of life itself. Having an abundant supply of heavy elements during the formation of the Solar System was also crucial; most regions of our Milky Way Galaxy were quite deficient in such elements billions of years ago, when the Sun formed. These arguments, and others, have led many astronomers to conclude that Earth-like planets capable of developing complex, technologically advanced, communicating civilizations really are rare in our Milky Way Galaxy, though primitive life such as bacteria might be very common indeed. However, other astronomers (such as those at the SETI Institute) argue that there are major potential flaws in the reasoning for this "rare Earth" hypothesis, and that communicating civilizations may be common.

SETI Institute

FIGURE 20–17 An artist's conception of the Allen Telescope Array, being built in California for the SETI Institute and the University of California, Berkeley.

It is not clear, for example, whether the apparently "special" conditions of Earth are essential to the development of intelligent life; maybe our view has been too highly skewed by the single example we know—ourselves. Moreover, is it necessarily true that if other intelligent creatures evolved, they would choose to colonize space, or have the ability to do so? The rather late appearance of technologically advanced life on Earth may also be a statistically unlikely fluke. Finally, some of the arguments relied on the unproven assumption that life elsewhere is quite similar to that on Earth in its properties and evolutionary path.

In any case, unless we make the effort to actually look for signs of intelligent extraterrestrial life, we might never know whether humans are indeed alone in our Galaxy, or simply one of many such creatures. Thus, many astronomers support the search for extraterrestrial intelligence, especially using telescopes that are simultaneously doing other, more conventional types of research projects.

Some major radio-telescope projects are being built for SETI purposes, with more ordinary astronomy to be carried out as a bonus (■ Fig. 20–17). Perhaps someday we will be scanning the skies with radio telescopes on the far side of the Moon, shielded from Earth's radio interference by the Moon's bulk. The chances for success might be slim, but all agree that the actual detection of extraterrestrial signals would be one of the most important and mind-blowing discoveries ever, if not the greatest discovery of all time.

AceAstronomy™ Log into AceAstronomy and select this chapter to see the Active Figure called "Drake Equation."

20.7 UFOs and the Scientific Method

If some or many astronomers believe that technologically advanced life exists elsewhere in our Galaxy, why do they not accept the idea that unidentified flying objects (UFOs) represent visitations from these other civilizations? The answer to this question leads us not only to explore the nature of UFOs but also to consider the nature of knowledge and truth. The discussion that follows is a personal view of the authors, but one that is shared to a greater or lesser extent by most scientists.

Jay M. Pasachoff

FIGURE 20–18 A Christmas-light UFO seen on a rooftop.

20.7a UFOs

The most common UFO is a light that appears in the sky that seems to the observer unlike anything made by humans or any commonly understood natural phenomenon. UFOs may appear as a point or extended, or may seem to vary in size or brightness.

THE FAR SIDE® By GARY LARSON

"YEEEEEHAAAAAAAAA!"

But the observations are usually anecdotal, are not controlled as in a scientific experiment, and are not accessible to study by sophisticated instruments.

Most of the sightings of UFOs that are reported can actually be explained in terms of natural phenomena. The Earth's atmosphere can display a variety of strange effects, and these can be used to explain many apparent UFOs. When Venus shines brightly near the horizon, for example, we sometimes get telephone calls from people asking us about the "UFO"—especially if the crescent moon happens to also be in that direction. It is not well known that a bright star or planet low on the horizon can seem to flash red or green because of atmospheric refraction. Atmospheric effects can affect radar (radio) waves as well as visible light.

For many of the effects that have been reported, the UFOs would have been defying well-established laws of physics. Why haven't we heard the expected sonic booms, for example, from rapidly moving UFOs? Scientists treat challenges to laws of physics very seriously, since our science and technology are based on these laws and they seem to work extremely well.

Most professional astronomers feel that UFOs can be so completely explained by natural phenomena that they are not worthy of more of our time. Although most of us do not categorically deny the possibility that UFOs exist (after all, the Voyager spacecraft might someday pass by a planet orbiting another star), the standard of evidence expected of all claims in science has not yet been met.

Some individuals may ask why we reject the identification of UFOs with flying saucers, when—they may say—that explanation is "just as good an explanation as any other." Let us go on to discuss what scientists mean by "truth" and how that applies to the above question.

ASIDE 20.4: Not UFOs

Sometimes other natural phenomena (flocks of birds, for example) are mistakenly reported as UFOs. Even the more mysterious sightings lack convincing evidence that rules out alternatives. One should not accept explanations that UFOs are flying saucers from other planets before more mundane explanations–including hoaxes, exaggeration, and fraud–are exhausted (Figs. 20–18, 20–19).

20.7b Of Truth and Theories

At every instant, we can explain what is happening in a variety of ways. When we flip a light switch, for example, we assume that the switch closes an electric circuit in the wall and allows the electricity to flow. But it is certainly possible, although not very likely, that the switch activates a relay that turns on a radio that broadcasts a message to the president of the United States. The president then might send back a telepathic message to the electricity to flow, making the light go on. The latter explanation sounds so unlikely that we don't seriously consider it. We would even call the former explanation "true," without qualification.

We usually regard as "true" the simplest explanation that satisfies all the data we have about any given thing. This principle is known as **Occam's Razor;** it is named after a 14th-century British philosopher who originally proposed it. Without this rule, we would always be subject to such complicated doubts that we would accept nothing as known to be true.

Science is based on Occam's Razor, though we don't usually bother to think about it. Sometimes something we call "true" might be more accurately described as a theory (see Chapter 1). An example of a theory is the Newtonian theory of gravitation, which for many years explained almost all of the planetary motions. Albert Einstein's 1916 theory of gravity, known as the general theory of relativity, provided an explanation for a nagging discrepancy in the orbit of Mercury, as we described in Chapter 10.

ASIDE 20.5: A cutting edge

Occam's Razor, sometimes called the Principle of Simplicity, is a "razor" in the sense that it is a cutting edge that allows a distinction to be made among theories.

Is Newton's theory "true"? Though we know it is "false," it is a good approximation of the truth in most regions of space; it is generally an accurate model for what we observe. Is Einstein's theory "true"? We may say so, although one day a newer theory may come along that is more general than Einstein's in the same way that Einstein's is more general than Newton's. Indeed, as we discussed in Chapter 19, superstring theory is a leading candidate for the unification of general relativity and quantum physics, and hence may someday be a more complete theory than Einstein's.

How does this view of truth tie in with the previous discussion of UFOs? Scientists have assessed the probability of UFOs being flying saucers from other worlds, and most have decided that the probability is so low that we have better things to do with our time and with our national resources. We have so many other, simpler explanations of the phenomena that are reported as UFOs that when we apply Occam's Razor, we call the identifications of UFOs with extraterrestrial visitation "false."

Jay M. Pasachoff

FIGURE 20–19 Though they may look like giant UFOs, these objects are really the planetarium and the Arts Palace in Valencia, Spain, projects of the architect Santiago Calatrava.

20.8 Conclusion

You have covered a lot of material in this book, and learned much about the Universe. We, the authors, hope that this new knowledge increases your awe and fascination for nature. The understanding of a phenomenon should enhance its beauty, not detract from it. The magnificence of the Universe comes in part from its logical structure—and the foundation is perhaps unexpectedly simple. Indeed, Einstein remarked that "The most incomprehensible thing about the Universe is that it's comprehensible."

The actual consequences of the basic laws can be extraordinarily complicated and varied. The best example is life itself: It is the most complex known structure. Even the simplest cell is more difficult to understand than the formation of galaxies or the structure of stars. Our highly advanced brains and dexterity are what sets humans apart from other forms of life. We are able to question, explore, and ultimately understand the inner workings of nature through a process of observation, experimentation, and logical thought.

It is almost as if the Universe has found a way to know itself, through us. We are the observers and explorers of the Universe; we are its brains and its conscience. This makes each one of us special to the Universe as a whole. Perhaps it need not have been this way: Alter the physical constants ever so slightly and the Universe may have been stillborn, with no such complexity. But here we are, enjoying life in this beautiful, amazing, mind-blowing Universe.

CONCEPT REVIEW

The study of life elsewhere than on Earth is called **exobiology** (Introductory paragraph). One difficulty in the search for life is that we don't have a clear definition of what life is, and we generally must restrict ourselves to life as we know it, which is based on **organic** compounds—those containing carbon (Section 20.1). Specifically, life on Earth is based on chains of carbon called **amino acids,** which link together many at a time to form **proteins,** chief among them DNA. Experiments in which sparks were passed through certain mixtures of gases showed that amino acids form easily, but it is now thought that these mixtures are not representative of Earth's primitive atmosphere. Nevertheless, amino acids seem to form in a variety of environments, including harsh ones.

Primitive life arose very quickly on Earth, suggesting that it is not highly improbable, though the discovery of indigenous life in at least one other location would provide much support for this hypothesis. Within the Solar System, Mars and Europa are the most likely candidates for life (Section 20.2). Extraterrestrial life outside our planetary system, especially intelligent life, is most likely to have formed on moons or planets orbiting Sun-like main-sequence stars; such stars have a long life and a relatively large **ecosphere,** or range of distances in which conditions suitable for life might be found (Section 20.3).

In searching for signs of extraterrestrial life, we can wait for their spacecraft to reach us, but the odds of success are low, even though Einstein's special theory of relativity in principle allows interstellar voyages to be completed in a reasonably short time (Section 20.4). A much more likely method of detection is by searching for patterns among electromagnetic signals, especially radio waves. Several such projects are being conducted, most notably the "search for extraterrestrial intelligence, at home" (SETI@home), in which the general public can participate with personal computers. Humans have sent radio signals into space, intentionally and unintentionally, which might someday be detected by extraterrestrials (Section 20.5).

Although we do not know how many communicating civilizations exist in the Milky Way Galaxy, an estimate can be attempted by using the **Drake equation** (Section 20.6a). Among the most uncertain factors are the probability that intelligence arises after the formation of primitive life and the typical lifetime of a communicating civilization. Some scientists think that intelligent life might be common in our Galaxy, but many others are concluding that intelligence and technological capability at the level of humans is rather rare (Section 20.6b). Unless we actually search for signals, however, the answer may always elude us.

Evidence that UFOs visit us is not convincing; most of the reports can be easily explained by atmospheric and other natural phenomena (Section 20.7a). Although it is possible that some of the sightings are of actual UFOs, **Occam's Razor** (the Principle of Simplicity) suggests otherwise, until the evidence becomes much stronger (Section 20.7b).

Having finished this textbook, you should now have a much deeper and broader perspective of the Universe and our place within it (Section 20.8). We, the authors, hope that this new knowledge increases your awe and fascination for nature.

QUESTIONS

1. Discuss the most likely places in the Solar System, other than Earth, where there might be primitive (microbial) life.
2. What is the significance of the Miller-Urey experiments?
3. What types of stars are most likely to have planets on which indigenous life developed? Explain.
4. Why do we think that indigenous intelligent life wouldn't evolve on planets orbiting stars much more massive than the Sun?
5. List two means by which we might detect extraterrestrial intelligence.
6. Why have astronomers generally used radio wavelengths in their search for life elsewhere?
7. In what way can optical (visible) signals be used in a search for extraterrestrial life?
8. Describe the radio signals humans have sent to outer space, both intentionally and unintentionally.
9. †Work out the binary value given for the population of Earth on the message we sent to M13, and compare it with the current actual value. (This question assumes knowledge of the binary system.)
10. †Work out the binary value given for the size of the Arecibo telescope on the message we sent to M13, and compare it with the actual value. (This question assumes knowledge of the binary system.)
11. Why can't we hope to carry on a conversation at a normal rate with extraterrestrials on a planet orbiting a distant star?
12. State the Drake equation and its significance. Discuss the importance of the value of L, the lifetime of the civilization.
13. †In the Drake equation, use your own preferences for each quantity to derive the number of intelligent, communicating civilizations in our Galaxy.
14. †If 10% of all stars are of suitable type for life to develop, 30% of all stars have planets, and 20% of planetary systems have a planet or moon at a suitable distance from the star, what fraction of stars have a planet suitable for life? Roughly how many such stars would there be in our Milky Way Galaxy?
15. †How long would it take a rocket ship travelling at 100 km/sec to reach a star that is 20 light-years away?
16. Explain how it is possible, in principle, to travel many light-years in a short time interval as measured by the traveller.
17. †Suppose you are in a rocket ship that is moving at 97% of the speed of light. If your journey to another star appeared to take 30 years from the perspective of an Earth-bound observer, how long did it take in your frame of reference?
18. †Referring back to question 17, how far did you travel, in your frame of reference?
19. Describe how Einstein's general theory of relativity serves as an example of the scientific method.
20. **True or false?** Unmistakable bacteria fossils have been found in at least one martian meteorite.
21. **True or false?** Biological processes on Earth are largely responsible for the free oxygen in its atmosphere.
22. **True or false?** Organic compounds can be created by sending electric discharges through mixtures of gases, some of which might be similar to the early atmospheres of Earth or some extra-solar planets.
23. **True or false?** Planets orbiting the most massive stars have the best chance of developing intelligent life, since these stars have long lives.
24. **True or false?** One of the most uncertain factors in estimating the number of communicating civilizations in our Galaxy is the lifetime of such civilizations.
25. **Multiple choice:** Around which one of the following objects is there thought to be the best chance of finding life, outside the Solar System? **(a)** A neutron star. **(b)** A main-sequence F-type star. **(c)** A red supergiant star. **(d)** A main-sequence O-type star. **(e)** A planetary nebula.
26. **Multiple choice:** Which one of the following is *not* a conclusion from Einstein's special theory of relativity? **(a)** It is possible to travel into the future, but not into

the past. **(b)** A moving clock runs more slowly according to a person at rest. **(c)** A moving ruler looks shorter according to a person at rest. **(d)** The speed of light can be exceeded, in principle, if an object is given a truly gargantuan amount of energy. **(e)** Interstellar travel is possible within a single human lifetime, in principle, if one travels sufficiently close to the speed of light.

27. **Multiple choice:** Which one of the following is *not* used in the Drake equation? **(a)** The fraction of Earth-like planets or moons on which life actually arises. **(b)** The fraction of Sun-like stars with planetary systems. **(c)** The average rate of star formation in the Milky Way Galaxy. **(d)** The fraction of stars that are on the main sequence. **(e)** The fraction of intelligent civilizations that develop electromagnetic communication techniques.
28. **Multiple choice:** Which one of the following is a reason to believe there might be life elsewhere in the Universe, outside our Solar System? **(a)** The presence of a large number of O-type and B-type stars in the Milky Way Galaxy. **(b)** The extra-solar planets recently discovered using Doppler techniques are a lot like the Earth. **(c)** The fact that life exists in many extreme environments on Earth, and seems to have arisen shortly after Earth formed. **(d)** The substantial quantities of free methane and oxygen (together) detected in spectra of some extra-solar planets, suggesting that the methane might have been produced by decaying organisms. **(e)** The many UFO sightings reported in the U.S. and elsewhere in the world.
29. †**Multiple choice:** According to the most recent data and the Drake equation, the odds that there is intelligent, communicating life outside our Solar System, but somewhere within 10,000 light-years of it, are roughly which one of the following? **(a)** 1 in 10,000. **(b)** 1 in 10^8 (i.e., $1/10{,}000^2$, according to the inverse-square law). **(c)** Virtually certain, since there are so many stars within 10,000 light-years. **(d)** 0, since intelligent life on Earth has been scientifically proven to be the one-time-only work of a special Creator. **(e)** Nobody really knows.
30. **Fill in the blank:** Jupiter's moon ______ probably has liquid water beneath its icy surface, and hence could be an abode for primitive life.
31. **Fill in the blank:** Searches for signals from extraterrestrial life have been conducted mainly at ______ wavelengths.
32. **Fill in the blank:** The volume around a star in which conditions may be suitable for life is called the ______.
33. **Fill in the blank:** The principle that the simplest explanation that satisfies all the data regarding a phenomenon is "true" is known as ______.
34. **Fill in the blank:** A well-known effort to find signals from intelligent life elsewhere in our Galaxy is called ______; it uses the computers of millions of people to process data.

†This question requires a numerical solution.

TOPICS FOR DISCUSSION

1. How restrictive do you think we are being when we consider only "life as we know it?"
2. How useful, in your opinion, is the Drake equation for estimating the number of intelligent, communicating civilizations in our Galaxy? Is it perhaps more useful as a tool for identifying the most uncertain aspects of the issue?
3. Do you think humans or their successors (machines?) will ever overcome the enormous barriers to rapid interstellar travel?

MEDIA

Virtual Laboratories

Extra-solar Planets

AceAstronomy™ Log into AceAstronomy at **http://astronomy.brookscole.com/cosmos3** to access quizzes and animations that will help you assess your understanding of this chapter's topics.

Log into the Student Companion Web Site at **http://astronomy.brookscole.com/cosmos3** for more resources for this chapter including a list of common misconceptions, news and updates, flashcards, and more.

EPILOGUE

We have completed our grand tour of the Universe. We have seen stars and planets, matter between the stars, giant collections of stars called galaxies and clusters of galaxies, and very distant objects such as quasars. We have witnessed the evolution of stars, in some cases ending with spectacular explosions with compact remnants the size of a city but half a million times more massive than Earth. We have pondered the properties of still more bizarre objects, black holes. We have learned how our Universe began in a hot, compressed state and has been expanding ever since—seemingly faster and faster during the past few billion years, perhaps driven by a cosmic antigravity effect. We have explored the origins of the Universe, galaxies, stars, the chemical elements, planets, and ultimately life itself. If you are thirsty for more information, as the authors would like you to be, you can consult the books listed in the Selected Readings.

Further, we have seen the vitality of contemporary science in general and astronomy in particular. The individual scientists who call themselves astronomers are engaged in fascinating studies, often pushing modern technologies to their limits. New telescopes on the ground and in space, new types of detectors, new computer capabilities for studying data and carrying out calculations, and new theoretical ideas are linked in research about the Universe.

Our views at the cutting edge of astronomy are changing so rapidly that within a few years some of what you read here will have been revised. Science is a dynamic process: New ideas are developed and tested, and modified when necessary. The authors hope that, over the years, you will keep up by following astronomical articles and stories in newspapers, magazines, and books, and on television. We would like you to consider the role of scientific research as you vote. And we hope you will remember the methods of science—the mixture of logic and standards of evidence by which scientists operate—that you have seen illustrated in this book.

Jay M. Pasachoff

The largest optical telescope in the world as of its completion in 2006, the Gran Telescopio Canarias (the Great Telescope of the Canary Islands), a 10.4-m reflector shown under construction on La Palma, Spain. We see the dome, the base, and an interior view that includes the back of the mirror cell, which contains 36 glass segments each 1.9-m across. (The name "Canaries" comes from the Latin for dog, as in the constellation Canis Major, given the many dogs that were on those islands hundreds of years ago.)

APPENDIX 1/2 Measurement Systems/ Basic Constants

Appendix 1A International System of Units (Système International d'Unités)

	SI Units	SI Abbrev.	Other Abbrev.
length	meter	m	
volume	liter	L	ℓ
mass	kilogram	kg	kgm
time	second	s	sec
temperature	kelvin	K	°K

Appendix 1B Conversion Factors

1 erg = 10^{-7} joule = 10^{-7} kg • m^2/s^2
1 joule = 6.2419 × 10^{18} eV
1 in = 25.4 mm = 2.54 cm
1 yd = 0.9144 m
1 mi = 1.6093 km ≃ 8/5 km
1 oz = 28.3 g
1 lb = 0.4536 kg

Appendix 2A Physical Constants (1998 CODATA Recommended Values)

Speed of light*	c	= 299 792 458 m/sec (exactly)
Constant of gravitation	G	= (6.673 90 ± 0.000 01) × 10^{-11} $m^3/kg \cdot sec^2$
Planck's constant	h	= (6.626 069 01 ± 0.000 000 34) × 10^{-34} J · sec
Boltzmann's constant	k	= (1.380 650 3 ± 0.000 002 4) × 10^{-23} J/K
Stefan-Boltzmann constant	σ	= (5.670 400 ± 0.000 040) × 10^{-8} $W/m^2 \cdot K^4$
Wien displacement constant	$\lambda_{max}T$	= (2.897 768 6 ± 0.000 005) × 10^7 Å · K

See http://physics.nist.gov/constants. Constant of universal gravitation and Planck's constant are 2005 values.

Appendix 2B Astronomical Constants

Astronomical Unit*	1 A.U.	= 1.495 978 70 × 10^{11} m
Parsec	1 pc	= 3.086 × 10^{16} m
		= 206 264.806 A.U.
		= 3.261 633 lt yr
Light-year	1 lt yr	= (9.460 530) × 10^{15} m
		= 6.324 × 10^4 A.U.
Mass of Sun*	$M_{\odot}^{**}$	= (1.988 843 ± 0.000 03) × 10^{30} kg
Radius of Sun*	$R_{\odot}$	= 696 000 km
Luminosity of Sun	$L_{\odot}^{**}$	= 3.827 × 10^{26} J/sec
Mass of Earth*	M_E^{**}	= (5.972 23 ± 0.000 08) × 10^{24} kg
Equatorial radius of Earth*	R_E	= 6 378.140 km

*Adopted as "IAU (1976) system of astronomical constants" at the General Assembly of the International Astronomical Union that year. The meter was redefined in 1983 to be the distance travelled by light in a vacuum in 1/299,792,458 second.
**2000 values

APPENDIX 3 Planets

Appendix 3A Our Solar System: Intrinsic and Rotational Properties

Name	Equatorial Radius km	Equatorial Radius ÷ Earth's	Mass ÷ Earth's	Mean Density (g/cm^3)	Oblateness	Surface Gravity (Earth = 1)	Sidereal Rotation Period	Inclination of Equator to Orbit	Apparent Magnitude During 2006
Mercury	2,439.7	0.3824	0.0553	5.43	0	0.378	58.646^{d}	0.0°	−2.1 to +0.9
Venus	6,051.8	0.9489	0.8150	5.24	0	0.894	243.02^{d}R	177.3°	−4.6 to −3.7
Earth	6,378.14	1	1	5.515	0.0034	1	$23^{h}56^{m}04.1^{s}$	23.45°	—
Mars	3,397	0.5326	0.1074	3.94	0.006	0.379	$24^{h}37^{m}22.662^{s}$	25.19°	−0.7 to +1.8
Jupiter	71,492	11.194	317.896	1.33	0.065	2.54	$9^{h}50^{m}$ to $> 9^{h}55^{m}$	3.12°	−2.5 to −1.7
Saturn	60,268	9.41	95.185	0.70	0.098	1.07	$10^{h}39.9^{m}$	26.73°	−0.2 to +0.6
Uranus	25,559	4.0	14.537	1.30	0.022	0.8	$17^{h}14^{m}$R	97.86°	+5.7 to +5.9
Neptune	24,764	3.9	17.151	1.76	0.017	1.2	$16^{h}7^{m}$	29.56°	+7.8 to +8.0
Pluto	1,195	0.2	0.0025	2.1	0	0.01	$6^{d}9^{h}17^{m}$R	120°	+13.9 to +14.0
2003 UB_{313}	1,200–1,500	0.2	—	—	—	—	—	—	—

R signifies retrograde rotation.

The masses and radii for Mercury, Venus, Earth, and Mars are the values recommended by the International Astronomical Union in 1976. The radii are from *The Astronomical Almanac 2006.* Surface gravities were calculated from these values. The length of the martian day is from G. de Vaucouleurs (1979). Most densities, oblatenesses, inclinations, and magnitudes are from *The Astronomical Almanac 2006.* Neptune data from *Science,* December 15, 1989, and August 9, 1991. Values for the masses of the giant planets are based on Voyager data for the mass of the Sun divided by the mass of the planet (E. Myles Standish, Jr., *Astronomical Journal* **105,** 2000, 1992); Jupiter: 1047.3486; Saturn: 3497.898; Uranus: 22 902.94; Neptune: 19 412.24.

Appendix 3B Our Solar System: Orbital Properties

Name	Semimajor Axis A.U.	Semimajor Axis 10^6 km	Sidereal Period Years	Sidereal Period Days	Synodic Period (Days)	Eccentricity	Inclination to Ecliptic
Mercury	0.387 099	57.909	0.240 84	87.96	115.9	0.205 63	7.004 87°
Venus	0.723 332	108.209	0.615 18	224.68	583.9	0.006 77	3.394 71°
Earth	1	149.598	0.999 98	365.25	—	0.016 71	0.000 05°
Mars	1.523 662	227.939	1.880 7	686.95	779.9	0.093 41	1.850 61°
Jupiter	5.203 363	778.298	11.857	4,337	398.9	0.048 39	1.305 30°
Saturn	9.537 070	1429.394	29.424	10,760	378.1	0.054 15	2.484 46°
Uranus	19.191 264	2875.039	83.75	30,700	369.7	0.047 168	0.769 86°
Neptune	30.068 963	4504.450	163.72	60,200	367.5	0.008 59	0.769 17°
Pluto	39.481 687	5915.799	248.02	90,780	366.7	0.248 81	17.141 75°
2003 UB_{313}	67.7	—	557	203,305	—	0.442	44.177°

Mean elements of planetary orbits for 2000, referred to the mean ecliptic and equinox of J2000 (E. M. Standish, X. X. Newhall, J. G. Williams, and D. K. Yeomans, *Explanatory Supplement to the Astronomical Almanac,* P. K. Seidelmann, ed., 1992). Periods are calculated from them.

For planetary satellites, see
http://ssd.jpl.nasa.gov/sat_props.html
http://ssd.jpl.nasa.gov/sat_elem.html

APPENDIX 4/5

Appendix 4 The Brightest Stars

Star		Name	Position (2000.0) r.a. h min sec	Position (2000.0) Decl. ° ′ ″	Apparent Magnitude (V)	Spectral Type	Absolute Magnitude (M_v)	Distance D (lt yr)	Proper Motion μ ″/yr	Proper Motion θ °	Radial Vel. (km/sec)
1.	α CMa A	Sirius	06 45 09	−16 42 58	−1.46	A1 V	+1.5	9	1.324	204	−8
2.	α Car	Canopus	06 23 57	−52 41 44	−0.72	A9 Ib	−5.4	313	0.034	50	+21
3.	α Boo	Arcturus	14 15 31	+19 10 57	−0.04	K2 IIIp	−0.6	37	2.281	209	−5
4.	α Cen A	Rigil Kentaurus	14 39 37	−60 50 02	0.00	G2 V	+4.2	4	3.678	28	−25
5.	α Lyr	Vega	18 36 56	+38 47 01	0.03	A0 V	+0.6	25	0.348	35	−14
6.	α Aur	Capella	05 16 41	+45 59 53	0.08	G6 +G2	−0.8	42	0.430	169	+30
7.	β Ori A	Rigel	05 14 32	−08 12 06	0.12	B8 Ia	−6.6	773	0.004	236	+21
8.	α CMi A	Procyon	07 39 18	+05 13 30	0.38	F2 IV-V	+2.8	11	1.248	214	−3
9.	α Eri	Achernar	01 37 43	−57 14 12	0.46	B3 V	−2.9	144	0.108	105	+19
10.	α Ori	Betelgeuse	05 55 10	+07 24 36	0.50	M2 Iab	−5.0	522	0.028	68	+21
11.	β Cen AB	Hadar	14 03 49	−60 22 22	0.61	B1 III	−5.5	526	0.030	221	−12
12.	α Aql	Altair	19 50 47	+08 52 06	0.77	A7 IV-V	+2.1	17	0.662	54	−26
13.	α Tau A	Aldebaran	04 35 55	+16 30 33	0.85	K5 III	−0.8	65	0.200	161	+54
14.	α Sco A	Antares	16 29 24	−26 25 55	0.96	M1.5 Iab	−5.8	604	0.024	197	−3
15.	α Vir	Spica	13 25 12	−11 09 41	0.98	B1 V	−3.6	262	0.054	232	+1
16.	β Gem	Pollux	07 45 19	+28 01 34	1.14	K0 IIIb	+1.1	34	0.629	265	+3
17.	α Ps A	Formalhaut	22 57 39	−29 37 20	1.16	A3 V	+1.6	25	0.373	116	+7
18.	α Cyg	Deneb	20 41 26	+45 16 49	1.25	A2 Ia	−7.5	1467	0.005	11	−5
19.	β Crucis		12 47 43	−59 41 19	1.25	B0.5 III	−4.0	352	0.042	246	+20
20.	α Leo A	Regulus	10 08 22	+11 58 02	1.35	B7 V	−0.6	77	0.264	271	+4
21.	α Cru A		12 26 35	−63 05 56	1.41	B0.5 IV	−4.0	321	0.030	236	−11
22.	ϵ CMa A	Adara	06 58 38	−28 59 20	1.50	B2 II	−4.1	431	0.002	27	+27
23.	λ Sco	Shaula	17 33 36	−37 06 14	1.63	B1.5 IV	−3.6	359	0.029	178	0
24.	γ Ori	Bellatrix	05 25 08	+06 20 59	1.64	B2 III	−2.8	243	0.018	221	+18
25.	β Tau	Alnath	05 26 18	+28 36 27	1.65	B7 III	−1.3	131	0.178	172	+8

Based on a table by Robert Garrison in the *Observers Handbook 2001 of the Royal Astronomical Society of Canada* and Robert Garrison and Toomas Karmo in *Observers Handbook 2005*, with the permission of the RASC.

Appendix 5 The Nearest Stars

	Name	Proper (2000.0) r.a. h	m	Decl. °	′	Parallax π ″	Distance lt yr	Proper Motion μ ″/yr	Proper Motion θ °	Radial Velocity (km/sec)	Spectral Type	V	B−V	M_V	Visual Luminosity ($L_\odot = 1$)
1.	Sun										G2 V	−26.75	0.65	4.82	1
2.	Proxima Cen	14	29.7	−62	41	0.772	4.21	3.86	282	−22	M5.5e	11.05	1.90	15.49	0.000 05
	α Cen A	14	39.6	−60	50	.742	4.40	3.71	278	−22	G2 V	.02	0.65	4.37	1.51
	α Cen B							3.69	281	−18	K0 V	1.36	0.85	5.71	0.44
3.	Barnard's star	17	57.8	+04	42	.549	5.94	10.37	356	−111	M4 V	9.54	1.74	13.24	0.000 4
4.	Wolf 359 (CN Leo)	10	56.5	+07	1	.419	7.80	4.69	235	+13	M6 V	13.45	2.0	16.56	0.000 02
5.	BD +36°2147 HD95735 (Lalande 21185)	11	03.4	+35	58	.392	8.32	4.81	187	−85	M2 V	7.49	1.51	10.46	0.006
6.	Sirius A	6	45.1	−16	43	.379	8.61	1.34	204	−8	A1 V	−1.45	0.00	1.44	22.49
	Sirius B							1.34	204		DA 2	8.44	−0.03	11.33	0.002 5
7.	L 726 − 8, BL Cet = A	1	39.0	−17	57	.374	8.74	3.37	81	+29	M5.5 V	12.41	1.87	15.27	0.000 07
	UV Cet = B							3.37	81	+28	M6 V	13.25		16.11	0.000 03
8.	Ross 154 (V1216 Sgr)	18	49.8	−23	50	.337	9.69	0.67	107	−12	M3.5 V	10.45	1.76	13.08	0.000 5
9.	Ross 248 (HH And)	23	41.9	+44	10	.316	10.31	1.62	177	−78	M5.5 V	12.29	1.91	14.79	0.000 1
10.	ϵ Eri	3	32.9	−09	27	.311	10.50	0.98	271	+16	K2 V	3.72	0.88	6.18	0.286

Parallaxes and distances are from the Hipparcos satellite (1997), courtesy of Hartmut Jahreiss, updated in 2000. One red dwarf, too faint for Hipparcos, was reported in 2003 to be close, but the accuracy of the measurement is insufficient to know if it will be in the top 10. See a discussion and an updated table at http://www.chara.gsu.edu/RECONS.

APPENDIX 6 Messier Catalogue

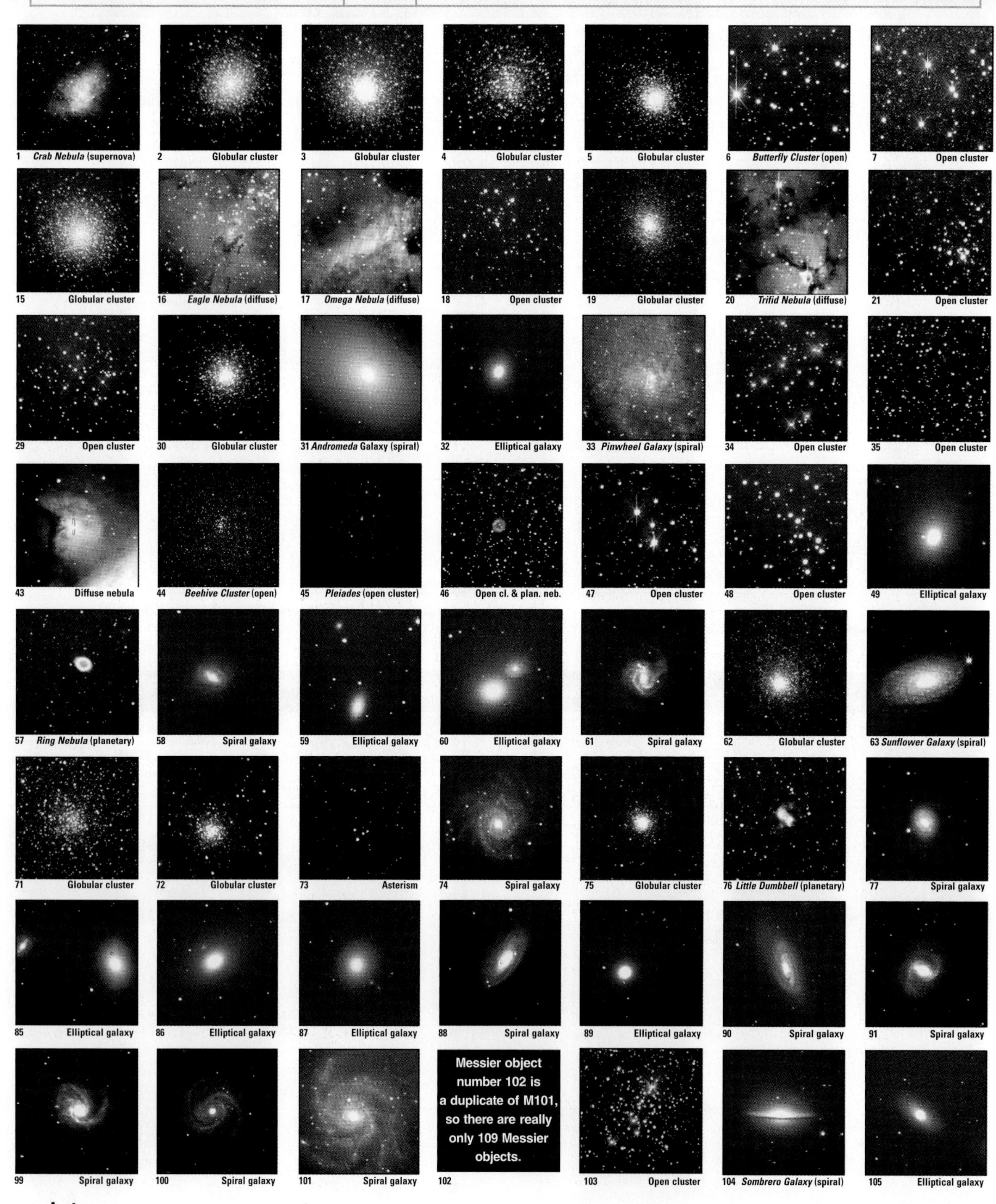

1 *Crab Nebula* (supernova) · 2 Globular cluster · 3 Globular cluster · 4 Globular cluster · 5 Globular cluster · 6 *Butterfly Cluster* (open) · 7 Open cluster

15 Globular cluster · 16 *Eagle Nebula* (diffuse) · 17 *Omega Nebula* (diffuse) · 18 Open cluster · 19 Globular cluster · 20 *Trifid Nebula* (diffuse) · 21 Open cluster

29 Open cluster · 30 Globular cluster · 31 *Andromeda* Galaxy (spiral) · 32 Elliptical galaxy · 33 *Pinwheel Galaxy* (spiral) · 34 Open cluster · 35 Open cluster

43 Diffuse nebula · 44 *Beehive Cluster* (open) · 45 *Pleiades* (open cluster) · 46 Open cl. & plan. neb. · 47 Open cluster · 48 Open cluster · 49 Elliptical galaxy

57 *Ring Nebula* (planetary) · 58 Spiral galaxy · 59 Elliptical galaxy · 60 Elliptical galaxy · 61 Spiral galaxy · 62 Globular cluster · 63 *Sunflower Galaxy* (spiral)

71 Globular cluster · 72 Globular cluster · 73 Asterism · 74 Spiral galaxy · 75 Globular cluster · 76 *Little Dumbbell* (planetary) · 77 Spiral galaxy

85 Elliptical galaxy · 86 Elliptical galaxy · 87 Elliptical galaxy · 88 Spiral galaxy · 89 Elliptical galaxy · 90 Spiral galaxy · 91 Spiral galaxy

99 Spiral galaxy · 100 Spiral galaxy · 101 Spiral galaxy · 102 Messier object number 102 is a duplicate of M101, so there are really only 109 Messier objects. · 103 Open cluster · 104 *Sombrero Galaxy* (spiral) · 105 Elliptical galaxy

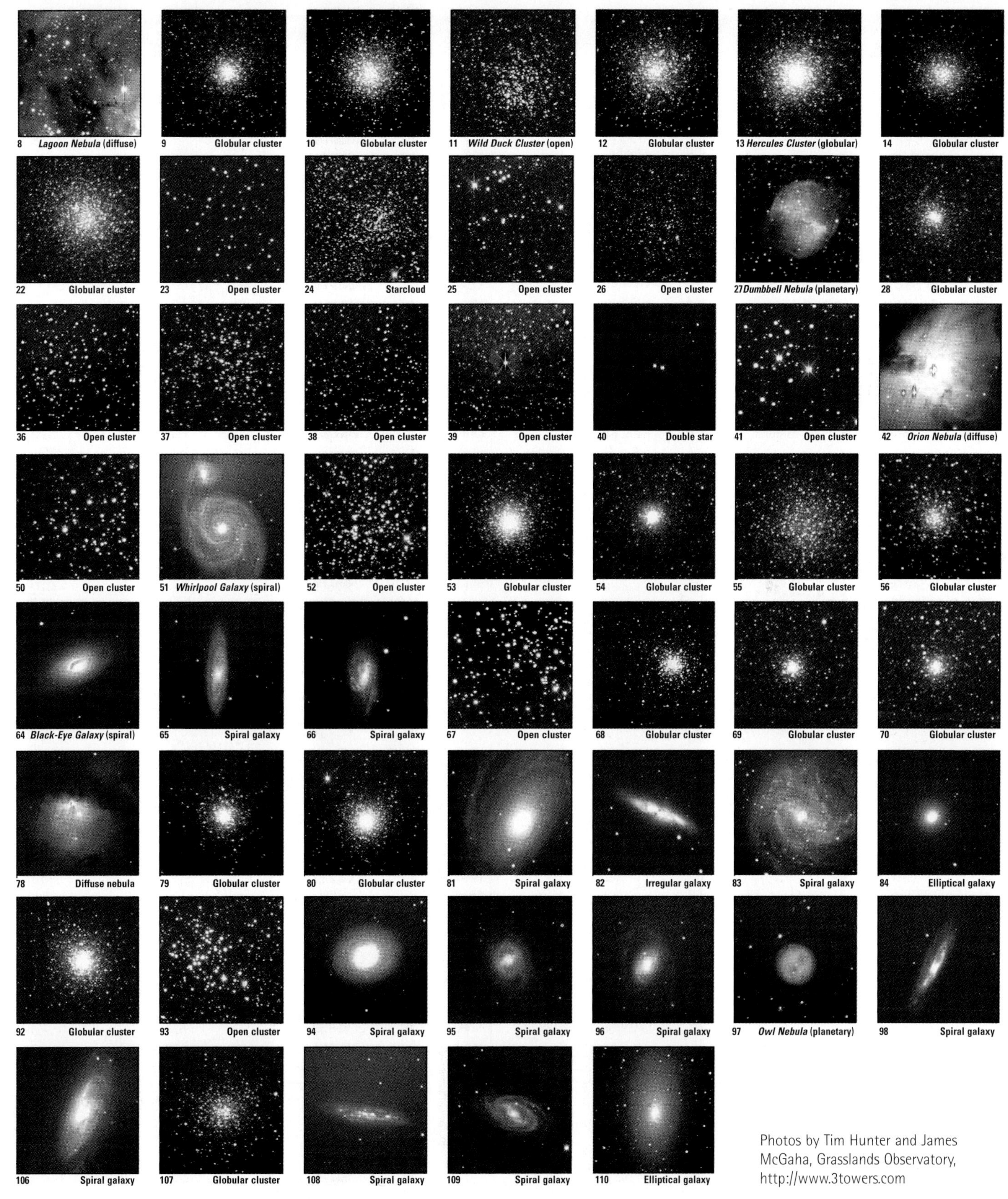

Photos by Tim Hunter and James McGaha, Grasslands Observatory, http://www.3towers.com

APPENDIX 7

The Constellations

Latin Name	Genitive	Abbre-viation	Translation	Latin Name	Genitive	Abbre-viation	Translation
Andromeda	Andromedae	And	Andromeda*	Lacerta	Lacertae	Lac	Lizard
Antlia	Antliae	Ant	Pump	Leo	Leonis	Leo	Lion
Apus	Apodis	Aps	Bird of Paradise	Leo Minor	Leonis Minoris	LMi	Little Lion
Aquarius	Aquarii	Aqr	Water Bearer	Lepus	Leporis	Lep	Hare
Aquila	Aquilae	Aql	Eagle	Libra	Librae	Lib	Scales
Ara	Arae	Ara	Altar	Lupus	Lupi	Lup	Wolf
Aries	Arietis	Ari	Ram	Lynx	Lyncis	Lyn	Lynx
Auriga	Aurigae	Aur	Charioteer	Lyra	Lyrae	Lyr	Harp
Boötes	Boötis	Boo	Herdsman	Mensa	Mensae	Men	Table (mountain)
Caelum	Caeli	Cae	Chisel	Microscopium	Microscopii	Mic	Microscope
Camelopardalis	Camelopardalis	Cam	Giraffe	Monoceros	Monocerotis	Mon	Unicorn
Cancer	Cancri	Cnc	Crab	Musca	Muscae	Mus	Fly
Canes Venatici	Canum Venaticorum	CVn	Hunting Dogs	Norma	Normae	Nor	Level (square)
Canis Major	Canis Majoris	CMa	Big Dog	Octans	Octantis	Oct	Octant
Canis Minor	Canis Minoris	CMi	Little Dog	Ophiuchus	Ophiuchi	Oph	Ophiuchus* (serpent bearer)
Capricornus	Capricorni	Cap	Goat	Orion	Orionis	Ori	Orion*
Carina	Carinae	Car	Ship's Keel**	Pavo	Pavonis	Pav	Peacock
Cassiopeia	Cassiopeiae	Cas	Cassiopeia*	Pegasus	Pegasi	Peg	Pegasus* (winged horse)
Centaurus	Centauri	Cen	Centaur*	Perseus	Persei	Per	Perseus*
Cepheus	Cephei	Cep	Cepheus*	Phoenix	Phoenicis	Phe	Phoenix
Cetus	Ceti	Cet	Whale	Pictor	Pictoris	Pic	Easel
Chamaeleon	Chamaeleonis	Cha	Chameleon	Pisces	Piscium	Psc	Fish
Circinus	Circini	Cir	Compass	Piscis Austrinus	Piscis Austrini	PsA	Southern Fish
Columba	Columbae	Col	Dove	Puppis	Puppis	Pup	Ship's Stern**
Coma Berenices	Comae Berencies	Com	Berenice's Hair*	Pyxis	Pyxidis	Pyx	Ship's Compass**
Corona Australis	Coronae Australis	CrA	Southern Crown	Reticulum	Reticuli	Ret	Net
Corona Borealis	Coronae Borealis	CrB	Northern Crown	Sagitta	Sagittae	Sge	Arrow
Corvus	Corvi	Crv	Crow	Sagittarius	Sagittarii	Sgr	Archer
Crater	Crateris	Crt	Cup	Scorpius	Scorpii	Sco	Scorpion
Crux	Crucis	Cru	Southern Cross	Sculptor	Sculptoris	Scl	Sculptor
Cygnus	Cygni	Cyg	Swan	Scutum	Scuti	Sct	Shield
Delphinus	Delphini	Del	Dolphin	Serpens	Serpentis	Ser	Serpent
Dorado	Doradus	Dor	Swordfish	Sextans	Sextantis	Sex	Sextant
Draco	Draconis	Dra	Dragon	Taurus	Tauri	Tau	Bull
Equuleus	Equulei	Equ	Little Horse	Telescopium	Telescopii	Tel	Telescope
Eridanus	Eridani	Eri	River Eridanus*	Triangulum	Trianguli	Tri	Triangle
Fornax	Fornacis	For	Furnace	Triangulum Australe	Trianguli Australis	TrA	Southern Triangle
Gemini	Geminorum	Gen	Twins	Tucana	Tucanae	Tuc	Toucan
Grus	Gruis	Gru	Crane	Ursa Major	Ursae Majoris	UMa	Big Bear
Hercules	Herculis	Her	Hercules*	Ursa Minor	Ursae Minoris	UMi	Little Bear
Horologium	Horologii	Hor	Clock	Vela	Velorum	Vel	Ship's Sails**
Hydra	Hydrae	Hya	Hydra* (water monster)	Virgo	Virginis	Vir	Virgin
Hydrus	Hydri	Hyi	Sea serpent	Volans	Volantis	Vol	Flying Fish
Indus	Indi	Ind	Indian	Vulpecula	Vulpeculae	Vul	Little Fox

*Proper names
**Formerly formed the constellation Argo Navis, the Argonauts' ship

SELECTED READINGS

Non-Technical Magazines

Sky and Telescope, P.O. Box 9111, Belmont, MA 02138, 800 253 0245, www.skyandtelescope.com.

Astronomy, 21027 Crossroads Circle, P.O. Box 1612, Waukesha, WI 53187, 800 533 6644, astronomy.com.

Mercury, Astronomical Society of the Pacific, 390 Ashton Ave., San Francisco, CA 94112, 800 962 3412, www.astrosociety.org.

StarDate, 2609 University, Rm. 3.118, University of Texas, Austin, TX 78712, 800 STARDATE, www.stardate.org.

The Griffith Observer, 2800 East Observatory Road, Los Angeles, CA 90027, www.griffithobs.org.

Science News, P.O. Box 1925, Marion, OH 43305, 800 552 4412, www.sciserv.org, published weekly.

Scientific American, P.O. Box 3186, Harlan, IA 51593-2377, 800 333 1199, www.scientificamerican.com.

National Geographic, P.O. Box 96583, Washington, DC 20078-9973, 800 NGC LINE, www.nationalgeographic.com.

Natural History, P.O. Box 5000, Harlan, IA 51593-5000, 800 234 5252, www.naturalhistorymag.com.

New Scientist, 151 Wardour St., London W1V 4BN, U.K., 888 822 4342, rbi.subscriptions@rbi.co.uk; www.newscientist.com.

Physics Today, American Institute of Physics, 2 Huntingdon Quadrangle, Melville, NY 11747, 800 344 6902, www.aip.org/pt; www.physicstoday.org.

Science Year, World Book Encyclopedia, Inc., Post Office Box 11207, Des Moines, IA 50340-1207, 800 504 4425, www.worldbook.com.

Smithsonian, P.O. Box 420311, Palm Coast, FL 32142-0311, Washington, DC 20560, 800 766 2149; www.smithsonianmag.si.edu.

Discover, P.O. Box 42105, Palm Coast, FL 34142-0105, 800 829 9132.

The Planetary Report, The Planetary Society, 65 North Catalina Avenue, Pasadena, CA 91106-2301, 818 793 5100, planetary.org/tps.

Observing Reference Books

Jay M. Pasachoff, *A Field Guide to the Stars and Planets,* 4th ed. (Boston: Houghton Mifflin Co., 2000; updated 2003). All kinds of observing information, including monthly maps and the 2000.0 sky atlas by Wil Tirion, and Graphic Timetables to locate planets and special objects like clusters and galaxies.

Jay M. Pasachoff, *Peterson's First Guide to Astronomy* (Boston: Houghton Mifflin Co., 1997). A brief, beautifully illustrated introduction to observing the sky. Tirion monthly maps.

Jay M. Pasachoff, *Peterson's First Guide to the Solar System* (Boston: Houghton Mifflin Co., 1997). Color illustrations and simple descriptions mark this elementary introduction. Tirion maps of Mars, Jupiter, and Saturn's positions through 2010.

H. J. P. Arnold, P. Doherty, and P. Moore, *The Photographic Atlas of the Stars* (Bristol, UK: Institute of Physics, 1997).

The Astronomical Almanac (yearly), U.S. Government Printing Office, Washington, DC 20402.

Michael Covington, *Astrophotography for the Amateur,* 2nd ed. (Cambridge University Press, 1999)

Terence Dickinson, *Nightwatch: A Practical Guide to Viewing the Universe,* 3rd ed. (Willowdale, Ontario: Firefly Books, 1998).

Philip S. Harrington, *The Deep Sky: An Introduction* (Cambridge, MA: Sky Publishing Co., 1997).

Alan Hirshfeld, Roger W. Sinnott, and François Ochsenbein, 1991 (vol. 1); Alan Hirshfeld and Roger W. Sinnott, 1985 (vol. 2). *Sky Catalogue 2000.0,* 2nd ed (Cambridge, Mass.: Sky Publishing Corp) Vol. 1 is stars, and Vol. 2 is full of tables of all other objects.

Chris Kitchen and Robert W. Forrest, *Seeing Stars: The Night Sky Through Small Telescopes* (New York: Springer-Verlag, 1998).

David H. Levy, *Observing Variable Stars: A Guide for the Beginner* (New York: Cambridge University Press).

Jim Mullaney, *Celestial Harvest: 3000-Plus Showpieces of the Heavens for Telescope Viewing & Contemplation* (self published, PO Box 1146, Exton, PA 1934, 1998); jimullaneysm@nospam.com; store.yahoo.com/doverpublications.

Observer's Handbook (yearly), Royal Astronomical Society of Canada, 136 Dupont Street, Toronto, Ontario M5R 1V2 Canada.

Stephen James O'Meara and David H. Levy, *The Messier Objects* (New York: Cambridge University Press, 1998).

Guy Ottewell, *Astronomical Calendar* (yearly) and *The Astronomical Companion,* Department of Physics, Furman University, Greenville, SC 29613, 864 294 2208, guyverno@aol.com; www.kalend.com.

Ian Ridpath, ed., *Norton's star atlas and reference handbook (epoch 2000.0),* 20th ed. (Pi Press/Longman, 2004). An updated old standard.

Ian Ridpath, *Eyewitness Handbooks: Stars and Planets* (DK Publishing, 1998).

Roger W. Sinnott, ed., *NGC 2000.0: the complete new general catalogue and index catalogues of nebulae and star clusters,* (Cambridge, Mass.: Sky Publishing Corp. and New York: Cambridge University Press, 1998). A centennial reissue of Dreyer's work with updated data.

Roger W. Sinnott and Michael A. C. Perryman, *Millennium Star Atlas* (Cambridge, MA: Sky Publishing Corp., 1997). Uses the Hipparcos data.

Wil Tirion, *Cambridge Star Atlas,* 3rd ed. (Cambridge University Press, 2001). A naked-eye star atlas in full color. A moon map, 24 monthly sky maps, 20 detailed star charts, and 6 all-sky maps.

Wil Tirion and Roger W. Sinnott, *Sky Atlas 2000.0,* 2nd ed (Cambridge, Mass.: Sky Publishing Corp. and New York: Cambridge University Press, 1998). Large-scale star charts to magnitude 8.5.

Wil Tirion, Barry Rappaport, and Will Remaklus, *Uranometria 2000.0,* 2nd ed. (Richmond, VA: Willmann-Bell, 2001). Vol. 1 covers +90° to –6°. Vol. 2 covers +6° to –90°. Star maps to magnitude 9.7.

Willmann-Bell catalogue, *Astronomy Books* (P.O. Box 35025, Richmond, VA 23235), 800 825 7827, www.willbell.com. They publish many observing guides and distribute all kinds of astronomy books.

For Information About Amateur Societies

American Association of Variable Star Observers (AAVSO), 25 Birch St., Cambridge, MA 02138, www.aavso.org.

American Meteor Society, Dept. of Physics and Astronomy, SUNY, Geneseo, NY 14454, www.amsmeteors.org.

Astronomical League, the umbrella group of amateur societies. For their newsletter, *The Reflector,* write The Astronomical League, Executive Secretary, c/o Science Service Building, 1719 N St., N.W., Washington, DC 20030, www.astroleague.org.

Astronomical Society of the Pacific, 390 Ashton Ave., San Francisco, CA 94112, www.astrosociety.org.

British Astronomical Association, Burlington House, Piccadilly, London W1V 0NL, England, www.britastro.org.

International Dark Sky Association, c/o David Crawford, 3545 N. Stewart, Tucson, AZ 85716, 877 600 5888. www.darksky.org.

The Planetary Society, 65 North Catalina Ave., Pasadena, CA 981106-2301, 818 793 5100, www.planetary.org.

Royal Astronomical Society of Canada, 124 Merton St., Toronto, Ontario M4S 2Z2, Canada, www.rasc.ca.

Careers in Astronomy

AAS Education Office, Adler Planetarium & Astronomy Museum, 2000 Florida Ave., N.W., Suite 400, Washington, DC 20009, aased@aas.org; www.aas.org/education. Information on careers in astronomy is available online.

Teaching

Jay M. Pasachoff and John R. Percy, *Teaching and Learning Astronomy: Effective Strategies for Educators Worldwide* (Cambridge Univ. Press, 2005). The proceedings of International Astronomical Union Special Session, held at the Sydney, Australia, General Assembly in 2003.

Jay M. Pasachoff and John R. Percy, *The Teaching of Astronomy* (Cambridge Univ. Press, 1990; paperbound, 1993). The proceedings of International Astronomical Union Colloquium #105, held in Williamstown in 1988. Available at adswww.harvard.edu.

Lucienne Gouguenheim, Derek McNally, and John R. Percy, *New Trends in Astronomy Teaching* (Cambridge Univ. Press, 1997). The proceedings of International Astronomical Union Colloquium #162, held in London in 1996.

John R. Percy, *Astronomy Education: Current Developments, Future Coordination* (San Francisco: Astronomical Society of the Pacific, 1996). An ASP conference, held in College Park, MD in 1995.

General Reading and Reference

Sybil P. Parker and Jay M. Pasachoff, *McGraw-Hill Encyclopedia of Astronomy* (McGraw-Hill, 1993).

Jean Audouze and Guy Israel, eds. *The Cambridge Atlas of Astronomy,* 3rd ed. (Cambridge Univ. Press, 1994). Coffee-table size with fantastic photos and authoritative text.

Marcia Bartusiak, *Archives of the Universe: A Treasury of Astronomy's Historic Works of Discovery* (Pantheon, 2004; paperback, Vintage, 2006).

Christopher De Pree and Alan Axelrod, *The Complete Idiot's Guide to Astronomy,* 3rd ed. (Indianapolis: Alpha Books, 2004). *The Complete Idiot's Guide* series has as a theme "You're no idiot, of course," at what you do best, but on this subject. . . .

Terence Dickinson, *The Universe and Beyond,* 4th ed. (Firefly, 2004).

Paul Murdin, ed., *Encyclopedia of Astronomy and Astrophysics* (Bristol, UK: Macmillan and Institute of Physics, 2000). See www.ency-astro.com.

Robert Nemiroff and Jerry Bonnell, *The Universe: 365 Days* (Abrams, 2003). Selections from Astronomy Picture of the Day; see antwrp.gsfc.nasa.gov.

Philip C. Plait, *Bad Astronomy: Misconceptions and Misuses Revealed, from Astrology to the Moon Landing "Hoax,"* www.badastronomy.com.

Martin Rees, ed., *Universe* (DK Publishing, 2005).

Neil deGrasse Tyson and Donald Goldsmith, *Origins: Fourteen Billion Years of Cosmic Evolution* (W. W. Norton, 2004).

History

James A. Connor, *Kepler's Witch* (San Francisco: Harper San Francisco, 2004).

Stillman Drake, *Galileo: A Very Short Introduction* (New York: Oxford University Press, 2001).

Kitty Ferguson, *The Nobleman and His Housedog: Tycho Brahe and Johannes Kepler: The Strange Partnership that Revolutionized Science* (Walker & Co, 2002), tychoandkepler.com.

Owen Gingerich and James MacLachlan, *Nicolaus Copernicus: Making the Earth a Planet* (New York: Oxford University Press, 2005).

Owen Gingerich, *The Book Nobody Read: Chasing the Revolutions of Nicolaus Copernicus* (Walker and Co., 2004, paperback, 2005). About the hunt for copies of Copernicus's book.

Owen Gingerich, *The Eye of Heaven: Ptolemy, Copernicus and Kepler* (New York: American Institute of Physics, 1993).

Michael Hoskin, ed., *The Cambridge Concise History of Astronomy* (Cambridge University Press, 1999).

Michael Hoskin, ed., *Cambridge Illustrated History: Astronomy* (Cambridge University Press, 1997).

George Johnson, *Miss Leavitt's Stars* (Atlas Books/W. W. Norton, 2005).

James MacLachlan, *Galileo Galilei: First Physicist* (New York: Oxford University Press, 1999).

Arthur I. Miller, *Empire of the Stars: Obsession, Friendship, and Betrayal in the Quest for Black Holes* (Houghton Mifflin, 2005).

Simon Mitton, *Conflict in the Cosmos: Fred Hoyle's Life in Science* (Joseph Henry Press, 2005).

Andrew Norton, ed., *Observing the Universe* (Cambridge University Press, 2004). How professional astronomers work.

Naomi Pasachoff, *World Book Biographies in Context: Newton, Einstein, and Oppenheimer* (Chicago: World Book, 2006).

James Voelkel, *Johannes Kepler and the New Astronomy* (New York: Oxford University Press, 1999). A short biography in the Oxford Portraits series.

Fred Watson, *Stargazer: The Life and Times of the Telescope* (Da Capo Press, 2004).

Art and Astronomy

Roberta J. M. Olson and Jay M. Pasachoff, *Fire in the Sky: Comets and Meteors, the Decisive Centuries, in British Art and Science* (Cambridge University Press, 1998, 1999). From the time of Newton and Halley to the present.

Solar System

J. Kelly Beatty, Carolyn Collins Petersen, and Andrew Chaikin, *The New Solar System,* 4th ed. (Cambridge, MA: Sky Publishing Corp. and Cambridge University Press, 1999). Each chapter is written by a different expert.

Jim Bell and Jacqueline Mitton, eds., *Asteroid Rendezvous: NEAR Shoemaker's Adventures at Eros* (Cambridge University Press, 2002).

Joseph M. Boyce, *The Smithsonian Book of Mars* (Smithsonian Institution Press, 2002).

Imke de Pater and Jack J. Lissauer, *Planetary Sciences* (Cambridge University Press, 2001).

Jeffrey S. Kargel, *Mars—A Warmer, Wetter Planet* (Springer Praxis Books, 2004).

Jeffrey Kluger, *Moon Hunters: NASA's Remarkable Expeditions to the Ends of the Solar System* (Simon and Schuster, 2001).

Kenneth Lang, *The Cambridge Guide to the Solar System* (Cambridge University Press, 2003).

Eli Maor, *June 8, 2004—Venus in Transit* (Princeton University Press, 2000).

Carl Sagan, *Pale Blue Dot* (New York: Random House, 1994). One of astronomy's most eloquent spokeperson's last works.

William Sheehan and John Westfall, *The Transits of Venus* (Prometheus Books, 2004).

Steven Squyres, *Roving Mars: Spirit, Opportunity, and the Exploration of the Red Planet* (2005). By the principal investigator of the Mars Exploration Rovers Spirit and Opportunity.

S. Alan Stern, *Our Worlds: The Magnetism and Thrill of Planetary Exploration: As Described by Leading Planetary Scientists* (Cambridge University Press, 1999).

S. Alan Stern and Jacqueline Mitton, *Pluto & Charon: Ice Worlds on the Ragged Edge of the Solar System* (New York: Wiley, 1997).

Ben Zuckerman and Michael Hart, eds., *Extraterrestrials—Where Are They?* 2nd ed. (Cambridge University Press, 1995).

The Sun

Arvind Bhatnagar and William C. Livingston, *Fundamentals of Solar Astronomy* (World Scientific, 2005). Comprehensive and phenomenological but relatively non-mathematical.

Michael J. Carlowicz and Ramon E. Lopez, *Storms from the Sun: The Emerging Science of Space Weather* (Joseph Henry Press, 2000).

Fred Espenak, *Fifty Year Canon of Solar Eclipses* (NASA Ref. Pub. 1178, Rev. 1987). Maps and tables.

Leon Golub and Jay M. Pasachoff, *Nearest Star: The Surprising Science of Our Sun* (Harvard University Press, 2001). A non-technical trade book.

Leon Golub and Jay M. Pasachoff, *The Solar Corona* (Cambridge Univ. Press, 1997). An advanced text.

Kenneth R. Lang, *Sun, Earth, and Sky* (Springer-Verlag, 1995).

Jay M. Pasachoff, *The Complete Idiot's Guide to the Sun* (Alpha Books, 2003).

Kenneth J. H. Phillips, *Guide to the Sun* (New York: Cambridge Univ. Press, 1992).

Peter O. Taylor and Nancy L. Hendrickson, *Beginner's Guide to the Sun* (Waukesha, WI: Kalmbach Books, 1995).

Jack B. Zirker, *Journey from the Center of the Sun* (Princeton University Press, 2001, paperback, 2004). A non-technical trade book.

Jack B. Zirker, *Sunquakes: Probing the Interior of the Sun* (Princeton University Press, 2003). A non-technical trade book.

Stars and the Milky Way Galaxy

Nigel Henbest and Heather Couper, *The Guide to the Galaxy* (Cambridge Univ. Press, 1994). Exciting, contemporary results; profusely illustrated.

Alan W. Hershfeld, *Parallax* (W. H. Freeman, 2001).

James B. Kaler, *The Hundred Greatest Stars* (New York: Copernicus, 2002).

Jonathan I. Katz, *The Biggest Bangs: The Mystery of Gamma-Ray Bursts, the Most Violent Explosions in the Universe* (New York: Oxford University Press, 2002).

Robert Kirshner, *The Extravagant Universe: Exploding Stars, Dark Energy, and the Accelerating Cosmos* (Princeton University Press, 2002, paperback 2004).

Alfred Mann, *Shadow of a Star* (W. H. Freeman, 1997). The story of the neutrinos from SN 1987A.

Laurence Marschall, *The Supernova Story*. Supernovae in general plus SN 1987A. (Princeton University Press, 1994).

Govert Schilling, *Flash! The Hunt for the Biggest Explosions in the Universe* (Cambridge University Press, 2002).

Kip Thorne, *Black Holes and Time Warps: Einstein's Outrageous Legacy* (Norton, 1994). An excellent account of general relativity.

Wallace H. Tucker and Karen Tucker, *Revealing the Universe: The Making of the Chandra X-ray Observatory* (Harvard University Press, 2001).

J. Craig Wheeler, *Cosmic Catastrophes: Supernovae, Gamma-Ray Bursts, and Adventures in Hyperspace* (Cambridge University Press, 2000).

Galaxies and Cosmology

Fred Adams and Greg Laughlin, *The Five Ages of the Universe* (New York: Free Press, 1999). Tells the story of what will happen as the Universe expands forever.

Mitchell Begelman and Martin Rees, *Gravity's Fatal Attraction: Black Holes in the Universe* (New York: Scientific American Library, 1996).

Richard Berendzen, Richard Hart, and Daniel Seeley, *Man Discovers the Galaxies* (New York: Neale Watson Academic Publications, 1976). A historical review.

Dennis Danielson, *The Book of the Cosmos* (New York: Helix Books, 2000). A compendium from millennia of writings about the cosmos, with useful commentaries.

Timothy Ferris, *The Whole Shebang: A State of the Universe(s) Report* (New York: Simon & Schuster, 1997, 1998). Contemporary cosmology, from a gifted journalist.

Michael Friedlander, *A Thin Cosmic Rain: Particles from Outer Space* (Harvard University Press, 2000).

Donald Goldsmith, *The Astronomers* (New York: St. Martin's Press, 1991). The companion to a PBS television series.

Don Goldsmith, *The Runaway Universe* (Cambridge, MA: Perseus Books, 2000). Describes the discovery of a nonzero cosmological constant.

Brian Greene, *The Elegant Universe* (New York: Norton, 1999). Superstrings and the Universe; Phi Beta Kappa book award winner.

Alan H. Guth, *The Inflationary Universe: The Quest for a New Theory of Cosmic Origins* (New York: Helix Books/Addison Wesley, 1997). By the originator of the inflationary theory.

Stephen W. Hawking, *A Brief History of Time, Updated and Expanded* (New York: Bantam Books, 1998). A best-selling discussion of fundamental topics. *The Illustrated Brief History of Time* (1997) is also available.

Stephen W. Hawking, *The Universe in a Nutshell* (New York: Bantam Books, 2001).

Craig Hogan, *The Little Book of the Big Bang: A Cosmic Primer* (New York: Copernicus, 1998).

Michio Kaku, *Hyperspace* (Doubleday, 1994). A very readable introduction to string theory and other dimensions.

Lawrence Krauss, *Quintessence* (New York: Basic Books, 2000). Discusses forms of repulsive dark energy that may be accelerating the Universe.

Leon M. Lederman and David N. Schramm, *From Quarks to the Cosmos,* 2nd ed. (New York: Scientific American Library, 1995).

Michael Lemonick, *Echo of the Big Bang* (Princeton University Press, 2003). About the WMAP satellite and its results.

Alan Lightman and Roberta Brawer, *Origins: The Lives and Worlds of Modern Cosmologists* (Cambridge, MA: Harvard Univ. Press, 1990). Interviews with 27 cosmologists.

Mario Livio, *The Accelerating Universe: Infinite Expansion, the Cosmological Constant, and the Beauty of the Cosmos* (New York: John Wiley & Sons, 2000). Contemporary cosmology by an expert at the Space Telescope Science Institute.

David Malin, *The Invisible Universe* (Boston: Bulfinch Press, 1999). By the master of ground-based astronomical color photography.

Fulvio Melia, *The Edge of Infinity: Supermassive Black Holes in the Universe* (Cambridge University Press, 2003).

Fulvio Melia, *The Black Hole at the Center of Our Galaxy* (Princeton University Press, 2003).

Dennis Overbye, *Lonely Hearts of the Cosmos: The Scientific Quest for the Secrets of the Universe* (New York: HarperCollins, 1991). Humanizing the cosmologists.

Jay M. Pasachoff, Hyron Spinrad, Patrick Osmer, and Edward S. Cheng, *The Farthest Things in the Universe* (Cambridge Univ. Press, 1995). The cosmic background of radiation, quasars, and distant galaxies.

Carolyn Collins Petersen and John C. Brandt, *Visions of the Cosmos* (Cambridge University Press, 2003). Highly illustrated with images from Hubble, Chandra, and other telescopes on the ground and in space.

Martin Rees, *Before the Beginning: Our Universe and Others* (Helix Books/Addison Wesley, 1997). By the Astronomer Royal, a major researcher in the field.

Martin Rees, *Our Cosmic Habitat* (Princeton University Press, 2001). By the Astronomer Royal, a major researcher in the field.

Michael Rowan-Robinson, *The Nine Numbers of the Cosmos* (Oxford, 1999).

Vera Rubin, *Bright Galaxies, Dark Matters* (Woodbury, New York: American Institute of Physics, 1997). Science and biography.

Allan Sandage and John Bedke, *The Carnegie Atlas of Galaxies* (Washington, D.C.: Carnegie Institution of Washington, 1994). Images to stare at and pore over.

Joseph Silk, *A Short History of the Universe* (New York: Scientific American Library, 1994).

Lee Smolin, *The Life of the Cosmos* (Oxford: Oxford Univ. Press, 1997). Suggests that the properties of our Universe resulted from natural selection.

George Smoot and Keay Davidson, *Wrinkles in Time* (New York: Morrow, 1993). A personal account of the discovery of ripples in the cosmic background radiation.

Ian Stewart, *Flatterland: Like Flatland, Only More So* (Perseus Books, 2001).

William H. Waller and Paul W. Hodge, *Galaxies and the Cosmic Frontier* (Harvard University Press, 2004).

Steven Weinberg, *The First Three Minutes,* 2nd ed. (New York: Basic Books, 1993). A readable discussion of the first minutes after the big bang, including a discussion of the cosmic background radiation.

John Noble Wilford, ed., *Cosmic Dispatches: The* New York Times *Reports on Astronomy and Cosmology* (W. W. Norton, 2002).

Richard Wolfson, *Simply Einstein: Relativity Demystified* (W. W. Norton, 2003). An excellent account of relativity.

Ben Zuckerman and Matthew A. Malkan, *The Origin and Evolution of the Universe* (Boston: Jones and Bartlett, 1996).

Books for the Vision-Impaired or Blind

Noreen Grice, *Touch the Universe: A NASA Braille Book of Astronomy* (Joseph Henry Press, 2002). Braille text and images whose outlines can be felt.

Noreen Grice, *Touch the Sun: A NASA Braille Book* (Joseph Henry Press, 2005). Braille text and images whose outlines can be felt.

Noreen Grice, *Touch the Stars II* (National Braille Press). Braille text and images whose outlines can be felt.

GLOSSARY

absolute magnitude The magnitude that a star would appear to have if it were at a distance of ten parsecs (32.6 lt yr) from us.
absorption line Wavelengths at which the intensity of radiation is less than it is at neighboring wavelengths.
absorption nebula Gas and dust seen in silhouette.
accelerating universe The model for the Universe based on observations late in the 1990s that the expansion of the Universe is speeding up over time, rather than slowing down in the way that gravity alone would modify its expansion.
accretion disk Matter that an object has taken up and that has formed a disk around the object.
active galactic nucleus (AGN) A galaxy nucleus that is exceptionally bright in some part of the spectrum; includes radio galaxies, Seyfert galaxies, quasars, and QSOs.
active galaxy A galaxy whose nucleus radiates much more than average in some part of the spectrum, revealing high-energy processes; radio galaxies and Seyfert galaxies are examples.
active optics Optical systems providing slow adjustments to components to keep them lined up properly.
active region One of the regions on the Sun where sunspots, plages, flares, etc., are found.
active Sun The group of solar phenomena that vary with time, such as active regions and their phenomena.
adaptive optics Optical systems providing rapid corrections to counteract atmospheric blurring.
AGN Active galactic nucleus.
albedo The fraction of incident light reflected by a body.
allowed states The energy values that atoms can have by laws of quantum theory.
alpha particle A helium nucleus; consists of two protons and two neutrons.
altitude (a) Height above the surface of a planet; (b) for a telescope mounting, elevation in angular measure above the horizon.
amino acid A type of molecule containing a chain of carbon atoms and the group NH_2 (the amino group). Amino acids are fundamental building blocks of life.
angstrom A unit of length equal to 10^{-8} cm.
angular momentum An intrinsic property of a system corresponding to the amount of its revolution or spin. The amount of angular momentum of a body orbiting around a point is the mass of the orbiting body times its (linear) velocity of revolution times its distance from the point. The amount of angular momentum of a spinning object is the amount of inertia, an intrinsic property of the distribution of mass, times the angular velocity of spin. The conservation of angular momentum is a law that states that the total amount of angular momentum remains constant in a system that is undisturbed from outside itself.
angular resolution See *resolution.*
angular velocity The rate at which a body rotates or revolves expressed as the angle covered in a given time (for example, in degrees per hour).
anisotropy Deviation from isotropy; changing with direction.
annular eclipse A type of solar eclipse in which a ring (annulus) of solar photosphere remains visible.
anthropic principle The idea that since we exist, the Universe must have certain properties or it would not have evolved so that life formed and humans evolved.
antimatter A type of matter in which each particle (antiproton, antineutron, etc.) is opposite in charge and certain other properties to a corresponding particle (proton, neutron, etc.) of the same mass of the ordinary type of matter from which the Solar System is made.
antiparticle The antimatter corresponding to a given particle.
aperture The diameter of the lens or mirror that defines the amount of light focused by an optical system.
aphelion For an orbit around the Sun, the farthest point from the Sun.
Apollo asteroids A group of asteroids, with semimajor axes greater than Earth's and less than 1.017 A.U., whose orbits overlap the Earth's.
apparent magnitude The brightness of a star as seen by an observer, given in a specific system in which a difference of five magnitudes corresponds to a brightness ratio of one hundred times; the scale is fixed by correspondence with a historical background.
archaebacteria A primitive type of organism, different from either plants or animals, perhaps surviving from billions of years ago.
association A physical grouping of stars; in particular, we talk of O and B associations.
asterism A special apparent grouping of stars, part of a constellation.
asteroid A "minor planet," a non-luminous chunk of rock smaller than planet-size but larger than a meteoroid, in orbit around a star.
asteroid belt A region of the Solar System, between the orbits of Mars and Jupiter, in which most of the asteroids orbit.
astrometric binary A system of two stars in which the existence of one star can be deduced by study of its gravitational effect on the proper motion of the other star.
astrometry The branch of astronomy that involves the detailed measurement of the positions and motions of stars and other celestial bodies.
Astronomical Unit The average distance from the Earth to the Sun.
astrophysics The science, now essentially identical with astronomy, applying the laws of physics to the Universe.
atom The smallest possible unit of a chemical element. When an atom is subdivided, the parts no longer have properties of any chemical element.
atomic clock A system that uses atomic properties to provide a measure of time.
atomic number The number of protons in an atom.
atomic weight The number of protons and neutrons in an atom, averaged over the abundances of the different isotopes.
aurora Glowing lights visible in the sky, resulting from processes in the Earth's upper atmosphere and linked with the Earth's magnitude field.
aurora australis The southern aurora.
aurora borealis The northern aurora.
autumnal equinox Of the two locations in the sky where the ecliptic crosses the celestial equator, the one that the Sun passes each year when moving from northern to southern declinations.
A.U. Astronomical Unit.
azimuth The angular distance around the horizon from the northern direction, usually expressed in angular measure from 0° for an object in the northern direction, to 180° for an object in the southern direction, back around to 360°.

background radiation See *cosmic background radiation.*
Baily's beads Beads of light visible around the rim of the Moon at the beginning and end of a total solar eclipse. They result from the solar photosphere shining through valleys at the edge of the Moon.
Balmer series The set of spectral absorption or emission lines resulting from a transition up from or down to (respectively) the second energy level (first excited level) of hydrogen.
bar The straight structure across the center of some spiral galaxies, from which the arms unwind.
baryons Nuclear particles (protons, neutrons, etc.) subject to the strong nuclear force; made of quarks.

basalt A type of rock resulting from the cooling of lava.
baseline The distance between points of observation when it determines the accuracy of some measurement.
beam The cone within which a radio telescope is sensitive to radiation.
belts Dark bands around certain planets, notably Jupiter.
big-bang theory A cosmological model, based on Einstein's general theory of relativity, in which the Universe was once compressed to nearly infinite density and has been expanding ever since.
binary pulsar A pulsar in a binary system.
binary star Two stars revolving around each other.
bipolar flow A phenomenon in young or forming stars in which streams of matter are ejected from the poles.
black body An object that absorbs all radiation that hits it and emits radiation that exactly follows Planck's law.
black-body curve A graph of brightness vs. wavelength that follows Planck's law; each such curve corresponds to a given temperature. Also called a Planck curve.
black-body radiation Radiation whose distribution in wavelength follows Planck's law, the black-body curve.
black dwarf A feebly radiating ball of gas that results when a white dwarf has radiated nearly all its energy.
black hole A region of space from which, according to the general theory of relativity, neither radiation nor matter can escape.
black-hole era The future era, following the degenerate era, when the only objects (besides photons and subatomic particles) the Universe will contain will be black holes.
blueshift A shift of optical wavelengths toward the blue or in all cases towards shorter wavelengths; when the shift is caused by motion, from a velocity of approach.
Bohr atom Niels Bohr's model of the hydrogen atom, in which the energy levels are depicted as concentric circles of radii that increase as (level number)^2.
bound-free transition An atomic transition in which an electron starts bound to the atom and winds up free from it.
brown dwarf A self-gravitating, self-luminous object, not a satellite and insufficiently massive for sustained nuclear fusion (less than 0.075 solar mass).

carbonaceous Containing a lot of carbon.
carbon-nitrogen-oxygen (CNO) cycle A chain of nuclear reactions, involving carbon as a catalyst and including nitrogen and oxygen as intermediaries, that transforms four hydrogen nuclei into one helium nucleus with a resulting release in energy. The CNO cycle is important only in stars hotter than the Sun.
Cassegrain (telescope) A type of reflecting telescope in which the light focused by the primary mirror is intercepted short of its focal point and refocused and reflected by a secondary mirror through a hole in the center of the primary mirror.
Cassini (a) Jean Dominique Cassini (1625–1712); (b) the NASA mission that began to orbit Saturn and sent a probe into Titan upon its arrival there in 2004.
Cassini's division The major division in the rings of Saturn.
catalyst A substance that speeds up a reaction but that is left over in its original form at the end.
CCD Charge-coupled device, a solid-state imaging device.
celestial equator The intersection of the celestial sphere with the plane that passes through the Earth's equator.
celestial poles The intersection of the celestial sphere with the axis of rotation of the Earth.
celestial sphere The hypothetical sphere centered at the center of the Earth to which it appears that the stars are affixed.
centaur A Solar-System object, with 2060 Chiron as the first example, intermediate between comets and icy planets or satellites and orbiting the Sun between the orbits of Jupiter and Neptune.
center of mass The "average" location of mass; the point in a body or system of bodies at which we may consider all the mass to be located for the purpose of calculating the gravitational effect of that mass or its mean motion when a force is applied.
central star The hot object at the center of a planetary nebula, which is the remaining core of the original star.
Cepheid variable A type of supergiant star that oscillates in brightness in a manner similar to the star δ Cephei. The periods of Cepheid variables, which are between 1 and 100 days, are linked to the average luminosity of the stars by known relationships; this allows the distances to Cepheids to be found.
Chandra X-ray Observatory The major NASA x-ray observatory, the former Advanced X-ray Astrophysics Facility, launched in 1999 to make high-resolution observations in the x-ray part of the spectrum.
Chandrasekhar limit The limit in mass, about 1.4 solar masses, above which electron degeneracy (a quantum-mechanical pressure) cannot support a star, and so the limit above which white dwarfs cannot exist.
chaos The condition in which very slight differences in initial conditions typically lead to very different results. Also, specifically, examples of orbits or rotation directions that change after apparently random intervals of time because of such sensitivity to initial conditions.
chromatic aberration A defect of lens systems in which different colors are focused at different points.
chromosphere The part of the atmosphere of the Sun (or another star) between the photosphere and the corona. It is probably entirely composed of spicules and probably roughly corresponds to the region in which mechanical energy is deposited.
circle A conic section formed by cutting a cone perpendicularly to its axis.
circumpolar stars For a given observing location, stars that are close enough to the celestial pole that they never set.
classical When discussing atoms, not taking account of quantum-mechanical effects.
closed universe A big-bang universe with positive curvature; it has finite volume and will eventually contract, assuming there is no repulsive cosmological constant.
cluster (a) Of stars, a physical grouping of many stars; (b) of galaxies, a physical grouping of at least a few galaxies.
CNO cycle See *carbon-nitrogen-oxygen cycle.*
COBE (Cosmic Background Explorer) A spacecraft launched in 1989 to study the cosmic background radiation.
cold dark matter Non-luminous matter that moves slowly through the expanding Universe, such as mini black holes and exotic nuclear particles.
color (a) Of an object, a visual property that depends on wavelength; (b) an arbitrary name assigned to a property that distinguishes three kinds of quarks.
coma (a) Of a comet, the region surrounding the head; (b) of an optical system, an off-axis aberration in which the images of points appear with comet-like asymmetries.
comet A type of object orbiting the Sun, often in a very elongated orbit, that when relatively near to the Sun shows a coma and may show a tail.
comparative planetology Studying the properties of Solar-System bodies by comparing them.
comparison spectrum A spectrum of known elements on Earth photographed in order to provide a known set of wavelengths for zero Doppler shift.
composite spectrum The spectrum that reveals a star is a binary system, since spectra of more than one object are apparent.
condensation A region of unusually high mass or brightness.
conic sections Geometric shapes obtained by slicing a cone.
conservation law A statement that the total amount of some property (angular momentum, energy, etc.) of a body or set of bodies does not change.
constellation One of 88 areas into which the sky has been divided for convenience in referring to the stars or other objects therein.
continental drift The slow motion of the continents across the Earth's surface, explained in the theory of plate tectonics as a set of shifting regions called plates.
continuous spectrum A spectrum with radiation at all wavelengths but with neither absorption nor emission lines.
continuum (plural: **continua**) The continuous spectrum that we would measure from a body if no spectral lines were present.

convection The method of energy transport in which the rising motion of masses from below carries energy upward in a gravitational field. Boiling is an example.
convection zone The subsurface zone in certain types of stars in which convection dominates energy transfer.
coordinate systems Methods of assigning positions with respect to suitable axes.
core The central region of a star or planet.
corona, solar or stellar The outermost region of the Sun (or of other stars), characterized by temperatures of millions of kelvins.
coronagraph A type of telescope with which the corona (and, sometimes, merely the chromosphere) can be seen in visible light at times other than that of a total solar eclipse by occulting (hiding) the bright photosphere. Also, any telescope in which a bright central object is occulted to reveal fainter things.
coronal holes Relatively dark regions of the corona having low density; they result from open field lines.
correcting plate A thin lens of complicated shape at the front of a Schmidt camera that compensates for spherical aberration.
cosmic abundances The overall abundances of elements in the Universe.
cosmic background radiation Isotropic millimeter and submillimeter radiation following a black-body curve for about 3 K; interpreted as a remnant of the big bang.
cosmic microwave radiation See *cosmic background radiation.*
cosmic rays Nuclear particles or nuclei travelling through space at high velocity.
cosmogony The study of the origin of the Universe, usually applied in particular to the origin of the Solar System.
cosmological constant A constant arbitrarily added by Einstein to an equation in his general theory of relatively in order to provide a solution in which the Universe did not expand or contract. It was only subsequently discovered that the Universe does expand after all, leading Einstein to renounce the cosmological constant. Recent evidence suggests, however, that its value is indeed nonzero, and is somewhat larger than originally postulated by Einstein, causing the Universe's expansion to currently accelerate.
cosmological principle The principle that on the whole the Universe looks the same in all directions and in all regions.
cosmology The study of the Universe as a whole.
critical density The density of matter that would allow the Universe to expand forever, but at a rate that would decrease to 0 at infinite time. The density of the Universe with respect to the critical density defines whether the Universe will expand forever or eventually contract, assuming zero cosmological constant..
crust The outermost solid layer of some objects, including neutron stars and some planets.

dark era The final era of the Universe, when only low-energy photons, neutrinos, and some elementary particles (the ones that did not find a partner to annihilate) remain.
dark matter Non-luminous matter. See *hot dark matter* and *cold dark matter.*
dark nebula Dust and gas seen in silhouette.
declination Celestial latitude, measured in degrees north or south of the celestial equator.
deferent In the Ptolemaic system of the Universe, the larger circle, centered at the Earth, on which the centers of the epicycles revolve.
degenerate era The future era when the Universe will be filled with cold brown dwarfs, white dwarfs, and neutron stars. (Black holes will also exist.)
density Mass divided by volume.
density wave A circulating region of relatively high density, important, for example, in models of spiral arms of galaxies.
density-wave theory The explanation of spiral structure of galaxies as the effect of a wave of compression that rotates around the center of the galaxy and causes the formation of stars in the compressed region.
deuterium An isotope of hydrogen that contains 1 proton and 1 neutron.
deuteron A deuterium nucleus, containing 1 proton and 1 neutron.
diamond-ring effect The last Baily's bead glowing brightly at the beginning of the total phase of a solar eclipse, or its counterpart at the end of totality.
differential force A net force resulting from the difference of two other forces; a tidal force.
differential rotation Rotation of a body in which different parts have different angular velocities (and thus different periods of rotation).
differentiation For a planet, the formation of layers of different structure or composition.
diffraction A phenomenon affecting light as it passes any obstacle, spreading it out in a complicated fashion.
diffraction grating A very closely ruled series of lines that, through diffraction of light, provides a spectrum of radiation that falls on it.
dirty snowball A theory explaining comets as amalgams of ices, dust, and rocks.
discrete Separated; isolated.
disk (a) Of a galaxy, the disk-like flat portion, as opposed to the nucleus or the halo; (b) of a star or planet, the two-dimensional projection of its surface.
dispersion Of light, the effect that different colors are bent by different amounts when passing from one substance to another.
D lines A pair of lines from sodium that appear in the yellow part of the spectrum.
DNA Deoxyribonucleic acid, a long chain of molecules that contains the genetic information of life.
Dobsonian An inexpensive type of large-aperture amateur telescope characterized by a thin mirror, composition tube, and Teflon bearings on an altitude-azimuth mount.
Doppler effect A change in wavelength that results when a source of waves and the observer are moving relative to each other.
Doppler shift The change in wavelength that arises from the Doppler effect, caused by relative motion toward or away from the observer.
double star A binary star; two or more stars orbiting each other.
double-lobed structure An object in which radio emission comes from a pair of regions on opposite sides.
Drake equation An equation advanced by Frank Drake and popularized by Carl Sagan that attempts to estimate the number of communicating civilizations by breaking the calculation down into a series of steps that can be assessed individually, such as the rate of star formation, the fraction of stars with planets, and the average lifetime of a civilization.
dust tail The dust left behind a comet, reflecting sunlight.
dwarf ellipticals Small, low-mass elliptical galaxies.
dwarf stars Main-sequence stars.
dwarfs Dwarf stars.

$E = mc^2$ Einstein's formula (special theory of relativity) for the equivalence of mass and energy.
early universe The Universe during its first minutes.
earthshine Sunlight illuminating the Moon after having been reflected by the Earth.
eccentric Deviating from a circle.
eccentricity A measure of the flatness of an ellipse, defined as the distance between the foci divided by the major axis.
eclipse The passage of all or part of one astronomical body into the shadow of another.
eclipsing binary A binary star in which one member periodically hides the other.
ecliptic The path followed by the Sun across the celestial sphere in the course of a year.
ecliptic plane The plane of the Earth's orbit around the Sun.
ecosphere The region around a star in which conditions are suitable for life, normally a spherical shell.
electric field A force set up by an electric charge.
electromagnetic force One of the four fundamental forces of nature, giving rise to electromagnetic radiation.
electromagnetic radiation Radiation resulting from changing electric and magnetic fields.
electromagnetic spectrum Energy in the form of electromagnetic waves, in order of wavelength.
electromagnetic waves Waves of changing electric and magnetic fields, travelling through space at the speed of light.

electromagnetism The combined force of electricity and magnetism, which follows the formulae unified by Maxwell.
electron A particle of one negative charge, 1/1830 the mass of a proton, that is not affected by the strong force. It is a lepton.
electron volt (eV) The energy necessary to raise an electron through a potential of one volt.
electroweak force The unified electromagnetic and weak forces, according to a recent theory.
element A kind of atom, characterized by a certain number of protons in its nucleus. All atoms of a given element have similar chemical properties.
elementary particle One of the constituents of an atom.
ellipse A curve with the property that the sum of the distances from any point on the curve to two given points, called the foci, is constant.
elliptical galaxy A type of galaxy characterized by elliptical appearance.
emission line Wavelengths (or frequencies) at which the brightness of radiation is greater than it is at neighboring wavelengths (or frequencies).
emission nebula A glowing cloud of interstellar gas.
energy A fundamental quantity usually defined in terms of the ability of a system to do something that is technically called "work," that is, the ability to move an object by application of force, where the work is the force times the displacement.
energy, law of conservation of Energy is neither created nor destroyed, but may be changed in form.
energy level A state corresponding to an amount of energy that an atom is allowed to have by the laws of quantum mechanics.
energy problem For quasars, how to produce so much energy in such a small emitting volume.
ephemeris A listing of astronomical positions and other data that change with time. From the same root as *ephemeral.*
epicycle In the Ptolemaic theory, a small circle, riding on a larger circle called the deferent, on which a planet moves. The epicycle is used to account for retrograde motion.
equal areas, law of Kepler's second law.
equant In Ptolemaic theory, the point equally distant from the center of the deferent as the Earth but on the opposite side, around which the epicycle moves at a uniform angular rate.
equator (a) Of the Earth, a great circle on the Earth, midway between the poles; (b) celestial, the projection of the Earth's equator onto the celestial spheres; (c) Galactic, the plane of the Milky Way Galaxy's disk as projected onto a map.
equinox An intersection of the ecliptic and the celestial equator. The center of the Sun is geometrically above and below the horizon for equal lengths of time on the two days of the year when the Sun passes the equinoxes; if the Sun were a point and atmospheric refraction were absent, then day and night would be of equal length on those days.
erg A unit of energy in the metric system, corresponding to the work done by a force of one dyne (the force that is required to accelerate one gram by one cm/sec^2) producing a displacement of one centimeter.
ergosphere A region surrounding a rotating black hole (or other system satisfying Kerr's solution) from which work can be extracted.
escape velocity The initial velocity that an object must have to escape the gravitational pull of a mass.
event horizon The sphere around a black hole from within which nothing can escape; the place at which the exit cones close.
evolutionary track The set of points on a temperature–luminosity diagram showing the changes of a star's temperature and luminosity with time.
excitation The raising of atoms to energy states higher than the lowest possible.
excited level An energy level of an atom above the ground level.
exit cone The cone that, for each point within the photon sphere of a black hole, defines the directions of rays of radiation that escape.
exobiology The study of life located elsewhere than Earth.
exoplanet A planet orbiting a star other than the Sun.
exponent The "power" representing the number of times a number is multiplied by itself.
exponential notation The writing of numbers as a power of 10 times a number with one digit before the decimal point.
extended objects Objects with detectable angular size.
extinction The dimming of starlight by scattering and absorption as the light traverses interstellar space.
extragalactic Exterior to the Milky Way Galaxy.
extra-solar planet A planet orbiting a star other than the Sun.
eyepiece The small combination of lenses at the eye end of a telescope, used to examine the image formed by the objective.

field of view The angular expanse viewable.
filament A feature of the solar surface seen in Hα as a dark wavy line; a prominence projected on the solar disk.
fireball An exceptionally bright meteor.
fission, nuclear The splitting of an atomic nucleus.
flare An extremely rapid brightening of a small area of the surface of the Sun, usually observed in Hα and other strong spectral lines and accompanied by x-ray and radio emission.
flash spectrum The solar chromospheric spectrum seen in the few seconds before or after totality at a solar eclipse.
flat (critical) universe A universe in which Euclid's parallel postulate holds. It barely expands forever, assuming there is no repulsive cosmological constant. It has infinite volume and its age is 2/3 the Hubble time (assuming zero cosmological contant).
flatness problem One of the problems solved by the inflationary theory, that the Universe is exceedingly close to being flat for no obvious reason.
flavors A way of distinguishing quarks; up, down, strange, charmed, truth (top), beauty (bottom).
flux The amount of something (such as energy) passing through a surface per unit time.
focal length The distance from a lens or mirror to the point to which rays from an object at infinity are focused.
focus (plural: **foci**) (a) A point to which radiation is made to converge; (b) of an ellipse, one of the two points the sum of the distances to which remains constant.
force In physics, something that can or does cause change of momentum, measured by the rate of change of momentum with time.
Fraunhofer lines The absorption lines of a solar or other stellar spectrum.
frequency The rate at which waves pass a given point.
full moon The phase of the Moon when the side facing the Earth is fully illuminated by sunlight.
fusion The amalgamation of nuclei into heavier nuclei.
fuzz The faint light detectable around nearby quasars.

galactic cannibalism The incorporation of one galaxy into another.
galactic year The length of time the Sun takes to complete an orbit of our Galactic Center.
Galilean satellites Io, Europa, Ganymede, and Callisto: the four major satellites of Jupiter, discovered by Galileo in 1610.
Galileo (a) The Italian scientist Galileo Galilei (1564–1642); (b) the NASA spacecraft in orbit around the planet Jupiter that sent a probe into Jupiter's clouds on December 7, 1995.
gamma rays Electromagnetic radiation with wavelengths shorter than approximately 0.1 Å.
gamma-ray burst A brief burst of gamma rays coming from specific locations uniformly distributed in the sky; discovered to come, at least in most cases, from exceedingly distant and thus extremely powerful sources of energy.
gas tail The puffs of ionized gas trailing a comet.
general theory of relativity Einstein's 1916 theory of gravity.
geocentric Earth-centered.
geology The study of the Earth, or of other solid bodies.
geothermal energy Energy from the Earth's surface.
giant A star that is larger and brighter than main-sequence stars of the same color.
giant ellipticals Elliptical galaxies that are very large.
giant molecular cloud A basic building block of our Galaxy, containing dust, which shields the molecules present.
giant planets Jupiter, Saturn, Uranus, and Neptune; or even larger extra-solar planets.

giant star A star more luminous than a main-sequence star of its spectral type; a late stage in stellar evolution.
gibbous moon The phases between quarter moon and full moon.
globular cluster A spherically symmetric type of collection of stars that shared a common origin.
gluon The particle that carries the color force (and thus the strong nuclear force).
grand unified theories (GUTs) Theories unifying the electroweak force and the strong force.
granulation Convection cells on the Sun about 1 arc sec across.
grating A surface ruled with closely spaced lines that, through diffraction, breaks up light into its spectrum.
gravitational force One of the four fundamental forces of nature, the force by which two masses attract each other.
gravitational instability A situation that tends to break up under the force of gravity.
gravitational lens In the gravitational lens phenomenon, a massive body changes the path of electromagnetic radiation passing near it so as to make more than one image (or a brightening) of an object. A double quasar was the first example to be discovered.
gravitationally Controlled by the force of gravity.
gravitational redshift A redshift of light caused by the presence of mass, according to the general theory of relativity.
gravitational waves Waves that most scientists consider to be a consequence, according to the general theory of relativity, of changing distributions of mass.
gravity The tendency for all masses to attract each other; described in a formula by Newton and more recently described by Einstein as a result of a warping of space and time by the presence of a mass.
gravity assist Using the gravity of one celestial body to change a spacecraft's trajectory.
grazing incidence Striking at a low angle.
great circle The intersection of a plane that passes through the center of a sphere with the surface of that sphere; the largest possible circle that can be drawn on the surface of a sphere.
Great Dark Spot A giant circulating region on Neptune seen by Voyager 2 in 1989; it has since disappeared.
Great Red Spot A giant circulating region on Jupiter.
greenhouse effect The effect by which the atmosphere of a planet heats up above its equilibrium temperature because it is transparent to incoming visible radiation but opaque to the infrared radiation that is emitted by the surface of the planet.
Gregorian calendar The calendar in current use, with normal years that are 365 days long, with leap years every fourth year except for years that are divisible by 100 but not by 400.
ground level An atom's lowest possible energy level.
ground state See *ground level.*
GUTs See *grand unified theories.*

H_0 Hubble's constant.
H I region An interstellar region of neutral hydrogen.
H II region An interstellar region of ionized hydrogen.
H line The spectral line of ionized calcium at 3968 Å.
Hα The first line of the Balmer series of hydrogen, at 6563 Å.
half-life The length of time for half a set of particles to decay through radioactivity or instability.
halo Of a galaxy, the region of the galaxy that extends far above and below the plane of the galaxy, containing globular clusters.
head Of a comet, the nucleus and coma together.
heat flow The flow of energy from one location to another.
heavyweight stars Stars of more than about 8 or 10 solar masses.
heliocentric Sun-centered; using the Sun rather than the Earth as the point to which we refer. A heliocentric measurement, for example, omits the effect of the Doppler shift caused by the Earth's orbital motion.
helium flash The rapid onset of fusion of helium into carbon through the triple-alpha process that takes place in most red-giant stars.
Herbig-Haro objects Blobs of gas ejected in star formation.
hertz The measure of frequency, with units of /sec (per second); formerly called cycles per second.
Hertzsprung–Russell diagram A graph of temperature (or equivalent) vs. luminosity (or equivalent) for a group of stars.
high-energy astrophysics The study of x-rays, gamma rays, and cosmic rays, and of the processes that make them.
highlands Regions on the Moon or elsewhere that are above the level that may have been smoothed by flowing lava.
homogeneous Uniform throughout.
horizon problem One of the problems of cosmology solved by the inflationary theory: why the Universe has the same average temperature in all directions, even though widely separated regions could never have been in thermal equilibrium with each other since they are beyond each other's horizons.
hot dark matter Non-luminous matter with large speeds, like neutrinos.
Hubble Deep Field A small part of the sky in Ursa Major extensively studied by the Hubble Space Telescope in December 1995 and thereafter also studied by a wide variety of telescopes on the ground and in space; the long exposures allowed astronomers to see "deep" into space, that is, to great distances and therefore far back in time. See http://www.stsci.edu/ftp/science/hdf.html. A Hubble Deep Field — South was observed subsequently (http://www.stsci.edu/ftp/science/hdfsouth/hdfs.html).
Hubble flow The assumed uniform expansion of the Universe, on which any peculiar motions of galaxies or clusters of galaxies is superimposed.
Hubble time The amount of time since the big bang, assuming a constant speed for any given galaxy; the Hubble time is calculated by tracing the Universe backward in time using the current Hubble's constant.
Hubble type Hubble's galaxy classification scheme; E0, E7, Sa, SBa, etc.
H–R diagram Hertzsprung–Russell diagram.
Hubble's constant (H_0) The constant of proportionality in Hubble's law linking the speed of recession of a distant object and its current distance from us.
Hubble's law The linear relation between the speed of recession of a distant object and its current distance from us, $v = H_0 d$.
hyperfine level A subdivision of an energy level caused by such relatively minor effects as changes resulting from the interactions among spinning particles in an atom or molecule.
hypothesis The first step in the traditional formulation of the scientific method; a tentative explanation of a set of facts that is to be tested experimentally or observationally.

igneous Rock cooled from lava.
inclination Of an orbit, the angle of the plane of the orbit with respect to the ecliptic plane.
inclined Tilted with respect to some other body, usually describing the axis of rotation or the plane of an orbit.
inferior planet A planet whose orbit around the Sun is within the Earth's, namely, Mercury and Venus.
inflation The theory that the Universe expanded extremely fast, by perhaps 10 to the 100th power, in the first fraction of a second after the big bang. The concept of inflation solves several problems in cosmology, such as the horizon problem.
inflationary universe A model of the expanding Universe involving a brief period of extremely rapid expansion.
infrared Radiation beyond the red, about 7000 Å to 1 mm.
interference The property of radiation, explainable by the wave theory, in which waves in phase can add (constructive interference) and waves out of phase can subtract (destructive interference); for light, this gives alternate light and dark bands.
interferometer A device that uses the property of interference to measure such properties of objects as their positions or structure.
interferometry Observations using an interferometer.
intergalactic medium Material between galaxies in a cluster.
interior The inside of an object.
international date line A crooked imaginary line on the Earth's surface, roughly corresponding to 180° longitude, at which, when crossed from east to west, the date jumps forward by one day.
interplanetary medium Gas and dust between the planets.
interstellar medium Gas and dust between the stars.

interstellar reddening The relatively greater extinction of blue light than of red light by interstellar matter.
inverse-square law Decreasing with the square of increasing distance.
ion An atom that has lost one or more electrons.
ionized Having lost one or more electrons.
ionosphere The highest region of the Earth's atmosphere.
ion tail See *gas tail.*
IRAS The Infrared Astronomical Satellite (1983).
iron meteorites Meteorites with a high iron content (about 90%); most of the rest is nickel.
irregular galaxy A type of galaxy showing no regular shape or symmetry.
ISO (Infrared Space Observatory) A European Space Agency project, 1995–1998.
isotope A form of chemical element with a specific number of neutrons.
isotropic Being the same in all directions.

joule The SI unit of energy, 1 kg $\cdot$ m^2/s^2.
Jovian planets Same as *giant planets.*
JPL The Jet Propulsion Laboratory in Pasadena, California, funded by NASA and administered by Caltech; a major space contractor.
Julian calendar The calendar with 365-day years and leap years every fourth year without exception; the predecessor to the Gregorian calendar.
Julian day The number of days since noon on January 1, 713 B.C.; used for keeping track of variable stars or other astronomical events. January 1, 2004, noon, begins Julian day 2,453,006.

K line The spectral line of ionized calcium at 3933 Å.
Keplerian Following Kepler's law.
Kerr's solution Equations describing rotating black holes.
Kuiper belt A reservoir of perhaps hundreds of thousands of Solar-System objects, each tens or hundreds of kilometers in diameter, orbiting the Sun outside the orbit of Neptune. It is the source of short-period comets.
Kuiper-belt object An object in the Kuiper belt; many comets and probably even the planet Pluto are examples.

large-scale structure The network of filaments and voids or other shapes distinguished when studying the Universe on the largest scales of distance.
laser An acronym for "**l**ight **a**mplification by **s**timulated **e**mission of **r**adiation," a device by which certain energy levels are populated by more electrons than normal, resulting in an especially intense emission of light at a certain frequency when the electrons drop to a lower energy level.
latitude Number of degrees north or south of the equator measured from the center of a coordinate system.
law of equal areas Kepler's second law.
leap year A year in which a 366th day is added.
lens A device that focuses waves by refraction.
lenticular A galaxy of type S0.
light Electromagnetic radiation between about 4000 and 7000 Å.
light curve The graph of the brightness of an object vs. time.
light pollution Excess light in the sky.
light-year The distance that light travels in a year.
lighthouse model The explanation of a pulsar as a spinning neutron star whose beam we see as it comes around.
lightweight stars Stars between about 0.075 and 10 solar masses.
limb The edge of a star or planet.
line profile The graph of the intensity of radiation vs. wavelength for a spectral line.
lithosphere The crust and upper mantle of a planet.
lobes Of a radio source, the regions to the sides of the center from which high-energy particles are radiating.
local In our region of the Universe.
Local Group The two dozen or so galaxies, including the Milky Way Galaxy, that form a small cluster.
Local Supercluster The supercluster of galaxies in which the Virgo Cluster, the Local Group, and other clusters reside.
logarithmic A scale in which equal intervals stand for multiplying by ten or some other base, as opposed to linear, in which increases are additive.
longitude The angular distance around a body measured along the equator from some particular point; for a point not on the equator, it is the angular distance along the equator to a great circle that passes through the poles and through the point.
long-period variable A Mira variable.
lookback time The duration over which light from an object has been travelling to reach us.
luminosity The total amount of energy given off by an object per unit time; its power.
luminosity class Different regions of the H–R diagram separating objects of the same spectral type: supergiants (I), bright giants (II), giants (III), subgiants (IV), dwarfs (V).
lunar eclipse The passage of the Moon into the Earth's shadow.
lunar occultation An occultation by the Moon.
lunar soils Dust and other small fragments on the lunar surface.
Lyman-alpha The spectral line (1216 Å) that corresponds to a transition between the two lowest major energy levels of a hydrogen atom.
Lyman-alpha forest The many Lyman-alpha lines, each differently Doppler-shifted, visible in the spectra of some quasars.
Lyman lines The spectral lines that correspond to transitions to or from the lowest major energy level of a hydrogen atom.

$M_\odot$ Solar mass; the mass of the Sun, used as a unit of measurement. For example, a star with $5M_\odot$ has 5 times the mass of the Sun.
MACHOs **M**assive **C**ompact **H**alo **O**bjects — like dim stars, brown dwarfs, or black holes — that could account for some of the dark matter. The name was chosen to contrast with WIMPs.
Magellanic Clouds Two small irregular galaxies, satellites of the Milky Way Galaxy, visible in the southern sky.
magnetic-field lines Directions mapping out the direction of the force between magnetic poles; the packing of the lines shows the strength of the force.
magnetic lines of force See *magnetic-field lines.*
magnetic monopole A single magnetic charge of only one polarity; may or may not exist.
magnetosphere A region of magnetic field around a planet.
magnification An apparent increase in angular size.
magnitude A factor of $\sqrt[5]{100} = 2.511886...$ in brightness. See *absolute magnitude* and *apparent magnitude.* An *order of magnitude* is a power of ten.
magnitude scale The scale of apparent magnitudes and absolute magnitudes used by astronomers, in which each factor of 100 in brightness corresponds to a difference of 5 magnitudes.
main sequence A band on a Hertzsprung–Russell diagram in which stars fall during the main, hydrogen-burning phase of their lifetimes.
major axis The longest diameter of an ellipse; the line from one side of an ellipse to the other that passes through the foci. Also, the length of that line.
mantle The shell of rock separating the core of a differentiated planet from its thin surface crust.
mare (plural: **maria**) One of the smooth areas of the Moon or on some of the other planets.
mascon A concentration of mass under the surface of the Moon, discovered from its gravitational effect on spacecraft orbiting the Moon.
mass A measure of the inherent amount of matter in a body.
mass-luminosity relation A well-defined relation between the mass and luminosity for main-sequence stars.
mass number The total number of protons and neutrons in a nucleus; also known as *atomic weight.*
Maunder minimum The period 1645–1715, when there were very few sunspots, and no periodicity, visible.
mean solar day A solar day for the "mean Sun," which moves at a constant rate during the year.
merging The interaction of two galaxies in space with a single galaxy as the result; the merging of two spiral galaxies to form an elliptical galaxy, or of two galaxies interacting with a ring galaxy resulting, are examples.

meridian The great circle on the celestial sphere that passes through the celestial poles and the observer's zenith.
Messier numbers Numbers of fuzzy-looking objects in the 18th-century list of Charles Messier.
metal (a) For stellar abundances, any element higher in atomic number than 2, that is, more massive than helium. (b) In general, neutral matter that is a good conductor of electricity.
meteor A track of light in the sky from rock or dust burning up as it falls through the Earth's atmosphere.
meteor shower The occurrence at yearly intervals of meteors at a higher-than-average rate as the Earth goes through a comet's orbit and the dust left behind by that comet.
meteorite An interplanetary chunk of rock after it impacts on a planet or the Moon, especially on the Earth.
meteoroid An interplanetary chunk of rock smaller than an asteroid.
micrometeorite A tiny meteorite. The micrometeorites that hit the Earth's surface are sufficiently slowed down that they can reach the ground without being vaporized.
midnight sun The Sun seen around the clock from locations sufficiently far north or south during the suitable season.
Milky Way The band of light across the sky from the stars and gas in the plane of the Milky Way Galaxy.
Milky Way Galaxy The collection of gas, dust, and perhaps 400 billion stars in which we live; the Milky Way is our view of the plane of the Milky Way Galaxy.
mini black hole A black hole the size of a pinhead (and thus the mass of an asteroid) or less, thought by some to be left over from the big bang; Stephen Hawking deduced that such mini black holes should seem to emit radiation at a rate that allows a temperature to be assigned to it.
minor axis The shortest diameter of an ellipse; the line from one side of an ellipse to the other that passes midway between the foci and is perpendicular to the major axis. Also, the length of that line.
minor planets Asteroids.
Mira variable A long-period variable star similar to Mira (omicron Ceti).
missing-mass problem The discrepancy between the mass visible and the mass derived from calculating the gravity acting on members of clusters of galaxies.
model A physical or mathematical equivalent to a situation; the former is typified by a model airplane that scales down a real airplane, while the latter is typified by a set of equations or of tables that describes, for example, the interior of Jupiter or processes that work there.
molecule Bound atoms that make the smallest collection that exhibits a certain set of chemical properties.
momentum A measure of the tendency that a moving body has to keep moving. The momentum in a given direction (the "linear momentum") is equal to the mass of the body times its speed in that direction. See also *angular momentum.*
mountain ranges Sets of mountains on the Earth, Moon, etc.
multiverse The set of parallel universes that may exist, with our observable universe as only one part.

naked singularity A singularity that is not surrounded by an event horizon and is therefore kept from our view.
neap tides The tides when the gravitational pulls of the Sun and Moon are perpendicular, making the tides relatively low.
Near-Earth Object A comet or asteroid whose orbit is sufficiently close to Earth's that a collision is possible or even likely in the long term. See *Apollo asteroids.*
nebula (plural: **nebulae**) Interstellar regions of dust or gas.
nebular hypothesis The particular nebular theory for the formation of the Solar System advanced by Laplace.
nebular theories The theories that the Sun and the planets formed out of a cloud of gas and dust.
neutrino A spinning, neutral elementary particle with little rest mass, formed in certain radioactive decays.
neutron A massive, neutral elementary particle, one of the fundamental constituents of an atom.
neutron star A star that has collapsed to the point where it is supported against gravity by neutron quantum mechanical pressure.
New General Catalogue *A New General Catalogue of Nebulae and Clusters of Stars* by J. L. E. Dreyer, 1888.
new moon The phase when the side of the Moon facing the Earth is the side that is not illuminated by sunlight.
Newtonian (telescope) A reflecting telescope in which the beam from the primary mirror is reflected by a flat secondary mirror to the side.
NGC *New General Catalogue.*
nonthermal radiation Radiation that cannot be characterized by a single number (the temperature). Normally, we derive this number from Planck's law, so that radiation that does not follow Planck's law is called nonthermal.
nova (plural: **novae**) A star that suddenly increases in brightness; an event in a binary system when matter from the giant component falls on the white dwarf component, or suddenly fuses on the surface of the white dwarf.
nuclear bulge The central region of spiral galaxies.
nuclear burning Nuclear fusion.
nuclear force The strong force, one of the fundamental forces.
nuclear fusion The amalgamation of lighter nuclei into heavier ones.
nucleosynthesis The formation of the elements.
nucleus (plural: **nuclei**) (a) Of an atom, the core, which has a positive charge, contains most of the mass, and takes up only a small part of the volume; (b) of a comet, the chunks of matter, no more than a few km across, at the center of the head; (c) of a galaxy, the innermost region.

O and B association A group of O and B stars close together.
objective The principal lens or mirror of an optical system.
oblate With equatorial greater than polar diameter.
Occam's razor The principle of simplicity, from the medieval philosopher William of Occam (approximately 1285–1349): The simplest explanation for all the facts will be accepted.
occultation The hiding of one astronomical body by another.
Olbers's paradox The observation that the sky is dark at night contrasted to a simple argument that shows that the sky should be uniformly bright.
one atmosphere The air pressure at the Earth's surface.
one year The length of time the Earth takes to orbit the Sun.
Oort comet cloud The trillions of incipient comets surrounding the Solar System in a 50,000 A.U. (radius) sphere.
open cluster A galactic cluster, a type of star cluster.
open universe A big-bang cosmology in which the Universe has infinite volume and will expand forever.
optical In the visible part of the spectrum, 4000–7000 Å, or having to do with reflecting or refracting that radiation.
optical double A pair of stars that appear extremely close together in the sky even though they are at different distances from us and are not physically linked.
organic Containing carbon in its molecular structure.
Orion Molecular Cloud The giant molecular cloud in Orion behind the Orion Nebula, containing many young objects.
Ozma A project that searched nearby stars for radio signals from extraterrestrial civilizations.
ozone layer A region in the Earth's upper stratosphere and lower mesosphere where O_3 absorbs solar ultraviolet radiation.

paraboloid A 3-dimensional surface formed by revolving a parabola around its axis.
parallax (a) Trigonometric parallax, half the angle through which a star appears to be displaced when the Earth moves from one side of the Sun to the other (2 A.U.); it is inversely proportional to the distance; (b) other ways of measuring distance, as in spectroscopic parallax.
parallel light Light that is neither converging nor diverging.
parsec The distance from which 1 A.U. subtends one second of arc; approximately 3.26 lt yr.
particle physics The study of elementary nuclear particles.
peculiar motion The motion of a galaxy with respect to the Hubble flow.
penumbra (a) For an eclipse, the part of the shadow from which the Sun is only partially occulted; (b) of a sunspot, the outer region, not as dark as the umbra.

perfect cosmological principle The assumption that on a large scale the Universe is homogeneous and isotropic in space and unchanging in time.

perihelion The near point to the Sun of the orbit of a body orbiting the Sun.

period The interval over which something repeats.

phase (a) Of a planet, the varying shape of the lighted part of a planet or moon as seen from some vantage point; (b) the relation of the variations of a set of waves.

phase transition Change from one state of matter to another, as from solid to liquid or liquid to gas; phase transitions in the early Universe marked the separation of the fundamental forces.

photometry The electronic measurement of the amount of light.

photomultiplier An electronic device that through a series of internal stages multiplies a small current that is given off when light is incident on it; a large current results.

photon A packet of energy that can be thought of as a particle travelling at the speed of light.

photon sphere The sphere around a black hole, $\frac{3}{2}$ the size of the event horizon, within which exit cones open and in which light can orbit.

photosphere The region of a star from which most of its light is radiated.

plage The part of a solar active region that appears bright when viewed in $H\alpha$.

Planck's constant The constant of a proportionality, h, between the frequency of an electromagnetic wave and the energy of an equivalent photon. $E = h\nu = hc/\lambda$.

Planck's law The formula that predicts, for an opaque object at a certain temperature, how much radiation there is at every wavelength.

Planck time The time very close to the big bang, 10^{-43} seconds, before which a quantum theory of gravity would be necessary to explain the Universe and which is therefore currently inaccessible to our computations.

planet A celestial body of substantial size (more than about 1000 km across), basically non-radiating and of insufficient mass for nuclear reactions ever to begin, ordinarily in orbit around a star.

planetary nebulae Shells of matter ejected by low-mass stars after their main-sequence lifetime, ionized by ultraviolet radiation from the star's remaining core.

planetesimal One of the small bodies into which the primeval solar nebula condensed and from which the planets formed.

plasma An electrically neutral gas composed of exactly equal numbers of ions and electrons.

plates Large, flat structures making up a planet's crust.

plate tectonics The theory of the Earth's crust, explaining it as plates moving because of processes beneath.

plumes Thin structures in the solar corona near the poles.

point objects Objects in which no size is distinguishable.

pole star A star approximately at a celestial pole; Polaris is now the pole star; there is no south pole star.

poor cluster A cluster of stars or galaxies with few members.

positive ion An atom that has lost one or more electrons.

positron An electron's antiparticle (charge of $+e$), where e is the unit of electric charge.

precession The very slowly changing position of stars in the sky resulting from variations in the orientation of the Earth's axis; a complete cycle takes 26,000 years.

precession of the equinoxes The slow variation of the position of the equinoxes (intersections of the ecliptic and celestial equator) resulting from variations in the orientation of the Earth's axis.

pre-main-sequence star A ball of gas in the process of slowly contracting to become a star as it heats up before beginning nuclear fusion.

pressure Force per unit area.

primary cosmic rays The cosmic rays arriving at the top of the Earth's atmosphere.

primary distance indicators Ways of measuring distance directly, as in trigonometric parallax.

prime focus The location at which the main lens or mirror of a telescope focuses an image without being reflected or refocused by another mirror or other optical element.

primeval solar nebula The early stage of the Solar System in nebular theories.

primordial nucleosynthesis The formation of the nuclei of isotopes of hydrogen (such as deuterium), helium, and lithium in the first 10 minutes of the Universe.

principal quantum number The integer n that determines the main energy levels in an atom.

prograde motion The apparent motion of the planets when they appear to move forward (from west to east) with respect to the stars; see also *retrograde motion.*

prolate Having the diameter along the axis of rotation longer than the equatorial diameter.

prominence Solar gas protruding over the limb, visible to the naked eye only at eclipses but also observed outside the eclipses by its emission-line spectrum.

proper motion Angular motion across the sky with respect to a framework of galaxies or fixed stars.

proteins Long chains of amino acids; fundamental components of all cells of life as we know it.

proton Elementary particle with positive charge $+e$, where e is the unit of electric charge, one of the fundamental constituents of an atom.

proton-proton chain A set of nuclear reactions by which four hydrogen nuclei combine one after the other to form one helium nucleus, with a resulting release of energy.

protoplanets The loose collections of particles from which the planets formed.

protostar A nebula of gas and dust that is on its way to becoming a star.

protosun The Sun in formation.

pulsar A celestial object that gives off pulses of radio waves; thought to be a rotating neutron star with a very strong magnetic field.

QSO Quasi-stellar objects, formally a radio-quiet version of a quasar, though now radio-loud quasars are also sometimes included.

quantized Divided into discrete parts.

quantum A bundle of energy.

quantum fluctuation The spontaneous formation and disappearance of virtual particles; it may be a quantum violation of the law of conservation of energy because of Heisenberg's Uncertainty Principle.

quantum mechanics The branch of 20th-century physics that describes atoms and radiation.

quantum theory The set of theories that evolved in the early 20th century to explain atoms and radiation as limited to discrete energy levels or quanta of energy, with quantum mechanics emerging as a particular version.

quark One of the subatomic particles of which modern theoreticians believe such elementary particles as protons and neutrons are composed. The various kinds of quarks have positive or negative charges of $\frac{1}{3}e$ or $\frac{2}{3}e$, where e is the unit of electric charge.

quasar A very-large-redshift object that is almost stellar (point-like) in appearance, but has a very non-stellar spectrum consisting of broad emission lines; thought to be the nucleus of a galaxy with an accreting supermassive black hole.

quiescent prominence A long-lived and relatively stationary prominence.

quiet Sun The collection of solar phenomena that do not vary with the solar activity cycle.

radar The acronym for **ra**dio **d**etection **a**nd **r**anging, an active rather than passive radio technique in which radio signals are transmitted and their reflections received and studied.

radial velocity The velocity of an object along a line (the radius) joining the object and the observer; the component of velocity toward or away from the observer.

radiant The point in the sky from which all meteors in a meteor shower appear to be coming.

radiation Electromagnetic radiation; sometimes also particles such as protons, electrons, and helium nuclei.
radiation belts Belts of charged particles surrounding planets.
radioactive Having the property of spontaneously changing into another isotope or element.
radio galaxy A galaxy that emits radio radiation orders of magnitude stronger than those from normal galaxies.
radio telescope An antenna or set of antennas, often together with a focusing reflecting dish, that is used to detect radio radiation from space.
radio waves Electromagnetic radiation with wavelengths longer than about one millimeter.
red giant A post-main-sequence stage of the lifetime of a star; the star becomes relatively luminous and cool.
reddened See *reddening.*
reddening The phenomenon by which the extinction of blue light by interstellar matter is greater than the extinction of red light so that the redder part of the continuous spectrum is relatively enhanced.
redshift A shift of optical wavelengths toward the red, or in all cases toward longer wavelengths.
red supergiant Extremely luminous, cool, and large stars; a post-main-sequence phase of evolution of stars of more than about 10 solar masses.
reflecting telescope A type of telescope that uses a mirror or mirrors to form the primary image.
reflection nebula Interstellar gas and dust that we see because it is reflecting light from a nearby star.
refracting telescope A type of telescope in which the primary image is formed by a lens or lenses.
refraction The bending of electromagnetic radiation as it passes from one medium to another or between parts of a medium that has varying properties.
refractory Having a high melting point.
relativistic Having a speed that is such a large fraction of the speed of light that the special theory of relativity must be applied.
resolution The ability of an optical system to distinguish detail; also called "angular resolution."
rest mass The mass an object would have if it were not moving with respect to the observer.
rest wavelength The wavelength radiation would have if its emitter were not moving with respect to the observer.
retrograde motion The apparent motion of the planets when they appear to move backward (from east to west) from the direction that they move ordinarily with respect to the stars.
retrograde rotation The rotation of a moon or planet opposite to the dominant direction in which the Sun rotates and the planets orbit and rotate.
revolution The orbiting of one body around another.
revolve To move in an orbit around another body.
rich cluster A cluster of many galaxies.
right ascension Celestial longitude, measured eastward along the celestial equator in hours of time from the vernal equinox.
rims The raised edges of craters.
Roche limit The sphere for each mass inside of which blobs of gas cannot agglomerate by gravitational interaction without being torn apart by tidal forces; normally about 2.5 times the radius of a planet.
rotate To spin on one's own axis.
rotation Spin on an axis.
rotation curve A graph of the speed of rotation vs. distance from the center of a rotating object like a galaxy.
RR Lyrae variable A short-period variable star. All RR Lyrae stars have approximately equal luminosity and so are used to determine distances.

S0 A transition type of galaxy between ellipticals and spirals; has a disk but no arms.
scarps Lines of cliffs; found on Mercury, Earth, the Moon, and Mars.
scattered Light absorbed and then reemitted in all directions.
Schmidt telescope (Schmidt camera) A telescope that uses a spherical mirror and a thin lens to provide photographs of a wide field.
Schwarzschild radius The radius that, according to Schwarzschild's solutions to Einstein's equations of the general theory of relativity, corresponds to the event horizon of a black hole.
scientific method No easy definition is possible, but it has to do with a way of testing and verifying hypotheses.
scientific notation Exponential notation.
secondary cosmic rays High-energy particles generated in the Earth's atmosphere by primary cosmic rays.
secondary distance indicators Ways of measuring distances that are calibrated by primary distance indicators.
sedimentary Formed from settling in a liquid or from deposited material, as for a type of rock.
seeing The steadiness of the Earth's atmosphere as it affects the resolution that can be obtained in astronomical observations.
seismic waves Waves travelling through a solid planetary body from an earthquake or impact.
seismology The study of waves propagating through a body and the resulting deduction of the internal properties of the body. "Seismo-" comes from the Greek for earthquake.
semimajor axis Half the major axis, that is, for an ellipse, half the longest diameter.
semiminor axis Half the minor axis, that is, for an ellipse, half the shortest diameter.
Seyfert galaxy A type of galaxy that has an unusually bright nucleus and whose spectrum shows broad emission lines that cover a wide range of ionization stages.
Shapley-Curtis debate The 1920 debate (and its written version) on a scale of our Galaxy and over "spiral nebulae."
shear wave A type of twisting seismic wave.
shock wave A front marked by an abrupt change in pressure caused by an object moving faster than the speed of sound in the medium through which the object is travelling.
shooting stars Meteors.
showers A time of many meteors from a common cause.
sidereal With respect to the stars.
sidereal day A day with respect to the stars.
sidereal year A circuit of the Sun with respect to the stars.
significant figure A digit in a number that is meaningful (within the accuracy of the data).
singularity A point in space where quantities become exactly zero or infinitely large; one is present in a black hole, according to classical general relativity.
slit A long, thin gap through which light is allowed to pass.
SOHO The Solar and Heliospheric Observatory, a joint NASA and European Space Agency mission to study the Sun.
solar-activity cycle The 11-year cycle with which solar activity like sunspots, flares, and prominences varies.
solar atmosphere The photosphere, chromosphere, and corona of the Sun.
solar constant The total amount of energy that would hit each square centimeter of the top of the Earth's atmosphere at the Earth's average distance from the Sun.
solar day A full rotation with respect to the Sun.
solar flares An explosive release of energy on the Sun.
solar mass The mass of the Sun, 1.99×10^{33} grams, about 330,000 times the Earth's mass.
solar rotational period The time for a complete rotation with respect to the Sun.
solar time A system of timekeeping with respect to the Sun such that the Sun is overhead of a given location at noon.
solar wind An outflow of particles from the Sun representing the expansion of the corona.
solar year (tropical year) An object's complete circuit of the Sun; a tropical year in between vernal equinoxes.
solstice The point on the celestial sphere of northernmost or southernmost declination of the Sun in the course of a year; colloquially, the time when the Sun reaches that point.
space velocity The velocity of a star with respect to the Sun.
spallation The break-up of heavy nuclei that undergo nuclear collisions.
special theory of relativity Einstein's 1905 theory of relative motion.
speckle interferometry A method obtaining higher resolution of an image by analysis of a rapid series of exposures that freeze atmospheric blurring.

spectral classes See *spectrum.*
spectral lines Wavelengths at which the brightness is abruptly different from the brightness at neighboring wavelengths.
spectral type One of the categories O, B, A, F, G, K, M, L, into which stars can be classified from study of their spectral lines, or extensions of this system. The above sequence of spectral types corresponds to a decreasing sequence of surface temperatures.
spectrograph A device to make and record a spectrum.
spectrometer A device to make and electronically measure a spectrum.
spectroscopic binary A type of binary star that is known to have more than one component because of the changing Doppler shifts of the spectral lines that are observed.
spectroscopic parallax The distance to a star derived by comparing its apparent brightness with its luminosity deduced from study of its position on a temperature-luminosity diagram (determined by observing its spectrum—spectral type and luminosity class).
spectroscopy The use of spectrum analysis.
spectrum (plural: **spectra**) A display of electromagnetic radiation spread out by wavelength or frequency.
speed of light By Einstein's special theory of relativity, the speed at which all electromagnetic radiation travels in a vacuum, and the largest possible speed of an object moving through space.
spherical aberration For an optical system, a deviation from perfect focusing by having a shape that is too close to that of a sphere.
spicule A small jet of gas at the edge of the quiet Sun, approximately 1000 km in diameter and 10,000 km high, with a lifetime of about 15 minutes.
spin-flip transition Transition in the relative orientation of the spins of an electron and the nucleus it is orbiting.
spiral arms Bright regions looking like a pinwheel.
spiral density wave A wave that travels around a galaxy in the form of a spiral, compressing gas and dust to relatively high density and thus beginning star formation at those compressions.
spiral galaxy A class of galaxy characterized by arms that appear as though they are unwinding like a pinwheel.
spiral nebula The old name for a shape in the sky, seen with telescopes, with arms spiralling outward from the center; proved in the 1920s actually to be a galaxy.
sporadic Not regular.
sporadic meteor A meteor not associated with a shower.
spring tides The tides at their highest, when the Earth, Moon, and Sun are in a line (from "to spring up").
stable Tending to remain in the same condition.
star A self-luminous ball of gas that shines or has shone because of nuclear reaction in its interior.
star clusters Groupings of stars of common origin.
stationary limit In a rotating black hole, the radius within which it is impossible for an object to remain stationary, no matter what it does.
steady-state theory The cosmological theory based on the perfect cosmological principle, in which the average properties of the Universe are unchanging over time.
Stefan-Boltzmann law The radiation law that states that the energy emitted by a black body varies with the fourth power of its surface temperature.
stellar atmosphere The outer layers of stars not completely hidden from our view.
stellar chromosphere The region above a photosphere that shows an increase in temperature.
stellar corona The outermost region of a star characterized by temperatures of 10^6 K and high ionization.
stellar evolution The changes of a star's properties with time.
stelliferous era The current era, with lots of stars in the Universe.
stones A stony type of meteorite.
stratosphere An upper layer of a planet's atmosphere, above the weather, where the temperature begins to increase. The Earth's stratosphere is at 20–50 km.
streamers Coronal structures at low solar latitudes.
string theories See *superstring theories.*
strong force The nuclear force, the strongest of the four fundamental forces of nature.
strong nuclear force The strong force.
subtend The angle that an object appears to take up in your field of view; for example, the full Moon subtends ½°.
sunspot A region of the solar surface that is dark and relatively cool; it has an extremely high magnetic field.
sunspot cycle The 11-year cycle of variation of the number of sunspots visible on the Sun.
supercluster A cluster of clusters of galaxies.
supercooled The condition in which a substance is cooled below the point at which it would normally make a phase change; for the Universe, the point in the early Universe at which it may have cooled below a certain temperature without breaking its symmetry; the strong and electroweak forces remained unified.
supergiant A post-main-sequence phase of evolution of stars of more than about 10 solar masses. They fall in the upper right of the temperature-luminosity diagram; luminosity class I.
supergravity A theory that attempts to unify the four fundamental forces.
superluminal speed An apparent speed greater than that of light.
supermassive black hole A black hole of millions or billions of times the Sun's mass, as is found in the centers of galaxies and quasars.
supernova (plural: **supernovae**) The explosion of a star with the resulting release of tremendous amounts of radiation.
supernova remnant The gaseous remainder of the star destroyed in a supernova.
superstring (string) theories A possible unification of quantum theory and general relativity in which fundamental particles are really different vibrating forms of a tiny, one-dimensional "string" instead of being localized at single points.
symmetric A correspondence of shape so that rotating or reflecting an object gives you back an identical form; symmetry of forces in the early Universe corresponds to forces acting identically that now act differently as the four fundamental forces: gravity, electromagnetism, weak nuclear force, and strong nuclear force.
synchronous orbit An orbit of the same period; a satellite in geostationary orbit has the same period as the Earth's rotation and so appears to hover.
synchronous rotation A rotation of the same period as an orbiting body.
synchrotron radiation Nonthermal radiation emitted by electrons spiralling at relativistic velocities in a magnetic field.

tail Gas and dust left behind as a comet orbits sufficiently close to the Sun, illuminated by sunlight.
temperature-luminosity diagram A diagram of a group of stars with temperatures on the horizontal axis and intrinsic brightness (luminosity) on the vertical axis; also called a temperature-magnitude diagram or a Hertzsprung–Russell diagram.
terminator The line between night and day on a moon or planet; the edge of the part that is lighted by the Sun.
terrestrial planets Mercury, Venus, Earth, and Mars.
theory A later stage of the traditional form of the scientific method, in which a hypothesis has passed enough of its tests that it is generally accepted.
thermal pressure Pressure generated by the motion of particles that can be characterized by temperature.
thermal radiation Radiation whose distribution of intensity over wavelength can be characterized by a single number (the temperature). Black-body radiation, which follows Planck's law, is thermal radiation.
thermosphere The uppermost layer of the atmosphere of the Earth and some other planets, the ionosphere, where absorption of high-energy radiation heats the gas.
3° background radiation The isotropic black-body radiation at 3 K, thought to be a remnant of the big bang.
tide A periodic variation of the force of gravity on a body, based on the difference in the strength of gravitational pull from one place on the body to another (the tidal force); on Earth, the

tide shows most obviously where the ocean meets the shore as a periodic variation of sea level.

time dilation According to relativity theory, the slowing of time perceived by an observer watching another object moving rapidly or located in a strong gravitational field.

TRACE The **T**ransition **R**egion **a**nd **C**oronal **E**xplorer, a solar satellite to study the solar corona and the transition region between the chromosphere and corona at high spatial resolution.

transit The passage of one celestial body in front of another celestial body. When a planet is *in transit*, we understand that it is passing in front of the Sun. Also, *transit* is the moment when a celestial body crosses an observer's meridian, or the special type of telescope used to study such events.

transition zone The thin region between a chromosphere and a corona.

Trans-Neptunian Object (TNO) One of the sub-planetary objects in the Kuiper belt; many people feel that Pluto is one, though it also retains its status as a planet, at least for the time being.

transparency Clarity of the sky.

transverse velocity Velocity along the plane of the sky.

trigonometric parallax See *parallax*.

triple-alpha process A chain of fusion processes by which three helium nuclei (alpha particles) combine to form a carbon nucleus.

tropical year The length of time between two successive vernal equinoxes.

troposphere The lowest level of the atmosphere of the Earth and some other planets, in which all weather takes place.

T Tauri star A type of irregularly varying star, like T Tauri, whose spectrum shows broad and very intense emission lines. T Tauri stars have presumably not yet reached the main sequence and are thus very young.

tuning-fork diagram Hubble's arrangement of types of elliptical, spiral, and barred spiral galaxies.

21-cm line The 1420-MHz line from neutral hydrogen's spin-flip.

twinkle A scintillation — rapid changing in brightness — and slight changing in position of stars as their light passes through the Earth's atmosphere.

Type Ia supernova A supernova whose distribution in all types of galaxies, and the lack of hydrogen in its spectrum, make us think that it is an event in low-mass stars, probably resulting from the incineration of a white dwarf in a binary system.

Type II supernova A supernova often associated with spiral arms, and that has hydrogen in its spectrum, making us think that it is the explosion of a massive star.

ultraviolet The region of the spectrum 100–4000 Å, also used in the restricted sense of ultraviolet radiation that reaches the ground, namely, 3000–4000 Å.

umbra (plural: **umbrae**) (a) Of a sunspot, the dark central region; (b) of an eclipse shadow, the part from which the Sun cannot be seen at all.

uncertainty principle Heisenberg's statement that the product of uncertainties of position and momentum is greater than or equal to Planck's constant. Consequently, both position and momentum cannot be known to infinite accuracy. Another version is that the product of uncertainties of energy and time is greater than or equal to Planck's constant.

universal gravitation constant The constant G of Newton's law of gravity: force = Gm_1m_2/r^2.

valleys Depressions in the landscapes of solid objects.

Van Allen belts Regions of high-energy particles trapped by the magnetic field of the Earth.

variable star A star whose brightness changes over time.

vernal equinox The equinox crossed by the Sun as it moves to northern declinations.

Very Large Array The National Radio Astronomy Observatory's set of radio telescopes in New Mexico, used together for interferometry.

Very Large Telescope The set of instruments, including four 8.2-m reflectors and several smaller telescopes, erected by the European Southern Observatory on a mountaintop in Chile through 2001.

Very-Long-Baseline Array The National Radio Astronomy Observatory's set of radio telescopes dedicated to very-long-baseline interferometry and spread over an 8000-km baseline across the United States.

very-long-baseline interferometry The technique using simultaneous measurements made with radio telescopes at widely separated locations to obtain extremely high resolution.

visible light Light to which the eye is sensitive, 4000–7000 Å.

visual binary A binary star that can be seen through a telescope to be double.

VLA See *Very Large Array*.

VLBA See *Very-Long-Baseline Array*.

VLBI See *very-long-baseline interferometry*.

VLT See *Very Large Telescope*.

void A giant region of the Universe in which few galaxies are found.

volatile Evaporating (changing to a gas) readily.

wave front A plane in which parallel waves are in step.

wavelength The distance over which a wave goes through a complete oscillation.

weak nuclear force One of the four fundamental forces of nature, weaker than the strong force and the electromagnetic force. It is important only in the decay of certain elementary particles, such as neutrons.

weight The force of the gravitational pull on a mass.

white dwarf The final stage of the evolution of a star initially between 0.075 and about 10 solar masses but with a final mass no larger than 1.4 $M_\odot$. It is supported by electron degeneracy. White dwarfs are found to the lower left of the main sequence of the temperature-luminosity diagram.

white light All the light of the visible spectrum together.

Wien's displacement law The expression of the inverse relationship of the temperature of a black body and the wavelength of the peak of its emission.

WIMP **W**eakly **I**nteracting **M**assive **P**article, an as yet undiscovered massive particle, interacting with other elementary particles only through the weak nuclear force, that may be a major type of cold dark matter.

winter solstice For northern-hemisphere observers, the southernmost declination of the Sun, and its date.

wormhole The hypothetical connection or bridge between two black holes; they probably do not exist except in idealized black holes.

x-rays Electromagnetic radiation between 0.1 and 100 Å.

year The period of revolution of a planet around its central star; more particularly, the Earth's period of revolution around the Sun.

zenith The point in the sky directly overhead an observer.

zero-age main sequence The curve of a temperature-luminosity diagram determined by the locations of stars at the time they begin nuclear fusion.

zodiac The band of constellations through which the Sun, Moon, and planets move in the course of a year.

zodiacal light A glow in the nighttime sky near the ecliptic from sunlight reflected by interplanetary dust.

INDEX

References to illustrations, either photographs or drawings, are in italics. References to Appendicies are prefixed by *App.* Significant initial numbers followed by letters are alphabetical under their spellings; for example, 21 cm is alphabetized as *twenty-one.* Less important initial numbers are ignored in alphabetizing. For example, 3C 273 appears at the beginning of the Cs. M4 appears at the beginning of the Ms. Greek letters are alphabetized under their English equivalents.

SPRING SKY

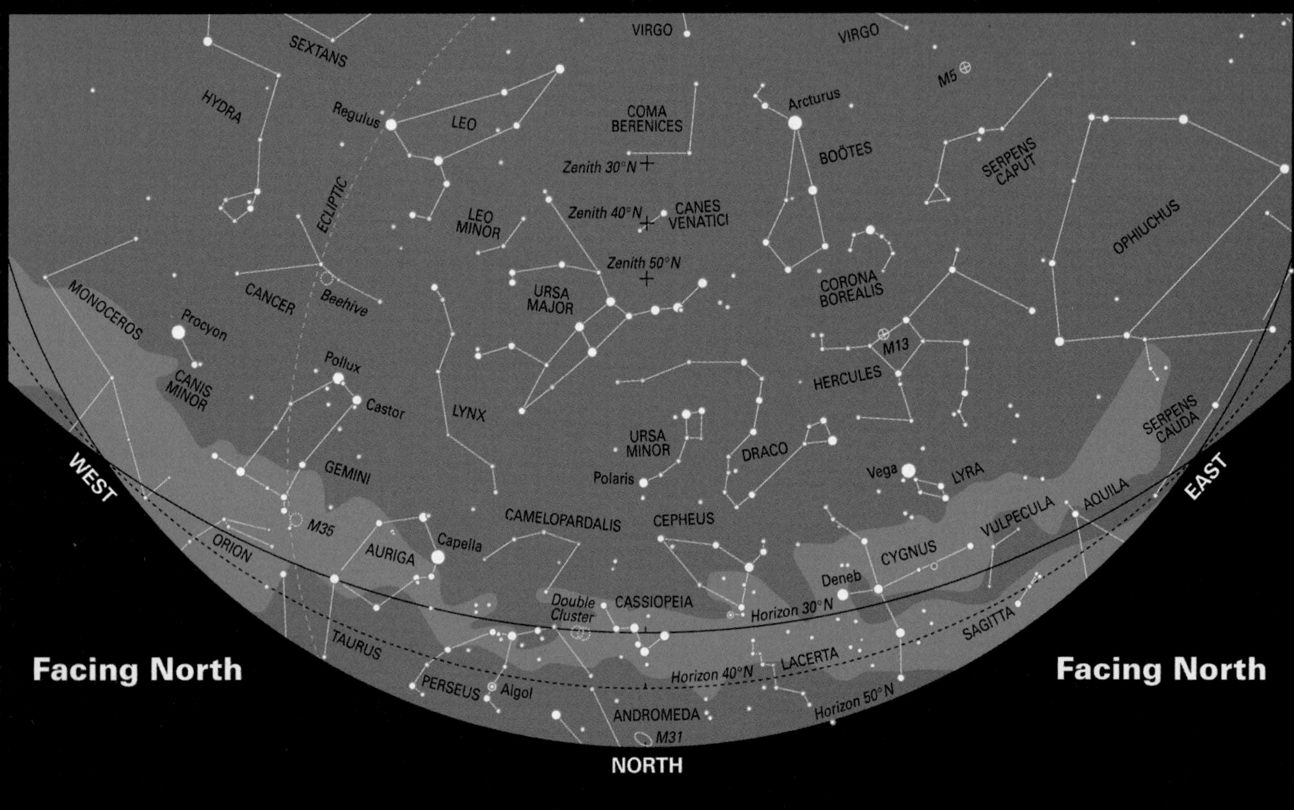

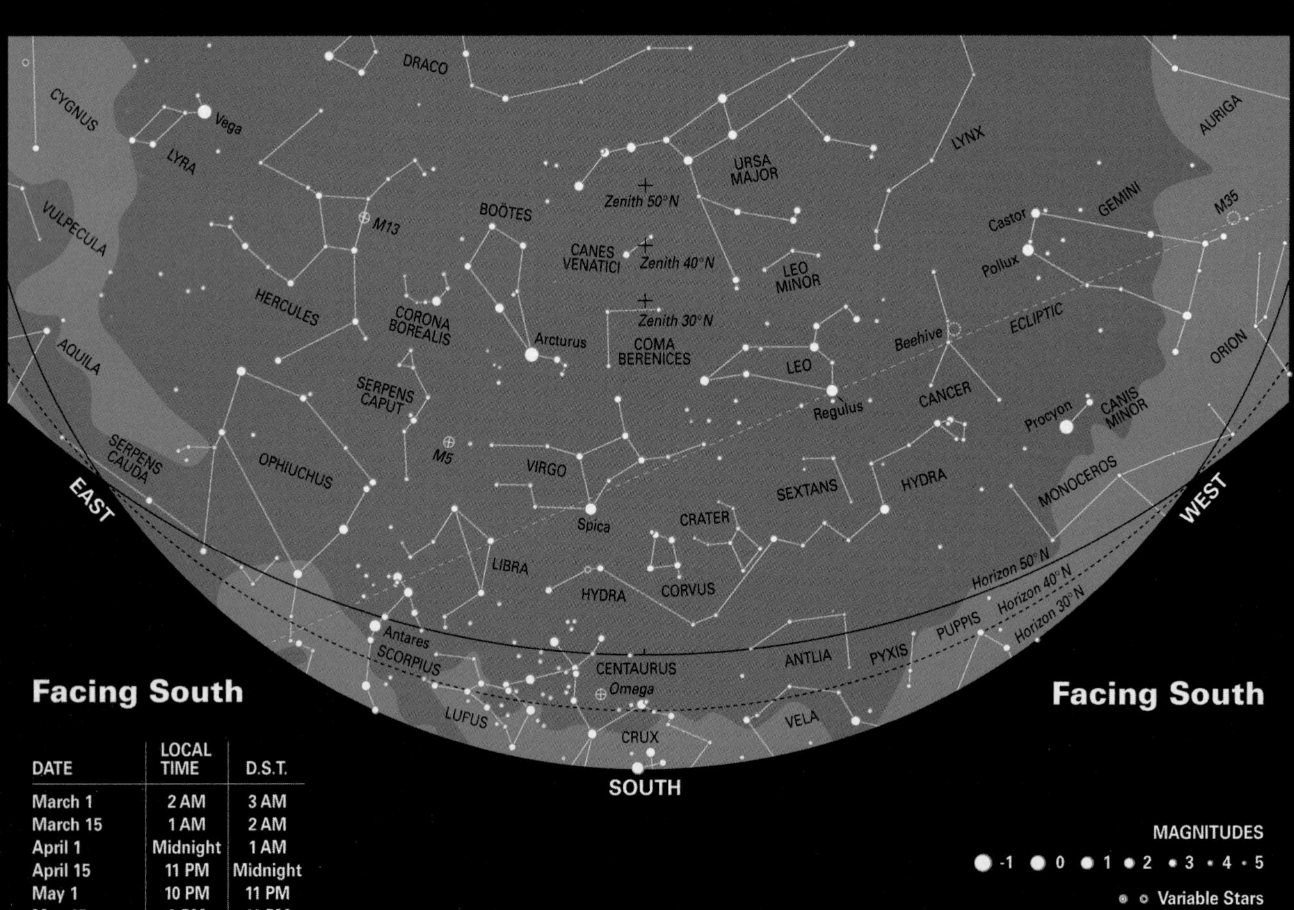

DATE	LOCAL TIME	D.S.T.
March 1	2 AM	3 AM
March 15	1 AM	2 AM
April 1	Midnight	1 AM
April 15	11 PM	Midnight
May 1	10 PM	11 PM
May 15	9 PM	10 PM
June 1	8 PM	9 PM
June 15	7 PM	8 PM

MAGNITUDES
-1 0 1 2 3 4 5
Variable Stars
Open Cluster
Globular Cluster
Nebula
Galaxy

MAP BY WIL TIRION; FOR JAY M. PASACHOFF